Handbook of
Chiral Chemicals

Second Edition

Handbook of Chiral Chemicals

Second Edition

edited by
David Ager

Taylor & Francis
Taylor & Francis Group
Boca Raton London New York

A CRC title, part of the Taylor & Francis imprint, a member of the
Taylor & Francis Group, the academic division of T&F Informa plc.

Chemistry Library

Published in 2006 by
CRC Press
Taylor & Francis Group
6000 Broken Sound Parkway NW, Suite 300
Boca Raton, FL 33487-2742

© 2006 by Taylor & Francis Group, LLC
CRC Press is an imprint of Taylor & Francis Group

International Standard Book Number-10: 1-57444-664-9 (Hardcover)
International Standard Book Number-13: 978-1-57444-664-7 (Hardcover)
Library of Congress Card Number 2005043716

Library of Congress Cataloging-in-Publication Data

Handbook of chiral chemicals / edited by David J. Ager.—2nd ed.
    p. cm. Includes bibliographical references and index.
    ISBN 1-57444-664-9 (alk. paper)
    1. Chemicals—Handbooks, manuals, etc. 2. Enantiomers—Handbooks, manuals, etc. I. Ager, David J.

TP201.H26 2005
661'.8--dc22   2005043716

informa
Taylor & Francis Group
is the Academic Division of Informa plc.

Visit the Taylor & Francis Web site at
http://www.taylorandfrancis.com

and the CRC Press Web site at
http://www.crcpress.com

Contents

Preface

The purpose of this book is to highlight the problems associated with the production of chiral compounds at a commercial scale. With the movement by pharmaceutical companies to develop single enantiomers as drug candidates, the problems associated with this subclass of organic synthesis are being highlighted. As the stereogenic center can be derived from nature through the use of "chiral pool" starting materials, the major classes of natural products are discussed.

Despite the explosion of asymmetric methods in the past 20 years, very few can be performed at scale due to cost, thermodynamic, or equipment limitations. The major reactions that have been used are covered. Resolution, whether chemical or enzymatic, still holds a key position. However, this is changing as highlighted by a short discussion of the best-selling compounds of 2002 compared to 1996.

The most mature chemical method for large-scale asymmetric synthesis is hydrogenation. This is highlighted by chapters on the uses of new ligands for hydrogenation as well as hydride reducing agents. Although we have made considerable advances in this area, the general catalyst is still elusive. Although the struggle goes on to identify the ultimate hydrogenation catalyst, for example, the use of enzymes and biological systems for the production of chiral compounds continues to increase at an incredible rate. Now that we have learned to manipulate nature's catalysts, this area will continue to grow and become more important in the production of fine chemicals.

The chapters have been grouped by topic, as shown in the table of contents. The first chapter under a topic heading is a general introduction to that topic. These chapters are not intended to be comprehensive reviews, but to critically discuss the options available at scale.

The first chapter of the book is a general introduction. Chapters 2 through 5 discuss how the key subclasses of the chiral pool are obtained. The amino acid chapters are more specific as there are other examples of amino acid syntheses contained within other chapters.

The next 23 chapters cover methods that can be used to introduce or control stereogenic centers. In some cases, such as asymmetric hydrogenations, the approach is well established and has been employed for the large-scale synthesis of a number of commercially important compounds. In other cases, such as pericyclic reactions, the potential is there—it just has not been used. Two of these chapters cover enzymatic methods, and this area seems to be more important as we understand how to manipulate enzymes to allow them to catalyze new reactions or take new substrates. The rush to market for pharmaceutical companies is forcing the chemical development time to be minimized. This is leading to large-scale usage of chiral auxiliaries. The chapters on resolution have a number of examples as illustrations. This methodology is still important to obtain chiral compounds. Although, ultimately, it may not be the most cost-effective method, it can provide material in a rapid manner, and can usually be scaled up. The introduction of large-scale chromatographic techniques, as well as the availability of a large number of enzymes that can be used to perform reactions on only one enantiomer, will ensure that this approach remains a useful tool in the future. The remaining chapters discuss various examples and topics to augment other chapters and provide a perspective of the different methods available. In two cases, ozonolysis and metathesis, the technique need not directly introduce a new stereogenic center. They are, however, still important reactions in the asymmetric context.

The final section of the book has three chapters that illustrate applications of the methodologies to prepare specific compounds. One chapter is from a small company and describes a number of projects; the second, from a large company for the synthesis of a relatively small volume product. The final chapter covers large monetary value products.

If the reader thinks that topics are missing, or if there seems to be a company bias, all major fine chemical manufacturing companies were invited to submit chapters. Even if a methodology has not been specifically covered in depth, the introduction chapter to the topic will most likely contain a reference to follow for further reading.

This is the second edition of this book. The publishers asked me whether they should print more of the first edition, or if I was willing to produce a second edition. I chose the latter course. All chapters have been updated, some very significantly illustrating the speed of development in this area. Additional chapters have also been added to highlight successes of the approaches in an industrial setting.

I would like to thank all the contributors to this book. Some have had to wait patiently for other contributors to catch up. I would especially like to thank my former colleagues from NSC Technologies for writing and updating many of the chapters that are overviews. They have also supplied numerous suggestions and ideas (and supported the hypothesis that the number of ideas generated is directly proportional to beer consumption). I would also like to make special mention of my new colleagues at DSM, who responded to my requests without hesitation and also provided a number of useful suggestions and ideas. To all the contributors, thanks; this book could not have been completed without you.

Dave Ager
Raleigh, North Carolina

The Editor

David Ager was born in Northampton, England, in 1953. He received a B.Sc. from Imperial College, London, and a Ph.D. from the University of Cambridge, working with Dr. Ian Fleming on organosilicon chemistry. In 1977 he was awarded a Science Research Council Postdoctoral Fellowship that allowed him to collaborate with Professor Richard Cookson FRS at the University of Southampton. In 1979, he joined the faculty of the University of Liverpool as a Senior Demonstrator. This was followed by an assistant professor position at the University of Toledo in Ohio. In 1986, he joined the NutraSweet Company's Research and Development group, and became responsible for the scale-up of new sweetener development candidates as a Monsanto Fellow. He is a founding member of NSC Technologies, which became an independent unit of Monsanto in 1995 as part of Monsanto Growth Enterprises. In 1999, NSC was sold to Great Lakes Fine Chemicals; Dr. Ager was a Fellow with GLFC, responsible for the development of new synthetic methodology. After leaving GLFC he worked as a consultant on chiral and process chemistry. He joined DSM at the beginning of 2002 as the competence manager for homogeneous catalysis. Dr. Ager is a member of a number of scientific advisory boards. He has over 80 publications including three books.

Contributors

David Ager
DSM Pharma Chemicals
Raleigh, North Carolina
and
PharQuest LLC
Raleigh, North Carolina

David Allen
Stepan Company
Northfield, Illinois

Nelson Barton
Diversa Corporation
San Diego, California

Colin Bayley
PharQuest LLC
Raleigh, North Carolina

Jens Beckmann
Chirogen Pty Ltd
Parkville, Victoria, Australia

Hans-Ulrich Blaser
Solvias AG
Basel, Switzerland

Wilhelmus H.J. Boesten
DSM Research
Advanced Synthesis, Catalysis and
  Development
Geleen, The Netherlands

Gary Bond
University of Central Lancaster
Centre for Material Science
Preston, UK

Quirinus Broxterman
DSM Research
Advanced Synthesis, Catalysis and
  Development
Geleen, The Netherlands

Michel Bulliard
SynKem
Chenôve Cedex, France

Mark Burk
Diversa Corporation
San Diego, California

Dainis Dakternieks
Chirogen Pty Ltd
Parkville, Victoria, Australia

Paul Dalby
University College London
Department of Biochemical Engineering
London, UK

Ben de Lange
DSM Research
Advanced Synthesis, Catalysis and
  Development
Geleen, The Netherlands

Johannes de Vries
DSM Research
Advanced Synthesis, Catalysis and
  Development
Geleen, The Netherlands

Grace DeSantis
Diversa Corporation
San Diego, California

Michael East
Uquifa Inc.
Golf, Illinois

Henk Elsenberg
DSM Research
Advanced Synthesis, Catalysis and
  Development
Geleen, The Netherlands

Karen Etherington
UltraFine (UFC Limited),
 a division of Sigma-Aldrich
Manchester Science Park, UK

Ian Fotheringham
Ingenza
Edinburgh, UK

Martin Fox
Chirotech Technology Ltd.,
 a subsidiary of The Dow Chemical Company
Cambridge, UK

Isabelle Gallou
Boehringer-Ingelheim Pharmaceuticals Inc.
Ridgefield, Connecticut

William Greenberg
Diversa Corporation
San Diego, California

Reinier Grimbergen
DSM Research
Centre for Particle Technology
Geleen, The Netherlands

Peter Haslam
Uquifa Laboratories
Barcelona, Spain

Ed Irving
UltraFine (UFC Limited),
 a division of Sigma-Aldrich
Manchester Science Park, UK

Mark Jackson
Chirotech Technology Ltd.,
 a subsidiary of The Dow Chemical Company
Cambridge, UK

Bernard Kaptein
DSM Research
Advanced Synthesis, Catalysis and
 Development
Geleen, The Netherlands

Richard Kellogg
Syncom BV
Groningen, The Netherlands

Dhileepkumar Krishnamurthy
Boehringer-Ingelheim Pharmaceuticals Inc.
Ridgefield, Connecticut

Scott Laneman
Digital Specialty Chemicals, Inc.
Toronto, Ontario, Canada

Ian Lennon
Chirotech Technology Ltd.,
 a subsidiary of The Dow Chemical Company
Cambridge, UK

Weiguo Liu
Merck Research Laboratories
Rahway, New Jersey

Matthias Lotz
Solvias AG
Basel, Switzerland

Gary Lye
University College London
Department of Biochemical Engineering
London, UK

Raymond McCague
Chirotech Technology Ltd.,
 a subsidiary of The Dow Chemical Company
Cambridge, UK

Janine McGuire
University of Central Lancaster
Centre for Material Science
Preston, UK

José Nieuwenhuijzen
Syncom BV
Groningen, The Netherlands

David Pantaleone
Baxter Healthcare Corporation
McGaw Park, Illinois

James Ramsden
Dowpharma, Chirotech Technology Ltd.
Cambridge, UK

David Schaad
Ferro Pfanstiehl, Ferro Corporation
Waukegan, Illinois

Feodor Scheinmann
UltraFine (UFC Limited),
  a division of Sigma-Aldrich
Manchester Science Park, UK

Carl Schiesser
Chirogen Pty Ltd
Parkville, Victoria, Australia

Chris Senanayake
Boehringer-Ingelheim Pharmaceuticals Inc.
Ridgefield, Connecticut

Yian Shi
Colorado State University
Department of Chemistry
Fort Collins, Colorado

Felix Spindler
Solvias AG
Basel, Switzerland

Martin Studer
Solvias AG
Basel, Switzerland

Paul Taylor
zuChem Inc.
Chicago, Illinois

Patrick Uiterweerd
Syncom BV
Groningen, The Netherlands

Marcel van der Sluis
Syncom BV
Groningen, The Netherlands

Ton Vries
Syncom BV
Groningen, The Netherlands

Basil Wakefield
UltraFine (UFC Limited),
  a division of Sigma-Aldrich
Manchester Science Park, UK

David Weiner
Diversa Corporation
San Diego, California

John Woodley
University College London
Department of Biochemical Engineering
London, UK

1 Introduction

David J. Ager

CONTENTS

This book discusses various aspects of chiral fine chemicals including their synthesis and uses at scale. There is an increasing awareness of the importance of chirality in biological molecules because the two enantiomers can sometimes have different effects.[1–4]

In many respects, chiral compounds have been regarded as special entities within the fine chemical community. As we will see, the possession of chirality does not, in many respects, make the compound significantly more expensive to obtain. Methods for the preparation of optically active compounds have been known for more than 100 years (many based on biological processes). The basic chemistry to a substrate on which an asymmetric transformation is then performed can offer more challenges in terms of chemistry and cost optimization than the "exalted" asymmetric step.

The book is organized into sections that relate either to a compound or reaction type, as outlined in the table of contents. When there are several chapters on a specific topic, the first chapter of that section is a general introduction.

1.1 CHIRALITY

The presence of a stereogenic center within a molecule can give rise to chirality. Unless a chemist performs an asymmetric synthesis, equal amounts of the two antipodes will be produced. To separate these, or to perform an asymmetric synthesis, a chiral agent has to be used. This can increase the degree of complexity in obtaining a chiral compound in a pure form. However, nature has been kind and does provide some chiral compounds in relatively large amounts. Chirality does provide an additional problem that is sometimes not appreciated by those who work outside of the field;* analysis of the final compound is often not a trivial undertaking. The various approaches to undertake an asymmetric synthesis are summarized in Figure 1.1; many of these methods are discussed later in this chapter.

The chiral drug market continues to grow at a significant rate (Chapter 31), partially as a result of the U.S. Food and Drug Administration's directive to have chiral drugs developed as single enantiomers. In addition to this, the introduction of a single enantiomer of a drug that has already been on the market in racemic form has been used to extend patent and product life. An example is AstraZeneca's Omeprazole, a proton pump inhibitor whose product life has been extended by the switch from racemic form to a single enantiomer (Chapter 31).

1.2 CHIRAL POOL

Nature has provided a wide variety of chiral materials, some in great abundance. The functionality ranges from amino acids to carbohydrates to terpenes (Chapters 2–5). All of these classes of compounds are discussed in this book. Despite the breadth of functionality available from natural sources, very few compounds are available in optically pure form on large scale. Thus, incorporation of a "chiral pool" material into a synthesis can result in a multistep sequence. However, with the advent of synthetic methods that can be used at scale, new compounds are being added to the chiral pool, although they are only available in bulk by synthesis. When a chiral pool material is available at large scale, it is usually inexpensive. An example is provided by L-aspartic acid, where the chiral material can be cheaper than the racemate.

How some of these chiral pool materials have been incorporated into syntheses of biologically active compounds is illustrated throughout this book. In addition, chiral pool materials are often incorporated, albeit in derivatized form, into chiral reagents and ligands that allow for the transfer of chirality from a natural source into the desired target molecule.

1.3 CHIRAL REAGENTS

Chiral reagents allow for the transfer of chirality from the reagent to the prochiral substrate. Almost all of these reactions involve the conversion of an sp^2 carbon to an sp^3 center. For example, reductions of carbonyl compounds (Chapters 16–19), asymmetric hydrogenations (Chapters 12–15), and asymmetric oxidations of alkenes (Chapters 9 and 10) are all of this type. The reagents can be catalytic for the transformation they bring about, or they can be stoichiometric. The former usually is preferred because it allows for chiral multiplication during the reaction—the original stereogenic center gives rise to many product stereocenters. This allows for the cost of an expensive catalyst to be spread over a large number of product molecules.

---

* For a review on synthetic strategies, see reference 5.

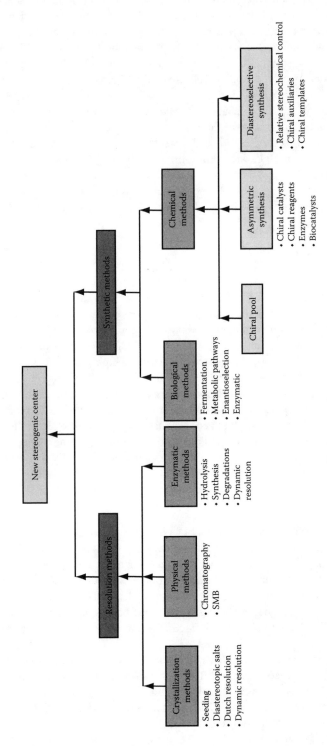

FIGURE 1.1 Approaches for the introduction of a new stereogenic center.

1.4 CHIRAL CATALYSTS

Considerable resources are being expended in the quest for new asymmetric catalysts for a wide variety of reactions (Chapter 12). In many cases, these catalysts are based on transition metals, where the ligands provide the chiral environment. However, as our understanding of biotransformations increases, coupled with our ability to produce mutant enzymes at scale, biocatalysts are beginning to become key components of our asymmetric synthetic toolbox (Chapter 19).

The move toward catalytic reactions is reflected in the increase in the number of chapters in this book on the topic compared to the first edition. The trend has been observed by noted chemists in the previous decade. Professor Seebach, for example, in 1990 stated "the primary center of attention for all synthetic methods will continue to shift toward catalytic and enantioselective variants: indeed, it will not be long before such modifications will be available for every standard reaction."[6] Professor Trost in 1995 was a little more specific with "catalysis by transition metal complexes has a major role to play in addressing the issue of atom economy—both from the point of view of improving existing processes, and, most importantly, from discovering new ones."[7] However, the concept can be extended to biological and organic catalysts and to those based on transition metals.

1.4.1 CHEMICAL CATALYSTS

The development of transition metal catalysts for the asymmetric reduction of functionalized alkenes allowed synthetic chemists to perform reactions with a stereochemical fidelity approaching that of nature (Chapters 12–18). We now have a number of reactions at our disposal that can be performed with chemical catalysts, and the number continues to grow. However, there are still problems associated with this approach because many catalysts have specific substrates requirements, often involving just one alkene isomer of the substrate. The chiral multiplication associated with use of a chiral catalyst often makes for attractive economic advantages. However, the discovery and development of a chemical catalyst to perform a specific transformation is often tedious, time consuming, and expensive. There are many reports of chiral ligands in the literature, for example, to perform asymmetric hydrogenation, yet very few have been used at scale.[8,9] This highlights the problem that there are few catalysts that can be considered general. As mentioned earlier, often the preparation of the substrate is the expensive part of a sequence, especially with catalysts that have high turnover numbers and can be recycled.

1.4.2 BIOLOGICAL CATALYSTS

Biological catalysts for asymmetric transformations have been used in specific cases for a considerable period of time, excluding the chiral pool materials. However, until recently, the emphasis has been on resolutions with enzymes rather than asymmetric transformations (Chapter 19; see also Chapters 20 and 21). With our increasing ability to produce mutant enzymes that have different or broad-spectrum activities compared to the wild types, the development of biological catalysts is poised for major growth. In addition to high stereospecificities, an organism can be persuaded to perform more than one step in the overall reaction sequence and may even make the substrate (Chapter 3).

Unlike the design of a chemical catalyst, which has to be semiempirical in nature and therefore is very difficult to apply to a completely different transformation, screening for an enzyme that performs a similar reaction is relatively straightforward and often gives the necessary lead for the development of a potent biological catalyst. The use of molecular biology, site-specific mutagenesis, and enzymology all contribute to the development of such a catalyst. This approach is often ignored because these methods are outside of traditional chemical methodologies.

There are a large number of reports of abzymes, or catalytic antibodies, in the literature.[10-15] Although catalysis has been observed in a large number of examples, the problems associated with the production of large amounts of abzymes compounded by the low turnover numbers often observed makes this technology only a laboratory curiosity, at present. The increasing use of mutant enzymes without isolation from the host organisms makes this latter approach economically more attractive.

1.5 STOICHIOMETRIC REAGENTS

To understand a specific transformation, chemists have often developed asymmetric synthetic methods in a logical, stepwise manner. Invariably, the mechanism of the reaction and the factors that control the stereochemical outcome of a transformation are paramount in the design of an efficient catalyst for use at scale with a wide variety of substrates. There are some noticeable exceptions to this approach, such as the empirical approach used to develop asymmetric hydrogenation catalysts. In other instances, an empirical approach provided sufficient insight to allow for the development of useful chiral catalysts, such as the empirical rule for the oxidation of alkenes[16,17] that led to the asymmetric hydroxylation catalysts.[18,19] Often the first generations of asymmetric reagents are chiral templates or stoichiometric reagents. These are then superseded by chiral auxiliaries, if the substrate has to be modified, or chiral catalysts in the case of external reagents.

1.5.1 CHIRAL AUXILIARIES

This class of compounds modifies the substrate molecule to introduce a stereogenic center that will influence the outcome of a reaction to provide an asymmetric synthesis. The auxiliary has to be put on to the substrate and removed. Although this involves two steps, concurrent protection of sensitive functionality can also take place, so that one inefficient sequence (protection and deprotection) is traded for another (auxiliary introduction and removal). A large number of asymmetric transformations have been performed with chiral auxiliaries, which has provided a wealth of literature. Thus, precedence for most reactions is available, providing for a large degree of certainty that a specific reaction, even with a new substrate, will work (see Chapters 23–25). As a result of the curtailed time lines for the development of new pharmaceutical products, coupled with the decrease in the costs of many auxiliaries, this approach is now being used at large scale.

Although an auxiliary is recovered intact after the asymmetric transformation and has the potential for recycle, there are often problems associated with the practical implementation of this concept.

1.5.1.1 Chiral Templates

Chiral templates can be considered a subclass of chiral auxiliaries. Unlike auxiliaries that have the potential for recycle, the stereogenic center of a template is destroyed during its removal. Although this usually results in the formation of simple by-products that are simple to remove, the cost of the template's stereogenic center is transferred to the product molecule. Under certain circumstances, chiral templates can provide a cost-effective route to a chiral compound (Chapter 25). Usually, the development of a template is the first step in understanding a specific transformation and the knowledge gained is used to develop an auxiliary or catalyst system.

1.6 RESOLUTION

The separation of enantiomers through the formation of derivative diastereoisomers and the subsequent separation of these by physical means has been practiced on large scale for many years (Chapters 6–8). In addition the racemic mixture can be reacted with a chiral reagent, where the rates of reaction are very different for the two enantiomers, allowing for a resolution. This approach is applicable to chemical agents, such as the Sharpless epoxidation procedure,[20–22] and biological agents, such as enzymes. Unless a *meso*-substrate is used, or the wrong isomer can be converted back to the racemate *in situ*, to provide a dynamic resolution, the "off-isomer" can present an economic problem. It either has to be disposed of—resulting in a maximum overall yield of 50% from the racemic substrate—or epimerized to allow for recycle through the resolution sequence. This latter approach often involves additional steps in a sequence that can prove to be costly. In either case, recovery of the resolving agent also has to be considered. As the development of robust, general, asymmetric methods to a class of compounds becomes available, it is becoming apparent that the more traditional resolution approaches are not economically viable. However, there is still a place for resolutions, especially when speed is involved. The use of chromatographic methods has also helped to decrease the time to perform a resolution, and some of these techniques, such as simulated moving bed (SMB) chromatography, can be performed at scale.

1.7 SYNTHESIS AT SCALE

There are many problems associated with carrying out asymmetric synthesis at scale. Many asymmetric transformations reported in the literature use the technique of low temperature to allow differentiation of the two possible diastereoisomeric reaction pathways. In some cases, the temperature requirements to see good asymmetric induction can be as low as –100°C. To obtain this temperature in a reactor is costly in terms of cooling and also presents problems associated with materials of construction and the removal of heat associated with the exotherm of the reaction itself. It is comforting to see that many asymmetric catalytic reactions do not require the use of low temperature. However, the small number of "robust" reactions often leads development chemists to resort to a few tried and tested approaches, namely chiral pool synthesis, use of a chiral auxiliary, or resolution. In addition, the scope and limitations associated with the use of a chiral catalyst often result in a less than optimal sequence either because the catalyst does not work well on the necessary substrate or the preparation of that substrate is long and costly. Thus, the availability of a number of different approaches helps to minimize these problems (Chapter 2).

A short overview of the syntheses of some of the large-scale and monetary value chiral products is given in Chapter 31. This illustrates to the reader the relative importance of some of the approaches discussed within this book, especially the power of biological approaches.

1.7.1 Reactions Amenable to Scale

When reactions that are "robust" are considered, only a relatively small number are available. Each of these reaction types are discussed within this book, although some do appear under the chiral pool materials that allowed for the development of this class of asymmetric reagent. Such an example is the use of terpenes that have allowed for the development of chiral boranes (Chapter 5).

1.7.1.1 Biological Methods

The biological methods class of reagents holds the most promise for rapid development in the near future because most reactions are asymmetric. The problems that are being overcome are the tight substrate specificity of many enzymes and the need for co-factor regeneration. Systems are now being developed for asymmetric synthesis rather than resolution approaches. Some of these reactions are discussed in Chapters 19 through 21, but see also Chapters 2 and 3.

1.7.1.2 Chemically Catalyzed Oxidations

The development of simple systems that allow for the asymmetric oxidation of allyl alcohols and simple alkenes to epoxides or 1,2-diols has had a great impact on synthetic methodology because it allows for the introduction of functionality with concurrent formation of one or two stereogenic centers. This functionality can then be used for subsequent reactions that usually fall into the substitution reaction class. Because the catalysts for these reactions do not require the use of low temperatures to ensure high degrees of induction, they can be considered robust. However, catalyst turnover numbers are sometimes low, and the synthesis of the substrate can still be a crucial economic factor. Aspects of asymmetric oxidations are discussed in Chapters 9 and 10, the latter discussing organocatalysis.

1.7.1.3   Transition Metal Catalyzed Reductions

As mentioned elsewhere, the development of transition metal catalysts that allowed for high enantioselectivity in reduction reactions showed that chemists could achieve comparable yields and ee's to enzymes. A large number of transition metal catalysts and ligands are now available. A number of reactions have been scaled up for commercial production that use an asymmetric hydrogenation (Chapters 12–18) or a hydride delivery (Chapters 12, 16, and 17) for the key step that forms the new stereogenic center.

1.7.1.4   Transition Metal Catalyzed Isomerizations

Transition metal catalyzed isomerization reactions are closely related to asymmetric hydrogenations, especially because similar catalysts are used. A number of these reactions are used at scale (Chapters 12 and 31).

1.7.1.5   Reductions

The reduction of a carbonyl group to a secondary alcohol with the generation of a new stereogenic center has been achieved at a laboratory scale with a large number of reagents, many stoichiometric with a ligand derived from the chiral pool. The development has led to boron-based reagents that perform this transformation with high efficiency (Chapters 16 and 17). The use of ruthenium-based catalysts that can be extremely efficient (Chapter 12) has opened up this methodology to exploitation at scale.

1.7.1.5.1   Hydroborations

In addition to being useful reagents for the reductions of carbonyl compounds, boron-based reagents can also be used for the conversion of an alkene to a wide variety of functionalized alkanes. Because the majority of these reagents carry a terpene substituent, they are discussed under these chiral pool materials (Chapter 5).

1.7.1.6   Pericyclic Reactions

Many pericyclic reactions are stereospecific and, because they have to be run at temperatures higher than ambient, are very robust. It is somewhat surprising that there are very few examples of pericyclic reactions being run at scale, especially in light of our understanding of the factors that control the stereochemical course of the reaction either through the use of a chiral auxiliary or catalyst (Chapter 26).

1.7.1.7   Substitution Reactions (S_N2)

The heading "substitution reactions" has been used to describe the conversion of a stereogenic center to another. Of course, this means that the substrate stereogenic center has had to be obtained by one of the reaction types outlined earlier, from the chiral pool, or by resolution. Reactions that fall into this category include epoxide and cyclic sulfate openings and iodolactonizations (Chapter 22). Perhaps the most important reaction of this type for asymmetric synthesis is allylic substitution in the presence of a transition metal catalyst.

1.7.1.8 Radical Reactions

Until recently, radical reactions were not considered for the generation of new stereogenic centers, especially if the reaction had to be performed at scale. As our understanding of the factors that control this class of reactions has increased, we are now in a position to use them in a synthetically useful manner. Some of these reactions are discussed in Chapter 27.

1.7.1.9 Other Reactions

Reactions that do not directly generate a stereogenic center have also been incorporated into this book. With these types of reactions, chiral molecules result if a stereocenter is already present in the substrate. Examples are ozonolysis (Chapter 11) and metathesis (Chapter 28).

1.8 ANALYSIS

The analysis of chiral compounds to determine their optical purity is still not a trivial task. The analysis method has to differentiate between the two antipodes and, thus, has to involve a chiral agent. However, the development of chiral chromatography, especially HPLC (high-performance liquid chromatography), has done a significant amount to relieve this problem. The purpose of this book is to discuss large-scale synthetic reactions, but the reader is reminded that the development of chiral analytic methods may not have been a trivial undertaking in many examples.

1.9 EXAMPLES

To illustrate and highlight methods that have been used in practice, the final section of this book contains examples of specific compound syntheses. A number of examples are given by a small company (Chapter 29); a larger company making a smaller volume product (Chapter 30); and large sales products in pharmaceutical, agrochemical, and food applications (Chapter 31).

1.10 SUMMARY

The development of optically active biological agents, such as pharmaceuticals, has led to the increase in large-scale chiral syntheses. The chirality may be derived from the chiral pool or a chiral agent, such as an auxiliary, template, reagent, or catalyst. There are, however, relatively few general asymmetric methods that can be used at scale.

REFERENCES

1. Blashke, G., Kraft, H. P., Fickenscher, K., Kohler, F. *Arzniem-Forsch/Drug Res.* 1979, *29*, 10.
2. Blashke, G., Kraft, H. P., Fickenscher, K., Kohler, F. *Arzniem-Forsch/Drug Res.* 1979, *29*, 1140.
3. Powell, J. R., Ambre, J. J., Ruo, T. I. In *Drug Stereochemistry,* Wainer, I. W., Drayer, D. E. Eds., Marcel Dekker: New York, 1988, p. 245.
4. Ariëns, E. J., Soudijn, W., Timmermas, P. B. M. W. M. *Stereochemistry and Biological Activity of Drugs*, Blackwell Scientific: Palo Alto, 1983.
5. Fuchs, P. L. *Tetrahedron* 2001, *57*, 6855.
6. Seebach, D. *Angew. Chem., Int. Ed.* 1990, *29*, 1320.
7. Trost, B. M. *Angew. Chem., Int. Ed.* 1995, *34*, 259.
8. Blaser, H.-U. *J. Chem. Soc., Chem. Commun.* 2003, 293.
9. Blaser, H.-U., Schmidt, E. Introduction. In *Asymmetric Catalysis on Industrial Scale,* Blaser, H.-U., Schmidt, E. Eds., Wiley-VCH: Weinheim, 2004, p. 1.
10. Lerner, R. A., Benkovic, S. J., Schultz, P. G. *Science* 1991, *252*, 659.

11. Blackburn, G. M., Kang, A. S., Kingsbury, G. A., Burton, D. R. *Biochem. J.* 1989, *262*, 381.
12. Schultz, P. G., Lerner, R. A. *Acc. Chem. Res.* 1993, *26*, 391.
13. Hilvert, D. *Acc. Chem. Res.* 1993, *26*, 552.
14. Stewart, J. D., Liotta, L. J., Benkovic, S. J. *Acc. Chem. Res.* 1993, *26*, 396.
15. Stewart, J. D., Benkovic, S. J. *Chem. Soc. Rev.* 1993, *22*, 213.
16. Cha, J. K., Christ, W. J., Kishi, Y. *Tetrahedron Lett.* 1983, *24*, 3943.
17. Cha, J. K., Christ, W. J., Kishi, Y. *Tetrahedron* 1984, *40*, 2247.
18. Sharpless, K. B., Behrens, C. H., Katsuki, T., Lee, A. W. M., Martin, V. S., Takatani, M., Viti, S. M., Walker, F. J., Woodard, S. S. *Pure Appl. Chem.* 1983, *55*, 589.
19. Sharpless, K. B., Verhoeven, T. R. *Aldrichimica Acta* 1979, *12*, 63.
20. Brown, J. M. *Chem. Ind. (London)* 1988, *612*.
21. Gao, Y., Hanson, R. M., Klunder, J. M., Ko, S. Y., Masamune, H., Sharpless, K. B. *J. Am. Chem. Soc.* 1987, *109*, 5765.
22. Ager, D. J., East, M. B. *Asymmetric Synthetic Methodology*, CRC Press: Boca Raton, 1995.

2 Amino Acids

David J. Ager

CONTENTS

2.1 INTRODUCTION

The 20 proteinogenic amino acids have become key building blocks as chiral pool materials. In addition to these 20 amino acids, there are many analogues with modified sidechains, backbones, or different stereochemistry. Some of these other amino acids, such as ᴅ-alanine, are found in nature, but in this chapter they will be collected under the descriptor of "unnatural" amino acids.

Although nature has been the primary source of the proteinogenic amino acids through extraction processes, many of the unnatural analogues have to be synthesized. Modern asymmetric synthetic methodology is now in the position to provide cheap, pure, chiral materials at scale. Some of the unnatural amino acids are now made at scale and have been used to extend the chiral pool. To avoid duplicating sections of this book, this chapter discusses the problems associated with the synthesis of unnatural amino acids at various scales. This illustrates that a single, cheap method need not fulfill all of the criteria to provide a chiral pool material to a potential customer, and a number of approaches are required.

2.1.1 UNNATURAL AMINO ACID SYNTHESIS

A variety of methods have been described in this book for the preparation of amino acids. They range from the use of chiral auxiliaries (Chapter 23), through the use of enzymes (Chapters 3 and

11

17) and asymmetric hydrogenation (Chapters 12–15), to total fermentation (Chapters 3 and 21). The aim of this chapter is to put into perspective the choice of the appropriate method.

The majority of chiral fine chemicals produced at present are used in pharmaceutical candidates and products. During the development of a compound as a drug candidate, the factors governing needs for intermediates change. Methodology has to be available that can be used to make material at a relatively small scale but that can be implemented rapidly. Also, there must be a low probability of failure because the change to a second method would be very time consuming. As the material needs increase, then cost becomes more of a factor, but the time scale lengthens allowing for the development of cheaper processes. However, there is still a reasonable chance that the drug candidate will not pass through the testing phase, and investment into method development must be made prudently. When commercialization of the compound is close, and requirements become large, then cost is the overriding factor and the manufacturing processes discussed elsewhere in this book and in other sources are paramount.[1,2]

As noted in a review, no economical chemical process has emerged for the asymmetric synthesis of amino acids.[3] This presumably refers to large-scale production where the research costs associated with the development of a biological approach can be justified. Chemical methods can produce racemic substrates, as illustrated later in this chapter with L-methionine and D-phenylglycine. With methods in place, however, biological methods can be implemented rapidly.[4]

2.2 CHOICE OF APPROACH

Unnatural amino acids are becoming more important as starting materials for synthesis as a result of their expanded usage in pharmaceutical drug candidates. This increase has been spurred by a number of factors. Rational drug design has allowed natural substrates of an enzyme to be structurally modified without the need for a large screening program. Because many of these substrates are polypeptides, substitution of unnatural amino acids for the native component is a relatively small step. Contrary to this, many combinatorial approaches rely on the production of a large number of potential compounds that are then screened. With a wide range of reactions available, the history of chemistry performed on solid supports, and unnatural amino acids extending the library of amino acids to almost boundless limits, these building blocks are being used in a wide range of applications.

Thus, the need has arisen for larger amounts of unnatural amino acids—those that contain unusual side chains or are in the D-series. Because α-amino acids contain an epimerizable center, D-amino acids are usually accessible through epimerization of the natural isomer, followed by a resolution. Of course, the racemic mixture can also be accessed by synthesis. Because resolution can be either wasteful—if the undesired isomer is discarded—or clumsy—when the other isomer is recycled through an epimerization protocol—many large-scale methods now rely on a dynamic resolution, where all of the starting material is converted to the desired isomer (*vide infra,* Chapters 6 and 7). With the advent of asymmetric reactions that can be performed at large scale, a substrate can now be converted to the required stereoisomer without the need for any extra steps associated with a resolution approach.

In addition to α-amino acids (**1**), β-amino acids (**2**) are beginning to find favor as building blocks. With β-amino acids, the extra carbon atom between the two functional groups increases the complexity. The stereogenic center can be at the β-position (**2**, R ≠ H), at the α-position (**2**, R¹ ≠ H), or at both. Of course, both of these positions have the potential to be differentially disubstituted (see Section 2.6).

The provision of intermediates to the pharmaceutical, agrochemical, and other industries requires a number of robust synthetic methods to be available for the preparation of a single class of compounds. The choice of which method to use is determined by the scale of the reaction, economics for a specific run or campaign, ease of operation, time frame, and availability of equipment. There are a large number of methodologies available in the literature for the preparation of "unnatural" amino acids.[3,5,6]

Chiral auxiliaries play a key role in the scale up of initial samples of materials and for small quantities. In addition, this method of approach can be modified to allow for the preparation of closely related materials that are invariably required for toxicologic testing during a pharmaceutical's development. There are a number of advantages associated with the use of an established auxiliary: The scope and limitations of the system are well defined; it is simple to switch to the other enantiomeric series (as long as mismatched pairs do not occur); concurrent protection of sensitive functionality can be achieved. This information can result in a short development time. The auxiliary's cost has the potential to be limited through recycles. However, the need to put on and take off the auxiliary unit adds two extra steps to a synthetic sequence that will reduce the overall yield. Most auxiliaries are not cheap, and this must be considered carefully when large amounts of material are needed. Finally, because the auxiliary has to be used on a stoichiometric scale, a by-product—recovered auxiliary—will be formed somewhere in the sequence. This by-product has to be separated from the desired product; sometimes, this is not a trivial task.

Chiral auxiliaries modify the structure of the precursor to allow induction at a new stereogenic center. An alternative approach uses a modifier group—a template—in a similar manner to an auxiliary, but this group is destroyed during removal. Because the modifier group cannot be recovered, the cost of the stereogenic center is incorporated into the product molecule (*cf.* Chapter 25). This type of approach is becoming less common just on the grounds of simple economics.

Chiral reagents allow an external chiral influence to act on an achiral precursor. Chiral reagents usually use chiral ligands that modify a more familiar reagent. Often the ligand has to be used in stoichiometric quantities, although it usually can be recovered at the end of the reaction. In contrast, a chiral catalyst allows many stereogenic centers to be derived from one catalyst molecule. Thus, the cost of the catalyst can be shared over a larger quantity of product, even when the potential for recycle is minimal, as in small production runs. The catalysts can be chemical or biological. The development of these catalyst systems is time consuming, and we look for general applicability so that a wide range of compounds can be prepared from a single system. The current trend is to use a chiral catalyst, be it chemical or biological, as early as possible.

Each of these approaches is discussed in detail in the following sections.

2.3 SMALL-SCALE APPROACHES

As noted earlier, the usual purpose of these methods is to provide relatively small amounts of material, in a rapid manner, with little risk of the chemistry failing. Cost is not of primary concern. One of the methods cited later in this chapter for larger scale approaches may be applicable for the product candidate, but experience with these methodologies has to be such that success can be attained in a synthesis campaign with little work involved in the transfer of technology to this project.

One of the most general approaches available is to use a chiral auxiliary because many examples are available in the literature and the scope and limitations of the various structures available as auxiliary units have been documented (see also Chapters 23–25).[6–8]

With high-throughput experimentation, it is possible to perform screening in a rapid manner to find a suitable resolution agent (*cf.* 2.3.3 and Chapters 6 and 7). However, with chromatographic equipment now being readily available, the use of this technique, either as the traditional application or SMB (simulated moving bed), is finding widespread usage to separate enantiomers. Of course, the racemic mixture still has to be prepared.

2.3.1 CHIRAL TEMPLATES

The use of phenylglycine amide (**3**) for the preparation of amino acids is discussed in Chapter 25.

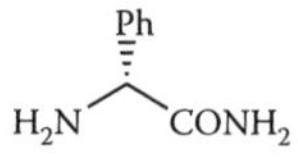

$$\text{H}_2\text{N} \overset{\text{Ph}}{\underset{\textbf{3}}{\bigwedge}} \text{CONH}_2$$

Dehydromorpholinones **4** can be used to prepare unnatural amino acids and α-substituted amino acids (Scheme 2.1).[9] The phenylglycinol unit is cleaved by hydrogenolysis.

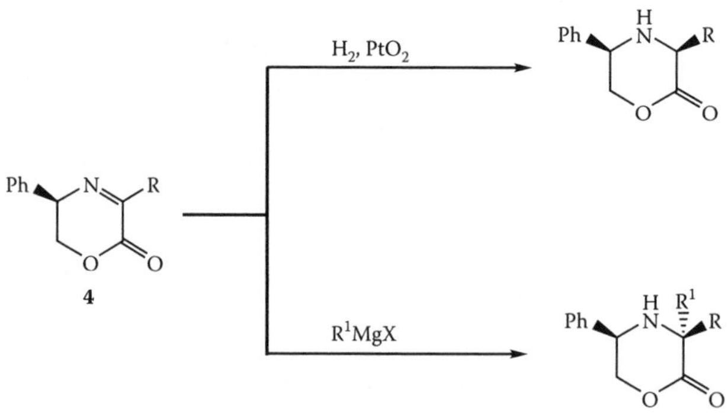

SCHEME 2.1

This class of compounds serves as a chiral template rather than a chiral auxiliary because the original stereogenic center is destroyed in the reaction sequence. Dehydromorpholinones have been prepared from the amino alcohol by a rather extended reaction sequence that uses expensive reagents,[9] but it has now been shown that the heterocycles can be made simply and cheaply in a one-pot reaction (Scheme 2.2).[9,10]

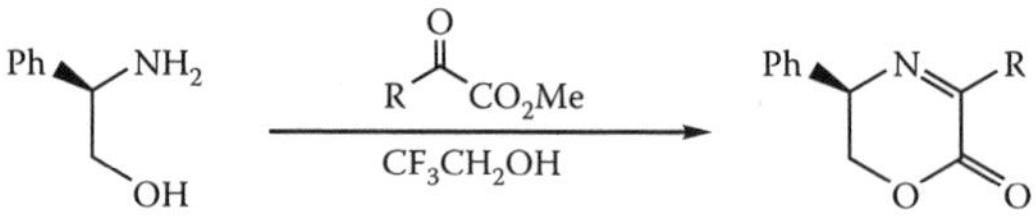

SCHEME 2.2

An example of the use of phenylglycinol as a template in the synthesis of an amino acid is given in Section 2.3.2.1 (Scheme 2.8).

2.3.2 CHIRAL AUXILIARIES

A wide variety of chiral auxiliaries are available to prepare amino acids (Chapter 23). The most popular are oxazolidinones, α-amino acids, aminoindanol (Chapter 24), imidazolidones, and pseudoephedrine.[7] Oxazolidinones **5** have found widespread usage as chiral auxiliaries. A wide range of reactions is available and well documented[7] and includes the following: aldol; alkylations;

α-substitution with a heteroatom such as halogenation, aminations, hydroxylations, and sulfenylations; Diels-Alder cycloadditions; and conjugate additions.

A method has been developed for a one-pot procedure from an amino acid to an oxazolidinone (Scheme 2.3). The sequence can be run at scale.[11]

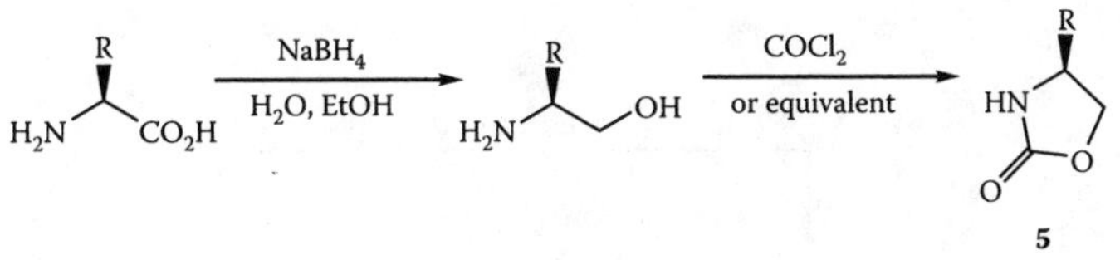

SCHEME 2.3

In addition to the availability issues of the oxazolidinone unit, there has been some reluctance to scale up the reactions because formation of the N-acyl derivatives usually uses n-butyl lithium as a base. This problem can be circumvented by the use of an acid chloride or anhydride (symmetric or mixed) with triethylamine as the base in the presence of a catalytic amount of DMAP (4-dimethylamino pyridine). The reaction is general and provides good yields even with α,β-unsaturated acid derivatives (Scheme 2.4).[12]

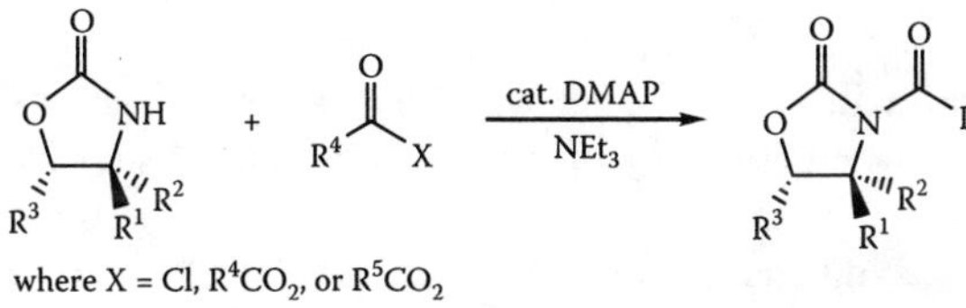

where X = Cl, R^4CO_2, or R^5CO_2

SCHEME 2.4

N-Acyloxazolidinones are, therefore, readily available. The use of this class of compounds for the preparation of unnatural amino acids has been well documented.[7] Di-tert-butyl azodicarboxylate (DBAD) reacts readily with the lithium enolates of N-acyloxazolidinones to provide hydrazides **6** in excellent yield and high diastereomeric ratio (Scheme 2.5); these adducts **6** can be converted to amino acids.[13,14]

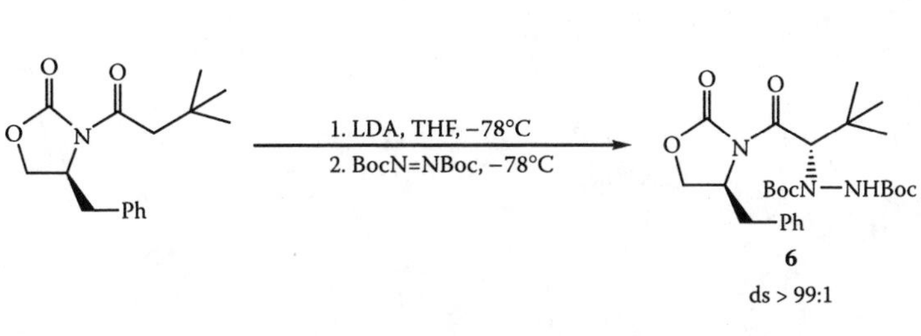

1. LDA, THF, −78°C
2. BocN=NBoc, −78°C

6

ds > 99:1

SCHEME 2.5

The electrophilic introduction of azide with chiral imide enolates has also been used to prepare α-amino acids with high diastereoselection (Scheme 2.6). The reaction can be performed with either the enolate directly,[15-19] or through a halo intermediate.[20] The resultant azide can be reduced to an amine.[21]

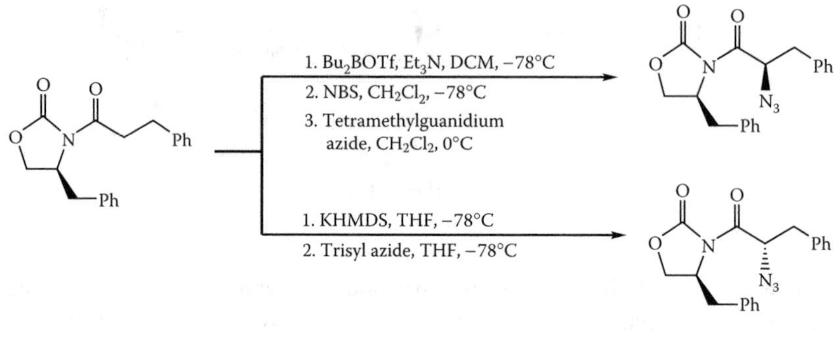

SCHEME 2.6

2.3.2.1 Vancomycin

As an example of amino acid synthesis, let us consider approaches to the constituents of the antibiotic vancomycin (**7**), where a variety of methods have been used. This section outlines some of the approaches that have used chiral auxiliaries and takes from a number of approaches rather than describing a complete synthetic sequence. One amino acid residue has been prepared by lactim chemistry, where an amino acid is used as the auxiliary (Scheme 2.7).[22]

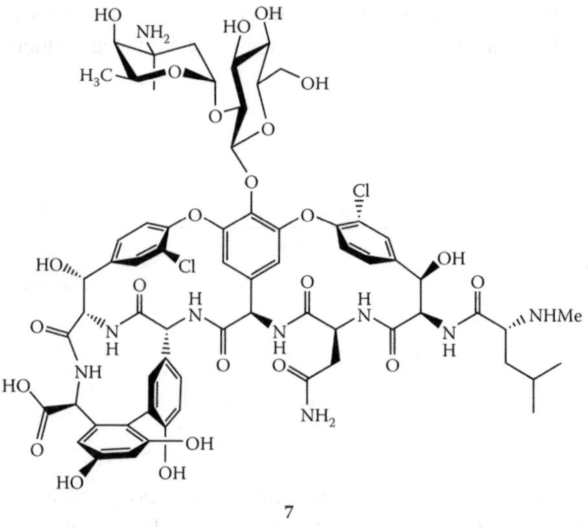

7

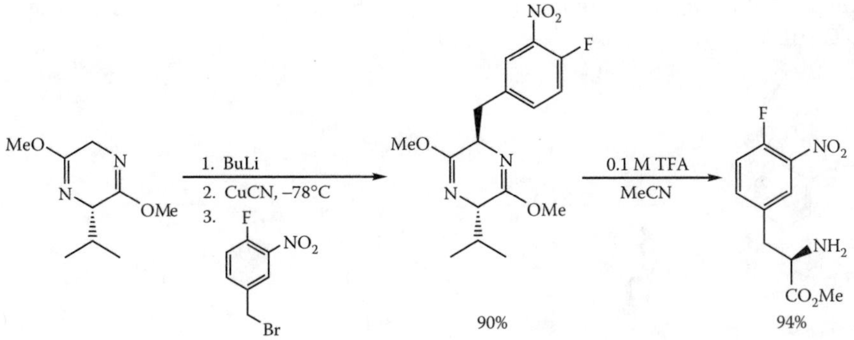

SCHEME 2.7

An important unit of the antibiotic vancomycin (**7**) can be synthesized using phenylglycinol as a chiral template (Scheme 2.8).[23,24]

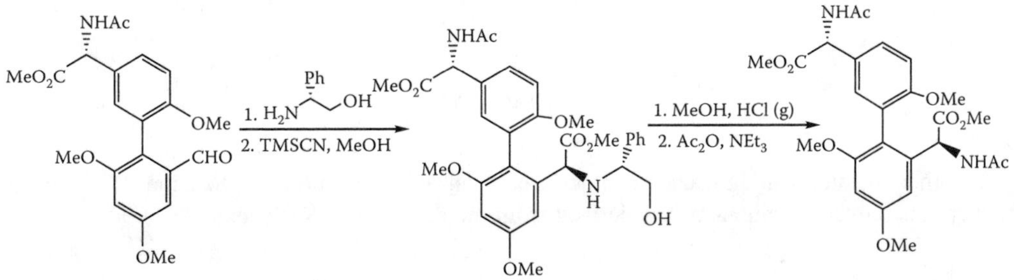

SCHEME 2.8

In another synthesis of the amino acid components, an asymmetric Strecker reaction with a 1,2-amino alcohol template was used where phenylglycinol was the nitrogen component (Scheme 2.9). Some epimerization occurred during the hydrolysis of the nitrile group.[25,26]

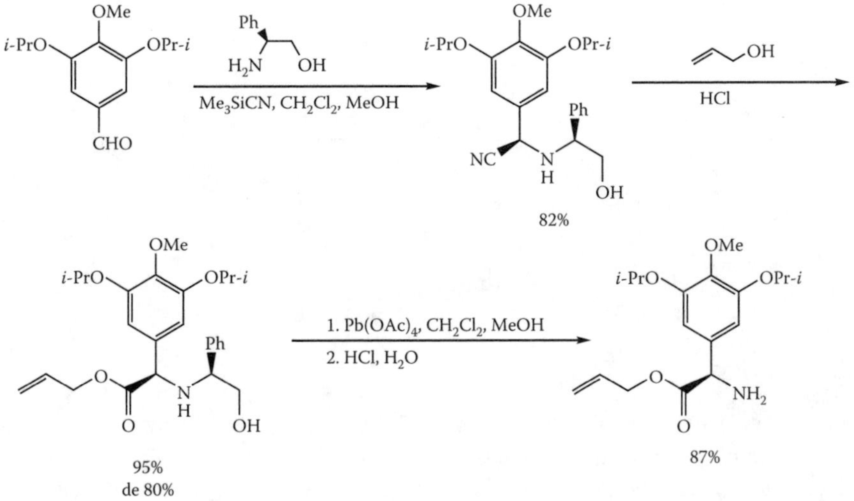

SCHEME 2.9

To circumvent the epimerization problem, oxazolidinone chemistry was used as an alternative (Scheme 2.10).[25]

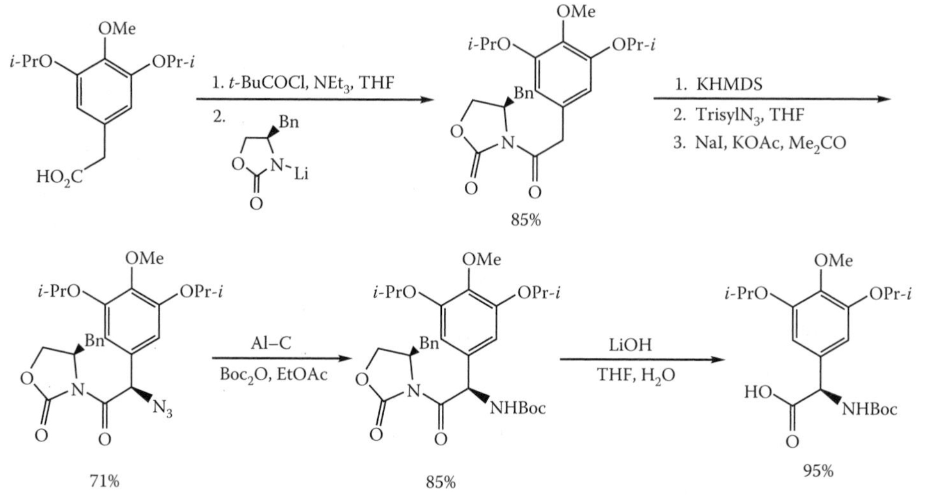

SCHEME 2.10

Another unnatural amino acids synthesis uses an azide reaction (*cf.* Scheme 2.6), whereas another vancomycin component was derived from the thiocyanate **8** (Scheme 2.11).[21,27,28]

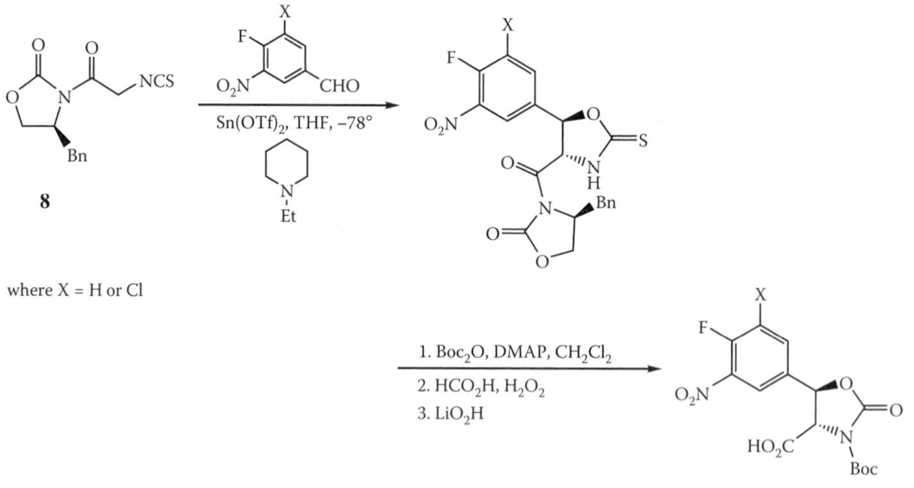

where X = H or Cl

SCHEME 2.11

2.3.3 RESOLUTIONS

There are occasions when a resolution method can be useful. On the chemical side, this approach usually comes into play when small amounts of material are required and alternative methodology is under development. However, if a dynamic kinetic resolution can be achieved, then the approach can be very cost effective, as illustrated by D-phenylglycine (Chapters 7 and 25). In contrast, biological approaches can be efficient if a dynamic resolution can be achieved or the starting materials are very cheap and readily available (*vide infra*).

Asymmetric hydrogenation methodology does not allow simple access to cyclic amino acid derivatives, at present. To obtain D-proline, a dynamic resolution with L-proline as the starting material can be performed. The presence of butyraldehyde allows racemization of the L-proline in solution, whereas the desired D-isomer is removed as a salt with D-tartaric acid (Scheme 2.12; see also Chapter 6).[29,30]

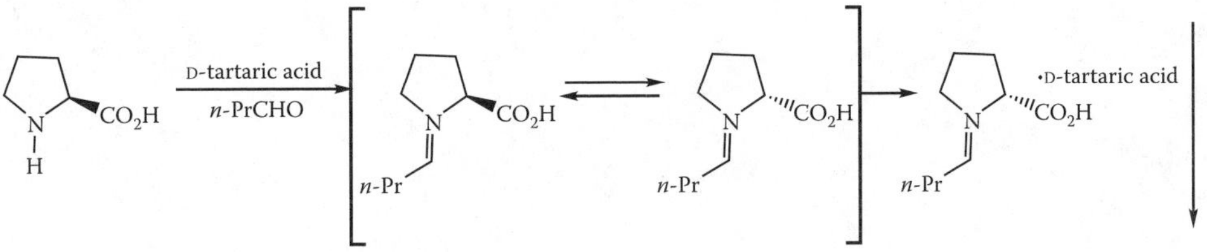

SCHEME 2.12

Another example is provided by D-*tert*-leucine (9) (Scheme 2.13), where an asymmetric approach cannot be used. This unnatural amino acid is prepared by a resolution method.[31,32]

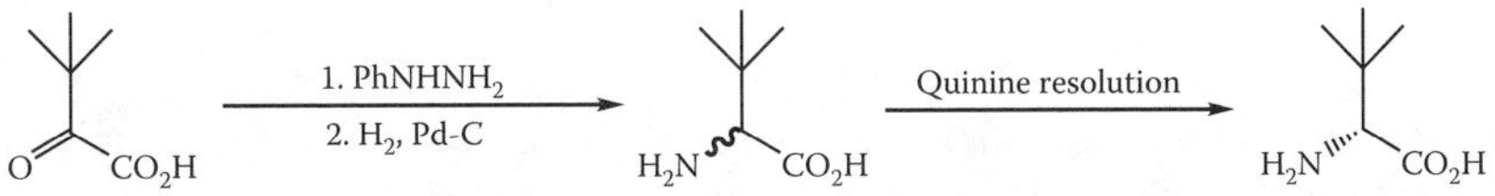

SCHEME 2.13

One resolution approach that is showing promise is the use of an enzyme, such as an amino acid oxidase (AAO) to convert one isomer of an amino acid to the corresponding achiral imine. A chemical reducing agent then returns the amino acid as a racemic mixture (Scheme 2.14). As the enzyme only acts on one enantiomer of the amino acid, the desired product builds up. After only seven cycles, the product ee is >99%.[33] The key success factor is to have the reducing agent compatible with the enzyme.[34,35] Rather than use the antipode of the desired product, the racemic mixture can be used as the starting material. The enzymes are available to prepare either enantiomeric series of amino acid.

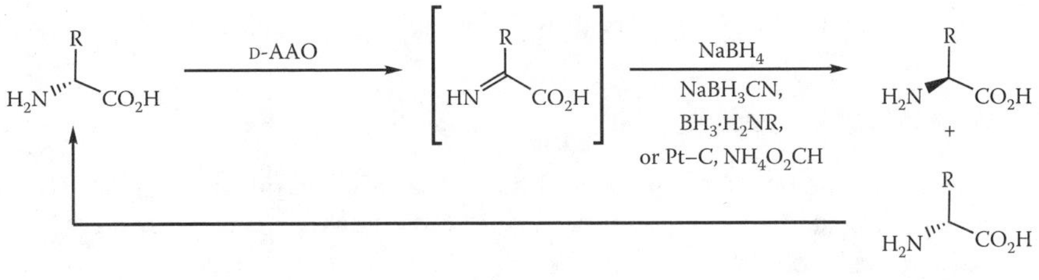

SCHEME 2.14

The approach is amenable to accessing β-substituted α-amino acids.[36] The methodology has culminated in a way to prepare all four possible isomers of β-aryl α-amino acids by a combination of asymmetric hydrogenation (Chapter 13) and the use of the deracemization process to invert the α-center (Scheme 2.15).[37]

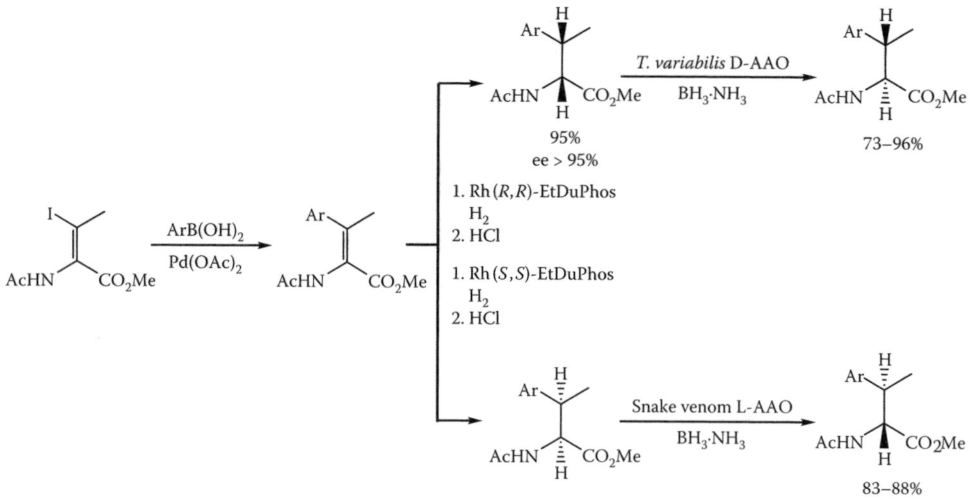

SCHEME 2.15

For other biological resolution methods see Section 2.5.2.3 and Chapter 19. Another deracemization method uses radical chemistry. This topic is discussed in detail in Chapter 27.

2.4 INTERMEDIATE-SCALE APPROACHES

Intermediate-scale reactions take the scale of reactions above laboratory and kilo lab scale to pilot plant. It is usually at this stage that cost becomes a factor, although the speed by which a material can be obtained may still be important. In certain cases, the intermediate may only be needed at this multi-kilo level even when the pharmaceutical end-product is commercialized.

Although the trend is changing, the use of chiral auxiliaries, such as oxazolidinones, is often thought of as too expensive. There are examples where this approach has been taken to large scale (Chapters 23 and 24). The use of a cheap chiral template can also have advantages (Chapter 25).

Asymmetric hydrogenations are often used at this scale because the methodology accommodates a wide variety of substrates. In addition, the technique has been scaled. In-depth discussions on the synthesis of amino acids by asymmetric hydrogenations can be found in Chapters 12–15. Resolution techniques, including SMB, are also viable options.

2.5 LARGE-SCALE METHODS

At larger scale, cost factors become important. It is no longer economical to use chiral auxiliaries, especially under good manufacturing practice (GMP) conditions, because they are not atom economical and so are costly for what they finally contribute to the product in terms of atoms. As volumes increase, research costs can be amortized over these larger amounts, and often routes are developed to specific target amino acids. The major approaches to amino acids at production scale are resolution, as with D-phenylglycine (see Chapters 7 and 25), asymmetric hydrogenation (Chapters 12–14), chemical sidechain manipulations, and biological methods (see also Chapter 3).

2.5.1 CHEMICAL METHODS

2.5.1.1 Asymmetric Hydrogenations

Today, a large number of catalysts and ligands have been used to prepare amino acids by asymmetric hydrogenation. For fine chemical companies and, indeed, some pharmaceutical companies, this plethora of ligands has arisen not because of chemical needs and limitations but because of freedom to operate issues. That is not to say that any catalyst can provide any desired amino acids; ligands systems do have limitations and so some complement others. Therefore some ligand systems do have advantages in certain applications. The method has become mature.[38] In the introduction to this chapter it was noted that there is no general asymmetric chemical solution for the synthesis of α-amino acids at scale. However, there are specific examples and solutions, as illustrated by the Monsanto L-Dopa process that is still operated today (*vide infra*).[39–41]

Whichever catalyst system is used to prepare an α-amino acid derivative at scale, the cost of the ligand can play a major economical role, often more than the metal. The cost of the catalyst can be offset by large substrate-to-catalysts ratios that can be improved by recycles and fast reactions. However, because metal hydride species are invariably involved in the catalytic cycle, recycles usually mean reuse in a short time span. The conclusion is that expensive ligands must be extremely good at the desired reduction. This is particularly true with amino acid derivatives because almost all are crystalline and offer the possibility of enantioenrichment during the purification process. Thus, high ee's may not be required in the reduction itself. In many cases, the synthesis of the substrate is the difficult part of the synthesis. This problem has been highlighted in the synthesis of β-amino acids (see Section 2.6).

The first catalyst to be used for the synthesis of α-amino acids was based on the *P*-chiral ligand, dipamp (Chapter 12). Knowles and Horner independently developed the ligand system, but Monsanto has been successful in the scale up to produce L-Dopa commercially.[42,43] The chemistry of the Rh(dipamp) system has been reviewed.[40,41,44–47] Although the catalyst system is more than 25 years old and in certain cases has been superseded by other systems that can provide higher ee's with certain substrates, it is still extremely useful. The turnover numbers are often high, the scope and limitations of the system are well defined, and the recycle of the catalyst has been practiced at scale (as part of the Dopa process).[40,48] The mechanism has been elucidated; it was found that the major product is derived from the minor, but less stable, organometallic intermediate and results from the latter's higher reactivity to hydrogen.[49]

The substrate requirements have been rationalized for Rh(dipamp) reductions. Thus, for an enamine they can be divided into four specific regions (Figure 2.1).[47] This has allowed the system to be used with confidence to prepare a wide variety of unnatural amino acids.[39]

The hydrogen region will only tolerate hydrogen and still give a reasonable rate. When other groups fall into this region, the stereospecificity decreases (*vide infra*). The aryl region will tolerate

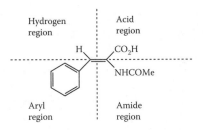

FIGURE 2.1 Regions of the enamide substrate for asymmetric hydrogenations.

a wide range of groups including heterocycles, alkyl groups, and hydrogen. In a similar manner, the amide region will allow a wide variety of amides to be used not only in terms of size but also in electronic features so that protected amino acids can be synthesized directly. The acid region is also forgiving as long as it contains an electronic withdrawing group.[48] Fortunately, the general method for the preparation of the enamides is stereoselective to the Z-isomer when an aryl aldehyde is used (Scheme 2.16).[50] Other approaches to the alkene are also available (cf. Scheme 2.15).

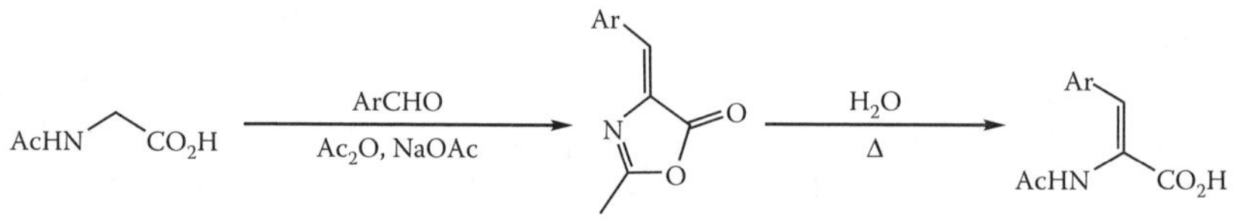

SCHEME 2.16

Although the chemistry of Knowles' catalyst system has been highlighted here, many other systems are now available. Some have distinct advantages, such as DuPhos and the ability to reduce enamides with two β-substituents with high enantioselectivity (Chapter 13; see also Scheme 2.15), whereas the monodentate ligands (Chapter 14) have distinct cost advantages.

2.5.1.2   Chiral Pool Approaches

Unnatural amino acids are available from more readily available amino acids. A wide variety of reactions are available in the literature. An example is provided by the synthesis of L-homoserine (**10**) from methionine (Scheme 2.17).[51] Of course, the problems associated with the formation of one equivalent of dimethyl sulfide have to be overcome.

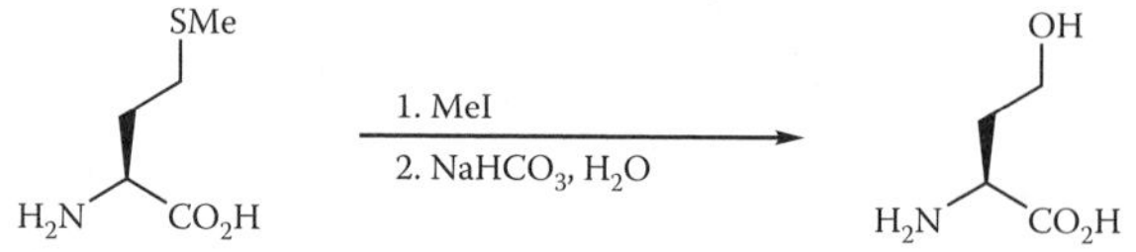

SCHEME 2.17

A wide variety of simple transformations on chiral pool materials can lead to unnatural amino acid derivatives. This is illustrated by the Pictet-Spengler reaction of L-phenylalanine, followed by amide formation and reduction of the aromatic ring (Scheme 2.18).[52] The resultant amide (**11**) is an intermediate in a number of commercial human immunodeficiency virus (HIV) protease inhibitors.

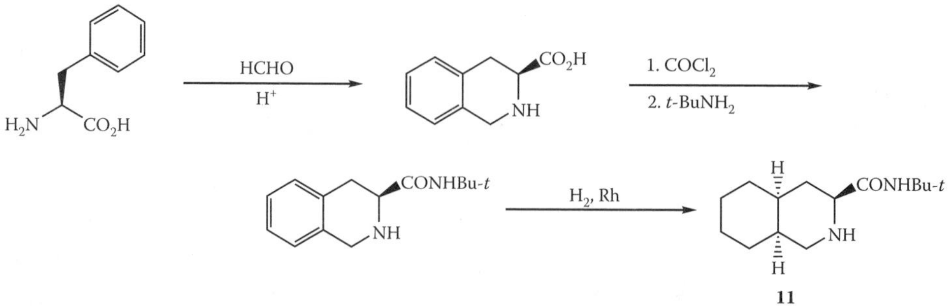

SCHEME 2.18

2.5.2 BIOLOGICAL APPROACHES

The fermentation methods used to prepare L-phenylalanine, threonine, lysine, and cysteine are discussed in detail in Chapter 3. The adaptation of these methods to prepare unnatural amino acids, such as the use of transaminases, is also discussed in that chapter. One of the large-scale amino acids, L-glumatic acid, which is often sold as its monosodium salt, is not covered because its preparation by fermentation is long established.[3]

In addition to resolution approaches, there are three main methods to prepare amino acids by biological methods: addition of ammonia to an unsaturated carboxylic acid; the conversion of an α-keto acid to an amino acid by transamination from another amino acid, and the reductive amination of an α-keto acid. These approaches are discussed in Chapter 19 and will not be discussed here to avoid duplication. The use of a lyase to prepare L-aspartic acid is included in this chapter as is the use of decarboxylases to access D-glutamic acid.

2.5.2.1 L-Aspartic Acid

L-Aspartic acid (**12**) is an industrially important, large volume, chiral compound. The primary use of aspartic acid in the fine chemical arena is in the production of aspartame, a high potency sweetener (Chapter 31). Other uses of L-aspartate include dietary supplements, pharmaceuticals, production of alanine (by decarboxylation), antibacterial agents, and lubricating compounds. There have been a number of reviews on L-aspartate production.[53-56]

Historically, L-aspartic acid was produced by hydrolysis of asparagine, by isolation from protein hydrolysates, or by the resolution of chemically synthesized D,L-aspartate. With the discovery of aspartase (L-aspartate ammonia lyase, EC 4.3.1.1),[57] fermentation routes to L-aspartic acid quickly superseded the initial chemical methods. These processes are far more cost effective than the fermentation routes, and aspartate is now made exclusively by enzymatic methods that use variations of the general approach outlined in Scheme 2.19.[53,57-65]

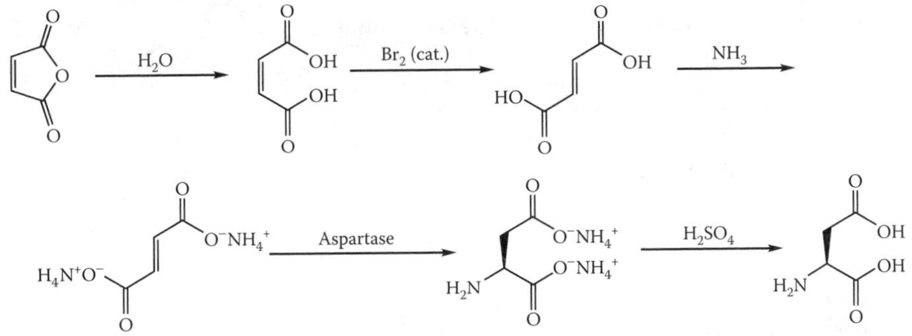

SCHEME 2.19

The basic enzymatic procedure usually involves immobilization of bacterial cells containing aspartase activity or of semi-purified enzyme, in either a fed batch or a continuous, packed-column process. The starting material is maleic anhydride, an inexpensive, high volume, bulk chemical.[56]

2.5.2.2 Decarboxylases

Amino acid decarboxylases can be used to catalyze the resolution of several amino acids, and for the most part their utility has been underestimated, especially with respect to the irreversible reaction equilibrium. The decarboxylases are ideally suited to large-scale industrial application as a result of their robust nature. D-Aspartate (**13**, n = 1) and D-glutamate (**13**, n = 2) can be made economically by use of the corresponding L-amino acids as starting material through an enzymatic racemization prior to the resolution with decarboxylase (Scheme 2.20).[66] The process for D-glutamate consists of racemization of L-glutamate at pH 8.5 with the cloned glutamate racemase enzyme from *Lactobacillus brevis*, expressed in *Escherichia coli* followed by adjustment of the pH to 4.2 to allow *E. coli* cells with elevated decarboxylase activity to decarboxylate the L-glutamate.[67] This reaction can be greatly improved by combination and overexpression of the glutamate racemase and glutamate decarboxylase in one bacterial strain. The combination of both activities is possible because glutamate decarboxylase is inactive at pH 8.5 during the racemization step, but it regains full activity when the pH is 4.2. Conversely, glutamate racemase is active at pH 8.5, but it is completely inhibited at pH 4.2.

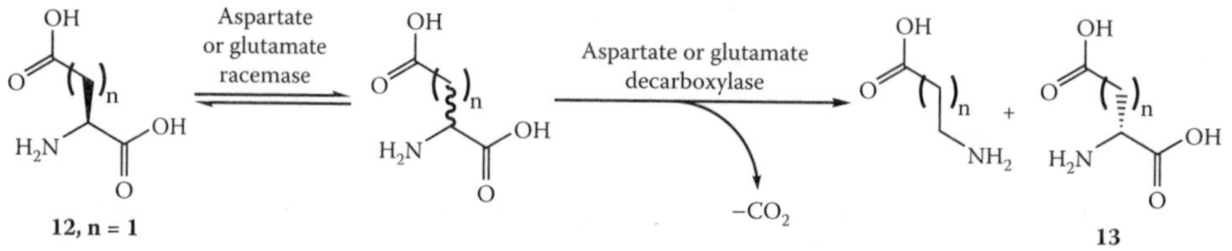

SCHEME 2.20

2.5.2.3 Biocatalyzed Resolution

2.5.2.3.1 Amino Acylases

The hydrolysis of an *N*-acylated amino acid by an enzyme provides a resolution method to amino acids. Because the starting materials are readily available in the racemic series by the Schotten-Baumann reaction, the method can be cost effective (Scheme 2.21).[68–71] The L-amino acid product can be separated by crystallization, whereas the D-amino acid, which is still *N*-acylated, can be recycled by being resubjected to the Schotten-Baumann conditions used for the next batch. Tanabe has developed a process with an immobilized enzyme,[72,73] whereas Degussa uses the method in a membrane reactor.[69,74] The process is used to make L-methionine.

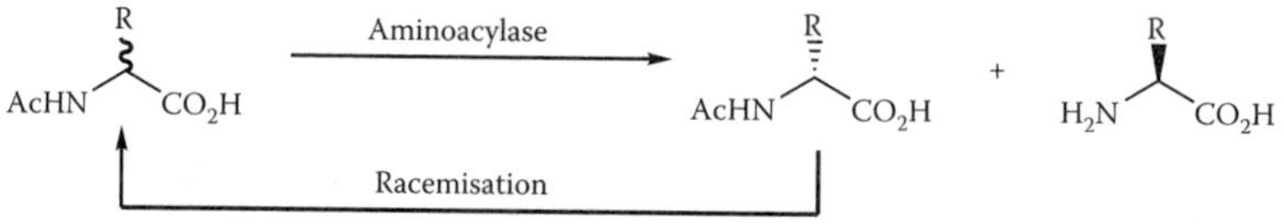

SCHEME 2.21

2.5.2.3.2 Amidases

DSM has developed a general, industrial-scale process for the production of either L- or D-amino acids through the hydrolysis of the amide (Scheme 2.22).[75–78] The product amino acid and untouched amide are easily separated. The amide can be recycled. The resolution method has been extended to α,α-disubstituted of which L-methyl-dopa (**14**) is an example.[79]

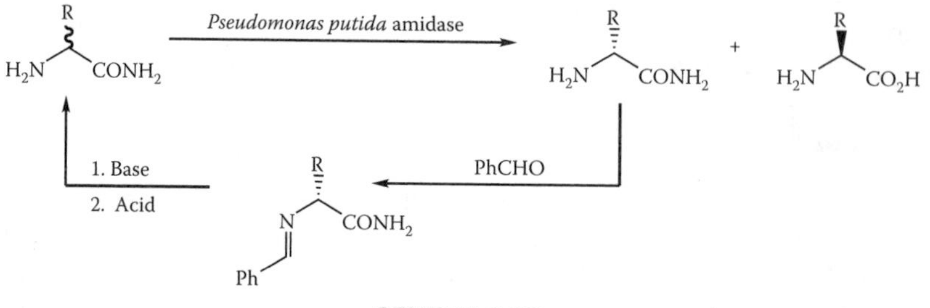

SCHEME 2.22

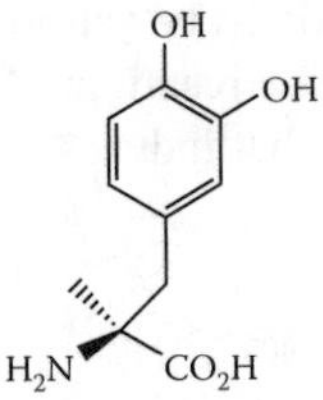

14

2.5.2.3.3 Hydantoins

Hydantoins (HD) are readily epimerized, which allows access to either D- or L-amino acids.[3,80] The coupling of enzymatic steps allows equilibria to be pushed in the desired direction. Degussa and DSM both use the approach, although they have slightly different variations.[69,81,82] Scheme 2.23 shows the DSM route to D-*p*-hydroxyphenylglycine.

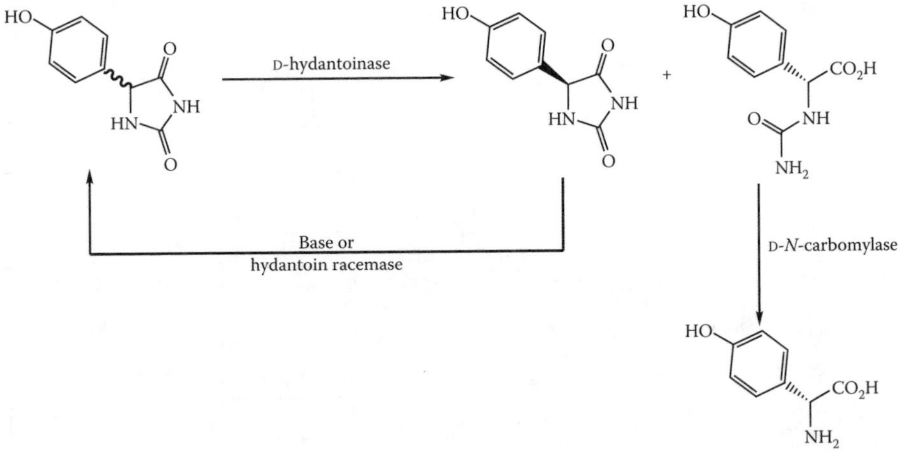

SCHEME 2.23

2.6 β-AMINO ACIDS

β-Amino acids (**2**) are now starting to enjoy use of building blocks within drug development candidates. There are a large number of approaches to this class of compounds, but most are amenable only to laboratory-scale synthesis. These methods have been reviewed.[83–86]

One example of the problems associated with finding a scaleable approach to β-amino acids is highlighted by methyl (*S*)-3-amino-3-(3′-pyridyl)propionate (**15**). Despite trying many asymmetric approaches, a resolution of the *N*-Boc protected amino acid with (−)-ephedrine was found to be the best method.[87]

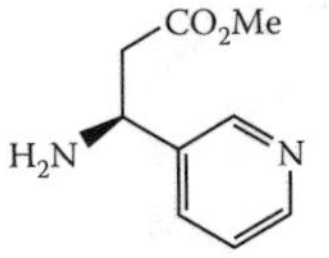

15

The Michael addition of a nitrogen nucleophile to an α,β-unsaturated system has been advocated as a method to prepare β-amino acid derivatives (Scheme 2.24).[88–91] However, the method is not atom economical with regard to nitrogen delivery and low temperatures are also required, making the approach a poor candidate for large-scale reactions. An additional problem can be encountered during the removal of the template group that induces the asymmetry.[92]

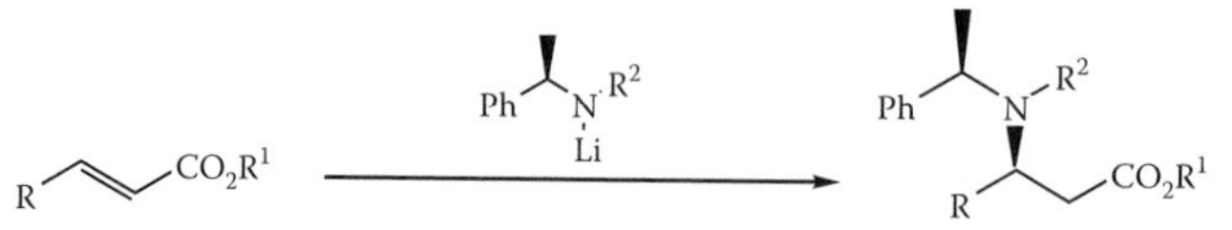

SCHEME 2.24

This template removal was encountered during studies to prepare the $\alpha_v\beta_3$ integrin antagonist (**16**). The route that emerged used phenylglycinol as a template in a Reformatsky approach (Scheme 2.25). Unfortunately, lead tetraacetate had to be used to remove this template, so there are still problems to be overcome in this area.[93]

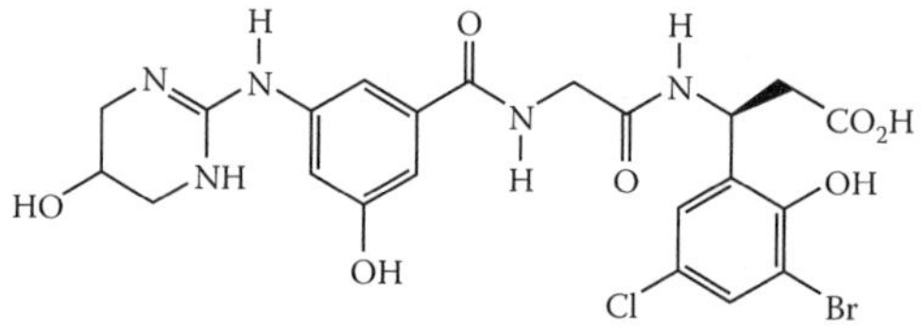

16

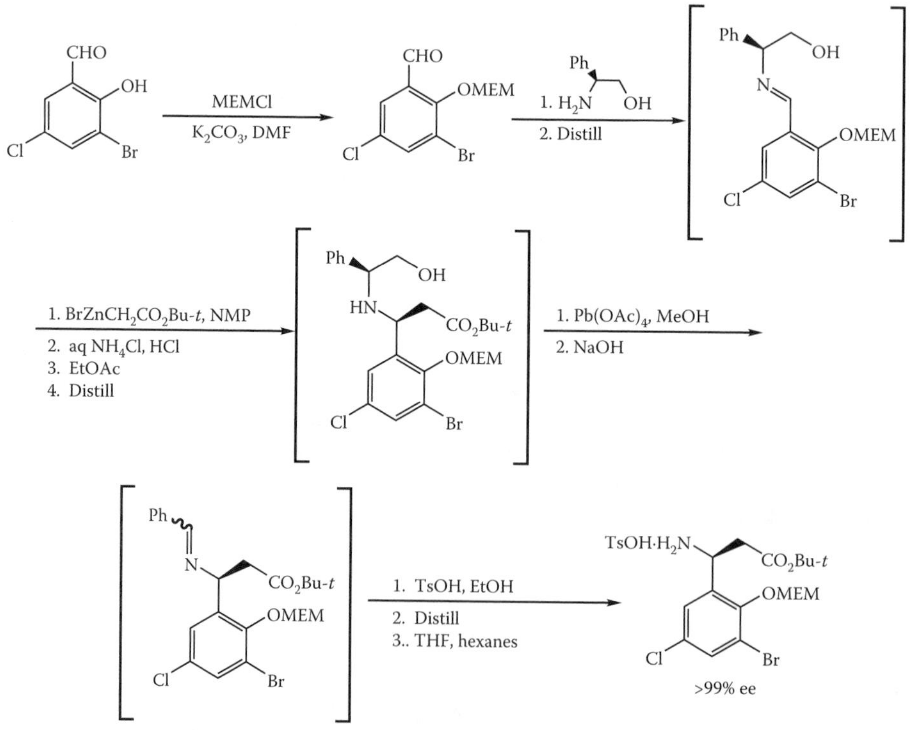

SCHEME 2.25

One method that shows great promise of being general and scaleable is an asymmetric hydrogenation approach. Although many attempts have been made to reduce enamines, such as **17**, to β-amino acid derivatives, the problems associated with obtaining just one isomer of **17** thwarted high ee's. It is now possible to prepare just one isomer of the enamine.[94] In addition, the MonoPhos family of ligands have been shown to provide good ee's with either isomer of **17** (Scheme 2.26).[95] The ability to prepare ligand libraries and screen them for asymmetric hydrogenations, such as to prepare β-amino acids,[96–98] makes this approach a powerful one. For a full discussion on monodentate ligands see Chapter 14.

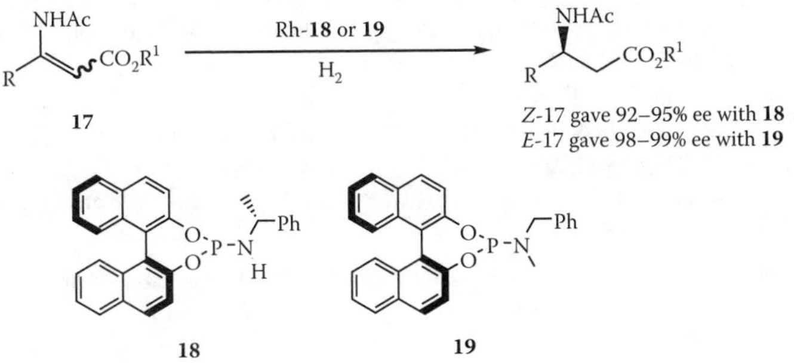

SCHEME 2.26

For β-amino acids, biological resolutions have been developed, but because the stereogenic center is not easily epimerized, as in α-amino acids, this approach will probably have to wait for a large value or volume product before it is exploited.

2.7 SUMMARY

As with another class of compounds, the scale of synthesis and time required at the research stage before product can be made influence which method is finally used. At small scale, a plethora of methods exist to prepare amino acids, in addition to isolation of the common ones from natural sources. The majority of these small-scale reactions rely on the use of a chiral auxiliary or template. At larger scale, asymmetric hydrogenation and biocatalytic processes come into their own. For the amino acids approaching commodity chemical scales, biological approaches, either as biocatalytic or total fermentation, provide the most cost-efficient processes.

REFERENCES

1. Collins, A. N., Sheldrake, G. N., Crosby, J., Eds. *Chirality in Industry: The Commercial Manufacture and Applications of Optically Active Compounds,* Wiley: Chichester, 1992.
2. Collins, A. N., Sheldrake, G. N., Crosby, J., Eds. *Chirality in Industry II: Developments in the Commercial Manufacture and Applications of Optically Active Compounds,* Wiley: Chichester, 1997.
3. Breuer, M., Ditrich, K., Habicher, T., Hauer, B., Keßeler, M., Stürmer, R., Zelinski, T. *Angew. Chem., Int. Ed.* 2004, 43, 788.
4. Schoemaker, H. E., Mink, D., Wubbolts, M. G. *Science* 2003, 299, 1894.
5. Williams, R. M. *Synthesis of Optically Active α-Amino Acids,* Pergamon Press: Oxford, 1989.
6. Ager, D. J., East, M. B. *Asymmetric Synthetic Methodology,* CRC Press: Boca Raton, 1995.
7. Ager, D. J., Prakash, I., Schaad, D. R. *Chem. Rev.* 1996, 96, 835.
8. Ager, D. J., Prakash, I., Schaad, D. R. *Aldrichimica Acta* 1997, 30, 3.
9. Cox, G. G., Harwood, L. M. *Tetrahedron: Asymmetry* 1994, 5, 1669.
10. Iyer, M. S., Yan, C., Kowalczyk, R., Shone, R., D., A., R., S. D. *Synth. Commun.* 1997, 27, 4355.
11. Schaad, D. R. Unpublished results.
12. Ager, D. J., Allen, D. R., Schaad, D. R. *Synthesis* 1996, 1283.
13. Evans, D. A., Britton, T. C., Dorow, R. L., Dellaria, J. F. *J. Am. Chem. Soc.* 1986, 108, 6395.

14. Trimble, L. A., Vederas, C. J. *J. Am. Chem. Soc.* 1986, 108, 6397.
15. Evans, D. A., Clark, J. S., Metternich, R., Novack, V. J., Sheppard, G. S. *J. Am. Chem. Soc.* 1990, 112, 866.
16. Evans, D. A., Britton, T. C. *J. Am. Chem. Soc.* 1987, 109, 6881.
17. Doyle, M. P., Dorow, R. L., Terpstra, J. W., Rodenhouse, R. A. *J. Org. Chem.* 1985, 50, 1663.
18. Evans, D. A., Lundy, K. M. *J. Am. Chem. Soc.* 1992, 114, 1495.
19. Evans, D. A., Britton, T. C., Ellman, J. A., Dorow, R. L. *J. Am. Chem. Soc.* 1990, 112, 4011.
20. Evans, D. A., Ellman, J. A., Dorow, R. L. *Tetrahedron Lett.* 1987, 28, 1123.
21. Evans, D. A., Everard, D. A., Rychnovsky, S. D., Fruh, T., Whittington, W. G., DeVries, K. M. *Tetrahedron Lett.* 1992, 33, 1189.
22. Bois-Choussy, M., Neuville, L., Beugelmans, R., Zhu, J. *J. Org. Chem.* 1996, 61, 9309.
23. Rao, A. V. R., Chakraborty, T. K., Joshi, S. P. *Tetrahedron Lett.* 1992, 33, 4045.
24. Chakraborty, T. K., Reddy, G. V., Hussain, K. A. *Tetrahedron Lett.* 1991, 32, 7597.
25. Vergne, C., Bouillon, J.-P., Chastanet, J., Bois-Choussy, M., Zhu, J. *Tetrahedron: Asymmetry* 1998, 9, 3095.
26. Zhu, J., Bouillon, J.-P., Singh, G. P., Chastanet, J., Beugelmans, R. *Tetrahedron Lett.* 1995, 36, 7081.
27. Pearson, A. J., Chelliah, M. V., Bignan, G. C. *Synthesis* 1997, 536.
28. Evans, D. A., Wood, M. R., Trotter, B. W., Richardson, T. I., Barrow, J. C., Katz, J. L. *Angew. Chem., Int. Ed.* 1998, 37, 2700.
29. Shiraiwa, T., Shinjo, K., Kurokawa, H. *Chem. Lett.* 1989, 1413.
30. Shiraiwa, T., Shinjo, K., Kurokawa, H. *Bull. Chem. Soc. Jpn.* 1991, 64, 3251.
31. Miyazawa, T., Takashima, K., Mitsuda, Y., Yamada, T., Kuwota, S., Watanabe, H. *Bull. Chem. Soc. Jpn.* 1979, 52, 1539.
32. East, M. B. Unpublished results.
33. Turner, N. J. *TIBTECH* 2003, 21, 474.
34. Beard, T. M., Turner, N. J. *J. Chem. Soc. Chem. Commun.* 2002, 246.
35. Alexandre, F.-R., Pantaleone, D. P., Taylor, P. P., Fotheringham, I. G., Ager, D. J., Turner, N. J. *Tetrahedron Lett.* 2002, 43, 707.
36. Enright, A., Alexandre, F.-R., Roff, G., Fotheringham, I. G., Dawson, M. J., Turner, N. J. *J. Chem. Soc. Chem. Commun.* 2003, 2636.
37. Roff, G., Lloyd, R. C., Turner, N. J. *J. Am. Chem. Soc.* 2004, 126, 4098.
38. Blaser, H.-U., Schmidt, E. Introduction. In *Asymmetric Catalysis on Industrial Scale,* Blaser, H.-U., Schmidt, E. Eds., Wiley-VCH: Weinheim, 2004, p. 1.
39. Ager, D. J., Laneman, S. A. The Synthesis of Unnatural Amino Acids. In *Asymmetric Catalysis on Industrial Scale,* Blaser, H.-U., Schmidt, E. Eds., Wiley-VCH: Weinheim, 2004, p. 259.
40. Knowles, W. S. Asymmetric Hydrogenations—The Monsanto L-Dopa Process. In *Asymmetric Catalysis on Industrial Scale,* Blaser, H.-U., Schmidt, E. Eds., Wiley-VCH: Weinheim, 2004, p. 23.
41. Knowles, W. S. *Angew. Chem., Int. Ed.* 2002, 41, 1998.
42. Knowles, W. S., Sabacky, M. J. *J. Chem. Soc. Chem. Commun.* 1968, 1445.
43. Horner, L., Siegel, H., Buthe, H. *Angew. Chem., Int. Ed.* 1968, 7, 942.
44. Knowles, W. S. *Acc. Chem. Res.* 1983, 16, 106.
45. Koenig, K. E., Sabacky, M. J., Bachman, G. L., Christopfel, W. C., Barnstorff, H. D., Friedman, R. B., Knowles, W. S., Stults, B. R., Vineyard, B. D., Weinkauff, D. *J. Ann. N. Y. Acad. Sci.* 1980, 333, 16.
46. Koenig, K. E. In *Catalysis of Organic Reactions,* Kosak, J. R. Ed., Marcel Dekker: New York, 1984, p. 63.
47. Koenig, K. E. In *Asymmetric Synthesis,* Morrison, J. D. Ed., Academic: Orlando, 1985, Vol. 5, p. 71.
48. Vineyard, B. D., Knowles, W. S., Sabacky, M. J., Bachman, G. L., Weinkauff, D. J. *J. Am. Chem. Soc.* 1977, 99, 5946.
49. Landis, C. R., Halpern, J. *J. Am. Chem. Soc.* 1987, 109, 1746.
50. Herbst, R. M., Shemin, D. *Org. Synth.* 1943, Coll. Vol. II, 1.
51. Boyle, P. H., Davis, A. P., Dempsey, K. J., Hoskin, G. D. *Tetrahedron: Asymmetry* 1995, 6, 2819.
52. Allen, D. R., Jenkins, S., Klein, L., Erickson, R., Froen, D. U.S. Pat. 5,587,481, 1996.
53. Terasawa, M., Yukawa, H. *Bioproc. Technol.* 1993, 16, 37.
54. Chibata, I., Tosa, T., Sato, T. *Appl. Biochem. Biotechnol.* 1986, 13, 231.
55. Carlton, G. J. The Enzymatic Production of L-Aspartic Acid. In *Biocatalytic Production of Amino Acids and Derivatives,* Rozzell, J. D., Wagner, F. Eds., Hanser Publishers, 1992, p. 4.

56. Taylor, P. P. Synthesis of L-aspartic acid. In *Handbook of Chiral Chemicals,* Ager, D. J. Ed., Marcel Dekker: New York, 1999, p. 317.
57. Quastel, J. H., Wolf, B. *J. Biochem.* 1926, 20, 545.
58. Terasawa, M., Yukawa, H., Takayama, Y. *Proc. Biochem.* 1985, 124.
59. Plachy, J., Sikyta, B. *Folia Microbiol.* 1977, 22, 410.
60. Laanbroek, H. J., Lambers, J. T., Vos, W. M. d., Veldkamp, H. *Arch. Microbiol.* 1978, 117, 109.
61. Suzuki, Y., Yasui, T., Mino, Y., Abe, S. *Eur. J. Appl. Microbiol. Biotechnol.* 1980, 11, 23.
62. Shiio, I., Ozaki, H., Ujigawa-Takeda, K. *Agric. Biol. Chem.* 1982, 46, 101.
63. Takagi, T., Kisumi, M. *J. Bacteriol.* 1985, 161, 1.
64. Takagi, J. S., Ida, N., Tokushige, M., Sakamoto, H., Shimura, Y. *Nucleic Acids Res.* 1985, 13, 2063.
65. Costa-Ferreira, M., Duarte, J. C. *Biotechnol. Lett.* 1992, 14, 1025.
66. Fotheringham, I. G., Pantaleone, D. P. Unpublished results.
67. Yagasaki, M., Azuma, M., Ishino, S., Ozaki, A. *J. Ferm. Bioeng.* 1995, 79, 70.
68. Sonntag, N. O. V. *Chem. Rev.* 1953, 52, 237.
69. Gröger, H., Drauz, K. Methods for the Enantioselective Biocatalytic Production of L-Amino Acids on an Industrial Scale. In *Asymmetric Catalysis on Industrial Scale,* Blaser, H.-U., Schmidt, E. Eds., Wiley-VCH: Weinheim, 2004, p. 131.
70. Bommarius, A. S. In *Enzyme Catalysis in Organic Synthesis,* Drauz, K. Ed., Wiley-VCH: Weinheim, 2002, Vol. 2, 2nd Ed., Chapter 12.3.
71. Bommarius, A. S., Drauz, K., Klenk, H., Wandrey, C. *Ann. N. Y. Acad. Sci.* 1992, 672, 126.
72. Chibata, I., Tosa, T., Sato, T., Mori, T. *Meth. Enzymol.* 1976, 746.
73. Tosa, T., Mori, T., Fuse, N., Chibata, I. *Agric. Biol. Chem.* 1969, 1047.
74. Wandrey, C., Flashel, E. *Adv. Biochem. Eng.* 1979, 147.
75. Kamphius, J., Boesten, W. H. J., Broxterman, Q. B., Hermes, H. F. M., van Balken, J. A. M., Meijer, E. M., Schoemaker, H. E. *Adv. Biochem. Eng. Biotechnol.* 1991, 42, 133.
76. Kamphius, J., Boestein, W. H. J., Kaptein, B., Hermes, H. F. M., Sonke, T., Broxterman, Q. B., van der Tweel, W. J. J., Schoemaker, H. E. In *Chirality in Industry,* Collins, A. N., Sheldrake, G. N., Crosby, J. Eds., Wiley: Chichester, 1992, p. 187.
77. Rutjes, F. P. J. T., Schoemaker, H. E. *Tetrahedron Lett.* 1997, 38, 677.
78. Schoemaker, H. E., Boesten, W. H. J., Broxterman, Q. B., Roos, E. C., Kaptein, B., van den Tweel, W. J. J., Kamphius, J., Meijer, E. M. *Chimica* 1997, 51, 308.
79. Kamphius, J., Hermes, H. F. M., van Balken, J. A. M., Schoemaker, H. E., Boesten, W. H. J., Meijer, E. M. In *Amino Acids: Chemistry, Biology, Medicine,* Lubec, G., Rosenthal, G. A. Eds., ESCOM Science: Leiden, 1990, p. 119.
80. Wegman, M. A., Janssen, M. H. A., van Rantwijk, F., Sheldon, R. A. *Adv. Synth. Catal.* 2001, 343, 559.
81. Syldatk, C., Müller, R., Siemann, M., Wagner, F. In *Biocatalytic Production of Amino Acids and Derivatives,* Rozzell, D., Wagner, F. Eds., Hanser: Munich, 1992, p. 75.
82. May, O., Nguyen, P. T., Arnold, F. H. *Nat. Biotechnol.* 2000, 18, 317.
83. Palomo, C., Aizpurua, J. M., Ganboa, I., Oiarbide, M. *Synlett* 2001, 1813.
84. Liu, M., Sibi, M. *Tetrahedron* 2002, 58, 7991.
85. Seward, N. *Angew. Chem., Int. Ed.* 2003, 42, 5794.
86. Ma, J.-A. *Angew. Chem., Int. Ed.* 2003, 42, 4290.
87. Boesch, H., Cesco-Cancian, S., Hecker, L. R., Hoekstra, W. J., Justus, M., Maryanoff, C. A., Scott, L., Shah, R. D., Solms, G., Sorgi, K. L., Stefanick, S. M., Thurnheer, U., Villani Jr., F. J., Walker, D. G. *Org. Proc. Res. Dev.* 2001, 5, 23.
88. Davies, S. G., Ichihara, O. *Tetrahedron: Asymmetry* 1996, 7, 1919.
89. Davies, S. G., Ichihara, O. *Tetrahedron: Asymmetry* 1991, 2, 183.
90. Bunnage, M. E., Burke, A. J., Davies, S. G., Goodwin, C. J. *Tetrahedron: Asymmetry* 1995, 6, 165.
91. Davies, S. G., Dixon, D. J. *J. Chem. Soc., Perkin Trans. I* 1998, 2629.
92. Bull, S. D., Davies, S. G., Delgrado-Ballester, S., Kelley, P. M., Kotchie, L. J., Gianotti, M., Laderas, M., Smith, A. D. *J. Chem. Soc., Perkin Trans. I* 2001, 3112.
93. Clark, J. D., Weisenburger, G. A., Anderson, D. K., Colson, P.-J., Edney, A. D., Gallagher, D. J., Klein, H. P., Knable, C. M., Lantz, M. K., Moore, C. M. V., Murphy, J. B., Rogers, T. E., Ruminski, P. G., Shah, A. S., Storer, N., Wise, B. E. *Org. Proc. Res. Dev.* 2004, 8, 51.
94. You, J., Drexler, H.-J., Zhang, S., Fischer, C., Heller, D. *Angew. Chem., Int. Ed.* 2003, 42, 913.

95. Peña, D., Minnaard, A. J., de Vries, J. G., Feringa, B. L. *J. Am. Chem. Soc.* 2002, 124, 14552.
96. Peña, D., Minnaard, A. J., Boogers, J. A. F., de Vries, A. H. M., de Vries, J. G., Feringa, B. L. *Org. Biomol. Chem.* 2003, 1, 1087.
97. de Vries, J. G., de Vries, A. H. M. *Eur. J. Org. Chem.* 2003, 799.
98. Lefort, L., Boogers, J. A. F., de Vries, A. H. M., de Vries, J. G. *Org. Lett.* 2004, 6, 1733.

3 Microbial Pathway Engineering for Amino Acid Manufacture

Ian Fotheringham and *Paul P. Taylor*

CONTENTS

3.1 INTRODUCTION

A number of amino acids are produced annually in large tonnage for industrial applications. In particular, L-glutamic acid (mainly as monosodium glutamate), L-phenylalanine, L-aspartic acid, and glycine are produced as human food additives and L-lysine, L-threonine, L-methionine, and L-tryptophan are manufactured as animal feed additives.[1-4] These and other amino acids are also used as pharmaceutical intermediates and in a variety of nutritional and health care industries. Although a number of amino acids are produced by the extraction of natural products such as hair, feathers, and plant hydrolysates, in several cases bacteria are used to produce amino acids through large-scale fermentation, particularly when high enantiomeric purity is required.

The development of microbial strains for fermentative production of amino acids initially relied heavily on classical methods of random mutation of the DNA of the organism using chemicals or ultraviolet irradiation, followed by screening for overproducing variants that showed resistance to toxic amino acid analogues. In this way commercially viable amino acid overproducing strains were obtained. However, during the 1980s this approach was complemented by the availability of molecular biological methods and a rapid increase in the understanding of the molecular genetics and enzymology of amino acid production, in particular bacterial species. The availability of plasmid vector systems, transformation protocols, and general recombinant DNA methodology enabled biochemical pathways to be altered in much more precise ways to further enhance amino acid titer and production efficiency in organisms such as *Escherichia coli* and the *Coryneform* bacteria, *Corynebacterium* and *Brevibacterium,* the most prominent amino acid–producing bacteria. During the 1990s the further advancement of gene manipulation protocols including increased use of the polymerase chain reaction (PCR)[5] and more powerful mutation and screening regimens enabled additional, more ambitious, and broad-reaching applications of microbial pathway engineering to be undertaken. These have resulted in the engineering of microbial strains to produce unnatural L- or D-amino acids in addition to the proteinogenic L-amino acids previously produced in this way.

This chapter focuses on the most significant aspects of biochemical pathway engineering such as the identification of rate-limiting enzymatic steps and the elimination of allosteric feedback inhibition of key enzymes. Emphasis will be placed on the shikimate and phenylalanine pathways of *E. coli,* which are among the best characterized biochemically and genetically. These pathways have been the subject of some of the most widespread efforts in amino acid production as a result of the application of L-phenylalanine in the sweetener aspartame (Chapter 31). The chapter also addresses aspects of metabolic engineering that have been successfully applied in the *Coryneform* organisms through increased understanding of the molecular genetics of those organisms and their importance in amino acid production. Additionally, it will cover the increasing potential to adapt existing biochemical pathways or introduce new combinations of enzymes, expressed from cloned heterologous genes, in organisms such as *E. coli,* so as to produce nonproteinogenic or unnatural amino acids.

3.2 BACTERIAL PRODUCTION OF L-PHENYLALANINE

Efforts to develop L-phenylalanine (**1**) overproducing organisms have been vigorously pursued by many companies including The NutraSweet Company, Ajinomoto, Kyowa Hakko Kogyo, and others. The focus has centered on bacterial strains that have previously demonstrated the ability to overproduce other amino acids. Such organisms include principally the *coryneform* bacteria, *Brevibacterium flavum*[6] and *Corynebacterium glutamicum*[7,8] used in L-glutamic acid production. In addition, *E. coli*[9] has been extensively studied in L-phenylalanine manufacture as a result of the detailed characterization of the molecular genetics and biochemistry of its aromatic amino acid pathways and its amenability to recombinant DNA methodology. The biochemical pathway, which results in the synthesis of L-phenylalanine from chorismate (**2**), is identical in each of these organisms (Figure 3.1), along with the common aromatic pathway to chorismate (Figure 3.2). In

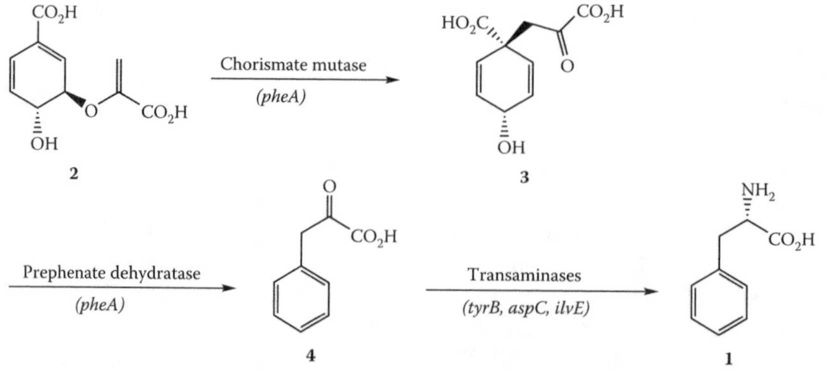

FIGURE 3.1 The biosynthetic pathway from chorismate to L-phenylalanine in *Escherichia coli* K12. The mnemonic of the genes involved are shown in parentheses below the enzymes responsible for each step. Compound **1** is L-phenylalanine, **2** is chorisimic acid, **3** is prephenic acid, and **4** is phenylpyruvic acid.

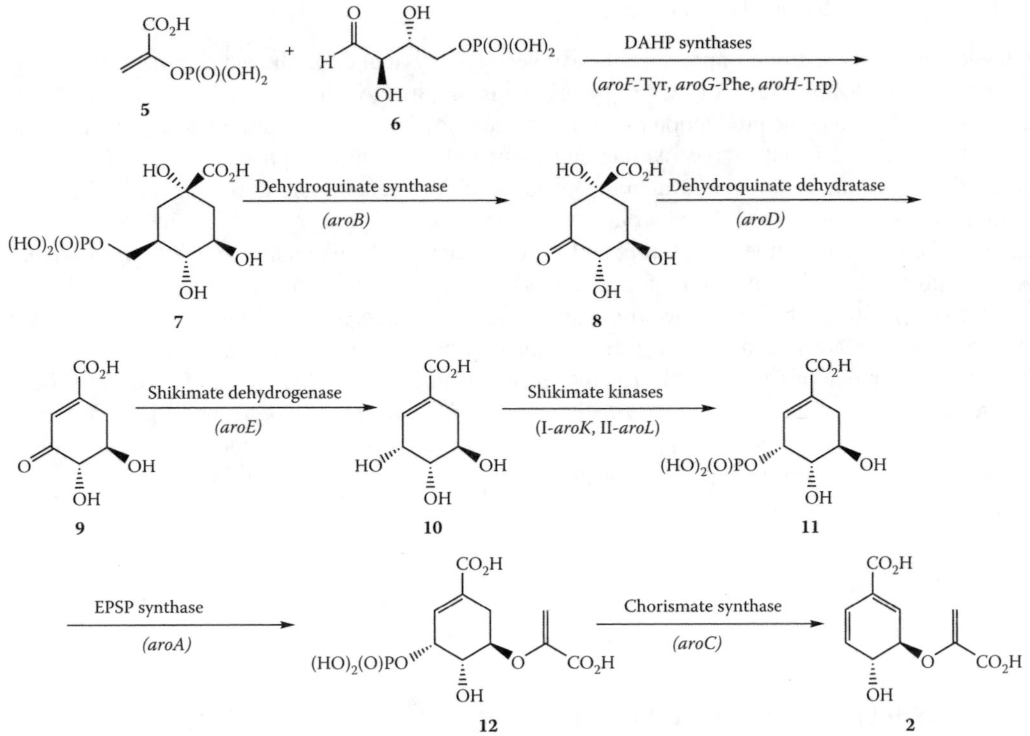

FIGURE 3.2 The common aromatic pathway to chorismate in *Escherichia coli* K12, where **5** is phosphoe-nolpyruvate, **6** is erythrose 4-phosphate, **7** is 3-deoxy-D-arabinoheptulose 7-phosphate, **8** is 3-dehydroquinic acid, **9** is 3-dehydroshikimic acid, **10** is shikimic acid, **11** is shikimic acid 3-phosphate, and **12** is 5-enolpyru-vylshikimic acid 3-phosphate.

each case the L-phenylalanine biosynthetic pathway comprises three enzymatic steps from choris-mate (**2**), the product of the common aromatic pathway. The precursors of the common aromatic pathway, phosphoenolpyruvate (PEP) (**5**) and erythrose-4-phosphate (E4P) (**6**), derive, respectively, from the glycolytic and pentose phosphate pathways of sugar metabolism. Although the intermediate compounds in both pathways are identical in each of the organisms, there are differences in the organization and regulation of the genes and enzymes involved.[6,10,11] Nevertheless, the principal points of pathway regulation are very similar in each of the three bacteria.[8,10,12] Conversely, tyrosine biosynthesis, which is also carried out in three biosynthetic steps from chorismate, proceeds through different intermediates in *E. coli* than do the *coryneform* organisms.[10,12]

Regulation of phenylalanine biosynthesis occurs both in the common aromatic pathway and in the terminal phenylalanine pathway, and the vast majority of efforts to deregulate phenylalanine biosynthesis have focused on two specific rate-limiting enzymatic steps. These are the steps of the aromatic and the phenylalanine pathways carried out respectively by the enzymes 3-deoxy-D-arabino-heptulosonate-7-phosphate (DAHP) synthase and prephenate dehydratase (PD). Classical mutagenesis approaches using toxic amino acid analogues and the molecular cloning of the genes encoding these enzymes have led to very significant increases in the capability of host strains to overproduce L-phenylalanine. This has resulted from the elimination of the regulatory mechanisms controlling enzyme synthesis and specific activity. The cellular mechanisms, which govern the activity of these particular enzymes, are complex and illustrate many of the sophisticated means by which bacteria control gene expression and enzyme activity. Chorismate mutase (CM), the first step in the phenylalanine specific pathway, and the shikimate kinase activity of the common aromatic pathway are subject to a lesser degree of regulation and have also been characterized in detail.[13,14]

3.2.1 CLASSICAL STRAIN IMPROVEMENT

Classical methods of strain improvement have been widely applied in the development of phenyl-alanine overproducing organisms.[9,15–17] Tyrosine auxotrophs have frequently been used in efforts to increase phenylalanine production through mutagenesis. These strains often already overproduce phenylalanine as a result of the overlapping nature of tyrosine and phenylalanine biosynthetic regulation.[7,18] Limiting tyrosine availability leads to partial genetic and allosteric deregulation of common biosynthetic steps.[19] Such strains have been subjected to a variety of mutagenesis proce-dures to further increase the overall titer and the efficiency of phenylalanine production. In general this has involved the identification of mutants that display resistance to toxic analogues of phenyl-alanine or tyrosine such as β-2-thienyl-L-alanine (13) and p-fluoro-L-phenylalanine (14).[7,9,20] Such mutants can be readily identified on selective plates where the analogue is present in the growth medium. The mutations responsible for the phenylalanine overproduction have frequently been located in the genes encoding the enzymatic activities DAHP synthase, CM, and PD. In turn this has prompted molecular genetic approaches to further increase phenylalanine production through the isolation and *in vitro* manipulation of these genes, as described later in this chapter.

13   14

3.2.2 DEREGULATION OF DAHP SYNTHASE

The activity of DAHP synthase commits carbon from intermediary metabolism to the common aromatic pathway converting equimolar amounts of PEP (5) and E4P (6) to DAHP.[21] In *E. coli* there are three isoenzymes of DAHP synthase of comparable catalytic activity encoded by the genes *aroF*, *aroG*, and *aroH*.[22] Enzyme activity is regulated respectively by the aromatic amino acids tyrosine, phenylalanine, and tryptophan.[19,23–26] In each case regulation is mediated both by repression of gene transcription and by allosteric feedback inhibition of the enzyme, although to different degrees.

The *aroF* gene lies in an operon with *tyrA*, which encodes the bifunctional protein chorismate mutase/prephenate dehydrogenase (CMPD). Both genes are regulated by the TyrR repressor protein complexed with tyrosine. The *aroF* gene product accounts for 80% of the total DAHP synthase activity in wild-type *E. coli* cells. The *aroG* gene is repressed by the TyrR repressor protein complexed with phenylalanine and tryptophan. Repression of *aroH* is mediated by tryptophan and the TrpR repressor protein.[23] The gene products of *aroF* and *aroG* are almost completely feedback inhibited respectively by low concentrations of tyrosine or phenylalanine,[2] whereas the *aroH* gene product is subject to maximally 40% feedback inhibition by tryptophan.[2] In *Brevibacterium flavum*, DAHP synthase forms a bifunctional enzyme complex with CM and is feedback inhibited by tyrosine and phenylalanine synergistically but not by tryptophan.[27,28] Similarly in *Corynebacterium glutamicum* DAHP synthase is inhibited most significantly by phenylalanine and tyrosine acting in concert,[29] but unlike *Brevibacterium flavum*, reportedly it does not show tyrosine-mediated repression of transcription.[27,30]

Many examples of analogue-resistant mutants of these organisms display reduced sensitivity of DAHP synthase to feedback inhibition.[7,9,11,27,31,32] In *E. coli* the genes encoding the DAHP synthase isoenzymes have been characterized and sequenced.[10,33,34] The mechanism of feedback inhibition of the *aroF*, *aroG*, and *aroH* isoenzymes has been studied in considerable detail[34] and variants of the *aroF* gene on plasmid vectors have been used to increase phenylalanine overpro-duction. Simple replacement of transcriptional control sequences with powerful constitutive or inducible promoter regions and the use of high copy number plasmids has readily enabled overproduction of the enzyme,[35,36] and reduction of tyrosine-mediated feedback inhibition has been described using resistance to the aromatic amino acid analogues β-2-thienyl-D,L-alanine (**13**, *ent*-**13**) and *p*-fluoro-D,L-phenylalanine (**14**, *ent*-**13**).[9,35]

3.2.3 DEREGULATION OF CHORISMATE MUTASE AND PREPHENATE DEHYDRATASE

The three enzymatic steps by which chorismate is converted to phenylalanine appear to be identical between *C. glutamicum, B. flavum,* and *E. coli,* although only in *E. coli* have detailed reports appeared on the characterization of the genes involved. In each case the principal regulatory step is the one catalyzed by PD. In *E. coli,* chorismate (**2**) first is converted to prephenate (**3**) and then to phenylpyruvate (**4**) by the action of a bifunctional enzyme (Figure 3.1), CMPD encoded by the *pheA* gene.[10] The final step, in which phenylpyruvate (**3**) is converted to L-phenylalanine (**1**), is carried out predominantly by the aromatic aminotransferase encoded by *tyrB*.[2] However, both the aspartate aminotransferase, encoded by *aspC,* and the branched chain aminotransferase, encoded by *ilvE,* can catalyze this reaction.[37] Phenylalanine biosynthesis is regulated by control of CMPD through phenylalanine-mediated attenuation of *pheA* transcription[10] and by feedback inhibition of the PD and CM activities of the enzyme. Inhibition is most pronounced on the PD activity, with almost total inhibition observed at micromolar phenylalanine concentrations.[38,39] CM activity, in contrast, is maximally inhibited by only 40%.[39] In *B. flavum,* PD and CM are encoded by distinct genes. PD is again the principal point of regulation with the enzyme subject to feedback inhibition, but not transcriptional repression, by phenylalanine.[40,41] CM, which in this organism forms a bifunctional complex with DAHP synthase, is inhibited by phenylalanine and tyrosine up to 65%, but this is significantly diminished by the presence of very low levels of tryptophan.[42] Expression of CM is repressed by tyrosine.[31,43] The final step is carried out by at least one transaminase.[44] Similarly, in *C. glutamicum,* the activities are encoded in distinct genes with PD again being the more strongly feedback inhibited by phenylalanine.[45–47] The only transcriptional repression reported is that of CM by phenylalanine. The *C. glutamicum* genes encoding PD and CM have been isolated and cloned from analogue-resistant mutants of *C. glutamicum*[11,48] and used along with the cloned DAHP synthase gene to augment L-phenylalanine biosynthesis in overproducing strains of *C. glutamicum*.[48]

Many publications and patents have described efforts to reduce and eliminate regulation of PD activity in phenylalanine overproducing organisms.[9,14,36] As with DAHP synthase, the majority of the reported work has focused on the *E. coli* enzyme encoded by the *pheA* gene, which is transcribed convergently with the tyrosine operon on the *E. coli* chromosome.[10] The detailed characterization of the *pheA* regulatory region has facilitated expression of the gene in a variety of transcriptional configurations leading to elevated expression of CMPD and a number of mutations in *pheA,* which affect phenylalanine-mediated feedback inhibition, have been described.[14,35,49] Increased expression of the gene is readily achieved by cloning *pheA* onto multi-copy plasmid vectors and deletion of the nucleotide sequences comprising the transcription attenuator.[14] Most mutations that affect the allosteric regulation of the enzyme by phenylalanine have been identified through resistance to phenylalanine analogues such as β-2-thienylalanine, but there are examples of feedback-resistant mutations arising through insertional mutagenesis and gene truncation.[35,49] Two regions of the enzyme in particular have been shown to reduce feedback inhibition to different degrees. Mutations at position Trp338 in the peptide sequence desensitize the enzyme to levels of phenylalanine in the 2- to 5-mM range but are insufficient to confer resistance to higher concentrations of L-phenylalanine.[35,49] Mutations in the region of residues 304–310 confer almost total resistance to feedback inhibition at L-phenylalanine concentrations of at least 200 mM.[14] Feedback-inhibition profiles of four such variants (JN305–JN308) are shown in Figure 3.3, in comparison to the profile of wild-type enzyme (JN302). It is not clear if the mechanism of resistance is similar in either case, but the difference is significant to commercial application because overproducing organisms readily achieve extracellular concentrations of L-phenylalanine of more than 200 mM.

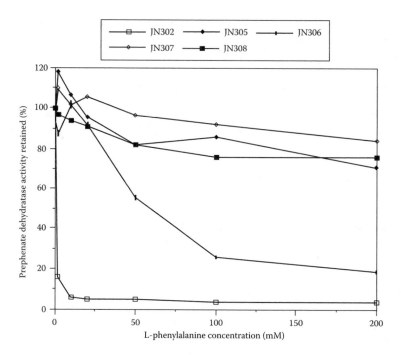

FIGURE 3.3 L-Phenylalanine–mediated feedback inhibition of wild-type *Escherichia coli* K12 prephenate dehydratase (JN302) and four feedback inhibition–resistant enzyme variants (JN305-JN308). Activity is expressed as a percentage of normal wild-type enzyme activity.

3.2.4 PRECURSOR SUPPLY

Rate-limiting steps in the common aromatic and phenylalanine biosynthetic pathways are obvious targets in the development of phenylalanine overproducing organisms. However, the detailed biochemical and genetic characterization of *E. coli* has enabled additional areas of its metabolic function to be specifically manipulated to determine their effect on aromatic pathway throughput. Besides efforts to eliminate additional points of aromatic pathway regulation, attempts have been made to enhance phenylalanine production by increasing the supply of aromatic pathway precursors and by facilitating exodus of L-phenylalanine from the cell. The precursors of the common aromatic pathway D-erythrose 4-phosphate (**6**) and phosphoenolpyruvate (**5**) are the respective products of the pentose phosphate and glycolytic pathways. Precursor supply in aromatic amino acid biosynthesis has been reviewed.[50,51] Theoretical analyses of the pathway and the cellular roles of these metabolites suggest that the production of aromatic compounds is likely to be limited by PEP availability[52,53] because phosphoenolpyruvate is involved in a number of cellular processes including the generation of metabolic energy through the citric acid cycle[54] and the transport of glucose into the cell by the phosphotransferase system.[55] Strategies to reduce the drain of PEP by these processes have included mutation of sugar transport systems to reduce PEP-dependent glucose transport[56] and modulation of the activities of pyruvate kinase, PEP synthase, and PEP carboxylase, which regulate PEP flux to pyruvate, oxaloacetate, and the citric acid cycle.[50,54,57] Similarly, the availability of E4P has been increased by altering the levels of transketolase, the enzyme responsible for E4P biosynthesis.[58] In general, these efforts have been successful in directing additional flux of PEP of E4P into the aromatic pathway, although their effect has not proved to be as predictable because the deregulation of rate-limiting pathway steps and their overall impact on L-phenylalanine overproduction has not been well characterized.

The incremental gains made from the various levels at which phenylalanine biosynthesis has been addressed have led to the high-efficiency production strains in use today. Almost all large-scale L-phenylalanine manufacturing processes in present operation are fermentations using bacterial strains such as those described earlier, where classical strain development and/or molecular genetics have been extensively applied to bring about phenylalanine overproduction. The low substrate costs and the economies of scale associated with this approach have resulted in significant economic advantages over alternative approaches (see Chapter 2).

3.3 BACTERIAL PRODUCTION OF L-LYSINE AND L-THREONINE

3.3.1 BIOCHEMICAL PATHWAY ENGINEERING IN CORYNEBACTERIA

The fermentative production of L-glutamate and amino acids of the aspartate family, such as lysine and threonine, has been dominated by the use of the *Coryneform* organisms, especially *C. glutamicum*. Several strains of the *corynebacteria* have been used for decades to produce L-glutamic acid.[1,2,59] These organisms have been found to secrete the amino acid at very high concentrations through changes in the cell membrane induced either by biotin limitation or by the addition of various surfactants.[60,61] The enzyme glutamate dehydrogenase encoded by the *gdh* gene is principally responsible for the biosynthesis of L-glutamate in the *Corynebacteria* from the keto acid precursor 2-ketoglutarate.[62] Although relatively few genetic studies have been published on this enzyme,[63] many advances in both the genetic organization and the technology to manipulate DNA in the *Corynebacteria* have been made in the last two decades.[64–66] This has enabled molecular genetic approaches to complement conventional strain development for the production of amino acids, such as L-threonine (**22**) and L-lysine (**19**) in *Corynebacterium*.[67–69] These advances have been thoroughly reviewed[67,70,71] and will be summarized here with only the key aspects relevant to microbial pathway engineering emphasized.

The use of *E. coli* has been a valuable tool to the researchers in the *Corynebacteria* because many of the genes encoding the key enzymes for amino acid biosynthesis were initially isolated through complementation of the corresponding mutants in *E. coli*.[67] Similarly, transconjugation of plasmids from *E. coli* to various *Corynebacterium* strains, facilitating gene disruption and replacement, has helped characterize the roles of specific genes, particularly in lysine and threonine biosynthesis.[72,73] The development of *Corynebacterium* cloning vectors has further facilitated the studies of gene regulation and expression in these strains.[64,74,75]

3.3.2 ENGINEERING THE PATHWAYS OF L-LYSINE AND L-THREONINE BIOSYNTHESIS

The biosynthesis of the aspartate-derived amino acids is significantly affected by the regulation of aspartokinase, the enzyme governing the flow of carbon entering the initial common pathway (Figure 3.4) and encoded by the *ask* gene in *C. glutamicum*. In the case of *C. glutamicum* and *B. flavum*, aspartokinase is tightly regulated by concerted feedback inhibition, mediated by threonine and lysine.[76] Through classical mutagenesis/selection using the toxic lysine analogue S-(α-aminoethyl)-D,L-cysteine (**15**), lysine overproducing strains of *C. glutamicum* were obtained, which frequently carried variants of aspartokinase resistant to feedback inhibition by lysine and/or threonine.[77,78] Sequence comparisons between the wild-type and mutant *ask* gene sequences located mutations conferring feedback resistance to the β-subunit of the enzyme,[78,79] although the mechanism of feedback inhibition and of resistance remains unknown. Similar mutations have been identified in the β-subunit of the *C. lactofermentum* and in *B. flavum*.[80,81]

15

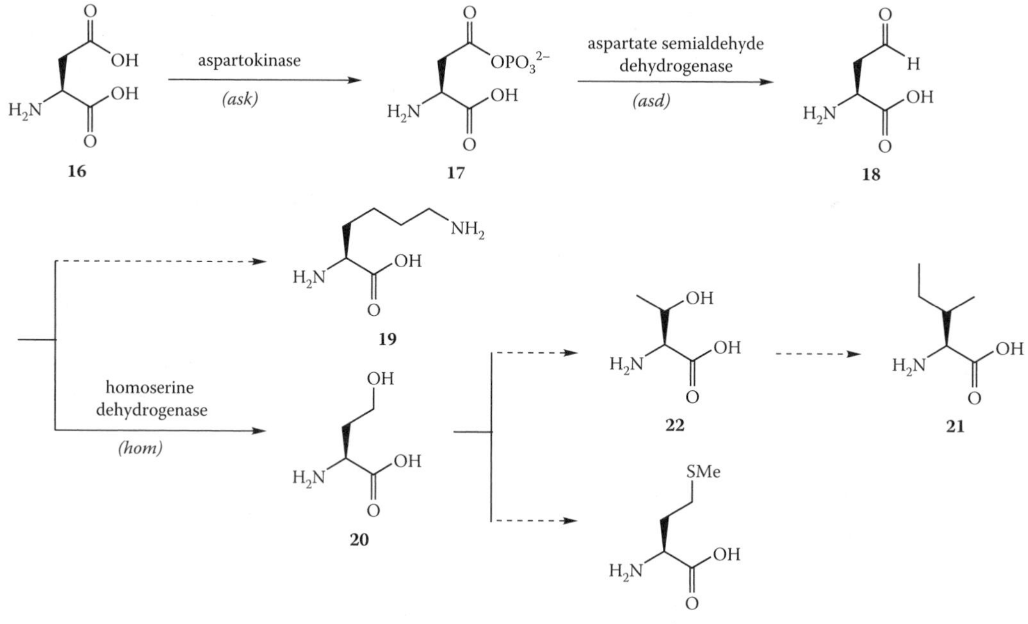

FIGURE 3.4 The common pathway of the aspartate-derived amino acids in *Corynebacteria*. The mnemonic of the genes involved are shown in parentheses below the enzymes responsible for each step. Dotted lines indicate multiple enzymatic steps, and **16** is L-aspartic acid, **17** is L-aspartyl phosphate, **18** is L-aspartate semialdehyde, **19** is L-lysine, **20** is L-homoserine, **21** is L-isoleucine, **22** is L-threonine, and **23** is L-methionine.

By analogy to the feedback-resistant mutations in PD and the DAHP synthases of the *E. coli* phenylalanine and common aromatic pathways, the identification of feedback-resistant variants of aspartokinase has contributed significantly to efforts to engineer lysine and threonine overproducing strains of *Corynebacterium*. Introduction of feedback-resistant *Ask* variants to lysine overproducing strains of *C. glutamicum* and *C. lactofermentum* enhanced extracellular lysine levels, the latter in a gene dosage–dependent manner.[82,83] Overexpression of the wild-type dihydropicolinate synthase, on a multi-copy plasmid, also leads to elevated lysine production,[83] as might be expected because this is the branch point of lysine and threonine biosynthesis in *Corynebacterium*. No increase in extracellular lysine was observed with the overexpression of each of the remaining enzymes of the primary lysine biosynthetic pathway. A secondary pathway of lysine biosynthesis exists in *C. glutamicum,* which allows complementation of mutations in the *ddh* gene encoding diaminopimelate dehydrogenase but was unable at least in the wild-type state to provide high throughput of precursors to D,L-diaminopimelate.[83,84]

As a complementary approach to the funneling of carbon to the lysine-specific pathways by overproduction of dihydropicolinate synthase, the elimination of this activity by mutation of the *dapA* gene led to significant overproduction of L-threonine[85] (**22**) to more than 13 g/L in a strain of *B. flavum,* which also carried a feedback inhibition–resistant aspartokinase. This productivity was determined to be comparable to the more typical overproducing strains of *B. flavum* carrying mutations in homoserine dehydrogenase, which eliminate the normal threonine-mediated feedback inhibition. This feedback inhibition usually is selected by the now-familiar mutagenesis/selection methods involving resistance to the toxic threonine analogue D,L-α-amino-β-hydroxyvaleric acid (**24**). Similar levels of threonine overproduction have been reported in recombinant *E. coli* strains where the threonine operon from a classically derived overproducing mutant was cloned on a multi-copy plasmid and reintroduced to the same host.[86] Significantly greater titers of threonine production have been reported by introducing the same cloned threonine operon into mutant strains of *B. flavum* grown with rigorous plasmid maintenance.[68]

24

The ability of the *Corynebacteria* to overproduce these amino acids sufficiently for commercial use is partly a result of the extraordinary efficiency with which this type of organism can secrete amino acids. For this reason it has been useful in the production of isoleucine[87] (**21**) and other amino acids[71] without necessitating the elimination of feedback inhibition of key enzymes in the biochemical pathways involved. Nevertheless, in the examples described earlier the availability of detailed information on the molecular genetic basis of metabolic and biosynthetic pathways is of fundamental importance in achieving optimal levels of amino acid production. Similarly, the application of recombinant DNA methodology and directed evolution to an ever wider array of microbes is enhancing the potential to effectively screen for mutations that relieve feedback inhibition, frequently the most crucial aspect of deregulating the biosynthesis of primary metabolites such as amino acids. As a result the potential to engineer microbial pathways is increasing significantly and includes opportunities to construct novel biochemical routes for the biosynthesis of nonproteinogenic amino acids. This broader application of pathway engineering to produce valuable unnatural amino acids has now been exploited by a number of groups, as discussed later in this chapter.

3.4 ENGINEERING NOVEL BIOCHEMICAL PATHWAYS FOR AMINO ACID PRODUCTION

3.4.1 ADAPTATION OF THE L-CYSTEINE PATHWAY TO PRODUCE UNNATURAL AMINO ACIDS

The Wacker-Chemie Group has described an extremely elegant approach[88] for the production of a family of unnatural amino acids from a common intermediate in the natural biochemical pathway to L-cysteine in *E. coli*. The method is based on earlier work by Wacker to develop efficient strains that produce natural L-cysteine by fermentation[89] but takes advantage of the naturally broad substrate specificity of *O*-acetylserine sulfhydrylase, the final enzyme in L-cysteine biosynthesis. This enzyme accepts a variety of alternative nucleophiles in addition to sulfide, which is normally required for cysteine biosynthesis. Accordingly, this enzyme can be provided with its substrate *O*-acetylserine (**25**) and compounds prepared using chemical synthesis, such as thiophenol, mercaptoethanol, or phenylselenol, resulting in the biosynthesis of *S*-phenyl-L-cysteine (**26**), *S*-hydroxyethyl-L-cysteine (**27**), or phenylseleno-L-cysteine (**28**), respectively. As might be expected, a major factor in the commercial viability of this process is the economic production of *O*-acetylserine (**25**), the precursor of each of these unnatural amino acids. The major regulatory point for the biosynthesis of this compound in *E. coli* is the very tight allosteric regulation of serine acetyltransferase, encoded by the *cysE* gene, which produces *O*-acetylserine from L-serine and acetylcoenzyme A. The strains used in the Wacker process overcome this natural regulation by incorporating a mutated *cysE* gene encoding a variant of serine acetyltransferase in which L-cysteine–mediated feedback inhibition is greatly reduced. Through the combined effect of this mutation and the overproduction of the feedback-resistant enzyme from a multi-copy plasmid significant overproduction of **25** is achieved when strains carrying this mutated allele of *cysE* are fermented. The approach used to isolate feedback inhibition–resistant variants of serine acetyltransferase was the screening and isolation of *E. coli* mutants[90] using methods quite analogous to those applied in the development strains, which overproduce natural amino acids. One key aspect of the success of this process is the interdisciplinary approach undertaken in which classical microbial genetics, molecular biology, and synthetic organic chemistry can be integrated to engineer and adapt a natural biochemical pathway for the manufacture of valuable unnatural compounds. This multidisciplinary approach is now increasingly apparent in applied biocatalysis.

25, R = OAc
26, R = SPh
27, R = S(CH$_2$)$_2$OH
28, R = SePh
29, R = Me

3.5   HYDANTOINASE WHOLE-CELL BIOCATALYSTS

Well-understood and readily engineered microbes such as *E. coli* can also serve as hosts for novel biosynthetic pathways composed entirely of heterologous enzymes producing commercially important amino acids. Such an example lies in the successful development of the hydantoinase-carbamoylase system for the production of D-amino acids and, more recently, L-amino acids (Chapter 2).[91,92] The hydantoinase-carbamoylase process is well established as an efficient means to produce D-amino acids in particular by the dynamic kinetic resolution of D,L-5-monosubstituted hydantoins. D-Amino acids are now increasingly in demand as valuable intermediates in the development of new pharmaceuticals including peptidomimetic drugs (Chapter 2). The efficiency and enantioselectivity of this process requires the concerted use of a hydantoinase biocatalyst with a carbamoylase and, depending on the substrate, a racemase biocatalyst. Efficient production of these biocatalysts is therefore very important to the overall process economics and the system is therefore highly suited to the application of a recombinant microbial strain that can produce each of the required enzymes from cloned genes. Such strains of *E. coli* have been constructed to produce the necessary biocatalysts required for the hydantoinase system.[92,93] This whole-cell biocatalytic system has been adapted for the production of L-amino acids using a directed evolution approach to alter the enantioselectivity of the hydantoinase enzyme from a D-selective to an L-selective biocatalyst.[92] Again, the integration of various technologies, including judicious and efficient screening regimens, molecular genetic and genomic methodologies, and the advantages of whole-cell biocatalysts, results in improved and competitive economic routes to amino acids.

3.6   NOVEL AMINOTRANSFERASE PATHWAYS

The biosynthesis of most proteinogenic amino acids including those described earlier directly or indirectly involves a transamination step.[2] Biochemical studies on a number of the bacterial L-amino acid transaminases (aminotransferases) responsible for this reaction have revealed that these enzymes frequently display broad substrate specificity (see also Chapter 19).[94] The relatively relaxed substrate specificity of microbial transaminases has been useful in the development of biotransformation approaches for the synthesis of nonproteinogenic L-amino acids, which are now in increasing demand as intermediates for the synthesis of peptidomimetic pharmaceuticals. Additionally the availability of a highly efficient screen has enabled the cloning of a number of genes encoding D-amino acid transaminase from a variety of microbial species.[95,96] In an example of the application of aminotransferases to the production of nonproteinogenic amino acids, both isomers of 2-aminobutyrate (**29**) have been produced.[97,98]

3.6.1 A Novel Biosynthetic Pathway to l-2-Aminobutyrate

In addition to the identification of the appropriate biosynthetic enzyme, the feasibility of all biotransformation processes depends heavily on other criteria, such as the availability of inexpensive starting materials, the reaction yield, and the complexity of product recovery. In the case of transaminase processes, the reversible nature of the reaction (Scheme 3.1) and the presence of a keto acid by-product is a concern that limits the overall yield and purity of product and has led to efforts to increase the conversion beyond the typical 50% yield of product.[99,100] Additionally, there are cost considerations in the large-scale preparation of keto acid substrates such as 2-ketobutyrate, which are not commodity chemicals.

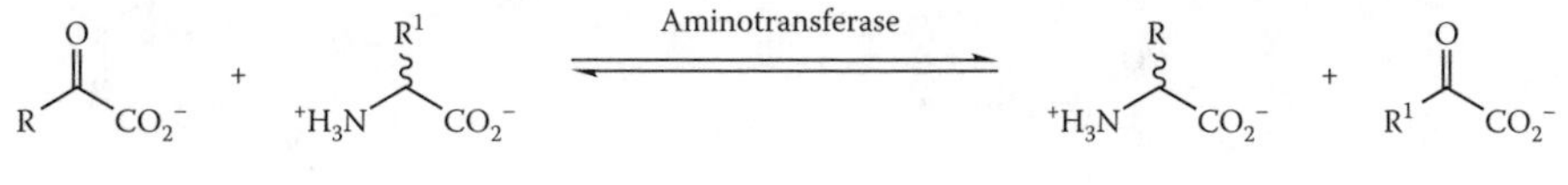

SCHEME 3.1

These issues are readily addressed in whole-cell systems because additional microbial enzymes can be used to generate substrates from inexpensive precursors and convert unwanted by-products to more easily handled compounds. For the biosynthesis of 2-aminobutyrate using the *E. coli* aromatic transaminase,[101] the engineered *E. coli* incorporates the cloned *E. coli* K12 *ilvA* gene encoding threonine deaminase to generate 2-ketobutyrate from the commodity amino acid l-threonine and the cloned *alsS* gene of *Bacillus subtilis* 168 encoding acetolactate synthase that eliminates pyruvate, the keto acid by-product of the reaction, through the formation of nonreactive acetolactate. The *B. subtilis* catabolic acetolactate synthase is preferable to the acetohydroxy acid synthases of *E. coli,* which also use 2-ketobutyrate as a substrate as a result of their overlapping roles in branched-chain amino acid biosynthesis. The reaction scheme for the combined use of acetolactate synthase with either a d- or l-amino acid aminotransferase is shown in Scheme 3.2. The l-2-aminobutyric acid (l-**29**) process is operated as a whole-cell biotransformation in a single strain, producing 27g/L of l-2-aminobutyrate. The benefit from the concerted action of these three enzymes in l-2-aminobutyrate biosynthesis is significant. Process economics benefit from the use of an inexpensive amino acid such as l-threonine (**22**) as the source of 2-ketobutyrate and l-aspartic acid as the amino donor. Second, with the additional acetolactate synthase activity present the ratio of l-2-aminobutyrate to l-alanine, the major amino acid impurity, increases to 22.5:1 from 2.4:1, reducing the complexity of the product recovery. Incremental improvements remain to be made in the prevention of undesired catabolism of substrates by metabolically active whole cells because the product yield is only 54%, although the substrates are almost entirely consumed. The detailed understanding of *E. coli* metabolism and genetics increases the likelihood of eliminating such undesirable catabolic side reactions.

3.7 SUMMARY

A variety of bacterial species have been used for almost 40 years in the large-scale fermentative production of amino acids for industrial uses. The *Corynebacteria* have been most widely used in this application as a result of their propensity to overproduce and excrete very high concentrations of amino acids under specific process conditions and their early development for monosodium glutamate production. However, other organisms including *E. coli* have also been successfully used in the production of amino acids such as phenylalanine.

The engineering of bacterial metabolic pathways to improve commercial fermentative production of amino acids has traditionally involved the use of relatively crude, nonspecific mutagenesis methods coupled to repeated rounds of arduous screening for resistance to toxic amino acid

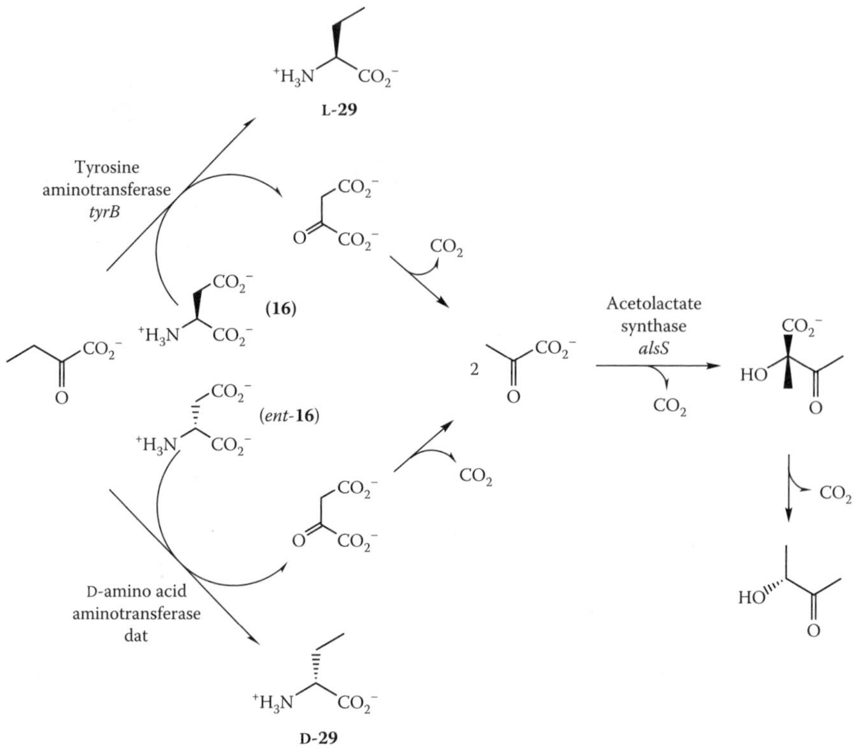

SCHEME 3.2

analogues. Although inelegant in execution, these methods nevertheless led to highly efficient and economically important organisms for large-scale production of many amino acids, although the approach becomes increasingly self-limiting as production organisms accumulate nonspecific mutations.

With the increasing understanding of the biochemistry and molecular genetics of amino acid production in bacteria and the availability of recombinant DNA methodology in the 1980s, the rate-limiting steps of biosynthetic pathways and precursor supply were addressed through the cloning and characterization of the genes encoding the key biosynthetic enzymes. This led to a greater understanding of the molecular mechanisms involved in biosynthetic pathway regulation and the characterization of mutations introduced previously through classical methods. Deregulation of gene expression, reintroduction of specific genes on multi-copy plasmids, and the relief of end-product–mediated feedback inhibition has led to even greater titers of product from amino acid–overproducing bacterial strains.

Consequently, many successes of amino acid overproduction in bacteria have come from the powerful hybrid approach of rational design and more traditional mutagenesis/selection methodology. As microbial DNA sequence information becomes increasingly available through genomic sequencing, the possibility to combine genes from diverse organisms in a single host is providing new opportunities to further engineer specific host bacterial strains to overproduce many unnatural amino acids. Accordingly, there are increasing examples of strains of E. coli or alternative hosts containing genes from multiple organisms that enable efficient whole-cell biosynthesis of various D-amino acids and nonproteinogenic L-amino acids.

This trend will surely continue as the application of whole-cell biocatalysis continues to grow, driven by improved tools now available to engineer microbial pathways and increasing accessibility of microbial genes from genomic sources. The introduction of the polymerase chain reaction (PCR), as a general method of gene cloning and manipulation, vastly reduced the time frame required to isolate genes encoding new enzymes of interest using DNA sequence information. With more than 150 complete microbial genomes now available from public sources such as GENBANK the accessibility of genes encoding new uncharacterized biocatalytic activities is unprecedented. More versatile gene mutation protocols and expanded screening regimens also allow the optimization of specific enzymatic activities to eliminate undesirable properties and rate-limiting steps in pathways of interest. Through these advances, the capability to engineer microbial biosynthetic pathways to produce industrially important amino acids will continue to grow and diversify.

REFERENCES

1. Abe, S., Takayama, K. *Amino acid-producing microorganisms. Variety and classification,* Kodansha Ltd, 1972.
2. Herrmann, K. M., Somerville, R. L., Eds., *Amino Acids: Biosynthesis and Genetic Regulation,* Addison-Wesley Publishing Co.: Reading, Mass., 1983.
3. Batt, C. A., Follettie, M. T., Shin, H. K., Yeh, P., Sinskey, A. J. *Trends Biotechnol.* 1985, *3*, 305.
4. Shiio, I. *Prog. Ind. Microbiol.* 1986, 24, 188.
5. Mullis, K. B., Ferre, F., Gibbs, R. A. *The Polymerase Chain Reaction,* Birkhaeuser: Cambridge, Mass., 1995.
6. Shiio, I., Sugimoto, S., Kawamura, K. *Agric. Biol. Chem.* 1988, *52*, 2247.
7. Hagino, H., Nakayama, K. *Agric. Biol. Chem.* 1974, *38*, 157.
8. Ikeda, M., Ozaki, A., Katsumata, R. *Appl. Microbiol. Biotechnol.* 1993, *39*, 318.
9. Tribe, D. E. U.S. Pat. 4,681,852, 1987.
10. Hudson, G. S., Davidson, B. E. *J. Mol. Biol.* 1984, 180, 1023.
11. Ozaki, A., Katsumata, R., Oka, T., Furuya, A. *Agric. Biol. Chem.* 1985, *49*, 2925.
12. Shiio, I., Sugimoto, S. *Agric. Biol. Chem.* 1981, *45*, 2197.
13. Millar, G., Lewendon, A., Hunter, M. G., Coggins, J. R. *Biochem. J.* 1986, *237*, 427.
14. Nelms, J., Edwards, R. M., Warwick, J., Fotheringham, I. *Appl. Environ. Microbiol.* 1992, *58*, 2592.
15. Shiio, I., Ishii, K., Yokozeki, K. *Agric. Biol. Chem.* 1973, 37, 1991.
16. Tsuchida, T., Kubota, K., Morinaga, Y., Matsui, H., Enei, H., Yoshinaga, F. *Agric. Biol. Chem.* 1987, *51*, 2095.
17. Tokoro, Y., Oshima, K., Okii, M., Yamaguchi, K., Tanaka, K., Kinoshita, S. *Agric. Biol. Chem.* 1970, *34*, 1516.
18. Hwang, S. O., Gil, G. H., Cho, Y. J., Kang, K. R., Lee, J. H., Bae, J. C. *Appl. Microbiol. Biotechnol.* 1985, *22*, 108.
19. Wallace, B. J., Pittard, J. *J. Bacteriol.* 1969, *97*, 1234.
20. Choi, Y. J., Tribe, D. E. *Biotechnol. Lett.* 1982, *4*, 223.
21. Haslam, E. *The Shikimate Pathway,* Halsted: New York, 1974.
22. Pittard, J., Gibson, F. *Curr. Top. Cell. Regul.* 1970, *2*, 29.
23. Brown, K. D. *Genetics* 1968, *60*, 31.
24. Brown, K. D., Somerville, R. L. *J. Bacteriol.* 1971, *108*, 386.
25. Camakaris, H., Pittard, J. *J. Bacteriol.* 1973, *115*, 1135.
26. Im, S. W. K., Davidson, H., Pittard, J. *J. Bacteriol.* 1971, *108*, 400.
27. Shiio, I., Sugimoto, S., Miyajima, R. *J. Biochem.* 1974, *75*, 987.
28. Sugimoto, S., Shiio, I. *J. Biochem.* 1980, *87*, 881.
29. Hagino, H., Nakayama, K. *Agric. Biol. Chem.* 1975, *39*, 351.
30. Hagino, H., Nakayama, K. *Agric. Biol. Chem.* 1974, *38*, 2125.
31. Shiio, I., Sugimoto, S. *J. Biochem.* 1979, *86*, 17.
32. Sugimoto, S., Makagawa, M., Tsuchida, T., Shio, I. *Agric. Biol. Chem.* 1973, *37*, 2327.
33. Davies, W. D., Davidson, B. E. *Nucleic Acids Res.* 1982, *10*, 4045.
34. Ray, J. M., Yanofsky, C., Bauerle, R. *J. Bacteriol.* 1988, *170*, 5500.
35. Edwards, M. R., Taylor, P. P., Hunter, M. G., Fotheringham, I. G. PCT Int. Appl. Pat. WO 8700202, 1987.
36. Foerberg, C., Eliaeson, T., Haeggstroem, L. *J. Biotechnol.* 1988, *7*, 319.
37. Fotheringham, I. G., Dacey, S. A., Taylor, P. P., Smith, T. J., Hunter, M. G., Finlay, M. E., Primrose, S. B., Parker, D. M., Edwards, R. M. *Biochem. J.* 1986, *234*, 593.

38. Gething, M. J., Davidson, B. E. *Eur. J. Biochem.* 1978, *86*, 165.
39. Dopheide, T. A. A., Crewther, P., Davidson, B. E. *J. Biol. Chem.* 1972, *247*, 4447.
40. Shiio, I., Sugimoto, S. *J. Biochem.* 1976, *79*, 173.
41. Sugimoto, S., Shiio, I. *J. Biochem.* 1974, *76*, 1103.
42. Shiio, I., Sugimoto, S. *J. Biochem.* 1981, *89*, 1483.
43. Sugimoto, S., Shiio, I. *Agric. Biol. Chem.* 1985, *49*, 39.
44. Shiio, I., Mori, M., Ozaki, H. *Agric. Biol. Chem.* 1982, *46*, 2967.
45. De Boer, L., Dijkhuizen, L. *Adv. Biochem. Eng. Biotechnol.* 1990, *41*, 1.
46. Hagino, H., Nakayama, K. *Agric. Biol. Chem.* 1974, *38*, 2367.
47. Hagino, H., Nakayama, K. *Agric. Biol. Chem.* 1975, *39*, 331.
48. Ikeda, M., Katsumata, R. *Appl. Environ. Microbiol.* 1992, *58*, 781.
49. Backman, K. C., Balakrishnan, R. U.S. Pat. 4,753,883, 1988.
50. Berry, A. *Trends Biotechnol.* 1996, *14*, 250.
51. Frost, J. W., Draths, K. M. *Annu. Rev. Microbiol.* 1995, *49*, 557.
52. Patnaik, R., Liao, J. C. *Appl. Environ. Microbiol.* 1994, *60*, 3903.
53. Patnaik, R., Spitzer, R. G., Liao, J. C. *Biotechnol. Bioeng.* 1995, *46*, 361.
54. Miller, J. E., Backman, K. C., O'Connor, M. J., Hatch, R. T. *J. Ind. Microbiol.* 1987, *2*, 143.
55. Postma, P. W. Phosphotransferase system for glucose and other sugars. In *Escherichia coli and Salmonella typhimurium,* Neidhardt, F. C., Ingraham, J. L., Brooks Low, K., Magasanik, B., Schaechter, M., and Umbarger, H. E., Eds., American Society for Molecular Biology: Washington, DC, 1987, p. 127.
56. Flores, N., Xiao, J., Berry, A., Bolivar, F., Valle, F. *Nat. Biotechnol.* 1996, *14*, 620.
57. Bledig, S. A. Ph. D. Thesis. University of Warwick, England. 1994.
58. Draths, K. M., Pompliano, D. L., Conley, D. L., Frost, J. W., Berry, A., Disbrow, G. L., Staversky, R. J., Lievense, J. C. *J. Am. Chem. Soc.* 1992, *114*, 3956.
59. Yamada, K., Kinoshita, S., Tsunoda, T., Aida, K. *The Microbial Production of Amino Acids,* Halsted: New York, 1973.
60. Clement, Y., Laneelle, G. *J. Gen. Microbiol.* 1986, *132*, 925.
61. Duperray, F., Jezequel, D., Ghazi, A., Letellier, L., Shechter, E. *Biochim. Biophys. Acta* 1992, *1103*, 250.
62. Tesch, M., Eikmanns, B. J., De Graaf, A. A., Sahm, H. *Biotechnol. Lett.* 1998, *20*, 953.
63. Boermann, E. R., Eikmanns, B. J., Sahm, H. *Mol. Microbiol.* 1992, *6*, 317.
64. Martin, J. F., Santamaria, R., Sandoval, H., Del Real, G., Mateos, L. M., Gil, J. A., Aguilar, A. *BioTechnol.* 1987, *5*, 137.
65. Santamaria, R. I., Martin, J. F., Gil, J. A. *Gene* 1987, *56*, 199.
66. Archer, J. A. C., Sinskey, A. J. *J. Gen. Microbiol.* 1993, *139*, 1753.
67. Jetten, M. S. M., Sinskey, A. J. *Crit. Rev. Biotechnol.* 1995, *15*, 73.
68. Ishida, M., Yoshino, E., Makihara, R., Sato, K., Enei, H., Nakamori, S. *Agric. Biol. Chem.* 1989, *53*, 2269.
69. Ishida, M., Sato, K., Hashiguchi, K., Ito, H., Enei, H., Nakamori, S. *Biosci., Biotechnol., Biochem.* 1993, *57*, 1755.
70. Jetten, M. S. M., Gubler, M. E., McCormick, M. M., Colon, G. E., Follettie, M. T., Sinskey, A. J. Molecular organization and regulation of the biosynthetic pathway for aspartate-derived amino acids in Corynebacterium glutamicum, *Ind. Microorg.* 1993, 97.
71. Nampoothiri, M., Pandey, A. *Process Biochem.* 1998, *33*, 147.
72. Puehler, A., Kassing, F., Winterfeldt, A., Kalinowski, J., Schaefer, A., Schwarzer, A., Seep-Feldhaus, A., Rossol, I. A novel system for genetic engineering of amino acid producing corynebacteria and the analysis of AEC-resistant mutants, *Eur. Congr. Biotechnol.,* 5th, 2, *975*, 1990.
73. Schwarzer, A., Puehler, A. *BioTechnol.* 1991, *9*, 84.
74. Miwa, K., Matsui, K., Terabe, M., Ito, K., Ishida, M., Takagi, H., Nakamori, S., Sano, K. *Gene* 1985, *39*, 281.
75. Jetten, M. S. M., Follettie, M. T., Sinskey, A. J. *Ann. N. Y. Acad. Sci.* 1994, *721*, 12.
76. Shiio, I., Miyajima, R. *J. Biochem.* 1969, *65*, 849.
77. Tosaka, O., Takinami, K. *Prog. Ind. Microbiol.* 1986, *24*, 152.

78. Kalinowski, J., Cremer, J., Bachmann, B., Eggeling, L., Sahm, H., Pruehler, A. *Mol. Microbiol.* 1991, *5*, 1197.
79. Follettie, M. T., Peoples, O. P., Agoropoulou, C., Sinskey, A. J. *J. Bacteriol.* 1993, *175*, 4096.
80. Shiio, I., Sugimoto, S., Kawamura, K. *Biosci. Biotechnol. Biochem.* 1993, *57*, 51.
81. Shiio, I., Yoshino, H., Sugimoto, S. *Agric. Biol. Chem.* 1990, *54*, 3275.
82. Jetten, M. S. M., Follettie, M. T., Sinskey, A. *J. Appl. Microbiol. Biotechnol.* 1995, *43*, 76.
83. Cremer, J., Eggeling, L., Sahm, H. *Appl. Environ. Microbiol.* 1991, *57*, 1746.
84. Schrumpf, B., Schwarzer, A., Kalinowski, J., Puehler, A., Eggeling, L., Sahm, H. *J. Bacteriol.* 1991, *173*, 4510.
85. Shiio, I., Toride, Y., Yokota, A., Sugimoto, S., Kawamura, K. Fr. Pat. 2626286, 1989.
86. Miwa, K., Tsuchida, T., Kurahashi, O., Nakamori, S., Sano, K., Momose, H. *Agric. Biol. Chem.* 1983, *47*, 2329.
87. Colon, G. E., Nguyen, T. T., Jetten, M. S. M., Sinskey, A. J., Stephanopoulos, G. *Appl. Microbiol. Biotechnol.* 1995, *43*, 482.
88. Maier, T. H. P. *Nature Biotechnol.* 2003, *21*, 422.
89. Maier, T. H. P., Winterhalter, C. PCT Pat. Appl. WO 01027307, 2001.
90. Denk, D., Boeck, A. J. *Gen. Microbiol.* 1987, *133*, 515.
91. May, O., Verseck, S., Bommarius, A., Drauz, K. *Org. Proc. Res. Dev.* 2002, *6*, 452.
92. May, O., Nguyen, P., Arnold, F. H. *Nature Biotechnol.* 2000, *18*, 317.
93. Wilms, B., Wiese, A., Syldatk, C., Mattes, R., Altenbuchner, J. J. Biotechnol. 2001, *86*, 19.
94. Christen, P., Metzler, D. E., Eds. *Transaminases,* John Wiley and Sons, New York: 1985, p. 463.
95. Fotheringham, I. G., Bledig, S. A., Taylor, P. P. *J. Bacteriol.* 1998, *180*, 4319.
96. Taylor, P. P., Fotheringham, I. G. *Biochim. Biophys. Acta* 1997, *1350*, 38.
97. Ager, D. J., Fotheringham, I. G., Laneman, S. A., Pantaleone, D. P., Taylor, P. P. *Chim. Oggi* 1997, *15* (3/4), 11.
98. Fotheringham, I. G., Pantaleone, D. P., Taylor, P. P. *Chim. Oggi* 1997, *15* (9/10), 33.
99. Crump, S. P., Rozzell, J. D. Biocatalytic production of amino acids by transamination. In *Biocatalytic Production of Amino Acid Derivatives,* Rozzell, J. D. Ed., Wiley: New York, 1992, p. 43.
100. Taylor, P. P., Pantaleone, D. P., Senkpeil, R. F., Fotheringham, I. G. *Trends Biotechnol.* 1998, *16*, 412.
101. Fotheringham, I. G, Grinter, N., Pantaleone, D. P., Senkpeil, R. F., Taylor, P. P. *Bioorg, Med. Chem.* 1999, *7*, 2209.

4 Carbohydrates in Synthesis

David J. Ager

CONTENTS

4.1 INTRODUCTION

Carbohydrates are extremely important in nature, being used as building blocks for structures such as cell walls, as modifiers of solubility, and as means for storing energy. Many carbohydrate derivatives, however, are not abundant in the host organism, which, despite the diverse array of functionality and stereochemistry known, only allows for a small number of these compounds to be available at a commercial scale.[1] This has to be counterbalanced against the breadth of sugar derivatives that are found in nature and, although only available at small scale (and often at a very high price), are cited as chirons.[2] In addition to sucrose, the carbohydrates produced at scale are D-glucose, D-sorbitol, D-lactose, D-fructose, D-mannitol, D-maltose, D-isomaltulose, D-gluconic acid, D-xylose, and L-sorbose.[3,4] Most of these are used as food additives rather than chemical raw materials. A comparison of volumes and price indications for sugars and some simple derivatives is given in Table 4.1.[5]

TABLE 4.1
Annual Production and Prices for Sugars and Derivatives[5]

		World Production (MT/year)	Price ($/Kg)
Sugars	Sucrose	130,000,000	0.35
	D-Glucose	5,000,000	0.72
	Lactose	295,000	0.72
	D-Fructose	60,000	1.20
	Isomaltulose	50,000	2.40
	Maltose	3,000	3.60
	D-Xylose	25,000	5.40
	L-Sorbose	60,000	9.00
Sugar alcohols	D-Sorbitol	650,000	2.15
	D-Xylitol	30,000	6.00
	D-Mannitol	30,000	9.60
Sugar acids	D-Gluconic acid	60,000	1.70
	L-Lactic acid	>100,000	2.10
	Citric acid	500,000	3.00
	L-Tartaric acid	35,000	7.20

The vast array of similar functionality often turns out to be a major problem to the synthetic chemist who needs to differentiate between them. This can result in long, tedious, protection–deprotection sequences.[1,6,7] Although enzymes can be of help, there is still a significant way to go. As carbohydrates are components of many pharmaceutical compounds, many methods have been developed for their synthesis.[8,9] This includes methodology that is amenable to use at scale, including the Sharpless epoxidation.[10–14]

Of course, carbohydrates are used and manipulated at commercial scales. The isolation of sucrose and the enzymatic formation of high fructose corn syrup are obvious examples. However, although sucrose is available from a wide variety of sources at very large scale, it has found relatively little application in chemical transformations.[15,16]

This chapter will cover the uses of carbohydrates as "chiral pool" materials. Because hydroxy acids are closely related, they are also discussed here.

4.2 DISACCHARIDES

4.2.1 SUCROSE

The disaccharide sucrose (**1**) is readily available in bulk quantities from sugar cane or beet. Its major use is in the food industry as a sweetening agent. Despite numerous publications, and a significant amount of research, sucrose has not found a place as a chiral chemical raw material. Sucrose has, however, been derivatized to provide useful food products that have become large-volume products (*vide infra*).

4.2.1.1 Sucrose Esters

Esters of this disaccharide, derived from fatty acids, have been developed as fat substitutes of which the Proctor and Gamble product, Olestra, has been commercialized.[17] Although the market is still small, there is hope that this will be a large-volume commercial product. The esters are prepared by transesterification reactions.[15]

4.2.1.2 Sucralose

Sucralose (2), 4,1′,6′-trichloro-4,1′,6′-trideoxy-*galacto*-sucrose (TGS), is a trichloro disaccharide nonnutritive sweetener.[18] This compound was discovered through a systematic study in which sucrose derivatives were prepared. It was found that substitution of certain hydroxy groups by a halogen increased the sweetness potency dramatically.[19,20] Sucralose was chosen as the development candidate by Tate and Lyle.[21–24]

4.2.1.2.1 Chemical Approaches

The synthesis of TGS (2) involves a series of selective protection and deprotection steps so that the 4-hydroxy group can be converted to chloro with inversion of configuration.[23,25] Differentiation between the primary hydroxy groups of the two sugar moieties is also required (Scheme 4.1).[26,27]

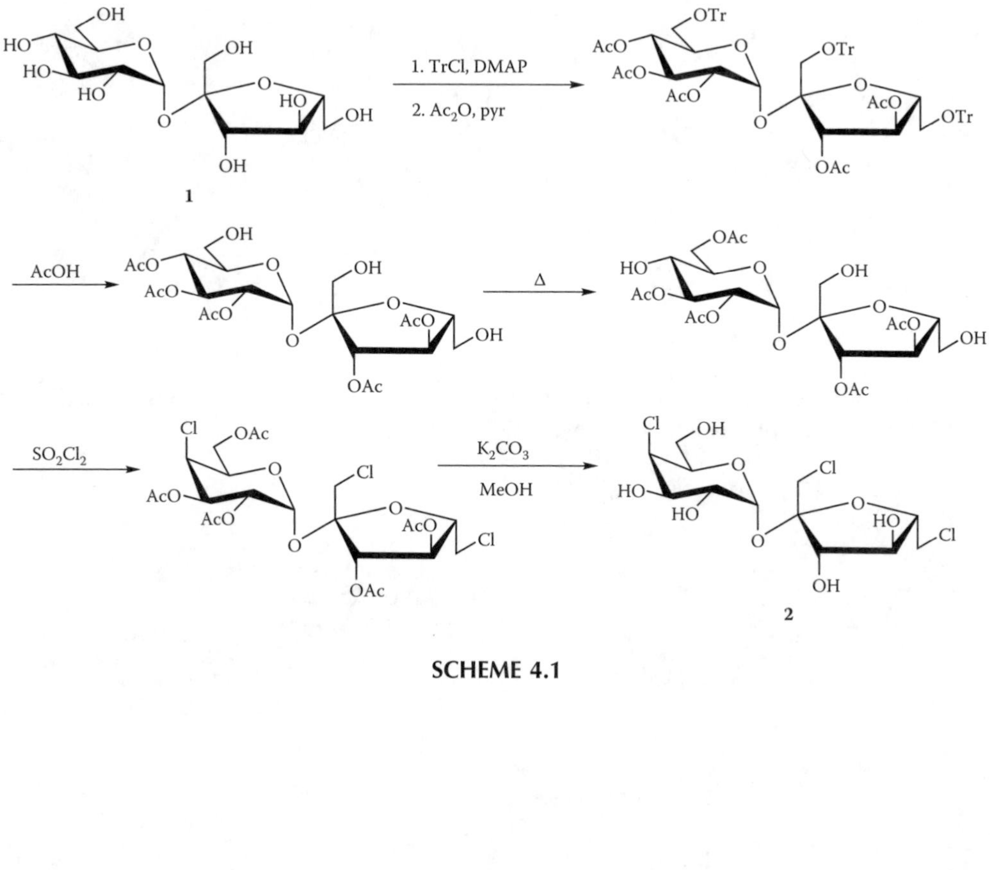

SCHEME 4.1

4.2.1.2.2 Enzymatic Approaches

Another route to **2** uses both chemical and biocatalytic transformations and starts from glucose (**3**) (Scheme 4.2).[28,29]

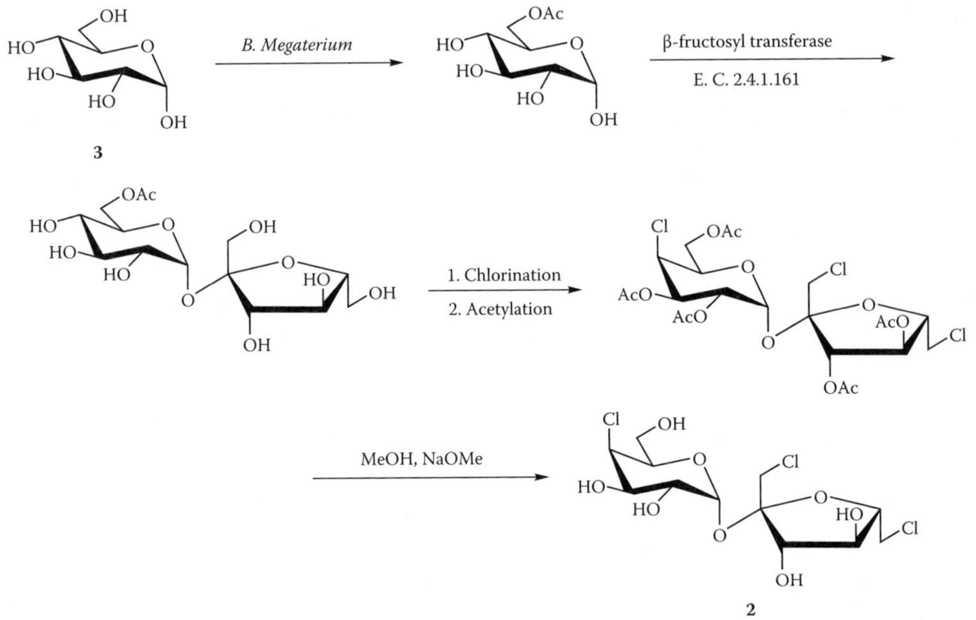

SCHEME 4.2

An enzymatic preparation to the useful intermediate **4** has been described from sucrose (Scheme 4.3).[30,31] Chlorination of this intermediate **4** yields TGS (**2**).

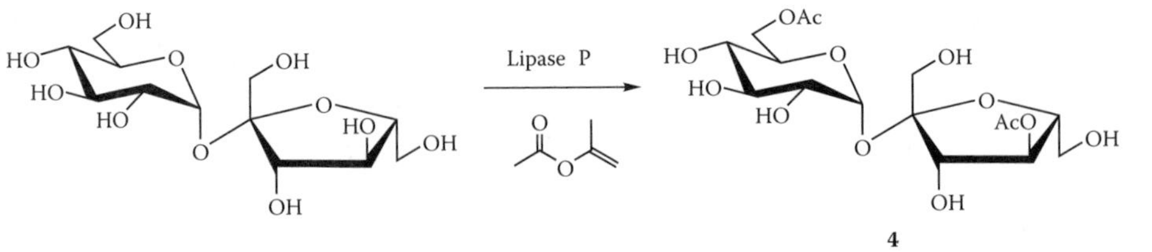

SCHEME 4.3

4.2.2 Isomalt

Isomalt (**5**), also called Palatinit, is used as a low-calorie sweetener in some countries. It is a mixture of two compounds obtained by the hydrogenation of isomaltulose (**6**) (Scheme 4.4).[15]

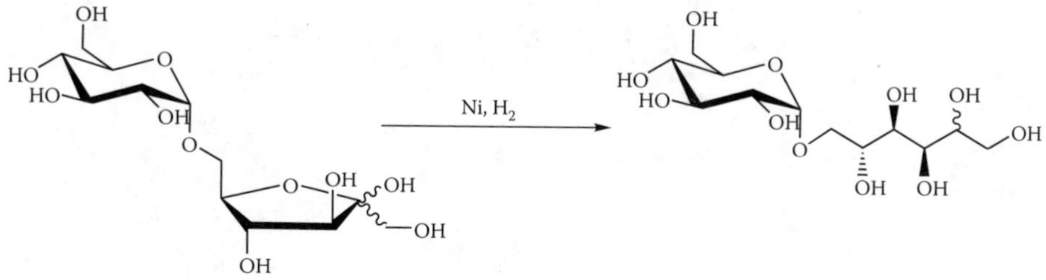

SCHEME 4.4

The disaccharide **6** is derived from sucrose by an enzymatic transformation catalyzed by *Protaminobacter rubrum* as well as other organisms.[15,32]

4.3 MONOSACCHARIDES AND RELATED COMPOUNDS

4.3.1 D-Gluconic Acid

D-Gluconic acid (**7**) is produced at scale by the fermentative oxidation of D-glucose (**3**) with *Acetobacter, Pseudomonas,* or *Penicillium sp.* (Scheme 4.5).[33] It is used as a processing aid.

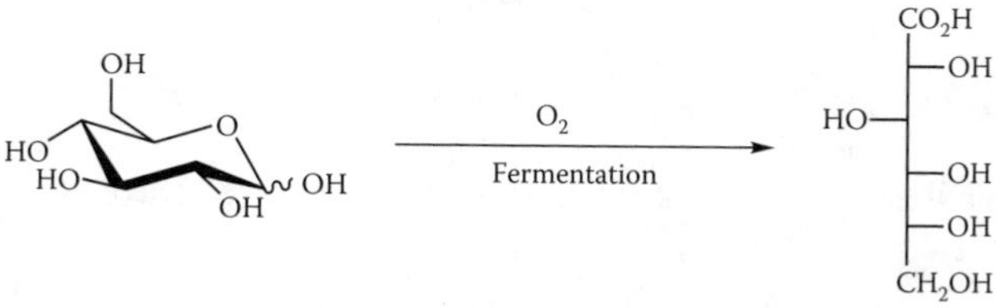

SCHEME 4.5

The same transformation also can be achieved by air oxidation in the presence of a number of metal catalysts and enzymes.[4,33]

4.3.2 D-Sorbitol

Sorbitol (**8**) is used at scale in the manufacture of resins and surfactants.[34] It is produced by the reduction of D-glucose (**3**) (Scheme 4.6).

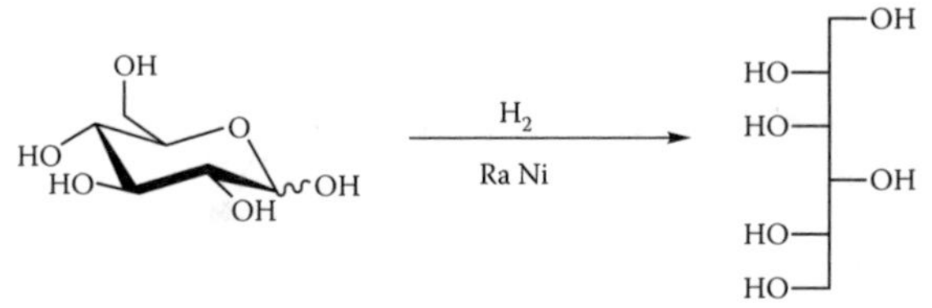

SCHEME 4.6

4.3.3 L-Ascorbic Acid

L-Ascorbic acid (**9**), vitamin C, is produced at large scale by the Reichstein-Grussner process that dates back to the 1920s (Scheme 4.7).[34] It involves the enzymatic oxidation of D-sorbitol (**8**) that is, in turn, obtained from D-glucose (Section 4.3.2).

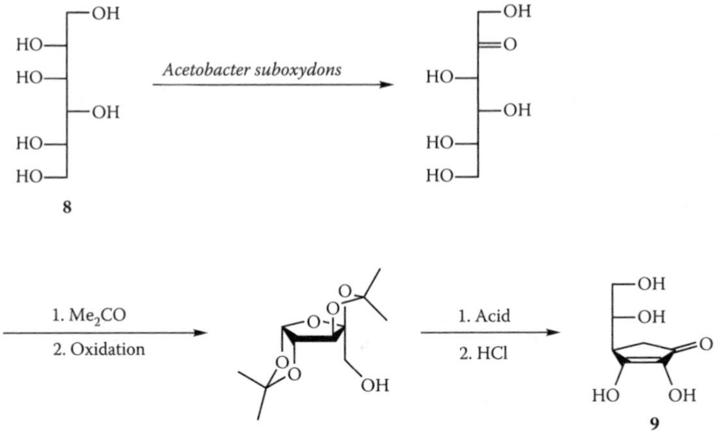

SCHEME 4.7

4.3.4 D-Erythrose

This valuable intermediate **10** is readily available by the oxidative cleavage of D-fructose (**11**) with hypochlorite (Scheme 4.8).[4]

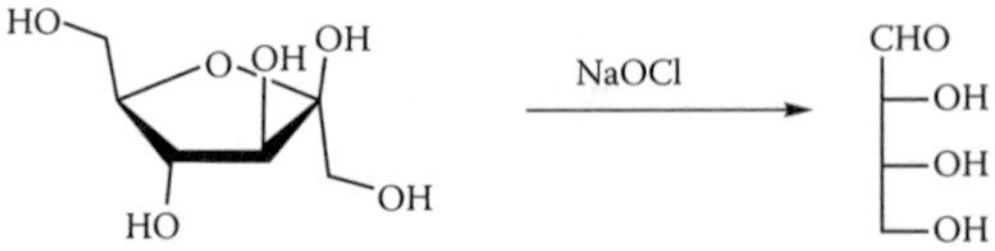

SCHEME 4.8

4.3.5 NUCLEOSIDES

Nucleosides are the building blocks of DNA and RNA and contain a carbohydrate unit together with a base. A number of derivatives have been prepared as pharmaceutical agents, perhaps the best known being AZT (azidothymidine, or zidovudine) (**12**). Some of these compounds may be accessed from ribose.[35] In many cases, a nucleoside is used as the starting material for modification; a number of routes to **12** use this approach.[36–38]

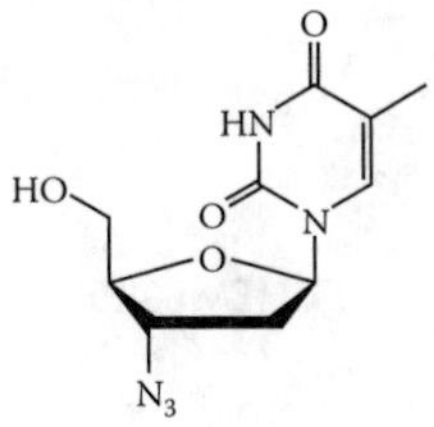

12

When the sugar has been modified, as in lamivudine (**13**) and GR95168 (**14**), most approaches do not use a carbohydrate precursor as the starting material; instead they rely on a resolution or asymmetric synthesis (Chapter 6).[39] Again, this illustrates the problems associated with the multitude of similar functional groups.[40–42]

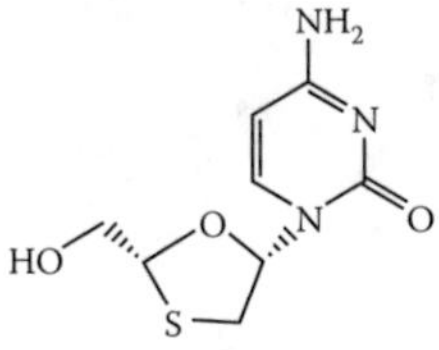

13

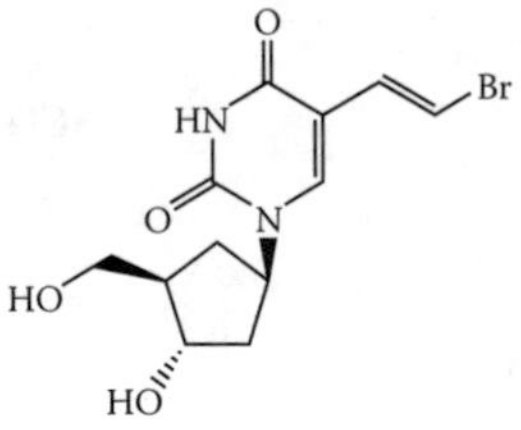

14

4.4 GLYCERALDEHYDE DERIVATIVES

4.4.1 S-SOLKETAL

The material S-solketal (**15**) can be obtained by the oxidative cleavage of D-mannitol (**16**) (Scheme 4.9). Of course, the use of heavy metals in the cleavage reaction does bring along environmental problems. The aldehyde **17**, although used for a wide variety of transformations, does suffer from epimerization problems; however, additions to the carbonyl group can be manipulated through the use of chelation control.[1]

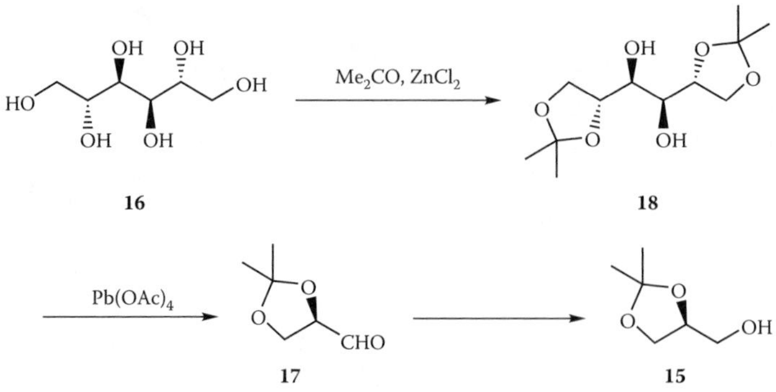

SCHEME 4.9

The alcohol **15** can also be prepared from protected glyceric acid.[43] This, in turn, can be obtained by cleavage of the protected diol **18** with bleach in the presence of a ruthenium catalyst (Scheme 4.10).[4,43]

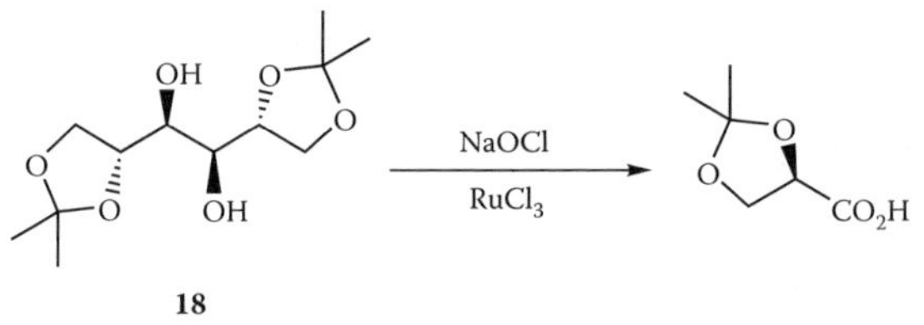

SCHEME 4.10

4.4.2 *R*-Solketal

This isomer of the aldehyde, **19**, can be accessed from L-ascorbic acid (**9**) (Scheme 4.11).[44]

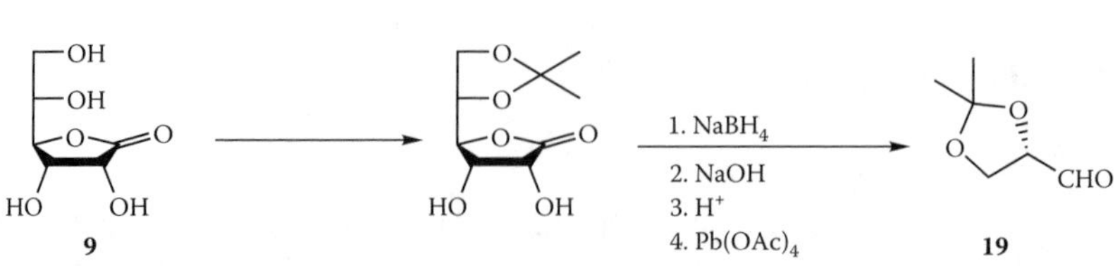

SCHEME 4.11

4.5 HYDROXY ACIDS

4.5.1 TARTARIC ACID

Both isomers of the compound tartaric acid are available at scale, yet it has not found large-scale application as a chiral building block. Tartaric acid can be used as a resolution agent (Chapter 6) and as a ligand for asymmetric catalysis. The *R,R*-isomer is available as a waste product of wine fermentation.[4] The *S,S*-isomer is a natural product but is produced by resolution and is much more expensive than its antipode.[4]

4.5.2 MALIC ACID

Again, the hydroxy acid malic acid is available by a wide variety of methods, including fermentation,[45] but it has not found a large-scale application as a building block.

4.5.3 LACTIC ACID

Both isomers of this hydroxy acid are available from fermentation.[45] Although *R*-lactic acid can be used for the synthesis of herbicides, the development of alternative methods to *S*-2-chloropropionic acid alleviates this need (Chapter 31).

4.5.4 3-HYDROXYBUTYRIC ACID

The hydroxy acid 3-hydroxybutyric acid is available as the *R*-isomer from the biodegradable polymer, Biopol (Chapter 19).[45] A number of derivatives are accessible from the polymer itself (Scheme 4.12).[46,47]

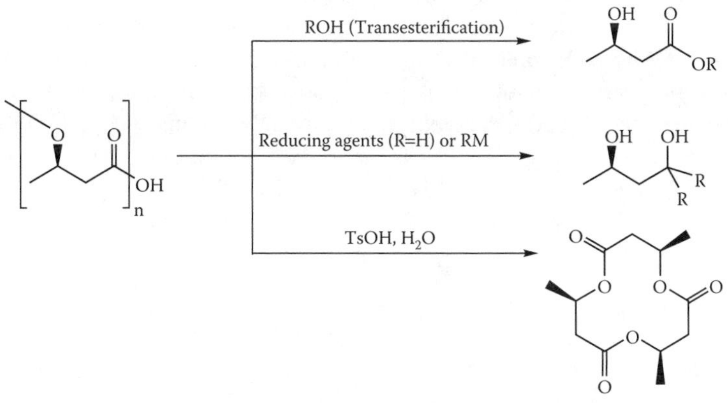

SCHEME 4.12

The *S*-isomer is available by yeast reduction of acetoacetate.[1,48–50] Both isomers are available by asymmetric hydrogenation of the β-keto ester (Chapters 12 and 13).[51]

The monomeric unit, as illustrated by 3*R*-hydroxybutyric acid, is a precursor to a wide range of β-hydroxy acid derivatives (Scheme 4.13).[52]

Although a wide range of β-hydroxy acids is available by reduction of the corresponding β-keto ester or acid (*vide supra*), and a number of schemes have been proposed to commercially important compounds, such as captopril, this methodology and class of compounds still have to wait for a new opportunity to show their versatility.[49,50]

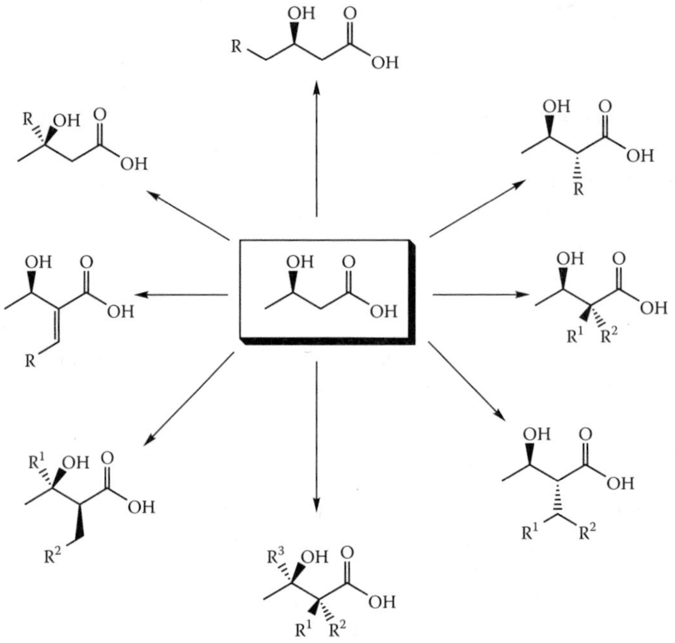

SCHEME 4.13

4.5.5 MACROLIDE ANTIBIOTICS

Macrolide antibiotics, in some instances, may be considered as hydroxy acid derivatives. In addition, many of them have carbohydrates attached, often with unique structural features, as illustrated by erythromycin (20).[53] Although some of these antibiotics are semi-synthetic, all are derived by a fermentation process where the antibiotic is formed as a secondary metabolite. This approach, of course, alleviates the need to perform complex carbohydrate chemistry (see Chapter 19).

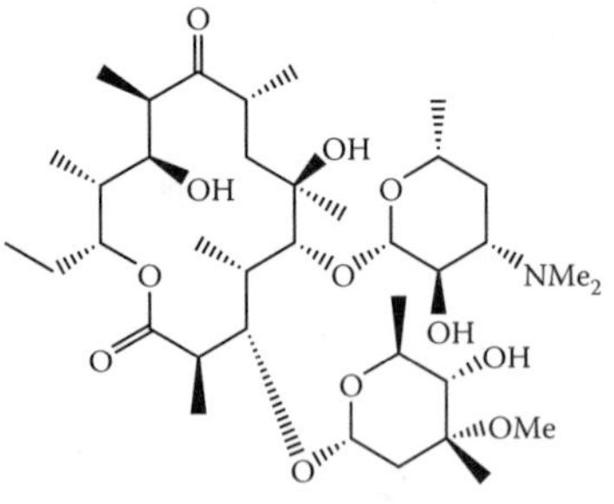

20

4.6 SUMMARY

Despite their widespread occurrence in nature, and despite some members of the class being available at scale, carbohydrates and hydroxy acids have not found widespread application as synthons in large-scale fine chemical synthesis with the exception of food ingredients. This is presumably a result of the problems associated with the differentiation of very similar functional groups, although enzymatic methods are known, and the low atom efficiency when incorporated into a synthesis.

REFERENCES

1. Ager, D. J., East, M. B. *Asymmetric Synthetic Methodology*, CRC Press: Boca Raton, 1995.
2. Scott, J. S. In *Asymmetric Synthesis,* Morrison, J. D. Ed., Academic Press: Orlando, 1984, Vol. 4, p. 1.
3. Lichtenthaler, F. W., Cuny, E., Martin, D., Rönninger, S., Weber, T. In *Carbohydrates as Organic Raw Materials,* Lichtenthaler, F. W. Ed., VCH: Weinheim, 1991, p. 207.
4. Sheldon, R. A. *Chirotechnology: Industrial Synthesis of Optically Active Compounds*, Marcel Dekker, Inc.: New York, 1993.
5. Lichtenhaler, F. W. *Acc. Chem. Res.* 2002, *35*, 728.
6. Hanessian, S., Banoub, J. *Synthetic Methods for Carbohydrates*, American Chemical Society: Washington, D.C., 1976.
7. Hanessian, S. *The Total Synthesis of Natural Products: The Chiron Approach*, Pergamon: Oxford, 1983.
8. Ager, D. J., East, M. B. *Tetrahedron* 1992, *48*, 2803.
9. Ager, D. J., East, M. B. *Tetrahedron* 1993, *49*, 5683.
10. Katsuki, T., Lee, A. W. M., Ma, P., Martin, V. S., Masamune, S., Sharpless, K. B., Tuddenham, D., Walker, F. J. *J. Org. Chem.* 1982, *7*, 1373.
11. Lee, A. W. M., Martin, V. S., Masamune, S., Sharpless, K. B., Walker, F. J. *J. Am. Chem. Soc.* 1982, *104*, 3515.
12. Ma, P., Martin, V. S., Masamune, S., Sharpless, K. B., Viti, S. M. *J. Org. Chem.* 1982, *47*, 1378.
13. Ko, S. Y., Lee, A. W. M., Masamune, S., Reed, L. A., Sharpless, K. B., Walker, F. J. *Science* 1983, *220*, 949.
14. Ko, S. Y., Lee, A. W. M., Masamune, S., Reed, L. A., Sharpless, K. B., Walker, F. J. *Tetrahedron* 1990, *46*, 245.
15. Schiweck, H., Munir, M., Rapp, K. M., Schneider, B., Vogel, M. In *Carbohydrates as Organic Raw Materials,* Lichtenthaler, F. W. Ed., VCH: Weinheim, 1991, p. 57.
16. BeMiller, J. N. In *Carbohydrates as Organic Raw Materials,* Lichtenthaler, F. W. Ed., VCH: Weinheim, 1991, p. 57.
17. van der Plank, P., Rozendaal, A. Eur. Pat. EP 256,585, 1988.
18. Ager, D. J., Pantaleone, D. P., Katritzky, A. R., Henderson, S. A., Prakash, I., Walters, D. E. *Angew. Chem., Int. Ed.* 1998, *37*, 1803.
19. Hough, L. *Chem. Soc. Rev.* 1985, *14*, 357.
20. Hough, L. *Nature* 1976, *263*, 500.
21. Jenner, M. R. In *Sweeteners: Discovery, Molecular Design, and Chemoreception,* Walters, D. E., Orthoefer, F. T., DuBois, G. E. Eds., American Chemical Society: Washington, DC, 1991, p. 68.
22. Jenner, M. R., Waite, D. U.S. Pat. 4,343,934, 1982.
23. Jenner, M. R., Waite, D., Jackson, G., Williams, J. C. U.S. Pat. 4,362,869, 1982.
24. Hough, L., Phandis, S. P., Khan, R. A. U.S. Pat. 4,435,440, 1984.
25. Fairclough, P. H., Hough, L., Richardson, A. R. *Carbohydr. Res.* 1975, *40*, 285.
26. Tully, W., Vernon, N. M., Walsh, P. A. Eur. Pat. EP 220907 A2, 1987.
27. Jackson, G., Jenner, M. R., Waite, D., Williams, J. C. U.K. Pat. 2065648, 1980.
28. Rathbone, E. B., Hacking, A. J., Cheetham, P. S. J. U.S. Pat. 4,617,269, 1986.
29. Jones, J. D., Hacking, A. J., Cheetham, P. S. J. *Biotechnol. Bioeng.* 1992, *39*, 203.
30. Dordick, J. S., Hacking, A. J., Khan, R. A. U.S. Pat. 5,270,460, 1993.
31. Dordick, J. S., Hacking, A. J., Khan, R. A. U.S. Pat. 5,128,248, 1992.
32. Crueger, W., Drath, L., Munir, M. Eur. Pat. 1,099, 1978.
33. Röper, H. In *Carbohydrates as Organic Raw Materials,* Lichtenthaler, F. W. Ed., VCH: Weinheim, 1991, p. 267.
34. Crosby, J. In *Chirality in Industry: The Commercial Manufacture and Applications of Optically Active Compounds,* Collins, A. N., Sheldrake, G. N., Crosby, J. Eds., Wiley: Chichester, 1992, p. 1.
35. Dekker, C. A., Goodman, L. In *The Carbohydrates: Chemistry and Biochemistry,* Pigman, W., Horton, D. Eds., Academic: New York, 1970, Vol. IIA, p. 1.
36. Chen, B.-C., Quinlan, S. L., Reid, G. J. *Tetrahedron Lett.* 1995, *36*, 7961.
37. Azhayev, A. V., Korpela, T. PCT Pat. Appl. WO 9307162, 1993.
38. Czernecki, S., Valery, J. M. *Synthesis* 1991, 239.

39. Bray, B. L., Goodyear, M. D., Partridge, J. J., Tapolczay, D. J. In *Chirality in Industry II: Developments in the Commercial Manufacture and Applications of Optically Active Compounds,* Collins, A. N., Sheldrake, G. N., Crosby, J. Eds., Wiley: Chichester, 1997, p. 41.

40. Paulsen, H., Pflughaupt, K.-W. In *The Carbohydrates: Chemistry and Biochemistry,* Pigman, W., Horton, D. Eds., Academic: New York, 1980, Vol. IB, p. 881.

41. Almond, M. R., Wilson, J. D., Rideout, J. L. U.S. Pat. 4,916,218, 1990.

42. Wilson, J. D. U.S. Pat. 4,921,950, 1990.

43. Emons, C. H. H., Kuster, B. F. M., Vekemans, J. A. J. M., Sheldon, R. A. *Tetrahedron: Asymmetry* 1991, *2*, 359.

44. Jung, M. E., Shaw, T. J. *J. Am. Chem. Soc.* 1980, *102*, 6304.

45. Crosby, J. In *Chirality in Industry,* Collins, A. N., Sheldrake, G. N., Crosby, J. Eds., Wiley: New York, 1992, p. 1.

46. Akita, S., Einaga, Y., Miyaki, Y., Fujita, H. *Macromolecules* 1976, *9*, 774.

47. Schnurrenberger, P., Seebach, D. *Helv. Chim. Acta* 1982, *65*, 1197.

48. Iimori, T., Shibasaki, M. *Tetrahedron Lett.* 1985, *26*, 1523.

49. Ohashi, T., Hasegawa, J. In *Chirality in Industry: The Commercial Manufacture and Applications of Optically Active Compounds,* Collins, A. N., Sheldrake, G. N., Crosby, J. Eds., Wiley: Chichester, 1992, p. 269.

50. Ohashi, T., Hasegawa, J. In *Chirality in Industry: The Commercial Manufacture and Applications of Optically Active Compounds,* Collins, A. N., Sheldrake, G. N., Crosby, J. Eds., Wiley: Chichester, 1992, p. 249.

51. Ager, D. J., Laneman, S. A. *Tetrahedron: Asymmetry* 1997, *8*, 3327.

52. Seebach, D. *Angew. Chem., Int. Ed.* 1990, *29*, 1320.

53. Hanessian, S., Haskell, T. H. In *The Carbohydrates: Chemistry and Biochemistry,* Pigman, W., Horton, D. Eds., Academic: New York, 1970, Vol. IIA, p. 139.

5 Terpenes: The Expansion of the Chiral Pool

Weiguo Liu

CONTENTS

5.1 INTRODUCTION

Nature has been, and continues to be, one of the greatest sources of chiral molecules that range from small amino acids to large proteins and nucleic acids to DNA. Synthetic organic chemists have long benefited from this resource by constantly being able to draw on members of these chiral natural products and converting them into more complex synthetic compounds, such as pharmaceuticals and other biologically active agents. During the past decade, an overwhelming amount of work has been done, with great success, on the use of amino acids and carbohydrates as chiral building blocks. In a similar manner, terpenes, especially monoterpenes, are also an important group of abundant natural products that contribute to nature's chiral pool. Many of these mono-terpenes are widely distributed in nature and can be isolated from a variety of plants in high yields.[1] They usually possess one or two stereogenic centers within their structures together with modest functionality, which allows convenient structural manipulation and transformations. Many terpenes have constrained asymmetric carbocyclic ring systems that can be incorporated into the frameworks of complex polycyclic target molecules or converted into chiral ligands for chiral auxiliaries, reagents, and resolving agents. Although amino acids and carbohydrates continue to be the major source of chiral building blocks in organic synthesis, terpenes are being used more and more frequently in asymmetric synthesis. In this chapter, we will discuss some of the applications in the use of terpenes, especially monoterpenes, in the synthesis of chiral organic molecules. The emphasis of the discussion will be on the use of natural or unnatural monoterpenes as building blocks in the synthesis of biologically active compounds and the stereoselective transformations and manipula-tions of these terpene molecules for asymmetric synthetic purposes.

5.2 ISOLATION

Some of the most abundant and readily available monoterpenes in nature's chiral pool include pinenes, menthones, menthols, carvones, camphor, and limonenes, which can be isolated in large quantities from various species and parts of plants found in different regions of the world.[2] These terpenes can also be interconverted through chemical and biological transformations and can be derivatized into a variety of oxygenated forms by simple chemical manipulations.[3] Current industrial production of many optically active monoterpenes is still through the extraction of plants, usually by steam distillation. Essential oils extracted from a variety of plants contain a mixture of a number of different monoterpenes, which are separated and purified by repeated distillation.[4] Because the essential oil industry is highly localized by region, and characterized by its diversity, the chemical and optical purities of commercial monoterpenes vary widely depending on region, the plant source, and production methods.[5] Unlike natural amino acids, which almost exclusively have the (S)-configuration, many monoterpenes exist in nature in the form of both enantiomers, which are often distributed in different plant species and in different quantities. The ee of a particular compound, therefore, can vary according to the region of origin. Many terpenes are "scalemic," which means that both antipodes are made by a single source, although one usually predominates.[6] Chemists who use these terpenes for chiral synthetic purposes should always be cognizant of the quality and producer of the product and, if possible, stay with just one for consistent, reliable product quality.

5.3 MONOTERPENES

5.3.1 α-Pinene

α-Pinene is perhaps the most abundant and chemically exploited monoterpene. It is isolated principally from a variety of pine trees, and both of its optical enantiomers occur in nature, although (+)-α-pinene (**1**) is the most abundant.[7] Bulk quantities of both optically active (+)- and (−)-α-pinenes are available. (For example, this is available from Penta Manufacturing Co., Fairfield, NJ., and from other fine chemical producers and dealers.) α-Pinenes obtained from natural sources have optical purities ranging from 80–90% ee as a result of the coexistence of the other enantiomer. The best (+)-α-pinene is about 91.3% ee, whereas the (−)-α-pinene (**2**) is only 81.3%.[8] This kind of optical purity is usually not good enough for many chiral synthetic purposes. For the preparation of optically pure chiral borane reagent, diisopinocampheylborane (Ipc$_2$BH) (**3**), a simple and practical procedure has been developed to upgrade the optical purity of commercial pinenes to 99% ee.[9] For (+)-α-pinene, the commercial material (91.3% ee) was treated with BH$_3$•THF (tetrahydro-furan) or BH$_3$•SMe$_2$ to produce the hydroboration product Ipc$_2$BH (**3**) as a crystalline solid. Digestion of the solid suspended in THF with 15% excess α-pinene caused the (+)-isomer to become incorporated into the crystalline Ipc$_2$BH, whereas the (–)-isomer accumulates in solution. Filtration removes the (–)-isomer to give crystalline **3** that contains 99% (+)-α-pinene. Elimination of the alkyl groups from the trialkylboranes by treatment of the Ipc$_2$BH (**3**) with benzaldehyde and a catalytic amount of BF$_3$ at 100°C, followed by distillation, gives (+)-α-pinene (99.5% ee) (Scheme 5.1).[10] The same process can be applied to upgrade (–)-α-pinene.

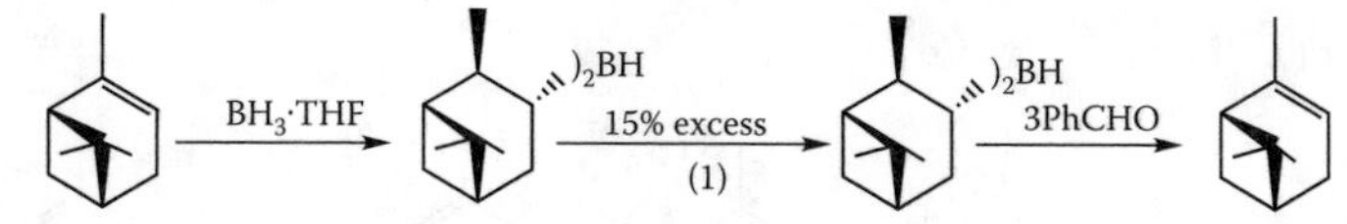

SCHEME 5.1

As a result of less abundant natural sources and its inferior optical purity, good-quality (–)-α-pinenes are usually obtained by chemical isomerization of (–)-β-pinenes. The isomerization is brought about by various catalysts, including acids,[11] bases,[12] and metals.[13] For example, (–)-α-pinene (**2**) (92% ee) was obtained from commercial (–)-β-pinene (**4**) of the same optical purity in high yield through isomerization catalyzed by potassium 3-aminopropylamide (KAPA) (**5**) (Scheme 5.2).[14] The (–)-α-pinene (**2**) obtained by this method can be further upgraded in optical purity by the hydroboration method mentioned earlier.

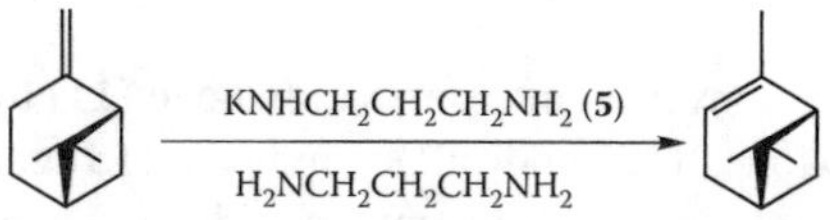

SCHEME 5.2

5.3.2 β-Pinene

β-Pinene is an important raw material for various perfumes and polyterpene resins.[7] In contrast to α-pinene, only the (–)-isomer of β-pinene is isolated from nature in large quantities. The (+)-isomer of β-pinene (**6**) is produced mainly by chemical means from (+)-α-pinene (**1**) through the use of various isomerization methods.[15–17] One of these methods is to heat Ipc$_2$BH (**3**) from **1** with high optical purity to about 130°C, followed by treatment with 1-hexene and benzaldehyde to release the (+)-β-pinene (**6**) (>99.5% ee) (Scheme 5.3).[10]

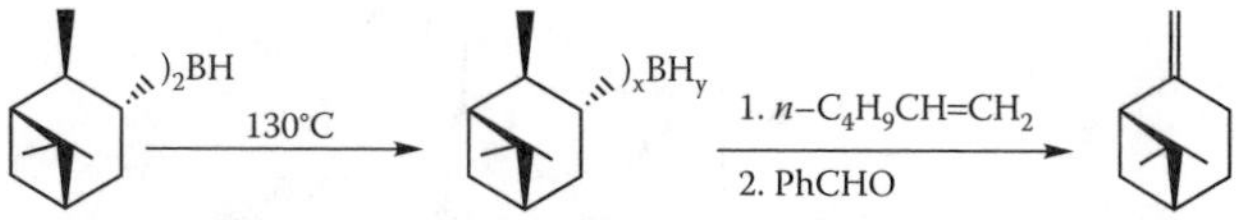

SCHEME 5.3

5.3.2.1 Other Monoterpenes Derived from Pinenes

Both α- and β-pinenes are popular starting materials for the synthesis of other monoterpene chiral synthons such as carvone, terpineol, and camphor (*vide infra*). Reactions leading to other monoterpenes are briefly summarized in Figure 5.1. Treatment of α-pinene with lead tetraacetate followed by rearrangement gives *trans*-verbenyl acetate (**7**), which is hydrolyzed to yield *trans*-verbenol (**8**).[18] Subsequent oxidation of **8** gives verbenone (**9**), which can be reduced to give *cis*-verbenol

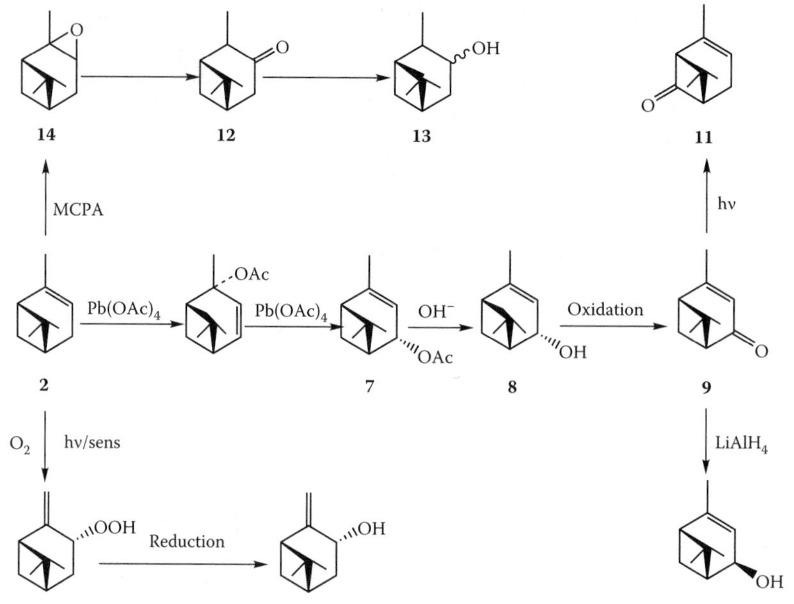

FIGURE 5.1 Conversion of α-pinene to other monoterpenes.

(**10**). Verbenone (**9**) can also be converted into chrysanthenone (**11**) by ultraviolet irradiation.[19] The naturally occurring pinocamphones (**12**) and pinocampheols (**13**) can be prepared from α-pinene by conventional ring-opening reactions of α-pinene epoxide (**14**) and the reduction.[20,21] One of the most effective methods for making *trans*-pinocarveol (**15**) is by sensitized photooxygenation of α-pinene followed by reduction of the resultant *trans*-pinocarveyl hydroperoxide (**16**).[22]

5.3.3 Limonene

Limonene is another inexpensive and abundantly available chiral pool material. Both *d*- and *l*-limonenes (**17** and **18**) are commercially available. *d*-Limonene (**17**) is the principal stereoisomer that can be isolated in large quantities from orange, caraway, dill, grape, and lemon oils, whereas *dl*- and *l*-limonenes are obtained from a variety of terpentine oils.[23] Racemic limonenes can be synthesized by the dimerization of isoprene in a Diels-Alder condensation. Also, a 5-carbon to 4-carbon coupling under Diels-Alder conditions followed by Wittig addition of the methylene group has been used to produce racemic limonene (Scheme 5.4).[24]

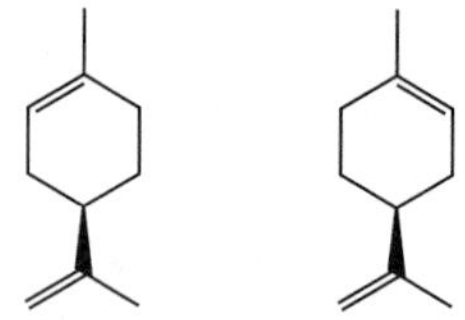

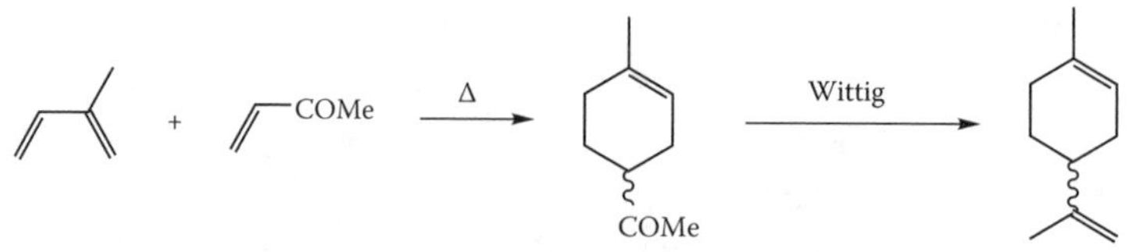

SCHEME 5.4

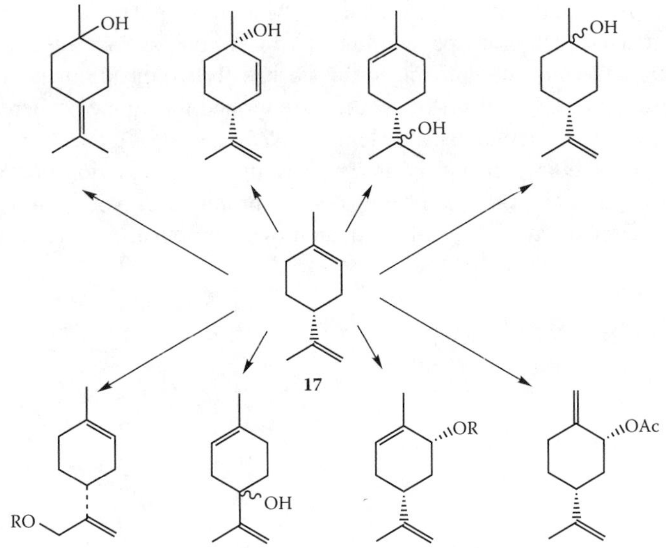

FIGURE 5.2 Monoterpenes accessible from *d*-limonene (**17**).

The degree of chemical and optical purities of limonenes, again, depends on the exact source of the oil. Bulk *d*-limonene currently available in the United States has optical purities in the range of 96–98% ee.

5.3.3.1 Monoterpenes Derived from Limonene

Although limonene is used directly in the constitution of natural flavors and perfumes, its principal value lies in its use as a starting material for chemical synthesis and production of a variety of oxygenated *p*-menthane monoterpenes, a group of important organoleptic compounds and chiral building blocks in organic synthesis.[1,25] Many of these compounds are derived from a common intermediate, limonene epoxide, which is obtained through peroxyacid oxidation of limonene.[26,27] For example, acid catalyzed ring opening of (+)-limonene epoxide (**19**) gives a 1,2-diol, which after acetylation, pyrolysis of the resultant diacetate, and hydrolysis yields carveol, which in turn can be oxidized to give (−)-carvone (**20**) (Scheme 5.5).[28] Other *p*-menthane type of monoterpenes that can be directly derived from limonene include α- and β-terpineol (**21**) and various mentha-di-enols and menthen-diols (Figure 5.2).[29]

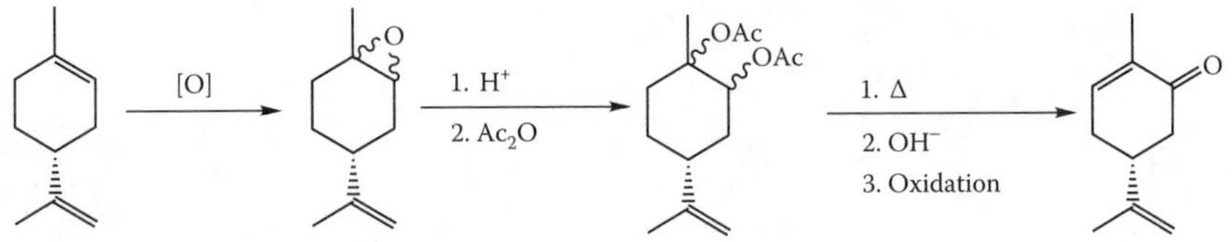

SCHEME 5.5

5.3.4 MENTHONES

p-Menthanes, which include menthol, menthone, terpineol, and carvone, are some of the best-known monoterpene-based chiral synthons in organic synthesis. They are all relatively inexpensive and are commercially available in bulk quantities. Menthols and their corresponding ketones, menthones, were first isolated from peppermint oils of various species of *Mentha piperita L.*[30] Menthones can exist in two diastereomeric forms, each with two enantiomers. The ones with the methyl and isopropyl groups in a *trans*-orientation are termed menthones, whereas those with the two in a *cis*-orientation are isomenthones. *l*-Menthone (**22**), also denoted as (*1R,4S*)-(–)-menthone, is the most abundant stereoisomer and is obtained by the dry distillation of the wood of *Pinus palustris Mell.*[31] It can also be produced chemically by chromic acid oxidation of *l*-menthol (**23**).[32] *l*-Menthone is commercially available in bulk quantities with optical purities of 90–98% ee.[33]

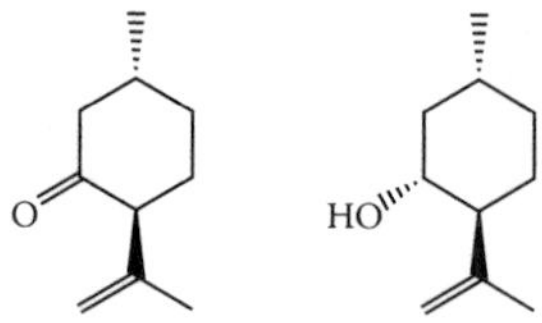

22   23

5.3.5 MENTHOLS

There are four diastereomeric menthols according to the relative orientations of the methyl, isopropyl, and hydroxyl groups. They are known as menthol, isomenthol, neomenthol, and neoisomenthol, each of which exists in two enantiomeric forms. The most abundant and readily available is *l*-menthol, or (*1R,2S,5R*)-(–)-menthol (**23**), whereas other stereoisomers, although available from nature, are more economically produced through chemical derivatizations from menthone, limonene, and α-pinene. Commercial (–)-menthol from natural sources can be as pure as 99% in both chemical and optical purities. Synthetic menthol is currently made by an asymmetric isomerization (see Chapters 12 and 31).[34]

5.3.6 CARVONES

Both enantiomeric forms of carvone are readily available. *l*-(–)-Carvone (**20**) is the major constituent of spearmint oil from *Mentha spicata*, whereas *d*-(+)-carvone is a major constituent of caraway oil from the fruit of *Carum carui* or dill from the fruit of *Anethum graveolens*. *d,l*-Carvone is found in gingergrass oils. It is interesting that the two enantiomers of carvone have quite different characteristic odors and flavors, suggesting that chirality plays an important role in organoleptic function.[1] Synthetic *l*-carvone (**20**) is produced on an industrial scale either from *d*-limonene (**17**) (*cf.* Scheme 5.5) or from α-pinene. The conversion of **17** into **20** is accomplished in 3 steps through the intermediates *d*-limonene nitrosochloride (**24**) and *l*-carvoxime (**25**), with greater than 60% overall yield (Scheme 5.6).[35]

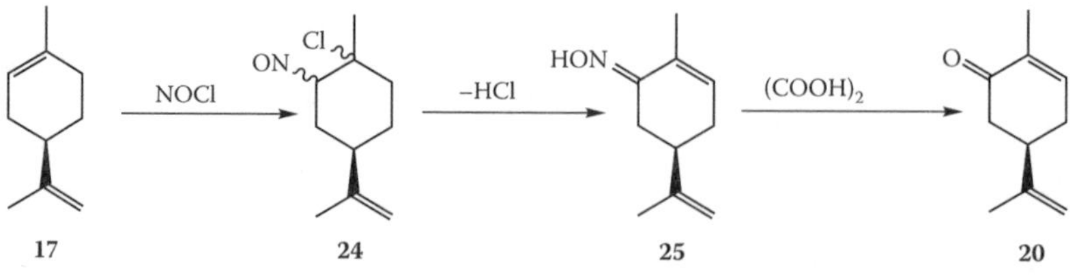

17   24   25   20

SCHEME 5.6

An alternative route to *l*-(–)-carvone (**20**) was developed from the more abundant (–)-α-pinene (**2**), albeit a somewhat lower overall yield (Scheme 5.7).[36] The process involves hydroboration of **2** followed by oxidation of the resultant alcohol. Subsequent treatment of the *l*-isopinocamphone

(**26**) with isopropenyl acetate then anodic oxidation gives almost pure *l*-(–)-carvone (34% overall). Both processes are being applied on industrial scale production, depending on the availability of the starting material. Optical purity of currently available commercial carvones ranges from 98–99% ee.

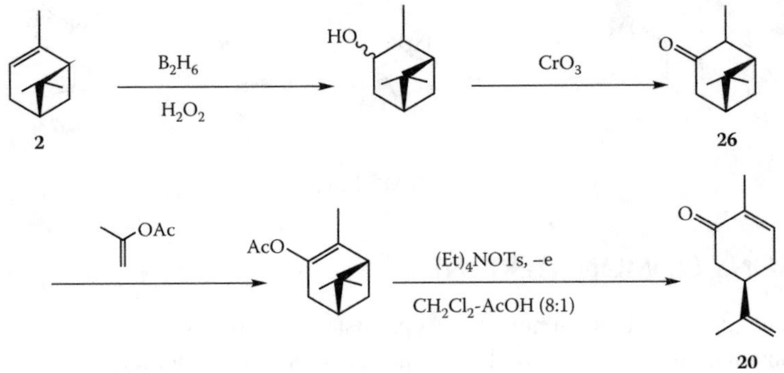

SCHEME 5.7

5.3.7 TERPINEOL

Terpineol is mainly isolated from plants of the Eucalyptus species and is a mixture of α-terpineol, β-terpineol, 4-terpineol, and 1,4- and 1,8-terpindiols. Both enantiomers of α-terpineol are commercially produced through hydration of α-pinenes.[37]

5.3.8 CAMPHOR

Camphor (**27**) is, perhaps, the most well-known monoterpene. It certainly has seen the most widespread commercial applications than any other single terpene, with a current 50–60 million pounds per year market. The most abundant natural source of camphor is the wood of various species of *Cinnamomum camphora,* a tree common to the Far East.[38] Although both *d-* and *l-* camphor have been isolated from a number of natural sources, most commercial camphor is now obtained through the chemical transformation of pinene (Figure 5.3).[39] Wagner-Meerwein rearrangement of α-pinene in HCl leads to bornyl chloride, which loses hydrogen chloride on heating to give camphene (**28**). The latter is then converted to an isobornyl ester, which is oxidized to yield camphor. There have been many variations of this pinene-to-camphor conversion process, such as direct catalytic conversion of pinene to camphene and methanolysis of camphene in the presence of strong cation exchange resin.

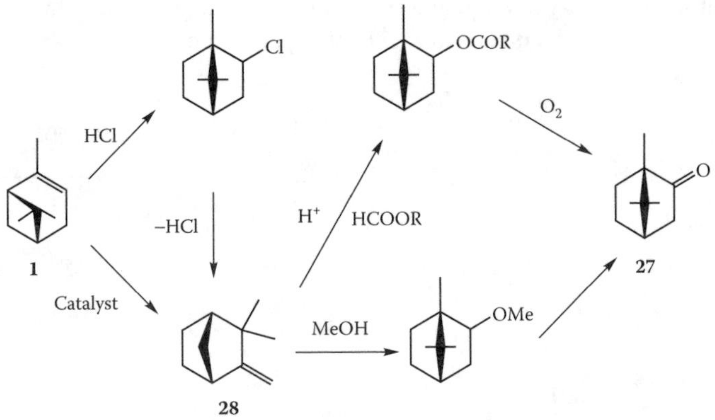

FIGURE 5.3 Routes for conversion of α-pinene to camphor.

Racemic camphor can be produced by total synthesis from 1,1,2-trimethylcyclopentadiene and vinyl acetate through a Diels-Alder reaction (Scheme 5.8).[40] Currently, about three-fourths of the camphor sold in the United States are produced synthetically, and most is in the racemic form.[41]

SCHEME 5.8

5.4 REACTIONS OF MONOTERPENES

The most fascinating aspect of monoterpene chemistry is perhaps the variety of stereoselective organic reactions that taken place around the chiral centers of these molecules. The architecture of many of the natural monoterpenes, especially cyclic and multi-cyclic monoterpenes, renders them unique steric features and imposes strong influences of steric interactions and facial selectivities. These differentiated steric interactions around the molecule allow distinct stereospecific or stereo-selective reactions to take place. Taking advantage of these steric features, synthetic organic chemists have designed various reactions to translate the chirality of these monoterpenes into various organic compounds generated in specific synthetic reactions. In these reactions, the chiral terpenes are used as chiral auxiliaries to direct the reaction to occur only in the preferred steric conformation resulting in stereochemically pure or enriched products. As a result of these stereoselective transformations, a great number of chiral organic compounds or building blocks have been derived directly or indirectly from this class of natural products. A few examples are given later in the chapter.

In addition, monoterpenes can provide some useful chiral reagents, such as the pinene-based organoborane reagents for chiral reductions, which have been reviewed extensively.[42] Camphor-derived organic acids such as camphenesulfonic acid can be used for the resolution of racemic bases and is a common practice in industry (Chapter 6).

5.4.1 CAMPHOR AS A CHIRAL AUXILIARY

5.4.1.1 Alkylations

Camphor and camphor-derived analogues are used frequently as chiral auxiliaries in asymmetric synthesis (*cf.* Chapter 23). There have been numerous reports in the use of camphor imine as templates to direct enantioselective alkylation for the synthesis of α-amino acids, α-amino phos-phonic acids, α-substituted benzylamines, and α-amino alcohols (e.g., Scheme 5.9).[43–47] Enantiomeric excesses of the products range from poor to excellent depending on the type of alkyl halides used.

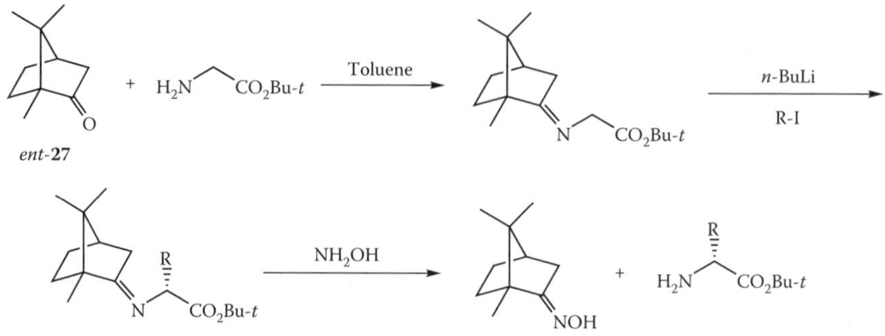

SCHEME 5.9

Other camphor analogues with the C–8 methyl group replaced by larger and more hindered groups can significantly increase the stereoselectivity of the alkylation and result in much higher enantiomeric purity of the products.[48]

5.4.1.2 Aldol Reaction

Camphor derivatives are also used as chiral auxiliaries in asymmetric aldol condensations (e.g., Scheme 5.10).[49-51]

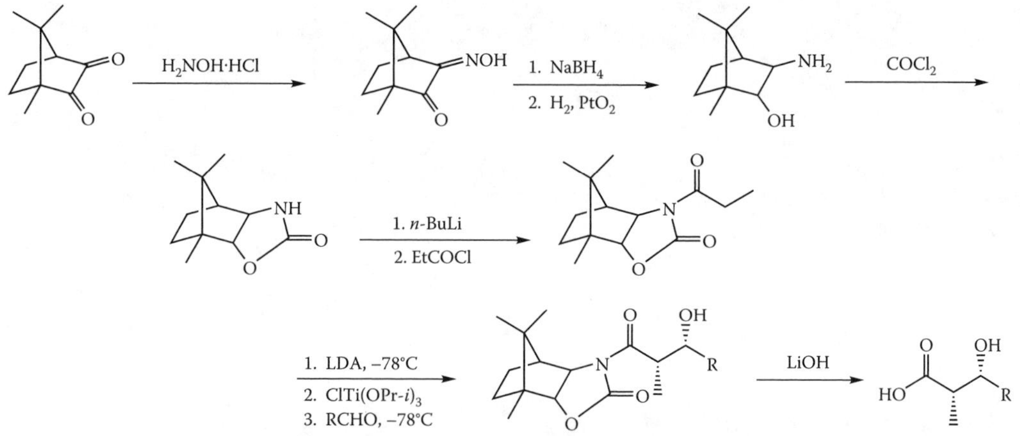

SCHEME 5.10

As with most camphor-based chiral auxiliaries, the size and steric congestion at the C–8 position of the camphor moiety can determine its efficiency. Through the formation of a connection between C–8 and C–2, a novel camphor-based oxazinone auxiliary **29**, which can be prepared from camphor in 3 steps,[52] becomes highly effective in directing stereoselective aldol reactions (Scheme 5.11).[53]

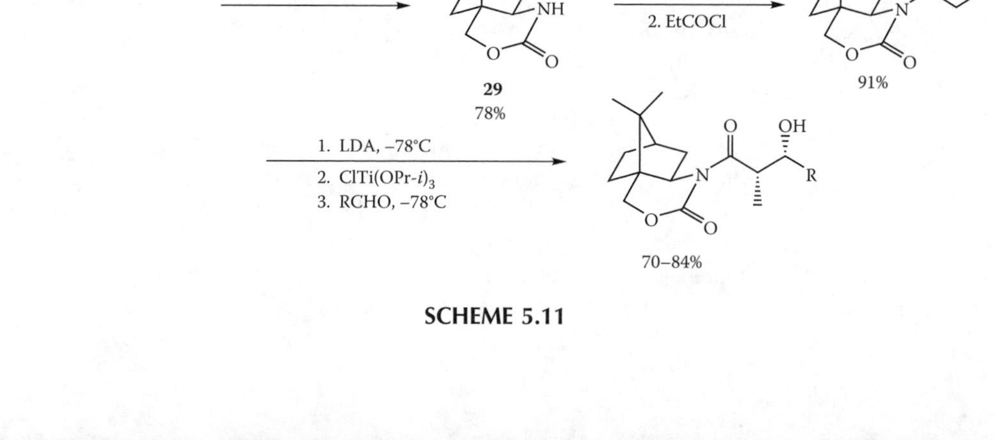

SCHEME 5.11

5.4.1.3 Diels-Alder Reaction

Another application of camphor-based chiral auxiliaries is the use in stereoselective Diels-Alder cycloadditions (Chapter 26), especially for the construction of quaternary carbon centers. One example, shown in Scheme 5.12, uses a camphor-derived lactam **30** as the auxiliary.[54]

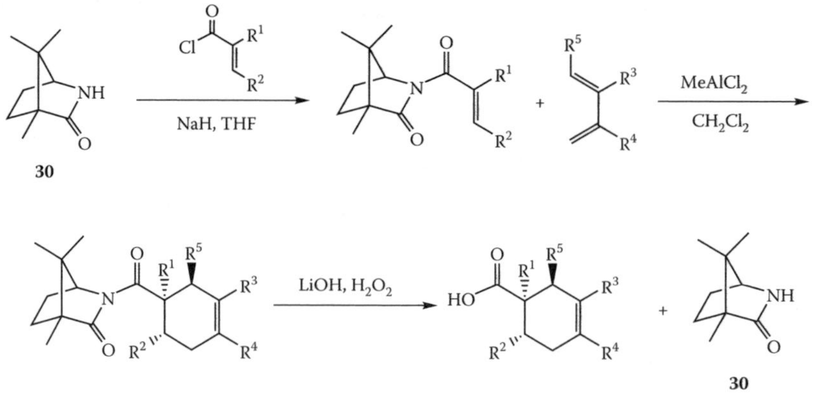

SCHEME 5.12

5.4.2 MENTHOL AS AN AUXILIARY

Optically active menthols are also commonly used as chiral auxiliaries in industrial production of fine chemicals. The most recent example of using this inexpensive and readily available monoterpene as chiral auxiliary for industrial-scale production of enantiomerically pure pharmaceuticals includes the synthesis of the anti-human immunodeficiency virus (HIV) drug 3TC (**31**) or Lamivudine marketed by Glaxo Wellcome. 3TC is an oxathiolane nucleoside with 2-hydroxymethyl-5-oxo-1,3-oxathiolane as the sugar portion and cytosine as the nucleobase. It contains two chiral centers and, thus, can exist in 4 stereoisomeric forms. The drug was originally developed as a racemic mixture of the β-anomer, but it later was converted to the optically pure form as a result of favorable toxicology and pharmacokinetic profiles. There have been several independent routes leading to the synthesis of optically active 3TC that involve classic resolution, or enzymatic resolution, or asymmetric synthesis. The current production route uses (–)-menthol (**23**) as the chiral auxiliary (Scheme 5.13).[55] The glycolate **32** is coupled with dithiodiene (**33**), the dimer of thioacetyl aldehyde, to produce the menthyl 4-oxo-1,3-oxathiolane-2-carboxylate (**34**). The presence of (–)-menthol as the chiral directing group results in the stereoselective formation of the 2R-isomer. The lactol **34** is then treated with thionyl chloride to generate, *in situ*, the chloro derivative **35**, which is coupled with TMS-cytosine to yield, predominately, the β-nucleoside analogue **36**. Again, the presence of the (–)-menthyl carboxylate group at the 5-position plays a major role in directing the stereoselectivity of this coupling reaction.

5.4.3 MONOTERPENES AS CHIRONS

Despite their low cost and abundant availability, the applications of monoterpenes as chiral synthons or building blocks for synthesis of chiral fine chemicals on an industrial scale have lagged far behind amino acids and carbohydrates. Most of the work in this area is related to multi-step total synthesis of complex natural products in laboratory scale. With the structures of new drug candidates in the research and development pipeline of pharmaceutical companies getting bigger and more complicated, the application of more sophisticated chiral building blocks such as the terpenes will

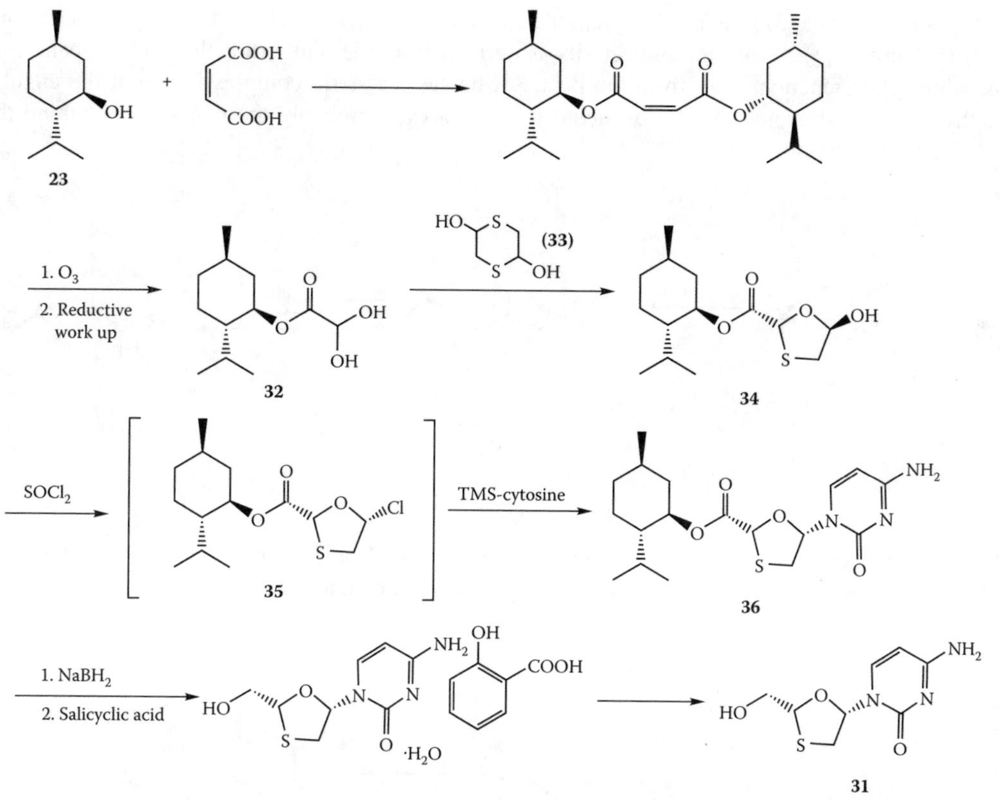

SCHEME 5.13

increase in the coming years. The major challenge lies in the design and development of practical chemical or biochemical reactions that effectively incorporate the simple monoterpenes into the target molecule of the desired products.

5.4.3.1 Robinson Annelation

One of the earliest industrial applications of monoterpenes is in the steroid synthesis. Robinson annelation is used frequently in these transformations as a key step. The first example of this reaction was by Robinson himself to synthesize α-cyperone from dihydrocarvone.[56] This synthesis has been shown to follow a stereospecific course to give (+)-α-cyperone (**37**) from (+)-dihydrocarvone (**38**) (Scheme 5.14).[57-59] Since then, numerous modifications and improvements have been made to apply this type of reaction for syntheses of a variety of natural and unnatural products.[60]

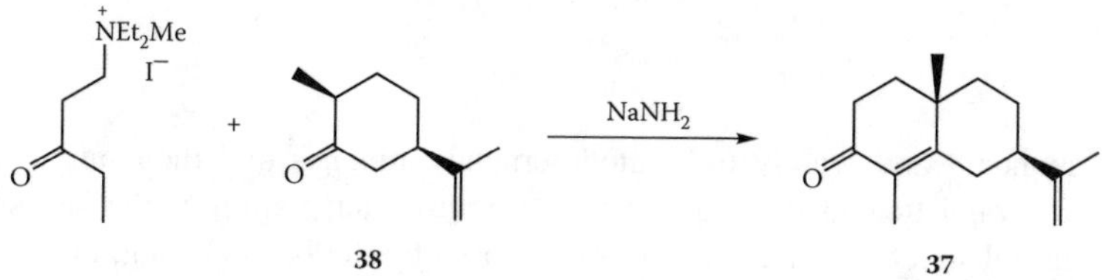

SCHEME 5.14

(+)-α-Cyperone (**37**) can also be prepared stereoselectively in good yields from (−)-3-caranone (**39**) (Scheme 5.15).[61] The stereoselectivity is derived from the bulky dimethylcyclopropane ring that directs the reaction away from the β-face, whereas the steric compression from the angular methyl group and avoidance of 1,3-diaxial interactions are probably the major force behind the selectivity of scission.

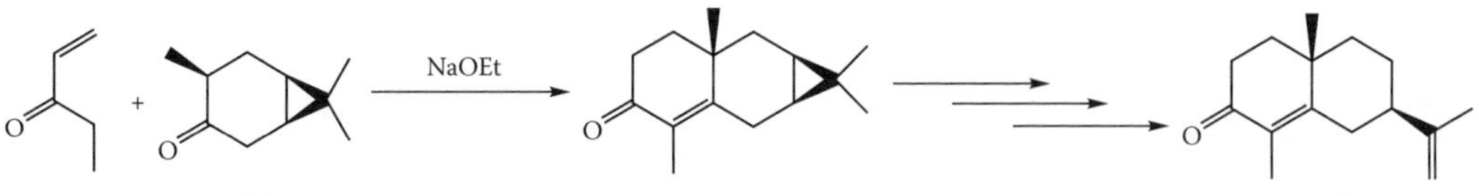

SCHEME 5.15

It is much easier to prepare (+)-β-cyperone (**40**) by the Robinson method because it has only one asymmetric center and one enol form for alkylation (Scheme 5.16).[62]

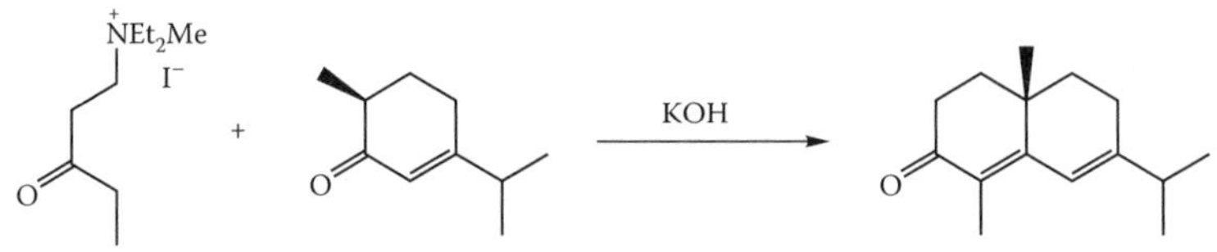

SCHEME 5.16

The annelation process in the Robinson reaction can also be carried out in a stepwise mode, so the alcohols from the aldol condensation step are often obtained (e.g., Scheme 5.17).[63]

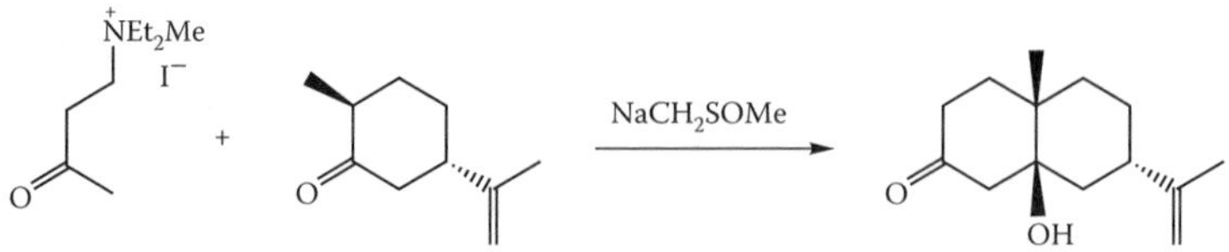

SCHEME 5.17

The most abundant and widely used monoterpenes, pinenes and their analogues, are often upgraded to more sophisticated structures with a Robinson annelation as the key step. However, as a result of unusual stereochemical constrains imposed by the bicyclic ring, only certain conformational isomers of the pinane skeleton undergo the annelation. Whereas the diketone (**41**) derived from *cis*-methylnopinone cyclized satisfactorily to the tricyclic ketone under Robinson conditions (Scheme 5.18), the epimeric diketone would not.[64]

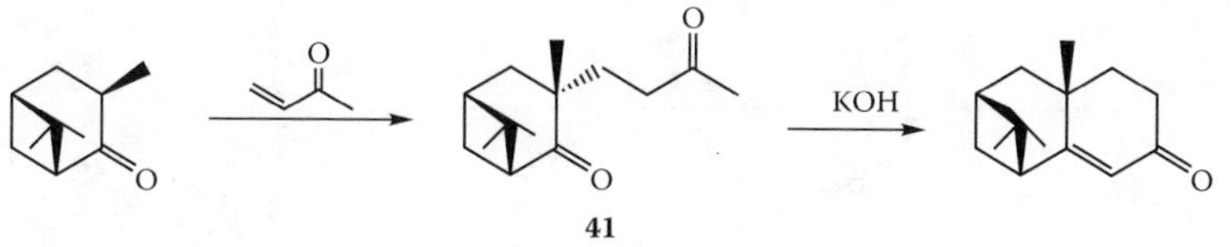

41

SCHEME 5.18

5.4.3.2 Steroid Synthesis

As a result of the biological importance and widespread medical and industrial utilities, estrogens and related hormones are among the earliest targets in steroid total synthesis. Major efforts in synthesis of these aromatic steroids were to set the correct chirality in the C,D rings because the A,B rings are mostly achiral. The following is an example of using monoterpenes as building blocks for synthesis of estrogen-type steroids (Scheme 5.19). Starting from the readily available *d*-(–)-camphor (**27**), direct functionalization at the C–9 methyl group was accomplished through 3 efficient steps to give (–)--bromocamphor (**42**) with complete retention of chirality. The bromo ketone was converted to the cyano ketone **43** by a number of transformations. The ketone **43** was then reacted with methyl vinyl ketone followed by POCl$_3$-pyridine treatment to give the tricyclenone **44**, a critical C,D intermediate in aromatic steroid synthesis. At this stage, all the chiral centers in the final product are correctly set, and the tricyclo-ring system is together with the appropriate functionalities to allow completion of the steroid synthesis.[65,66] The construction of the A ring on the C,D intermediate can be affected by many alternative methods.[67,68]

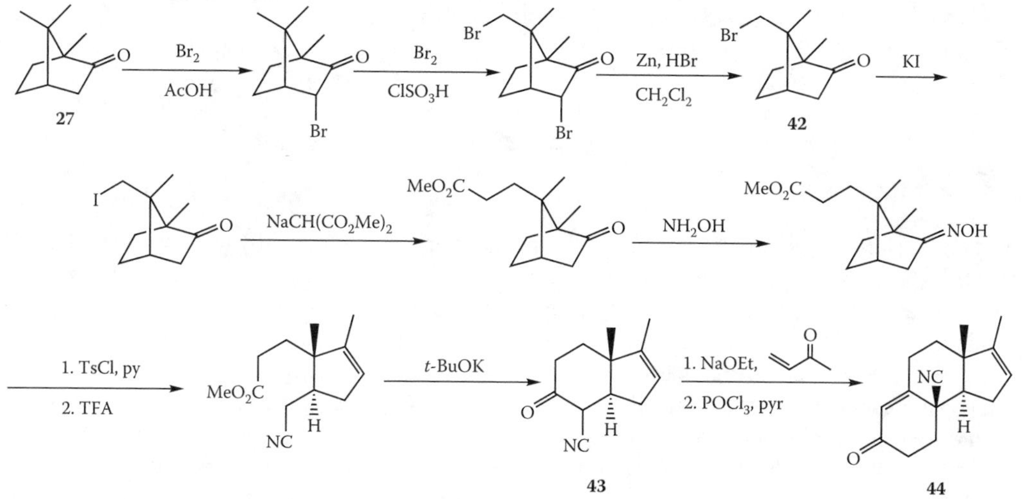

43

44

SCHEME 5.19

5.4.3.3 Cinmethylin

Finally, an example of using monoterpenes as building blocks for production of chiral fine chemicals is demonstrated in the synthesis of cinmethylin (**45**), a herbicide developed by Shell Oil Company (Scheme 5.20).[69] Enantiomerically pure *S*-terpinene-4-ol (**46**) was subjected to diastereoselective epoxidation by *tert*-butyl hydroperoxide in the presence of vanadium(III) acetylacetonate to generate the *cis*-epoxy-alcohol (**47**).[70] The diastereoselectivity of this epoxidation is effected through chelation of vanadium with the 4-hydroxyl group to deliver the oxygen from the top face. The chiral epoxide **47** was opened by sulfuric acid followed by dehydration of the resultant 1,4-diol to give 2-(*R*)-*exo*-1,4-cineole (**48**), which reacts with *o*-methylbenzylbromide to give cinmethylin (**45**).

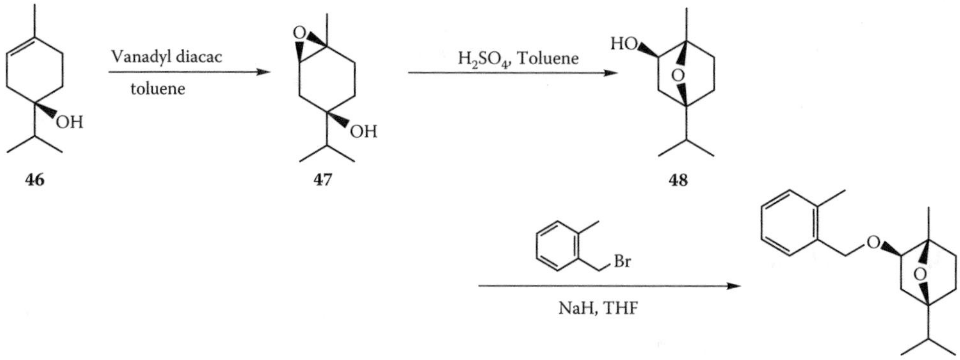

SCHEME 5.20

The chiral epoxide intermediate **47** can also be used for the synthesis of dihydropinol (**49**) (Scheme 5.21)[71] and an analogue of sobrerol,[72] which is a mucolytic drug[73] marketed as a racemate, although differences in activity between the two enantiomers have been reported.[74]

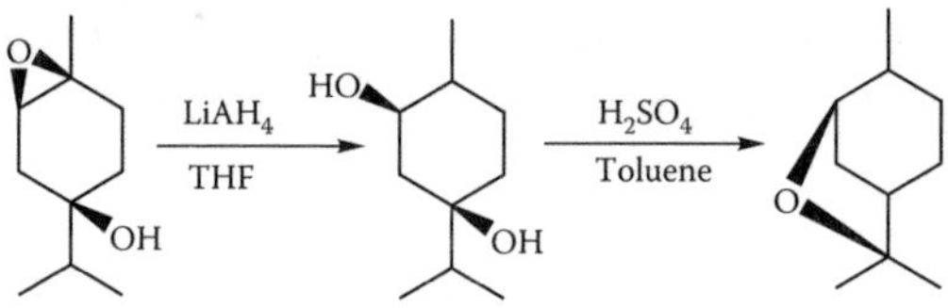

SCHEME 5.21

5.5 SUMMARY

All the optically active terpenes mentioned in this chapter are commercially available in bulk (>kg) quantities and are fairly inexpensive. Although many of them are isolated from natural sources, they can also be produced economically by synthetic methods. Actually, two thirds of these monoterpenes sold in the market today are manufactured by synthetic or semi-synthetic routes. These optically active molecules usually possess simple carbocyclic rings with one or two stereo-genic centers and have modest functionality for convenient structural manipulations. These unique features render them attractive as chiral pool materials for synthesis of optically active fine chemicals or pharmaceuticals. Industrial applications of these terpenes as chiral auxiliaries, chiral synthons, and chiral reagents have increased significantly in recent years. The expansion of the chiral pool into terpenes will continue with the increase in complexity and chirality of new drug candidates in the research and development pipeline of pharmaceutical companies.

REFERENCES

1. Erman, W. E. *Chemistry of Monoterpenes*, Marcel Dekker: New York, 1985.
2. Gildemeister, E., Hoffmann, F. *Die Atherischen Ole*, Vol. III, Akademie-Verlag: Berlin, 1960.
3. Newman, A. A. *Chemistry of Terpenes and Terpenoids*, Academic Press: New York, 1972.
4. Finnemore, H. *The Essential Oils*, Ernest Benn Ltd: London, 1927.
5. Guenther, E. *The Essential Oils*, Vol. 1–7, van Nostrand: New York, 1948–1952.
6. Pinder, A. R. *The Chemistry of Terpenes*, Wiley: New York, 1960.

7. Banthorpe, D. V., Whittaker, D. *Chem. Rev.* 1966, *66*, 643.
8. Bir, G., Kaufmann, D. *Tetrahedron Lett.* 1987, *28*, 777.
9. Brown, H. C., Jadhav, P. K., Desai, M. C. *J. Org. Chem.* 1982, *47*, 4583.
10. Brown, H. C., Joshi, N. N. *J. Org. Chem.* 1988, *53*, 4059.
11. Settine, R. L. *J. Org. Chem.* 1970, *35*, 4266.
12. Bank, S., Rowe, C. A., Schriesheim, A., Naslund, C. A. *J. Am. Chem. Soc.* 1967, *89*, 6897.
13. Richter, F., Wolff, W. *Chem. Ber.* 1926, *59*, 1733.
14. Brown, C. A. *Synthesis* 1978, 754.
15. Lavalee, H. C., Singaram, B. *J. Org. Chem.* 1986, *51*, 1362.
16. Andrianone, M., Delmoond, B. *J. Chem. Soc., Chem. Commun.* 1985, 1203.
17. Zaidlewicz, M. *J. Organometal. Chem.* 1985, *239*, 139.
18. Whitham, G. H. *J. Chem. Soc.* 1961, 2232.
19. Hurst, J. J., Whitham, G. H. *J. Am. Chem. Soc.* 1960, *82*, 2864.
20. Nigam, I. C., Levi, L. *Can. J. Chem.* 1968, *46*, 1944.
21. Chloupek, F. J., Zweifel, G. *J. Org. Chem.* 1964, *29*, 2062.
22. Schenck, G. O., Eggert, H., Denk, W. *Annalen* 1953, *584*, 177.
23. Simonsen, J. L. *The Terpenes*, Cambridge University Press: Cambridge, 1974, *Vol. I*, p. 143.
24. Vig, O. P., Matta, K. L., Lal, A., Raj, I. *J. Ind. Chem. Soc.* 1964, *41*, 142.
25. Bates, R. B., Caldwell, E. S., Klein, H. P. *J. Org. Chem.* 1969, *34*, 2615.
26. Klein, E., Ohleff, G. *Tetrahedron* 1963, *19*, 1091.
27. Royals, E. E., Leffingwil, J. C. *J. Org. Chem.* 1966, *31*, 1937.
28. Linder, S. M., Greenspan, F. P. *J. Org. Chem.* 1957, *22*, 949.
29. ApSimon, J. W. *The Total Synthesis of Natural Products*, Wiley: New York, 1973, *Vol. 2*, p. 88.
30. Simonsen, J. L., Owen, L. N. *The Terpenes*, Cambridge University Press: Cambridge, 1953, *Vol. I*, p. 230.
31. *J. Pharm. Chim.* 1918, *18*, 139.
32. Hussey, A. S., Baker, R. H. *J. Org. Chem.* 1960, *25*, 1434.
33. Brown, H. C., Garg, C. P. *J. Am. Chem. Soc.* 1961, *83*, 2952.
34. Akutagawa, S., Tani, K. *Catalytic Asymmetric Synthesis,* Ojima, I. Ed., VCH Publishers, Inc.: New York, 1993, p. 41.
35. Royals, E. E., Hornes, S. E. *J. Am. Chem. Soc.* 1951, *73*, 5856.
36. Shono, T., Nishiguchi, I., Yokoyama, T., Nitta, N. *Chem. Lett.* 1975, 433.
37. Merkel, D. *Die Ätherischen Ölen (Gildermeister Hoffmann)*, Vol. IIIb, Akademie-Verlag: Berlin, 1962, p. 70.
38. Simonsen, J. L. *The Terpenes*, Cambridge University Press: Cambridge, 1957, *Vol. II*, p. 373.
39. Alder, K. *New Methods of Preparative Organic Chemistry*, Interscience: New York, 1948.
40. Vaughan, W. R., Perry, R. *J. Am. Chem. Soc.* 1953, *75*, 3168.
41. Budavar, S. E. *Merck Index*, Twelfth ed., Merck: Whitehouse Station, NJ, 1996.
42. Srebnik, M., Ramachandran, P. V. *Aldrichimica Acta* 1987, *20*, 9.
43. McIntosh, J. M., Mishra, P. *Can. J. Chem.* 1986, *64*, 726.
44. Schöllköpf, U., Schütze, R. *Liebigs Ann. Chem.* 1987, 45.
45. Yaozhong, J., Guilan, L., Jinchu, L., Changyou, Z. *Synth. Commun.* 1987, *17*, 1545.
46. Yaozhong, J., Guilan, L., Jingen, D. *Synth. Commun.* 1988, *18*, 1291.
47. McIntosh, J. M., Cassidy, K. C., Matassa, L. *Tetrahedron* 1989, *45*, 5449.
48. Yaozhong, J., Peng, G., Guilan, L. *Synth. Commun.* 1990, *20*, 15.
49. Oppolzer, W., Blagg, J., Rodriguez, I., Walther, W. *J. Am. Chem. Soc.* 1990, *112*, 2767.
50. Boeckman, R. K., Johnson, A. T., Musselman, R. A. *Tetrahedron Lett.* 1994, *35*, 8521.
51. Kelly, T. R., Arvanitis, A. *Tetrahedron Lett.* 1984, *25*, 39.
52. Bartlett, P. D., Knox, L. H. *Org. Synth.* 1973, *Coll. Vol. 5*, 689.
53. Anh, K. H., Lee, S., Lim, A. *J. Org. Chem.* 1992, *57*, 5065.
54. Boeckman, R. K., Nelson, S. G., Gaul, M. D. *J. Am. Chem. Soc.* 1992, *114*, 2258.
55. Goodyear, M. D., Dwyer, P. O., Hill, M. L., Whitehead, A. J., Hornby, R., Hallett, P. PCT Pat. Appl. WO 9529174, 1995.
56. Adamson, P. S., McQuillin, F. C., Robinson, R., Simonsen, J. L. *J. Chem. Soc.* 1937, 1576.
57. Tanaka, A., Kamata, H., Yamashita, K. *Agric. Biol. Chem.* 1988, *52*, 2043.

58. McQuillin, F. J. *J. Chem. Soc.* 1955, 528.
59. Howe, R., McQuillin, F. J. *J. Chem. Soc.* 1955, 423.
60. Ho, T.-L. *Carbocycle Construction in Terpene Synthesis*, VCH: New York, 1988.
61. Fringuelli, F., Taticchi, A., Trazerso, G. *Gazz. Chim. Ital.* 1969, *99*, 231.
62. Roy, J. K. *Chem. Ind. (London)* 1954, 1393.
63. Humber, D. C., Pinder, A. R., Williams, R. A. *J. Org. Chem.* 1967, *32*, 2335.
64. Thomsa, A. F. *Pure Appl. Chem.* 1990, *62*, 1369.
65. Velluz, L., Nominé, G., Bucourt, R., Pierdet, A., Dufay, P. *Tetrahedron Lett.* 1961, 127.
66. Stevens, R. V., Caeta, C. A. *J. Am. Chem. Soc.* 1977, *99*, 6105.
67. Danishefsky, S., Cain, P. *J. Am. Chem. Soc.* 1975, *97*, 5282.
68. Danlewski, A. R. *J. Org. Chem.* 1975, *40*, 3135.
69. Payne, G. B. U.S. Pat. 4,487,945, 1984.
70. Ohloff, G., Uhde, G. *Helv. Chim. Acta* 1965, *48*, 10.
71. Liu, W.-G., Rosazza, J. P. N. *Synth. Commun.* 1996, *26*, 2731.
72. Bovara, R., Carrea, G., Ferrara, L., Riva, S. *Tetrahedron: Asymmetry* 1991, *2*, 931.
73. Braga, P. C., Allegra, L., Bossi, R., Scuri, R., Castiglioni, C. L., Romandini, S. *Int. J. Clin. Pharm. Res.* 1987, *7*, 381.
74. Klein, E. A. U.S. Pat. 2,815,378, 1957.

6 Resolutions at Large Scale: Case Studies

Weiguo Liu

CONTENTS

6.1 INTRODUCTION

Resolution and chirality are like twins born on the day Louis Pasteur separated crystals of salts of D- and L-tartaric acid under his microscope. Since then, the separation of each enantiomer from a racemic mixture has been the primary means to obtain optically actively organic compounds. Only recently, the fast and explosive new developments in asymmetric synthesis involving the use of organometallic catalysts, enzymes, and chiral auxiliaries have begun to challenge the resolution approach. Even so, owing to its simplicity, reliability, and practicality, resolution is so far still the most widely applied method for the production of optically pure fine chemicals and pharmaceuticals.[1] Through many years of evolution, the art of resolution has become a multi-disciplinary science that includes diastereomeric, kinetic, dynamic, chromatographic, and enzymatic components. In this chapter, we will describe some examples of applications of the resolution approach for the production of chiral fine chemicals on an industrial scale. The following chapter has more examples of resolutions.

The rapid development of chiral technology in recent years has provided process chemists an array of methodologies to choose from when considering the synthesis of optically active organic compounds. The choice may be a resolution, but, generally speaking, the nature of a 50% maximum theoretical yield makes the process less desirable when an alternative, asymmetric synthesis is available. In practice, however, and especially in industry, the major issue, and often the deciding factor behind the selection of the technology, is the practicality of the process for large-scale production. Also, at different stages of drug development there are different needs and priorities. In the early development stages, the quantity of material needed is relatively small and is often required on short notice. In these cases, simple resolution processes often win out owing to their swiftness and ease of scale up. For well-established bulk actives, lower costs become the primary

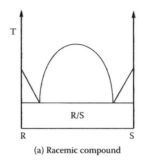
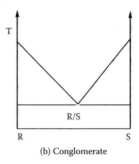

FIGURE 6.1 Melting point diagrams of racemates.

goal of process development, and 100% theoretical yield reactions are almost a necessity. In these situations, resolution is less attractive but can still be the method of choice if the unwanted enantiomer can be recycled in the process, or the starting material is cheap and the resolution is performed very early in the synthesis, or both. The method of choice to synthesize optically active fine chemicals or pharmaceuticals is entirely dependent on the specific case of that project and therefore should be discussed on a case-by-case basis.

6.2 CHEMICAL RESOLUTION

Classical resolution generally refers to the processes involving preferential crystallization of one enantiomer from a racemic mixture or the treatment of a racemate with a chiral reagent followed by crystallization of one of the resultant diastereoisomers. A large number of commercially successful pharmaceuticals or their intermediates are primarily manufactured by these processes.[2,3] On the basis of their melting point diagram, about 5–10% of racemates are referred to as conglomerate or racemic mixtures that consist of a mechanical mixture of crystals of the two enantiomers in equal amounts—each individual crystal contains only one enantiomer (Figure 6.1). The remaining 90% of racemates are called racemic compounds and consist of crystals of an ordered array of R- and S-enantiomers, where each individual crystal contains equal amounts of both enantiomers. Racemic mixtures of conglomerates may be separated by direct crystallization that usually involves seeding of supersaturated solution of the racemate with the desired enantiomer.[4] The resolution by direct crystallization method has been used for industrial-scale manufacturing of several major pharmaceuticals such as α-methyl-L-dopa[5] and chloramphenicol.[6] The Merck process of direct crystallization for α-methyl-L-dopa production has been carried on the scale of several hundred tons per year.[7]

An overwhelming majority of classic resolutions still involve the formation of diastereomeric salts of the racemate with a chiral acid or base (Table 6.1). These chiral-resolving agents are relatively inexpensive and readily available in large quantities (Table 6.2). They also tend to form salts with good crystalline properties.[8]

Chiral acids or bases are tested for salt formation with the target racemate in a variety of solvents. The selected diastereomeric salt must crystallize well, and there must be an appreciable difference in solubility between the two diastereoisomers in an appropriate solvent. This type of selection process is still a matter of trial and error, although knowledge about the target racemate and the available chiral-resolving agents and experience with the art of crystallization does provide significant help.[9–11]

6.2.1 NAPROXEN

Naproxen was introduced to the market by Syntex in 1976 as a nonsteroidal antiinflammatory drug (NSAID) in an optically pure form. The original manufacturing process (Scheme 6.1) before product launch started from β-naphthol (1), which was brominated in methylene chloride to produce 1,6-dibromonaphthol (2). The labile bromine at the 1-position was removed with bisulfite to give

TABLE 6.1
Examples of Pharmaceuticals Resolved by Classic Resolution*

Pharmaceutical	Resolving Agent
Ampicillin	D-Camphorsulfonic acid
Ethambutol	L-(+)-Tartaric acid
Chloramphenicol	D-Camphorsulfonic acid
Dextropropoxyphene	D-Camphorsulfonic acid
Dexbrompheniramine	D-Phenylsuccinic acid
Fosfomycin	R-(+)-Phenethylamine
Thiamphenicol	D-(−)-Tartaric acid
Naproxen	Cinchonidine
Diltiazem	R-(+)-Phenethylamine

* See Bayley and Vaidya.[2]

TABLE 6.2
Commonly Used Resolving Agents*

Chiral Bases	Chiral Acids
α-Methylbenzylamine	l-Camphor-10-sulfonic acid
α-Methyl-p-nitrobenzylamine	Malic acid
α-Methyl-p-bromobenzylamine	Mandelic acid
2-Aminobutane	α-Methoxyphenylacetic acid
N-Methylglucamine	α-Methoxy-α-trifluoromethylphenylacetic acid
Cinchonine	2-Pyrrolidone-5-carboxylic acid
Cinchonidine	Tartaric acid
Ephedrine	2-Ketogulonic acid
Quinine	
Brucine	

* See Bayley and Vaidya.[2]

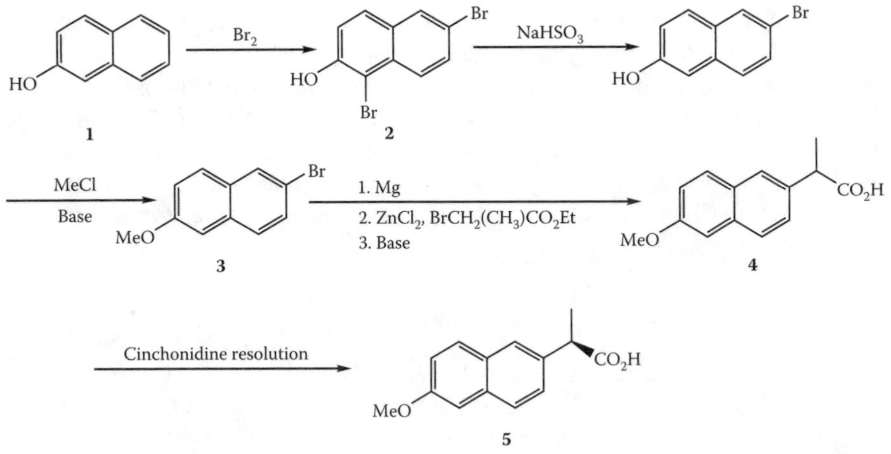

SCHEME 6.1

2-bromo-6-hydroxynaphthalene, which was then methylated with methyl chloride in water-isopropanol to obtain 2-bromo-6-methoxynaphthalene (3) in 85–90% yield from β-naphthol. The bromo compound was treated with magnesium followed by zinc chloride. The resultant naphthylzinc was coupled with ethyl bromopropionate to give naproxen ethyl ester, which was hydrolyzed to afford the racemic acid 4. The final optically active naproxen (5) was obtained by a classic resolution process. The racemic acid 4 was treated with cinchonidine to form diastereomeric salts. The *S*-naproxen-cinchonidine salt was crystallized and then released with acid to give *S*-naproxen (5) in 95% of the theoretical yield (48% chemical yield).[12,13]

Shortly after the product launch, the original process was modified (Scheme 6.2). Zinc chloride was removed from the coupling reaction, and ethyl bromopropionate was replaced with bromopropionic acid magnesium chloride salt. Most importantly, the resolution agent cinchonidine was replaced with *N*-alkylglucamine, which can be readily obtained from the reductive amination of D-glucose. This new resolution process was equally efficient, with greater than 95% theoretical yield, in addition to being much more cost effective as a result of the low cost of the resolving agent.[14,15]

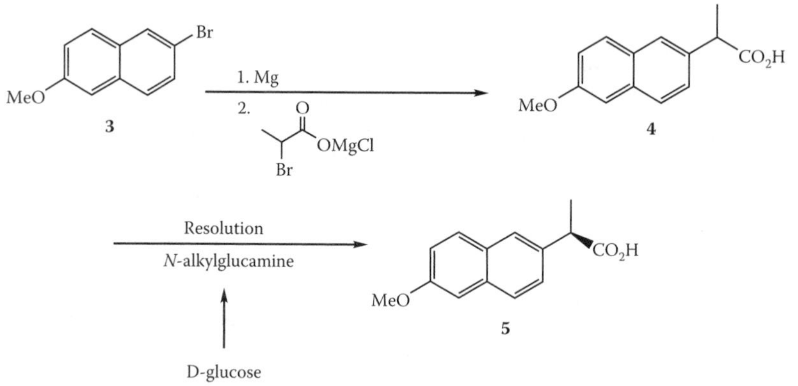

SCHEME 6.2

Twenty years has passed since then. During this period, naproxen has become a billion dollar drug, lost its patent protection, and seen numerous alternative routes and improvements to its synthesis. Significant progress in the field of chiral technology created a long list of methods for generation of optically pure organic compounds. There are several asymmetric synthetic processes specifically designed for production of *S*-naproxen,[16] including the well-known Zambon process (see Chapter 23)[17] and asymmetric hydrogenations (Chapter 12).[18–21]

Other approaches that have been suggested include catalytic asymmetric hydroformylation of 2-methoxy-6-vinylnaphthalene (6) using a rhodium catalyst on BINAPHOS ligand followed by oxidation of the resultant aldehyde 7 to yield *S*-naproxen (Scheme 6.3).[22] However, the tendency of the aldehyde to racemize and the co-generation of the linear aldehyde isomer make the process less attractive. Other modifications related to this process include catalytic asymmetric hydroesterification,[23] hydrocarboxylation,[24] and hydrocyanation.[25]

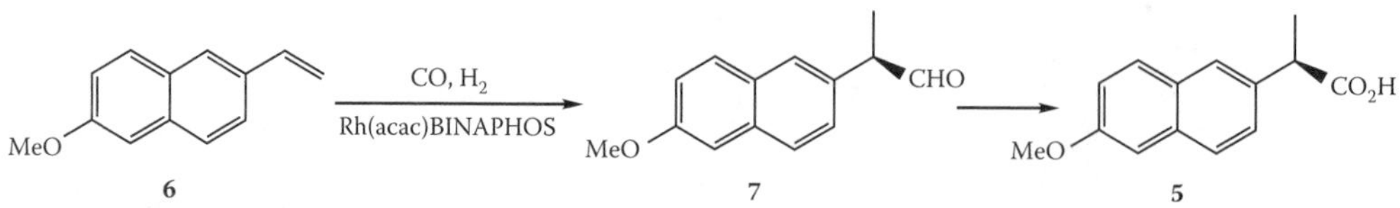

SCHEME 6.3

One of the alternative routes of using the chiral pool approach for production of *S*-naproxen was developed at Syntex in the 1980s (Scheme 6.4).[26,27] The process starts with inexpensive ethyl L-lactate (**8**). Conversion of the 2-hydroxyl group to the mesylate followed by hydrolysis of the ethyl ester and conversion of the acid to acid chloride yield the chiral acylating agent **9** for the Grignard reagent derived from 2-bromo-6-methoxynaphthalene to provide the optically pure ketone **10**. Protection of the ketone with dimethyl-1,3-propanediol, followed by rearrangement at 115–120°C, gives the naproxen ester **11**. Hydrolysis of this ester yields *S*-naproxen (**5**) with a 75% overall yield from 2-bromo-6-methoxynaphthalene.

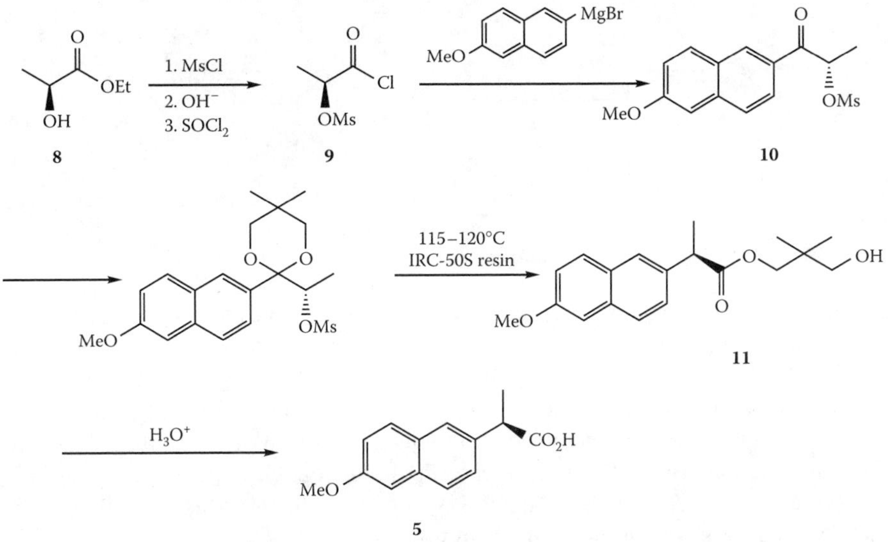

SCHEME 6.4

Several biocatalytic processes for the production of (*S*)-(+)-naproxen (**5**) have also been developed (see Chapter 19). Direct isomerization of racemic naproxen (**4**) by a microorganism catalyst, *Exophialia wilhansil,* was reported to give the (*S*)-isomer **5** (92%, 100% ee) (Scheme 6.5).[28] A 1-step synthesis of (*S*)-(+)-naproxen (**5**) by microbial oxidation of 6-methoxy-2-isopropylnaphthalene (**12**) was developed by IBIS (Scheme 6.6).[29] In both cases, typical bioprocess-related issues such as productivity, product isolation, and biocatalyst production have apparently prevented them from rapid commercialization.

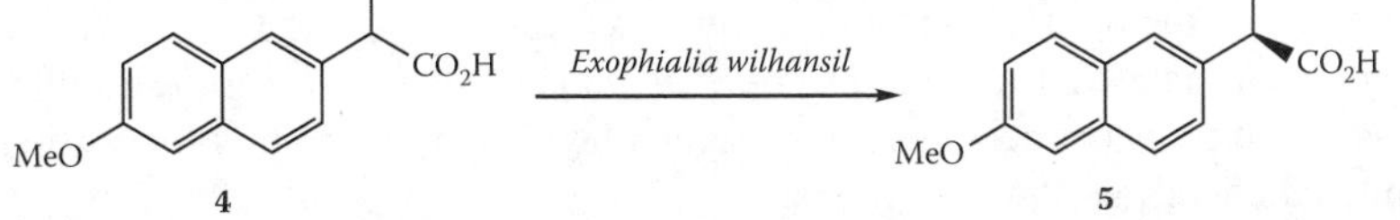

SCHEME 6.5

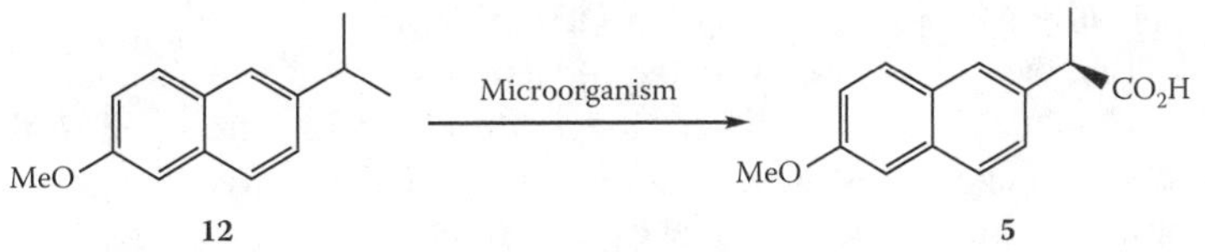

SCHEME 6.6

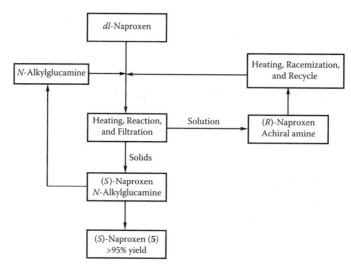

FIGURE 6.2 Block flow diagram of the naproxen resolution process.

During the same period, continuous process improvements on the resolution route to S-naproxen have achieved dramatic cost reductions. The breakthrough comes from in-process racemization and recycle of the R-naproxen by-product and the recovery of the resolving agent (Figure 6.2). In this novel and efficient resolution process, the racemic naproxen is reacted with half an equivalent of the chiral N-alkylglucamine and another half an equivalent of an achiral amine.[13] Theoretically, there should be an equilibrium of four different salts in this mixture: the S-acid chiral amine salt, S-acid achiral amine salt, R-acid chiral amine salt, and R-acid achiral amine salt. However, only the S-acid chiral amine salt is insoluble in the system and crystallizes out. This process drives the equilibrium toward complete formation of S-naproxen N-alkylglucamine that is collected by filtration; acidification liberates S-naproxen (**5**). The mother liquor, which contains the unwanted acid and achiral amine, is heated. The amine base catalyzes the racemization of the R-acid; the resultant salt of the racemic naproxen and the achiral amine is then recycled to the resolution loop in a continuous operation. Although each individual cycle of the resolution yields the diastereomeric salt of 45–46%, the overall result from continuous operation produces S-naproxen of 99% ee in greater than 95% yield from the racemic mixture. The recovery of the resolving agent, the N-alkylglucamine, is greater than 98% per cycle. This effective combination of resolution, racemization, and recycle of both unwanted enantiomer and resolving agent dramatically increases the yield, reduces the cost, and cuts the wastestream, resulting in an ideal case for a Pope Peachy resolution.[1] Despite the availability of so many elegant asymmetric and chiral pool synthetic processes that we have discussed earlier, the optimized resolution process still turns out to be the most cost-effective route for S-naproxen manufacture. This example demonstrates that classic resolution technology when combined with proper process engineering can be a superior method for the manufacturing of chiral industrial chemicals.[13]

6.2.2 IBUPROFEN

Manufacture of optically pure (S)-(+)-ibuprofen (**13**), an NSAID similar to naproxen, is another example demonstrating the role of resolution in production of chiral fine chemicals, although from a somewhat different angle. Unlike naproxen, ibuprofen (**14**) was introduced to the market as a racemate almost 30 years ago.[30,31] At the time of the introduction, it was thought that both R- and S-isomers of ibuprofen had the same *in vivo* activity.[32] It has been demonstrated that the R-isomer is converted to the S-isomer *in vivo*[29] by a unique enzyme system called invertase.[34] Based on these data, ibuprofen has since been marketed as a racemate and has achieved sales of more than a billion dollars per year. The results from certain animal and human studies have begun to reveal the possible superior performance of pure (S)-isomer (13) in comparison with the racemate 14.[35] These findings have prompted several companies to invest in the research and development of 13 as a "chiral switch" to replace the racemate. As a result, a large number of synthetic processes have been developed for the production of (S)-(+)-ibuprofen (13). Many of these processes are the variations of those discussed in the synthesis of (S)-(+)-naproxen (5). For example, the electrocarboxylation and catalytic asymmetric hydrogenation process developed at Monsanto can be applied to the synthesis of (S)-(+)-ibuprofen.[20,21,36]

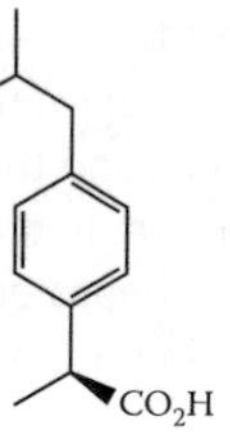

13 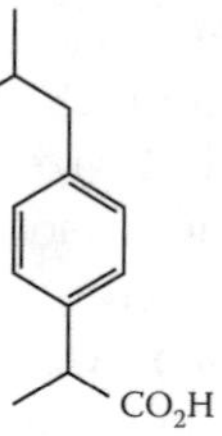   14

Other asymmetric synthetic processes used for the manufacturing of (S)-(+)-naproxen can also be applied to the production of (S)-(+)-ibuprofen; these include the Rh-phosphite catalyzed hydroformylation,[37] hydrocyanation,[25] and hydrocarboxylation reactions.[24]

Because ibuprofen has been a successful drug on the market for almost 30 years with no patent protection since 1985, there is a widespread competition for commercial production of this product throughout the world. As a result, several practical and economical industrial processes for the manufacture of racemic ibuprofen (14) have been developed and are in operation on commercial scales.[38] Most of these processes start with isobutylbenzene (15) and go through an isobutylstyrene[39–41] or an acetophenone intermediate.[42] The most efficient route is believed to be the Boots-Hoechst-Celanese process, which involves 3 steps from isobutylbenzene, all catalytic, and is 100% atom-efficient (Scheme 6.7).[43,44]

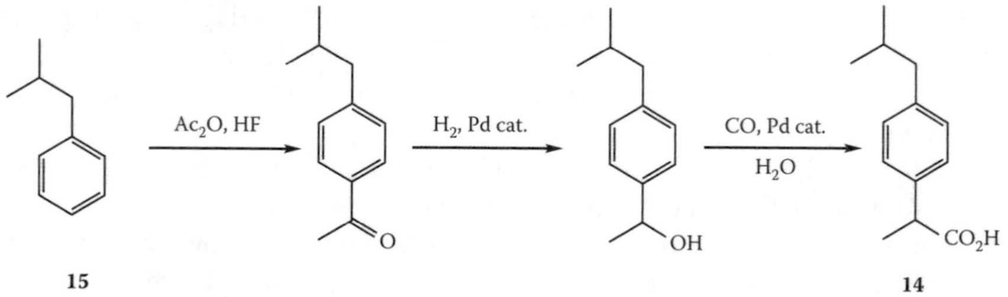

15   14

SCHEME 6.7

Taking advantage of the ready availability of racemic ibuprofen, the resolution approach for production of (S)-(+)-ibuprofen becomes an attractive alternative. Merck's resolution process involves the formation of a diastereomeric salt of ibuprofen with (S)-lysine, an inexpensive and readily available natural amino acid.[45] The racemic ibuprofen is mixed with 1.0 equivalent of (S)-lysine in aqueous ethanol. The slurry is agitated to allow full dissolution. The supernatant, which is a supersaturated solution of ibuprofen-lysine salt, is separated from the solid and seeded with (S)-ibuprofen-(S)-lysine to induce crystallization. The precipitated solid is collected by filtration, and the mother liquor is recycled to the slurry of racemic ibuprofen and (S)-lysine. This process is continued until essentially all (S)-ibuprofen in the original slurry is recovered, resulting in the formation of greater than 99% ee (S)-ibuprofen-(S)-lysine. Subsequent acidification of the salt followed by crystallization gives enantiomerically pure (S)-ibuprofen. The undesired (R)-ibuprofen isomer can be recycled through racemization. A variety of methods have been reported for the racemization of ibuprofen, such as heating (R)-ibuprofen sodium salt in aqueous sodium hypochloride,[46] the formation of anhydrides in the presence of thionyl chloride or acetic anhydride followed by heating and hydrolysis,[47] or heating (R)-ibuprofen with a catalytic amount of palladium on carbon in a hydrogen atmosphere.[48]

Another classic resolution process developed by Ethyl Corp. for (S)-ibuprofen production uses (S)-(−)-α-methylbenzylamine (MAB) as the chiral base for diastereomeric salt formation.[49] The difference in solubility between (S)- and (R)-ibuprofen MAB salts is so substantial that only half an equivalent of MAB is used for each mole of racemic ibuprofen, and no seeding is needed. The process can also be performed in a wide range of solvents, and the unwanted (R)-ibuprofen can be recycled conveniently by heating the mother liquor in sodium hydroxide or hydrochloric acid. Other designer amines have been developed for resolution of ibuprofen with good stereoselectivities,[50] but these chiral amines were prepared specifically for ibuprofen resolution and are thus unlikely to be economical for industrial production.

It is worth mentioning that ibuprofen is a racemic compound. Its eutectic composition determined from the binary melting point phase diagram requires that the enantiomeric purity has to be 90% and higher to achieve optical purification by crystallization.[51] Even above this composition limit, crystallization is still rather inefficient.[52] However, the sodium salt of ibuprofen exhibits a much more favorable phase diagram in which the eutectic composition is about 60%. This means that ibuprofen of at least 60% optical purity can be upgraded through the sodium salt. A one-step crystallization of 76% (S)-ibuprofen after treatment with sodium hydroxide in acetone gave 100% ee (S)-ibuprofen.[53]

Preparation of (S)-ibuprofen by enzyme-catalyzed enantioselective hydrolysis of racemic ibuprofen esters has been investigated by several companies, such as Sepracor,[54,55] Rhone-Poulenc (now Sanofi Aventis),[56,57] Gist-Brocades (now DSM),[58,59] and the Wisconsin Alumni Research Foundation.[60] These processes are usually performed under mild reaction conditions, yield highly optically pure product, and can be readily scaled up for industrial production. The disadvantages of these processes for (S)-ibuprofen production are the extra step needed to produce the corresponding ester of racemic ibuprofen in addition to the cost for producing the enzyme and microorganism catalysts.

From comparison of all the available processes for the manufacturing of (S)-ibuprofen, it was found that many of them are highly efficient and cost competitive as far as the chemistry is concerned. For example, the RuBINAP catalyzed asymmetric hydrogenation process is a superior route to (S)-ibuprofen to a resolution approach. However, in the broader economic environment other nonprocess–related factors come into play. The risks associated with the unknowns of a new asymmetric catalysis technology and the capital investment for high-pressure hydrogenation vessels make potential producers reluctant to implement the new methodology. However, the existing, well-established processes and production facilities for racemic ibuprofen allow a step of classic resolution to be added with minimal front-end investment therefore giving the racemic producers a competitive edge for production of the optically active form. Even the new players can easily purchase the bulk racemate and perform a simple resolution to obtain the optically pure form for packaging and sales. It is widely anticipated that classic resolution will be the ultimate method of choice for commercial manufacturing of (S)-(+)-ibuprofen if it becomes a successful product in the marketplace.

6.2.3 D-PROLINE

D-Proline (16) is an unnatural amino acid and an important chiral synthon for the synthesis of a variety of biologically active compounds. There are few chemical methods for asymmetric synthesis of this compound. Almost all processes for the production of D-proline at scale are based on resolution of dl-proline, and most of them involve the racemization of L-proline (17).

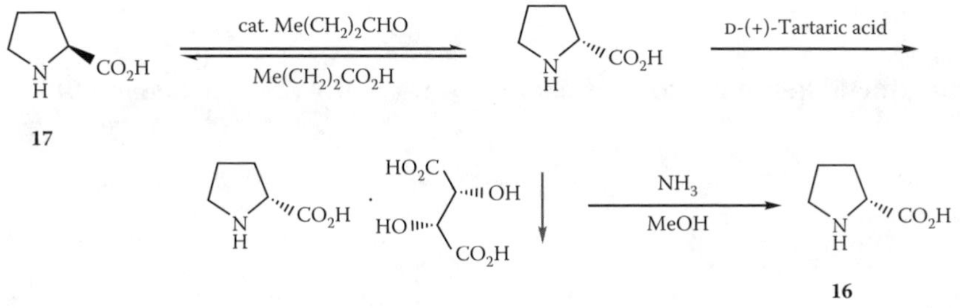

SCHEME 6.8

It is well-known that catalytic amounts of aldehyde can induce racemization of α-amino acids through the reversible formation of Schiff bases.[61] Combination of this technology with a classic resolution leads to an elegant asymmetric transformation of L-proline to D-proline (Scheme 6.8).[62,63] When L-proline is heated with one equivalent of D-tartaric acid and a catalytic amount of *n*-butyraldehyde in butyric acid, it first racemizes as a result of the reversible formation of the proline-butyraldehyde Schiff base. The newly generated D-proline forms an insoluble salt with D-tartaric acid and precipitates out of the solution, whereas the soluble L-proline is continuously being racemized. The net effect is the continuous transformation of the soluble L-proline to the insoluble D-proline-D-tartaric acid complex, resulting in near-complete conversion. Treatment of the D-proline-D-tartaric acid complex with concentrated ammonia in methanol liberates the D-proline (**16**) (99% ee, with 80–90% overall yield from L-proline). This is a typical example of a dynamic resolution where L-proline is completely converted to D-proline with simultaneous *in situ* racemization. As far as the process is concerned, this is an ideal case because no extra step is required for recycle and racemization of the undesired enantiomer and a 100% chemical yield is achievable. The only drawback of this process is the use of stoichiometric amount of D-tartaric acid, which is the unnatural form of tartaric acid and is relatively expensive. Fortunately, more than 90% of the D-tartaric acid is recovered at the end of the process as the diammonium salt that can be recycled after conversion to the free acid.[64]

6.2.4 HYDROLYTIC KINETIC RESOLUTION OF EPOXIDES

Chiral epoxides are extremely useful intermediates and building blocks for synthesis of a variety of optically active organic compounds including pharmaceuticals and industrial fine chemicals. There have been several methods developed in recent years for asymmetric synthesis of chiral epoxides.[65–67] However, as with other reactions, these chiral epoxidations are restricted to certain types of substrates. Optically active terminal epoxides are not effectively prepared by currently available methodologies[68] but are one of the most important and widely sought after group of chiral compounds. In 1996 Jacobsen discovered a (salen)Co(III) catalyzed enantioselective opening of racemic terminal epoxides by water.[69] This chemistry has since been applied successfully to the hydrolytic kinetic resolution (HKR) of terminal epoxides such as propylene oxide. The neat oil of racemic propylene oxide (1.0 mol) containing 0.2 mol % of (salen)Co(III)(OAc) complex (**18**) was treated with 0.55 equivalent of water at room temperature for 12 hours to afford a mixture of unreacted epoxide and propylene glycol (Scheme 6.9; R = Me). This mixture was then separated by fractional distillation to provide both compounds in high chemical and enantiomeric purity

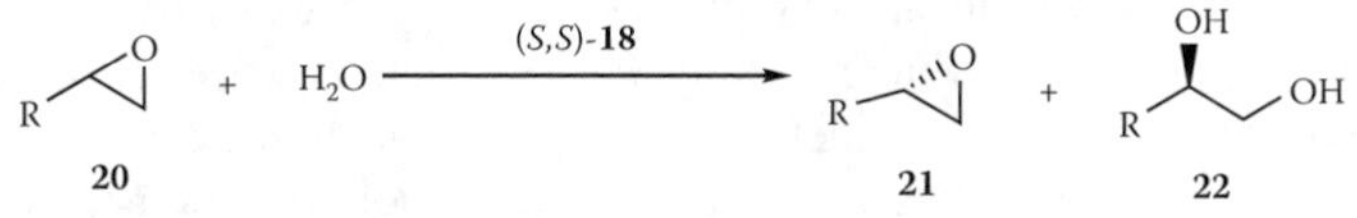

SCHEME 6.9

TABLE 6.3
Hydrolytic Kinetic Resolution of Terminal Epoxides with Water (Scheme 6.9)

Entry	20, R=	18 (mol %)	Water Equivalent	Time (hour)	21 (%)	21 (ee%)	22 (%)	22 (ee%)	krel
1	Me	0.2	0.55	12	44	>98	50	98	>400
2	CH$_2$Cl	0.3	0.55	8	44	98	38	86	50
3	(CH$_2$)$_3$Me	0.42	0.55	5	46	98	48	98	290
4	(CH$_2$)$_5$Me	0.42	0.55	6	45	99	47	97	260
5	Ph	0.8	0.70	44	38	98	39*	98*	20
6	CH=CH$_2$	0.61	0.50	20	44	84	49	94	30
7	CH=CH$_2$	0.85	0.70	68	29	99	64	88	30

(>98% ee) with nearly quantitative yield. The critical element in this resolution process is the chiral catalyst (**18**). At the end of the process, after removing products, the catalyst was found reduced to the Co(II) complex (**19**), which can be recycled back to the active catalyst by treatment with acetic acid in air with no observable loss in activity or stereoselectivity (Scheme 6.10). This highly efficient hydrolytic kinetic resolution procedure has been applied to the preparation of a series of chiral terminal epoxides that have previously not been readily accessible in optically pure form (Table 6.3).

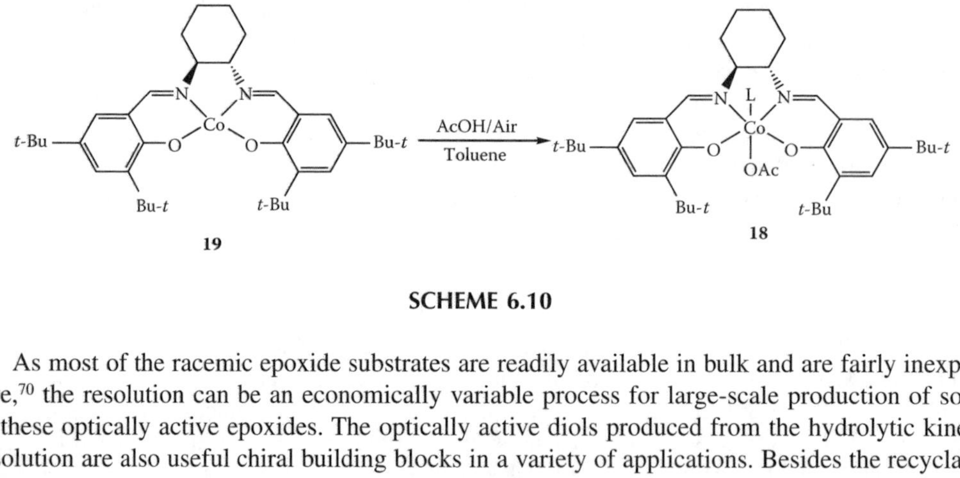

SCHEME 6.10

As most of the racemic epoxide substrates are readily available in bulk and are fairly inexpensive,[70] the resolution can be an economically variable process for large-scale production of some of these optically active epoxides. The optically active diols produced from the hydrolytic kinetic resolution are also useful chiral building blocks in a variety of applications. Besides the recyclable catalyst, water is the only reagent used in the reaction, and no solvent was added. There is virtually no wastestream from the process, and the operation is simple. Rhodia Chirex (now Rhodia Pharma Solutions) has licensed this technology and started producing some of the chiral epoxides and diols at scale.[71,72] The catalyst (salen)Co(III)(OAc) has also been manufactured in multi-kilogram quantities. This simple, environmentally friendly, and highly efficient resolution process has a potential to offer a wide range of cost-effective chiral fine chemicals for existing and new products.

6.3 ENZYMATIC RESOLUTIONS

The use of enzymes and microorganisms in organic synthesis, especially in the production of chiral organic compounds, has grown significantly in recent years and has been accepted as an effective and practical alternative for certain synthetic organic transformations. There have been plenty of excellent review articles,[73,74] including Chapter 19 of this book, describing recent advances and various aspects of applications in this field. In general, biotransformations in organic synthesis have advanced from the stage of exploratory laboratory research to industrial-scale production and applications. Certain enzymes such as lipases and esterases are now considered as standard and routine reagents in organic synthesis. Using these enzymes for resolution of racemic alcohols, esters, amides, amines, and acids through enantioselective hydrolysis, esterification, or acylation are routinely practiced by scientists in both academic and industrial laboratories. Compared with classic diastereomeric resolutions, enzymatic resolutions are catalytic, highly selective, and environmentally benign. Their mild reaction conditions allow complex target compounds with multifunctional groups to be resolved effectively. However, in large-scale productions, process economics is often the overriding factor for the determination of the practicality of a technology. For an enzymatic resolution to be performed effectively on large scale, issues such as solvent selection, volume throughput, product isolation, enzyme recycle, and the recycle of unwanted enantiomer have to be dealt with to achieve the acceptable economy. Very often, these evaluations are carried out in comparison with the results obtained using other alternative technologies such as classic resolution and asymmetric synthesis. Traditionally, research and development work on biocatalysis and biotransformations are carried out in a bioprocess laboratory setting because the isolation of new enzymes and handling of microorganisms and fermentation requires special sets of skills. With the rapid advances in this field, a very large number of industrial enzymes have become commercially available, and they are affordable in large quantities. This change has prompted many companies to set up biocatalysis groups within the chemical process development laboratories, so that all available technologies relating to the preparation of chiral organic compounds are directly accessible for a specific development project and are coordinated based on the needs of the project.

6.3.1 α,α-Disubstituted α-Amino Acids

α,α-Disubstituted α-amino acids have attracted increased attention in recent years. This group of nonproteinogenic amino acids induces dramatic conformational change when incorporated into peptides[75–78] and renders them more resistant to protease hydrolysis. They are found in natural peptide antibiotics,[79–82] and some are potent inhibitors of amino acid decarboxylases.[83–86] Chiral α,α-disubstituted α-amino acids are also important building blocks for the synthesis of pharmaceuticals and other biological agents.[87–91] For the synthesis of an anticholesterol drug under development at the then Burroughs Wellcome Co. as 2164U90 (**22**), optically pure (*R*)-2-amino-2-ethylhexanoic acid (**23**) was needed as a critical building block (Scheme 6.11), and a campaign was launched for the development of a cost-effective process for the large-scale preparation of this chiral disubstituted amino acid and its analogues.

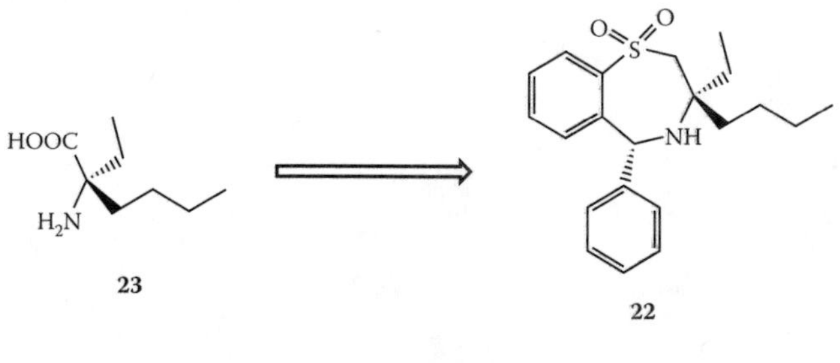

23

22

SCHEME 6.11

A variety of methods exists for the synthesis of optically active amino acids including asymmetric synthesis[92–100] and classic and enzymatic resolutions.[101–104] However, most of these methods are not readily applicable to the preparation of α,α-disubstituted amino acids as a result of poor stereoselectivity and lower activity at the α-carbon. Attempts to resolve the racemic 2-amino-2-ethylhexanoic acid and its ester through classic resolution failed. Several approaches for the asymmetric synthesis of the amino acid were evaluated including alkylation of 2-aminobutyric acid using a camphor-based chiral auxiliary and chiral phase-transfer catalyst. A process based on Schöllkopf's asymmetric synthesis was developed (Scheme 6.12).[105] Formation of piperazinone **24** through dimerization of methyl (*S*)-(+)-2-aminobutyrate (**25**) was followed by enolization and methylation to give (3*S*,6*S*)-2,5-dimethoxy-3,6-diethyl-3,6-dihydropyrazine (**26**) (Scheme 6.12). This dihydropyrazine intermediate is unstable in air and can be oxidized by oxygen to pyrazine **27**, which was isolated as a major impurity.

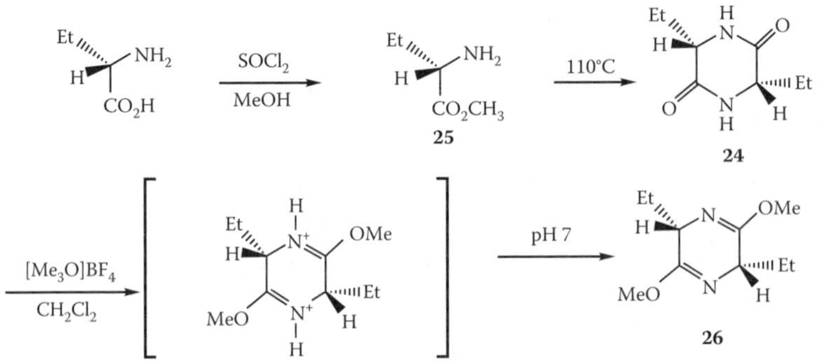

SCHEME 6.12

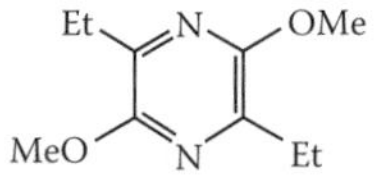

27

Operating carefully under nitrogen, compound **26** was used as a template for diastereoselective alkylation with a series of alkyl bromides to produce derivatives **28**, which were then hydrolyzed by strong acid to afford the dialkylated (*R*)-amino acids **29** (~90% ee), together with partially racemized 2-aminobutyric acid, which was separated from the product by ion-exchange chromatography (Scheme 6.13). This process provided gram quantities of the desired amino acids as analytical markers and test samples but was neither practical nor economical for large-scale production.

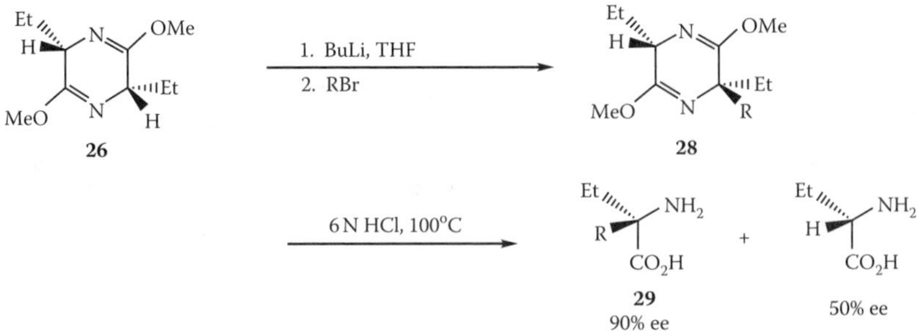

SCHEME 6.13

Facing difficulties with chemical methods, development work was aimed at finding a biocatalytic process to resolve the amino acids. Typical commercial enzymes reported for resolution of amino acids were tested. Whole-cell systems containing hydantoinase were found to produce only α-monosubstituted amino acids;[106–112] the acylase catalyzed resolution of *N*-acyl amino acids had extremely low rates (often zero) of catalysis toward α-dialkylated amino acids;[113,114] and the nitrilase system obtained from Novo Nordisk showed no activity toward the corresponding 2-amino-2-

ethylhexanoic amide.[115,116] Finally, a large-scale screening of hydrolytic enzymes for enantio-selective hydrolysis of racemic amino esters was carried out. Racemic α,α-disubstituted α-amino esters were synthesized by standard chemistry through alkylation of the Schiff's base of the corresponding natural amino esters (Scheme 6.14)[117] or through formation of hydantoins.[118]

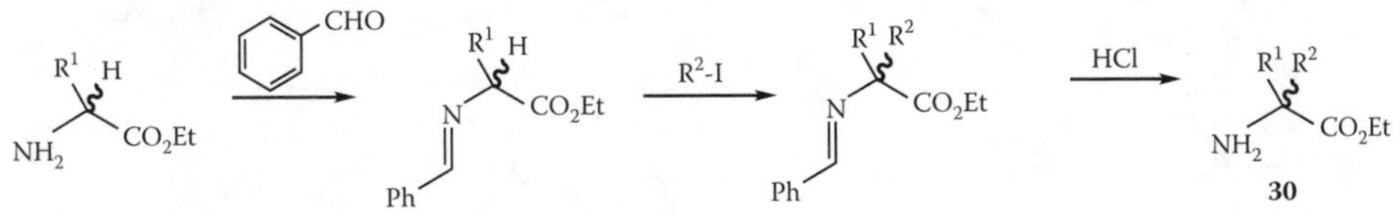

SCHEME 6.14

Initial enzyme screening was aimed at obtaining optically active 2-amino-2-ethylhexanoic acid **23** or the corresponding amino alcohol. Enzymes reported for resolving α-H or α-Me analogues of amino acids failed to catalyze the corresponding reaction of this substrate,[119] primarily as a result of the presence of the α-ethyl group that causes a critical increase in steric hindrance at the α-carbon. Out of 50 different enzymes and microorganisms screened, pig liver esterase and *Humicola langinosa* lipase (Lipase CE, Amano) were the only ones found to catalyze the hydrolysis of the substrate.

Both enzymes catalyze the hydrolysis of the amino ester **30** enantioselectively (Scheme 6.15). At about 60% substrate conversion, the enantiomeric excess of recovered ester **32** from both reactions exceeds 98%. In addition, the acid product **31** (96–98% ee) was obtained by carrying the hydrolysis of the ester to 40%. The rates of hydrolysis become significantly slower when conversion approaches 50%, allowing a wide window for kinetic control of the resolution process. Both enzymes function well in a concentrated water/substrate (oil) two-phase system while maintaining high enantioselectivity, making this system very attractive for industrial processes.

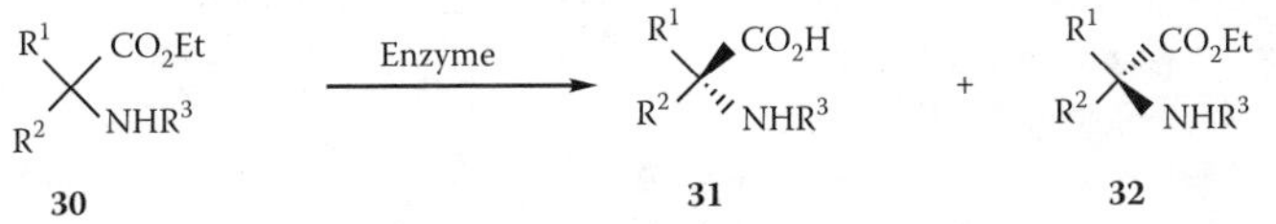

SCHEME 6.15

Although pig liver esterase (PLE) catalyzes the hydrolysis of all amino esters tested in this work, it was only enantioselective toward esters **30** (R^1 = ArCH$_2$, or n-Bu, R^2 = Et). This lack of correlation between enantioselectivity and substrate structure has been reported for many PLE catalyzed reactions.[120,121] The lipase CE was obtained as a crude extract containing about 10% of total protein. The active enzyme catalyzing the amino ester hydrolysis was isolated from the mixture and partially purified. This new enzyme, *Humicola* amino esterase, is present as a minor protein component, has a molecular weight of about 35,000 daltons, and shows neither esterase activity toward *o*-nitrophenol butyrate nor lipase activity to olive oil. It is, however, highly effective in catalyzing the amino ester hydrolysis with very broad substrate specificity and high enantioselec-tivity. Various substrates including aliphatic, aromatic, and cyclic amino esters were resolved into optically active esters and acids (Table 6.4) with good E values.[122] Aliphatic amino esters with alkyl or alkenyl sidechains as long as 10 carbon atoms were well-accepted by the enzyme. Substrates with longer chain length were limited by the solubility of the compounds rather than their binding with the enzyme. Resolutions were also extended to α,α-disubstituted amino esters in which the two alkyl groups differ in length by as little as a single carbon atom. The fact that the enzyme successfully catalyzes the resolution of straight-chain aliphatic amino esters with two α-alkyl groups both larger than a methyl group is unique. These amino esters and their acids have been difficult to resolve by chemical and biochemical means as a result of the increased flexibility of the two

TABLE 6.4
Enantioselective Hydrolysis of α-Amino Esters Catalyzed by *Humicola* Amino Esterase

30, R¹ =	R² =	R³ =	%Conv.	31 (%ee)	32 (%ee)	E value
$Me(CH_2)_8^-$	Et	H	50	91 (R)[d]	—	58
$Me(CH_2)_5^-$	Et	H	50	90 (R)[d]	99 (S)	58
$Me(CH_2)_5^-$	$Me(CH_2)_2^-$	H	37	92 (R)[c]	60 (S)	41
$ArCH_2^-$	Et	H	50	85 (R)[a]	—	35
$ArCH_2^-$	Me	H	74	32 (R)[a]	88 (S)	4.4
$ArCH_2^-$	H	H	68	—	72 (S)[b]	4.0
Ar^-	H	H	62	—	78 (S)[b]	6.3
$Me(CH_2)_3^-$	Et	H	46	92 (R)[d]	—	58
			58	—	98 (S)[e]	
$Me(CH_2)_3^-$	Et	–OCOMe	NR			
MeCH=CH–	Et	H	42	94 (R)[d]	66 (S)	66
$Me(CH_2)_2^-$	Et	H	65	26 (R)[d]	53 (S)	2.6
Me	Et	H	72	36 (S)[a]	100 (R)	20
H	$Me(CH_2)_3^-$	H	40	72 (S)[b]	42 (R)	9.6
H	Et	H	55	53 (S)[b]	42 (R)	6.1
H	Et	–OCOMe	NR			

Absolute configurations were assigned by the following:
[a] Comparison of optical rotation with literature data;
[b] Comparison of high-performance liquid chromatography (HPLC) retention time with authentic samples;
[c] Analogy to the results obtain from d;
[d] Comparison of HPLC retention time of its diastereomeric derivative with those obtained through Schollkopf's asymmetric synthesis;
[e] X-ray crystallographic analysis.

large alkyl groups, which become indistinguishable to most resolving agents. Unsubstituted amino esters underwent significant chemical hydrolysis under the experimental conditions, resulting in relatively lower E values.

When some amino esters are protected by *N*-acetylation, they become resistant to hydrolysis by the enzyme. Replacement of the α-amino group of the amino ester with a hydroxyl group also changes it from substrate to nonsubstrate. The enzyme showed high catalytic activity toward hydrolysis of phenylglycine ethyl ester but was found incapable of catalyzing the hydrolysis of mandelic ethyl ester in which the amino group had been replaced with an α-hydroxyl group. The *N*-acetyl compounds and mandelic ester were not inhibitors, indicating that the free amino group is necessary for binding between the enzyme and substrates. It is most likely that the substrate is protonated at the amino group. The ammonium cation then binds to an anion on the enzyme's active site forming a strong ionic bonding to facilitate the catalysis. This mechanism was further supported by the strong pH dependence of the enzymatic activity toward amino ester hydrolysis. Stability and relative activity of the amino esterase were measured over a range of pHs with ethyl 2-amino-2-ethylhexanoate (**30**, R¹ = Et, R² = *n*-Bu) as substrate, and the data were compared as in Figure 6.3. The enzyme has an optimum pH of 7.5 and is most stable at pH 8. At low pHs, both the stability and activity of the enzyme suffered a gradual loss; this is consistent with the effect of gradual protein denaturation. At high pHs, however, the enzyme remained relatively stable but its activity decreased sharply. The most significant decrease of activity occurs when the medium's pH is greater than 9.6, which coincides with the substrate's pKa. Clearly when the substrate exists

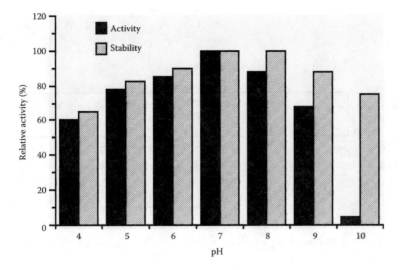

FIGURE 6.3 Activity and stability of *Humicula* amino esterase at different pHs. Activity was determined as % conversion of the amino ester to the amino acid in 0.1 M phosphate buffer at 25°C for 4 hours. Stability was measured by incubating the enzyme at the given pH for 24 hours and then assaying for activity at pH 7.5.

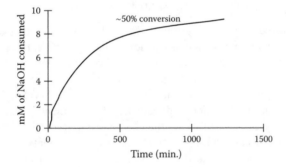

FIGURE 6.4 Profile of the enantioselective hydrolysis of ethyl 2-amino-2-ethylhexanoate catalyzed by *Humicola* amino esterase.

mostly in the form of a free base, its affinity with the enzyme dramatically diminishes at the active site where a charged ammonium cation is required for binding.

Initial scale up of the enzymatic resolution for production of kilogram quantities of (R)-2-amino-2-ethylhexanoic acid was performed in a batch process. The oil of ethyl 2-amino-2-ethyl-hexanoate was suspended in an equal volume of water containing the enzyme. The enantioselective hydrolysis of the ester proceeded at room temperature with titration of the produced acid by NaOH through a pH stat (Figure 6.4).

It is significant that the reaction mixture was worked up by removal of the unreacted ester by hexane extraction and concentration of the aqueous layer to obtain the desired (R)-amino acid. The process has a high throughput and was easy to handle on a large scale. However, because of the nature of a batch process, the enzyme catalyst could not be effectively recovered, adding significantly to the cost of the product. In the further scale up to 100-kg quantity productions, the resolution process was performed using Sepracor's membrane bioreactor module. The enzyme was immobilized by entrapment into the interlayer of the hollow-fiber membrane. Water and the substrate amino ester as a neat oil or hexane solution were circulated on each side of the membrane. The ester was hydrolyzed enantioselectively by the enzyme at the membrane interface, and the chiral acid product

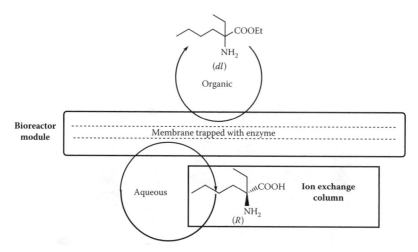

FIGURE 6.5 Membrane bioreactor for enzyme catalyzed enantioselective hydrolysis of racemic ethyl 2-amino-2-ethylhexanoate.

was diffused to the aqueous phase to be recovered by simple concentration or ion-exchange chromatography (Figure 6.5). Optical purity of (*R*)-2-amino-2-ethylhexanoic acid obtained from this process ranged from 98–99% ee.[123]

6.3.2 PREPARATION OF D-AMINO ACIDS USING HYDANTOINASE

D-Amino acids have played an increasingly important role as building blocks for pharmaceuticals and other biologically active agents such as unnatural therapeutic peptides,[124] ACE (angiotensin-converting enzyme) inhibitors,[125] and semi-synthetic β-lactam antibiotics.[126] D-Phenylglycine and D-*p*-hydroxyphenylglycine are used as raw materials in the production of some well-known anti-biotics such as ampicillin, amoxicillin, cephalexin, and cefadroxil. D-Valine is a building block for the synthesis of the pyrethroid insecticide, fluvalinate.[127] Most manufacturers of D-phenylglycine and D-*p*-hydroxyphenylglycine are still using the classic resolution process using camphorsulfonic acid and bromocamphorsulfonic acid, respectively, as resolving agents (but see Chapters 2, 3, 7, and 25).[92] Both compounds are produced in thousands of tonnes per annum. Although industrial productions of most of the D-amino acids are still through classic resolution, enzymatic methods have begun to catch up.[128] At present, the most promising enzymatic processes for production of D-amino acids seem to be those catalyzed by D-specific hydantoinases.

D-Hydantoinases exist at different levels in many microorganism strains, and many of them have been cloned and overexpressed.[110,129,130] This group of enzymes catalyzes enantioselective ring cleavage of 5-substituted hydantoins to give the corresponding D-*N*-carbamylamino acids, which can be further hydrolyzed chemically or enzymatically to the free D-amino acids. The chemical process for removal of the carbamoyl group is often performed under diazotation conditions using sodium nitrite and sulfuric or hydrochloric acid. Enzymatic cleavage of the carbamoyl residue has been reported.[131] Actually, the enzyme, *N*-carbamoyl D-amino acid amidohydrolase, can be incor-porated with hydantoinase to produce D-amino acids in 1 step from the corresponding racemic hydantoins.[132] Because many of the 5-substituted hydantoins undergo spontaneous racemization under the enzymatic reaction conditions (pH >8), no extra steps are needed for racemization of the unwanted enantiomer and complete conversion from racemic hydantoin to the D-amino acid can be achieved. The starting hydantoins are readily prepared according to the procedure of Bucherer and Bergs (Scheme 6.16),[133] by simple condensation of potassium cyanide and ammonium carbonate with the corresponding aliphatic or aromatic aldehydes, which are usually also the starting materials for synthesis of the corresponding racemic amino acids. Many of the D-hydantoinases have broad substrate specificities and can be used for production of a variety of D-amino acids.[109,134]

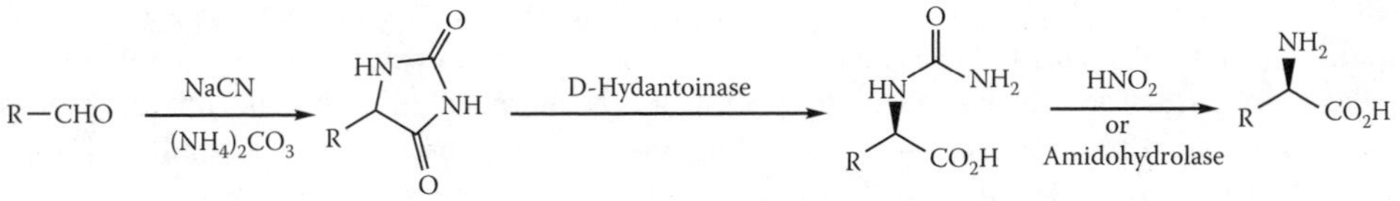

SCHEME 6.16

D-Valine is produced from isobutyraldehyde through a Bucherer-Bergs reaction followed by a 1-step enzymatic conversion of the hydantoin to the free D-amino acid.[135] The biocatalyst is *Agrobacterium radiobacter,* which contains high levels of both D-hydantoinase and *N*-carbamoyl D-amino acid amidohydrolase.[132] A typical run of the biotransformation involves a 10% solution of the hydantoin with a weight ratio of hydantoin versus resting cells (dry wt.) of 5. The incubation is usually performed at 40°C for 48 hours with continuous addition of 1N NaOH to maintain a constant pH of 7.5. The final broth is worked up by centrifugation or ultrafiltration followed by simple concentration and precipitation with ethanol or ethanol/pyridine to give the optically pure D-valine with an overall yield of about 60% from the starting aldehyde.

In the case of D-phenylglycine and D-*p*-hydroxphenylglycine manufacture, several producers are believed to have started the hydantoinase route (Scheme 6.17). Both enzymatic and classic resolution use, as the starting material, benzaldehyde for D-phenylglycine and phenol for D-*p*-hydroxphenylglycine. The route that will eventually become the most cost-effective is very much dependent on the process steps, cost of resolving agent or biocatalyst, and the final yield of product. The hydantoinase process has the advantage of not requiring a racemization step. If the step of chemical hydrolysis of the carbamoyl residue can be completely replaced by the *in situ* enzymatic hydrolysis, the enzymatic route will be much more simple and streamlined than the classic resolution. A few other issues to be worked out are to increase the activity and efficiency of the biocatalyst and to recover the product effectively from the aqueous solution. At least in the manufacture of D-*p*-hydroxyphenylglycine, where the classic resolution using bromocamphorsulfonic acid is not as well-established as in the case of phenylglycine, the hydantoinase route has shown a competitive edge. Kanegafuchi uses a bacterial strain, *Bacillus brevis,* as the biocatalyst for enantioselective conversion of the racemic hydantoin to the *N*-carbamoyl-D-amino acid, which is then diazotized to give D-phenylglycine. Recordati applies a strain of *Agrobacterium radiobacter* that contains both hydantoinase and amidohydrolyase to convert the racemic hydantoin to D-phenylglycine in one step.[7] With continuous strain improvement and genetic engineering, the gap between enzymatic and classic resolution for the production of D-phenylglycine and D-*p*-hydroxyphenylglycine is getting closer.

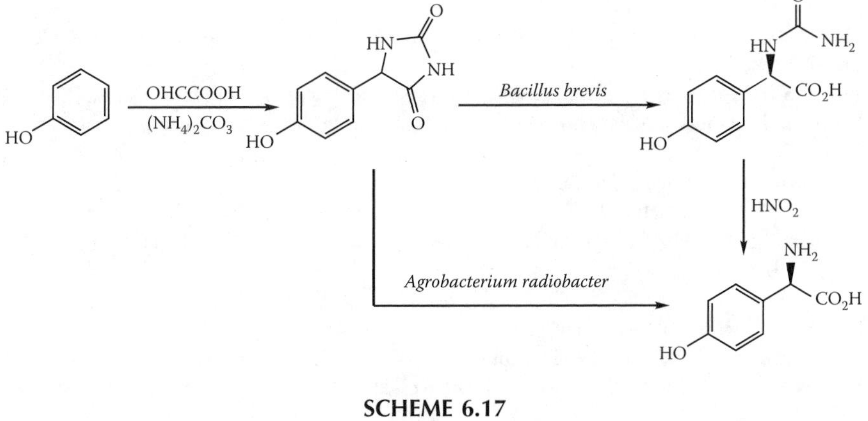

SCHEME 6.17

Beside being used in the whole-cell systems, D-hydantoinases are also commercially available as purified or partially purified enzymes in both free and immobilized forms. A chemist can easily test these enzymes for preparation of a specific target amino acid and scale up the process to produce large quantity materials. Boehringer Mannheim offers two types of immobilized D-hydantoinases, D-Hyd1 and D-Hyd2, in both research and bulk quantities. The enzymes are cloned from thermophilic microorganisms and expressed and overproduced in *Escherichia coli*. Both of them have broad substrate specificities, although D-Hyd1 is more active toward aromatic substituted hydantoins, whereas D-Hyd2 shows better activity to substrates with aliphatic substitution.[136] The combination of the two enzymes allows quick synthesis of a variety of D-amino acids in high chemical and optical yields and relatively large quantities. Some D-amino acids reportedly prepared by these two enzymes include alanine, norleucine, homophenylalanine, phenylalanine, valine, isoleucine, methionine, serine, threonine, 2-thienylglycine, phenylglycine, and *p*-hydroxyphenylglycine. However, unlike the whole cells, which could produce both D-hydantoinase and *N*-carbamoyl-amino acid amidohydrolase, these commercial hydantoinases only catalyze the cleavage of hydantoins to the *N*-carbamoyl-amino acid. A separate chemical hydrolysis step is necessary to obtain the free D-amino acids.

6.4 SUMMARY

Despite the revolutionary advances achieved in the field of catalytic asymmetric synthesis, resolution methods both chemical and enzymatic are still probably the most used methods for preparation of optically pure organic compounds. This is especially true on large scale for the production of industrial fine chemicals. A very large number of chiral pharmaceuticals and pharmaceutical intermediates are manufactured by the process involving resolution. The reason behind the continued dominance of resolution in industrial production of optically pure fine chemicals is perhaps the reliability and scalability of these processes.

REFERENCES

1. Jacques, J., Collet, A., Wilen, S. *Enantiomers, Racemates, and Resolutions*, Wiley: New York, 1981.
2. Bayley, C. R., Vaidya, N. A. in *Chirality in Industry,* Collins, A. N., Sheldrake, G. N., Crosby, J., Eds., Wiley: New York, 1992, p. 71.
3. *CRC Handbook of Optical Resolutions via Diastereomeric Salt Formation,* Kozma, D., Ed., CRC Press: Boca Raton, 2002.
4. Kinbara, K., Tagawa, Y., Saigo, K. *Tetrahedron: Asymmetry* 2001, *12*, 2927.
5. Reinhold, D. F., Firestone, R. A., Gaines, W. A., Chemerda, J. M., Sletzinger, M. *J. Org. Chem.* 1968, *33*, 1209.
6. Amiard, G. *Experientia* 1959, *15*, 1.
7. Sheldon, R. A. *Chirotechnology: Industrial Synthesis of Optically Active Compounds*, Marcel Dekker, Inc.: New York, 1993.
8. Wilen, S. H. *Tables of Resolving Agents*, University of Notre Dame: Notre Dame, IN, 1972.
9. Newman, P. *Optical Resolution Procedures*, Resolution Information Center, Manhattan College: New York, 1981.
10. Eames, J. *Angew. Chem., Int. Ed.* 2000, *39*, 885.
11. Johnson, D. W., Singleton, D. A. *J. Am. Chem. Soc.* 1999, *121*, 9307.
12. Harrison, I. T., Lewis, B., Nelson, P., Rooks, W., Roszkowski, A., Tomolonis, A., Fried, J. H. *J. Med. Chem.* 1970, *13*, 203.
13. Harrington, P. J., Lodewijk, E. *Org. Proc. Res. Dev.* 1997, *1*, 72.
14. Holton, P. G. U.S. Pat. 4,246,193, 1981.
15. Felder, E., San Vitale, R., Pitre, D., Zutter, H. U.S. Pat. 4,246,164, 1981.
16. Sonawane, H. R., Bellur, N. S., Ahuja, J. R., Kulkarni, D. G. *Tetrahedron: Asymmetry* 1992, *3*, 163.

17. Giordano, C., Castaldi, G., Cavicchioli, S., Villa, M. *Tetrahedron* 1998, *45*, 4243.
18. Ohta, T., Takaya, H., Kitamura, M., Nagai, K., Noyori, R. *J. Org. Chem.* 1987, *52*, 3174.
19. Chan, A. S. C., Laneman, S. A. U.S. Pat. 5,202,473, 1993.
20. Chan, A. S. C., Laneman, S. A. U.S. Pat. 5,144,050, 1992.
21. Chan, A. S. C. *Chemtech.* 1993, *23*, 46.
22. Stille, J. K., Su, H., Brechot, P., Parinello, G., Hegedus, L. S. *Organometallics* 1991, *10*, 1183.
23. Hiyama, T., Wakasa, N., Kusumoto, T. *Synlett* 1991, 569.
24. Alper, H., Hamel, N. *J. Am. Chem. Soc.* 1990, *112*, 2803.
25. RajanBabu, T. V., Casalnuovo, A. L. *J. Am. Chem. Soc.* 1992, *114*, 6265.
26. Piccolo, O., Azzena, U., Melloni, G., Delogu, G., Valoti, E. *J. Org. Chem.* 1991, *56*, 183.
27. Piccolo, O., Speafico, F., Visentin, G. *J. Org. Chem.* 1985, *50*, 1946.
28. Reid, A. J., Phillips, G. T., Marx, A. F., Desmet, M. J. Eur. Pat. 0 338 645-A, 1989.
29. Phillips, G. T., Robertson, B. W., Watts, P. D., Matcham, G. W. J., Bertola, M. A., Marx, A. F., Koger, H. S. Eur. Pat. Appl. 205 215, 1986.
30. Nicholson, J. S., Adams, S. S. U.K. Pat. 971 700, 1964.
31. Nicholson, J. S., Adams, S. S. U.S. Pat. 3,228,831, 1966.
32. Adams, S. S., Bresloff, P., Mason, G. *Pharm. Pharmacol.* 1967, *28*, 256.
33. Adams, S. S., Cliffe, E. E., Lessel, B., Nicholson, J. S. *J. Pharm. Sci.* 1967, *56*, 1686.
34. Wechter, W. J., Loughhead, D. G., Reischer, R. J., Van Giessen, G. J., Kaiser, D. G. *Biochem. Biophys. Res. Commun.* 1974, *61*, 833.
35. Sunshine, A., Laska, E. M. U.S. Pat. 4,851,444, 1989.
36. Chan, A. S. C. U.S. Pat. 4,994,607, 1991.
37. Fowler, R., Connor, H., Bachl, R. A. *Chemtech* 1976, 722.
38. Rieu, J.-P., Boucherle, A., Cousse, H., Mouzin, G. *Tetrahedron* 1986, *42*, 4095.
39. Shimizu, I., Matsumura, Y., Tokumoto, Y., Uchida, K. U.S. Pat. 4,694,100, 1987.
40. Shimizu, I., Matsumaura, Y., Tokumoto, Y., Uchida, K. U.S. Pat. 5,097,061, 1992.
41. Tokumoto, Y., Shimizu, I., Inooue, S. U.S. Pat. 5,166,419, 1992.
42. Boots Pure Drug Co. *Fr. Pat.* 1,545,270, 1968.
43. *Chem. Mkt. Rep.* 1993, 20.
44. *Chem. Mkt. Rep.* 1993, 41.
45. Tung, H.-H., Waterson, S., Reynolds, S. D. U.S. Pat. 4,994,604, 1991.
46. Trace, T. L. U.S. Pat. 5,278,338, 1994.
47. Larsen, R. D., Reider, P. U.S. Pat. 4,946,997, 1990.
48. Lin, R. W. U.S. Pat. 5,278,334, 1994.
49. Manimaran, T., Impastato, F. J. U.S. Pat. 5,015,764, 1991.
50. Nohira, H. U.S. Pat. 5,321,154, 1994.
51. Dwivedi, S. K., Sattari, S., Jamali, F., Mitchell, A. G. *Int. J. Pharm.* 1992, *87*, 95.
52. Manimaran, T., Stahly, P. G. *Tetrahedron: Asymmetry* 1993, *4*, 1949.
53. Manimaram, T., Stahly, P. G. U.S. Pat. 5,248,813, 1993.
54. Matson, S. L. U.S. Pat. 4,800,162, 1989.
55. Wald, S. A., Matson, S. L., Zepp, C. M., Dodds, D. R. U.S. Pat. 5,057,427, 1991.
56. Cobbs, C. S., Barton, M. J., Peng, L., Goswani, A., Malick, A. P., Hamman, J. P., Calton, G. J. U.S. Pat. 5,108,916, 1992.
57. Goswani, A. U.S. Pat. 5,175,100, 1992.
58. Bertola, M. A., Marx, A. F., Koger, H. S., Quax, W. J., Van der Laken, C. J., Phillips, G. T., Robertson, B. W., Watts, P. O. U.S. Pat. 4,886,750, 1989.
59. Bertola, M. A., De Smet, M. J., Marx, A. F., Phillips, G. T. U.S. Pat. 5,108,917, 1992.
60. Sih, C. J. Eur. Pat. Appl. 227 078, 1986.
61. Yamada, S., Hongo, C., Yoshioka, R., Chibata, I. *J. Org. Chem.* 1983, *48*, 843.
62. Shiraiwa, T., Shinjo, K., Kurokawa, H. *Chem. Lett.* 1989, 1413.
63. Shiraiwa, T., Shinjo, K., Kurokawa, H. *Bull. Chem. Soc., Jpn.* 1991, *64*, 3251.
64. Huang, T., Qian, X. *Shipin Yu Fajiao Gongye* 1987, *6*, 98.
65. Kolb, H. C., VanNieuwenhze, M. S., Sharpless, K. B. *Chem. Rev.* 1994, *94*, 2483.
66. Jacobsen, E. N. in *Comprehensive Organometallic Chemistry II,* Wilkinson, G., Stone, F. G. A., Abel, E. W., Hegedus, L. S., Eds., Pergamon: New York, 1995, *Vol. 12*, Chapter 11.1.

67. Aggarwal, V. K., Ford, J. G., Thompson, A., Jones, R. V. H., Standen, M. *J. Am. Chem. Soc.* 1996, *118*, 7004.
68. Sinigalia, R., Michelin, R. A., Pinna, F., Strukul, G. *Organometallics* 1987, *6*, 728.
69. Tokunaga, M., Larrow, J. F., Kakiuchi, F., Jacobsen, E. N. *Science* 1997, *277*, 936.
70. Sato, K., Aoki, M., Ogawa, M., Hashimoto, T., Noyori, R. *J. Org. Chem.* 1996, *61*, 8310.
71. Pettman, R. B. *Chemspec Europe BACS Symposium* 1997, Manchester, 37.
72. Aouni, L., Hemberger, K. E., Jasmin, S., Kabir, H, Larrow, J. F., Le-Fur, I., Morel, P., Schlama, T. In *Asymmetric Catalysis on Industrial Scale,* Blaser, H. U., Schmidt, E., Eds., Wiley-VCH: Weinheim, 2004, p. 165.
73. Sticher, H., Faber, K. *Synthesis,* 1997, 1.
74. Asymmetric Catalysis on Industrial Scale, Blaser, H. U., Schmidt, E., Eds., Wiley-VCH, Weinheim, 2004.
75. Paul, P. K. C., Sukumar, M., Blasio, B. D., Pavone, V., Pedone, C., Balaram, P. *J. Am. Chem. Soc.* 1986, *108*, 6363.
76. Bardi, R., Piazzesi, A. M., Toniolo, C., Sukumar, M., Balaram, P. *Biopolymers* 1986, *25*, 1635.
77. Mutter, M. *Angew. Chem.* 1985, *97*, 639.
78. Barone, V., Lelj, F., Bavoso, A., Blasio, B. D., Grimaldi, P., Pavone, V., Pedone, C. *Biopolymers* 1985, *24*, 1759.
79. Khosla, M. C., Stackiwiak, K., Smeby, R. R., Bumpus, F. M., Piriou, F., Lintner, K., Fermandijan, S. *Proc. Natl. Acad. Sci., USA* 1981, *78*, 757.
80. Turk, J., Panse, G. T., Marshall, G. R. *J. Org. Chem.* 1975, *40*, 953.
81. Deeks, T., Crooks, T. P. A., Waigh, R. D. *J. Med. Chem.* 1983, *26*, 762.
82. Christensen, H. N., Handlogten, M. E., Vadgama, J. V., de la Cuesta, E., Ballesteros, P., Trigo, G. C., Avendano, C. *J. Med. Chem.* 1983, *26*, 1374.
83. Zembower, D. E., Gilbert, J. A., Ames, M. M. *J. Med. Chem.* 1993, *36*, 305.
84. Kollonitsch, J., Patchett, A. A., Marburg, S., Maycock, A. L., Perkins, L. M., Doldouras, G. A., Duggan, D. E., Aster, S. D. *Nature* 1978, *274*, 906.
85. Bhattacharjee, M. K., Snell, E. E. *J. Biol. Chem.* 1990, *265*, 6664.
86. Schirlin, D., Gerhart, F., Hornsperger, J. M., Hamon, M., Wagner, J., Jung, M. J. *J. Med. Chem.* 1988, *31*, 30.
87. Coppola, G. M., Schuster, H. F. *Asymmetric Synthesis: Construction of Chiral Molecules Using Amino Acids*, Wiley: New York, 1987.
88. Evans, D. A. in *Asymmetric Synthesis,* Morrison, J. D., Ed., Academic Press: Orlando, 1984, *Vol. 3*, p. 1.
89. Mzengeza, S., Yang, C. M., Whitney, R. A. *J. Am. Chem. Soc.* 1987, *109*, 276.
90. Knudson, C. G., Palkowitz, A. D., Rappaport, H. *J. Org. Chem.* 1985, *50*, 325.
91. Lubell, W. D., Rappaport, H. *J. Am. Chem. Soc.* 1987, *109*, 236.
92. Williams, R. M., Im, M.-N. *J. Am. Chem. Soc.* 1991, *113*, 9276.
93. Schollkopf, U. *Pure Appl. Chem.* 1983, *55*, 1799.
94. Schollkopf, U., Busse, U., Lonsky, R., Hinrichs, R. *Liebigs Ann. Chem.* 1986, 2150.
95. Seebach, D., Boes, M., Naef, R., Schweizer, W. B. *J. Am. Chem. Soc.* 1983, *105*, 5390.
96. Seebach, D., Aebi, J. D., Naef, R., Weber, T. *Helv. Chim. Acta* 1985, *65*, 144.
97. O'Donnell, M. J., Bennett, W. D., Bruder, W. A., Jacobson, W. N., Knuth, K., LeClef, B., Polt, R. L., Bordwell, F. G., Mrozack, S. R., Cripe, T. A. *J. Am. Chem. Soc.* 1988, *110*, 8520.
98. Kolb, M., Barth, J. *Angew. Chem.* 1980, *92*, 753.
99. Goerg, G. I., Guan, X., Kant, J. *Tetrahedron Lett.* 1988, *29*, 403.
100. Obrecht, D., Spiegler, C., Schönholzer, P., Müller, K., Heimgartner, H., Stierli, F. *Helv. Chim. Acta* 1992, *75*, 1666.
101. Bosch, R., Brückner, H., Jung, G., Winter, W. *Tetrahedron* 1982, *38*, 3579.
102. Anantharamaiah, G. M., Roeske, R. W. *Tetrahedron Lett.* 1982, *23*, 3335.
103. Berger, A., Smolarsky, M., Kurn, N., Bosshard, H. R. *J. Org. Chem.* 1973, *38*, 457.
104. Kamphuis, J., Boesten, W. H. J., Broxterman, Q. B., Hermes, H. F. M., Meijer, E. M., Schoemaker, H. E. in *Adv. Biochem. Engineering Biotechnol.,* Flechter, A., Ed., Springer-Verlag: Berlin, 1990, *Vol. 42*.
105. Schollkopf, U., Hartwig, W., Groth, U., Westphalen, K.-O. *Liebigs Ann. Chem.* 1981, 696.
106. Gross, C., Syldatk, C., Mackowiak, V., Wagner, F. J. *Biotechnology* 1990, *14*, 363.

107. Chevalier, P., Roy, D., Morin, A. *Appl. Microbiol. Biotechnol.* 1989, *30*, 482.
108. West, T. P. *Arch. Microbiol.* 1991, *156*, 513.
109. Shimizu, S., Shimada, H., Takahashi, S., Ohashi, T., Tani, Y., Yamada, H. *Agric. Biol. Chem.* 1980, *44*, 2233.
110. Syldatk, C., Wagner, F. *Food Biotechnol.* 1990, *4*, 87.
111. Runser, S., Chinski, N., Ohleyer, E. *Appl. Microbiol. Biotechnol.* 1990, *33*, 382.
112. Nishida, Y., Nakamicho, K., Nabe, K., Tosa, T. *Enzyme Micro. Technol.* 1987, *9*, 721.
113. Chenault, H. K., Dahmer, J., Whitesides, G. M. *J. Am. Chem. Soc.* 1989, *111*, 6354.
114. Keller, J. W., Hamilton, B. J. *Tetrahedron Lett.* 1986, *27*, 1249.
115. Kruizinga, W. H., Bolster, J., Kellogg, R. M., Kamphuis, J., Boesten, W. H. J., Meijer, E. M., Schoemaker, H. E. *J. Org. Chem.* 1988, *53*, 1826.
116. Schoemaker, H. E., Boesten, W. H. J., Kaptein, B., Hermes, H. F. M., Sonke, T., Broxterman, Q. B., van den Tweel, W. J. J., Kamphuis, J. *Pure Appl. Chem.* 1992, *64*, 1171.
117. Stein, G. A., Bronner, H. A., Pfister, K. *J. Am. Chem. Soc.* 1955, *77*, 700.
118. Ware, E. *Chem. Rev.* 1950, *46*, 403.
119. Yee, C., Blythe, T. A., McNabb, T. J., Walts, A. E. *J. Org. Chem.* 1992, *57*, 3525.
120. Toone, E. J., Jones, J. B. *Tetrahedron: Asymmetry* 1991, *2*, 201.
121. Lam, L. K. P., Brown, C. M., Jeso, B. D., Lym, L., Toone, E. J., Jones, J. B. *J. Am. Chem. Soc.* 1988, *110*, 4409.
122. Chen, C. S., Fujimoto, Y., Girdaukas, G., Sih, C. J. *J. Am. Chem. Soc.* 1982, *104*, 7294.
123. Liu, W., Ray, P., Beneza, S. A. *J. Chem. Soc., Perkin Trans. 1* 1995, 553.
124. Roth, H. J., Fenner, H. *Arzneistoffe-Pharmazeutische Chemie III*, Georg Thieme: Stuttgart, 1988.
125. Patchett, A. A. *Nature* 1980, *288*, 280.
126. Dürekheimer, W., Blumbach, J., Lattreil, R., Scheunemann, K. H. *Angew. Chem.* 1985, *97*, 183.
127. Kamphuis, J., Boesten, W. H. J., Schoemaker, H. E., Meijer, E. M. In *Conference Proceedings on Pharmaceutical Ingredients and Intermediates,* Reuben, B. G., Ed., Expoconsult: Maarsen, 1991, p. 28.
128. Williams, R. M. *Synthesis of Optically Active α-Amino Acids*, Pergamon Press: Oxford, 1989.
129. Morin, A., Hummel, W., Kula, M.-R. *Appl. Microbiol. Biotechnol.* 1986, *25*, 91.
130. Morin, A., Touzel, J.-P., Lafond, A., Leblanc, D. *Appl. Microbiol. Biotechnol.* 1991, *35*, 536.
131. Olivieri, R., Fascetti, E., Angelini, L., Degen, L. *Enzyme Microb. Technol.* 1979, *1*, 201.
132. Olivieri, R., Fascetti, E., Angelini, L., Degen, L. *Biotechnol. Bioeng.* 1981, *23*, 2173.
133. Bucherer, H. T., Steiner, W. *J. Prakt. Chem.* 1934, *140*, 291.
134. Takahashi, S., Ohashi, T., Kii, Y., Kumagai, H., Yamada, H. *J. Ferment. Technol.* 1979, *57*, 328.
135. Battilott, M., Barberini, U. *J. Mol. Catal.* 1988, *43*, 343.
136. Keil, O., Schneider, M. P., Rasor, J. P. *Tetrahedron: Asymmetry* 1995, *6*, 1257.

7 New Developments in Crystallization-Induced Resolution

Bernard Kaptein, Ton R. Vries, José W. Nieuwenhuijzen, Richard M. Kellogg, Reinier F. P. Grimbergen, and Quirinus B. Broxterman

CONTENTS

7.1 INTRODUCTION

The cost-effective synthesis of enantiopure compounds is and will continue to be a challenge for the fine chemical industry (see Chapter 1). For 2003, the revenues generated by fine chemical companies via application of chiral technology for pharmaceutical and agrochemical intermediates was estimated to be $8 billion with an annual growth of approximately 11%.[1] Of the drugs currently under development, approximately 80% are chiral. Although in recent years great progress in asymmetric synthesis[2] and (enantioselective) biocatalysis[3] has been realized, often separation of enantiomers is still the method of choice on industrial scale. Although on a laboratory scale the application of preparative high-performance liquid chromatography (HPLC) methods is becoming increasingly important, large-scale preparation of single enantiomers by diastereoisomeric salt formation is still the preferred method.[4] The process of crystallization-driven resolution is usually referred to as a "classical" resolution. Our estimation based on current use of technology is that 30–50% of single enantiomers are obtained by resolution of the corresponding racemate (*cf.* Chapter 31).

In this chapter some of the recent developments in crystallization-induced resolution are presented. After an introduction to the state-of-the-art of resolution by diastereoisomeric salt formation and some recent developments in identifying new resolving agents, the new and coming generation of automated screening and speed in development will be discussed.

A new concept in classical resolution is the application of families of structurally and stereo-chemically related resolving agents. This technique now goes under the name of "Dutch Resolution" (DR). This concept, with a resulting high chance of success, is described together with some recent modifications of the DR concept, new families of resolving agents, and the latest insights in understanding the process.

The general drawback of all these resolution concepts is the maximum yield of 50%. In the last part of this chapter the state-of-the-art and recent developments in 100% yield concepts in crystallization-induced resolution are discussed. This so-called crystallization-induced asymmetric transformation combines classical resolution with *in situ* racemization. Most examples in this chapter originate from day-to-day research efforts at DSM Pharma Chemicals and at Syncom.

7.2 RESOLUTION BY DIASTEREOISOMERIC SALT FORMATION

Since its discovery by Pasteur in 1853,[5] classical resolution by selective crystallization of diastereo-isomers, despite wide and frequent use, remains to a large degree a method of trial and error. Various attempts to rationalize classical resolutions and predict a successful combination of race-mate and resolving agent by computational approaches so far have not been crowned with remark-able success.[6] Even when the crystal structures of both diastereoisomeric salts are known, molecular modeling calculations do not provide a basis for a reliable prediction. Only recently has some progress been made in the calculation of the relative thermodynamic stability of ephedrine–cyclic phosphoric acid **4** diastereoisomers,[7] a diastereoisomeric salt frequently used as a model system (*vide infra*).

Discovery of an efficient resolution, even today, still often relies on screening of a large number of resolving agents. Few systematic studies on the average chance of finding a suitable resolution have been performed; Collet estimated the chance of success as 20–30% for a typical resolution experiment.[8] For an ideal resolution two factors are of importance: a large solubility difference between the diastereoisomers—reflected in the position of the eutecticum E (Figure 7.1)—and the

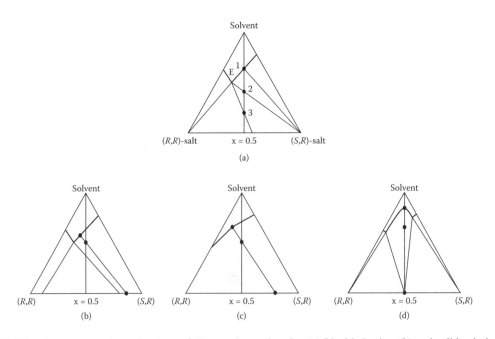

FIGURE 7.1 Solubility phase diagrams of diastereoisomeric salts. (a) Ideal behavior; (b) end solid-solution behavior; (c) full solid-solution behavior; and (d) double salt formation.

absence of formation of (end) solid solutions. A typical solubility phase diagram describing such an ideal resolution (where the resolving agent has the R-configuration in the example) is depicted in Figure 7.1a. At a starting concentration (2) the highest yield of the least soluble diastereoiso-merically pure salt (S,R in this example) is obtained, leaving the mother liquor saturated in both diastereoisomers (E). At lower concentrations, between points 1 and 2, a lower yield of pure salt is obtained, whereas at higher concentration (point 3) also the other diastereoisomer will start to crystallize, resulting in a lower diastereoisomeric purity.

Frequently deviations from this ideal behavior are observed—for instance, as a result of end or even full solid-solution behavior (Figures 7.1b and 7.1c) or double salt formation (Figure 7.1d). In the former situation repeated (or fractional) crystallization is needed to obtain a diastereoiso-merically pure salt, whereas in the latter it is virtually impossible to obtain a pure salt. In addition to these physical constraints, cost and availability play a determining role in the actual application of resolving agents on an industrial scale.

Many resolving agents are available nowadays, and a good overview has been given by Kozma.[9] Some, such as brucine, strychnine, and amphetamine, cannot be used because they are poisonous. However, new resolving agents steadily become available from various synthetic endeavors in the fine chemical industry. Some recent examples from our own efforts are the basic resolving agents (*1S,2R*)-1-amino-2-indanol (**1**) (see also Chapter 8),[10] (*R*)-phenylglycine amide (**2**) (see Chapter 25), and (*R*)-α-methylphenylglycine amide (**3**) and the acidic resolving agents phencyphos **4**[11] and chalconesulfonic acid **5**.[12]

Other resolving agents are readily prepared from inexpensive chiral starting materials such as glucose, aspartic acid, or glutamic acid. A literature example is the use of *N*-methylglucamine **6**, obtained by reductive amination of D-glucose, in the resolution of naproxen (Chapter 6).[13]

At DSM Pharma Chemicals a series of (substituted) *N*-benzoylated L-aspartic and L-glutamic acids has been developed for use as families of resolving agents in a DR (see Section 7.4). Particularly, the *N*-benzoyl-L-glutamic acid series has resulted in some efficient resolving agents for the resolution of chiral amino nitriles. An example of a very efficient resolution is given in Scheme 7.1.

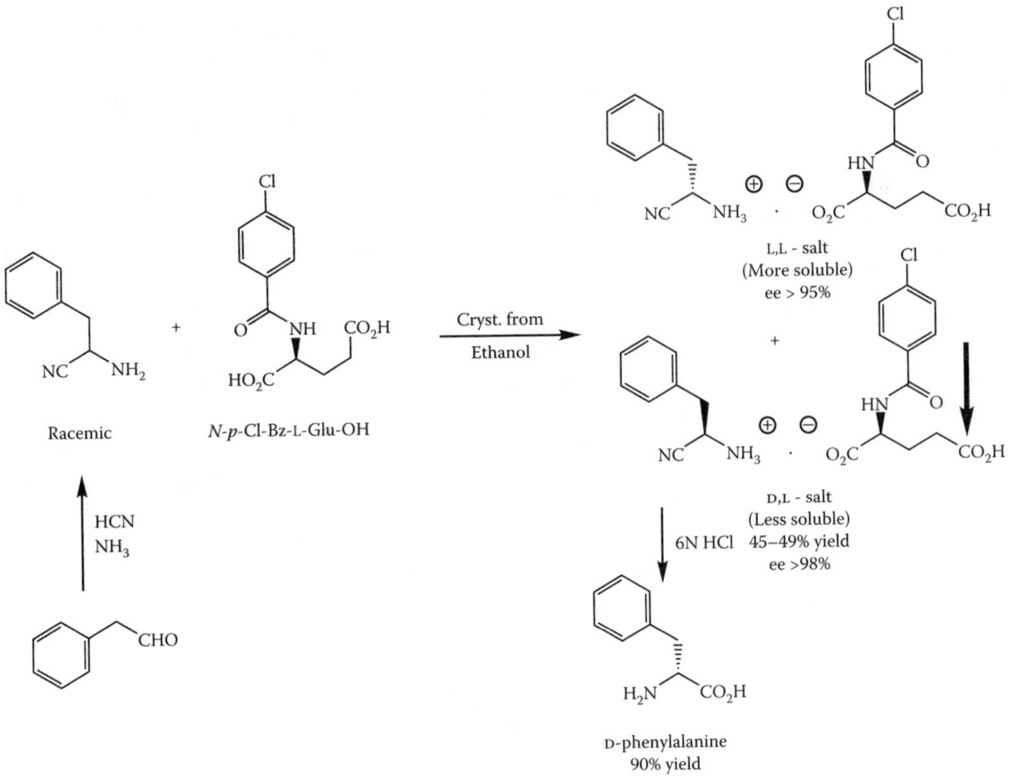

SCHEME 7.1

In this example the resolution of racemic phenylalanine nitrile, readily obtained by a Strecker reaction with phenylacetaldehyde, is performed with *N-p*-chlorobenzoyl-L-glutamic acid. The yield of this resolution nearly reaches the theoretic 50% maximum yield of enantiopure D-amino nitrile. After acidic workup of the crystalline diastereoisomeric salt, the (*R*)-amino nitrile is easily hydrolyzed to D-phenylalanine.

7.3 AUTOMATED SCREENING OF RESOLVING AGENTS

Efficient and fast screening procedures are required to circumvent the necessity of time-consuming trial and error for identification of suitable resolving agents. Modern high-throughput screening procedures allow for the fast determination of phase diagrams, which are indicators of the ideal resolution conditions. Various methods have been tested for their applicability. One of these methods consists of an automated solubility determination of mixtures of both diastereoisomeric salts in various ratios. To this end the Chemspeed ASW2000™ equipment[14] combined with an automated HPLC analysis was used to determine the dilution at which these mixtures become soluble. From this information the solubility lines and the eutectic composition for various resolving agents can be determined. An example of this procedure is shown in Figure 7.2 for the resolution of racemic *cis*-1-amino-2-indanol with L-(+)-tartaric acid.

This method gives a good fit with the solubilities known, especially if applied with high starting concentrations. At lower starting concentrations the solubility of the more soluble diastereoisomer is underestimated, probably because of slow dissolution (thermodynamic equilibrium not being reached during the time period of the experiment). The position of the eutectic point E is indicative of a highly effective resolution. Obviously, this method does not identify the existence of end solid-solution behavior in the crystalline state.

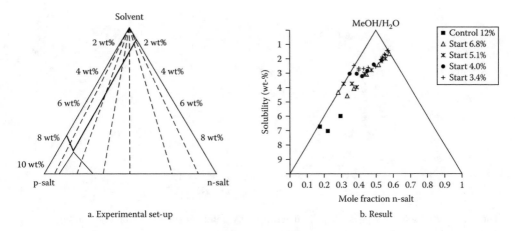

FIGURE 7.2 Automated determination of a solubility phase diagram of a classical resolution. (a) Experimental setup with different ratios of p- and n-salt (dilution along the dashed lines); (b) solubilities at different starting concentrations.

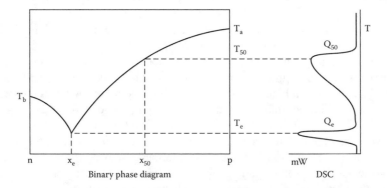

FIGURE 7.3 Determination of the fusion phase diagram by DSC measurement in a 1:1 composition.

In general, however, the pure enantiomers of the racemate under study are not available to prepare the diastereoisomerically pure salts. Therefore, the automated screening described earlier has limited application.

An alternative screening method, using a 1:1 mixture of the diastereoisomeric salts (prepared directly from the racemate), is highly favorable. One such procedure suggested by Kozma[15] and Bruggink[16] has been used by Mitchell[17] as a screening tool for the identification of resolving agents. Differential scanning calorimetry (DSC) is used to determine the melting point thermogram of the 1:1 diastereoisomeric mixture. The heat of fusion ΔH_f of the less soluble (i.e., higher melting) diastereoisomer and the eutectic composition can be calculated from the thermogram (Figure 7.3) by means of the Schröder-Van Laar equation (Equation 7.1). Conglomerate melting behavior is a prerequisite for a successful application of this procedure.

$$\ln(x_{50}) = \frac{\Delta H_a}{R} \cdot \left(\frac{1}{T_a} - \frac{1}{T_{50}} \right) \tag{7.1}$$

We tested this procedure on 10 different known resolutions, some of which were run on large scale. However, in only three cases was a clear result obtained in line with the known resolution efficiency. For all other examples no workable DSC thermogram was obtained. Phase transitions, loss of solvates

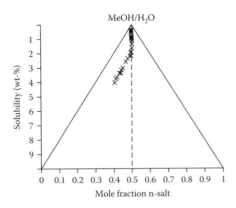

FIGURE 7.4 Partial solubility phase diagram determined by dilution experiments on a 1:1 diastereoisomeric mixture (*dashed line*).

or decomposition of the salt during heating, and incomplete separation of the DSC melting curves are some of the causes for the disparities. False-negative results were obtained in 70% of the cases studied.[18] Therefore, this method is not suitable as a general screening tool for resolving agents.

A better screening procedure is based on the differences in solubility behavior of the 1:1 diastereoisomeric mixture (i.e., racemic starting material). A suitable screening method was developed by performing automated dilution experiments in the HEL automate™ equipment.[19] During the dilution experiment, automated samples of the solution were taken and analyzed by HPLC for enantiomeric excess and concentration. Initially, at high starting concentrations between point 3 and 2 (see the phase diagram in Figure 7.1) both diastereoisomeric salts will be saturated in solution (eutectic composition with constant ee). On further dilution beyond point 2, the more soluble diastereoisomer will be fully dissolved, whereas the concentration of the less soluble diastereoisomer is still beyond saturation. The concentration in solution will change from the eutectic point E to point 1. Analysis of the remaining solid at these dilutions can give information about the nature of the diastereoisomeric salt (solid-solution behavior, solvates, etc.). At point 1 all of the solids are dissolved and only dilution of the 1:1 composition is observed on addition of more solvent. In this way a partial solubility line is determined, as shown in Figure 7.4, again for the previously mentioned racemic *cis*-1-amino-2-indanol – L-(+)-tartaric acid salt.

7.4 DUTCH RESOLUTION

In 1998 we described a new concept in classical resolution at Syncom, whereby a mixture of structurally closely related resolving agents (families) was used.[20] By this method, the chance of success in finding a suitable resolution significantly increases. Since the initial discovery of this method in 1996, about 1000 racemates have been examined. The success rate to effect a resolution has been >95%, and usually only a few families of resolving agents need to be tested.[21] Obviously, many of these examples fall under customer confidentiality. A selection of nonconfidential structures is shown in Table 7.1. Other examples can be found in the original DR publication.[20] By use of this method the development time for a resolution process on lab scale can be sharply reduced compared to the standard trial and error method. We coined the name "Dutch Resolution" (or DR) for this concept of a family approach to the resolution of racemates. Two typical examples are shown in Figure 7.5 and Scheme 7.2.

With the use of a family of three structurally related cyclic phosphoric acids (P-mix) as resolving agent, a diastereoisomerically pure salt containing the (*2S,3R*)-enantiomer in 99% ee precipitates. The resolved *threo*-(4-methylthiophenyl)serine amide (Figure 7.5) can be used as an intermediate in the synthesis of thiamphenicol.[22] Note that the individual phosphoric acids give salts of lower

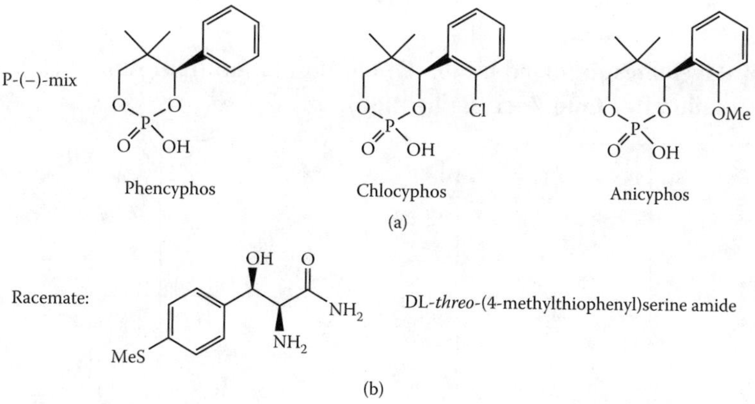

P-(–)-mix

Phencyphos

Chlocyphos

Anicyphos

(a)

Racemate:

DL-*threo*-(4-methylthiophenyl)serine amide

(b)

Resolving Agent	Yield (%)	ee (%)	Mix-Ratio
(–)-Phencyphos	47	52 (*2R,3S*)	—
(–)-Chlocyphos	55	17 (*2R,3S*)	—
(–)-Anicyphos	41	67 (*2R,3R*)	—
(–)-P-mix (1:1:1)	25	98.8 (*2S,3R*)	12:35:53 (P:C:A)

FIGURE 7.5 Dutch Resolution of D,L-*threo*-(4-methylthiophenyl)serine amide with P-(–)-mix.

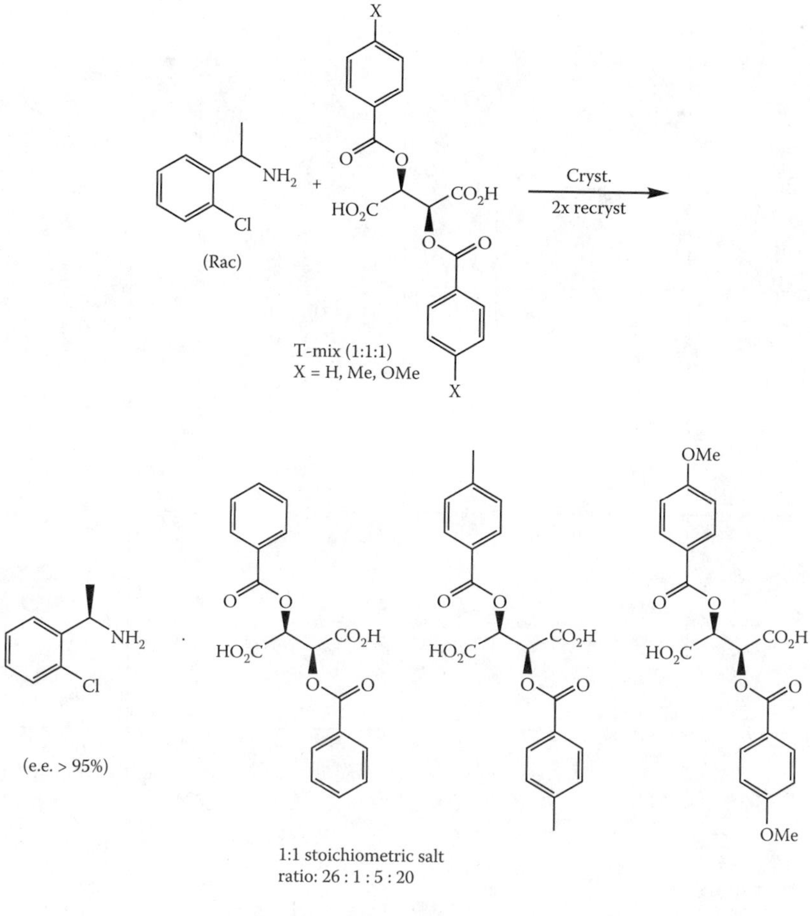

(Rac)

Cryst.
2x recryst

T-mix (1:1:1)
X = H, Me, OMe

(e.e. > 95%)

1:1 stoichiometric salt
ratio: 26 : 1 : 5 : 20

SCHEME 7.2

TABLE 7.1
Selection of Racemates Resolved by the Dutch Resolution (DR) Approach: ee and Mix (Ratio after One Recrystallization)

Structure	ee (%)	DR Family[a]	Ratio	Structure	ee (%)	DR Family[a]	Ratio
(structure)	90	M-mix	1:1[b]	(structure, Cl)	99	P-mix	20:0:1
(structure, Br, NH₂)	90	M-mix	1:1[b]	(structure, NH₂)	99	P-mix	5:5:1
(structure, Cl, NH₂)	99	M-mix	1:1[b]	(structure, Br, NH₂)	99	M-mix	1:5[b]
(structure, H₃CO, NH₂)	94	M-mix	1:10[b]	(structure, Ph, NH₂)	86	P-Mix	–[c]
(structure, NH₂)	98	P-Mix	–[c]	(structure, Cl, NH₂)	94	M-mix	1:5:0
(structure, Br, NH₂)	96	M-mix	1:5:0				
(structure, H₂N, OH, O)	92	P-mix	4:1:2	(structure, H₂N, NH₂, O)	98	P-mix	2:1:1
(structure, HO, OH, NH₂, O)	95	P-mix	1:2:3	(structure, F, OCH₃, NH₂, O)	94	P-mix	15:3:2
(structure, F, CN, NH₂)	95	P-mix	20:2:1	(structure, CN, NH₂)	95[d]	P-mix	10:1:1
(structure, cis/trans)	98	P-mix	4:1:0				
(structure, OH, NH₂, O)	98[d]	P-Mix	5:4:1	(structure, NH₂, N)	89[e]	P-mix	4:1:0
(structure, NH₂, OH)	98	M-mix	1:3:1	(structure, NHMe, OH) [f]	98	P-mix	1:1:1
(structure, NH₂, OH[f])	96	PGA-mix	1:2:1	(structure, NH₂, OH)	99[d]	M-mix	1:4:1

TABLE 7.1 (continued)
Selection of Racemates Resolved by the Dutch Resolution (DR) Approach: ee and Mix (Ratio after One Recrystallization)

Structure	ee	Mix	Ratio	Structure	ee	Mix	Ratio
Br–C₆H₄–CH(NH₂)CH₂OH	99[d]	P-mix	1:0:0	CH₃–C₆H₄–CH(NH₂)CH₂OH	91	P-mix	10:1:1
MeO–C₆H₄–CH(NH₂)CH₂OH	97[d]	P-mix	7:1:1				
quinuclidine–OH	66	P-mix	1:1:2	quinuclidine–O–C(=O)–Ph	98	T-mix	1:10:4
3-ethylmorpholine	96[e]	T-mix	0:2:1	2-phenyl-3-hydroxypiperidine	99	T-mix	5:1:1
2-phenylpyrrolidine	93	P-mix	–[c]	2-(4-chlorophenyl)pyrrolidine	96	P-mix	2:7:4
3,4-diphenylpyrrolidine	99	T-mix	1:3:3	3,4-diphenyl-1-benzylpyrrolidine[f]	97	T-mix	6:1:1
Ph–CH(Et)COOH	93	PG-mix	–[c]	Br–C₆H₄–CH(Et)COOH	95	PE-II-mix	1:0:0
Cl–C₆H₄–O–CH(CH₃)COOH	88	PE-I-mix	–[c]	F–C₆H₄–CH(OH)COOH	99[d]	PE-II-mix	–[c]
Cl–C₆H₄–CH(OH)COOH	95	PE-II-mix	35:1:0	Br–C₆H₄–CH(OH)COOH	95	PE-II-mix	15:1:0
CH₃–C₆H₄–CH(OH)COOH	95	PE-I-mix	–[c]	C₆H₄–C(OH)(COOH)	97	PE-I-mix	–[c]
Ph–CH(COOH)CH₂COOH	97	PG-mix	–[c]	Ph–CH₂–CH(COOH)CH₂COOH	88	PE-II-mix	2:2:1

[a] Structures of the mixtures are given in Figure 7.6.
[b] Only with 1:1 family of X = H and X = Me.
[c] Mix composition could not be accurately determined, but all three components were present.
[d] No recrystallization.
[e] Two recrystallizations.
[f] Racemic *anti*-isomer.

ee and in two of the three cases with the mirror image isomer of the amino acid amide. Moreover, most of the time the DR salt contains all three of the resolving agents in a nonstoichiometric ratio. Solid-solution behavior of the resolving agents is a general phenomenon observed in almost all DR experiments and can also be seen in the example in Scheme 7.2 for the T-mix.

At the time of the initial DR work, various families of acidic and basic resolving agents were available (Figure 7.6).[21]

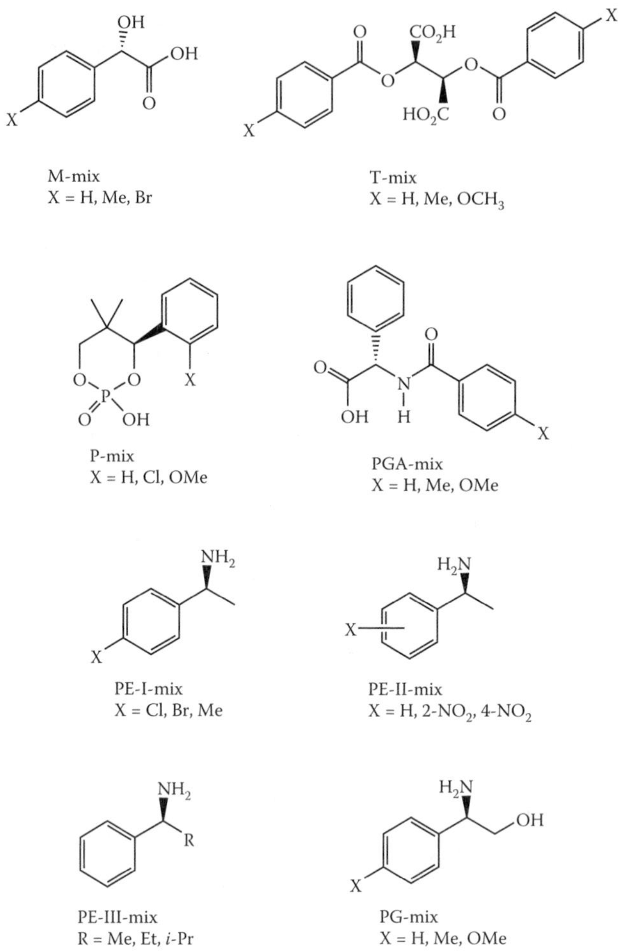

FIGURE 7.6 Families of resolving agents.

For practical reasons most families in general consist of a set of closely related homochiral resolving agents that differ in substituents on the aromatic group. However, family behavior is not limited to this, as can be seen from results with the PE-III mix. Although solid-solution behavior of the family of resolving agents is a general phenomenon, it does not constitute the (thermodynamic) explanation for the improved resolution (*vide infra*). The solubility of the mixed crystals often is higher than that of the diastereoisomerically pure salt of one of the resolving agents. Repeated crystallization of the (diastereoisomerically pure) mixed crystals results in a gradual shift of the mix-composition to one of the single resolving agents. This is the resolving agent with the lowest solubility, initially present in the highest fraction in the precipitated mixed salt. An explanation for the higher efficiency of DR as compared to the classical resolution approach might find its origin in this change in composition of the resolving agents. In general, crystals of the less soluble diastereoisomeric salt are obtained that are enriched in one or more of the resolving agents. The solubility of the diastereoisomeric salt with this new family composition is reduced, compared with the initial family composition. As a result, the composition of the family mix remaining in solution is shifted in the opposite direction. In general, this results in a higher solubility of the more soluble diastereoisomer with the remaining family mix. This can be considered as a special modification of a resolution according to the method of Pope and Peachey.[23]

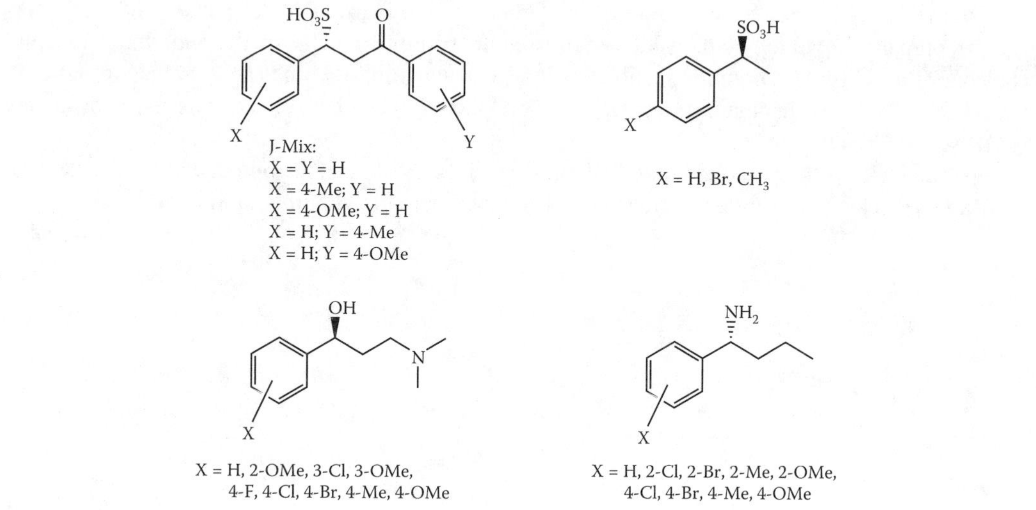

FIGURE 7.7 New families of resolving agents.

Other explanations of the DR effect can be offered, as has been suggested by Collet.[24] It is also not unlikely that various effects, likely complementary, are involved. For instance, nucleation inhibition and crystal growth inhibition can play an important role during many DR experiments (*vide infra*).

7.4.1 NEW FAMILIES OF RESOLVING AGENTS

Various efficient families of resolving agents are available (Figure 7.6) and have been described before.[20] The selection is, however, not all that large. To find new efficient (families of) resolving agents we have developed in recent years several synthetic approaches to new families.[12,21,25] Four of these are depicted in Figure 7.7.

It should be realized that the preparation of a new family takes a considerable amount of time. Two general approaches can be followed. The first consists of the synthesis of the racemic family members, followed by the individual resolution of each member and combining them to the preferred combination of a 1:1:1 mix. This general procedure is used, for example, for the preparation of the J-mix (Scheme 7.3).[12,25]

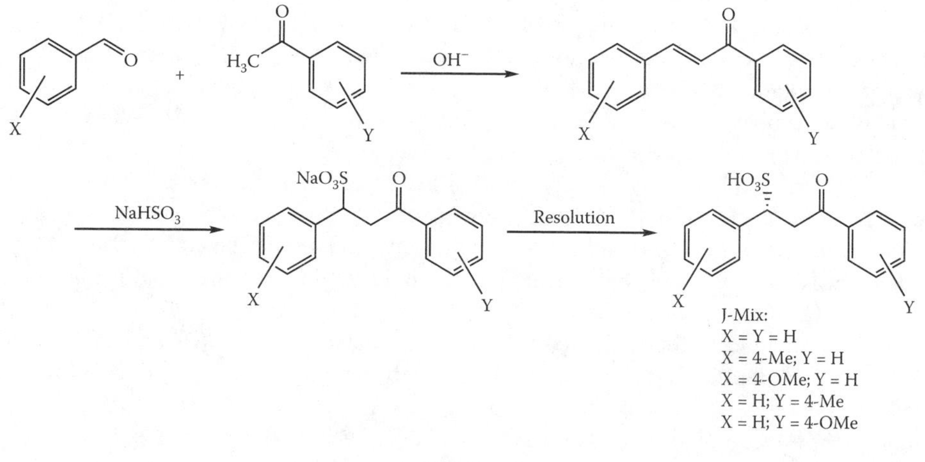

SCHEME 7.3

Starting with a range of substituted acetophenones and benzaldehydes substituted chalcones can be prepared in high yield. Addition of sodium bisulfite results in the individual racemic chalconesulfonic acids. These are resolved using (S)-leucine to yield the (R)-(+)-chalconesulfonic acids. Also (R)-phenylglycinamide and (R)-4-methylphenylglycinol are effective as resolving agents for these chalonesulfonic acids.

Asymmetric synthesis of the individual family members is a second approach to the preparation of a family.[26] This is illustrated in Scheme 7.4 for the phenylbutylamine family.

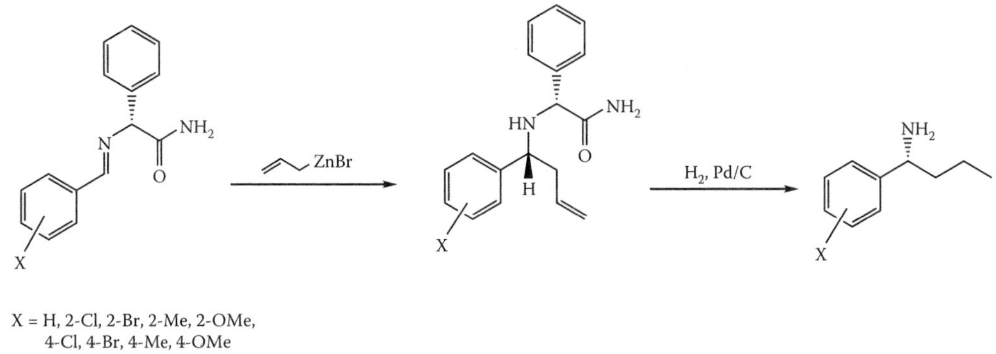

X = H, 2-Cl, 2-Br, 2-Me, 2-OMe,
    4-Cl, 4-Br, 4-Me, 4-OMe

SCHEME 7.4

Obviously, instead of preparing each family member individually, combinatorial approaches can also be applied. Also the resolution of the various racemic family members can be combined. For example, the resolution of the chalconesulfonic acids and of the members of the PE-I mix can be combined. A disadvantage is that the families are obtained in nonstoichiometric ratio.

Most of these families are still under investigation and are routinely taken along in trial DR experiments aimed at resolving new racemates.

7.4.2 Reciprocal and Reverse Dutch Resolution

Instead of resolving a racemate with a mixture of resolving agents, a mixture of racemates can be resolved with a single resolving agent (or even with a mixture of resolving agents). Some examples were mentioned earlier. A special application of this reciprocal approach is the use of an enantiopure family member of the racemate. If one of the enantiomers of the racemate and its enantiopure family member show solid-solution behavior (like the resolving agents in a normal DR), co-resolution is possible. This offers the possibility to resolve compounds with a resolving agent that would not be effective under normal resolution conditions.

Such an example is formed by the resolution of 4-hydroxyphenylglycine (Hpg) (Figure 7.8). Hpg cannot be resolved by (+)-camphor-10-sulfonic acid [(+)-Csa] and is, therefore, resolved on industrial scale with the more expensive and difficult to recycle (1R)-(+)-(endo,anti)-3-bromo-camphor-8-sulfonic acid. However, if racemic or (R)-phenylglycine (Phg) is added to the resolution of Hpg with (+)-Csa, co-resolution of both phenylglycines is possible. (R)-(–)-Hpg is incorporated in the crystal lattice of the (R)-(–)-Phg—(+)-Csa salt by partial replacement of (R)-(–)-Phg.[27]

Analogously, para-fluorophenylglycine, for which no single resolving agent is known, can be resolved with (R)-(–)-Phg and (+)-Csa.

A second example is the reciprocal DR of (R,S)-alaninol. Both (R)- and (S)-alaninol are interesting chiral intermediates used in recent drug developments[28] but are difficult to obtain by classical resolution.[29] Alaninol is also one of the few examples that resisted the standard DR procedure. However, resolution of (R,S)-alaninol could be achieved with mandelic acid in isopropanol/water (19:1) in the presence of enantiopure 2-amino-1-butanol.[30] Whereas the latter amine

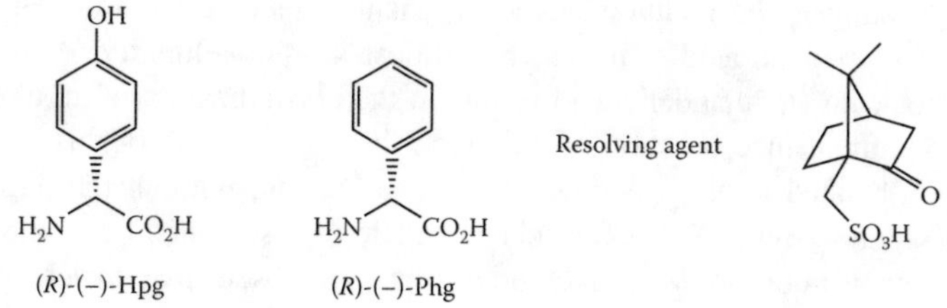

Starting Ratio Hpg:Phg	Crystal. Yield (%)	Ratio Hpg:Phg	ee (%) (–)-Phg	ee (%) (–)-Hpg
With (±)-Phg				
1:3	35	13:87	94 (>99)	94 (100)
1:1	17	25:75	95 (>99)	89 (>95)
With (–)-Phg				
2:1	26	26:74	–	87
1:1	53	19:81	–	67

Resolutions on 40-46 wt% solutions with 0.75 eq. Of (+)-CSA in 1N HCl (ee after recrystallization). In parentheses, ee after second crystallization.

FIGURE 7.8 Dutch Resolution of (R,S)-p-hydroxyphenylglycine with (+)-camphor-10-sulfonic acid and (R,S)- or (R)-phenylglycine as a family member.

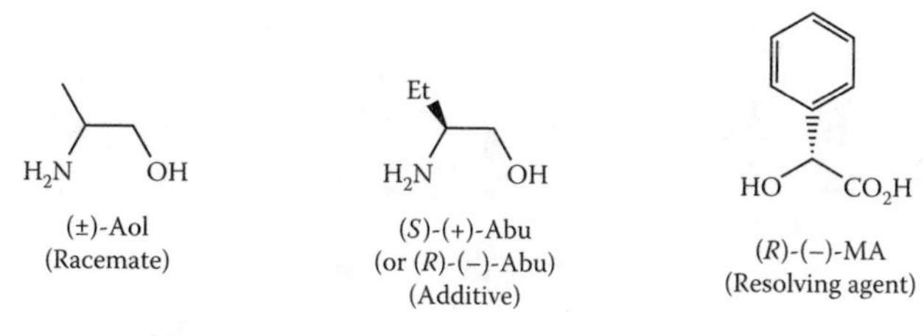

	Starting ratio Abu		(R)-(–)-alaninol in salt		
(±)-Aol	(additive)	(R)-(–)-MA	Yield (%)	ee (%)	Aol/Abu
100	—	100	43	13	—
50	50 (R)-(–)	100	38	94	25:75
75	25 (R)-(–)	100	34	92	57:43
90	10 (R)-(–)	100	37	88	76:24
67	33 (S)-(+)	100	24	94	98:2

FIGURE 7.9 Dutch Resolution of (R,S)-alaninol with (R)-mandelic acid using (R)-, or (S)-2-amino-1-butanol as a family member.

can be effectively resolved with mandelic acid, the resolution of alaninol with mandelic acid proceeds poorly. Addition of racemic or enantiopure (R)-2-amino-1-butanol (forming the less soluble diastereoisomer) to the resolution of alaninol with (R)-mandelic acid results in co-crystallization of a mixed salt of (R)-alaninol, (R)-2-amino-1-butanol, and (R)-mandelic acid (Figure 7.9). Depending on the starting ratio of (R,S)-alaninol and (R)-2-amino-1-butanol, more or less of the (R)-alaninol is incorporated in the diastereoisomeric salt.

Even more interesting is the addition of the (*S*)-enantiomer of 2-amino-1-butanol, which itself forms a salt with (*R*)-mandelic acid. This salt has almost infinite solubility. With this additive, the resolution of alaninol with (*R*)-mandelic acid results in the crystallization of an almost diastereoisomerically pure salt, without incorporation of 2-amino-1-butanol. On use of this method enantiopure (*R*)-alaninol can be obtained after an additional recrystallization, without the need to separate the alaninol from the 2-amino-1-butanol. We call this procedure reverse Dutch Resolution, to indicate that the family mix remains in solution. Most likely in this case stereoselective nucleation inhibition plays a role by suppressing the nucleation of the (*S*)-alaninol-(*R*)-mandelic acid diastereoisomer.

7.4.3 The Role of Nucleation Inhibition

On reexamination of the solid-solution behavior of the resolving agents in the various DR experiments, our attention was drawn to the following set of results. Analogous to the case of the (*R*)-alaninol-(*R*)-mandelic acid salt, in 46 other examples the crystalline salts lacked one or more of the resolving agents of the family mix (or resolving agents were incorporated only in a low percentage) but still gave efficient resolution. Further investigation revealed that the poorly incorporated resolving agent acts as an effective nucleation inhibitor, even at lower mol-fractions.[31] As an example of what can be accomplished, the resolution of mandelic acid with (*S*)-1-phenylethylamine can be significantly improved by replacement of 10 mol% of the resolving agent by a mixture of 2- and 4-nitro substituted (*S*)-1-phenylethylamine (Scheme 7.5).

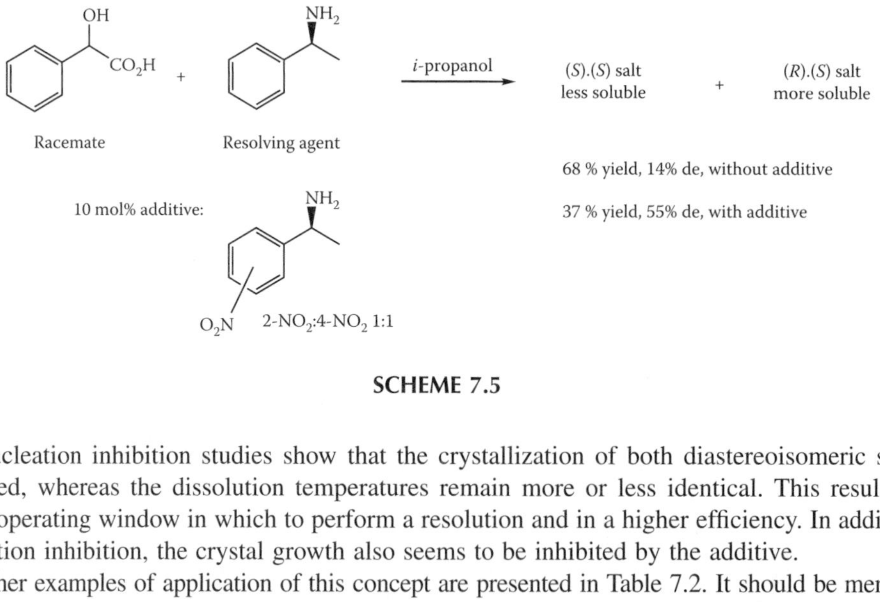

SCHEME 7.5

Nucleation inhibition studies show that the crystallization of both diastereoisomeric salts is inhibited, whereas the dissolution temperatures remain more or less identical. This results in a larger operating window in which to perform a resolution and in a higher efficiency. In addition to nucleation inhibition, the crystal growth also seems to be inhibited by the additive.

Other examples of application of this concept are presented in Table 7.2. It should be mentioned that the resolutions described in Table 7.2 intentionally are performed in nonoptimized conditions. Therefore, further improvement of the resolution efficiency should be possible when nucleation inhibitors are applied. The procedure clearly offers the opportunity to improve the outcome of a classical resolution without the need for stoichiometric mixtures of resolving agents. The original DR procedure as well as molecular modeling calculations[32] can help in identifying efficient nucleation inhibitors.

7.5 CRYSTALLIZATION-INDUCED ASYMMETRIC TRANSFORMATION

Combination of (classical) resolution with *in situ* racemization is a powerful but highly underestimated technology. This technology is often referred to as crystallization-induced asymmetric

TABLE 7.2
Resolution in the Presence of 10 mol% of a Family Member of the Resolving Agent

Racemate	Resolving agent	Additive[a]	Solvent	Yield(%)	ee(%)
		–	i-PrOH	58	22
		–	EtOH	23	68
			EtOH	29	86
			EtOH	30	88
			i-PrOH	36	86
		–	i-PrOH	40	74
			i-PrOH	39	93
			i-PrOH	36	89
		–	i-PrOH	58	49
			i-PrOH	45	63

[a] 10 mol%.

transformation or second order asymmetric transformation.[33] Whereas the resolution processes described earlier are limited to 50% yield of the desired pure enantiomer, asymmetric transformation allows a yield of nearly 100%. It is good to keep in mind that the outcome of such a process is thermodynamically determined. A small solubility difference between the two diastereoisomers (i.e., a poor eutectic composition) can lead to a high yield of one diastereoisomer, whereas in a standard classical resolution this would either lead to a poor yield or a poor ee.

As described earlier for classical resolutions (Figure 7.1), asymmetric transformations can also be described by phase diagrams. However, because the overall composition of the ternary mixture (2 diastereoisomers + solvent) is not constant, only starting and end composition can be visualized (Figure 7.10).

At the start of an asymmetric transformation the high concentration 3 results in the crystallization of both diastereoisomers in an almost 50:50 ratio. As a result of racemization the solution composition E will tend to reach composition 2, becoming supersaturated in the (S,R)-diastereoisomer and subsaturated in the (R,R)-diastereoisomer. This process results in a continuous dissolution of the (R,R)-diastereoisomer and crystallization of the (S,R)-diastereoisomer until equilibrium is

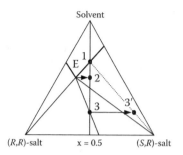

FIGURE 7.10 Description of a crystallization-induced asymmetric transformation by a ternary phase diagram.

reached. At equilibrium most of the diastereoisomeric mixture is transferred into the solid (S,R)-salt, leaving only the equimolar composition 1 in solution. The overall composition of the mixture by that time has been changed from 3 to 3'. Note that the outcome of an asymmetric transformation is fully determined by the tie line at composition 1 and not by the position of the eutecticum. If (end) solid-solution behavior occurs, the composition of the solid will deviate from pure (S,R)-diastereoisomers as a result of the incorporation of the (R,R)-diastereoisomer (a solid-solution). Thus, an ideal asymmetric transformation is defined by the crystallization of pure (less soluble) diastereoisomer and not by the efficiency of the resolution.

Crystallization-induced asymmetric transformation has already been described by Leuchs in 1913 during the resolution of 2-(2-carboxybenzyl)-1-indanone with brucine.[34] In this case spontaneous racemization occurred. More recently researchers at Sanofi observed spontaneous racemization during the resolution of 3-cyano-3-(3,4-dichlorophenyl)propionic acid (**7**), most likely as a result of the basic resolving agent D-(−)-N-methylglucamine [D-(−)-MGA] (**8**) (Scheme 7.6).[35] The enantiopure cyano acid, obtained in 91% overall yield, is subsequently reduced to (+)-4-amino-3-(3,4-dichlorophenyl)-1-butanol (**9**), a key intermediate in the phase 2 synthesis of tachykinin antagonists.

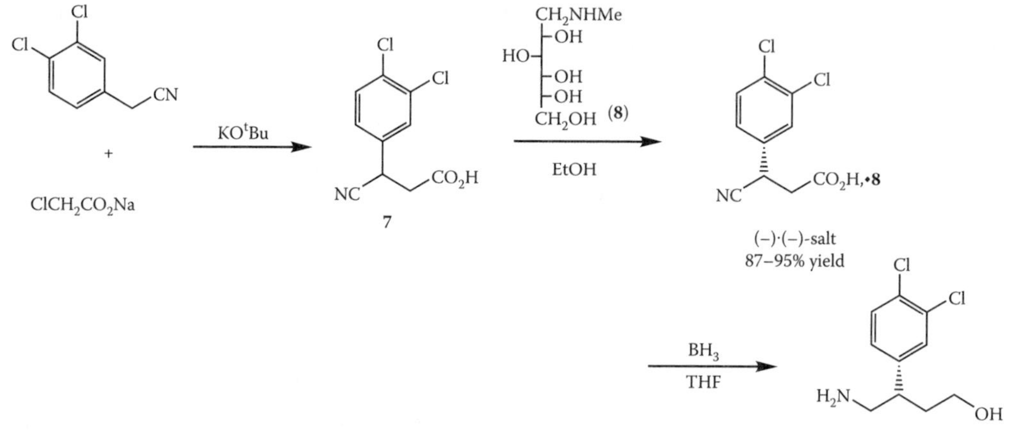

SCHEME 7.6

In the past decade many new examples have been described, mostly for amino acid derivatives using (substituted) benzaldehyde as a racemization catalyst. In Scheme 7.7 an application of this procedure on multi-kilogram scale from Merck & Co is described for the asymmetric transformation of a benzodiazepinone in the synthesis of the CCK antagonist, L-364,718 (**10**).[36] More recently, the Merck group also applied this method to other bendiazepinones.[37]

Other examples of crystallization-induced asymmetric transformation have been described in the literature and have been reviewed.[38,39]

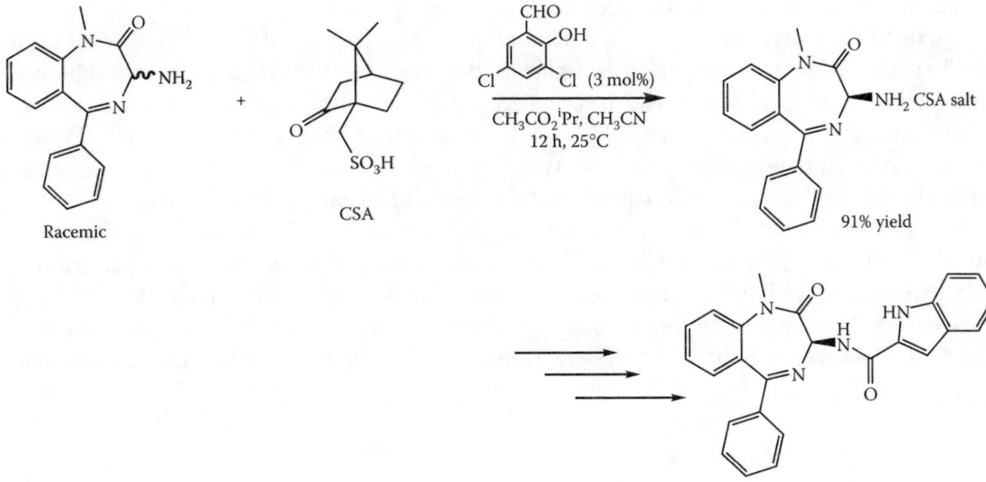

10
79% yield
99.8% e.e.

SCHEME 7.7

An example of a very efficient asymmetric transformation is the preparation of (*R*)-phenylglycine amide (Scheme 7.8) (see also Chapter 25).[40] This offers a good alternative to the enzymatic resolution of (*R,S*)-phenylglycine amide with the (*S*)-specific amidase from *Pseudomonas putida.*[41] This amide is used in a coupling process for semi-synthetic antibiotics.[42]

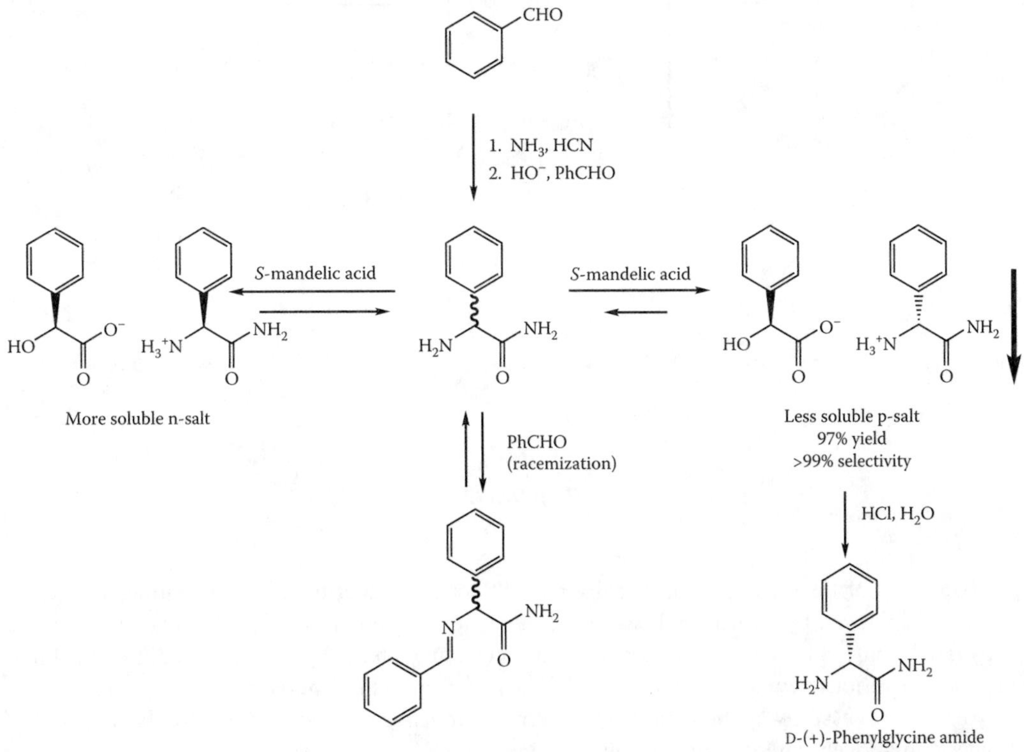

SCHEME 7.8

The high-yield synthesis of the racemate via a Strecker synthesis is elegantly combined with the asymmetric transformation process. Addition of the resolving agent (*S*)-mandelic acid results in the formation of both diastereoisomeric salts. In the presence of benzaldehyde these salts are in equilibrium with the Schiff base, which racemizes readily. The low solubility of the diastereoisomeric salts (in apolar solvents) eventually allows obtainment of a >95% yield of the (*R*)·(*S*)-salt in more than 99% diastereoisomeric excess. After decomposition of this salt by hydrochloric acid, pure (*R*)-phenylglycine amide is obtained, and the resolving agent can be recycled.

As a second example, we recently used the readily available (*R*)-phenylglycine amide as a "Chirality Transfer" reagent (see Chapter 25).[43] In this case, a Strecker reaction is performed on pivaldehyde under equilibrium conditions resulting in two covalent diastereoisomeric products (Scheme 7.9). The lower solubility of the (*S*,*R*)-diastereoisomer results in transformation of the (*R*,*R*)-diastereoisomer into the (*S*,*R*)-diastereoisomer in 93% yield and >99% diastereoselectivity.

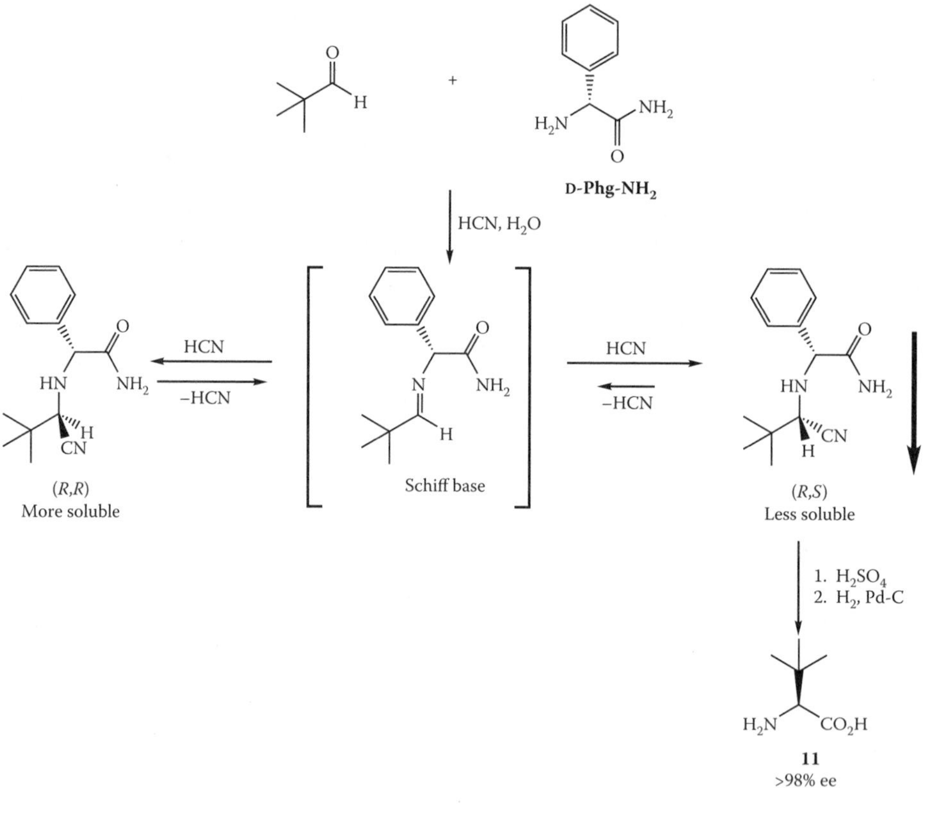

SCHEME 7.9

Hydrolysis of the diastereoisomerically pure Strecker product under racemization-free acidic conditions, followed by hydrogenolysis of the benzylic C–N bond, results in a 73% overall yield of (*S*)-*tert*-leucine (**11**). The amino acid **11** is an interesting, sterically constrained amino acid that is applied in various new antiviral and anti-HIV (human immunodeficiency virus) agents.

We have successfully applied the same asymmetric transformation procedure for the preparation of other α-hydrogen- and α,α-disubstituted amino acids.

7.6 SUMMARY

On an industrial scale, classical resolution using single resolving agents is still preferred over the use of a family of resolving agents (as applied in the DR approach). The effort to synthesize and recycle mixtures of resolving agents on an industrial scale should not be underestimated. However, the DR approach is a fast and reliable method for the generation of enantiopure material in the (early) development phase. Once a successful resolution is identified in this manner, usually the resolving agent of the family present in the highest fraction in the diastereoisomeric salt can be selected as the single resolving agent. The other family members can then be tested as additive to increase the resolution efficiency by nucleation inhibition. Of course, all resolution processes are by necessity held to a maximum yield of 50%, but the unwanted enantiomer can often be racemized and recycled.

Alternatively, to increase the yield to more than 50%, a classical resolution can be further developed into a crystallization-induced asymmetric transformation.

REFERENCES AND NOTES

1. Frost & Sullivan Market Report, 2002.
2. Jacobsen, E. N., Pfalz, A., H. Yamamoto, H., Eds., *Comprehensive Asymmetric Catalysis,* Vol 1–3, Springer: Berlin, 1999.
3. Patel, R. N., Ed., *Stereoselective Biocalysis,* Marcel Dekker: New York, 2000.
4. Lough, W. J. In Lough, W. J., Wainer, I. W., Eds., *Chirality in Natural and Applied Science,* Blackwell Science: Oxford, 2002, pp. 179–202.
5. Pasteur, L. *C. R. Acad. Sci.* 1853, *37,* 162.
6. Bruggink, A. Collins, A. N., Sheldrake, G. N., Crosby, J., Eds., In *Chirality in Industry II,* Wiley & Sons: Chichester, 1997, pp. 81–98.
7. Leusen, F. J. J. *Cryst. Growth & Design* 2003, *2,* 189.
8. Jacques, J., Collet, A., Wilen, S. H. *Enantiomers, Racemates, and Resolution,* Wiley & Sons: New York, 1981, pp. 380–383.
9. Kozma, D., Ed., *CRC Handbook of Optical Resolution via Diastereoisomeric Salt Formation,* CRC Press: Boca Raton, 2001.
10. For the synthesis and use of (*1S,2R*)-1-amino-2-indanol in the HIV protease inhibitor Crixivan® see Maligres, P. E., Upadhyay, V., Rossen, K., Ciancosi, S. J., Pucick, R. M., Eng, K. K., Reamer, R. A., Askin, D., Volante, R. P., Reider, P. J. *Tetrahedron Lett.* 1995, *36,* 2195.
11. ten Hoeve, W., Wynberg, H. *J. Org. Chem.* 1985, *50,* 4508.
12. Nieuwenhuijzen, J. W. *PhD Thesis,* University of Groningen, 2002.
13. Holton, P. G., U.S. Pat. 4,246,193, 1981; Felder, E., San Vitale, R., Pitre, D., Zutter, H. U.S. Pat. 4,246,164, 1981.
14. The Chemspeed ASW2000™ equipment consists of vortex stirred units of 16 vessels of 10 ml, designed for automated screening and high throughput experimentation.
15. Kozma, M., Polok, G., Acs, M. *J. Chem. Soc., Perkin Trans. II* 1992, 435.
16. Ebbers, E., Ariaans, G. J. A., Zwanenburg, B., Bruggink, A. *Tetrahedron: Asymmetry* 1998, *9,* 2745.
17. Dyer, U. C., Henderson, D. A., Mitchell, M. B. *Org. Proc. Res. Dev.* 1999, *3,* 161.
18. Verhey M., Hulshof, L. A. Unpublished results, DSM Fine Chemicals, 2000.
19. The HEL autoMate™ equipment consists of a multiple (four) reactor system of 50–100 ml, equipped with automated sampling and analysis. The autoMate™ is specially suited for route-scouting and process development.
20. Vries, T. R., Wynberg, H., van Echten, E., Koek, J., ten Hoeve, W., Kellogg, R. M., Broxterman, Q. B., Minnaard, A. J., Kaptein, B., van der Sluis, S., Hulshof, L. A., Kooistra, J. *Angew. Chem., Int. Ed.* 1998, *37,* 2349, Hulshof, L. A., Broxterman, Q. B., Vries, T. R., Wynberg, H., van Echten, E. Eur. Pat. Appl. 0.838.448, 1998, Broxterman, Q. B., van Echten, E., Hulshof, L. A., Kaptein, B., Kellogg, R. M., Minnaard, A. J., Vries, T. R., Wynberg, H. *Chimica Oggi* 1998, *16,* Sept. issue 34.

21. Kellogg, R. M., Nieuwenhuijzen, J. W., Pouwer, K., Vries, T. R., Broxterman, Q. B., Grimbergen, R. F. P., Kaptein, B., La Crois, R. M., de Wever, E., van der Laan, A. C., Zwaagstra, K. *Synthesis* 2003, 1626.

22. Kaptein, B., van Dooren, T. J. G. M., Boesten, W. H. J., Sonke, T., Duchateau, A. L. L., Broxterman, Q. B., Kamphuis, J. *Org. Proc. Res. Dev.* 1998, *2*, 10.

23. Ref 8, pp. 307–317.

24. Collet, A. *Angew. Chem., Int. Ed.* 1998, *37*, 3239.

25. Wynberg, H., Pouwer, K., Nieuwenhuijzen, J., Vries, T. R. PCT Int. Appl. WO 01/014327, 2001.

26. van der Sluis, M., Dalmolen, J., de Lange, B., Kaptein, B., Kellogg R. M., Broxterman. Q. B. *Org. Lett.* 2001, *3*, 3943.

27. Kaptein, B., Elsenberg, H., Grimbergen, R. F. P., Broxterman, Q. B., Hulshof, L. A., Pouwer, K. L., Vries, T. R. *Tetrahedron: Asymmetry* 2000, *11*, 1343.

28. a) Mitscher, L. A., Sharma, P. N., Chu, D. T. W., Shen, L. L., Pernet, A. G. *J. Med. Chem.* 1987, *30*, 2283; b) Musso, D. L., Kelley, J. L. *Tetrahedron: Asymmetry* 1995, *6*, 1841; c) Kelley, K. L., Musso, D. L., Boswell, G. E., Sokoro, F. E., Cooper, B. R. *J. Med. Chem.* 1996, *39*, 347; d) Carocci, A., Frannchini, C., Lentini, G., Loiodice, F., Tortorella, V. *Chirality* 2000, *12*, 103; e)Randolph, J. T., Haviv, F., Sauer, D., Waid, P., Nichols, C. J., Dalton, C. R., Greer, J. U.S. Pat. 6,020,521, 2000; f) Noguchi, S., Yokoyama, Y. Korean Pat. 078.569, 2001; g) Lee, B. S., Shin, S. H. Jap. Pat. 172.283, 2001.

29. a) Stoll, A., Peyer, J., Hofmann, A. *Helv. Chim. Acta* 1943, *26*, 929, b) Den Hollander, C. W., Leimgruber, W., Mohacsi, E. U.S. Pat. 3.682.925, 1972, for other methods see c) Boesten, W. H. J., Schepers, C. H. M., Roberts, M. J. A. Eur. Pat. 322.982, 1989, d) Abiko, A., Masamune, S. *Tetrahedron Lett.* 1992, *33*, 5517, e) Anand, R. C., Vimal, *Tetrahedron Lett.* 1998, *39*, 917, f) Kottenhahn, M., Drauz, K., Hilpert, H. Eur. Pat. 778.823, 1999, g) Studer, M., Burkhardt, S., Blaser, H.-U. *Adv. Synth. Cat.* 2001, *343*, 802, h) Leisinger, T., Van Der Ploeg, J., Kiener, A. M., De Azevedo Wäsch, S. I. PCT Int. Appl. WO 01/73038, 2001, i) Francalanci, F., Cesti, P., Cabri, W., Bianchi, D., Martinengo, T., Foa, M. *J. Org. Chem.* 1987, *52*, 5079, j) Fernandez, S., Brieva, R., Rebolledo, F., Gotor, V. *J. Chem. Soc., Perkin Trans. I*, 1992, 2885, k) Nikaido, T., Kawada, N. Jap. Pat. 245.692, 2001.

30. Kaptein, B., Pouwer, K.L., Vries, T. R., Grimbergen, R. F. P., Grooten, H. M. J., Elsenberg, H. L. M., Nieuwenhuijzen, J., Kellogg, R. M., Broxterman, Q. B. Manuscript in preparation.

31. Nieuwenhuijzen, J. W., Grimbergen, R. F. P., Koopman, C., Kellogg, R. M., Vries, T. R., Pouwer, K., van Echten, E., Kaptein, B., Hulshof L. A., Broxterman, Q. B. *Angew. Chem., Int. Ed.*, 2002, *41*, 4281.

32. Gervais, C., Grimbergen, R. F. P., Markovits, I., Ariaans, G. J. A., Kaptein, B., Bruggink, A., Broxterman, Q. B. *J. Am. Chem. Soc.*, 2004, *126*, 565.

33. Ref 8, pp. 369–377.

34. Leuchs, H., Wutke, J. *Chem. Ber.* 1913, *46*, 2420.

35. Deschamps, M., Raddison, J. Eur. Pat. 612,716, 1994.

36. Reider, P. J., Davies, P., Hughes, D. L., Grabowski, E. J. J. *J. Org. Chem.* 1987, *52*, 957.

37. Shi, Y.-J., Wells, K. M., Pye, P. J., Choi, W.-B., Churchill, H. R. O., Lynch, J. E., Maliakal, A., Sager, J. W., Rossen, K., Volante, R. P., Reider, P. J. *Tetrahedron* 1999, *55*, 909.

38. Ebbers, E., Ariaans, G. J. A., Houbiers, J. P. M., Bruggink, A., Zwanenburg, B. *Tetrahedron* 1997, *53*, 9417.

39. For one of the most recent publications see Shiraiwa, T., Katayama, T., Ishikawa, J., Asai, T., Kurokawa, H. *Chem. Pharm. Bull.* 1999, *47*, 1180 and references cited therein.

40. Boesten, W. H. J. Eur. Pat. 0,442,584, 1991.

41. Sonke, T., Kaptein, B., Boesten, W. H. J., Broxterman, Q. B., Kamphuis, J., Formaggio, F., Toniolo, C., Rutjes, F. P. J. T., Schoemaker, H. E. in Ref. 2, pp. 23–58.

42. Bruggink, A., Roos, E. C., de Vroom, E. *Org. Proc. Res. Dev.* 1998, *2*, 128.

43. Boesten, W. H. J., Seerden, J. P. G., de Lange, B., Dielemans, H. J. A., Elsenberg, H. L. M., Kaptein, B., Moody, H. M., Kellogg, R. M., Broxterman, Q. B. *Org. Lett.* 2001, *3*, 1121, Boesten, W. H. J., Moody, H. M., Kaptein, B., Seerden, J. P. G., van der Sluis, M., de Lange, B., Broxterman, Q. B. PCT Int. Appl. WO 01/42173, 2001.

8 The Role of cis-1-Amino-2-Indanol in Resolution Processes

Chris H. Senanayake, Dhileepkumar Krishnamurthy, and Isabelle Gallou

CONTENTS

8.1 INTRODUCTION

The amino alcohol *cis*-1-amino-2-indanol (**1**) has been shown to be an extremely versatile reagent in asymmetric synthesis. It has been used as a chiral auxiliary. This chemistry and the synthesis of **1** and its uses in biologically active agents are discussed in Chapter 24. Reactions where the amino alcohol **1** is used as a ligand in catalytic reactions will be found in Chapter 17. This chapter discusses reactions where **1** is used as a resolving agent. Other resolution methods can be found in Chapters 6 and 7.

(1*S*,2*R*)-**1**

8.2 CHIRAL DISCRIMINATION OF 2-ARYLALKANOIC ACIDS

Production of enantiomerically pure α-arylpropanoic acids, also known as profens, is of critical importance to the pharmaceutical industry because they constitute a major class of antiinflammatory agents. One of the most practical approaches to preparing optically pure α-arylpropanoic acids is by resolution with chiral amines. Notable examples include brucine, quinidine, cinchonidine, morphine, ephedrine, and α-(1-naphthyl)ethylamine. For instance, (*S*)-α-methylbenzylamine and

(–)-cinchonidine have been reported to resolve racemic ibuprofen[1] and ketoprofen,[2] respectively. However, most of these chiral amines are either expensive, difficult to recover, available in only one enantiomeric form, or substrate-specific.

Enantiopure *cis*-1-amino-2-indanols have been demonstrated to effectively overcome these limitations. Both enantiomers of *cis*-1-amino-2-indanols are rapidly accessible via an inexpensive process, and they also have been shown to be extremely effective in the resolution of a number of 2-arylalkanoic acids,[3,4] including ketoprofen, flurbiprofen, and ibuprofen.[3] In the case of ketoprofen,[3] use of (1R,2S)-1-amino-2-indanol allowed for the selective crystallization of (S)-ketoprofen in the presence of (R)-ketoprofen (Scheme 8.1). Similarly, (R)-ketoprofen was shown to selectively precipitate with (1S,2R)-1-amino-2-indanol in the presence of (S)-ketoprofen. Extensive solubility studies led to the following observations: (a) diastereomeric salts of (R)- and (S)-ketoprofen with aminoindanols exhibited larger differences in solubility than diastereomeric salts of (R)- and (S)-ketoprofen with other chiral amines; (b) diastereomeric salts of (R)- and (S)-acids with aminoindanol displayed large differences in solubility for a wide range of chiral acids; and (c) a catalytic amount of water in acetonitrile had a significant impact on the rate of crystallization, leading to higher yields of the precipitating diastereoisomer. Finally, the undesired isomer [(R)-ketoprofen in Scheme 8.1] can be racemized and recycled, and enantiopure *cis*-aminoindanol can be easily recovered for reuse.[3]

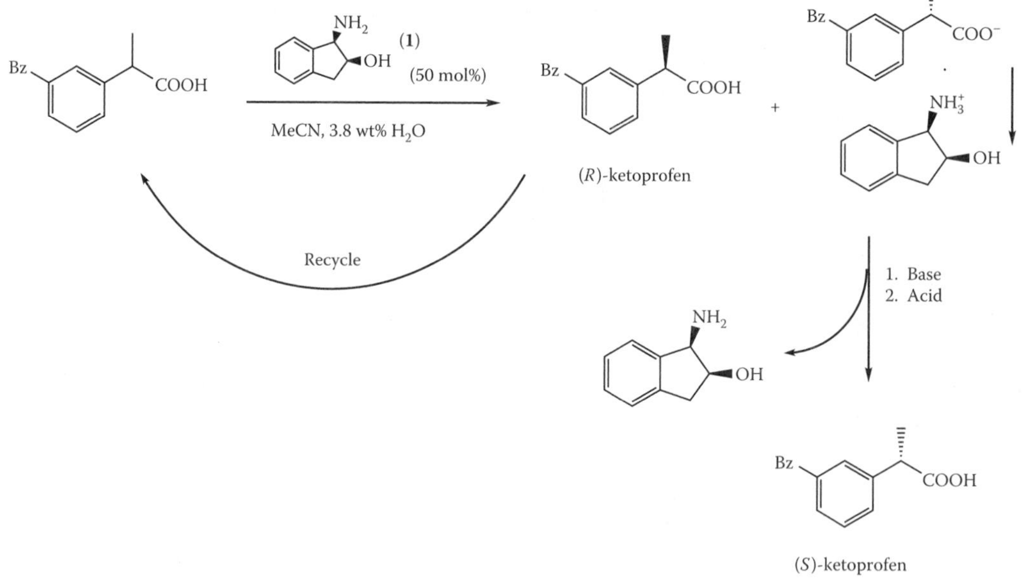

SCHEME 8.1

8.3 ENANTIOSELECTIVE ACETYLATION OF RACEMIC SECONDARY ALKYL AMINES

A new *cis*-aminoindanol–derived (*N*-cyanoimino)oxazolidine acetylating agent was developed by Tanaka and co-workers, which was applied to the kinetic resolution of secondary alkyl amines.[5] The chiral acetylating agents **2** and **3** were readily prepared from (1R,2S)-1-amino-2-indanol by reaction with S,S′-dimethyl *N*-cyanodithioiminocarbonate followed by acetylation (78–85% yield, 2 steps). The kinetic resolutions of racemic 1-phenylethylamine **5** using 10 mol% of **2** at –70°C led to (R)-N-benzoyl-1-phenylethylamine **6** in 87% yield and 83% ee, along with the recovered

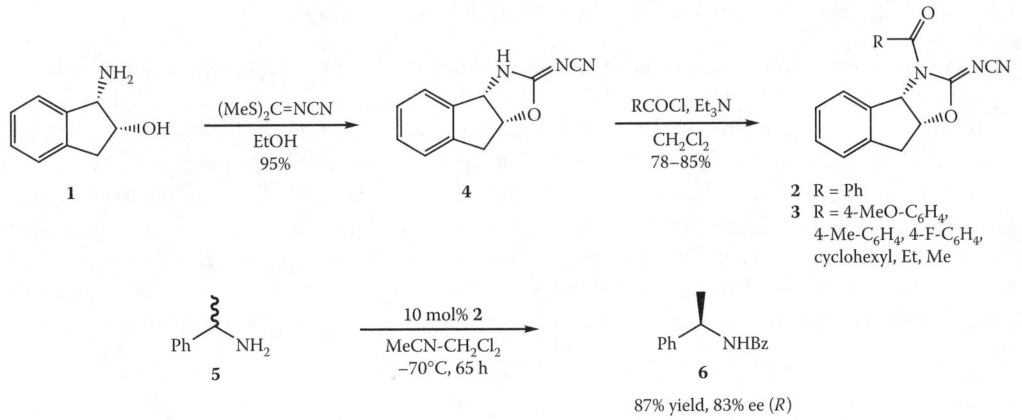

SCHEME 8.2

chiral **4**, which was easily separated by column chromatography for reuse (Scheme 8.2). The transfer of the *N*-acyl group was activated by the strong electron-withdrawing *N*-cyanoimino moiety of **2**. It is interesting that, no aminolysis of the oxazolidine ring of **2** was observed. Other acetylating agents **3** (R = 4-MeO-C$_6$H$_4$, 4-Me-C$_6$H$_4$, 4-F-C$_6$H$_4$, Me) were also shown to efficiently resolve **5** at low temperature (–70°C for R = aryl and –20°C for R = Me), leading to protected secondary alkyl amines in moderate to good enantioselectivities (65–78% ee). Kinetic resolution of a variety of racemic secondary alkyl amines using a catalytic amount of **2** was also examined. It was rapidly established that a phenyl group in amines was essential to achieve high selectivities. Replacement of the methyl group of **5** by larger groups decreased the reaction rate but led to similar selectivities.[5]

To explain the stereochemical outcome of the reaction, Tanaka and co-workers speculated that the amide carbonyl and the *N*-cyanoimino groups in **2** would prefer opposite orientation to minimize unfavorable dipole–dipole interaction. In this conformation, the *Re*-face of the carbonyl group would be shielded by the large and conformationally rigid indane ring, thus favoring the approach of the amine from the *Si*-face. Overlap of the amine phenyl ring with the carbonyl aryl or alkyl group of **2** would create a stabilizing – or CH– interaction, respectively, in which case the approach of the (*S*)-isomer would suffer from steric interaction between R^1 and *N*-cyanoimino groups (Figure 8.1).[5]

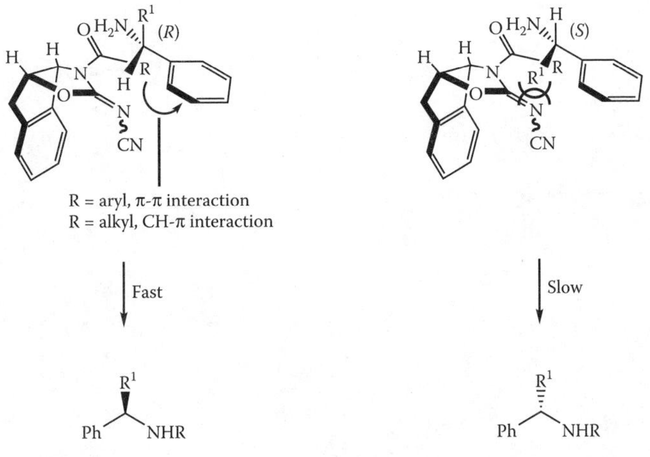

FIGURE 8.1 Postulated model of asymmetric acetylation of racemic phenylalkylamine.

8.4 KINETIC RESOLUTION OF SECONDARY ARYL ALCOHOLS

A novel kinetic resolution of secondary aryl alcohols using an enantiopure *cis*-1-amino-2-indanol-derived catalyst has been disclosed by Faller.[6] (For a full discussion of catalysts with **1** and related compounds as ligands see Chapter 17.) The investigation was based on earlier findings by Palmer[7] that stereochemically rigid (1*R*,2*S*)-1-amino-2-indanol was a powerful ligand for the asymmetric ruthenium-catalyzed transfer hydrogenation of aryl ketones using 2-propanol, leading to enantioselectivities up to 98%. The stereoselectivity of the reaction was obtained primarily by kinetic discrimination of enantiofaces of the prochiral ketone, but thermodynamic factors favoring the reverse process were not negligible (Scheme 8.3).[8,9] Thus, it was argued that an asymmetric transfer hydrogenation catalyst should be capable of dehydrogenation as well as hydrogenation.

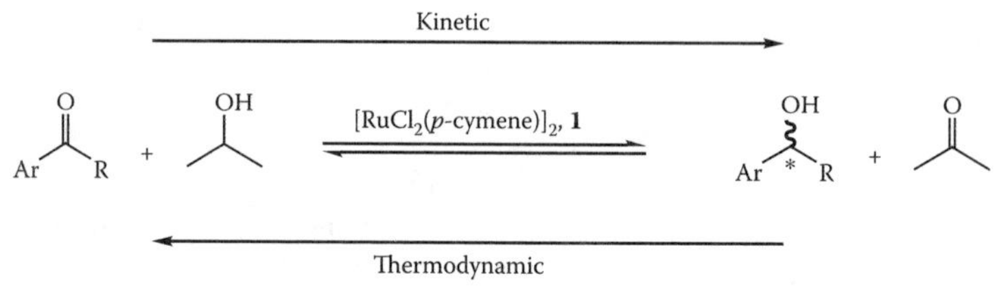

SCHEME 8.3

Because Palmer's hydrogenation catalyst provided an excellent route to (*S*)-aryl alcohols from aryl ketones, Faller decided to expand the scope of the same catalyst to kinetic resolution of racemic aryl alcohols to prepare (*R*)-aryl alcohols. Results showed that products can be obtained with enantiomeric purities in the 90–99% range. In particular, when applied to the kinetic resolution of (±)-tetralol (**7**) and (±)-indanol, the [RuCl$_2$(*p*-cymene)]$_2$-(1*R*,2*S*)-1-amino-2-indanol-KO-*t*-Bu combination in acetone yielded the corresponding (*R*)-alcohols in 99% ee, arylketones, and 2-propanol. In turn, reduction of 1′-tetralone and 1′-indanone using the same catalytic system in 2-propanol led to (*S*)-alcohols in 97% ee (Scheme 8.4). These studies provide a compelling demonstration of the generally unappreciated notion that a single enantiomer of a catalyst can lead to both antipodes of the same product with high levels of stereoselectivities.[6]

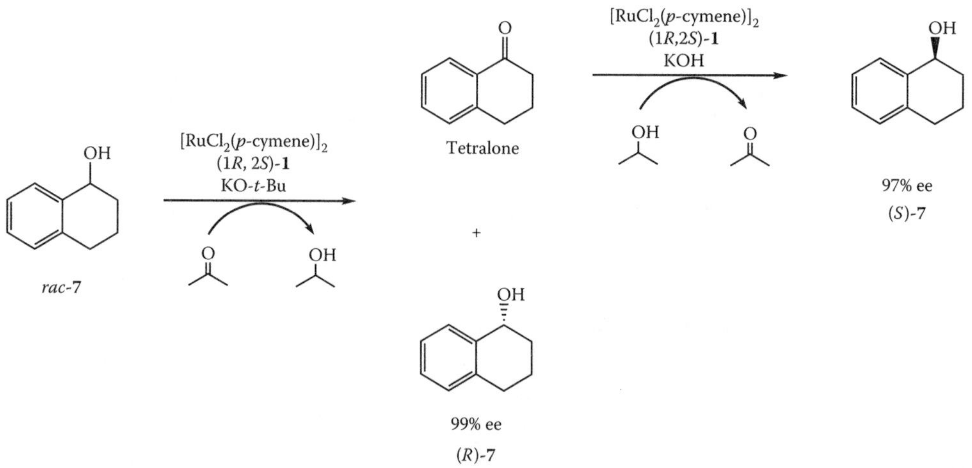

SCHEME 8.4

8.5 RELATED AMINO ALCOHOLS AND THEIR APPLICATIONS IN ASYMMETRIC SYNTHESIS

8.5.1 *Trans*-6-Nitro-1-Amino-2-Indanol

Researchers at Eli Lilly have disclosed the significance of the *trans*-6-nitro-1-amino-2-indanol template in a new class of resistant neoplasms inhibitors of type **8**. Current effective treatments of several forms of cancer include surgery, radiotherapy, and chemotherapy. Unfortunately, multi-drug resistance remains an important issue in chemotherapy, where cells may become cross resistant to a wide range of drugs with different structures and cellular targets. Compounds of type **8** were shown to selectively inhibit intrinsic and/or acquired resistance conferred in part or in total to the 190 kDa multi-drug resistance protein, commonly referred to as MRP1.[10]

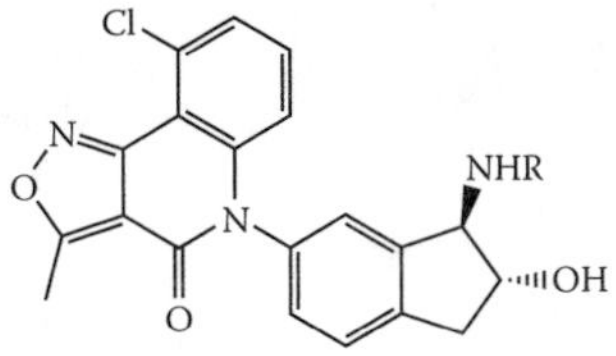

8　　　　R = Hydrogen
　　　　　　　ketone
　　　　　　　sulfone
　　　　　　　thioamide

Racemic substituted aminoindanol **9** was synthesized in a 5-step sequence by nitration of 1-indanone, followed by ketone reduction and dehydration to give 6-nitro-1-indene and subsequent epoxidation of the olefin and final regioselective amination (Scheme 8.5). Optically pure (1*R*,2*R*)- and (1*S*,2*S*)-6-nitro-1-amino-2-indanol **9** were eventually obtained by resolution with mandelic acid.[10,11]

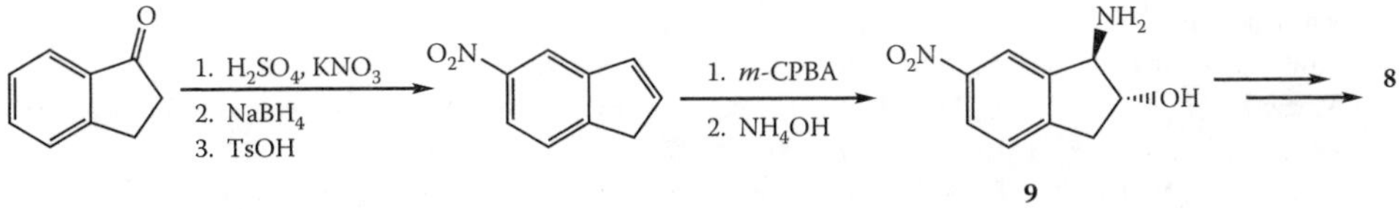

SCHEME 8.5

Compound **9** has been identified as a valuable chiral agent in the preliminary studies on the crystallization-induced dynamic resolution (CIDR) of imines.[11] This resolution–deracemization process provided a practical access to large quantities of nonracemic, α-epimerizable, unfunctionalized ketones and aldehydes. For example, crystalline imines of epimerizable 2-methylcyclohexanone **10** were formed and **11** was found to exist as a 3:1 solution mixture of *E*:*Z* imine isomers with a 1:1 diastereoselective ratio. In the case of ketone **10**, equilibration in methanol afforded the best results, where imine (*E*,*R*)-**11** is the least soluble crystalline product. Filtration and hydrolysis under acidic conditions gave (*R*)-1-methylcyclohexanone in 97% yield and 92% ee, and the chiral agent was recovered for reuse (Scheme 8.6).[11]

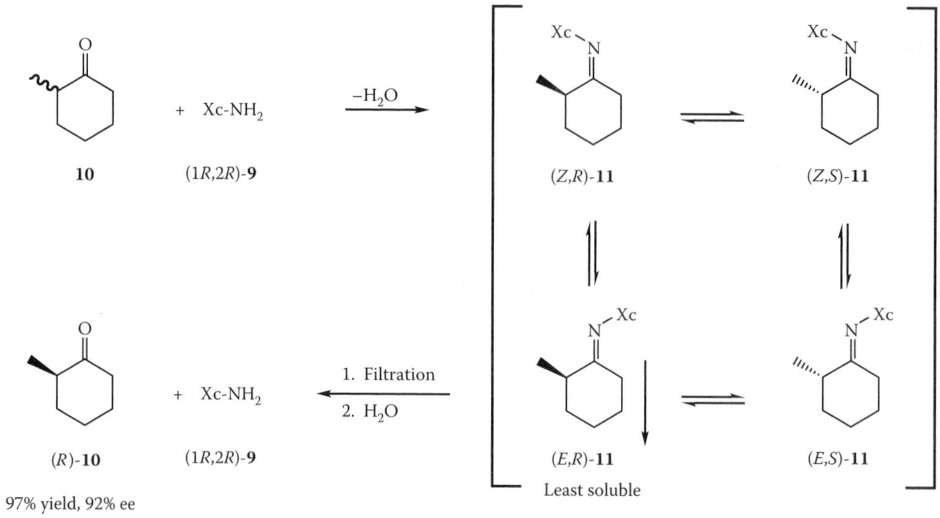

SCHEME 8.6

8.6 SUMMARY

Since the discovery of *cis*-1-amino-2-indanol as a ligand for human immunodeficiency virus protease inhibitors and the development of a practical industrial process for the synthesis of either *cis*-isomers in enantiopure form, the remarkable properties of the rigid indane platform have been used extensively in an ever-increasing number of asymmetric methodologies. In addition to the use of this amino alcohol as a chiral auxiliary and ligand for asymmetric synthesis, it has found application as a useful resolution agent. Applications include amines, carboxylic acids, and alcohols.

REFERENCES

1. Manimaran, T, Impastato, F. J. U.S. Pat. 5,015,764, 1991.
2. Manimaran, T., Potter, A. A. U.S. Pat. 5,162,576, 1992.
3. van Eikeren, P., McConville, F. X., López, J. L. U.S. Pat. 5,677,469, 1997.
4. Kinbara, K., Kobayashi, Y., Saigo, K. *J. Chem. Soc., Perkin Trans. II* 2000, 111.
5. Maezaki, N., Furusawa, A., Uchida, S., Tanaka, T. *Tetrahedron* 2001, *57*, 9309 and references therein.
6. Faller, J. W., Lavoie, A. R. *Org. Lett.* 2001, *3*, 3703.
7. Palmer, M., Walsgrove, T., Wills, M. *J. Org. Chem.* 1997, *62*, 5226 and references therein.
8. Ohkuma, T., Noyori, R. *Comprehensive Asymmetric Catalysis* 1999, *1*, 199–246.
9. Palmer, M. J., Wills, M. *Tetrahedron: Asymmetry* 1999, *10*, 2045.
10. Gruber, J. M., Hollinshead, S. P., Norman, B. H., Wilson, J. W. PCT Appl. WO 99/51236, 1999.
11. Kosmrlj, J., Weigel, L. O., Evans, D. A., Downey, C. W., Wu, J. *J. Am. Chem. Soc.* 2003, *125*, 3208.

9 Asymmetric Oxidations

David J. Ager and David R. Allen

CONTENTS

9.1 INTRODUCTION

Considerable advances have been made in catalytic methodologies to perform asymmetric oxidations.[1–3] Although no large-scale processes for commercial chiral pharmaceuticals currently use the technology, the methods are relatively new compared to catalytic asymmetric hydrogenation. The approach, however, is now in the synthetic arsenal, and it is surely just a matter of time before it comes to fruition as many drug candidates use asymmetric oxidations. An example of an up and coming asymmetric oxidation is the epoxidation method based on carbohydrate ketones (Chapter 10).

Asymmetric oxidations have followed the usual development pathway where face selectivity was observed through the use of chiral auxiliaries and templates. The breakthrough came with the Sharpless asymmetric epoxidation method, which, although stoichiometric, allowed for a wide range of substrates and the stereochemistry of the product to be controlled in a predictable manner.[4]

The need for a catalytic reaction was very apparent, but this was developed and now the Sharpless epoxidation is a viable process at scale, although it is subject to the usual economic problems of a cost-effective route to the substrate (*vide infra*).[5] The Sharpless epoxidation has been joined by other methods, and a wide range of products is now available. The power of these oxidations is augmented by the synthetic utility of the resultant epoxides or diols that can be used for further transformations, especially those that use a substitution reaction (Chapter 22).[4] However, asymmetric oxidation is still an area where significant basic advances still need to be made to have workable, scaleable reactions. The Shi epoxidation (Chapter 10) is beginning to be used at scale. An example where further research is needed is the allylic oxidation of an alkene to provide an allyl alcohol.[6–8] Hypervalent iodine reagents that have been used for a wealth of achiral oxidations are another example, yet progress is only just being made with chiral analogues.[9]

Oxidations of carbon–heteroatom species often results in the destruction of a stereogenic center, as in the oxidation of a secondary alcohol to a ketone. In some instances, this reaction can be coupled with another to provide a chiral product (see Chapter 21). One example is the enzymatic acetylation of one enantiomer of a secondary alcohol, where a redox reaction with a transition metal catalyst equilibrates the unreactive isomer of the alcohol (Scheme 9.1).[10–12] The redox reaction can also be performed by an enzyme.[13]

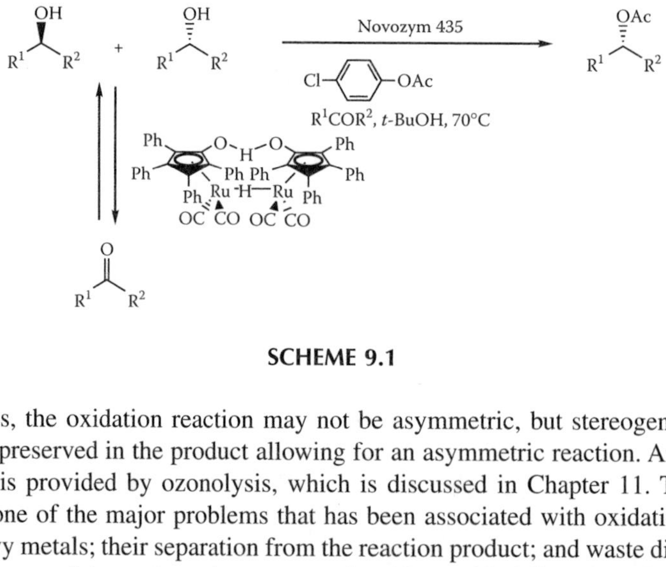

SCHEME 9.1

In other cases, the oxidation reaction may not be asymmetric, but stereogenic centers within the substrate are preserved in the product allowing for an asymmetric reaction. An example of this type of reaction is provided by ozonolysis, which is discussed in Chapter 11. The use of ozone also overcomes one of the major problems that has been associated with oxidations at scale—the use of toxic, heavy metals; their separation from the reaction product; and waste disposal. However, there are still some useful reactions that use metals without chiral ligands and provide stereodifferentiation. An example is provided by the manganese oxide oxidation of ferrocenyl amino alcohols (Scheme 9.2).[14]

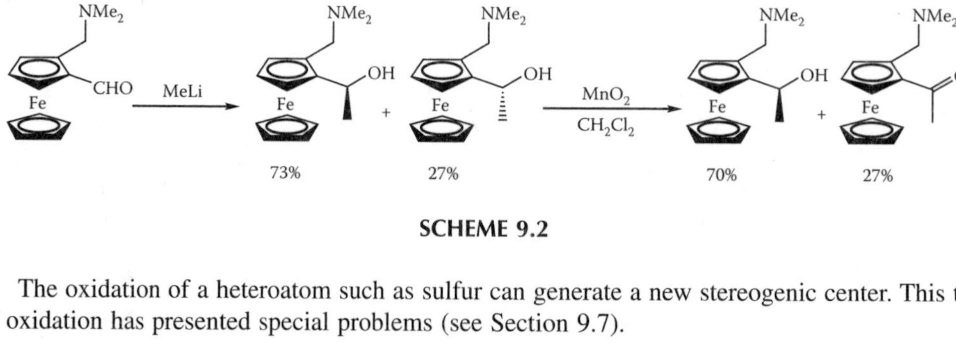

SCHEME 9.2

The oxidation of a heteroatom such as sulfur can generate a new stereogenic center. This type of oxidation has presented special problems (see Section 9.7).

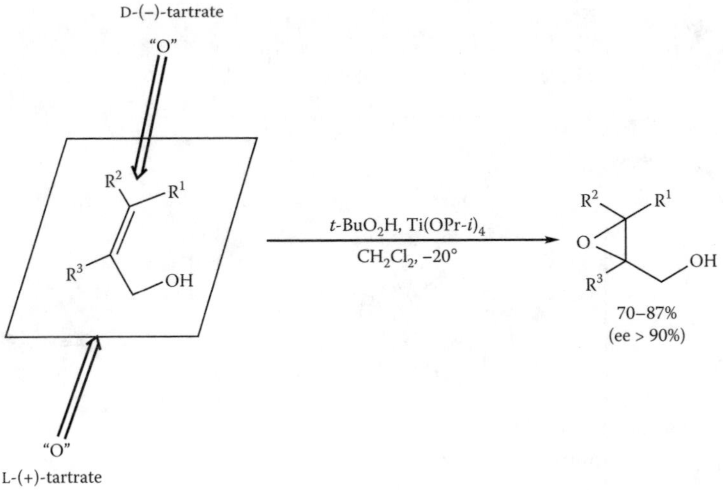

FIGURE 9.1 Face selectivity for the Sharpless epoxidation of allyl alcohols.

9.2 EPOXIDATIONS

9.2.1 SHARPLESS EPOXIDATION

One of the major advantages of the Sharpless[15,16] titanium asymmetric epoxidation is the simple method by which the stereochemical outcome of the reaction can be predicted.[17] The other powerful feature is the ability to change this selectivity to the other isomer by simple means (Figure 9.1).[18–20]

Although the original Sharpless epoxidation method was stoichiometric, the development of a catalytic method has allowed the reaction to be amenable to scale up. The addition of molecular sieves for the removal of trace amounts of water is important in the catalytic procedure.[5,21–23]

In addition to the reagent's ingredients being commercially available, the reaction is promiscuous and proceeds in good chemical yield with excellent enantiomeric excesses. The reaction, however, does suffer when bulky substituents are *cis* to the hydroxymethyl functionality (R^1 in Figure 9.1). For prochiral alcohols, the absolute stereochemistry of the transformation is predictable, whereas for a chiral alcohol, the diastereofacial selectivity of the reagent is often sufficient to override those preferences inherent in the substrate. When the chiral atom is in the E-β-position of the allyl alcohol (R^2), then the epoxidation can be controlled to access either diastereoface of the alkene. In contrast, when the chirality is at either the α- or Z-β-positions (R^1 or R^3), the process is likely to give selective access of the reagent from only one of the two diastereotopic faces.[18,24] Many examples of substrates for the epoxidation protocol are known.[4,25,26]

An improved workup procedure increases the yield for allyl alcohols containing a small number of carbon atoms.[27–29] Furthermore, less reactive substrates provide the epoxide readily.[30]

The structure of the titanium–tartrate derivatives has been determined,[25,26,31–37] and based on these observations together with the reaction selectivity, a mechanistic explanation has been proposed (Scheme 9.3).[38] The complex **1** contains a chiral titanium atom through the appendant tartrate ligands. The intramolecular hydrogen bond ensures that internal epoxidation is only favored at one face of the allyl alcohol. This explanation is in accord with the experimental observations that substrates with an α-substituent (b = alkyl; a = alkyl or hydrogen) react much slower than when this position is not substituted (b = hydrogen).

The reaction time has been reduced dramatically by the addition of calcium hydride, silica gel, or montmorillonite catalysts.[30,39–41]

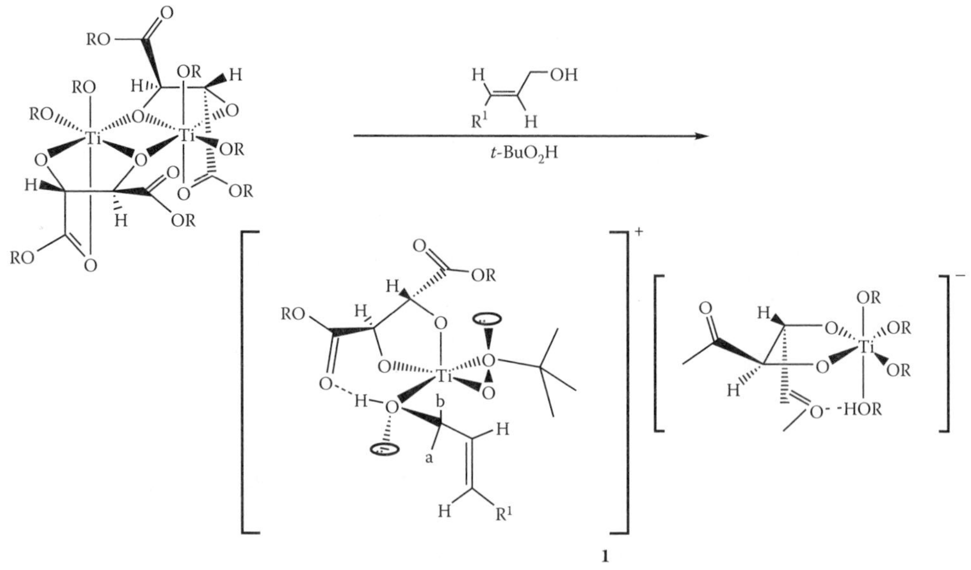

SCHEME 9.3

The power of the Sharpless epoxidation method is augmented by the versatility of the resultant 2,3-epoxy alcohols[4] and the development of the catalytic variation.[5]

Although glycidols are important chemical intermediates, it is unfortunate that the Sharpless approach gives lower enantiomeric purities in this simple case. However, if the arylsulfonates are prepared from the oxidation products, enantioenrichment during crystallization is observed.[42] There is an additional advantage as the simple glycidyl epoxide undergoes attack at C–1 and C–3, but the presence of the sulfonate leaving the group gives rise to preponderance of C–1 attack (*vide infra*).

9.2.1.1 Kinetic Resolutions

The ability of the Sharpless epoxidation catalyst to differentiate between the two enantiomers of an asymmetric allyl alcohol affords a powerful synthetic tool to obtain optically pure materials through kinetic resolution.[43] Because the procedure relies on one enantiomer of a secondary allyl alcohol undergoing epoxidation at a much faster rate than its antipode, reactions are usually run to 50–55% completion.[22] In this way, resolution can often be impressive.[18,25,26,44–46] An increase in steric bulk at the olefin terminus increases the rate of reaction.[46,47]

This resolution method has been used to resolve furfuryl alcohols—the furan acts as the alkene portion of the allyl alcohol (Scheme 9.4).[48–50]

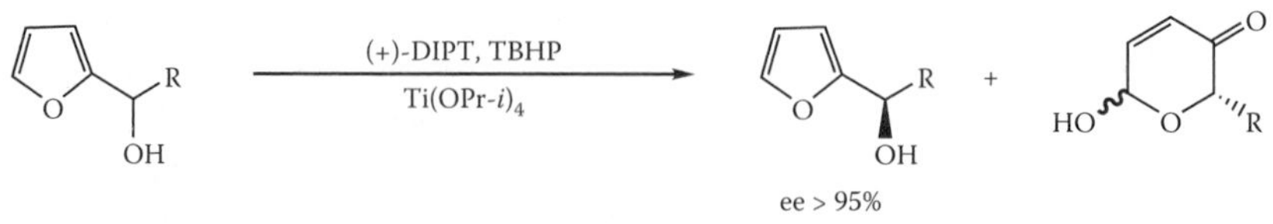

ee > 95%

SCHEME 9.4

9.2.1.2 Reactions of 2,3-Epoxy Alcohols

As noted earlier, one reason for the powerful nature of the Sharpless epoxidation is the ability of the resultant epoxy alcohols to undergo regioselective and stereoselective reaction with nucleo-philes.[4,18–20,51–55] Often the regiochemistry is determined by the functional group within the substrate.[56–83]

A key reaction of 2,3-epoxy alcohols is the Payne rearrangement, an isomerization that produces an equilibrium mixture. This rearrangement then allows for the selective reaction with a nucleophile at the most reactive, primary position (Scheme 9.5).[18,84]

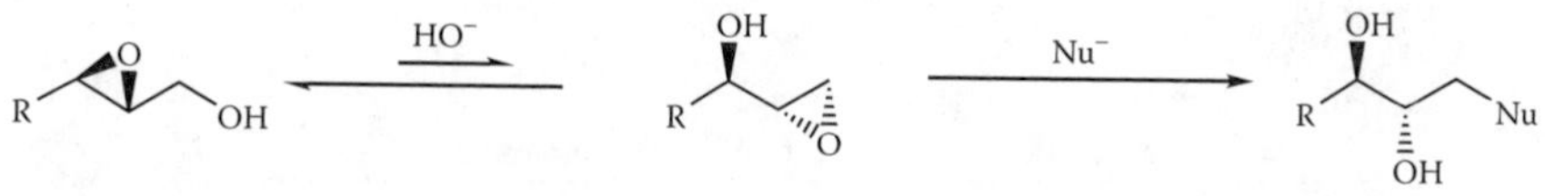

SCHEME 9.5

Under Payne rearrangement conditions, sodium t-butylthiolate provides 1-t-butylthio-2,3-diols with very high regioselectivity. The selectivity is affected, however, by many factors including reaction temperature, base concentration, and the rate of addition of the thiol. These sulfides can then be converted to the 1,2-epoxy-3-alcohols, which in turn react with a wide variety of nucleophiles specifically at the 1-position (Scheme 9.6). This methodology circumvents the problems associated with the instability of many nucleophiles under "Payne" conditions.[85]

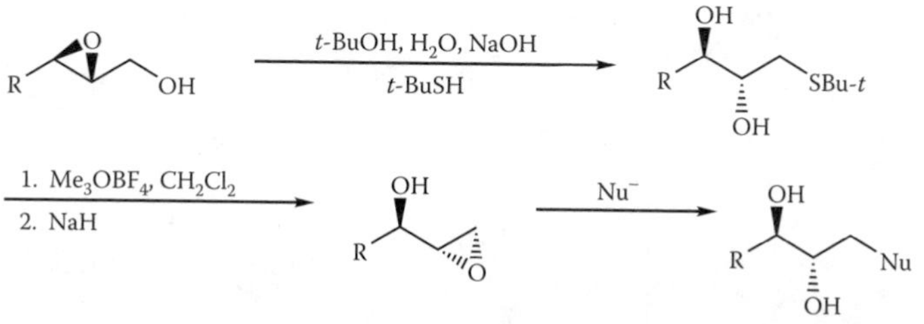

SCHEME 9.6

With good nucleophiles, under relatively mild conditions, 2,3-epoxy alcohols will undergo epoxide ring opening at C–2 or C–3. In simple cases, nucleophilic attack at C–3 is the preferred mode of reaction. However, as the steric congestion at C–3 is increased, or if substituents play a significant electronic role, attack at C–2 can predominate.[86]

9.2.1.3 Commercial Applications

Commercial applications include the synthesis of both isomers of glycidols (ARCO), oct-3-en-1-ol (Upjohn), and the synthesis of disparulene (**2**) (Scheme 9.7).[21,87–91]

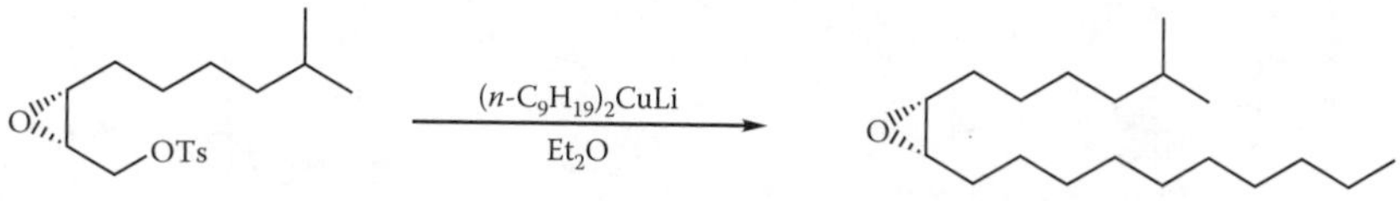

SCHEME 9.7

Although no specific synthesis uses a carbohydrate-type synthon that has been prepared by an asymmetric oxidation, the methodology offers many advantages over a synthesis starting from a carbohydrate (Chapter 4). The useful building block **3** is available by a Sharpless protocol (Scheme 9.8).[92]

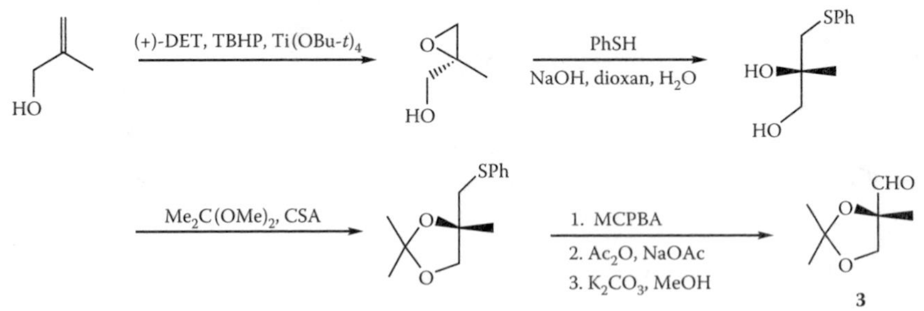

SCHEME 9.8

This methodology has been used for the preparation of L-threitol (**4**) (Scheme 9.9) and erythritol derivatives.[93] This simple iterative process has been used for the simple alditols,[93,94] deoxyalditols,[95] and aldoses[93] and for all of the L-hexoses.[96,97]

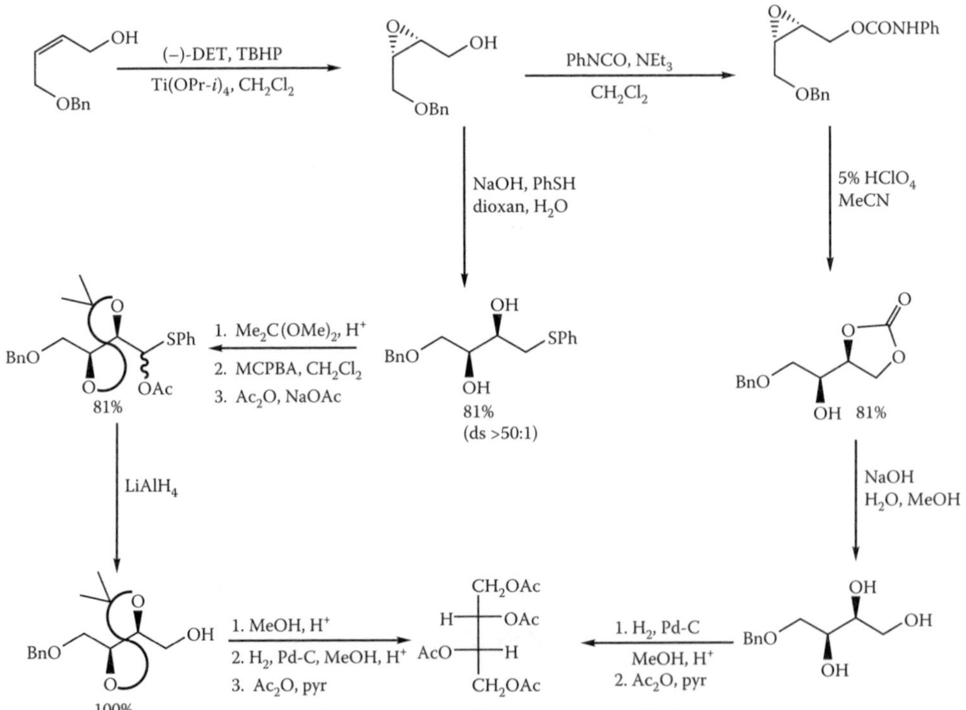

SCHEME 9.9

9.2.2 JACOBSEN EPOXIDATION

The facial selectivity required for an asymmetric epoxidation can be achieved with manganese complexes to provide sufficient induction for synthetic utility (Scheme 9.10).[98–103] This manganese(III) salen complex 5 can also use bleach as the oxidant rather than an iodosylarene.[104,105] The best selectivities are seen with *cis*-alkenes.

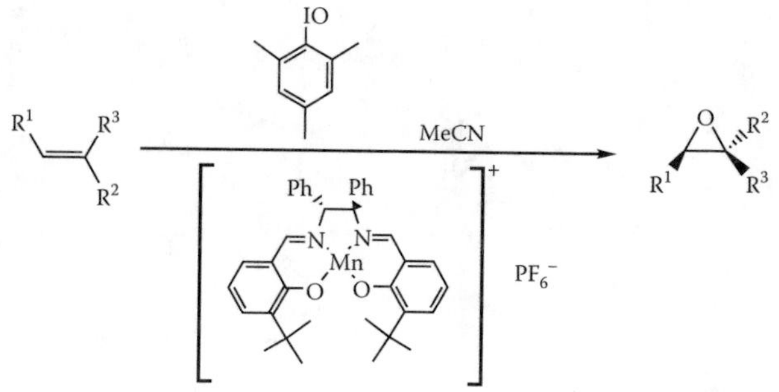

SCHEME 9.10

In effect the metal is planar, but the ligands are slightly buckled and direct the approach of the alkene to the metal reaction site in a particular orientation. The enantioselectivity is highest for Z-alkenes and correlates directly with the electronic properties of the ligand substituents (Scheme 9.10).[98,106,107] The use of blocking groups other than *t*-butyl provides similar selectivities.[99,104,108,109] The chemistry of metallosalen complexes has been reviewed.[110]

9.2.2.1 Resolutions

Although terminal epoxides are not yet accessible with high enantiomeric excess by an asymmetric oxidation methodology, a resolution method has been developed based on the catalytic opening with water (Scheme 9.11; see also Chapter 6).[111,112]

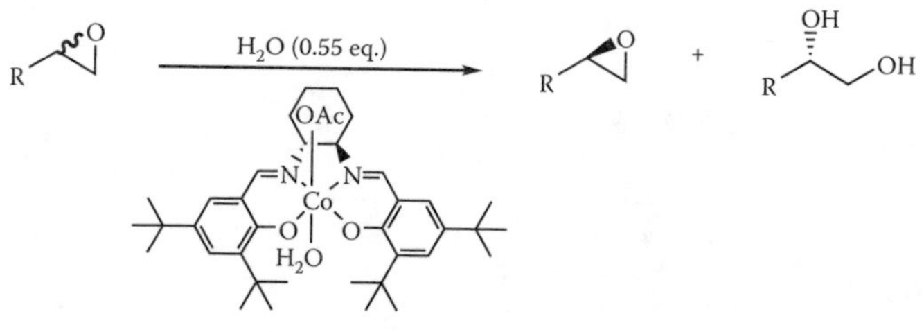

SCHEME 9.11

Although this is a resolution approach, racemic epoxides are readily available, and other than the catalysts, the substrate and water are the only materials present. The method can be made more amenable to scale up by the use of oligomer catalysts.[113]

The general approach has been used to access a wide variety of chiral epoxides. One example is in the synthesis of β-adrenergic blocking agents (Scheme 9.12).[114]

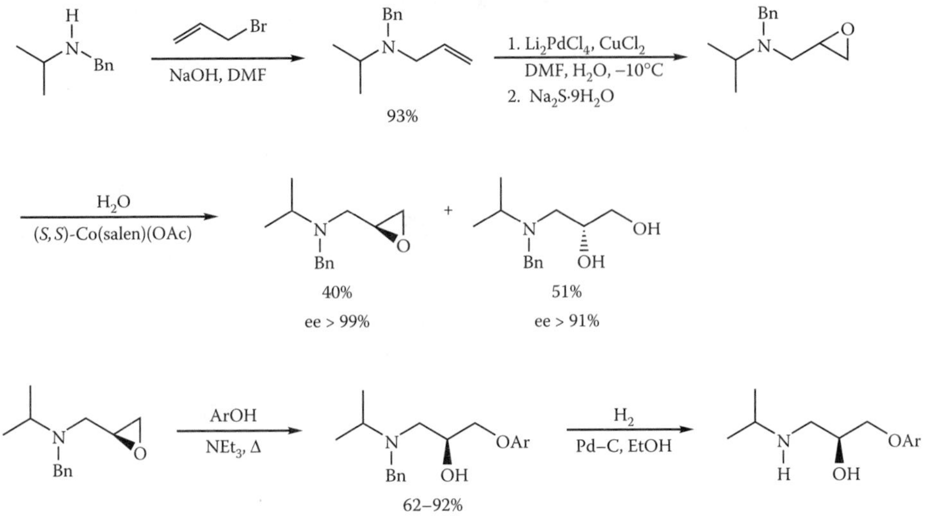

SCHEME 9.12

Another example that uses the hydrolytic kinetic resolution (HKR) methodology is the syntheses of HETE derivatives and LTB$_4$ (**6**) (Scheme 9.13).[115]

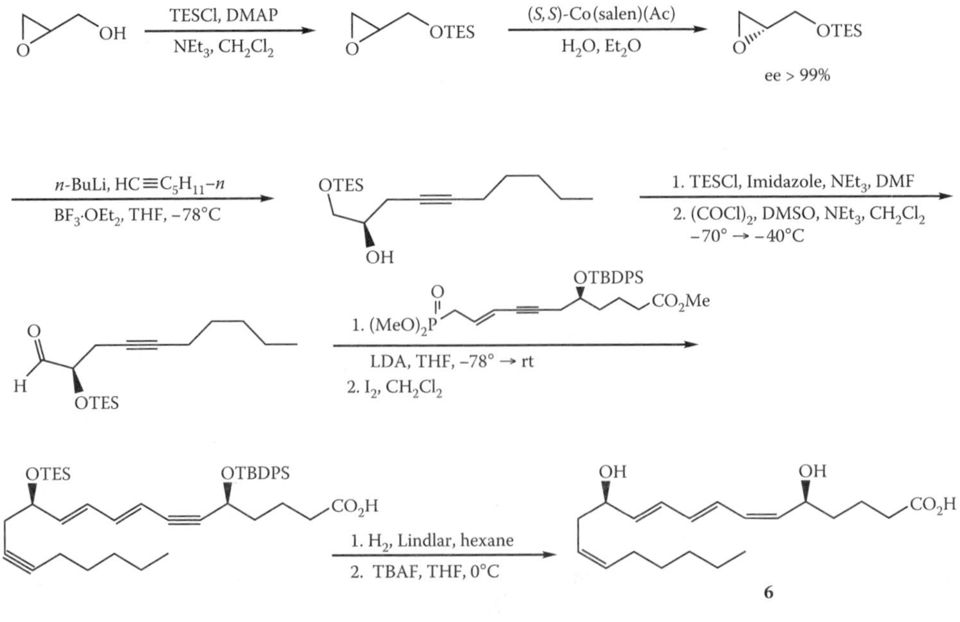

SCHEME 9.13

In a comparison study for the synthesis of the arrhythmia and hypertension drug candidate **7**, the intermediate epoxide **8** was prepared by a Sharpless dihydroxylation and a Jacobsen HKR. The latter HKR method gave the highest ee's.[116]

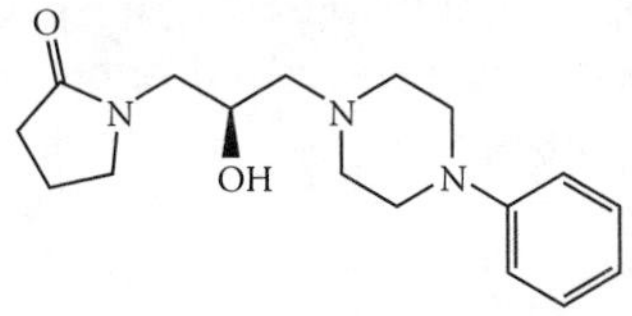

HKR has been used in a short synthesis of (R)-(–)-phenylephrine hydrochloride (**9**), an adrenergic agent and β-receptor sympathomimetic drug (Scheme 9.14).[117]

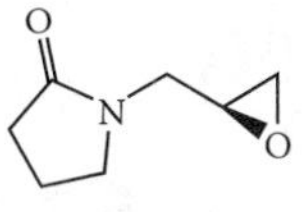

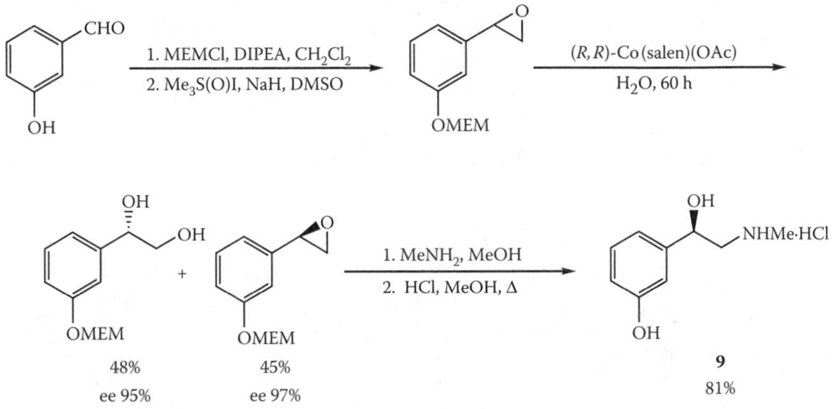

SCHEME 9.14

A similar approach to the hydrolytic methodology uses a chromium-salen complex to open an epoxide with trimethylsilyl azide, as illustrated by the synthesis of the antihypertensive agent, (S)-propranolol (**10**) (Scheme 9.15).[118,119]

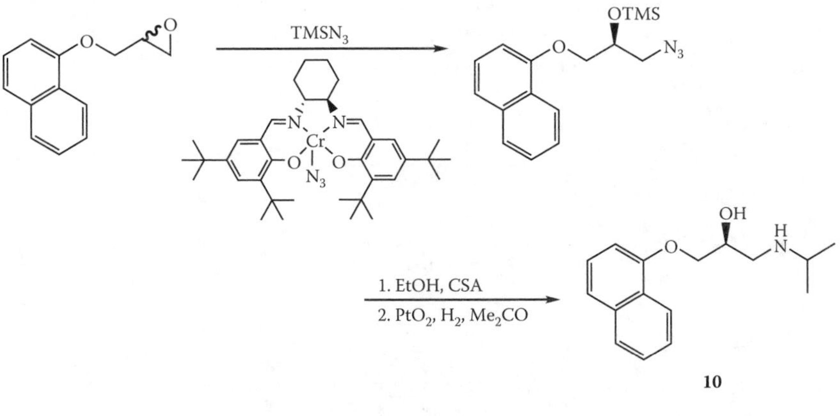

SCHEME 9.15

The methodology, including the use of other nucleophiles, can be used to produce chiral compounds from *meso*-epoxides in high yield.[120–122]

9.2.3 OTHER CHEMICAL METHODS

Simple alkenes can be epoxidized by dioxiranes in an asymmetric manner. This chemistry is discussed in detail in Chapter 10.

In an approach to the potent calcium antagonist, diltiazem (**11**), a BINOL-derived ketone (**12**) was found to provide higher ee than when a carbohydrate-derived ketone was used to prepare the epoxide **13** (Scheme 9.16).[123]

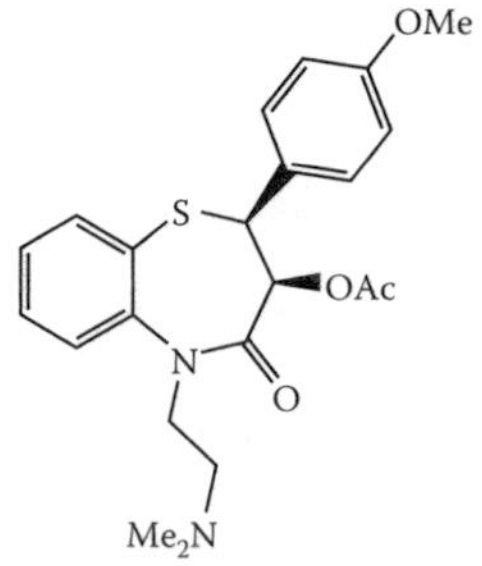

11

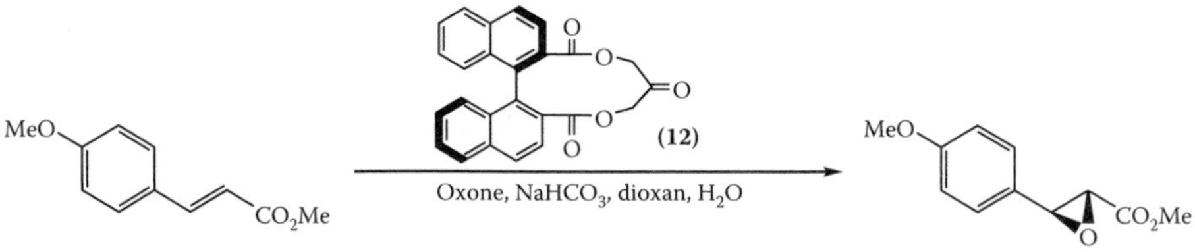

Oxone, NaHCO$_3$, dioxan, H$_2$O

13
80%
>76% ee

SCHEME 9.16

Indeed, a wide variety of BINOL derivatives can be used for the epoxidation of enones in the presence of a lanthanide group salt.[124,125]

The Julia-Colonna method, which uses polyleucine, can form an epoxide from a chalcone (Scheme 9.17).[126–132] However, the method is limited to aryl-substituted enones and closely related systems, and even then scale up of the procedure has been found to be problematic.[133] The product of the epoxidation **14** has been used in a synthesis of (+)-clausenamide (**15**).[134]

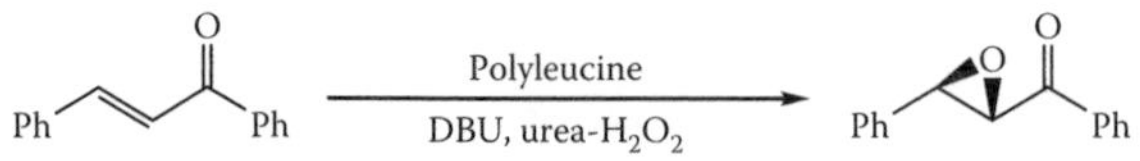

Polyleucine
DBU, urea-H$_2$O$_2$

14

SCHEME 9.17

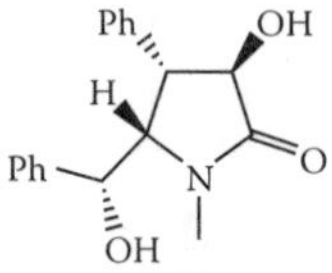

15

9.2.4 BIOLOGICAL METHODS

Epoxidation of alkenes can be achieved by chloroperoxidases (see also Section 9.6. and Chapter 19) (Scheme 9.18).[135–141]

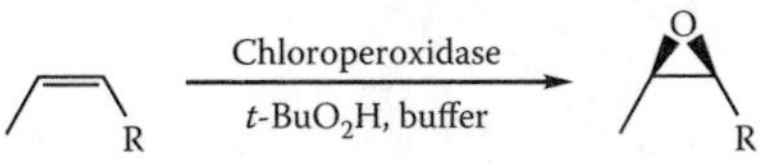

SCHEME 9.18

9.3 ASYMMETRIC DIHYDROXYLATION

This asymmetric dihydroxylation problem was first solved by the use of cinchona alkaloid esters (**16** and **17**; R = p-ClC$_6$H$_4$) together with a catalytic amount of osmium tetroxide.[142,143] The alkaloid esters act as pseudoenantiomeric ligands (Scheme 9.19).[144–148] They can also be supported on a polymer.[149,150]

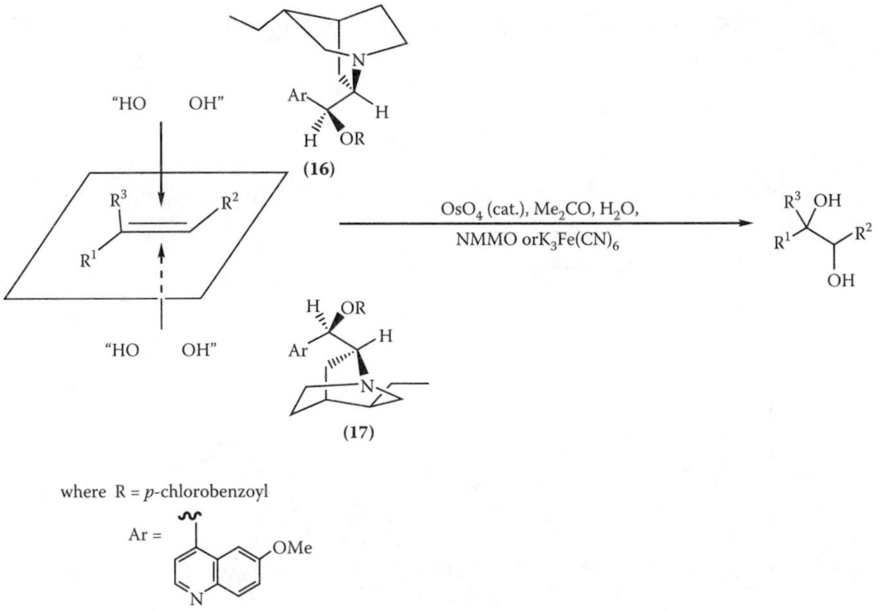

SCHEME 9.19

The original procedure has been modified by the use of a slow addition of the alkene to afford the diol in higher optical purity, and ironically this modification results in a faster reaction. This behavior can be rationalized by consideration of two catalytic cycles operating for the alkene (Scheme 9.20); the use of low alkene concentrations effectively removes the second, low enantioselective cycle.[145,151] The use of potassium ferricyanide in place of N-methylmorpholine-N-oxide (NMMO) as oxidant also improves the level of asymmetric induction.[152,153]

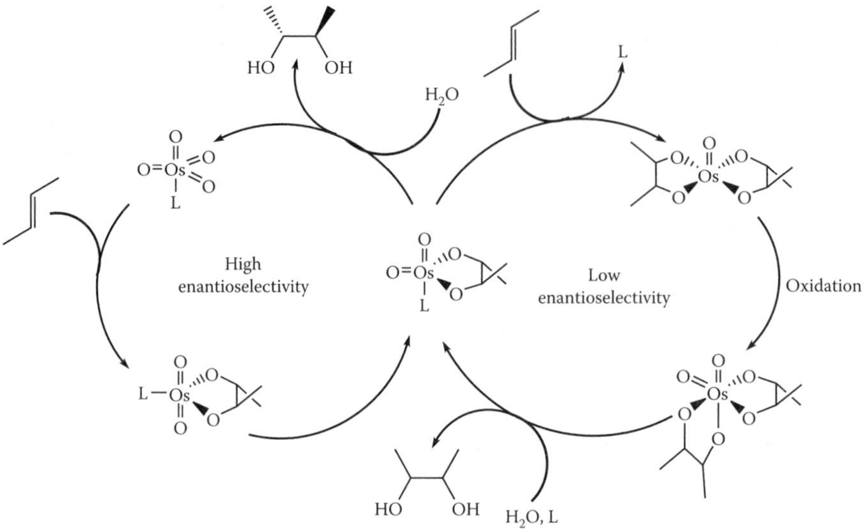

SCHEME 9.20

Although the use of cinchona alkaloids as chiral ligands does provide high asymmetric induction with a number of types of alkene, the search for better systems has resulted in better catalysts (Scheme 9.21).[154–158]

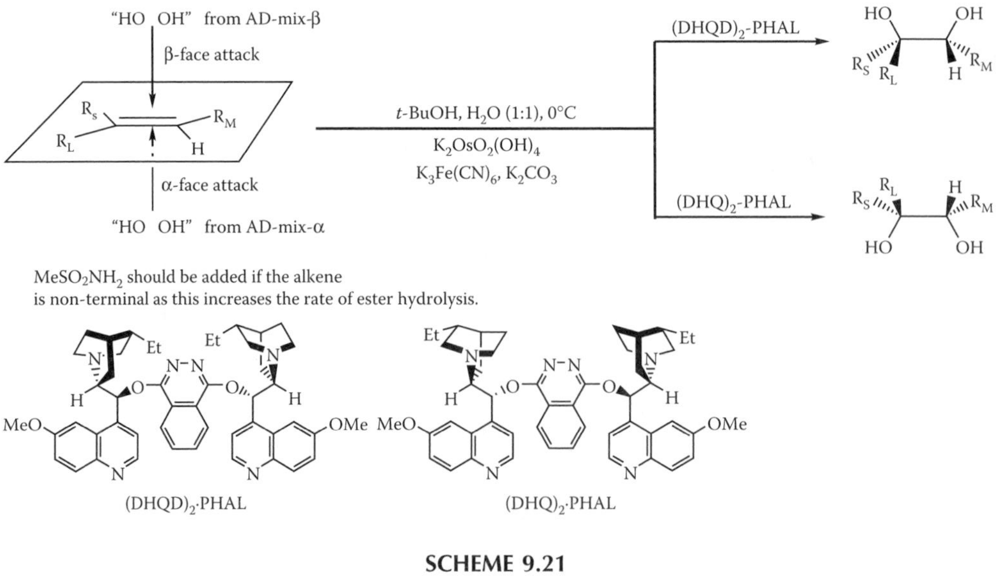

SCHEME 9.21

Kinetic resolutions can be achieved by the dihydroxylation approach.[159]

The Sharpless dihydroxylation method has been run at scale by Pharmacia-Upjohn with *o*-isopropoxy-*m*-methoxysytrene as substrate and *N*-methylmorpholine *N*-oxide as oxidant.[160]

9.3.1 Reactions of Diols

The 1,2-diols formed by the asymmetric oxidation can be used as substrates in a wide variety of transformations. Conversion of the hydroxy groups to *p*-toluenesulfonates then allows nucleophilic displacement by azide at both centers with inversion of configuration (Scheme 9.22).[161]

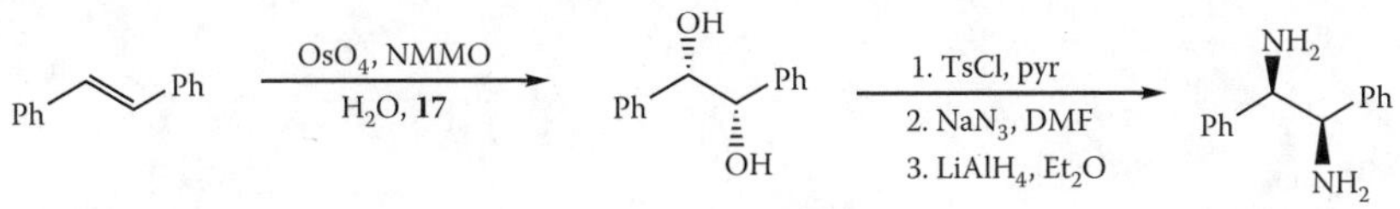

SCHEME 9.22

The diol **18**, prepared by a Sharpless hydroxylation, was converted to the epoxide **19** through the mesylate (Scheme 9.23).[162]

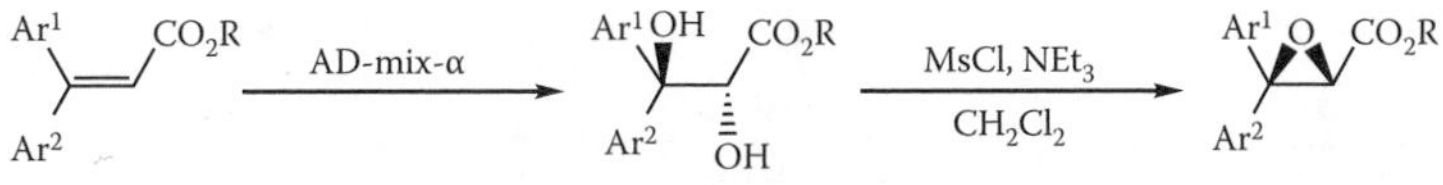

SCHEME 9.23

This conversion of a 1,2-diol to an epoxide has been used as an approach to 2-arylpropanoic acids, members of the nonsteroidal antiinflammatory drugs (NSAIDS) family of drugs.[163] However, the sequence can be shortened by a selective hydrogenolysis, as illustrated for naproxen (**20**) (Scheme 9.24).[164]

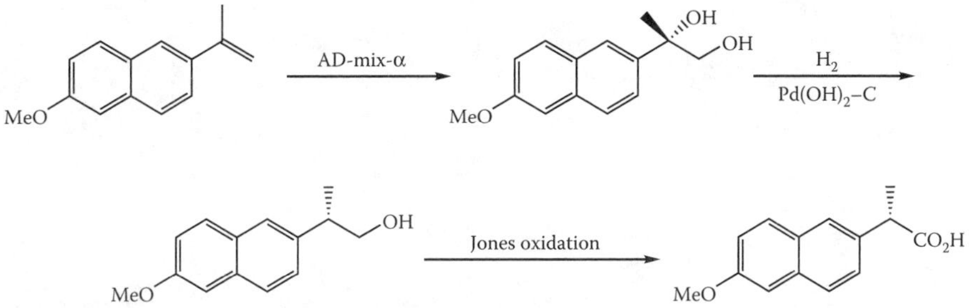

SCHEME 9.24

Another useful variant is to oxidize the primary hydroxy group of the diol leaving the tertiary hydroxy untouched (Scheme 9.25). This reaction sequence can be performed as a one-pot reaction.[165]

SCHEME 9.25

Cyclic sulfates provide a useful alternative to epoxides now that it is viable to produce a chiral diol from an alkene. These cyclic compounds are prepared by reaction of the diol with thionyl chloride, followed by ruthenium-catalyzed oxidation of the sulfur (Scheme 9.26).[166] This oxidation has the advantage over previous procedures because it only uses a small amount of the transition metal catalyst.[167,168]

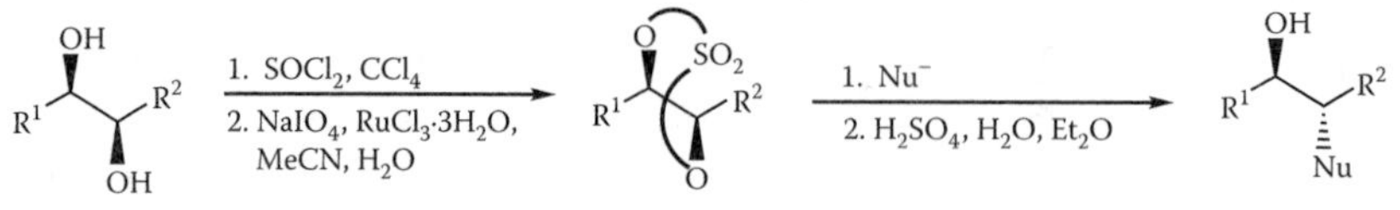

SCHEME 9.26

The cyclic sulfates undergo ring opening with a wide variety of nucleophiles, such as hydride, azide, fluoride, benzoate, amines, and Grignard reagents. The reaction of an amidine with a cyclic sulfate provides an expeditious entry to chiral imidazolines **21** and 1,2-diamines (Scheme 9.27).[169]

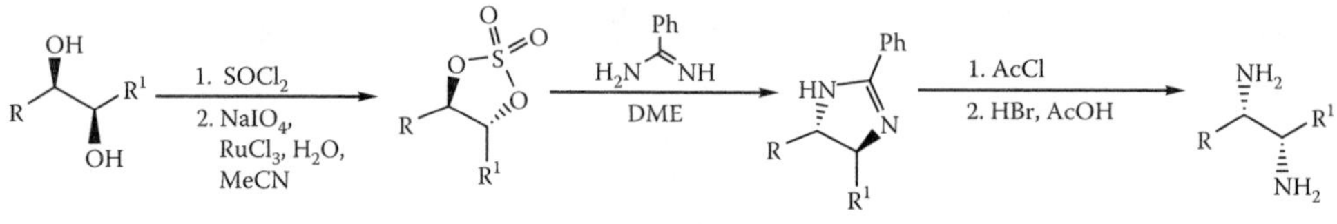

SCHEME 9.27

In the case of an ester ($R^2 = CO_2Me$) the addition occurs exclusively at C–2 (Scheme 9.28); the analogous epoxide does not demonstrate such selectivity.[166,171–173] Terminal cyclic sulfates open in a manner completely analogous to the corresponding epoxide.[170]

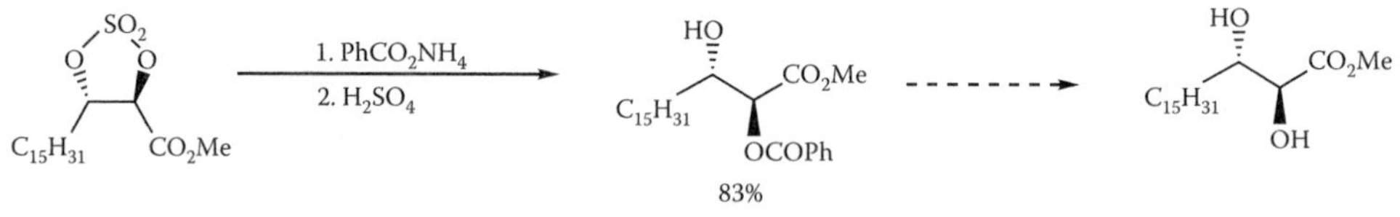

SCHEME 9.28

The resultant sulfate ester can be converted to the alcohol by acid hydrolysis. If an acid-sensitive group is present, this hydrolysis is still successful through use of a catalytic amount of sulfuric acid in the presence of 0.5–1.0 equivalents of water with tetrahydrofuran as solvent. The use of base in the formation of the cyclic sulfates themselves can also alleviate problems associated with acid-sensitive groups.[174,175]

An alternative to the use of cyclic sulfates is the use of cyclic carbonates, as illustrated for the synthesis of 1,2-amino alcohols (Scheme 9.29).[176]

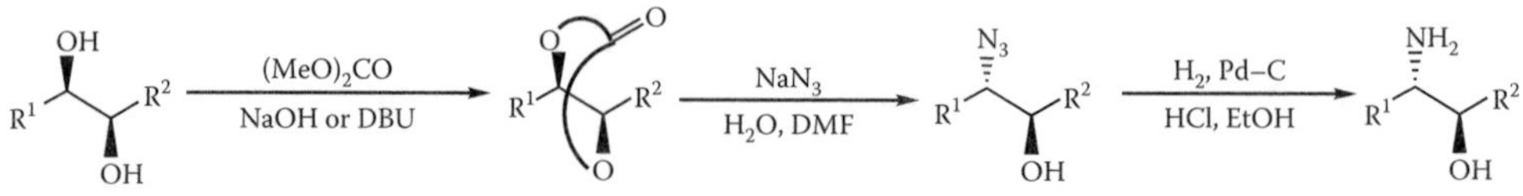

SCHEME 9.29

A method to 1,2-amino alcohols through the dihydroxylation of an enol ether has been reported by Merck (Scheme 9.30).[177] The ee fell off if shorter chain alcohols were used.

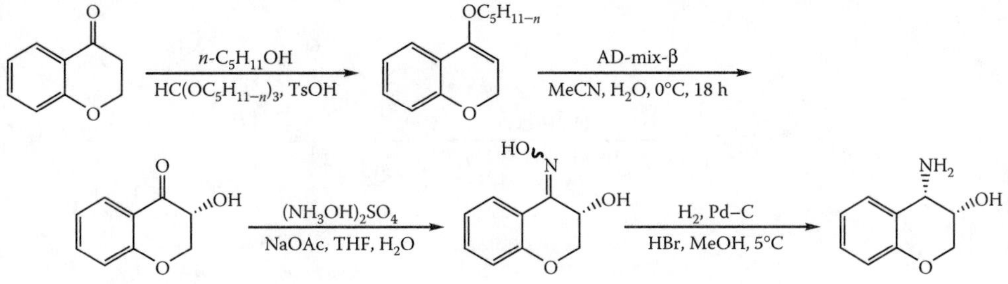

SCHEME 9.30

Another example, again by Merck, is in the synthesis of the cyclooxygenase-2 (COX-2) inhibitor L-784,512 (**22**) (Scheme 9.31).[178]

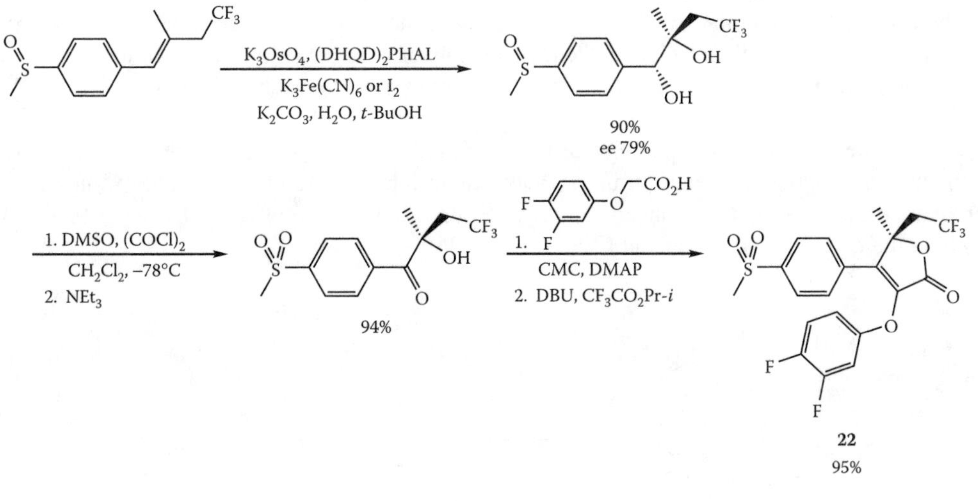

SCHEME 9.31

9.4 AMINOHYDROXYLATION

A variation on the Sharpless dihydroxylation methodology allows for the preparation of amino alcohols. As the groups are introduced in a *syn* manner, this complements an epoxide opening or a similar reaction that uses substitution at one center (Chapter 22). Higher selectivity is observed when one of the alkene substituents is electron withdrawing, as in an ester group (Scheme 9.32).[179-181]

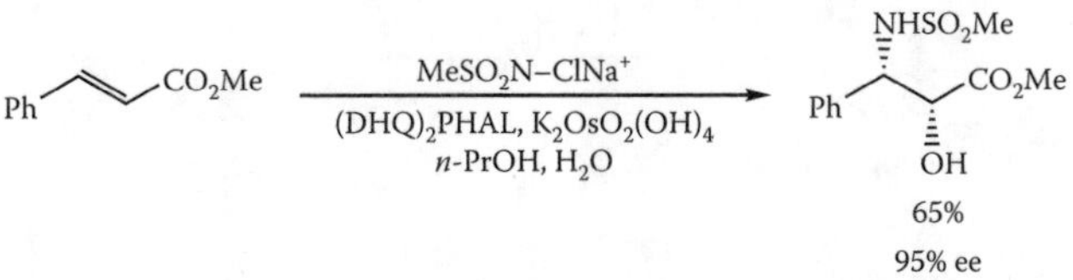

SCHEME 9.32

The methodology has been extended to provide α-arylglycinols from styrenes.[182] The amino-hydroxylation has also been used to access the sidechain of paclitaxel from cinnamate (Scheme 9.33, *cf* Scheme 9.32).[183]

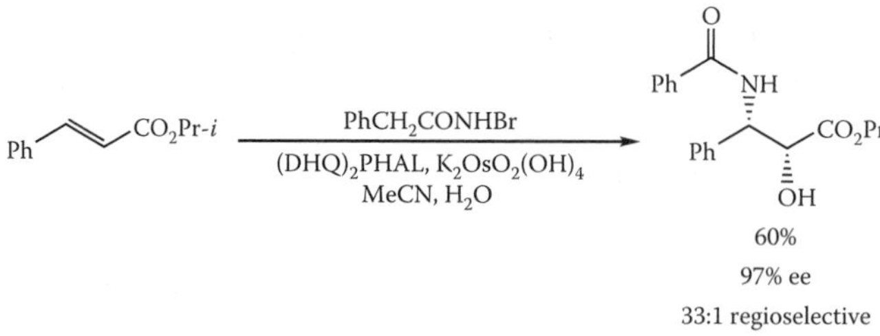

60%
97% ee
33:1 regioselective

SCHEME 9.33

9.5 HALOHYDROXYLATIONS

The conversion of an alkene to a halohydrin can also be considered as an epoxidation because this can be achieved by a simple ring closure.[184] Although no reagent is yet available to perform an asymmetric conversion of an isolated alkene to a halohydrin, the reaction can be controlled through diastereoselection. One such case is the halolactonization of γ,δ-unsaturated carboxylic acids, *N,N*-dialkylamides.[185–189]

One of the few industrial examples of the asymmetric synthesis of halohydrins is in a process to the human immunodeficiency virus (HIV) protease inhibitor, Indinavir.[190] The γ,δ-unsaturated carboxamide **23** is smoothly converted into iodohydrin **24** (92%, 94% de) (Scheme 9.34). (For more on the chemistry of the indanol, see Chapter 24.)[191]

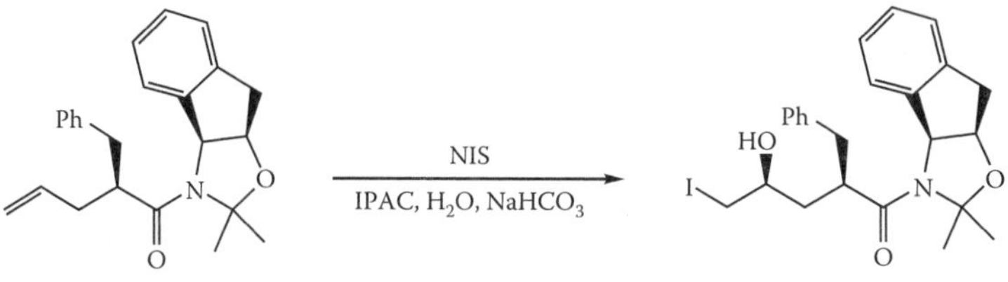

SCHEME 9.34

The reaction presumably proceeds by attack of the amide carbonyl on iodonium ion, followed by collapse of tetrahedral intermediate (Scheme 9.35).[192]

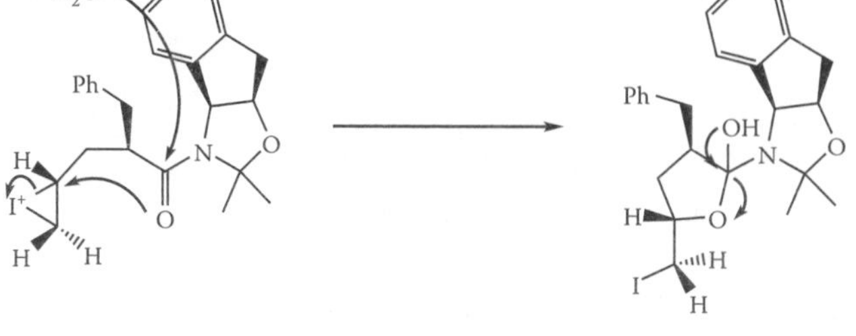

SCHEME 9.35

N-Iodosuccinimide, iodine, diiodohydantoin,[193] or electrochemical means[194] can be used as the oxidant. Other *N,N*-dialkyl-2-substituted-4-enamides work equally well.[191,193]

9.6 BIOLOGICAL METHODS

The epoxidation of alkenes was discussed in Section 9.2.4.

The oxidation of an aromatic substrate has provided a synthesis of 2,3-isopropylidene L-ribonic γ-lactone (**25**) (Scheme 9.36).[195] Indeed, this dihydroxylation of aryl substrates provides a wealth of synthetic intermediates and reaction pathways.[195–206]

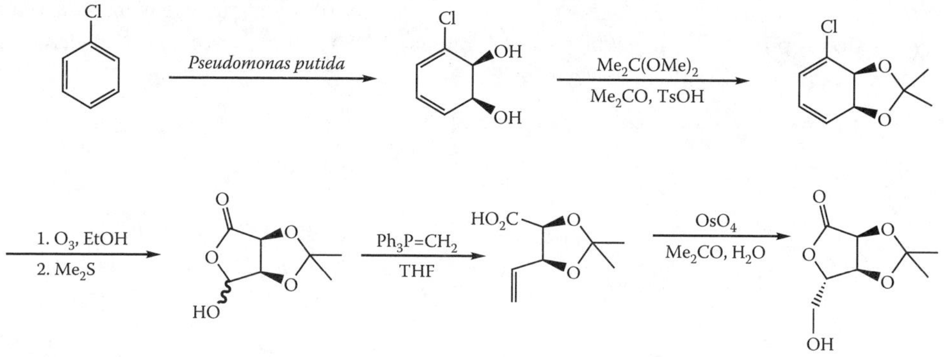

SCHEME 9.36

Other oxidations catalyzed by enzymes are known (Chapter 19). As our understanding of the processes coupled with schemes to alleviate the need for co-factors develop, we will no doubt see more examples of the use of enzymes to achieve oxidations.

9.7 HETEROATOM OXIDATIONS

A number of pharmaceuticals now contain stereogenic heteroatoms where one enantiomer has the desired biological activity. An example is a sulfoxide, and perhaps the best-known example is esomeprazole (see Chapter 31).[207] Access to these compounds by asymmetric synthesis has not been trivial. Although biological methodology continues to advance in terms of ee and substrate promiscuity, there is still considerable progress to be made before a general method is available.

A variant of the Sharpless epoxidation methodology, introduced by Kagan, does provide a useful chemical method.[4,208,209] The method is susceptible to the reaction conditions as shown by nonlinear effects,[210] and the mechanism is not clear.[211,212]

9.8 OTHER OXIDATIONS

9.8.1 CARBONYL COMPOUNDS

9.8.1.1 α-Oxidations

A number of methods have been used to stereoselectively introduce a hydroxy group adjacent to a ketone. Epoxidation of an enol ether or ester can be used (see Chapter 10). Chiral auxiliaries have also been used to lead to the introduction of the hydroxy group.[213] Another method is the reaction of nitrosobenzene with an aldehyde in the presence of L-proline as catalyst (Scheme 9.37).[214]

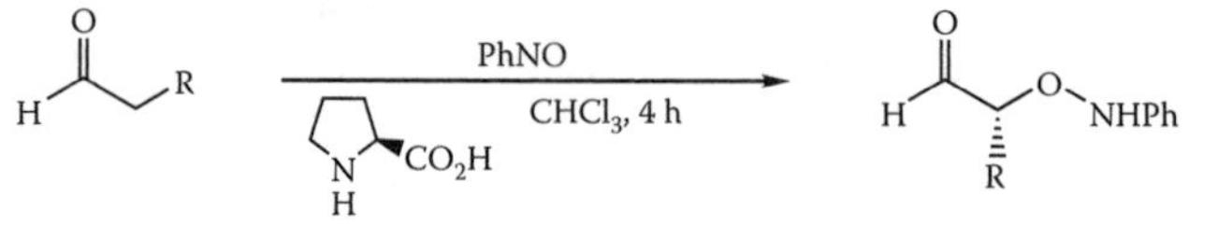

SCHEME 9.37

9.8.1.2 Baeyer-Villiger Oxidations

A Baeyer-Villiger oxidation has the potential of being asymmetric if a regioselective reaction is conducted on a *meso*-substrate or if a stereodifferentiated stereogenic center is already present. The oxidation can be chemical or enzymatic.[213,215,216] An example of a chemical reagent in a Baeyer-Villiger reaction is provided in the synthesis of fragment A **26** from *R*-carvone (**27**) (Scheme 9.38) as part of a synthesis of Cryptophycin A (**28**).[217]

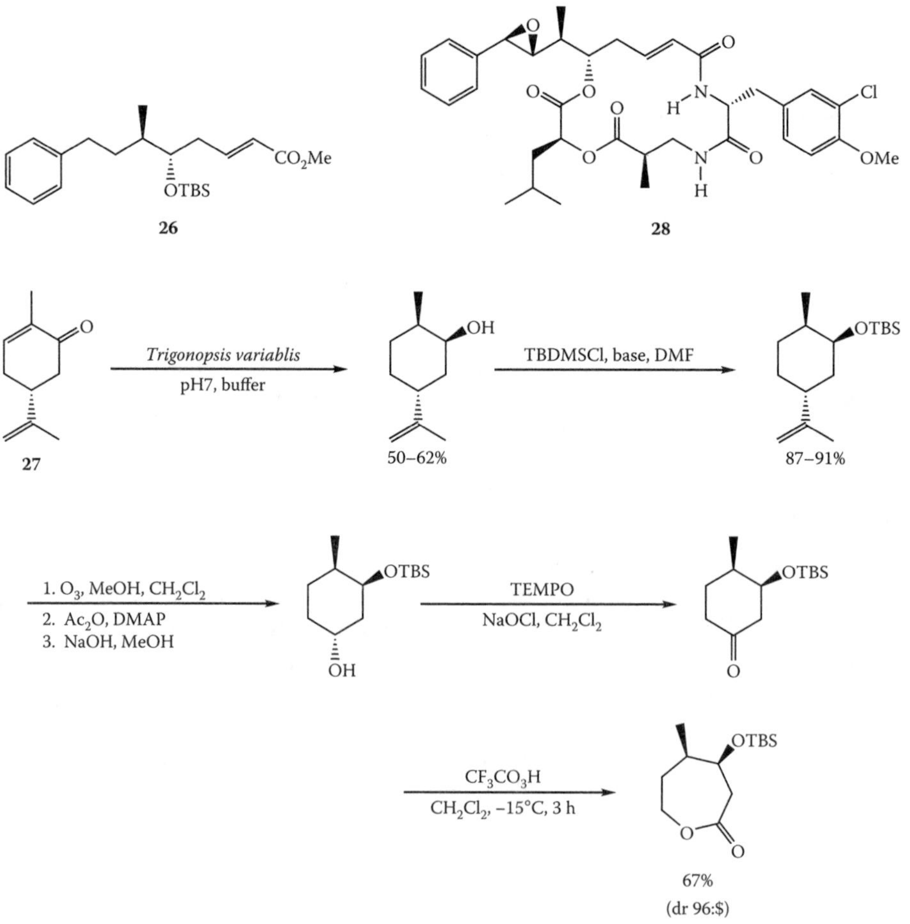

SCHEME 9.38

9.8.2 Silyl Compounds

A reaction that is related to the Baeyer-Villiger reaction is that associated with the use of a silyl group as a masked hydroxy moiety. This approach allows for the wealth of organosilicon chemistry to be used in a synthesis and provides protection for the hydroxy group.[218,219] An example of this conversion is provided in a synthesis of ebelactone A (**29**) where most stereochemical operations were controlled through organosilicon chemistry. The diol **30,** which constitutes the C8–C12 portion of **29**, was prepared by an allene condensation with an aldehyde. The hydroxy group was unmasked by oxidation (Scheme 9.39).[220]

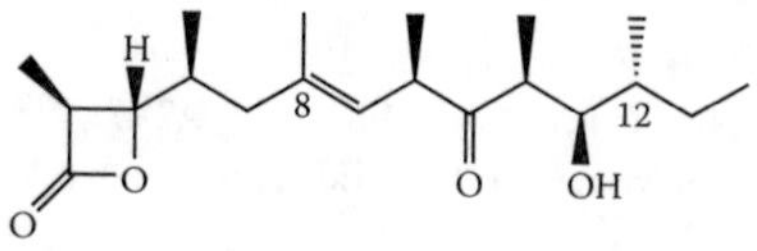

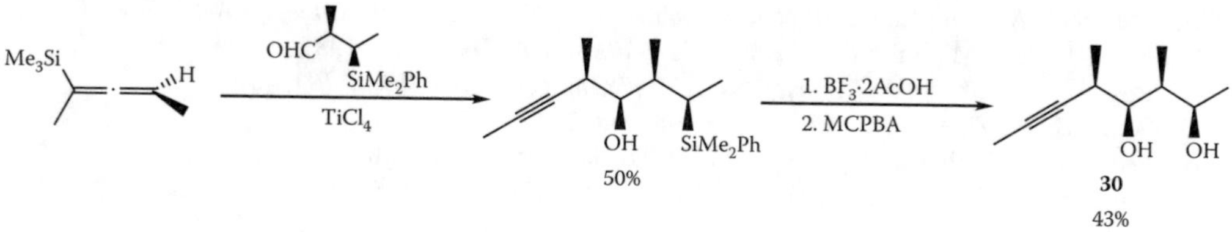

SCHEME 9.39

9.9 SUMMARY

Methodology has been found that allows for the asymmetric oxidation of alkenes and allyl alcohol to the corresponding epoxides or diols. The HKR method is beginning to be used to access chiral epoxides. Although asymmetric oxidations are now being used at scale, they are not used nearly as often as asymmetric reductions.

REFERENCES

1. Tye, H. *J. Chem. Soc., Perkin Trans. 1* 2000, 275.
2. Bonini, C., Righi, G. *Tetrahedron* 2002, *58*, 4981.
3. Bolm, C. *Coord. Chem. Rev.* 2003, *237*, 245.
4. Ager, D. J., East, M. B. *Asymmetric Synthetic Methodology,* CRC Press: Boca Raton, 1995.
5. Shum, W. P., Cannarsa, M. J. In *Chirality in Industry II: Developments in the Commercial Manufacture and Applications of Optically Active Compounds,* Collins, A. N., Sheldrake, G. N., Crosby, J. Eds., Wiley: Chichester, 1997, p. 363.
6. Trost, B. M., Tang, W. *Angew. Chem., Int. Ed.* 2002, *41*, 2795.
7. Eames, J., Watkinson, M. *Angew. Chem., Int. Ed.* 2001, *40*, 3567.
8. Malkov, A. V., Baxendale, I. R., Bella, M., Langer, V., Fawcett, J., Russell, D. R., Mansfield, D. J., Valko, M., Kocovsky, P. *Organometallics* 2001, *20*, 673.
9. Hirt, U. H., Spingler, B., Wirth, T. *J. Org. Chem.* 1998, *63*, 7674.
10. Nishibayashi, Y., Takei, I., Uemura, S., Hida, M. *Organometallics* 1999, *18*, 2291.
11. Persson, B. A., Larsson, A. L. E., Le Ray, M., Bäckvall, J.-E. *J. Am. Chem. Soc.* 1999, *121*, 1645.
12. Verzijl, G. K. M., de Vries, J. G., Broxterman, Q. B. PCT Appl. Pat. WO 0190396, 2001.
13. Stampfer, W., Kosjek, B., Moitzi, C., Kroutil, W., Faber, K. *Angew. Chem., Int. Ed.* 2002, *41*, 1014.
14. Delacroix, O., Picart-Goetgheluck, S., Maciejewski, L., Brocard, J. *Tetrahedron: Asymmetry* 1999, *10*, 4417.
15. Katsuki, T. J. *Syn. Org. Chem., Japan* 1987, *45*, 90.
16. Jorgensen, K. A. *Chem. Rev.* 1989, *89*, 431.
17. Katsuki, T., Sharpless, K. B. *J. Am. Chem. Soc.* 1980, *103*, 5974.
18. Sharpless, K. B., Behrens, C. H., Katsuki, T., Lee, A. W. M., Martin, V. S., Takatani, M., Viti, S. M., Walker, F. J., Woodard, S. S. *Pure Appl. Chem.* 1983, *55*, 589.

19. Pfenninger, A. *Synthesis* 1986, 89.
20. Sato, F., Kobayashi, Y. *Synlett* 1992, 849.
21. Sheldon, R. A. *Chirotechnology: Industrial Synthesis of Optically Active Compounds,* Marcel Dekker, Inc.: New York, 1993.
22. Gao, Y., Hanson, R. M., Klunder, J. M., Ko, S. Y., Masamune, H., Sharpless, K. B. *J. Am. Chem. Soc.* 1987, *109*, 5765.
23. Hansen, R. M., Sharpless, K. B. *J. Org. Chem.* 1986, *51*, 1922.
24. Kende, A. S., Rizzi, J. P. *J. Am. Chem. Soc.* 1981, *103*, 4247.
25. Finn, M. F., Sharpless, K. B. In Morrison, J. D. Ed., *Asymmetric Synthesis,* Academic Press: Orlando, 1986, Vol. 5, p. 247.
26. Rossiter, B. E. In Morrison, J. D. Ed., *Asymmetric Synthesis,* Academic Press: Orlando, 1985, Vol. 5, p. 193.
27. Mulzer, J., Angermann, A., Munch, W., Schlichthorl, G., Hentzschel, A. *Liebigs Ann. Chem.* 1987, 7.
28. Kang, J., Park, M., Shin, H. T., Kim, J. K. *Bull. Korean Chem. Soc.* 1985, *6*, 376.
29. Rossiter, B. E., Katsuki, T., Sharpless, K. B. *J. Am. Chem. Soc.* 1981, *103*, 464.
30. Wang, Z.-M., Zhou, W.-S. *Tetrahedron* 1987, *43*, 2935.
31. Pedersen, S. F., Dewan, J. C., Eckman, R. R., Sharpless, K. B. *J. Am. Chem. Soc.* 1987, *109*, 1279.
32. Burns, C. J., Martin, C. A., Sharpless, K. B. *J. Org. Chem.* 1989, *54*, 2826.
33. Carlier, P. R., Sharpless, K. B. *J. Org. Chem.* 1989, *54*, 4016.
34. Woodard, S. S., Finn, M. G., Sharpless, K. B. *J. Am. Chem. Soc.* 1991, *113*, 106.
35. Finn, M. G., Sharpless. *J. Am. Chem. Soc.* 1991, *113*, 113.
36. Hawkins, J. M., Sharpless, K. B. *Tetrahedron Lett.* 1987, *28*, 2825.
37. Potvin, P. G., Bianchet, S. *J. Org. Chem.* 1992, *57*, 6629.
38. Corey, E. J. *J. Org. Chem.* 1990, *55*, 1693.
39. Wang, Z.-M., Zhou, W.-S., Lin, C. q. *Tetrahedron Lett.* 1985, *26*, 6221.
40. Choudary, B. M., Valli, V. L. K., Prasad, A. D. *J. Chem. Soc., Chem. Commun.* 1990, 1186.
41. Whang, Z.-M., Zhou, W.-S. *Synth. Commun.* 1989, *19*, 2627.
42. Klunder, J. M., Onami, T., Sharpless, K. B. *J. Org. Chem.* 1989, *54*, 1295.
43. Brown, J. M. *Chem. Ind. (London)* 1988, 612.
44. Schweiter, M. J., Sharpless, K. B. *Tetrahedron Lett.* 1985, *26*, 2543.
45. Martin, V. S., Woodard, S. S., Katsuki, T., Yamada, Y., Ikeda, M., Sharpless, K. B. *J. Am. Chem. Soc.* 1981, *103*, 6237.
46. Carlier, P. R., Mungall, W. S., Schroder, G., Sharpless, K. B. *J. Am. Chem. Soc.* 1988, *110*, 2978.
47. Kitano, Y., Matsumoto, T., Wakasa, T., Okamoto, S., Shimazaki, T., Kobayashi, Y., Sato, F., Miyaji, K., Arai, K. *Tetrahedron Lett.* 1987, *28*, 6351.
48. Kobayashi, Y., Kusakabe, M., Kitano, Y., Sato, F. *J. Org. Chem.* 1988, *53*, 1586.
49. Kusakabe, M., Kitano, Y., Kobayashi, Y., Sato, F. *J. Org. Chem.* 1989, *54*, 2085.
50. Martin, S. F., Zinke, P. W. *J. Am. Chem. Soc.* 1989, *111*, 2311.
51. Behrens, C. H., Sharpless, K. B. *Aldrichimica Acta* 1983, *16*, 67.
52. Hanson, R. M. *Chem. Rev.* 1991, *91*, 437.
53. Ward, R. S. *Chem. Soc. Rev.* 1990, *19*, 1.
54. Ager, D. J., East, M. B. *Tetrahedron* 1992, *48*, 2803.
55. Ager, D. J., East, M. B. *Tetrahedron* 1993, *49*, 5683.
56. Marshall, J. A., Trometer, J. D. *Tetrahedron Lett.* 1987, *28*, 4985.
57. Marshall, J. A., Trometer, J. D., Blough, B. E., Crute, T. D. *Tetrahedron Lett.* 1988, *29*, 913.
58. Chakraborty, T. K., Joshi, S. P. *Tetrahedron Lett.* 1990, *31*, 2043.
59. Meyers, A. I., Comins, D. L., Roland, D. M., Henning, R., Shimizu, K. *J. Am. Chem. Soc.* 1979, *101*, 7104.
60. Trost, B. M., Angle, S. R. *J. Am. Chem. Soc.* 1985, *107*, 6124.
61. Hungerbuhler, E., Deebach, D., Wasmuth, D. *Angew. Chem., Int. Ed.* 1979, *18*, 958.
62. Askin, D., Volante, R. P., Ryan, K. M., Reamer, R. A., Shinkai, I. *Tetrahedron Lett.* 1988, *29*, 4245.
63. Howe, G. P., Wang, S., Procter, G. *Tetrahedron Lett.* 1987, *28*, 2629.
64. Bernet, B., Vasella, A. *Tetrahedron Lett.* 1983, *24*, 5491.
65. Roush, W. R., Adam, M. A., Peseckis, S. M. *Tetrahedron Lett.* 1983, *24*, 1377.
66. Isobe, M. I., Kitamura, M., Mio, S., Goto, T. *Tetrahedron Lett.* 1982, *23*, 221.

67. Ohfune, Y., Kurokawa, N. *Tetrahedron Lett.* 1984, *25*, 1587.
68. Solladie, G., Hamouchi, C., Vicente, M. *Tetrahedron Lett.* 1988, *29*, 5929.
69. Page, P. C. B., Rayner, C. M., Sutherland, I. O. *J. Chem. Soc., Perkin Trans. I* 1990, 1375.
70. Mulzer, J., Schollhorn, B. *Angew. Chem., Int. Ed.* 1990, *29*, 1476.
71. Rao, A. V. R., Bose, D. S., Gurjar, M. K., Ravindranathan, T. *Tetrahedron* 1989, *45*, 7031.
72. White, J. D., Bolton, G. L. *J. Am. Chem. Soc.* 1990, *112*, 1626.
73. Rao, A. V. R., Dhar, T. G. M., Bose, D. S., Chakraborty, T. K., Gurjar, M. K. *Tetrahedron* 1989, *45*, 7361.
74. Marshall, J. A., Blough, B. E. *J. Org. Chem.* 1991, *56*, 2225.
75. Jung, M. E., Jung, Y. H. *Tetrahedron Lett.* 1989, *30*, 6637.
76. Wang, Z., Schreiber, S. L. *Tetrahedron Lett.* 1990, *31*, 31.
77. Mori, Y., Kuhara, M., Takeuchi, A., Suzuki, M. *Tetrahedron Lett.* 1988, *29*, 5419.
78. Murphy, P. J., Procter, G. *Tetrahedron Lett.* 1990, *31*, 1059.
79. Hummer, W., Gracza, T., Jager, V. *Tetrahedron Lett.* 1989, *30*, 1517.
80. Mori, Y., Takeuchi, A., Kageyama, H., Suzuki, M. *Tetrahedron Lett.* 1988, *29*, 5423.
81. Mori, Y., Suzuki, M. *J. Chem. Soc., Perkin Trans. I* 1990, 1809.
82. Mori, Y., Kohchi, Y., Ota, T., Suzuki, M. *Tetrahedron Lett.* 1990, *31*, 2915.
83. Williams, N. R. *Adv. Carbohydr. Chem. Biochem.* 1970, *25*, 109.
84. Payne, G. B. *J. Org. Chem.* 1962, *27*, 3819.
85. Behrens, C. H., Ko, S. Y., Sharpless, K. B., Walker, F. J. *J. Org. Chem.* 1985, *50*, 5687.
86. Behrens, C. H., Sharpless, K. B. *J. Org. Chem.* 1985, *50*, 5696.
87. Mori, K., Kbata, T. *Tetrahedron* 1986, *42*, 3471.
88. Bell, T. W., Clacclo, J. A. *Tetrahedron Lett.* 1988, *29*, 865.
89. Kurth, M. J., Abreo, M. A. *Tetrahedron* 1990, *46*, 5085.
90. Marczak, S., Masnyk, M., Wicha, J. *Tetrahedron Lett.* 1989, *30*, 2845.
91. Kang, S.-K., Kim, Y.-S., Lim, J.-S., Kim, K.-S., Kim, S.-G. *Tetrahedron Lett.* 1991, *32*, 363.
92. Dung, J.-S., Armstrong, R. W., Anderson, O. P., Williams, R. M. *J. Org. Chem.* 1983, *48*, 3592.
93. Katsuki, T., Lee, A. W. M., Ma, P., Martin, V. S., Masamune, S., Sharpless, K. B., Tuddenham, D., Walker, F. J. *J. Org. Chem.* 1982, *47*, 1373.
94. Lee, A. W. M., Martin, V. S., Masamune, S., Sharpless, K. B., Walker, F. J. *J. Am. Chem. Soc.* 1982, *104*, 3515.
95. Ma, P., Martin, V. S., Masamune, S., Sharpless, K. B., Viti, S. M. *J. Org. Chem.* 1982, *47*, 1378.
96. Ko, S. Y., Lee, A. W. M., Masamune, S., Reed, L. A., Sharpless, K. B., Walker, F. J. *Science* 1983, *220*, 949.
97. Ko, S. Y., Lee, A. W. M., Masamune, S., Reed, L. A., Sharpless, K. B., Walker, F. J. *Tetrahedron* 1990, *46*, 245.
98. Zhang, W., Loebach, J. L., Wilson, S. R., Jacobsen, E. J. *J. Am. Chem. Soc.* 1990, *112*, 2801.
99. Irie, R., Noda, K., Ito, Y., Matsumoto, N., Katsuki, T. *Tetrahedron Lett.* 1990, *31*, 7345.
100. Van Draanen, N. A., Arseniyadis, S., Crimmins, M. T., Heathcock, C. H. *J. Org. Chem.* 1991, *56*, 2499.
101. Okamoto, Y., Still, W. C. *Tetrahedron Lett.* 1988, *29*, 971.
102. Schwenkreis, T., Berkessel, A. *Tetrahedron Lett.* 1993, *34*, 4785.
103. Quan, R. W., Li, Z., Jacobsen, E. N. *J. Am. Chem. Soc.* 1996, *118*, 8156.
104. Zhang, W., Jacobsen, E. N. *J. Org. Chem.* 1991, *56*, 2296.
105. Deng, L., Jacobsen, E. N. *J. Org. Chem.* 1992, *57*, 4320.
106. Palucki, M., Finney, N. S., Pospisil, P. J., Güler, M. L., Ishida, T., Jacobsen, E. N. *J. Am. Chem. Soc.* 1998, *120*, 948.
107. Finney, N. S., Pospisil, P. J., Chang, S., Palucki, M., Konsler, R. G., Hansen, K. B., Jacobsen, E. N. *Angew. Chem., Int. Ed.* 1997, *36*, 1720.
108. Irie, R., Noda, K., Ito, Y., Katsuki, T. *Tetrahedron Lett.* 1991, *32*, 1055.
109. O'Connor, K. J., Wey, S.-J., Burrows, C. J. *Tetrahedron Lett.* 1992, *33*, 1001.
110. Katsuki, T. *Synlett* 2003, 281.
111. Tokunaga, M., Larrow, J. F., Kakiuchi, F., Jacobsen, E. N. *Science* 1997, *277*, 936.
112. Brandes, B. D., Jacobsen, E. N. *Tetrahedron: Asymmetry* 1997, *8*, 3927.
113. Ready, J. M., Jacobsen, E. N. *Angew. Chem., Int. Ed.* 2002, *41*, 1374.
114. Hou, X.-L., Li, B.-F., Dai, L. X. *Tetrahedron: Asymmetry* 1999, *10*, 2319.

115. Rodríguez, A., Nomen, M., Spur, B. W., Godfried, J. J., Lee, T. H. *Tetrahedron* 2001, *57*, 25.

116. Kulig, K., Holzgrabe, U., Malawska, B. *Tetrahedron: Asymmetry* 2001, *12*, 2533.

117. Gurjar, M., Krishna, L. M., Sarma, B. V. N. B. S., Chorghade, M. S. *Org. Proc. Res. Dev.* 1998, *2*, 422.

118. Larrow, J. F., Schaus, S. E., Jacobsen, E. N. *J. Am. Chem. Soc.* 1996, *118*, 7420.

119. Schaus, S. E., Jacobsen, E. N. *Tetrahedron Lett.* 1996, *37*, 7937.

120. Jacobsen, E. N., Kakiuchi, F., Konsler, R. G., Larrow, J. F., Tokunaga, M. *Tetrahedron Lett.* 1997, *38*, 773.

121. Schaus, S. E., Larrow, J. F., Jacobsen, E. N. *J. Org. Chem.* 1997, *62*, 4197.

122. Hansen, K. B., Leighton, J. L., Jacobsen, E. N. *J. Am. Chem. Soc.* 1996, *118*, 10924.

123. Imashiro, R., Seki, M. *J. Org. Chem.* 2004, *69*, 4216.

124. Chen, R., Qian, C., de Vries, J. G. *Tetrahedron Lett.* 2001, *42*, 6919.

125. Daikai, K., Kamaura, M., Inanaga, J. *Tetrahedron Lett.* 1998, *39*, 7321.

126. Banfi, S., Colonna, S., Molinari, H., Juliá, S., Guixer, J. *Tetrahedron* 1984, *40*, 5207.

127. Lasterra-Sánchez, M. E., Roberts, S. M. *Curr. Org. Chem.* 1997, 187.

128. Ebrahim, S., Wills, M. *Tetrahedron: Asymmetry* 1997, *8*, 3163.

129. Baars, S., Drauz, K.-H., Krimmer, H.-P., Roberts, S. M., Sander, J., Skidmore, J., Zanardi, G. *Org. Proc. Res. Dev.* 2003, *7*, 509.

130. Lauret, C., Roberts, S. M. *Aldrichimica Acta* 2002, *35*, 47.

131. Porter, M. J., Roberts, S. M., Skidmore, *J. Bioorg. Med. Chem.* 1999, 7, 2145.

132. Allen, J. V., Bergeron, S., Griffiths, M. J., Mukherjee, S., Roberts, S. M., Williamson, N. M. *J. Chem. Soc., Perkin Trans. I* 1998, 3171.

133. Senn-Bilfinger, J., *FAST,* 2004, Cambridge, U.K.

134. Cappi, M., Chen, W.-P., Flood, R. W., Liao, Y.-W., Roberts, S. M., Skidmore, J., Smith, J. A., Williamson, N. M. *J. Chem. Soc., Chem. Commun.* 1998, 1159.

135. Hu, S., Hager, L. P. *Tetrahedron Lett.* 1999, *40*, 164.

136. Koch, A., Reymond, J., Lerner, R. A. *J. Am. Chem. Soc.* 1994, *116*, 803.

137. Zaks, A., Dodds, D. R. *J. Am. Chem. Soc.* 1995, *117*, 10419.

138. Dexter, A. F., Lakner, F. J., Campbell, R. A., Hager, L. P. *J. Am. Chem. Soc.* 1995, *117*, 6412.

139. Lakner, F. J., Hager, L. P. *J. Org. Chem.* 1996, *61*, 3923.

140. Lakner, F. J., Cain, K. P., Hager, L. P. *J. Am. Chem. Soc.* 1997, *119*, 443.

141. Lakner, F. J., Hager, L. P. *Tetrahedron: Asymmetry* 1997, *8*, 3547.

142. Kolb, H. C., VanNieuwenhze, M. S., Sharpless, K. B. *Chem. Rev.* 1994, *94*, 2483.

143. Sharpless, K. B. *Angew. Chem., Int. Ed.* 2002, *41*, 2024.

144. Jacobsen, E. N., Markó, I., Mungall, W. S., Schröder, G., Sharpless, K. B. *J. Am. Chem. Soc.* 1988, *110*, 1968.

145. Jacobsen, E. N., Marko, I., France, M. B., Svendsen, J. S., Sharpless, K. B. *J. Am. Chem. Soc.* 1989, *111*, 737.

146. Lohray, B. B., Kalantar, T. H., Kim, B. M., Park, C. Y., Shibata, T., Wai, J. S. M., Sharpless, K. B. *Tetrahedron Lett.* 1989, *30*, 2041.

147. Shibata, T., Gilheany, D. G., Blackburn, B. K., Sharpless, K. B. *Tetrahedron Lett.* 1990, *31*, 3817.

148. Jorgensen, K. A. *Tetrahedron Lett.* 1990, *31*, 6417.

149. Kim, B. M., Sharpless, K. B. *Tetrahedron Lett.* 1990, *31*, 3003.

150. Han, H., Janda, K. D. *J. Am. Chem. Soc.* 1996, *118*, 7632.

151. Wai, J. S. M., Marko, I., Svendsen, J. S., Finn, M. G., Jacobsen, E. N., Sharpless, K. B. *J. Am. Chem. Soc.* 1989, *111*, 1123.

152. Kwong, H.-L., Sorato, C., Ogino, Y., Chen, H., Sharpless, K. B. *Tetrahedron Lett.* 1990, *31*, 2999.

153. Minato, M., Yamamoto, K., Tsuji, J. *J. Org. Chem.* 1990, *55*, 766.

154. Sharpless, K. B., Amberg, W., Bennani, Y. L., Crispino, G. A., Hartung, J., Jeong, K.-S., Kwong, H.-L., Morikawa, K., Wang, Z.-M., Xu, D., Zhang, X.-L. *J. Org. Chem.* 1992, *57*, 2768.

155. Vidari, G., Giori, A., Dapiaggi, A., Lanfranchi, G. *Tetrahedron Lett.* 1993, *34*, 6925.

156. Arrington, M. P., Bennani, Y. L., Göbel, T., Walsh, P., Zhao, S.-H., Sharpless, K. B. *Tetrahedron Lett.* 1993, *34*, 7375.

157. Morikawa, K., Park, J., Andersson, P. G., Hashiyama, T., Sharpless, K. B. *J. Am. Chem. Soc.* 1993, *115*, 8463.

158. Wang, L., Sharpless, K. B. *J. Am. Chem. Soc.* 1992, *114*, 7568.

159. VanNieuwenhze, M. S., Sharpless, K. B. *J. Am. Chem. Soc.* 1993, *115*, 7864.
160. Ahrgren, L., Sutin, L. *Org. Proc. Res. Dev.* 1997, *1*, 425.
161. Pini, D., Iuliano, A., Rosini, C., Salvadori, P. *Synthesis* 1990, 1023.
162. Jansen, R. PCT Appl. Pat. WO 0153281, 2001.
163. Griesbach, R. C., Hamon, D. P. G., Kennedy, R. J. *Tetrahedron: Asymmetry* 1997, *8*, 507.
164. Ishibashi, H., Maeki, M., Yagi, J., Ohba, M., Kanai, T. *Tetrahedron* 1999, *55*, 6075.
165. Aladro, F. J., Guerra, F. M., Moreno-Dorado, F. J., Bustamante, J. M., Jorge, Z. D., Massanet, G. M. *Tetrahedron Lett.* 2000, *41*, 3209.
166. Gao, Y., Sharpless, K. B. *J. Am. Chem. Soc.* 1988, 110, 7538.
167. Denmark, S. E. *J. Org. Chem.* 1981, *46*, 3144.
168. Lowe, G., Salamone, S. J. *J. Chem. Soc., Chem. Commun.* 1983, 1392.
169. Oi, R., Sharpless, K. B. *Tetrahedron Lett.* 1991, *32*, 999.
170. Gao, Y., PhD Thesis Massachusetts Institute of Technology 1988.
171. Lohray, B. B., Gao, Y., Sharpless, K. B. *Tetrahedron Lett.* 1989, *30*, 2623.
172. Berridge, M. S., Franceschini, M. P., Rosenfeld, E., Tewson, T. J. *J. Org. Chem.* 1990, *55*, 1211.
173. Shao, H., Goodman, M. *J. Org. Chem.* 1996, *61*, 2582.
174. Kim, B. M., Sharpless, K. B. *Tetrahedron Lett.* 1989, *30*, 655.
175. Pearlstein, R. M., Blackburn, B. K., Davis, W. M., Sharpless, K. B. *Angew. Chem., Int. Ed.* 1990, *29*, 639.
176. Chang, H.-T., Sharpless, K. B. *Tetrahedron Lett.* 1996, *37*, 3219.
177. Marcune, B. F., Karady, S., Reider, P. J., Miller, R. A., Biba, M., DiMichele, L., Reamer, R. A. *J. Org. Chem.* 2003, *68*, 8088.
178. Tan, L., Chen, C.-Y., Larsen, R. D., Verhoeven, T. R., Reider, P. J. *Tetrahedron Lett.* 1998, *39*, 3961.
179. Rudolph, J., Sennhenn, P. C., Vlaar, C. P., Sharpless, K. B. *Angew. Chem., Int. Ed.* 1996, *35*, 2810.
180. Li, G., Chang, H.-T., Sharpless, K. B. *Angew. Chem., Int. Ed.* 1996, *35*, 451.
181. Rubin, A. E., Sharpless, K. B. *Angew. Chem., Int. Ed.* 1997, *36*, 2637.
182. Reddy, K. L., Sharpless, K. B. *J. Am. Chem. Soc.* 1998, *120*, 1207.
183. Song, C. E., Oh, C. R., Roh, E. J., Lee, S.-G., Choi, J. H. *Tetrahedron: Asymmetry* 1999, *10*, 671.
184. Ng, J. S., Przybyla, C. A., Liu, C., Yen, J. C., Muellner, F. W., Weyker, C. L. *Tetrahedron* 1995, *51*, 6397.
185. Fuji, K., Node, M., Naniwa, Y., Kawabata, T. *Tetrahedron Lett.* 1990, *31*, 3175.
186. Hart, D. J., Huang, H. C., Krishnamurthy, R., Schawrtz, T. *J. Am. Chem. Soc.* 1989, *111*, 7507.
187. Kurth, M. J., Brown, E. G. *J. Am. Chem. Soc.* 1987, *109*, 6844.
188. Moon, H.-S., Eisenberg, S. W. E., Wilson, M. E., Schore, N. E., Kurth, M. J. *J. Org. Chem.* 1994, *59*, 6504.
189. Tamaru, Y., Mizutani, M., Furukawa, Y., Kawamura, S., Yoshida, Z., Yanagi, K., Minobe, M. *J. Am. Chem. Soc.* 1984, *106*, 1079.
190. Dorsey, B. D., Levin, R. B., McDaniel, J. P., Vacca, J. P., Guare, J. P., Darke, P. L., Zugay, J. A., Emini, E. A., Schleif, W. A., Quintero, J. C., Lin, J. H., Chen, I.-W., Holloway, M. K., Fitzgerald, P. M. D., Axel, M. G., Ostovic, D., Anderson, P. S., Huff, J. R. *J. Med. Chem.* 1994, *37*, 3443.
191. Maligres, P. E., Upadhyay, V., Rossen, K., Cianciosi, S. J., Purick, R. M., Eng, K. K., Reamer, R. A., Askin, D., Volante, R. P., Reider, P. J. *Tetrahedron Lett.* 1995, *36*, 2195.
192. Rossen, K., Reamer, R. A., Volante, R. P., Reider, P. J. *Tetrahedron Lett.* 1996, *38*, 6843.
193. Maligres, P. E., Weissman, S. A., Upadhyay, V., Cianciosi, S. J., Reamer, R. A., Purick, R. M., Sager, J., Rossen, K., Eng, K. K., Askin, D., Volante, R. P., Reider, P. J. *Tetrahedron* 1996, *52*, 3327.
194. Rossen, K., Volante, R. P., Reider, P. J. *Tetrahedron Lett.* 1997, *38*, 777.
195. Hudlicky, T., Price, J. D. *Synlett* 1990, 159.
196. Ley, S. V., Sternfeld, F., Taylor, S. *Tetrahedron Lett.* 1987, *28*, 225.
197. Hudlicky, T., Luna, H., Price, J. D., Rulin, F. *Tetrahedron Lett.* 1989, *30*, 4053.
198. Ley, S. V., Sternfeld, F. *Tetrahedron* 1989, *45*, 3463.
199. Amici, M. D., Micheli, C. D., Carrea, G., Spezia, S. *J. Org. Chem.* 1989, *54*, 2646.
200. Ley, S. V., Sternfeld, F. *Tetrahedron Lett.* 1988, *29*, 5305.
201. Hudlicky, T., Entwistle, D. A., Pitzer, K. K., Thorpe, A. J. *Chem. Rev.* 1996, *96*, 1195.
202. Carless, H. A. J. *Tetrahedron: Asymmetry* 1992, *3*, 795.
203. Ley, S. V., Yeung, L. L. *Synlett* 1992, 997.
204. Dumortier, L., Liu, P., Dobbelaere, S., Van der Eycken, J., Vandewalle, M. *Synlett* 1992, 243.

205. Ley, S. V., Yeung, L. L. *Synlett* 1992, 291.
206. Hudlicky, T., Gonzalez, D., Gibson, D. T. *Aldrichimica Acta* 1999, *32*, 35.
207. Federsel, H.-J. *Chirality* 2003, *15*, 5128.
208. Scettri, A., Bonadies, F., Lattanzi, A. *Tetrahedron: Asymmetry* 1996, *7, 629.*
209. Brunel, J.-M., Diter, B., Duetsch, M., Kagan, H. B. *J. Org. Chem.* 1995, 60, 8086.
210. Brunel, J.-M., Luukas, T. O., Kagan, H. B. *Tetrahedron: Asymmetry* 1998, *9*, 1941.
211. Potvin, P. G., Fieldhouse, B. G. *Tetrahedron: Asymmetry* 1999, *10*, 1661.
212. Kagan, H. B., Dunach, E., Nemecek, C., Pitchen, P., Samuel, O., Zhao, S. *Pure Appl. Chem.* 1985, *57*, 1911.
213. Ager, D. J., Prakash, I., Schaad, D. R. *Chem. Rev.* 1996, *96*, 835.
214. Brown, S. P., Brochu, M. P., Sinz, C. J., MacMillan, D. W. C. *J. Am. Chem. Soc.* 2003, *125*, 10808.
215. Stewart, J. D., Reed, K. W., Martinez, C. A., Zhu, J., Chen, G., Kayser, M. M. *J. Am. Chem. Soc.* 1998, *120*, 3541.
216. Stewart, J. D. *Curr. Org. Chem.* 1998, *2*, 195.
217. Varie, D. L., Brennan, J., Biggs, B., Cronin, J. S., Hay, D. A., Rieck III, J. A., Zmijewski, M. J. *Tetrahedron Lett.* 1998, *39*, 8405.
218. Fleming, I., Henning, R., Parker, D. C., Plaut, H. E., Sanderson, P. E. I. *J. Chem. Soc., Perkin Trans. I* 1995, 317.
219. Fleming, I., Barbero, A., Walter, D. *Chem. Rev.* 1997, *97*, 2063.
220. Archibald, S. C., Barden, D. J., Bazin, J. F. Y., Fleming, I., Foster, C. F., Mandal, A. K., Mandal, A. K., Parker, D., Takaki, K., Ware, A. C., Williams, A. R. B., Zwicky, A. B. *Org. Biomol. Chem.* 2004, *2*, 1051.

10 Asymmetric Epoxidation of Olefins by Chiral Ketones

Yian Shi

CONTENTS

10.1 INTRODUCTION

Asymmetric epoxidation of olefins is a powerful method for the generation of optically active epoxides, which are very useful intermediates for the synthesis of a wide variety of enantiomerically enriched molecules. A number of efficient epoxidation methods have been developed,[1,2] including the epoxidation of allylic alcohols,[3,4] the metal-catalyzed epoxidation of unfunctionalized olefins (particularly for conjugated *cis-* and trisubstituted olefins),[5–8] and the nucleophilic epoxidation of electron-deficient olefins.[9–11] Chiral dioxiranes generated *in situ* from ketones and Oxone™ (potassium peroxymonosulfate) have been shown to be a class of effective agents for the epoxidation of a wide variety of olefins (Scheme 10.1), particularly for unfunctionalized *trans*-olefins, a long-standing challenge.[12–20] Since the first chiral ketone was reported by Curci in 1984,[21] a variety of chiral ketone catalysts have been investigated in various laboratories (Figure 10.1). This chapter will briefly summarize some of our studies in this area.

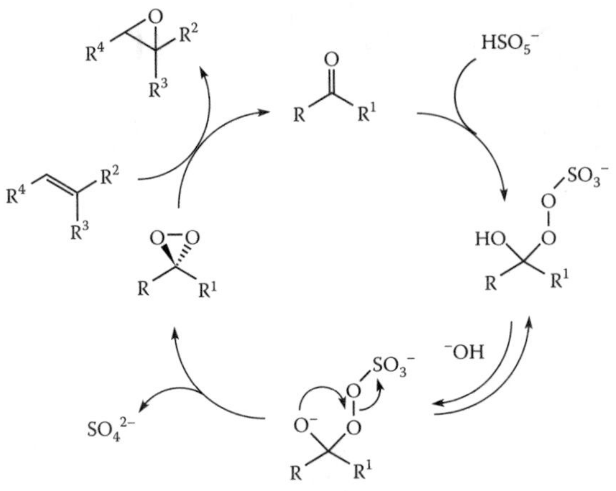

SCHEME 10.1

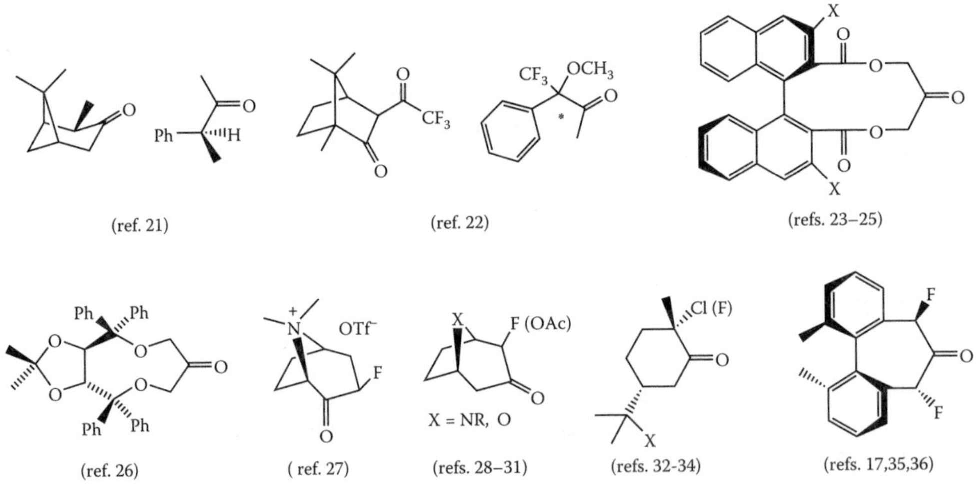

(ref. 21)   (ref. 22)   (refs. 23–25)

(ref. 26)   (ref. 27)   (refs. 28–31)   (refs. 32-34)   (refs. 17,35,36)

FIGURE 10.1 Some examples of reported chiral ketones.

10.2 KETONE CATALYSTS FOR *TRANS-* AND TRISUBSTITUTED OLEFINS

10.2.1 SYNTHESIS OF FRUCTOSE-DERIVED KETONE

In our search for ketone catalysts, fructose-derived ketone **1** has been discovered and developed to be a highly effective epoxidation catalyst. Ketone **1** is a member of a general class of ketone catalysts (**2** and **3**) (Scheme 10.2) designed to maximize the stereochemical interaction between substrate and catalyst by bringing the chiral control element close to the reacting carbonyl, to minimize the potential epimerization of the stereogenic centers at the α position of the carbonyl using fused ring(s), to control the approach of an olefin to the reacting dioxirane by sterically blocking one face or by a C_2 or pseudo C_2 symmetric element, and to activate the carbonyl by introducing electron-withdrawing substituents. Fructose-derived ketone **1** is readily prepared in two steps from D-fructose by ketalization and oxidation (Scheme 10.2).[37–40] Both alcohol **5** and ketone **1** are stable crystalline compounds and can be purified by recrystallization. Simple acids such as H_2SO_4 can also be used for the ketalization.[40] For the oxidation, many oxidants such as PCC, PDC-Ac_2O, DMSO-Ac_2O, DMSO-DCC, DMSO-$(COCl)_2$, $RuCl_3$-$NaIO_4$, Ru-TBHP, and so on can be used.[40] Sheldon and co-workers have shown that inexpensive NaOCl can act as an effective oxidant using 1 mol% TPAP as catalyst, and the Ru catalyst can be recycled a number of times.[41] Ketone **1** has been produced at DSM-Catalytica (now DSM Pharma Chemicals) in multi-kilogram quantities and is currently commercially available. The enantiomer of ketone catalyst **1** (ketone *ent-***1**) can be synthesized similarly from L-fructose, which can be prepared from readily available L-sorbose based on a literature procedure[39,42] Ketone *ent-***1** prepared in this way showed the same enantioselectivity for epoxidation as ketone **1**.[39]

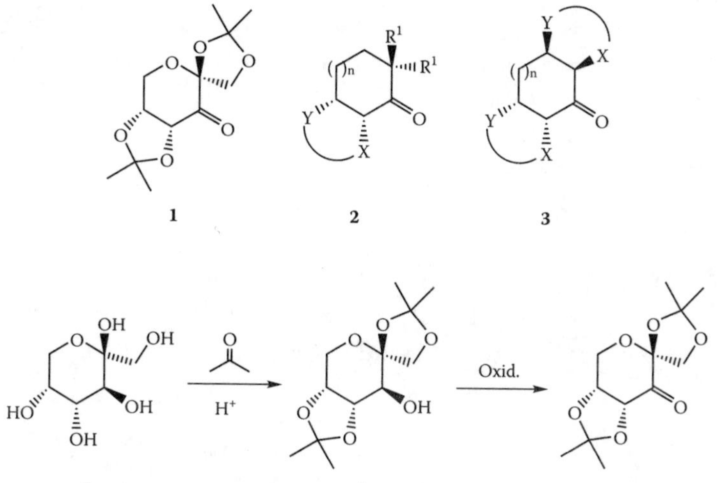

SCHEME 10.2

10.2.2 THE pH EFFECT ON EPOXIDATION

Controlling the reaction pH is crucial for the epoxidation with *in situ*–generated dioxiranes.[43,44] Ketone-mediated epoxidations are usually performed at pH 7–8 since poor conversions are often obtained at higher pH because Oxone can autodecompose rapidly at these pHs.[45–47] However, studies showed that a higher pH is beneficial for the epoxidation with ketone **1**.[38,39,48,49] As shown in Figure 10.2, with 20 mol% ketone **1**, the conversion of *trans*-β-methylstyrene increased from <10% at pH 7–8 to around 80% at pH >10, while maintaining high enantioselectivity (90–92% ee) at high pH. Running the reaction at higher pH greatly reduces the amount of catalyst needed and leads to a catalytic asymmetric epoxidation process. The epoxidation usually is performed around pH 10.5, and the reaction pH can be controlled by adding either K_2CO_3 or KOH along with Oxone. Whereas a clear mechanistic understanding of the pH effect on ketone-catalyzed epoxidation requires further studies, a higher pH possibly facilitates the formation of anion **7** and the subsequent formation of dioxirane **8**, thus reducing the unproductive Baeyer-Villiger oxidation (Scheme 10.3).

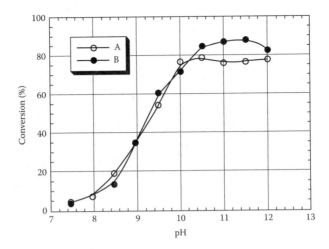

FIGURE 10.2 Plot of the conversion of *trans*-β-methylstyrene against pH using ketone **1** (0.2 eq.) as catalyst in two solvent systems, H_2O-MeCN (1:1.5, v/v) (**A**), H_2O-MeCN-DMM (2:1:2, v/v) (**B**).[39]

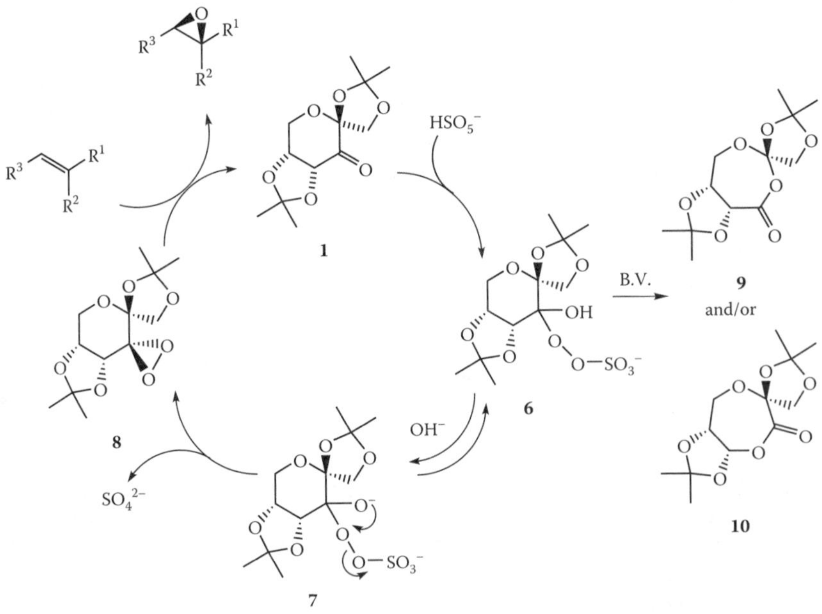

SCHEME 10.3

10.2.3 SUBSTRATE SCOPE

The substrate scope of this epoxidation was subsequently investigated using a variety of olefins with a catalytic amount of ketone **1** (usually 20–30 mol%). A variety of *trans*-substituted and trisubstituted olefins have been shown to be effective substrates (Table 10.1),[39] and the high ee obtained with *trans*-7-tetradecene suggests that this epoxidation is quite general for simple *trans*-olefins (Table 10.1, Entry 5). Various functional groups such as ethers, ketals, esters, and so on are compatible with the epoxidation conditions (Table 10.1). A variety of 2,2-disubstituted vinylsilanes

TABLE 10.1
Asymmetric Epoxidation of Representative *trans*-Substituted and Trisubstituted Olefins by Ketone 1

Entry	Epoxide	Yield (ee) (%)	Entry	Epoxide	Yield (ee) (%)
1	Ph—epoxide—Ph	85 (98)	10	Ph cyclohexene oxide	94 (98)
2	Ph—epoxide	94 (95)	11	Ph/Ph—epoxide—C₁₀H₂₁	92 (97)
3	Ph—epoxide—OTBS	87 (94)	12	Ph—epoxide	89 (97)
4	—epoxide—OTBS	83 (94)	13	Ph—epoxide	93 (76)
5	C₆H₁₃—epoxide—C₆H₁₃	89 (95)	14	C₁₀H₂₁—epoxide	97 (86)
6	(structure)	92 (92)	15	cyclohexyl—epoxide—CO₂Me	89 (94)
7	Ph—epoxide—C(O)OMe	68 (92)	16	Ph—epoxide—SiMe₃	74 (94)
8	Ph—epoxide—Ph	89 (95)	17	Ph—epoxide—SiMe₃	66 (93)
9	Ph/Ph—epoxide—Ph	54 (97)	18	—epoxide—SiMe₃	51 (90)

TABLE 10.2
Asymmetric Epoxidation of Representative Hydroxyalkenes and Enol Esters by Ketone 1

Entry	Epoxide	Yield (ee) (%)	Entry	Epoxide	Yield (ee) (%)
1	Ph—epoxide—OH	85 (94)	7	Ph—epoxide—OH	75 (74)
2	Ph—epoxide—OH	45 (91)	8	—epoxide—OH	82 (90)
3	—epoxide—OH	68 (91)	9	BzO—epoxide	79 (80)
4	Ph/Ph—epoxide—OH	87 (94)	10	BzO—epoxide	82 (93)
5	cyclohexane—epoxide—OH	93 (94)	11	BzO—epoxide	87 (91)
6	—epoxide—OH	85 (92)	12	BzO—epoxide	82 (95)

can also be enantioselectively epoxidized (Table 10.1, Entries 16–18).[50] From the resulting epoxy-silanes, optically active 1,1-disubstituted terminal epoxides can be obtained by desilylation with TBAF. The epoxidation can also be extended to hydroxyalkenes such as allylic and homoallylic alcohols (Table 10.2, Entries 1–8)[51] and enol esters with good enantioselectivity (Table 10.2, Entries 9–12).[52–54] Conjugated dienes can be regioselectively epoxidized to provide vinyl epoxides with high ee's (Table 10.3, Entries 1–8),[55] and conjugated enynes can be chemoselectively and enantio-selectively epoxidized to produce optically active propargyl epoxides (Table 10.3, Entries 9–16).[56]

TABLE 10.3
Asymmetric Epoxidation of Representative Conjugated Dienes and Enynes by Ketone 1

Entry	Epoxide	Yield (ee)(%)	Entry	Epoxide	Yield (ee)(%)
1		77 (97)	9		78 (93)
2		54 (95)	10		71 (93)
3		41 (96)	11		97 (77)
4		68 (96)	12		98 (96)
5		68 (90)	13		59 (96)
6		82 (95)	14		71 (89)
7		89 (94)	15		84 (95)
8		60 (92)	16		60 (93)

TABLE 10.4
Asymmetric Epoxidation of Representative Olefins by Ketone 1 Using H_2O_2

Entry	Epoxide	Yield (ee)(%)	Entry	Epoxide	Yield (ee)(%)
1		90 (98)	7		90 (96)
2		93 (92)	8		88 (89)
3		75 (93)	9		77 (92)
4		71 (89)	10		93 (95)
5		97 (92)	11		75 (96)
6		94 (95)	12		76 (95)

10.2.4 Hydrogen Peroxide as Primary Oxidant

Further studies have shown that asymmetric epoxidation with ketone **1** can also be effective using hydrogen peroxide (H_2O_2), a highly desirable oxidant as a result of its high active oxygen content with its reduction product being water, as oxidant in the presence of a nitrile.[57–59] Among nitriles investigated, MeCN and EtCN were found to be the most effective for the epoxidation.[58] A variety of olefins can be epoxidized in good yields and high enantioselectivity with CH_3CN-H_2O_2 (Table 10.4).[58] This epoxidation system is carried out under mild conditions with relatively high reaction concentrations. It is found that the choice of solvents is also very important for the reaction. A mixed solvent such as MeCN-EtOH-CH_2Cl_2 is beneficial for olefins with poor solubility.[58] Peroxyimidic acid **11** is

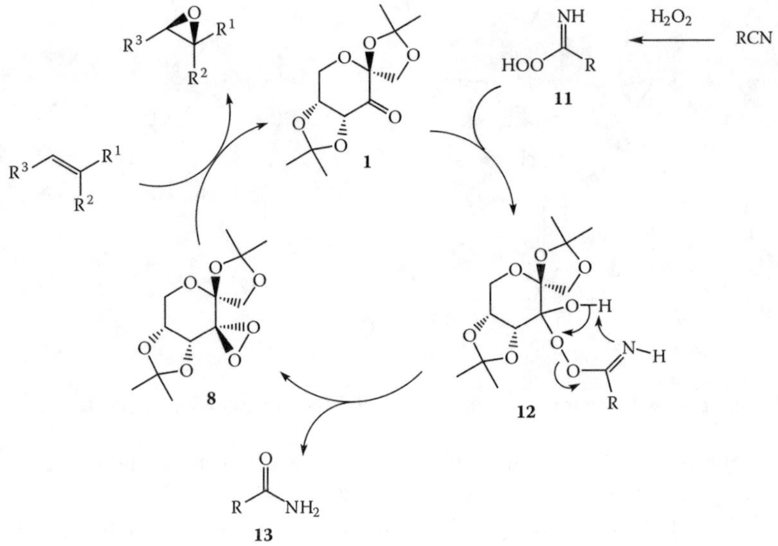

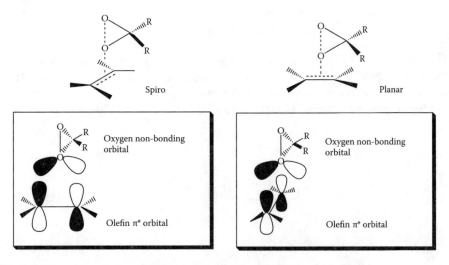

SCHEME 10.4

likely to be the active oxidant that reacts with the ketone to generate the dioxirane (Scheme 10.4). Compound **11** itself has been found to be a good epoxidation reagent to produce racemic epoxides.[60,61] However, our studies have shown that the background epoxidation can be minimized by control of the reaction pH. Additional work revealed that some other ketones can also be effective epoxidation catalysts using this RCN-H_2O_2 system. For example, a variety of olefins can be epoxidized in high yields using 10–30% trifluoroacetone as catalyst.[49]

10.2.5 TRANSITION STATE MODEL

The two extreme transition state geometries (spiro and planar) for the epoxidation with dioxiranes are shown in Figure 10.3. Both experimental[14,15,24,25,37,39,62] and computational[63–67] studies suggest that the spiro transition state is generally favored, presumably as a result of the stabilizing interaction of the oxygen lone pair with the π^* orbital of the alkene (Figure 10.3).[63–67] Our studies show that

FIGURE 10.3 The spiro and planar transition states for the epoxidation with dioxiranes.

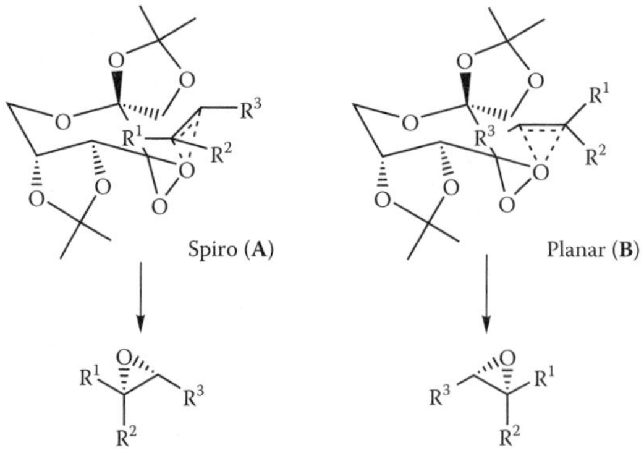

FIGURE 10.4 The spiro and planar transition states for the epoxidation with ketone **1**.

the epoxidation of *trans*-substituted and trisubstituted olefins with ketone **1** proceeds mainly through spiro **A**, whereas planar **B** is the major competing transition state.[37–39,50–56] Spiro **A** and planar **B** provide opposite enantiomers of the epoxide product; thus, factors influencing the competition of these two transition states would consequently affect the ee of the epoxide (Figure 10.4). Studies show that the competition between these two transition states is dependent on the electronic and steric nature of the substituents on the olefins. Spiro **A** can be further favored if the reacting alkenes are conjugated with groups that can enhance the stabilizing secondary orbital interaction by lowering the energy of the π^* orbital of the reacting alkene, such as phenyl, alkene, alkyne, and so on, thus giving higher enantioselectivity for the epoxidation. With regard to the steric effect, generally higher ee can be obtained with a smaller R^1 (favoring spiro **A**) and/or a larger R^3 (disfavoring planar **B**). This transition state model is also validated by a kinetic resolution study of racemic olefins with ketone **1**.[68]

10.2.6 STRUCTURAL EFFECT OF KETONES ON CATALYSIS

A variety of analogues of ketone **1** were prepared to further understand the structural requirements for ketone catalysis. It was found that the 5-membered spiro ketal[69] and the pyranose oxygen[70] of ketone **1** are important for both the activity and enantioselectivity. However, when the fused ketal of ketone **1** was replaced with a more electron-withdrawing oxazolidinone, the resultant ketone (**14**) was found to be highly active, giving good yields and enantioselectivities for a variety of olefin substrates.[71] For example, the epoxidation of *trans*-β-methylstyrene with 5 mol% of **14** gave 100% conversion and 88% ee. In the case of 1-phenylcyclohexene, the epoxidation with 2 mol% of the ketone provided the epoxide in 93% yield and 97% ee. The enhanced stability and activity of ketone **14** could result from the fact that the presumed Baeyer-Villiger decomposition was reduced by the more electron-withdrawing oxazolidinone. Further studies showed that diacetate ketone **15,** readily available from **1,** was effective for the epoxidation of α,β-unsaturated esters.[72] High ee's and good yields can be obtained for a variety of α,β-unsaturated esters using 20–30 mol% ketone **15** (Table 10.5).

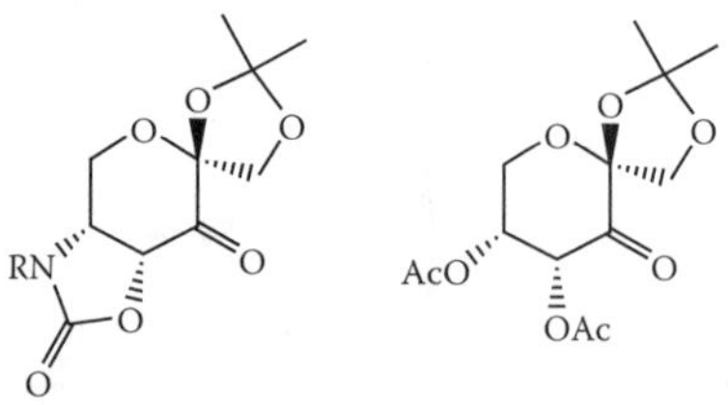

14   15

TABLE 10.5
Asymmetric Epoxidation of Representative Olefins by Ketone 15

Entry	Epoxide	Yield (ee) (%)	Entry	Epoxide	Yield (ee) (%)
1	Ph—CO$_2$Et	73 (96)	6	Ph—CO$_2$Et	93 (96)
2	—CO$_2$Et	91 (97)	7	CO$_2$Et	77 (93)[a]
3	MeO—CO$_2$Et	57 (90)	8	CO$_2$Et	64 (82)[a]
4	Cl—CO$_2$Et	64 (97)	9	Ph—CO$_2$Et	96 (94)[a]
5	F—CO$_2$Et	77 (96)	10	n-Bu—CO$_2$Et	74 (98)[a]

[a] The configuration was not determined.

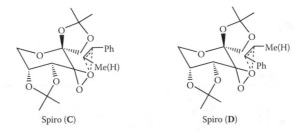

Spiro (**C**)   Spiro (**D**)

FIGURE 10.5 Spiro transition states for the epoxidation of *cis*-β-methylstyrene with ketone **1**.

10.3 KETONE CATALYSTS FOR *CIS*- AND TERMINAL OLEFINS

Fructose-derived ketone **1** is an effective epoxidation catalyst for a wide variety of *trans*-substituted and trisubstituted olefins, but it does not give high enantioselectivity for *cis*- and terminal olefins.[39,73] For example, the epoxidations of *cis*-β-methylstyrene and styrene with ketone **1** gave only 39% ee and 24% ee, respectively, suggesting that the phenyl and methyl (or hydrogen) groups of the olefin cannot be effectively differentiated in spiro transition states **C** and **D,** thus giving poor enantioselectivity (Figure 10.5).

During our further studies of ketone catalysts, ketone **16** was found to be highly enantioselective for a number of acyclic and cyclic *cis*-olefins (Table 10.6).[73,74] It is important to note that the epoxidation is stereospecific with no isomerization observed in the epoxidation of acyclic systems. Ketone **16** also provides encouragingly high ee's for certain terminal olefins, particularly styrenes.[74,75] In general, ketones **1** and **16** have complementary substrate scopes. In our subsequent study of the conformational and electronic effects of ketone catalysts on epoxidation, ketone **17**, a carbocyclic analog of **16**, was found to be highly enantioselective for various styrenes (Table 10.7).[76]

16   17   18

TABLE 10.6
Asymmetric Epoxidation of Representative *cis*- and Terminal Olefins by Ketone 16

Entry	Epoxide	Yield (ee)(%)	Entry	Epoxide	Yield (ee)(%)
1		87 (91)	7		61 (97)[a]
2		91 (92)	8		88 (94)[a]
3		88 (83)	9		74 (83)
4		61 (91)	10		90 (85)
5		77 (91)	11		86 (84)
6		82 (91)	12		93 (71)[a]

[a] The configuration was not determined.

TABLE 10.7
Asymmetric Epoxidation of Styrenes by Ketone 17

Entry	Epoxide	Yield (ee)(%)	Entry	Epoxide	Yield (ee)(%)
1		63 (90)	7		76 (91)
2		84 (89)	8		67 (90)
3		91 (90)	9		81 (90)
4		71 (93)	10		69 (93)
5		62 (89)	11		56 (93)
6		64 (90)			

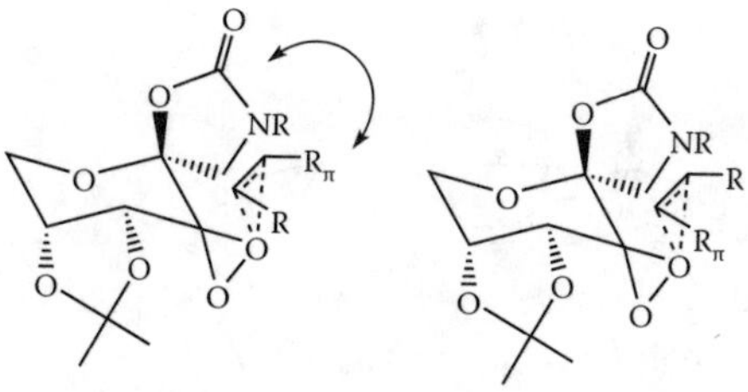

FIGURE 10.6 Transition states for the epoxidation of *cis*-alkenes with ketone **1**.

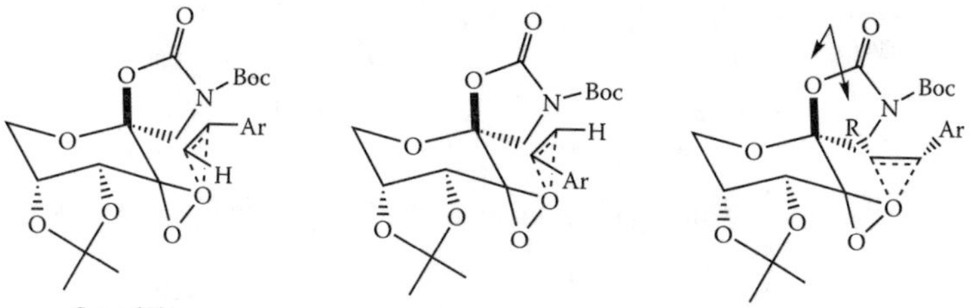

FIGURE 10.7 Transition states for the epoxidation of styrenes with ketone **16**.

The enantioselectivity obtained with ketone **16** for *cis*- and terminal olefins is likely to result from electronic interactions. Studies show that an attraction may exist between the R_π group of the olefin and the oxazolidinone of the ketone (Figure 10.6).[73–75] As a result, transition state spiro **E** is favored over spiro **F**, giving high enantioselectivity. Studies with *N*-aryl-substituted oxazolidinone-containing ketones **18** show that the attractive interaction is generally enhanced by a group that withdraws electron density from the oxazolidinone through conjugation.[77] The attractive interaction could cause a change of the transition state from the normally favored spiro to planar in some cases.[74,77]

Similar to *cis*-olefins, the epoxidation of styrenes with ketone **16** proceeds mainly through transition state spiro **G**. One difference for styrenes is that planar transition state **I** could also be competing (R = H) in addition to spiro **H**, whereas the corresponding planar **I** for *cis*-olefins would be less competitive because of the steric effect (Figure 10.7). As a result, the enantioselectivities obtained for *cis*-olefins with **16** are usually higher than styrenes.[74,75] Reducing the competition from planar **I** would be important to further improve the enantioselectivity for terminal olefins.

The higher ee's obtained for styrenes with ketone **17** compared to **16** suggest that the replacement of the pyranose oxygen with a carbon has a significant effect on the competition between these transition states.[76] Studies showed that ketones **16** and **17** gave similar enantioselectivities for *cis*-β-methylstyrene. However, the opposite enantiomers were obtained for the epoxidation of 1-phenylcyclohexene with these two ketones (Figure 10.8). The X-ray structures show that ketones **16** and **17** have similar conformations, suggesting that the pyranose oxygen influences the transition states via an electronic effect rather than a conformational effect. The switch of the epoxide configuration observed for the two structurally similar ketones for 1-phenylcyclohexene (Figure 10.8) could result from the fact that the replacement of the pyranose oxygen in ketone **16** with a

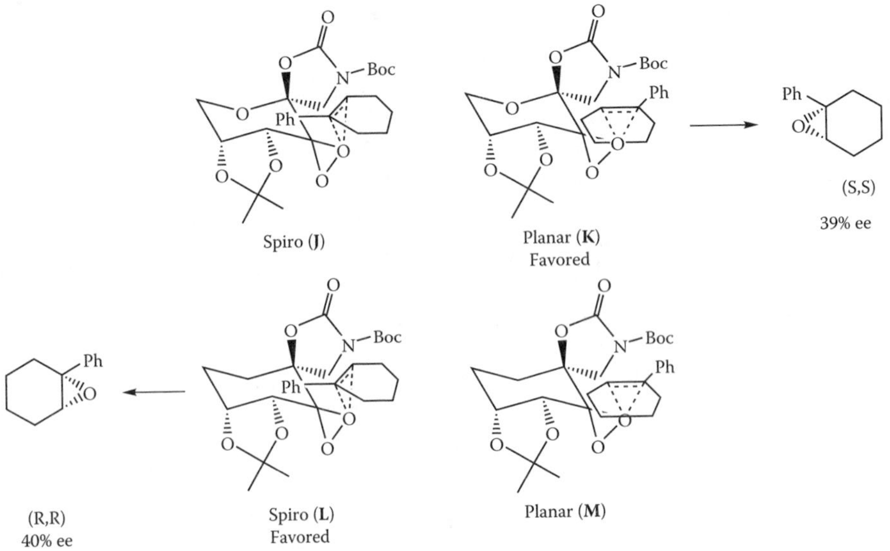

FIGURE 10.8 Transition states for the epoxidation of 1-phenylcyclohexene with ketones **16** and **17**.

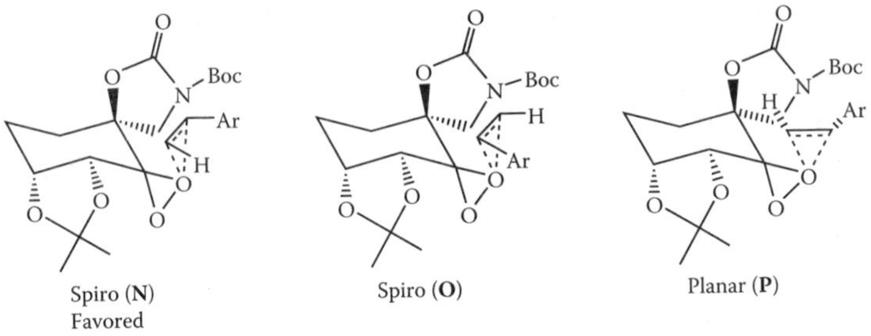

FIGURE 10.9 Transition states for the epoxidation of styrenes with ketone **17**.

carbon in ketone **17** increases the interaction of the nonbonding orbital of the dioxirane with the π* orbital of the alkene by increasing the energy of the nonbonding orbital of the dioxirane, consequently favoring spiro **L** over planar **M**. In the case of styrene, the change from an oxygen to a carbon leads to the favoring of desired spiro **N** and undesired spiro **O** over undesired planar **P** (Figure 10.9), thus reducing the minor enantiomer generated via planar pathway **P** and enhancing the enantioselectivity of the reaction overall.[76]

Although an electron-withdrawing substituent may frequently increase the reactivity and stability of a ketone catalyst, such a substituent at the same time may have an unfavorable effect on the enantioselectivity of the reaction by decreasing the energy of the nonbonding orbital of the dioxirane, thus disfavoring the major spiro transition state. An effective catalyst requires delicate balance between reactivity and enantioselectivity. The comparative studies between ketones **16** and **17** provide extremely valuable insight for future catalyst design.

Ketone **16** can be prepared from glucose in 6 steps without extensive chromatography purification (Scheme 10.5).[78] N-Aryl-substituted oxazolidinone-containing ketone **18** can be synthesized

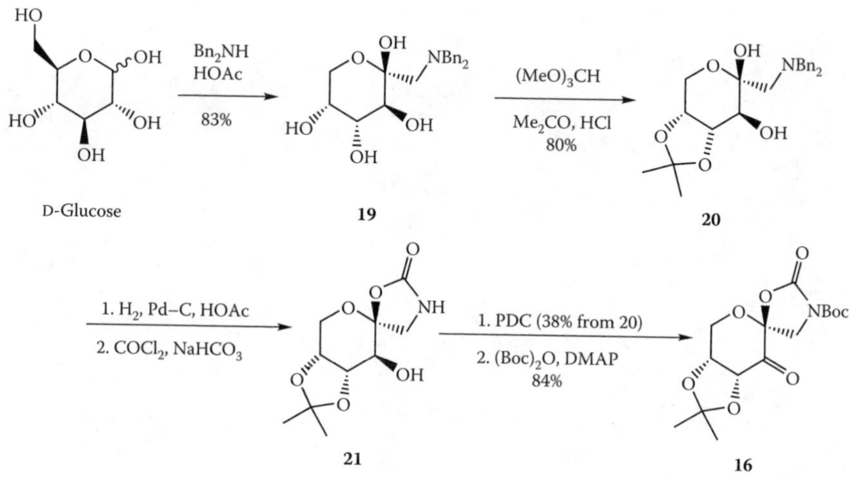

SCHEME 10.5

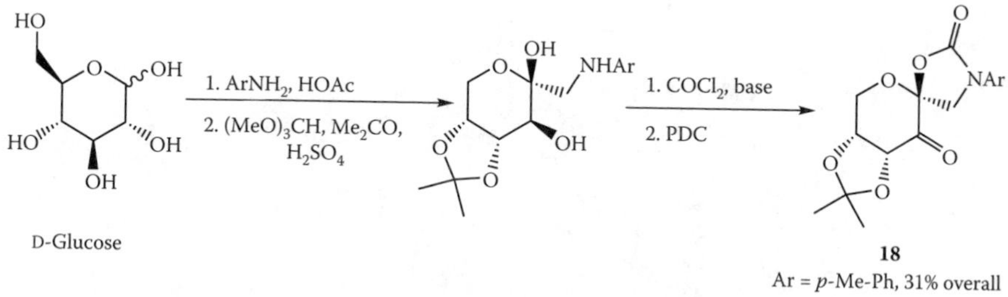

SCHEME 10.6

from D-glucose in 4 steps by Amadori rearrangement, ketalization, oxazolidinone formation, and oxidation (Scheme 10.6).[77] The substrate scope of ketone 18 is under investigation.

10.4 SYNTHETIC APPLICATIONS

Ketone 1 is highly general and enantioselective for the epoxidation of *trans*-substituted and trisubstituted olefins. Its ready availability and predictability potentially make this ketone useful. Its utilization in synthesis has been reported by other researchers.[79–100] For example, Corey and coworkers have reported that pentaoxacyclic compound 24 can be obtained in 31% overall yield by enantioselective epoxidation of (R)-2,3-dihydroxy-2,3-dihydrosqualene (22) and subsequent cyclization of pentaepoxide 23 (Scheme 10.7).[82] Eli Lilly has also used the epoxidation to introduce the epoxide on the sidechain of Cryptophycin 52 (Scheme 10.8).[90] The epoxidation has been carried out at multi-kilogram scale at DSM-Catalytica. Around 100 kg of lactone 29 was prepared by the epoxidation of olefin 27 and subsequent *in situ* cyclization of epoxide 28 (Scheme 10.9).[101]

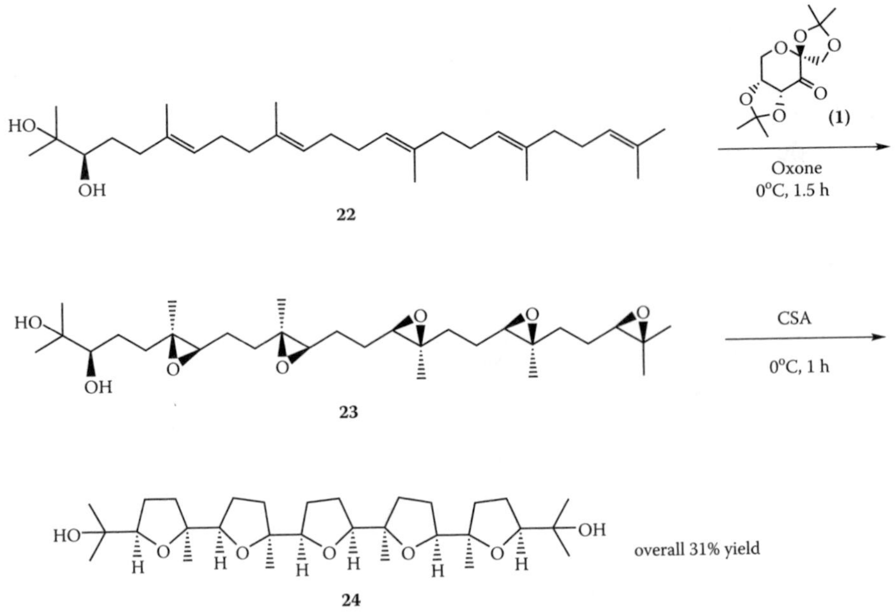
(1)
Oxone
0°C, 1.5 h
22
CSA
0°C, 1 h
23
overall 31% yield
24

SCHEME 10.7

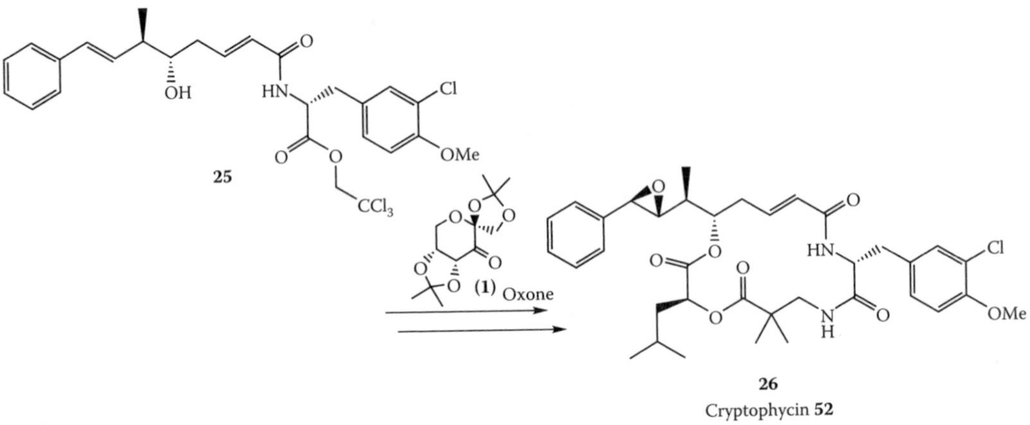
25
(1) Oxone
26
Cryptophycin 52

SCHEME 10.8

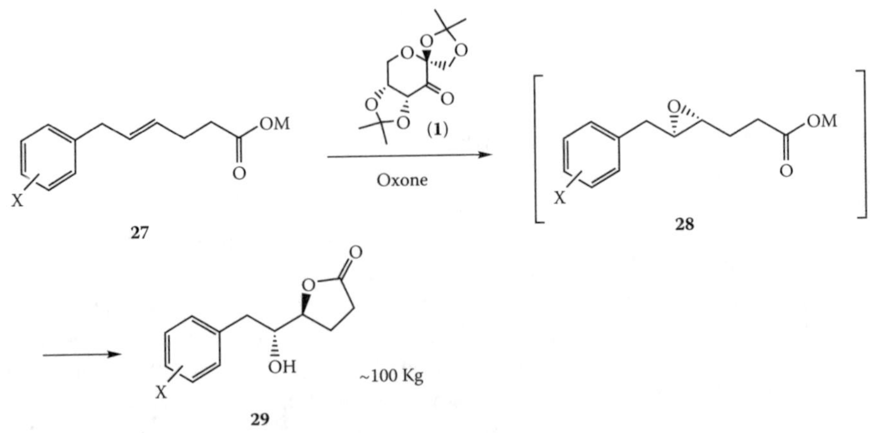
(1)
Oxone
27
28
~100 Kg
29

SCHEME 10.9

10.5 SUMMARY

During the past few years, chiral ketones with various structures have been investigated for asymmetric epoxidation of olefins by a number of laboratories, and significant progress has been made in the field. Because of various undesired processes that compete with the catalytic cycle of the epoxidation (for a detailed discussion see reference 18 and references cited therein) the development of an efficient ketone catalyst requires a delicate balance of the steric and electronic properties of the chiral control elements around the carbonyl group, which has proved to be a challenging task. The studies in our laboratories have shown that the fructose-derived ketone (**1**) is a highly enantioselective catalyst for the epoxidation of *trans*-substituted and trisubstituted olefins with broad substrate scope. The stereochemical outcome of the epoxidation for various olefin systems can be rationalized and predicted with a reasonable level of confidence by a simple spiro transition state model. In addition, the epoxidation conditions are mild and environmentally friendly, and the workup is straightforward. In many cases, the epoxide can be obtained by simple extraction of the reaction mixture with hexane, leaving the ketone catalyst in the aqueous phase. It is important to note that the epoxidation is also amenable to a large scale. Studies with ketones **16–18** show that the dioxirane-mediated asymmetric epoxidation can also be expanded to *cis*- and terminal olefins. The substrate scope of these ketones is expected to be broadened in the near future.

ACKNOWLEDGMENT

We are grateful to the generous financial support from the General Medical Sciences of the National Institutes of Health and the National Science Foundation. I am extremely indebted to my co-workers for their contributions to our program and those who have been identified in the references. I also thank Dr. David Ager for his kind suggestions and input for the writing of this chapter.

REFERENCES

1. For a leading review see: Besse, P., Veschambre, H. *Tetrahedron* 1994, *50*, 8885.
2. For asymmetric epoxidation of aldehydes by chiral ylides see: Li, A-H., Dai, L-X., Aggarwal, V. K. *Chem. Rev.* 1997, *97*, 2341.
3. Johnson, R. A., Sharpless, K. B. In *Catalytic Asymmetric Synthesis*, Ojima, I. Ed., VCH: New York, 1993, p. 103.
4. Katsuki, T., Martin, V. S. *Org. React.* 1996, *48*, 1.
5. Jacobsen, E. N. In *Catalytic Asymmetric Synthesis*, Ojima, I. Ed., VCH: New York, 1993, p. 159.
6. Collman, J. P., Zhang, X., Lee, V. J., Uffelman, E. S., Brauman, J. I. *Science* 1993, *261*, 1404.
7. Mukaiyama, T. *Aldrichimica Acta* 1996, *29*, 59.
8. Katsuki, T. In *Catalytic Asymmetric Synthesis*, Ojima, I. Ed., VCH: New York, 2000, p. 287.
9. Porter, M. J., Skidmore, J. *J. Chem. Soc., Chem. Commun.* 2000, 1215.
10. Lauret, C., Roberts, S. M. *Aldrichimica Acta* 2002, *35*, 47.
11. Nemoto, T., Ohshima, T., Shibasaki, M. *J. Synth. Org. Chem. Jpn.* 2002, *60*, 94.
12. Adam, W., Curci, R., Edwards, J. O. *Acc. Chem. Res.* 1989, *22*, 205.
13. Murray, R. W. *Chem. Rev.* 1989, *89*, 1187.
14. Curci, R., Dinoi, A., Rubino, M. F. *Pure Appl. Chem.* 1995, *67*, 811.
15. Adam, W., Smerz, A. K. *Bull. Soc. Chim. Belg.* 1996, *105*, 581.
16. Adam, W., Saha-Möller, C. R., Zhao, C-G. *Org. React.* 2002, *61*, 219.
17. Denmark, S. E., Wu, Z. *Synlett* 1999, 847.
18. Frohn, M., Shi, Y. *Synthesis* 2000, 1979.
19. Shi, Y. *J. Synth. Org. Chem. Jpn.* 2002, *60*, 342.
20. Shi, Y. In *Modern Oxidation Methods*, Bäckvall, J-E. Ed., Wiley-VCH, in press.
21. Curci, R., Fiorentino, M., Serio, M. R. *J. Chem. Soc., Chem. Commun.* 1984, 155.
22. Curci, R., D'Accolti, L., Fiorentino, M., Rosa, A. *Tetrahedron Lett.* 1995, *36*, 5831.

23. Yang, D., Yip, Y.-C., Tang, M-W., Wong, M.-K., Zheng, J.-H., Cheung, K-K. *J. Am. Chem. Soc.* 1996, *118*, 491.
24. Yang, D., Wang, X.-C., Wong, M.-K., Yip, Y.-C., Tang, M-W. *J. Am. Chem. Soc.* 1996, *118*, 11311.
25. Yang, D., Wong, M.-K., Yip, Y.-C., Wang, X.-C., Tang, M-W., Zheng, J.-H., Cheung, K-K. *J. Am. Chem. Soc.* 1998, *120*, 5943.
26. Adam, W., Zhao, C-G. *Tetrahedron: Asymmetry* 1997, *8*, 3995–3998.
27. Denmark, S. E., Wu, Z., Crudden, C. M., Matsuhashi, H. *J. Org. Chem.* 1997, *62*, 8288.
28. Armstrong, A., Hayter, B. R. *J. Chem. Soc., Chem. Commun.* 1998, 621.
29. Armstrong, A., Hayter, B. R., Moss, W. O., Reeves, J. R., Wailes, J. S. *Tetrahedron: Asymmetry* 2000, *11*, 2057.
30. Armstrong, A., Moss, W. O., Reeves, J. R. *Tetrahedron: Asymmetry* 2001, *12*, 2779.
31. Armstrong, A., Ahmed, G., Dominguez-Fernandez, B., Hayter, B. R., Wailes, J. S. *J. Org. Chem.* 2002, *67*, 8610.
32. Yang, D., Yip, Y.-C., Chen, J., Cheung, K-K. *J. Am. Chem. Soc.* 1998, *120*, 7659.
33. Solladié-Cavallo, A., Bouérat, L. *Org. Lett.* 2000, 2, 3531.
34. Solladié-Cavallo, A., Bouérat, L., Jierry, L. *Eur. J. Org. Chem.* 2001, 4557.
35. Denmark, S. E., Matsuhashi, H. *J. Org. Chem.* 2002, *67*, 3479.
36. Stearman, C. J., Behar, V. *Tetrahedron Lett.* 2002, *43*, 1943.
37. Tu, Y., Wang, Z-X., Shi, Y. *J. Am. Chem. Soc.* 1996, *118*, 9806.
38. Wang, Z-X., Tu, Y., Frohn, M., Shi, Y. *J. Org. Chem.* 1997, *62*, 2328.
39. Wang, Z-X., Tu, Y., Frohn, M., Zhang, J-R., Shi, Y. *J. Am. Chem. Soc.* 1997, *119*, 11224.
40. Tu, Y., Wang, Z-X., Frohn, M., Shi, Y. *Org. Synth.* 2003, *80*, 1 and references cited therein.
41. Gonsalvi, L., Arends, I. W. C. E., Sheldon, R. A. *Org. Lett.* 2002, *4*, 1659.
42. Chen, C-C., Whistler, R. L. *Carbohydr. Res.* 1988, *175*, 265.
43. Edwards, J. O., Pater, R. H., Curci, R., Di Furia, F. *Photochem. Photobiol.* 1979, *30*, 63.
44. Denmark, S. E., Forbes, D. C., Hays, D. S., DePue, J.S., Wilde, R. G. *J. Org. Chem.* 1995, *60*, 1391.
45. Ball, D. L., Edwards, J. O. *J. Am. Chem. Soc.* 1956, *78*, 1125.
46. Goodman, J. F., Robson, P. *J. Chem. Soc.* 1963, 2871.
47. Montgomery, R. E. *J. Am. Chem. Soc.* 1974, *96*, 7820.
48. Frohn, M., Wang, Z-X., Shi, Y. *J. Org. Chem.* 1998, *63*, 6425.
49. Shu, L., Shi, Y. *J. Org. Chem.* 2000, *65*, 8807.
50. Warren, J. D., Shi, Y. *J. Org. Chem.* 1999, *64*, 7675.
51. Wang, Z-X., Shi, Y. *J. Org. Chem.* 1998, *63*, 3099.
52. Zhu, Y., Tu, Y., Yu, H., Shi, Y. *Tetrahedron Lett.* 1998, *39*, 7819.
53. Zhu, Y., Manske, K. J., Shi, Y. *J. Am. Chem. Soc.* 1999, *121*, 4080.
54. Zhu, Y., Shu, L., Tu, Y., Shi, Y. *J. Org. Chem.* 2001, *66*, 1818.
55. Frohn, M., Dalkiewicz, M., Tu, Y., Wang, Z-X., Shi, Y. *J. Org. Chem.* 1998, *63*, 2948.
56. Wang, Z-X., Cao, G-A., Shi, Y. *J. Org. Chem.* 1999, *64*, 7646.
57. Shu, L., Shi. Y. *Tetrahedron Lett.* 1999, *40*, 8721.
58. Shu, L., Shi, Y. *Tetrahedron* 2001, *57*, 5213.
59. Wang, Z-X., Shu, L., Frohn, M., Tu, Y., Shi, Y. *Org. Synth.* 2003, *80*, 9.
60. Payne, G. B., Deming, P. H., Williams, P. H. *J. Org. Chem.* 1961, *26*, 659.
61. Arias, L. A., Adkins, S., Nagel, C. J., Bach, R. D. *J. Org. Chem.* 1983, *48*, 888.
62. Baumstark, A. L., McCloskey, C. J. *Tetrahedron Lett.* 1987, *28*, 3311.
63. Bach, R. D., Andres, J. L., Owensby, A. L., Schlegel, H. B., McDouall, J. J. W. *J. Am. Chem. Soc.* 1992, *114*, 7207.
64. Houk, K. N., Liu, J., DeMello, N. C., Condroski, K. R. *J. Am. Chem. Soc.* 1997, *119*, 10147.
65. Jenson, C., Liu, J., Houk, K. N., Jorgensen, W. L. *J. Am. Chem. Soc.* 1997, *119*, 12982.
66. Armstrong, A., Washington, I., Houk, K. N. *J. Am. Chem. Soc.* 2000, *122*, 6297.
67. Deubel, D. V. *J. Org. Chem.* 2001, *66*, 3790.
68. Frohn, M., Zhou, X., Zhang, J-R., Tang, Y., Shi, Y. *J. Am. Chem. Soc.* 1999, *121*, 7718.
69. Tu, Y., Wang, Z-X., Frohn, M., He, M., Yu, H., Tang, Y., Shi, Y. *J. Org. Chem.* 1998, *63*, 8475.
70. Wang, Z-X., Miller, S. M., Anderson, O. P., Shi, Y. *J. Org. Chem.* 2001, *66*, 521.
71. Tian, H., She, X., Shi, Y. *Org. Lett.* 2001, *3*, 715.
72. Wu, X-Y., She, X., Shi, Y. *J. Am. Chem. Soc.* 2002, *124*, 8792.

73. Tian, H., She, X., Shu, L., Yu, H., Shi, Y. *J. Am. Chem. Soc.* 2000, *122*, 11551.
74. Tian, H., She, X., Yu, H., Shu, L., Shi, Y. *J. Org. Chem.* 2002, *67*, 2435.
75. Tian, H., She, X., Xu, J., Shi, Y. *Org. Lett.* 2001, *3*, 1929.
76. Hickey, M., Goeddel, D., Crane, Z., Shi, Y. *Proc. Natl. Acad. Sci. USA* 2004, *101*, 5794.
77. Shu, L., Wang, P., Gan, Y., Shi, Y. *Org. Lett.* 2003, *5*, 293.
78. Shu, L., Shen, Y-M., Burke, C., Goeddel, D., Shi, Y. *J. Org. Chem.* 2003, *68*, 4963.
79. Bluet, G., Campagne, J-M. *Synlett* 2000, 221.
80. Tokiwano, T., Fujiwara, K., Murai, A. *Synlett,* 2000, 335.
81. Hioki, H., Kanehara, C., Ohnishi, Y., Umemori, Y., Sakai, H., Yoshio, S., Matsushita, M., Kodama, M. *Angew. Chem., Int. Ed.* 2000, *39*, 2552.
82. Xiong, Z., Corey, E. J. *J. Am. Chem. Soc.* 2000, *122*, 4831.
83. Xiong, Z., Corey, E. J. *J. Am. Chem. Soc.* 2000, *122*, 9328.
84. McDonald, F. E., Wang, X., Do, B., Hardcastle, K. I. *Org. Lett.* 2000, *2*, 2917.
85. Morimoto, Y., Iwai, T., Kinoshita, T. *Tetrahedron Lett.* 2001, *42*, 6307.
86. Guz, N. R., Lorenz, P., Stermitz, F. R. *Tetrahedron Lett.* 2001, *42*, 6491.
87. Shen, K-H., Lush, S-F., Chen, T-L., Liu, R-S. *J. Org. Chem.* 2001, *66*, 8106.
88. McDonald, F. E., Wei. S. *Org. Lett.* 2002, *4*, 593.
89. McDonald, F. E., Bravo, F., Wang, X., Wei, X., Toganoh, M., Rodriguez, J. R., Do, B., Neiwert, W. A., Hardcastle, K. I. *J. Org. Chem.* 2002, *67*, 2515.
90. Hoard, D. W., Moher, E. D., Martinelli, M. J., Norman, B. H. *Org. Lett.* 2002, *4*, 1813.
91. Kumar, V. S., Aubele, D. L., Floreancig, P. E. *Org. Lett.* 2002, *4*, 2489.
92. Morimoto, Y., Takaishi, M., Iwai, T., Kinoshita, T., Jacobs, H. *Tetrahedron Lett.* 2002, *43*, 5849.
93. Olofsson, B., Somfai, P. *J. Org. Chem.* 2002, *67*, 8574.
94. Altmann, K-H., Bold, G., Caravatti, G., Denni, D., Flörsheimer, A., Schmidt, A., Rihs, G., Wartmann, M. *Helv. Chim. Acta* 2002, *85*, 4086.
95. Madhushaw, R. J., Li, C-L., Su, H-L., Hu, C-C., Lush, S-F., Liu, R-S. *J. Org. Chem.* 2003, *68*, 1872.
96. Olofsson, B., Somfai, P. *J. Org. Chem.* 2003, *68*, 2514.
97. Bravo, F., McDonald, F. E., Neiwert, W. A., Do, B., Hardcastle, K. I. *Org. Lett.* 2003, *5*, 2123.
98. Heffron, T. P., Jamison, T. F. *Org. Lett.* 2003, *5*, 2339.
99. Zhang, Q., Lu, H., Richard, C., Curran, D. P. *J. Am. Chem. Soc.* 2004, *126*, 36.
100. Smith, A. B. III, Fox, R. J. *Org. Lett.* 2004, *6*, 1477.
101. Ager, D., DSM Pharma Chemicals, private communication.

11 Ozonolysis in the Production of Chiral Fine Chemicals

Janine McGuire, Gary Bond, and Peter J. Haslam

CONTENTS

11.1   INTRODUCTION

Ozonolysis is a convenient and highly effective method of oxidatively cleaving a double bond to give an array of compounds including carboxylic acids, ketones, aldehydes, and alcohols. The final product depends on whether the intermediate ozonide is exposed to oxidative or reducing conditions.[1,2] Ozonolysis reactions are used during a synthetic process because they can proceed under remarkably mild conditions and display a high level of selectivity. For chemical efficiency, there are few reactions that can match ozonolysis. The reaction, however, does not usually introduce new stereogenic centers and these have to be present within the substrate. There are a few exceptions, as illustrated in Section 11.4. A number of examples in which ozonolysis is a key transformation in the synthesis of pharmaceutically important compounds are highlighted in this chapter. The examples have been collected into structural types.

11.2 AMINO ACIDS AND DERIVATIVES

Over recent years amino acids have enjoyed unprecedented renaissance in virtually all disciplines. A number of nonproteinogenic amino acids have demonstrated biological and pharmacologic activity,[3] including their incorporation in semi-synthetic penicillins, cephalosporins, and biologically active peptides. Other chapters in this book cover other approaches to amino acids (see Chapters 2, 3, 12–16, 19, and 25).

11.2.1 L-ISOXAZOLYLALANINE

One approach to unnatural amino acids is to use a readily available amino acid, such as L-phenylalanine, as the starting material. The Birch reduction of L-phenylalanine (1) was carried out with lithium in ammonia, followed by acylation of the amino group to produce compound 2, which was further esterified to produce the cyclohexa-1,4-dienyl-L-alanine derivative 3 (Scheme 11.1). The ozonolysis step of the reaction was carried out at –78°C in a dichloromethane solution presaturated with ozone to reduce the extent of oxidation of the diene 3 to produce 4. Cyclization was then carried out by the introduction of either hydroxylamine hydrochloride to produce the isoxazol-5-ylalanine derivative 5 or phenylhydrazine to give a 1:1 mixture of (1-phenylpyrazol-3-yl)alanine derivative 6 and the (1-phenylpyrazol-2-yl)alanine derivative 7.[4,5]

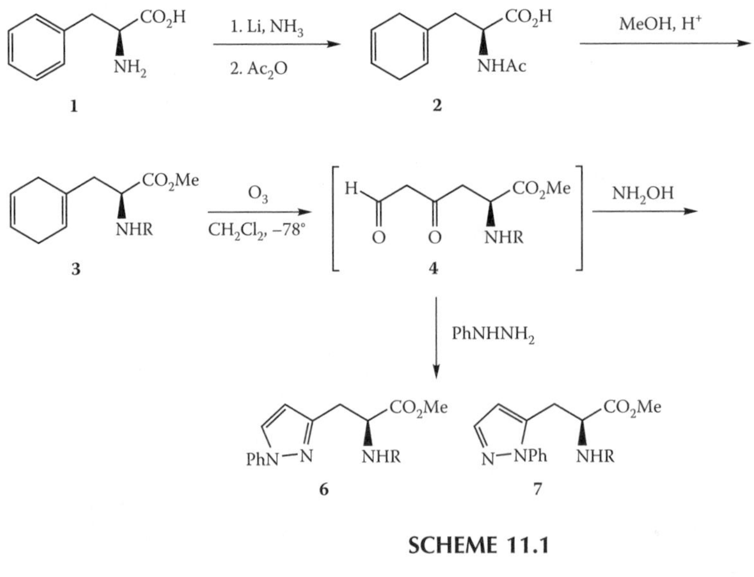

SCHEME 11.1

This approach provides a convenient route for the production of new optically active amino acids from readily available amino acids by a relatively simple reduction–oxidation reaction.

11.2.2 α-AMINO ALDEHYDES

α-Amino aldehydes derived from natural amino acids have also shown great potential as enzyme inhibitors and peptide isosteres.[6] However, although theoretically the synthesis of α-amino aldehydes is relatively straightforward, their strong tendency to polymerize, even in acidic solution, can make their isolation problematic. Ozonolysis of α-vinyl amines leads to α-amino aldehydes, particularly when the molecule carries a sidechain incorporating a functional group capable of participating in ring formation.[7] Thus, ozonolysis of 3-amino-3-vinylpropanoic acid hydrochloride (**8**), in 2M HCl at 0°C, resulted in an ozonide **9** that underwent further hydrolysis, followed by cyclization, under the reaction conditions to produce the cyclic form of aspartate 1-semialdehyde, 4-amino-5-hydroxy-4,5-dihydrofuran-2-one hydrochloride (HAD) (**10**) (Scheme 11.2).

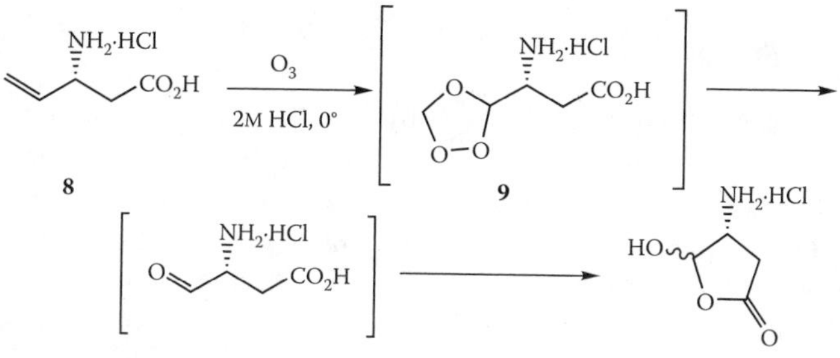

SCHEME 11.2

11.2.3 β-HYDROXY-γ-AMINO ACIDS

β-Hydroxy-γ-amino acids **11** (statines) have received much attention in recent years (see Chapter 31) because they are key structural units in peptidometric protease inhibitors.[8] These γ-amino acid compounds are increasing in interest because aspartic proteases, including pepsin, renin, human immunodeficiency virus-1 (HIV-1) and HIV-2 proteases, plasmepsin, cathepsin D, and β-secretase, are targets for peptidodometric inhibitors.[9] The statine structure acts as an isostere, restricting the conformation of a dipeptide unit within the protease.[10,11] Although synthesis of statines such as **12** is well-documented,[12,13] there are few examples of α-substituted statines, which would be of more interest with respect to restricting the conformation of a dipeptide unit. The majority of routes described to date are via ring opening of α,β-epoxy esters[14] and aldol-type reactions.[15]

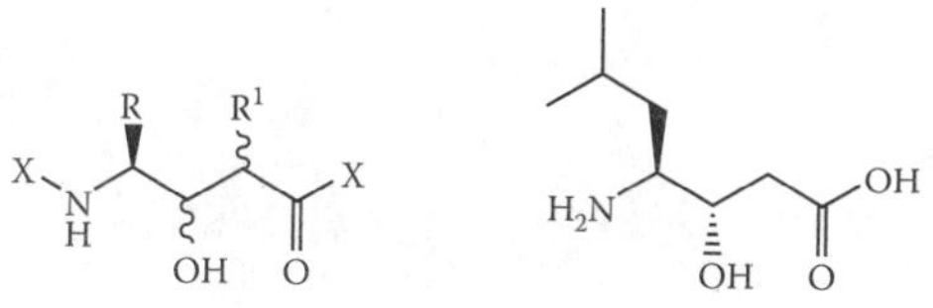

Le Carrer-Le Goff and co-workers have reported an expeditious synthetic route to α-substituted statines by the well-defined metal-mediated allylation of N-protected α-amino aldehydes followed by the ozonolysis of the double bond (Scheme 11.3).[16] Ozonolysis of the double bond in **13**, carried out in dichloromethane/methanol, at −78°C in the presence of sodium hydroxide permits the intermediate ozonide to be converted directly to the statine methyl ester **14**.[17]

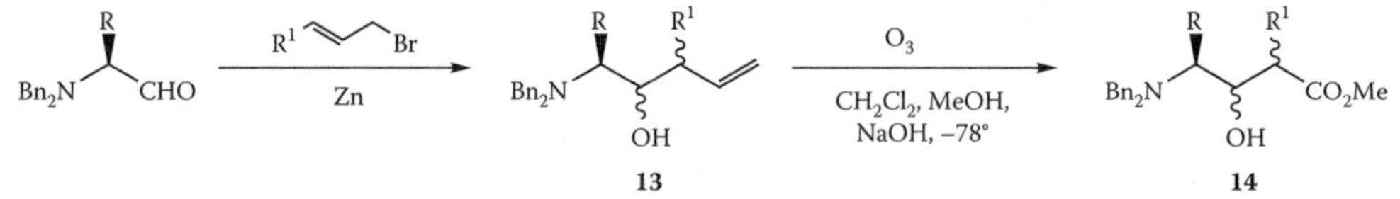

SCHEME 11.3

Amino aldehydes provide useful starting materials for the sequence of Scheme 11.3 because they can easily be prepared from α-amino acids in enantiomerically pure forms and are, therefore, useful chiral building blocks.[18] The addition of organometallic allylic reagents (including Zn, used in the reaction described) to such compounds can produce homoallylic alcohols **13** with a high *anti*-diastereoselectivity.

11.2.4 β-Amino-γ-Butyrolactones

β-Amino-γ-butyrolactones are components of many biologically active natural products and pharmaceuticals, such as antibiotic and antifungal peptides,[19,20] antimalarial alkaloids,[21] gastroprotective drugs,[22] inhibitors of phosphodiesterase,[23] and HIV-1 protease.[24] Additionally, β-amino-γ-butyrolactones are important intermediates in the production of a variety of β-amino acids[25] and β-lactam antibiotics.[26] Because there are only a few general stereoselective syntheses to β-amino-γ-butyrolactones, the development by Roos and colleagues[27] of a synthetic route not reliant on optically active compounds derived from the "chiral pool" as a starting material is a welcome addition to the literature. The route exploits the use of the silylated optically active cyanohydrins **16**, derived from aldehydes **15**. The cyanohydrin **16** is reacted with allylGrignard reagents to produce an imino intermediate that is hydrogenated *in situ* with NaBH$_4$ to produce the amino alcohol **17**. After acylation with acetic anhydride, ozonolysis in methanol at −78°C is then used to oxidatively cleave the terminal double bond resulting in the production of an aldehyde **18**, which can be further reacted to the desired β-amino-γ-butyrolactone (Scheme 11.4).

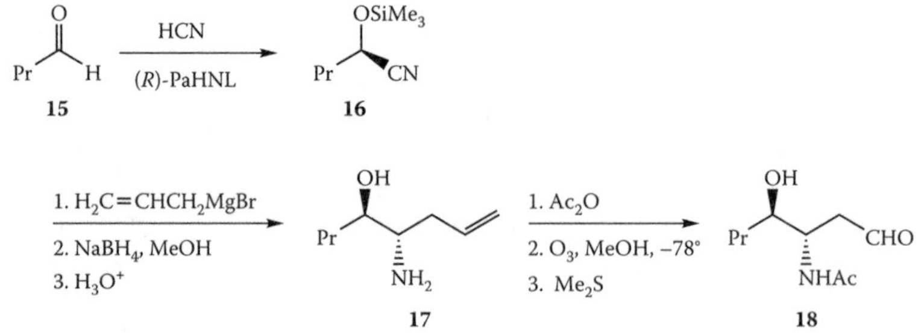

SCHEME 11.4

11.3 TERPENES, DITERPENES, AND SESQUITERPENES

The general chemistry of these classes of compounds is discussed in Chapter 5.

11.3.1 Drimanes

Drimanes, including the sesquiterpenoids polygodial and warburganal, have shown interesting biological properties that have stimulated a number of total and semi-synthetic pathways.[28–30] Semi-synthetic routes have an advantage of using a compound with a relatively advanced structure that simply requires modification, as opposed to the long, and often complicated, procedure of a fully synthetic route. Semi-synthetic routes offer the added advantage of producing only one stereoisomer of the desired product.

Barrero and colleagues have developed a synthetic route to the drimanic sesquiterpenoids **19**, from which the enal-aldehyde grouping is chemically easy to obtain, that incorporates two ozonolysis steps in the reaction.[31] Owing to its functionality and chirality, *trans*-communic acid **20**, found in important quantities in species belonging to the *cupresaceas* family and easily obtained using apolar extraction, was used as a starting material. Selective ozonolysis of the double bond Δ^{12} could be carried out at –78°C in dichloromethane to produce the aldehyde **21** (Scheme 11.5). Acylation of the aldehyde **21**, followed by epoxidations of the Δ^{17} double bond, gave **22** that permits the second ozonolysis reaction to afford the drimane **19**.

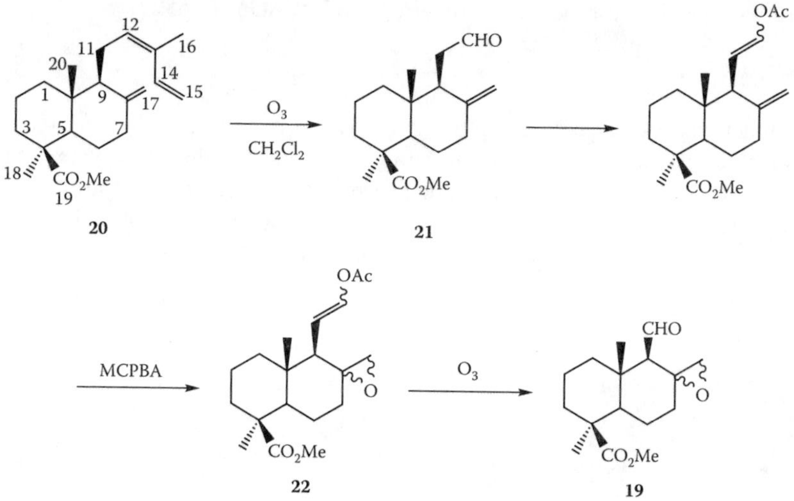

SCHEME 11.5

11.3.2 IMMUNOSUPPRESSANTS

Immunosuppressants, such as FK506, repamycinans cyclosporine A (CsA), attracted intensive research interests during the late 1980s and early 1990s,[32,33] with the aim to understand the mechanism of action.[34] Renewed interest in these compounds developed from the discovery that some of the compounds, such as FKBP-12, have the potential to penetrate the blood–brain barrier and have also demonstrated a capacity for nerve regeneration activity.

Synthetic routes to this class of compounds have often been long and complicated; the original route for AG5473/5507 required 20 steps. Therefore, the ability to reduce the number of steps has considerable advantage. Guo and colleagues have developed an 11-step synthetic route to AG5473/5507.[35] This sequence incorporates an effective ozonolysis protocol for the introduction of an aldehyde moiety **23** that is an essential step to the production of the tricyclic diazamide core **24** (Scheme 11.6). In the reaction sequence, a protected acetylene is added to acetoxyamide **25**, to

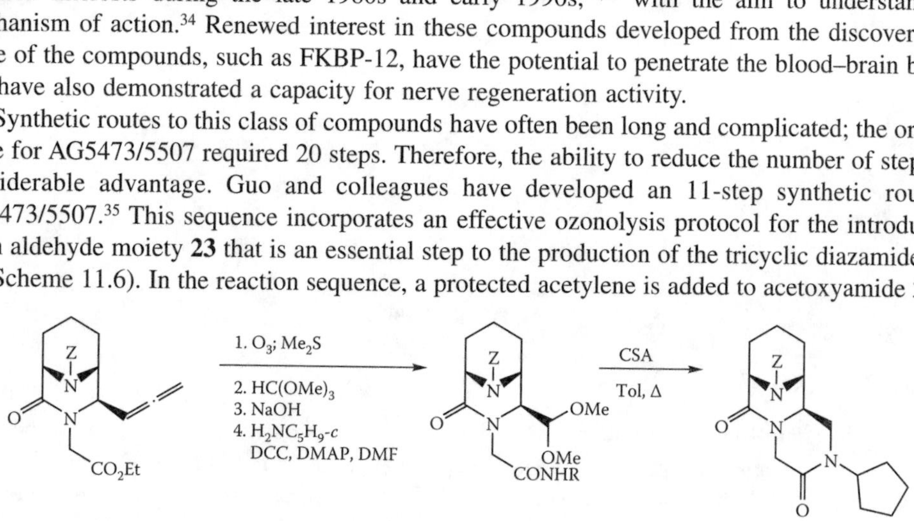

SCHEME 11.6

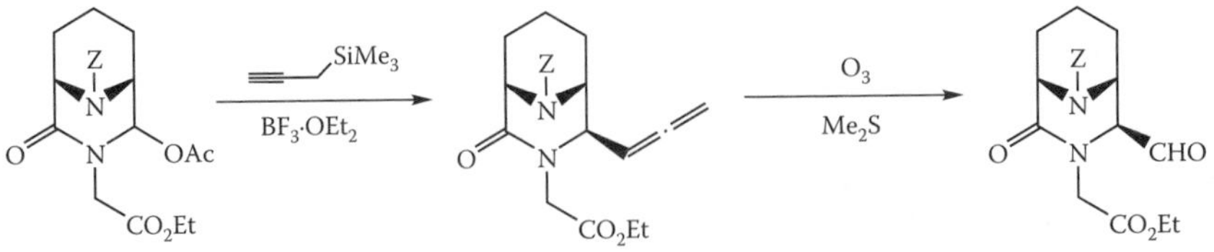

SCHEME 11.7

give the required allene **26,** which underwent smooth ozonolysis to the aldehyde **27** (Scheme 11.7), permitting further reaction to produce the required product AG5473/5507.[35]

11.3.3 Camptothecin

The antitumor activity displayed by camptothecin (**28**) and its analogues has led to a great deal of interest in such compounds.[36] Camptothecin is a complicated alkaloid to isolate, and analogues, which are of biological interest, are synthesized from the natural product itself.[37] This need to access a wide range of analogues has led to a need for synthetic studies directed toward the development of a concise, yet inexpensive, total synthesis of camptothecin.

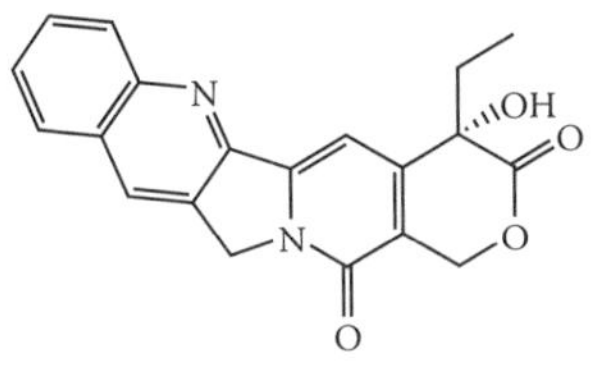

One synthetic approach has been developed by Shen and colleagues.[38] Modification of the readily available pyridone **29** to form the benzylidene acid **30** permitted ozonolysis to be carried out, in a methanol/dichloromethane mixed solvent, followed by esterification; this resulted in the formation of **31** (Scheme 11.8). This compound **31** could then be further transformed into camptothecin (**28**).

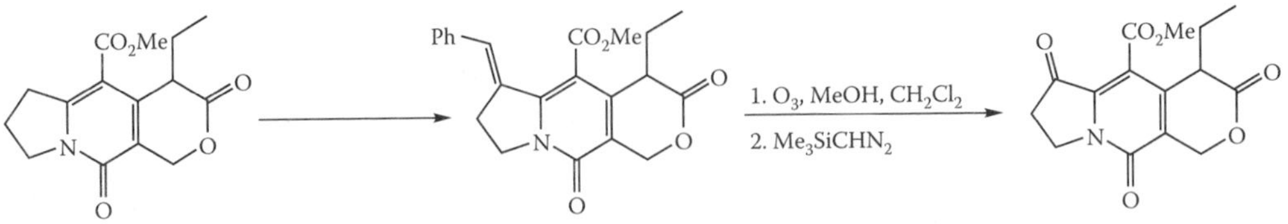

SCHEME 11.8

11.3.4 Tachykinin Antagonists

The ability to control pain and inflammation is an area of continued interest within the pharmaceutical industry. Sensoneuropeptide tachykinins, distributed in the peripheral and central nervous systems, are known to be involved in neurologic inflammation, pain transmission, and bronchoconstriction.[39] The activation of NK-1 (neurokinin-1), NK-2, and NK-3 receptors can mediate the effects of these tachykinins, resulting in a search for new antagonists as a means of controlling pain and inflammation. Pfizer has developed a class of nonpeptidic NK-2 antagonists consisting of a δ-lactam **32**.[40] As reported by Allan and colleagues, the synthesis of such tachykinin antagonists can be produced in a short and efficient 4-step sequence starting from a chiral enol acetate **33**.[41] Oxidative cleavage of the enol acetate **33** by ozonolysis, in a mixed methanol/dichloromethane solvent, followed by reductive workup produced an aldehyde ester **34**, which could be further reacted to produce the required NK-2 antagonist **32** (Scheme 11.9).

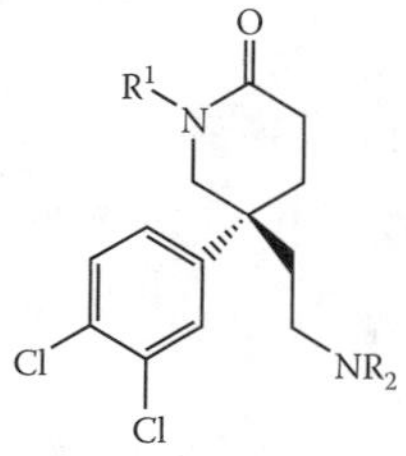

SCHEME 11.9

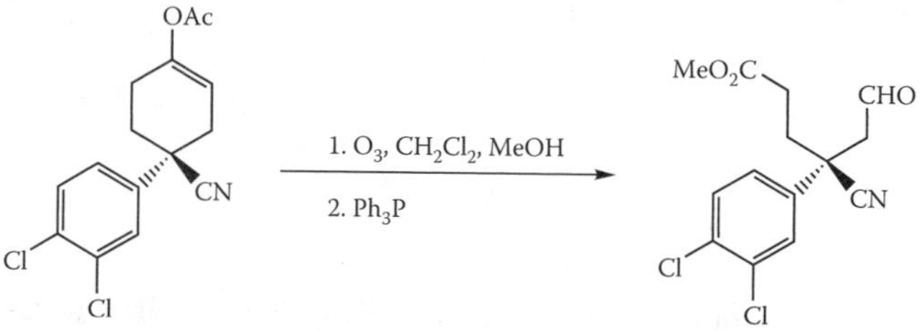

11.3.5 ANTIINFLAMMATORY DRUGS

Naturally occurring as well as synthetic enantiomerically pure 2-arylalkanoic acids are playing an increasing role as therapeutic agents. Of particular interest has been their emergence as an important class of nonsteroidal antiinflammatory agents dominated by α-arylpropionic acids[42] that includes ibuprofen, naproxen, ketoprofen, and flurbiprofen. Of significance in the production of many of these compounds is the role played by ozonolysis.

Since its introduction in 1969,[43] ibuprofen has been marketed in more than 120 countries. However, manufacturing routes used in its production lead to a racemic mixture, although it is known that the more potent form is the *S*-enantiomer (see Chapter 6). Acemoglu and colleagues have developed a route to (*S*)-ibuprofen (**35**) that incorporated a palladium-catalyzed allylic substitution reaction followed by ozonolysis.[44] A substituted allyl acetate **36** can undergo palladium-catalyzed allylic substitution with bis(phenylsulfonyl)methane, with good asymmetric induction, to afford the product **37** (see Chapter 22). Conversion of the alkene into a carboxylic acid, via an ozone-induced oxidative cleavage reaction, followed by removal of the sulfonyl groups provides a route to the required α-arylpropionic acids. Simple modification of the enantiomerically enriched substitution products provides a convenient route to a range of analogous compounds in this class of compounds. In the synthesis of ibuprofen (**35**) (Scheme 11.10), the enantioselectivity of the reaction is dependent on the reaction conditions. During the oxidative cleavage of **37** even ozonolysis followed by an oxidative workup lead to an unsatisfactory mixture of products. However, ozonolysis in a dichloromethane/methanol mixed solvent system at −78°C, followed by a reductive workup with sodium borohydride, produces the alcohol **38**, which, following the removal of the sulfonyl groups to give **38**, could be subsequently oxidized to the required carboxylic acid **35**.

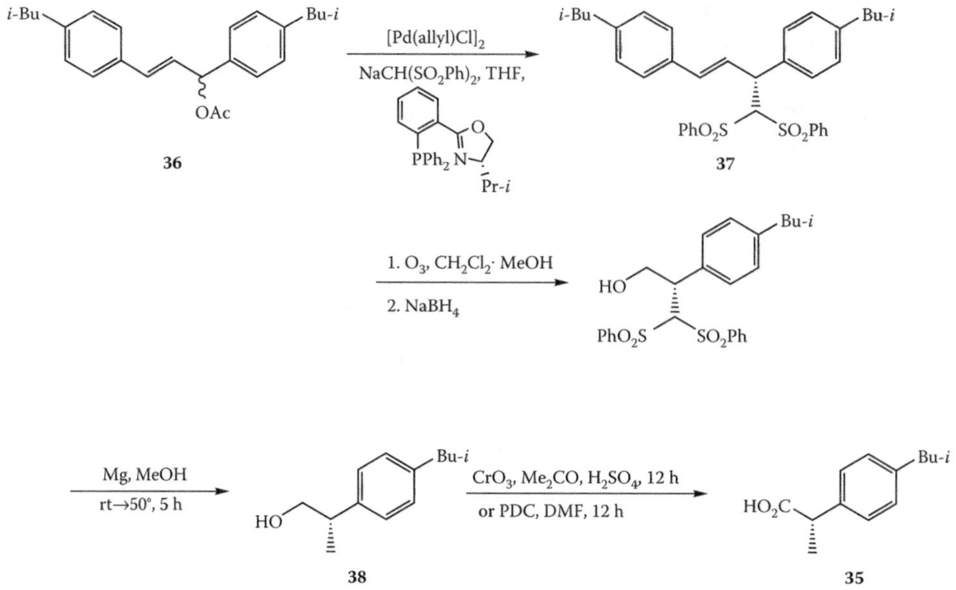

SCHEME 11.10

This reaction once again demonstrates the versatility that can be displayed by ozonolysis reactions by a simple variation in the experimental conditions used.

11.3.6 Prostaglandin Endoperoxides

The prostaglandin endoperoxides **39** (PGG$_2$) and **40** (PGH$_2$) are central to the biosynthesis of the primary prostaglandins, leukotrienes, and thromboxanes. These classes of compounds have a wide range of important biological activities. Because they also exhibit extreme instability, more stable analogues are required. Although most common routes involve either modification of naturally occurring prostaglandins or Diels-Alder approaches, Larock and colleagues have developed an allyl palladium reaction coupled with an ozonolysis to synthesize compound **41** that incorporates the cis-5,6-double bond found in naturally occurring prostaglandins.[45]

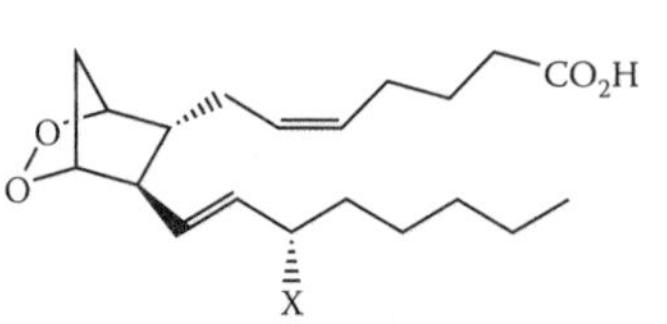

The two sidechains of the prostaglandin are readily introduced by the reaction of norbornene with an allylpalladium compound followed by reaction within an optically active lithium acetylide. The key step of the overall reaction depends on the selective oxidation of the allylic derivative **42** to an aldehyde **43** (Scheme 11.11).[46] The reaction exploits the fact that ozone is able to selectively oxidize a carbon–carbon double bond in the presence of an acetylene.[46] This is achieved by reaction at −78°C in dichloromethane in the presence of pyridine to moderate the reactivity of ozone.

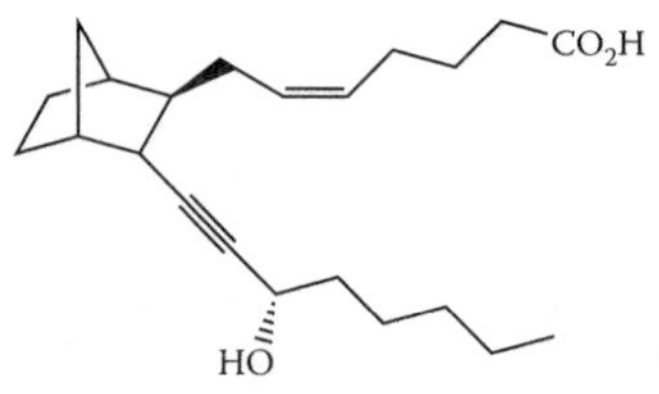

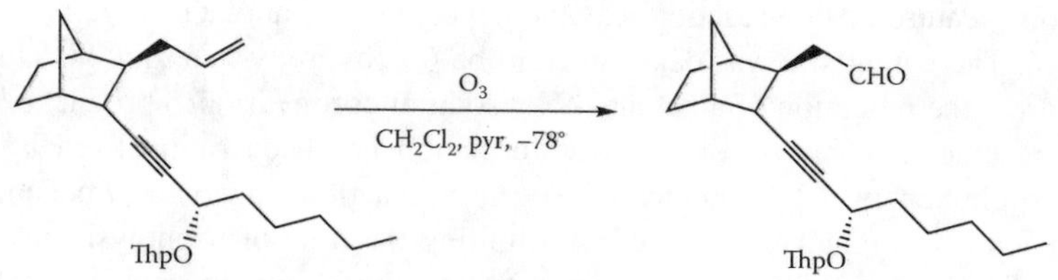

SCHEME 11.11

11.3.7 EPOXYCHOLESTEROL

Increasing evidence has shown that 24(S),25-epoxycholesterol (**44**) participates in the natural regulation of cholesterol metabolism.[47] Escalating evidence of the biological importance associated with **44** has made access to substantial quantities of this epoxide necessary, facilitating the development of a stereoselective synthetic route. A number of synthetic routes to **44** have been described[48,49] including one from the least costly of the steroidal starting materials stigmasterol (**45**).[50] The method involves the preparation of *i*-steroid **46** and ozonolysis in a mixed dichloromethane/methanol solvent mixture, at –78°C followed by treatment with sodium borohydride, to give compound **47** (Scheme 11.12). Further synthetic steps can then be permitted to produce epoxycholesterol (**44**).

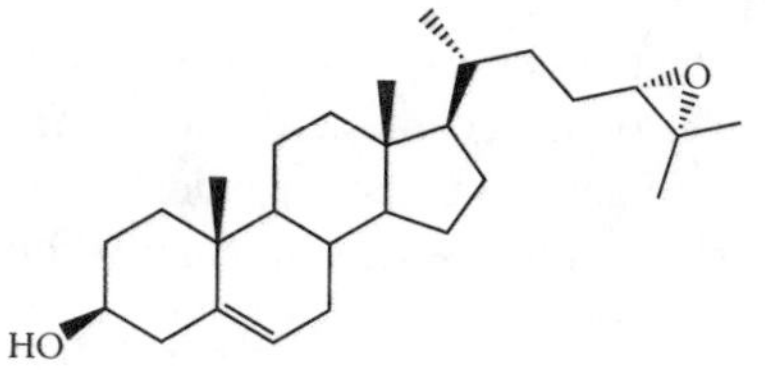

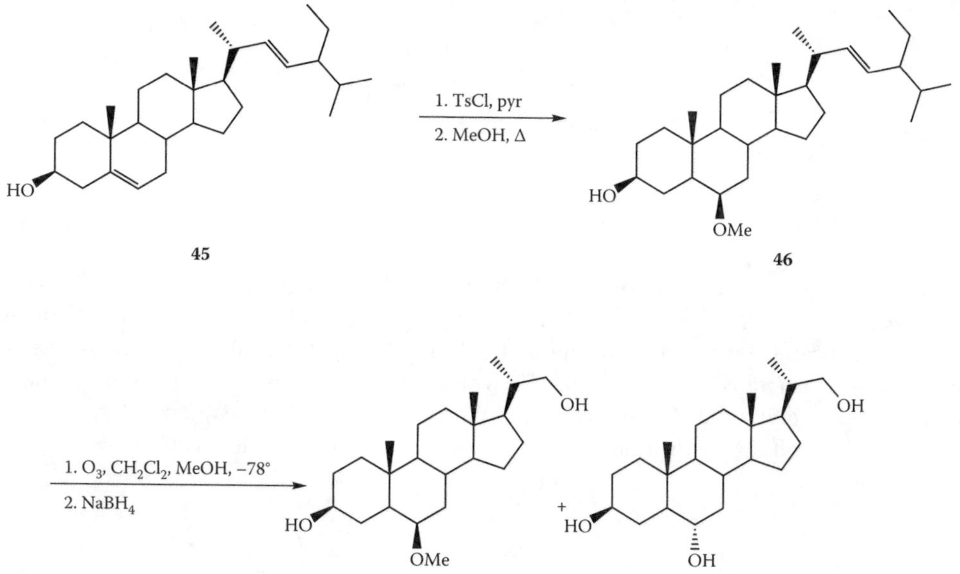

SCHEME 11.12

An interesting feature of this reaction was the appearance of another, more polar, product on ozonolysis of **46**. The compound was determined to be 6-α-hydroxy-*i*-steroid **48**. This compound can be formed from the reduction of a ketone **49** by sodium borohydride, a product often observed on ozonolysis of ethers.[51] This type of ketone formation has been rationalized by the type of mechanism shown in Scheme 11.13 and tends to occur in reactions in which carbocation formation at the site of oxidation is relatively favorable.[51] Limiting the time of ozonolysis has the effect of suppressing the formation of the ketone, demonstrating the ability to control the formation of products by controlling the time of ozonolysis.

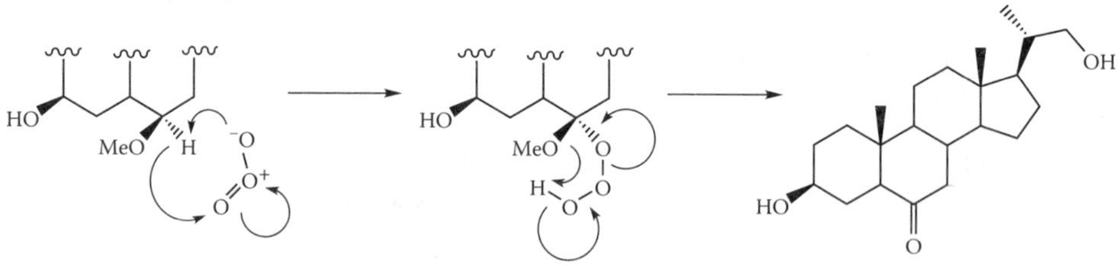

SCHEME 11.13

11.3.8 PRAVASTATIN

Another approach to the reduction in blood cholesterol levels has been the development of prav-astatin (**50**) and its congeners as potent inhibitors of 3-hydroxy-3-methylglutaryl coenzyme A (HMG Co-A) reductase, the rate-limiting enzyme in cholesterol biosynthesis.[52] It has also been discovered that the biological activity of such compounds is dependent largely on the β-hydroxy-δ-lactone moiety within the compound.[53] This led to the development of more potent compounds, such as NK-104 (**51**), a congener of pravastatin.

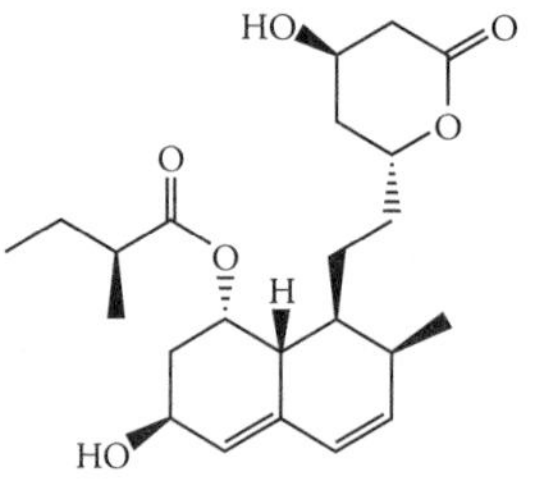

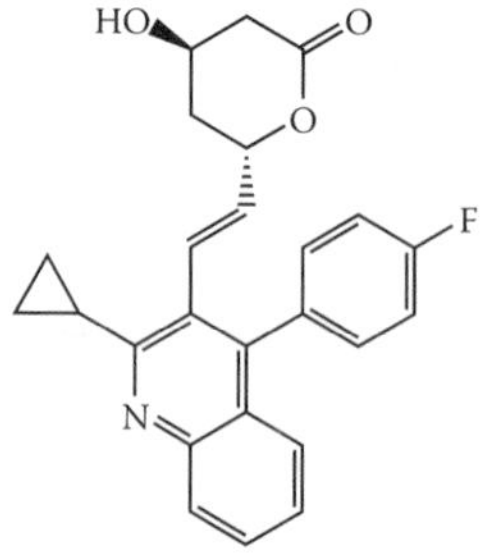

The major synthetic problem associated with the synthesis of compounds such as NK-104 is the stereoselective construction of the lactone moiety and its connection to an aromatic or ethylenic core. This has been overcome by Minami and colleagues with the enantioselective synthesis of methyl 6-oxo-3,5-isopropylidenehexanoate (**52**) by ozonolysis of the ester **53**, followed by reductive workup using dimethyl sulfide (Scheme 11.14).[54] With the required aldehyde **52** now formed, further reaction could permit the production of NK-104 **51** to take place enantioselectively.

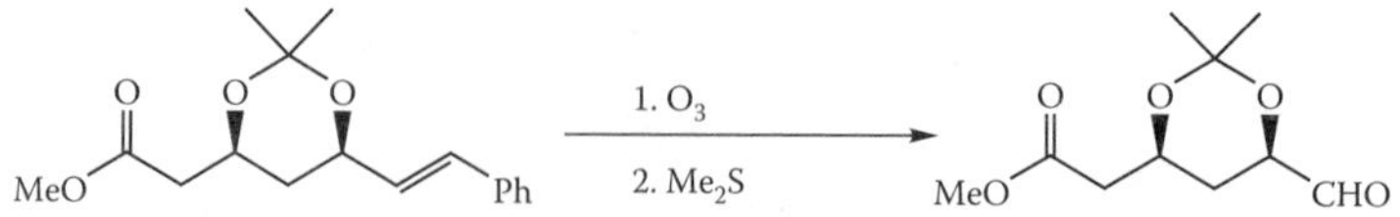

SCHEME 11.14

Although just the synthesis of **51** was demonstrated here, the method is equally applicable for the synthesis of a variety of analogues of this compound, ensuring an effective route to a number of potentially useful inhibitory drugs.

11.4 ANTIMALARIAL COMPOUNDS

11.4.1 ARTEMISININ

Malaria is one of the most prevalent diseases in the world. Although there are numerous drugs on the market for both the treatment and prevention of the disease, multiple drug resistance by parasites is now ubiquitous in all parts of the world where malaria is endemic.[55] Therefore, there is an increasing need for the development of new antimalarial drugs, especially ones that possess novel modes of action. The antimalarial properties of nonalkaloidal compounds such as artemisinin (**54**) and yingzhaosu A (**55**) have attracted considerable attention because of the limited availability of the natural product by extraction from the Chinese *Artemisia annua* plant.[56–58]

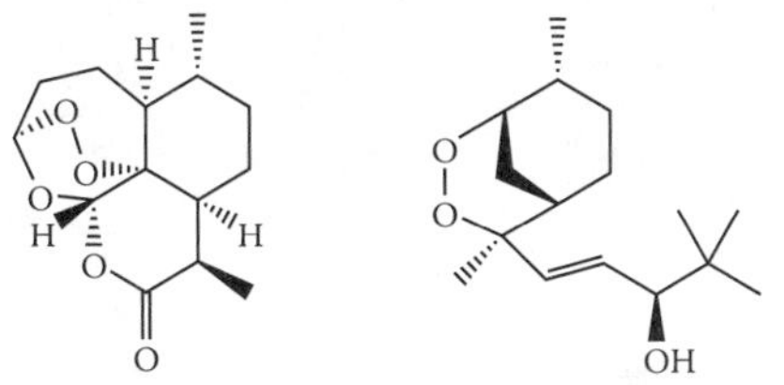

A number of total synthetic routes to artemisinin, and hence analogues of the compound, have been reported,[59–63] a key step in many of these chemical syntheses being the ozonolysis of a vinylsilane that leads, transiently, to an α-hydroperoxycarbonyl moiety[64] and, hence, to the desired product. The synthesis can proceed from a number of starting materials including the 10-step stereoselective transformation[61] of compounds such as (*R*)-(+)-pulegone to the 2β,3α-hexanone **56**. In turn, the ketone **56** can be converted to the substituted vinylsilane **57** (Scheme 11.15). Subsequent ozonolysis of **57** produces the desired polyoxatetracyclic compound **54** in a one-pot procedure that involves the sequential treatment with ozone followed by wet acidic silica gel to affect a complex process of deprotection and multiple cyclizations.

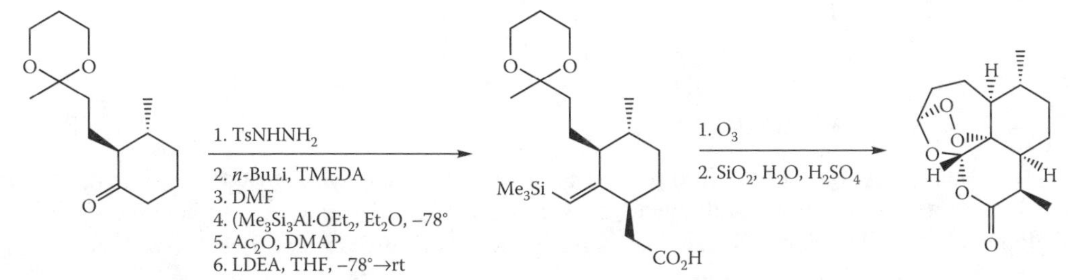

SCHEME 11.15

Variation in substituents generates a variety of analogues of artemisinin.[59,62] Ozonolysis is carried out at low temperatures (–70°C to –80°C) in a reaction solvent chosen to assure compatibility with both ozone and the vinylsilane. Generally, lower alcohols such as methanol, or haloalkanes such as dichloromethane, or a combination of both types of solvent are used. The reaction is carried out by mixing the vinylsilane with the reaction mixture followed by controlled addition of ozone to avoid an excess. Good results are obtained when the amount of ozone added is limited to less than 1.25 equivalents based on the amount of vinylsilane added.

A second synthetic route to artemisinin (**54**) is characterized by the inclusion of two ozonolysis steps within the sequence of reactions.[59] The route uses the unsaturated bicyclic ketone **58**, which again is converted to a vinylsilane **59** (Scheme 11.16). The subsequent ozonolytic cleavage of its olefinic bond yields a member of a family of unique carboxyl and carbonyl-substituted vinylsilanes **60**. A final one-pot ozonolysis–acidification step closes the oxygen-containing ring structure, producing the required compound **54**. Vinylsilanes **60** can be subjected to a wide range of reactions prior to the final ozonolysis step to afford a number of artemisinin analogues. Ozonolysis in both synthetic steps is carried out under similar conditions, with the temperature preferentially being in the range of –70°C to –80°C and the solvent used being methanol, dichloromethane, or a combination of both. Controlling the flow of ozone and maintaining a limit of not more than 1.25 equivalents, based on the amount of vinylsilane present, gives higher yields.

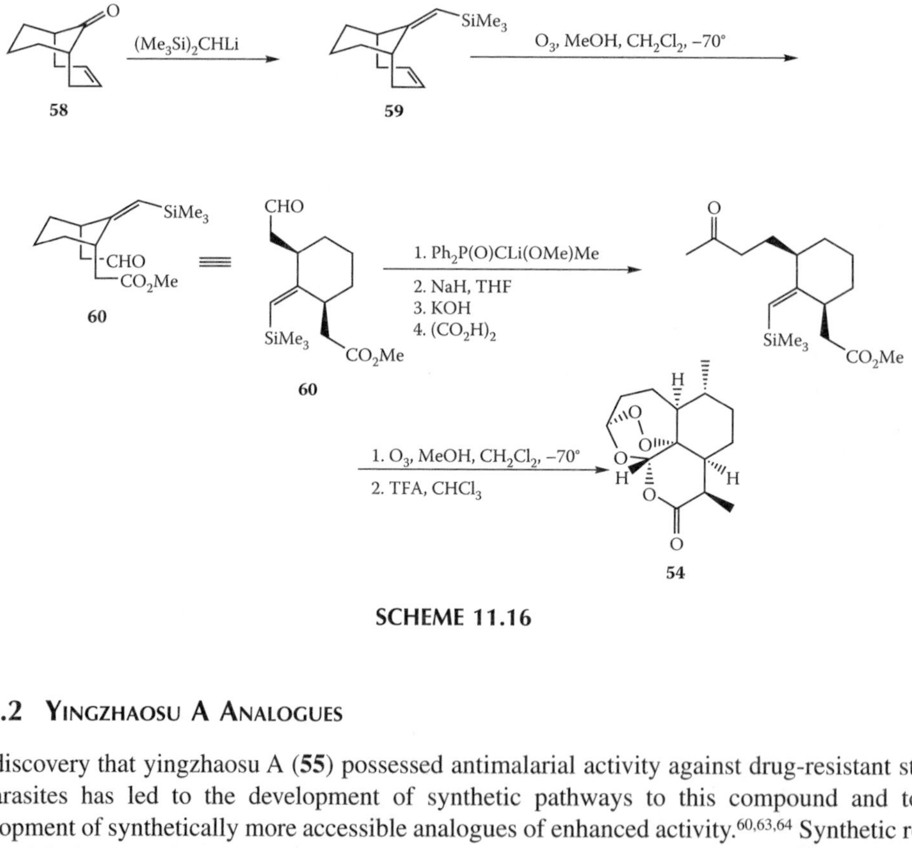

SCHEME 11.16

11.4.2 Yingzhaosu A Analogues

The discovery that yingzhaosu A (**55**) possessed antimalarial activity against drug-resistant strains of parasites has led to the development of synthetic pathways to this compound and to the development of synthetically more accessible analogues of enhanced activity.[60,63,64] Synthetic routes that exploit the ozonolysis of unsaturated hydroperoxy compounds such as the ketals[57] **61** and the acetals[63] **62** have been developed (Schemes 11.17 and 11.18). Using the commercially available derivatives of *R*-carvone **63** and **64**, regioselective mono-ozonolysis produces the desired hydroperoxides **61** and **62**. The hydroperoxide ketal **61** can be converted to analogue **65** of yingzhaosu A by a base-catalyzed intramolecular cyclization reaction. Ozone-mediated cyclization of the hydroperoxide acetal **62**, in 2,2,2-trifluoroethanol (TFE) at 0°C, affords the hydroperoxy-substituted yingzhaosu A analogue **66**.[65]

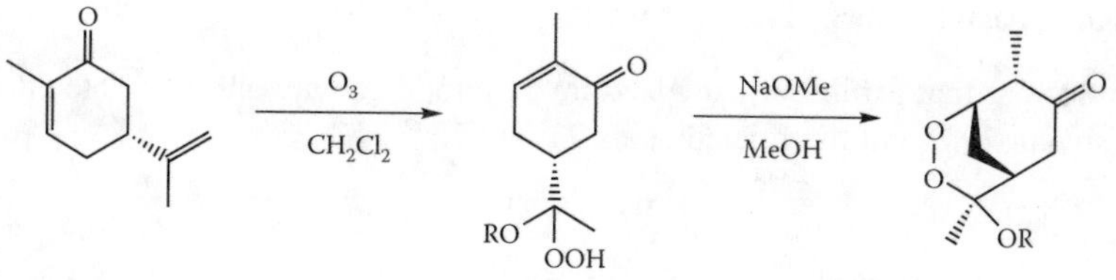

SCHEME 11.17

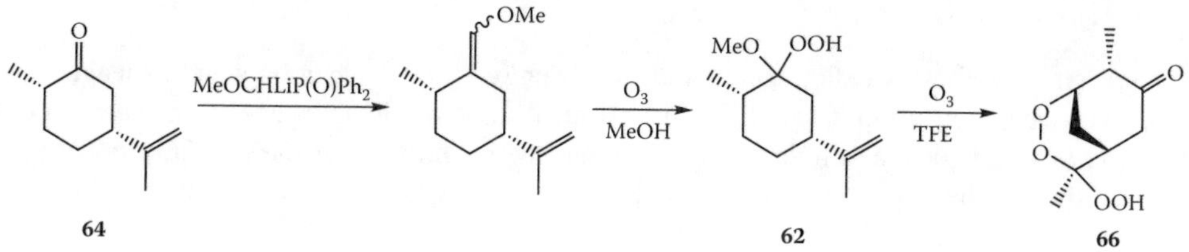

SCHEME 11.18

11.4.3 OZONIDES

Another source of potential antimalarial compounds has been the isolation of stable ozonides **67**, produced via the ozonolytic cleavage of the bicyclic ketones **68** (Scheme 11.19).[66] Variation of the groups R within **68** allows for formation of a wide range of stable ozonides **67** as exemplified by R = *n*-heptyl that has a melting point of 75°C to 76°C. The ozonolysis can be carried out in dichloromethane at –5°C to 0°C, and the ozonides **67** usually crystallize directly from the reaction mixture.

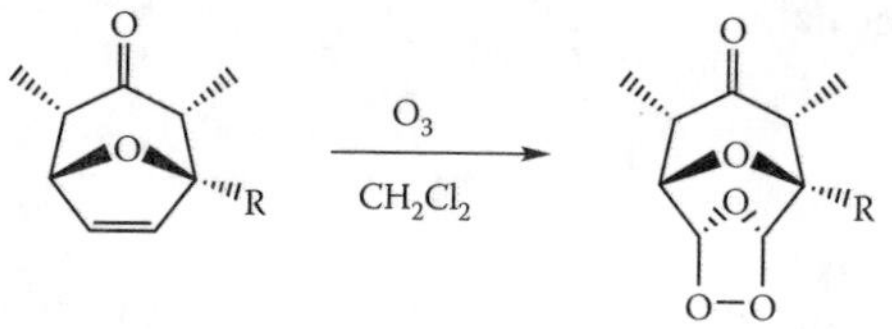

SCHEME 11.19

11.5 ANTIBIOTICS

Antibiotics have proved to be invaluable for the treatment of disease and infection, although new antibiotics require development constantly as a result of resistance. Because antibiotics are commonly complex compounds, it is often extremely difficult to structurally modify their natural configuration. Therefore, the synthesis of such compounds proves to be the only route of the production of potentially important derivatives.

One common structural element of natural products exhibiting antibiotic properties is the incorporation of a β-lactam moiety. This structural unit is somewhat labile. Because ozonation is generally recognized as a mild, clean, and selective method for oxidation of a compound, it can be considered an attractive synthetic transformation during the synthesis of antibiotics and their analogues.

11.5.1 Thiolactomycin and Thiotetramycin

A group of molecules that exhibit broad-spectrum antibiotic activity are the related thiolactomycin **69**,[67] thiotetramycin **70**,[68] and the related acids **71** and **72**.[69]

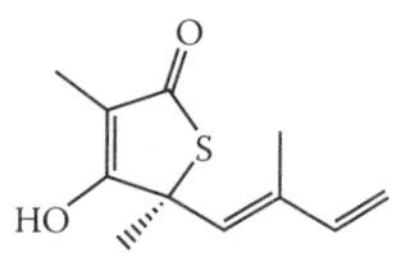

The racemic synthesis of such compounds has been reported.[70] Chambers and colleagues have developed an asymmetric synthesis of 5,5-disubstituted thiotetronic acids and the production of an asymmetric synthesis of (5S)-thiolactomycin *ent*-**69**, the enantiomer of the natural product.[71] The key step in this synthetic pathway is the ozonolysis of a thioether **73** to the aldehyde **74** (Scheme 11.20).

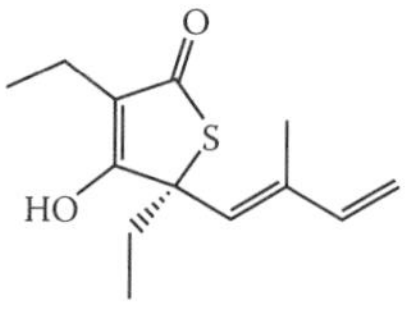

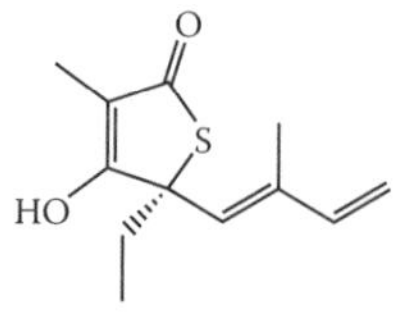

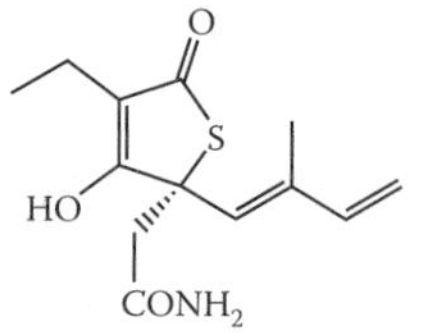

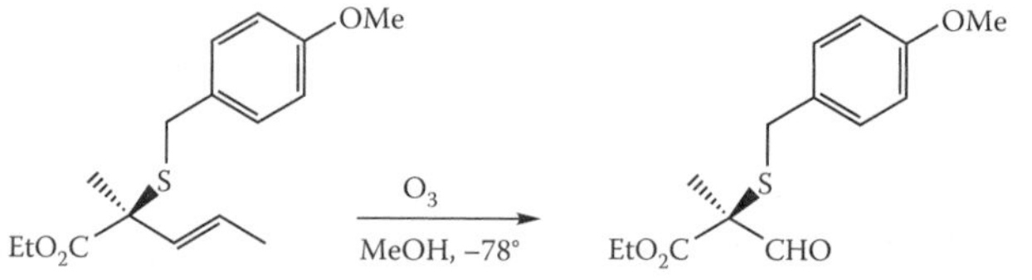

SCHEME 11.20

The ability for selective ozonolysis to occur is verified here. The reaction, carried out in methanol at −78°C, does not exhibit oxidation of the thioether functional group but selectively oxidizes the alkene group.

11.5.2 Acetomycin

Ozonolysis also plays a key role in the stereoselective synthesis of acetomycin antibiotics. The densely functionalized acetomycin **75** has other stereochemical arrangements **76–78** of interest.[72]

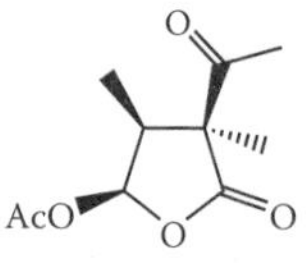

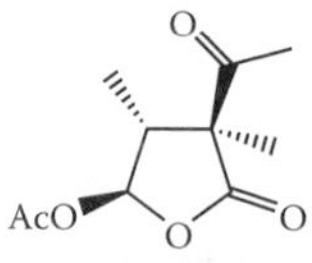

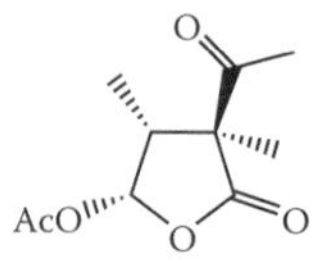

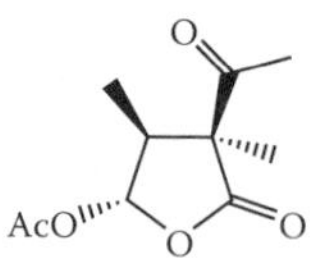

Acetomycin **75** is an antibiotic that also shows activity against bacteria, fungi, and protozoa[73] as well as antitumor activity.[74] These biological properties have prompted the development of synthetic routes to acetomycin and analogues.[75,76] The stereochemistry of the molecule may be of use in influencing its activity and, therefore, the ability to stereoselectively synthesize a particular epimer of acetomycin is of significant use. An example of this is the stereoselective synthesis of the 4-*epi*-acetomycin **76**, described by Echavarren and colleagues.[72] The sequence involves the conversion of an acetoacetate **79** into a dienolate **80** that smoothly undergoes a Carroll rearrangement[77] to a β-ketoacid **81**. The synthesis is completed with an ozonolysis reaction, in dichloromethane at –78°C, which enables spontaneous cyclization to occur concurrently with *in situ* acetylation, to give the desired 4-*epi*-acetomycin **76** (Scheme 11.21).

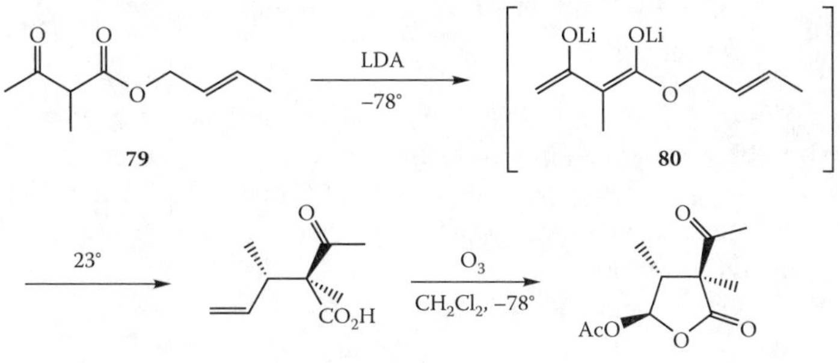

SCHEME 11.21

The use of ozonolysis during this reaction provides not only an efficient route to the oxygenated compound that can undergo cyclization but also permits the acetylation reaction to be carried out *in situ* because both reactions can be performed under the same conditions, therefore eliminating expensive separation an purification.

11.5.3 SULFENIC ACID DERIVATIVES

The use of ozonolysis to provide analogues of penicillin and dihydroxacetoxycephalosporin in the development of new antibiotics has generally been hindered by the fact that these compounds tend to produce high yields of diastereoisomeric sulfoxide mixtures that do not undergo further reaction with ozone.[78] The Δ²-cephem **82** derivatives, however, have been shown, by Botta and colleagues[79] to be reactive enough toward ozone to compete with the sulfur atom in the oxidative transformation. This leads to selective ozonolysis, by the use of ozone in dichloromethane at –20°C, to produce the highly functionalized sulfenic acid derivatives **83**, instead of the sulfoxide **84** usually produced (Scheme 11.22).[80]

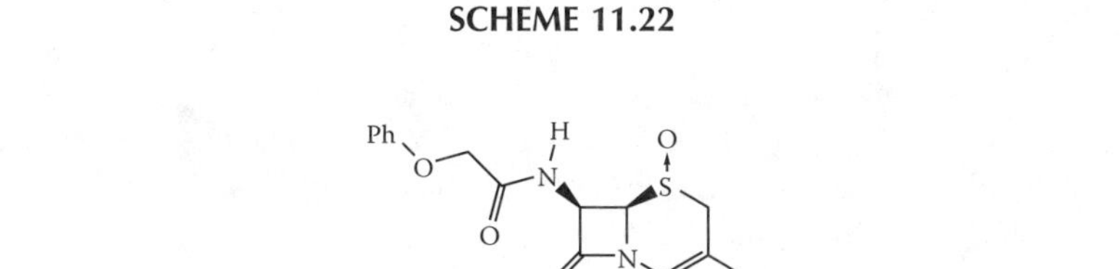

SCHEME 11.22

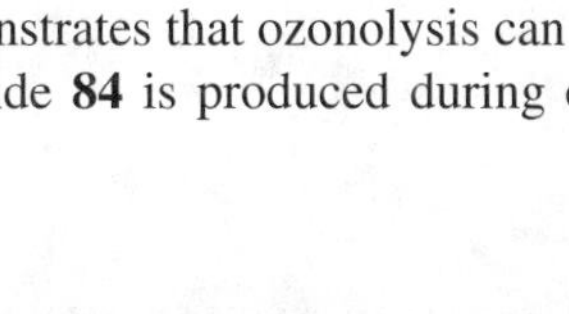

The previous explanation demonstrates that ozonolysis can be a highly selective reaction. Only a very small amount of the sulfoxide **84** is produced during ozonolysis, showing that the sulfur atom oxidation[81] and the double-bond isomerization that occurs are uncompetitive reactions under these conditions. Compounds such as **83** can be used as useful intermediates for the synthesis of penems, a useful class of antibiotics.

11.5.4 PENEM-TYPE ANTIBIOTICS

Another example of ozonolysis being used in the production of penem-type antibiotics, which are regarded as a hybrid between penicillins and cephalosporins, has been demonstrated by Osborne and colleagues[82] who describe the convenient and chiral synthesis of the triazolymethylene penem **85**, from 6-aminopenicillanic acid (6-APA) (**86**), an inexpensive and readily available chiral synthon. The multi-step synthesis incorporates a key synthetic step involving the ozonolytic cleavage of a double bond in **87** to produce an amide **88** (Scheme 11.23).

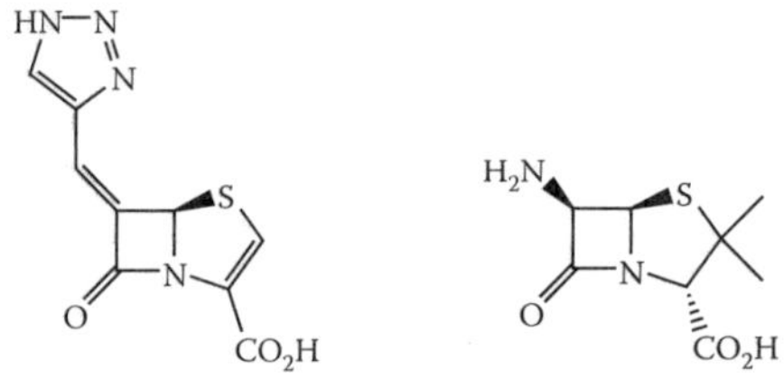

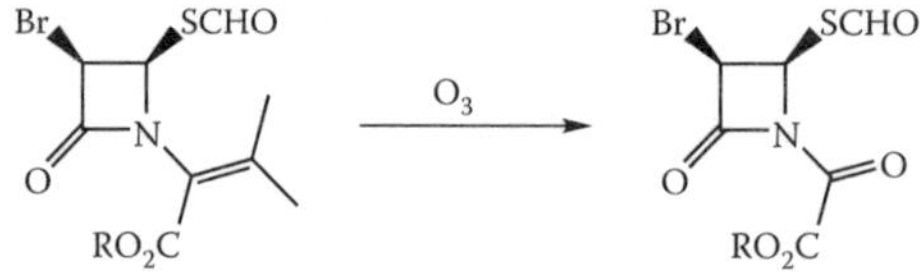

SCHEME 11.23

The reaction has the advantage of utilizing much of the original penem framework from 6-APA, retaining the stereogenic centers of 6-APA and providing a versatile synthetic route to a variety of analogues of penem compounds.

11.5.5 β-LACTAM ANTIBIOTICS

The ability to produce synthetic β-lactam antibiotics that have not come from fermentation is of considerable advantage industrially because microbiological spores are not present, negating the use of a dedicated contained plant. Evans and colleagues have described the synthesis of 1-carbacephalosporins (**89**), which are useful intermediates in the production of monocyclic β-lactam antibiotics.[83] A significant step during this procedure is the conversion of an acylaminoazetidinone **90** to a β-keto ester **91** by ozonolysis (Scheme 11.24).

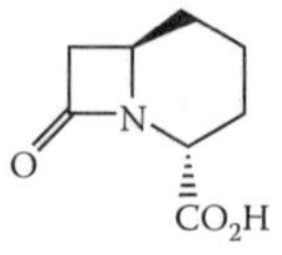

89

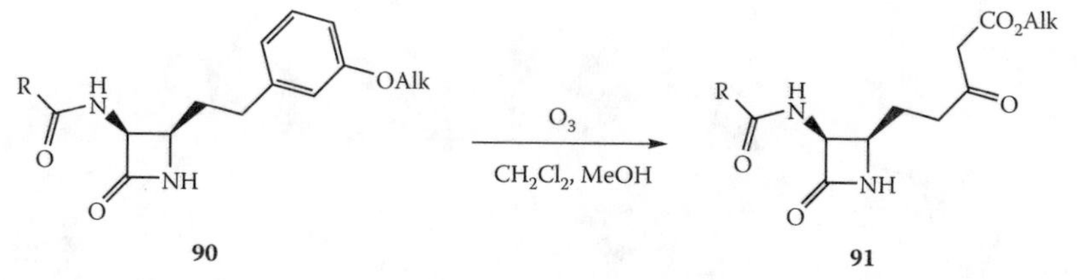

SCHEME 11.24

Ozonolysis is carried out preferentially in a methanol–dichloromethane mixed solvent system, followed by a reductive workup to produce the desired β-keto ester **91**, which is then further reacted to the 1-carbacephalosporin **89**, providing a versatile and viable route to an important intermediate compound for monocyclic β-lactam antibiotics.

11.5.6 HYDROXYL-CEPHAM SULFOXIDE ESTERS

Ozonolysis has also been demonstrated to play a role in the production of another important intermediate in β-lactam synthesis, hydroxyl-cepham sulfoxide esters **92**.[84] Ozonolysis plays a role in the conversion of a methylene group in compound **93** into the required hydroxy group in compound **92** (Scheme 11.25).

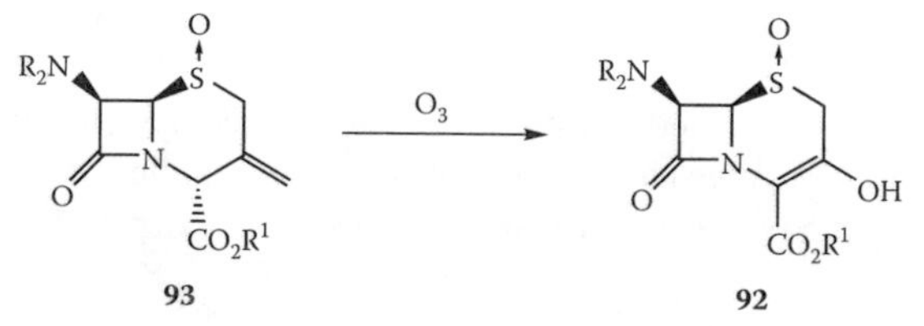

SCHEME 11.25

11.5.7 *TRANS*-CONFIGURATED β-LACTAMS

The use of *trans*-configured β-lactam moieties has led to drugs displaying much higher stability toward resistant bacteria. Examples of such compounds are thienamycin **94** and the recently discovered trinems **95**,[85] which also contain a hydroxyalkyl substituent at the C–3 position.

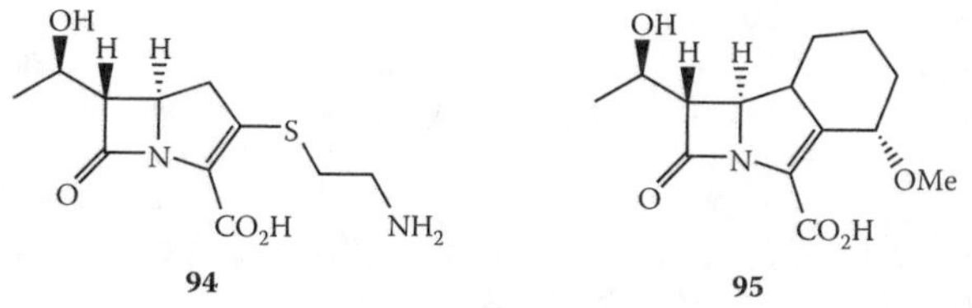

One method for the synthesis of hydroxyalkyl-substituted β-lactams is by the Staudinger reaction, the most frequently used method for the synthesis of β-lactams.[86] This method for the preparation of 4-acetoxy- and 4-formyl-substituted β-lactams involves the use of diazoketones prepared from amino acids. These diazoketones are precursors for ketenes, in a diastereoselective, photochemically induced reaction to produce exclusively *trans*-substituted β-lactams. The use of cinnamaldimines **96**, considered as vinylogous benzaldimines, resulted in the formation of styryl-substituted β-lactams. Ozonolysis, followed by reductive workup with dimethyl sulfide, led to the formation of the aldehyde **97**, whereas addition of trimethyl orthoformate permitted the production of the dimethyl acetal **98** (Scheme 11.26).

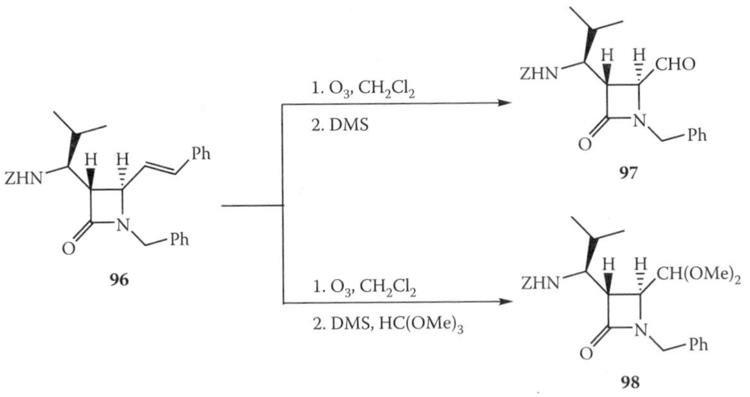

SCHEME 11.26

This synthesis leads to the production of new β-lactam analogues with the potential for exhibiting antibiotic activity and supplies useful intermediates in the synthesis of bicyclic β-lactam antibiotics; it also demonstrates the versatility of ozonolysis workup leading to the preferential selection of desired products.

11.6 SUMMARY

Ozonolysis provides a mild method for the cleavage of alkenes that does not compromise a stereogenic center in the substrate. The methodology is powerful because a variety of methods are available to further react the ozonide that results in access to a variety of functionality.

REFERENCES

1. Bailey, P. S. *Chem. Rev.* 1958, *58*, 925.
2. Criegee, R. In *Oxidation in Organic Chemistry,* Winberg, K. B., Ed., Academic Press: New York, 1965.
3. Johnson R., Koerner, J. F. *J. Med. Chem.* 1988, *31*, 2057.
4. Zvilichovsky G., Gurvich, V. *J. Chem. Soc. Perkin Trans. I* 1995, 2509.
5. Zvilichovsky G., Gurvich, V. *Tetrahedron* 1997, *53*, 4457.
6. Saino, T., Someno, T., Miyazaki H., Ishii, S. *Chem. Pharm. Bull.* 1982, *30*, 2319.
7. Cheung, K.-M., Shoolingin-Jordan, P. M. *Tetrahedron* 1997, *53*, 15807.
8. Mishi, T., Saito, F., Magahori, H., Kataoka, M., Morisowa, Y., Yabe, Y., Sakurai, M., Higashida, S., Shoji, M., Matsushita, Y., Iijima, Y., Ohizumi, K., Koike, H. *Chem. Pharm. Bull.* 1990, *38*, 103.
9. Maly, D. J., Huang, L., Ellman, J. A. *Chem. Biochem.* 2002, *3*, 16.
10. Rich, D. H. *J. Med. Chem.* 1985, *28*, 263.
11. Alemany, C., Bach, J., Farras, J., Garcia, J. *Org. Lett.* 1999, *1*, 1831.
12. Maibaum J., Rich, D. H. *J. Org. Chem.* 1988, *53*, 869.
13. Kwon, S. J., Ko, S. Y. *Tetrahedron Lett.* 2002, *43*, 639.
14. Reetz, M. T., Lauterbach, E. H. *Tetrahedron Lett.* 1991, *32*, 4477.
15. Piveteau, N., Audin, P., Paris, J. *Synlett* 1997, 1269.
16. Le Carrer-Le Goff, N., Audin, P., Paris, J., Cazes, B. *Tetrahedron Lett.* 2002, *43*, 6325.
17. Marshall, J. A., Garofalo, A. W., Sedrani, R. C. *Synlett* 1992, 643.
18. Reetz, M. T. *Chem. Rev.* 1999, *99*, 1121.
19. Matsunaga, S., Fusetani, N., Hashimoto, K., Walchli, M. *J. Am. Chem. Soc.* 1989, *111*, 2582.
20. Harding, K. E., Nam, D. *Tetrahedron Lett.* 1988, *29*, 4263.
21. Barringer, D. F., Berkelhammer, G., Wayne, R. S. *J. Org. Chem.* 1973, *38*, 1191.
22. Shimojima, Y., Hayashi, H., Ooka, T., Shibukawa, M., Iitaka, Y. *Tetrahedron* 1994, *40*, 2519.

23. Hoff, H., Drautz, H., Fiedler, H.-P., Zahner, H., Shultz, J. E., Keller-Schierlein, W., Philipps, S., Ritzau, M., Zeeck, A, *J. Antibiot.* 1992, *45*, 1096.
24. Roggo, B. E., Petersen, F., Delmendo, R., Jenny, H.-B., Peter, H. H., Roesel, J. *J. Antibiot.* 1994, *47*, 136.
25. Jefford, C. W., Wang, J. *Tetrahedron Lett.* 1993, *34*, 1111.
26. Collis, M. P., Perlmutter, P. *Tetrahedron Asymmetry* 1996, *7*, 2117.
27. Roos, J., Effenberger, F. *Tetrahedron: Asymmetry* 2002, *13*, 1855.
28. Okawara, H., Nakai, H., Ohno, M. *Tetrahedron Lett.* 1982, *23*, 1087.
29. Ayer, W. A., Talamas, F. X. *Can. J. Chem.* 1988, *66*, 1675.
30. Akita, H., Oishi, T. *Tetrahedron Lett.* 1978, *39*, 3733.
31. Barrero, A. F., Sanchez, J. F., Altarejos, J. *Tetrahedron Lett.* 1989, *30*, 5515.
32. Shreiber, S. L. *Science* 1991, *251*, 283.
33. Rosen, M. K., Shreiber, S. L. *Angew. Chem., Int. Ed.* 1992, *31*, 384.
34. Liu, J., Farmer, J. D., Lane, W. S., Friedman, J., Weissman, I., Shreiber, S. L. *Cell* 1991, *66*, 807.
35. Guo, C., Reich, S., Showalter, R., Villafranca, E., Dong, L. *Tetrahedron Lett.* 2000, *41*, 5307.
36. Cai, J. C., Huchinson, C. R. *Chem. Heterocycl. Compd.* 1983, *25*, 753.
37. Sawada, S., Matsuoka, S., Nokata, K., Nagata, H., Furuta, T., Yokokura, T., Miyasaka, T. *Chem. Pharm. Bull.* 1991, *39*, 183.
38. Shen, W., Coburn, C. A., Bornmann, W. G., Danishefsky, S. J. *J. Org. Chem.* 1993, *58*, 611.
39. Fauchere, J. *J. Pharmacol. Rev.* 1994, *46*, 551.
40. Meadows, S. D., Middleton, D. S. PCT Pat. WO 96/05193-A1, 1996.
41. Allan, G., Carnell, A. J., Hernandez, M. L. E., Pettman, A. *Tetrahedron* 2001, *57*, 8193.
42. Sonawane, H., Bellur, N. S., Ahuja, J. R., Kulkarni, D. G. *Tetrahedron: Asymmetry* 1992, *3*, 163.
43. Buckler, J. W., Adams, S. S. *Med. Proc.* 1968, *14*, 574.
44. Acemoglu, L., Williams, J. M. J. *J. Mol. Cat. A: Chem.* 2003, *196*, 1.
45. Larock, R. C., Narayanan, K. *Tetrahedron* 1988, *44*, 6995.
46. Bailey, P. S. In *Ozonation in Organic Chemistry,* Vol 2, Trahanovsky, W. Ed., Academic Press: New York, 1982.
47. Spencer, T. A., Gayen, A. K., Phirwa, S., Nelson, J. A., Taylor, F. R., Kandutsch, A. A., Erickson, S. K. *J. Biol. Chem.* 1985, *260*, 13391.
48. Tomkinson, N. O. C., Willson, T. M., Russel, J. S., Spencer, T. A. *J. Org. Chem.* 1998, *63*, 9919.
49. Corey, E. J., Grogan, M. J. *Tetrahedron Lett.* 1998, *39*, 9351.
50. Spencer, T. A., Li, D., Russel, J. S. *J. Org. Chem.* 2000, *65*, 1919.
51. Bailey, P. S., Lerdal, D. A. *J. Am. Chem. Soc.* 1978, *100*, 5820.
52. Endo, A. *J. Med. Chem.* 1985, *28*, 401.
53. Stokker, G. E., Hoffman, W. F., Alberts, A. W., Cragoe Jr., E. J., Deana, A. A., Gilifillan, J. L., Huff, J. W., Novello, F. C., Prugh, J. D., Smith, R. L., Willard, A. K. *J. Med. Chem.* 1985, *28*, 347.
54. Minami, T., Hiyama, T. *Tetrahedron Lett.* 1992, *33*, 7525.
55. Marshall, E. *Science* 1990, *247*, 399.
56. Meshnick, S. R., Jefford, C. W., Posner, G. H., Avery, M. A., Peters, W. *Parasitol. Today* 1996, *12*, 79.
57. O'Neill, P. M., Searle, N. L., Raynes, K. J., Maggs, J. L., Ward, S. A., Storr, R. C., Park, B. K., Posner, G. H. *Tetrahedron Lett.* 1998, *39*, 6065.
58. Cazelles, J., Robert, A., Meunier, B. *J. Org. Chem.* 1999, *64*, 6776.
59. Avery, M. A., Chong, W. K. M. U.S. Pat. 5,019,590, 1991.
60. Avery, M. A., Chong, W. K. M. U.S. Pat. 5,180,840, 1993.
61. Avery, M. A., Chong, W. K. M., Jennings-White, C. *J. Am. Chem. Soc.* 1992, *114*, 974.
62. Avery, M. A., Jennings-White, C. U.S. Pat. 5,225,554, 1993.
63. Tokuyasu, T., Masuyama, A., Nojima, M., Kim, H., Watya, Y. *Tetrahedron Lett.* 2000, *41*, 3145.
64. Buchi, G., Wuest, H. *J. Am. Chem. Soc.* 1978, *100*, 294.
65. Tokuyasu, T., Masuyama, A., Nojima, M., McCullough, K. J., Kim, H., Watya, Y. *Tetrahedron* 2001, *57*, 5979.
66. De Almeida-Barbosa, L. C., Cutler, D., Mann, J., Crabbe, M. J., Kirby, G. C., Warhurst, D. C. *J. Chem. Soc. Perkin Trans. I* 1996, 1101.
67. Hayashi, T., Yamamoto, O., Sasaki, H., Okazaki, H. *J. Antibiot.* 1984, *37*, 1456.
68. Omura, S., Nakagawa, A., Iwata, R., Hatano, A, *J. Antibiot.* 1983, *36*, 1781.

69. Sato, T., Suzuki, K., Kadota, S., Abe, K., Takamura, S., Iwanami, M. *J. Antibiot.* 1989, *42*, 890.
70. Wang, C.-L. J., Salvino, J. M. *Tetrahedron Lett.* 1984, *25*, 5243.
71. Chambers, M. S., Thomas, E. J. *J. Chem. Soc. Perkin Trans. I* 1997, 417.
72. Echavarren, A, M., de Mendoza, J., Prados, P., Zapata, A. *Tetrahedron Lett.* 1991, *32*, 6421.
73. Uhr, H., Zeeck, A., Clegg, W., Egert, E., Fuhrer, H., Peter, H. H. *J. Antibiot.* 1985, *38*, 1684.
74. Mamber, S. W., Mitulski, J. D., Borondy, P. E., Tunac, J. B. *J. Antibiot.* 1987, *40*, 77.
75. Tadano, K., Ishihara, J., Ogawa, S. *Tetrahedron Lett.* 1990, *31*, 2609.
76. Uenishi, J., Okadai, T., Wakabayashi, S. *Tetrahedron Lett.* 1991, *32*, 3381.
77. Wilson, S. R., Price, M. F. *J. Org. Chem.* 1994, *49*, 722.
78. Spry, D. O. *J. Org. Chem.* 1972, *37*, 793.
79. Botta, M., Crucianelli, M., Saladino, R., Mozzetti, C., Nicoletti, R. *Tetrahedron* 1996, *52*, 10205.
80. Bailey, P. S., Khashab, A. Y. *J. Org. Chem.* 1978, *43*, 675.
81. Corey, E. J., Ouannes, C. *Tetrahedron Lett.* 1976, 4263.
82. Osborne, N. F., Atkins, R. J., Broom, N. J. P., Coulton, S., Harbridge, J. B., Harris, M. A., Sterlig–Francois, I., Walker, G. *J. Chem. Soc. Perkin Trans. I* 1994, 179.
83. Evans, D. A., Sjogren, E. B. U.S. Pat. 4,775,752, 1988.
84. Brown Jr., F. Eur. Pat. EP0565351 A2, 1993.
85. Ghiron, C., Rossi, T., Thomas, R. J. *Tetrahedron Lett.* 1997, *38*, 3569.
86. Linder, M. R., Frey, W. U., Podlech, J. *J. Chem. Soc. Perkin Trans. I* 2001, 2566.

12 Transition Metal Catalyzed Hydrogenations, Isomerizations, and Other Reactions

Scott A. Laneman

CONTENTS

12.1 INTRODUCTION

The desire to produce enantiomerically pure pharmaceuticals and other fine chemicals has advanced the field of asymmetric catalytic technologies. Since the independent discoveries of Knowles and Horner,[1,2] the number of innovative asymmetric catalysis for hydrogenation and other reactions has mushroomed. Initially, nature was the sole provider of enantiomeric and diastereoisomeric compounds; these form what is known as the "chiral pool." This pool is comprised of relatively inexpensive, readily available, optically active natural products, such as carbohydrates, hydroxy acids, and amino acids, that can be used as starting materials for asymmetric synthesis.[3,4] Prior to 1968, early attempts to mimic nature's biocatalysis through noble metal asymmetric catalysis primarily focused on heterogeneous catalyst that used chiral supports,[5] such as quartz, natural fibers, and polypeptides. An alternative strategy was hydrogenation of substrates modified by a chiral auxiliary.[6]

Knowles[1] and Horner[2] independently discovered homogeneous asymmetric catalysts based on rhodium complexes bearing a chiral monodentate tertiary phosphine. Continued efforts in this field have produced hundreds of asymmetric catalysts with a plethora of chiral ligands,[7] dominated by chelating bisphosphines, which are highly active and enantioselective. These catalysts are beginning to rival biocatalysis in organic synthesis. The evolution of these catalysts has been chronicled in several reviews.[8–15]

Asymmetric catalysis possesses many advantages over stoichiometric methodologies, of which the most important is chiral multiplication. A single chiral catalyst molecule can generate thousands of new stereogenic centers. Stoichiometric methods use resolution of racemates or start from chiral pool materials. Resolution methods require the use of a resolving agent to form diastereoisomers that are then separated (see Chapters 6–8). This process can be wasteful because the undesired diastereoisomer has to be either racemized or discarded. The same arguments can be applied to the recovery of the resolving agent. Utilization of the chiral pool in asymmetric synthesis can be limited by the availability of an inexpensive reagent that possesses the correct stereochemistry and structure similarity to the final target.

Enzymes cannot perform the range of reactions that organometallic catalysts can and may be susceptible to degradation caused by heat, oxidation, and pH. Substrates not recognized by enzymes can be used in asymmetric catalysts, in which optimization of the enantioselectivities and overall chirality can be easily modified by change of the chiral ligands on the catalyst.[11]

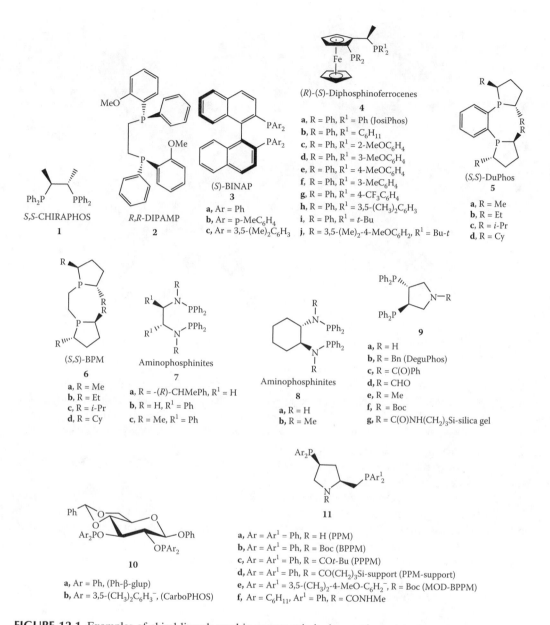

(R)-(S)-Diphosphinoferrocenes

4

a, R = Ph, R^1 = Ph (JosiPhos)
b, R = Ph, R^1 = C_6H_{11}
c, R = Ph, R^1 = 2-MeOC$_6$H$_4$
d, R = Ph, R^1 = 3-MeOC$_6$H$_4$
e, R = Ph, R^1 = 4-MeOC$_6$H$_4$
f, R = Ph, R^1 = 3-MeC$_6$H$_4$
g, R = Ph, R^1 = 4-CF$_3$C$_6$H$_4$
h, R = Ph, R^1 = 3,5-(CH$_3$)$_2$C$_6$H$_3$
i, R = Ph, R^1 = t-Bu
j, R = 3,5-(Me)$_2$-4-MeOC$_6$H$_2$, R^1 = Bu-t

(S)-BINAP

3

a, Ar = Ph
b, Ar = p-MeC$_6$H$_4$
c, Ar = 3,5-(Me)$_2$C$_6$H$_3$

Ph_2P PPh_2

S,S-CHIRAPHOS

1

R,R-DIPAMP

2

(S,S)-DuPhos

5

a, R = Me
b, R = Et
c, R = i-Pr
d, R = Cy

(S,S)-BPM

6

a, R = Me
b, R = Et
c, R = i-Pr
d, R = Cy

Aminophosphinites

7

a, R = -(R)-CHMePh, R^1 = H
b, R = H, R^1 = Ph
c, R = Me, R^1 = Ph

Aminophosphinites

8

a, R = H
b, R = Me

9

a, R = H
b, R = Bn (DeguPhos)
c, R = C(O)Ph
d, R = CHO
e, R = Me
f, R = Boc
g, R = C(O)NH(CH$_2$)$_3$Si-silica gel

11

a, Ar = Ar^1 = Ph, R = H (PPM)
b, Ar = Ar^1 = Ph, R = Boc (BPPM)
c, Ar = Ar^1 = Ph, R = COt-Bu (PPPM)
d, Ar = Ar^1 = Ph, R = CO(CH$_2$)$_3$Si-support (PPM-support)
e, Ar = Ar^1 = 3,5-(CH$_3$)$_2$-4-MeO-C$_6$H$_2$$^-$, R = Boc (MOD-BPPM)
f, Ar = C$_6$H$_{11}$, Ar^1 = Ph, R = CONHMe

10

a, Ar = Ph, (Ph-β-glup)
b, Ar = 3,5-(CH$_3$)$_2$C$_6$H$_3$$^-$, (CarboPHOS)

FIGURE 12.1 Examples of chiral ligands used in asymmetric hydrogenation catalysts.

The chiral ligand, when coordinated to the metal, plays an important role in the control of enantioselectivity during the course of the catalysis reaction. These ligands can contain chirality in the backbone (e.g., CHIRAPHOS, **1**), at the phosphorus atom (DIPAMP, **2**), atropisomerically from C_2 symmetric axial configurations (e.g., BINAP, **3a**), or in highly unsymmetric environments (e.g., ferrocenyl bisphosphines, **4**). Usually, most chiral ligands possess diphenylphosphine moieties (Figure 12.1). A new generation of ligands has been developed within these classes that displaces high catalytic activity and enantioselectivities for once-problematic substrates.

The asymmetric outcome of the reaction results from a combination of steric repulsion imposed by the ligand and complex reaction kinetics. It is imperative that the ligands remain coordinated to the metal during the formation of the new stereogenic center to achieve high enantioselectivities. Chiral catalysis begins by enantiofacial differentiation of the substrate in the initial coordination

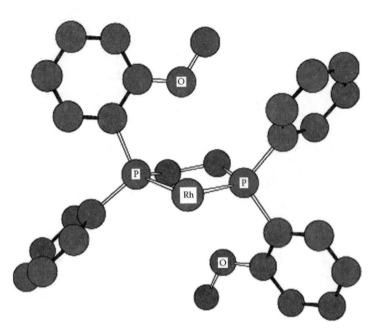

FIGURE 12.2 From the X-ray structure of [Rh(dipamp)COD]. Hydrogens and COD (1,5-cyclooctadiene) ligand are not shown.[17]

to the metal center. A chiral environment about the metal is created by the positioning of the groups, usually phenyl, that pucker in the phosphine-metallocycle and act as a template for the coordinating substrate. The metal can form two diastereoisomeric complexes on coordination of the prochiral substrate that are in rapid equilibrium. Enantioselectivity of the overall product is governed by the competing reaction rates at the metal center for each diastereoisomer.[16] The pucker and aryl array are illustrated by Rh(dipamp) (Figure 12.2) where the anisyl groups are showing their faces and the two phenyl groups are on edge.[17]

Asymmetric catalysis has been most prevalent in the area of homogeneous hydrogenations. As previously stated, producing considerable amounts of a single enantiomer or diastereoisomer from a small amount of chiral catalyst has a huge industrial impact. Natural and unnatural amino acids, particularly L-dopa (**12**), have been produced by this method.[17–21] Catalysts based on rhodium and ruthenium have enjoyed the most success.

12

Despite the explosion of asymmetric homogeneous catalysis technologies, few have made the jump from laboratory to manufacturing.[10,11,22–25] The focus of this chapter is to highlight and discuss the industrial feasibility of both old and new asymmetric hydrogenation technologies, as well as other transition metal catalyzed reactions, based on their respective chiral ligands for the synthesis of chiral intermediates. Several criteria must be filled for an asymmetric hydrogenation technology to be industrially feasible: the cost of catalyst preparation (cost of metal and chiral bisphosphine), catalyst efficiencies, catalyst selectivity expressed as enantiomeric excess (% ee), catalyst separation or recycle, reaction and operation conditions, and the need for specialized equipment. In each case the feasibility of an asymmetric hydrogenation technology is also critical in the cost of the prochiral substrate to be reduced and isolation issues.

12.2 HOMOGENEOUS CATALYSIS: CHIRAL LIGANDS IMPLEMENTED IN INDUSTRIAL PROCESSES AND POTENTIAL PROCESSES

12.2.1 DIPAMP

As stated earlier, Knowles and co-workers developed an efficient asymmetric catalyst based on the chiral bisphosphine, R,R-DIPAMP (**2**), that has chirality at the phosphorus atoms and can form a 5-membered chelate ring with rhodium. (Digital Specialty Chemical has optimized this process, and both antipodes of DIPAMP are available in kilogram quantities.) [Rh(COD)(R,R-DIPAMP)]$^+$BF$_4^-$ (**13**) has been used by Monsanto for the production of L-dopa (**12**), a drug used for the treatment of Parkinson's disease, by an asymmetric reduction of the Z-enamide, **14a**, in 96% ee (Scheme 12.1). The pure isomer of the protected amino acid intermediate, **15a**, can be obtained on crystallization from the reaction mixture because it is a conglomerate.[17]

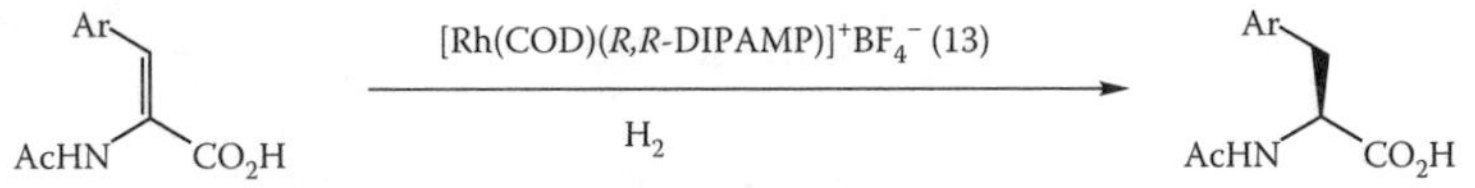

a, Ar = 3-AcO-4-MeOC$_6$H$_3$
b, Ar = Ph

SCHEME 12.1

The chemistry of Rh(DIPAMP) and mechanism has been reviewed.[9,14,16,17,26–28] Marginally higher catalyst efficiencies are observed with higher alcohols compared to methanol, whereas the presence of water can result in the reduction of slurries. Filtration of the product can improve the % ee while the catalyst and D,L-product remain in the mother liquor. The catalyst stereoselectivities decrease as hydrogen pressure increases. [Rh(COD)(R,R-DIPAMP)]$^+$BF$_4^-$ (**13**) affords the S-configuration of amino acids on reduction of the enamide substrate. Reduction of enamides in the presence of base eliminates the pressure variances on the stereoselectivities, but the rate of reaction under these conditions is slow.[17]

The advantages and disadvantages of Rh(DIPAMP) are summarized in Table 12.1. The catalyst precursor, **13,** is air-stable, which simplifies handling operations on a manufacturing scale. Despite these advantages, ligand synthesis is very difficult. After the initial preparation of menthylmethyl-phenylphosphinate (**16**), the $(R)_P$-isomer is separated by two fractional crystallizations (Scheme

---

TABLE 12.1
Advantages and Disadvantages of Rh(DIPAMP)

Advantages	Disadvantages
High enantioselectivity for various aromatic enamides	Lower enantioselectivities of alkyl enamides
Rapid rates	Ligand synthesis difficult and in low yields
High catalyst turnovers	Free ligand racemizes at low temperature
Low hydrogen pressures	
Mild reaction temperatures	
Air-stable catalyst precursor	

12.2). The yield of R,R-DIPAMP (**2**) is 18% yield based on **17**.[26] Another disadvantage is that the R,R-DIPAMP can racemize at 57°C, which can be problematic because the last step is performed at 70°C. Once the ligand is coordinated to rhodium, however, racemization is no longer a problem.

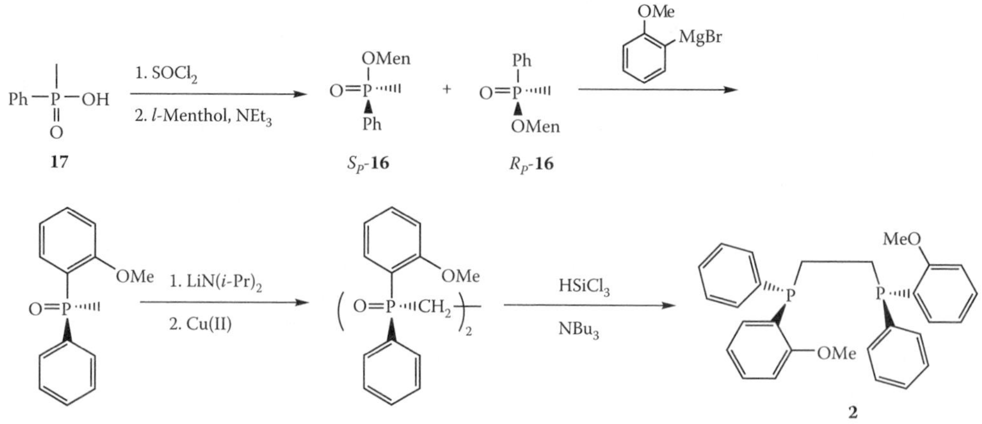

SCHEME 12.2

Other routes to R,R-DIPAMP (**2**) have been reported.[29–31] At present, the most practical synthesis of DIPAMP involves the formation of a single diastereoisomer of **18** by the combination of $PhP(NEt_2)_2$ and (–)-ephedrine followed by formation of the borane adduct (Scheme 12.3).[29,32,33]

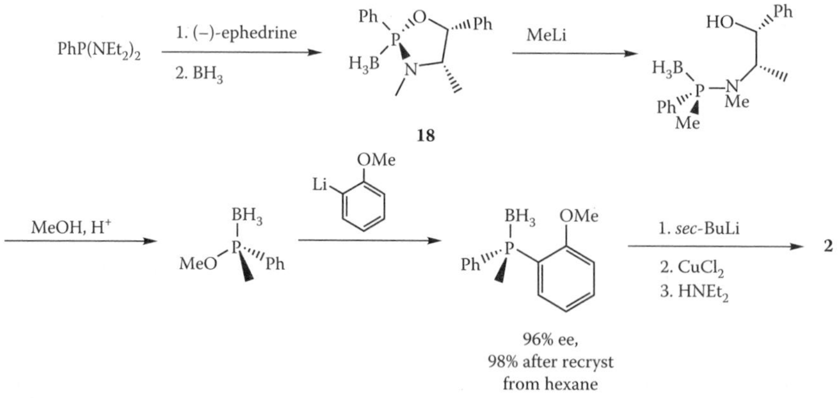

SCHEME 12.3

The cost of the ligand and catalyst preparation can be critical in the decision to pursue technical transfer to manufacturing stage. Fortunately, the high reactivity of $[Rh(COD)(R,R\text{-DIPAMP})]^+BF_4^-$ (**13**) in most enamide reductions can offset the high price, low yield synthesis of DIPAMP.

$[Rh(COD)(R,R\text{-DIPAMP})]^+BF_4^-$ (**13**) has shown remarkable activity and enantioselectivity in the asymmetric hydrogenation of various enamides, enol acetates, and unsaturated olefins.[20,26] For the past 20 years, **13** has been the premier asymmetric homogeneous catalyst for amino acid synthesis, but a new generation of catalysts based on novel bisphosphines (*vide infra*) has surpassed it in versatility.

Use of $[Rh(COD)(R,R\text{-DIPAMP})]^+BF_4^-$ (**13**) has continued in the industrial preparation of substituted phenylalanines. Bromo-, fluoro-, chloro-, and cyano-substituted phenylalanines have been produced in >99% ee at 10–300-kg scale via asymmetric hydrogenation.[20,21,34]

12.2.2 BINAP

Asymmetric catalysis undertook a quantum leap with the discovery of ruthenium and rhodium catalysts based on the atropisomeric bisphosphine, BINAP (**3a**). These catalysts have displayed remarkable versatility and enantioselectivity in the asymmetric reduction and isomerization of α,β- and γ-keto esters; functionalized ketones; allylic alcohols and amines; α,β-unsaturated carboxylic acids; and enamides. Asymmetric transformation with these catalysts has been extensively studied and reviewed.[8,13,15,35,36] The key feature of BINAP is the rigidity of the ligand during coordination on a transition metal center, which is critical during enantiofacial selection of the substrate by the catalyst. Several industrial processes currently use these technologies, whereas a number of other opportunities show potential for scale up.

Each industrial or potential scale up process that uses BINAP-ligated asymmetric homogeneous catalysts has a common major drawback despite their respective high reactivities and enantioselectivities—the cost to produce the chiral ligand and limited availability. Until 1995 Takasago Co. had been the only industrial producer of chiral BINAP as a result of the troublesome 6-step procedure with an overall yield of 14% from 2-naphthol (**19**) (Scheme 12.4),[37] until significant improvement in the synthesis was developed by Merck (Scheme 12.5).[38,39] The latter method has been improved further through modification of reaction parameters by the use of diphenylchlorophosphine rather than the pyrophoric and expensive diphenylphosphine.[40] Today, the availability of BINAP has significantly improved. The technology for the chlorophosphine method, developed by NSC Technologies (later purchased by Great Lakes Chemicals Co.), has been licensed to Rhodia.[41]

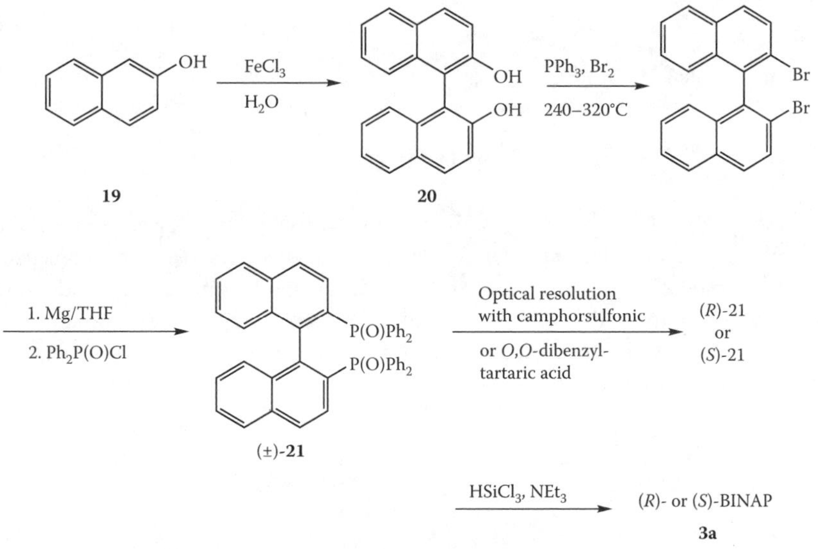

SCHEME 12.4

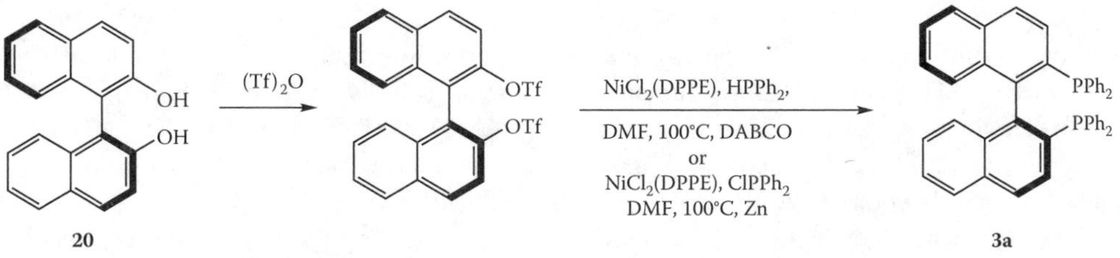

SCHEME 12.5

12.2.2.1 Menthol

Menthol is used in many consumer products, such as toothpaste, chewing gum, cigarettes, and pharmaceutical products, with an estimated worldwide consumption estimated at 4500 tons per year (Chapter 31).[42,43] (–)-Menthol (**22**) is manufactured by Takasago Co. from myrcene (**23**), which is available from the cracking of inexpensive β-pinene (Scheme 12.6).[42,44] The key step in the process is the asymmetric isomerization of *N,N*-diethylgeranylamine (**24**) catalyzed by either [Rh(L$_2$)(*S*-BINAP)]$^+$BF$_4^-$ (where L is diene or solvent) or [Rh(*S*-BINAP)$_2$]$^+$BF$_4^-$ to the diethyl enamine intermediate **25** in 96–99% ee.[36,45] Citronellal (**26**) is obtained in 100% ee after hydrolysis of the enamine intermediate; natural citronellal has an optical purity of 80%.[35] A stereospecific acid-catalyzed cyclization followed by reduction produces **22**.[42]

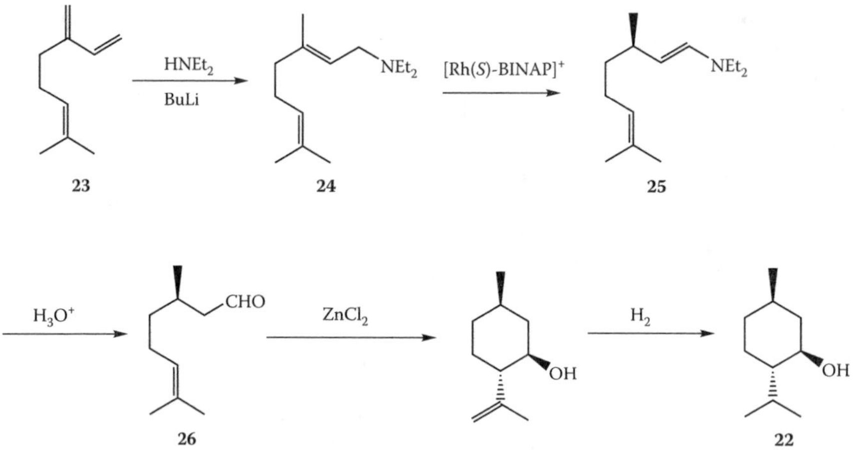

SCHEME 12.6

The catalysts used in the asymmetric isomerization of allylamines are very susceptible to water, oxygen, and carbon dioxide, and significant deactivation is observed by the presence of donor substances that include NEt$_4$, COD, **27**, and **28**. Unfortunately, commercial production of **24** is usually accompanied by formation of 0.5–0.7% of **28** and a thorough pretreatment of the substrate **24** is required for the reaction system to attain high turnover numbers (TON), especially when Rh(L$_2$)(*S*-BINAP)]$^+$BF$_4$ is used as the catalyst.[42]

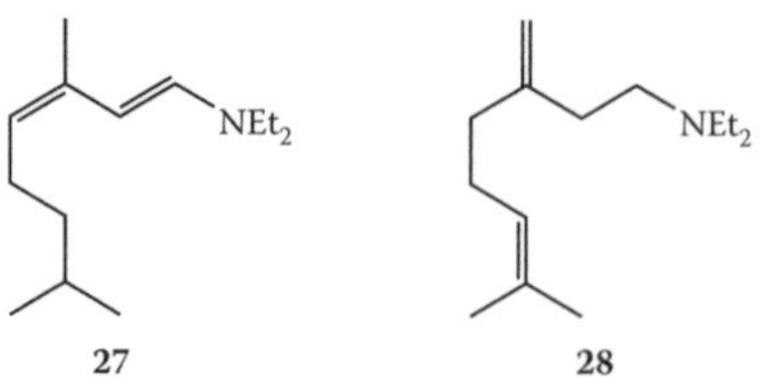

A significant improvement in the asymmetric isomerization came on the discovery of [Rh(*S*-BINAP)$_2$]$^+$BF$_4^-$. This catalyst shows remarkable catalytic activity at high temperature (>80°C) without deterioration of stereoselectivity (>96% ee). The TON is improved to 400,000 through catalyst recycle (2% loss of catalyst per reuse). Additional improvements in the catalyst have been achieved by modification of the BINAP to *p*-TolBINAP (**3b**).[42]

The stereochemical outcome of the enamine can be controlled by the configuration of the olefin geometry of the allylamine and the configuration of the chiral ligand (Scheme 12.7).[36,42]

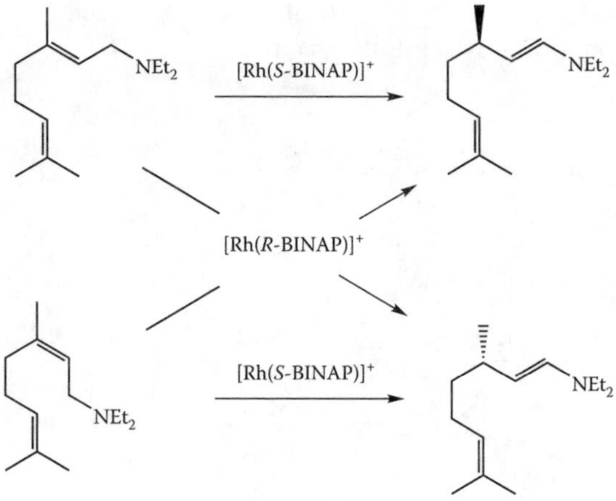

SCHEME 12.7

Takasago has implemented their asymmetric isomerization technology to produce a variety of optically active terpenoids from allyl amines at various manufacturing scales, as described in Chapter 31.[42,43]

12.2.2.2 Carbapenem Intermediates

Carbapenem antibiotics (**29**) can be manufactured from intermediates obtained by Ru(BINAP)-catalyzed reduction of α-substituted β-keto esters by a dynamic kinetic resolution (Scheme 12.8). 4-Acetoxy azetidinone (**30**) is prepared by a regioselective RuCl$_3$-catalyzed acetoxylation reaction of **31** with peracetic acid.[46] This process has been successful in the industrial preparation of the azetidinone **30** in a scale of 120 tons per year.[47] The current process has changed ligands to 3,5-Xyl-BINAP (**3c**), and **31** is obtained in 98% ee and >94% de (substrate-to-catalyst ratio, or S/C ratio = 1,000).[23]

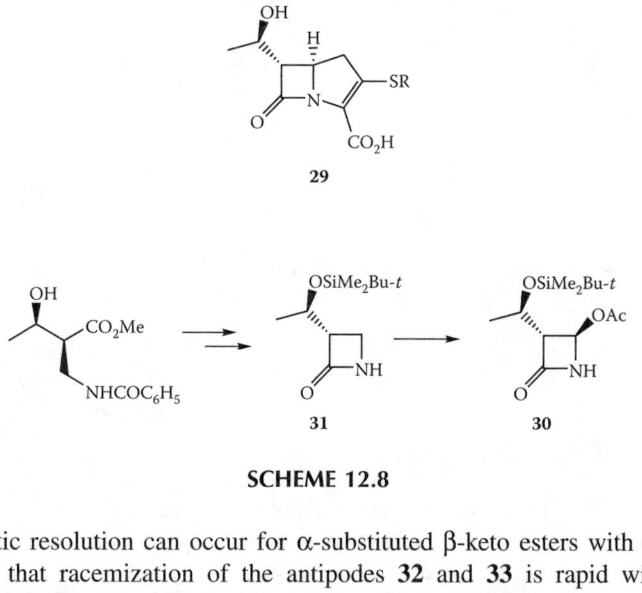

29

31   **30**

SCHEME 12.8

Dynamic kinetic resolution can occur for α-substituted β-keto esters with epimerizable substituents provided that racemization of the antipodes **32** and **33** is rapid with respect to the Ru(BINAP)-catalyzed reduction, thereby potentially allowing the formation of a single diastereoisomer (Scheme 12.9). Deuterium labeling experiments have confirmed the rapid equilibrium of epimers at C–2.[48] Generally, the configuration of C–3 is governed by the chirality of BINAP, whereas the configuration at C–2 is substrate specific.

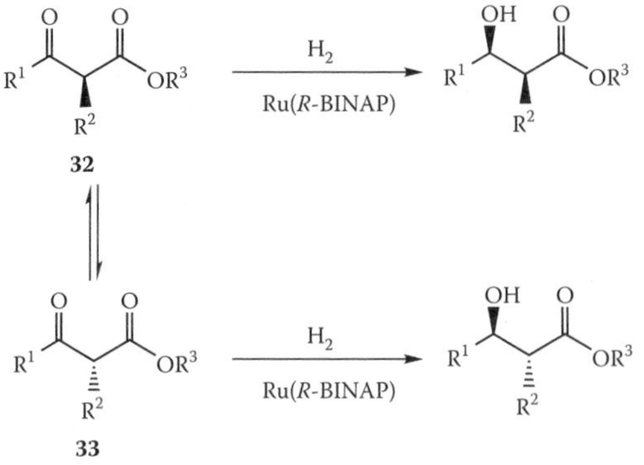

eg., R^1 = Me, R^2 = CH$_2$NHCOPh, R^3 = Et

SCHEME 12.9

Catalyst optimization has been reported only for the [RuX(arene)(BINAP)]$^+$X$^-$ series by variation of the bisphosphines, halogen, and coordinated arene ligands. The reductions of β-dicarbonyl compounds with Ru(BINAP-type) catalysts have been reviewed.[49]

12.2.2.3 Other Intermediates

Asymmetric hydrogenation of racemic 2-substituted β-keto esters to produce 2-substituted β-hydroxy esters with two new chiral centers is a powerful method, and it is useful in the production of other pharmaceutical intermediates. The methodology can be used in the preparation of protected threonine derivatives **34,** where **34d** and **34e** are key intermediates for the anti-Parkinsonian agent, L-Dops (**35**).

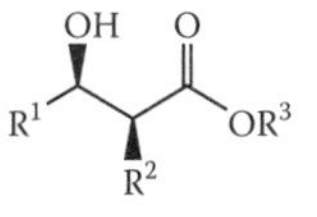

a, R^1 = Me, R^2 = NHCOMe, R^3 = Me
b, R^1 = Me, R^2 = NHCOMe, R^3 = t-Bu
c, R^1 = Me, R^2 = NHCOi-Pr, R^3 = Me
d, R^1 = 3,4-methylenedioxyphenyl, R^2 = NHCOMe, R^3 = Me
e, R^1 = 3,4-methylenedioxyphenyl, R^2 = NHCOPh, R^3 = Me
f, R^1 = Me, R^2 = Cl, R^3 = Me
g, R^1 = Ph, R^2 = Cl, R^3 = Et
h, R^1 = p-MeOPh, R^2 = Cl, R^3 = Me
i, R^1 = Me, R^2 = Me, R^3 = Me

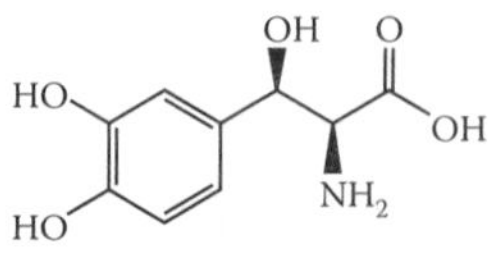

12.2.2.4 Nonsteroidal Antiinflammatory Drugs

(S)-Naproxen (**36**), a potent nonsteroidal antiinflammatory drug (NSAID), is a best-selling agent for arthritis and represents a billion dollar a year market.[50] Extensive research is aimed at the formation of the active S-enantiomer because the R-isomer is a potent liver teratogen. An asymmetric reduction has been proposed for the manufacture of **36** but has yet to be put into practice. Monsanto has reported that a 4-step manufacture process was in development for naproxen (Scheme 12.10).[51]

Several parameters of the hydrogenation affect the enantioselectivities and include temperature, hydrogen pressure, solvent, base, and catalyst.[50–54] However, a significant increase of the S/C ratio, decrease of the hydrogen pressure from 2000 psig, and reduction in the cost of production of the alkene **37** are required for this technology to become economically feasible.[50] Catalyst cost could also be reduced by use of solid supports that have the potential to improve TONs through catalyst recycle.[55]

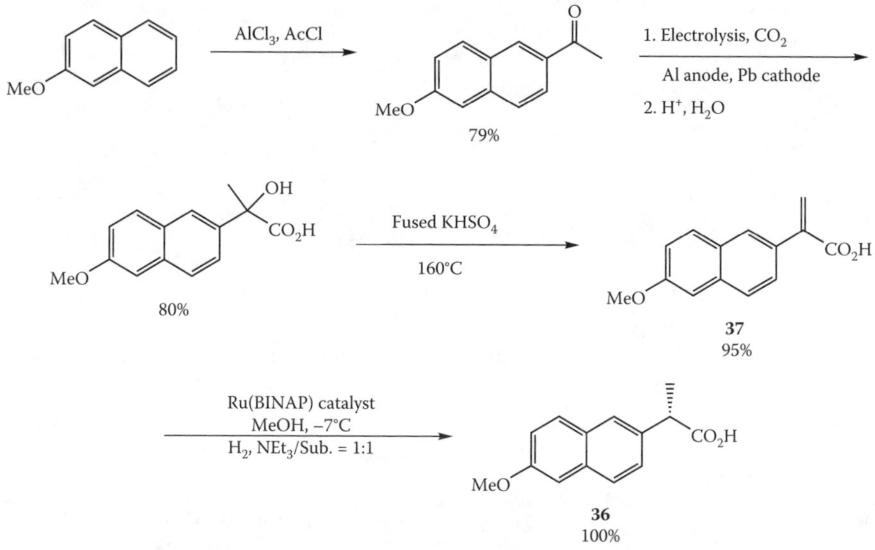

SCHEME 12.10

The naproxen technology can be applied to ibuprofen, another popular analgesic and antiin-flammatory drug sold over the counter with use volumes several times larger than naproxen. Ibuprofen is sold as a racemate, although extensive pharmacologic studies have shown that only the S-isomer (**38**) has significant therapeutic effect.[50] For this reason, several pharmaceutical companies have become interested in marketing S-ibuprofen as a premium analgesic.[50] The racemate of ibuprofen can be prepared by substitution of a heterogeneous catalyst (Pd–C) for the chiral catalyst.

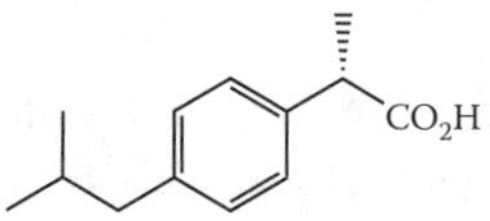

12.2.2.5 Levofloxacin

Another Ru(BINAP)-catalyzed asymmetric hydrogenation that has been performed at manufacturing scale involves the reduction of a functionalized ketone. The reduction of hydroxyacetone catalyzed by $[NH_2Et_2]^+[\{RuCl(p\text{-tol-BINAP})\}_2 (\mu\text{-Cl})_3]^-$ (**39**) proceeds in 94% ee (Scheme 12.11).[46] The chiral diol (**40**) is incorporated into the synthesis of levofloxacin (**41**), a quinolinecarboxylic acid that exhibits marked antibacterial activity. Current production of **40** is 40 tons per year by Takasago International Corp.[46]

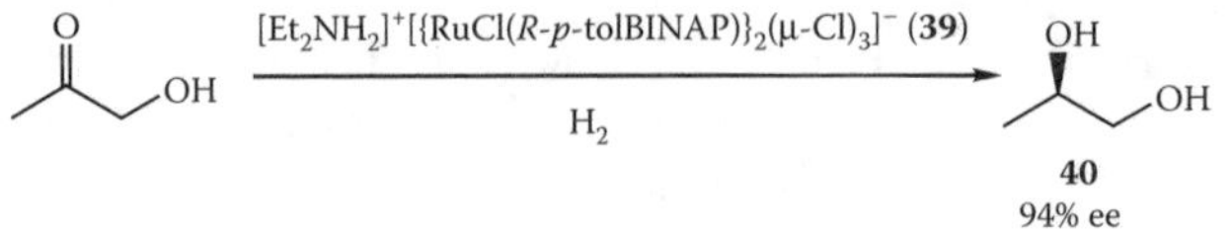

SCHEME 12.11

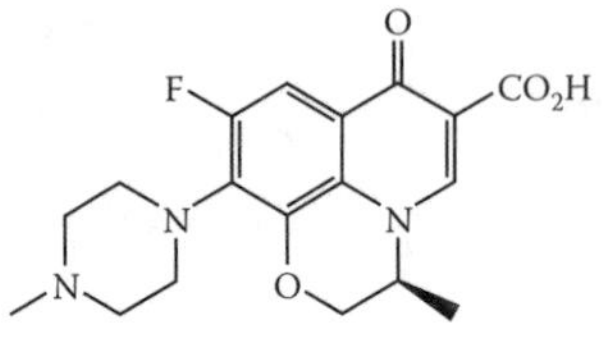

41

12.2.2.6   Latest and Potential Ru(BINAP) Technologies

The asymmetric hydrogenation of allylic alcohols has high potential for industrial scale up. $Ru(O_2CCF_3)_2(S\text{-BINAP})$ and $Ru(O_2CCF_3)_2(S\text{-}p\text{-tolBINAP})$ can catalyze the reduction of allylic alcohols in high enantioselectivities and reactivities.[8] Geraniol (**42**) and nerol (**43**) are hydrogenated in MeOH to give citronellol (**44**) in 96–99% ee with an S/C ratio of 50,000 (Scheme 12.12).[56] In a similar manner to the asymmetric isomerization of allylic amines, the new stereogenic center can be controlled by either the substrate or the configuration of the ligand on ruthenium.[56]

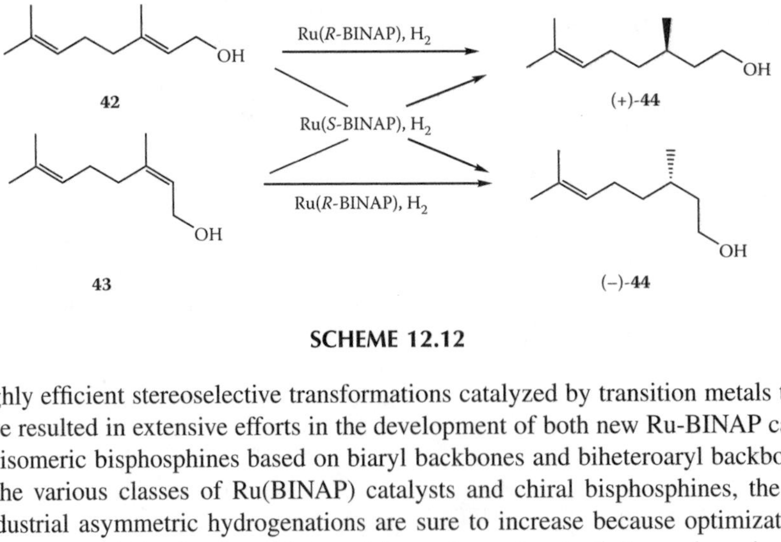

SCHEME 12.12

The highly efficient stereoselective transformations catalyzed by transition metals that contain BINAP have resulted in extensive efforts in the development of both new Ru-BINAP catalysts and chiral atropisomeric bisphosphines based on biaryl backbones and biheteroaryl backbones.[49] Coupled with the various classes of Ru(BINAP) catalysts and chiral bisphosphines, the number of efficient industrial asymmetric hydrogenations are sure to increase because optimization for fine precision and easy optimization in catalyst activity and enantioselectivity made easier.[57–66]

Asymmetric hydrogenation of proprietary β-keto esters with $[RuCl_2(BINAP)]_n$ in 98–99% ee had been performed by NSC Technologies (Scheme 12.13). Substrate-to-catalyst loadings of 10,000–20,000 have been achieved. These reductions have been performed at small-scale production to ton scale.[23] Phoenix Chemicals, whose specialty is in continuous processes, has developed methodology and equipment that can perform this type of transformation up to ~100 tons/year for a proprietary β-hydro ester with enantioselectivities of 98–99% ee (see Chapter 31).[67,68]

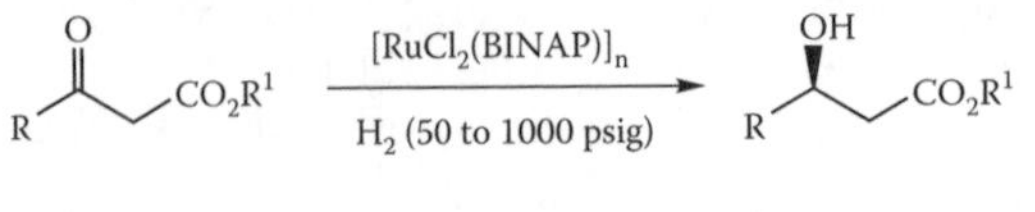

SCHEME 12.13

Ruthenium-BINAP complexes are capable of kinetic resolution of racemic allylic alcohols under hydrogenation conditions. Prostaglandin intermediate (R)-**45** has been prepared on a multi-kilogram scale by Takasago International Corp. and Teijin Ltd. Japan with this process (Scheme 12.14), which was used to prepare prostaglandin E (**46**) and prostaglandin F (**47**) derivatives.[43] Slow-reacting (R)-4-hydroxy-2-cyclopentenone, (R)-**45**, is obtained by kinetic resolution of rac-**45** with Ru(OAc)$_2$(S-BINAP) in 98% ee along with the R-enriched hydrogenation product 3-hydroxy-cyclopentanone (**48**).[69] 3-Hydroxycyclopentanone undergoes rapid dehydration during the silylation procedure to give the volatile 2-cyclopentenone (**49**) so that enantiomerically pure crystalline (R)-**50** is easily separated from the product mixture.[43,69]

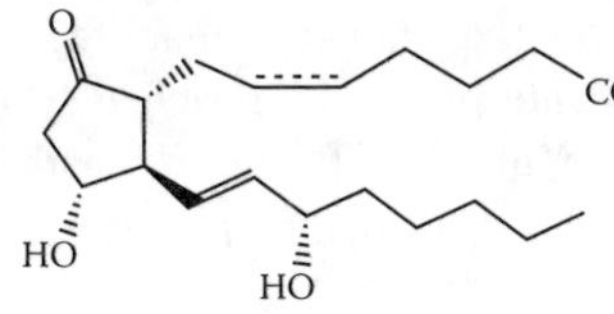

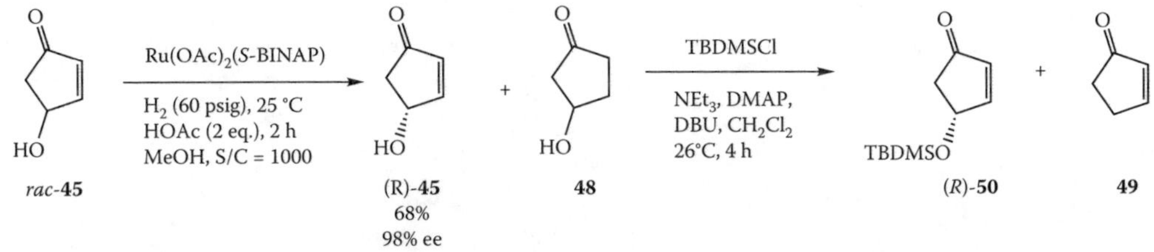

SCHEME 12.14

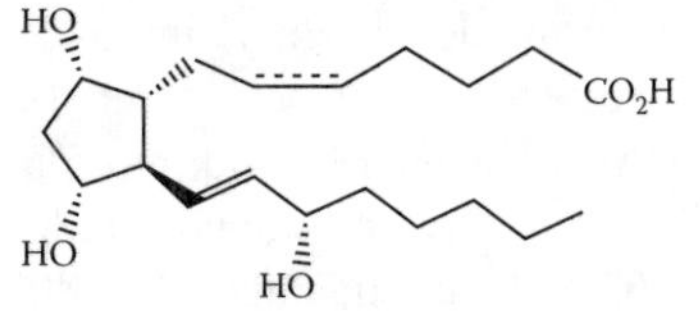

12.2.3 CHIRAL AMINOPHOSPHINES

A class of chiral bisphosphines that has not received much attention is aminophosphines (e.g., **7** and **8**) known as PNNP.[7] The chirality is derived from a 1,2-diamine. One advantage of amino-phosphines is that they can be easily prepared by the reaction of a chiral amine and Ph$_2$PCl in the presence of a base.[70] However, the rhodium-catalyzed reductions must be performed with care because the P–N bonds are easily hydrolyzed or solvolyzed.[17,71]

Rhodium complexes that contain these ligands have demonstrated moderate to high enantio-selectivities (24–96%) in the reduction of enamides to protected amino acids (cf. Scheme 12.1).[7,70] Despite the moderate degree of asymmetric induction, ANIC S.p.A. (EniChem) developed an industrial process with this catalyst system for the production of (S)-phenylalanine for the synthesis of aspartame.[11,45] The process uses cationic Rh-**7a** for the reduction of **14b** at 28 psig H$_2$ and 22°C for 3 hours (S/C = 15,000) to give **15b** in 83.3% ee that is enriched to 98.3% by recrystallization.[72]

12.2.4 FERROCENYLPHOSPHINES

Most catalysts that have been developed for asymmetric catalysis contain chiral C$_2$-symmetric bisphosphines.[7] The development of chiral ferrocenylphosphines ventures away from this conventional wisdom. Chirality in this class of ligands can result from planar chirality due to 1,2-unsymmetrically ferrocene structure as well as from various chiral substituents. Two classes of ferrocenylbisphosphines exist: two phosphino groups substituted at the 1,1′-position about the ferrocene backbone (**51**) and both phosphino groups contained within a single cyclopentadienyl (Cp) ring of ferrocene (**4**).[9]

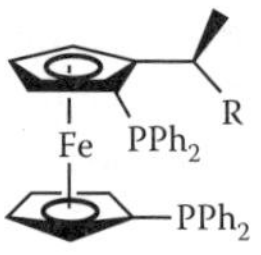

51

a, R = NMe$_2$
b, R = OH
c, R = OAc
d, R = H

Novartis (Ciba Geigy) has developed ferrocenyldiphosphines, **4**, for the asymmetric hydrogenation of sterically hindered *N*-aryl imines, specifically for the industrial manufacture of Metolachlor (**52**), a pesticide sold since March 1997 as the enantiomerically enriched form under the trade name DUAL MAGNUM™ and, more recently, Dual Gol (see Chapters 15 and 31).[73–76] The synthesis of **52** involves the asymmetric reduction by a catalyst formed *in situ* from [Ir(COD)Cl]$_2$ and **4h** to give **53** and subsequently **52** in 80% ee (Scheme 12.15).[74]

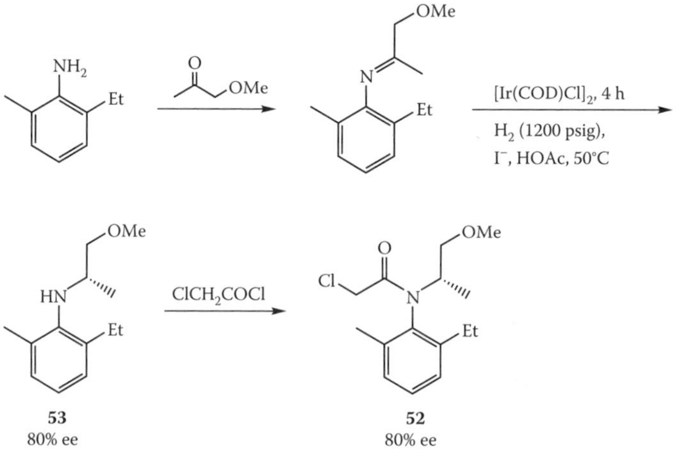

53
80% ee

52
80% ee

SCHEME 12.15

The enantioselectivity and catalytic activity are highly influenced by the purity of imine, electronic effects of the ligand, and additives. In the early stage of process development, iridium catalysts with a variety of known chiral ligands (DIPAMP, DIOP, BPPM, BDPP, and BPPFOH) would deactivate if the purity of the imine was below 99.1%, but deactivation is not observed with Ir-**4h**. The addition of both iodide anion and acetic acid additives is critical to the catalytic activity. High catalytic activities are not observed if either one of the additives is not present during the reduction.[73] The high catalyst cost can be offset by high catalytic activity (S/C = 2,000,000 and turnovers of >600,000/hr)[77] and low catalyst usage. Until recently, these are the highest S/C ratios and catalytic rates ever reported for a catalytic process.[73] An insightful personal account of the research and development of the iridium catalyzed asymmetric hydrogenation to Metachlor has been published by Blaser.[77]

The syntheses of **4** from the Ugi amine (**54**) are summarized in Scheme 12.16.[78–80]

Rhodium and palladium catalysts that contain **4** display high enantioselectivities for the asymmetric hydrogenation of enamides, itaconates, β-keto esters, asymmetric hydroboration, and asymmetric allylic alkylation,[80–82] but this ligand system distinguishes itself from other chiral bisphosphines in the asymmetric reduction of tetrahydropyrazines and tetrasubstituted olefins (see also Chapter 15). The reduction of tetrahydropyrazines produces the piperazine-2-carboxylate core,

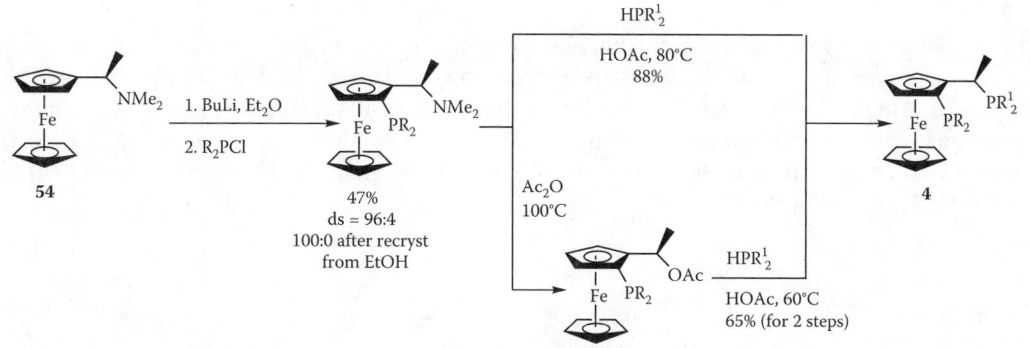

SCHEME 12.16

which is a common intermediate in the pharmaceuticals D-CPP-ene (**55**), an NMDA antagonist; Draflazine (**56**), a nucleoside transport blocker; and Indinavir (**57**), the well-known human immunodeficiency virus protease inhibitor. The chiral piperazine cores (**59**) can be prepared by the asymmetric reduction of tetrahydropyrazines (**58**) (Scheme 12.17).[83] The preparation of the 2-substituted piperazine **59b** has been performed at a scale of nearly 1 ton.[23,84]

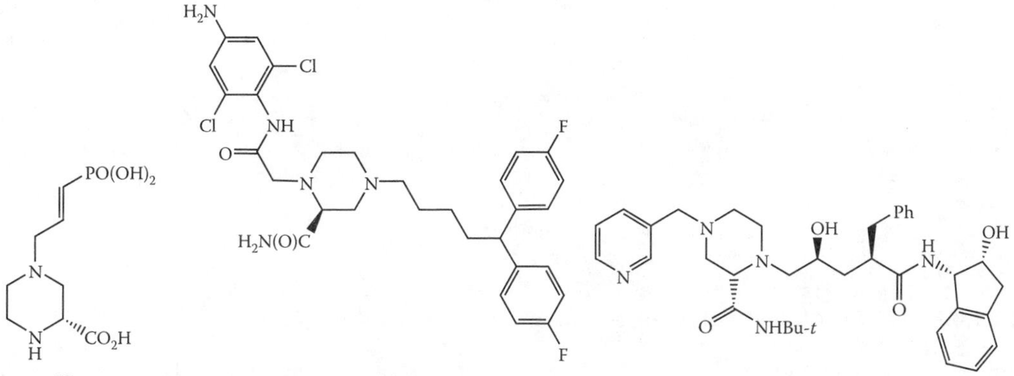

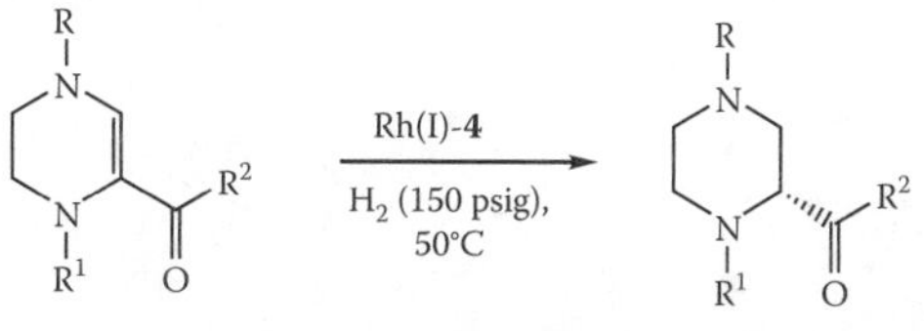

a, R = CHO, R^1 = C(O)Et, R^2 = NHBu-*t*
b, R = CHO, R' = C(O)CF$_3$, R^2 = NHBu-*t*
c, R = H, R^1 = C(O)Me, R^2 = OMe
d, R = H, R^1 = C(O)Me, R^2 = NHtBu
e, R = C(O)Me, R^1 = C(O)Me, R^2 = OMe
f, R = C(O)Me, R^1 = C(O)Me, R^2 = NHBu-*t*
g, R = Boc, R^1 = C(O)Me, R^2 = OMe
h, R = Boc, R^1 = C(O)Me, R^2 = NHBu-*t*
i, R = Boc, R^1 = Cbz, R^2 = OMe
j, R = Boc, R^1 = Cbz, R^2 = NHBu-*t*
k, R = Boc, R^1 = Boc, R^2 = OMe

SCHEME 12.17

12.2.4.1 Biotin

Biotin (**60**), a water-soluble vitamin with widespread application in the growing market for health and nutrition, acts as a co-factor for carboxylase enzymes and its essential fatty acid synthesis. The key step in the chemical synthesis of biotin is the asymmetric reduction of the tetrasubstituted olefins **61** by *in situ* Rh(I)-**4i** catalyst (Scheme 12.18).[79,83,85,86] Substrate-to-catalyst ratios of 2000 with diastereoselectivities of 99% de were achieved with Rh-**4i** at the multi-ton scale before production was terminated.[87]

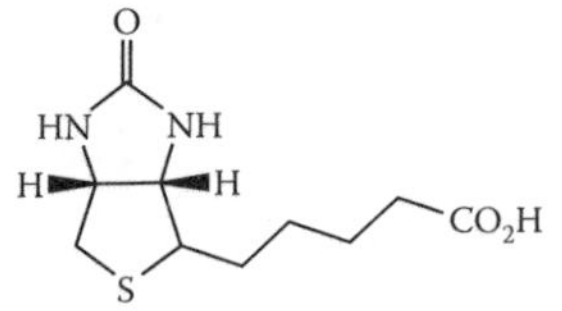

60

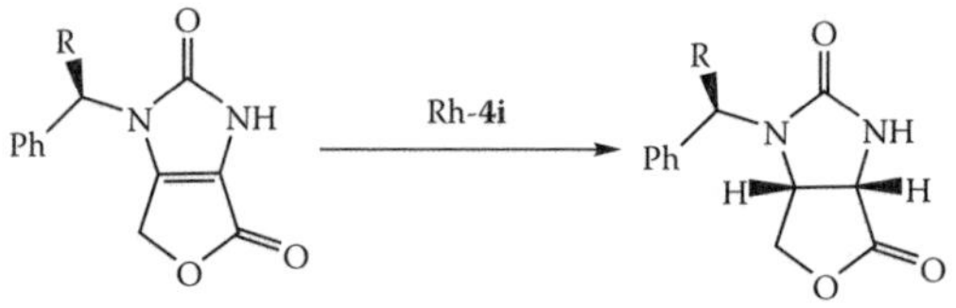

61

a, R = Me
b, R = H

a ds >99:1
b ds >90:10

SCHEME 12.18

12.2.4.2 (+)-*Cis*-Methyl Dihydrojasmonate

Firmenich has reported the asymmetric hydrogenation of an α,β-unsaturated ketone (**62**) with rhodium catalysts that contains **4b** to give (+)-*cis*-methyl dihydrojasmonate (**63**) in 90% ee (Scheme 12.19). The product, which is a fragrance, has been performed at the multi-100-kg scale with increased production planned.[87,88]

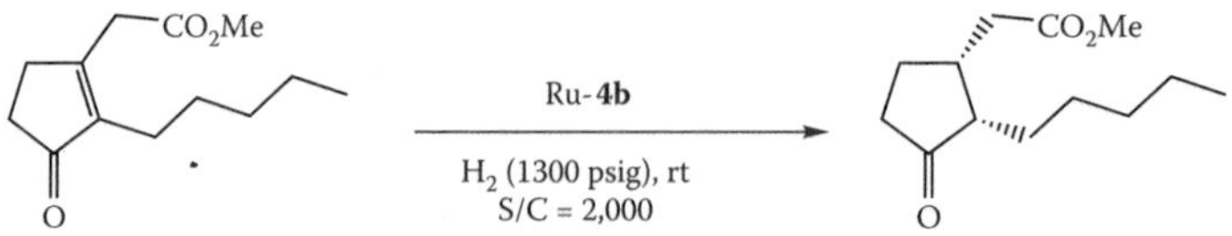

62

63
90% ee

SCHEME 12.19

12.2.4.3 Dextromethorphan

Lonza has performed the asymmetric hydrogenation of a cyclic imine **64** with iridium catalysts that contain **4j** to form dextromethorphan (**65**), an antitussive, in 90% ee (Scheme 12.20) at the multi-kilogram scale (>100 kg produced).[87]

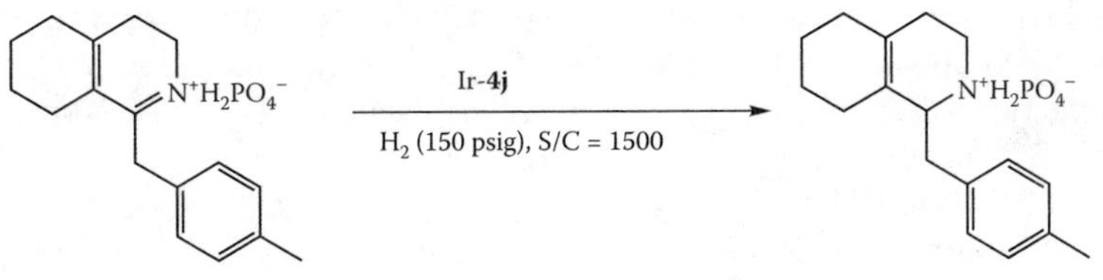

SCHEME 12.20

12.2.5 CARBOHYDRATE PHOSPHONITES

Carbohydrate backbones have been used as a chiral source in a series of phosphonite ligands. Several key features of these ligands are that the D-glucopyranoside backbone is very rigid and an inexpensive chiral source. The benzylidene protection of 4- and 6-hydroxy groups in **10a** is not critical for high enantioselectivity, but protection of the 1-hydroxy is required. Rhodium complexes that contain these ligands are enantioselective catalysts in the asymmetric hydrogenation of enamides.[89]

Rhodium catalysts that contain **10a** have been reported for the industrial preparation of L-dopa (*cf.* Scheme 12.1). Similar to the Monsanto Rh(DIPAMP) process, the enamide **14a** has been reduced with [Rh(COD)(**10a**)]⁺BF₄⁻ in >90% ee by VEB Isis-Chemie Zwickau.[90]

Electronic amplification of the enantioselectivity is observed by modification of the aryl moieties of the phosphonite. Although [Rh(COD)(**10a**)]⁺BF₄⁻ is an efficient catalyst for the asymmetric reduction of α-acetamidocinnamate esters to protected phenylalanine amino acid esters, enantioselectivities decrease in the asymmetric hydrogenation of substituted phenylalanines. Substitution of the diphenylphosphinyl moieties for a more electron-donating 3,5-dimethyphenylphosphinyl produce ligands (**10b**) can efficiently hydrogenate a wide variety of phenyl-substituted dehydroamino acid derivatives in high enantioselectives.[91]

The ligands **10** produce L-amino acid derivatives by rhodium-catalyzed reduction of the corresponding enamide.[91] It is amazing that D-amino acid derivatives can be produced in the hydrogenation with ligands that contain the same D-glucopyranoside backbone, only the O-diarylphosphino moieties are on the 3- and 4-hydroxy groups (**66** and **67**). The production of both D- and L-amino acids from the same inexpensive chiral source (D-glucose and D-glucopyranoside) is very advantageous.

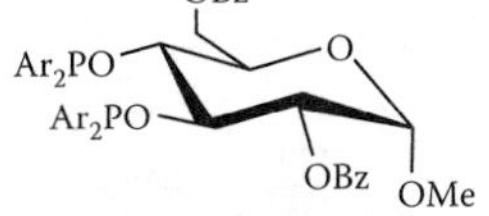

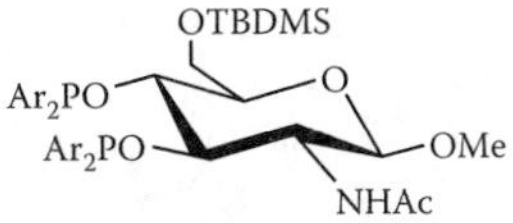

12.2.6 PYRROLIDINE-BASED BISPHOSPHINES

12.2.6.1 3,4-Bisphosphinopyrrolines (DeguPHOS)

A class of chiral bisphosphines based on 3,4-bis(diphenylphosphino)pyrrolidines (**9**) has been developed by Degussa and the University of Munich. Rhodium–bisphosphine catalysts of this class can reduce a variety of enamides to chiral amino acid precursors with high enantioselectivities. These catalysts are extremely rapid and can operate with high S/C ratios (10,000–50,000) under moderately high hydrogen pressure (150–750 psig). Contrary to other rhodium catalysts that contain chiral bisphosphines, the enantioselectivity of the pyrrolidine–bisphosphines catalysts are independent of the hydrogen pressure, although the rate of hydrogenation increases.[92]

The asymmetric hydrogenation to N-acetyl-L-phenylalanine (S/C = 10,000) is catalyzed by [Rh(COD)(**9b**)]$^+$BF$_4^-$ at 40 kg scale in 99.5% ee and 100% conversion. The reduction is completed in 3–4 hours at 50°C under 150–225 psig hydrogen pressure. The TON can be further increased by the recycle of catalyst and ligand or by the attachment of the ligand to a solid support (**9g**).[92–94] The asymmetric hydrogenation catalyst can be fine-tuned by the variation of the R group on the pyrrolidine backbone. The synthesis is summarized in Scheme 12.21.[92,93]

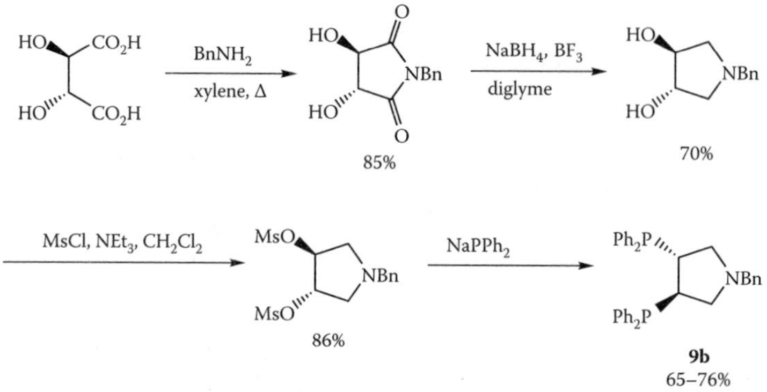

SCHEME 12.21

12.2.6.2 4-Phosphino-2-(phosphinomethyl)pyrrolidines (PPM and BPPM)

Another variation of the bisphosphinopyrrolidine ligand is chirality in the 2 and 4 positions (**11**).[95] The synthesis of these ligands begins with readily available L-4-hydroxyproline (**68**) (Scheme 12.22).[95,96] The synthesis of the antipode of BPPM requires a lengthy 9-step process from L-4-hydroxyproline that contains three stereocenter inversions.[96]

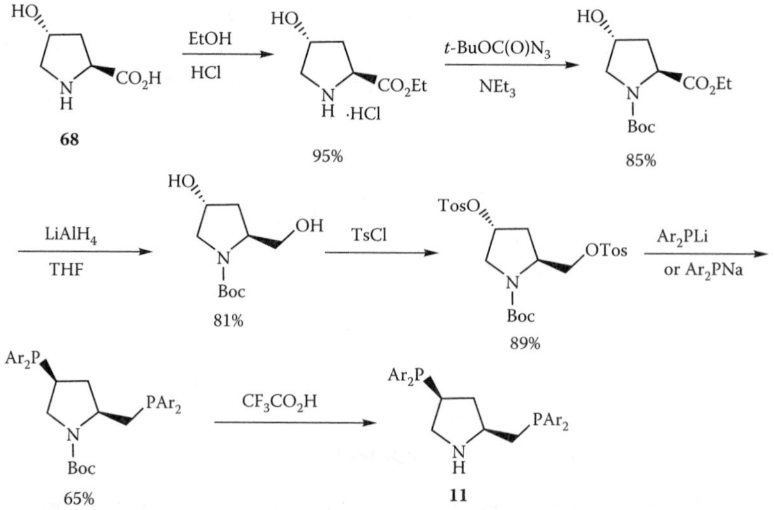

SCHEME 12.22

Rhodium complexes that contain (2S,4S)-PPM ligands efficiently catalyze the asymmetric reduction of enamides to form protected amino acids with the R-configuration.[95,97] However, no large-scale process has been realized to date, although the potential exists because high S/C ratios of 100,000 have been obtained in the hydrogenation of **69** (Scheme 12.23).[8,98]

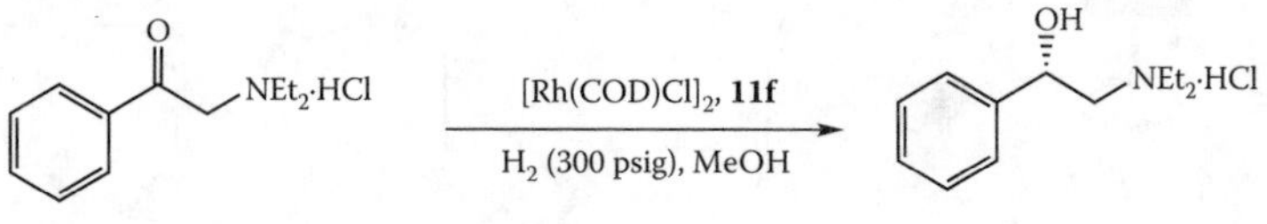

69 95% ee

SCHEME 12.23

12.2.7 Bis(phospholanes) (DuPhos)

A chiral ligand system based on C_2-symmetric chiral bis(phospholanes) (**5** and **6**) has shown to be a powerful ligand in the asymmetric hydrogenation of various substrates, which include enamides to α-amino ester and β-amino esters, enamides and imines to chiral amines, enol acetates, and β-keto esters. The development of the DuPhos ligands has inspired generations of new phospholane ligands (Figure 12.3). Detailed discussions of this class of ligands are in Chapter 13.

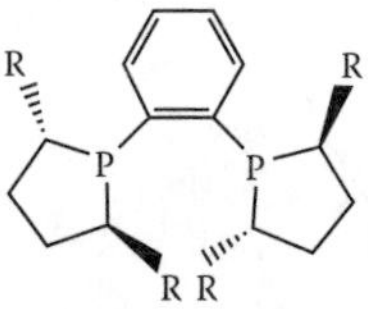

5
DuPhos family
R = Me, Et, Pr, *i*Pr, Ph

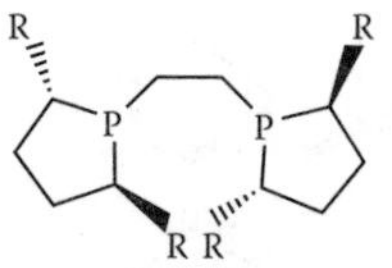

6
BPE family
R = Me, Et

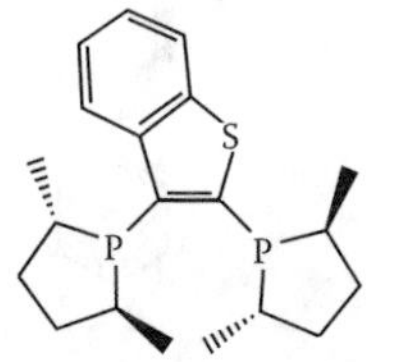

70
Butiphane

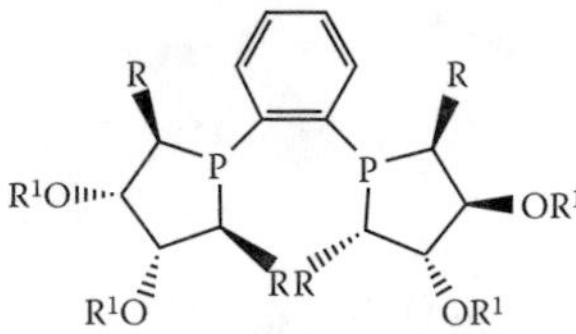

71
RoPhos family
R = Me, Et
R^1 = H, Bn, *t*-Bu

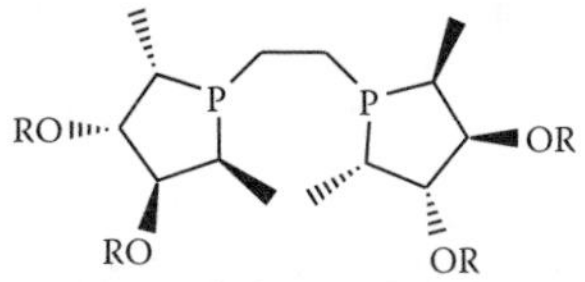

72
RoPhos family
R = Bn, *t*-Bu

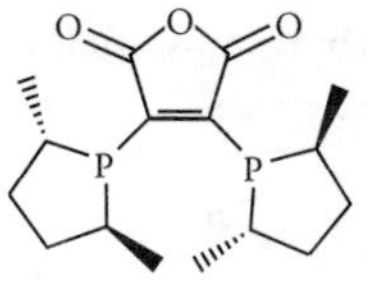

73
Malphos

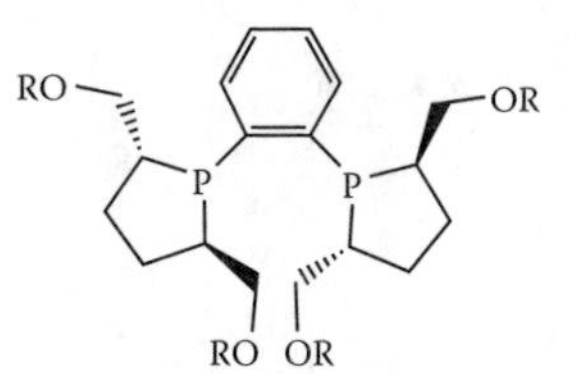

74
BasPhos family
R = H, Me, Bn

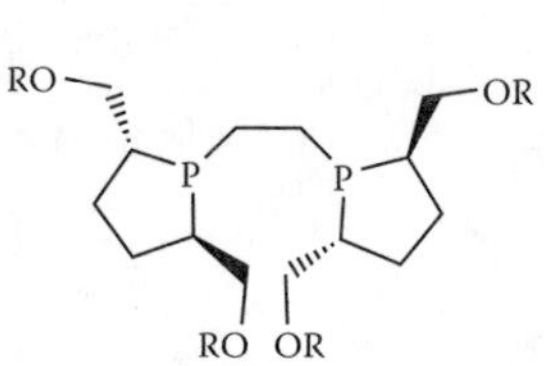

75
BasPhos family
R = H, Me, Bn

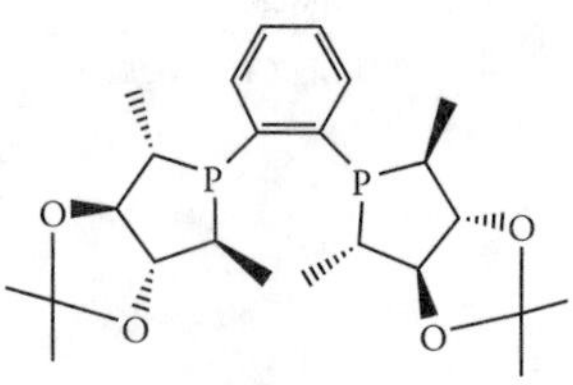

76
Me-KetalPhos

FIGURE 12.3 Structure and abbreviations of phospholane ligands.[23,99]

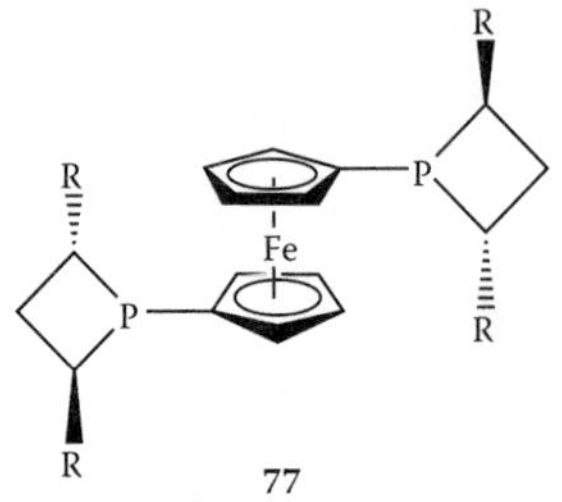

77
Ferrotane family

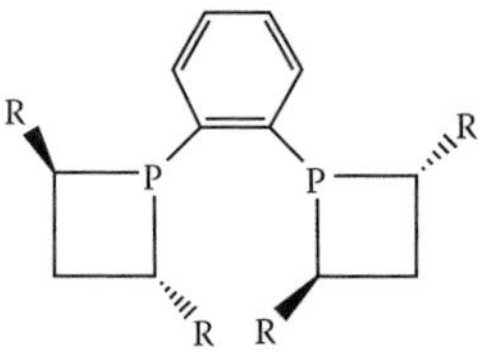

78
CnrPHOS family

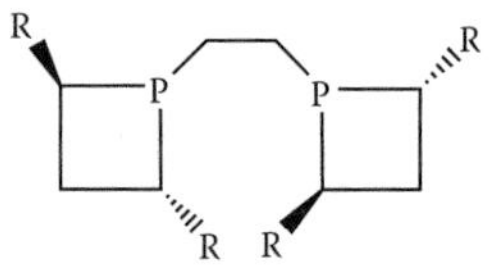

79
BPE-4 family

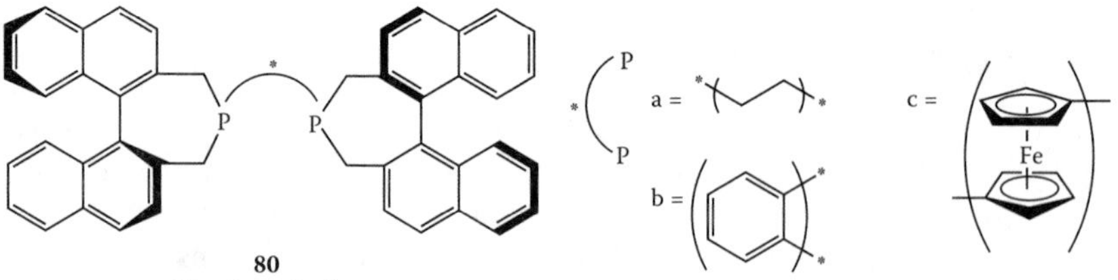

80
Binaphane family

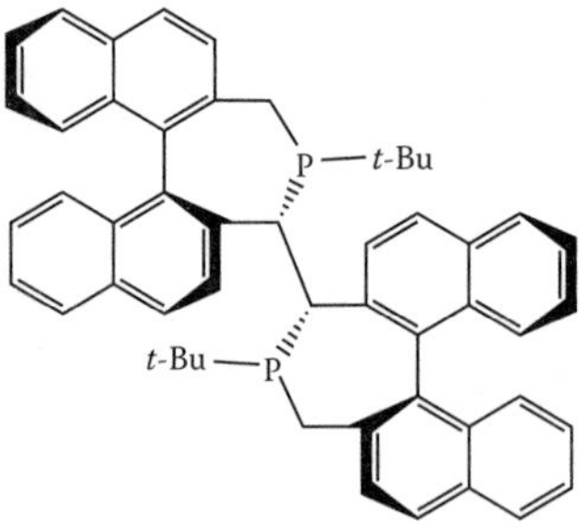

81
Binapine

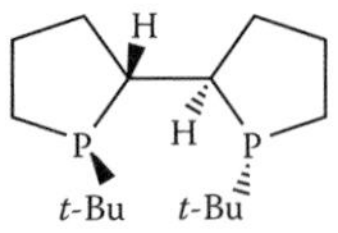

82
TangPhos

FIGURE 12.3 (continued)

12.2.8 TANGPHOS

TangPhos (**82**), developed by Zhang and Tang at Pennsylvania State University, is a small and rigid bisphosphine that can be classified as a phospholane and a *P*-chiral ligand. TangPhos is synthesized from PCl_3 (Scheme 12.24).[100,101] (This ligand is available from Chiral Quest.)

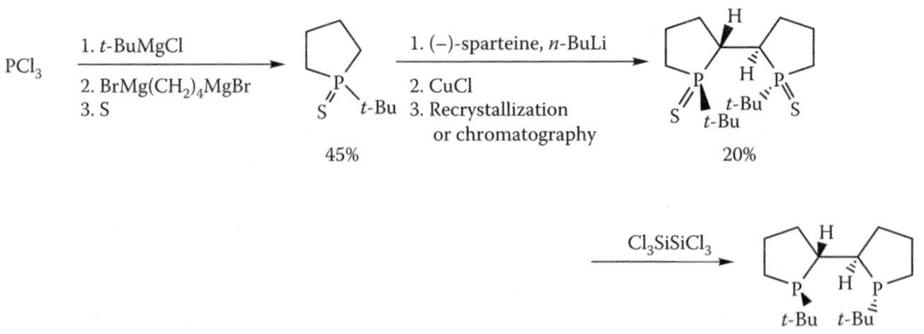

PCl_3

1. *t*-BuMgCl
2. $BrMg(CH_2)_4MgBr$
3. S

45%

1. (−)-sparteine, *n*-BuLi
2. CuCl
3. Recrystallization or chromatography

20%

$Cl_3SiSiCl_3$

82
88%

SCHEME 12.24

TangPhos (**82**) has shown remarkable versatility in the asymmetric hydrogenation of enamides (>99% ee),[100] α-arylenamides (97–99% ee),[100] *E/Z*-β-(acylamino)acrylates (83–99.5% ee),[102] itaconates (95–995 ee),[103] aromatic enol acetates (93–99% ee),[103] and *E/Z-o*-alkoxy-substituted arylenamides[104] (91–95% ee) to protected amino alcohols with high enantioselectivities.

12.2.9 MALPHOS AND CATASIUM FAMILY

Variation of bite angles has been studied in chiral phospholanes ligands. Holz and Boerner investigated the effect of the bite angles in the bis(phospholanes) ligands shown in Figure 12.4, which contain rigid, unsaturated backbones, on the asymmetric hydrogenation of various substrates.[105]

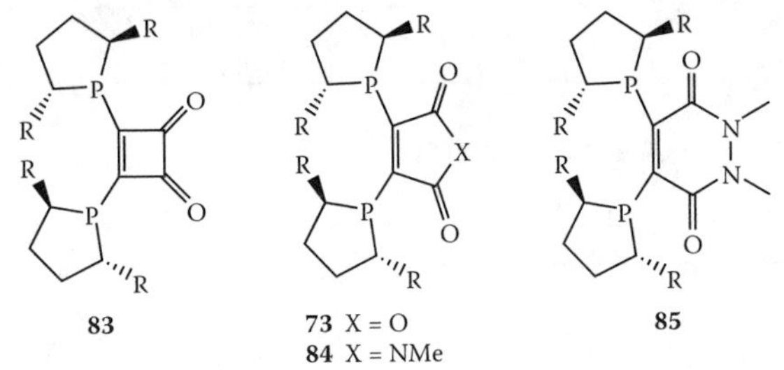

83   **73** X = O   **85**
                84 X = NMe

FIGURE 12.4 Various CatASium ligands.

This class of ligands has been prepared by the addition of a chiral silylphospholane reagent (**86**) to an unsaturated dichlorocyclic reagent to give bis(phospholane) ligands (Scheme 12.25). [Rh(COD)L]⁺BF₄⁻ complexes have been prepared and used in the asymmetric hydrogenation of various substrates (Table 12.2).[105] Asymmetric hydrogenations with MalPhos (**73**) and CatASium **83-85** ligands have been scaled to multi-kilogram pilot scale.[106]

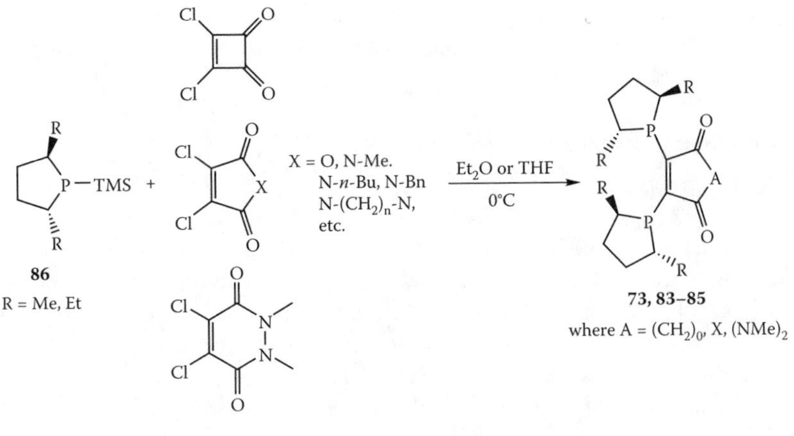

86
R = Me, Et

X = O, N-Me.
N-*n*-Bu, N-Bn
N-(CH₂)ₙ-N,
etc.

Et₂O or THF
0°C

73, 83–85
where A = (CH₂)₀, X, (NMe)₂

SCHEME 12.25

TABLE 12.2
Comparison of Enantioselectivity and Dihedral Angle in the Asymmetric Hydrogenation of Various Compounds[a]

	83	73	84	85	Me-DuPHOS
Bite angle[b]	87.0°	87.1°	86.3°	85.0°	84.6°
Ph $\diagup$ NHAc / CO_2Me	85%	98%	97%	98%	98%
MeO$_2$C $\diagup$ CO$_2$Me	93%	99%	99%	95%	97%[c]
MeO$_2$C $\diagup$ NHAc	98%	97%	99%	>99%	99%[d]
MeO$_2$C $\diagup$ NHAc	86%	86%	82%	78%	88%[d]

[a]Hydrogenations performed in CH_2Cl_2.[105,107]
[b]Obtained bite angles from crystal structure of [Rh(COD)L]$^+$BF$_4^-$.
[c]Reduction performed in THF.
[d]Reduction performed in MeOH.

12.3 NEW GENERATIONS OF CHIRAL PHOSPHINE LIGANDS AND CATALYSTS

This section will concentrate on the next generation of ligands that have been developed over the last 5 years. As stated previously, the investigation of new chiral ligands for asymmetric catalysis continues to draw considerable interest from both industry and academia. Whether the reason is the avoidance of intellectual property, cost and availability at large scale, or lack of efficiency in a particular asymmetric transformation, these new ligands have been successfully implemented in a variety of asymmetric hydrogenations that rivals the mature ligands described in Figure 12.1.

Several ligands developed have shown promise or have been used in large-scale production. Other ligands that have yet to be implemented in large-scale may soon be because catalyst and custom manufacturing companies have begun to assemble portfolios of chiral phosphines by development of their own technologies or by licensing technologies from industry and/or academia. This section will focus on those new ligands that have either been used at large-scale production or have been licensed (and poised to be used for large-scale production).

Because each generation learns from the last, this generation of ligands has focused on use of highly electron-rich phosphines to increase catalyst activity, increased backbone rigidity or the minute detail of the biaryl backbones to enhance enantioselectivities, or the development of a new class of asymmetric hydrogenation catalysts.

12.3.1 *P*-Chirogenic Phosphines

12.3.1.1 BisP* and MiniPHOS

The *P*-chirogenic nature of DIPAMP has paved the way to the development of new families of chiral bisphosphines ligands, BisP* (**87a**) and MiniPHOS (**88a**), by Imamoto and co-workers.[108,109] These bisphosphines ligands are very electron-rich at the phosphorus atom compared to DIPAMP as a result of the presence of alkyl groups on the phosphorus. The strategy is to use a small and very large alkyl group on each phosphorus atom. The electron-rich character of the ligand has produced active asymmetric hydrogenation catalysts. The electron-rich character of BisP* compared to DIPAMP has extended the applicability to the asymmetric hydrogenation of enamides to chiral amines.

BisP* (**87a**) and analogues have been prepared in 3 steps overall from PCl_3 (Scheme 12.26).[108,109] The metal complexes are prepared by the addition of $[Rh(nbd)_2]BF_4$ to form $[Rh(nbd)(BisP*)]BF_4$ and a minor amount of $[Rh(BisP*)_2]BF_4$. The use of $[Rh(COD)_2]BF_4$ produces $[Rh(BisP*)_2]BF_4$ as the major product. (−)-Sparteine produces the (*S,S*)-isomer of BisP*. Unfortunately, the preparation of (*R,R*)-BisP* is difficult because (+)-sparteine is not readily available. (*R,R*)-BisP* has been prepared by an alternate route from *R-tert*-butyl(hydroxymethyl)methylphosphine.[110]

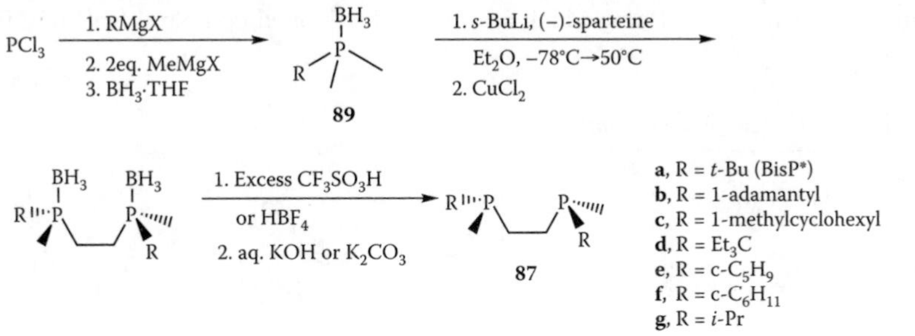

SCHEME 12.26

The synthesis of MiniPHOS (**88a**) and its analogues is similar to that of **87** (Scheme 12.27).[108,109] Reaction with $[Rh(nbd)_2]X$ (X = BF_4, PF_6) only produced $[Rh(MiniPHOS)_2]X$.

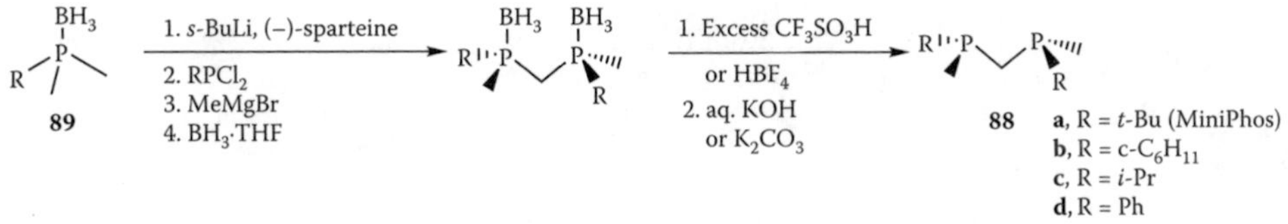

SCHEME 12.27

Rhodium-BisP* and –MiniPHOS catalysts are capable of high enantioselective reductions of dehydroamino acids in 96–99.9% ee.[109] A variety of aryl enamides give optically active amides with 96–99% ee with the exception of *ortho*-substituted substrates.[111] Despite the high enantioselectivity, the rate of reaction in this transformation is slow. Rhodium-BisP* and -MiniPHOS catalysts perform excellently in the asymmetric reduction of (*E*)-β-(acylamino)acrylates to the corresponding protected-β-amino esters in 95–99% ee.[112] Within the family of BisP* and MiniPHOS, the ligands that contain *t*-Bu groups were found to be the most effective in a variety of asymmetric hydrogenations.

These ligands are very air-sensitive as a result of high electron density on the phosphorus. Treatment of BisP* (**87a**) and MiniPHOS (**88a**) with strong Brønsted acids, such as HBF_4 aqueous or TfOH, produces air-stable *P*-chirogenic trialkylphosphonium salts that can be use to generate the metal catalysts (base required in some cases) without racemization about the phosphorus. The asymmetric hydrogenation of (*E*)-β-(acylamino)acrylates with rhodium catalyst prepared *in situ* with these phosphonium salts does not affect the high enantioselective.[113]

12.3.1.2 Trichickenfootphos

Pfizer developed a new ligand with an unforgettable name, Trichickenfootphos (TCFP) (**90**), for possible production of Pregabalin (**91**) by asymmetric hydrogenation. Pregabalin has indications as an anticonvulsant related to the inhibitory neurotransmitter γ-aminobutyric acid.

Trichickenfootphos is a C_1-symmetric ligand with a single *P*-chiral center. The ligand **90** was originally prepared by deprotonation of dimethyl-*t*-butylphosphine boronate (**92**, X = BH$_3$) (Scheme 12.28). The synthesis was improved by substitution of the boronato-group with sulfur.[114] The intermediate disulfide **93** can be monitored by ultraviolet light during chiral high-performance liquid chromatography (HPLC) separation, unlike the boronato complex. Sulfur deprotection was accomplished with Si$_2$Cl$_6$ (90% yield based on *rac*-**93**).[115]

Metal complexation with [Rh(COD)$_2$]$^+$BF$_4^-$ only gives the mono-ligated catalyst (*S*)-**94,** unlike metal complexations with MiniPhos, which affords the bis-ligated complex [Rh(MiniPhos)$_2$]$^+$BF$_4^-$.[114]

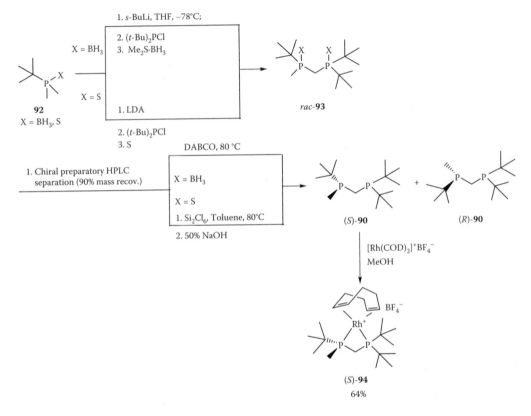

SCHEME 12.28

91

[Rh(TCFP)]$^+$ complex **94** has demonstrated high enantioselectivities in the asymmetric hydrogenation of standard benchmark enamide acids and esters as well as a β,β-disubstituted enamide. [Rh(TCFP)]$^+$ excels in the asymmetric hydrogenation of pregabalin precursor **95** (a 3.5:1 mixture of Z/E isomers). Enantioselectivities of 98% ee have been achieved in 40 hours with an S/C ratio of 27,000 (Scheme 12.29).[114] The previous best catalytic reduction was with Rh(*R*,*R*)-Me-DuPHOS, which could only achieve 97% ee and an S/C ratio of 2700.[116]

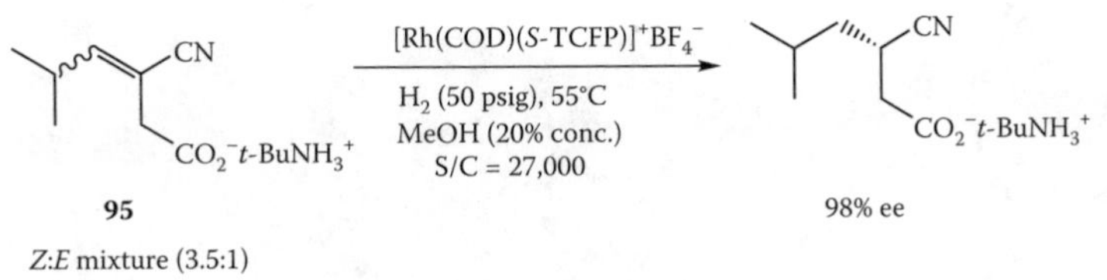

SCHEME 12.29

12.3.2 ATROPISOMERIC BIARYL BISPHOSPHINES

12.3.2.1 Biphenyl Backbone Series

12.3.2.1.1 MeO-BIPHEP

The binaphthyl backbone of BINAP has inspired many variations of atropisomeric biaryl bisphosphines. One approach by Roche was to substitute the binaphthyl backbone with a 6,6'-dimethoxybiphenyl backbone. MeO-Biphep (**96a**) was synthesized in approximately 26% yield in 6 steps from 3-bromoanisole (**97a**) (Scheme 12.30). MeO-Biphep can also be synthesized in 5 steps from 2-iodo-3-nitroanisole in approximately 18% yield. Several phosphine analogues can be prepared by the addition of R_2PCl to the lithio intermediate.[117]

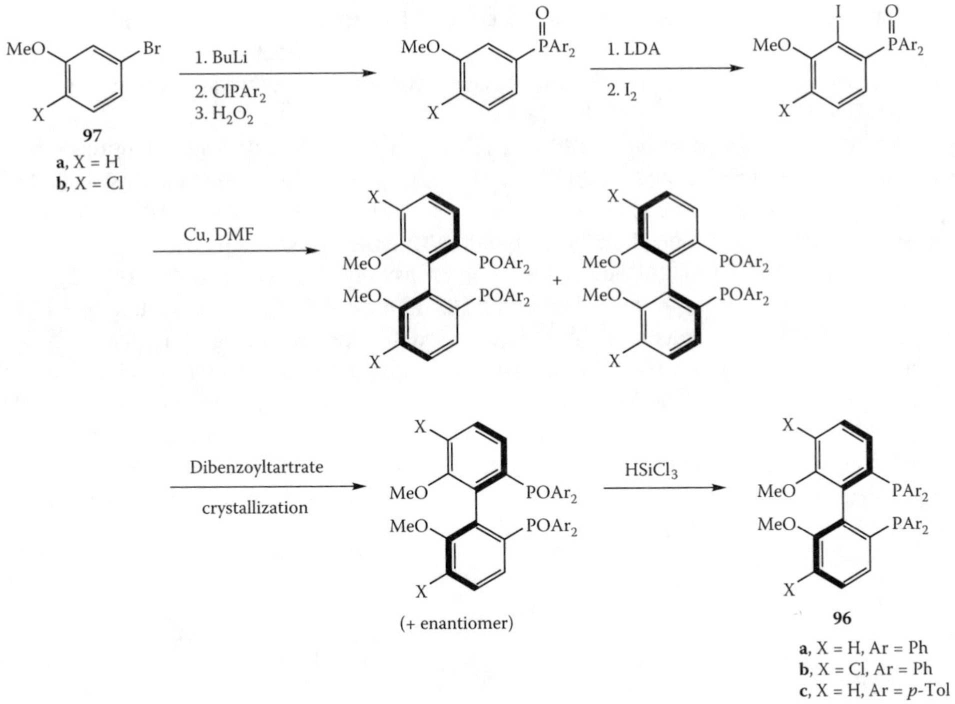

SCHEME 12.30

This ligand, MeO-BIPHEP (**96a**), has shown similar reactivities and enantioselectivities to catalysts that contain BINAP.[117] Ruthenium catalysts that contain MeO-BIPHEP have been used in several asymmetric hydrogenations from bench scale to multi-ton scale, which include the large-scale preparation of a β-keto ester, an aryl ketone, allylic alcohol, and several α,β-unsaturated carboxylic acid substrates, which are shown in Figure 12.5.

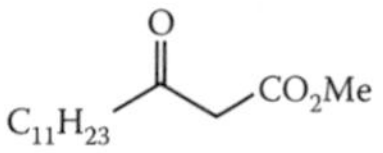

98
>98% ee

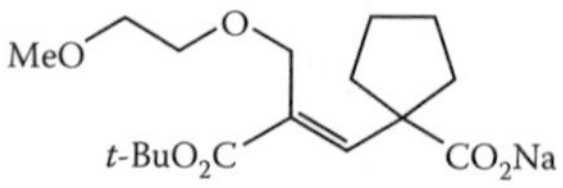

99
>99% ee

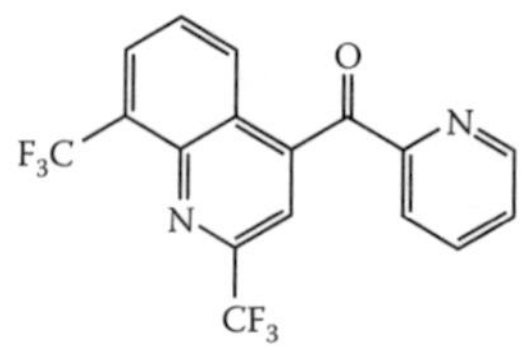

100
92% ee

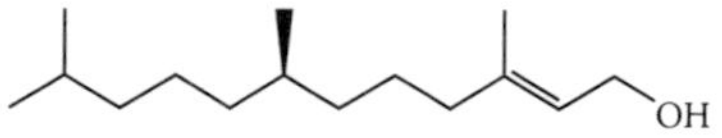

101
>98% de

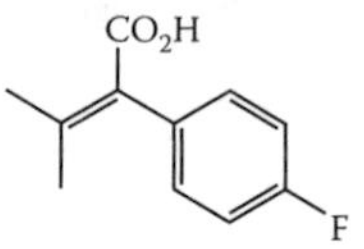

102
94% ee

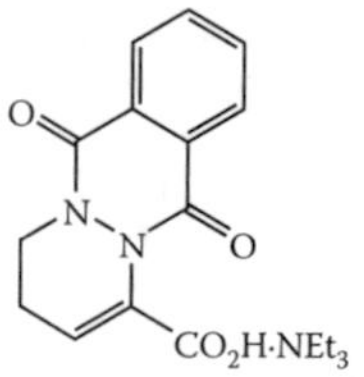

>99% ee

FIGURE 12.5 Examples of substrates of Ru/MeO-Biphep catalysts in industry.

Ru-MeO-Biphep (Ru-**96a**) was used by Roche to reduce **98** to the corresponding β-hydroxy ester with >98% ee at 240-kg scale. Turnover numbers of 50,000 were achieved in this reduction. A process was developed by PPG-Sipsy to reduce **99** with Ru-MeO-Biphep for Pfizer in an approach to candoxatril. The olefin was reduced in >99% ee at 230-kg scale (S/C = 1000–2000).[118] Although catalysts that contain DuPhos had been determined to be more effective based on overall yields and isomerization to an enol by-product, the Ru-MeO-Biphep catalyst was preferred as a result of catalyst availability at scale and more favorable licensing agreements.[119]

The asymmetric hydrogenation of **100** was piloted by Roche in kilogram quantities (>10 kg) with Ru-MeO-Biphep to give >99% ee (S/C = 20,000). Two industrial applications with Ru-**96c** catalysts were performed by Roche. The asymmetric hydrogenation of **101** and **102** with Ru-MeO-Biphep at kilogram scale produced the corresponding saturated products in >98% de and 94% ee, respectively. The TONs of 100,000 were achieved in the asymmetric hydrogenation of **101,** whereas 1000 turnovers were achieved in the reduction of **102**. Compound **102** is a key building block for a novel class of calcium antagonists: mibefradil used for the treatment of hypertension.[43] An asymmetric hydrogenation of a α-keto ester has been reported by Solvias for the preparation of a fungicidal agent (**103**) on behalf of agrochemical company Syngenta. The preparation of (*S*)-*p*-chloromandelate (**104**) from **105** by asymmetric reduction with a ruthenium-(*R*)-Biphep catalyst proceeded with 94% ee (Scheme 12.31).[76] Recrystallization of the free acid of **104** affords enantiomeric enhancements to >99% ee. This process has not been fully optimized but has been performed at the kilogram scale.

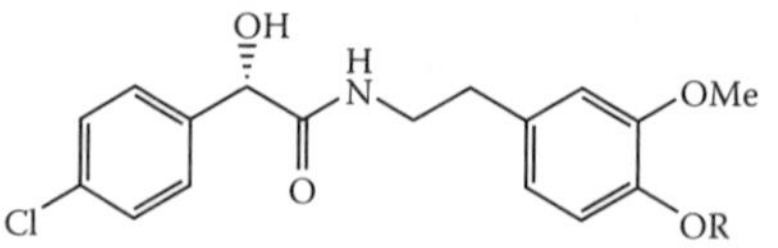

103

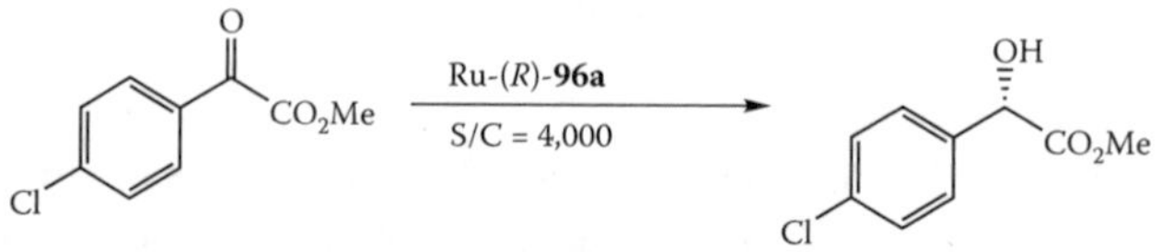

105

(*S*)-**104**
94% ee

SCHEME 12.31

12.3.2.1.2 Biphemp

The bisphosphine ligand Biphemp (**106a**) has been used in rhodium catalysts in the asymmetric isomerization of *N,N*-diethylnerylamine with enantioselectivities up to 99.5% ee. [Rh(Biphemp)]+ catalysts compared favorably to [Rh(BINAP)]+ catalysts in rate and enantioselectivities.[120]

A ruthenium catalyst that contained a mixed bisphosphine analogue of Biphemp (**106b**) was used in the asymmetric hydrogenation of **100**. The asymmetric hydrogenation of the bisaryl ketone proceeded with 92% ee at bench scale by Roche.[23]

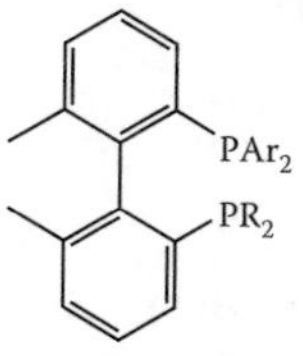

106

a, Ar = R = Ph
b, Ar = Ph, R = Cy

12.3.2.1.3 Cl-MeO-Biphemp

Bayer entered the asymmetric hydrogenation field with the development of a biphenyl ligand similar to MeO-Biphemp called Cl-MeO-biphemp (**96b**). This ligand has been prepared at kilogram scale by two procedures. One approach is an analogue of that used for MeO-Biphep (**96a**) (Scheme 12.30).[121–124] The second method proceeds through a tricyclic intermediate (Scheme 12.32).[122,125]

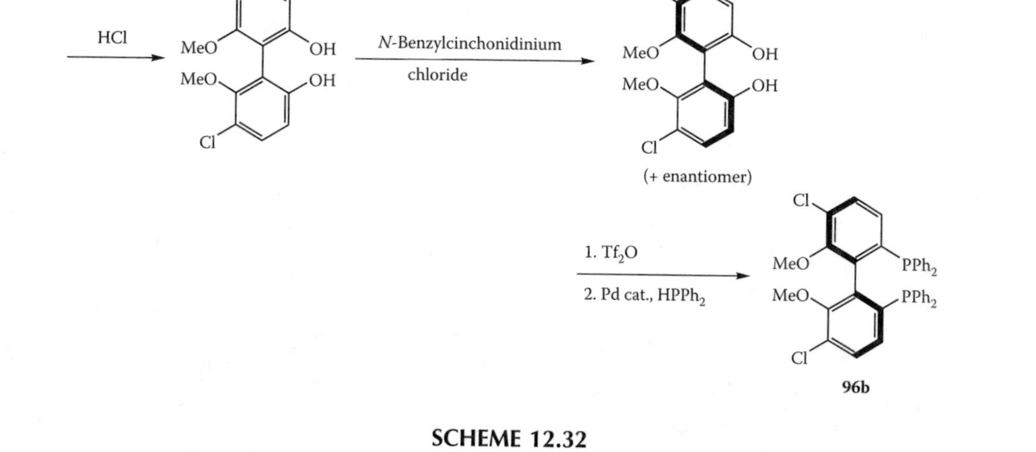

SCHEME 12.32

Ruthenium catalysts that contain Cl-MeO-BIPHEMP have been used in the asymmetric hydrogenation of β-keto esters (99% ee)[126] and the dynamic kinetic resolution of substituted β-keto esters (Scheme 12.33).[121] The asymmetric hydrogenation of methyl 3,3-dimethyl-2-oxobutyrate to the corresponding α-hydroxy ester has been reported with ruthenium catalyst, {RuBr$_2$[(−)-Cl-MeO-BIPHEMP]}$_2$ (Scheme 12.34).[121]

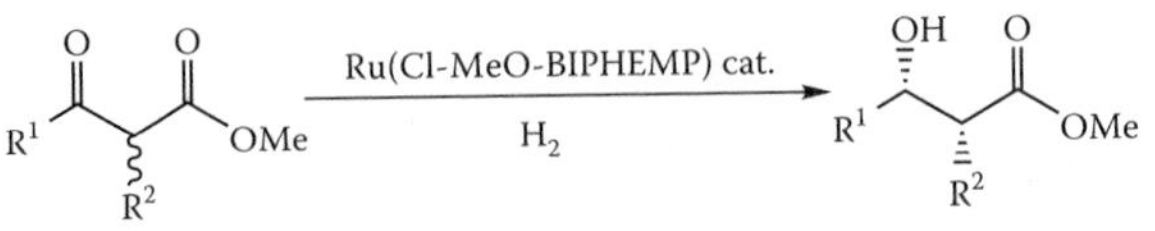

SCHEME 12.33

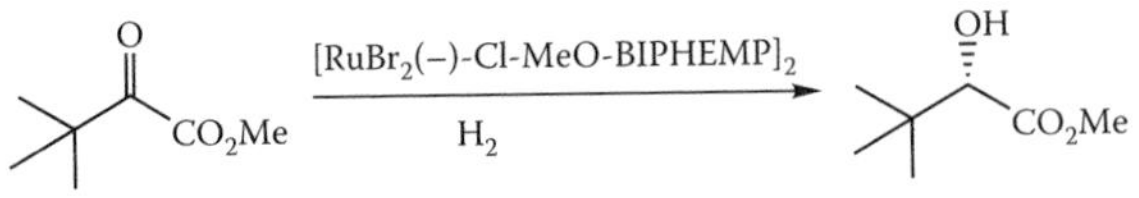

SCHEME 12.34

Pfizer reported an alternative route to atorvastatin (Lipitor®) (see Chapter 31) that used an asymmetric hydrogenation of a β,δ-diketo ester with the intent of reducing the cost of goods.[127,128] The asymmetric hydrogenation of the β,δ-diketo ester proceeds with 98% ee at C–5 and a mixture at C–3 (Scheme 12.35), which is overcome by a dehydration to an α,β-unsaturated lactone followed by a stereoselective oxy-Michael addition to give the benzyl ether of atorvastatin. Hydrogenolysis and subsequent lactone opening with base afforded the calcium salt of atorvastatin. As the alternate process was reaching the cost target, the price of the current starting material had significantly dropped, which shelved the alternate process.

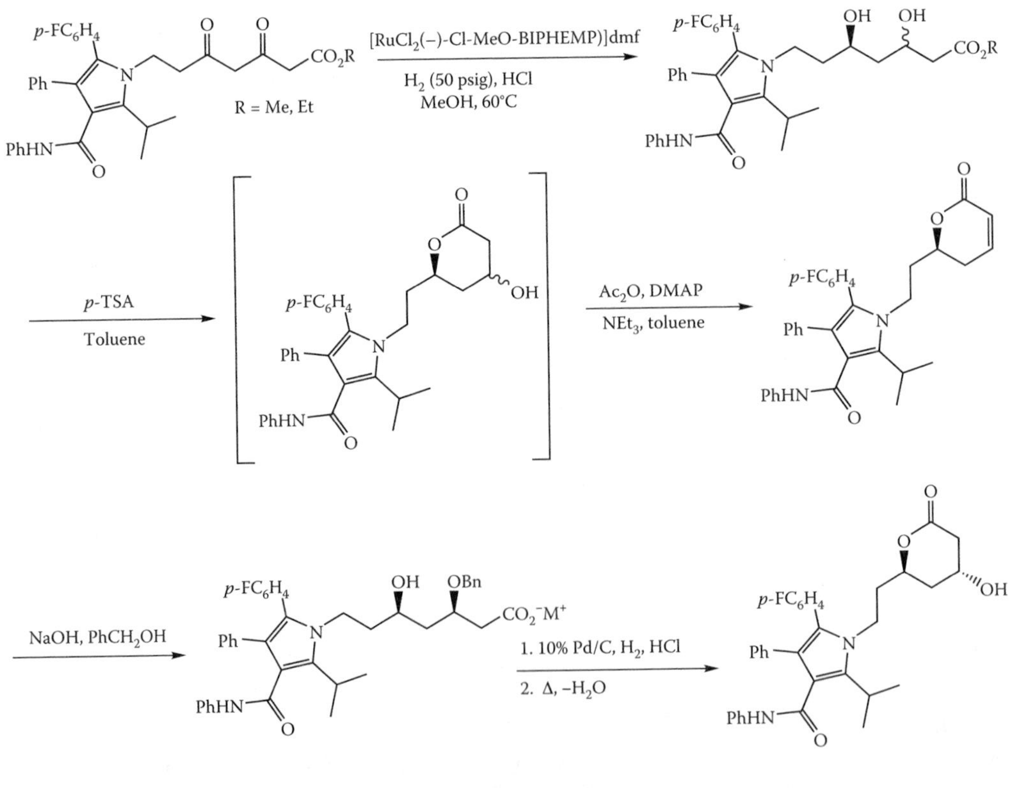

SCHEME 12.35

12.3.2.1.4 SegPhos

It has been well-established that small changes in the steric and/or electronic properties of a chiral ligand can have tremendous effect on both substrate selectivity and enantioselectivities. Takasago theorized that small changes in the dihedral angle of the biaryl backbone could change the enantioselectivity during asymmetric hydrogenation. Catalysts that contain similar biaryl bisphosphines ligands, such as **96** and **106**, produced high enantioselectivities in the asymmetric reduction of 2-oxo-1-propanol to (2R)-1,2-propanediol. Takasago observed a correlation between the dihedral angle of the biaryl backbone versus enantioselectivity. The enantioselectivity of this reduction increased in the following order: BINAP (**3a**) (89.0% ee), BIPHEMP (**106a**) (92.5% ee), and MeO-BIPHEP (**96a**) (96.0% ee). The dihedral angles of the Ru-complexes were estimated by a CAChe MM2 calculation to be 73.49 degrees (BINAP), 72.07 degrees, (BIPHEMP), and 68.56 degrees (MeO-BIPHEP). This correlation between dihedral angle and enantioselectivity suggested that higher enantioselectivities might be obtained by narrowing the dihedral angle of the biaryl backbone. This approach was pursued in the design of SegPHOS (**107**). SegPHOS is an atopisomeric ligand that contains 1,3-benzodioxole rings in the biaryl backbone.[129] The dihedral angle of a Ru-SegPHOS complex was estimated to be 64.99 degrees by the molecular mechanics calculations.[130]

107

a, Ar = Ph (SegPHOS)
b, Ar = 3,5-Me$_2$C$_6$H$_3$ (DM-SegPHOS)
c, Ar = 3,5-(t-Bu)-4-MeOC$_6$H$_2$ (DTBM-SegPHOS)

The decrease in the dihedral angle of the biaryl backbone had a profound effect on both reactivity and enantioselectivity. It was determined that the catalyst [NH$_2$Me$_2$]$^+$[{RuCl(R-SegPHOS)}$_2$(μ-Cl)$_3$]$^-$ can hydrogenate 2-oxo-1-propanol to (2R)-1,2-propanediol in 98.5% ee and with an S/C ratio of 10,000. (2R)-1,2-propanediol is used in the production of Levofloxacin (see Section 12.2.2.5). The current catalyst used in production is a R-Tol-BINAP-Ru(II) complex.[130]

[NH$_2$Me$_2$]$^+$[{RuCl(R-SegPHOS)}$_2$(μ-Cl)$_3$]$^-$ has been found to be very active and stereoselective compared to catalysts that contain BINAP and other biaryl ligands previously mentioned in the asymmetric hydrogenation α-keto esters, such as 3-oxo-3-phenylpropionate (SegPHOS 97.6% ee, BINAP 87.0% ee); β-keto esters, such as ethyl 4-chloroacetoacetate (98.5% ee SegPHOS, 95.9% ee BINAP) at significantly lower hydrogen pressures (~350 psig); and γ-keto esters.[130]

Several variations of SegPHOS (**107a**) have been prepared, in which the diarylphosphine moiety is varied as with, for example, DM-SegPHOS (**107b**) and DTBM-SegPHOS (**107c**). [NH$_2$Me$_2$]$^+$ [{RuCl((–)-DTBM-SegPHOS)}$_2$(μ-Cl)$_3$]$^-$ can be used to perform the kinetic dynamic resolution described earlier in the carbapenem production (Scheme 12.8).[130]

Takasago has patented an alternate route to l-menthol (**22**) (Scheme 12.36).[131] Chirality is set by the rhodium catalyzed asymmetric hydrogenation of piperitenone (**108**). Although many chiral biaryl bisphosphines catalysts have been used, DTMB-SegPHOS (**107c**) produced pulegone (**109**) in 90% yield and 98% ee with an S/C ratio of 50,000.[131]

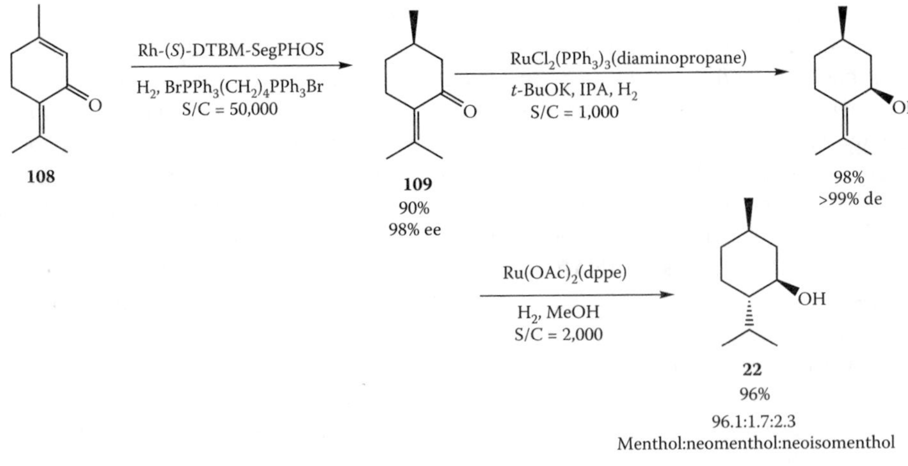

SCHEME 12.36

12.3.2.1.5 TunePhos Family

In keeping with the relationship between the dihedral angle/bite angle of the diaxial bisphosphines and enantioselectivities, two groups independently developed new chiral bisphosphines derived from MeO-BIPHEP (**96a**), which contain a linkage between the two aromatic rings that restrict rotation within the ligand. Zhang's group and Takasago filed patents on **110c** (C_3-TunePhos) at almost the same time.[132,133] Whereas Takasago only reported on the preparation and application of **110c**, Zhang prepared a family of chiral atropisomeric bisphosphines **110** with tunable bite angles and investigated the enantioselectivities in the asymmetric hydrogenation of β-keto esters,[134] enol acetates,[135] α-phthalamide ketones,[136] and tetrasubstituted cyclic β-(acylamino)acrylates.[137] This class of ligands has been named TunePhos (**110**), and the asymmetric hydrogenations of three model substrates are summarized in Table 12.3 and Schemes 12.37–12.39. In each case, the tunable bite angles of the bisphosphines allowed for successful screening of the optimal ligand in any particular asymmetric transformation.

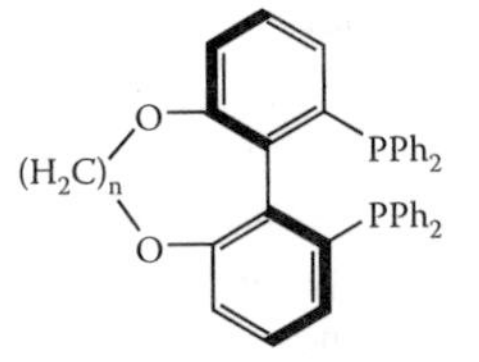

110

(S)-TunePhos

a, n = 1; C_1-TunePhos   **b**, n = 2; C_2-TunePhos
c, n = 3; C_3-TunePhos   **d**, n = 4; C_4-TunePhos
e, n = 5; C_5-TunePhos   **f**, n = 6; C_6-TunePhos

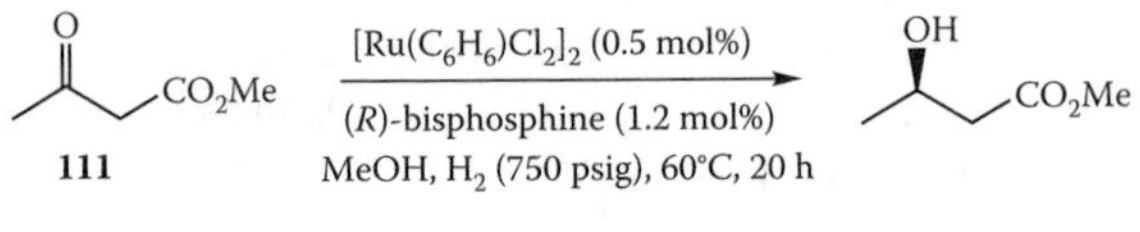

SCHEME 12.37

TABLE 12.3
Comparison of Enantioselectivity and Dihedral Angle in the Asymmetric Hydrogenation of Various Compounds

Phosphine	110a	110b	110c	110d	110e	110f	96a	3a
Dihedral angle (degree)[a]	60	74	77	88	94	106	87 (69)[b]	87 (73)[b]
Compound 111[c]	90.9	90.8	97.7	99.1	97.1	96.5	97.9	98.4
Compound 112[c]	95.9	95.9	92.1	88.9	91.9	92.3		
Compound 113[c]	98	99	99	99	99	97	99	99

[a] Calculated dihedral angles from CAChe MM2 program.
[b] CAChe MM2 calculation of dihedral angle in the Ru-complexes that contain these ligands.[138]
[c] Reaction conditions shown in Schemes 12.37–12.39.

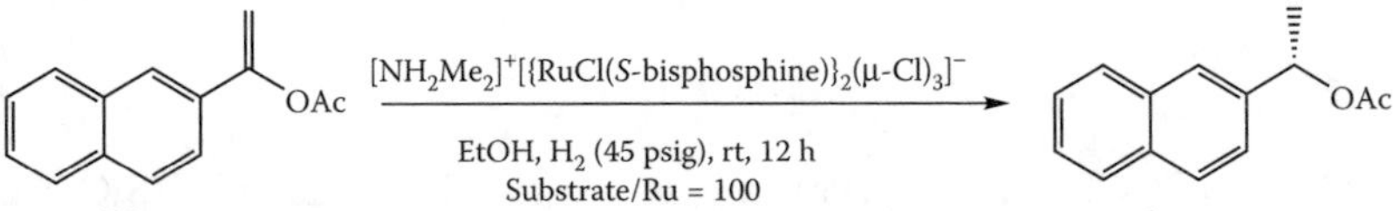

SCHEME 12.38

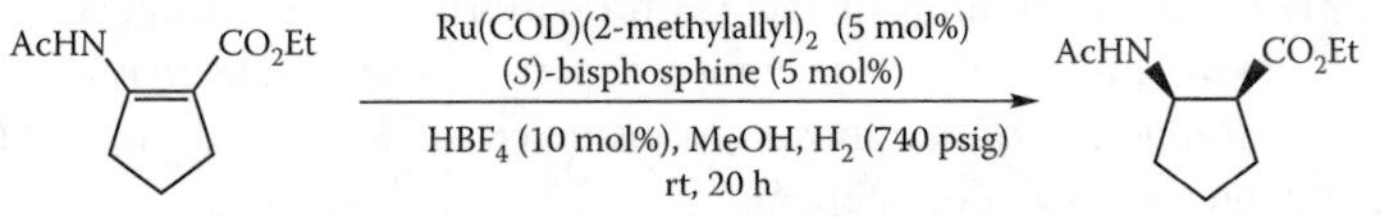

SCHEME 12.39

These ligands can be prepared by two methods: demethylation of enantiomerically pure MeO-BIPHEP (**96a**) followed by alkylation with the corresponding dihalide to form the cyclic diether backbone (Scheme 12.40)[134] or an intramolecular Fe(III)-promoted oxidative coupling of a *meta*-substituted bis(arylphosphine oxide) connected by a 1,3-propoxyl unit followed by resolution with DBTA and reduction to the bisphosphines (Scheme 12.41).[133]

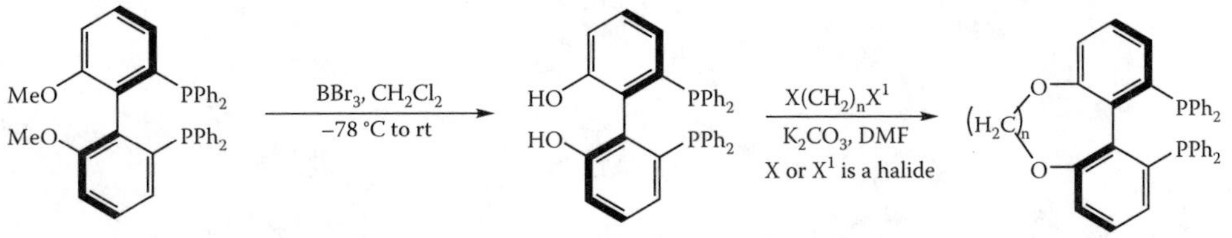

110a: 80%
110b: 61%
110c: 82%
110d: 84%
110e: 55%
110f: 55%

SCHEME 12.40

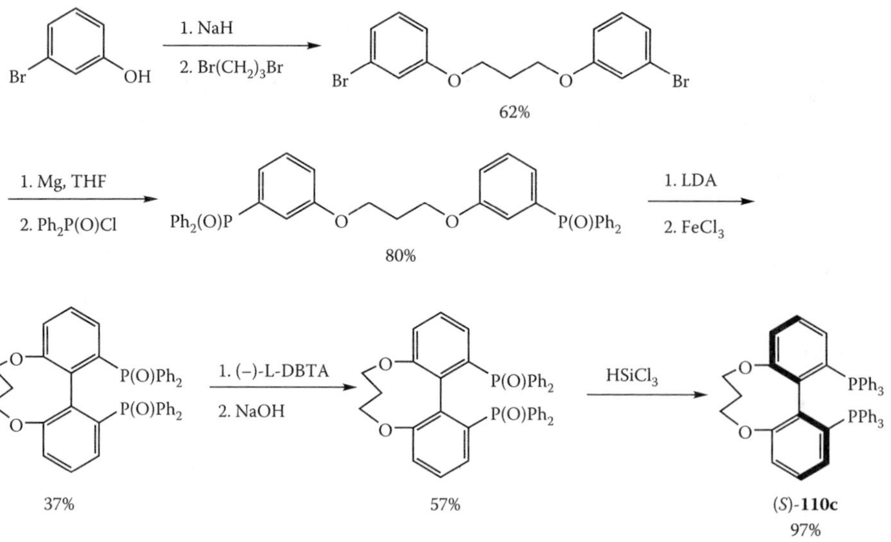

SCHEME 12.41

Metal complexes of C₃-TunePhos (**110c**) have demonstrated high activities (S/C = 45,000) and enantioselectivities (98–99%) in the reduction of ethyl 4-chloroacetoacetate (ECAA).[139] Chiral Quest disclosed that this catalyst has been supplied in kilogram quantities to a client for the commercial production of ECHB, which is an intermediate in the production of Lipitor (see Chapter 31).[76]

Chan has reported that substitution of the 1,3-propa-dioxy ether linkage with a (2S,3S)-buta-dioxy ether linkage, as in **114**, resulted in complete atropdiastereoselective coupling of the aryl units (Scheme 12.42). Reduction of the phosphine oxide (R)-**115** resulted in (R)-**116** in 96% yield. This procedure eliminates a resolution step.[140]

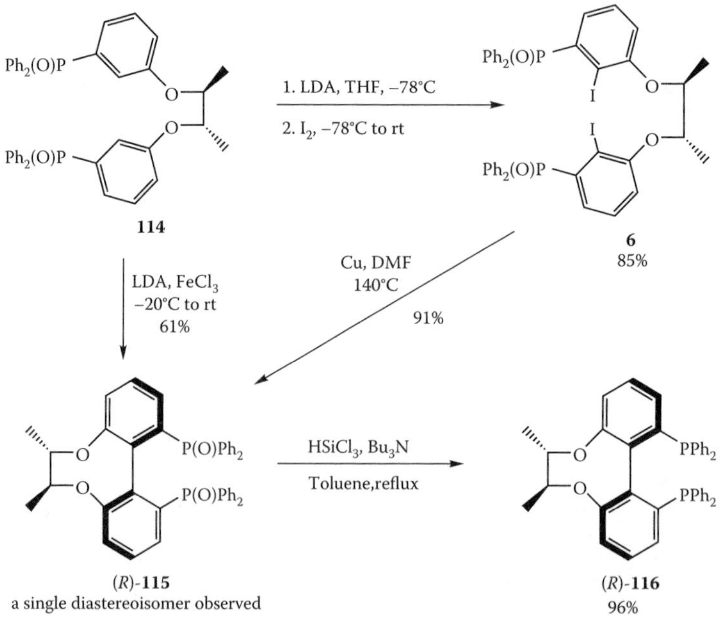

SCHEME 12.42

Comparisons in the asymmetric hydrogenation of methyl acetoacetate (111) with ruthenium catalysts that contain C_2-Tunephos (110b) and 116 indicated an improvement in enantiomeric excess when the ether linkage contained additional chirality.[134,140]

12.3.2.1.6 Synphos and Difluorphos

The success of Takasago's SegPHOS (107a) in various asymmetric hydrogenations, which spawn from variation of the dihedral angle, inspired two independent groups to develop a bisphosphphine that contain a bis-benzodioxane structure named SYNPHOS (117a). Genet's group (in collaboration with SYNKEM)[141] and Chan's group[142] reported nearly identical synthetic procedures (Scheme 12.43).[141–144] This chemistry has been performed at multi-kilogram scale after extensive optimization.[145]

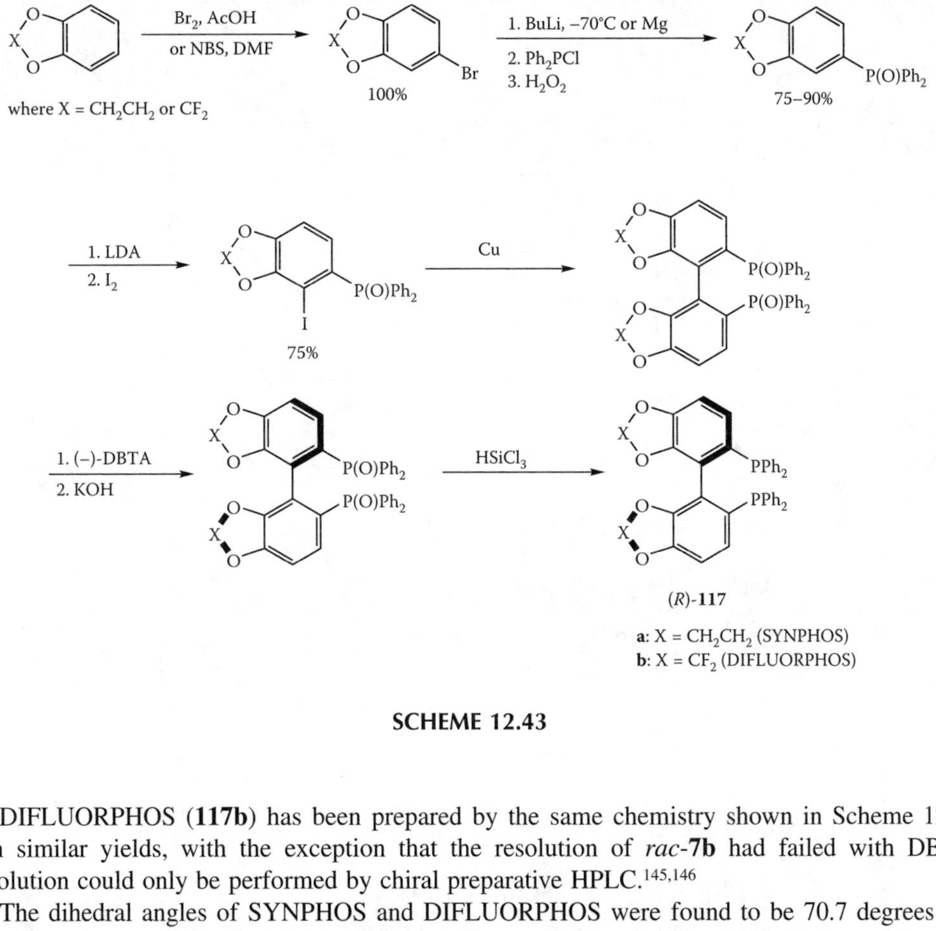

SCHEME 12.43

DIFLUORPHOS (117b) has been prepared by the same chemistry shown in Scheme 12.43 with similar yields, with the exception that the resolution of rac-7b had failed with DBTA. Resolution could only be performed by chiral preparative HPLC.[145,146]

The dihedral angles of SYNPHOS and DIFLUORPHOS were found to be 70.7 degrees and 67.6 degrees, respectively, by CAChe MM2 calculations (for comparison, SegPHOS was calculated to be 67.2 degrees) and some of the narrowest dihedral angles reported in the biphenyl backbone class of atropisomeric ligands (with the exception of C_1-TunePHOS).[144,145]

Metal complexes of 117a and 117b produced high enantioselectivities (98–99% ee) in the asymmetric hydrogenation of β-keto esters[141,142,146] with S/C ratios of 10,000.[145] The production of ethyl (R)-4-hydroxybutyrate (ECHB) from the corresponding β-keto esters (ECAA) has been achieved with 97% ee.[142] Deviation of enantioselectivities was more prevalent in the asymmetric hydrogenation of α-keto esters and fluorinated β-keto esters and diketones.[145] [Ru(117a)Br₂] reduced α-keto esters with 92–94% ee, whereas the 117b catalyst achieved only 67–87% ee. [Ru(DIFLUORPHOS)Br₂] reduced 118 to anti-(R,R)-119 in 98% ee and 86% de compared to 85% ee and 67% de for the SYNPHOS catalyst (Scheme 12.44).

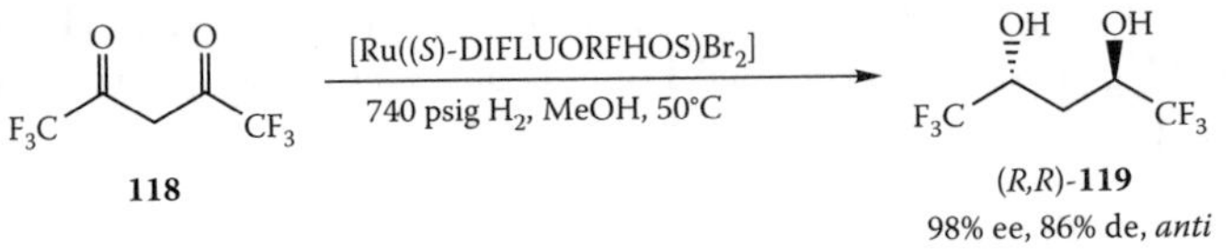

SCHEME 12.44

12.3.2.1.7 P-Phos Ligands

In this class of ligands, the biphenyl backbone has been replaced with a dipyridylphosphine and called P-Phos (**120**). (These ligands have been licensed to Johnson Matthey.) The synthesis is shown in Scheme 12.45.[147-149]

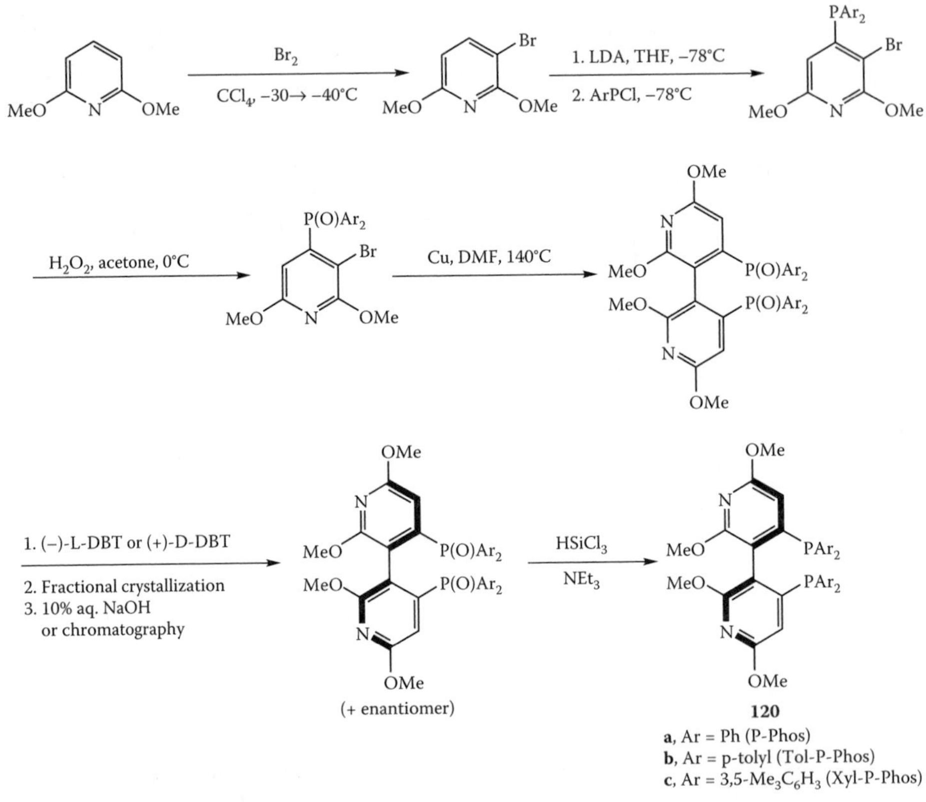

a, Ar = Ph (P-Phos)
b, Ar = p-tolyl (Tol-P-Phos)
c, Ar = 3,5-Me$_3$C$_6$H$_3$ (Xyl-P-Phos)

SCHEME 12.45

This class of ligand has been used in the asymmetric hydrogenations of enamides to form α-amino acids (73–97% ee),[150] E-β-alkyl-substituted β-(acylamino)acrylates to β-amino acids (97.0–99.7% ee from the E-β-(acylamino)acrylates; 55–83% from the Z-isomer),[151] α,β-unsaturated carboxylic acids, and β-keto esters (91–98% ee).[147–149] The enantiomeric ratios and activities usually are lower that those of other catalysts derived from BINAP. Implementation of hydrogenation conditions developed by Japan Science and Technology (JST) (see Section 12.3.4), addition of a chiral diamine to the ruthenium complex,[152] improved the enantioselectivities of β-keto ester reductions up to 99% ee with S/C ratios as high as 100,000.[153] Naproxen has been produced in 87–96% ee from the corresponding olefin with ruthenium catalysts.[147] In all cases, the higher enantioselectivities were obtained with phosphines with larger substitutes (Xyl-P-Phos>Tol-P-Phos>P-Phos).

The asymmetric 1,4-addition of aryl boronic acids to α,β-unsaturated ketones has been reported with rhodium catalysts that contained P-Phos (**120a**) (Scheme 12.46).[154]

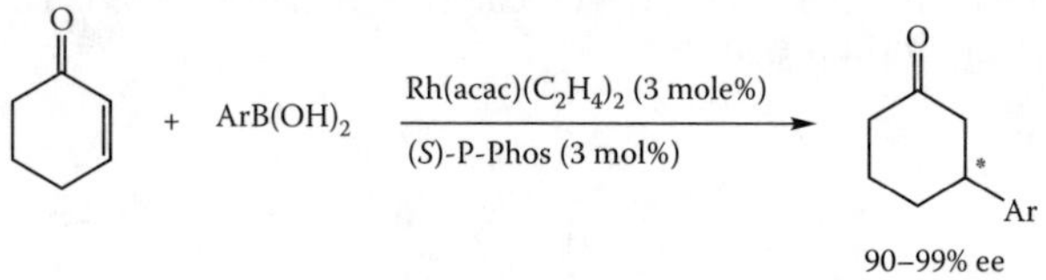

SCHEME 12.46

Despite the lower catalytic activities and enantioselectivities to other axial bisphosphines, the greatest advantage of P-Phos is its stability to air. All ligands, intermediates, and catalysts have shown a remarkable tolerance to air,[148–150,153] which can greatly simplify production operations that need to perform time-consuming air-exclusion procedures.

12.3.2.1.8   4,4-Bis(diphenylphosphino)-2,2′,5,5′-tetramethyl-3,3′-dithiophene

Another example of a chiral axial ligand with a heteroaromatic backbone is 4,4-bis(diphenylphosphine)-2,2′5,5′-tetramethyl-3,3′-dithiophene (**121**) (TMBTP), which has been used in asymmetric hydrogenations at production scale. (This ligand system is used by Chemi SpA.) In the past, most hydrogenation processes tended to shy away from any sulfur-containing compounds because they can poison these reactions. This ligand, among others in the thiophene class, has demonstrated high catalytic activity and enantioselectivities. The synthesis is shown in Scheme 12.47.[155]

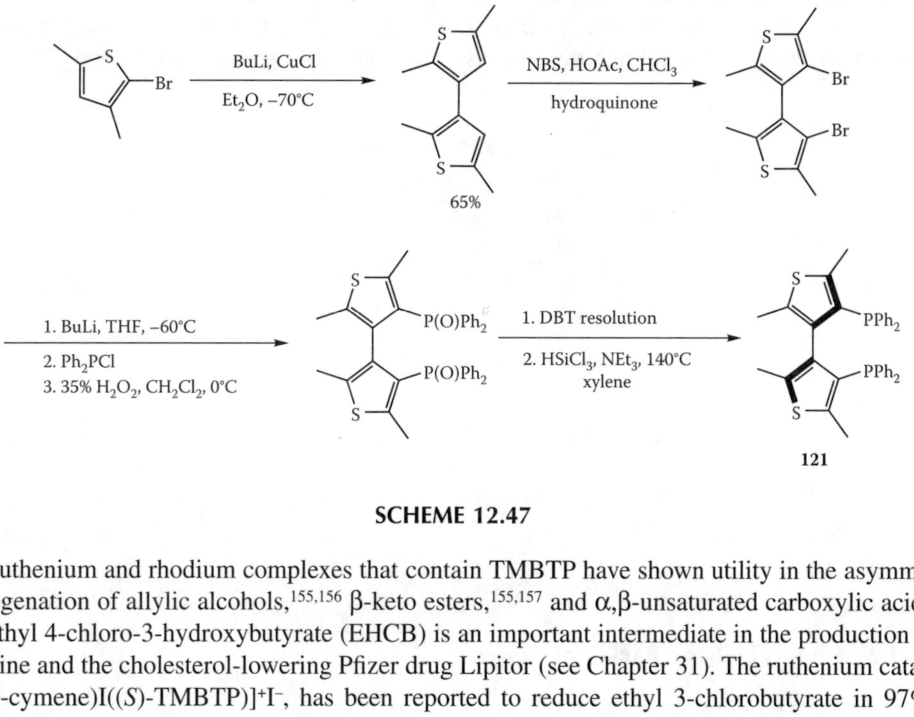

SCHEME 12.47

Ruthenium and rhodium complexes that contain TMBTP have shown utility in the asymmetric hydrogenation of allylic alcohols,[155,156] β-keto esters,[155,157] and α,β-unsaturated carboxylic acids.[155]

Ethyl 4-chloro-3-hydroxybutyrate (EHCB) is an important intermediate in the production of L-carnitine and the cholesterol-lowering Pfizer drug Lipitor (see Chapter 31). The ruthenium catalyst, [Ru(p-cymene)I((S)-TMBTP)]⁺I⁻, has been reported to reduce ethyl 3-chlorobutyrate in 97% ee with S/C ratios of 20,000 (Scheme 12.48).[87,157] This chemistry has been reported at scales >100 kg.[87]

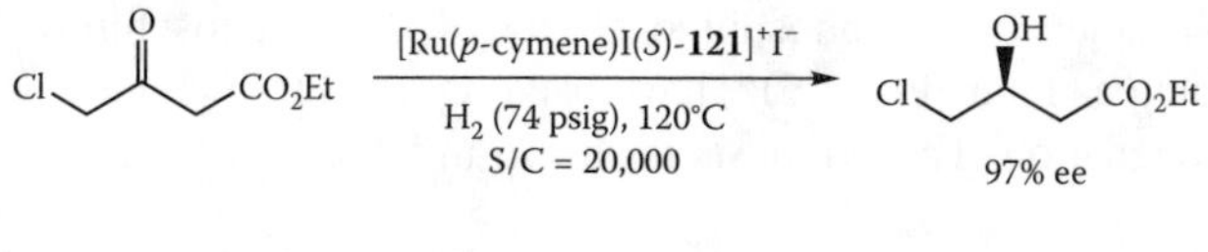

SCHEME 12.48

Astra Zeneca's ZD 3523 (**122**),[158] a potential leukotriene antagonist for the treatment of asthma, was prepared by an asymmetric hydrogenation of a 96:4 mixture of *E:Z*-4,4,4-trifluoro-2-methyl-2-butenoic acid (**123**) with a ruthenium/TMBTP catalyst (Scheme 12.49). This chemistry has been demonstrated at the multi-100-kg scale.[87]

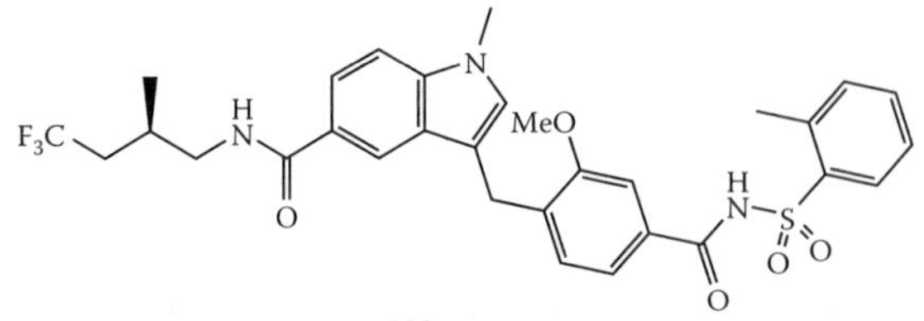

122

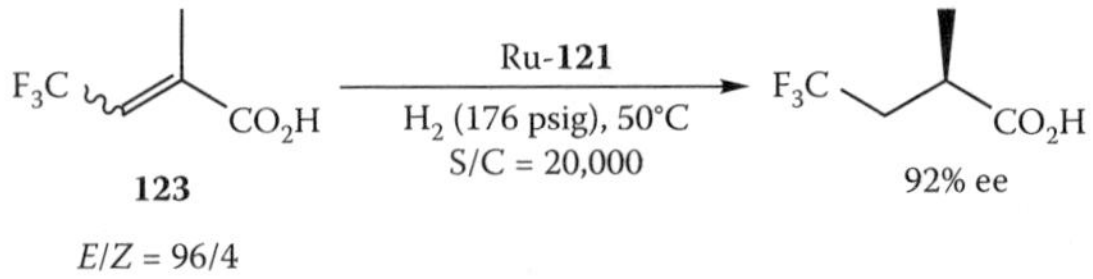

SCHEME 12.49

The decahydroisoquinoline derivative NVP-ACQ090 (**124**) is a potent and selective antagonist at the somatostatin sst$_3$ receptor. The asymmetric hydrogenation of allylic alcohol **125** with a rhodium catalyst that contained (*S*)-TMBTP produced (*R*)-**126** (97.5% ee) (Scheme 12.50). The authors indicated that enantiomerically pure **124** could be obtained from material acquired by asymmetric hydrogenation and that the process is suitable for large-scale production.[156]

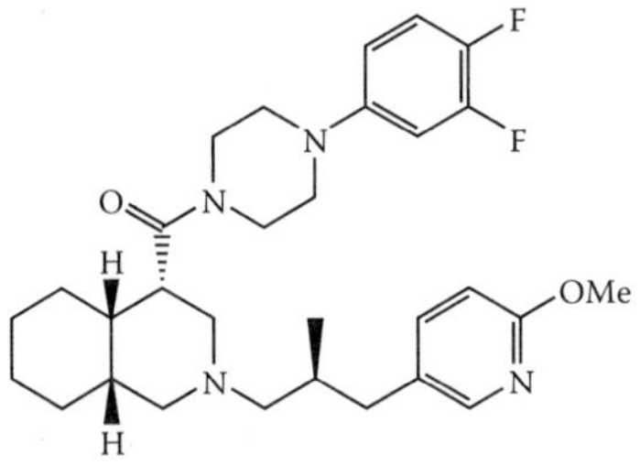

124

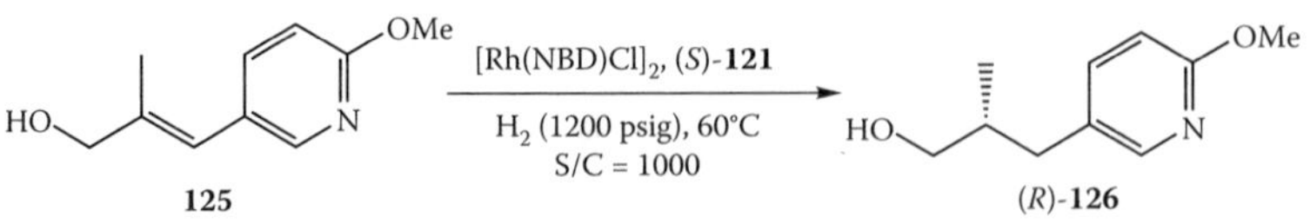

SCHEME 12.50

12.3.2.2 Paracyclophane Backbone Series

12.3.2.2.1 PhanePhos

A unique alternative to the traditional C_2 symmetric atopisomeric motif uses a paracyclophane backbone for the placement of the phosphino groups. 4,12-Bis(phosphino)-[2.2]-paracyclophane complexes, abbreviated as PhanePhos (**5**)*, have been reported to be highly active in a few classes of asymmetric hydrogenation. The synthesis is shown in Scheme 12.51.[159,160]

---

* This ligand system was developed at Merck and has been licensed by Dow Pharma and Johnson Matthey.

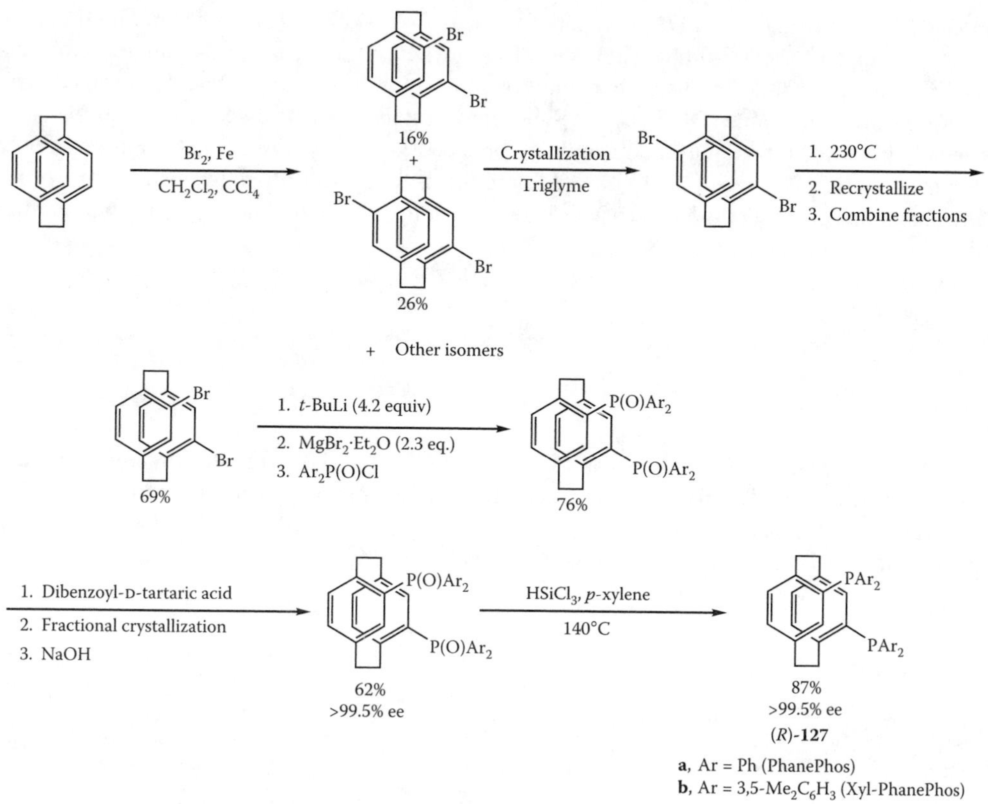

SCHEME 12.51

High catalytic activities and enantioselectivities are observed in the asymmetric reduction of enamides esters to N-protected α-amino esters. As a result of the high catalytic activity, hydrogenations can be routinely accomplished at –40°C and 22 psig H_2 without detriment to the enantioselectivity (94–99.6% ee).[159] Catalyst preparation is critical for obtaining the high activity. [Rh(COD)(PhanePhos)]$^+$OTf$^-$ is subjected to H_2 prior to addition of the substrate to produce [Rh(MeOH)$_2$(PhanePhos)]$^+$OTf$^-$.[159] The free carboxylic acid form of the enamides did not give high enantioselectivities on reduction.

Ruthenium complexes of PhanePhos have been used in the asymmetric hydrogenation of β-keto esters. Similar to the enamides reduction example, catalyst preparation is critical to the activity and consistency of the reduction. Initial catalyst preparation formed the ruthenium complex from Ru(2-methylallyl)$_2$(COD) followed by acidification with HBr in MeOH/acetone. The catalyst activity decreased rapidly on storage of catalyst solution and less rapidly in the solid state. Substitution of trifluoroacetic acid (TFA) in acetone instead of HBr produced Ru(PhanePhos)(CF$_3$CO$_2$$^-$)$_2$, which exhibited moderate enantioselectivity and activity in the asymmetric hydrogenation of ethyl isobutyrylacetate (75% ee).[161] Treatment of Ru(PhanePhos)(CF$_3$CO$_2$$^-$)$_2$ with excess Bu$_4$N$^+$I$^-$ produces a catalyst that exhibits high activity and enantioselectivities (95–96% ee) over a wide temperature range (–10°C to 50°C) for a variety of substrates. This catalyst can be stored under argon for weeks without loss of activity or enantioselectivity.

Preparation of the JST class (see Section 12.3.4) of the ruthenium–diphosphine–diamine complex, [(PhanePhos)Ru(diamine)Cl$_2$], produced highly active and enantioselective catalysts in the reduction of aryl methyl ketones (**128**, R = Me), as well as α,β-unsaturated ketones.[162,163] Higher enantioselectivities and catalyst activities (albeit small) are routinely observed with catalysts that contained Xylyl-PhanePhos (**127b**) compared to PhanePhos (**127a**). The choice of chiral diamine can be expanded to use 1,2-diphenylethylenediamine (DPEN) and *trans*-1,2-diaminocyclohexane (DACH), which are less expensive and more readily available at scale than 1,1-dianisyl-2-isopropyl,1,2-ethylenediamine (DIAPEN).[162] Substrate-to-catalyst ratios ranged from 3000 to 40,000. Extremely high S/C ratios of 100,000 could be obtained in the asymmetric reduction of 4-fluoroacetophenone (**128**, Ar = *p*-FC$_6$H$_4$, R = Me) if the ketone is purified by distillation prior to hydrogenation. This observation has been reproduced at the kilogram scale with preparation of the corresponding chiral alcohol in 98.3% ee.[163]

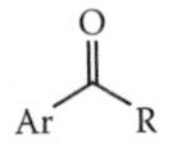

128

A potential industrial example with [((*S*)-**127a**)Ru((*R*,*R*-DPEN)Cl$_2$] is shown in Scheme 12.52 for the preparation of Eli Lilly's Duloxetine® (**129**), an antidepressant used for the treatment of urinary incontinence. Asymmetric hydrogenation of **130** produced **131** with 93.4% ee at kilo scale. This material has been taken onward to **129** with enantiomeric enhancement to >99%.[164]

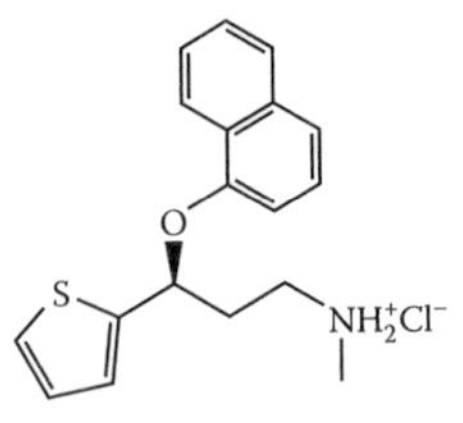

129

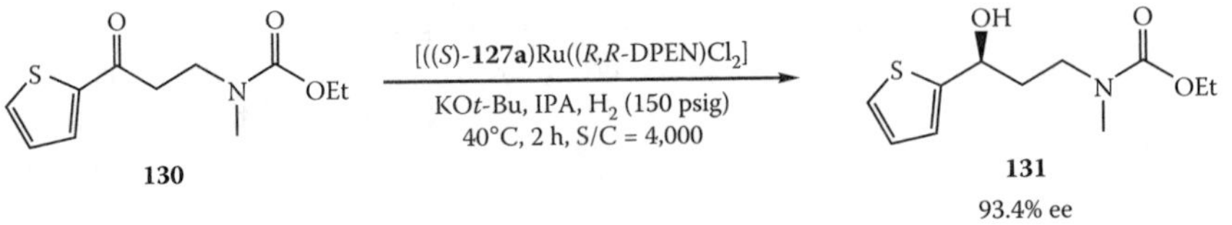

SCHEME 12.52

12.3.3 New Generation Ferrocenylbisphosphines

12.3.3.1 BoPhoz

BoPhoz (**132**) can be classified as an unsymmetric ferrocenyl aminophosphine ligand and has demonstrated hydrolytic and air-stability (see also Chapter 15). (This ligand system was developed at Eastman Kodak and has been licensed by Johnson Matthey.) The ligand has been successfully used in the asymmetric hydrogenation of enamide esters, itaconates, and α-keto esters.[165] BoPhoz can be easily prepared in two steps from the Ugi amine (**54**), of which synthesis has been further improved for the preparation of BoPhoz (Schemes 12.53 and 12.54).[165,166] The *S*,*R*-isomer of BoPhoz begins with (*S*)-**133**, which is converted to the acetate with Ac$_2$O and reacted with H$_2$NR^1 to give (*S*,*R*)-**134**. (*S*,*R*)-BoPhoz (**132**) is produced upon addition of the chlorophosphine, R$_2$PCl.[166]

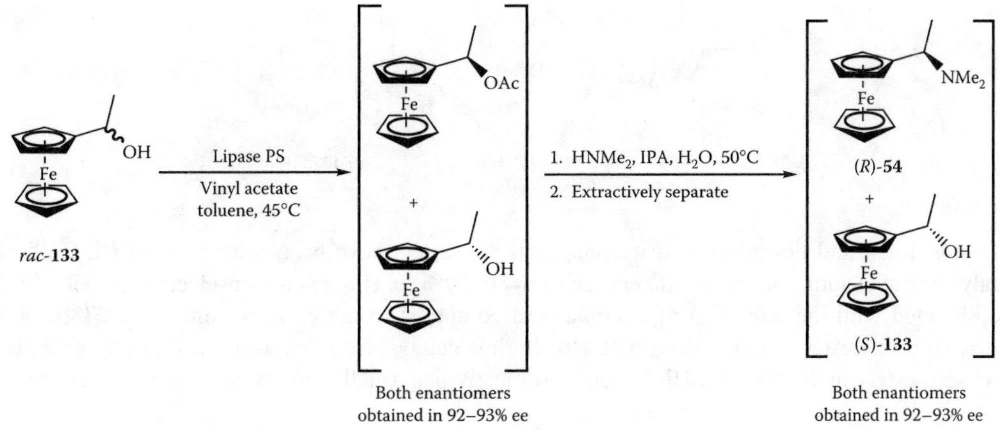

SCHEME 12.53

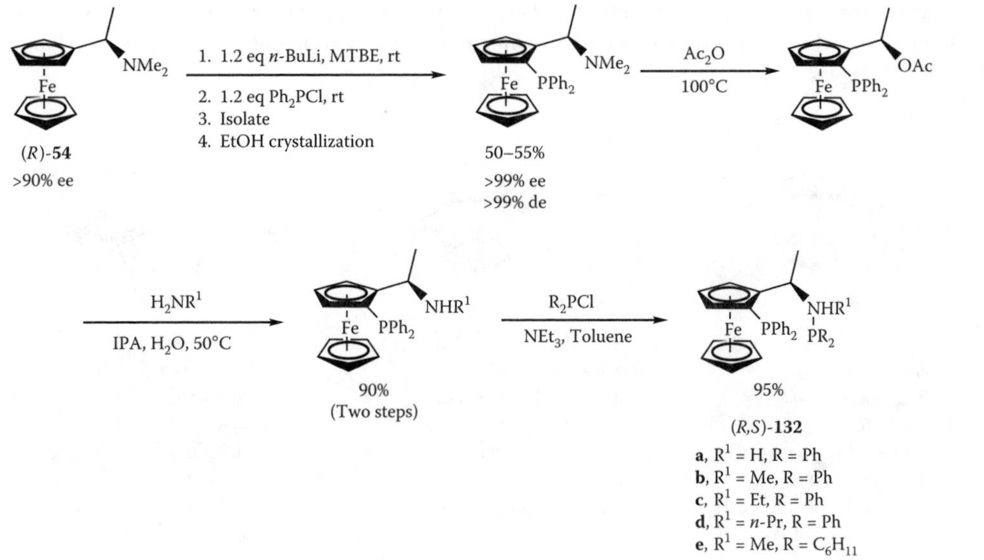

SCHEME 12.54

Rhodium catalysts with BoPhoz ligands have demonstrated high enantioselectivities in the asymmetric hydrogenation of various enamides esters to protected α-amino esters (93.0–99.5% ee). Enamide esters were reduced with consistently higher enantioselectivities compared to the corresponding acids. High S/C ratios of 10,000 have been reported along with catalyst activities of 30,000 turnovers/hour. High enantioselectivities are observed with enamide substrates that contain Ac, Cbz, and Boc groups on the amido group (Scheme 12.55).[165,167–169] Two examples of amino esters that have potential industrial application are cyclopropylalanine (**135**), which has been used in a variety of pharmaceutical agents in renin inhibitors[170] and inflammatory disorders (which include rheumatoid arthritis),[171,172] and ethyl 2-(2-oxopyrrolin-1-yl)butyrate (**136**), a potential intermediate in the drug Levetiracetam (**137**) (which is on the market for the treatment of epilepsy).[173]

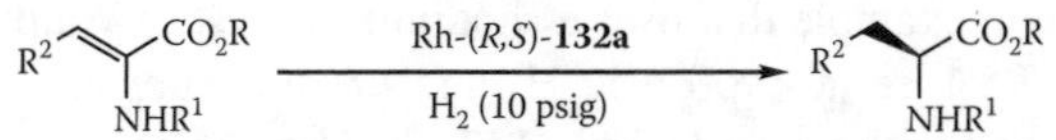

SCHEME 12.55

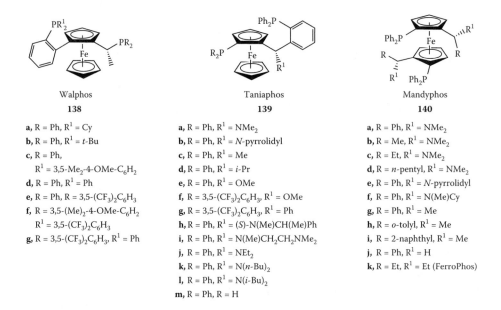

Substituted and nonsubstituted itaconic acid and esters have been reduced with Rh-BoPhoz catalysts with enantioselectivities that range from 80–99% ee. Higher enantioselectivities (89–99%) are obtained with the reduction of itaconic acid compared to the corresponding esters (80–94% ee). α-Hydroxy esters are produced with Rh-BoPhoz catalyst from the asymmetric hydrogenation of α-keto esters in 46–97% ee. Higher enantioselectivities usually are observed with the electron-rich dicyclohexylphosphino ligand (**132e**).

12.3.3.2 Miscellaneous Ferrocenylphosphines

A series of uniquely shaped chiral ferrocenyl-based bisphosphines are available from Solvias. These ligand families (Figure 12.6) exploit a highly unsymmetric environment on the ferrocenyl backbone (see also Chapter 15).

Walphos **138**	**Taniaphos** **139**	**Mandyphos** **140**

a, R = Ph, R^1 = Cy
b, R = Ph, R^1 = t-Bu
c, R = Ph,
  R^1 = 3,5-Me$_2$-4-OMe-C$_6$H$_2$
d, R = Ph, R^1 = Ph
e, R = Ph, R = 3,5-(CF$_3$)$_2$C$_6$H$_3$
f, R = 3,5-(Me)$_2$-4-OMe-C$_6$H$_2$
  R^1 = 3,5-(CF$_3$)$_2$C$_6$H$_3$
g, R = 3,5-(CF$_3$)$_2$C$_6$H$_3$, R^1 = Ph

a, R = Ph, R^1 = NMe$_2$
b, R = Ph, R^1 = N-pyrrolidyl
c, R = Ph, R^1 = Me
d, R = Ph, R^1 = i-Pr
e, R = Ph, R^1 = OMe
f, R = 3,5-(CF$_3$)$_2$C$_6$H$_3$, R^1 = OMe
g, R = 3,5-(CF$_3$)$_2$C$_6$H$_3$, R^1 = Ph
h, R = Ph, R^1 = (S)-N(Me)CH(Me)Ph
i, R = Ph, R^1 = N(Me)CH$_2$CH$_2$NMe$_2$
j, R = Ph, R^1 = NEt$_2$
k, R = Ph, R^1 = N(n-Bu)$_2$
l, R = Ph, R^1 = N(i-Bu)$_2$
m, R = Ph, R = H

a, R = Ph, R^1 = NMe$_2$
b, R = Me, R^1 = NMe$_2$
c, R = Et, R^1 = NMe$_2$
d, R = n-pentyl, R^1 = NMe$_2$
e, R = Ph, R^1 = N-pyrrolidyl
f, R = Ph, R^1 = N(Me)Cy
g, R = Ph, R^1 = Me
h, R = o-tolyl, R^1 = Me
i, R = 2-naphthyl, R^1 = Me
j, R = Ph, R^1 = H
k, R = Et, R^1 = Et (FerroPhos)

FIGURE 12.6 Various classes of ferrocenyl ligands offered by Solvias.

Walphos (**138**), developed by Sturm at the University of Vienna and optimized at Solvias, is derived from the Ugi amine **54**.[174,175] Walphos can be electronically fine-tuned as various phosphine groups are introduced in separate steps of the synthesis. Walphos catalysts have been used to reduce enamide esters to α-amino esters, β-keto esters, α,β-unsaturated carboxylic acids, and itaconate esters with enantioselectivities >90%.[176]

A pilot plant production example that uses a rhodium catalyst of Walphos is shown in Scheme 12.56 to produce **141** in 95% ee at <900 psig H$_2$ and <80°C (Sub/cat = 5700).[175,176] Compound **141** has been converted to Synthon A alcohol (**142**, 200-kg scale), which is an intermediate to SPP100 (**143**), a renin inhibitor from Speedel Pharma AG.[175]

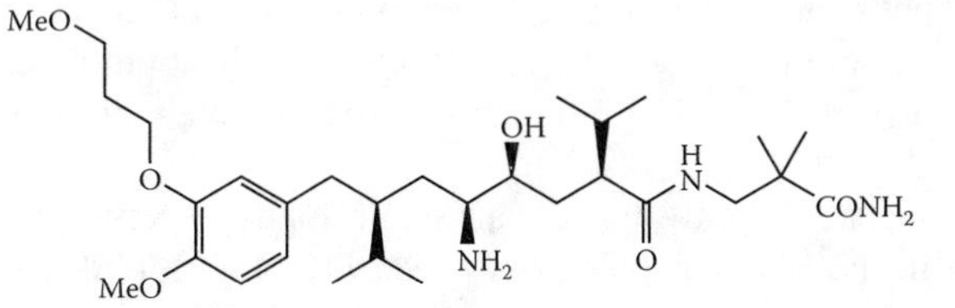

143

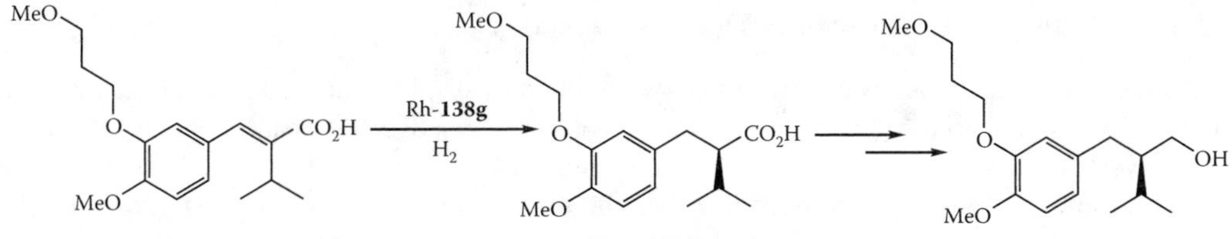

141
100% conversion
95% ee

142

SCHEME 12.56

Taniaphos (**139**), developed by Knochel (with cooperation with Degussa), can be electronically fine-tuned by variation of the phosphine groups and the groups on the chiral sidechain (R^1).[177,178] Taniaphos catalysts have been used to reduce enamide esters to α-amino esters,[178–181] β-acylamidoacrylates esters to β-amino esters,[175] 1,3-diketones,[179,180] β-keto esters,[179,180] enol acetate,[179,180] enamides to amide,[179] and itaconate esters[179,180] with enantioselectivities >90%.

The synthesis is outlined in Scheme 12.57.[178,180] Syntheses of other groups on the chiral sidechain (R^1 = alkyl or OMe) are reported elsewhere in literature.[179,181]

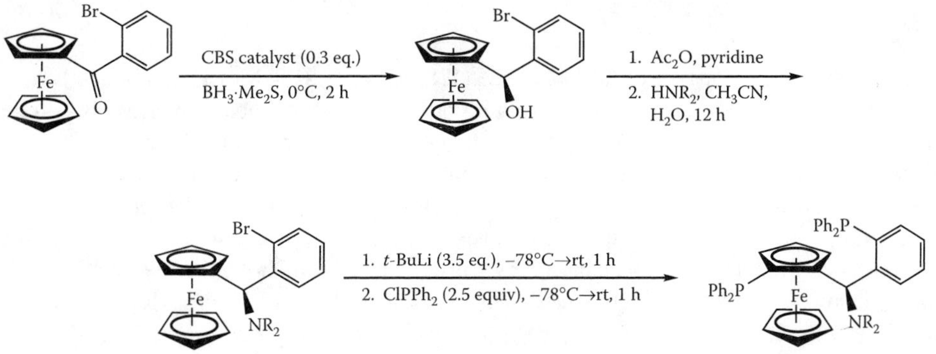

139

SCHEME 12.57

Mandyphos (**140**), previously known as Ferriphos) was originally synthesized by Hayashi[182] and further developed by Knochel (with the cooperation of Degussa).[183,184] The synthesis is similar to that of Walphos such that each cyclopentadiene unit contains a chiral alcohol unit prepared by CBS-catalyst reduction of 1,1'-diacetylferrocene. The chiral diol intermediate is converted to the bis-Ugi amine complex by acetylation and reaction with an amine.[183–185] Stereoselective dilithiation and chlorophosphine quench produced Mandyphos with diamine groups (**140a–g**). Preparation of the bisphosphines **140h–j** with chiral (and achiral) alkyl appendages required additional manipulations.[183]

Rhodium Mandyphos catalysts have been used to reduce enamide esters and acids with enantioselectivities that range from 95% to 99%.[175,183–185] Other applications reported are the asymmetric hydrogenation of tiglic acid and ethyl 3,3-dimethyl-2-oxobutyrate in 97% ee and >97% ee, respectively.[175]

Walphos, Taniaphos, and Mandyphos are air-stable solids.[175] Kang prepared an air-stable bisphosphine that falls in the Mandyphos class (named FerroPhos, **140k**). Rhodium complexes of **140k** have produced N-acetyl amino acids and esters in 96–99% ee from the corresponding enamide acid or ester by asymmetric hydrogenation.[186]

12.3.4 RUTHENIUM/BISPHOSPHINES/DIAMINE CATALYSTS

The discovery of Japan Science & Technology's ruthenium–bisphosphines–diamine catalyst (**144**) (JST catalysts) marked the dawning of a new class of highly active and enantioselective homogeneous hydrogenation catalysts.[187] In the past, ruthenium-biaxial bisphosphines catalysts have been excellent catalysts for a wide variety of prochiral substrates except aryl ketones **128**.[43,188] Aryl ketones could be reduced with good to high enantioselectivities under harsh reaction conditions (high hydrogen pressures and temperatures). The addition of a diamine and an inorganic base (t-BuOK, KOH, or K_2CO_3) had significantly improved the reactivity and enantioselectivity of Ru(binap)-like catalysts.[187]

Ar = Ph, p-tolyl, 3,5-xyl
Diamine unit = **145** or **146**

Asymmetric hydrogenation of acetophenone (**128**, Ar = Ph, R = Me) with **144** can be achieved with 96–99% ee and 100% conversions at S/C ratios of 100,000. An example of high activity is observed with an S/C ratio of 2,400,000! The enantioselectivities in this example had dropped to 80%.[187]

The scope of ketone substrates that have been reduced with high enantioselectivities includes acetophenones with various ring substitutions[189] and electron-rich or deficient heteroaryl ketones, such as thienyl, furanyl, pyrolyl, and thiazolyl rings.[190] Cyclic and acyclic α,β-unsaturated ketones can be selectively reduced in high enantiomeric purity to allylic alcohols.[189,190] The catalyst tolerates α-amino ketones[191] and ortho-substituted benzophenones.[192] rac-2-Isopropylcyclohexanone can undergo dynamic kinetic resolution with Ru(S,S-DPEN)(R-BINAP)Cl₂ to give (S,S)-2-isopropylcyclohexanol in 93% ee and 99.8:0.2 cis-selectivity.[193,194]

Base sensitive ketones can be successfully reduced with high enantioselectivities treatment of **144** with $NaBH_4$ to give **147**, which no longer requires a strong base to be present in the asymmetric reduction.[195]

147

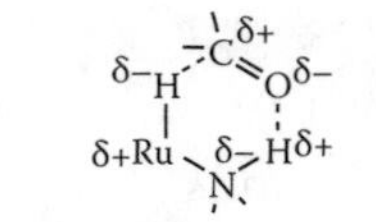

FIGURE 12.7 Transition state for carbonyl reductions with JST **144**.

The presence of an NH_2 group in the diamine is crucial. Mechanistic studies have shown, unlike other asymmetric homogeneous hydrogenation catalysts, that the ketone substrate does not coordinate to the metal but interacts with the NH_2 group and the metal hydride, as shown in the transition state (Figure 12.7). The steric environment of the overall catalyst provides excellent facial stereo-differentiation.[194]

The diamine ligands that have shown the most consistent usage are DPEN (**145**) and diapen (**146**). The best bisphosphines ligands used to date have been axial bisphosphines ligands, such as BINAP (**3a**), tol-BINAP (**3b**), xyl-BINAP (**3c**),[23,188] xyl-PhanePhos (**127b**),[162,163] and xyl-P-Phos (**120c**).[153]

A remarkable match–mismatch effect is observed. The difference in reactivity of the matched catalyst—for example, Ru(*R*-BINAP)(*S,S*-DPEN)Cl₂—can be 120 times more reactive than the mismatched catalyst, Ru(*S*-BINAP)(*S,S*-DPEN)Cl₂.[196,197]

The use of an activation/deactivation protocol with a chiral poison, (*R*)-DM-DABN (**148**), has been achieved with ruthenium catalysts that contained *rac*-xyl-BINAP and *rac*-tol-BINAP with chiral diamine (*S,S*)-DPEN. Asymmetric hydrogenation of 2-napthyl methyl ketone (**128**, Ar = 2-Naph, R = Me) without **148** gave the alcohol with 41% ee, whereas an enantioselectivity of 91% ee is obtained with deactivator **148** present (Scheme 12.58).[197]

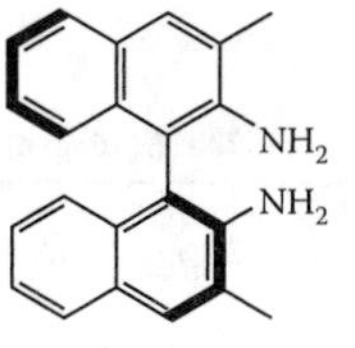

148

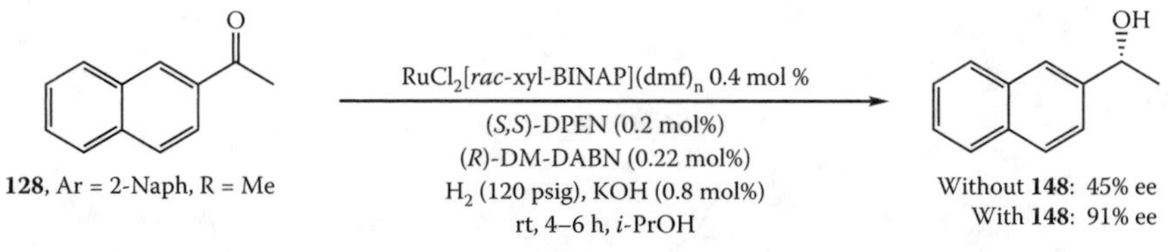

128, Ar = 2-Naph, R = Me

RuCl₂[*rac*-xyl-BINAP](dmf)ₙ 0.4 mol %

(*S,S*)-DPEN (0.2 mol%)
(*R*)-DM-DABN (0.22 mol%)
H₂ (120 psig), KOH (0.8 mol%)
rt, 4–6 h, *i*-PrOH

Without **148**: 45% ee
With **148**: 91% ee

SCHEME 12.58

Besides the production of (*R*)-1-phenylethanol as a fragrance,[198] various pharmaceutically important chiral compounds have been produced at various lab scales by asymmetric hydrogenation with JST catalysts. These compounds include a β^1-receptor antagonist denopamine hydrochloride (**149**),[191] antidepressant fluoxetine hydrochloride (**150**),[191] antipsychotic BMS 181100 (**151**),[191] serotonin and norepinephrine inhibitor duloxetine (**129**),[164] antihistaminic and anticholinergic orphenadrine (**152**),[192] and antihistaminic neobenodine (**153**).[192]

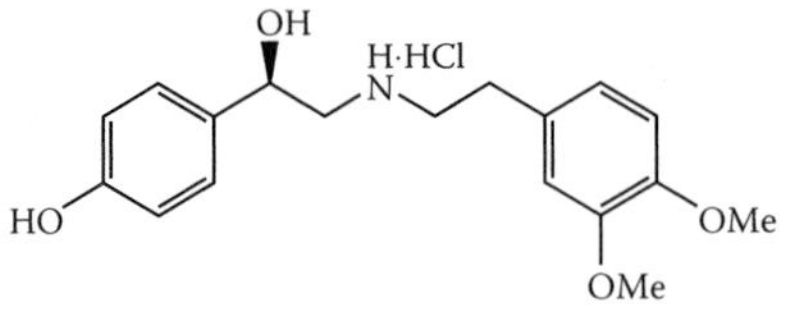

149

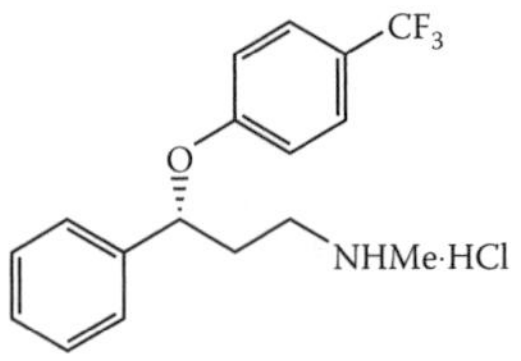

150

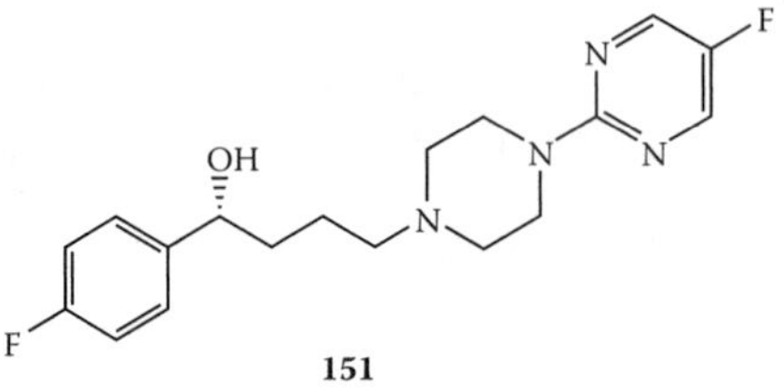

151

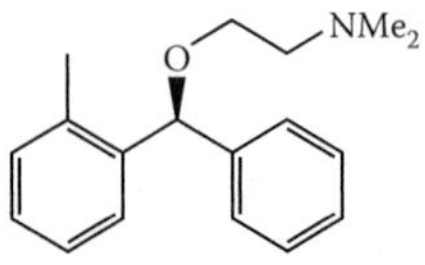

152 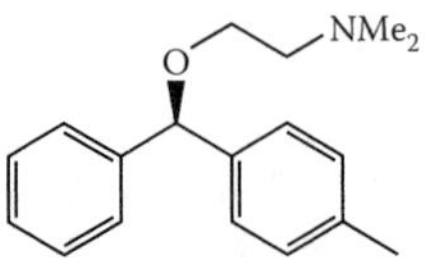   153

The enantioselective reduction of 5-benzoyl thioazoles with 2-substituents (**154**) with Ru(*R*-xyl-BINAP)(*R*-diapen)Cl$_2$ produced **155** in >99% ee, which could be converted to a potent DPE-IV inhibitor **156** (Scheme 12.59).[199]

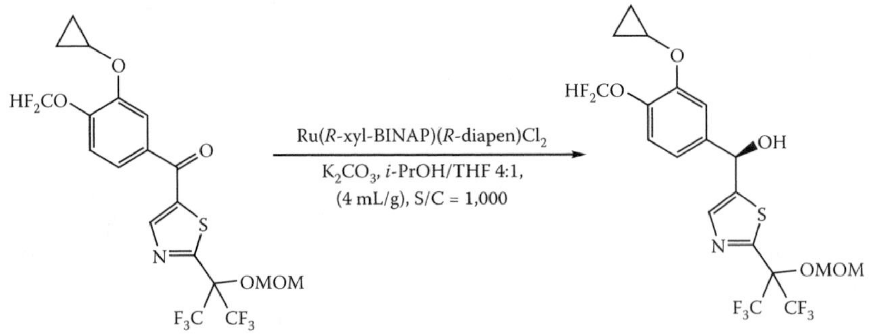

$$Ru(R\text{-xyl-BINAP})(R\text{-diapen})Cl_2$$
$$K_2CO_3, i\text{-PrOH/THF 4:1,}$$
$$(4\ mL/g),\ S/C = 1,000$$

154

155
>99% ee

SCHEME 12.59

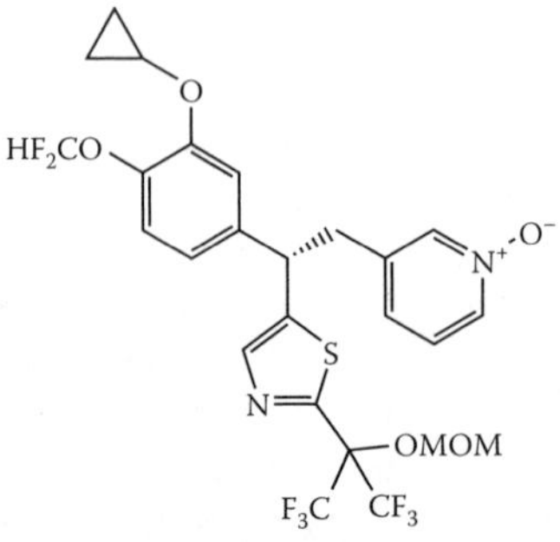

156

JST technology has shown large potential for many industrial processes, which prompted two companies to license the technology (Johnson Matthey and Dow Pharma).

12.3.5 MONODENTATE LIGANDS

Considerable success has been realized for asymmetric hydrogenations of carbon–carbon unsaturation with monodentate ligands, especially those derived from BINOL. The most popular class is the phosphoramidites (**157**) as used by DSM.[200,201] Phosphites (**158**) have also been successfully used in a wide range of asymmetric hydrogenations,[202] as have phosphonites (**159**).[203,204] These ligands are discussed in depth in Chapter 14.

| 157 | 158 | 159 |

12.4 ASYMMETRIC HETEROGENEOUS CATALYSTS IMPLEMENTED IN INDUSTRY

In the first forays into asymmetric organic synthesis the catalysts used were dominated by heterogeneous systems, which can boast a few catalytic systems that undergo asymmetric hydrogenation transformations with good to high enantioselectivities, but these have been overshadowed by the highly active and enantioselective homogeneous counterparts. Informative historical backgrounds and overviews on asymmetric heterogeneous catalytic systems have been published.[5,6]

Strategies to induce chirality in a prochiral substrate included modification of existing heterogeneous catalysts by addition of a naturally occurring chiral molecules, such as tartaric acid, natural amino acids, or alkaloids, and the implementation of chiral supports, which include quartz or natural fibers, for metallic catalysts. Both strategies have been successful on a limited basis.

Despite the convenience of handling and separation in heterogeneous catalysis, many other parameters have a strong influence on the stereochemical outcome: pressure; temperature; modifier; purity of substrates; high substrate specificity; catalyst preparation, which includes type, texture, and porosity of the support; dispersion; impregnation; reduction; and pretreatment of the metal.[22] Reproducibility of catalyst activities and enantioselectivities can often be a problem attributed to variations in catalyst preparations and purity of the substrate.[5,22]

The transformations that use asymmetric heterogeneous catalysis will be highlighted: β-keto esters and diketone reductions by Raney nickel catalyst modified with R,R-tartaric acid and NaBr. α-Keto acid reductions with cinchona modified Pt catalysts are discussed in Chapter 18.

12.4.1 NI/R,R-TARTARIC ACID/NABR

The Ni/tartaric acid/NaBr catalyst system has been extensively studied. A variety of ketone substrates have been reduced with Ni/tartaric acid/NaBr catalysts with variable enantioselectivities, but the highest (>85% ee) are obtained for the reductions of β-keto esters and β-diketones (Schemes 12.60 and 12.61).[5] Asymmetric reduction of diketones results in the formation of *meso* and chiral diols. The highest *meso*:chiral diol ratio of 2:98 and enantioselectivities of 98% ee are obtained with modified Raney nickel catalysts treated by sonication.[5]

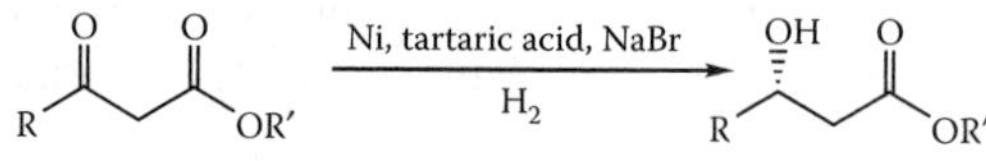

SCHEME 12.60

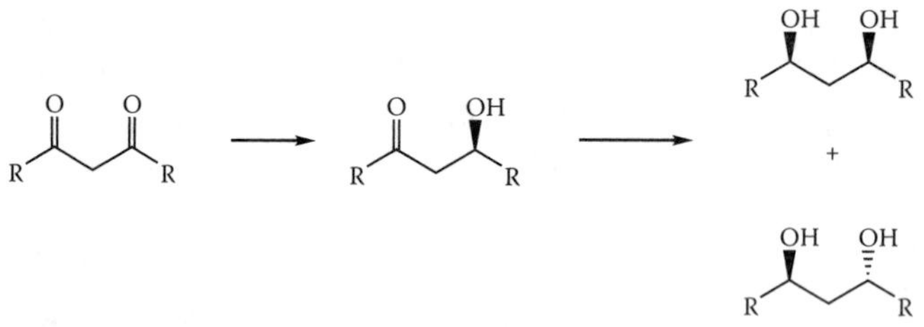

SCHEME 12.61

Hoffmann-LaRoche has developed a process that uses R,R-tartaric acid/NaBr modified Raney nickel catalyst in the asymmetric hydrogenation of **160** to give **161** in 100% yield and 90–92% ee (6–100-kg scale) for the synthesis of the intermediate of tetrahydrolipstatin (**162**), a pancreatic lipase inhibitor (Scheme 12.62).[5]

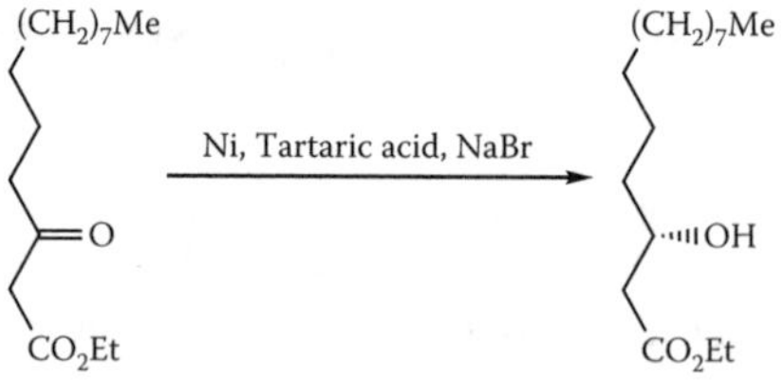

SCHEME 12.62

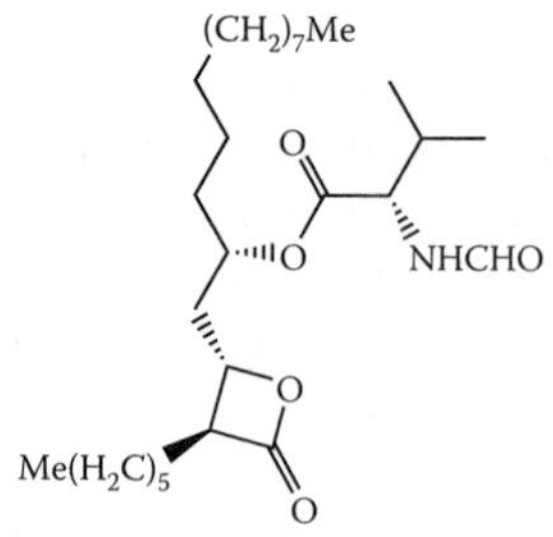

162

12.4.2 CINCHONA-MODIFIED PLATINUM

The stereoselective reduction of α-keto acid derivatives at a preparative scale is performed with cinchona-modified Pt catalyst (see Chapter 18). Enantioselectivities range from 57–95% for the reduction of α-keto acid derivatives[5] and is dependent on the preparation of the Pt catalyst.[5,22,205,206]

Novartis (Ciba Geigy) has reported the synthesis of Benazepril (**163**) (Scheme 12.63), an angiotensin-converting enzyme inhibitor, via an intermediate prepared by cinchona-modified Pt asymmetric hydrogenation (10–200-kg scale, >98%, 79–82% ee).[5] The low optical purities can be tolerated because enantio-enrichment is relatively easy in the latter stages of the synthesis.[22]

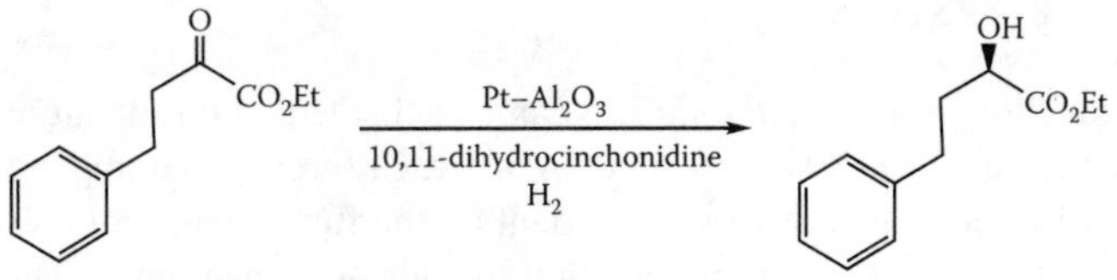

SCHEME 12.63

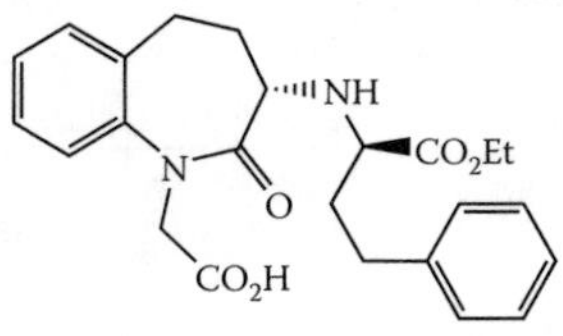

163

12.5　ASYMMETRIC HYDROGEN TRANSFER

Asymmetric hydrogen transfer shows promise for use at industrial scale because ruthenium complexes that contain chiral vicinal diamino **164** or amino alcohol **165** ligands allow the reductions of substrates such as aryl ketones and imines to be achieved under mild conditions.[13,207]

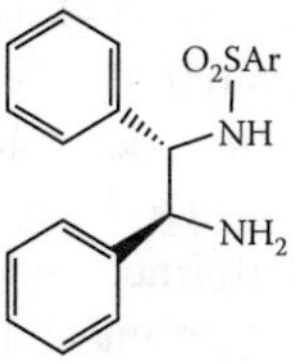

164

a, Ar = 4-CH$_3$C$_6$H$_4$ (TsDPEN)
b, Ar = 2,4,6-(CH$_3$)$_3$C$_6$H$_2$
c, Ar = 1-naphthyl

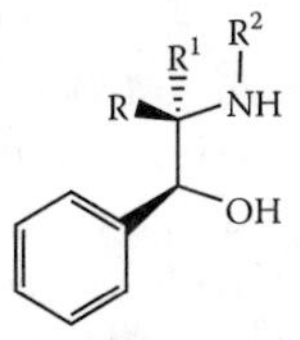

165

a, R = Ph, R^1 = H, R^2 = H
b, R = Ph, R^1 = H, R^2 = Me
c, R = Me, R^1 = H, R^2 = Me
d, R = Ph, R^1 = Me, R^2 = H
e, R = Me, R^1 = Me, R^2 = H

The progress in development and application of asymmetric hydrogen transfer has continued, albeit not at the same pace as reductions with molecular hydrogen.[23,188,208,209] Several examples of asymmetric hydrogen transfer have been reported at large scale, such as the preparation of (*R*)-1-tetralol, (*S*)-4-fluorophenylethanol, and (*R*)-1-methylnaphthylamine with CATHy based catalysts (**166–168**), whereas (*R*)-3,5-bistrifluoromethylphenylethanol has been made with *cis*-aminoindanol-Ru(*p*-cymene) complex (**169**) (see also Chapter 17).

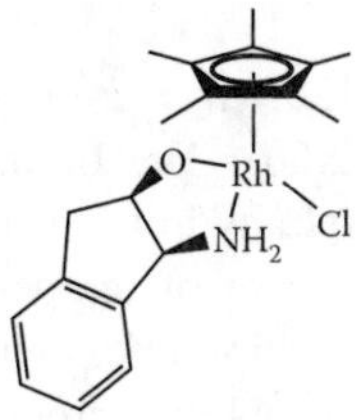

166 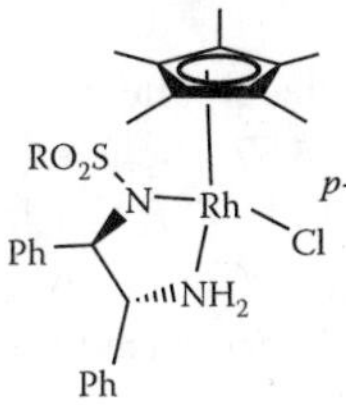　　　　　**167** 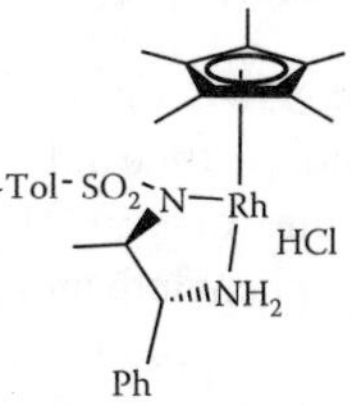　　　　　**168** 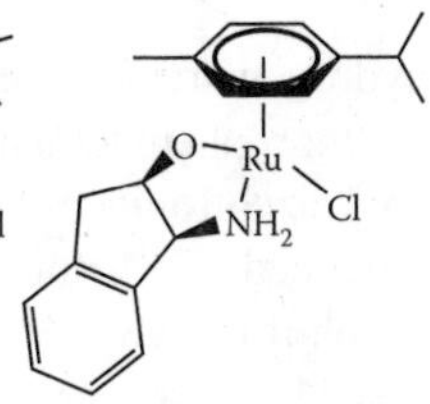　　　　　**169**

12.5.1 REDUCTIONS OF ARYL KETONES

The asymmetric reduction of an aryl ketone **128** can be achieved with ruthenium catalysts (Scheme 12.64), prepared separately or *in situ* by formation of [RuCl$_2$(arene)]$_2$ and ligand, in *i*-PrOH.[207] The high enantioselectivities and rate are very dependent on the functionality of the substrate, η^6-arene and *N*-substitution of the diamino or amino alcohol ligands on ruthenium.[207] The hydrogen transfer reaction in *i*-PrOH is reversible, necessitating low concentrations, whereas extensive reaction times degrade the enantioselectivities of the alcohol.[207] This limitation can be overcome by the use of an azeotropic mixture of formic acid and NEt$_3$ (5:2) as the reaction is now irreversible.[210]

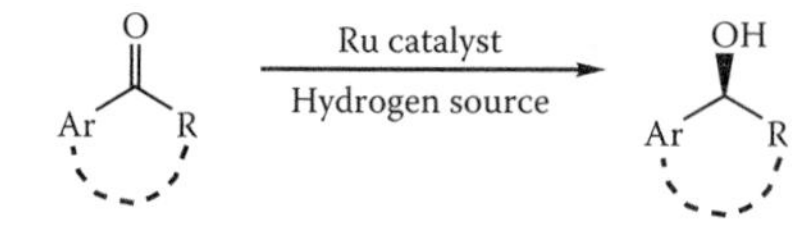

SCHEME 12.64

The asymmetric hydrogen transfer of aryl ketones can be accomplished with ruthenium catalysts that contain amino alcohols **165** in modest to high enantioselectivities.[211] With amino alcohol ligands, the optimal rate and stereoselectivities are produced from catalysts prepared *in situ* with [RuCl$_2$(η^6-C$_6$Me$_6$)]$_2$.

12.5.1.1   (*R*)-1-Tetralol

1-Tetralone has been reduced to (*R*)-1-tetralol with **166** at the 200-L scale (Scheme 12.65). The hydrogen source was *i*-PrOH. Initially, the maximum concentration to achieve high conversion was 0.05M in a closed system, but the ee decreased near the latter conversions. The key discovery in process optimization was removal of acetone by-product during the reaction. The catalyst is not stable at higher than 40°C; however, the reaction can be performed under a slight vacuum (10–50 mbar) to remove acetone. Fresh IPA is added to the reaction to maintain constant volume. Under these conditions coupled with efficient agitation, substrate concentrations of 0.5M are achieved with complete conversions (TOF = 500–2500 h^{-1}) and reproducible enantioselectivity.[212]

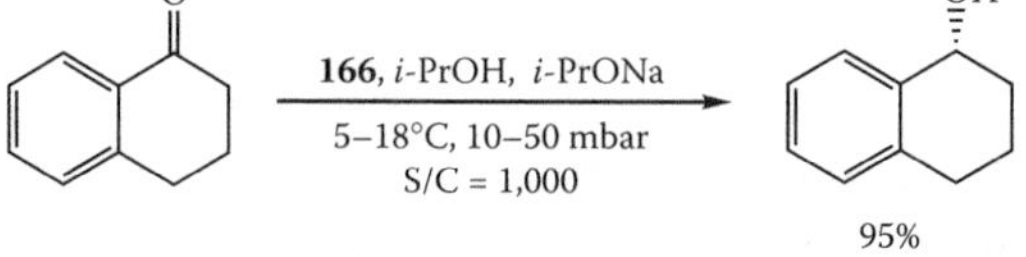

SCHEME 12.65

12.5.1.2   (*S*)-4-Fluorophenylethanol

Asymmetric hydrogen transfer of *p*-4-fluorophenylacetophenone (**128**, Ar = *p*-FC$_6$H$_4$, R = Me) produced (*S*)-4-fluorophenylethanol with catalyst **167** (Scheme 12.66). The optimal hydrogen source was a 5:2 azeotropic mixture of NEt$_3$:HCO$_2$H (TEAF) in tetrahydrofuran (THF). Initial experiments produced (*S*)-4-fluorophenylethanol in 96% ee. A decrease of temperature to 0°C produced the alcohol in >98.5% ee but significantly decreased the reaction rate. Optimal reaction rates were achieved by separate addition of TEAF and **167**. Nitrogen gas bubbling through the reaction solution and high agitation enabled successful operation at 50-L scale. The substrate concentration of 3.6M was most effective in the reduction with respect to reproducibility of the product stream.[212]

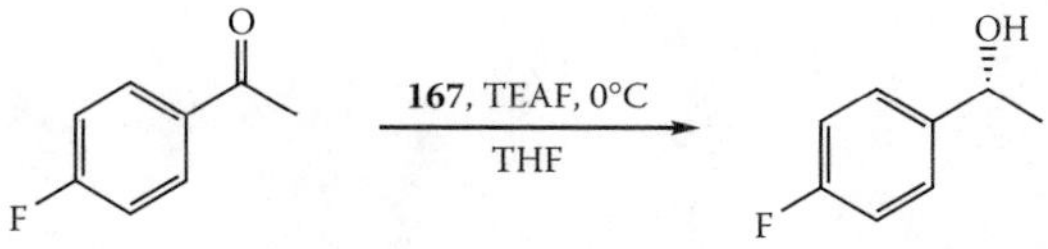

128, Ar = *p*-FC_6H_4, R = Me

85%
98.4% ee

SCHEME 12.66

12.5.1.3 (*R*)-3,5-Bistrifluoromethylphenylethanol

Merck reported the synthesis and isolation of (*R*)-3,5-bistrifluoromethylphenylethanol (**170**) in high yields and enantiomeric excess by asymmetric hydrogen transfer. Reduction of 3,5-bistrifluoroacetophenone (**128**, Ar = 3,5-$(CF_3)_2C_6H_3$, R = Me) with catalyst **169,** prepared *in situ* from [$RuCl_2$(*p*-cymene)]$_2$ and (1*S*,2*R*)-*cis*-1-aminoindan-2-ol, produced the chiral alcohol **170** in 91–93% ee (Scheme 12.67).[213]

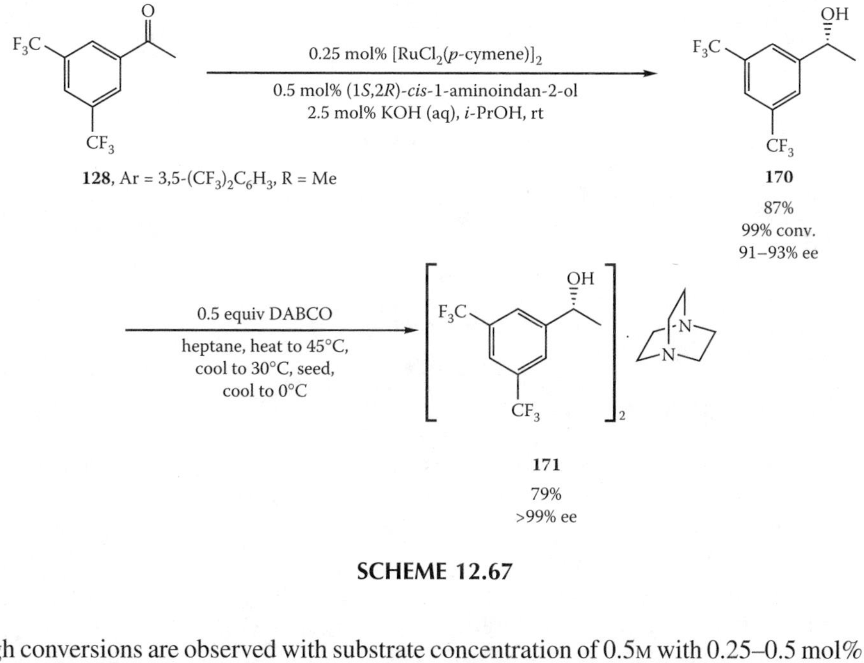

128, Ar = 3,5-$(CF_3)_2C_6H_3$, R = Me

170

87%
99% conv.
91–93% ee

171

79%
>99% ee

SCHEME 12.67

High conversions are observed with substrate concentration of 0.5M with 0.25–0.5 mol% catalyst loadings. Inconsistent reaction completions were observed at the lower catalyst loadings. Therefore, the researchers choose 0.5 mol% catalyst loading. Several process discoveries that contributed to the success of scale up included the fact that rigorous degassing of all reagents was not necessary until activation of the catalyst solution with base. Anhydrous solvents and bases are not required. Anhydrous *t*-BuOK and KOH can be replaced with aqueous NaOH and KOH (5M) without any decrease in performance compared to bases generated under anhydrous conditions.

The optical purity of 91–93% was too low for downstream chemistry. The physical properties of **170** made enantiomeric enhancement by crystallization. The formation of a DABCO inclusion complex **171** in heptane and crystallization under thermodynamic control provided material that was 99% ee, 98% purity, and 79% recovery. This procedure produced 90 kg of **170**, which was used to prepare an NK-1 receptor antagonist, Aprepitant (**172**), used for the treatment of chemotherapy-induced emesis.[213]

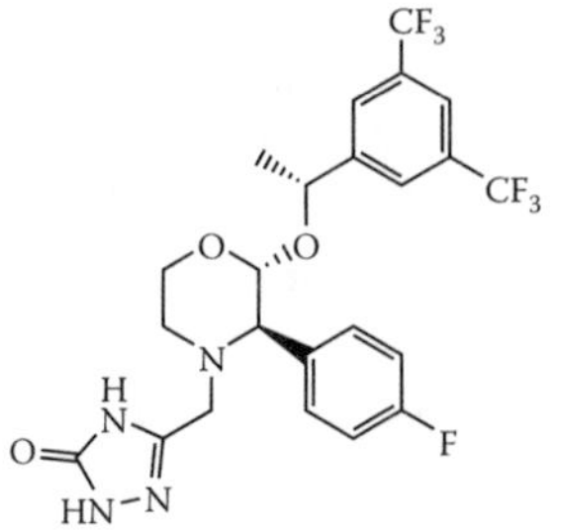

172

12.5.2 Reductions of Imines

Asymmetric reduction of imines by classic hydrogenation or hydrogen transfer has proved difficult, but selected imines have been reduced by chiral phosphine-Rh or -Ir to give secondary amines in moderate ee or by a chiral *ansa*-titanocene catalyst in 95–100% ee (S/C = 20).[214-220] A series of preformed ruthenium catalysts [Ru(**164**)(arene)Cl] have successfully reduced a variety of cyclic imines **173** with high enantioselectivities in a 5:2 mixture of HCO_2H-NEt_3 (Scheme 12.68). These reductions can be performed in acetone, although the ruthenium catalyst is capable of catalyzing the hydrogen transfer of ketones.[221]

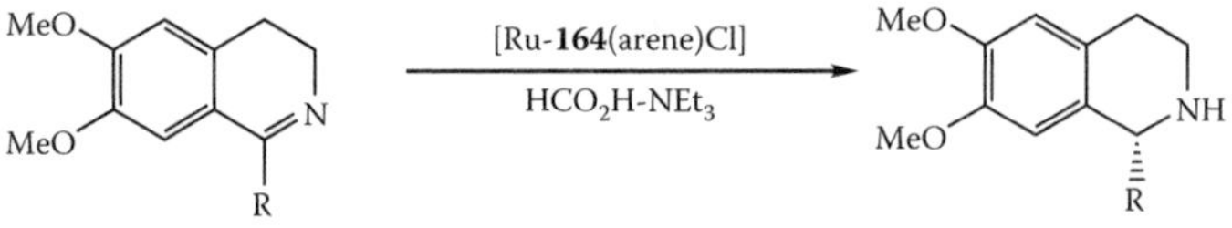

173

SCHEME 12.68

12.5.2.1 (*R*)-1-Methylnaphthylamine

Asymmetric hydrogen transfer of *N*-diphenylphosphinyl-1-naphthyl imine (**174**) produced (*R*)-*N*-4-diphenylphosphinyl-1-methylnaphthylamine (**175**) with catalyst **168** in 99% ee (Scheme 12.69). The optimal hydrogen source was a TEAF. The initial S/C ratio was 200, but separate addition of TEAF and catalyst to the imine solution produced increased rates without affecting the enantio-selectivities. Similar to the *p*-4-fluoroacetophenone process, bubbling of nitrogen and high agitation afforded the best catalyst activity (1000 turnovers/hour). This process was scaled up to an undisclosed size and produced product in 99% ee and 95% yield. The diphenylphosphinyl group can be removed by acid hydrolysis with acidic EtOH to give the primary amine.[212]

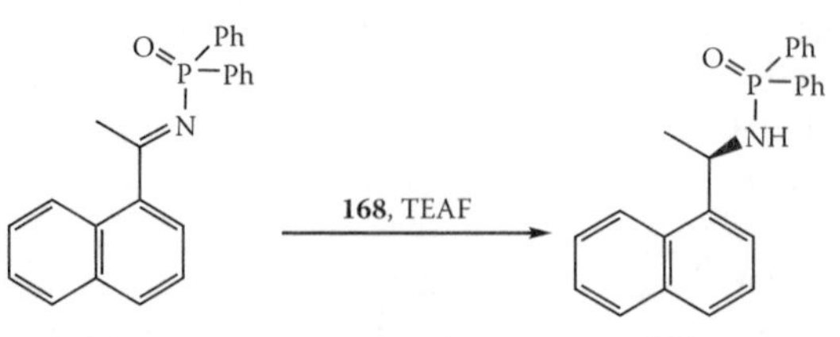

174

168, TEAF

175
95%
99% ee

SCHEME 12.69

12.6 HYDROFORMYLATION

Hydroformylation, also known as the oxo process, is the transition metal catalyzed conversion of olefins into aldehydes by the addition of synthesis gas (H_2/CO) and is the second largest homogeneous process in the world in which more than 12 billion pounds of aldehydes are produced each year.[222] Aldehydes are important intermediates in a variety of processes such as the production of alcohols, lubricants, detergents, and plasticizers. Several transition metals can catalyze the hydroformylation of olefins, but rhodium and cobalt have been used extensively.

Despite high volumes of aldehyde products that are manufactured and a deep understanding of the catalytic mechanistic process, no successful industrial asymmetric hydroformylation process has been achieved. Rhodium and platinum catalysts that contain chiral bisphosphine and bisphosphinite ligands have provided the highest enantioselectivies in the study of asymmetric hydroformylation and are chronicled in several reviews.[10,222–224]

From 1995 to the present, several advances in asymmetric hydroformylations have been reported. Ligands that have been used with reasonable success with benchmark substrates, such as styrene and vinyl acetate (Schemes 12.70 and 12.71), are shown in Figure 12.8.

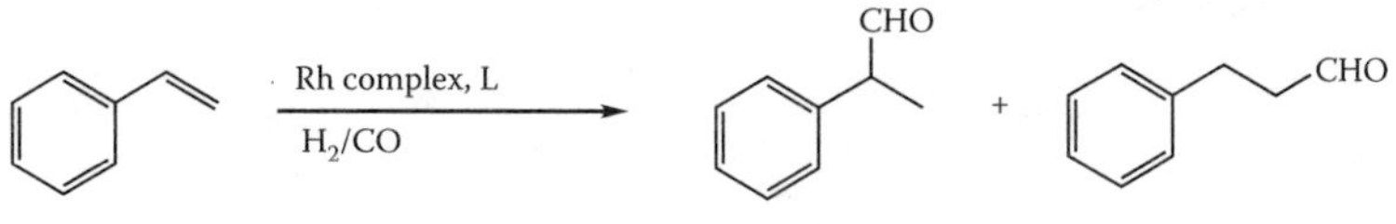

SCHEME 12.70

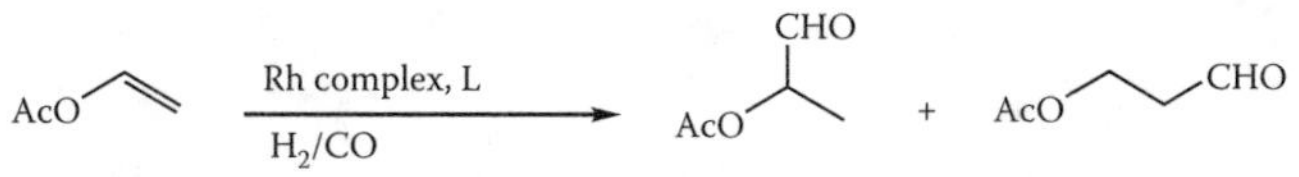

SCHEME 12.71

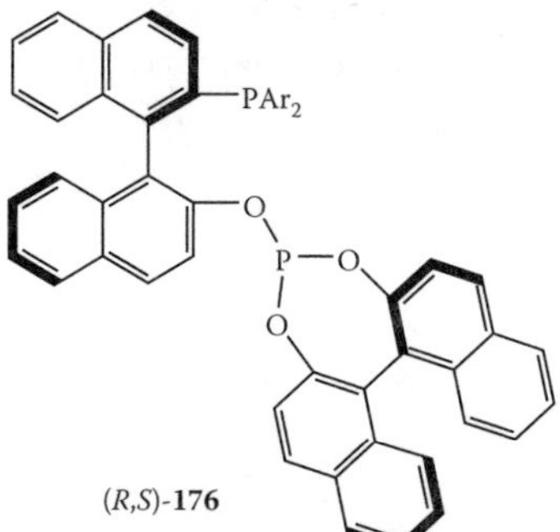

(R,S)-**176**

a, Ar = Ph (BINAPHOS)
b, Ar = 3-MeO-C$_6$H$_4$–
c, Ar = 3-iPro-C$_6$H$_4$–
d, Ar = 3-C$_6$F$_{13}$CH$_2$CH$_2$-C$_6$H$_4$–
e, Ar = 3,5-(MeO)$_2$C$_6$H$_3$–
f, Ar = 4-MeO-C$_6$H$_4$–
g, Ar = 3-Me-C$_6$H$_4$–

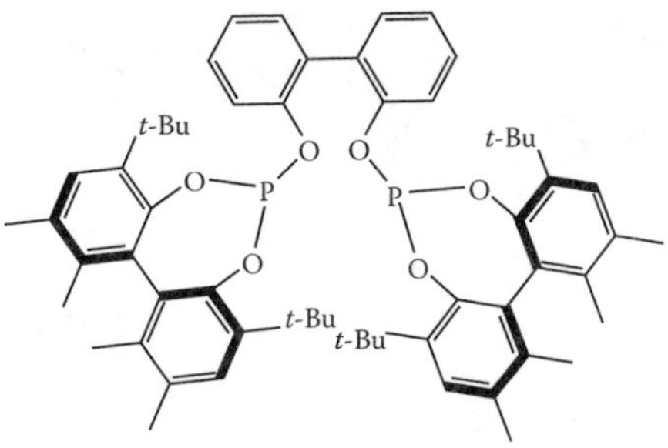

177
(S,S)-Kelliphite

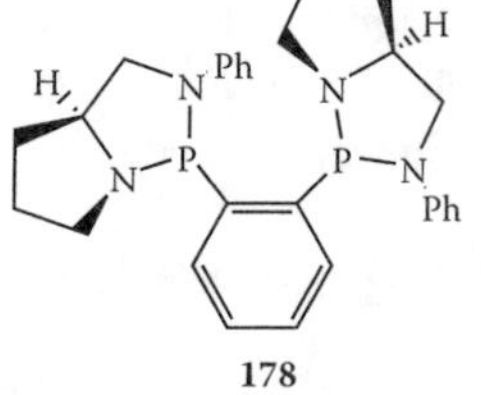

178
ESPHOS

FIGURE 12.8 Ligands used in asymmetric hydroformylations.

BINAPHOS (**176a**), developed by Takasago, has been used in the asymmetric hydroformylation of styrene and vinyl acetate with enantioselectivites of 94% ee and 90% ee, respectively. The regioselectivities (branch/normal ratios) ranged 88/12 to 85/15 in both transformations.[225,226]

Variations of phenyl substituents on the phosphorus produced enantioselectivities between 95–98% ee for the asymmetric hydroformylations of styrene. Ligand **176b** produced the best result with regard to enantioselectivities (95.0% ee), conversions (>99%), and regioselectivity (93:7).[226]

The asymmetric hydroformylation of vinyl arenes can provide a route to the preparation of the profen class of drugs. Naproxen and ibuprofen, two examples in the profen class, are NSAIDs on the market.[50]

DSM reported that the asymmetric hydroformylation of unsaturated nitriles (Scheme 12.72) to provide a potential route to (*R*)-4-amino-2-methyl-1-ol (**180**), which is a key intermediate for TAK-637 (**181**), a new Tachykinin NK$_1$ receptor antagonist, developed by Takeda (Scheme 12.73). Rhodium-BINAPHOS (4 equiv) has catalyzed the asymmetric hydroformylation of but-3-enenitrile (allyl cyanide) with regioselectivity of 72/28 (b/l) and 66% ee at 73% conversions.[227]

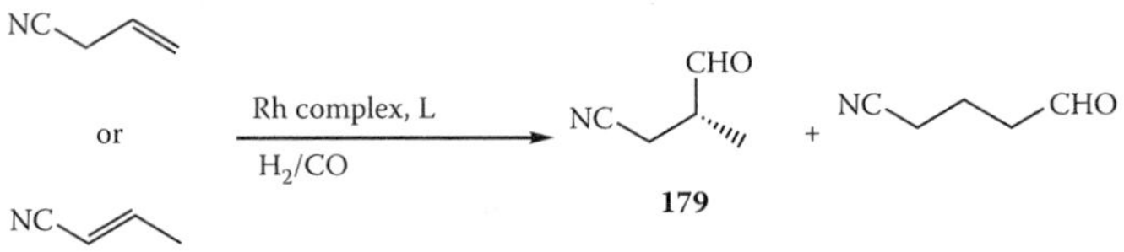

SCHEME 12.72

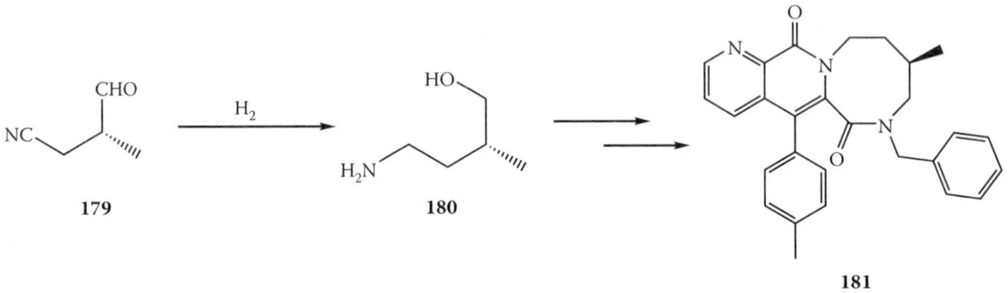

SCHEME 12.73

DowPharma reported improvements in the asymmetric hydroformylation of allyl cyanide with the development of a new ligand, Kelliphite (**177**), which proceeds with 78–80% ee and a regioselectivity of 15–23:1.[228,229] Under reaction conditions developed by DowPharma, an enantioselectivity of 76% ee and a regioselectivity of 73/27 are observed with the rhodium-BINAPHOS catalyst.[229] An S/C loading of 10,000 can be achieved without affecting conversions (100%), regioselectivities (b/l = 95/5), and enantioselectivities (79% ee) at low gas pressure (150 psig) and temperatures (35°C). The process runs neat in allyl cyanide, and the rhodium can be recovered by partitioning into hexane.[228,229]

DowPharma identified that selective hydrogenation of **179** with Pt-C will reduce the aldehyde moiety to **182**, which is an intermediate for a potent nonpeptide gonadotropin-releasing hormone antagonist **183** by Merck (Scheme 12.74). Addition of an acid, either oxalic acid or D-tartaric acid, to the Pt-C hydrogenation will reduce the nitrile to **169**, which is isolated as the salt.[228,229]

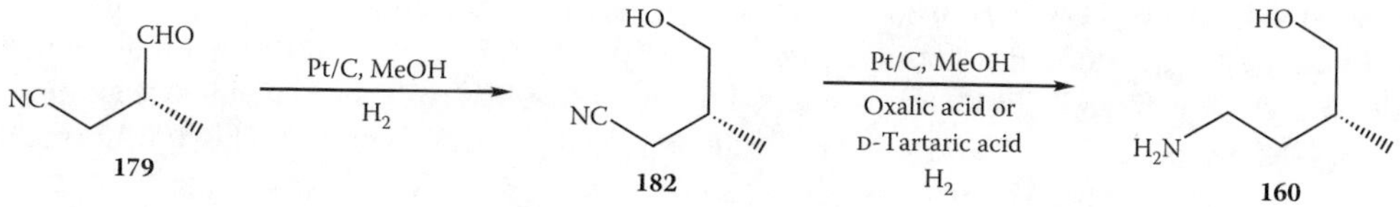

SCHEME 12.74

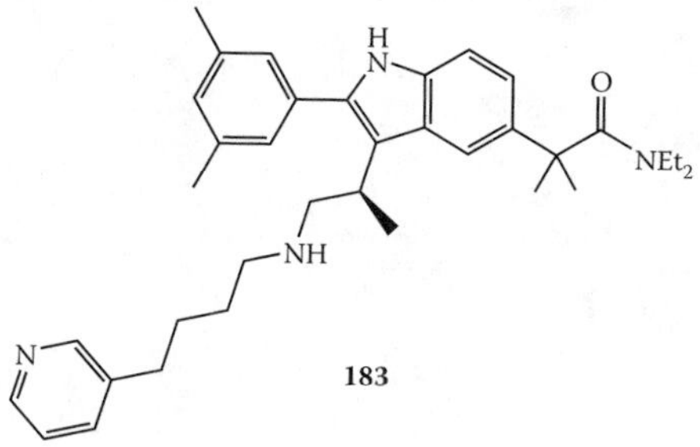

183

Rhodium catalysts with Kelliphite (**177**) have been tested in the asymmetric hydroformylation of vinyl acetate. Enantioselectivities of 90% ee and regioselectivities of >99.5/0.5 (b/l) have been achieved in initial screening experiments.[230]

ESPHOS (**178**), a chiral bis(diazaphospholidine) ligand developed by the Wills group, has been found to be active in the asymmetric hydroformylation of vinyl acetate.[231] (This ligand system is being commercialized by Stylacats.) Enantioselectivities of 87–89% ee and regioselectivities of 95/5 (b/l) have been obtained at 120 psig pressure of syn gas.[232] Unfortunately, the substrate scope is not as broad as BINAPHOS.

The production of 1β-methylcarbapenem (**184**), which has antibacterial activities and enhanced chemical and metabolic stability, has been reported by asymmetric hydroformylation of 4-vinyl-β-lactams **185** catalyzed by Rh-BINAPHOS complexes (Scheme 12.75). Under optimized conditions, the observed regioselectivity was 55/45 (b/l), enantioselectivity was 93/7 (**186β**:**186α**) at 95% conversion, and S/C = 1000.[233]

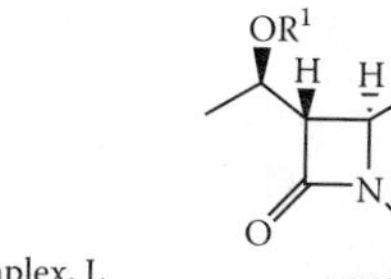

184

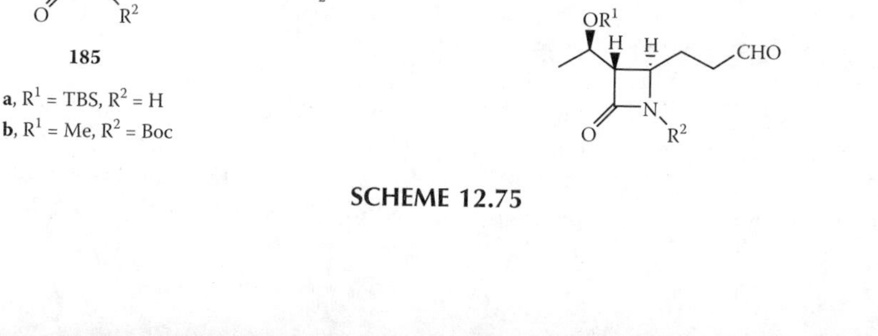

185

a, R^1 = TBS, R^2 = H
b, R^1 = Me, R^2 = Boc

SCHEME 12.75

Alper and co-workers improved the asymmetric hydroformylation of vinyl β-lactams. The substitution of a Zwitterionic rhodium catalyst (**187**) and (*S,S*)-BDPP (**188**) in the asymmetric hydroformylation of **185b** produced regioselectivities of >99/1 (b/l) and enantioselectivities of >99/1 (**186β**:**186α**) with 70% conversion at 1200 psig syn gas (S/C = 20).[234] This transformation has been performed at the kilogram scale.[235]

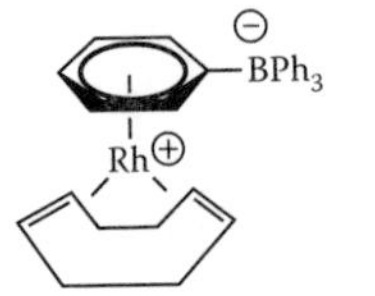

187 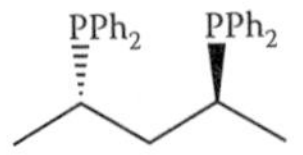   188

12.7 HYDROSILYLATION

Optically active alcohols, amines and alkanes can be prepared by the metal-catalyzed asymmetric hydrosilylation of ketones, imines, and olefins.[97,224,236] Several catalytic systems have been successfully demonstrated, such as the asymmetric silylation of aryl ketones with rhodium and Pybox ligands; however, there are no industrial processes that use asymmetric hydrosilylation. The asymmetric hydrosilyation of olefins to alkylsilanes (and the corresponding alcohol) can be accomplished with palladium catalysts that contain chiral monophosphines with high enantioselectives (up to 96% ee) and reasonably good turnovers (S/C = 1000).[237] Unfortunately, high enantioselectivities are only limited to the asymmetric hydrosilylation of styrene derivatives.[238] Hydrosilylation of simple terminal olefins with palladium catalysts that contain the monophosphine, MeO-MOP (**189**), can be obtained with enantioselectivities in the range of 94–97% ee and regioselectivities of the branched to normal of the products of 66:43 to 94:6 (Scheme 12.76).[239,240]

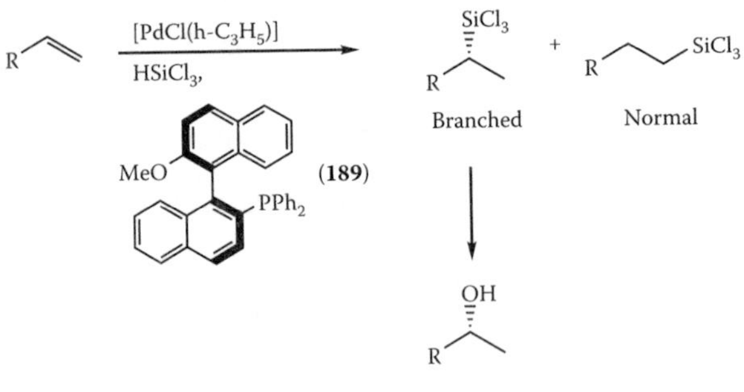

R=	Branched/Normal	%ee (alcohol)
n-C$_4$H$_9$	89/11	94% (*R*)
n-C$_6$H$_{13}$	93/7	95% (*R*)
n-C$_{10}$H$_{21}$	94/6	95% (*R*)
CH$_2$CH$_2$Ph	81/19	97% (*S*)
c-C$_6$H$_{11}$	66/43	96% (*R*)

SCHEME 12.76

12.8 ASYMMETRIC CYCLOPROPANATIONS

Insecticides of the pyrethroid class, such as *trans*-chrysanthemic acid (**190**), have significant commercial value (see Chapter 31).[241] An asymmetric synthesis of **190** has been achieved through the use of a chiral copper carbenoid reaction (Scheme 12.77).[242,243] With ethyl diazoacetate, equal amounts of the *cis*- and *trans*-cyclopropanes were formed. However, when the size of the alkyl group in the diazoester was increased, the geometric and enantioselectivity increased. The *l*-menthyl group was selected because this gave 93% of the *trans*-isomer with 94% ee.

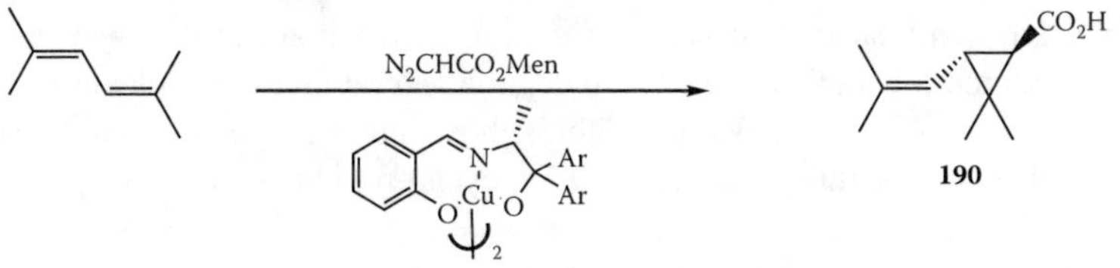

Where Ar = 5-*t*-butyl-2-octyloxyphenyl

SCHEME 12.77

12.9 SUMMARY

The preparation of chiral material via asymmetric hydrogenation has come a long way and continues to evolve. Asymmetric hydrogenation is a powerful and clean method to prepare chiral intermediates and products. The number of asymmetric hydrogenation processes continues to grow at the manufacture scale, although that number is still considered small compared with the burgeoning research efforts. Since Monsanto's pioneering L-dopa process, which key step is the asymmetric hydrogenation of an enamide to the protected amino acid, several other industrial asymmetric hydrogenation processes have been developed. L-Dopa, L-phenylalanine, and a variety of unnatural amino acids have been prepared at the manufacture scale from the asymmetric reduction of the corresponding enamide in moderate to high enantioselectivities by rhodium catalysts based on phosphines ligands of various classes: *P*-chiral, ferrocenyl, chiral pool (diol, diamines, and carbohydrates), and phospholanes.

The discovery of the chiral atropisomeric ligand, BINAP, greatly expanded the number of asymmetric homogeneous hydrogenation catalysis. Rhodium and ruthenium complexes that contain BINAP and similar ligands systems have demonstrated an amazing versatility in the reduction of a wide variety of substrate classes in excellent stereoselectivities and reactivities. (−)-Menthol, a variety of terpenoids, carbapenem intermediates, 1,2-propanediol, 4-hydroxy-2-cyclopentenone, and citranellol are manufactured by transition metal catalysis that contain BINAP. There are a number of potential industrial processes that could implement Rh- and Ru(BINAP) catalysts, for example, naproxen, and β-keto esters, such as ethyl 4-chloroacetoacetate. Several new ligands similar to BINAP have been synthesized, demonstrated excellent catalytic reactivities and selectivities, and provided a means of catalytic "fine-tuning" a specific process.

An area of "fine-tuning" that has been a major focus in ligand design is the dihedral angle of the biaryl backbone. Fine adjustment to the dihedral angle has resulted in the development of highly active and enantioselective catalysts that contain new ligands: BIPHEMP, MeO-BIPHEP, MeO-Cl-BIPHEP, Segphos, SYNPHOS, and TMBTP. Catalysts that have contained these ligands have been used in a wide range of asymmetric processes, which include reductions of β-keto esters (ethyl 4-chloroacetoacetate and dynamic kinetic resolutions for carbapenem production), β,γ-diketo esters (alternate Lipitor process), asymmetric allylic alcohols (NVP-ACQ090), α,β-unsaturated esters (Candoxatril), ketones (alternate menthol production) and acids (mibefradil and Zeneca ZD 3523), and α-keto esters (antifungal product).

Variation in electronic effects within the biaryl backbone has produced the P-Phos family of ligands, which contain a hindered dipyridinyl backbone. These ligands have demonstrated high catalytic activities and enantioselectivities with JST catalysts.

A novel series of atropisomeric ligands uses a paracyclophane backbone. Rhodium and ruthenium-PhanePhos catalysts have performed well in the asymmetric hydrogenation of enamide esters, β-keto esters, and especially arylketones with JST catalysts (Duloxetine).

Phosphines ligands that have chirality from ferrocenes have been implemented in the iridium-catalyzed asymmetric hydrogenation of imine with moderate enantioselectivities for Novartis's manufacture of metolachlor. Electronic modifications of these ferrocenyl ligands have increased the enantioselectivity and catalyst reactivity for Lonza's asymmetric hydrogenation processes of biotin and 2-substituted piperazines, intermediates for several pharmaceutical drugs. New ferrocenylbisphosphines, such as BoPhoz, Walphos, Taniaphos, and Mandyphos, have been developed and could be used in the manufacture of Levetiracetam, cyclopropylalanine, and Synthon A alcohol for SPP100.

Several asymmetric hydrogenation technologies show high potential for use at an industrial scale. Electronic modifications of Ph-β-glup have produced a rhodium catalyst that can reduce a wide variety of enamides to protected amino acids with high enantioselectivities. Alterations of the phosphinate positions of the same D-glucopyranoside backbone can produce rhodium catalysts that can produce the opposite antipode of protected amino acids. Noyori and co-workers have developed ruthenium catalysts that contain chiral diamino or amino alcohol ligands, which reduce aryl ketones with high stereoselectivities under both hydrogen transfer and hydrogenation conditions. Chiral bisphosphoranes ligands, such as DuPHOS, have been developed, whose transition metal complexes are very efficient in the asymmetric reductions of enamides, enol acetates, and β-keto esters. DuPHOS has become the new benchmark to which all new asymmetric homogeneous hydrogenations are compared. Like the atropisomeric ligands, the development of a second generation has mushroomed with the focus on dihedral angles (MalPhos, CatASium families, Butiphane), electronic effects on the phospholane ring (RoPhos, Me-KetalPhos, and BasPhos), phospholane ring size (Ferrotane, CnrPHOS, and BPE-4), and novel ligand design (Binaphane, Binapine, and Tangphos). Many of these ligands display similar enantioselectivities and activities in the production of β-amino esters, chiral succinates, chiral benzyl amines, and unnatural amino acids.

Several new ligands that possess chirality at the phosphorus center have been developed and shown to be excellent catalysts for the asymmetric reduction of enamides to amino acids and chiral amines and β-enamides to β-amino esters. One ligand that shows great promise to be used at the manufacture scale is Trichickenfootphos, which has demonstrated high activity and enantioselectivity in the production of a chiral intermediate for Pregabalin.

The next quantum leap in catalysis after DuPhos has been the development of the ruthenium–diamine–bisphosphine catalysts (JST). This combination of ligands on ruthenium produces a highly active and enantioselective catalyst for the reduction of aryl ketones at mild conditions. Although this technology is relatively young, the potential is strong for many industrial processes that use this catalyst system.

Another area that has received considerable attention and was not chronicled in this chapter (but see Chapter 14) is the development of monodentate phosphoramidites. Once thought to be unstable in reaction media and not highly enantioselective, monodentate phosphoramidites have changed the minds of the scientific field. These ligands have been used in the manufacture of unnatural α-amino and β-amino acids.

Hydrogenation, by far, has garnished the best results in homogeneous asymmetric catalysis, but efforts continue in the development of asymmetric hydroformylation. Several new ligands (Binaphos, Kelliphite, and ESPHOS) have been designed that have raised the level of enantioselectivities to 90–94% ee and regioselectivities to 88–99% in favor of the branched isomer.

Although homogeneous catalysts have overshadowed asymmetric heterogeneous hydrogenation catalysts, a few industrial processes have been developed, but the scope of these catalytic systems is narrow and highly substrate specific. The enantioselectivities that are obtained with heterogeneous systems are usually lower in comparison to homogeneous or biocatalysis, with some notable exceptions. Although heterogeneous systems are preferred to homogeneous systems in a commercial setting because of convenience with handling and separation, catalyst performance needs to be more reproducible. Raney nickel modified with R,R-tartaric acid and NaBr has been used by Hoffmann LaRoche in the asymmetric hydrogenation of β-keto esters in moderately high ee's.

Novartis (Ciba Geigy) has implemented a cinchona-modified Pt/Al_2O_3 catalyst in the asymmetric hydrogenation of α-keto esters for the industrial preparation of Benzaprin.

The stereoefficiencies of an asymmetric hydrogenation catalyst are not the sole criteria for a industrially feasible process. Many economic factors of a process that govern a successful industrial scale include the cost of catalyst and ligands, substrate, operation, product throughput, equipment, catalyst activities, and stereoselectivities. Ideally, a low-cost process would have low catalyst loadings (high catalyst activity and TON) and high stereoselectivities in the reduction of an inexpensive substrate at convenient hydrogen pressures and high concentrations. Unfortunately, this is not always the case because some processes are only stereoselective at high hydrogen pressures (2000 psig), which can be expensive to implement at scale. Other examples of high-cost scenarios include expensive catalysts that are only stereoselective at high catalyst loadings or low substrate concentrations. In addition, the cost and availability of the substrate needed for the asymmetric hydrogenation can be a key factor.

The past several years have witnessed an explosion of creativity in ligand design and asymmetric synthetic organic chemistry used in ligand preparation. This has fueled the development of new catalysts, ligands, and asymmetric transformations that should further increase the number of asymmetric hydrogenation processes at industrial scale in the years to come.

REFERENCES

1. Knowles, W. S., Sabacky, M. J. *J. Chem. Soc., Chem. Commun.* 1968, 1445.
2. Horner, L., Siegel, H., Buthe, H. *Angew. Chem., Int. Ed.* 1968, *7*, 942.
3. Crosby, J. *Tetrahedron* 1991, *47*, 4789.
4. Ager, D. J., East, M. B. *Asymmetric Synthetic Methodology*, CRC Press: Boca Raton, 1995.
5. Blaser, H. *Tetrahedron: Asymmetry* 1991, *2*, 843.
6. Harada, K. Asymmetric Heterogeneous Catalytic Hydrogenation. In *Asymmetric Synthesis,* Morrison, J. D. Ed., Academic Press, Inc.: Orlando, FL, 1985, *Vol. 5*, p. 345.
7. Kagan, H. B. Chiral Ligands for Asymmetric Catalysis. In *Asymmetric Synthesis,* Morrison, J. D. Ed., Academic Press, Inc.: Orlando, FL, 1985, *Vol. 5*, p. 1.
8. Takaya, H., Ohta, T., Noyori, R. Asymmetric Hydrogenation. In *Catalytic Asymmetric Synthesis,* Ojima, I. Ed., VCH Publishers, Inc.: New York, NY, 1993, p. 1.
9. Koenig, K. E. The Applicability of Asymmetric Homogeneous Catalytic Hydrogenation. In *Asymmetric Synthesis,* Morrison, J. D. Ed., Academic Press, Inc.: Orlando, FL, 1985, *Vol. 5*, p. 71.
10. Ojima, I., Clos, N., Bastos, C. *Tetrahedron* 1989, *45*, 6901.
11. Nugent, W. A., RajanBabu, T. V., Burk, M. J. *Science* 1993, *259*, 479.
12. Noyori, R. *Science* 1990, *248*, 1194.
13. Noyori, R. *Tetrahedron* 1994, *50*, 4259.
14. Knowles, W. S. *Angew. Chem., Chem. Int. Ed.* 2002, *41*, 1998.
15. Noyori, R. *Angew. Chem., Int. Ed.* 2002, *41*, 2008.
16. Halpern, J. Asymmetric Catalytic Hydrogenation: Mechanism and Origin of Enantioselection. In *Asymmetric Synthesis,* Morrison, J. D. Ed., Academic Press: Orlando, 1985, *Vol. 5*, p. 41.
17. Knowles, W. S. *Acc. Chem. Res.* 1983, *16*, 106.
18. Knowles, W. S. *J. Chem. Ed.* 1986, *63*, 222.
19. Vineyard, B. D., Knowles, W. S., Sabacky, M. J., Bachman, G. L., Weinkauff, D. J. *J. Am. Chem. Soc.* 1977, *99*, 5946.
20. Ager, D. J., Laneman, S. A. The Synthesis of Unnatural Amino Acids. In *Asymmetric Catalysis on Industrial Scale,* Blaser, H.-U., Schmidt, E. Eds., Wiley-VCH: Weinheim, 2004, p. 259.
21. Laneman, S. A., Froen, D. E., Ager, D. J. The Preparation of Amino Acids via Rh(DIPAMP)-Catalyzed Asymmetric Hydrogenations. In *Catalysis of Organic Reactions,* Herkes, F. E. Ed., Marcel Dekker: New York, 1998, p. 525.
22. Blaser, H.-U., Jalett, H.-P., Spindler, F. *J. Mol. Catalysis A: Chemical* 1996, *107*, 85.
23. Blaser, H.-U., Malan, C., Pugin, B., Spindler, F., Steiner, H., Studer, M. *Adv. Synth. Catal.* 2003, *345*, 103.

24. Blaser, H.-U., Schmidt, E. Introduction. In *Asymmetric Catalysis on Industrial Scale,* Blaser, H.-U., Schmidt, E. Eds., Wiley-VCH: Weinheim, 2004, p. 1.
25. Blaser, H.-U. *J. Chem. Soc., Chem. Commun.* 2003, 293.
26. Vineyard, B. D., Knowles, W. S., Sabacky, M. J., Bachman, G. L., Weinkauff, D. J. *J. Am. Chem. Soc.* 1977, *99,* 5946.
27. Koenig, K. E., Sabacky, M. J., Bachman, G. L., Christopfel, W. C., Barnstorff, H. D., Friedman, R. B., Knowles, W. S., Stults, B. R., Vineyard, B. D., Weinkauff, D. J. *Ann. N.Y. Acad. Sci.* 1980, *333,* 16.
28. Knowles, W. S. Asymmetric Hydrogenations—The Monsanto L-Dopa Process. In *Asymmetric Catalysis on Industrial Scale,* Blaser, H.-U., Schmidt, E. Eds., Wiley-VCH: Weinheim, 2004, p. 23.
29. Juge, S., Stephan, M., Laffitte, J. A., Genet, J. P. *Tetrahedron Lett.* 1990, *31,* 6357.
30. Schmidt, U., Riedl, B., Griesser, H., Fitz, C. *Synthesis* 1991, 655.
31. Juge, S., Genet, J. P. *Tetrahedron Lett.* 1989, *30,* 2783.
32. Juge, S., Stephan, M., Merdes, R., Genet, J. P., Halut-Desportes, S. *J. Chem. Soc., Chem. Commun.* 1993, 531.
33. Juge, S., Genet, J. P. U.S. Pat. 5,043,465, 1991.
34. Ager, D. J., Fotheringham, I. F., Laneman, S. A., Pantaleone, D. P., Taylor, P. P. *Chimica Oggi* 1997, *15 (3/4),* 11.
35. Noyori, N. *Chem. Soc. Rev.* 1989, *18,* 187.
36. Noyori, R., Takaya, H. *Acc. Chem. Res.* 1990, *23,* 345.
37. Takaya, H., Akutagawa, S., Noyori, R. *Org. Synth.* 1989, *67,* 20.
38. Cai, D., Payack, J. F., Bender, D. R., Hughes, D. L., Verhoeven, T. R., Reider, P. J. *J. Org. Chem.* 1994, *59,* 7180.
39. Cai, D., Payack, J. F., Verhoeven, T. R. U.S. Pat. 5,399,771, 1995.
40. Ager, D. J., East, M. B., Eisenstadt, A., Laneman, S. A. *J. Chem. Soc., Chem. Commun.* 1997, 2359.
41. Strong, J. G. *PharmaChem* 2003, *2 (June),* 20.
42. Akutagawa, S., Tani, K. Asymmetric Isomerization of Allylamines. In *Catalytic Asymmetric Synthesis,* Ojima, I. Ed., VCH Pulishers, Inc.: New York, NY, 1993, p. 41.
43. Noyori, R., Hasiguchi, S., Yamano, T. Asymmetric Synthesis. In *Applied Homogeneous Catalysis with Organic Compounds,* Herrmann, B. C. W. A. Ed., Wiley-VCH: Weinheim, 2002, *Vol. 1,* p. 557.
44. Otsuka, S., Tani, K. Asym. Cat. Isom. of Func. Olefin. In *Asymmetric Catalytic Isomerization of Functionalized Olefins,* Morrison, J. D. Ed., Academic Press, Inc.: Orlando, FL, 1985, *Vol. 5,* p. 171.
45. Kagan, H. B. *Bull. Soc. Chim. Fr.* 1988, 846.
46. Kumobayashi, H., Miura, T. *Chiral USA '96,* Chiral USA'96, 1996, Boston, MA.
47. Akutagawa. *Appl. Catal. A: General* 1995, *128,* 171.
48. Noyori, R., Ikeda, T., Ohkuma, T., Widhalm, M., Kitamura, M., Takaya, H., Akutagawa, S., Sayo, N., Saito, T., Taketomi, T., Kumobayashi, H. *J. Am. Chem. Soc.* 1989, *111,* 9134.
49. Ager, D. J., Laneman, S. A. *Tetrahedron: Asymmetry* 1997, *8,* 3327.
50. Chan, A. S. C. *CHEMTECH* 1993, *March,* 46.
51. Chan, A. S. C. U.S. Pat., 4,994,607, 1991.
52. Chan, A. S. C., Laneman, S. A. *Unpublished results.*
53. Ohta, T., Takaya, H., Kitamura, M., Nagai, K., Noyori, R. *J. Org. Chem.* 1987, *52,* 3176.
54. Chan, A. S. C., Laneman, S. A. U.S. Pat. 5,202,473, 1993.
55. Wan, K. T., Davis, M. E. *Nature* 1994, *370,* 449.
56. Takaya, H., Ohta, T., Sayo, N., Kumobayashi, H., Akutagawa, S., Inoue, S., Kasahara, I., Noyori, R. *J. Am. Chem. Soc.* 1987, *109,* 1596.
57. Noyori, R., Ohkuma, T., Kitamura, M. *J. Am. Chem. Soc.* 1987, *109,* 5856.
58. Ikariya, T., Ishii, Y., Kawano, H., Arnai, T., Saburi, M., Yoshikawa, S., Akutagawa, S. *J. Chem Soc., Chem. Commun.* 1985, 922.
59. Mashima, K., Kusano, K., Ohta, T., Noyori, R., Takaya, H. *J. Chem. Soc., Chem. Commun.* 1989, 1208.
60. Kitamura, M., Tokunaga, M., Ohkuma, T., Noyori, R. *Tetrahedron Lett.* 1991, *32,* 4163.
61. Heiser, B., Broger, E. A., Crameri, Y. *Tetrahedron: Asymmetry* 1991, *2,* 51.
62. Hoke, J. B., Hollis, L. S., Stern, E. W. *J. Organometal. Chem.* 1993, *455,* 193.
63. Genet, J. P., Pinel, C., Ratovelomanana-Vidal, V., Mallart, S., Pfister, X., Bischoff, L., De Andrade, M. C. C., Darses, S., Galopin, C., Laffitte, J. A. *Tetrahedron: Asymmetry* 1994, *5,* 675.
64. Chan, A. S. C., Laneman, S. A. U.S. Pat. 5,144,050, 1992.

65. Shao, L., Takeuchi, K., Ikemoto, M., Kawai, T., Ogasawara, M., Takeuchi, H., Kawano, H., Saburi, M. *J. Organometal. Chem.* 1992, *435*, 133.
66. Noyori, N., Kitamura, M., Sayo, N., Kumobayashi, H., Giles, M. F. U.S. Pat. 5,198,562, 1993.
67. Proctor, L. *Chiral USA '02,* Chiral USA '02, 2002, Boston, MA.
68. Leece, C., Personal communication, 2004.
69. Kitamura, M., Kasahara, I., Manabe, K., Noyori, R., Takaya, H. *J. Org. Chem.* 1988, *53*, 708.
70. Fiorini, M., Giongo, G. M. *J. Mol. Catal.* 1979, *5*, 303.
71. Pracejus, G., Pracejus, H. *Tetrahedron Lett.* 1977, *39*, 3497.
72. Fiorini, M., Riocci, M., Giongo, M. Eur. Pat. EP 77099, 1983.
73. Spindler, F., Pugin, B., Jalett, H.-P., Buser, H.-P., Pittelkow, U., Blaser, H.-U. Catalysis of Organic Reations. In *Chem. Ind.,* Malz, J. Ed., Dekker: New York, 1996, *Vol. 68*, p. 153.
74. Stinson, S. C. *Chem. Eng. News*, 1997, *75*, 34.
75. Blaser, H.-U., Spindler, F. *Chimia* 1997, *51*, 297.
76. Rouhi, A. M. *Chem. Eng. News* 2004, *82*, 47.
77. Blaser, H.-U. *Adv. Synth. Catal.* 2002, *344*, 17.
78. Hayashi, T., Mise, T., Fukushima, M., Kagotani, M., Nagashima, N., Hamada, Y., Matsumoto, A., Kawakami, S., Monishi, M., Yamomoto, K., Kumada, M. *Bull. Chem. Soc. Jpn.* 1980, *53*, 1138.
79. Brieden, W. PCT Pat. Appl. WO 9616971, 1996.
80. Togni, A., Breutel, C., Schnyder, A., Spindler, F., Landert, H., Tijani, A. *J. Am. Chem. Soc.* 1994, *116*, 4062.
81. Barbaro, P., Pregosin, P. S., Salzmann, R., Albinati, A., Kunz, R. *Organometallics* 1995, *14*, 5160.
82. Breutel, C., Pregosin, P. S., Salzmann, R., Togni, A. *J. Am. Chem. Soc.* 1994, *116*, 4067.
83. Brieden, W. *Chiral USA '97,* Chiral USA '97, 1997, Boston, MA.
84. Brieden, W. Personal communication, 1997.
85. Bader, R. R., Baumeister, P., Blaser, H.-U. *Chimia* 1996, *30*, 9.
86. Imwinkelried, R. *Chimia* 1997, *51*, 300.
87. Blaser, H.-U., Spindler, F., Studer, M. *Appl. Catal. A: General* 2001, *221*, 119.
88. Rautenstrauch, V., *Proceedings of the International Symposium on Chirality,* 1999, Cambridge, UK.
89. Selke, R. *J. Organomet. Chem.* 1989, *370*, 249.
90. Vocke, W., Hanel, R., Flother, F.-U. *Chem. Techn.* 1987, *39*, 123.
91. RajanBabu, T. V., Ayers, T. A., Casalnuovo, A. L. *J. Am. Chem. Soc.* 1994, *116*, 4101.
92. Andrade, J. G., Prescher, G., Schaefer, A., Nagel, U. In *Chem. Ind.,* Kosik, J. Ed., Marcel Dekker, 1990, p. 33.
93. Nagel, U., Kinzel, E., Andrade, J., Prescher, G. *Chem. Ber.* 1986, *119*, 3326.
94. Pugin, B., Muller, M., Spindler, S. U.S. Pat. 5,306,853, 1993.
95. Achiwa, K. *J. Am. Chem. Soc.* 1976, *98*, 8265.
96. Baker, G. L., Fritschel, S. J., Stille, J. R., Stille, J. K. *J. Org. Chem.* 1981, *46*, 2954.
97. Ojima, I., Kogure, T., Yoda, N. *J. Org. Chem.* 1980, *45*, 4728.
98. Takeda, H., Tachinami, T., Aburatani, M., Takahashi, H., Morimoto, T., Achiwa, K. *Tetrahedron Lett.* 1989, *30*, 363.
99. Tang, W., Zhang, X. *Chem. Rev.* 2003, *103*, 3029.
100. Tang, W., Zhang, X. *Angew. Chem., Int. Ed.* 2002, *41*, 1612.
101. Zhang, X., Tang, W. PCT Pat. Appl. WO 0342135, 2003.
102. Tang, W., Zhang, X. *Org. Lett.* 2002, *4*, 4159.
103. Tang, W., Zhang, X. *Org. Lett.* 2003, *5*, 205.
104. Cong-Dung Le, J., Pagxenkopf, B. L. *J. Org. Chem.* 2004, *69*, 4177.
105. Holz, J., Monsees, A., Jiao, H., You, J., Komarov, I. V., Fischer, C., Drauz, K., Borner, A. *J. Org. Chem.* 2003, *68*, 1701.
106. Borner, A., Holz, J., Personal communication 2004.
107. Holz, J., Riermeier, R., Monsees, A., Borner, A., *16th International Conference on Phosphorus Chemistry,* 2004, Birmingham, UK.
108. Imamoto, T., Watanabe, J., Wada, Y., Masuda, H., Yamada, H., Tsuruta, H., Matsukawa, S., Yamaguchi, K. *J. Am. Chem. Soc.* 1998, *120*, 1635.
109. Gridnev, I. D., Yamanoi, Y., Higashi, N., Tsuruta, H., Yasutake, M., Imamoto, T. *Adv. Synth. Catal.* 2001, *343*, 118.

110. Crepy, K. V. L., Imamoto, T. *Tetrahedron Lett.* 2002, *43*, 7735.
111. Gridnev, I. D., Yasutake, M., Higashi, N., Imamoto, T. *J. Am. Chem. Soc.* 2001, *123*, 5268.
112. Yasutake, M., Gridnev, I. D., Higashi, N., Imamoto, T. *Org. Lett.* 2001, *3*, 1701.
113. Danjo, H., Sasaki, W., Miyazaki, T., Imamoto, T. *Tetrahedron Lett.* 2003, *44*, 3467.
114. Hoge, G., Wu, H.-P., Kissel, W. S., Pflum, D. A., Greene, D. J., Bao, J. *J. Am. Chem. Soc.* 2004, *126*, 5966.
115. Hoge, G., *16th International Symposium on Chirality*, 2004, New York.
116. Burk, M. J., De Konig, P. D., Grote, T. M., Hoekstra, M. S., Hoge, G., Jennings, R. A., Kissel, W. S., Le, T. V., Lennon, I. C., Mulhern, T. A., Ramsden, J. A., Wade, R. A. *J. Org. Chem.* 2003, *68*, 5731.
117. Gautier, I., Ratovelomanana-Vidal, V., Savignac, P., Genet, J.-P. *Tetrahedron Lett.* 1996, *37*, 7721.
118. Bulliard, M., Laboue, B., Lastennet, J., Roussiasse, R. *Org. Proc. Res. Dev.* 2001, *5*, 438.
119. Rouhi, A. M. *Chem. Eng. News* 2003, *81 (18)*, 45.
120. Schmid, R., Foricher, J., Cereghetti, M., Schonholzer, P. *Helv. Chim. Acta* 1991, *74*, 370.
121. Schlummer, B., *7th International Conference on Organic Process Research and Development*, 2003, New Orleans.
122. Gerlach, A., Scholz, U. *Spec. Chem. Magazine* 2004, 37.
123. Pohl, T., Prinz, T., Giffels, G., Sirges, W. Eur. Pat. EP 1205486, 2001.
124. Schroder, G., Arlt, D. Eur. Pat. EP 749973, 1996.
125. Kralik, J., Ritzkopf, I., Steffens, C., Giffels, G., Dreisbach, C., Prinz, T., Lange, W. Eur. Pat. EP 1186609, 2001.
126. Neugebauer, T. *Chimica Oggi* 2002, 2.
127. Zeller, J., *Chiral USA '03*, 2003, Chicago, IL.
128. Butler, D. E., DeJong, R. L., Nelson, J. D., Pamment, M. G., Stuk, T. L. PCT Appl. Pat. WO 0255519, 2002.
129. Saito, T., Yokozawa, T., Zhang, X., Sayo, N. U.S. Pat. 5,872,273, 1999.
130. Saito, T., Yokozawa, T., Ishizaki, T., Moroi, T., Sayo, N., Miura, T., Kumobayashi, H. *Adv. Synth. Catal.* 2001, *343*, 264.
131. Sayo, N., Matsumoto, T. U.S. Pat. 6,342,644, 2002.
132. Zhang, X. PCT Pat. Appl. WO 0121625, 2001.
133. Yokozawa, T., Sayo, N., Saito, T., Ishizaki, T. Eur. Pat. EP 95946, 2001.
134. Zhang, Z., Qian, H., Longmire, J., Zhang, X. *J. Org. Chem.* 2000, *65*, 6223.
135. Wu, S., Wang, W., Tang, W., Lin, M., Zhang, X. *Org. Lett.* 2002, *4*, 4495.
136. Lei, A., Wu, S., He, M., Zhang, X. *J. Am. Chem. Soc.* 2004, *126*, 1626.
137. Tang, W., Wu, S., Zhang, X. *J. Am. Chem. Soc.* 2003, *125*, 9570.
138. Saito, T., Yokozawa, T., Ishizaki, T., Moroi, T., Sayo, N., Miura, T., Kumobayashi, H. *Adv. Synth. Catal.* 2001, *343*, 264.
139. Challener, C. Private communication.
140. Qui, L., Wu, J., Chan, S., Au-Yeung, T. T.-L., Ji, J.-X., Guo, R., Pai, C.-C., Zhong, Y. Z., Li, X., Fan, Q.-H., Chan, A. S. C. *Proc. Nat. Acad. Sci. USA* 2004, *101*, 5815.
141. Duprat de Paule, S., Jeulin, S., Ratovelomanana-Vidal, V., Genet, J.-P., Champion, N., Dellis, P. *Tetrahedron Lett.* 2003, *44*, 823.
142. Pai, C.-C., Li, Y.-M., Zhong, Y. Z., Chan, A. S. C. *Tetrahedron Lett.* 2002, *43*, 2789.
143. Duprat de Paule, S., Champion, N., Vidal, V., Genet, J.-P., Dellis, P. PCT Pat. Appl. WO 03029259, 2003.
144. Duprat de Paul, S., Jeulin, S., Ratovelomanana-Vidal, V., Genet, J.-P., Champion, N., Dellis, P. *Eur. J. Org. Chem.* 2003, 1931.
145. Jeulin, S., Duprat de Paul, S., Ratovelomanana-Vidal, V., Genet, J.-P., Champion, N., Dellis, P. *Proc. Nat. Acad. Sci. USA* 2004, *101*, 5799.
146. Jeulin, S., Duprat de Paul, S., Ratovelomanana-Vidal, V., Genet, J.-P., Champion, N., Dellis, P. *Angew. Chem., Int. Ed.* 2004, *43*, 320.
147. Pai, C.-C., Lin, C.-W., Lin, C.-C., Chen, C.-C., Chan, A. S. C. *J. Am. Chem. Soc.* 2000, *122*, 11513.
148. Wu, J., Wai, H. K., Kim, H. L., Zhong, Y. Z., Yeung, C. H., Chan, A. S. C. *Tetrahedron Lett.* 2002, 1539.
149. Wu, J., Chen, H., Zhou, Z.-Y., Yeung, C. H., Chan, A. S. C. *Synlett* 2001, 1050.
150. Wu, J., Pai, C.-C., Kwok, W., Guo, R., Au-Yeung, T. T.-L., Yeung, C. H., Chan, A. S. C. *Tetrahedron: Asymmetry* 2003, *14*, 987.

151. Wu, J., Chen, X., Guo, R., Yeung, C. H., Chan, A. S. C. *J. Org. Chem.* 2003, *68*, 2490.
152. Noyori, R., Ohkuma, T. *Pure Appl. Chem.* 1999, *71*, 1493.
153. Wu, J., Chen, H., Kwok, w., Guo, R., Zhou, Z.-Y., Yeung, C. H., Chan, A. S. C. *J. Org. Chem.* 2002, *67*, 7908.
154. Shi, Q., Xu, L., Jia, X., Wang, R., Au-Yeung, T. T.-L., Chan, A. S. C., Hayashi, T., Cao, R., Hong, M. *Tetrahedron Lett.* 2003, 6505.
155. Beninicori, T., Cesarotti, E., Piccolo, O., Sannicolo, F. *J. Org. Chem.* 2000, *65*, 2043.
156. Banzinger, M., Cercus, J., Hirt, H., Laumen, K., Malan, C., Spindler, F., Struber, F., Troxler, T. *Tetrahedron: Asymmetry* 2003, *14*, 3469.
157. Tinti, M. O., Piccolo, O., Bonifacio, O., Crescenzi, C., Penco, S. U.S. Pat. 6,566,552, 2003.
158. Jacobs, R. T., Bernstein, P. R., Cronk, L. A., Vacek, E. P., Newcomb, L. F., Aharony, D., Buckner, C. K., Kusner, E. J. *J. Med. Chem.* 1994, *37*, 1282.
159. Pye, P. J., Rossen, K., Reamer, R. A., Tsou, N. N., Volante, R. P., Reider, P. J. *J. Am. Chem. Soc.* 1997, *119*, 6207.
160. Pye, P. J., Rossen, K., Volante, R. P. PCT Pat. Appl. WO 9747632, 1997.
161. Pye, P. J., Rossen, K., Reamer, R. A., Volante, R. P., Reider, P. J. *Tetrahedron Lett.* 1998, *39*, 4441.
162. Burk, M. J., Hems, W., Herzberg, D., Malan, C., Zanotti-Gerosa, A. *Org. Lett.* 2000, *2*, 4173.
163. Chaplin, D., Harrison, P., Henschke, J. P., Lennon, I. C., Meek, G., Moran, P., Pilkington, C. J., Ramsden, J. A., Watkins, S., Zanotti-Gerosa, A. *Org. Proc. Res. Dev.* 2003, *7*, 89.
164. Hems, W., Rossen, K., Reichert, D., Kohler, K., Perea, A.-M. PCT Pat. Appl. WO 0411452, 2004.
165. Boaz, N. W., Debenham, S. D., Mackenzie, E. B., Large, S. E. *Org. Lett.* 2002, *14*, 2421.
166. Boaz, N. W. BoPhoz Ligands for Asymmetric Catalysis, *7th International Conference on Organic Process Research and Development,* 2003, New Orleans, USA.
167. Debenham, S. D., Boaz, N. W. PCT Pat. Appl. WO 0226695, 2002.
168. Boaz, N. W., Debenham, S. D., Large, S. E., Moore, M. K. *Tetrahedron: Asymmetry* 2003, *14*, 3575.
169. Boaz, N. W., Debenham, S. D. PCT Pat. Appl. WO 0226750, 2002.
170. Hanson, G. J., Chen, B. B., Baran, J. S. U.S. Pat. 5,268,391, 1993.
171. Wehner, V., Stilz, H., Schmidt, W., Seiffge, D. PCT Pat. Appl. WO 0069831, 2000.
172. Wehner, V., Flohr, S., Blum, H., Rutten, H., Stilz, H. PCT Pat. Appl. WO 0311288, 2003.
173. Surtees, J., Marmon, V., Differding, E., Zimmermann, V. PCT Pat. Appl. WO 0164637, 2001.
174. Weissensteiner, W., Sturm, T., Spindler, F. PCT Pat. Appl. WO 0202578, 2002.
175. Thommen, M. *Chiral USA '03,* 2003, Chicago, USA.
176. Sturm, T., Weissensteiner, W., Spindler, F. *Adv. Synth. Catal.* 2003, *345*, 160.
177. Knochel, P., Ireland, T., Grossheimann, G., Drauz, K., Klement, I. U.S. Pat., 6 191,284, 2001.
178. Ireland, T., Grossheimann, G., Wieser-Jeunesse, C., Knochel, P. *Angew. Chem., Int. Ed.* 1999, *38*, 3212.
179. Lotz, M., Polborn, K., Knochel, P. *Angew. Chem., Int. Ed.* 2002, *41*, 4708.
180. Ireland, T., Tappe, K., Grossheimann, G., Knochel, P. *Chem. Eur. J.* 2002, *8*, 843.
181. Tappe, K., Knochel, P. *Tetrahedron: Asymmetry* 2004, *15*, 91.
182. Hayashi, T., Yamamoto, A., Hojo, M., Ito, Y. *J. Chem. Soc., Chem. Commun.* 1989, 495.
183. Drauz, K., Klement, I., Knochel, P., Alema, P. J. J. U.S. Pat. 6,284,925, 2001.
184. Schwink, L. *Tetrahedron Lett.* 1996, *37*, 25.
185. Perea, J. J. A., Lotz, M., Knochel, P. *Tetrahedron: Asymmetry* 1999, *10*, 375.
186. Kang, J., Lee, J. H., Ahn, S. H., Choi, J. S. *Tetrahedron Lett.* 1998, *39*, 5523.
187. Doucet, H., Ohkuma, T., Murata, K., Yokozawa, T., Kozawa, M., Katayama, E., England, A. F., Ikariya, T., Noyori, R. *Angew. Chem., Int. Ed.* 1998, *37*, 1703.
188. Ohkuma, T., Kitamura, M., Noyori, R. Asymmetric Hydrogenation. In *Catalytic Asymmetric Synthesis, 2nd Edition,* Ojima, I. Ed., Wiley-VCH: New York, 2000, p. 1.
189. Ohkuma, T., Koizumi, M., Doucer, H., Pham, T., Kozawa, M., Murata, K., Katayama, E., Yokozawa, T., Ikariya, T., Noyori, R. *J. Am. Chem. Soc.* 1998, *120*, 13529.
190. Ohkuma, T., Koizumi, M., Yoshida, M., Noyori, R. *Org. Lett.* 2000, 2.
191. Ohkuma, T., Ishii, D., Takeno, H., Noyori, R. *J. Am. Chem. Soc.* 2000, *122*, 6510.
192. Ohkuma, T., Koizumi, M., Ikehira, H., Yokozawa, T., Noyori, R. *Org. Lett.* 2000, *2*, 659.
193. Ohkuma, T., Ooka, H., Yamakawa, M., Ikariya, T., Noyori, R. *J. Org. Chem.* 1996, *61*, 4872.
194. Noyori, R., Kitamura, M., Ohkuma, T. *Proc. Nat. Acad. Sci. USA* 2004, *101*, 5356.

195. Ohkuma, T., Koizumi, M., Muniz, K., Hilt, G., Kabuto, C., Noyori, R. *J. Am. Chem. Soc.* 2002, *124*, 6508.
196. Mikami, K., Korenaga, T., Ohkuma, T., Noyori, R. *Angew. Chem., Int. Ed.* 2000, *39*, 3707.
197. Mikami, K., Korenaga, T., Yusa, Y., Yamanaka, M. *Adv. Synth. Catal.* 2003, *345*, 246.
198. Kumobayashi, H., Miura, T., Sayo, N., Saito, T. *Synlett* 2001, 1055.
199. Chen, C.-Y., Reamer, R. A., Chilenski, J. R., McWilliams, C. J. *Org. Lett.* 2003, *5*, 5039.
200. de Vries, J. G., de Vries, A. H. M. *Eur. J. Org. Chem.* 2003, 799.
201. van den Berg, M., Minnaard, A. J., Schudde, E. P., van Esch, J., de Vries, A. H. M., de Vries, J. G., Feringa, B. L. *J. Am. Chem. Soc.* 2000, *122*, 11539.
202. Reetz, M. T., Mehler, G. *Angew. Chem., Int. Ed.* 2000, *39*, 3889.
203. Reetz, M. T., Sell, T. *Tetrahedron Lett.* 2000, *41*, 6333.
204. Claver, C., Fernandez, E., Gillon, A., Heslop, K., Hyett, D. J., Martorell, A., Orpen, A. G., Pringle, P. G. *J. Chem. Soc., Chem. Commun.* 2000, 961.
205. Garland, M., Blaser, H.-U. *J. Am. Chem. Soc.* 1990, *112*, 7048.
206. Augustine, R. L., Tanielyan, S. K., Doyle, L. K. *Tetrahedron: Asymmetry* 1993, *4*, 1803.
207. Hashiguchi, S., Fujii, A., Takehara, J., Ikariya, T., Noyori, R. *J. Am. Chem. Soc.* 1995, *117*, 7562.
208. Noyori, R., Hashiguchi, S. *Acc. Chem. Res.* 1997, *30*, 97.
209. Palmer, M. J., Wills, M. *Tetrahedron: Asymmetry* 1999, *10*, 2045.
210. Fujii, A., Hashiguchi, S., Uematsu, N., Ikariya, T., Noyori, R. *J. Am. Chem. Soc.* 1996, *118*, 2521.
211. Takehara, J., Hashiguchi, S., Fujii, A., Inoue, S., Ikariya, T., Noyori, R. *J. Chem. Soc., Chem. Commun.* 1996, 233.
212. Blacker, J., Martin, J. Scale-up Studies in Asymmetric Transfer Hydrogenation. In *Asymmetric Catalysis on Industrial Scale: Challenges, Approaches and Solutions,* H.U. Blaser, E. S. Ed., Wiley-VCH Verlag GmbH & Co.: Weinheim, 2004, p. 201.
213. Hansen, K. B., Chilenski, J. R., Desmond, R., Devine, P. N., Grabowski, E. J. J., Heid, R., Kubryk, M., Mathre, D. J., Varsolona, R. *Tetrahedron: Asymmetry* 2003, *14*, 3581.
214. Lee, N. E., Buchwald, S. L. *J. Am. Chem. Soc.* 1994, *116*, 5985.
215. Bakos, J., Toth, I., B., H., Marko, L. *J. Organometal. Chem.* 1985, *279*, 23.
216. Becalski, A. G., Cullen, W. R., Fryzuk, M. D., James, B. R., Kang, G.-J., Rettig, S. J. *Inorg. Chem.* 1991, *30*, 5002.
217. Bakos, J., Orosz, A., Heil, B., Laghmari, M., Lhoste, P., Sinou, D. *J. Chem. Soc., Chem. Commun.* 1991, 1684.
218. Chan, Y., Ng, C., Osborn, J. A. *J. Am. Chem. Soc.* 1990, *112*, 9600.
219. Morimoto, T., Achiwa, K. *Tetrahedron: Asymmetry* 1995, *6*, 2661.
220. Hoveyda, A. H., Morken, J. P. *Angew. Chem., Int. Ed.* 1996, *35*, 1262.
221. Uematsu, N., Fujii, A., Hashiguchi, S., Ikariya, T., Noyori, R. *J. Am. Chem. Soc.* 1996, *118*, 4916.
222. Stanley, G. G. Carbonylation Processes by Homogeneous Catalysis. In *Encyclopedia of Inorganic Chemistry,* King, R. B. Ed., J. Wiley, 1994, Vol. 2, p. 575.
223. Consiglio, G. Asymmetric Carbonylation. In *Catalytic Asymmetric Synthesis,* Ojima, I. Ed., VCH Publishers, Inc.: New York, 1993, p. 273.
224. Ojima, I., Hirai, K. Asymmetric Hydrosilylation and Hydrocarbonylation. In *Asymmetric Synthesis,* Morrison, J. D. Ed., Academic Press, Inc.: Orlando, 1985, *Vol. 5*, p. 103.
225. Sakai, N., Mano, S., Nozaki, K., Takaya, H. *J. Am. Chem. Soc.* 1993, *115*, 7033.
226. Nozaki, K., Matsuo, T., Shibahara, F., Hiyama, T. *Adv. Synth. Catal.* 2001, *343*, 61.
227. Lambers-Verstappen, M. M. H., de Vries, J. G. *Adv. Synth. Catal.* 2003, *345*, 478.
228. Rouhi, A. M. *Chem. Eng. News* 2003, *81*, 52.
229. Cobley, C. J., Gardner, K., Klosin, J., Praquin, C., Hill, C., Whiteker, G. T., Zanotti-Gerosa, A. *J. Org. Chem.* 2004, *69*, 4031.
230. Houlton, S. *Manufacturing Chemist* 2003.
231. Breeden, S., Cole-Hamilton, D. J., Foster, D. F., Schwarz, G. J., Wills, M. *Angew. Chem., Int. Ed.* 2000, *39*, 4106.
232. Clarkson, G. J., Ansell, J. R., Cole-Hamilton, D. J., Pogorzelec, P. J., Whittell, J., Wills, M. *Tetrahedron: Asymmetry* 2004, *15*, 1787.
233. Nozaki, K., Li, W., Horiuchi, T., Takaya, H. *J. Org. Chem.* 1996, *61*, 7658.
234. Park, H. S., Alberico, E., Alper, H. *J. Am. Chem. Soc.* 1999, *121*, 11697.

235. Alper, H. Asymmetric Synthesis Using Chiral Phosphine Ligands, *16th International Conference on Phosphorus Chemistry,* 2004, Birmingham, UK.
236. Brunner, H., Nishiyama, H., Itoh, K. Asymmetric Hydrosilylation. In *Catalytic Asymmetric Synthesis,* Ojima, I. Ed., VCH Publishers, Inc.: New York, 1993, p. 303.
237. Kitayama, K., Uozumi, Y., Hayashi, T. *J. Chem. Soc., Chem. Commun.* 1995, 1533.
238. Kitayama, K., Tsuji, H., Uozumi, Y., Hayashi, T. *Tetrahedron Lett.* 1996, *37,* 4169.
239. Uozumi, Y., Hayashi, T. *J. Am. Chem. Soc.* 1991, *113,* 9887.
240. Uozumi, Y., Kitayama, K., Hayashi, T., Yanagi, K., Fukuyo, E. *Bull. Chem. Soc. Jpn.* 1995, *68,* 713.
241. Matsui, M., Yamamoto, I. In *Naturally Occurring Insecticides,* Jacobsen, M., Crosby, G. D. Eds., Marcel Dekker: New York, 1971, p. 3.
242. Aranati, T. *Pure Appl. Chem.* 1985, *57,* 1839.
243. Nozaki, H., Moriuti, S., Takaya, H., Noyori, R. *Tetrahedron Lett.* 1966, 5239.

13 Modular, Chiral P-Heterocycles in Asymmetric Catalysis

Mark J. Burk and James A. Ramsden

CONTENTS

13.1 INTRODUCTION

An ever-increasing number of pharmaceutical, flavor and fragrance, animal health, polymer, and agrochemical products rely on the availability of enantiomerically pure building blocks.[1] More than 50% of the world's top-selling drugs are single enantiomers,[2] and it has been estimated that up to 80% of all drugs currently entering development are chiral and will be marketed as single enantiomer entities.[3] On the basis of this need, asymmetric catalysis has established itself as one of the most cost-effective and environmentally responsible methods for production of a large array of structurally diverse, enantiomerically pure compounds.

An important area of asymmetric catalysis research involves the design of chiral ligands and transition metal catalysts that can lead to a desired transformation with both high efficiency and high selectivity. Whereas selectivity in asymmetric catalysis often refers to control of absolute stereochemistry, other types, such as diastereoselectivities, chemoselectivities, and regioselectivities, can also play a crucial role in the development of a viable synthetic method. Because of the catalytic nature of the system, the intrinsic chirality of an asymmetric catalyst can be used effectively through many cycles, allowing many moles of a desired product to be generated from scant quantities of catalyst. This singular property engenders the economically and environmentally attractive features associated with asymmetric catalysis.

A blueprint for the design of effective chiral ligands and asymmetric catalysts is not available, and the factors that govern the efficiency and selectivity remain generally obscure. Accordingly, a well-conceived ligand class should possess one or more structural and/or electronic features (modules) that may be varied readily in a systematic fashion to optimize the design for a given purpose. The principle of modular design has been used extensively in nature to access proteins and secondary metabolites of specific function. In the present context, we borrow this principle for the conception and design of new ligand and catalyst systems. A design is informative to the extent that variations in the ligand modules can be correlated to changes in the reactivity or selectivity of the catalyst.

In this chapter, we review a growing family of modular phosphorus heterocycles that have been found broadly useful as chiral ligands in asymmetric catalysis.[4] Specifically, the utility of catalysts based on 4-membered phosphetane and 5-membered phospholane ligands will be the focus herein. A description of other types of chiral phosphorus heterocycles can be found in recent review articles[4] as well as in Chapters 12 and 15.

13.2 PHOSPHOLANE LIGANDS

Auxiliaries, reagents, and catalysts based on C_2-symmetric *trans*-2,5-disubstituted 5-membered nitrogen and boron heterocycles are well-documented to render very high levels of absolute stereocontrol in numerous reactions.[5] These systems furnished the original inspiration for the design of analogous phosphorus-based ligands. Brunner and Sievi had prepared enantiomerically pure *trans*-3,4-disubstituted phospholanes, but these ligands proved relatively ineffectual in asymmetric catalysis.[6] In this case, presumably the chiral environment was too distant from the metal coordination sphere to exert significant influence. It was reasoned, by contrast, that 5-membered *trans*-2,5-disubstituted phospholanes would position the chirality proximal to a metal's coordination sphere. Moreover, phospholanes possessing dialkyl- or trialkyl-substituted phosphorus atoms would represent a new series of electron-rich ligands that potentially could be used more effectively in certain asymmetric catalytic processes.[7]

Ligand electronic properties can dramatically influence the reactivity and selectivity of transition metal catalysts, and the electron-rich nature of phospholanes (Figure 13.1) is a unique feature that appears to differentiate these systems from many other available chiral ligands. Another important attribute of phospholanes is associated with the modularity of these systems. The ability to vary the phospholane R-substituents in a systematic fashion allows valuable information to be gathered concerning the steric requirements of the catalytic process. In this manner, the steric environment imposed by the ligand can be tuned to ideally accommodate the steric demands of the reactants and thus facilitate optimization of catalyst efficacy.

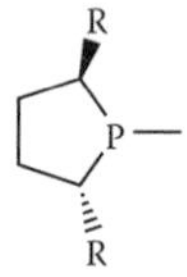

FIGURE 13.1 *trans*-2,5-Disubstituted phospholanes allow systematic variation of R-substituents.

Entry to *trans*-2,5-disubstituted phospholanes **3** was first achieved conveniently through the use of chiral 1,4-diol intermediates **1** (Scheme 13.1).[8] Originally, a series of 1,4-diols was prepared via electrochemical Kolbe coupling of enantiomerically pure α-hydroxy acids.[9] Commercially, the Kolbe procedure was not practical and more attractive routes involving biocatalytic methodologies currently are used to produce the requisite 1,4-diols.[10]

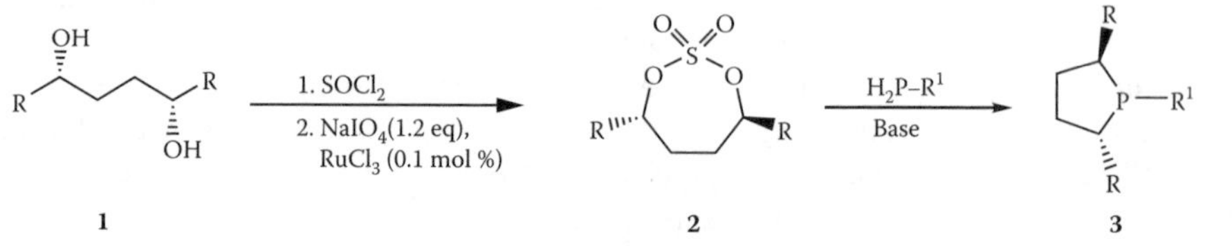

SCHEME 13.1

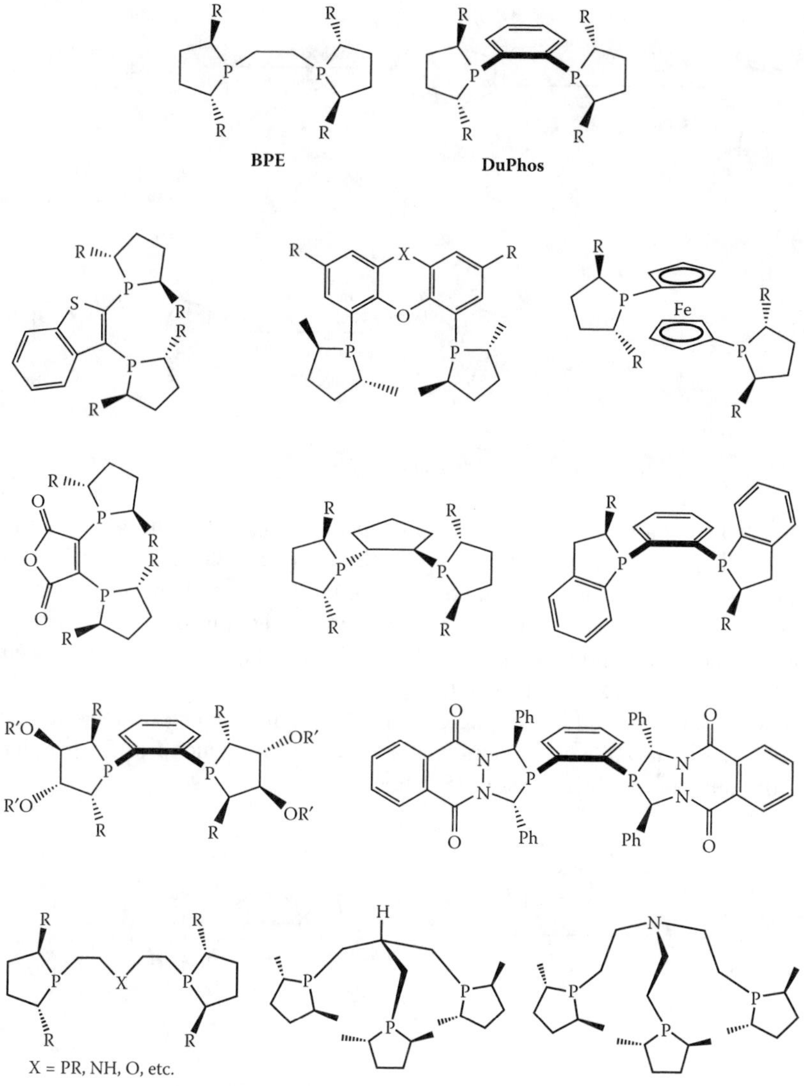

FIGURE 13.2 Chiral phospholane ligand family showing backbone diversity.

The availability of enantiomerically pure 1,4-diols **1** has allowed assembly of phospholanes possessing a range of R-substituents through base-induced reaction between 1,4-diol cyclic sulfates **2** and primary phosphines. Because the chirality of phospholanes resides within the phosphorus heterocycle, the R^1 group of **3** represents another module that may be varied to create a diverse collection of chiral phospholane ligands. In the event that R^1 possesses one or more additional primary phosphine groups ($-PH_2$ units), chelating ligand structures are possible.[9,11] In this case, the backbone unit of multidentate ligands has been explored as an element of diversity in an effort to control ligand electronic and conformational properties. More recently the range of both monocyclic and bicyclic phospholane-type ligands has been extended.[12,13] Figure 13.2 displays representative examples of this growing family of ligands.

The use of phosphide nucleophiles in an S_N2 attack on alkyl-substituted cyclic sulfates allows access to a wide range of phospholane derivatives. However, if aryl-substituted cyclic sulfates are used, elimination and racemization processes compete with substitution. To access bis(2,5-diphenylphospholano)ethane (**4**), an alternative strategy was required (Scheme 13.2).[14]

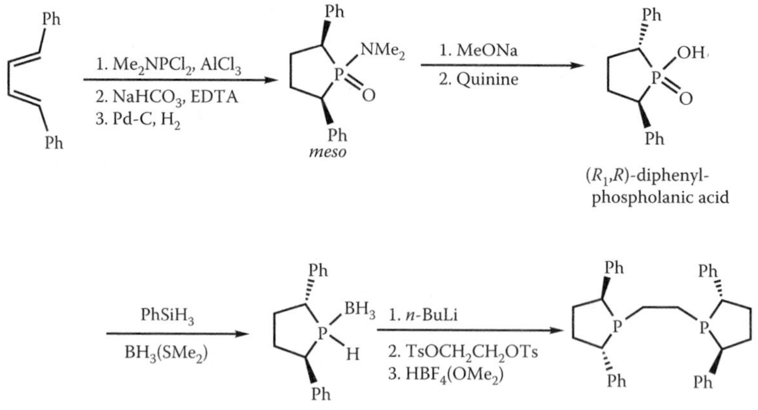

SCHEME 13.2

The *meso*-diphenylphospholanomide was obtained via a cycloaddition reaction of *N,N*-di-methylaminophosphinodichloride and 1,4-diphenylbutadiene followed by hydrogenation. The *meso*-compound was transformed into an equilibrium mixture containing predominately the *rac*-compound on treatment with base. After hydrolysis, the phospholanic acid was kinetically resolved using quinine. Reduction and protection of the phospholane as the borane adduct allowed the double displacement of 1,2-ethyleneglycol ditosylate.

A similar approach was used in the synthesis of 3,4-diazphospholanes. In this case, the addition of a bisphosphine to a diazine gave predominantly the *rac*-form, which was further elaborated and resolved via diastereomeric salt formation (Scheme 13.3).[12]

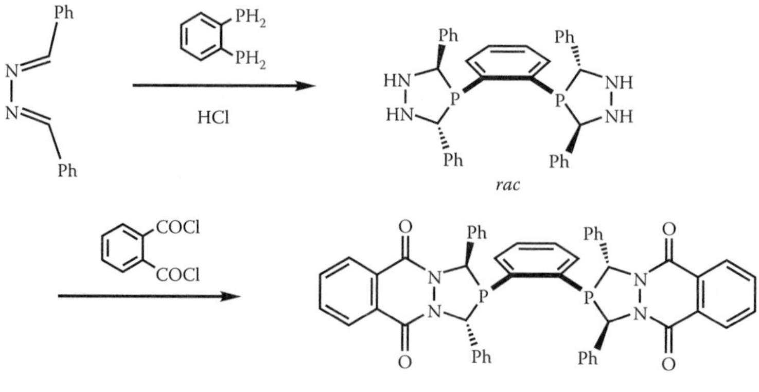

SCHEME 13.3

13.3 PHOSPHETANE LIGANDS

Success with phospholanes led to a logical exploration of other chiral phosphorus heterocycles. Four-membered phosphetanes appeared to offer a more rigid butterfly ring structure that was hoped would afford elevated enantioselectivities in certain reactions. A series of these ligands was prepared by an analogous route involving reaction between primary phosphines and chiral 1,3-diol cyclic sulfates in the presence of base (Scheme 13.4).[15] The preparation of phosphetanes was facilitated by ready access to enantiomerically pure 1,3-diols through asymmetric hydrogenation of a range of corresponding 1,3-diones using biaryldiphosphine-Ru catalysts.[16]

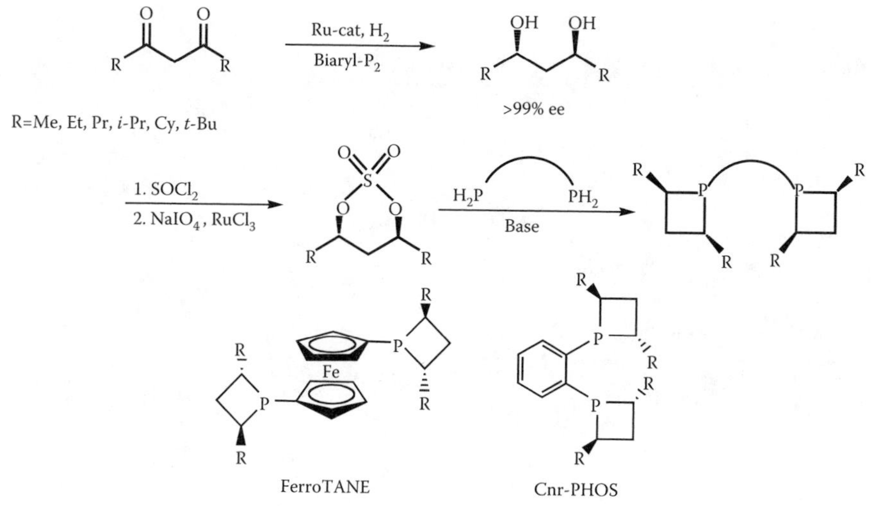

SCHEME 13.4

Reaction between 1,3-diol cyclic sulfates and bis(phosphines) in the presence of bases such as LDA afforded a family of chiral bis(phosphetane) ligands with a range of R groups. The first two members of this family were the FerroTANE[15a] and Cnr-PHOS[15c] ligands bearing 1,1′-ferrocenyl and 1,2-benzenyl backbones, respectively.

13.4 ASYMMETRIC CATALYTIC HYDROGENATIONS

Asymmetric hydrogenation reactions are ideal for commercial manufacture of single-enantiomer compounds because of the ease with which these robust processes are scaled up and because of the cleanliness of these transformations—few by-products are generated. Exceedingly high enantioselectivities and catalytic efficiencies have been realized in enantioselective hydrogenation reactions.[17] The versatility of phospholane and phosphetane ligands in asymmetric catalytic hydrogenations is highlighted in the following.

13.4.1 AMINO ACID DERIVATIVES

The advent of processes based on asymmetric catalytic homogeneous hydrogenation may be traced to the pursuit of economic routes to unnatural α-amino acids more than 30 years ago.[18] Impressive initial success was achieved with rhodium complexes bearing chiral diphosphines such as DIOP[19] and DIPAMP,[20] although general catalysts for the enantioselective hydrogenation of a range of different α-enamide substrates **5** had remained elusive, even in 1990. In the past 10 years, a number of useful catalysts systems have been discovered.

Cationic rhodium complexes of the type [(COD)Rh(DuPhos)]⁺X⁻ (X = weakly or noncoordinating anion) have been developed as one of the most general classes of catalyst precursors for efficient, enantioselective, low-pressure hydrogenation of α-enamides of type **5** (Scheme 13.5).[21] For substrates that possess a single β-substituent (e.g., R^1 = H), the Me-DuPhos-Rh and Et-DuPhos-Rh catalysts were found to render a multitude of amino acid derivatives with enantioselectivities of 95–99%. A variety of N-acyl protecting groups such as acetyl, Cbz, or Boc may be used, and enamides **5** may be used either as carboxylic esters or acids. Moreover, the substrates may be present as mixtures of E- and Z-geometric isomers with little detrimental effect. The commercial viability of these robust catalysts is revealed by the high catalyst activities (turnover frequencies >5000 h⁻¹) and catalyst productivities (substrate-to-catalyst [S/C] ratios up to 50,000) displayed by these systems.[22,23]

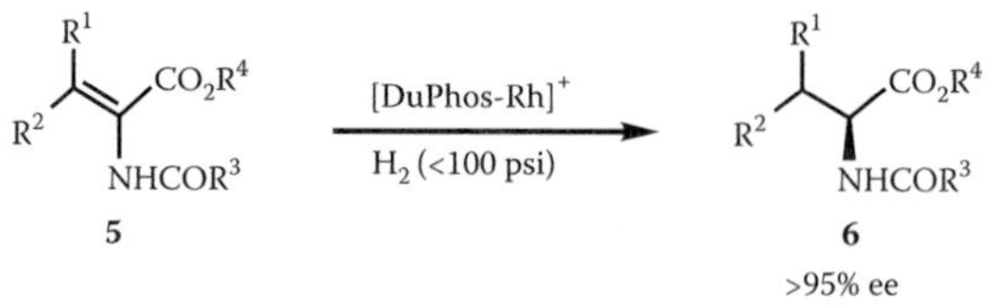

SCHEME 13.5

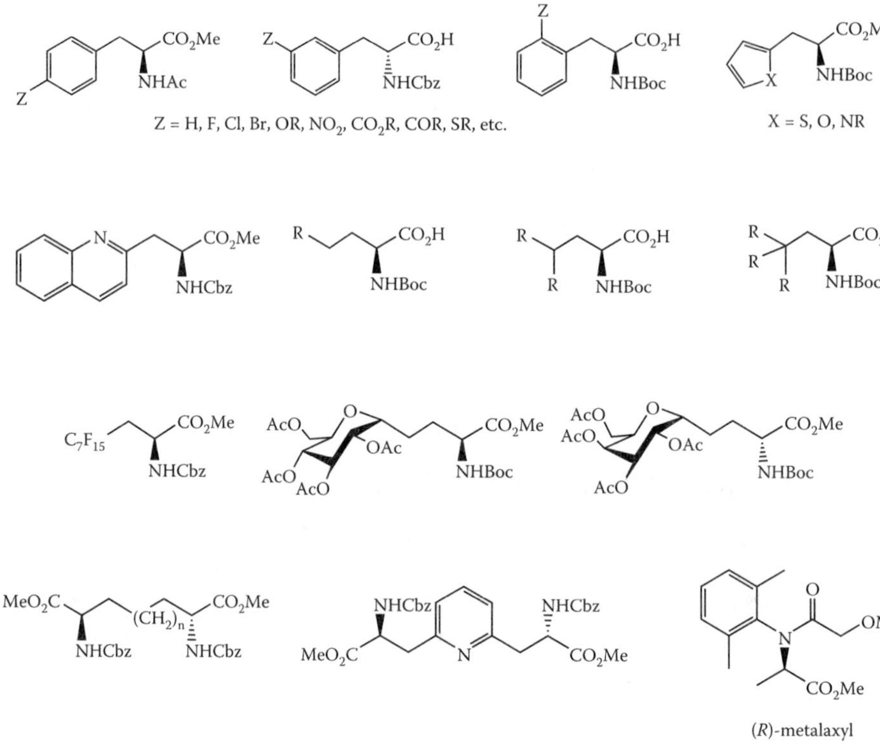

$Z = H, F, Cl, Br, OR, NO_2, CO_2R, COR, SR, etc.$

$X = S, O, NR$

(*R*)-metalaxyl

FIGURE 13.3 Amino acid derivatives produced via DuPhos-Rh catalysts.

Figure 13.3 highlights a representative selection of β-substituents and organic functional groups that are tolerated by these catalysts in the production of novel amino acids. Virtually any substituted aromatic, heteroaromatic, alkyl, fluoroalkyl, or other functionalized organic group may be incorporated into the amino acid product.[22–26] Assorted C-glycopeptide intermediates were generated in reactions whereby the catalyst displayed high levels of reagent control.[27] Polyamino acids of different sorts have been synthesized with very high selectivities.[28] Finally, the highly active fungicide, (*R*)-metalaxyl, has been produced efficiently using this technology.[17d] It is important to note that hydrogenation of haloaromatic enamides has provided facile access to a broad range of highly functionalized aromatic amino acids and peptides through subsequent Pd-catalyzed coupling processes.[29]

Other work has focused on the synthesis of more elaborate amino acids such as (*S*)-acromelobinic acid,[30] (*R*)-4-piperidinylglycine,[31] and L-azatyrosine[32] (Figure 13.4). Et-DuPhos-Rh specifically has become a routine catalyst used in the synthesis of structurally diverse amino acids required in peptide design (e.g., **7** and **8**).[33] The tolerance of these catalysts for a wide range of structural elements and functional groups is further emphasized by the ease with which these catalysts can be used in the synthesis of complex amino acid derivatives required in total synthesis (e.g., **9**).[34]

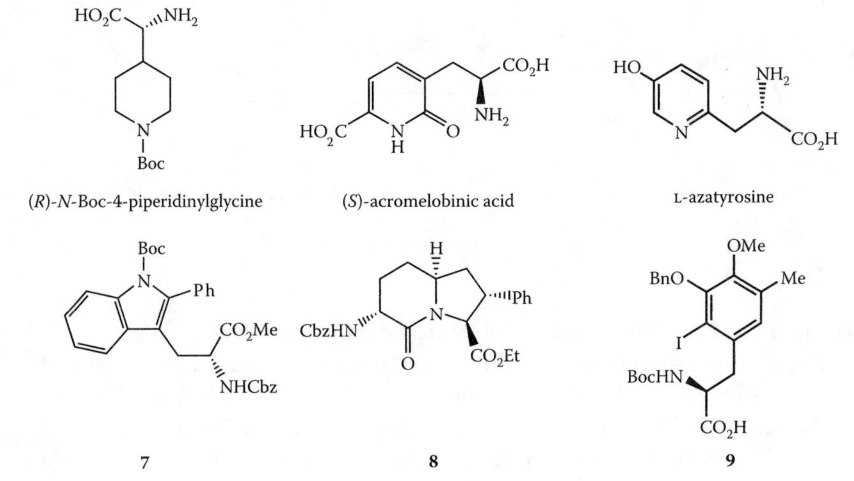

FIGURE 13.4 Elaborate amino acids accessible with DuPhos or BPE rhodium catalysts.

The advantages of modularity inherent to the DuPhos ligand series became apparent during attempts to hydrogenate enamides **5** possessing two β-substituents (R^1, $R^2 \neq H$). Enantioselective hydrogenation of tetrasubstituted alkene units of this type had presented a significant challenge for all known catalysts, and the highest selectivity reported prior to our work was 55% ee using a DIPAMP-Rh catalyst.[35] The rates and enantioselectivities were found to be dramatically dependent on the properties of the phospholane ligand.[36] In contrast to the preeminence of Et-DuPhos-Rh for enamides containing a single β-substituent, sterically less encumbered catalysts were required for hydrogenation of sterically congested β,β-disubstituted enamides. In particular, the Me-DuPhos-Rh and more electron-rich Me-BPE-Rh catalysts have been found superior for the production of multifarious β-branched amino acids.[37] Hence, both symmetric and dissymmetric β-substituted enamides were found to hydrogenate with very high enantioselectivites. In the latter case, hydrogenation of *E*- and *Z*-enamide isomers separately allowed the generation of a second new β-stereogenic center and production of both *threo*- and *erythro*-diastereomers in stereoisomerically pure form. For example, D-*allo*-isoleucine and D-isoleucine were each readily prepared in >98% ee through hydrogenation of the corresponding *Z*- and *E*-enamides using the (*R,R*)-Me-BPE-Rh catalyst. Other examples are shown in Figure 13.5.

The DuPhos-Rh catalysts also offer excellent regioselectivity in asymmetric hydrogenation reactions.[22,38] Selective hydrogenation of one alkene function within substrates possessing two or more different alkene groups can be a significant challenge, particularly if the most highly substituted alkene is targeted for reduction (Scheme 13.6).

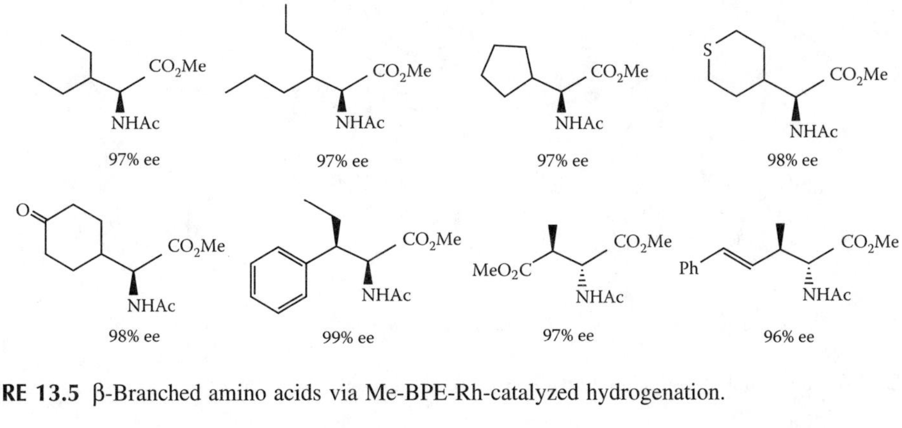

FIGURE 13.5 β-Branched amino acids via Me-BPE-Rh-catalyzed hydrogenation.

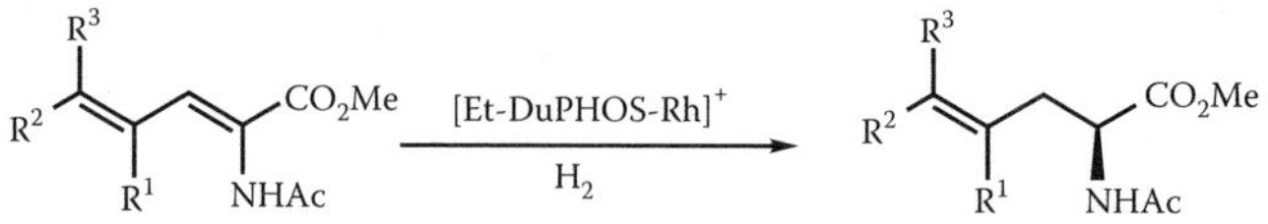

10 → 11

Regioselectivity: >98%
Enantioselectivity: >98%

SCHEME 13.6

A screen of myriad chiral diphosphine rhodium and ruthenium catalysts showed that the cationic Et-DuPhos-Rh catalysts are uniquely suited for hydrogenation of dienamides **10** with both high enantioselectivities (>98%) and high regioselectivities (>98%) to produce allylglycine derivatives **11**.[38a] No other catalyst, including the closely related Me-DuPhos-Rh and *i*-Pr-DuPhos-Rh, has been found as effective for this transformation, again demonstrating the benefits of modular ligand design. Enamide substrates are known to chelate to cationic diphosphine-Rh catalysts through the alkene and the carbonyl oxygen atom of the *N*-acyl group. Apparently, such chelation directs the hydrogenation to occur preferentially at the enamide alkene unit of **10**. In the case of the Et-DuPhos-Rh catalyst system, hydrogenation of other alkene groups within the substrate is particularly unfavorable, regardless of whether the alkenes are conjugated, as in **10**, or more remote from the enamide function.[22]

Allylglycine derivatives are functionalized amino acids that can serve as valuable synthetic intermediates. For instance, allylglycine **12** was prepared in 99% ee via the (*R*,*R*)-Et-DuPhos-Rh catalyst and then conveniently converted to the complex natural product (+)-bulgecinine (**13**) (Figure 13.6).[38a] In a further example, allyl glycine has been further elaborated by a hydroformylation. The resultant aldehydes are amenable to cyclization to furnish the piperidinecarboxylic acid derivative **14**.[38b] Through use of the Me-BPE-Rh catalysts, hydrogenation of dienamides bearing an additional β-substituent has allowed rapid access to β-branched allylglycines (**15–17**) possessing contiguous stereogenic centers.[39] These challenging examples demonstrate the powerful influence of substrate chelation, whereby the tetrasubstituted enamide alkene unit is reduced in preference to a disubstituted alkene function.

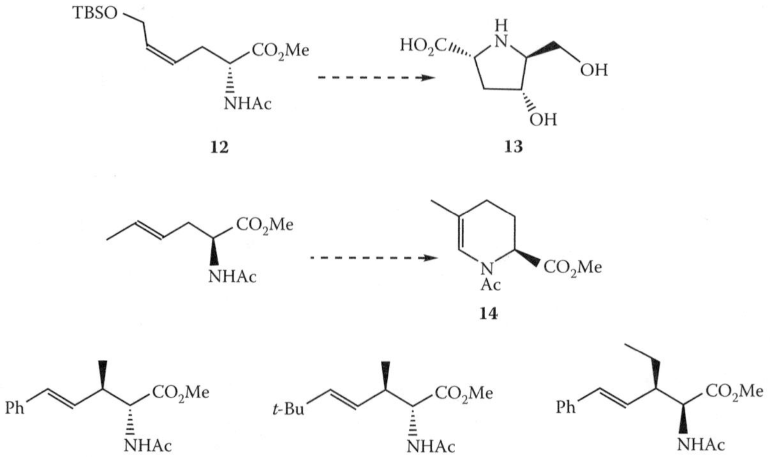

12, **13**, **14**

15, 96% ee; **16**, 96% ee; **17**, 93% ee

FIGURE 13.6 Functional allylglycine derivatives prepared through hydrogenation.

Me-DuPhos-Rh catalysts also have been shown to be very effective in the enantioselective and diastereoselective hydrogenation of dehydroaminoacids immobilized on polymer supports.[40]

Reduction of β-enamides can provide straightforward access to β-amino acids. It has been demonstrated that the Me-DuPhos-Rh catalysts are very effective for hydrogenation of a series of (E)-β-enamides **18** to afford the desired β-amino acid derivatives **19** with very high enantioselectivities (>97%) (Scheme 13.7).[41] The Me-and Et-FerroTANE-Rh catalysts also have been found to hydrogenate (E)-β-enamides with very high enantioselectivities (>98%) and rates.[42] Unfortunately, neither catalyst system was able to reduce the corresponding (Z)-β-enamides with the same high level of absolute stereocontrol, thus requiring use of geometrically pure β-enamide substrates. The asymmetric hydrogenation of (Z)-β-enamides appears to be peculiarly sensitive to temperature and pressure effects. Lower pressure generally favors higher enantioselectivity, and increasing the temperature from 15°C to 40°C also significantly enhances selectivity.[43] Work has shown that the use of i-Pr-DuPhos-Rh delivers good enantioselectivity at higher pressures and lower catalyst loading with either Z- or E-enamide.[43d] For example, the hydrogenation of (Z)-3-acetamido-2-butenoate (**18**, R = Me) at a molar S/C = 1000 gave the corresponding β-amino acid derivative **19** in 92% ee. Hydrogenation of the E-enamide gave the same product in >99% ee.

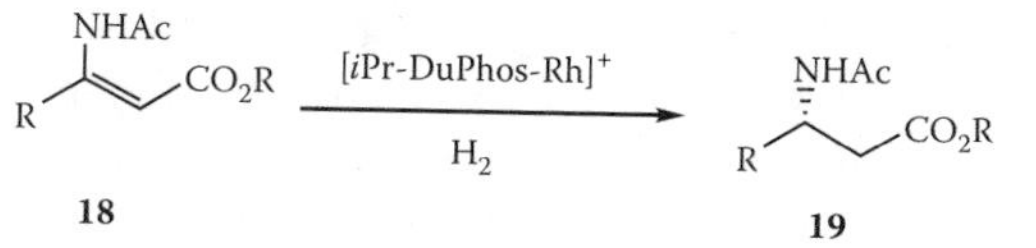

SCHEME 13.7

13.4.2 Amines and Amino Alcohols

Hydrogenation of enamides using bis(phospholane)-based rhodium catalysts has been extended to include substrates that previously had been considered intractable. Many enamides of general structure **20** may be prepared readily from ketones—for example, through reaction with hydroxylamine, followed by reduction of the oxime with iron powder in the presence of an acylating agent (Scheme 13.8).[44] Alternatively, the enamide can be accessed by a Grignard addition to the corresponding nitrile and trapping the intermediate magnesium salt with acetic anhydride.[45] Subsequent hydrogenation of enamides of type **20** with the DuPhos-Rh or BPE-Rh catalysts provides a simple 3-step route to a range of valuable amine derivatives **21**.

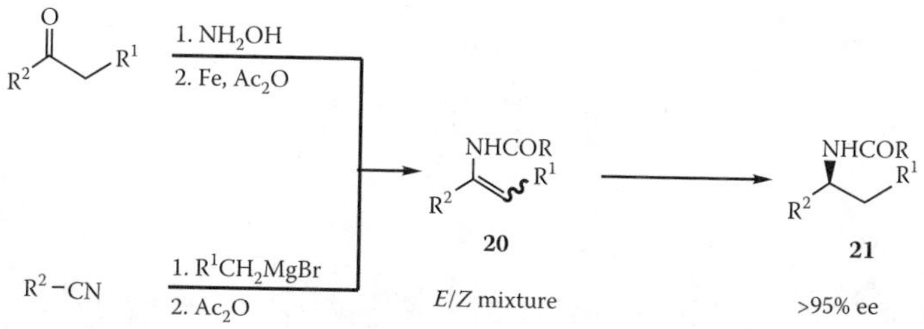

SCHEME 13.8

As shown in Figure 13.7, the Me-DuPhos-Rh and Me-BPE-Rh catalysts were found particularly effective for reduction of α-aryl enamides (**20**, R² = aromatic group) to provide an array of α-1-arylalkylamine derivatives **22** (95–99% ee).[46] In an analogous fashion, arylglycinol derivatives **23** were synthesized conveniently and with high enantioselectivity (>97%) using the Me-DuPhos-Rh catalysts.[47] As previously noted, the DuPhos-Rh and BPE-Rh catalysts tolerate β-substituents in either the E- or Z-position of enamides **20**, converging the geometric mixtures to products with

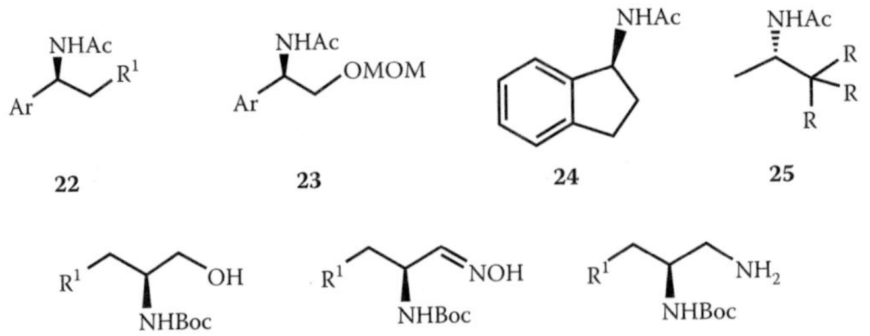

FIGURE 13.7 Amines, amino alcohols, and diamines available via DuPhos-Rh-catalyzed enamide hydrogenation.

high ee's. Cyclic amines, such as 1-aminoindane (**24**), have been prepared with enantioselectivities >99%.[44] Enamides **20** possessing tertiary-alkyl R^2-substituents are hydrogenated efficiently with the Me-DuPhos-Rh catalysts to yield enantiomerically pure dialkylamine derivatives of structure **25**, albeit with opposite absolute stereochemistry to that predicted based on other enamide hydrogenations.[44] Finally, hydrogenation of enamides **20** supporting different substitution patterns using the Me-DuPhos-Rh and Et-DuPhos-Rh catalysts has permitted preparation of a wide variety of amino alcohols, amino oximes, and diamines (**26-28**) with enantioselectivities up to >99% (see Figure 13.7).[48]

The preparation of amines also has been explored through enantioselective hydrogenation of imines using pholospholane- and phosphetane-based Rh and Ru catalysts. In general, lower rates and selectivities have been achieved with these substrates. The best results were obtained in hydrogenation of the *N*-benzylimine of acetophenone (**29**), whereby the corresponding amine **30** was obtained in up to 94% ee using the diphosphine–ruthenium–diamine catalyst derived from {RuCl₂[(*R,R*)-Et-DuPhos)((*R,R*)-Diaminocyclohexane)]} (Scheme 13.9).[49]

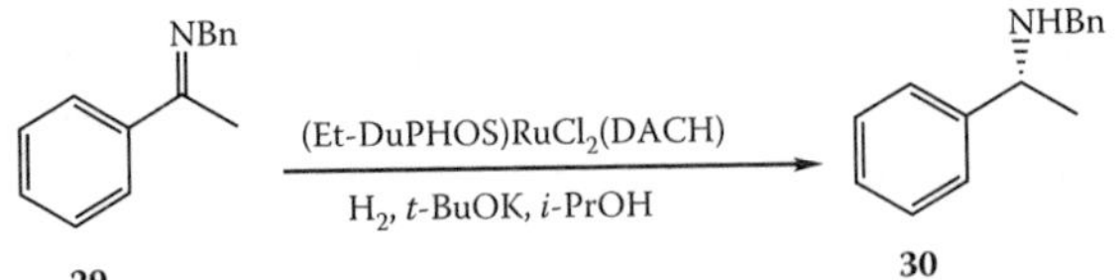

SCHEME 13.9

13.4.3 HYDRAZINES

As noted earlier, rhodium catalysts bearing the DuPhos and BPE ligands have been found generally ineffective for hydrogenation of the C=N group of imines and oximes. However, hydrogenation of *N*-acylhydrazones of structure **31** was shown to occur readily and rendered a broad range of hydrazine derivatives of structure **32** (Scheme 13.10). The Et-DuPhos-Rh catalysts were found to be consummate for this process, and a variety of hydrazines and hydrazino acids were prepared with high enantioselectivities (>90%).[50] It is interesting that solvents had a dramatic effect on selectivities in this process and the highest ee's were observed in isopropanol, similar to that observed earlier for hydrogenation of the imine **29**. Optically active amines also could be generated through this method by reductive cleavage of the N–N bond of hydrazines **32** with reagents such as samarium diiodide.

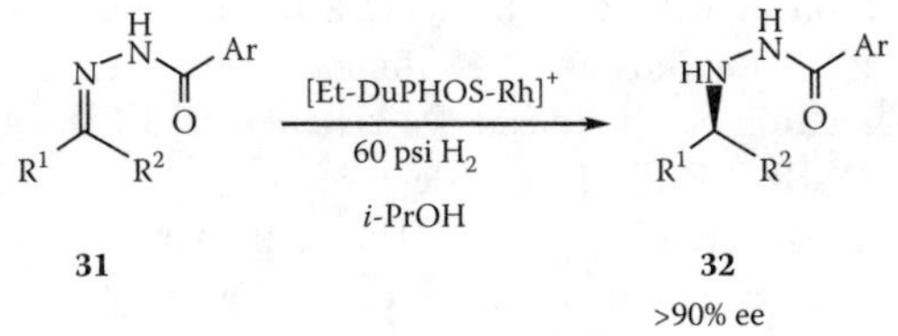

SCHEME 13.10

13.4.4 ALCOHOLS

Two different methods for production of chiral alcohols have been explored—one entailing alkene reduction and one involving keto group hydrogenation. Hydrogenation of enol acylates, formed from ketones and having the generic form **33**, proceeds efficiently and with very high enantio-selectivities using the DuPhos-Rh and BPE-Rh catalyst systems (Scheme 13.11 and Figure 13.8). Again, use of substrates **33** as *E/Z*-isomeric mixtures does not debase enantioselectivities. Acid- or base-catalyzed hydrolysis of the acyl group of **34** renders simple access to the desired alcohol products.

SCHEME 13.11

This strategy has been used to prepare numerous different types of alcohols, including aromatic alcohols **35** and trifluoromethyl-substituted derivatives **36**.[21] Alcohols of type **37** and **38**, respectively, were produced through highly enantioselective and regioselective hydrogenation of allylic and propargylic enol acetates using the same catalysts.[51] Moreover, a wide range of α-hydroxy phosphonate esters **39**[52] and valuable α-hydroxy carboxylic esters **40**[53] have been prepared with high enantioselectivities (>95% ee) by this method. Rather than hydrolysis to **40**, direct hydride-mediated reduction of the initially formed acetate products (i.e., **34**; R^2 = CO$_2$R) affords chiral 1,2-diols in high yield.[53] Cyclic enol acetates also may be hydrogenated with high enantioselectivity. For instance, multi-kilogram quantities of the intermediate (*S*)-phorenol (**41**) have been manufactured through efficient (S/C = 20,000; 4 hours) and highly enantioselective (98%) hydrogenation of the corresponding enol acetate using the (*R,R*)-Et-DuPhos-Rh catalyst.[54]

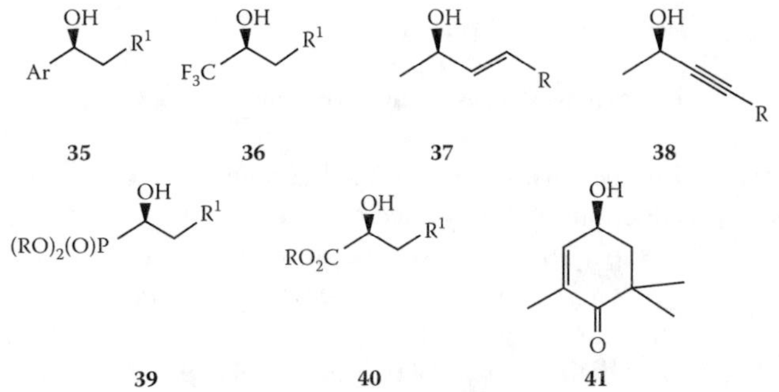

FIGURE 13.8 Alcohols produced through asymmetric hydrogenation of enol acylates **33**.

Asymmetric hydrogenation of ketones represents a more direct approach to chiral alcohols. Ruthenium catalysts bearing the DuPhos and BPE ligands were found to be effective for the low-pressure (60 psi H_2) hydrogenation of β-keto esters **42** (Scheme 13.12).[55] Using a catalyst precursor of general composition [i-Pr-BPE-RuBr$_2$], a wide selection of β-hydroxy esters **43** have been attained with very high enantioselectivities (>98%). The easily handled complex, [η^6-(C$_6$H$_6$)Ru(Me-DuPhos)Cl]OTf, also may serve as a comparable precursor for enantioselective β-keto ester hydrogenations.

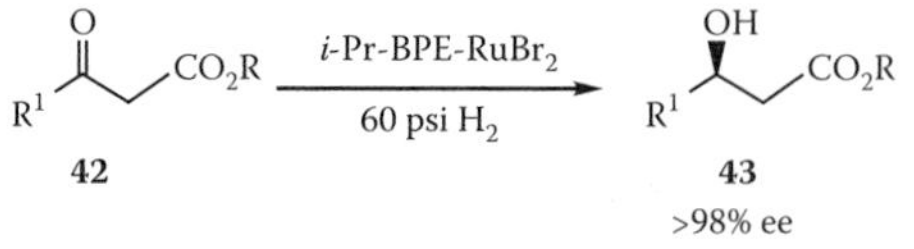

SCHEME 13.12

13.4.5 Carboxylic Acids

The creation of C–C stereogenic centers is an important and challenging area of asymmetric synthesis. The apparent need to synthesize geometrically pure alkenes that are devoid of attached heteroatoms has served as a specious impediment to the development of practical hydrogenation processes for this purpose. Unsaturated carboxylic acid derivatives of general structure **44** can be hydrogenated with a high degree of stereoselection to furnish products **45** with new C–C stereogenic centers at either the α- or β-carbon (Scheme 13.13 and Figure 13.9). For example, carboxylic acid products **46** and **47** have been obtained with high enantioselectivities using an *in situ*–generated *i*-Pr-DuPhos-Ru catalyst.[56] Unfortunately, this process is not general because substitution with larger groups at either α- or β-positions led to serious diminution of selectivity and rates. Functionalized carboxylic acids, such as **48**, were prepared directly and with high ee's through hydrogenation of the corresponding vinyl sulfones using Et-DuPhos-Rh.[57]

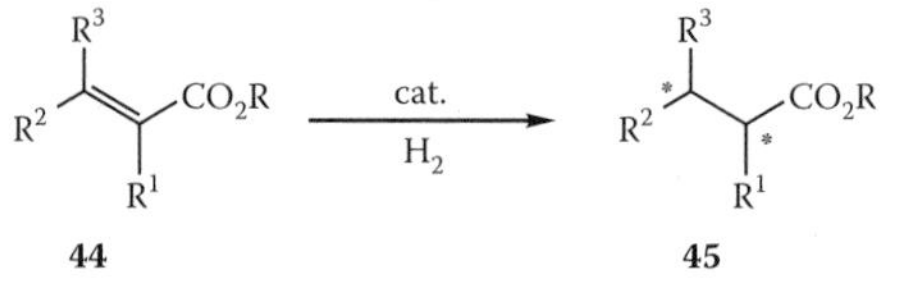

SCHEME 13.13

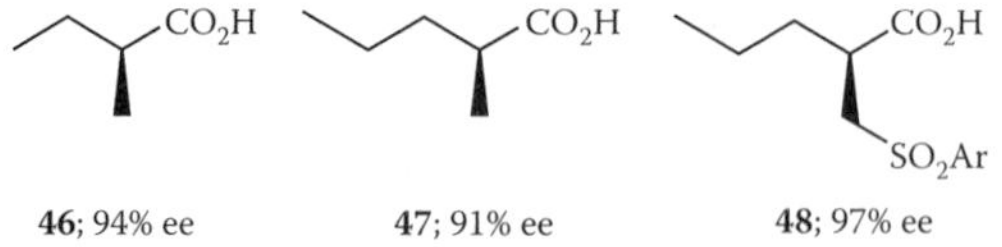

FIGURE 13.9 Carboxylic acid derivatives accessed by asymmetric hydrogenation.

Considerably improved scope was realized in the hydrogenation of itaconates **49**, which serve as useful substrates for production of 2-alkylsuccinate peptidomimetics **50** (Scheme 13.14). The Et-DuPhos-Rh catalysts are superlative for asymmetric hydrogenation of a very broad range of itaconate substrates with exceedingly high enantioselectivities (>97%) and high efficiencies (S/C >5000).[58] Itaconates may be prepared effectively as *E/Z*-isomeric mixtures through Stobbe condensation between aldehydes and dialkyl succinates. Fortunately, the Et-DuPhos-Rh catalysts once again were found to tolerate itaconates synthesized as geometric isomers and converged the mixture to the desired product with high levels of absolute stereocontrol.

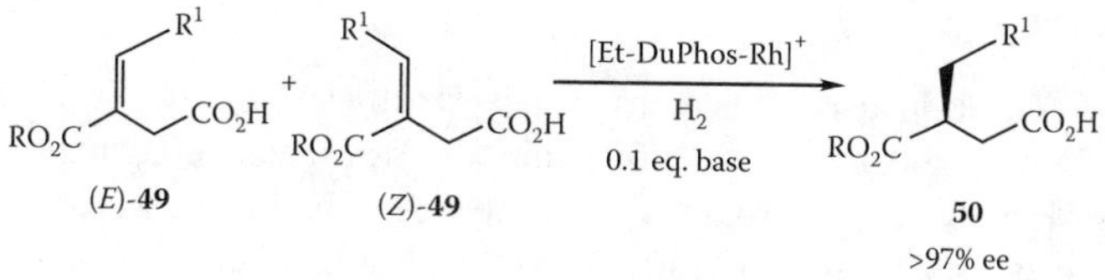

SCHEME 13.14

The Et-DuPhos-Rh catalysts were found superior for itaconates **49**; however, inverting the protecting group positions led to a decrease in enantioselectivity and catalytic rates, especially when amido itaconates of type **51** were used. It was here that the Et-FerroTANE-Rh catalysts first demonstrated their potential value for the production of useful peptidomimetic intermediates **52** (Scheme 13.15). Thus, a range of amido itaconates bearing different R and amide groups were reduced with very high enantioselectivities (>95%) and high catalytic efficiency (e.g., 3-hour reaction time at S/C = 20,000).[15a] An even higher S/C ratio (S/C = 100,000) has been demonstrated for the unsubstituted 2-methylenesuccinamic acid (**51**, R = H).[59] It is noteworthy that in the hydrogenation of this substrate, trace amounts of chloride bound to and inhibited the catalyst such that at S/C = 1000 the reaction time was reduced from ~2 hours to 4 minutes when the chloride was removed.

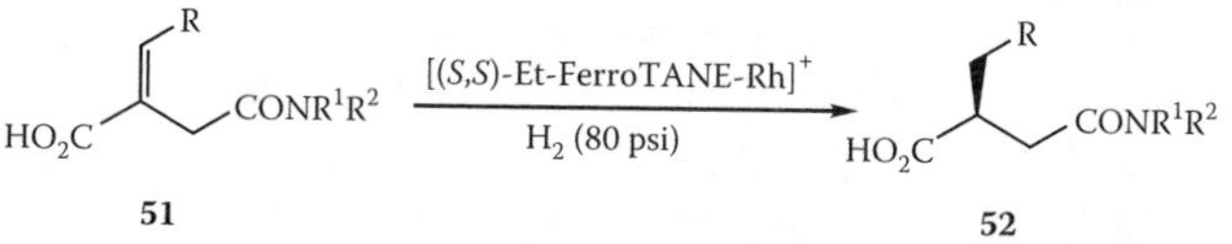

SCHEME 13.15

The unique itaconate-like substrate **53** has been hydrogenated efficiently and with high selectivity (>99% ee) using the (R,R)-Me-DuPhos-Rh catalyst to afford **54**, an important intermediate for the drug candoxatril (Scheme 13.16).[60]

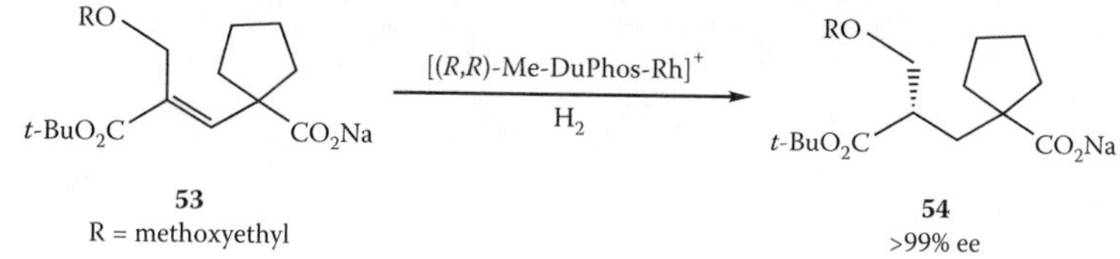

SCHEME 13.16

Hydrogenation of cyano-itaconate derivative **55** with Me-DuPhos-Rh catalyst was found to proceed in high enantioselectivity (98%) when an E/Z mixture of the carboxylate salt (M = K or t-BuNH$_3$) was used; **56** was formed in only 19% ee when the ethyl ester was used (M = Et). Subsequent hydrogenation of the cyano group of **55** using Ni sponge provided straightforward and practical access to pregabalin (**57**), a potent anticonvulsant drug under development by Pfizer (Scheme 13.17).[61] It is of interest to note that the (R,R)-Me-FerroTANE-Rh catalyst also provided the product **57** with high enantioselectivity (95%).

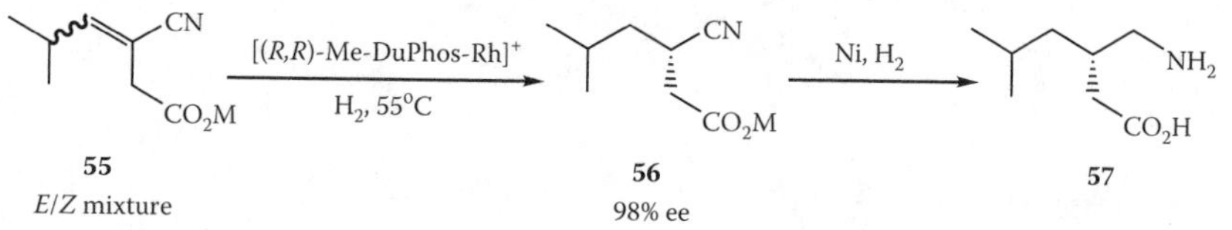

SCHEME 13.17

13.4.6 Dihydropyrones

Stereoselective hydrogenation also has been used successfully to introduce C–C stereogenic centers into 2 related dihydropyrone derivatives. Both enantiomers of the anticoagulant warfarin (**58**) have been prepared in 89% ee through hydrogenation of the corresponding α,β-unsaturated ketone using the Et-DuPhos-Rh catalysts.[62] More recently, the (*R,R*)-Me-DuPhos-Rh catalyst furnished a key intermediate **59** for production of tipranavir, a new nonpeptidic human immunodeficiency virus protease inhibitor originally discovered by Pharmacia & Upjohn (now Pfizer).[63]

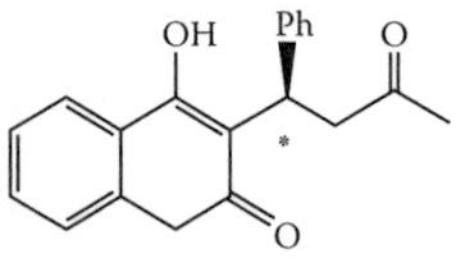

58
89% ee

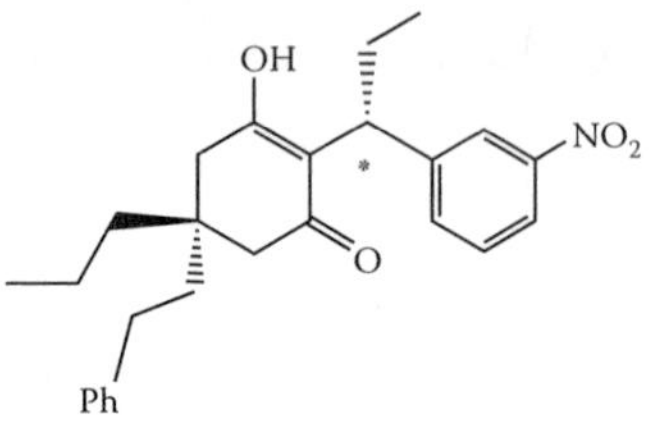

59
93% de

13.4.7 Cyclopentanones

A Me-DuPhos-ruthenium catalyst has found application in the industrial synthesis of the fragrance (+)-*cis*-methyl dihydrojasmonate **60** (Scheme 13.18).[64] The catalyst gave excellent *cis*-selectivity, and although the enantiomeric excess was modest (76%), it was acceptable for use of the product in perfumery.

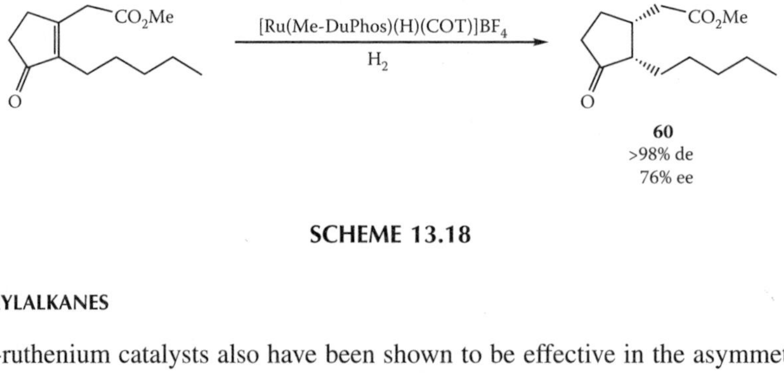

SCHEME 13.18

13.4.8 Arylalkanes

Me-DuPhos-ruthenium catalysts also have been shown to be effective in the asymmetric hydrogenation of α-ethylstyrenes **61** giving some of the highest enantioselectivities thus far recorded for this difficult class of substrate (Scheme 13.19).[65]

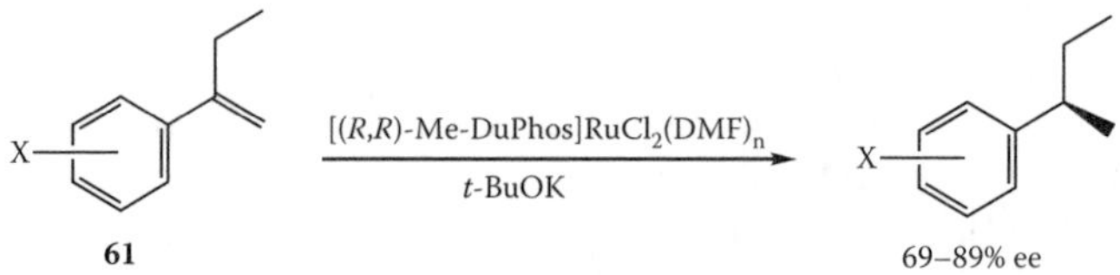

SCHEME 13.19

13.4.9 Cyclopentanes

Rhodium-bisphospholane catalysts have been used effectively in directed hydrogenation. For example, careful choice of catalyst in the hydrogenation of the cyclopent-1-enecarboxylic ester **62** allows access to either diastereoisomer of the hydrogenation product **63** (Scheme 13.20). Crabtree's catalyst apparently coordinates to the hydroxyl oxygen of **62**, leading to the (*S,R,R*) stereoisomer, whereas the Rh-Me-BPE catalyst delivers hydrogen from above by binding to the carbamate group to afford (*R,R,R*)-**63**.[66]

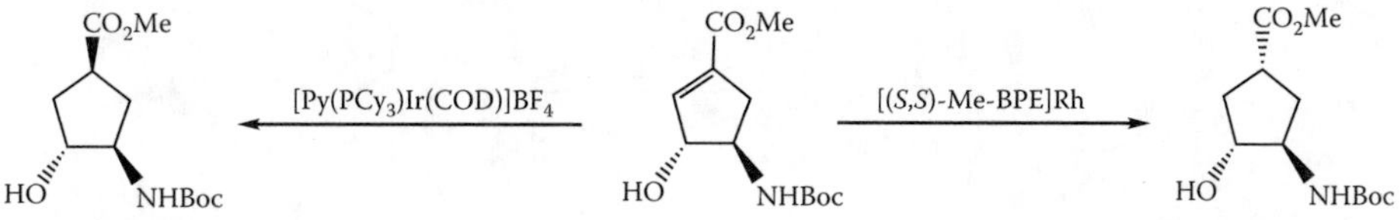

CO₂Me

[Py(PCy₃)Ir(COD)]BF₄

[(*S,S*)-Me-BPE]Rh

(*S,R,R*)-**63**
94% de

(*R,R*)-**62**

(*R,R,R*)-**63**
94% de

SCHEME 13.20

13.5 OTHER APPLICATIONS OF PHOSPHOLANE LIGANDS

In addition to hydrogenation reactions, modular phospholane ligands are being applied in a growing rank of other useful asymmetric catalytic transformations. For instance, Jiang and Sen reported the discovery of a dicationic Me-DuPhos-Pd catalyst for the alternating copolymerization of aliphatic α-olefins and carbon monoxide (Scheme 13.21).[67]

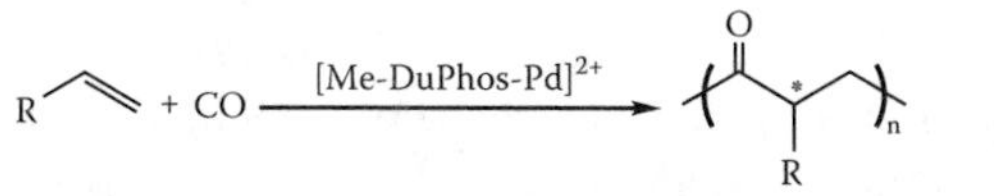

$R \diagdown\!\!\!\diagup$ + CO $\xrightarrow{\text{[Me-DuPhos-Pd]}^{2+}}$

Enantioselectivity: >90%

SCHEME 13.21

Analysis with chiral nuclear magnetic resonance shift reagents revealed that the isotactic poly(1,4-ketone) products were formed with an average or overall degree of enantioselectivity that was >90%. Using the same catalyst, Jiang and Sen also described the first example of alternating co-polymerization between an internal alkene (2-butene) and carbon monoxide to form an isotactic, optically active poly(1,5-ketone).

Murakami and Ito have highlighted the utility of cationic Me-DuPhos-Rh catalysts for novel asymmetric (4 + 1) cycloaddition reactions between vinylallenes and carbon monoxide.[68] Complex cyclopentenone derivatives such as **64** have been constructed in a single step and with enantioselectivities up to 95% in this process (Scheme 13.22).

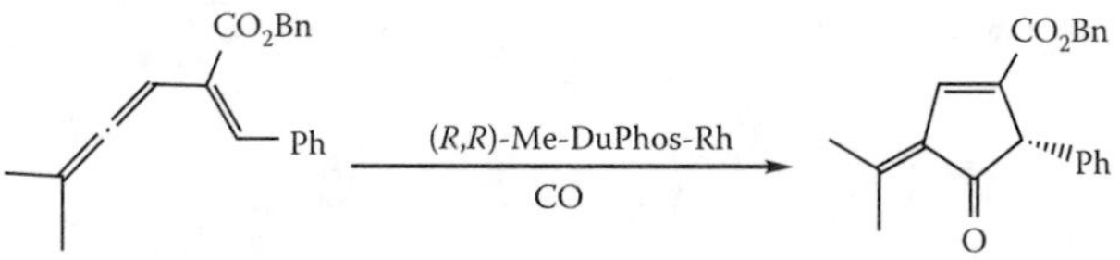

CO₂Bn

Ph

(*R,R*)-Me-DuPhos-Rh

CO

64
95% ee

SCHEME 13.22

The modular nature of the phospholane series of ligands allows different backbone structures to be explored. Osborn and van Leeuwen have reported bis(phospholane) ligands **65** possessing dibenzo[*b,d*]pyran and dibenzo-1,4-thioxin backbones that have large P–M–P bite angles.[12b] Preliminary studies using ligands possessing 2,5-dimethylphospholane units have affirmed that these systems are useful in asymmetric Pd-catalyzed allylic substitution reactions. Even notoriously difficult substrates, such as *O*-acetylcyclohexenol **66**, reacted with dimethylmalonate to yield allylic substitution products (e.g., **67**) with high enantioselectivities (Scheme 13.23).

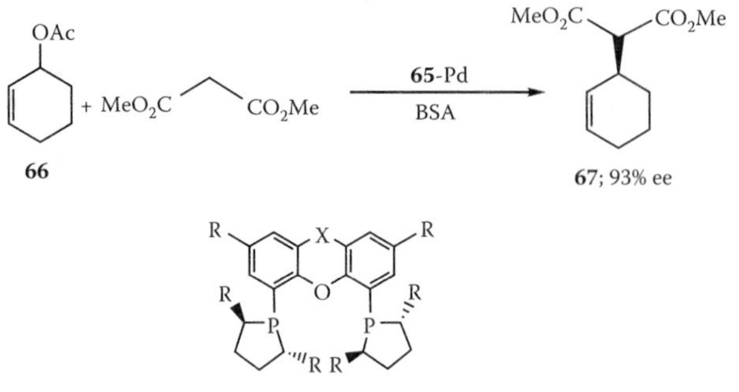

66 **67**; 93% ee

65

SCHEME 13.23

Two different reports have illustrated that cationic Me-DuPhos-Rh complexes serve as excellent catalysts for asymmetric C–C bond-forming cyclization reactions. In the first example, Bosnich and co-workers discovered that valuable 3-substituted cyclopentanones can be prepared simply by treatment of 4-pentenal derivatives **68** with the Me-DuPhos-Rh catalyst.[69] Asymmetric intramolecular hydroacylation furnished the product cyclopentanones **69** in high yield and with enantioselectivities ranging from 93% to 98% (Scheme 13.24).

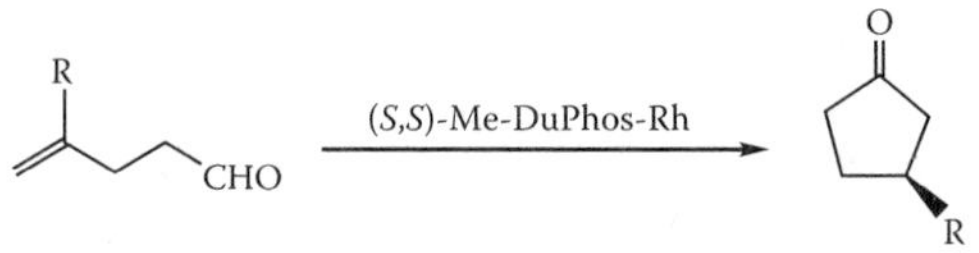

68 **69**

93–98% ee

SCHEME 13.24

In the second example, Gilbertson and colleagues have revealed that cationic Me-DuPhos-Rh catalysts effect asymmetric [4 + 2] cycloisomerization of standard dieneynes such as **70** to afford bicyclic products **71**, whereupon two stereogenic centers are established with high levels of absolute stereocontrol (Scheme 13.25).[70]

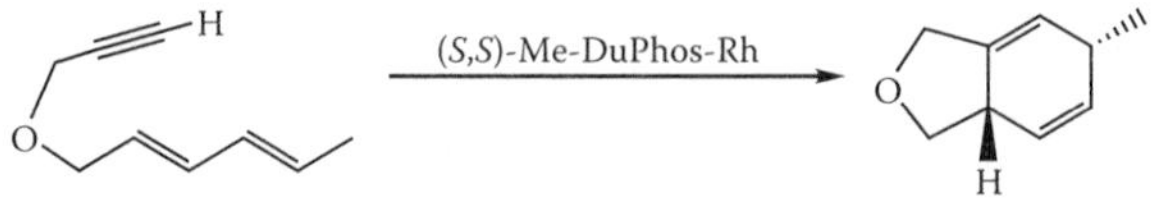

70 **71**

95% ee

SCHEME 13.25

A nickel catalyst containing Me-DuPhos has been shown to be effective in the enantioselective isomerization of 4,7-dihydro-1,3-dioxepins **72**.[71] Diastereoselective oxidation of the resulting 4,5-dihydro-1,3-dioxepins **73** allowed access to a range of C–4 chiral synthons (Scheme 13.26).

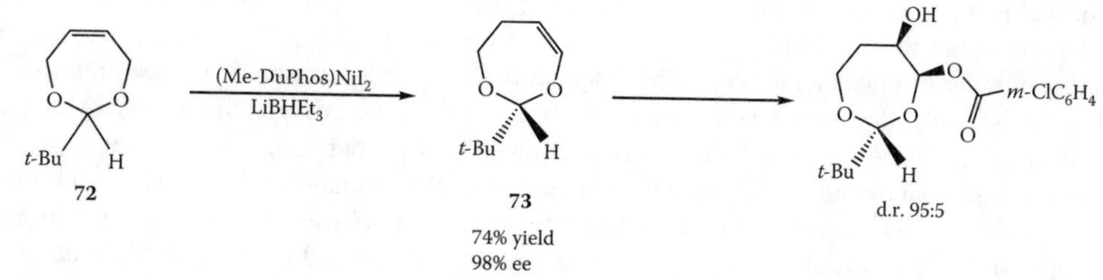

SCHEME 13.26

In a radically different approach, Alper has discovered that Me-DuPhos may be used in the synthesis of α-amino acids through a palladium-catalyzed double carbohydroamination whereby aryl iodides **74** can be converted to the corresponding arylglycine amides **75** in the presence of excess primary amine (Scheme 13.27).[72]

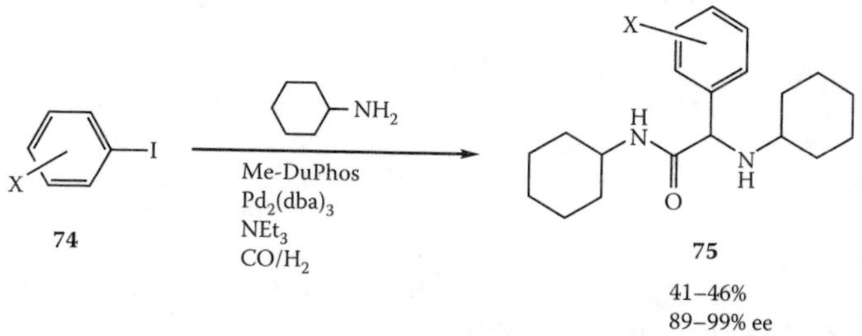

SCHEME 13.27

Finally, Boezio and Charette have shown that Me-DuPhos in combination with Cu(OTf)$_2$ is effective in the enantioselective addition of dialkylzinc reagents to *N*-diphenylphosphinoylimines **76** to provide phosphinoylamines **77** with high selectivities (Scheme 13.28).[73] Further study revealed that a bis(ligand)-Cu catalyst containing the mono-oxide of Me-DuPhos, and not Me-DuPhos itself, was responsible for the high enantioselectivities observed in this process.[74]

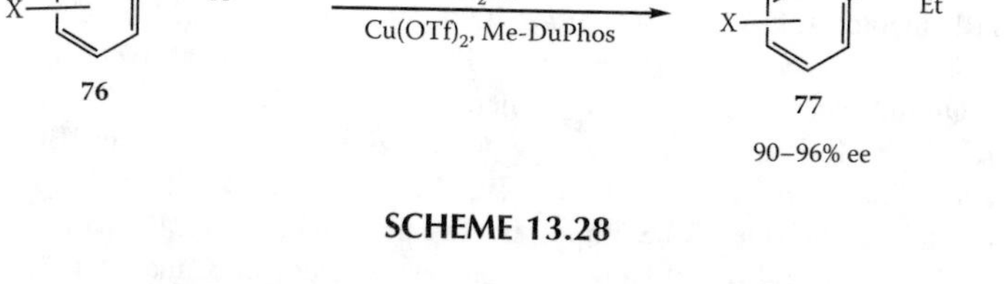

SCHEME 13.28

13.6 SUMMARY

Modular pholane and phosphetane ligands are finding useful application in a growing number of catalytic reaction types and are well-suited for commercial process development. Catalysts derived from the bidentate DuPhos and BPE ligands exhibit particularly high efficiency and selectivity in a wide spectrum of asymmetric hydrogenation reactions, providing access to a diverse range of valuable chiral compounds. The versatility of these catalyst systems emanates from the modular nature of the ligands, which allows ready adjustment of their steric, electronic, and conformational properties. Optimal results have been achieved through the ability to systematically vary both the phospholane or phosphetane R-substituents and the ligand backbones. Because only a limited number of phospholane- and phosphetane-based ligands have been prepared and studied to date, it is likely that many new and interesting catalytic processes will be discovered and developed as this series of novel ligands is extended. It is encouraging to note that a growing number of these ligands and catalysts are being made generally available to the research community. These ligands represent a family with individual members that are sterically and conformationally differentiated. It often is difficult or impossible to predict which ligand or catalyst might be ideally suited for any specific process. Hence, it is important to recognize that a series of ligand–catalyst combinations must be screened to identify the optimum system for a given application. As more of these ligands become available, we would anticipate that their scope and applicability will continue to expand into new and exciting areas. Moreover, the utility of these ligands and catalysts in industrial processes is only beginning to be realized, and many large-scale applications are anticipated.

REFERENCES AND NOTES

1. a) Sheldon, R. A. *Chirotechnology,* Marcel Dekker Inc.: New York, 1993; b) Collins, A. N., Sheldrake, G. N., Crosby, J., Eds., *Chirality in Industry,* John Wiley & Sons: New York, 1992.
2. Stinson, S. C. *Chem. Eng. News* September 21, 1998, pp. 83–104.
3. Richards, A., McCague, R. *Chem. Ind.* June 2, 1997, pp. 422–425.
4. a) For a tabulation of chiral phosphines, see Brunner, H., Zettlmeier, W. *Handbook of Enantioselective Catalysis with Transition Metal Compounds,* VCH: New York, 1993, Volume 2; b) For a recent review on the scope and applicability of chiral phosphorus ligands in asymmetric hydrogenation, see Tang, W., Zhang, X. *Chem. Rev.* 2003, *103*, 3029.
5. Whitesell, J. K. *Chem. Rev.* 1989, *89*, 1581.
6. Brunner, H., Sievi, R. *J. Organometal. Chem.* 1987, *328*, 71.
7. Burk, M. J. *ChemTracs-Org. Chem.* 1998, *11*, 787.
8. Burk, M. J., Feaster, J. E., Harlow, R. L. *Organometal.* 1990, *9*, 2653.
9. Burk, M. J., Feaster, J. E., Harlow, R. L. *Tetrahedron: Asymmetry* 1991, *2*, 569.
10. Taylor, S. J. C., Holt, K. E., Brown, R. C., Keene, P. A., Taylor, I. N. In *Stereoselective Biocatalysis,* Patel, R., Ed., Marcel Dekker: New York, 2000, Chapter 15, pp. 397–413.
11. Burk, M. J., Harlow, R. L. *Angew. Chem., Int. Ed.* 1990, *29*, 1462.
12. For additional examples of phospholane ligands, see a) Morimoto, T., Ando, N., Achiwa, K. *Synlett* 1996, 1211; b) Dierkes, P., Ramdeehul, S., Barloy, L., De Cian, A., Fischer, J., Kamer, P. C. J., van Leeuwen, P. W. N. M., Osborn, J. A. *Angew. Chem., Int. Ed.* 1998, *37*, 3116; c) Schmid, R., Broger, E. A., Cereghetti, M., Crameri, Y., Foricher, J., Lalonde, M., Muller, R. K., Scalone, M., Schoettel, G., Zutter, U. *Pure App. Chem.* 1996, *68*, 131; d) Burk, M. J., Pizzano, A., Martin, J. A., Liable-Sands, L., Rheingold, A. L. *Organometallics* 2000, *19*, 250; e) Burk, M. J., Gross, M. F. *Tetrahedron Lett.* 1994, *35*, 9363; f) Shimizu, H., Saito, T., Kumobayashi, H. *Adv. Synth. Catal.* 2003, *345*, 185; g) Holz, J., Quirmbach, M., Schmidt, U., Heller, D., Sturmer, R., Borner, A. *J. Org. Chem.* 1998, *63*, 8031; h) Yan, Y.-Y., RajanBabu, T. V. *J. Org. Chem,* 2000, *65*, 900; i) RajanBabu, T. V., Yan, Y.-Y., Shin, S. *J. Am. Chem. Soc.* 2001, *123*, 10207; j) Li, W., Zhang, Z., Zhang, X. *J. Org. Chem.* 2000, *65*, 3489; k) Holz, J., Stürmer, R., Schmidt, U., Drexler, H.-J., Heller, D., Krimmer, H.-P., Börner, A. *Eur. J. Org. Chem.* 2001, 4615; l) Holz, J., Monsees, A., Jiao, J., Komarov, I. V., Fischer, C., Drauz, K., Börner, A. *J. Org. Chem.* 2003, *68*, 1701; m) Fenandez, E., Gillon, A., Heslop, K., Horwood, E., Hyett, D. J., Orpen, A. G., Pringle, P. G. *J. Chem. Soc., Chem. Commun.* 2000, 1663; n) Berens, U. Life Science Molecules at Ciba SC- Selected Examples. Chiral Europe 2003 12–14 May 2003; o) Landis, C. R., Wiechang, J., Owen, J. S., Clark, T. P. *Angew. Chem., Int. Ed.* 2001, *40*, 3432.
13. a) Carmichael, D., Doucet, H., Brown, J. M. *J. Chem. Soc., Chem. Comm.* 1999, 261; b) Matsumura, K., Shimizu, H., Saito, T., Kumobayashi, H. *Adv. Synth. Catal.* 2003, *345*, 180; c) Hoge, G. S., Goel, O. P. U.S. Pat. Appl. 277125 P 19.03.2001.
14. Pilkington, C. J., Zanoti-Gerosa, A. *Org. Lett.,* 2003, *5*, 1273.

15. a) Berens, U., Burk, M. J., Gerlach, A., Hems, W. *Angew. Chem., Int. Ed.* 2000, *39*, 1981; b) Marinetti, A., Labrue, F., Genet, J.-P. *Synlett* 1999, 1975; c) Marinetti, A., Genet, J.-P., Jus, S., Blanc, D., Ratovelomanana-Vidal, V. *Chem. Eur. J.* 1999, *5*, 1160.

16. Kitamura, M., Ohkuma, T., Inoue, S., Sayo, N., Kumobayashi, H., Akutagawa, S., Ohta, T., Takaya, H., Noyori, R. *J. Am. Chem. Soc.* 1988, *110*, 629.

17. a) Ojima, I. *Catalytic Asymmetric Synthesis,* VCH: New York, 1993, Chapter 1; b) Noyori, R. *Asymmetric Catalysis in Organic Synthesis,* Wiley & Sons: New York, 1993, Chapter 2; c) Doucet, H., Ohkuma, T., Murata, K., Yokozawa, T., Kozawa, M., Katayama, E., England, A.F., Ikariya, T., Noyori, R. *Angew. Chem., Int. Ed.* 1998, *37*, 1703; d) Blaser, H.-U., Spindler, F. *Topics in Catalysis* 1997, *4*, 275.

18. a) Knowles, W. S., Sabacky, M. J. *J. Chem. Soc., Chem. Commun.* 1968, 1445; b) Horner, L., Siegel, H., Büthe, H. *Angew. Chem., Int. Ed.* 1968, *7*, 942.

19. Kagan, H. Dang, T.-P. *J. Am. Chem. Soc.* 1972, *94*, 6429.

20. Vineyard, B. D., Knowles, W. S., Sabacky, M. J., Bachman, G. L., Weinkauff, D. J. *J. Am. Chem. Soc.* 1977, *99*, 5946.

21. Burk, M. J. *J. Am. Chem. Soc.* 1991, *113*, 8518.

22. Burk, M. J., Feaster, J. E., Nugent, W. A., Harlow, R. L. *J. Am. Chem. Soc.* 1993, *115*, 10125.

23. For a review comparing DuPhos with different ligands in enamide hydrogenations, see Burk, M. J., Bienewald, F. In *Transition Metals for Organic Synthesis and Fine Chemicals,* Bolm, C., Beller, M., Eds., VCH Publishers: Wienheim, Germany, 1998, *Vol. 2*, pp. 13–25.

24. Stammers, T. A., Burk, M. J. *Tetrahedron Lett.* 1999, *40*, 3325.

25. Masquelin, T., Broger, E., Mueller, K., Schmid, R., Obrecht, D. *Helv. Chim. Acta* 1994, *77*, 1395.

26. Jones, S. W., Palmer, C. F., Paul, J. M., Tiffin, P. D. *Tetrahedron Lett.* 1999, *40*, 1211.

27. a) Debenham, S. D., Debenham, J. S., Burk, M. J., Toone, E. J. *J. Am. Chem. Soc.* 1997, *119*, 9897; b) Debenham S. D., Cossrow, J., Toone, E. J. *J. Org. Chem.* 1999, *64*, 9153; c) Xu, X., Fakha, G., Sinou, D. *Tetrahedron* 2002, *58*, 7539.

28. a) Rizen, A., Basu, B., Chattopadhyay, S. K., Dossa, F., Frejd, T. *Tetrahedron: Asymmetry* 1998, *9*, 503; b) Hiebl, J., Kollmann, H., Rovenszky, F., Winkler, K. *J. Org. Chem.* 1999, *64*, 1947; c) Maricic, S., Ritzén, A., Berg, U., Frejd, T. *Tetrahedron,* 2001, *57*, 6523.

29. Burk, M. J., Lee, J. R., Martinez, J. P. *J. Am. Chem. Soc.* 1994, *116*, 10847.

30. Adamczyk, M., Akireddy, S. R., Reddy, R. E. *Tetrahedron: Asymmetry* 2001, *12*, 2385.

31. Shieh, W.-C., Xue, S., Reel, N., Wu, R., Fitt, J., Repic, O. *Tetrahedron: Asymmetry* 2001, *12*, 2421.

32. Adamczyk, M., Akireddy, S. R., Reddy, R. E. *Org. Lett.* 2001, *3*, 3157.

33. a) Wang, W., Yang, J., Ying, J., Xiong, C., Zhang, J., Cai, C., Hruby, V. J. *J. Org. Chem.* 2002, *67*, 6353; b) Wang, W., Cai, M., Xiong, C., Zhang, J., Trivedi, D., Hruby, V. J. *Tetrahedron* 2002, *58*, 7365.

34. Endo, A., Yanagisawa, A., Abe, M., Tohma, S., Kan, T., Fukuyama, T. *J. Am. Chem. Soc.* 2002, *124*, 6552.

35. Scott, J. W., Keith, D. D., Nix, G., Parrish, D. R., Remington, S., Roth, G. M., Townsend, J. M., Valentine, D., Jr., Yang, R. *J. Org. Chem.* 1981, *46*, 5086.

36. Burk, M. J., Gross, M. F., Harper, T. G. P., Kalberg, C. S., Lee, J. R. Martinez, J. P. *Pure Appl. Chem.* 1996, *68*, 37.

37. a) Burk, M. J., Gross, M. F., Martinez, J. P. *J. Am. Chem. Soc.* 1995, *117*, 9375; b) Hoerrner, R. S., Askin, D., Volante, R. P., Reider, P. J. *Tetrahedron Lett.* 1998, *39*, 3455.

38. a) Burk, M. J., Allen, J. G., Kiesman, W. F. *J. Am. Chem. Soc.* 1998, *120*, 657; b) Teoh, E., Campi, E. M., Jackson, W. R., Robinson, A. J. *J. Chem. Soc., Chem. Commun.* 2002, 978.

39. Burk, M. J., Bedingfield, K. M., Kiesman, W. F., Allen, J. G. *Tetrahedron Lett.* 1999, *40*, 3093.

40. Doi, T., Fujimoto, N., Watanabe, J., Takahashi T. *Tetrahedron Lett.,* 2003, *44*, 2161.

41. Zhu, G., Chen, Z., Zhang, X. *J. Org. Chem.* 1999, *64*, 6907.

42. You, J., Drexler, H.-J., Zhang, S., Fischer, C., Heller, D. *Angew. Chem., Int. Ed.* 2003, *42*, 913.

43. a) Heller, D., Holz, J., Drexler, H.-J., Lang, J., Drauz, K., Krimmer, H.-P., Börner, A. *J. Org. Chem.* 2001, *66*, 6816; b) Jerphagnon, T., Renaud, J.-C., Demonchaux, P., Ferreira, A., Bruneau, C. *Tetrahedron: Asymmetry* 2003, *14*, 1973; c) Tang, W., Zhang, X. *Org. Lett.* 2002, *4*, 4159; d) Malan, C. G., Cobley, C. PCT Pat. Appl. WO 03/016264, 2003.

44. Burk, M. J., Casy, G., Johnson, N. B *J. Org. Chem.* 1998, *63*, 6084.

45. Storace, L., Anzalone, L., Confalone, P. N., Davis, W. P., Fortunak, J. M., Giangiordano, M., Haley Jr., J. J., Kamholz, K., Li, H.-Y., Ma, P., Nugent, W. A., Parsons Jr, R. L., Sheeran, P. J., Silverman, C. E., Waltermire, R. E., Wood, C. C. *Org. Proc. Res. Dev.* 2002, *6*, 54.

46. Burk, M. J., Wang, Y. M., Lee, J. R. *J. Am. Chem. Soc.* 1996, *118*, 5142.

47. Zhu, G., Casalnuovo, A. L., Zhang, X. *J. Org. Chem.* 1998, *63*, 8100.

48. Burk, M. J., Johnson, N. B., Lee, J. R. *Tetrahedron Lett.* 1999, *40*, 6685.

49. Cobley, C., Henschke, J. *Adv. Synth. Catal.* 2003, *345*, 195.

50. a) Burk, M. J., Feaster, J. E. *J. Am. Chem. Soc.* 1992, *114*, 6266; b) Burk, M. J., Martinez, J. P., Feaster, J. E., Cosford, N. *Tetrahedron* 1994, *50*, 4399.

51. Boaz, N. W. *Tetrahedron Lett.* 1998, *39*, 5505.

52. Burk, M. J., Stammers, T. A., Straub, J. A. *Org. Lett.* 1999, *1*, 387.

53. Burk, M. J., Kalberg, C. S., Pizzano, A. *J. Am. Chem. Soc.* 1998, *120*, 4345.

54. a) Scalone, M., Schmid, R., Broger, E., Burkart, W., Cereghetti, M., Crameri, Y., Foricher, J., Hennig, M., Kienzle, F., Montavon, F., Schoettel, G., Tesauro, D., Wang, S., Zell, R., Zutter, U. *Twenty Years of Chiral Catalysis—The Roche Experience.* ChiraTech '97 Proceedings, 1997, Session II; b) Broger, E. A., Crameri, Y., Schmid, R., Siegfried, T., U.S. Pat. 5543559, 1996.

55. Burk, M. J., Harper, T. G. P., Kalberg, C. S. *J. Am. Chem. Soc.* 1995, *117*, 4423.

56. Nugent, W. A., RajanBabu, T. V., Burk, M. J. *Science*, 1993, *259*, 479.

57. Paul, J. M., Palmer, C., PCT Pat. Appl. WO 9915481, 1999.

58. Burk, M. J., Bienewald, F., Harris, M., Zanotti-Gerosa, A. *Angew. Chem., Int. Ed.* 1998, *37*, 1931.

59. Cobley, C. J., Lennon, I. C., Praquin, C., Zanotti-Gerosa, A., Appell, R. B., Goralski, C. T., Sutterer, A. C. *Org. Proc. Res. Dev.* 2003, *7*, 407.

60. Burk, M. J., Bienewald, F., Challenger, S., Derrick, A., Ramsden, J. A. *J. Org. Chem.* 1999, *64*, 3290.

61. Burk, M. J., de Koning, P. D., Grote, T. M., Hoekstra, M. S., Hoge, G., Jennings, R. A., Kissel, W. S., Le, T. V., Lennon, I. C., Mulhern, T. A., Ramsden, J. A., Wade, R. A. *J. Org. Chem.* 2003, *68*, 5731.

62. Robinson, A., Li, H.-Y., Feaster, J. *Tetrahedron Lett.* 1996, *37*, 8321.

63. a) Fors, K. S., Gage, J. R., Heier, R. F., Kelly, R. C., Perrault, W. R., Wicnienski, N. *J. Org. Chem.* 1998, *63*, 7348; b) Gage, J. R., Kelly, R. C., Hewitt, B. D., PCT Pat. Appl. WO 9912919, 1999.

64. a) Dobbs, D. A., Vanhessche, K. P. M., Brazi, E., Rautenstrauch, V., Lenoir, J.-Y., Genêt, J.-P., Wiles, J., Bergens, S. H. *Angew. Chem., Int. Ed.* 2000, *39*, 1992; b) Wiles, J., Bergens, S. H., Vanhessche, K. P. M., Dobbs, D. A., Rautenstrauch, V. *Angew. Chem., Int. Ed.* 2001, *40*, 914.

65. Forman, G. S., Ohkuma, T., Hems, W. P., Noyori, R. *Tetrahedron Lett.* 2000, *41*, 9471.

66. Smith, M. E. B., Derrien, N., Lloyd, M. C., Taylor, S. J. C., Chaplin, D. A., McCague, R. *Tetrahedron Lett.*, 2001, *42*, 1347.

67. Jiang, Z., Sen, A. *J. Am. Chem. Soc.* 1995, *117*, 4455.

68. Murakami, M., Itami, K., Ito, Y. *J. Am. Chem. Soc.* 1999, *121*, 4130.

69. Barnhart, R. W., McMorran, D. A., Bosnich, B. *J. Chem. Soc., Chem. Comm.* 1997, 589.

70. Gilbertson, S. R., Hoge, G. S., Genov, D. G. *J. Org. Chem.* 1998, *63*, 10077.

71. Frauenrath, H., Brethauer, D., Reim, S., Mayrer, M., Raabe, G. *Angew. Chem., Int. Ed.* 2001, *40*, 177.

72. Nanayakkara, P., Alper, H. *J. Chem. Soc., Chem. Commun.* 2003, 2384.

73. Boezio, A. A., Charette, A. B. *J. Am. Chem. Soc.* 2003, *125*, 1692.

74. Côté, A., Boezio, A. A., Charette, A.B. *Angew. Chem., Int. Ed.* 2004, *43*, 6525.

14 Asymmetric Olefin Hydrogenation Using Monodentate BINOL- and Bisphenol-Based Ligands: Phosphonites, Phosphites, and Phosphoramidites

Johannes G. de Vries

CONTENTS

14.1 INTRODUCTION

The development of rhodium-catalyzed asymmetric olefin hydrogenation historically started with the work of Knowles[1] and Horner[2] who used *P*-chiral monodentate phosphines as ligands. Initial enantioselectivities were low, which was attributed to the many degrees of freedom of the rhodium complexes, in particular the free rotation around the Rh-P bond. Thus, it was a logical step when Dang and Kagan synthesized the first chiral bisphosphine ligand (DIOP) (**1**), which in principal should form a much more rigid complex. Their analysis was rewarded by the remarkably high ee at that time of 70% in the hydrogenation of 2-acetamidocinnamic acid.[3] Knowles subsequently showed that dimerization of the first-generation ligand PAMP (**2**) to DIPAMP (**3**) (Figure 14.1) raised the enantioselectivity of the rhodium-catalyzed hydrogenation of methyl 2-acetamido-cinnamate (**5**) from 55% to 95% (Scheme 14.1).[4]

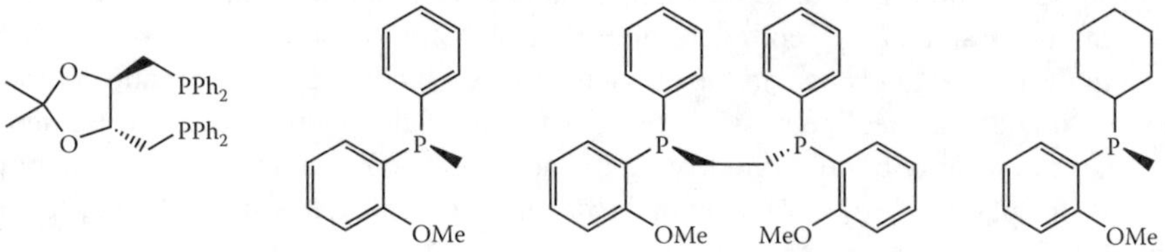

FIGURE 14.1 Structures of early chiral ligands.

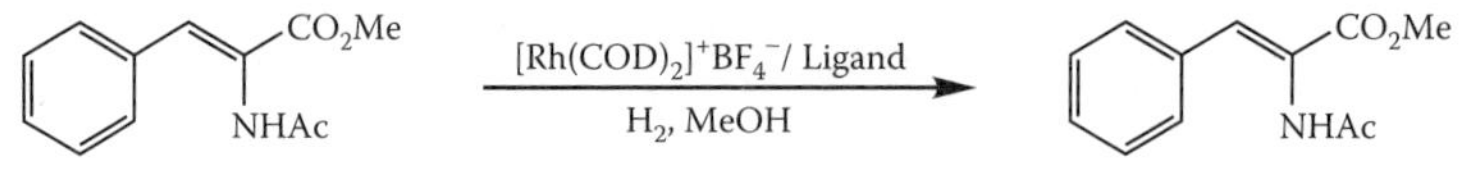

SCHEME 14.1

The application of DIPAMP in the commercial process for L-dopa,[1b] and the simplification of the synthesis by using non-*P*-chiral ligands, such as DIOP (**1**), has lead to the discovery of several dozens of families of chiral bisphosphines.[5] Despite the fact that Knowles already had shown that use of monodentate CAMP (**4**) led to formation of *N*-acetyl-phenylalanine (**6**) in up to 88% ee,[6] the dogma of the superiority of bidentate phosphorus ligands was well-established and remained unchallenged for many years.[7] Thus, it was very surprising that in a very short period 3 new classes of monodentate ligands were reported that showed remarkably high enantioselectivities, comparable with the best bidentate bisphosphines in the rhodium-catalyzed asymmetric hydrogenation of a range of substituted olefinic substrates. These monophosphonites, monophosphites, and monophosphoramidites are all based on the BINOL skeleton (Figure 14.2).

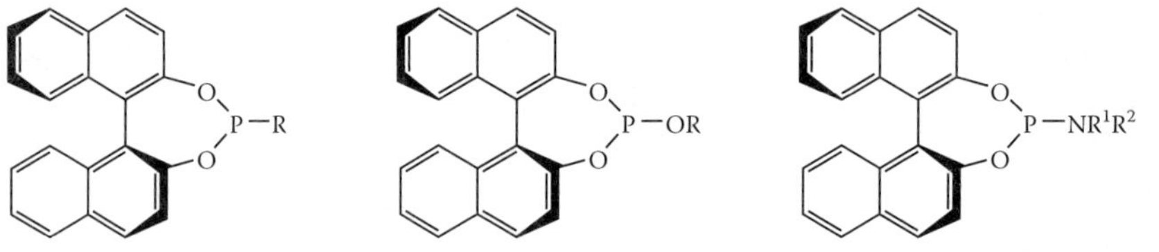

| Phosphonites | Phosphites | Phosphoramidites |
| Pringle *et al.* | Reetz *et al.* | Feringa, de Vries *et al.* |

FIGURE 14.2 Monodentate ligand families based on BINOL.

They also share the feature that the enantioselectivity of the hydrogenations in which they are used is strongly solvent dependent. In particular, methanol, the workhorse solvent of asymmetric hydrogenations with bisphosphine ligands, gives very poor results. These ligands give optimal results in nonprotic solvents such as CH_2Cl_2 (DCM), tetrahydrofuran (THF), acetone, or EtOAc.

This chapter summarizes the remarkable results that were obtained with these ligands in asymmetric olefin hydrogenation.

14.2 MONODENTATE PHOSPHONITES

Asymmetric hydrogenation using monodentate phosphonites was first reported by Pringle[8] and later by Reetz and colleagues.[9] These ligands are easily prepared in a single step from $RPCl_2$ and 2,2′-bisnaphthol or 9,9′-bisphenanthrol (Figure 14.3).

Surprisingly high enantioselectivities up to 94% are obtained in the asymmetric hydrogenation of methyl 2-acetamido-cinnamate (**5**), methyl 2-acetamido-acrylate (**9**), and dimethyl itaconate (**11**). The rate of these reactions is also fairly high; even at a substrate-to-catalyst (S/C) ratio of 1000, most reactions on **9** and **11** are complete after 3 to 4 hours. Based on the X-ray structures of platinum complexes of ligands **8c** and the ethylene bridged bidentate bis-phenanthrol based ligand (not shown), Pringle analyzed the possible reasons for the high enantioselectivity. He showed that in the complexes containing the monodentate ligands, the bisphenanthryl units actually protrude further outside the plane of projection than in the bidentate complex, leading to 2 blocked quadrants. Also, with these complexes, rotation around the Rh–P bond seems hindered.

These monodentate ligands are interesting for high-throughput experimentation (HTE) because of their ease of synthesis, which allows the easy assembly of a library of ligands. Moreover, it is

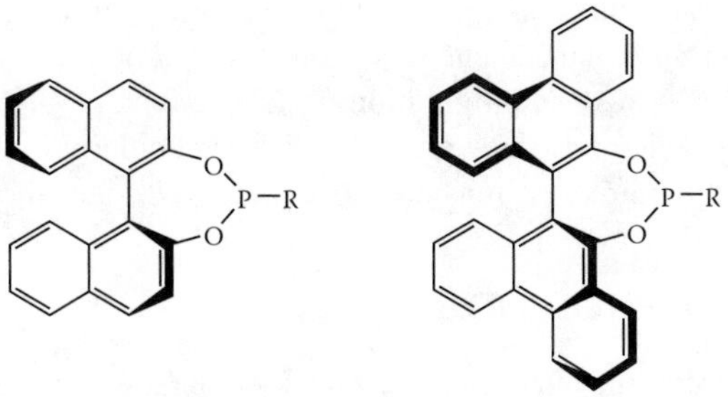

7 a R = Me
b R = *t*-Bu
c R = Ph
d R = Et
e R = Cy

8 a R = Me
b R = *t*-Bu
c R = Ph

FIGURE 14.3 Phosphonite ligands.

TABLE 14.1
Rhodium/Phosphonite-Catalyzed
Asymmetric Hydrogenation of 5, 9, and 11[a]

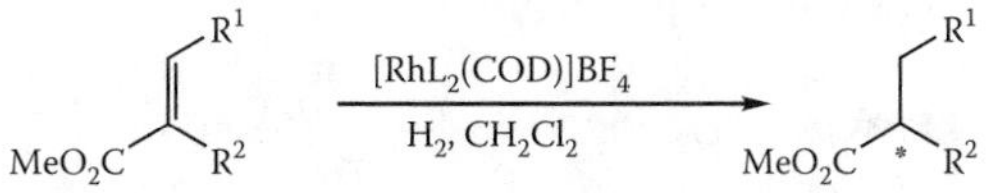

5 R^1 = Ph, R^2 = NHAc   **6**
9 R^1 = H, R^2 = NHAc   **10**
11 R^1 = H, R^2 = $CH_2CO_2CH_3$ **12**

Entry	Ligand	ee 6 (%)	ee 10 (%)	ee 12 (%)
1	**7a**	80 (R)	78 (R)[b]	90 (R)
2	**7b**	10 (R)	92 (R)	57 (R)
3	**7c**	63 (R)	73 (R)	29 (R)
4	**7d**	—	94 (R)	71 (R)
5	**8a**	14 (S)	29 (R)	—
6	**8b**	49 (R)	70 (R)	—
7	**8c**	59 (S)	78 (R)	—

[a] Conditions: Catalyst:S = 1:500, [S] = 0.47M in CH_2Cl_2, 25°C, H_2 1.5 bar, 3h (**9**) 20h (**5, 11**).
[b] This is the ee value reported by Pringle; Reetz reports 92%.

possible to further expand the available diversity by testing mixtures of ligands. Reetz and co-workers showed that several 1:1 combinations of different monophosphonites lead to higher enantioselectivities than those obtained with the "homo" catalysts.[10] For instance, in the rhodium-catalyzed hydrogenation of **9** using a 1:1 mixture of **7a** and **7b**, the product **10** was obtained in 97.8% ee, whereas they found 92% ee with each of the "homo" catalysts. The ratio of ligands does not need to be exactly 1:1. In the asymmetric hydrogenation of aromatic enamides (*vide infra*), it was shown that the ligand ratio could vary between 1:5 and 5:1 without much difference in the enantioselectivity. It is clear that this combinatorial approach has much potential. Indeed, a library of 10 ligands allows 55 combinations to be tested.[11] Using combinations of phosphonites and nonchiral phosphines or phosphites also gave interesting results in the rhodium-catalyzed hydrogenation of **9.**[12] Although no enhancement was found in enantioselectivity, there were profound changes, including a reversal of the configuration of the product.

In conclusion, these new ligands show great promise, particularly in view of the relatively high reactivity of their rhodium complexes in asymmetric olefin hydrogenation.

14.3 MONODENTATE PHOSPHITES

The monophosphites were discovered by Reetz and colleagues springing from a program on bidentate bisphosphites based on dianhydro-D-mannite and 2 equivalents of BINOL. It turned out that the analogous monophosphites (one BINOL replaced by MeOH) gave surprisingly high enantioselectivities in the rhodium-catalyzed hydrogenation of **11**. Comparison of matched and mismatched ligands revealed that the chirality of BINOL had the largest influence on the enantioselectivity. To test this hypothesis a number of simple BINOL-based monophosphites were synthesized, which showed excellent properties in the asymmetric hydrogenation.[13] Chiral monophosphites had been reported before by Union Carbide (now Dow) as ligands for rhodium-catalyzed asymmetric hydroformylation, although in this application they led to much poorer enantioselectivity than those obtained with the bidentate bisphosphites.[14] The monophosphites can be easily synthesized in two steps (Scheme 14.2). Although most ligands were initially made in a sequence that starts with the phosphinylation of BINOL (Scheme 14.2a), this procedure results in oils or foams that are not easily purified. Millitzer and colleagues showed that the reverse preparation (Scheme 14.2b) actually gives a much purer product.[15] In the case of the isopropyl ligand, they were able to obtain the stable $RhL_2(COD)BF_4$ complex in crystalline form. Monophosphites based on BINOL and *l*-menthol (**13v,w**) were reported by Xiao.[16] More recently, the octahydro-derivatives **14** were reported by Bakos and colleagues.[17]

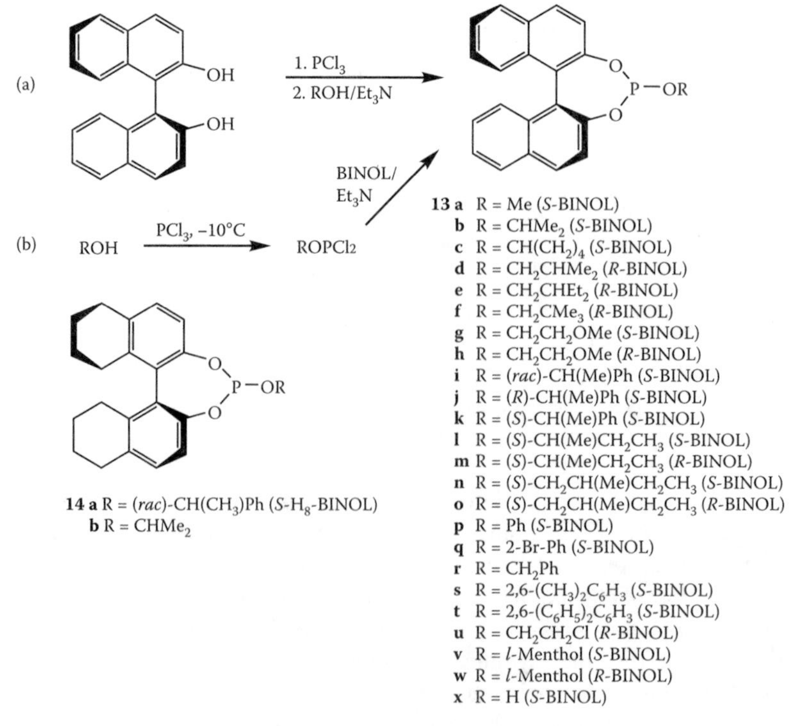
(a)
OH
OH
1. PCl3
2. ROH/Et3N
O
P—OR
O
BINOL/
Et3N
(b)
ROH
PCl3, –10°C
ROPCl2
O
P—OR
O
14 a R = (rac)-CH(CH3)Ph (S-H8-BINOL)
b R = CHMe2
13 a R = Me (S-BINOL)
b R = CHMe2 (S-BINOL)
c R = CH(CH2)4 (S-BINOL)
d R = CH2CHMe2 (R-BINOL)
e R = CH2CHEt2 (R-BINOL)
f R = CH2CMe3 (R-BINOL)
g R = CH2CH2OMe (S-BINOL)
h R = CH2CH2OMe (R-BINOL)
i R = (rac)-CH(Me)Ph (S-BINOL)
j R = (R)-CH(Me)Ph (S-BINOL)
k R = (S)-CH(Me)Ph (S-BINOL)
l R = (S)-CH(Me)CH2CH3 (S-BINOL)
m R = (S)-CH(Me)CH2CH3 (R-BINOL)
n R = (S)-CH2CH(Me)CH2CH3 (S-BINOL)
o R = (S)-CH2CH(Me)CH2CH3 (R-BINOL)
p R = Ph (S-BINOL)
q R = 2-Br-Ph (S-BINOL)
r R = CH2Ph
s R = 2,6-(CH3)2C6H3 (S-BINOL)
t R = 2,6-(C6H5)2C6H3 (S-BINOL)
u R = CH2CH2Cl (R-BINOL)
v R = l-Menthol (S-BINOL)
w R = l-Menthol (R-BINOL)
x R = H (S-BINOL)

SCHEME 14.2

TABLE 14.2
Rh-Catalyzed Asymmetric Hydrogenation of 9
and 11 Using BINOL-Based Phosphite Ligands[a]

$$\underset{MeO_2C}{\overset{}{\bigvee}} R \quad \xrightarrow[\substack{H_2, CH_2Cl_2}]{[RhL_2(COD)]BF_4} \quad MeO_2C \overset{*}{\underset{}{\bigvee}} R$$

9 R = NHAc 10
11 R = CH₂CO₂Me 12

Entry	Ligand	ee 10 (%)	ee 12 (%)
1	13a	73 (R)	89 (S)
2	13b	93 (R)	97 (S)
3	13c	97 (R)	99 (S)
4	13f	93 (S)	99 (R)
5	13h	88 (S)	95 (R)
6	13i	92 (R)	99 (S)
7	13j	96 (R)	99 (S)
8	13k	—	97 (S)
9	13l	96 (R)	99 (S)
10	13m	96 (S)	99 (R)
11	13n	—	64 (S)
12	13o	—	97 (R)
13	13p	81 (R)	97 (S)
14	13q	70 (R)	90 (S)
15	13s	32	39 (S)
16	13t	—	29 (S)
17	13v	75 (R)	95 (S)
18	13w	85 (S)	91 (R)
19	13x	—	85 (S)
20	14a	—	98 (S)
21	14b	—	99 (S)
22[b]	15	93 (S)	96 (R)
23[b]	16a	94 (S)	97 (S)
24[b]	16b	94 (S)	94 (S)
25[b]	16c	83 (S)	99 (S)
26[b]	16d	77 (S)	91 (S)
27[b]	16e	18 (S)	74 (S)

[a] Conditions: Rh:S:L = 1:2:1000, [S] = 0.5M in CH₂Cl₂ or ClCH₂CH₂Cl, 25°C, H₂ 1.3 bar, 20 h.
[b] Pressure is 3 bar.

Table 14.2 shows the results of the rhodium-catalyzed hydrogenation of **9** and **11** at 1.3 bar in DCM with these ligands. In the initial Reetz communication, a Rh/L ratio of 1:1 was used, but later experiments were done at a Rh/L of 1:2. Enantioselectivities are unaffected within this range and are generally excellent with these two substrates. Rhodium/**13**-catalyzed hydrogenation of **5** gave **6** with enantioselectivities ranging from 53–92%. Best results were obtained with **13j**. Similar to the phosphonites, these hydrogenations need nonprotic solvents for highest enantioselectivities. Reetz reported inferior results in isopropanol. Even at 1.3 bar, these reactions are fast at around 300 mol⁻¹ h⁻¹. At 20 bar, turnover frequencies (TOFs) up to 120,000 were obtained with **14b** in the hydrogenation of **11**. The enantioselectivity of **12** decreases only slightly (1–2%) when pressure

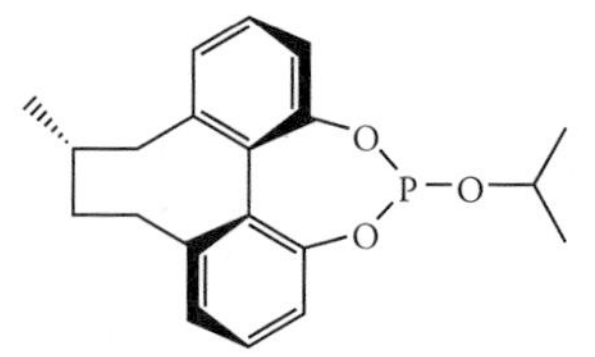

15

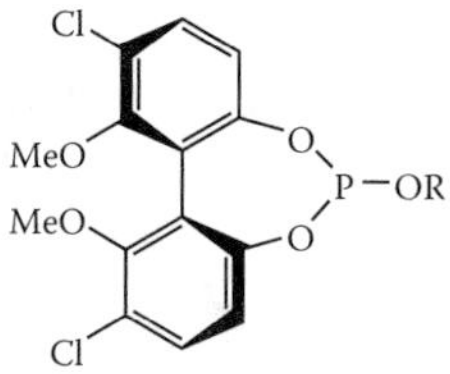

16a (S) R = CH(CH₃)₂
   b (S) R = Cyclohexyl
   c (S) R = (R)-Phenethyl
   d (S) R = Ph
   e (S) R = 2,6-(CH₃)₂C₆H₃

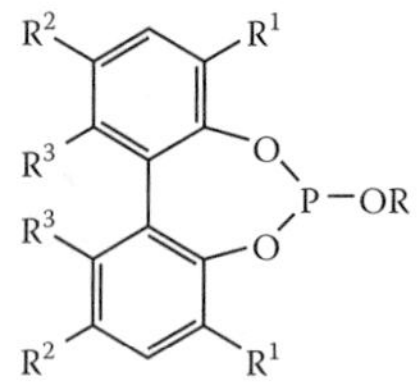

17 a-e R¹ = R² = R³ = H, R see below
   f-j R¹ = R² = t-Bu, R³ = H, R see below
   k   R¹ = H, R² = R³ = Me, R = Ph
   l   R¹ = t-Bu, R² = R³ = Me, R = Ph
   m   R¹ = H, R² = R³ = Me, R = 2-naphthyl
   n   R¹ = t-Bu, R² = R³ = Me, R = 2-naphthyl
   o   R¹ = H, R² = R³ = Me, R see below
   p   R¹ = t-Bu, R² = R³ = Me, R see below
   q   R¹ = H, R² = R³ = Me, R = (1S, 2R)-2-Ph-Cy
   r   R¹ = t-Bu, R² = R³ = Me R = (1S, 2R)-2-Ph-Cy
   s   R¹ = H, R² = R³ = Me, R = (1R, 2S)-2-Ph-Cy
   t   R¹ = t-Bu, R² = R³ = Me R = (1R, 2S)-2-Ph-Cy
   u   R¹ = Ph, R² = R³ = Me R = (1R, 2S)-2-Ph-Cy
   v   R¹ = Me, R² = R³ = Me R = (1R, 2S)-2-Ph-Cy
   w   R¹ = Br, R² = R³ = Me R = (1R, 2S)-2-Ph-Cy

ROH in 17 =

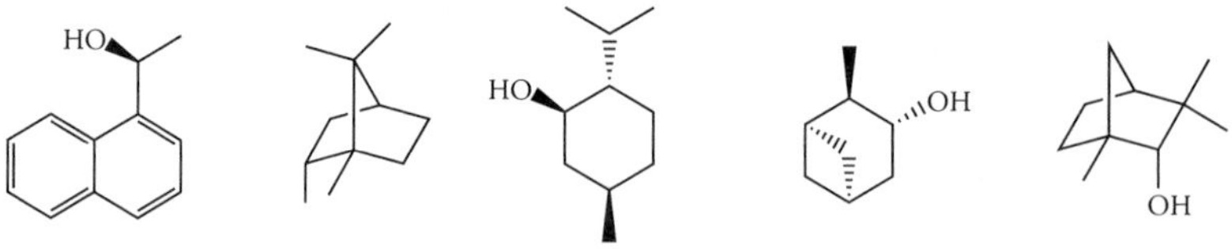

a, f   b, g   c, h, o, p   d, i   e, j

FIGURE 14.4 Monodentate phosphite ligands based on bisphenol.

is increased from 1 to 20 bar. The parent compound **13x** has been reported by Reetz and leads to the formation of **12** in 85% ee, which is only marginally less than the result with **13a**.[18]

Chemists from Bayer and the MPI reported a monodentate phosphite based on a biphenyl backbone locked in one conformation by using a chiral diol to bridge the 6,6′-positions (**15**).[19] In another cooperation between Bayer chemists and Driessen-Hölscher a series of 5-Cl-6-MeO-bisphenol–based monophosphites (**16**) were prepared.[13,20] Asymmetric hydrogenation results with these ligands are also listed in Table 14.2. Results are comparable with those obtained with the BINOL-based ligands. Although biphenyls without substituents on all the 2,2′-, 6,6′-positions are conformationally flexible, it was hoped that in the monophosphites based on biphenyl and readily available chiral alcohols (**17a–j**) such as menthol, a preponderance of one biphenyl conformation would be induced. In practice 1:1 diastereoisomeric mixtures were formed. However, on complexation of **17e** to half an equivalent of rhodium precursor a 5:1 diastereoisomeric mixture of RhL₂(COD)BF₄ complexes was obtained. It is surprising that the biphenyls of both ligands have the same conformation within each complex; no mixed complexes were found. Application of these complexes in the asymmetric hydrogenation of **11** gave up to 60% ee (best results with **17e**), which could be increased to 75% at −15°C.[21] Axially chiral biphenyl-based phosphites **17k–t** have been reported by Ojima and colleagues. Results with these ligands in the rhodium-catalyzed hydrogenation of **9–11** are listed in Table 14.3.

In these experiments, a few remarkable trends were found. It is well-known that bulky substituents in the *ortho*-positions confer hydrolytic stability to phosphite ligands. However, the effect on

TABLE 14.3
Rhodium-Catalyzed Asymmetric
Hydrogenation of Dimethyl Itaconate 11
Using Biphenyl-Based Phosphite Ligands[a]

Entry	Ligand	ee 12 (%) $[Rh(COD)_2]BF_{24}$	ee 12 (%) $[Rh(COD)_2]SbF_6$
1	17k	97 (S)	—
2	17l	14 (R)	16 (R)
3	17m	96 (S)	—
4	17n	19 (R)	23 (R)
5	17o	92 (S)	—
6	17p	25 (R)	15 (R)
7	17q	(Conv. < 1%)	—
8	17r	(Conv. < 1%)	—
9	17s	—	81 (S)
10	17t	(Conv. < 1%)	99 (R)
11	17u	—	98 (R)
12	17v	—	77 (R)
13	17w	—	97 (R)

[a] Conditions: Rh:S:L = 1:2:200, [S] = 0.5M in CH_2Cl_2 or $ClCH_2CH_2Cl$, 23°C (BF$_4$ exps.) 50°C (SbF$_6$ exps.), H$_2$ 6.8 bar, 20 h.

enantioselectivity is negative and even leads to a reversal of the configuration of the product, although the rate is not influenced much. The presence of a phenyl group on remote positions in the sidechain as in 17q–w had a disastrous effect on the rate. It was a surprise to find that switching the anion from BF$_4$ to SbF$_6$ restored activity and led to product with high enantioselection, even with ligand 17t containing t-Bu-groups in the *ortho* positions. Hydrogenations with 17k–w are extremely solvent dependent and are best performed in the chlorinated solvents CH_2Cl_2 or $ClCH_2CH_2Cl$. When THF, MeOH, EtOAc, or even CHCl$_3$ was used, no enantioselectivity was observed. Similar to the BINOL-based ligands, the biaryl configuration is the determining factor in the enantioselection. Change of configuration of the biaryl group in 17o and 17p led to inversion of configuration in the product 12.

Chiral amines can be prepared via hydrogenation of the enamides followed by hydrolysis. The starting enamides can be prepared from ubiquitous ketones via the oxime[22] or alternatively from the nitriles via Grignard addition followed by acylation.[23] Rh/13 is a good catalyst for the hydrogenation of aromatic enamides giving the *N*-acyl-amines with enantioselectivities up to 95% at ambient temperature (Table 14.4).[24] However, these hydrogenations are not very fast and need 20 hours at 60 bar and S/C =500 for full conversion. The stereochemistry of the double bond becomes important if the enamide has a third substituent on the double bond. Whereas *Z*-olefin 18d was hydrogenated smoothly with 97% ee, *E*-olefin 18e was only slowly hydrogenated with an enantioselectivity of 76%.

The asymmetric hydrogenation of enol esters is an alternative to asymmetric ketone hydrogenation. The precursors can be prepared from the ketones but also via ruthenium-catalyzed addition of the carboxylic acids to the 2-postion of terminal alkynes. This latter method allows the study of the effect of the carboxylate on the enantioselectivity of the asymmetric hydrogenation. A remarkable study by Reetz and colleagues established that it is possible to hydrogenate enolate

TABLE 14.4
Rh/13-Catalyzed Hydrogenation of Aromatic Enamides[a]

18 a Ar = Ph; R = H
   b Ar = *p*-Cl-Ph; R = H
   c Ar = 2-naphthyl; R = H
   d Ar = Ph; R = Me (95% *Z*)
   e Ar = Ph; R = Me (84% *E*)

Entry	Ligand	ee 19a (%)	ee 19b (%)	ee 19c (%)	ee 19d (%)	ee 19e (%)
1	13a	76	—	—	—	—
2	13b	89	—	—	—	—
3	13c	94	—	—	—	—
4	13d	93	—	92	—	—
5	13e	94	—	93	—	—
6	13f	95	—	94	—	—
7	13g	86	—	—	—	—
8	13j	95	96	—	97	76[b]
9	13l	92	—	91	—	—
10	13m	94	—	93	—	—
11	13r	94	—	94	—	—
12	13u	93	—	—	—	—

[a] [Rh(COD)$_2$]BF$_4$:**13**:**18** = 1:2:500 in CH$_2$Cl$_2$, 60 bar, 30°C, 20 h.
[b] 69% conversion.

esters of aliphatic ketones such as 2-butanone and 2-hexanone in good enantioselectivities, provided the carboxylate is aromatic.[25] Best results were obtained with enol esters based on 2-furanoic acid. Eight monodentate phosphites **13** were tested in conjunction with [Rh(COD)$_2$]BF$_4$, but no good enantioselection was obtained.[26] Eight new monodentate phosphites were synthesized based on (*R*)- or (*S*)-BINOL and 4 commercially available protected carbohydrate alcohols. Of these ligands, **13x** gave the best results (Scheme 14.3). Hydrogenation of **22e** using Rh/**13y** at −20°C gave the acylated alcohol in 94% ee. For comparison, the best result so far reported in the asymmetric hydrogenation of 2-hexanone was 75% ee (Ru/Pennphos).[27] The enol ester hydrogenations are not fast: the reaction took 20 hours to go to completion at S/C of 200 and 60 bar of hydrogen pressure.

The monophosphite ligands **13** have also been used in a combinatorial approach using mixtures of two different ligands, as described earlier for the monodentate phosphonites. Most combinations of phosphite ligands led to lower or comparable enantioselectivities as those obtained with the homo-catalysts; best results were obtained with a mixture of **13a** and **13s** in the rhodium-catalyzed hydrogenation of **9** (85% vs. 77% and 32% for the respective homo-catalysts). However, combinations of phosphonites 7 with phosphites **13** led to more spectacular hits. In the hydrogenation of **9**, the combination of **7b** with **13a** resulted in **10** with 98% enantioselectivity (vs. 93% and 77% with the respective homo-catalysts). Using mixtures of **13** and nonchiral phosphines or phosphites in the rhodium-catalyzed hydrogenation of **9** led to profound changes in enantioselectivity, even to a change in configuration of the product **10**.[12] However, no improved enantioselection has been found so far.

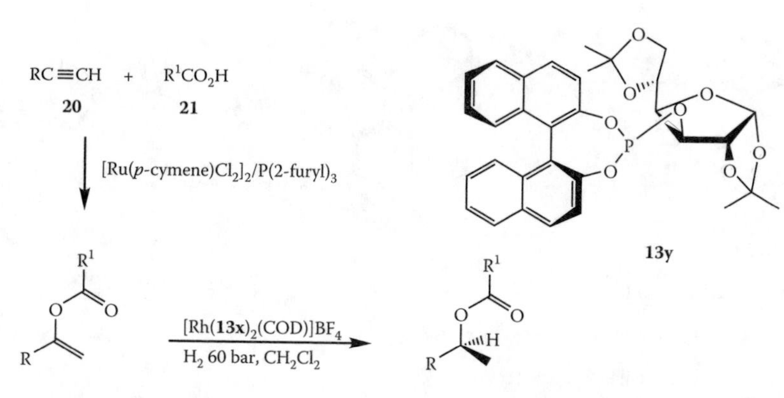

$$RC\equiv CH \quad + \quad R^1CO_2H$$
$$\textbf{20} \qquad\qquad \textbf{21}$$

[Ru(p-cymene)Cl$_2$]$_2$/P(2-furyl)$_3$

22 a R = n-Bu, R^1 = Ph
b R = n-Bu, R^1 = Me
c R = n-Bu, R^1 = Et
d R = n-Bu, R^1 = t-Bu
e R = n-Bu, R^1 = 2-furyl
f R = Et, R^1 = Ph
g R = Et, R^1 = 2-N-Me-Pyrrolyl
h R = Et, R^1 = 2-furyl

[Rh(13x)$_2$(COD)]BF$_4$
H$_2$ 60 bar, CH$_2$Cl$_2$

23 a 86% e.e.
b 73% e.e.
c 74% e.e.
d 42% e.e.
e 90% e.e. (94% at −20°C)
f 80% e.e.
g 72% e.e.
h 84% e.e. (88% at −20°C)

13y

SCHEME 14.3

14.4 MONODENTATE PHOSPHORAMIDITES

The use of monodentate phosphoramidite ligands in asymmetric hydrogenation[28] was borne out of the desire of DSM to have available an easy-to-synthesize ligand library that could be used in robotic screening of catalytic procedures for the production of chiral pharma intermediates.[29] Monodentate phosphoramidites had been developed by Feringa and co-workers as ligands for the highly enantioselective copper-catalyzed 1,4-addition of Et$_2$Zn to enones.[30] Thus, there was an excellent basis for a joint development. The ligands are easily prepared in a 1- or 2-step procedure, as illustrated in Scheme 14.4.[31]

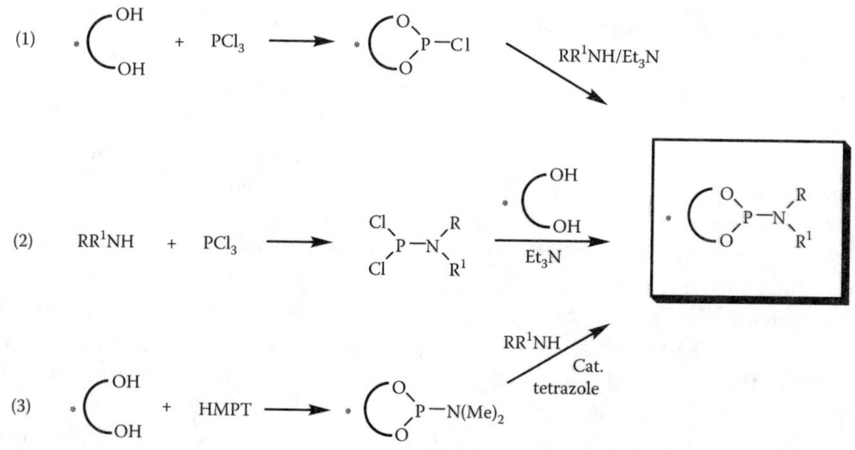

(1)   *—(OH)(OH) + PCl$_3$ ⟶ *—O$_2$P—Cl   RR^{1}NH/Et$_3$N

(2)   RR^{1}NH + PCl$_3$ ⟶ Cl$_2$P—N(R)(R^1) ⟶ (with *—(OH)(OH), Et$_3$N)

(3)   *—(OH)(OH) + HMPT ⟶ *—O$_2$P—N(Me)$_2$ RR^{1}NH, Cat. tetrazole

SCHEME 14.4

The first method gives relatively pure phosphoramidites with most substrates. The second, reversed method is preferred with hindered amines.[32] MonoPhos™, the dimethyl analogue **24a**, is made in a single step from BINOL and HMPT (hexamethylphosphoric triamide) in toluene.[33] It

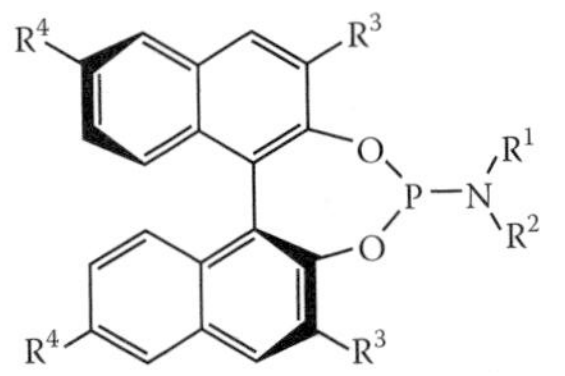

24 a $R^1 = R^2 = Me, R^3 = R^4 = H$
b $R^1 = R^2 = Et, R^3 = R^4 = H$
c $R^1 = R^2 = i\text{-}Pr, R^3 = R^4 = H$
d $R^1 = R^2 = \text{-}(CH_2)_5\text{-}, R^3 = R^4 = H$
e $R^1 = Me, R^2 = CH_2Ph, R^3 = R^4 = H$
f $R^1 = R^2 = (R)\text{-}CH(Me)Ph, R^3 = R^4 = H$
g $R^1 = R^2 = (S)\text{-}CH(Me)Ph, R^3 = R^4 = H$
h $R^1 = H, R^2 = (R)\text{-}CH(Me)Ph, R^3 = R^4 = H$
i $R^1 = H, R^2 = (S)\text{-}CH(Me)Ph, R^3 = R^4 = H$
j $R^1 = R^2 = Me, R^3 = Me, R^4 = H$
k $R^1 = R^2 = Me, R^3 = H, R^4 = Br$

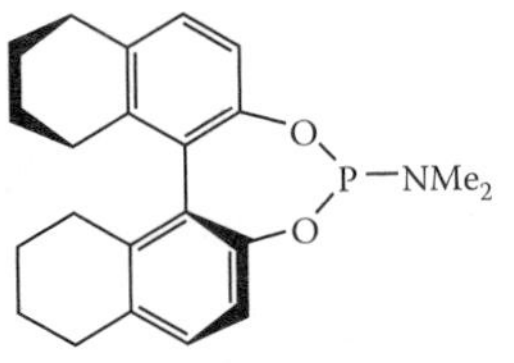

25

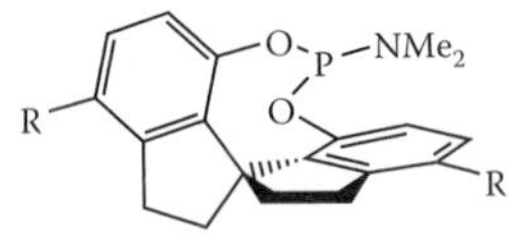

26 a R = H
b R = Br
c R = Ph
d R = OMe

FIGURE 14.5 Monodentate phosphoramidite ligands.

can be reacted with other secondary or primary amines in the presence of a catalyst such as tetrazole.[34] This latter method is highly suitable for the preparation of more labile ligands. A large library of ligands has been synthesized over the years with most diversity stemming from the readily available amines. Nevertheless, some substituted BINOLs have also been used as building blocks. In Figure 14.5, some of the most important ligands are depicted. (Members of the MonoPhos ligand family are available from STREM either separately or as a ligand-kit.) Ligand **25** based on octahydro-BINOL has been developed by both the Feringa group[31] and by Chan.[35] The *spiro-biindane* based phosphoramidite **26** was developed by Zhou and co-workers.[36] Phosphoramidite ligands based on TADDOL[31] and on D-mannitol[37] have also been reported but generally led to lower enantioselectivity than the BINOL-based ligands in the hydrogenation of α-dehydro-amino acids and itaconate. Moreover, these diols were prepared by lengthy synthetic sequences.

In the initial rhodium-catalyzed hydrogenation experiments on substituted olefins with Mono-Phos (**24a**) as ligand, it was found that the reaction is strongly solvent dependent. Very good enantioselectivities were obtained in the rhodium-catalyzed hydrogenation of **5** in nonprotic solvents (Table 14.5).[28,31]

The phosphoramidite ligands are surprisingly stable, even in protic solvents. Whereas in general P–N bonds are not stable in the presence of acids, the hydrogenation of *N*-acetyldehydroamino acids proceeded smoothly. A large range of substituted *N*-acetyl phenylalanines and their esters could thus be prepared by asymmetric hydrogenation with Rh/MonoPhos with enantioselectivities ranging from 93–99% (Table 14.6).[31,38] The configuration of the α-amino acid products always is the opposite of that of the BINOL in the ligand used in the hydrogenation.

The hydrogenations catalyzed by rhodium/**24a** are not fast at 1 bar pressure. However, at 5 bar in a well-stirred autoclave TOFs of 200–600 h^{-1} were reached, which is quite satisfactory even for industrial applications. Even faster reactions can be induced by further increasing the pressure. At elevated pressures the amount of catalyst can be substantially reduced to 0.01–0.1mol%. In all these hydrogenations, the enantioselectivity was not affected in any way by pressures up to 60 bar, in contrast to the deterioration of ee that occurs with increase of pressure when using bisphoshines.[31]

An interesting aspect in these hydrogenations is the effect of the L/Rh ratio on rate and enantioselectivity.[31] No reaction occurs at L/Rh 3. Surprisingly, it was found that reducing the L/Rh ratio from 2 to 1.5 or even 1 results in an increased rate, whereas the enantioselectivity was not affected. If a mixture of active catalysts is present, one would expect the enantioselectivity to be influenced by the L/Rh ratio. Thus, a single catalytic species must be responsible for the catalysis. Initially, these results suggested the possibility that the active catalyst species may contain only a single phosphoramidite ligand. However, results obtained with mixtures of different ligands (*vide*

TABLE 14.5

Solvent Dependency of Enantioselectivity in Rh/MonoPhos™-Catalyzed Asymmetric Hydrogenation of 5[a]

Entry	Solvent	Temp.	ee
1	CH_3OH	RT	70%
2	CH_2Cl_2	RT	95%
3	CH_2Cl_2	5°C	97%
4	THF	RT	93%
5	Acetone	RT	92%
6	$BuOCH_2CH_2OH$[b]	RT	77%
7	EtOAc	RT	97%

[a] Conditions: Substrate: $[Rh(COD)_2]BF_4$: MonoPhos 100:5:10 at 1 bar H_2 pressure in a Schlenck tube.

[b] This solvent has the same polarity as CH_2Cl_2.

infra) clearly show that the active catalyst must contain 2 ligands. Inspection of the electrospray mass spectrum (ES-MS) of a hydrogenation mixture revealed the presence of RhL_1, RhL_2, RhL_3, and even RhL_4, In addition, substrate complexes were found of the first 3. The RhL_4 is surprisingly stable and could be isolated in crystalline form. Thus, the increased rate at the lower L/Rh ratios can be explained by a shift away from the higher ligated catalytically inactive species.

Whereas bidentate phosphites and phosphonites are excellent ligands for rhodium-catalyzed asymmetric hydrogenation, bidentate phosphoramidites (not shown) gave very poor results with low reaction rates and enantioselectivities. Results of a number of other monodentate phosphoramidites in the asymmetric hydrogenation of *N*-acetyl dehydrophenylalanine derivatives are shown in Table 14.7.[31,35,36,39,40]

The best ligands in terms of enantioselectivity are **24b**[39] and **24d**.[40] The rate of the hydrogenation is retarded by bulky groups on the nitrogen atom and by electron withdrawing substituents on the BINOL. SIPHOS (**26a**) leads to excellent results in these hydrogenations but needs 7 synthetic steps, versus 1 for MonoPhos.[36,41]

Excellent results have been obtained in the asymmetric hydrogenation of itaconic acid (97% ee) and dimethyl ester (94% ee) using Rh/MonoPhos.[31] The latter substrate could be hydrogenated in 99% ee using the piperidine ligand **24d**.[40] These hydrogenations are relatively fast and have been carried out on a 100-g scale with an S/C of 10,000. Preliminary results with some alkylidene and benzylidene succinates were also very promising.

The monodentate phosphoramidites have also been used as ligands for rhodium-catalyzed enamide hydrogenation. Similar to the hydrogenations with the phosphites, these reactions are relatively slow, needing 1–20 hours at 10–20 bar of hydrogen pressure with S/C ratios of 50–100. The enantioselectivity is strongly influenced by the nitrogen substituents on the ligand. Table 14.8 shows results of the hydrogenation of a range of enamide substrates.[36a,c,42,43]

The results in Table 14.8 show that very high enantioselectivities can be obtained using the diethyl[39] or piperidinyl[40] MonoPhos ligands or with any member of the SIPHOS family.[36a,c]

An important class of compounds that are frequently used as building blocks in drugs is the β-amino acids. Here, the precursors are made from the β-keto esters by reaction with NH_4OAc, followed by acylation with Ac_2O. These enamides can be made predominantly in either the *E*-[44]

TABLE 14.6
Asymmetric Hydrogenation of Dehydroamino Acids and Esters with Rhodium/24[a]

$$R-C(=CH-CO_2R^1)(NHAc) \xrightarrow[\text{H}_2,\ \text{Solvent}]{[Rh(COD)_2]BF_4/\textbf{24a}} R-CH_2-CH(CO_2R^1)(NHAc)$$

Entry	R	R[1]	Mol% Rh	24a:Rh	Solvent	pH$_2$ (bar)	Time	ee (%)	TOF (h^{-1}) [a]
1	Phenyl	Me	5.0	2.2	CH$_2$Cl$_2$	1	3 h	95	6.7
2	Phenyl	Me	0.5	2.2	CH$_2$Cl$_2$	5	40 min	95	300
3	Phenyl	Me	0.1	2.2	CH$_2$Cl$_2$	15	2 h	95	500
4	Phenyl	Me	0.015	2.0	CH$_2$Cl$_2$	5	16 h	95	337[b] (81%)
5	Phenyl	Me	0.9	2.2	EtOAc	60	4 min	97	1667
6	Phenyl	H	0.2	2.2	CH$_2$Cl$_2$	15	1 h	97	500
7	Phenyl	H	0.1	2.2	CH$_2$Cl$_2$	5	3 h	97	333
8	Phenyl	Me	5.0	1.1	EtOAc	1	2 h	96	10
9	Phenyl	Me	0.015	1.0	CH$_2$Cl$_2$	5	16 h	95	375[c] (90%)
10	3-methoxyphenyl	Me	1.0	1.1	CH$_2$Cl$_2$	5	2 h	97	50
11	4-methoxyphenyl	Me	1.0	1.1	CH$_2$Cl$_2$	5	2 h	94	50
12	4-AcO-3-MeO-phenyl	Me	0.5	2.2	EtOAc	5	nd[d]	96	—
13	4-fluorophenyl	Me	1.0	1.1	CH$_2$Cl$_2$	5	25 min	96	240
14	4-fluorophenyl	H	2.0	2.2	CH$_2$Cl$_2$	27	10 min	93	300
15	3-fluorophenyl	Me	1.0	1.1	CH$_2$Cl$_2$	5	30 min	95	200
16	3-fluorophenyl	H	0.1	2.2	CH$_2$Cl$_2$	10	2 h	96	500
17	2-fluorophenyl	Me	1.0	1.1	CH$_2$Cl$_2$	5	15 min	95	400
18	4-chlorophenyl	Me	1.0	1.1	CH$_2$Cl$_2$	5	20 min	94	300
19	3,4-dichlorophenyl	H	0.1	1.1	CH$_2$Cl$_2$	5	2 h	97	500
20	3,4-dichlorophenyl	Me	1.0	1.1	CH$_2$Cl$_2$	5	30 min	99	200
21	3-nitrophenyl	Me	4.0	1.1	CH$_2$Cl$_2$	5	2 h	95	13
22	4-nitrophenyl	Me	1.0	1.1	CH$_2$Cl$_2$	5	nd	95	—
23	4-fluoro-3-nitrophenyl	Me	1.0	1.1	CH$_2$Cl$_2$	5	2 h	95	50
24	4-biphenyl	Me	1.0	1.1	CH$_2$Cl$_2$	5	25 min	95	240
25	3-fluoro-4-biphenyl	Me	1.0	1.1	CH$_2$Cl$_2$	5	25 min	93	240
26	4-acetylphenyl	Me	1.0	1.1	CH$_2$Cl$_2$	5	15 min	99	400
27	4-benzoylphenyl	Me	1.0	1.1	CH$_2$Cl$_2$	5	30 min	94	168
28	4-cyanophenyl	Me	1.0	1.1	CH$_2$Cl$_2$	5	18 h	92	(70%)
29	1-naphthyl	Me	1.0	1.1	CH$_2$Cl$_2$	5	10 min	93	600
30	H	Me	1.0	2.2	CH$_2$Cl$_2$	5	2 h	>99	50
31	H	H	5.0	2.2	EtOAc	1	nd	>99	—

[a] Yields are quantitative, unless noted otherwise. Because reaction times are unoptimized, turnover frequencies (TOFs) are indicative.
[b] Turnover number (TON) = 5400.
[c] TON = 6000.
[d] Not determined.

TABLE 14.7
Ligand Variations in the Rh-Catalyzed Hydrogenation of Dehydro-Amino Acids and Esters

R–CH=C(CO$_2$Me)(NHAc) $\xrightarrow[\text{H}_2,\ \text{CH}_2\text{Cl}_2]{[\text{Rh(COD)}_2]\text{BF}_4/\text{Ligand}}$ R–CH$_2$–CH(CO$_2$Me)(NHAc)

Entry	R = Ligand	Ph ee (%)	4-MeOPhe ee (%)	4-NO$_2$Ph ee (%)	4-ClPh ee (%)	H ee (%)
1	24a	95	94	95	94	99.7
2	24b[a]	98	99.0	99.7	99.1	97
3	24c	93	—	—	—	90
4	24d	>99	—	—	—	>99
5	24f	42	—	—	—	—
6	24j	89	—	—	—	—
7	24k	94	—	—	—	—
8	25	94	94[b]	96	98	99.9[b]
9	26a[c]	98	96	99	99	97
10	26b[c]	98	95	99	99	—
11	26c[c]	98	95	99	98	—
12	26d[c]	97	94	97	98	—

[a] Solvent is THF.
[b] Solvent is acetone.
[c] Solvent is toluene.

TABLE 14.8
Rhodium-Catalyzed Hydrogenation of Aromatic Enamides Using Monodentate Phosphoramidite Ligands

Ar–C(=CH$_2$)(NHAc) $\xrightarrow[\text{H}_2,\ \text{Solvent}]{[\text{Rh(COD)}_2]\text{BF}_4/\text{Ligand}}$ Ar–CH(CH$_3$)(NHAc)

Entry	Ar = Ligand	Ph ee (%)	4-ClPh ee (%)	4-CF$_3$Ph ee (%)	4-MeOPh ee (%)	2-Furyl ee (%)	2-Thienyl ee (%)
1	24a[a]	86	89	—	86	85	90
2	24b[b]	95	90	97	93		
3	24d[a]	97	99	—	99		
4	24j[a]	—	44[a]	—	—		
5	24k[a]	—	89	—	83		
6	25a[a]	—	86	—	62		
7	26a[c]	98	99	99	—	99	96
8	26b[c]	97	99	—	—		
9	26c[c]	98	98	—	—		
10	26d[c]	95	94	—	—		

[a] Solvent = CH$_2$Cl$_2$.
[b] Solvent = THF.
[c] Solvent = toluene.

TABLE 14.9
Rh–Phosphoramidite-Catalyzed Hydrogenation
of β-Dehydroamino Acid Esters

Entry	Substrate	Ligand	Solvent	ee (%)
1	E- R = Me, R^1 = Me	**24a**	CH_2Cl_2	95
2	E- R = Me, R^1 = Me	**24e**	CH_2Cl_2	99
3	Z- R = Me, R^1 = Me	**24h**	i-PrOH	95
4	Z- R = Ph, R^1 = Et	**24h**	i-PrOH	92

Conditions: The reaction was performed at room temperature by dissolving substrate, $Rh(COD)_2BF_4$, and ligand (100:1:2) in the suitable solvent and applying 10 bar of H_2 to the stirred reaction.

or the Z-[45] forms and can be purified by crystallization or chromatography. There is a marked difference in hydrogenation behavior between the E- and the Z-precursors. Whereas the E-β-dehydroamino acid esters are smoothly hydrogenated with many rhodium bisphosphine-type catalysts, the Z-isomers are much harder to hydrogenate; generally, both reaction rates and enantioselectivities are much lower. Presumably, this is the result of the presence of an intramolecular hydrogen bond between the amide NH and the ester group. Using rhodium with MonoPhos as ligand gave quite satisfactory results in the asymmetric hydrogenation of aliphatic E-β-dehydroamino acid esters in CH_2Cl_2 (Table 14.9, Entry 1).[46] Use of the slightly modified ligand **24e** even gave the product in 99% enantioselectivity (Entry 2). Use of protic solvents again led to a sharp decrease in enantioselectivity.

For the Z-isomers, a different approach was necessary. Most importantly, the use of isopropanol as a solvent is mandatory. Presumably, this solvent aids in breaking up the internal hydrogen bond, which is an essential step preceding the bidentate binding of the substrate to the metal complex. In view of the low reaction rates that are associated with the hydrogenation of the Z-substrate, a ligand was needed that induces higher hydrogenation rates. From earlier work, it had become clear that the rate of the reaction is very dependent on the steric bulk present at the nitrogen atom of the ligand. Because MonoPhos already has the smallest possible substituents, ligands based on primary amines were investigated. They were found to be quite stable as long as they can be obtained in crystalline form. Ligand **24h** could be made in good yield and performed beyond expectation in the asymmetric hydrogenation of both aliphatic and aromatic Z-β-dehydroamino acid esters (Entries 3 and 4).[46] In particular, the 92% ee obtained with the Z-aromatic substrate is quite remarkable and belongs to the highest values ever reached for this substrate. In addition, the catalyst made from $[Rh(COD)_2]BF_4$ and 2eq. of **24h** turned out to be very fast. Figure 14.6 shows a comparison of hydrogenation rates in the rhodium-catalyzed hydrogenation of ethyl Z-3-N-acetamidocrotonate (**27a**).[47] The performance of ligand **24h** in terms of rate is only surpassed by DUPHOS and comparable to PHANEPHOS. However, enantioselectivity with these ligands was only 67% and 11%, respectively.

The combinatorial approach was also investigated successfully with the phosphoramidite ligands **24** in the rhodium-catalyzed asymmetric hydrogenation of **27a** and **27b** (Scheme 14.5).[48]

The hydrogenation of the Z-dehydroamino acids was carried out in a solvent known not to give the best results to create room for improvement: CH_2Cl_2. The best homo-catalyst was Rh/**24h**. Most ligand combinations gave rise to lower enantioselectivities, but all combinations with ligand **24h**

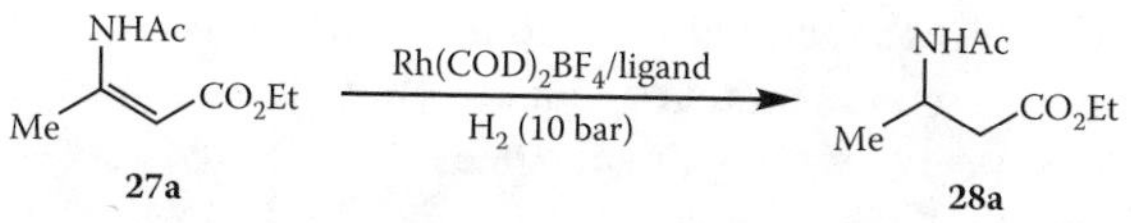

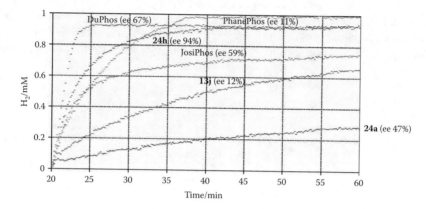

FIGURE 14.6 Comparison of hydrogenation rates in the asymmetric hydrogenation of *Z*-ethyl 3-*N*-acetamido-crotonate.

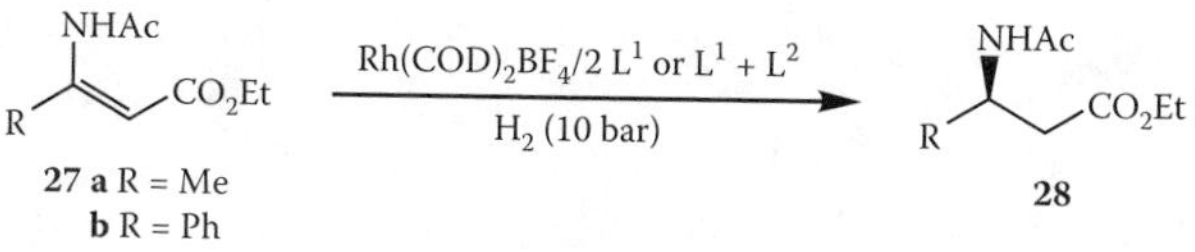

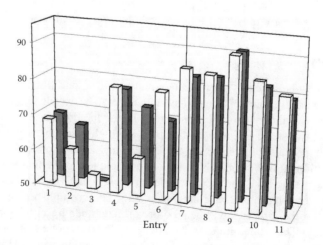

SCHEME 14.5

FIGURE 14.7 Combinatorial hydrogenation of **27a** (*light bars*) and **28b** (*dark bars*) with [Rh(COD)₂]BF₄/L¹ + L². Legend: 1 = **24a**, 2 = **25**, 3 = **24j**, 4 = **24k**, 5 = **24e**, 6 = **24 h**, 7 = **24h** + **24a**, 8 = **24h** + **25**, 9 = **24h** + **24j**, 10 = **24h** + **24k**, 11 = **24h** + **24e**.

gave better results than those obtained with the homo-catalysts. Most spectacular is the result of Entry 9, which shows that the combination of ligand **24h** with the worst performer in the "homo" series **24j** leads to the best results with both substrates.

Immobilized phosphoramidites **29** and **30** were made by Doherty and co-workers.[49] When loaded with [Rh(COD)$_2$]BF$_4$, material was obtained with a P/Rh ratio of 7 and 9.6, respectively. Hydrogenation of N-acetyl-dehydroamino acids and esters as well as dimethyl itaconate was performed at 1 bar H$_2$ in CH$_2$Cl$_2$. Products were obtained in ee's ranging from 49% to 80%. In general, enantioselectivities were lower than those obtained with the monomeric analogues. The catalyst made from **29** could be recycled 4 times, although only with substrates that gave relatively fast reactions.

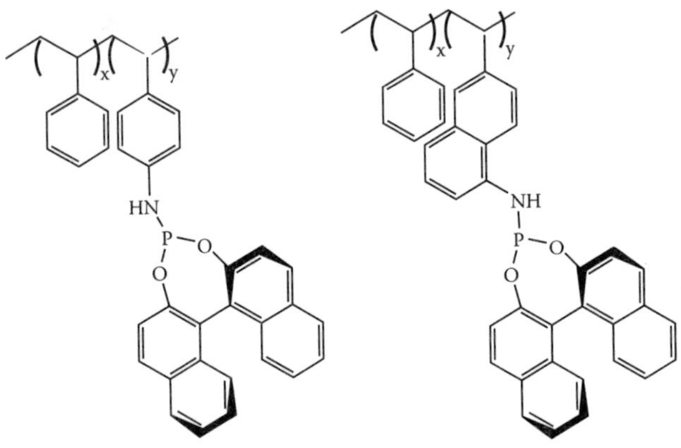

FIGURE 14.8 Immobilized phosphoramidite ligands.

Huttenloch and co-workers prepared an entire library of phosphoramidite ligands on solid phase.[50] However, these were only used for conjugate addition reactions.

13.5 SUMMARY

Quite contrary to long-held beliefs, monodentate phosphonites, phosphites and phosphoramidites based on BINOL, or axially chiral bisphenol have turned out to be excellent ligands for the rhodium-catalyzed asymmetric hydrogenation of substituted olefins. In addition, they are readily prepared in one or two synthetic steps. Because of this, they are highly suitable for a library approach.[51] These findings will have a strong impact on the applicability of asymmetric hydrogenation for the production of enantiopure fine chemicals. It will be possible to rapidly find the right catalyst for a given transformation using HTE and these libraries of monodentate ligands, and also the synthesis on kilogram scale for ton-scale production can readily be accomplished in a short time. Also important: the cost of these ligands will be an order of magnitude lower than that of conventional bisphosphine ligands. Thus, three important bottlenecks for the application of this technology in production have been removed. Indeed, the first ton-scale production using a MonoPhos-type ligand has just taken place.[52] The conclusion seems justified that asymmetric hydrogenation has undergone a rejuvenation.

REFERENCES AND NOTES

1. a) Knowles, W. S., Sabacky, M. J. *J. Chem. Soc. Chem. Commun.* 1968, 1445; b) Knowles, W. S. *Acc. Chem. Res.* 1983, *16*, 106.

2. Horner, L., Siegel, H., Büthe, H. *Angew. Chem., Int. Ed.* 1968, *7*, 942.

3. a) Dang, T. P., Kagan, H. B. *J. Chem. Soc., Chem. Commun.* 1971, 481; b) Kagan, H. B., Dang, T. P. *J. Am. Chem. Soc.* 1972, *94*, 6429.

4. Vineyard, B. D., Knowles, W. S., Sabacky, M. J., Bachman, G. L., Weinkauff, D. J. *J. Am. Chem. Soc.* 1977, *99*, 5946.

5. a) Brunner, H., Zettlmeier, W. *Handbook of Enantioselectieve Catalysis,* VCH: Weinheim, 1993; b) Noyori, R. *Asymmetric Catalysis in Organic Synthesis,* Wiley: New York, 1993; c) Kagan, H. B. In *Asymmetric Synthesis*, Morrison, J. D., Ed., Academic Press, Inc.: Orlando, 1985, Vol. 5, p. 1; d) Brunner, H. *Top. Stereochem.* 1988, *18*, 129; e) Blaser, H.-U., Malan, C., Pugin, B., Spindler, F., Steiner, H., Studer, M. *Adv. Syn. Catal.* 2003, *345*, 103.

6. Knowles, W. S., Sabacky, M. J., Vineyard, B. D. *J. Chem. Soc. Chem. Commun.* 1972, 10.

7. For a historical review on the use of chiral monodentate phosphorus ligands in enantioselectieve olefin hydrogenations, see Komarov, I. V., Börner, A. *Angew. Chem., Int. Ed.* 2001, *40*, 1197.

8. Claver, C., Fernandez, E., Gillon, A., Heslop, K., Hyett, D. J., Martorell, A., Orpen, A. G., Pringle, P. G. *J. Chem. Soc., Chem. Commun.* 2000, 961.

9. Reetz, M. T., Sell, T. *Tetrahedron Lett.* 2000, *41*, 6333.

10. Reetz, M. T., Sell, T., Meiswinkel, A., Mehler, G, *Angew. Chem., Int. Ed.*, 2003, *42*, 790.

11. For a library of n ligands, this number can be calculated according to the formula (n+1)!/2(n–1)!

12. Reetz, M. T., Mehler, G. *Tetrahedron Lett.* 2003, *44*, 4593.

13. a) Reetz, M. T., Mehler, G. *Angew. Chem., Int. Ed.*, 2000, *39*, 3889; b) Reetz, M. T., Mehler, G, Meiswinkel, A., Poster P2-3 at ESOC-12, Groningen, The Netherlands, 2001; c) Reetz, M.T., Mehler, G., Meiswinkel, A. PCT Pat. Appl. WO01/94278, 2001.

14. a) Babin, J. E., Whiteker, G., PCT Pat. Appl. WO93/03839, 1993; b) See also Dussault, P. H., Woller, K. R. *J. Org. Chem.* 1997, *62*, 1556.

15. Agel, F. Driessen-Hölscher, B., Meseguer, B., Scholz, U. Vogl, E. M., Gerlach, A., Millitzer, H.-C., Poster P.039 at ISHC-13, Tarragona, 2002.

16. Chen, W., Xiao J., *Tetrahedron Lett.*, 2001, *42*, 2897-2899.

17. Gergely, I., Hegedüs, C, Gulyás, H., Szöllsy, Á., Monsees, A., Riermeier, T., Bakos, J., *Tetrahedron: Asymmetry*, 2003, *14*, 1087.

18. Reetz, M. T., Sell, T., Goddard, R. *Chimia* 2003, *57*, 290.

19. Hannen, P., Millitzer, H.-C., Vogl, E. M., Rampf, F.A. *J. Chem. Soc., Chem. Commun.*, 2003, 2210.

20. Meseguer, B., Prinz, T., Scholz, U., Militzer, H.-C., Agel, F., Driessen-Hölscher, B. Eur. Pat. EP 1 298 136, 2003.

21. Chen, W., Xiao J., *Tetrahedron Lett.*, 2001, *42*, 8737.

22. a) Barton, D. H. R., Zard, S. Z. *J. Chem. Soc. Perkin. Trans. I* 1985, 2191; b) Burk, M. J., Casy, G., Johnson, N. B. *J. Org. Chem.* 1998, *63*, 6084; c) Zhang, Z., Zhu, G., Jiang, Q., Xiao, D., Zhang, X. *J. Org. Chem.* 1999, *64*, 1774.

23. a) Kagan, H. B., Langlois, N., Dang, T. P. *J. Organomet. Chem.* 1975, *96*, 353; b) Sinou, D., Kagan, H. B. *J. Organomet. Chem.* 1976, *114*, 325; c) Burk, M. J., Wang, Y. M., Lee, J. R. *J. Am. Chem. Soc.* 1996, *118*, 5142.

24. Reetz, M. T., Mehler, G., Meiswinkel, A., Sell, T. *Tetrahedron Lett.* 2002, *43*, 7941.

25. Reetz, M.T., Goossen, L.J., Meiswinkel, A., Paetzold, J., Feldthusen Jensen, J. *Org. Lett.* 2003, *5*, 3099.

26. The following ligands were tested: **13a, b, d, f, p, r** and the ligands **13** with R = Cy and with R = CH$_2$Cy. Enantioselectivities ranged from 21% to 65% with these ligands.

27. Jiang. Q., Xiao, D., Cao, P., Zhang, X. *Angew. Chem., Int. Ed.* 1998, *37*, 1100.

28. van den Berg, M., Minnaard, A. J., Schudde, E. P., van Esch, J., de Vries, A. H. M., de Vries, J. G., Feringa, B. L. *J. Am. Chem. Soc.* 2000, *122*, 11539.

29. de Vries, J. G., de Vries, A. H. M. *Eur. J. Org. Chem.*, 2003, 799.

30. a) de Vries, A.H.M., Meetsma, A., Feringa, B.L. *Angew. Chem., Int. Ed.*, 1996, *35*, 2374; b) Feringa, B.L. *Acc. Chem. Res.* 2002, *33*, 346.

31. van den Berg, M., Minnaard, A. J., Haak, R. M., Leeman, M., Schudde, E. P., Meetsma, A., Feringa, B. L., de Vries, A. H. M., Maljaars, C. E. P., Willans, C. E., Hyett, D., Boogers, J. A. F., Henderick, H. J. W., de Vries, J. G. *Adv. Synth. Catal.* 2003, *345*, 308.

32. van Rooy, A., Burgers, D., Kamer, P.C.J., van Leeuwen, P.W.N.M. *Recl. Trav. Chim. Pays-Bas* 1996, *115*, 492.

33. R. Hulst, N.K. de Vries and B.L. Feringa, *Tetrahedron: Asymmetry* 1994, *5*, 699.

34. It is not recommended to scale this method up to large scale because of shock sensitivity of tetrazole. An alternative catalyst is benzimidazolium triflate. See also reference 31 for more alternatives and references.

35. a) Zeng, Q., Liu, H, Cui, X., Mi, A., Jiang, Y., Li, X., Choi, M. C. K., Chan, A. S. C. *Tetrahedron: Asymmetry* 2002, *13*, 115; b) Zeng, Q., Liu, H., Mi, A., Jiang, Y., Li, X., Choi, M. C. K., Chan, A. S. C. *Tetrahedron* 2002, *58*, 8799.

36. a) Hu, A.-G., Fu, Y, Xie, J.-H., Zhou, H., Wang, L.-X., Zhou, Q.-L. *Angew. Chem., Int. Ed.* 2002, *41*, 2348; b) Fu, Y., Xie, J.-H., Hu, A.-G., Zhou, H., Wang, L.-X., Zhou, Q.-L. *J. Chem. Soc., Chem. Commun.* 2002, 480; c) Zhu, S.-F., Fu, Y., Xie, J.-H., Liu, B., Xing, L., Zhou, Q.-L. *Tetrahedron: Asymmetry* 2003, *14*, 3219.

37. Bayer, A., Murszat, P., Thewalt, U., Rieger, B. *Eur. J. Inorg. Chem.* 2002, 2614.

38. Willans, C. E., Mulders, J. M. C. A., de Vries, J. G., de Vries, A. H.M. *J. Organomet. Chem.*, 2003, *687*, 494.

39. Jia, X., Li, X., Xu, L., Shi, Q., Yao, X., Chan, A. S. C. *J. Org. Chem.* 2003, *68*, 4539.

40. Bernsmann, H., van den Berg, M., Hoen, R., Minnaard, A. J., Mehler, G., Reetz, M. T., de Vries, J. G., Feringa, B. L., *J. Org. Chem.* 2005, *70*, 943.

41. Birman, V. B., Rheingold, A. L., Lam, K.-C. *Tetrahedron: Asymmetry* 1999, *10*, 125.

42. van den Berg, M., Haak, R. M., Minnaard, A.J., de Vries, A. H. M., de Vries, J. G., Feringa, B.L. *Adv. Synth. Catal.* 2002, *344*, 1003.

43. Jia, X., Guo, R., Li, X. Yao, X. Chan, A. S. C. *Tetrahedron Lett.* 2002, *43*, 5541.

44. You, J., Drexler, H.-J., Zhang, S., Fischer, C., Heller, D., *Angew. Chem., Int. Ed.* 2003, *42*, 913.

45. Lubell, W. D., Kitamura, M., Noyori, R. *Tetrahedron: Asymmetry* 1991, *2*, 543.

46. Peña, D., Minnaard, A. J., de Vries, J. G., Feringa, B. L. *J. Am. Chem. Soc.* 2002, *124*, 14552.

47. Peña, D., Minnaard, A.J., de Vries, A.H.M., de Vries, J.G., Feringa, B.L. *Org. Lett.* 2003, 5, 475.

48. Peña, D., Minnaard, A. J., Boogers, J. A. F., de Vries, A. H. M., de Vries J. G., Feringa, B. L. *Org. Biomol. Chem.* 2003, *1*, 1087.

49. Doherty, S., Robins, E. G., Pál, I., Newman, C. R., Hardacre, C., Rooney, D., Mooney, D. A. *Tetrahedron: Asymmetry* 2003, *14*, 1517.

50. Huttenloch, O., Laxman, E., Waldmann, H. *Chem. Eur. J.* 2002, *8*, 4767.

51. Lefort, L., Boogers, J. A. F., de Vries, A. H. M., de Vries, J. G., Feringa, B. L. *Org. Lett.* 2004, *6*, 1733.

52. de Vries, A. H. M., Lefort, L., Boogers, J. A. F., de Vries, J. G., Ager, D. J. *Chimica Oggi* 2005, *23(2)* *Supplement on Chiral Technologies,* 18.

15 Asymmetric Catalytic Hydrogenation Reactions with Ferrocene-Based Diphosphine Ligands

Hans-Ulrich Blaser, Matthias Lotz, and Felix Spindler

CONTENTS

15.1 INTRODUCTION

Among the factors that control the chemical properties of a chiral metal complex (Figure 15.1), the choice of the ligand is usually the most crucial for its catalytic performance. The nature and structure of the ligand significantly influence how the catalyst transforms the substrate(s) to the desired product(s) via a number of well-understood elementary steps such as oxidative addition, insertion, and reductive elimination. During these transformations the ligand must be able to stabilize various oxidation states and coordination geometries of the intermediary complexes. For many enantioselective reactions an impressively large number of chiral ligands is recorded in the

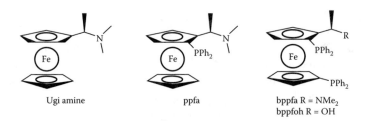

FIGURE 15.1 Design elements for chiral metal complexes and some privileged ligand classes (with selected examples).

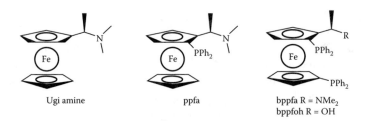

FIGURE 15.2 Starting compound and structures of the first ferrocene-based chiral ligands.

literature affording high enantioselectivities for a variety of catalytic reactions.[1] However, if one has a closer look at the ligands really used by the synthetic organic chemist in academia, and even more so in industry, very few chiral ligands are applied on a regular basis for the synthesis of commercially relevant target molecules. Jacobsen has coined the term "privileged ligands," and some of these are depicted in Figure 15.1.[2]

Ferrocene as a (at the time rather exotic) backbone for diphosphine ligands was introduced by Kumada and Hayashi based on Ugi's pioneering work related to the synthesis of enantiopure ferrocenes (Figure 15.2).[3] Ppfa as well as bppfa and bppfoh proved to be effective ligands for a variety of asymmetric transformations. From this starting point, several ligand families with an assortment of structural variations have been developed in the last few years. In this chapter we will describe effective structures developed over time, the main focus being on diphosphine deriv-atives (Figure 15.3) and their application as hydrogenation catalysts. Two published reviews cover much of the same area but from slightly different points of view. Colacot[4] presented a general overview on ferrocene-based chiral ligands and the application of the corresponding catalysts to all important transformations, whereas Blaser and colleagues[5] reviewed the recent progress in selective hydrogenation. These two papers can serve to put the present account into a broader perspective.

15.2 CATALYTIC TEST RESULTS

When assessing the results reported for new ligands, one has to keep in mind that the quality and relevance differ widely. For most new ligands only experiments with selected model test substrates were carried out under standard conditions (Figure 15.4). The test substrates for C=C functions used most frequently are acetamido cinnamic acid (ACA) (**1**) or its methyl esters (MAC) **2**, methyl

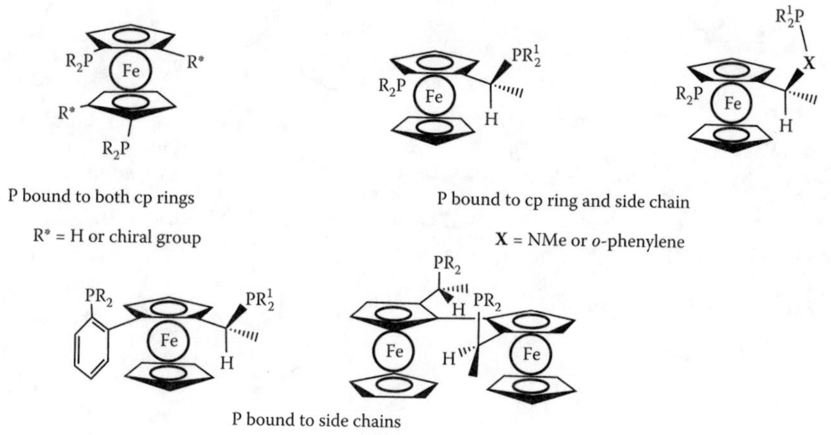

FIGURE 15.3 Subclasses of ferrocenyl diphosphines.

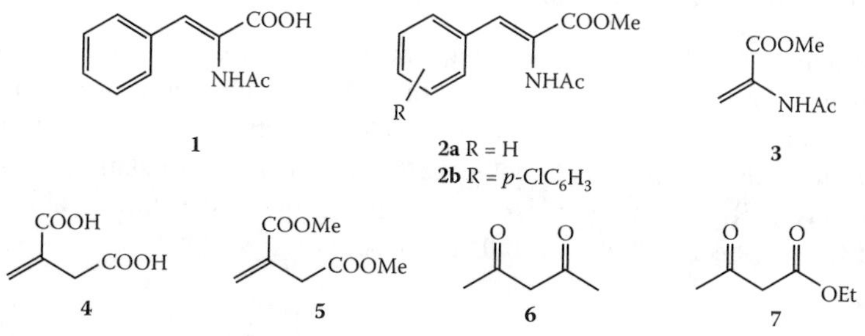

FIGURE 15.4 Structures and numbering of frequently used model test substrates.

acetamido acrylate (MAA) (**3**), itaconic acid (ITA) (**4**), or dimethyl itaconate (DMIT) (**5**), and selected aryl enamides. Test substrates for C=C functions are often acetoacetone (AcAc) (**6**) and alkyl acetyl acetates (AcAcOR) **7** and acetophenone, for the C=N function various derivatives of acetophenone.

Especially for new ligands, reaction conditions are usually optimized for enantioselectivity, whereas catalyst productivity (given as TON, turnover number, or S/C, substrate-to-catalyst ratio) and catalyst activity (given as TOF, turnover frequency, h^{-1}, at high conversion) are often only a first indication of the potential of the ligand. The decisive test, namely the application of a new ligand to "real world problems," which will tell about the scope and limitations of a ligand (family) concerning tolerance to changes in the substrate structure and/or the presence of functional groups, will often come much later. For this reason we will point out which ligands have been successfully applied on a technical scale.

15.3 LIGANDS WITH PHOSPHINE SUBSTITUENTS BOUND TO BOTH CYCLOPENTADIENE RINGS

As mentioned earlier, the first effective ligands were prepared by Kumada and Hayashi in the 1970s starting from Ugi's amine. Depending on the reaction conditions, phosphine substituents were introduced either in one or in both cyclopentadiene rings. It turned out that only the diphosphines bppfa (**8**) and bppfoh (**9**) proved to be useful for hydrogenation reactions. New ligand families have been prepared with excellent enantioselectivities for a variety of hydrogenation reactions (Figure 15.5).

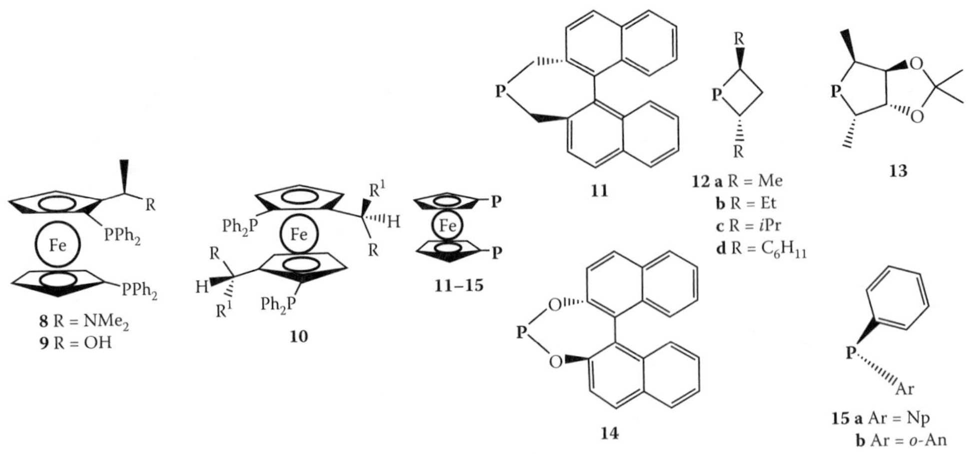

FIGURE 15.5 Structures of ferrocene-based diphosphines with P bound to both cp rings.

15.3.1 Bppfoh and Related Ligands

Bppfoh (**9**) and bppfa (**8**) derivatives have been successfully applied for the Rh-catalyzed hydrogenation of functionalized olefins and ketones (Table 15.1). The nature of auxiliary group has a significant effect on the enantioselectivity and often also on activity and is used to tailor the ligand for a particular substrate. These effects could be the result of electrostatic interactions between substrate and catalyst. Rh-bppfa complexes were among the first catalysts able to hydrogenate tetrasubstituted C = C bonds, albeit with rather low activity.

TABLE 15.1
Selected Results for Rh-Catalyzed Hydrogenations
Using bppf Derivatives

Ligand	Substrate	$p(H_2)$	TON	TOF (h^{-1})	ee (%)	Comments	Ref.
9	19	80	2000	125	97	*c.f.* Scheme 15.2	6
16	19	80	20[a]	<1[a]	99.8	—	6
9	20	50	100[a]	~2[a]	95	—	7
8	1	50	200[a]	n.a.	93	—	8
17	21	50	200[a]	7[a]	98.4	—	9
17	22	50	200[a]	2[a]	97	*cis/trans* 97/3	9
18	23	50	100[a]	2[a]	87	*cis/trans* >99/1	9

[a] Standard test results, not optimized.

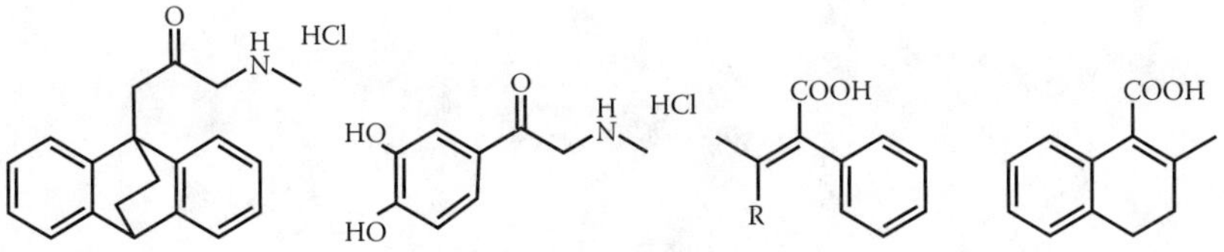

Somewhat surprisingly, the Pd-bppfa complex **24** tethered to a macroporous silica was shown to be moderately active for the enantioselective hydrogenation of ethyl nicotinate (**25**) (a rare example of a homogeneous Pd-catalyzed hydrogenation), but ee's were very low (Scheme 15.1).[10a]

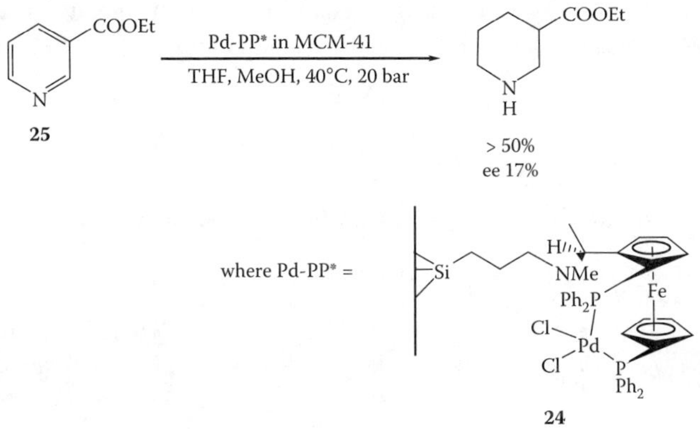

SCHEME 15.1

15.3.1.1   Industrial Applications

Rh-bppfoh complexes (Rh-**9**) have been used successfully for the technical synthesis of pharma intermediates. A pilot process to an intermediate for levoprotiline has been developed by the former Ciba-Geigy/Solvias for the reduction of an α-amino ketone moiety with very good enantioselectivity and acceptable catalyst activity (Scheme 15.2).[6] The final active compound was obtained in >99% ee and with <5 ppm Rh residue after one crystallization. The asymmetric hydrogenation of an unsaturated α-keto acid was carried out on bench scale by Roche with moderate ee and activity (Scheme 15.3).[11]

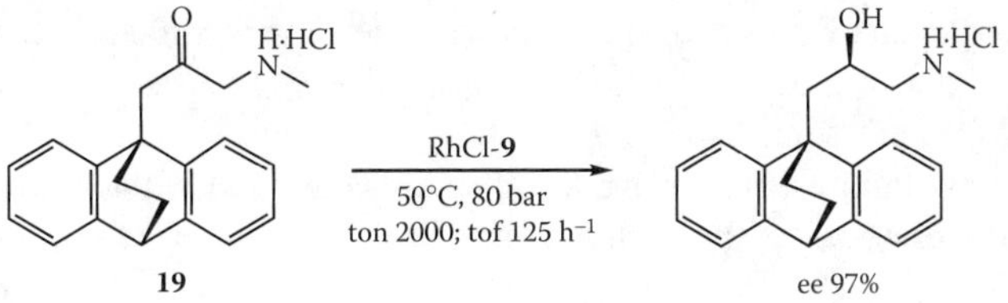

SCHEME 15.2

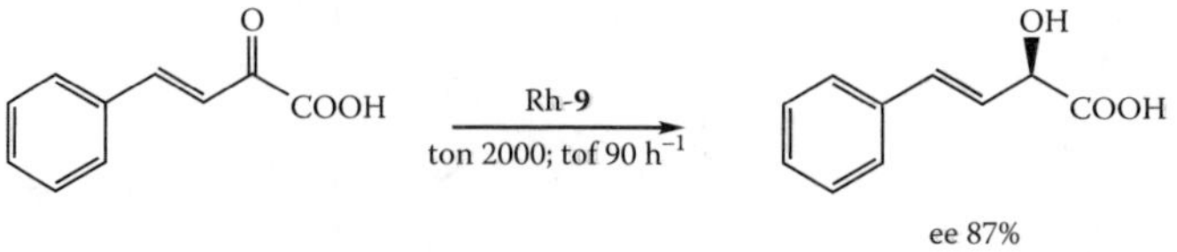

SCHEME 15.3

15.3.2 MANDYPHOS

Mandyphos[12] (**10**) are bidentate analogues of ppfa with C_2 symmetry, where in addition to the phosphine moieties, R and R^1 can also be used for fine-tuning purposes. Although several ligands have been prepared and tested, the scope of this family is not yet fully explored, but preliminary results indicate high enantioselectivities for the Rh-catalyzed hydrogenation of dehydroamino acid derivatives and enol acetate **26** (Table 15.2). Both enantiomers of the mandyphos family are equally well-accessible,[12a] and selected derivatives are now commercialized by Solvias in collaboration with Umicore (formerly OMG).[13]

TABLE 15.2
Selected Results for Rh-Mandyphos-Catalyzed Hydrogenation

Entry	R^1 in 10	Substrate	$p(H_2)$	TON[a]	TOF (h^{-1})[a]	ee (%)	Comments	Ref.
1	Me	**2a**	1	100	≥600	98.6	97.9% ee for **3**	12b
2	NMe_2	**2b**	1	100	n.a.	>99	98% ee for **2a**	12c
3	Me	**26**	1	100	8	95	—	12a

[a] Standard test results, not optimized

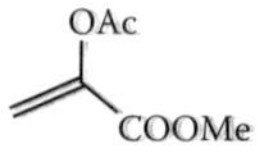

26

15.3.3 MISCELLANEOUS DIPHOSPHINES

A number of C_2 symmetric diphosphine ligands with a ferrocenyl backbone (see Figure 15.5) have been described and tested with sometimes very good results. Interesting examples are f-binaphane (**11**),[14] ferrotane (**12**),[15] the sugar-based phospholane **13**,[16] the ferrocenyl bisphosphonite **14**,[17] and the *P*-chiral phosphine **15**.[18]

Ir–f-binaphane complexes show good to excellent enantioselectivities but modest TONs and low TOFs for the hydrogenation of *N*-aryl imines with the general structure **27** (Table 15.3).[14] The reaction has to be performed in poorly coordinating solvents such as dichloromethane and at a relatively high hydrogen pressure. As with the Ir–Josiphos catalysts, the best ee's are obtained with 2,6-disubstituted *N*-aryl imines (Entries 1 and 2), whereas alkyl ketimines give low enantioselectivities (Entry 3). In some cases, the addition of I_2 has a beneficial effect on enantioselectivity (Entries 4 and 5).

Rh complexes of ferrotanes (**12**) show good performance for amido itaconates (Table 15.4, Entry 1), and Ru-ferrotane catalyzes the hydrogenation of β-diketones with high stereoselectivity

27 a R = Ph   R^1 = Ph
b R = Ph   R^1 = 2,6-xyl
c R = p-MeOC$_6$H$_4$ R^1 = 2,6-xyl
d R = p-CF$_3$C$_6$H$_4$ R^1 = 2,6-xyl
e R = iPr   R^1 = 2,6-xyl
f R = t-Bu   R^1 = 2,6-xyl
g R = 2,6-xyl   R^1 = Ph
h R = Ar   R^1 = P(O)Ph$_2$

TABLE 15.3
Selected Results for Ir–f-Binaphane (Ir-11)-Catalyzed Hydrogenation of Imines 27

Entry	Imine	p(H$_2$)	TON[a]	TOF (h^{-1})[a]	ee (%)	Comments
1	**27b**	70	180	4	>99	—
2	**27c,d**	70	200	4	98–99	—
3	**27e,f**	70	40–80	4	8–23	—
4	**27a**	70	250	6	84	With I$_2$ addition ton 10, ee 89%
5	**27a**	70	250	10	94	With I$_2$, at –5°C

[a] Standard test results, not optimized.

(Entry 2). The sugar-based ligand **13** is excellent for dehydroamino and itaconic acid derivatives with good TONs and very high ee's (Entries 3–5). Bisphosphonite **14** based on a binol or related moiety achieved very high ee's and respectable TONs for the hydrogenation of itaconates catalyzed by cationic Rh complexes (Entry 6). Rh-**15** complexes reduce MAC analogues **2** with ee's >96% (results not shown). In contrast to many other ligands, Rh-**15** catalysts are quite tolerant toward changes in the structure of the amide moiety showing high ee's for the N-methyl benzoyl (R = Ph) derivatives **29** (Entry 7).

28   **29** R = Me or Ph

TABLE 15.4
Selected Results for Rh-Catalyzed Hydrogenations Using Ligands 12–15

Entry	Ligand	Substrate	p(H$_2$)	TON	TOF (h^{-1})	ee (%)	Comments	Ref.
1	**12a**	**28**	5	1000[a]	1–6000[a]	92–99	—	15
2	**12c,d**	β-diketone	70	50[a]	<1[a]	95–98	**6**, de >95%	15
3	**13**	**4**	5	100[a]	<10[a]	99.5	ee 90% for **5**	16
4	**13**	**3**	3	10000	850	99.9	Best solvent THF[b]	16
5	**13**	**2**[c]	1	100[a]	100[a]	>99.9	Best solvent THF	16
6	**14**	**5**	1.3	5400	270	>99.5	—	17
7	**15**	**1,29**	1	200[a]	15[a]	96–98	—	18c,d

[a] Standard test results, no optimized.
[b] THF.
[c] Various substituted analogues were tested.

Josiphos

30 a	R = Ph	R^1 = Ph
b	R = Ph	R^1 = 2,6-xyl
c	R = Ph	R^1 = *p*-$CF_3C_6H_4$
d	R = Ph	R^1 = *i*-Bu
e	R = Ph	R^1 = *c*-C_6H_{11}
f	R = 2,6-xyl	R^1 = 2,6-xyl
g	R = *c*-C_6H_{11}	R^1 = *c*-C_6H_{11}

Taniaphos

31 a	R = H	R^1 = Me	Ar = Ph	Ar^1 = Ph
b	R = H	R^1 = H	Ar = Ph	Ar^1 = Ph
c	R = H	R^1 = Pr-*i*	Ar = Ph	Ar^1 = Ph
d	R = H	R^1 = NMe_2	Ar = Ph	Ar^1 = Ph
e	R = H	R^1 = N(Pr-*i*)$_2$	Ar = Ph	Ar^1 = Ph
f	R = H	R^1 = OMe	Ar = Ph	Ar^1 = Ph
g	R = MeO	R^1 = H	Ar = Ph	Ar^1 = Ph
h	R = H	R^1 = OMe	Ar = Ph	Ar^1 = 2,6-xyl
i	R = MeO	R^1 = H	Ar = Ph	Ar^1 = 2,6-xyl
j	R = H	R^1 = N(Pr-*i*)$_2$	Ar = Ph	Ar^1 = Ph
k	R = H	R^1 = NEt_2	Ar = Ph	Ar^1 = Ph

Bophoz

32

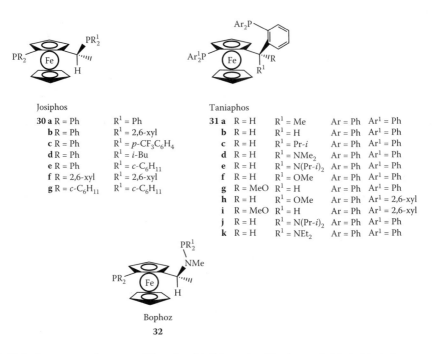

FIGURE 15.6 Structures and names of ligands with P bound to Cp ring and chiral sidechain.

15.4 LIGANDS WITH PHOSPHINE SUBSTITUENTS BOUND TO BOTH CYCLOPENTADIENE RING AND SIDECHAIN

The structures of ligands with phosphine substituents bound both to a cyclopentadiene ring and sidechain are shown in Figure 15.6.

15.4.1 Josiphos Family

The first successful variation of the ppfa structure was carried out by Togni and Spindler replacing the amino group at the stereogenic center of the sidechain by a second phosphino moiety. The resulting Josiphos ligands **30** (that are now owned by Solvias AG) represent arguably the most versatile and successful ferrocenyl ligand family. Because the two phosphine groups are introduced in consecutive steps with very high yields (Scheme 15.4) a variety of ligands are readily available with widely differing steric and electronic properties. The ligands are technically developed; available in commercial quantities from Solvias;[13] and have already been applied in three production processes, several pilot processes (see Figure 15.7) including the process for (*S*)-metolachlor (Scheme 15.5;[19] see also Chapters 5, 12, and 31),[19] and many other syntheses.[2b,20]

30

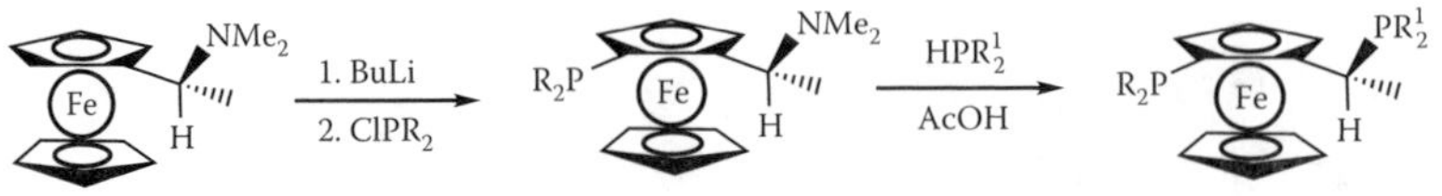

SCHEME 15.4

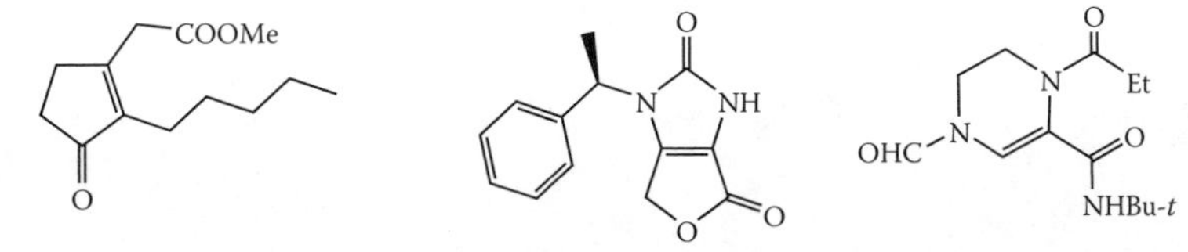

33a
Ru-Josiphos or Duphos; ee 90%
ton 2000; tof 200 h^{-1}
medium scale production
Firmenich

33b
Rh-Josiphos; de 99%
ton 2000; tof n.a.
medium scale production
Lonza

Rh-Josiphos; ee 97%
ton 1000; tof 450 h^{-1}
pilot process, >200 kg
Lonza

FIGURE 15.7 Industrial applications of Josiphos ligands.[2b]

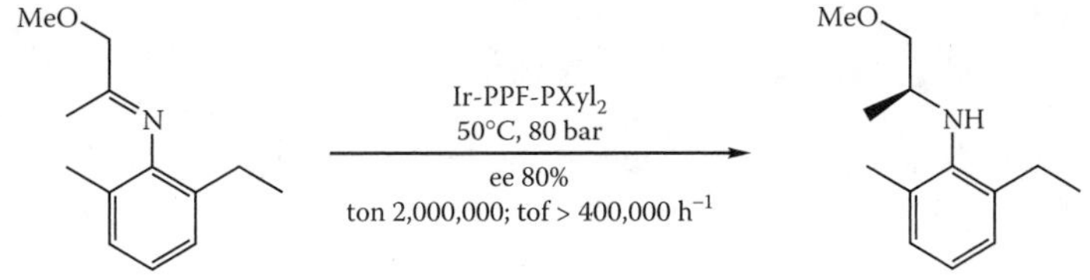

Ir-PPF-PXyl$_2$
50°C, 80 bar

ee 80%
ton 2,000,000; tof > 400,000 h^{-1}

34

SCHEME 15.5

Because a comprehensive review on the catalytic performance of Josiphos ligands has been published,[20] we restrict ourselves to a short overview on the most important fields of applications. Up to now, only the (*R*)-(*S*)-family (and its enantiomers) but not the (*R*)-(*R*) diastereoisomers have led to high enantioselectivities (the first descriptor stands for the stereogenic center, and the second stands for the planar chirality). The most important application is undoubtedly the hydrogenation of C = N functions, where the effects of varying R and R^1 have been extensively studied (for the most pertinent results see Table 15.5, Entries 1–4). Outstanding performances are also observed for tetrasubstituted C = C bonds (Entry 5) and itaconic and dehydroamino acid derivatives (Entries 6 and 7). A rare example of an asymmetric hydrogenation of a heteroaromatic compound **36** with a respectable ee is depicted in Scheme 15.6.[10b]

TABLE 15.5
Selected Results for Rh- and Ir-Catalyzed Hydrogenation Using Josiphos Ligands

Entry	Substrate	M–Ligand	p(H$_2$)	TON	TOF (h^{-1})	ee (%)	Ref.
1	**34**	Ir-**30b**	50	2 Mio	>400,000	80	19
2	**27g**	Ir-**30c**	80	200[a]	n.a.	96	21a
3	**27h**	Rh-**30g**	70	500	500	99	21b
4	**35**	Ir-**30f**	40	250[a]	56[a]	93	21a
5	**33a,b**	Rh-**30d**	—	2000	200	>90	20
6	**3**	Rh-**30e**	1	100[a]	330[a]	97	20
7	**5**	Rh-**30e**	1	100[a]	200[a]	90–99.9	20

[a] Standard test results, not optimized.

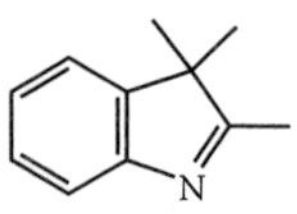

35

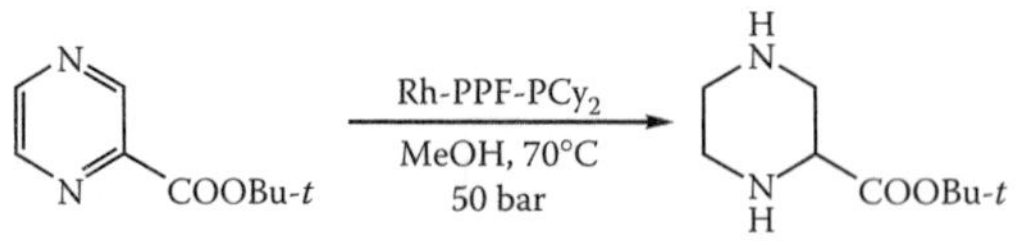

$$\text{Rh-PPF-PCy}_2$$
$$\text{MeOH, 70°C}$$
$$\text{50 bar}$$

36   ee 78%

SCHEME 15.6

15.4.1.1 Immobilized Catalysts

Several Josiphos ligands were functionalized at the lower Cp-ring and grafted to silica gel or a water-soluble group (Figure 15.8)[22a] to give very active catalysts for the Ir-catalyzed MEA imine **34** reduction; a Rh–Josiphos complex grafted to a dendrimer hydrogenated DMIT (**5**) with ee's up to 98.6%.[22b]

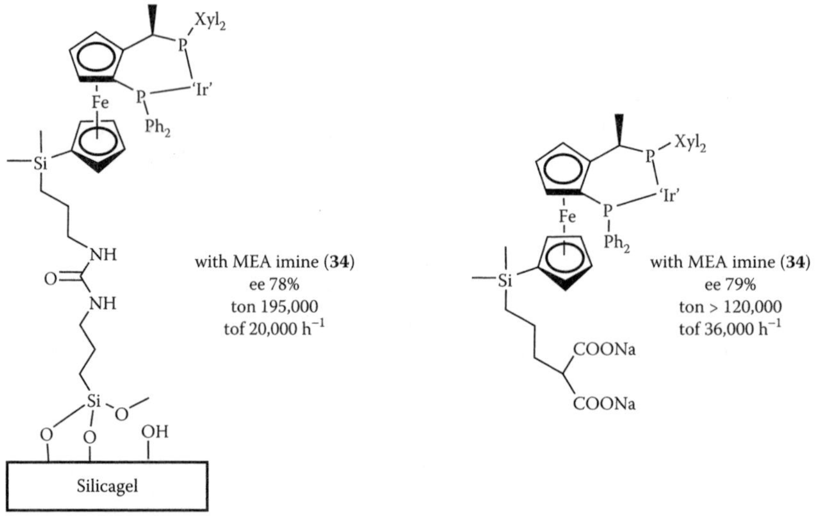

with MEA imine (**34**)
ee 78%
ton 195,000
tof 20,000 h⁻¹

with MEA imine (**34**)
ee 79%
ton > 120,000
tof 36,000 h⁻¹

FIGURE 15.8 Structure of immobilized catalysts; substrate, ee, TON, and TOF.

15.4.2 TANIAPHOS

Compared to the Josiphos ligands, Taniaphos ligands (**31**) have an additional phenyl ring inserted at the sidechain of the Ugi amine.[23] Whereas the effect of changing the two phosphine moieties has not yet been investigated in much detail (Table 15.6, Entries 3, 4, 7, and 9), the nature of the substituent at the stereogenic center has a strong effect on the stereoinduction for the Rh-catalyzed hydrogenation of MAC (**2**) and DMIT (**5**). It is rather surprising that a change of the substituent can even lead to a different sense of induction. For **2**, methyl or methoxy substituents lead to the opposite absolute configuration of the product compared to R = NMe₂, *i*-Pr or H (Entries 1–4). Similar effects are also observed for **5** (Entries 5–8) and for the hydrogenation of enol acetate **26** where ee's up to 99% but low activities are achieved (Entries 9 and 10). It is interesting that changing the absolute configuration of the stereogenic center has only an effect on the level of the ee but not on the sense of induction (compare Entries 3/4 and 7/8). Enamides **37** are hydrogenated with high ee's but low TOFs (Entries 11 and 12). β-Functionalized ketones are reduced using Ru–Taniaphos (R = dialkylamine), but catalytic activities are lower than for the state-of-the-art Ru–binap catalysts. Also for this reaction a dependence of the sense of induction on the nature of the substituent is observed with the bulkier N(*i*-Bu)$_2$ inducing opposite configuration (Entries 13 and 14). Dynamic kinetic resolution of α-substituted β-keto esters and of β-diketones is achieved with de's between 80–99% and ee's of up to >99%, comparable to Ru-binap catalysts (for an example see Scheme 15.7). Several Taniaphos ligands are being marketed by Solvias in collaboration with Umicore (formerly OMG).[13]

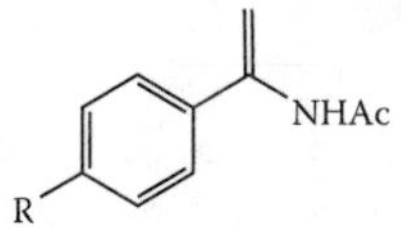

37 a R = H
b R = Br
c R = Me

TABLE 15.6
Selected Results for Rh- and Ru-Catalyzed Hydrogenation Using Taniaphos (31)

Entry	M - Ligand	Substrate	p(H$_2$)	TON[a]	TOF (h^{-1})[a]	ee (%)	Comments	Ref.
1	Rh-**31c**	**2a**	1	100	25	97 (R)	52% (S) for **31a**!	23a
2	Rh-**31d**	**2a**	1	100	200	95 (R)	77% (R) for **31b**	23a
3	Rh-**31f**	**2a**	1	100	50	94 (S)	92% (S) for **31h**	23b
4	Rh-**31g**	**2a**	1	100	67	99 (S)	99% (S) for **31i**	23b
5	Rh-**31c**	**5**	1	100	25	98 (S)	19% (R) for **31a**!	23a
6	Rh-**31d**	**5**	1	100	7	91 (S)	75% (S) for **31b**[b]	23a
7	Rh-**31g**	**5**	1	100	200	98 (R)	90% (R) for **31i**	23b
8	Rh-**31f**	**5**	1	100	40	95 (R)	—	23b
9	Rh-**31g**	**26**	1	100	5	98 (S)	99% (S) for **31i**	23b
10	Rh-**31f**	**26**	10	100	5	80 (S)	low conversion at 1 bar	23b
11	Rh-**31g**	**37a**	1	100	7	96	97% ee for **37b**	23b
12	Rh-**31i**	**37a**	1	100	67	92	95% ee for **37c**	23b
13	Ru-**31e**	**7** R = Et	100	200	25	98.6 (S)	70% (R) for **37j**!	23a
14	Ru-**31d**	**7** R = Et	100	200[c]	25	96 (R)	85% (R) for **31k**	23a

[a] Standard test results, not optimized.
[b] At 10 bar, low conversion at 1 bar.
[c] At S/C 5000 ee 93.2%, TOF 80 h^{-1}.

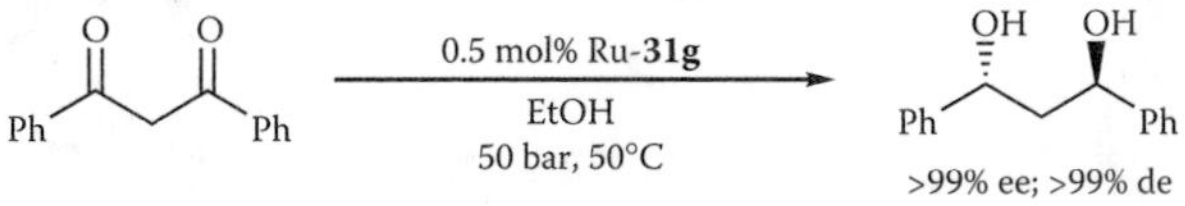

SCHEME 15.7

15.4.3 Bophoz

Bophoz (**32**) is a combination of a phosphine and an aminophosphine and is prepared in 4 steps from ppfa with high overall yields.[24] The ligand is air-stable and effective for the hydrogenation of enamides, itaconates, and α-keto acid derivatives (Table 15.7). As observed for several ligands

TABLE 15.7
Selected Results for Rh-Catalyzed Hydrogenation Using Bophoz

Ligand	Substrate	$p(H_2)$	TON[a]	TOF (h^{-1})[a]	ee (%)	Comments
32a	2a	1	100	≥20	99.1	ee 98.5% for **3**
32a	4	1	100	≥20	97	ee 94% for **5**
32b	38	20	100	≥15	97	—
32b	39	20	100	≥15	90–92	—

[a] Standard test results, not optimized.

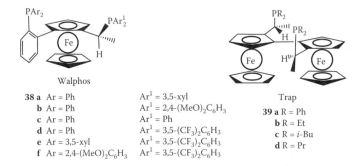

38

39 a R = Me
b R = PhCH₂CH₂

forming 7-membered chelates, high activities can be reached and TONs up to 10,000 have been claimed. The full scope of this modular ligand class has not yet been explored, and depending on R and R¹, the stability of the N-PR₂ bond might be a critical issue.

15.5 LIGANDS WITH PHOSPHINE SUBSTITUENTS BOUND ONLY TO SIDECHAINS

The structures of the class of compounds ligands with phosphine substituents bound only to sidechains are shown in Figure 15.9.

FIGURE 15.9 Structures and names of ferrocene-based diphosphines with P bound only to chiral sidechains.

15.5.1 WALPHOS FAMILY

The starting point for walphos (**38**) is also the Ugi amine. Like Josiphos, Walphos is modular but forms 8-membered metallocycles as a result of the additional phenyl ring attached to the cyclopentadiene ring.[25] It shows promise for the enantioselective hydrogenation of dehydroamino and itaconic acid derivatives (Table 15.8, Entries 1 and 2) and β-functionalized ketones (Entries 3 and 4) with noticeable electronic effects, but its scope is still under investigation. Several members of this ligand class are available on a technical scale.[13] The first synthetic application for the hydrogenation of SPP100-SyA, a sterically demanding α,β-unsaturated acid intermediate of the renin inhibitor SPP100, has just been realized in collaboration with Speedel/Novartis (Scheme 15.8).[26]

TABLE 15.8
Selected Results for Rh- and Ru-Catalyzed Hydrogenation Using the Walphos Ligand (38)[25]

Entry	M - (Ar, Ar[1])[a])	Substrate	p(H_2)	TON	TOF (h^{-1})	ee (%)	Comments
1	Rh-**38a–d**	**2a**	1	200[a]	≥10[a]	95	ee 94% for **38c**
2	Rh-**38a–d**	**5**	1	200[a]	≥10[a]	92	—
3	Ru-**38a–f**	**40**	5	1000	≥60[a]	95	ee 85% for **38c**
4	Ru-**38a–f**	**6**[b]	20	1000	≥60[a]	96	ee 97% for **38c**[c]

[a] Standard test results, not optimized.
[b] *dl*: meso: >99:1 for resulting diol.
[c] At 80 bar.

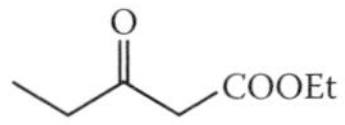
O
COOEt

40

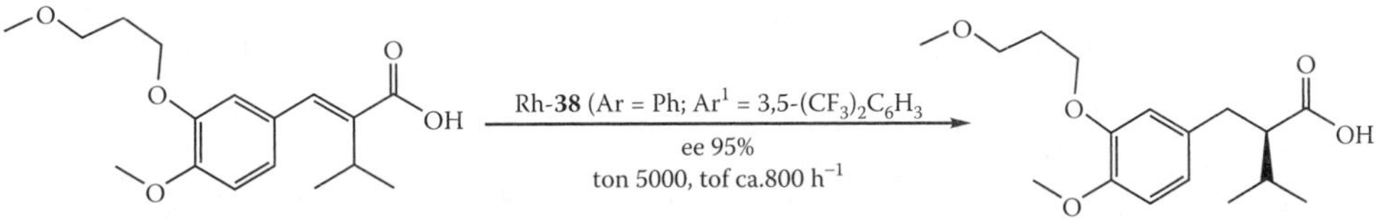
O
O
O
O
OH
Rh-38 (Ar = Ph; Ar¹ = 3,5-(CF₃)₂C₆H₃
ee 95%
ton 5000, tof ca.800 h⁻¹
O
O
O
O
OH

SCHEME 15.8

15.5.2 Trap (*Trans*-Chelating) Ligands

The trap ligands (**34**) developed by Kuwano and colleagues[27] form 9-membered metallocycles for which *trans*-chelation is possible. However, it is not clear whether the *cis*-isomer that has been shown to be present in small amounts or the major *trans*-isomer is responsible for the catalytic activity. Up to now only a few different PR_2 fragments have been tested, but it is clear that the choice of R strongly affects the level of enantioselectivity and sometimes even the sense of induction (e.g., see Table 15.9, Entries 3 and 4). The Rh complexes work best at very low pressures of 0.5–1 bar but often need elevated temperatures. Effectively reduced are indole-derivatives **41**[27a] (Entries 1 and 2; the first example of a heteroaromatic substrate with high ee's; note the effect of Cs_2CO_3) and dehydroamino[27b] (Entries 3 and 4; best ligand PEt_2-trap for **3**; unusual P and T effects are seen) and itaconic[27c] acid derivatives (Entry 5). β-Hydroxy-α-amino acids and α,β-diamino acids can be prepared by asymmetric hydrogenation of tetrasubstituted alkenes **43** and **44**[27e,f] with respectable de's of 99–100% and ee's of 97% and 82%, respectively, but low catalyst activities (entry 6). Also described was the hydrogenation of an indinavir intermediate **44**[27d] (Entry 7).

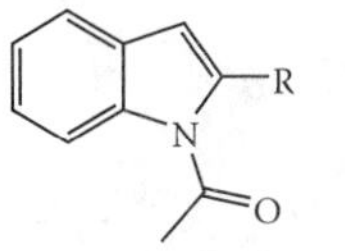
R
N
O

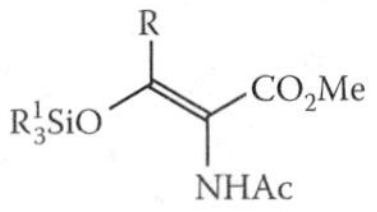
R
CO₂Me
R¹₃SiO
NHAc

41 a R = M
b R = COOEt

42

43 a X = OOCtBu
   b X = NHCbz

44

TABLE 15.9
Selected Results for Rh-Catalyzed Hydrogenation Using Trap Ligands (39)

Entry	Ligand (R)	Substrate	p(H$_2$)	TON[a]	TOF (h^{-1})[a]	ee (%)	Comments
1	Rh-**39a**[b]	**41a**	50	100	50[c]	94 (R)	ee 7% (S) without Cs2CO3
2	Rh-**39a**[b]	**41b**	50	100	200	95	ee 78% for Boc derivative
3	Rh-**39b**[d]	**3**	0.5	100	50[c]	96 (R)	ee 70/2%(!) at 1/100 bar
4	Rh-**39c**	**2a**	1	100	4	92 (S)	R = Et[c] ee 77% (R)!
5	Rh-**39b**[e]	**4**	1	200	30	96	ee 68% for **5**
6	Rh-**39d**	**42,43**	1	100	4	97	ee 82% for **43b**
7	Rh-**39c**	**44**	1	100	5	97	Intermediate for indinavir

[a] Standard test results, not optimized.
[b] In presence of Cs$_2$CO$_3$.
[c] At 60°C.
[d] For R = Pr/Ph/i-Pr, ee = 85%(R)/21%(S)/5%(S), respectively.
[e] For R = i-Bu, ee = 17%.

15.6 SUMMARY OF THE MAJOR CATALYTIC APPLICATIONS OF FERROCENE-BASED CATALYSTS

As can be seen in the preceding chapters, ferrocene-based complexes are very versatile ligands for the enantioselective hydrogenation of a variety of functional groups. One reason for this is undoubtedly the modularity of most of the described ligand families that allows influence on the activity and enantioselectivity in an extraordinarily broad range. In the following, we give a short overview for substrates where ferrocene-based ligands are defining the state of the art. A comparison with other classes of ligands can be found in the already mentioned review (see also Chapter 12).[5]

15.6.1 HYDROGENATION OF SUBSTITUTED OLEFINS

Rh complexes of ferrocene-based ligands are very effective for the hydrogenation of several types of dehydroamino (**2,3,29,41,42,44**) and itaconic acid derivatives (**4,5,28**) as well as for enamide **45**, enol acetate **26**, and a tetrasubstituted C = C-COOH **21**. Of particular interest are substrates that have unusual substituents (**41,42,44**) at the C = C moiety or are more sterically hindered than the usual model compounds (**21,42**). Table 15.10 lists typical examples with very high ee's and often respectable TONs and TOFs. Several industrial applications have already been reported using Rh–Josiphos and Ru-Josiphos (see Figure 15.7) as well as Rh–Walphos (Scheme 15.8).

45

TABLE 15.10
Best New Catalysts for the Hydrogenation
of Selected Functionalized Olefins

Substrate	Metal–Ligand	TON	TOF (h^{-1})	ee (%)
2,3	Rh–**10,32**	100[a]	8–600[a]	98–>99
42	Rh–**39**	100[a]	4[a]	97
44	Rh–**39**	100[a]	5[a]	97
41	Ru–**31**, Rh–**39**	100–200[a]	25–200[a]	94–95.5
4,5	Rh–**30**	100[a]	200[a]	97–99.9
28	Rh–**12**	1000[a]	1–6000[a]	92–99
29	Rh–**15**	200[a]	15[a]	96–98
45	Rh–**14**	1000	6000	97–98
26	Rh–**10**	100[a]	8[a]	95
21	Rh–**8**	200[a]	7[a]	98

[a] Standard test results, not optimized

15.6.2 HYDROGENATION OF C=O AND C=N FUNCTIONS

Compared to some other ligand classes, ferrocene-based complexes have a relatively limited potential for the enantioselective reduction of ketones.[5] Rh complexes of bppfa (**8**), Bophoz (**32**), and josiphos (**30**) are among the most effective catalysts for the hydrogenation of α-functionalized ketones (Table 15.11; **46**,**47**; see also Scheme 15.2). Ru complexes of walphos (**38**) and ferrotane (**12**) are quite effective for β-keto esters and diketones **6** and **7**, usually the domain of Ru–binap type catalysts. Josiphos and f-binaphane, however, are the ligands of choice for the Ir-catalyzed hydrogenation of *N*-aryl imines. Special mention should be made of the Ir–Josiphos catalyst system, which is able to hydrogenate MEA imine **34** with TONs up to 2 million (Scheme 15.5).

46   **47**

TABLE 15.11
Best Catalysts for the Hydrogenation of C=O and C=N Functions

Substrate	Metal–Ligand (Modifier)	TON	TOF (h^{-1})	ee (%)
27g	Ir–**30c** (I$^-$,H$^+$)	200[a]	n.a.	96
27b–e	Ir–**11**	100[a]	2[a]	>99
27h	Rh–**30g**	500	500	99
35	Ir–**30f** (I$^-$,H$^+$)	250[a]	56[a]	93
46	Rh–**9,16**	200–2000	2–125	95–>99
47	Rh–**30,32**	100–200[a]	1[a]–>1000	97–99
6,7	Ru–**12,38**	5–200[a]	<1–25[a]	95–99
34	Ir–**30b** (I$^-$,H$^+$)	2,000,000	>400,000	80

[a] Standard test results, not optimized.

REFERENCES

1. a) Brunner, H., Zettlmeier, W. In *Handbook of Enantioselective Catalysis,* VCH: Weinheim, 1993; b) Brown J. M. In *Comprehensive Asymmetric Catalysis,* Jacobsen, E. N., Pfaltz, A., Yamamoto, H., Eds. Springer: Berlin, 1999, p. 121; c) Ohkuma, T., Kitamura, M., Noyori R. In *Catalytic Asymmetric Synthesis,* 2nd Ed., Ojima, I., Ed., Wiley-VCH: New York, 2000, p. 1.

2. a) Blaser, H. U., Pugin, B., Spindler F. In *Applied Homogeneous Catalysis by Organometallic Complexes,* 2nd Ed., Cornils, B., Herrmann W. A., Eds., Wiley-VCH: Weinheim, 2002, p. 1131; b) Blaser, H. U., Spindler, F., Studer, M. *Applied Catal. A: General* 2001, *221*, 119; c) Yoon, T. P., Jacobsen, E.N. *Science*, 2003, *299*, 1691.

3. For an account see Hayashi T. In *Ferrocenes,* Togni, A., Hayashi T., Eds., VCH: Weinheim, 1995, p. 105.

4. Colacot, T. J. *Chem. Rev.* 2003, *103*, 3101.

5. Blaser, H. U., Malan, Ch., Pugin, B., Spindler, F., Steiner, H., Studer, M. *Adv. Synth. Catal.* 2003, *345*, 103.

6. Blaser, H. U., Gamboni, R., Rihs, G., Sedelmeier, G., Schaub, E. Schmidt, E., Schmitz, B., Spindler, F., Wetter Hj. In *Process Chemistry in the Pharmaceutical Industry,* Gadamasetti, K. G., Ed., Marcel Dekker, Inc: New York, 1999, p. 189.

7. Hayashi, T., Kanehira, K., Kumada, M. *Tetrahedron Lett.* 1979, 425.

8. Hayashi, T., Mise, T., Mitachi, S., Yamamoto, K., Kumada, M. *Tetrahedron Lett.* 1976, 1133.

9. Hayashi, T., Kawamura, N., Ito, Y. *J. Am. Chem. Soc.* 1987, *109*, 7876, Hayashi, T., Kawamura, N., Ito, Y. *Tetrahedron Lett.* 1988, *29*, 5969.

10. a) Raynor, S. A., Thomas, J. M., Raja, R., Johnson, B. F. G., Bell, R. G., Mantle, M. D. *J. Chem. Soc., Chem. Commun.* 2000, 1925; b) Fuchs, R. Eur. Patent 1997, 0803502 A2.

11. Schmid, R., Broger, E. A. In *Proceedings of the Chiral Europe '94 Symposium,* Spring Innovations, Stockport, U.K., 1994, p. 79.

12. a) Lotz, M., Ireland, T., Almena Perea, J., Knochel, P. *Tetrahedron: Asymmetry* 1999, *10*, 1839; b) Almena Perea, J., Börner, A., Knochel, P. *Tetrahedron Lett.* 1998, *39*, 8073; c) Almena Perea, J., Lotz, M., Knochel, P. *Tetrahedron: Asymmetry* 1999, *10*, 375.

13. For more information see www.solvias.com/ligands, Thommen, M., Blaser, H. U. *PharmaChem*, July/August, 2002, 33.

14. Xiao, D., Zhang, X. *Angew. Chem.* 2001, *113*, 3533.

15. a) Berens, U., Burk, M. J., Gerlach, A., Hems, W. *Angew. Chem.* 2000, *112*, 2057; b) Marinetti, A., Genet, J-P., Jus, S., Blanc, D., Ratovelamanana-Vidal, V. *Chem. Eur. J.* 1999, *5*, 1160, Marinetti, A., Carmichael, D. *Chem. Rev.* 2002, *102*, 201.

16. Liu, D., Li, W., Zhang, X. *Org. Lett.* 2002, *4*, 4471.

17. Reetz, M. T., Gosberg, A., Goddard, R., Kyung, S.-H. *J. Chem. Soc., Chem. Commun.* 1998, 2077, Reetz, M. T., Gosberg, A. PCT Patent, 2000, WO 0014096.

18. a) For overviews see Ohff, M., Holz, J., Quirmbach, M., Börner, A. *Synthesis*, 1998, 1391, Maienza, F., Spindler, F., Thommen, M., Pugin, B., Mezzetti, A. *Chimia* 2001, *55*, 694; b) Maienza, F. *Thesis ETH Zürich*, 2001; c) Stoop, R. M., Mezzetti, A., Spindler, F. *Organometallics* 1998, *17*, 668; d) Maienza, F., Wörle, M., Steffanut, P., Mezzetti, A., Spindler, F. *Organometallics* 1999, *18*, 1041.

19. Blaser, H. U., Buser, H. P., Coers, K., Hanreich, R., Jalett, H. P., Jelsch, E., Pugin, B., Schneider, H. D., Spindler, F., Wegmann, A. *Chimia* 1999, *53*, 275.

20. For an overview see Blaser, H. U., Brieden, W., Pugin, B., Spindler, F., Studer, M., Togni, A. *Topics in Catalysis* 2002, *19*, 3.

21. a) Blaser, H. U., Buser, H. P., Häusel, R., Jalett, H. P., Spindler, F. *J. Organomet. Chem.* 2001, *621*, 34; b) Spindler, F., Blaser, H. U. *Adv. Synth. Catal.* 2001, *343*, 68.

22. a) Pugin, B., Landert, H., Spindler, F., Blaser, H.-U. *Adv. Synth. Catal.* 2002, *344*, 974; b) Köllner, C., Pugin, B., Togni, A. *J. Am. Chem. Soc.* 1998, *120*, 10274.

23. a) Ireland, T., Tappe, K., Grossheimann, G., Knochel, P. *Chem. Eur. J.* 2002, *8*, 843; b) Lotz, M., Polborn, K., Knochel, P. *Angew. Chem., Int. Ed.* 2002, *41*, 4708.

24. Boaz, N. W., Debenham, S. D., Mackenzie, E. B., Large, S. E. *Org. Lett.* 2002, *4*, 2421.

25. a) Sturm, T., Xiao, L., Weissensteiner, W. *Chimia* 2001, *55*, 688; b) Weissensteiner, W., Sturm, T., Spindler, F. *Adv. Synth. Catal.* 2003, *345*, 160.

26. Herold, P., Stutz, S., Sturm, T., Weissensteiner, W., Spindler, F. PCT Patent 2002, WO 02/02500.

27. a) Kuwano, R., Sato, K., Kurokawa, T., Karube, D., Ito, Y. *J. Am. Chem. Soc.* 2000, *122*, 7614; b) Kuwano, R., Sawamura, M., Ito, Y. *Bull. Chem. Soc. Jpn.* 2000, *73*, 2571; c) Kuwano, R., Sawamura, M., Ito, Y. *Tetrahedron: Asymmetry* 1995, *6*, 2521; d) Kuwano, R., Ito, Y. *J. Org. Chem.* 1999, *64*, 1232; e) Kuwano, R., Okuda, S., Ito, Y. *J. Org. Chem.* 1998, *63*, 3499; f) Kuwano, R., Okuda, S., Ito, Y. *Tetrahedron: Asymmetry* 1998, *9*, 2773.

16 Asymmetric Reduction of Prochiral Ketones Catalyzed by Oxazaborolidines

Michel Bulliard

CONTENTS

16.1 INTRODUCTION

A number of the new drugs under development are chiral. The two main reasons for this choice by pharmaceutical companies are the regulatory constraints to assess the properties of single enantiomers, even if the racemic mixture is developed, and the scientific evidence that enantiomers often have very different and, sometimes, opposite pharmacologic effects. Among the numerous techniques available today to industrial chemists, asymmetric synthesis has been used successfully to obtain chiral compounds. From an industrial point of view, asymmetric catalysis is becoming the preferred approach because of its low environmental impact and high potential productivity.

In particular, reduction of unsymmetric ketones to alcohols has become one of the more useful reactions. To achieve the selective preparation of one enantiomer of the alcohol, chemists first modified the classical reagents with optically active ligands; this led to modified hydrides. The second method consisted of reaction of the ketone with a classical reducing agent in the presence of a chiral catalyst. The aim of this chapter is to highlight one of the best practical methods that could be used on an industrial scale: the oxazaborolidine catalyzed reduction.[1-4] This chapter gives an introductory overview of oxazaborolidine reductions and covers those of proline derivatives in-depth. For the oxazaborolidine derivatives of 1-amino-2-indanol for ketone reductions see Chapter 17.

16.2   THE STOICHIOMETRIC REACTIONS

The first solution to the problem of reducing a prochiral ketone was to use a conventional reducing agent modified with stoichiometric amounts of chiral ligand. Among the reagents that illustrate this approach, one has to mention first those prepared from lithium aluminum hydride (LAH). Weigel at Eli Lilly[5] demonstrated the scope and limitation of the Yamagushi-Mosher reagent prepared from LAH and Chirald® [(2S,3R)-(+)4-dimethylamino-1,2-diphenyl-3-methyl-2-butanol] or its enantiomer, ent-Chirald®, in the course of the synthesis of an analogue of the antidepressant fluoxetine (Prozac®).

Another method has been described by Noyori. Optically active binaphthol reacts with LAH to afford BINAL-H (1).[6] This reagent has proved to be an extremely efficient hydride donor achieving very high enantiomeric excesses at low temperature.

The Yamaguchi-Mosher reagent and BINAL-H (1) have been most used over the last few years. Although these reagents have proved to be quite efficient, their manipulation is not easy and requires very low temperatures, sometimes below –78°C, to secure good stereoselectivity.

Another excellent way to produce enantiomerically pure secondary alcohols uses chiral organoboranes. Following the work pioneered by Brown, many reagents have been developed, such as β-3-pinanyl-9-borabicyclo-[3.3.1]-nonane (Alpine-Borane) (2) and chlorodiisopinocampheylborane (Ipc$_2$BCl) (3).[7] The latter is now commercially available at scale and is a useful reagent to reach high chemical and optical yields. A good example of its efficacy has been reported in the preparation of the antipsychotic BMS 181100 from Bristol-Myers.

Many reagents such as glucoride, other modified borohydrides, or Masamune's borolanes have also been tried successfully.[8]

16.3   THE CATALYTIC APPROACH

All of the previously mentioned reagents, despite their effectiveness, have an important drawback. They are used in stoichiometric amounts and, accordingly, are often costly. Their use involves separation and purification steps that can be troublesome. Very often, efficient recycling of the ligand cannot be achieved. Both researchers and industrial chemists have designed catalytic reagents to overcome this limitation.

The most well-known technology is probably homogeneous catalytic hydrogenation. Numerous articles on asymmetric hydrogenation have appeared over the last few years, and a whole chapter of this book refers to this technology. It constitutes one of the best techniques available nowadays, although it is sometimes more elegant than useful as a result of its technical or commercial drawbacks. These processes can be particularly difficult to develop on an industrial scale and require a long chemical development time that hampers their use for the rapid preparation of quantities for clinical trials. Although enzymes and yeasts have been shown to be quite useful on a laboratory scale, they suffer the one major disadvantage that limits their usefulness on a large scale as a result of the high dilution required and troublesome workups. Moreover, enzymes can lack broad application, and the outcome of an enzymatic resolution is often difficult to predict.

Apart from these techniques, the most interesting method recently described is the Itsuno-Corey reduction. This catalytic reaction is general, is highly predictable, and can rapidly be scaled up.

16.3.1 THE CATALYST STRUCTURE AND THE EFFECT ON ENANTIOMERIC EXCESS

Building on the excellent work of Itsuno,[9,10] who first described the use of oxazaborolidine as a chiral ligand, and of Kraatz,[11] Corey was the first to report the enantioselective reduction of ketones to chiral secondary alcohols in the presence of an oxazaborolidine in substoichiometric amounts.[12,13] This general method was named the CBS method (Scheme 16.1).

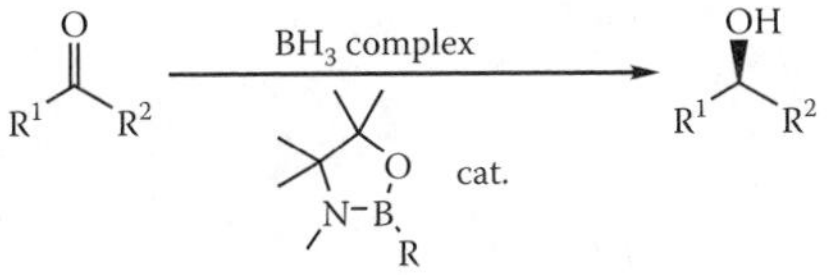

SCHEME 16.1

Since that time, numerous articles dealing with this versatile reagent have appeared in the literature. Many new catalysts have been designed, usually by replacement of the diphenylprolinol moiety with other derivatives of various natural or unnatural amino acids. A large number of new oxazaborolidines have been reported. The enantiomeric excess varies substantially from one amino alcohol to another. Generally, the carbon adjacent to the nitrogen bears the chirality simply because amino alcohols are easily prepared from chiral amino acids belonging to the chiral pool. There are only a few examples of catalysts in which the carbon adjacent to the oxygen is the stereogenic center.

In the case of acetophenone reduction, it appears that amino alcohols that are sterically hindered at the carbon adjacent to the alcohol lead to much better results. A dramatic effect has been found in the case of prolinol and diphenylprolinol: the enantiomeric excess increases from 50% to 97%. Cyclic amino alcohols where the nitrogen is in a 4- or 5-membered ring have exceptional catalytic properties and lead to very good enantiomeric excesses (Tables 16.1 and 16.2). Diphenylprolinol (**4**) is a very good choice because of its availability and performance.

Diphenyl oxazaborolidine, reported by Quallich at Pfizer, is possibly a good alternative because it has the advantage that both enantiomers of the *erythro* aminodiphenylethanol are inexpensive and commercially available.[27] Its performance, however, is somewhat lower than diphenylprolinol's. On the industrial side, the discovery of this technique prompted many groups to investigate its versatility. A group of chemists at Merck has investigated the use of oxazaborolidines to prepare various optically pure pharmaceutical intermediates. In parallel with Corey's work, they studied the mechanistic aspects of this reaction and clearly demonstrated the pivotal role of the borane adduct CBS-B.[29,30] Their research culminated in the isolation of this stable complex described as a "free-flowing crystalline solid." They were the first to report its single-crystal X-ray structure. (For other references on the mechanism of the reaction, see references 31 and 32 and Chapter 17.)

16.3.2 CATALYST PREPARATION

Two different approaches can be followed to prepare and use the catalyst. The first is to prepare it *in situ* by mixing (*R*)- or (*S*)-diphenylprolinol (DPP) (**4**) and a borane complex (Scheme 16.2). This route is advantageous because there is no need to use boronic acids (or boroxines) and to remove water to form the catalyst. Another possible way is to use preformed catalysts, some of which are commercially available from suppliers such as Callery.

Diphenylprolinol (**4**) itself is now commercially available at scale, or it can be prepared several ways (Scheme 16.2): Direct addition of an aryl Grignard reagent to a proline ester leads to the diarylprolinol with a low yield in the range of 20–25%.[33] A more efficient route is based on an

TABLE 16.1
Effect of Catalyst Structure in the Reduction of Acetophenone

	ee (%)	Reference		ee (%)	Reference
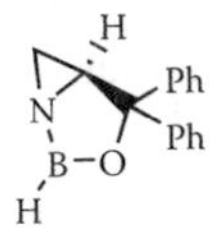	94	14	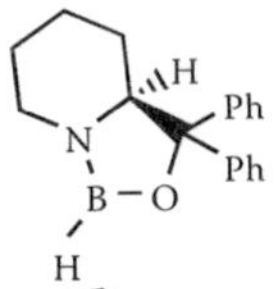	87	15
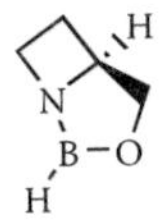	0	16	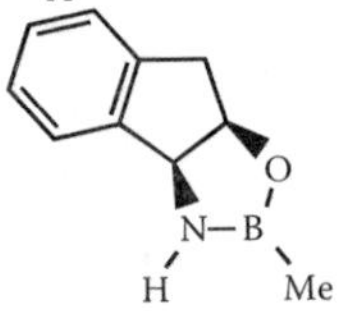	86 87 (1 eq.)	17 18
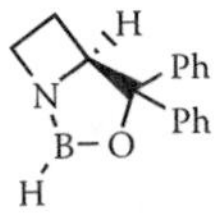	95–98	19,20	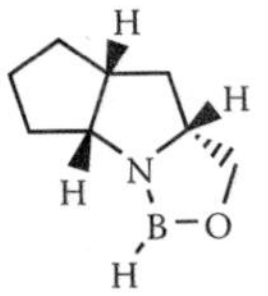	33	
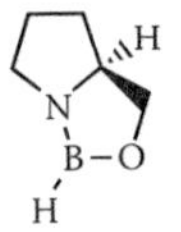	48	16	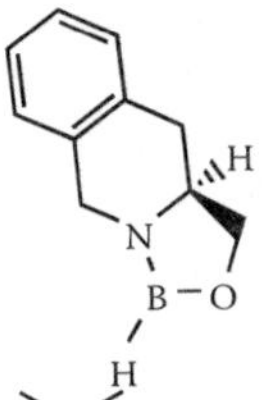	71	
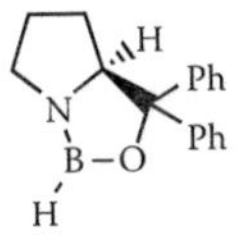	97	12	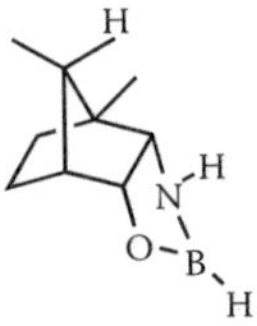	93	21
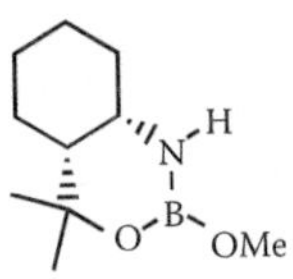	99	22	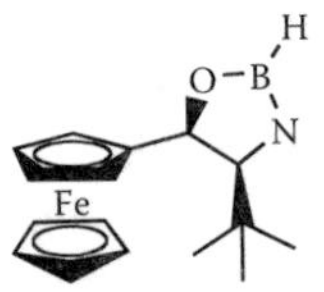	90	23

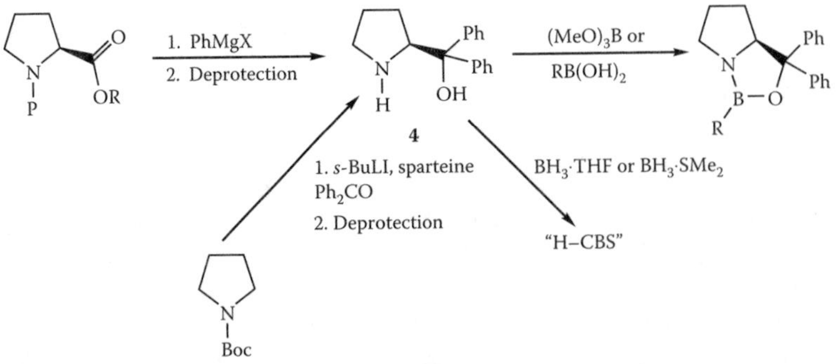

SCHEME 16.2

TABLE 16.2
Effect of Catalyst Structure in the Reduction of Acetophenone

(structure)	ee (%)	Reference	(structure)	ee (%)	Reference
MeS-substituted oxazaborolidine	79	24	oxazaborolidine	84	25
isopropyl oxazaborolidine	94 (1 eq.)	9	Ph-substituted oxazaborolidine	94	25
cyclopropyl oxazaborolidine	80	26	Me-B oxazaborolidine	92	27
tert-butyl oxazaborolidine	89	20	Ph,Ph oxazaborolidine	96	28

earlier publication from Corey. Suitably protected proline is reacted with phenylmagnesium halide, and then, after deprotection, diphenylprolinol is obtained.[13,34,35] An alternative 2-step enantioselective synthesis of diphenylprolinol from proline, based on the addition of proline-N-carboxyanhydride to 3 equivalents of phenylmagnesium chloride, has been reported by the Merck group.[36,37] An elegant asymmetric preparation of **4** has been reported by Beak.[38] The enantioselective deprotonation of Boc-pyrrolidine is achieved with *sec*-butyllithium in the presence of sparteine as chiral inducer. Subsequent quench with benzophenone gives N-Boc-DPP in 70% yield and 99% ee after one recrystallization. To cope with the rather high price of (*R*)-proline, Corey developed an alternative route to this ligand from racemic pyroglutamic acid.[39]

The oxazaborolidine is finally made by heating diphenylprolinol (**4**) under reflux with a suitable alkyl(aryl)boronic acid or, better, with the corresponding boroxine in toluene in the presence of molecular sieves—the water can also be removed by azeotropic distillation. According to the literature, methyl-oxazaborolidine (Me-CBS) can be either distilled or recrystallized. The key point is that the catalyst must be free from any trace of water or alkyl(aryl)boronic acid because those impurities decrease enantioselection.

In the case of H-CBS, this catalyst could be prepared by mixing diphenylprolinol with borane–THF (tetrahydrofuran) or borane dimethylsulfide. Despite numerous efforts in many groups to isolate or even characterize H-CBS by nuclear magnetic resonance spectroscopy, all attempts have been unsuccessful. H-CBS is used *in situ*, and good results can be obtained in many cases.

16.3.3 REDUCING AGENT

Borane complexes, borane–THF and borane dimethylsulfide, are generally the appropriate reducing agents in this reaction. Borane thioxane has been used, but it suffers from the inconvenience that thioxane is quite expensive. Diborane gas has been tried by Callery and has proved to be very efficient. The use of this reagent could open new industrial developments for this technology. Catecholborane has shown some advantages at low temperature when selectivity is needed.[40]

16.3.4 MAIN PARAMETERS OF THE REACTION

The best reaction conditions depend both on the nature of the substrate and the catalyst. A wide range of conditions have been studied by Mathre at Merck,[41–45] by Quallich and co-workers at Pfizer,[25,46–48] by Stone at Sandoz,[49] and by the group at Sipsy. The reaction must be carried out under anhydrous conditions. Traces of moisture have a dramatic effect on enantiomeric excesses.[46] The catalyst loading is usually 1–20%, depending on the nature of the ketone to be reduced. Diphenylprolinol can be recovered at the end of the reaction and recycled in some cases. The temperature is the most important parameter of the reaction. Although a clear effect has been shown by Stone, no good mechanistic explanations have been presented. The effect of temperature could be interpreted by assessing the accumulation of an oxazadiborane intermediate at low temperature, which would reduce the ketone with a low enantioselectivity.[50]

When one plots enantiomeric excess versus temperature, all reductions show the same shape of curve. At low temperature, lower enantiomeric excesses are obtained. When the temperature is increased, the enantiomeric excess reaches a maximum value that depends on the reducing agent, the structure of the amino alcohol and of the catalyst itself (methyl, butyl, and phenyl oxazaborolidines do not give the same result), and the substrate. Then, as the temperature increases further, the enantiomeric excess decreases once again. This general behavior was experienced in the reduction of p-chloro-α-chloroacetophenone (5) where the range of optimal reaction temperature is broad and very practical. Indeed, a very high enantiomeric excess could be obtained at 20°C (Scheme 16.3).

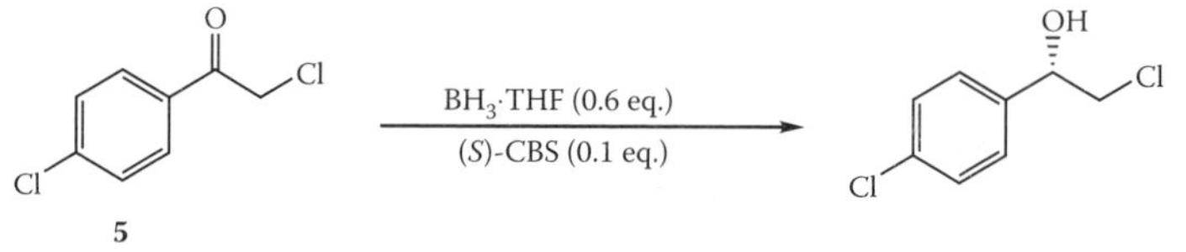

SCHEME 16.3

It is more interesting that reduction of complex ketones could be dramatically improved by optimization of the reaction temperature. In the case of the phenoxyphenylvinyl methyl ketone (6), a 5-lipoxygenase inhibitor synthesis intermediate, we were able to improve the enantiomeric excess from the 80–85% range up to 96% by selection of the optimal temperature for the reduction. In this case, the temperature range for an acceptable enantiomeric excess is very narrow. Generally, Me-CBS is the best catalyst with borane complexes as reducing agent.

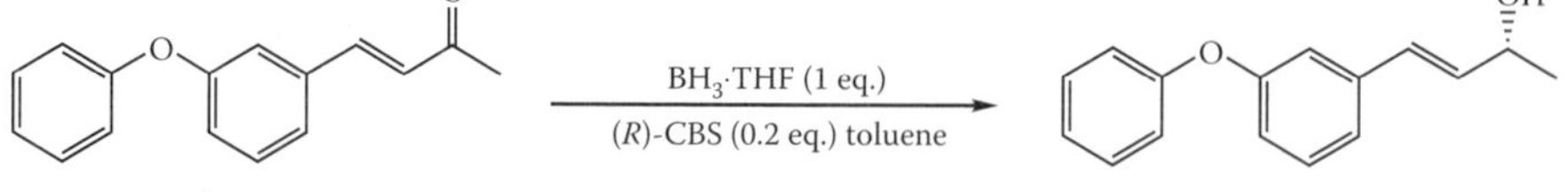

SCHEME 16.4

16.3.5 VERSATILITY OF THE REACTION

The enantioselectivity versus the nature of the substrate is highly predictable. The best results are obtained with aryl ketones where the difference between the large and the small substituent, as well as the basicity of the ketone, are favorable.

Besides the preparation of aryl alkyl carbinol, many other secondary alcohols have been prepared in good to excellent optical purity through the use of this method, although improvements

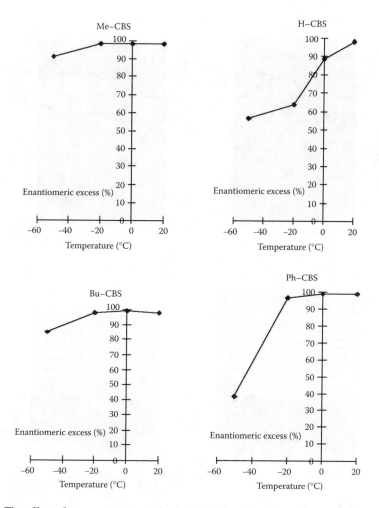

FIGURE 16.1 The effect of temperature on ee for the reduction of *p*-chlorochloroacetophenone (**5**).

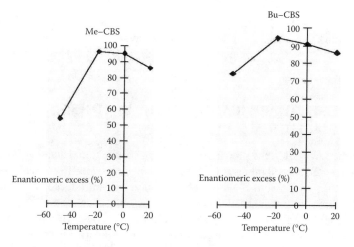

FIGURE 16.2 The reduction of the ketone (**6**).

TABLE 16.3
Versatility of the Reduction

Starting material	Product	ee or de (%)	Reference
		96	51
		98	41
		97	52
		90	53
		93	54
		94	55
		95	56

of the catalytic system have to be found to obtain good results with all types of ketones, especially dialkyl ketones. Numerous complex target molecules have been prepared where the asymmetric reduction step is key.[57] The Corey reduction is now a standard reaction for laboratory-scale preparation.

16.4 THE INDUSTRIAL APPLICATION IN THE SYNTHESIS OF PHARMACEUTICALS

The first to become involved in the preparation of pharmaceutical molecules using this technology was Corey, who reported several preparations of important drugs. His first success was the synthesis of a pure enantiomer of fluoxetine, the Lilly antidepressant introduced under the brand name Prozac.[58] Shortly after, he published the preparation of both enantiomers of isoproterenol, a β-adrenoreceptor agonist and denopamine, a useful drug for congestive heart failure.[59] From a pharmaceutical point of view, the structure of isoproterenol is remarkable because its basic structure is encountered in a variety of drugs such as those that act at the adrenoreceptor site. Therapeutic indications are broad, ranging from asthma (β$_2$-agonist), to obesity and diabetes (β$_3$-agonist). Within this frame, halo aryl ketones have been shown to be valuable intermediates, giving rise after reduction and basic cyclization to the corresponding epoxystyrenes (Scheme 16.5). These are important intermediates for the preparation of β_3-agonists.

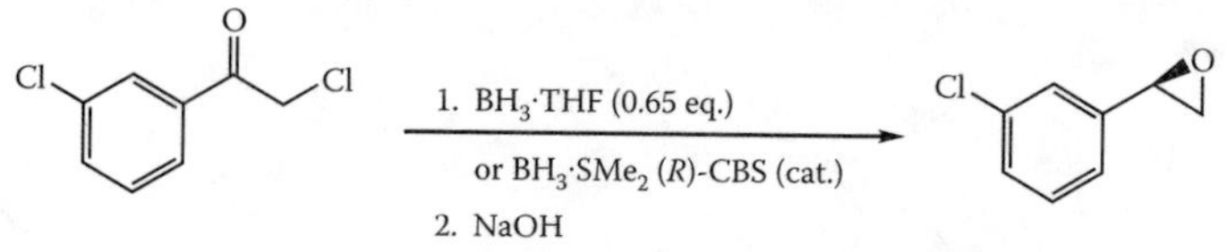

1. BH$_3$·THF (0.65 eq.)

or BH$_3$·SMe$_2$ (R)-CBS (cat.)

2. NaOH

SCHEME 16.5

The regiospecific ring opening of these epoxides is achieved with the appropriate amine to give advanced intermediates possessing the right configuration at the carbinol stereocenter. Sipsy has demonstrated the industrial feasibility of this approach and produced hundreds of kilos of related molecules.

Along the same lines, several groups of chemists at Merck successfully produced substantial quantities of various important pharmaceutical intermediates used in the preparation of drugs under development. Among these we can mention MK-0417 (**7**), a soluble carbonic anhydrase inhibitor useful to treat glaucoma,[41] the LTD4 antagonist L-699392 (**8**),[42] and the antiarrhythmic MK-499 (**9**).[43,44]

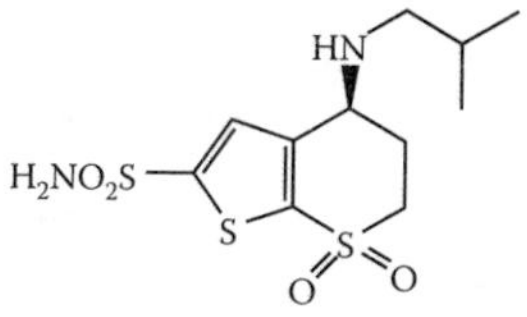

7

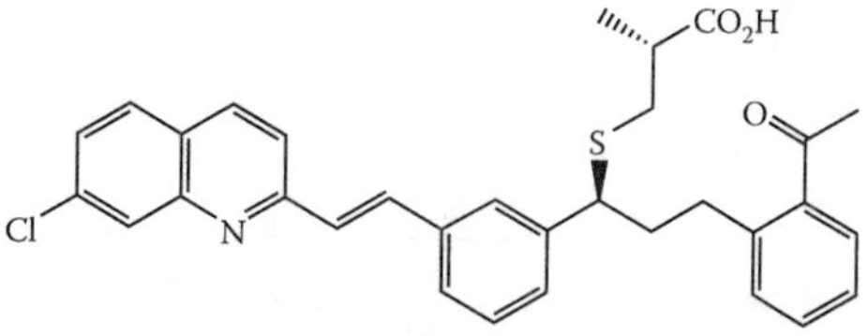

8

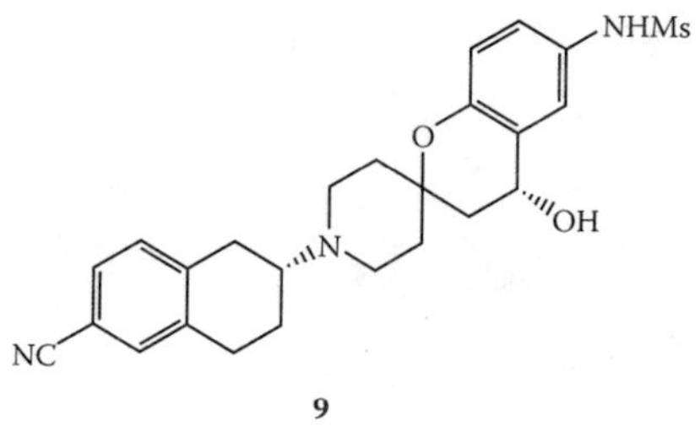

9

Equally successful has been the group of Quallich at Pfizer[46] who reported the preparation of a substituted dihydronaphthalenone, a pivotal intermediate in the preparation of the antidepressant sertraline (Lustral®) (**10**) (Scheme 16.6). In this synthesis, the new chiral center created subsequently determines the absolute stereochemistry of the final product (see Chapter 31).

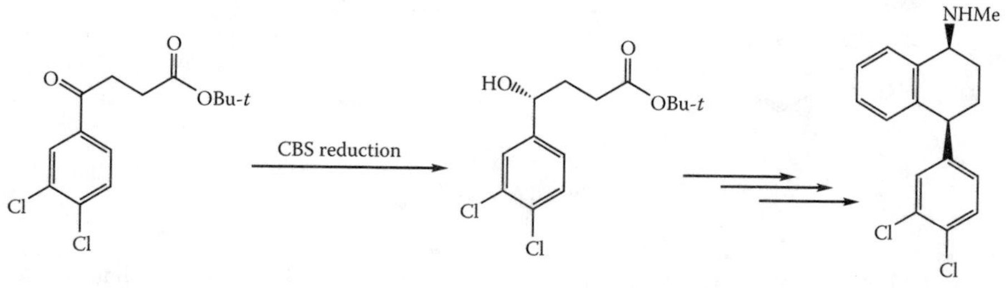

CBS reduction

10

SCHEME 16.6

Corey reduction is a valuable tool to introduce the chirality in a complex molecule because of its high compability with other reactive centers. Quallich demonstrated that high enantiomeric excesses also were obtained when other heteroatoms were present in the starting material. Specific chelation/protection of nitrogen functions can occur besides the reduction without affecting it (Scheme 16.7).[47]

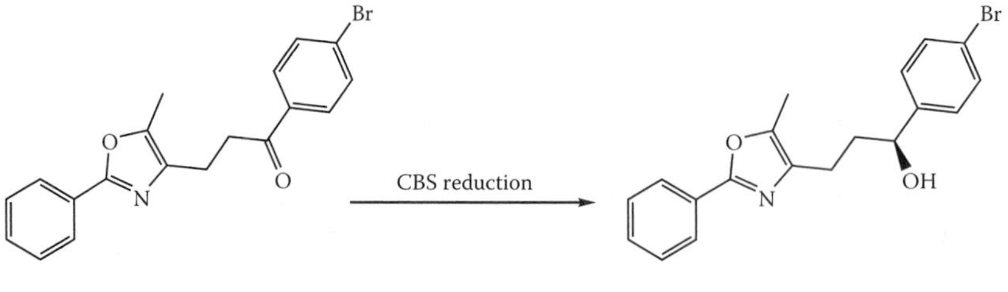

SCHEME 16.7

Moreover, reduction of alkyl aryl ketones can be used to access optically pure secondary aryl alkyl amines, as illustrated in an enantioselective synthesis of SDZ-ENA-713 (**11**)[60] and as we have demonstrated in a related process (Scheme 16.8).

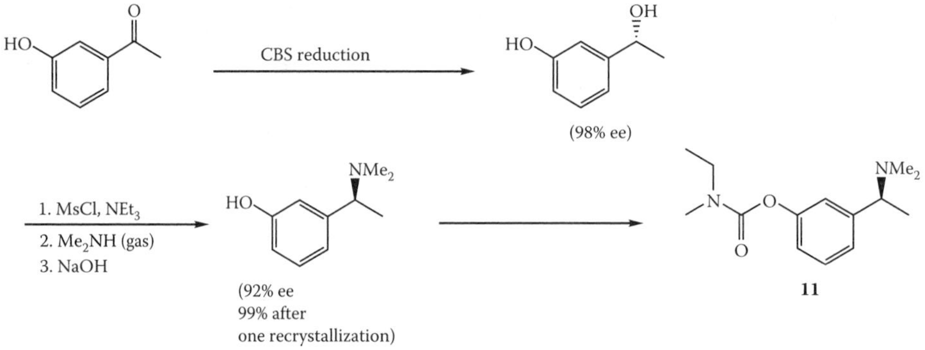

SCHEME 16.8

In the synthesis of another carbonic anhydrase inhibitor, we have successfully performed the following reduction with a very high enantioselectivity.

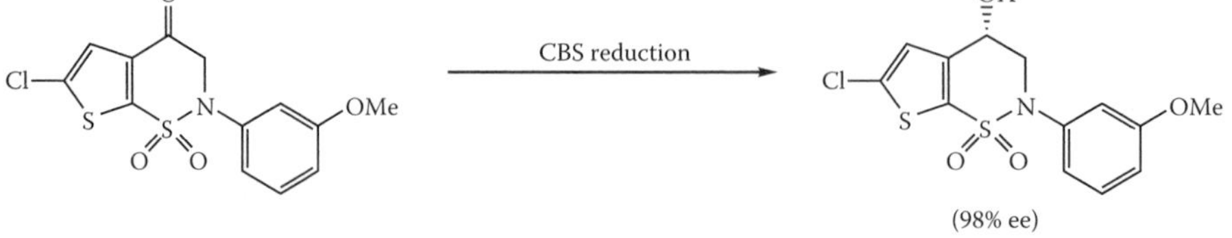

SCHEME 16.9

Finally, Corey reduction was used to solve the particular problem of the diastereoselective reduction of the ketone in C–15 in a prostaglandin synthesis. Corey reported a selectivity of 90:10.[39] We have investigated the reduction of a modified prostaglandin and, despite considerable efforts, we were not able to reach excesses greater than 72% (Scheme 16.10) (*cf.* Chapter 30). The yield of stereochemically pure product was 60% after 1 recrystallization. This yield is higher than that obtained by reduction with sodium borohydride, followed by chromatography and recycle of the undesired isomer by oxidation. Even in difficult cases where stereoselectivity is good but not excellent, Corey reduction can still be used to advantage.

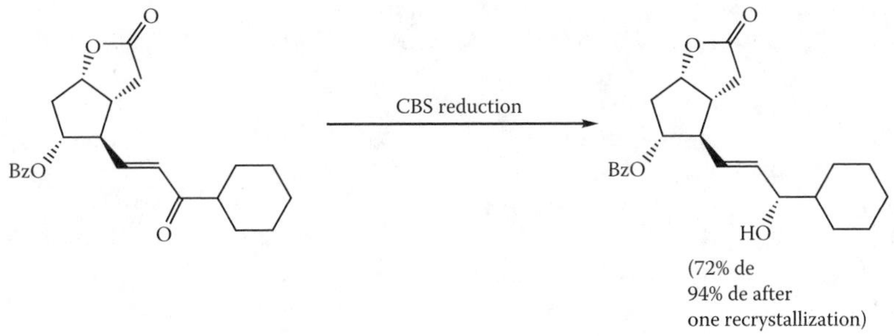

(72% de
94% de after
one recrystallization)

SCHEME 16.10

A number of methods have been proposed to circumvent the use of the borane complex.[61–64]

16.5 SUMMARY

The asymmetric reduction of ketones by borane catalyzed by oxazaborolidines has been widely studied since the beginning of the 1980s. Despite the use of borane complexes, which are hazardous chemicals, this reaction is an excellent tool to introduce the chirality in a synthesis and has demonstrated its usefulness in industrial preparation of chiral pharmaceutical intermediates. As a result of its performance, versatility, predictability, and scale up features, this method is particularly suitable for the rapid preparation of quantities of complex chiral molecules for clinical trials.

ACKNOWLEDGMENT

I would like to thank all the people at Callery and Sipsy, and especially Dr. J. M. Barendt, Dr. J-C. Caille, Dr. B. Laboue, and S. Frein for their contribution to the development of CBS technology.

REFERENCES

1. Singh, V. K. *Synthesis* 1992, 605.
2. Caille, J. C., Bulliard, M., Laboue, B. *Chimica Oggi* 1995, *13(4-5)*, 17.
3. Wills, M. *Curr. Opin. Drug Discov. Develop.* 2002, 5, 881.
4. Corey, E. J., Helal, C. J. *Angew. Chem., Int. Ed.* 1998, *37*, 1986.
5. Weigel, L. O. *Tetrahedron Lett.* 1990, 7101.
6. Noyori, R., Tomino, I., Tanimoto, Y., Nishizawa, M. *J. Am. Chem. Soc.* 1984, *106*, 6709.
7. Brown, H. C. In *Organic Synthesis via Boranes*, Wiley-Interscience: New York, 1975.
8. For a critical examination of the different stoichiometric chiral reducing agents, see Brown, H. C., Park, W. S., Cho, B. T., Ramachandran, P. V. *J. Org. Chem.* 1987, *52*, 5406.
9. Itsuno, S., Hirao, A., Nakahama, S., Yamazaki, N. *J. Chem. Soc. Perkin Trans. I* 1983, 1673.
10. Itsuno, S., Nakano, M., Miyazaki, K., Masuda, H., Ito, K., Mhizao, A., Nakahama, S. *J. Chem. Soc. Perkin Trans. I* 1985, 2039.
11. Kraatz, U. Ger. Pat. DE3609152, 1987.
12. Corey, E. J., Bakshi, R. K., Shibata, S. *J. Am. Chem. Soc.* 1987, *109*, 5551.

13. Corey, E. J. U.S. Pat. 4,943,635, 1990.
14. Willems, J. G. H., Dommerholt, F. J., Hamminck, J. B., Vaarhorst, A. M., Thÿss, L., Zwanenburg, B. *Tetrahedron Lett.* 1995, *36*, 603.
15. Rao, A. V. R., Gurjar, M. K., Sharma, P. A., Kaiwar, V. *Tetrahedron Lett.* 1990, *31*, 2341.
16. Mehler, T., Behnen, W., Wilken, J., Martens, J. *Tetrahedron: Asymmetry* 1994, *5*, 185.
17. Huong, Y., Gao, Y., Nie, X., Zepp, C. M. *Tetrahedron Lett.* 1994, *35*, 6631.
18. Didier, E., Loubinoux, B., Tombo, G. M. R., Rihs, G. *Tetrahedron* 1991, *47*, 4941.
19. Rao, A. V. R., Gurjar, M. K., Kaiwar, V. *Tetrahedron: Asymmetry* 1992, *3*, 859.
20. Behnen, W., Dauelsberg, C., Wallbaum, S., Martens, J. *Synth. Comm.* 1992, *22*, 2143.
21. Santhi, V., Rao, J. M. *Tetrahedron: Asymmetry* 2000, *11*, 3553.
22. Draper, T. R. PCT Int. Appl. WO 0222623, 2002.
23. Brunin, T., Cabou, J., Bastin, S., Brocard, J., Pelinski, L. *Tetrahedron: Asymmetry* 2002, *13*, 1241.
24. Mehler, T., Martens, J. *Tetrahedron: Asymmetry* 1993, *4*, 1983.
25. Quallich, G. J., Woodall, T. M. *Synlett* 1993, 929.
26. Wallbaum, S., Martens, J. *Tetrahedron: Asymmetry* 1992, *3*, 1475.
27. Quallich, G. J., Woodall, T. M. *Tetrahedron Lett.* 1993, *34*, 4145.
28. Berenguer, R., Garcia, J., Vilarrasa, J. *Tetrahedron: Asymmetry* 1994, *5*, 165.
29. Thompson, A. S., Douglas, A. W., Hoogsteen, K., Carroll, J. D., Corley, E. G., Grabowski, E. J. J., Mathre, D. J. *J. Org. Chem.* 1993, *58*, 2880.
30. Douglas, A. W., Tschaen, D. M., Reamer, R. A., Shi, Y.-J. *Tetrahedron: Asymmetry* 1996, *7*, 1303.
31. Zhoa, J., Bao, X., Liu, X., Wan, B., Han, X., Yang, C., Hang, J., Feng, Y., Jiang, B. *Tetrahedron: Asymmetry* 2000, *11*, 3351.
32. Alagona, G., Ghio, C., Persico, M., Tomasi, S. *J. Am. Chem. Soc.* 2003, *125*, 10027.
33. Kraatz, U. Eur. Pat. EP 0237902, 1987.
34. Dauelsberg, C., Behnen, W. Ger. Pat. DE 4,341,605, 1995.
35. Klinger, F. D., Sobotta, R. Eur. Pat. EP 0682013, 1995.
36. Blacklock, T. J., Jones, T. K. U.S. Pat. 5,039,802, 1991.
37. Mathre, D. J., Jones, T. K., Xavier, L. C., Blacklock, T. J., Reamer, R. A., Mohan, J. J., Turner-Jones, E. T., Hoogsteen, K., Baum, M. W., Grabowski, E. J. J. *J. Org. Chem.* 1991, *56*, 751.
38. Beak, P., Kerrick, S. T., Wu, S., Chu, J. *J. Am. Chem. Soc.* 1994, *116*, 3231.
39. Corey, E. J., Bakshi, R. K., Shibata, S., Chn C. P., Singh, V. K. *J. Am. Chem. Soc.* 1987, *109*, 7925.
40. See also Manju, K., Trehan, S. *Tetrahedron: Asymmetry* 1998, *9*, 3365.
41. Jones, T. K., Mohan, J. J., Xavier, L. C. C., Blacklock, T. J., Mathre, D. J., Sohar, P., Jones, E. T. T., Reaner, R. A., Robers, F. E., Grabowski, E. J. J. *J. Org. Chem.* 1991, *56*, 763.
42. King, A. O., Corley, E. G., Anderson, R. K., Larsen, R. D., Verhoeven, T. R., Reider, P. J. *J. Org. Chem.* 1993, *58*, 3731.
43. Shi, Y. J., Cai, D., Dolling, U. M., Douglas, A. W., Tschaen, D., Verhoeven, T. R. *Tetrahedron Lett.* 1994, *35*, 6409.
44. Cai, D., Tschaen, D., Shi, Y. J., Verhoeven, T. R., Reamer, R. A., Douglas, A. W. *Tetrahedron Lett.* 1993, *34*, 3243.
45. Tschaen, D., Abramson, M. L., Cai, D., Desmond, R., Dolling, U. M., Frey, L., Karady, S., Shi, Y. J., Verhoeven, T. R. *J. Org. Chem.* 1995, *60*, 4324.
46. Quallich, G. J., Woodall, T. M. *Tetrahedron* 1992, *48*, 10239.
47. Quallich, G. J., Woodall, T. M. *Tetrahedron Lett.* 1993, *34*, 785.
48. Quallich, G. J., Woodall, T. M. *Synlett* 1993, 929.
49. Stone, G. B. *Tetrahedron: Asymmetry* 1994, *5*, 465.
50. Mathre, D. J. *Proceedings of the Chiral USA'96 Symposium*, Spring Innovation Ltd, 1996, p. 69.
51. Corey, E. J., Shibata, S., Bakshi, R. K. *J. Org. Chem.* 1988, *53*, 2861.
52. Denmark, S. E., Schnute, M. E., Marcin, L. R., Thorarensen, A. *J. Org. Chem.* 1995, *60*, 3205.
53. Corey, E. J., Rao, K. S. *Tetrahedron Lett.* 1991, *32*, 4623.
54. Corey, E. J., Jardine, P. D. S. *Tetrahedron Lett.* 1989, *30*, 7297.
55. Dumartin, H., Le Floc'h, Y., Gree, R. *Tetrahedron Lett.* 1994, *35*, 6681.
56. Meier, C., Laux, W. H. G. *Tetrahedron: Asymmetry* 1995, *61*, 1089.
57. Boudreau, C., Tillyer, R. D., Tschaen, D. M. PCT Int. Appl. WO 9627581, 1996.
58. Corey, E. J., Reichard, G. A. *J. Org. Chem.* 1989, *30*, 5207.

59. Corey, E. J., Link, J. O. *J. Org. Chem.* 1991, *56*, 442.
60. Chen, C.-P., Prasad, K., Repic, O. *Tetrahedron Lett.* 1991, *32*, 7175.
61. Salunkhe, A. M., Burkhardt, E. R. *Tetrahedron Lett.* 1997, *38*, 1523.
62. Matos, K., Corella, J. A., Burkhardt, E. R., Nettles, S. M. U.S. Pat. 6,218,585, 2001.
63. Nettles, S. M., Matos, K., Burkhardt, E. R., Rouda, D. R., Corella, J. A. *J. Org. Chem.* 2002, *67*, 2970.
64. Huertas, R. E., Corella, J. A., Soderquist, J. A. *Tetrahedron Lett.* 2003, *44*, 4435.

17 *cis*-1-Amino-2-Indanol-Derived Ligands in Asymmetric Catalysis

Chris H. Senanayake, Dhileepkumar Krishnamurthy, and Isabelle Gallou

CONTENTS

17.1   INTRODUCTION

Enantiomerically pure *cis*-1-amino-2-indanol and its derivatives have been used as ligands in numerous catalytic asymmetric carbon–hydrogen, carbon–carbon, and carbon–heteroatom bond formation reactions. The conformationally constrained indanyl platform has emerged as a particularly valuable backbone in a variety of catalytic processes leading to high levels of asymmetric induction. The aminoindanol **1** has also been used as a resolution agent (Chapter 8) as well as a chiral auxiliary (Chapter 24). For the synthesis of **1** see Chapter 24.

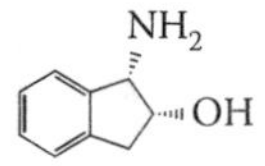

(1*S*,2*R*)-1

cis-1-Amino-2-indanol **1** itself was shown to be an efficient ligand in the reduction of carbonyl groups.[1–3] (The ligand **1** has been used in a kinetic resolution of secondary aryl alcohols; see Chapter 8.) Ligand derivatives of **1** include oxazaborolidines, bis(oxazolines)[4] **2** and pyridine bis(oxazoline) **3**, phosphinooxazoline **4**, Schiff bases, aryl phosphite **5**, benzoquinone **6**, and phosphaferrocene-oxazoline **7** (Figure 17.1).[5]

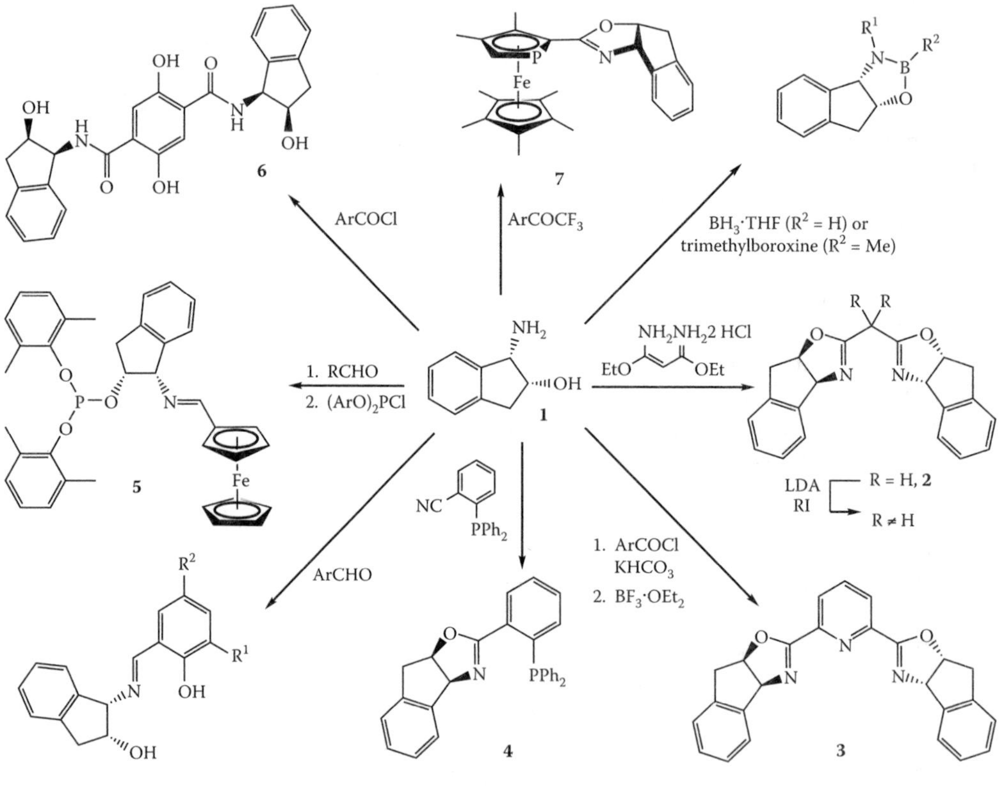

FIGURE 17.1 Ligands derived from *cis*-1-amino-2-indanol (**1**).

Oxazaborolidines were developed for reduction of carbonyls and were shown to be of value in the reduction of imines.[6–8] *B*-Hydrogen oxazaborolidines were prepared *in situ* from **1** and BH$_3$·THF (tetrahydrofuran), whereas stock solutions of *B*-methyl oxazaborolidines were obtained by reaction with trimethylboroxine.[6–9]

The extremely versatile class of aminoindanol-derived chiral bis(oxazoline) ligands were developed by Davies, Senanayake, and co-workers and Ghosh and co-workers independently to study the effect of conformational rigidity of ligands in the catalytic asymmetric Diels-Alder reaction.[4,9–16] These ligands were later found equally efficient for asymmetric hetero Diels-Alder reactions,[17,18] conjugate additions,[19] and conjugate radical additions.[20] Inda-box ligand **2** (R = H) was obtained by condensation of **1** with the appropriate amide enol ether dihydrochlorate.[11] More constrained inda-box ligands (R ≠ H) were prepared either by Ritter type reaction of **1** with the corresponding dinitriles in the presence of trifluoromethanesulfonic acid[21] or by dialkylation of **2** with the corresponding alkyl iodides.[14]

Chiral pybox ligands were synthesized as ligands for the asymmetric cyclopropanation of styrene.[10] In-pybox ligand **3** was prepared by reaction of **1** with 2,6-pyridine dicarbonyl dichloride in the presence of potassium hydrogen carbonate in isopropyl acetate followed by cyclization of the *bis*-hydroxyamide with $BF_3 \cdot OEt_2$ at 120°C.[22]

In light of the remarkable results obtained with the ligands described previously, several new classes of aminoindanol-containing ligands were disclosed that take advantage of the rigidity and steric bulk of the indanyl platform. Among those, phosphinooxazoline **4**, prepared by reaction of **1** with diphenylphosphinobenzonitrile, was studied as a possible ligand for the asymmetric allylic alkylation of small acyclic allyl acetate substrates.[23]

Aminoindanol-derived Schiff bases were developed as tridentate ligands for the chromium-catalyzed hetero Diels-Alder reaction between weakly nucleophilic dienes and unactivated aldehydes.[24] The generality of the utility of these Schiff bases, readily obtained by condensation of **1** with the corresponding aldehyde, was later demonstrated in the hetero Diels-Alder reaction between Danishefsky's diene and chiral aldehydes,[25] in the inverse electron-demand hetero Diels-Alder reaction of α,β-unsaturated aldehydes with alkyl vinyl ethers,[26] and in hetero-ene reactions.[27]

Aryl phosphite **5** was studied as a possible ligand in the palladium-catalyzed allylic alkylation and allylic sulfonation reactions. This P,N-ligand was easily accessed by condensation of ferrocenealdehyde followed by reaction with bis(2,6-dimethylphenyl)chlorophosphite.[28]

Benzoquinone **6** was reported as one of a new class of amino alcohol-derived benzoquinones tested in the palladium-catalyzed 1,4-dialkylation of 1,3 dienes. These ligands were prepared by reaction of **1** with C_2-symmetric 1,4-diallyloxy-2,5-benzenedicarboxylic acid chloride followed by allyl deprotection.[29]

Bidentate phosphaferrocene-oxazoline **7**, generated by acylation with the corresponding phosphaferrocene trifluoroacetate followed by oxazoline formation, has proved to be a highly efficient ligand for asymmetric induction in the copper-catalyzed conjugate addition of diethylzinc to α,β-unsaturated ketones.[30]

Considering the wide variety of *cis*-1-amino-2-indanol–containing ligands synthesized to date and the numerous applications reported in the literature and detailed in the present section, it is to be expected that novel ligands and methodologies will continue to emerge featuring the remarkable characteristics of *cis*-1-amino-2-indanol.

17.2 REDUCTION OF CARBONYLS

17.2.1 TRANSFER HYDROGENATION

As mentioned earlier, *cis*-aminoindanols have been shown to be excellent ligands for asymmetric transfer hydrogenation (see also Chapter 8).[1,2] Reduction of acetophenone in isopropanol in the presence of 0.25 mol% of $[RuCl_2(p\text{-cymene})]_2$, 1 mol% (1R,2S)-*cis*-1-amino-2-indanol, and 2.5 mol% of potassium hydroxide proceeded in 70% yield and 91% ee to give the *S*-alcohol. Enantioselectivities were considerably lower with phenylglycinol (23% ee) or *N*-methyl-*cis*-aminoindanol (27%) as ligands. These results strongly suggested that a stereochemically rigid backbone enhanced the degree of asymmetric induction and that a primary amine function in the ligand was essential (Scheme 17.1).

Mechanistic studies[3] tended to demonstrate that the hydrogen transfer proceeds via a 6-center transition state similar to that proposed by Noyori with monotosylated diamine complexes of ruthenium(II) (Figure 17.2).[31]

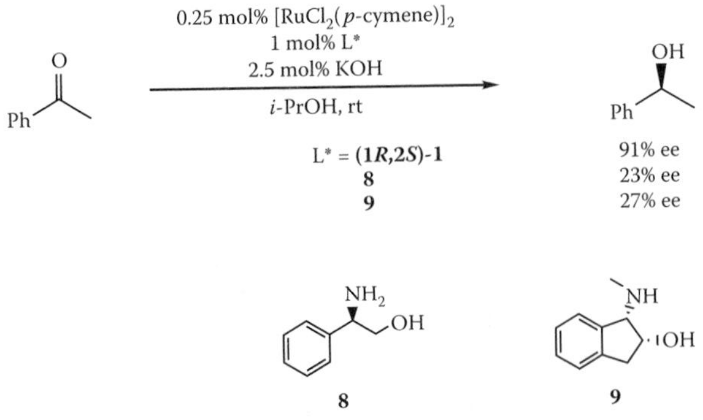

SCHEME 17.1

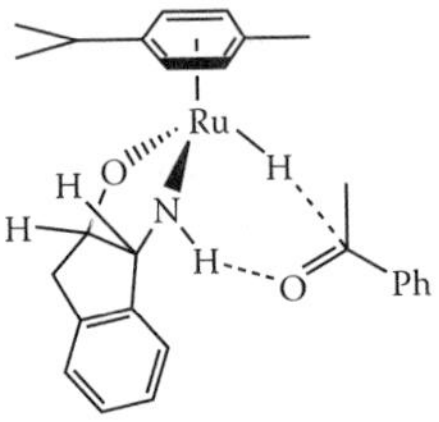

FIGURE 17.2 Proposed transition state for the transfer hydrogenation to carbonyls.

17.1.2 BORANE REDUCTIONS

Since the discoveries of Itsuno[32] and Corey,[33] remarkable advances have been made in the enantio-selective reduction of prochiral ketones using amino alcohol–derived oxazaborolidines (see Chapter 16).[34,35] In most cases, these amino alcohols were obtained from chiral pool sources. Consequently, extensive synthetic manipulations were often necessary to access their unnatural antipode. Didier and co-workers were first to examine the potential of *cis*-aminoindanol as a ligand for the asymmetric oxazaborolidine reduction of ketones.[36] Several acyclic and cyclic amino alcohols were screened for the reduction of acetophenone (Scheme 17.2), and *cis*-aminoindanol led to the highest enantioselectivity (87% ee).

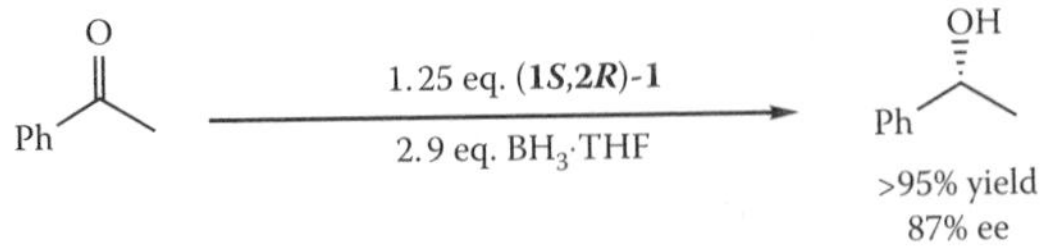

SCHEME 17.2

It is interesting that Didier also studied the reduction of *anti*-acetophenone oxime methyl ether (Scheme 17.3)[36] and again observed that *cis*-aminoindanol yielded one of the highest selectivity (95% ee). However, a stoichiometric amount of amino alcohol was needed to achieve high degrees of asymmetric induction. Use of a catalytic amount of amino alcohol resulted in a considerable decrease in product enantioselectivity.

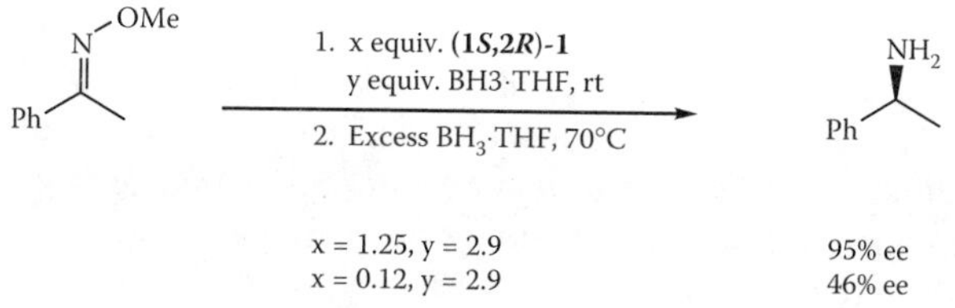

SCHEME 17.3

Researchers at Sepracor later disclosed the use of a new class of chiral oxazaborolidines derived from *cis*-aminoindanol in the enantioselective borane reduction of α-haloketones.[6,7] The *B*-hydrogen oxazaborolidine ligand **10** was prepared *in situ* from *cis*-aminoindanol **1** and BH$_3$•THF.[8] Stock solutions of *B*-methyl oxazaborolidine **11–16** were obtained by reaction of the corresponding *N*-alkyl aminoindanol with trimethyl boroxine.[6,7] *B*-Methyl catalyst **11** was found to be more selective (94% ee at 0°C) than the *B*-hydrogen catalyst **10** (89% ee at 0°C), and enantioselectivities with **11** increased at lower temperatures (96% ee at –20°C). The catalyst structure was modified by introduction of *N*-alkyl substituents. As a general trend, reactivities and selectivities decreased as the steric bulk or the chelating ability of the *N*-alkyl substituent increased (Scheme 17.4).

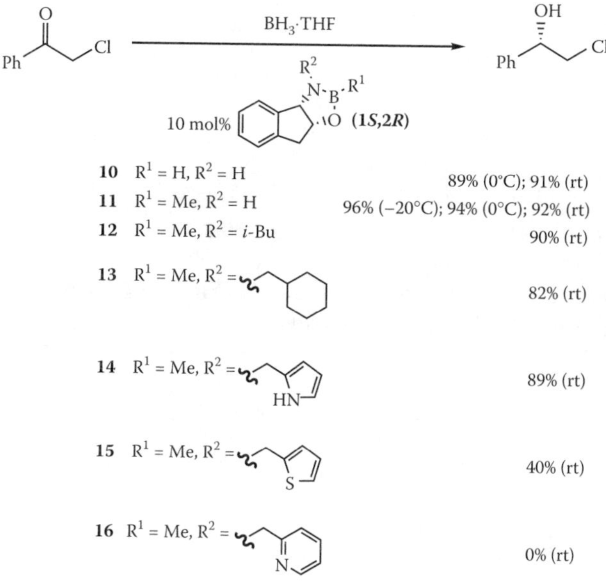

SCHEME 17.4

Borane reduction of a variety of aromatic ketones using 5–10 mol% of *B*-methyl catalyst **11** proceeded in >95% yield and in 80–97% ee. α-Haloketones were generally more reactive (90–97% ee) than simple ketones, which required higher temperatures (0°C compared to –20°C) to react to completion and led to lower enantioselectivities (80–90% ee).[118] A complementary study by Umani-Ronchi and co-workers[37] described the borane reduction of cyclic and acyclic ketones using catalyst **10**. All products were obtained in >89% yield and >85% ee. Cyclic and hindered ketones led to the highest enantioselectivities (up to 96% ee) at room temperature.

This methodology was successfully applied to the asymmetric synthesis of (*R,R*)-formoterol (**17**), a potent β$_2$-agonist for the treatment of asthma and bronchitis.[38–40] Reduction of bromo ketone

18 was a key step in the synthesis (Scheme 17.5). An extensive study, involving varying temperatures, boron sources, and phenylglycinol catalyst backbones, was undertaken to determine the
optimal conditions for the reaction.[40] Experimental data clearly demonstrated that each catalyst had
its own optimal conditions with respect to temperature, boron source, and additives. Conformationally rigid oxazaborolidine catalysts, containing the tetraline or indane backbone, proved to be
the most effective in the reduction of **18**. The highest selectivity was achieved using a stock solution
of catalyst **11** at –10°C (96% ee). Although lower selectivity was obtained with the *in situ* prepared
10 at 0°C (93% ee), this catalyst was chosen for the large-scale process because its preparation
was easier, it was less time-consuming, and it involved inexpensive reagents. On a multi-kilogram
scale, bromohydrin **19** could be isolated in 85% yield by crystallization, which directly enriched
its enantiopurity to >99% ee.

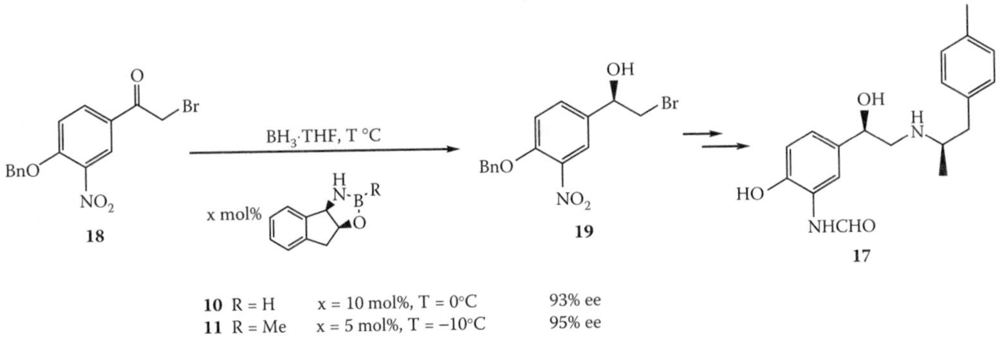

10 R = H	x = 10 mol%, T = 0°C	93% ee	
11 R = Me	x = 5 mol%, T = –10°C	95% ee	

SCHEME 17.5

The proposed mechanism of reduction[33,37] involves the coordination of a borane molecule to
the nitrogen of the oxazaborolidine in a *trans*-relationship to the indanyl substituent. The ketone
coordinates to the boron center of the oxazaborolidine, *cis* to the coordinated borane, in a possible
boatlike transition state **20** or chairlike transition state **21** (Figure 17.3). In both cases, intramolecular
hydride attack occurs from the *Re*-face of the carbonyl.

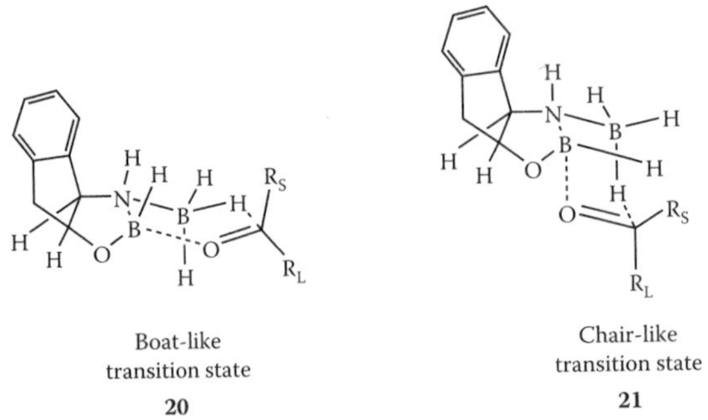

Boat-like transition state **20**

Chair-like transition state **21**

FIGURE 17.3 Proposed transition states for borane reduction.

Mechanistic studies[41–43] by Dixon and Jones excluded the possibility of dimeric catalytic species
because a linear dependence was observed between the catalyst's enantiopurity and the reaction's
enantioselectivity.[43] The test reaction was the desymmetrization of *meso*-imide **22** using chiral
oxazaborolidine catalysts. The sense of the enantioselectivity of the reduction was established by
conversion of hydroxy lactam **23** to the known ethoxy lactam **24** (Scheme 17.6).

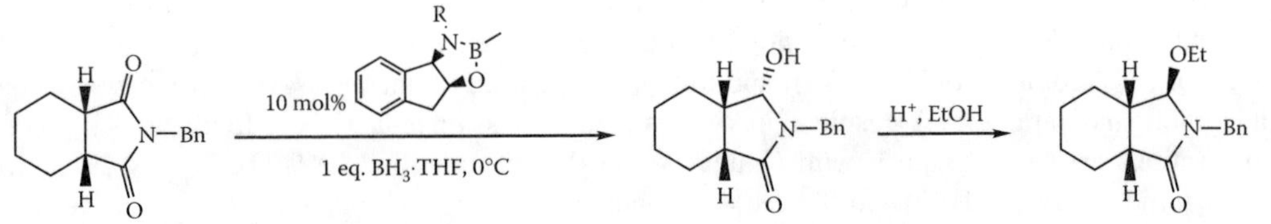

22 23 24

R = H — 75% yield (2 steps), 86% ee
R = Me — 52% yield, 55% ee
R = Et — 28% yield, 2% ee

SCHEME 17.6

The decreasing enantioselectivity with increasing *N*-alkyl steric bulk was rationalized by assuming the model described in Figure 17.3. As mentioned earlier, the key feature of this model is the coordination of borane on the least-hindered *exo*-face of the oxazaborolidine. This coordination is reversible, and large substituents on the nitrogen can displace the equilibrium to *endo*-face coordination, therefore diminishing the sense of asymmetric induction (Scheme 17.7).[43]

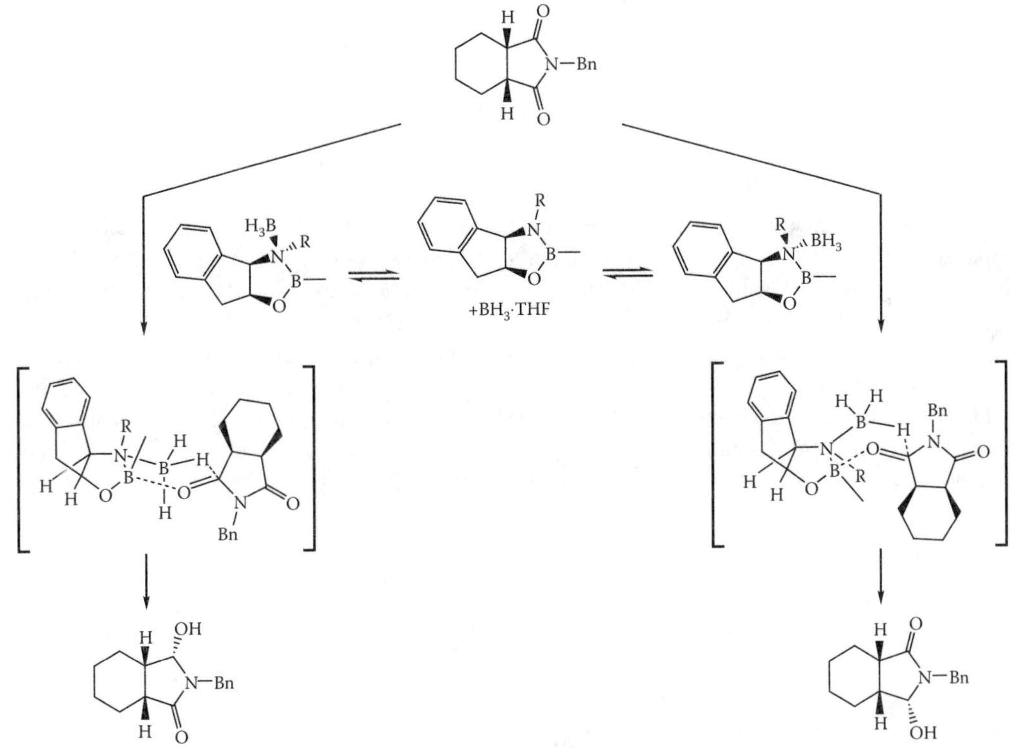

SCHEME 17.7

17.3 ADDITION TO CARBONYLS

17.3.1 TRIMETHYLSILYLCYANIDE ADDITIONS

Tridentate salen ligands derived from enantiopure *cis*-aminoindanol were prepared in an effort to explain the relationship between the ligand structure and the enantioselectivity in the Ti(IV)-Schiff base catalyzed asymmetric addition of trimethylsilylcyanide to benzaldehyde.[44] The nature and position of substituents on the Schiff base phenyl ring were shown to have a great influence on the reactivity and selectivity of the reaction. An enantiomeric excess of 85% was reached using 20 mol% of a 1:1 complex of Ti(O-i-Pr)$_4$ and tridentate ligand **25** (Scheme 17.8). Modification of the titanium:ligand ratio to 1:2 resulted in a considerable loss in enantioselectivity (19%). These observations are in accordance with Oguni's studies[45] that identified L*Ti(O-i-Pr)$_2$ as the active species and determined *bis*(Schiff base) compounds L*$_2$Ti to be inactive toward the catalytic hydrocyanation of aldehydes.

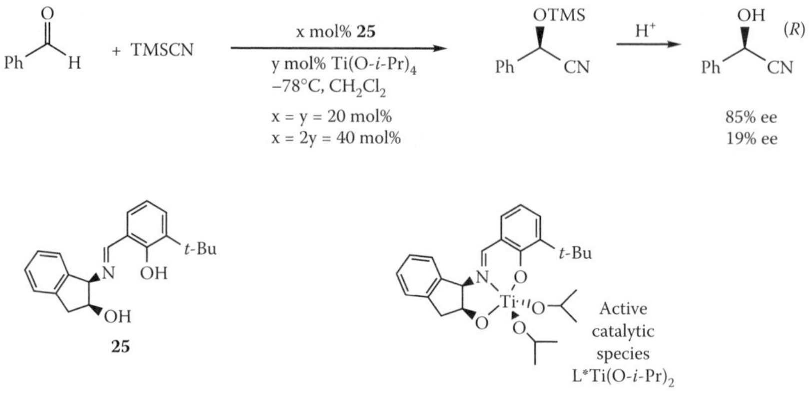

SCHEME 17.8

17.3.2 DIETHYLZINC ADDITIONS

In parallel to their investigations on the asymmetric reduction of ketones, Umani-Ronchi and co-workers examined the utility of *cis*-aminoindanol derivatives as catalysts in the addition of diethylzinc to aldehydes.[37] Using *N*-dibutyl or *N*-diallylaminoindanol as catalysts, secondary alcohols could be obtained in high yields but the enantioselectivities remained low, in the 40–50% range (Scheme 17.9). It is interesting that several studies[46,47] have taken advantage of alternative isomers of aminoindanol: both *cis*- and *trans*-*N*-disubstituted-2-amino-1-indanol were found to give high yields but moderate enantiomeric excesses (56–80% ee).[46] High degrees of enantioselection in the diethylzinc addition to aliphatic and aromatic aldehydes were eventually achieved by introducing an alkyl or aryl substituent at C–1 of *trans*-*N*-disubstituted-2-amino-1-indanol (up to 93% ee).[47]

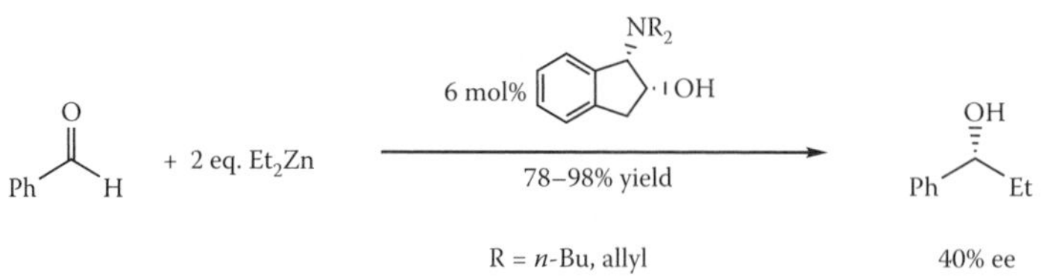

SCHEME 17.9

17.4 CONJUGATE ADDITIONS

A new class of phosphaferrocene-oxazoline ligands has been disclosed by Fu and co-workers[30] and applied to the copper-catalyzed asymmetric conjugate addition of diethylzinc to acyclic enones with good enantioselectivity. The substitution pattern on the phosphoryl ring as well as on the oxazoline was shown to have an enormous impact on the selectivity. Ligands **26** and **27**, which share the same absolute configuration in the oxazoline, provide the (*S*)-1,4-adduct preferentially. This observation led to the conclusion that the stereochemistry at the oxazoline, and not the planar chirality of the phosphaferrocene, was responsible for the stereochemistry of the conjugate addition. Introduction of a phenyl group at C–5 of the phosphoryl ring further enhanced the enantioselectivity of the reaction. The best result was obtained with the *cis*-aminoindanol derived ligand **7** (Scheme 17.10). The reaction conditions were optimized by using CuOTf instead of $Cu(OTf)_2$, and diethylzinc addition to chalcone with **7** as ligand gave the 1,4-adduct in 87% ee.

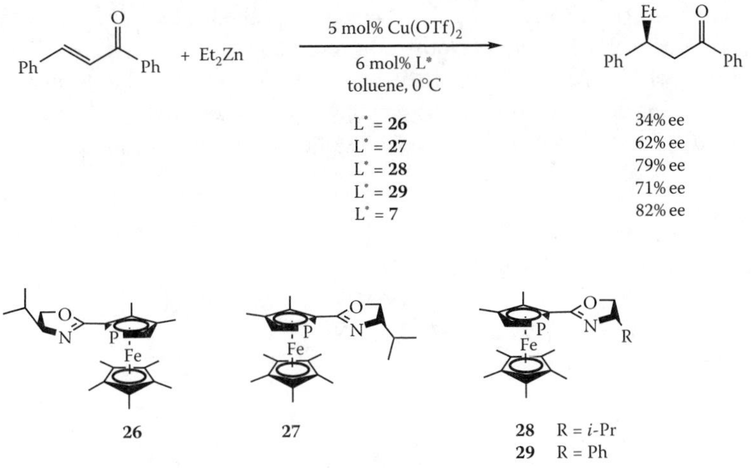

Ph Ph O
+ Et2Zn
5 mol% Cu(OTf)2
6 mol% L*
toluene, 0°C
Et O
Ph Ph
L* = 26 34% ee
L* = 27 62% ee
L* = 28 79% ee
L* = 29 71% ee
L* = 7 82% ee
26
27
28 R = i-Pr
29 R = Ph

SCHEME 17.10

Sibi and co-workers have studied the copper-catalyzed conjugate addition of silylketene acetals to β-enamidomalonates using chiral bisoxazolines as ligands.[19] Addition of neutral nucleophile *O*,*S*-ketene silyl acetal **30** to malonate **31** was performed with 10 mol% of Lewis acid $Cu(OTf)_2$ and 10 mol% of bisoxazolines **2** and **32–35** as chiral source. Using ligand **33**, β-amino acid derivative **37** was obtained in 96% yield and 89% ee (Scheme 17.11). A working model for the conjugate addition was proposed where the copper metal coordinates malonate carbonyls and the addition occurs from the *Si*-face of the olefinic bond. However, other chelation modes can be speculated, involving the amido functional group in 6- or 8-membered rings.[19]

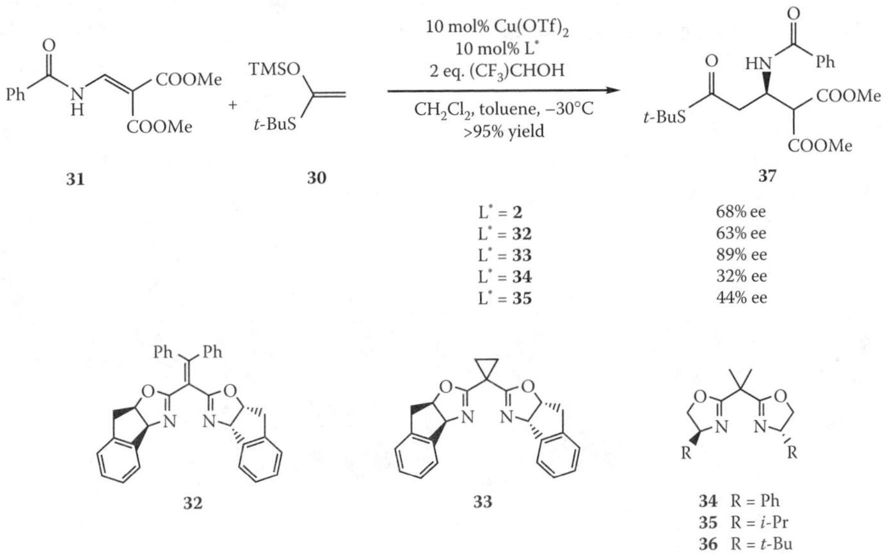

COOMe
COOMe
TMSO
t-BuS
10 mol% Cu(OTf)2
10 mol% L*
2 eq. (CF3)CHOH
CH2Cl2, toluene, −30°C
>95% yield
HN Ph
COOMe
COOMe
t-BuS
31
30
37
L* = 2 68% ee
L* = 32 63% ee
L* = 33 89% ee
L* = 34 32% ee
L* = 35 44% ee
32
33
34 R = Ph
35 R = i-Pr
36 R = t-Bu

SCHEME 17.11

The enantioselective conjugate radical addition using chiral *bis*(oxazoline)-based Lewis acid was also studied by Sibi, Porter, and co-workers.[20] Stoichiometric amounts of metal-ligand complexes derived from a number of box-ligands, and a variety of Lewis acids including MgI$_2$ and Zn(OTf)$_2$ were examined in the addition to β-substituted-α,β-unsaturated-*N*-oxazolidinone compounds. The enantioselectivity of product **38**, which ranged from 37% to 82% ee, was dramatically improved (up to 93% ee) using rigid aminoindanol-derived inda-box ligands (Scheme 17.12).[48] Catalytic amounts (5–30 mol%) of the MgI$_2$-**33** catalyst retained excellent enantioselectivity levels (90–97% ee). The reaction could be performed at room temperature with little loss of selectivity (93% ee using 30 mol% of **33**). In comparison, conjugate radical addition to pyrazole derivatives using Zn(OTf)$_2$-box-ligand complexes led to moderate enantioselectivities.[49] It is still unclear why pyrazole and oxazolidinone templates gave products of opposite configuration using the same chiral Lewis acid.

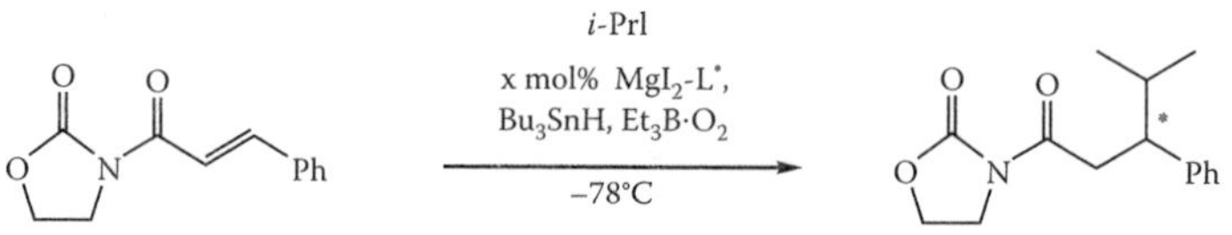

L* = **36**	x = 100 mol%	61% ee (*R*)
L* = **34**	x = 100 mol%	47% ee (*S*)
L* = **33**	x = 100 mol%	93% ee (*R*)
L* = **33**	x = 30 mol%	97% ee (*R*)

SCHEME 17.12

17.5 DIELS-ALDER AND RELATED REACTIONS

17.5.1 DIELS-ALDER REACTIONS

Inda-box ligands were applied to the metal-catalyzed asymmetric Diels-Alder reaction (see Chapter 26) of cyclopentadiene with α,β-unsaturated-*N*-oxazolidinones by Davies and Senanayake[9,10] and Ghosh[11] independently. Conformationally constrained inda-box ligand **39** displayed excellent selectivity level (82% ee) compared to phe-box ligand **34** (30% ee) in the copper-catalyzed Diels-Alder reaction.[9] In addition, Ghosh has shown that the use of magnesium as Lewis acid led to moderate and reversed enantioselectivities compared to copper Lewis acid (Scheme 17.13).[11] These results have been rationalized using a square planar conformation of copper and *s-cis* conformation of the dienophile proposed by Evans,[12] which results in a preferential *endo-Si*-face attack of the diene. For magnesium Lewis acids, a Corey-Ishihana transition state would explain the reversal of selectivity as magnesium adopts a tetrahedral geometry, favoring the diene attack from the less-hindered *endo-Re*-face (Figure 17.4).[13]

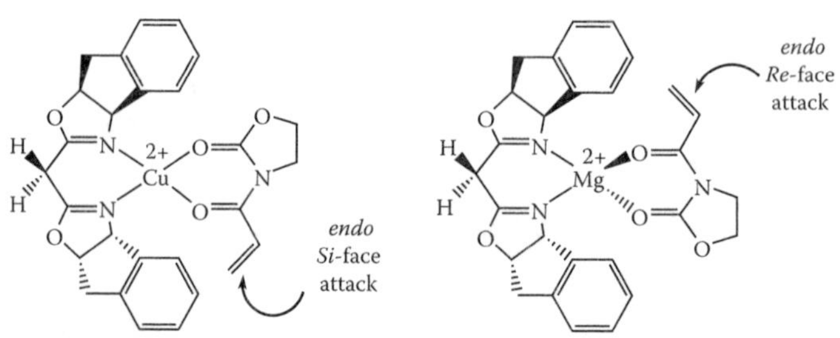

FIGURE 17.4 Proposed transition states for the copper- and magnesium-catalyzed asymmetric Diels-Alder reaction.

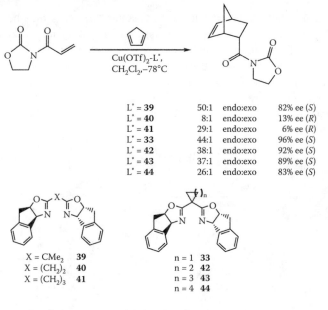

L* = *ent*-34	M = Cu (10 mol%)	T = –50°C		30% ee
L* = **39**	M = Cu (10 mol%)	T = –50°C	50:1 *endo:exo*	82% ee (*S*)
L* = **39**	M = Cu (10 mol%)	T = –65°C	130:1 *endo:exo*	92% ee (*S*)
L* = **2**	M = Cu (8 mol%)	T = –78°C	>99:1 *endo:exo*	94% ee (*S*)
L* = *ent*-**2**	M = Cu (4 mol%)	T = –78°C	>99:1 *endo:exo*	99% ee (*R*)
L* = *ent*-**2**	M = Mg (10 mol%)	T = –78°C	95:5 *endo:exo*	34% ee (*S*)

39

SCHEME 17.13

The ligand bite angle and its impact on the enantioselectivity was examined by Davies, Senan-ayake, and co-workers.[9,10,14] In a first series of experiments, such ligands as **39**, which form a 6-membered copper chelate, were found to be the most selective. It was postulated that ligands **40** and **41** were far less selective because of the increased flexibility of the 7- and 8-membered metal chelates formed with copper.[9] Modification of ligand **39** to increase the bite angle (calculated in the increasing order **44**<**39**<**43**<**42**<**33**) resulted in higher *endo:exo* selectivities as well as *endo*-enantioselectivities of the cycloadduct (Scheme 17.14).[10,14] The orientation of the C–8 proton of the aminoindanol moieties in the "chiral pocket" of the copper complex was believed to play an important role on the stereoselectivity of the Diels-Alder reaction.[14] Another explanation raised the possibility that a larger bite angle would altogether change the coordination around the copper center, away from the idealized square planar model.[14]

L* = **39**	50:1	endo:exo	82% ee (*S*)
L* = **40**	8:1	endo:exo	13% ee (*R*)
L* = **41**	29:1	endo:exo	6% ee (*R*)
L* = **33**	44:1	endo:exo	96% ee (*S*)
L* = **42**	38:1	endo:exo	92% ee (*S*)
L* = **43**	37:1	endo:exo	89% ee (*S*)
L* = **44**	26:1	endo:exo	83% ee (*S*)

X = CMe₂ **39**	n = 1 **33**
X = (CH₂)₂ **40**	n = 2 **42**
X = (CH₂)₃ **41**	n = 3 **43**
	n = 4 **44**

SCHEME 17.14

Ghosh and co-workers improved the utility of inda-box ligands in the Diels-Alder reaction by using cationic aqua complex derived from **2** and Cu(ClO₄)•6H₂O. The complex had the considerable advantage of being air-stable, and reaction of *N*-acryoyloxazolidinone and cyclopentadiene at −78°C for 11 hours using 10 mol% of aqua complex afforded the cycloadduct in 88% yield, >99:1 *endo:exo* selectivity, and 98% ee. Cryogenic temperatures were not necessary to achieve high levels of enantioselectivity because the same reaction could be run at 0°C, which allowed for completion of the reaction within 1 hour, to give the cycloadduct in similar yield (91%) with little loss of selectivity (98:2 *endo:exo*, 95% ee).[15]

The importance of the rigid aminoindanol backbone in asymmetric catalytic Diels-Alder reactions is a subject of continued interest.[16,50] One example immobilized the copper-inda-box complex onto mesoporous silica in the context of continuous large-scale production of chiral compounds.[16] Using 10 mol% of this catalyst (Figure 17.5), the Diels-Alder reaction between *N*-acryoyloxazolidinone and cyclopentadiene proceeded in 99% yield, 17:1 *endo:exo* selectivity, and 78% ee of *endo* cycloadduct. The catalyst could easily be recovered and reused several times without significant loss of diastereoselectivity (15:1 *endo:exo* selectivity after the fifth reuse) or enantioselectivity (72% ee after the fifth reuse).[16] The same remarkable reactivity was observed with a number of diene–dienophile partners.

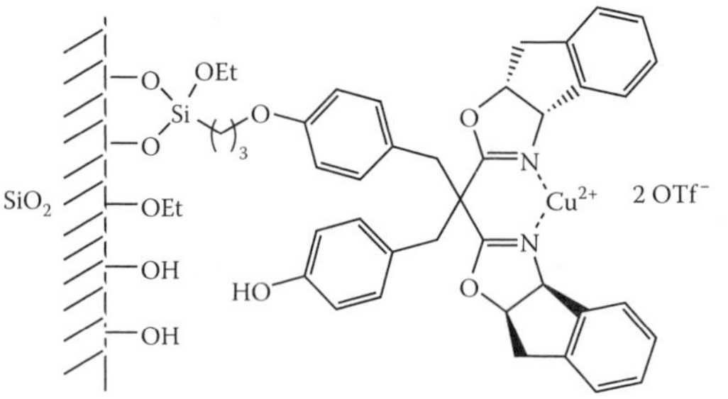

FIGURE 17.5 Immobilized copper-inda-box catalyst.

17.5.2 HETERO DIELS-ALDER REACTIONS

The asymmetric hetero Diels-Alder reaction (see Chapter 26) using inda-box ligands was first investigated by Ghosh and co-workers.[17] Danishefsky's diene (**45**) and glyoxylate esters **46** were reacted using copper triflate as Lewis acid. The reaction gave a mixture of Mukayama aldol **47** and pyranone derivative **48** after standard workup. Treatment of the crude mixture with trifluoroacetic acid allowed for ring-closing of **47** and led to **48** as the sole product. Aminoindanol-derived ligand *ent*-**2** afforded the highest yields and enantioselectivities (Scheme 17.15). As observed for the Diels-Alder reaction, use of magnesium triflate as Lewis acid resulted in largely decreased and reversed enantioselectivities.[17]

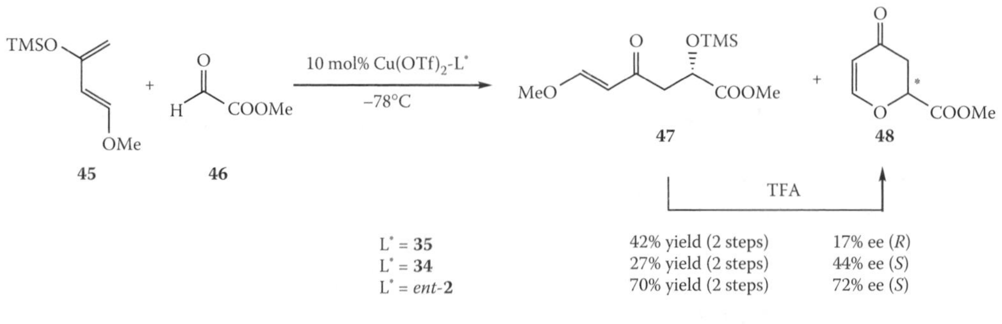

L* = **35**	42% yield (2 steps)	17% ee (*R*)
L* = **34**	27% yield (2 steps)	44% ee (*S*)
L* = *ent*-**2**	70% yield (2 steps)	72% ee (*S*)

SCHEME 17.15

The synthetic utility of this methodology was demonstrated in the asymmetric construction of the C3–C14 segment of antitumor macrolide laulimalide (**50**).[18] Using Cu(OTf)$_2$-*ent*-**2** as catalyst, Danishefsky's diene reacted with benzyloxyacetaldehyde **51** to provide cycloadduct **52** in 76% yield and 85% ee. Standard synthetic manipulations on dihydropyran **52** led to the C$_3$–C$_{14}$ segment **49** of laulimalide (Scheme 17.16).

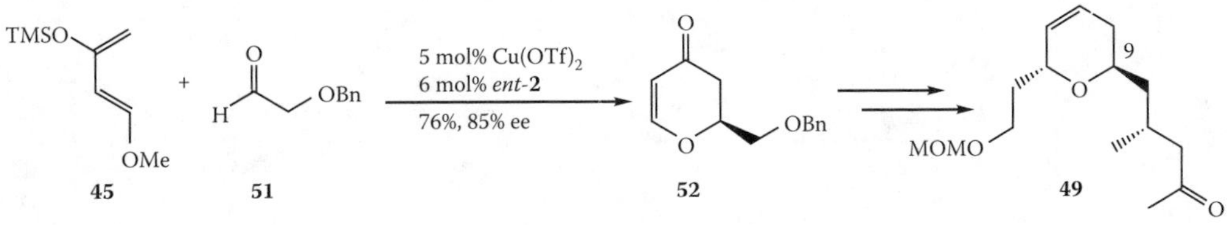

SCHEME 17.16

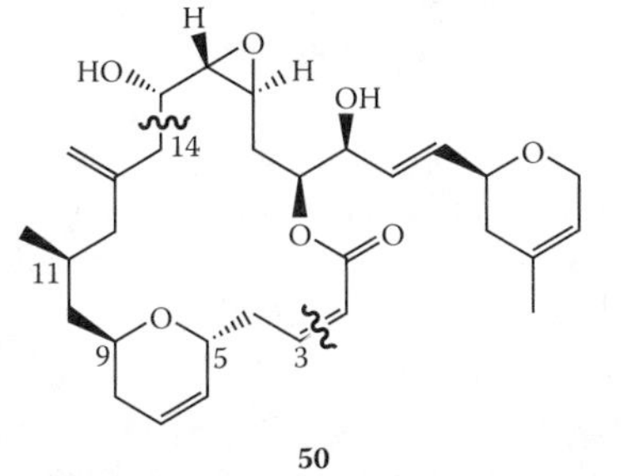

Jacobsen and co-workers investigated the diastereoselective hetero Diels-Alder reaction between Danishefsky's diene and chiral aldehydes catalyzed by chromium-Schiff base complexes.[25] A variety of chiral aldehydes underwent the doubly diastereoselective reaction using catalysts **53** and **54** in good yield (up to 99%), satisfactory diastereoisomeric ratio (~10:1 for unhindered aldehydes), and excellent enantiomeric excess of the major diastereoisomer (>97% ee). In the case of congested aldehydes, tridentate indane-derived catalysts were far less effective, leading to dihydropyrans in moderate yields (44–58%) and selectivities (<4:1 dr). Whereas the reaction of **45** and **55**, using achiral complex **56** as catalyst, gave **57** in a relatively high 1:4.5 diastereomeric ratio, the matched catalyst-to-substrate system using (1*S*,2*R*)-**53** provided the highest level of diastereoselectivity (1:33) of the study (Scheme 17.17). With judicious choice of chiral aldehyde and catalyst enantiomer, any of the four possible diastereoisomers of dihydropyranone can be synthesized with high selectivity.

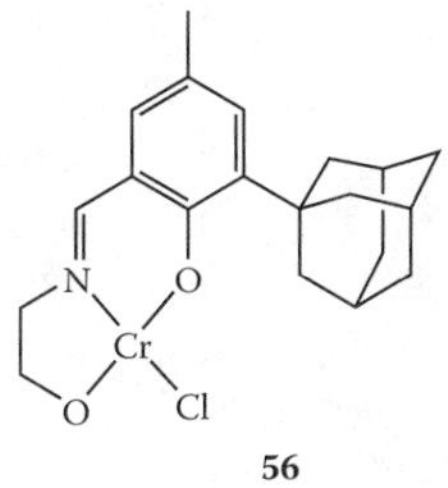

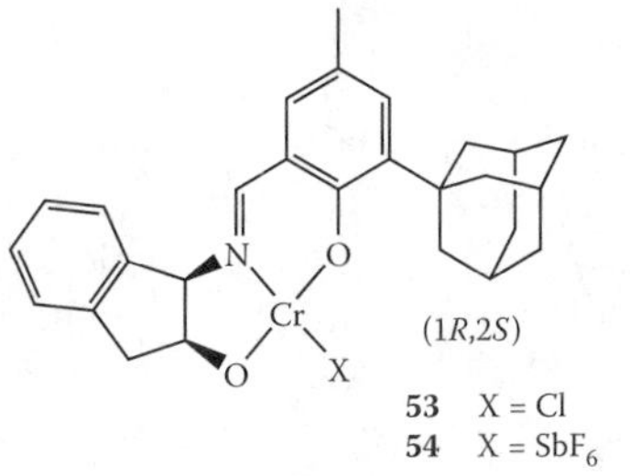

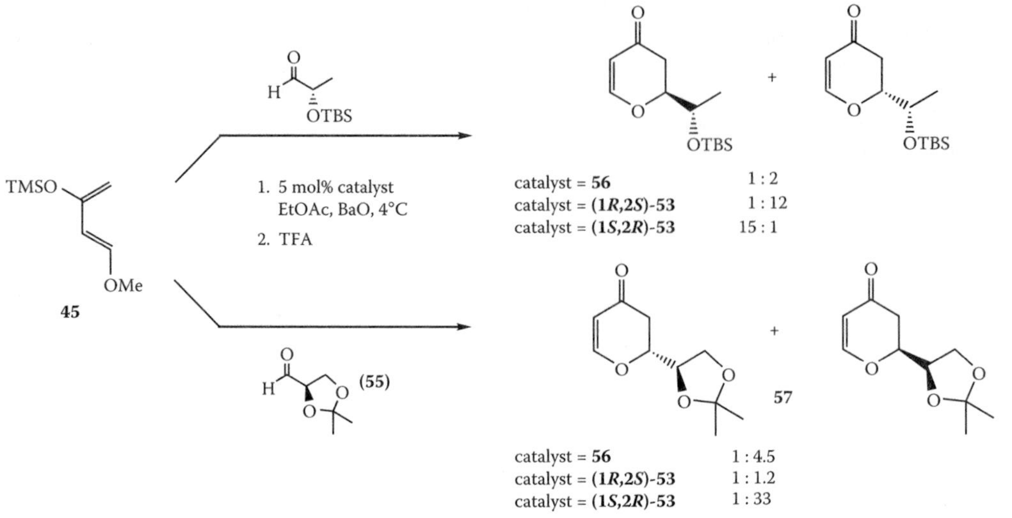

1. 5 mol% catalyst
   EtOAc, BaO, 4°C
2. TFA

catalyst = 56	1 : 2
catalyst = (1R,2S)-53	1 : 12
catalyst = (1S,2R)-53	15 : 1

45

(55)

57

catalyst = 56	1 : 4.5
catalyst = (1R,2S)-53	1 : 1.2
catalyst = (1S,2R)-53	1 : 33

SCHEME 17.17

In a complementary study, Jacobsen and co-workers used their new chromium complex of aminoindanol-derived Schiff bases to perform an efficient hetero Diels-Alder reaction between less-nucleophilic monooxygenated dienes and unactivated achiral aldehydes.[24] This reaction provided enantiomerically enriched dihydropyrans with three defined stereogenic centers in one step (Scheme 17.18). The best results were obtained using catalysts **53** and **54,** which bear a very large adamantyl group. In all cases, reaction of dienes **58** with various aldehydes gave excellent all-*cis* diastereo-selectivities (>95% de). In addition, all aliphatic aldehydes led to remarkable enantioselectivities (>94%). The tetrahydropyranone formed from benzaldehyde was obtained with a lower level of enantioselectivity (65% ee using **53** and 81% using **54**). In general, reactions with hexafluoroanti-monate catalyst **54** were faster and more enantioselective than with chloride **53**. With the exception of aromatic aldehydes, where use of acetone as solvent was critical, solvent-free conditions gave satisfactory results. This method provided highly efficient access to several interesting synthons, which, by further elaboration of the double bond, would ultimately lead to tetrahydropyran deriv-atives with 5 defined stereogenic centers.

1. 3 mol% **53** or **54**, sieves, rt
2. TFA

> 95% de (all-*cis*)
> 94% ee

58

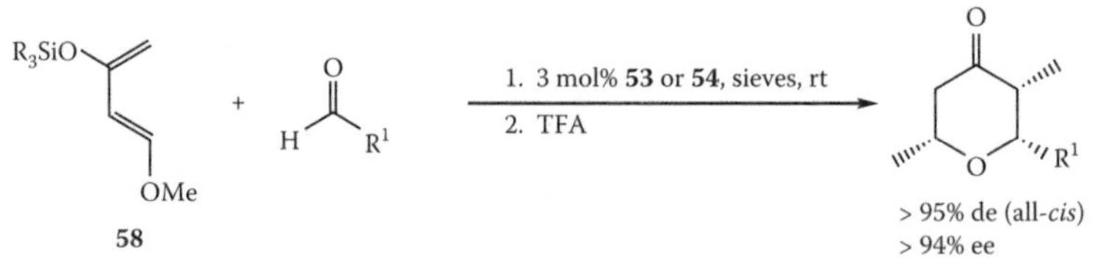

SCHEME 17.18

Chromium complex **53** was also shown to efficiently catalyze the inverse electron-demand hetero Diels-Alder reaction of α,β-unsaturated aldehydes with alkyl vinyl ethers (Scheme 17.19).[26] Although the uncatalyzed process required elevated temperatures and pressures to give dihydropy-rans in good yields but poor *endo:exo* selectivities, the reaction proceeded at room temperature in the presence of 5 mol% of *ent*-**53** and 4Å molecular sieves in dichloromethane of *tert*-butyl methyl ether with excellent diastereoselectivity (*endo:exo* >96:4) and promising enantioselectivities (72–78% ee). Optimal results were achieved using a solvent-free system and excess vinyl ether.

Both reactivity and selectivity decreased with increasing steric bulk of alkyl group on the vinyl ether (Et>*n*-Pr>*n*-Bu~*i*-Bu), and *tert*-butyl vinyl ether was completely unreactive. The cycloaddition of ethyl vinyl ether with a wide variety of α,β-unsaturated aldehydes bearing aliphatic and aromatic β-substituents proceeded with high selectivity (>95% de, 89–98% ee). Only 5 mol% of catalyst were necessary except in the case of sterically more demanding substituents (R = *i*-Pr, Ar) that required 10 mol% of catalyst loading. Substitution could also be introduced in the α-position of the unsaturated aldehyde, and cycloadducts were obtained with similar high selectivities.[26]

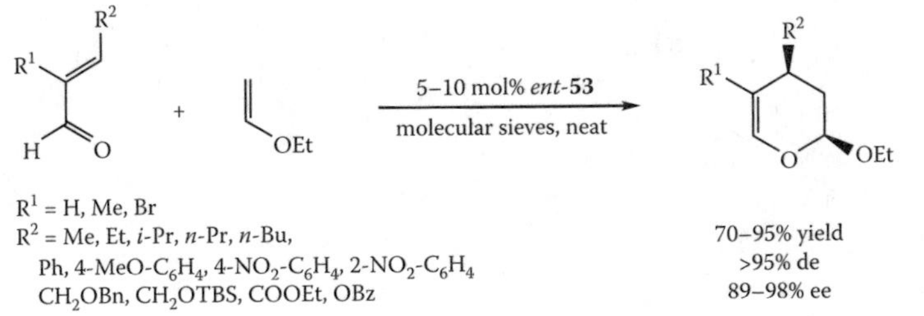

R^1 = H, Me, Br
R^2 = Me, Et, *i*-Pr, *n*-Pr, *n*-Bu,
  Ph, 4-MeO-C_6H_4, 4-NO_2-C_6H_4, 2-NO_2-C_6H_4
  CH_2OBn, CH_2OTBS, COOEt, OBz

5–10 mol% *ent*-**53**
molecular sieves, neat

70–95% yield
>95% de
89–98% ee

SCHEME 17.19

These results were particularly remarkable in light of the few precedents that only involved chelation of oxabutadienes bearing oxygen-containing electron-withdrawing groups by a 2-point binding catalyst, which would easily explain the enantioface discrimination. In contrast, Jacobsen's tridentate chromium catalyst **53** is a one-point binding complex, which activated the unsaturated aldehyde and discriminated its enantiotopic faces through simple chelation to the carbonyl. In efforts to understand the source of stereoinduction, the crystal structure of **53** was analyzed and the catalyst was shown to exist in the solid state as a dimeric structure bridged through a single molecule of water and bearing a terminal water ligand on each chromium center. Based on preliminary solution molecular weight and kinetic studies, it was proposed that the dimeric species was maintained in the catalytic cycle and that dissociation of one terminal water ligand would open one coordination site for substrate binding. These mechanistic studies not only explained the crucial need for molecular sieves in the reaction but also clearly indicated that the role of **53** is comparable to activation of the aldehyde by a Lewis acid.[26]

17.5.3 HETERO-ENE REACTIONS

The generality of the activation of aldehydes for reaction with weak nucleophiles, using tridentate Schiff base chromium(III) complexes, was further demonstrated by the successful and highly selective ene-reaction of alkoxyalkenes and silyloxyalkenes with aromatic aldehydes.[27] The ene-reaction of 2-methoxypropene or 2-trimethylsilyloxypropene with a number of substituted benzaldehydes catalyzed by **59** in acetone or ethyl acetate proceeded in high yields (75–97%) and good to excellent enantioselectivities (70–96% ee, with >85% ee in the majority of cases). Catalysts **53** and **60** were also effective in the ene-reaction, but enantioselectivities were 2–5% lower. It is interesting that chiral (salen)CrCl complexes afforded good yields of ene-reaction but considerably poorer selectivities (<30% ee). β-Hydroxy enol ethers thus obtained were readily transformed into β-hydroxy ketone and β-hydroxy ester derivatives (Scheme 17.20).[27]

Crystal structure of **60** revealed that the complex existed as a dimeric species where the two chromium centers are bridged through the indane oxygen and each chromium metal bears one molecule of water (Figure 17.6). It was therefore proposed, as for the inverse electron-demand hetero Diels-Alder reaction, that the barium oxide desiccant removed one molecule of bound water from the catalyst dimer, which opened one coordination site for binding of the substrate carbonyl.[26]

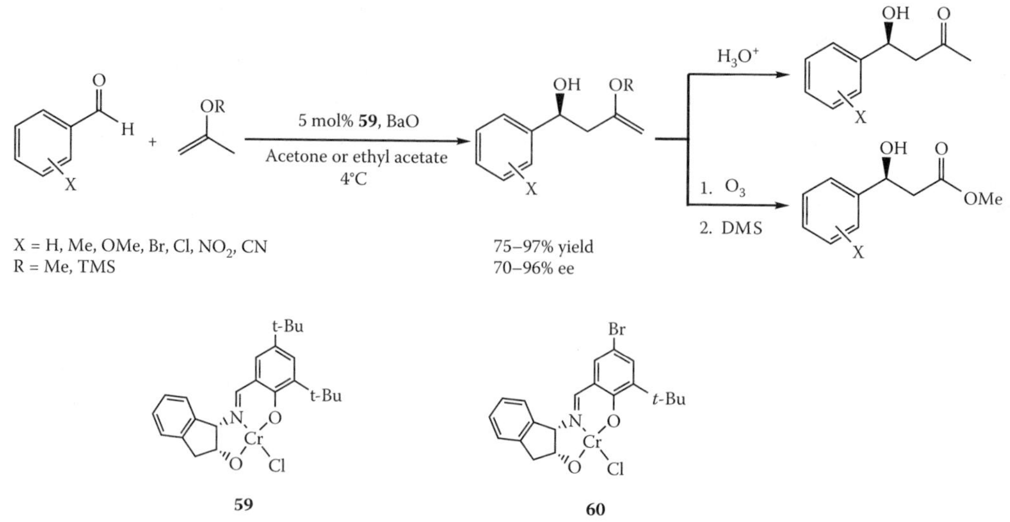

SCHEME 17.20

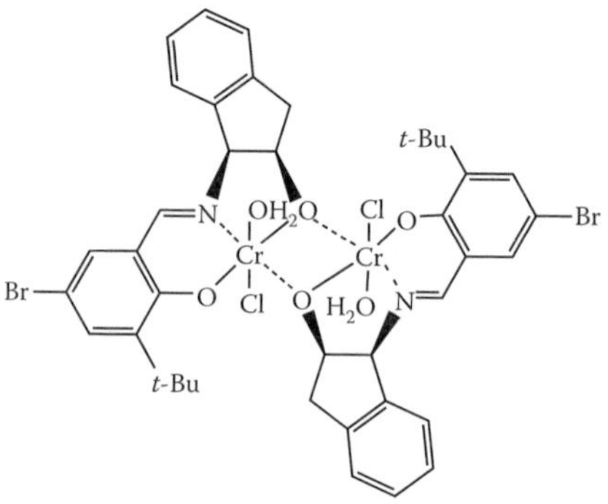

FIGURE 17.6 X-ray crystal structure of catalyst **60**.

17.6 AZIDE ADDITIONS

Tridentate Schiff base chromium(III) complexes were identified as the optimal catalysts for the enantioselective ring opening of *meso*-aziridines by TMSN$_3$.[51] Indeed, preliminary studies have shown that, although the (salen)chromium complexes catalyzed the reaction to some extent, they consistently led to low enantioselectivities (<14% ee). It was rationalized that the diminished reactivity and selectivity of the salen complexes with aziridines compared to epoxides was a result of the steric hindrance created by the *N*-substituent of the coordinated aziridine. As expected, improved results were observed using tridentate ligands on the chromium center because they offer a less-hindered coordination environment (Figure 17.7).[51]

Extensive optimization studies identified highly electron-deficient 2,4-dinitrobenzyl-substituted aziridines as the most reactive substrates, chromium as the metal of choice, and indanol-derived Schiff bases as the most effective ligands. In this ring-opening process, catalyst **61** provided the highest selectivities. Using these optimized conditions, a variety of aziridines were selectively opened in a very efficient manner (Scheme 17.21).[51] This reaction can provide an easy access to C_2-symmetric 1,2-diamines, a valuable class of chiral auxiliaries, and even to less accessible non-C_2-symmetric 1,2-diamines because of the differentially protected amines of the ring-opened products.

O
N—R
N—R
Cr
L

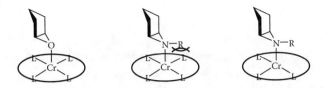

FIGURE 17.7 Schematic illustration of the possible advantage of tridentate ligands over tetradentate ligands for the activation of aziridines.

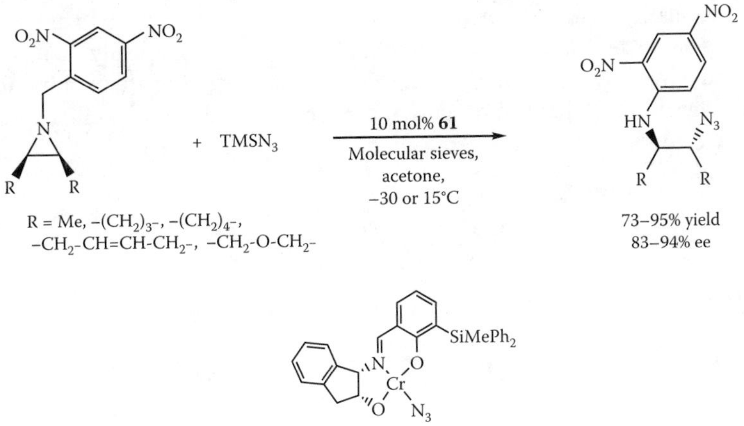
O_2N
NO_2
N
R
+ $TMSN_3$
10 mol% **61**
Molecular sieves,
acetone,
−30 or 15°C
R = Me, $-(CH_2)_3-$, $-(CH_2)_4-$,
$-CH_2-CH=CH-CH_2-$, $-CH_2-O-CH_2-$
HN
N_3
73–95% yield
83–94% ee
$SiMePh_2$
Cr
N_3

61

SCHEME 17.21

17.7 ASYMMETRIC ALKYLATIONS

17.7.1 Catalytic Asymmetric Cyclopropanations

Davies and co-workers have explored the role of ligand conformation in the ruthenium(II)-catalyzed cyclopropanation of styrene.[10] This study was based on results reported by Nishiyama in which the catalyst prepared *in situ* from pyridine-bis(oxazoline) **62** and $[RuCl_2(p\text{-cymene})]_2$ was found to be highly active and selective in the reaction of ethyl diazoacetate with styrene (66% yield, 84% de, and 89% ee of major *trans*-isomer).[52] Several ligands hindered on the oxazoline ring, including **3**, were tested and poorer yields and selectivities were obtained (for **3**, 50% yield, 81% de, and 59.5% ee of major *trans*-isomer), which indicated unfavorable steric interactions between styrene and the Ru(in-pybox) carbene complex (Scheme 17.22).[10]

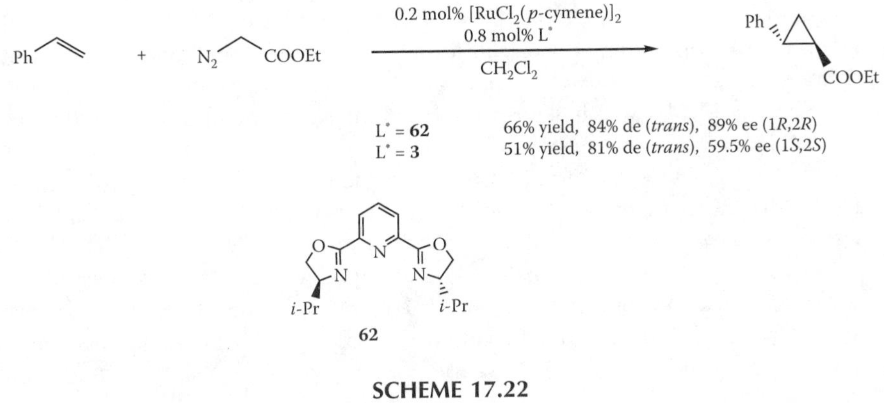
Ph
+
N_2
COOEt
0.2 mol% $[RuCl_2(p\text{-cymene})]_2$
0.8 mol% L*
CH_2Cl_2
Ph
COOEt
L* = **62** 66% yield, 84% de (*trans*), 89% ee (1*R*,2*R*)
L* = **3** 51% yield, 81% de (*trans*), 59.5% ee (1*S*,2*S*)
i-Pr

62

SCHEME 17.22

17.7.2 ASYMMETRIC 1,4-DIALKYLATION OF 1,3-DIENES

The potential of asymmetric induction of chiral benzoquinones containing β-amino alcohols as ligands in palladium(II) catalysis was demonstrated by Bäckvall and co-workers, although enantioselectivities remained low.[29] A variety of β-amino alcohols was tested as ligand components in the 1,4-dialkylation of 1,3-dienes, and it became clear that introducing a substituent on the β-carbon resulted in decreased selectivities, whereas introduction of bulky groups on the α-carbon enhanced enantioselectivities (Scheme 17.23). It was hypothesized that the hydroxy group was coordinating to the palladium center, which brought the sterically demanding group at the π-position of the amide closer to the π-allyl. This intermediate **63** is in equilibrium with the noncoordinated complex **64**, which should result in lower levels of asymmetric induction. Sterically demanding groups at the β-carbon would indeed favor the noncoordinated intermediate. The effect of the solvent seemed to confirm this suggested mechanism because the use of nonpolar noncoordinating solvents led to higher enantioselectivities.[29]

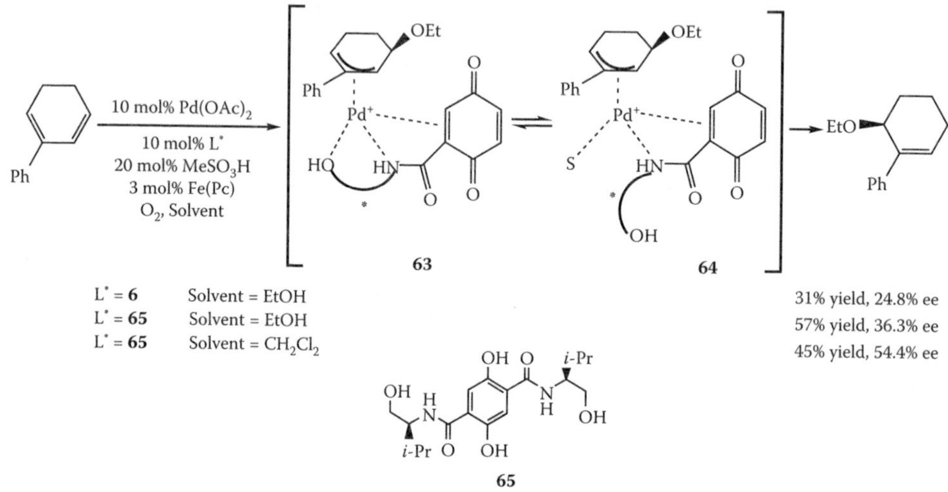

L* = **6**	Solvent = EtOH	31% yield, 24.8% ee
L* = **65**	Solvent = EtOH	57% yield, 36.3% ee
L* = **65**	Solvent = CH₂Cl₂	45% yield, 54.4% ee

SCHEME 17.23

17.7.3 ALLYLIC ALKYLATIONS

Helmchen and co-workers reported phosphinooxazoline (PHOX) ligands to be particularly efficient for the palladium-catalyzed allylic alkylations.[23,53] Although P,N-chelate ligand **64** furnished excellent results with large acyclic substrates (e.g., 1,3-diphenylallyl acetate) up to 99% ee, reaction on small acyclic substrates [e.g., 1,3-dimethylallyl acetate (**67**)] gave low enantioselectivities. Taking into account the many aspects of the allylic alkylation mechanism, it was argued that bulkier substituents on the oxazoline ring would lead to improved enantiofacial discrimination of small acyclic substrates. Indeed, constrained ligands **4** and **68** led to considerably increased enantioselectivities (Scheme 17.24), which were further optimized by running the reaction at lower temperatures (**68** led to 89.5% ee at –40°C).[23]

A series of chiral P,N-bidentate aryl phosphite ligands were studied for the allylic alkylation of small acyclic substrates (Figure 17.8). In this case, the rigid indanyl backbone led to moderate enantioselectivities (50% ee using **5** in THF) compared to ligand **69** (69% in THF and 82% in CH₂Cl₂).[28] The fact that the allylic alkylations could be performed at room temperature and did not necessitate lower reaction temperatures to achieve high selectivities is particularly noteworthy. These ligands were also tested in the allylic sulfonation reaction, and the same trends in enantioselectivities were observed.[28]

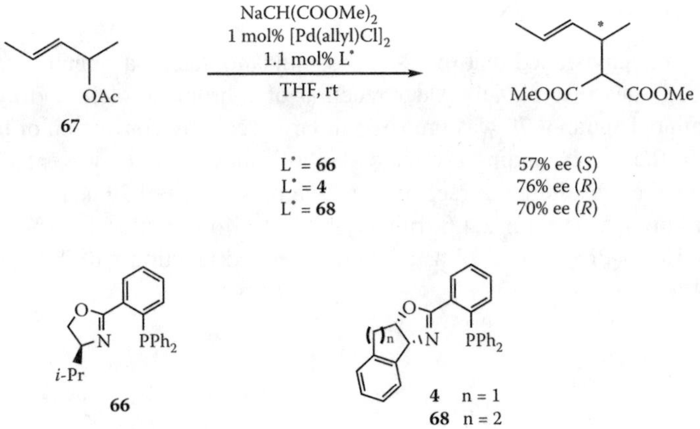

L* = 66	57% ee (S)
L* = 4	76% ee (R)
L* = 68	70% ee (R)

66

i-Pr

4 n = 1
68 n = 2

SCHEME 17.24

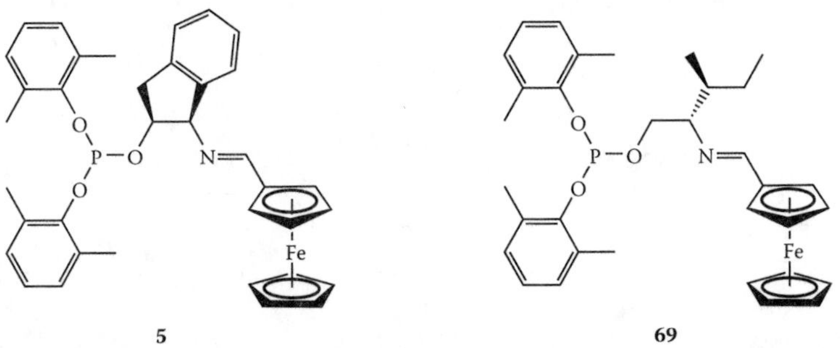

5

69

FIGURE 17.8 P,N-Bidentate aryl phosphite ligands.

17.8 RELATED AMINO ALCOHOLS AND THEIR APPLICATIONS IN ASYMMETRIC SYNTHESIS

Isomers of *cis*-1-amino-2-indanol have attracted considerably less attention, although improved asymmetric inductions have been reported on several occasions. Diethylzinc addition to aldehydes with *cis*-*N*-disubstituted-1-amino-2-indanols as catalysts yielded secondary alcohols with low enantiomeric excesses (40–50%),[37] whereas *cis*-*N*-disubstituted-2-amino-1-indanols led to increased selectivities (up to 80% ee) (see Section 17.3.2).[46] High degrees of enantioselection were eventually achieved in the addition of diethylzinc to aliphatic and aromatic aldehydes with *trans*-*N*-dialkyl-1-substituted-2-amino-1-indanols as catalysts (Scheme 17.25).[47] Optimal results were obtained with bulky groups at the hydroxy-bearing carbon and at the nitrogen (R = Ph, R¹ = *n*-Bu), which led to the formation of (*R*)-1-phenylpropanol in 90% yield and 93% ee.

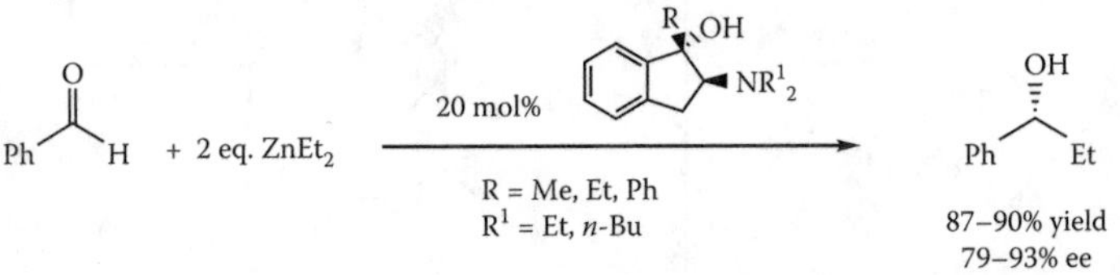

| 20 mol% |
| R = Me, Et, Ph |
| R¹ = Et, *n*-Bu |

87–90% yield
79–93% ee

SCHEME 17.25

17.8.1 *cis*-*N*-Sulfonyl-2-Amino-1-Indanol

Corey and co-workers envisioned that the 2-amino-1-indanol–derived titanium complex **74** could provide high selectivities in the Diels-Alder reaction of 2-bromoacrolein and cyclopentadiene.[54] Racemic *cis*-2-amino-1-indanol **70** was prepared in large scale by conversion of the racemic *trans* bromohydrin **71** to the corresponding *cis*-azido alcohol followed by hydrogenation. The optically pure **72** was obtained by resolution of racemic *cis*-2-amino-1-indanol **70** using (–)-tartaric acid and subsequent amine protection with 2,4,6-trimethylbenzylsulfonyl chloride. The sulfonamide was then treated with Ti(O-*i*-Pr)$_4$ to give **73** as an aggregate, and reaction with SiCl$_4$ finally led to the active catalyst **74** (Scheme 17.26).[54]

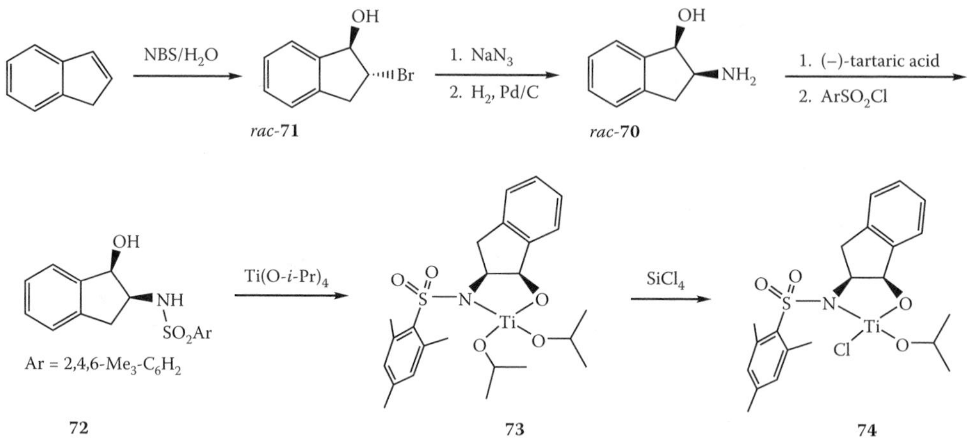

SCHEME 17.26

Diels-Alder reaction of 2-bromoacrolein and cyclopentadiene using 10 mol% of titanium catalyst **74** gave the synthetically versatile (*R*)-bromoaldehyde adduct **75** in 94% yield, 67:1 *exo:endo* diastereoselectivity, and 93% ee. The absolute stereochemical outcome of the reaction is consistent with the proposed transition state assembly **76** in which the dienophile coordinates at the axial site of the metal, proximal to the indane moiety through π-attractive interactions. In this complex, the π-basic indole and the π-acidic dienophile can assume a parallel orientation facilitated by the octahedral geometry of the transition metal. The aldehyde would then react through a preferential *s-cis* conformation (Scheme 17.27).[54]

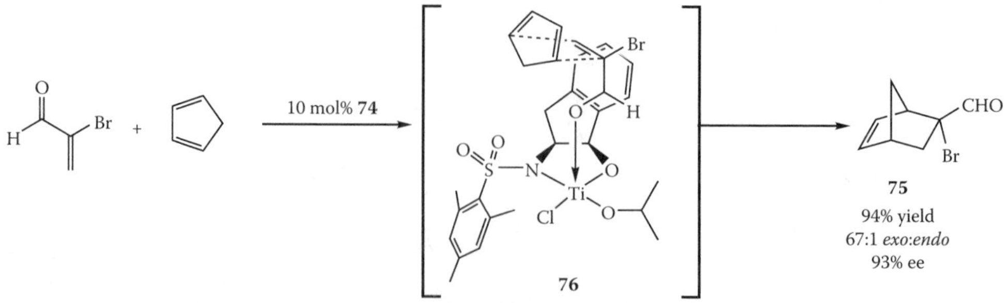

SCHEME 17.27

17.8.2 *CIS*-2-AMINO-3,3-DIMETHYL-1-INDANOL

cis-2-Amino-3,3-dimethyl-1-indanol (**77**) can be used as the basis for a ligand as well as a chiral auxiliary (see Section 24.5.1). Thus, Saigo and co-workers synthesized phosphorous-containing oxazoline **78** derived from **77** and studied the utility of this new ligand in palladium- and rhodium-catalyzed asymmetric processes.[55–57] The chiral ligand was rapidly prepared by condensation of optically pure (1*R*,2*S*)-**77** with 2-fluoronitrile in the presence of a catalytic amount of zinc chloride followed by reaction with potassium diphenylphosphide (Scheme 17.28).[55]

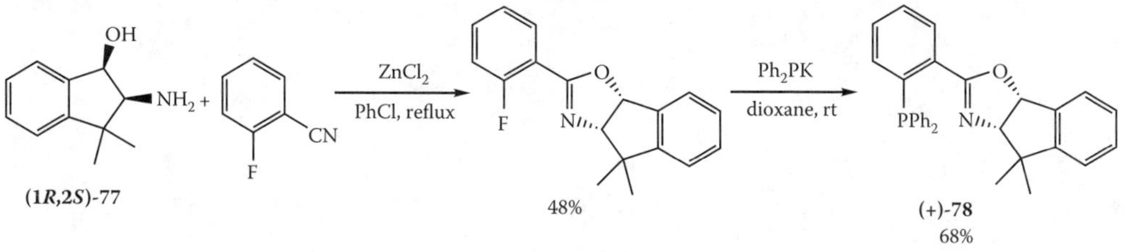

SCHEME 17.28

Ligand (+)-**78** was first tested in the palladium-catalyzed allylic amination reaction. Preliminary studies on 1,3-bisphenyl-2-propen-1-yl acetate (**79**) showed that phosphine-oxazoline **78** was a more effective ligand than the parent ligands **80** and **81** with shorter reaction time and better enantioselectivity (Scheme 17.29).[55,58] Other 1,3-bis(*p*-substituted-aryl)-2-propen-1-yl acetates were also converted to the corresponding amines with excellent selectivity (>95% ee). The amination reaction of 1-alkyl-3,3-diphenyl-2-propen-1-yl acetates was examined, and addition of acetic acid to the reaction system was found to be necessary to minimize the competing elimination reaction and to achieve high enantioselectivities.[55] In the case of **82** (alkyl = Me), the allylic amination with 2 equivalents of benzylamine led to a mixture of allyl amine **83** (38% yield, 92% ee) and elimination product **84** (54% yield). Addition of 10 equivalents of acetic acid resulted in a faster reaction, decreased elimination product formation (13%), and higher enantiomeric excesses of **83** (98% ee). Increased amounts of benzylamine (10 equiv.) were necessary to achieve high chemical yields of allylic aromatic product, probably to balance the acidity of the medium, which may lead to ligand decomposition.

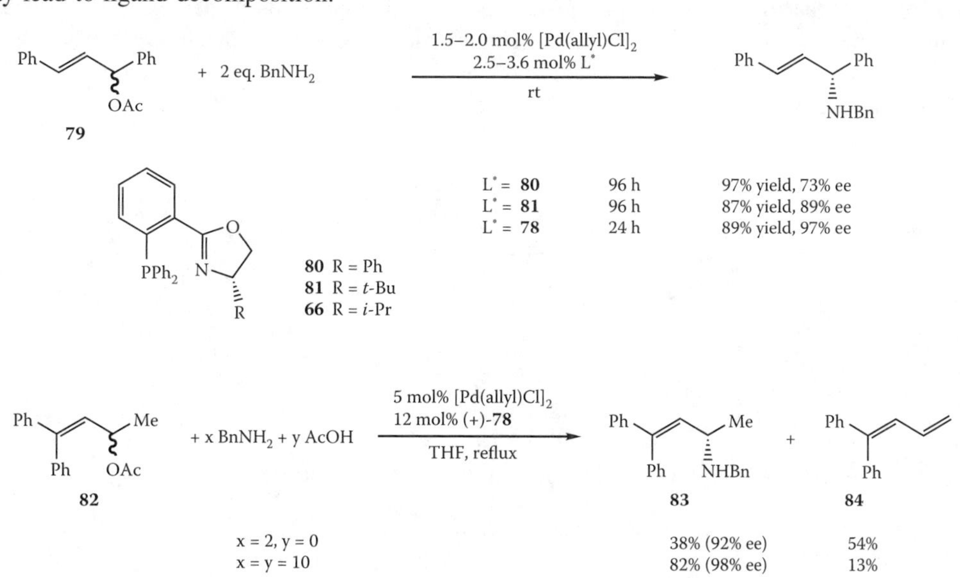

SCHEME 17.29

Encouraged by these successful results, Saigo and co-workers tested ligand **78** in the rhodium-catalyzed hydrosilylation of ketones.[56] Indeed, asymmetric hydrosilylation of acetophenone and tetralone using **78** as a chiral source led to considerably improved enantioselectivities (94% and 89% ee, respectively) compared to reactions performed with valinol-derived phosphorous-containing oxazoline **66** (82% and 59%, respectively).[59,60] The equal accessibility of the two enantiomers of the *cis*-2-amino-3,3-dimethyl-1-indanol backbone in **78** represented an additional advantage over oxazoline **66**, which is derived from an amino alcohol of the chiral pool because (*S*)-tetralol could easily be obtained using (–)-**78** in 97% yield and 92% ee (Scheme 17.30).[56]

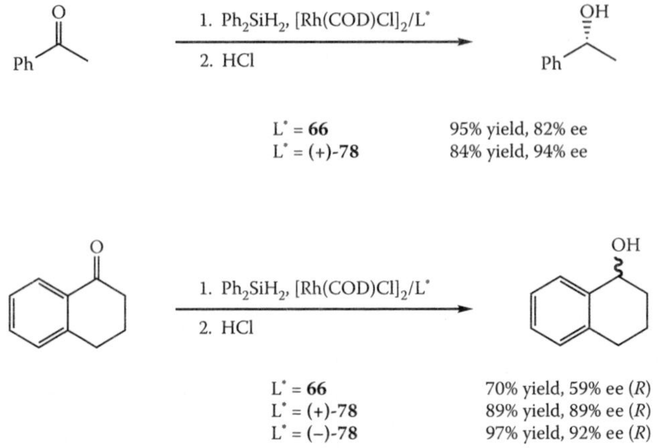

L· = **66**   95% yield, 82% ee
L· = (+)-**78**   84% yield, 94% ee

L· = **66**   70% yield, 59% ee (*R*)
L· = (+)-**78**   89% yield, 89% ee (*R*)
L· = (–)-**78**   97% yield, 92% ee (*R*)

SCHEME 17.30

Phosphine-oxazoline **78** also proved a valuable ligand in the palladium-catalyzed asymmetric Heck reaction.[57] Although slightly less selective than ligand **81**,[61] the Heck coupling of aryl and alkenyl triflates using **78** as ligand proceeded with high asymmetric induction (>90% ee) in all cases. Furthermore, both enantiomers of Heck reaction product can easily be obtained by proper choice of the ligand enantiomer (Scheme 17.31). Low yields were observed with bulky alkenes and were attributed to the rigid methyl groups of **78**, which ensure high stereoselection but may also hinder the approach of the alkene in the coordination sphere of the metal. Finally, in contrast to BINAP ligand, which led to a 2,3- and 2,5-dihydro-2-phenylfuran mixture, phosphine-oxazolines were shown to have a low tendency to promote carbon–carbon bond migration. In the case of phosphine-oxazoline **78**, no migration was observed.[57]

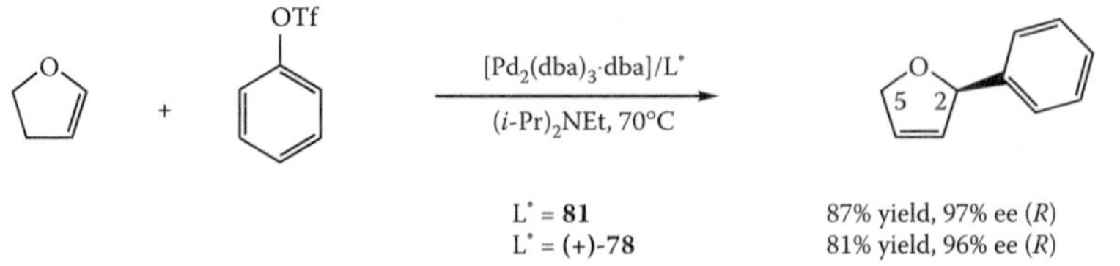

L· = **81**   87% yield, 97% ee (*R*)
L· = (+)-**78**   81% yield, 96% ee (*R*)

SCHEME 17.31

17.8.3 *cis*-1-Amino-2-Hydroxy-1,2,3,4-Tetrahydronaphthalene

Another promising conformationally constrained amino alcohol is *cis*-1-amino-2-hydroxy-1,2,3,4-tetrahydronaphthalene. Early studies using the tetrahydronaphthol backbone as a ligand in catalysis often showed that the relative flexibility of the 6-membered ring was highly detrimental to the achievement of good levels of asymmetric induction, such as in the Diels-Alder reaction[62] or in the addition of diethylzinc to benzaldehyde.[63]

Jørgensen and co-workers have described the first enantioselective Lewis acid–catalyzed Mannich reaction of imino glycine alkyl esters with imines as a new approach to optically active α,β-diamino acid derivatives.[66] In the course of the study, copper(I) complexes of phosphine-oxazoline ligands were found to be the most effective catalysts for the transformation. Among the P,N-ligands tested, those derived from (1*R*,2*S*)-1-amino-2-hydroxy-1,2,3,4-tetrahydronaphthalene gave the most encouraging results in terms of *syn:anti* selectivities. Although **68** led to improved diastereoselectivities compared to **4**, enantioselectivities were much poorer. Rapid steric and electronic tuning of the phosphine aryl substituents yielded the novel ligand **85**, which gave the diamine adduct **86** in a remarkable 79:21 *syn:anti* selectivity and 97% ee for the *syn*-isomer (Scheme 17.32).[66]

L* = *ent*-**66**	81% yield	67:33 *syn:anti*	64% ee (syn)/68% ee (*anti*)
L* = **87**	>10%	–	—
L* = **4**	>95%	63:37	84% ee/82% ee
L* = **68**	76%	72:28	14% ee/46% ee
L* = **85**	94%	79:21	97% ee/94% ee

66 R = *i*-Pr
87 R = Ph

4 n = 1, Ar = Ph
68 n = 2, Ar = Ph
85 n = 2, Ar = 2,4,6-Me-C$_6$H$_2$

SCHEME 17.32

Spectroscopic ^{1}H nuclear magnetic resonance investigations revealed that only **88** coordinated to the Lewis acid. It was therefore assumed that the catalyst activates **88** by coordination followed by deprotection by the base to give the chiral ligand Cu(I)-stabilized imino glycine alkyl ester anion. Mechanistic studies gave the geometry of **90–93** derived from ligands **66** and **85** as tetrahedral around the copper center (Figure 17.9).[66] Semiempirical PM3 calculations showed that intermediate

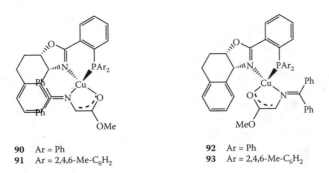

90 Ar = Ph
91 Ar = 2,4,6-Me-C$_6$H$_2$

92 Ar = Ph
93 Ar = 2,4,6-Me-C$_6$H$_2$

FIGURE 17.9 Possible coordination modes.

90 is <1 kcal/mol less stable than **92**, whereas **91** was found 12.4 kcal/mol more stable than **93**. The remarkable gap in energy differences correlates and accounts for the difference in enantioselectivities obtained experimentally with ligands **66** and **85** (20% ee and 97% ee, respectively). The relative stability of **91** compared to **93** was attributed to steric repulsion in **93** between the 2- and 6-methyl substituents of the phosphine aryl groups of the chiral ligand with the phenyl groups of the benzophenone imine. The same methyl substituents of the phosphine aryl groups apparently shield the *Re*-face of the carbon atom of **88**, which acts as a nucleophile. Indeed, the proposed preferred approach of imine **89** from the *Si*-face of benzophenone imine **88** accounts for the diastereoselectivities and enantioselectivities observed (Figure 17.9).[66]

17.9 SUMMARY

Since the discovery of *cis*-1-amino-2-indanol as a ligand for human immunodeficiency virus protease inhibitors and the development of a practical industrial process for the synthesis of either *cis*-isomers in enantiopure form, the remarkable properties of the rigid indane platform have been extensively used in an ever-increasing number of asymmetric methodologies.

The explosion of new aminoindanol-derived ligands disclosed in the past few years is the unmistakable mark of recognition and appreciation of the indanyl platform as a valuable backbone in the field of catalytic asymmetric synthesis. The emergence of related and novel structural designs and their success in achieving improved selectivities clearly reveal the continuing interest in the synthetic potential of conformationally constrained amino alcohols.

REFERENCES

1. Faller, J. W., Lavoie, A. R. *Org. Lett.* 2001, *3*, 3703.
2. Faller, J. W., Lavoie, A. R. *Organometallics* 2002, *21*, 3493.
3. Kenny, J. A., Versluis, K., Heck, A. J. R., Walsgrove, T., Wills, M. *J. Chem. Soc., Chem. Commun.* 2000, 99 and references therein.
4. Ghosh, A. K., Mathivanan, P., Cappiello, J. *Tetrahedron: Asymmetry* 1998, *9*, 1.
5. Fache, F., Schulz, E., Tommasino, M. L., Lemaire, M. *Chem. Rev.* 2000, *100*, 2159.
6. Hong, Y, Gao, Y., Nie, X., Zepp, C. M. *Tetrahedron Lett.* 1994, *35*, 6631.
7. Gao, Y., Hong, Y, Zepp, C. M. PCT Appl. WO 95/32937, 1995.
8. Mathre, D. J., Thompson, A. S., Douglas, A. W., Hoogsteen, K., Carroll, J. D., Corley, E. G., Grabowski, E. J. J. *J. Org. Chem.* 1993, *58*, 2880.
9. Davies, I. W., Senanayake, C. H., Larsen, R. D., Verhoeven, T. R., Reider, P. J. *Tetrahedron Lett.* 1996, *37*, 1725 and references therein.
10. Davies, I. W., Gerena, L., Cai, D., Larsen, R. D., Verhoeven, T. R., Reider, P. J. *Tetrahedron Lett.* 1997, *38*, 1145 and references therein.
11. Ghosh, A. K., Mathivanan, P., Cappiello, J. *Tetrahedron Lett.* 1996, *37*, 3815 and references therein.
12. Evans, D. A., Miller, S. J., Lectka, T. *J. Am. Chem. Soc.* 1993, *115*, 6460.
13. Corey, E. J., Ishihara, K. *Tetrahedron Lett.* 1992, *33*, 6807.
14. Davies, I. W., Gerena, L., Castonguay, L., Senanayake, C. H., Larsen, R. D., Verhoeven, T. R., Reider, P. J. *J. Chem. Soc., Chem. Commun.* 1996, 1753.
15. Ghosh, A. K., Cho, H., Cappiello, J. *Tetrahedron: Asymmetry* 1998, *9*, 3687 and references therein.
16. Park, J. K., Kim, S.-W., Hyeon, T., Kim, B. M. *Tetrahedron: Asymmetry* 2001, *12*, 2931.
17. Ghosh, A. K., Mathivanan, P., Cappiello, J., Krishnan, K. *Tetrahedron: Asymmetry* 1996, *7*, 2165.
18. Ghosh, A. K., Mathivanan, P., Cappiello, J. *Tetrahedron Lett.* 1997, *38*, 2427.
19. Sibi, M. P., Chen, J. *Org. Lett.* 2002, *4*, 2933.
20. Sibi, M. P., Ji, J., Wu, J. H., Gürtler, S., Porter, N. A. *J. Am. Chem. Soc.* 1996, *118*, 9200.
21. Davies, I. W., Senanayake, C. H., Larsen, R. D., Verhoeven, T. R., Reider, P. J. *Tetrahedron Lett.* 1996, *37*, 813.
22. Davies, I. W., Gerena, L., Lu, N., Larsen, R. D., Reider, P. J. *J. Org. Chem.* 1996, *61*, 9629.

23. Wiese, B., Helmchen, G. *Tetrahedron Lett.* 1998, *39*, 5727.
24. Dossetter, A. G., Jamison, T. F., Jacobsen, E. N. *Angew. Chem., Int. Ed.* 1999, *38*, 2398.
25. Joly, G. D., Jacobsen, E. N. *Org. Lett.* 2002, *4*, 1795.
26. Gademann, K., Chavez, D. E., Jacobsen, E. N. *Angew. Chem., Int. Ed.* 2002, *41*, 3059.
27. Ruck, R. T., Jacobsen, E. N. *J. Am. Chem. Soc.* 2002, *124*, 2882.
28. Gavrilov, K. N., Bondarev, O. G., Lebedev, R. V., Shiryaev, A. A., Lyubimov, S. E., Polosukhin, A. I., Grintselev-Knyazev, G. V., Lyssenko, K. A., Moiseev, S. K., Ikonnikov, N. S., Kalinin, V. N., Davankov, V. A., Korostylev, A. V., Gais, H.-J. *Eur. J. Inorg. Chem.* 2002, 1367.
29. Itami, K., Palmgren, A., Thorarensen, A., Bäckvall, J.-E. *J. Org. Chem.* 1998, *63*, 6466.
30. Shintani, R., Fu, G. *Org. Lett.* 2002, *4*, 3699.
31. Noyori, R., Yamakawa, M., Hashiguchi, S. *J. Org. Chem.* 2001, *66*, 7931.
32. Itsuno, S., Sakurai, Y., Ito, K., Hirao, A., Nakahama, S. *Bull. Chem. Soc. Jpn.* 1987, *60*, 395.
33. Corey, E. J., Bakshi, R. K., Shibata, S. *J. Am. Chem. Soc.* 1987, *109*, 5551.
34. Wallbaum, S., Martens, J. *Tetrahedron: Asymmetry* 1992, *3*, 1475.
35. Deloux, L., Srebnik, M. *Chem. Rev.* 1993, *93*, 763.
36. Didier, E., Loubinoux, B., Ramos Tombo, G. M., Rihs, G. *Tetrahedron Lett.* 1991, *47*, 4941.
37. Di Simone, B., Savoia, D., Tagliavini, E., Umani-Ronchi, A. *Tetrahedron: Asymmetry* 1995, *6*, 301.
38. Hett, R., Fang, Q. K., Gao, Y., Hong, Y., Butler, H. T., Nie, X., Wald, S. A. *Tetrahedron Lett.* 1997, *38*, 1125.
39. Hett, R., Fang, Q. K., Gao, Y., Wald, S. A., Senanayake, C. H. *Org. Proc. Res. Dev.* 1998, *2*, 96.
40. Hett, R., Senanayake, C. H., Wald, S. A. *Tetrahedron Lett.* 1998, *39*, 1705.
41. Kaptein, B., Elsenberg, H., Minnaard, A. J., Broxterman, Q. B., Hulshof, L. A., Koek, J., Vries, T. R. *Tetrahedron: Asymmetry* 1999, *10*, 1413.
42. Jones, S., Atherton, J. C. C. *Tetrahedron: Asymmetry* 2000, *11*, 4543.
43. Dixon, R. A., Jones, S. *Tetrahedron: Asymmetry* 2002, *13*, 1115.
44. Flores-Lopéz, L. Z., Parra-Hake, M., Somanathan, R., Walsh, P. J. *Organometallics* 2000, *19*, 2153 and references therein.
45. Hayashi, M, Miyamoto, Y., Inoue, T., Oguni, N. *J. Org. Chem.* 1993, *58*, 1515.
46. Solà, L., Vidal-Ferran, A., Moyano, A., Pericàs, M. A., Riera, A. *Tetrahedron: Asymmetry* 1997, *8*, 1559.
47. Xu, Q., Yang, H., Pan, X., Chan, A. S. C. *Tetrahedron: Asymmetry* 2002, *13*, 945.
48. Sibi, M. P., Ji, J. *J. Org. Chem.* 1997, *62*, 3800.
49. Sibi, M. P., Shay, J. J., Ji, J. *Tetrahedron Lett.* 1997, *38*, 5955.
50. Davenport, A. J., Davies, D. L., Fawcett, J., Garratt, S. A., Russell, D. R. *J. Chem. Soc., Dalton Trans.* 2000, 4432.
51. Li, Z., Fernández, M., Jacobsen, E. N. *Org. Lett.* 1999, *1*, 1611.
52. Nishiyama, H., Itoh, Y., Matsumoto, H., Park, S.-B., Itoh, K. *J. Am. Chem. Soc.* 1994, *116*, 2223.
53. Helmchen, G., Pfaltz, A. *Acc. Chem. Res.* 2000, *33*, 336 and references therein.
54. Corey, E. J., Roper, T. D., Ishihara, K., Sarakinos, G. *Tetrahedron Lett.* 1993, *34*, 8399 and references therein.
55. Sudo, A., Saigo, K. *J. Org. Chem.* 1997, *62*, 5508.
56. Sudo, A., Yoshida, H., Saigo, K. *Tetrahedron: Asymmetry* 1997, *8*, 3205.
57. Hashimoto, Y., Horie, Y., Hayashi, M., Saigo, K. *Tetrahedron: Asymmetry* 2000, *11*, 2205.
58. von Matt, P., Loiseleur, O., Koch, G., Pfaltz, A. *Tetrahedron: Asymmetry* 1997, *5*, 573.
59. Langer, T., Janssen, J., Helmchen, G. *Tetrahedron: Asymmetry* 1996, *7*, 1599.
60. Newman, L. M., Williams, J. M. J. *Tetrahedron: Asymmetry* 1996, *7*, 1597.
61. Loiseleur, O., Hayashi, M., Schmees, N., Pfaltz, A. *Synthesis* 1997, 1338.
62. Davies, I. W., Senanayake, C. H., Castonguay, L., Larsen, R. D., Verhoeven, T. R., Reider, P. J. *Tetrahedron Lett.* 1995, *36*, 7619.
63. Bellucci, C. M., Bergamini, A., Cozzi, P. G., Papa, A., Tagliavini, E., Umani-Ronchi, A. *Tetrahedron: Asymmetry* 1997, *8*, 895.
64. Higashijima, S., Itoh, H., Senda, Y., Nakano, S. *Tetrahedron: Asymmetry* 1997, *8*, 3107.
65. Senanayake, C. H., Fang, K., Grover, P., Bakale, R. P., Vandenbossche, C. P., Wald, S. A. *Tetrahedron Lett.* 1999, *40*, 819.
66. Bernardi, L., Gothelf, A. S., Hazell, R. G., Jørgensen, K. A. *J. Org. Chem.* 2003, *68*, 2583.

18 Enantioselective Hydrogenation of Activated Ketones Using Heterogeneous Pt Catalysts Modified with Cinchona Alkaloids

Martin Studer and Hans-Ulrich Blaser

CONTENTS

18.1   INTRODUCTION

The enantioselective reduction of ketones in general and of functionalized ketones in particular is of high interest, both from an academic and industrial point of view.[1] The most effective methods available today are hydride reductions using both stoichiometric and catalytic amounts of chiral auxiliaries,[2] various biocatalytic methodologies,[3] catalytic transfer hydrogen,[4,5] and catalytic hydrogenation.[5] Although the recent activities in the enantioselective hydrogenation of ketones were predominantly in the field of homogeneous catalysis, highlighted by the Nobel Prize for Ryoji Noyori, the use of chirally modified heterogeneous catalyst has a much longer tradition and has been very successful for several classes of activated ketones.[6] Here we will summarize the application of heterogeneous platinum catalysts modified with cinchona alkaloids (for structures see Figure 18.1)[7] to the enantioselective hydrogenation of a number of ketones carrying an activating substituent in the α-position of the carbonyl group (Figure 18.2). We will organize the review with respect to classes of ketones and will restrict our discussion to aspects important for the practical

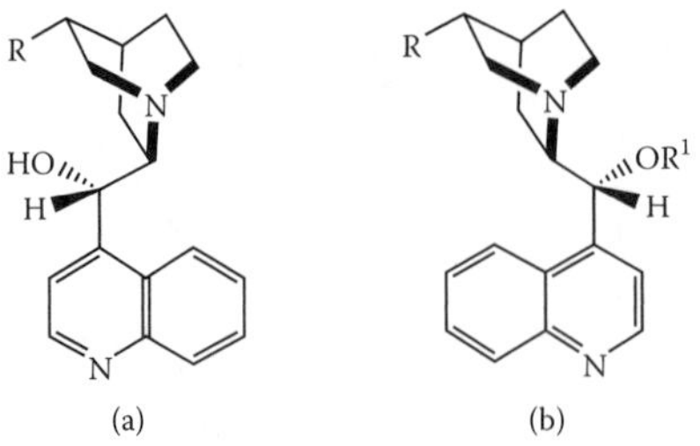

	R	R¹		
Cinchonine	(Cn)	Vinyl	H	(Cd)	Cinchonidine
10,11-Dihydrocinchonine	(HCn)	Ethyl	H	(HCd)	10,11-Dihydrocinchonidine
		Ethyl	OMe	(MeOHCd)	Methoxy-HCd

FIGURE 18.1 Structure, names, and abbreviations of the applied cinchona alkaloids. The Cn modifiers depicted on the left preferentially lead to (S)-alcohols, members of the Cd family to (R)-alcohols.

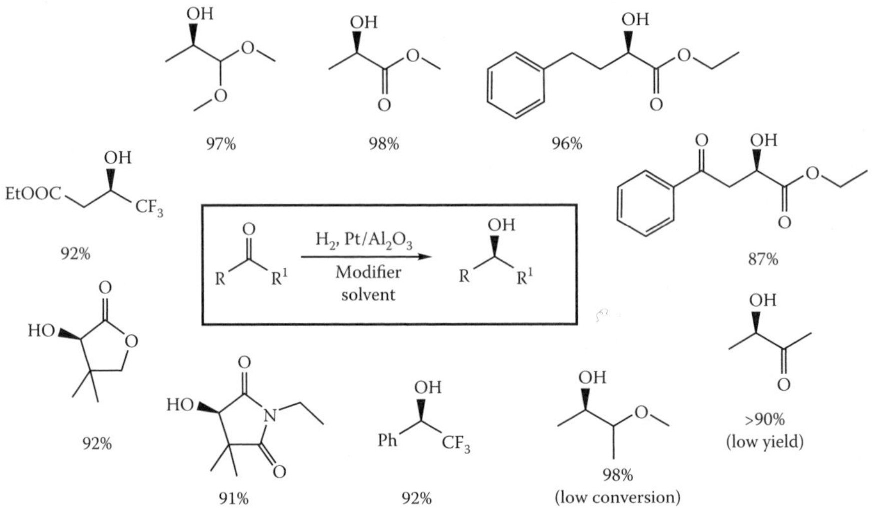

FIGURE 18.2 Alcohols prepared with the Pt-cinchona system and best ee's reported.

application of these catalysts such as scope and limitations concerning substrate and modifier structure or the influence of catalyst and reaction conditions.

18.2 α-KETO ACID DERIVATIVES

α-Keto esters such as methyl and ethyl pyruvate and phenyl glyoxylic acid esters are the substrates giving the highest ee's (Table 18.1) and activities. The following strategies were applied for improving the ee: addition of trifluoroacetic acid[8] and amines,[9] use of special colloidal catalysts,[10] use of ultrasound,[11] slow addition of modifier,[12] and use of solvent mixtures.[13,14] It is not surprising that the reaction conditions have to be optimized for every substrate. An analysis of the results shows that AcOH in combination with MeOHCd or HCd gave the best results for most aliphatic keto ester derivatives. For aromatic and conjugated systems, the combination of HCd and toluene was usually optimal. In most cases, the question of activity and productivity of the catalytic systems was not addressed.

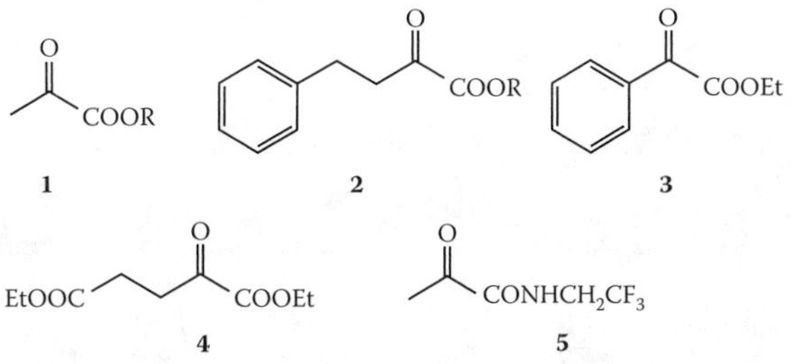

FIGURE 18.3 α-Keto acid derivatives listed in Table 18.1.

TABLE 18.1
Best ee for Various α-Keto Acid Derivatives
(For Substrate Structures see Figure 18.3)

Substrate	ee (%)	Catalyst, Modifier, Solvent, Reaction Conditions, Remarks	Ref.
1 R = Me	98	Pt colloids, Cd, AcOH, 40 bar, 25°C	10
1 R = Et	97	5% Pt/Al$_2$O$_3$ (E 4759), MeOHCd, AcOH, 10 bar, 25°C, ultrasound	11a
2 R = Et	96	5% Pt/Al$_2$O$_3$ (E 4759), MeOHCd, AcOH, 10 bar, 25°C, ultrasound	11a
2 R = Et	94	1% Pt/Al$_2$O$_3$ (Aldrich), HCd, AcOH, 5.8 bar, 17°C, dosing of modifier	12
2 R = H	85	5% Pt/Al$_2$O$_3$ (JMC 94), MeOHCd, EtOH/H2O 9:1, 100 bar, 20–30°C	13
3	98	5% Pt/Al$_2$O$_3$ (E 4759), HCd, AcOH / toluene mixture, 25 bar, 0°C	14
4	96	5% Pt/Al$_2$O$_3$ (JMC 94), MeOHCd, AcOH, 20 bar, 20°C	15
5	60	5% Pt/Al$_2$O$_3$ (E 4759), Cd, AcOH, 60 bar, room temperature	16

Szöri and colleagues[17] investigated a series of α-keto ester RCOCOOR[1] and found that bulky R and to a lower extent R[1] groups caused a significant decrease in ee as well as rate. Similar results were reported for various R[1] with Pt-colloids.[10] For the corresponding α-keto acids, much less work has been carried out; a preliminary study found somewhat lower ee's and different optimal solvent systems than for the corresponding esters.[13] Keto amides related to **5** also showed significantly lower ee's.[16] The hydrogenation of **2** (R = Et) was scaled up to several 100-kg scale for the production of (R)-2-hydroxy-4-phenylbutyric acid ethyl ester [(R)-HPB ester], an important intermediate for the synthesis of angiotensin-converting enzyme (ACE) inhibitors (*vide infra*).[18a]

18.3 α,γ-DIKETO ESTERS

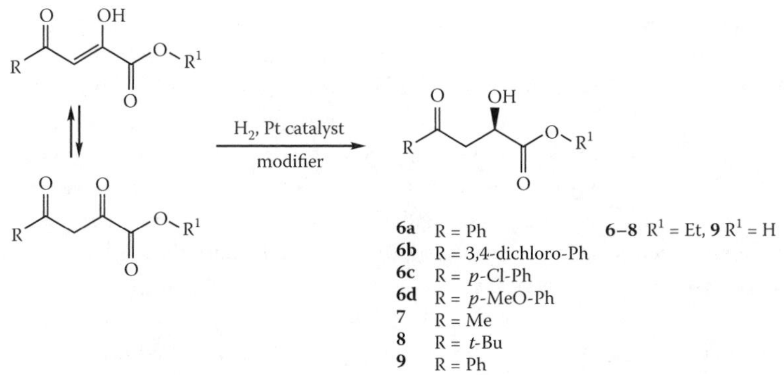

SCHEME 18.1

The hydrogenation of 2,4-diketo acid derivatives **6–9** to the corresponding 2-hydroxy compounds with cinchona-modified Pt catalysts as depicted in Scheme 18.1 can be carried out with chemoselectivities

TABLE 18.2
Hydrogenation of Various α,γ-Diketo Esters[18c]

Substrate	Solvent	Catalyst	Pressure (bar)	Modifier	ee (%)	Rate (mmol g^{-1} min^{-1})
6a	Toluene	E 4759	5	HCd	80 (R)	1.5
6a	Toluene	E 4759	60	HCd	86 (R)	3.4
6a	Toluene	E 4759	135[a]	HCd	87 (R)	4.0
6a	Toluene	E 4759	60	HCn	59 (S)	0.9
6a	AcOH	E 4759	60	HCn	48 (S)	0.8
9	Toluene	JMC 94	60	MeOHCd	23 (R)	0.24
9	EtOH	JMC 94	60	MeOHCd	52 (R)	3.4
9	THF	JMC 94	60	MeOHCd	56 (R)	0.88
6b	Toluene	E 4759	60	HCd	84 (R)	n.d.
6b	Toluene	E 4759	60	HCn	68 (S)	1.3
6c	Toluene	E 4759	60	HCd	82 (R)	5.0
6c	Toluene	E 4759	60	HCn	64 (S)	n.d.
6d	Toluene	E 4759	60	HCd	79 (R)	5.0
6d	Toluene	E 4759	60	HCn	49 (S)	n.d.
7	Toluene	JMC 94	60[b]	HCd	78 (R)	3.5
7	AcOH	JMC 94	60	MeOHCd	65 (R)	7.5
8	Toluene	JMC 94	60	HCd	63 (R)	0.5
8	AcOH	JMC 94	60	MeOHCd	74 (R)	1.5

Reaction conditions: Magnetically stirred 50 ml autoclave, 1–2 g substrate, 5% Pt/Al$_2$O$_3$ catalyst (2 h pretreated with H$_2$ at 400°C), 5–50 mg modifier, 15–20 mL solvent, 25°C. Content of diol always <5%.

[a] 0°C.
[b] 5°C.

of >99% and enantioselectivities up to 87% (R) and 68% (S), respectively.[18b] Enrichment to >98% ee was possible for substrates **6a–d** by crystallization, opening up an efficient technical synthesis of (R)-HPB ester, a building block for several ACE inhibitors.[18c]

For all substrates tested, the modified systems proved to be highly chemoselective. This is remarkable, considering that only a few homogenous catalysts show high chemoselectivity for **7** and **8** and even fewer show both high chemoselectivity and enantioselectivity.[19] As shown for **6a**, the highest ee's (70–80%) were usually obtained in toluene. The only exception was **8**, where AcOH gave 74% and toluene gave only 63% ee. For all esters, the reactions in EtOH proceeded with the lowest ee's but often with the highest rates (results not shown). Comparing the results for **6a**, **6b**, **6c**, and **6d**, it seems that electron-withdrawing groups on the phenyl groups lead to slightly higher ee's and rates. For all these substrates, HCn was also investigated as modifier to give the (2S)-4-keto esters, but compared with the (R)-series, the ee's were always lower by 20–30%.

An improved new synthetic route to (R)-HPB ester with the selective hydrogenation of an α,γ-diketo ester as a key step has been developed and is feasible on large scale (Scheme 18.2).[18b,c] The following aspects were the keys to success: a) the low price of the diketo ester (prepared via Claisen condensation of acetophenone and diethyl oxalate); b) the high chemoselectivity in the Pt-cinchona catalyzed hydrogenation, and c) the possibility to enrich the hydroxy ketone intermediate with ee's as low as 70% to >99% ee in one crystallization step. The removal of the second keto group via Pd-catalyzed hydrogenolysis did not lead to any racemization. A whole range of chiral building blocks derived from the keto hydroxy intermediate depicted in Figure 18.4 is now available in laboratory quantities from Fluka both in the (R)- and in the (S)-form.[18d,e]

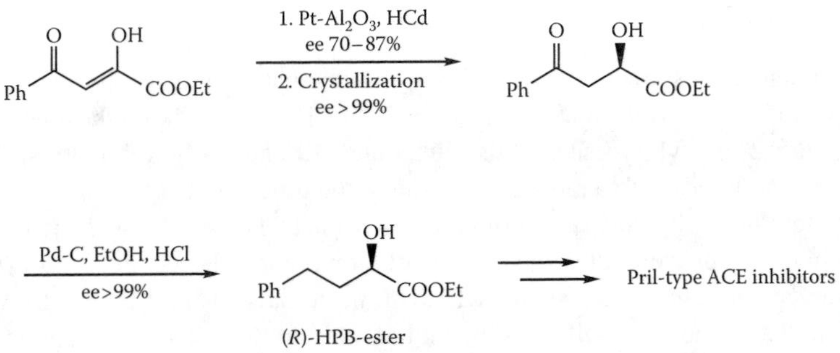

SCHEME 18.2

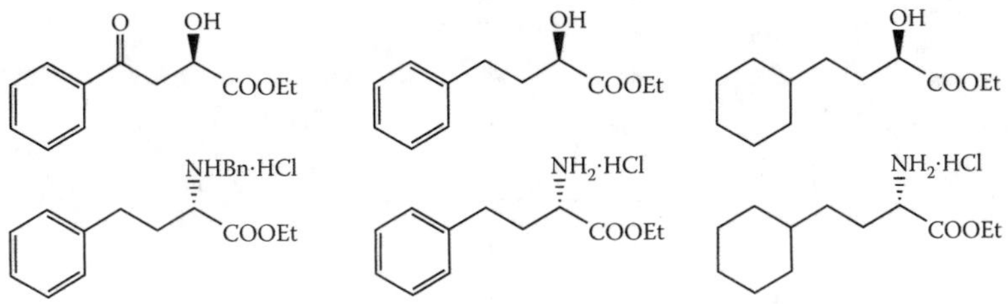

FIGURE 18.4 HPB ester–related compounds available from Fluka.[18d,e]

18.4 TRIFLUOROMETHYL KETONES

The trifluoromethyl group was identified as having similar activating properties to an ester group in 1997, and several papers by Baiker and colleagues elaborate on this topic.[20] Methyl, ethyl, and isopropyl esters of 4,4,4-trifluoroacetoacetate gave ee's of 90–93% in AcOH or trifluoroacetic acid/THF mixtures, and MeOHCd was often significantly more efficient than HCd (see Table 18.3). Various trifluoroacetophenone derivatives with additional CF_3 or $N(Et)_2$-substitutents on the aromatic ring gave ee's between 36% and 81%, but ee's and turnover frequencies (TOFs) were highest without any substituent. The same substrates showed significantly lower ee's in o-dichlorobenzene/EtOH with Pt-PVP colloids[21] or in o-dichlorobenzene using Pt/Al$_2$O$_3$.[11b,20d] Other CF_3-substituted ketones were tested, but except for 2-trifluoroacetylpyrrole (63% ee), none of them gave ee's significantly higher than 20%.[11b,20d] In contrast to α-keto esters, trifluoroketones easily form hemiketals that can negatively affect ee's during the course of the reaction.[20c]

TABLE 18.3
Best ee's for Trifluoromethyl Ketones

R	ee (%)	Catalyst, Modifier, Solvent, Reaction Conditions, Remarks	Ref.
CH$_2$COOEt	93	5% Pt/Al$_2$O$_3$ (E 4759), MeOHCd, THF/TFA mixture, 10 bar, 20°C	20a
Ph	92	5% Pt/Al$_2$O$_3$ (E 4759), toluene, Cd, 10 bar, 0°C	20e
m-CF$_3$C$_6$H$_3$	81	5% Pt/Al$_2$O$_3$ (E 4759), toluene, Cd, 10 bar, room temperature	20e
p-CF$_3$C$_6$H$_3$	60	5% Pt/Al$_2$O$_3$ (E 4759), Cd, 1,2-dichlorobenzene, 2 bar, room temperature, mass transport limitations	20e
p-Tol	46	5% Pt/Al$_2$O$_3$ (E 4759), Cd, AcOH, 10 bar, room temperature	20e
2-pyrryl	63	5% Pt/Al$_2$O$_3$ (E 4759), Cd, toluene, 2 bar, room temperature	20d

18.5 α-KETO ACETALS

It has been simultaneously shown by two different groups that α-keto acetals can be hydrogenated with high rates and ee's up to 97% (Scheme 18.3; Table 18.4).[22,23] The highest ee and rate values were obtained with 2-methylglyoxal acetals (substrates **12** and **13** in Table 18.4). In these cases the addition of the modifier led to a rate acceleration in the order of 10, comparable to that observed for ethyl pyruvate. Other aliphatic and aromatic α-keto acetals **14–18**, **20–22** with relatively low bulkiness gave high ee's as well but with significantly lower rates. Lower ee's and very much lower rates were observed for keto acetals such as **19** with more bulky R and especially with larger R[1]. Aromatic and aliphatic ethers as well as esters and amides are tolerated as functional groups in the R residue and do not seem to affect the enantioselectivity very much (**21–24**). α-Keto ketals **25** and **26** are hydrogenated very slowly and with negligible induction.

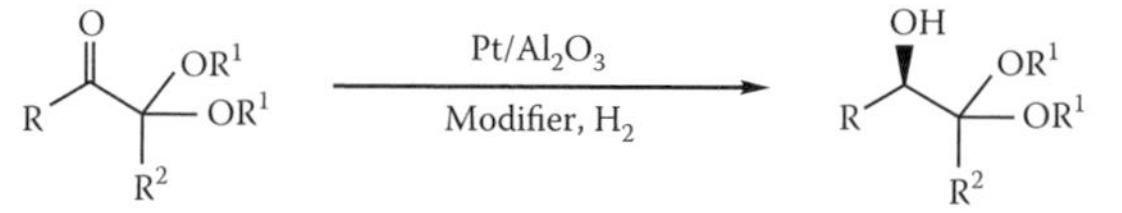

SCHEME 18.3

TABLE 18.4
Effect of Substrate Structure in the Hydrogenation of α-Keto Acetals[22]

Substrate	R	R¹	R²	ee (%)	Rate (mmol/g*min)
12	Me	Me	H	96	53
13	Me	-(CH$_2$)$_3$-	H	97	42
14	Me	Et	H	91	5.8
15	Me	n-Bu	H	85	1.8
16	Ph	Me	H	89	1.5
17	Ph	Et	H	81	0.5
18	CH$_3$CH$_2$CH$_2$	Me	H	93	4
19	(CH$_3$)$_2$CHCH$_2$	Me	H	62	<0.1
20	PhCH$_2$CH$_2$	Me	H	93	15
21	Ph-O-(CH$_2$)$_3$	Me	H	93	5
22	CH$_3$CH$_2$O(CH$_2$)$_3$	Me	H	92	5
23	(CH$_3$)$_2$NOCCH$_2$CH$_2$	Me	H	80	0.7
24	CH$_3$OOCCH$_2$CH$_2$	Me	H	50	<0.1
25	Me	Me	Me	<1	<0.1
26	Ph	Me	Ph	<1	<0.1

Reaction conditions: Magnetically stirred 50 mL autoclave, 1–2 g substrate, 5% Pt/Al$_2$O$_3$ (JMC type 94, 2 h pretreated with H$_2$ at 400°C), 5–50 mg MeOHCd, 15–20 mL AcOH, 60 bar, 25°C.

Enantiomerically enriched α-hydroxy acetals are interesting synthons and can be transformed to a variety of chiral building blocks such as 1,2-diols, α-hydroxy acids, or 1,2-amino alcohols (Scheme 18.4). Whereas the oxidation to (R)-ethyl lactate was rather difficult and required the protection of the OH group, the reduction could be easily accomplished after hydrolysis of the acetal. No significant racemization was observed. With a boronic acid derivative and a secondary amine as described by Petasis and Zavialov,[24] it was also possible to synthesize an amino alcohol with high diastereoselectivity.

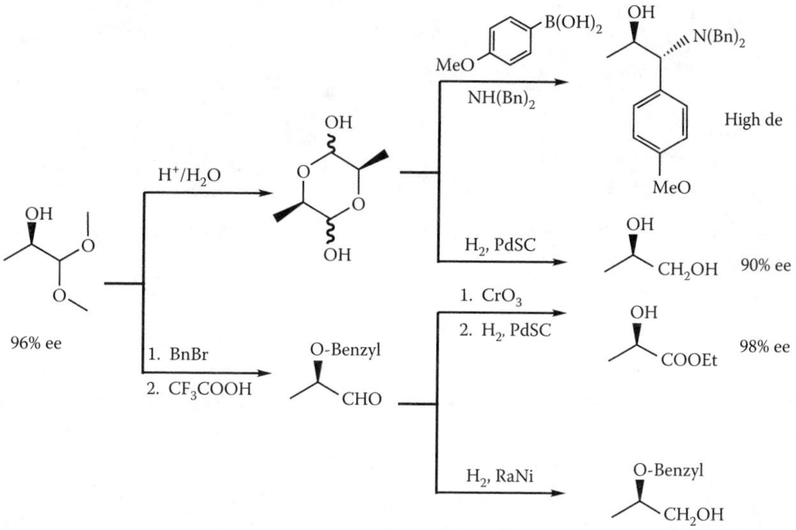

SCHEME 18.4

18.6 α-KETO ETHERS

Substituted aliphatic and aromatic α-keto ethers (Scheme 18.5) are also amenable to enantioselective hydrogenation catalyzed by cinchona-modified Pt catalysts.[25] However, as opposed to the prochiral ketones discussed earlier, kinetic resolution is observed for these chiral substrates. At conversions of 20–42%, ee's of 91–98% were obtained when starting with a racemic substrate (see Table 18.5). It is somewhat surprising that α-keto ethers without substituent in the α-position, such as methoxy acetone, reacted very slowly or not at all and led to very low enantioselectivities,[6] and from the results described earlier for α-ketoacetals, the same is expected if 2 substituents are present.

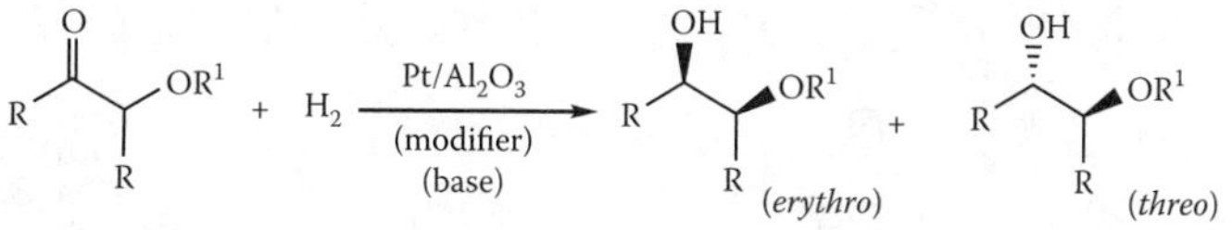

a	R = Me	R^1 = Me
b	2-methoxycyclohexanone	
c	R = Me	R^1 = $CH_2C_6H_5$
d	R = C_6H_5	R^1 = Me
e	R = C_6H_5	R^1 = Et

SCHEME 18.5

TABLE 18.5
Kinetic Resolution of Various α-Keto Ethers

Substrate	Solvent	Modifier	Time (min)	Conversion (%)	Yield of 28 (%)	ee of 28 (%)	Yield of 29 (%)
27a	AcOH	MeOHCd	12	22	20	98	<1
27b	Toluene	—	145	33	25	—	7
27b	Toluene	HCd	24	44	42	92	<1
27b	i-PrOH	HCd	6	38	38	79	<1
27b	AcOH	MeOHCd	16	53	52	88	2
27c	Toluene	HCd	53	7	5	75	2
27d	Toluene	—	23	34	29	—	5
27d	Toluene	HCd	63	42	42	91	<1
27e	Toluene	HCd	13	10	10	90	<1

Reaction conditions: 1–2 g substrate, 20–25 mL solvent, 50–100 mg 5% Pt/Al$_2$O$_3$ (2 h pretreated with H$_2$ at 400°C), 5–10 mg modifier, 25°C, 60 bar.

Although the very high initial ee's were impressive, it was also clear that this method with yields of <50% and gradually decreasing ee's is of little preparative value. The obvious solution would be dynamic kinetic resolution as reported for some homogeneous systems.[26] In fact, with >5 mM KOH in iPrOH a very high rate was obtained for **27b** (100% conversion in 6 min!), but the ee was 0%, most likely because the chirally modified surface was disturbed. To get dynamic kinetic resolution the OH$^-$ ions had to be immobilized on a solid ion exchanger. With OH$^-$-activated Amberlite IRA-402 (gel type) and Amberlite IRA-900 (macroreticular), dynamic kinetic resolution was indeed observed in iPrOH, and (R,S)-**28b** was obtained with ee's of up to 56% at >95% conversion. In toluene, even higher ee's of >80% at >95% conversion with less than 1% of **29b** were obtained, but only macroreticular ion exchangers showed the desired effect. A kinetic analysis of the system is depicted in Figure 18.5 for **27b.** Similar experiments were conducted with **27d,** and an ee of 90% was achieved at 88% conversion.

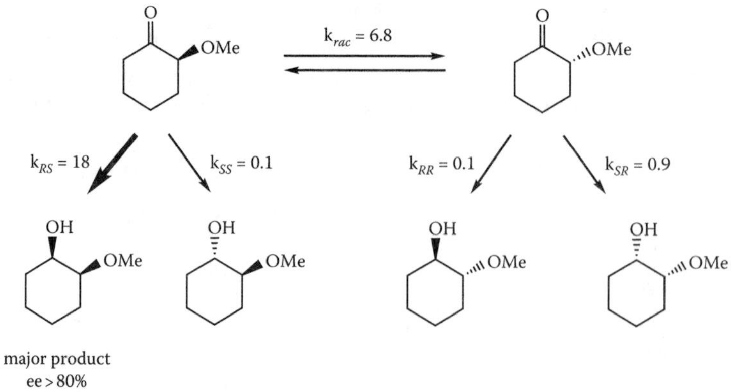

FIGURE 18.5 Reaction pathways, rates, and product isomers for the hydrogenation of cyclic α-keto ether **27b.**[25]

18.7 MISCELLANEOUS KETONES

Under optimized conditions both keto pantolactone (**30**)[27] and a cyclic imidoketone **31**[28] could be hydrogenated with ee's 90%. The hydrogenation of 1,2-butanedione (**32a**)[29,30] and 1-phenyl-1,2-propanedione (**32b**)[31] gave significantly lower ee's. For both **32a** and **32b**, the ee's of the hydroxy ketones increased slowly during the reaction because in a second step the minor enantiomer reacted preferentially to the corresponding diol (kinetic resolution).[29–31] Ketones such as acetophenone,[32] keto isophorone,[33] or various β-keto esters[34] all gave ee's of 20%.

| 30 | 31 | 32a R = Me |
| | | 32b R = Ph |

TABLE 18.6
Best ee for Miscellaneous Activated Ketones

Substrate	ee (%)	Catalyst, Modifier, Solvent, Reaction Conditions	Ref.
30	92	5% Pt/Al$_2$O$_3$ (E 4759), Cd, toluene, 70 bar, –13°C	27
31	91	5% Pt/Al$_2$O$_3$ (E 4759), Cd, toluene, 70 bar, 17°C	28
32a	65 (>90)	5% Pt/Al$_2$O$_3$ (Strem), Cd, CH2Cl2, 5 bar, 25°C.[a]	31
32b	50 (90)	5% Pt/Al$_2$O$_3$ (JMC 94), HCd, toluene, 107 bar, 25°C.[b]	29,30

[a] Main product was 1-hydroxy-1-phenyl propanone. The value in parentheses was obtained by kinetic resolution in EtOAc from enriched hydroxy ketone.
[b] The value in parentheses was obtained by kinetic resolution.

18.8 EXPERIMENTAL PROCEDURES

18.8.1 CATALYSTS

The catalyst of choice is usually 5% Pt/Al$_2$O$_3$. Two commercially available 5% Pt/Al$_2$O$_3$ were until now considered to be "standard" catalysts: E 4759 from Engelhard and JMC 94 from Johnson Matthey.[35] However, both have the disadvantage that reduction in hydrogen at 300–400°C before use is necessary to obtain good performance.[36] A new 5% Pt/Al$_2$O$_3$ (catASium® F214) catalyst has become available from Degussa, where this tedious pretreatment is no longer necessary because high ee's and rates are obtained with the "as received" commercial catalyst.[37]

18.8.2 MODIFIERS

Several types of modifiers have been shown to be successful. Common features are a basic nitrogen atom close to one or more stereogenic centers connected to an extended aromatic system. By far the best overall results are obtained with cinchonidine derivatives for an excess of the (*R*)- and with cinchonine derivatives for the (*S*)-alcohols.[7]

18.8.3 REACTION CONDITIONS

The preferred solvents are acetic acid or toluene; base and acid additives can be effective. As pointed out in several papers,[6] high pressure (20–100 bar) is usually beneficial for good ee's and rates; a temperature range of 0–50°C is suitable.

18.8.4 HYDROGENATION PROCEDURE FOR α-KETO ESTERS AND ACETALS

For a detailed description of a low pressure procedure of the reaction depicted in Scheme 18.6, see reference 38.

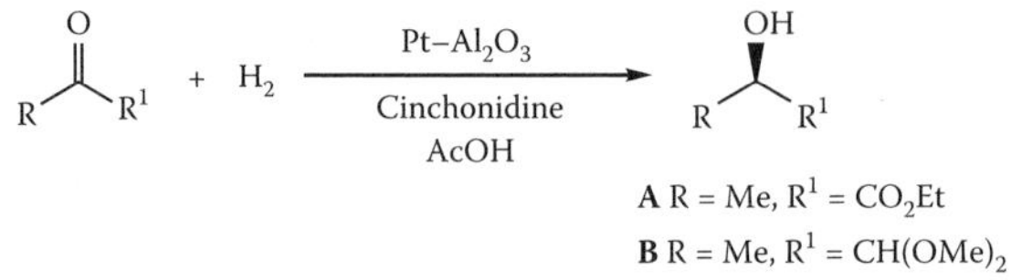

$$
\underset{R}{\overset{O}{\|}}\underset{R^1}{\big\backslash} + H_2 \xrightarrow[\substack{\text{Cinchonidine}\\\text{AcOH}}]{\text{Pt–Al}_2\text{O}_3} \underset{R}{\overset{OH}{\big|}}\underset{R^1}{\big\backslash}
$$

A R = Me, R^1 = CO$_2$Et
B R = Me, R^1 = CH(OMe)$_2$

SCHEME 18.6

Considerably higher ee's (up to 95% for **A** and up to 97% for **B**), with higher reaction rates and, therefore, shorter reaction times can be obtained by the following measures. The use of higher hydrogen pressure (e.g., 60 bar) leads to a significant increase of both rates and the ee's (especially for **A**). The replacement of Cd by HCd or MeOHCd often leads to higher ee's. Generally, acetic acid with MeOHCd gives the best results, but other solvents such as toluene or ethanol are also suitable. Distillation of the starting material often gives higher rates and ee's, especially for **A**. Under optimized conditions, much lower catalyst and modifier loadings for both **A** and **B** are possible (10-mg catalyst and 2-mg modifier for the example described previously).

18.8.5 HYDROGENATION PROCEDURE FOR α-KETO ETHERS[25]

For a detailed description of a low pressure procedure for the reaction depicted in Scheme 18.7, see reference 39. The reaction can also be used in a dynamic kinetic resolution mode if a basic ion-exchange resin is added to the reaction medium.[40]

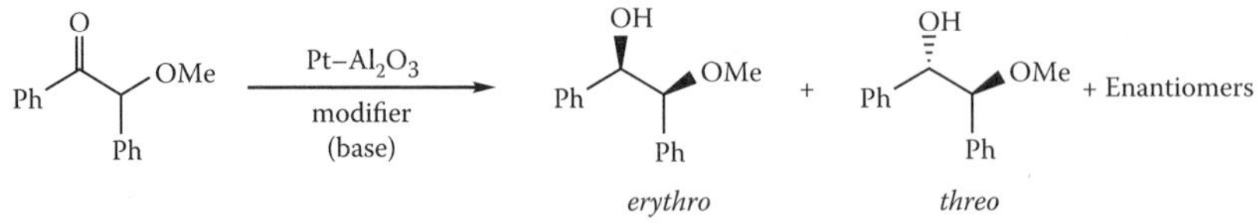

$$
\underset{\substack{Ph\\Ph}}{\overset{O}{\big|}}\text{OMe} \xrightarrow[\substack{\text{modifier}\\\text{(base)}}]{\text{Pt–Al}_2\text{O}_3} \underset{\substack{Ph\\Ph}}{\overset{OH}{\big|}}\text{OMe} + \underset{\substack{Ph\\Ph}}{\overset{OH}{\big|}}\text{OMe} + \text{Enantiomers}
$$

erythro   *threo*

SCHEME 18.7

18.9 SUMMARY AND OUTLOOK

Significant progress in the substrate scope of the Pt-cinchona systems has been made in the last 5 years. Besides α-keto acids and esters, α-keto acetals, α-keto ethers, and some trifluoromethyl ketones have been shown to give high ee's. It is now possible to classify ketones concerning their suitability as substrates for the Pt-cinchona catalyst system, as depicted in Figure 18.6. Nevertheless, for the synthetic chemist, the substrate scope is still relatively narrow, and it is not expected that new important substrate classes will be found in the near future. However, the chemoselectivity of this system has not yet been exploited to its full value, and this might be a potential for future synthetically useful applications.

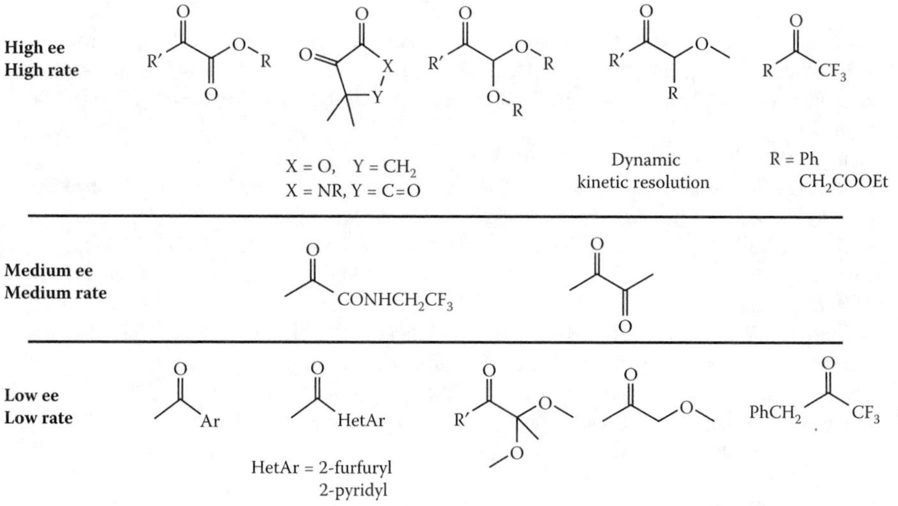

FIGURE 18.6 Structures of "good," "medium," and "bad" substrates for cinchona-modified Pt catalysts.

ACKNOWLEDGMENTS

The authors would like to thank Stephan Burkhardt and Pavel Kukula for their skillful experimental work.

REFERENCES AND NOTES

1. Wills, M., Hannedouche, J. *Curr. Opin. Drug Discov. Develop.* 2002, *5*, 881.
2. a) Cho, B. T. *Aldrichimia Acta* 2002, 35, 3; b) Daverio, P., Zanda, M. *Tetrahedron: Asymmetry* 2001, *12*, 2225; c) Itsuno, S. *Org. React.* 1998, *52*, 395.
3. Csuk, R., Glänzer, B. I. *Chem. Rev.* 1991, *91*, 49.
4. a) Noyori R., Hashiguchi, S. *Acc. Chem. Res.* 1997, *30*, 97; b) Palmer, M. J., Wills, M. *Tetrahedron: Asymmetry* 1999, *10*, 2045.
5. Blaser, H. U., Malan, C., Pugin, B., Spindler, F., Steiner, H., Studer, M. *Adv. Synth. Catal.* 2003, *345*, 103 and references therein.
6. Studer, M., Blaser, H. U., Exner, C. *Adv. Synth. Catal.* 2003, *345*, 45 and references therein.
7. For recent reviews on the effect of modifier structure see a) Pfaltz, A., Heinz, T. *Topics Catal.* 1997, *4*, 229; b) Blaser, H. U., Jalett, H. P., Lottenbach, W., Studer, M. *J. Am. Chem. Soc.* 2000, *122*, 12675; c) Exner, C., Pfalz, A., Studer, M., Blaser, H. U. *Adv. Synth. Catal.* 2003, *345*, 1253.
8. Török, B., Balázsik, K., Felföldi, K., Bartók, M. Stud. *Surf. Sci. Catal.* 2000, *130*, 3381.
9. Margitfalvi, J. L., Tálas, E., Hegedüs, M. *J. Chem. Soc., Chem. Commun.* 1999, 645.
10. Zuo, X., Liu, H., Guo, D., Yang, X. *Tetrahedron* 1999, *55*, 7787.
11. a) Török, B., Balázsik, K., Török, M., Szöllösi, G., Bartók, M. *Ultrasonics Sonochemistry* 2000, *7*, 151; b) Balázsik, K., Török, B., Felföldi, K., Bartók, M. *Ultrasonics Sonochemistry* 1999, *5*, 149; c) Török, B., Felföldi, K., Szakonyi, G., Bartók, M. *Ultrasonics Sonochemistry* 1997, *4*, 301; d) Török, B., Felföldi, K., Szakonyi, G., Balázsik, K., Bartók, M. *Catal. Lett.* 1998, *52*, 81.
12. a) LeBlond, C., Wang, J., Liu, J., Andrews, A. T., Sun, Y.-K. *J. Am. Chem. Soc.* 1999, *121*, 4920; b) LeBlond, C., Wang, J., Andrews, A. T., Sun, Y.-K. *Topics Catal.* 2000, *13*, 169.
13. Blaser, H. U., Jalett, H. P. Stud. *Surf. Sci. Catal.* 1993, *78*, 139.
14. Sutyinszki, M., Szöri, K., Felföldi, K., Bartók, M. *Catal. Commun.* 2002, *3*, 125.
15. Balázsik, K., Szöri, K., Felföldi, K., Török, B., Bartók, M. *J. Chem. Soc., Chem. Commun.* 2000, 555.
16. Wang, G. Z., Mallat, T., Baiker, A. *Tetrahedron: Asymmetry* 1997, *8*, 2133.
17. Szöri, K., Török, B., Felföldi, K., Bartók, M. *Chem. Ind. (Dekker)* 2001, *82*, 489.

18. a) Sedelmeier, G. H., Blaser, H. U., Jalett, H. P., Eur. Pat. 1986, 206993; b) Studer, M., Burkhardt, S., Indolese, A. F., Blaser, H. U. *J. Chem. Soc., Chem. Commun.* 2000, 1327; c) Herold, P., Indolese, A. F., Studer, M., Jalett, H. P., Blaser, H. U. *Tetrahedron* 2000, *56*, 6497; d) Fluka laboratory chemicals catalogue, 2001/2002; e) Blaser, H.U., Burkhardt, S., Kirner, H.-J., Mössner, T., Studer, M. *Synthesis* 2003, 1679.

19. Blandin, V., Carpentier, J. F., Mortreux, A. *Eur. J. Org. Chem.* 1999, 1787.

20. a) von Arx, M., Bürgi, T., Mallat, T., Baiker, A. *Chem. Eur. J.* 2002, *8*, 1430 and references cited therein; b) Bodmer, M., Mallat, T., Baiker, A. *Chem. Ind. (Dekker)* 1998, *75*, 75; c) von Arx, M., Mallat, T., Baiker, A. *J. Catal.* 2001, *202*, 169; d) von Arx, M., Mallat, T., Baiker, A. *Spec. Pub., Royal Chemical Society* 2001, *266*, 247; e) von Arx, M., Mallat, T., Baiker, A. *Tetrahedron: Asymmetry* 2001, *12*, 3089.

21. a) Zhang, J., Yan, X., Liu, H. *J. Mol. Catal. A: Chemical* 2001, *175*, 125; b) Zhang, J., Yan, X., Liu, H. *J. Mol. Catal. A: Chemical* 2001, *176*, 281; c) Zuo, X., Liu, H., Yue, C. *J. Mol. Catal. A: Chemical* 1999, *147*, 63; d) Yan, X., Liu, H., Zhao, *J. Catal. Lett.* 2001, *74*, 81.

22. Studer, M., Burkhardt, S., Blaser, H. U. *J. Chem. Soc., Chem. Commun.* 1999, 1727.

23. Török, B., Felföldi, K., Balázsik, K., Bartók. M. *J. Chem. Soc., Chem. Commun.* 1999, 1725.

24. Petasis, N. A., Zavialov, I. A. *J. Am. Chem, Soc.* 1998, *120*, 11798.

25. Studer, M., Blaser, H. U., Burkhardt, S. *Adv. Synth. Catal.* 2002, *344*, 511.

26. Matsumoto, T., Murayama, T., Mitsuhashi, S., Miura, T. *Tetrahedron Lett.* 1999, *40*, 5043, Murata, K., Okano, K., Miyagi, M., Iwane, H, Noyori, R., Ikariya, T. *Org. Lett.* 1999, *1*, 1119.

27. Wandeler, R., Künzle, N., Schneider, M. S., Mallat, T., Baiker, A. *J. Chem, Soc., Chem. Commun.* 2001, 673.

28. Künzle, N., Szabó, A., Schürch, M., Wang, G., Mallat, T., Baiker, A. *J. Chem. Soc., Chem. Commun.* 1998, 1377.

29. Studer, M., Blaser, H. U., Okafor, V. *J. Chem. Soc., Chem. Commun.* 1998, 1053.

30. Slipszenko, J. A., Griffith, S. P., Johnston, P., Simons, K. E., Vermeer, W. A., Wells, P. B. *J. Catal.* 1998, *179*, 267.

31. Toukoniitty, E., Mäki-Arvela, P., Kuzma, M., Villela, A., Neyestanaki, A. K., Salmi, T., Sjöholm, R., Leino, R., Laine, E., Murzin D.Y. *J. Catal.* 2001, *204*, 281; Toukoniitty, E., Mäki-Arvela, P., Wärnå, J., Salmi, T. *Catal. Today* 2001, *66*, 411; Toukoniitty, E., Mäki-Arvela, P., Sjöholm, R., Leino, R., Salmi, T., Murzin, D.Y. *React. Kinet. Catal. Lett.* 2002, *75*, 21.

32. Perosa, A., Tundo, P., Selva, M. *J. Mol. Catal. A: Chemical* 2002, *180*, 169.

33. von Arx, M., Mallat, T., Baiker, A. *J. Mol. Catal. A: Chemical* 1999, *148*. 275.

34. Török, B., Balázsik, K., Szöllsi, G., Felföldi, K., Bartók, M. *Chirality* 1999, *11*, 470.

35. Blaser, H. U., Jalett, H. P., Monti, D. M., Baiker. A., Wehrli, J. T. *Stud. Surf. Sci. Catal.* 1991, *67*, 147.

36. Blaser, H. U., Jalett, H. P., Monti, D. M., Wehrli, J. T. *Appl. Catal.* 1989, *52*, 19.

37. See http://www.degussa-catalysts.com/catalysts/technical/special/enantioselective.html, Kranter, J., *Speciality Chem.* 2004, *24*(4), 26.

38. Cinchonidine (20 mg) was placed in a 100-mL magnetically stirred glass vessel, glass shaker, or autoclave. A slurry of 200 mg of the catalyst (5% Pt/Al$_2$O$_3$ [F 214 VHA/D]) in 10 mL AcOH was added, followed by 2 g ethyl pyruvate A or pyruvaldehyde dimethyl acetal B dissolved in 8 mL AcOH. Then, the vessel was closed, evacuated, and purged 3 times with hydrogen (1.1 bar). The reaction began after the agitation was set into motion under 1.1 bar hydrogen at room temperature. The hydrogen uptake was measured by the pressure drop in a reservoir equipped with a pressure regulator (alternatively, a hydrogen-filled balloon can be used). After 75 min and 120 min (for A and B, respectively), the reaction was stopped. The pressure was released and the vessel was purged with nitrogen. The catalyst was filtered off and the reaction mixtures were analyzed by GLC (Beta-dex 110, length 30 m, Supelco Art 2-4301). Using helium as carrier gas at 75°C, the retention times for A were 6.0 min for the starting material, 7.7 min for the (*R*)-product, and 8.3 min for the (*S*)-product. With hydrogen as carrier gas at 60°C, the retention times for B were 4.1 min for the staring material, 7.1 min for the (*R*)-product, and 7.7 min for the (*S*)-product. Results: 99% conversion and 64% ee were observed for A, and 99% conversion and 91% ee were observed for B. The products could be isolated by evaporation to dryness and distillation. Care has to be taken because the products are quite volatile.

39. Dihydrocinchonidine (10 mg) was placed in a 50-mL pressure autoclave equipped with a magnetic stirring bar and baffles. Catalysts (100 mg) (5% Pt/Al$_2$O$_3$, JMC 94, Batch 14017/01, Supplier: Johnson Matthey, pretreated 2 h at 400°C under a flow of hydrogen) were suspended in 2 mL toluene and transferred to the autoclave. One g 2-methoxy-2-phenylacetophenone was dissolved in 18 mL toluene and also transferred to the autoclave. After sealing, the autoclave was purged 3 times with argon and 3 times with hydrogen and then pressurized with hydrogen to 60 bar. The reaction was started by turning the magnetic stirrer on, and the temperature was kept constant at 25°C with the help of a cryostat. The pressure in the autoclave was kept constant at 60 bar during the reaction by the use of a pressure transducer. The hydrogen consumption was measured by the pressure drop in a reservoir with a known volume. After 128 min, the reaction was stopped, the pressure was released, and the autoclave was purged with argon 3 times. The catalyst was filtered off, and the reaction mixture was evaporated to dryness. Yield: 0.85 g (85%). HPLC analysis was carried out on an HP 1100 with a Chiracel® OD (Daicel) column of 0.46 × 25 cm and hexane–isopropanol (98:2) as eluent and detection at 210 nm. Retention time: substrate (S) 12.2 min (R) 20.3 min; *erythro* product 24.3 min for (S,R) and 30.6 min for (R,S); *threo* diastereoisomers 15.8 and 18.7 min, respectively, but no assignment of the absolute configuration was possible for the *threo* diastereoisomers (less than 2% for all systems containing modifier). Sample: 91% ee, 42% conversion; final product 88% ee, 54% conversion; *threo* product always <2%.

40. Amberlite IRA-900 (1.6 g) (strongly basic anion exchanger, converted to the OH⁻ form by washing with 0.1M NaOH until chloride free) was added after the substrate. The rest of the experiment was carried out as described in reference 39 except that 2 g substrate were used. After 203 min, 90% ee at 92% conversion was measured, and the concentration of *threo* product was <2%.

19 Biotransformations: "Green" Processes for the Synthesis of Chiral Fine Chemicals

David P. Pantaleone

CONTENTS

19.1 INTRODUCTION

The need for chiral fine chemical intermediates has become commonplace today in light of the trend toward single enantiomer drugs manufactured by pharmaceutical companies.[1-6] Also, as waste disposal costs skyrocket, processes developed today must emphasize environmental safety. Although, at present, chemical approaches account for the majority of processes to produce fine chemicals, biotransformation steps are making more of an impact and are being integrated into chemical process sequences, especially where the chirality of the target compound must be maintained.[7,8] New biocatalysts are being isolated and developed by companies such as Dowpharma (Midland, MI), Diversa, (San Diego, CA), BioCatalytics (Pasadena, CA), Protéus (Nimes, France), and Codexis (Redwood City, CA). Many of these biocatalysts have been engineered using directed evolution approaches to have altered substrate specificity, enhanced thermal stability, and greater organic solvent tolerance. In addition, commercial enzyme suppliers such as Amano Enzyme Co. (Nagoya, Japan), Novozymes A/S (Bagsvaerd, Denmark), and Genencor International (Palo Alto, CA) are promoting their enzymes for specific biotransformations. Regardless of the source of these enzymes, because they are chiral catalysts and operate under mild reaction conditions, their utility to produce chiral molecules is being exploited more and more.

The term "biotransformation" means many things to many people. Throughout this chapter, it is defined as the use of a biological catalyst from a bacterial, fungal, plant, or mammalian source to carry out a desired transformation; the discussion, however, will be limited to the formation of a single enantiomer or diastereoisomer. This biocatalyst could be a commercial enzyme (often an impure mixture of many different activities), whole cell, or cell lysate. These biocatalysts are sometimes immobilized onto solid supports that can provide added benefits such as greater thermal stability or reusability, despite an increase in catalyst cost. Enzymatic resolutions are a specific type of biotransformation. They will not be discussed here because they do not result in the creation of an asymmetric center; instead they convert one isomer of a mixture to a compound that will generally allow easy separation of the two isomers. Because this field is extremely large, only selected examples of some industrially important biotransformations will be discussed along with certain biotransformations that offer potential to become significant routes to chiral fine chemicals. The reader is directed to the numerous textbooks, review articles, and symposia proceedings that have been published that exemplify this rapidly developing field.[9–21]

19.2 BIOCATALYST CLASSIFICATION

Enzymes are classified by a 4-digit number referred to as the Enzyme Commission number or "EC number," which is assigned according to the enzyme's function, based on recommendations of the Nomenclature Committee of the International Union of Biochemistry and Molecular Biology (IUBMB).[22,23] There are six general groups into which enzymes are classified, and the first digit of the EC number corresponds to the following general categories: (1) oxidoreductases; (2) transferases; (3) hydrolases; (4) lyases; (5) isomerases; and (6) ligases. The other three digits are then assigned based on further specifics of the type of reaction catalyzed. The EC number will be used in this chapter to conveniently categorize the biocatalysts of industrial importance. Only the first five classes will be discussed because the ligase class currently does not represent any significant number of commercial bioprocesses. Although this EC number is assigned to a specific enzyme, reference to microorganisms catalyzing the same reaction as the purified enzyme will also be cited. A publication of sources of enzymes has been compiled that allows one to locate a commercial supplier of a particular enzyme using the EC number.[24] The reader is also directed to the many electronic databases available both commercially and on the Internet to locate a commercial supplier of the desired enzyme.

19.2.1 OXIDOREDUCTASES (E.C. 1.x.x.x)

The oxidoreductase class of biocatalysts is one of the most common of all biological reactions, comprising dehydrogenases, oxidases, and reductases. All these enzymes act on substrates through the transfer of electrons with various co-factors or co-enzymes serving as acceptor molecules. Only a select group of reactions will be discussed because of space limitations, so the reader is referred to other texts for more in-depth discussions of other oxidation-reduction reactions.[17,20,25–28]

19.2.1.1 Dehydrogenases

There are a number of different types of dehydrogenases that catalyze redox reactions and have been used to synthesize chiral molecules.[29,30] Some are co-factor dependent, generally requiring $NAD^+/NADH$ or $NADP^+/NADPH$ such as the alcohol dehydrogenases (AdHs), and some are co-factor independent, such as the AdH from *Gluconobacter suboxydans*.[31] The dependent enzymes use the co-factor to supply a hydride ion stereospecifically to reduce the substrate, whereas the independent dehydrogenases are often membrane bound and thus associated with the cytochromes and quinoproteins that facilitate hydrogen removal and ultimately reduce oxygen to water. For AdHs, the hydride can be delivered to either the *si* or *re* face, which will depend on how the substrate binds in the enzyme active site based on the steric bulk of the R^1 and R^2 substituents (Scheme 19.1). Enzymes such as yeast (YAdH), horse liver (HLAdH), and *Thermoanaerobium brockii* AdHs deliver the *pro-R* hydrogen to the *re*-face, thus forming the (*S*)-alcohol. Another enzyme from *Mucor javanicus* delivers the *pro–S* hydrogen to the *si*-face to yield the (*R*)-alcohol. An enzyme isolated from *Pseudomonas* sp. strain PED (ATCC 49794) has also been shown to form (*R*)-alcohols from a wide variety of substrates.[32] Further mechanistic details, discussion of Prelog's rule to predict the stereochemical outcome of a particular reduction, and listings of other enzymes with different stereospecificities are discussed elsewhere.[27] Some specific examples of dehydrogenases and their use to produce chiral alcohols and amino acids are given later in this chapter.

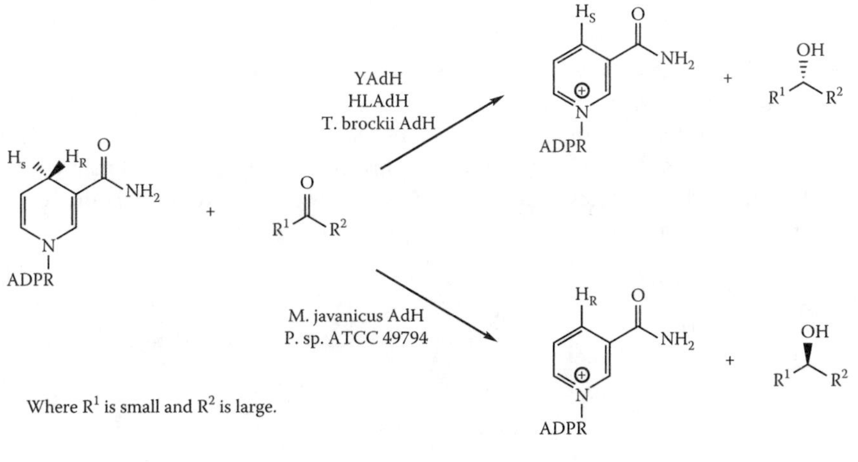

SCHEME 19.1

The AdHs have been further subdivided into primary (PAdH) and secondary (SAdH) enzymes, based on their preference for primary or secondary alcohols. Substrate specificity and site-directed mutagenesis studies with the SAdH from *Thermoanaerobacter ethanolicus* have shown this enzyme to have a very broad profile, such that certain alkynyl ketones and ketoesters serve as substrates.[33–35]

One of the major disadvantages of using these enzymes to reduce ketones is generally the very poor aqueous solubility of the substrates (usually <5–10 mM) and the fact that the co-factor regeneration system is sensitive to organic solvents. To address this limitation, biphasic and membrane bioreactors have been used that allow for moderate to high conversions with often >99% ee for selected products. This has been demonstrated with the (*S*)-AdH from *Rhodococcus erythropolis*.[36,37]

19.2.1.1.1 Reductions of 2-Oxo Acids

There are two lactate dehydrogenases (LdHs) with different stereospecificities that have been very useful for the preparation of chiral 2-hydroxy acids.[38–41] L-LdH (E.C. 1.1.1.27) and D-LdH (E.C. 1.1.1.28) are obtained from various microbial sources (Table 19.1). Along with hydroxyisocaproate

TABLE 19.1
Selected Substrates and Stereochemical Product Configuration
with Lactate (L) and Hydroxyisocaproate (Hic) Dehydrogenases
(dH) from Various Sources that Catalyze the Reaction Shown
in Scheme 19.2

R	Enzyme	Source	Product Configuration	Ref.
Me	L-LdH	Rabbit muscle (iso M)	S	38
Me	D-LdH	*L. mesenteroides*	R	38
Me$_3$C	L-2-HicdH	*L. confusus*	S	43
Me$_3$C	D-2-HicdH	*L. casei*	R	43
C$_6$H$_5$(CH$_2$)$_2$	D-LdH	*L. mesenteroides*	R	40
C$_6$H$_5$(CH$_2$)$_2$	D-LdH	*S. epidermidis*	R	41
ZHN(CH$_2$)$_4$	L-2-HicdH	*S. epidermidis*	R	44
E-Me(CH)$_2$	L-LdH	*B. stearothermophilus*	S	45
E-Me(CH)$_2$	D-LdH	*S. epidermidis*	R	45
c-(CH$_2$)$_2$CH	D-LdH	*S. epidermidis*	R	39
p-F-C$_6$H$_4$CH$_2$	D-LdH	*S. epidermidis*	R	46
FCH$_2$	L-LdH	Rabbit muscle (iso M)	S	47

dehydrogenase (HicdH), these enzymes have been used in conjunction with a co-factor regeneration system, usually the formate dehydrogenase (FdH) (E.C. 1.2.1.2) system from *Candida boidinii,* to generate important chiral intermediates and synthons (Scheme 19.2 and Table 19.1),[42] illustrating the broad range of substrates that these enzymes work on. In all cases, the enantiomeric excess and chemical yields of the products were high. The reader is referred to the specific references for additional substrates tested with the respective enzymes.

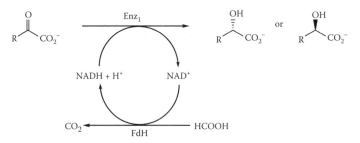

For Enz$_1$ see Table 19.1

SCHEME 19.2

Because the commercial availability of 2-oxo acids is somewhat limited and their chemical stability is relatively poor, a system was developed to generate these *in situ* using an amino acid deaminase (amino acid oxidase, E.C. 1.4.3.x) (Scheme 19.3).[48] This keto acid was then reduced to the chiral 2-hydroxy acid with either L-LdH or D-LdH to produce the respective enantiomer in moderate to high yield with high enantiomeric purity.[49]

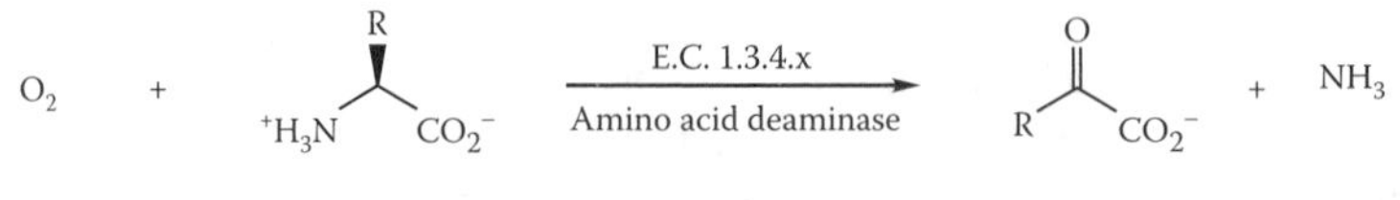

$$O_2 \quad + \quad \underset{\substack{+ \\ H_3N \quad CO_2^-}}{\overset{R}{\diagup\hspace{-0.3em}\diagdown}} \quad \xrightarrow[\text{Amino acid deaminase}]{\text{E.C. 1.3.4.x}} \quad \underset{R \quad CO_2^-}{\overset{O}{\parallel}} \quad + \quad NH_3$$

SCHEME 19.3

19.2.1.1.2 Reductive Amination

Natural and unnatural amino acids have been prepared through reductive amination reactions catalyzed by amino acid dehydrogenases (dHs) (E.C. 1.4.1.x) (Scheme 19.4) in combination with the formate dH co-factor recycling system from *Candida boidinii*.[29,50] Another co-factor recycling system that is sometimes used with certain amino acid dHs is glucose dehydrogenase from *Bacillus megaterium* that oxidizes glucose to gluconate with the concomitant reduction of $NADP^+$ to NADPH. Those enzymes, whose natural substrates are alanine (R = Me) (x = 1), leucine (R = Me_2CHCH_2) (x = 9), and phenylalanine (R = $C_6H_5CH_2$) (x = 20), are the most well-studied, although others exist.[51,52] For example, cloned, thermostable alanine dH has been used with a coupling enzyme system to prepare D-amino acids, and the alanine dH gene has been incorporated into selected bacterial strains to enhance L-alanine production.[53,54] In addition, halogenated derivatives of L-alanine have also been prepared using alanine dH.[55,56]

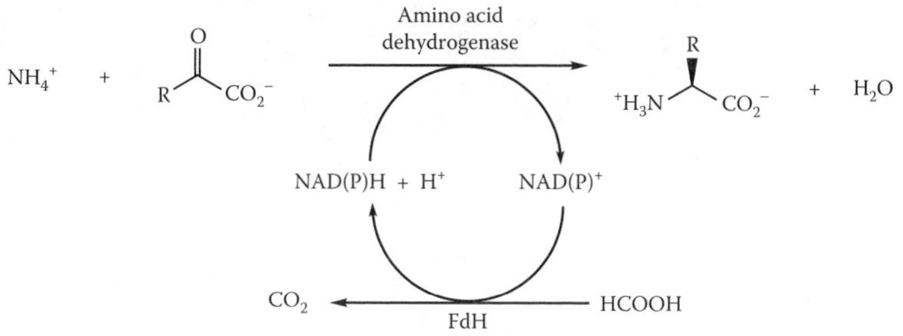

SCHEME 19.4

In the mid to late 1980s, many research groups focused on methods and processes to prepare L-phenylalanine (Chapter 3). This was a direct result of the demand for the synthetic, artificial sweetener aspartame. One of the many routes studied was the use of phenylalanine dH (Scheme 19.4, R = $C_6H_5CH_2$) with phenylpyruvate (PPA) as substrate.[57,58] This enzyme from *Bacillus sphaericus* shows a broad substrate specificity and, thus, has been used to prepare a number of derivatives of L-phenylalanine.[59] A phenylalanine dH isolated from a *Rhodococcus* strain M4 has been used to make L-homophenylalanine [(*S*)-2-amino-4-phenylbutanoic acid], a key, chiral component in many angiotensin-converting enzyme (ACE) inhibitors.[40] More recently, that same phenylalanine dH has been used to synthesize a number of other unnatural amino acids (UAAs) that do not contain an aromatic sidechain.[43]

Leucine dH is the enzyme used as the biocatalyst in the process commercialized by Degussa AG (Hanau, Germany) to produce L-*tert*-leucine (L-Tle).[28,60] This UAA has found widespread use in peptidomimetic drugs in development, and the demand for this unique amino acid continues to increase.[61,62] This process, which has been the subject of much study, requires a co-factor recycling system (Scheme 19.4, R = Me_3C).[63,64] Similar to phenylalanine dH, leucine dH has been used to prepare numerous UAAs because of its broad substrate specificity.[43,65,66]

19.2.1.2 Microbial Reductions

Yeast, bacteria, and fungi, often used as whole-cell preparations, are known to reduce a variety of substrates. Following are selected examples of these types of reductions. The reader is referred to the literature for additional details and examples of this technology.[67,68]

19.2.1.2.1 Yeast Reductions

Yeast reductions have provided the synthetic organic chemist with highly versatile methods to prepare chiral alcohols from prochiral ketones of which *Saccharomyces cerevisiae* (baker's yeast) is the most commonly used biocatalyst. In addition to prochiral ketone reductions, hundreds of other examples of yeast reductions have been cited in the literature using a variety of substrates such as β-keto esters, β-diketones, and analogues such as sulfur and nitrogen-containing compounds.[69–75] Examples of yeast reductions used to prepare some important chiral intermediates as reported by several pharmaceutical companies and academia are presented in this chapter.

On the synthetic route to a novel class of orally active 2,3-benzodiazepines, scientists at Lilly have described a yeast reduction using *Zygosaccharomyces rouxii* (ATCC 14462) (Scheme 19.5).[76–78] This intermediate is used in the synthesis of a novel class of benzodiazepines such as LY-300164 (**1**). The choice of this yeast strain came from a screen of numerous microbes. The product was isolated in >95% yield and >99.9% ee by use of an adsorbant polymeric resin.[76–78]

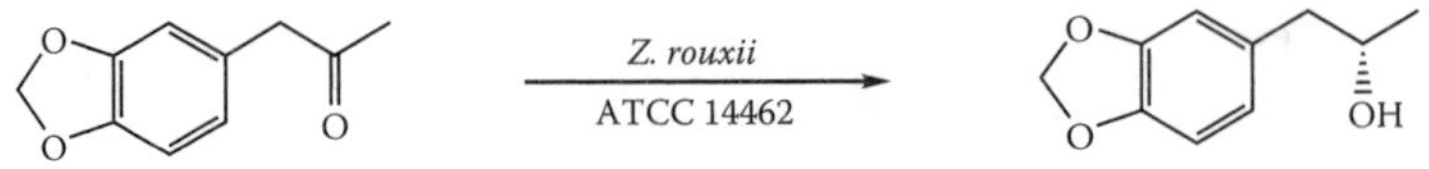

SCHEME 19.5

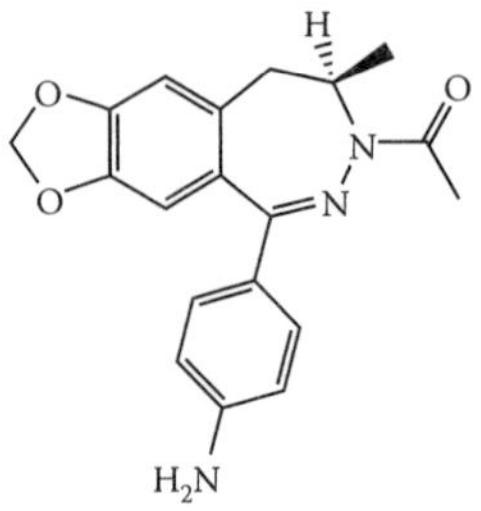

1

Scientists at Merck have reported a number of biocatalytic routes derived from screening various microorganisms targeted to produce key intermediates that are then combined with chemical reactions to prepare the target molecule. The biocatalytic step was often carried out by necessity as a result of poor chemical yield, low optical purity, or both. 6-Bromo-β-tetralone (**2**) was reduced to (*S*)-6-bromo-β-tetralol (**3**) by the yeast *Trichosporon capitatum* MY 1890 (Scheme 19.6).[79] The tetralol **3** is a key intermediate for the synthesis of MK-0499 (**4**), a potassium channel blocker. The (*S*)-β-tetralol **3** was produced in gram quantities with an ee of >99% to support further development of MK-0499. Baker's yeast was tested for its ability to carry out this reduction but showed insignificant product formation.

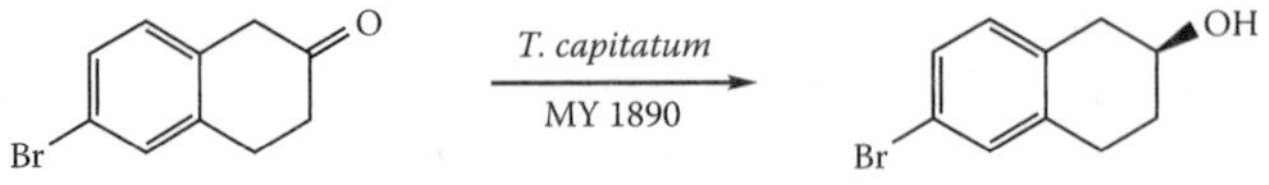

2   **3**

SCHEME 19.6

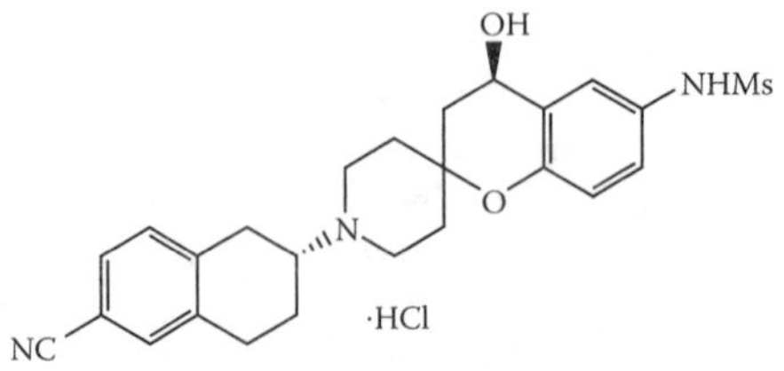

4

Benzylacetoacetate (**5**) was reduced to benzyl-(*S*)-(+)-3-hydroxybutyrate (**6**) by the yeast *Candida schatavii* MY 1831 (Scheme 19.7), which is considered a possible intermediate in the synthesis of L-734,217 (**7**), an experimental fibrinogen receptor antagonist.[80] Initially, 8 microorganisms were tested for their reduction ability using **5** as substrate. The best culture, *C. schatavii* MY 1831, yielded product with an ee of 93% and an estimated conversion yield of >95% (by thin layer chromatography [TLC]).

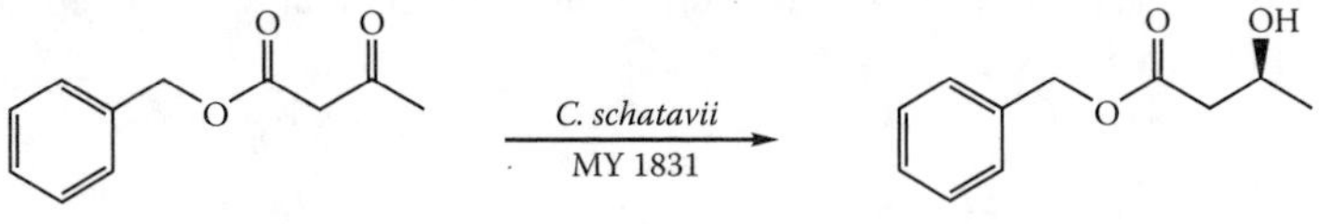

SCHEME 19.7

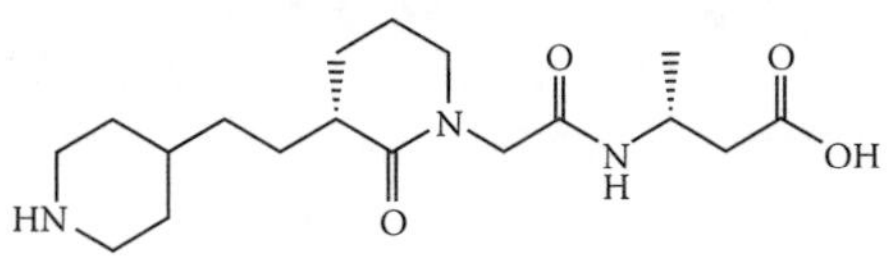

Another yeast, *Rhodotorula rubra* MY 2169, has been shown to reduce a ketosulfone **8** to the corresponding *trans*-hydroxysulfone **9** (Scheme 19.8). This hydroxysulfone is an intermediate in the drug candidate L-685,393 (**10**), a carbonic anhydrase inhibitor.[81] Results of this biotransformation yielded gram quantities of product with a de of >96%. Studies by Zeneca discuss additional screening experiments aimed at finding microorganisms to reduce a similar ketosulfone.[82]

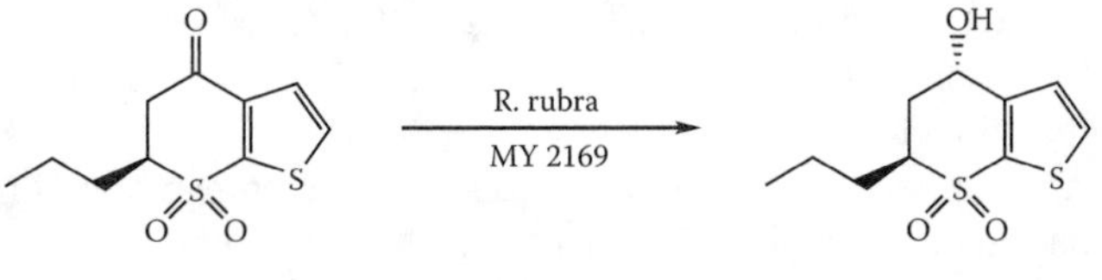

SCHEME 19.8

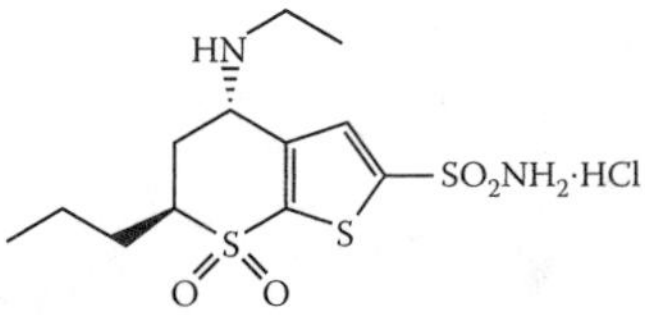

Another example of a β-keto ester reduction was studied by scientists at Zeneca with collaborators from the University of Warwick (Coventry, UK). They have reported on the reduction of ethyl 4-chloroacetoacetate (**11**) after screening a number of yeast strains.[83,84] The reduction of this substrate has been studied previously because it has been realized for a number of years to be a potential intermediate for a number of pharmaceuticals, including the lipid carrier molecule, L-carnitine (see Chapter 31),[85–87] and an intermediate in 3-hydroxy-3-methylglutaryl coenzyme A (HMG-CoA) reductase inhibitors such as atorvastatin (Lipitor®) (see Chapter 31).[88–90] It is interesting that the Zeneca group and colleagues found that a co-substrate (e.g., glucose, 2-propanol, xylose, glycerol) could dramatically affect the chirality of the stereogenic center of the product. Two different yeast strains, *Zygosaccharomyces rouxii* (NCYC 564) and *Pichia capsulata* (CBS 837), catalyzed the reduction, forming the (*S*)-3-hydroxy and the (*R*)-3-hydroxy isomers, respectively, depending on which co-substrate was added (Scheme 19.9). The authors suggested that these co-substrates inhibit certain enzymes possibly involved in the NADH regeneration rather than the (*R*)-specific enzymes themselves. Nevertheless, this reversal in enantioselectivity depending on co-substrate indicates the complexity of yeast reductions and the fact that there are competing enzymes that produce both antipodes.

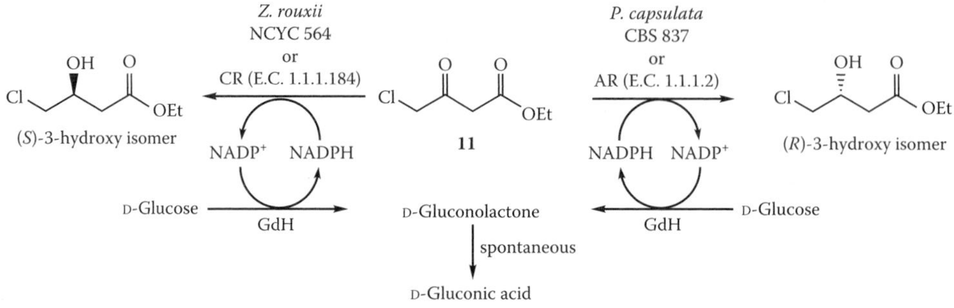

SCHEME 19.9

One way of circumventing these competing reactions encountered when using whole yeast cells for reductions is to isolate the specific enzymes responsible for a particular transformation. This has been demonstrated for the reduction of **11** whereby an isolated aldehyde reductase (AR) (E.C. 1.1.1.2) from *Sporobolomyces salmonicolor* forms the (*R*)-isomer and an isolated carbonyl reductase (CR) (E.C. 1.1.1.184) from *Candida magnoliae* forms the (*S*)-isomer (Scheme 19.9).[91–94] Both are NADPH-dependent reductases, and a glucose dehydrogenase (GdH) recycling system was used to regenerate the co-factor.

An example of a 2-step reduction system developed for the preparation of (4*R*,6*R*)-4-hydroxy-2,2,6-trimethylcylcohexanone (actinol) (**12**), a useful chiral building block, has been described (Scheme 19.10).[95–97] The first step was a yeast reduction with *S. cerevisiae* old yellow enzyme (E.C. 1.6.99.1) cloned into *Escherichia coli* that stereospecifically reduced 2,6,6-trimethyl-2-cyclo-hexen-1,4-dione (ketoisophorone) (**13**) to (6*R*)-2,2,6-trimethyl-cyclohexane-1,4-dione [(6*R*)-levo-dione] (**14**). The second step used levodione reductase, whose gene was cloned from *Corynebacterium aquaticum* M-13 into *E. coli,* to reduce **14** stereoselectively and produce **12**. Both enzymes required the co-factor NADH for reducing equivalents, which was supplied as a glucose dehydrogenase/glucose/NAD+ regeneration system.

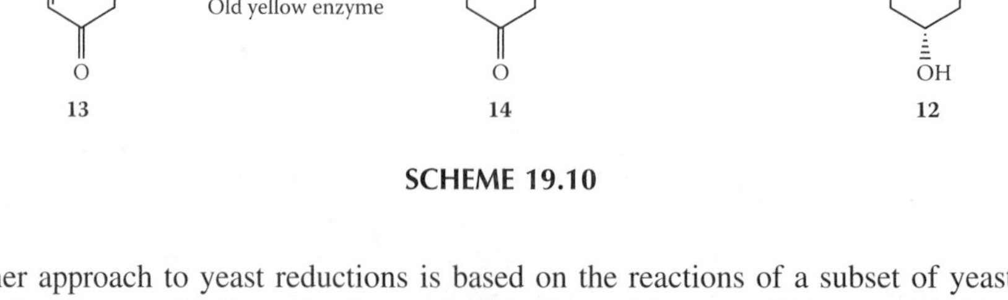

SCHEME 19.10

Another approach to yeast reductions is based on the reactions of a subset of yeast proteins, which catalyze stereoselective reductions of aldehydes and ketones. This was aided by the yeast genome sequence and the design of 6144 individual yeast strains, each containing a different open reading frame (ORF) fused to glutathione *S*-transferase.[98] Based on proteins capable of reducing carbonyl compounds, Kaluzna and co-workers described a subset of fusion proteins comprised of 24 individual ORFs that was cloned into *E. coli* and evaluated for reduction of the α-ketoester (**15**) (Scheme 19.11).[99] Reduction of **15** to the (*R*)-isomer is desirable because this is an intermediate used in the synthesis of a number of ACE inhibitors. Two individual yeast fusion proteins, Ypr1p and Gre2p, were shown to catalyze the reduction of α-keto ester **15** to afford the (*R*)-hydroxy ester (**16**) and the (*S*)-hydroxy ester (**17**), respectively.[99] Both of these enzymes require NADPH as a co-factor. Other yeast proteins out of this subset of 24 ORFs have been described for their utility to reduce α-keto esters.[100,101]

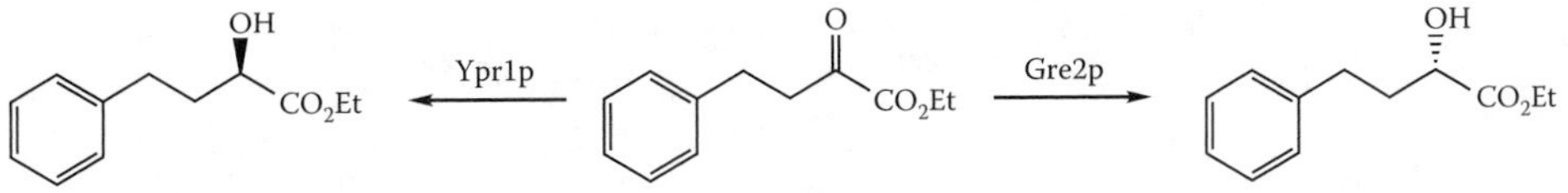

SCHEME 19.11

19.2.1.2.2 Bacterial and Fungal Reductions

In addition to yeast described previously, bacterial and fungal cultures also possess enzymes that will carry out synthetically useful biotransformations *en route* to a particular chiral intermediate. Some examples of these types of bioconversions are given in the following.

In the route to BO-2727 (**18**), a broad-spectrum β-methyl carbapenem being developed by Merck, a bioreduction catalyzed by the fungus *Mortierella alpina* MF 5534 is used to form a precursor (*R*)-β-hydroxy ester **19** (Scheme 19.12).[102] This fungal culture was a result of screening approximately 260 strains of microorganisms and resulted in the production of gram quantities of product with a de of >98%.

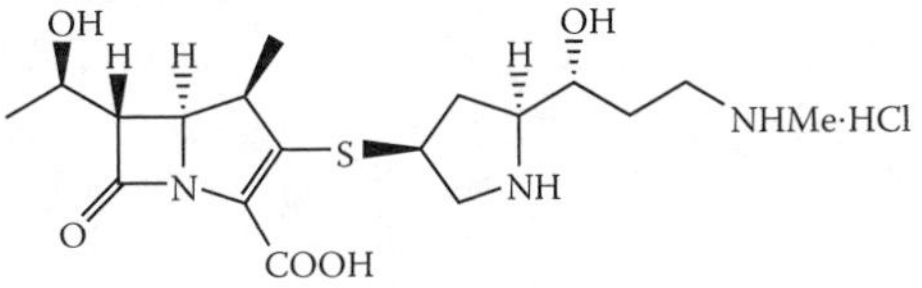

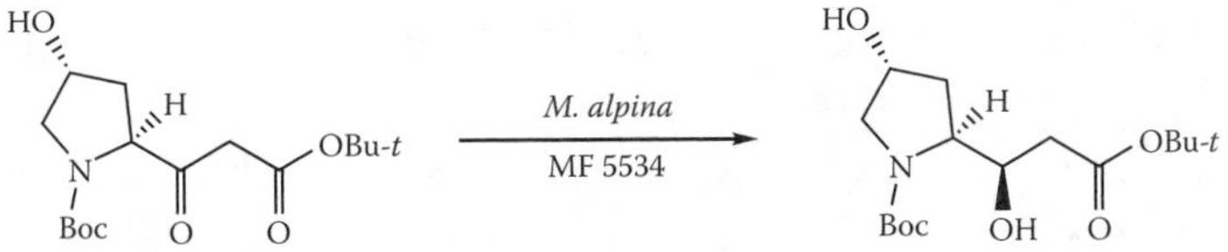

SCHEME 19.12

Scientists at Bristol-Myers Squibb, after screening various microorganisms, selected a bacterial strain of *Acinetobacter calcoaceticus* SC 13876 to reduce a 3,5-dioxo ester **20** to the dihydroxy ester **21** (Scheme 19.13).[103] The diol **21** is a key intermediate in the synthesis of an HMG-CoA reductase inhibitor (**22**). In a 1-L batch reaction, a yield of 92% was obtained with an optical purity of 99%.

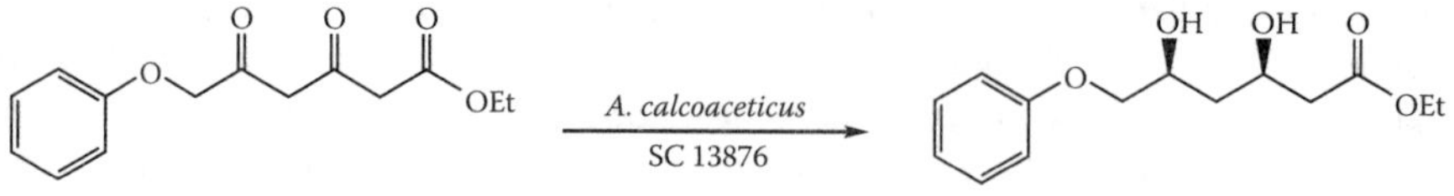

SCHEME 19.13

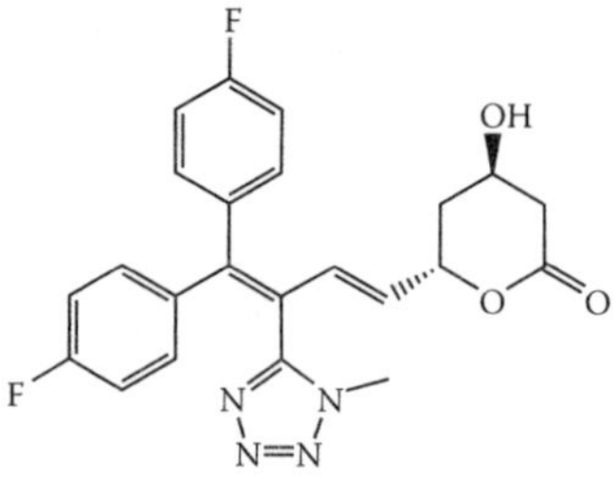

In another study, screening was carried out for reduction of substituted benzazepin-2,3-dione **23** to a 3-hydroxy derivative **24** (Scheme 19.14). This was accomplished by a bacterial strain of *Rhodococcus fascians* ATCC 12975 (*Norcardia salmonicolor* SC 6310) with a conversion of 97% and an optical purity of >99.9%. This reaction product **24** is a key intermediate in the synthesis of the calcium antagonist SQ 31,765 (**25**).[104,105] The Bristol-Myers Squibb group has also shown the selective reduction of the β-keto ester, methyl-4-chloro-3-oxobutanoate, by the fungus *Geotrichum candidum* SC 5469 to the corresponding (*S*)-hydroxy ester.[106]

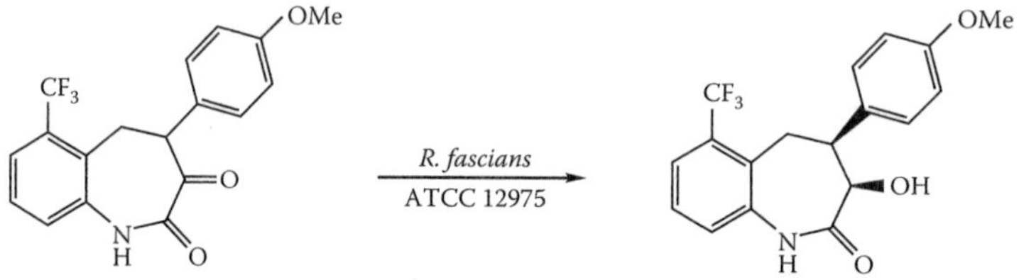

SCHEME 19.14

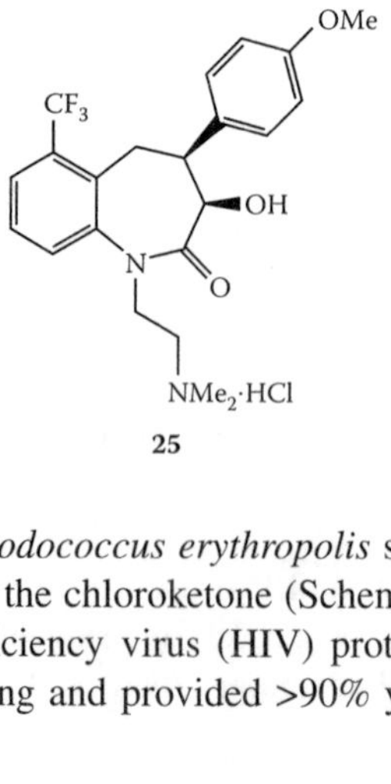

A microbial reduction with a *Rhodococcus erythropolis* strain was used by BMS scientists to prepare the chiral chlorohydrin from the chloroketone (Scheme 19.15). This intermediate is incorporated into the human immunodeficiency virus (HIV) protease inhibitor Atazanavir (**26**). This strain was identified through screening and provided >90% yield with a diastereomeric purity of >98% and an ee of >99%.[62]

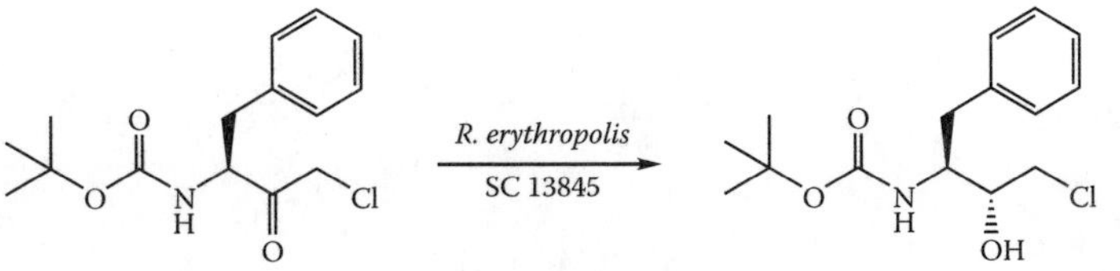

SCHEME 19.15

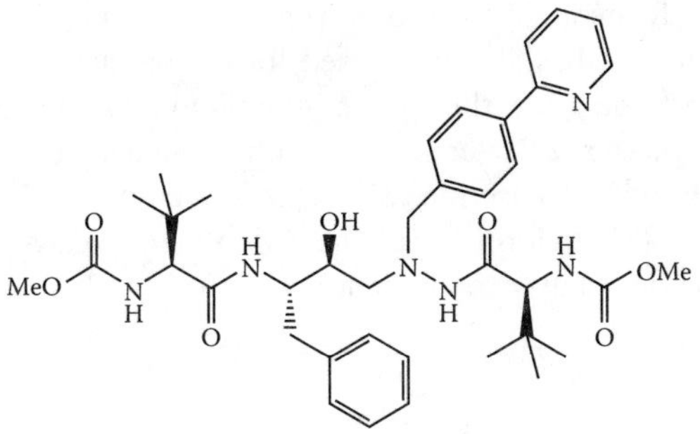

26

The reactions shown here illustrate the versatility of yeast, bacterial, and fungal reductions. In all these cases, screening studies were usually conducted first followed by refinement of the bioreduction using the best microorganism. Other examples where reductions are used to produce pharmaceutical intermediates can be found in a review by Patel.[107]

19.2.2 Transferases (E.C. 2.x.x.x)

The transferase class of biocatalysts is one of the most common of all biological reactions, catalyzing the general reaction shown in Scheme 19.16. This type of reaction uses a wide variety of substrates such as amino acids, keto acids, nucleotides, and carbohydrates, to name a few. In general, the carbohydrate transferases do not result in the generation of a new stereocenter but have been used most extensively to generate novel saccharides.[108]

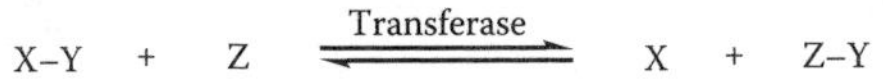

SCHEME 19.16

A transferase that also has aldolase activity and has been used to prepare a number of chiral compounds is the enzyme serine hydroxymethyltransferase (SHMT) (E.C. 2.1.2.1). This enzyme, also known as threonine aldolase, catalyzes the physiologic reaction of the interconversion of serine and glycine with pyridoxal phosphate (PLP) and tetrahydrofolate (FH_4) as the shuttling co-factor of the C–1 unit. It also catalyzes a number of other reactions, some of which are independent of PLP and FH_4.[109] The SHMT-catalyzed aldolase reaction generates two stereocenters; that at the α-carbon is formed with a high degree of stereospecificity, whereas there is a less strict stereo-preference at the β-carbon (Scheme 19.17). Nevertheless, this enzyme from porcine liver, *E. coli*, and *Candida humicola* (threonine aldolase) has been used to prepare a number of β-hydroxy-α-amino acids.[110–113] A chiral intermediate used in the synthesis of the carbacephem antibiotic lora-carbef (Lorabid®), L-*erythro*-2-amino-3-hydroxy-6-heptenoic acid has been synthesized using the recombinant *E. coli* SHMT enzyme.[114,115]

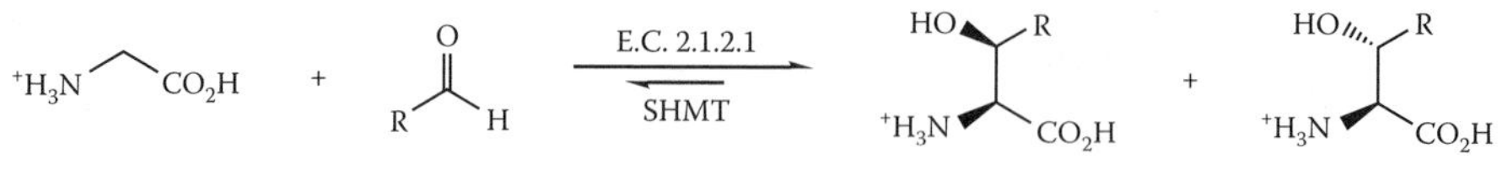

SCHEME 19.17

The enzyme has also been used in the production of several natural amino acids such as L-serine from glycine and formaldehyde and L-tryptophan from glycine, formaldehyde, and indole.[116–118] In addition, SHMT has also been used for the production of a precursor, **27**, to the artificial sweetener aspartame (**28**) through a nonphenylalanine-requiring route (Scheme 19.18).[119–122] Glycine methyl ester (**29**) is condensed with benzaldehyde under kinetically controlled conditions to form L-*erythro*-β-phenylserine (**30**). This is then coupled enzymatically using thermolysin with Z-aspartic acid (**31**) to form *N*-carbobenzyloxy-L-α-aspartyl-L-*erythro*-β-phenylserine (**27**) and affords aspartame on catalytic hydrogenation.

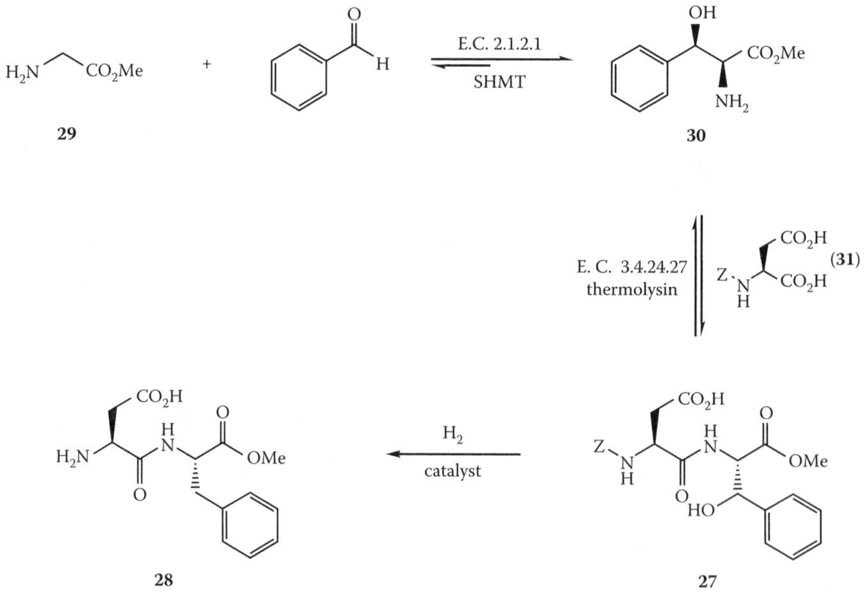

SCHEME 19.18

19.2.2.1 Aminotransferases

The aminotransferase class of enzymes (E.C. 2.6.1.x), also known as transaminases, are ubiquitous, PLP-requiring enzymes that have been used extensively to prepare natural L-amino acids and other chiral compounds.[30,123,124] The L-aminotransferases catalyze the general reaction shown in Scheme 19.19 where an amino group from one L-amino acid is transferred to an α-keto acid to produce a new L-amino acid and the respective α-keto acid (see also Chapter 3). Those enzymes most commonly used as industrial biocatalysts have been cloned, overexpressed, and generally used as whole-cell or immobilized preparations. These include branched chain aminotransferase (BCAT) (E.C. 2.6.1.42), aspartate aminotransferase (AAT) (E.C. 2.6.1.1), and tyrosine aminotransferase (TAT) (E.C. 2.6.1.5).

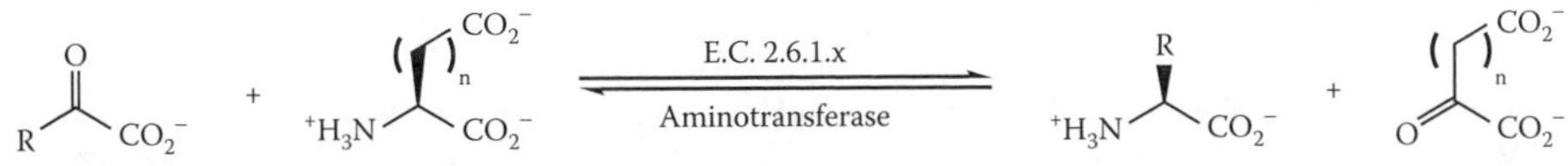

where n = 1 or 2

SCHEME 19.19

A transaminase patented by Celgene Corporation (Warren, NJ), called an ω-aminotransferase [(ω–AT) E.C. 2.6.1.18] does not require an α-amino acid as amino donor; instead it requires a primary amine and hence has the ability to produce chiral amines.[125,126] A similar ω-AT from *Vibrio fluvialis* has been described for the production of chiral amines along with chiral alcohols when coupled with AdH or chiral amino acids when coupled with an α-amino acid aminotransferase.[127–130] Another ω-AT, ornithine (lysine) aminotransferase (E.C. 2.6.1.68), has been described for the preparation of a chiral pharmaceutical intermediate used in the synthesis of Omapatrilat, a vasopeptidase inhibitor developed by Bristol-Myers Squibb, as well as the UAA Δ^1-piperidine-6-carboxylic acid.[131,132]

Another useful transaminase, D-amino acid transaminase (DAT) (E.C. 2.6.1.21), has been the subject of much study.[53,133,134] This enzyme catalyzes the reaction using a D-amino acid donor, either alanine, aspartate, or glutamate (Scheme 19.20), to produce another D-amino acid.

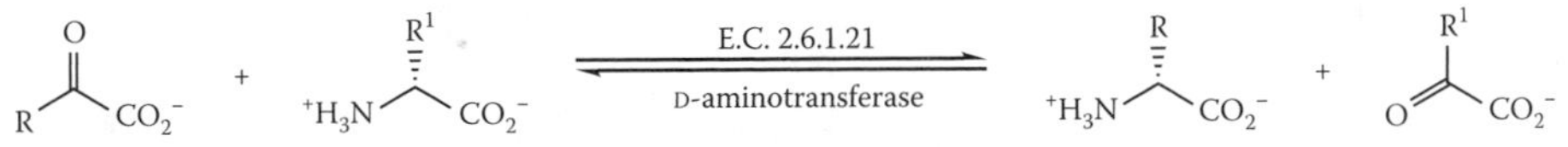

where R^1 = Me, $CH_2CO_2^-$ or $(CH_2)_nCO_2^-$

SCHEME 19.20

Although the utility of transaminases has been widely examined, one such limitation is the fact that the equilibrium constant for the reaction is near unity. Therefore, a shift in this equilibrium is necessary for the reaction to be synthetically useful. A number of approaches to shift the equilibrium can be found in the literature.[53,124,135] Another method to shift the equilibrium is a modification of that previously described. Aspartate, when used as the amino donor, is converted into oxaloacetate (**32**) (Scheme 19.21). Because **32** is unstable, it decomposes to pyruvate (**33**) and thus favors product formation. However, because pyruvate is itself an α-keto acid, it must be removed, or it will serve as a substrate and be transaminated into alanine, which could potentially cause downstream processing problems. This is accomplished by including the *alsS* gene encoding for the enzyme acetolactate synthase (E.C. 4.1.3.18), which condenses two moles of pyruvate to form (*S*)-acetolactate (**34**). The (*S*)-acetolactate undergoes decarboxylation either spontaneously or by the enzyme acetolactate decarboxylase (E.C. 4.1.1.5) to the final by-product, (*R*)-acetoin (**35**), which is metabolically inert. This process, for example, can be used for the production of both L- and D-2-aminobutyrate (**36** and **37**, respectively) (Scheme 19.21).[8,132,136,137]

In addition to UAAs and nonproteinogenic amino acids such as L- and D-2-aminobutyrate (**36, 37**), other examples of the use of transaminases to synthesize UAAs have also been described in the literature including L-*tert*-leucine (L-Tle), L-2-amino-4-(hydroxymethylphosphinyl)butanoic acid (phosphinothricin), and L-thienylalanines.[8,138–141] These UAAs are becoming increasingly important for their use in peptidomimetics, especially with the application of rational drug design being used by many pharmaceutical companies.

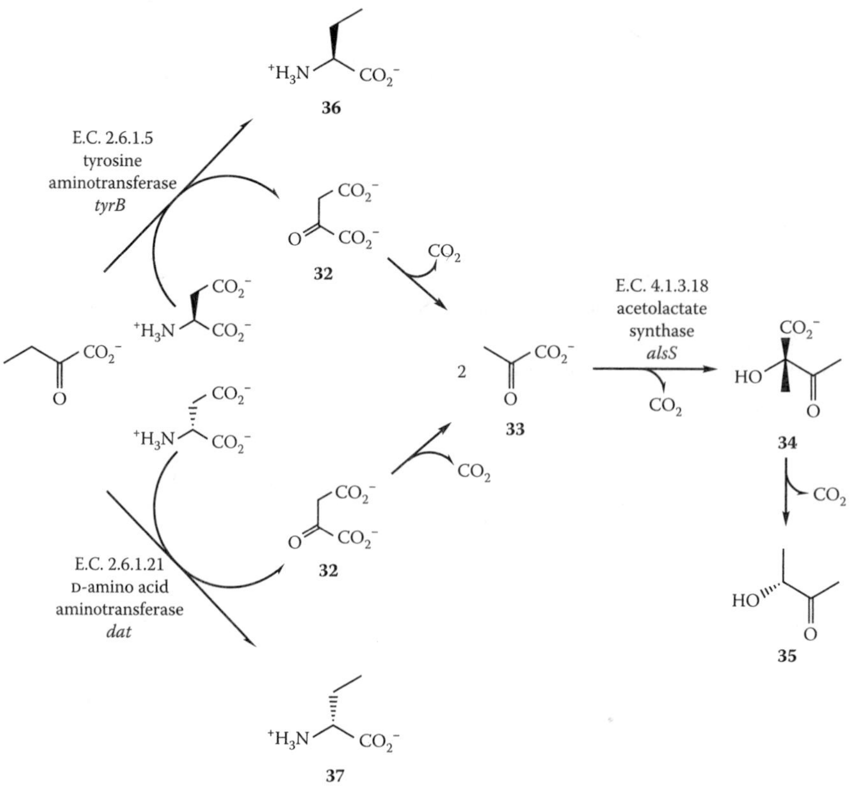

SCHEME 19.21

19.2.3 HYDROLASES (E.C. 3.x.x.x)

The hydrolase class of enzymes is the largest, representing about 75% of all the industrial enzymes where their use ranges from the hydrolysis of polysaccharides (carbohydrases) and nitriles (nitrilases) to proteins (proteases).[142] Most of these industrial enzymes are used in processing-type reactions to reduce protein, carbohydrate (i.e., viscosity), and lipid content in the detergent and food industries. Some of the more common reactions catalyzed by hydrolases are shown in Figure 19.1. The glycosidases (Figure 19.1, Reaction 1) will not be discussed here because they do not result in the preservation or formation of a chiral center. Nevertheless, their industrial use in the corn wet milling industry for starch conversion is of utmost significance.[143] Reactions catalyzed by esterases and lipases, epoxidases,[144] and proteases (Figure 19.1, Reactions 2–4, respectively) have most frequently been used to resolve substrates into their desired enantiomers.

The degradation of nitriles by nitrilases (E.C. 3.5.5.1) has been the subject of intense study, especially as it relates to the preparation of the commodity chemical acrylamide. Nitrilases catalyze the hydrolysis of nitriles to the corresponding acid plus ammonia (Figure 19.1, Reaction 5), whereas nitrile hydratases (E.C. 4.2.1.84) add water to form the amide.[145] Strains such as *Rhodococcus rhodochrous* J1, *Brevibacterium* sp., and *Pseudomonas chlororaphis* have been used to prepare acrylamide from acrylonitrile, which contain the hydratase and not nitrilase activity.[9] A comparison of these strains has been discussed.[146] Other uses of nitrilases, however, have primarily been directed at resolution processes to stereoselectively hydrolyze one enantiomer over another or regioselectively hydrolyze dinitriles.[147–149] The development of a nitrilase library and mutagenesis through gene site saturation have produced biocatalysts used to prepare (*R*)-4-cyano-3-hydroxybutyric acid, an intermediate in the synthesis in the cholesterol-lowering drug Lipitor® (see Chapter 20).[150,151]

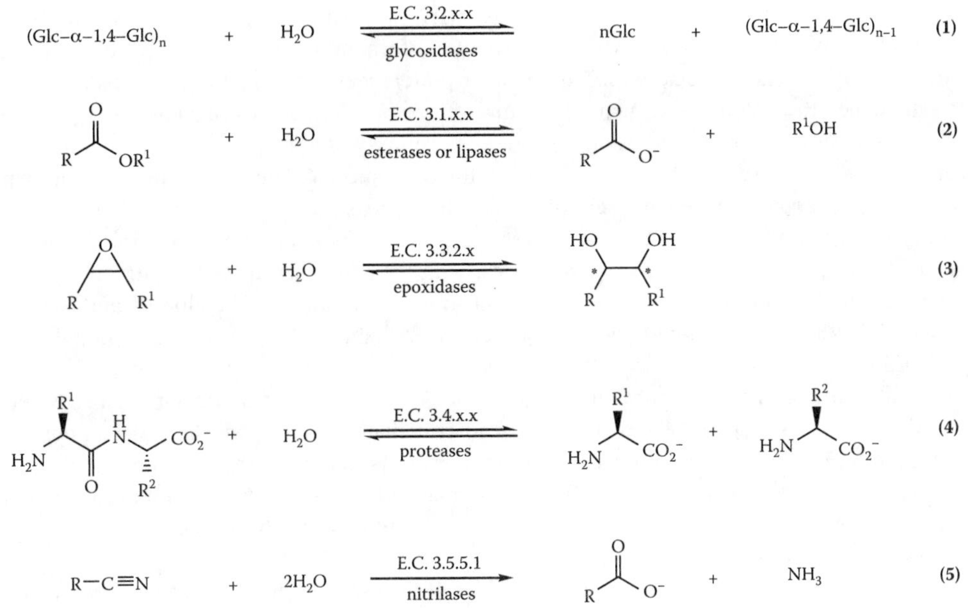

FIGURE 19.1 General reactions of hydrolase enzymes.

In the same group as nitrilase enzymes are the amidases. This includes amino acid amidase (E.C. 3.5.1.4) used to prepare amino acids, usually through resolution, and also penicillin G acylase (penicillin G amidohydrolase) (E.C. 3.5.1.11), used in the manufacture of semisynthetic penicillins.[152,153] Immobilized penicillin G acylase has most recently been used to catalyze the formation of N-α-phenylacetyl amino acids, which can then be used in peptide coupling reactions (see Section 19.2.3.2).[154] Bacterial aminoacylase I (N-acyl-L-amino acid amidohydrolase, E.C. 3.5.1.14) has also been used to acylate chiral amines with poor to moderate enantioselectively.[155]

The esterases, lipases, and proteases are often used to prepare chiral intermediates when the reactions are carried out in the synthetic mode. Selected examples of these enzymatic biotransformations will be discussed in the respective sections later in this chapter. The reader is directed to the following reviews and textbooks for additional information, especially regarding resolutions.[9,19,26,72,156–159]

19.2.3.1 Esterases and Lipases

One of the reactions catalyzed by esterases and lipases is the reversible hydrolysis of esters (Figure 19.1, Reaction 2). These enzymes also catalyze transesterifications and the desymmetrization of *meso*-substrates (*vide infra*). Many esterases and lipases are commercially available, making them easy to use for screening desired biotransformations without the need for culture collections and/or fermentation capabilities.[160] In addition, they have enhanced stability in organic solvents, require no co-factors, and have a broad substrate specificity, which make them some of the most ideal industrial biocatalysts. Alteration of reaction conditions with additives has enabled enhancement and control of enantioselectivity and reactivity with a wide variety of substrate structures.[159,161–164]

As increasing research has been carried out with these enzymes, a less empirical approach has been taken as a result of the different substrate profiles that have been compiled for various enzymes in this class. These profiles have been used to construct active site models for such versatile enzymes as the carboxylester hydrolase, pig liver esterase (PLE) (E.C. 3.1.1.1), and the microbial lipases (E.C. 3.1.1.3) from *Burkholderia cepacia* (formerly *Pseudomonas cepacia*) lipase (PCL), *Candida rugosa* (formerly *C. cylindracea*) lipase (CRL), lipase SAM-2 from *Pseudomonas* sp., *Chromobacterium viscosum* (CVL), *Aspergillus niger* lipase (ANL, Amano lipase AP), and *Rhizopus oryzae* lipase (ROL).[165–175] In addition, X-ray crystal structure information on PCL and CRL has been most helpful at predicting substrate activities and isomer preferences.[176–178] Furthermore, a database has been established to exploit the sequence structure–function relationships of lipases.[179]

Any discussion of lipases and esterases would be remiss without specifically mentioning PLE. Because of its extensive biocatalytic versatility, this enzyme has been the subject of numerous reviews and has been screened for stereoselective hydrolysis using hundreds of substrates.[180,181] Suffice it to say that PLE has been the workhorse of enzymes for many organic chemists catalyzing both enantioselective and regioselective reactions, with the active site model constructed by Jones being extremely helpful to allow one to predict substrate reactivities.[169] The cloning and functional expression of PLE will undoubtedly aid in a better understanding of the mechanistic details of this important industrial biocatalyst.[182]

A number of the major pharmaceutical companies have used biocatalytic approaches based on esterases and lipases to prepare target drugs or intermediates.[107,183-187] Most of these approaches involve resolutions that use a racemic ester or amide as substrate, whereby yields of 50% can only be realized. Examples of resolutions applied to pharmaceutical intermediates such as the paclitaxel (Taxol®) sidechain and *R*-(+)-BMY-14802, an antipsychotic agent, have been described by the Bristol-Myers Squibb group.[107]

The following sections discuss selected examples of the use of esterases and lipases to hydrolyze prochiral or *meso*-substrates, where yields of 100% can theoretically be attained, followed by a brief discussion of dynamic kinetic resolution (DKR) where reaction yields of 100% can also potentially be achieved.

19.2.3.1.1 Prochiral and meso-Substrates

Selected prochiral and *meso*-substrates (Figure 19.2) have been used with various esterases and lipases, which illustrates the wide variety of structural elements that can be used with these enzymes to afford desymmetrization. A more complete listing of substrates used in these types of reactions can be found in other sources.[9,156,158,159,180,181,188–190] Reaction conditions often use organic solvents, which can have a profound effect as to what product isomer is formed.[191] The use of porcine pancreatic lipase (PPL) to carry out an asymmetric hydrolysis of a *meso*-diacetate for the production of an intermediate in pheromone synthesis has been reported[192] as well as the desymmetrization of prochiral 2-benzyl-1,3-propanediol derivatives for the preparation of phosphorylated tyrosine analogues.[193] Both (*R*)- and (*S*)-enantiomers of the UAA 1-amino-2,2-difluorocyclopropane-1-carboxylic acid were also prepared from prochiral starting precursors using lipases.[194] Two specific examples are discussed that used this approach directed at key chiral pharmaceutical intermediates.

Dodds and colleagues at Schering-Plough described the acylation of a prochiral diol **38** on the synthetic route to a potential antifungal drug, SCH 51048 (**39**) (Scheme 19.22). Numerous commercial

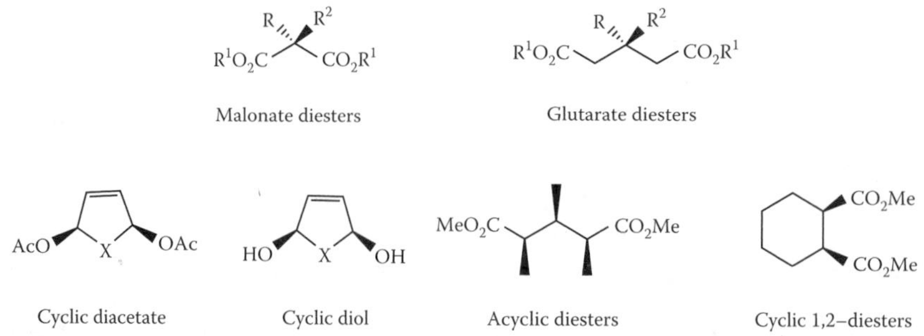

FIGURE 19.2 Examples of prochiral and *meso*-substrates (hydrolytic and synthetic) for esterases and lipases.

enzymes were screened for their ability to carry out the stereoselective acylation of the diol using many different conditions of solvents and temperature as well as enzyme, substrate, and acylating agent concentrations. Among the better enzymes for monoacylation were Amano CE (*Humicola lanuginosa*) and Biocatalyst (*Pseudomonas fluorescens*). Yields using both enzymes were 100% with ee's of 97 and 99, respectively. The enzyme Novo SP435 (*Candida antarctica*) gave an 70% yield of (*S*)-monoacetate **40** with good ee (>99%) but with the remaining product (30%) being the diacetate.[183,185,187]

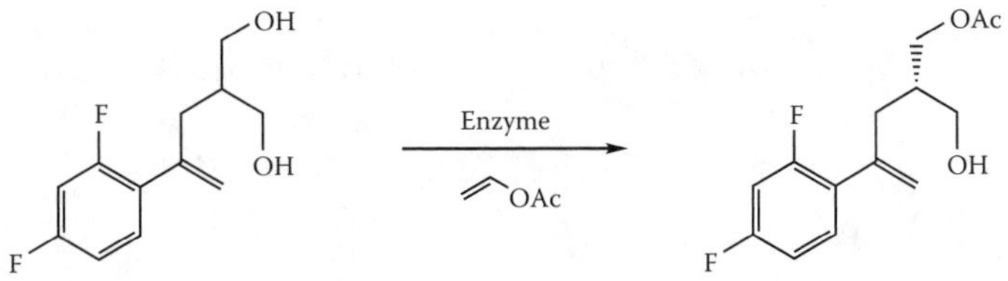

SCHEME 19.22

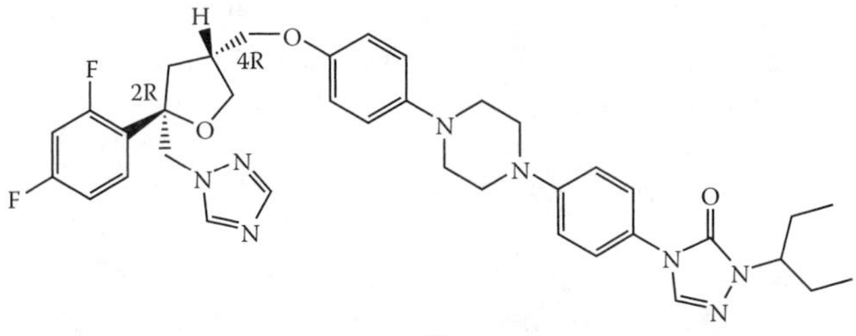

39

Another example using a prochiral acetate and asymmetric hydrolysis was described by the Bristol-Myers Squibb group for an intermediate in the synthesis of Monopril® (fosinopril sodium) (**41**), an ACE inhibitor (Scheme 19.23). The prochiral substrate **42** was hydrolyzed both when R = phenyl or cyclohexyl to the corresponding (*S*)-(−)-monoacetate **43**. The reaction was carried out in a 10% toluene biphasic system with either PPL or *Chromobacterium viscosum* lipase. The cyclohexyl monoacetate was obtained in 90% yield with an optical purity of 99.8%.[107,195]

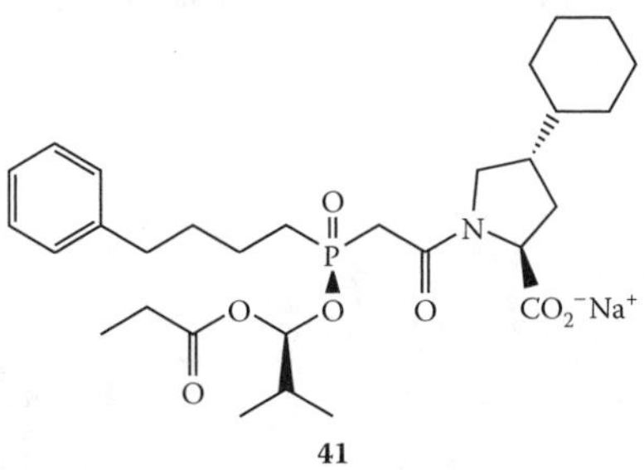

41

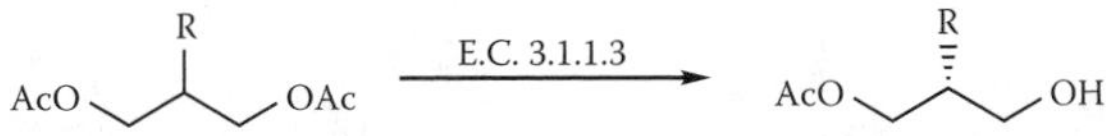

R = C_6H_5 or C_6H_{11}

SCHEME 19.23

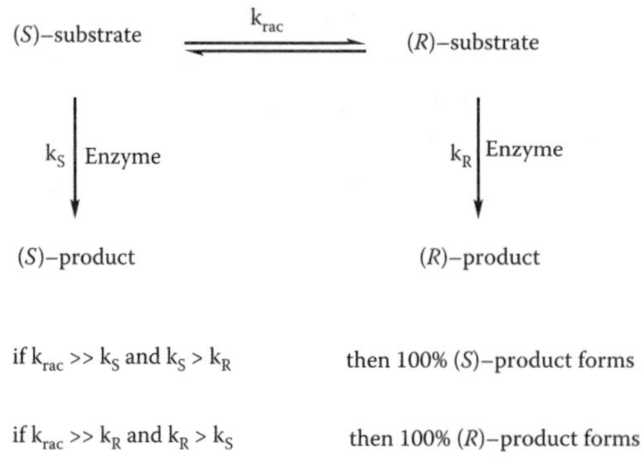

$$(S)-\text{substrate} \quad \underset{}{\overset{k_{rac}}{\rightleftharpoons}} \quad (R)-\text{substrate}$$

k_S | Enzyme k_R | Enzyme

$(S)-\text{product}$ $(R)-\text{product}$

if $k_{rac} \gg k_S$ and $k_S > k_R$ then 100% (S)–product forms

if $k_{rac} \gg k_R$ and $k_R > k_S$ then 100% (R)–product forms

FIGURE 19.3 Dynamic kinetic resolution (DKR).

19.2.3.1.2 Dynamic Kinetic Resolution

A method that has been used to approach 100% theoretical yield in asymmetric syntheses is dynamic kinetic resolution, or DKR. Although this method has been practiced based on strictly chemical reactions, only those chemoenzymatic DKR reactions will be discussed here. Most often, the enzyme used by this method is a hydrolase (lipase, esterase, protease), but other enzymes such as hydantoinases, N-acylamino acid racemases, and dehydrogenases have also been exploited to effectively carry out DKR reactions.[196] For additional details the reader is directed to the many review articles written on DKR.[197–206]

If the interconversion between the two enantiomeric substrates is rapid and the product is relatively stable, and thus irreversibly formed, then the magnitude of the rate constants, k_R and k_S, will dictate which product isomer is formed (Figure 19.3). The racemization reaction is most often catalyzed by metal ion complexes containing Ru or Pd, but silica, ion-exchange resins, or enzymes can also be used. The addition of aldehydes can also be used to facilitate the racemization process through the formation of a Schiff base with primary amines and amino acids.

An example is described for the UAA (S)-tert-leucine (**44**) (Scheme 19.24).[207] It uses the commercially available Lipozyme® (*Mucor miehei*) from Novozymes to hydrolyze the racemic 2-phenyl-4-tert-butyl-oxazolin-5-(4H)-one (**45**) to the (S)-N-benzoyl-tert-leucine butyl ester (**46**) followed by Alcalase® (subtilisin, *Bacillus licheniformis* from Novozymes) treatment to hydrolyze the butyl ester, which on debenzoylation yields (S)-tert-leucine (**44**) with an ee of 99.5% and a yield of 74%.

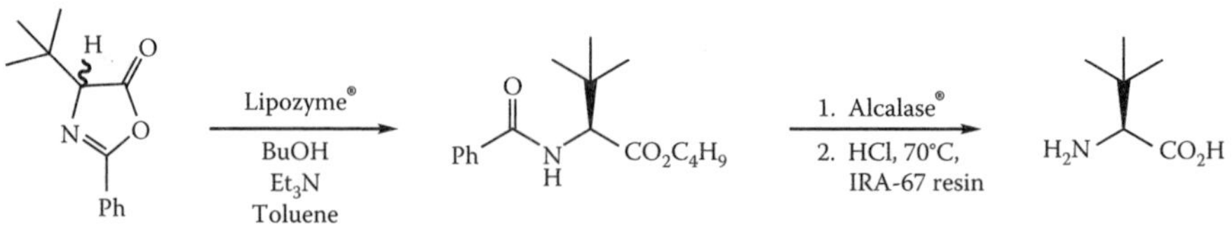

SCHEME 19.24

The antipode, (R)-tert-leucine (**47**), was synthesized using DKR with an enantioselective (R)-hydantoinase. Here the racemic 5-tert-butyl-hydantoin (**48**), which racemizes *in situ* at a pH >8, produces the N-carbamoyl-(R)-tert-leucine (**49**) in >99% yield through the action of the (R)-hydantoinase. Decarbamoylation of intermediate **49** produced enantiomerically pure **47** in 85.5%

isolated yield with an ee of 99.5% (Scheme 19.25).[61] This is an example where the instability of the stereogenic center in **48** at slightly alkaline pH results in near-quantitative conversions to a single isomer.

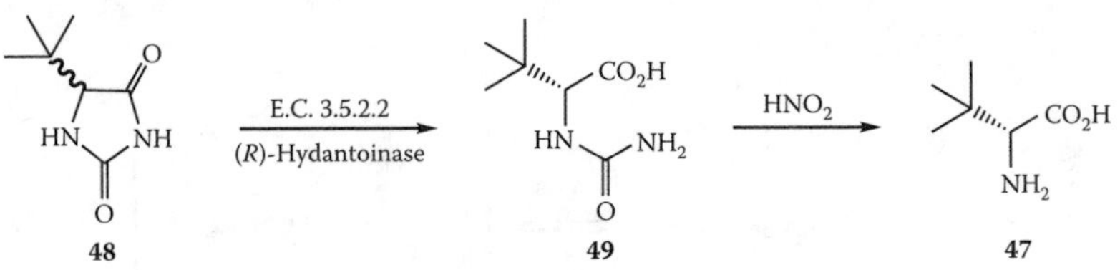

SCHEME 19.25

An example where a transition metal catalyst is used in combination with an enzyme has been described (Scheme 19.26).[207] The racemic alcohol **50** was converted to the (*R*)-acetate **51**, using a ruthenium catalyst along with Novozym 435® (immobilized Lipase B from *Candida antarctica*), 3 equivalents of *p*-chlorophenylacetate in *t*-BuOH, and 1 equivalent of 1-indanone. The reaction yield was 81% with an optical purity of >99.5% ee.

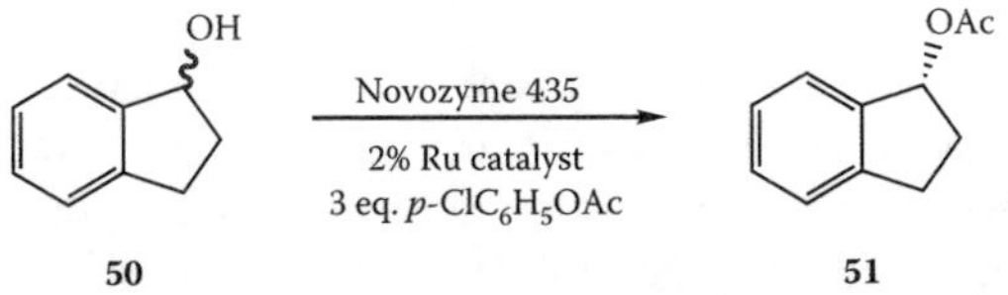

SCHEME 19.26

There are many other examples of DKR reports that have been published.[208–210] This approach to asymmetric synthesis, especially combining chemical and biochemical regimens, is developing into a powerful tool for synthesizing asymmetric molecules with high yields and optical purities.

19.2.3.2 Proteases

As mentioned earlier, proteases have been extensively studied to carry out resolutions both of natural and unnatural amino acids. These useful reactions provide chiral compounds, which very often cannot be obtained by chemical means. The use of proteases to carry out enantioselective and regioselective coupling reactions, especially incorporating the use of organic solvents, has provided numerous examples of synthetic and semi-synthetic peptides.[211–218]

With regard to the use of protease in the synthetic mode, the reaction can be carried out using a kinetic or thermodynamic approach. The kinetic approach requires a serine or cysteine protease that forms an acyl-enzyme intermediate, such as trypsin (E.C. 3.4.21.4), α-chymotrypsin (E.C. 3.4.21.1), subtilisin (E.C. 3.4.21.62), or papain (E.C. 3.4.22.2), and the amino donor substrate must be "activated" as the ester (Scheme 19.27) or amide (not shown). Here the nucleophile R^3–NH_2 competes with water to form the peptide bond. Besides amines, other nucleophiles such as alcohols or thiols can be used to compete with water to form new esters or thioesters. Reaction conditions such as pH, temperature, and organic solvent modifiers are manipulated to maximize synthesis. Examples of this approach using carboxypeptidase Y (E.C. 3.4.16.5) from baker's yeast have been described.[219]

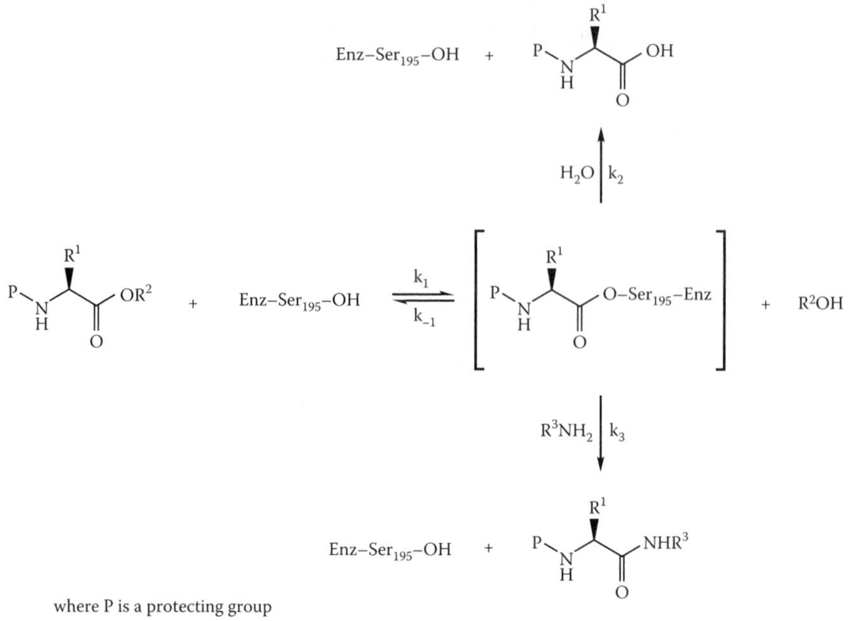

where P is a protecting group

SCHEME 19.27

The thermodynamic approach is shown whereby the uncharged amino acids are coupled to form the peptide bond (Scheme 19.28). This requires suppression of the substrate charges, which again can be accomplished by altering reaction conditions similar to that described for the kinetic approach. Although protease activity generally decreases with increasing organic solvent, water miscible solvents such as DMSO, DMF, methanol, or acetonitrile have been used in small to moderate amounts with some success in both synthetic approaches. Other methods such as *in situ* product removal, solid-to-solid synthesis, and frozen state reactions have been conducted to maximize yields and alter substrate specificity.[220–223]

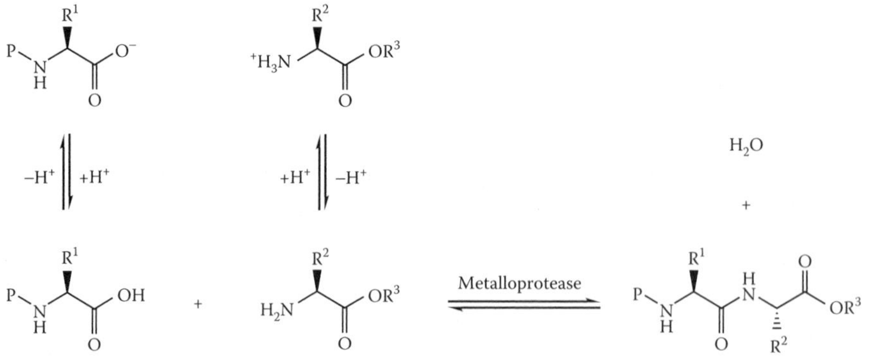

where P is a protecting group

SCHEME 19.28

A prominent example of peptide coupling is used for the production of the synthetic dipeptide sweetener aspartame (α-APM) (**28**) (Chapter 31). The chemical coupling method yields approximately an 80:20 mixture of the α- and β-isomers, whereas the regioselectivity of the enzymatic method produces 100% of the desired α-APM isomer. The thermodynamic approach is the basis of the TOSOH process for α-APM synthesis, which is practiced by the Holland Sweetener Company. Carbobenzyloxy-L-aspartic acid (Z-L-asp, **31**) is coupled with L–phenylalanine methyl ester (L-PM, **52**) using the enzyme thermolysin (E.C. 3.4.24.27) from *Bacillus thermoproteolyticus* to form Z-α-APM (**53**), which then on catalytic hydrogenation forms α-APM (Scheme 19.29). Because of the enantioselectivity of thermolysin, D/L–PM can also be used with only the L–isomer being coupled. The D-isomer does not react and forms a salt with Z-α-APM (Z-α-APM•D-PM); the D-PM can be recycled on acidification. Many variations have been explored for the enzymatic preparation of α-APM as a result of its high economic value.[224-227]

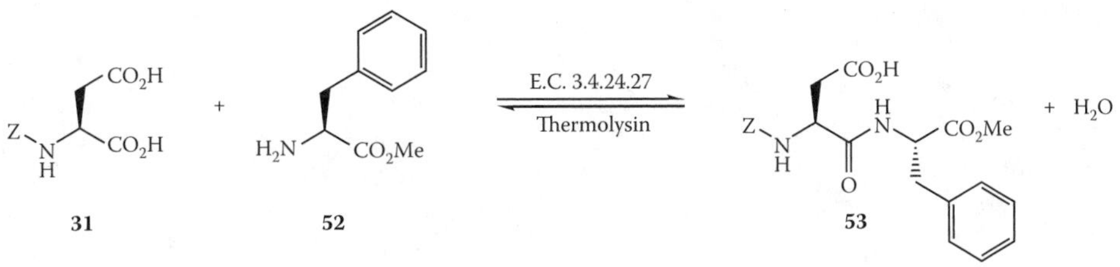

SCHEME 19.29

With the increased efforts to use peptidomimetics as selective enzyme inhibitors in a number of disease states such as hypertension (renin), cancer (matrix metalloproteases), and acquired immune deficiency syndrome (AIDS; HIV protease), the incorporation of UAAs into peptides is becoming necessary. The use of proteases such as subtilisin, trypsin, α-chymotrypsin, and thermolysin have been used with or without organic solvents to synthesize peptides containing UAAs. In addition, protein engineering has redesigned certain proteases to effect thermostability and organic solvent tolerance.[228] As shown in Table 19.2, N-protecting groups such as Z (benzyloxycarbonyl), Boc (t-butyloxycarbonyl), and Moz [(p-methoxybenzyl)oxy]carbonyl have been used with a variety of solvent conditions and enzymes to prepare the final dipeptide or tripeptide, in most cases, with high yield. A number of different acyl donors and acceptors containing UAAs are shown to function very well in the synthesis of coupled peptides. These include D-amino acids (Leu, Phe, Ala, Trp); an α,α-disubstituted amino acid (Aib, α-aminobutyrate); a statine-type isostere; and two noncoded amino acids, norvaline (Nvl) and homophenylalanine (hPhe).

Synthesis of polypeptides containing all D-amino acids has also been demonstrated using the enzyme clostripain (E.C. 3.4.22.8).[229] Halogenated phenylalanines can be used in peptide coupling reactions with thermolysin, but the coupling was conducted with a racemic derivative.[230] Another reaction where UAAs have been incorporated into dipeptides is with the enzyme penicillin acylase (E.C. 3.5.1.11), using the kinetic approach with both D- and L-phenylglycine.[231] The approach using proteases to synthesize peptides containing UAAs is likely to become a key technology as additional peptidomimetics are discovered and become approved as drugs.

19.2.4 LYASES (E.C. 4.x.x.x)

Lyases are an attractive group of enzymes from a commercial perspective, as demonstrated by their use in many industrial processes.[240] They catalyze the cleavage of C–C, C–N, C–O, and other bonds by means other than hydrolysis, often forming double bonds. For example, two well-studied ammonia lyases, aspartate ammonia lyase (aspartase) (E.C. 4.3.1.1) and phenylalanine ammonia lyase (PAL) (E.C. 4.3.1.5), catalyze the *trans*-elimination of ammonia from the amino acids, L-aspartate and L-phenylalanine, respectively. Most commonly used in the synthetic mode, the reverse reaction has been used to prepare the L-amino acids at the ton scale (Schemes 19.30 and 19.31).[240–242] These reactions are conducted at very high substrate concentrations such that the equilibrium is shifted, resulting in very high conversion to the amino acid products.

TABLE 19.2
Peptide Coupling Reactions Using Unnatural Amino Acids in Both the Acyl Donor and Acceptor Sites (the Arrow Indicates the Peptide Bond That Was Synthesized)

Product	Enzyme	Solvent	Yield (%)	Ref.
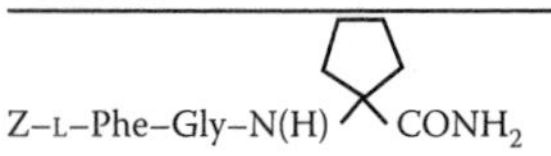 Z–L–Phe–Gly–N(H)⁀CONH₂ (cyclopentyl)	subtilisin BPN	DMF/H₂O (8/3)	89	232
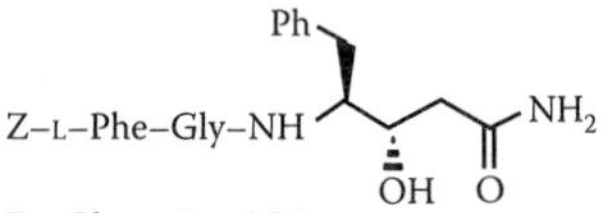 Z–L–Phe–Gly–NH (Ph, OH, O, NH₂)	subtilisin BPN	DMF/H₂O (8/3)	70	232
Z-L-Phe-D-Leu-NH₂	subtilisin CLEC	MeCN	96	233
Z-L-Phe-D-Phe-NH₂	subtilisin CLEC	MeCN	95	233
Boc-Aib-L-Ala-pNA	trypsin (*S. griseus*)	DMSO/buffer (1/1)	96	234
Boc–L–Phe–D–Ala–L–Ala–pNA (↑)	trypsin (bovine pancreas)	DMSO/buffer (1/1)	80	235
Z–L–Val–L–Tyr–D–Phe–NH₂ (↑)	α-chymotrypsin (bovine pancreas)	isooctane/THF (7/3)	68	236
Moz-L-Phe-D-Ala-NH₂	subtilisin [Alcalase] (*B. licheniformis*)	*t*-butanol	87	237
Boc-L-Phe-D-TrpOMe	subtilisin [Alcalase] (*B. licheniformis*)	EtOH	60	238
Z-D-Phe-L-LysOMe	subtilisin [Alcalase] (*B. licheniformis*)	EtOH	53	238
Z-L-Phe-L-Nvl-NH₂	thermolysin (*B. thermoproteolyticus*)	buffer	92	239
Z-L-hPhe-L-Leu-NH₂	thermolysin (*B. thermoproteolyticus*)	buffer	91	239

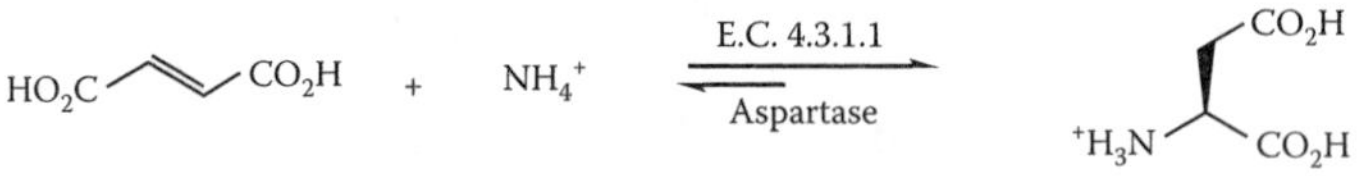

$$HO_2C\diagup\diagdown CO_2H \quad + \quad NH_4^+ \quad \underset{Aspartase}{\overset{E.C.\ 4.3.1.1}{\rightleftharpoons}} \quad {}^+H_3N\diagdown{}^{CO_2H}_{CO_2H}$$

SCHEME 19.30

$$\text{(PhCH=CH)}CO_2H \quad + \quad NH_4^+ \quad \underset{PAL}{\overset{E.C.\ 4.3.1.5}{\rightarrow}} \quad {}^+H_3N\diagdown CO_2H \text{ (CH}_2\text{Ph)}$$

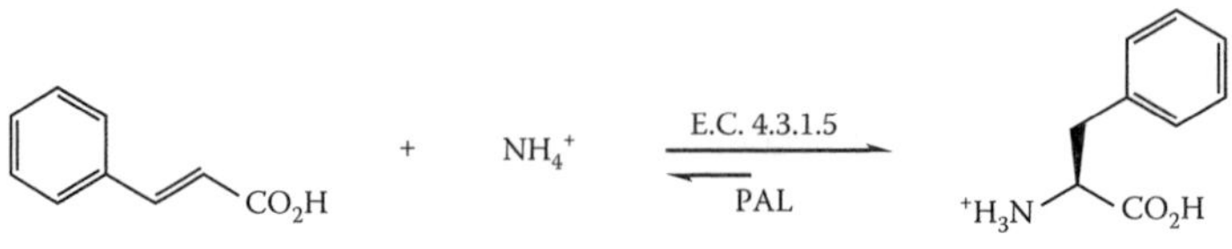

SCHEME 19.31

Aspartase exhibits incredibly strict substrate specificity and thus is of little use in the preparation of L-aspartic acid analogues. However, a number of L-phenylalanine analogues have been prepared with various PAL enzymes from the yeast strains *Rhodotorula graminis*, *Rhodotorula rubra*, *Rhodoturula glutinis*, and several other sources that have been cloned into *E. coli*.[243–247] Future work in this area will likely include protein engineering to design new enzymes that offer a broader substrate specificity such that additional L-phenylalanine analogues could be prepared.

19.2.4.1   Aldolases

A specific type of lyase, the aldolase class of enzymes catalyzes the formation of a C–C bond with control over the newly created stereogenic centers, which is a most useful reaction to the synthetic organic chemist.[248–252] These enzymes are most often used in the synthesis of novel saccharides, such as aminosugars, thiosugars, and disaccharide mimetics, because they use these types of substrates in nature and their substrate specificity is quite broad.[253–255] Aldolases are classified as Type I or Type II enzymes depending on their source and mechanism of action.[256,257] Type I aldolases are from higher plants and mammalian sources and require no metal ion co-factor and use a Schiff-base intermediate in catalysis. Type II enzymes are from bacterial and fungal sources and require a metal ion, usually Zn^{2+}, for catalysis. In addition to different mechanisms, depending on the specific aldolase, the stereochemistry of the two new stereogenic centers can be controlled (Scheme 19.32).[9,14]

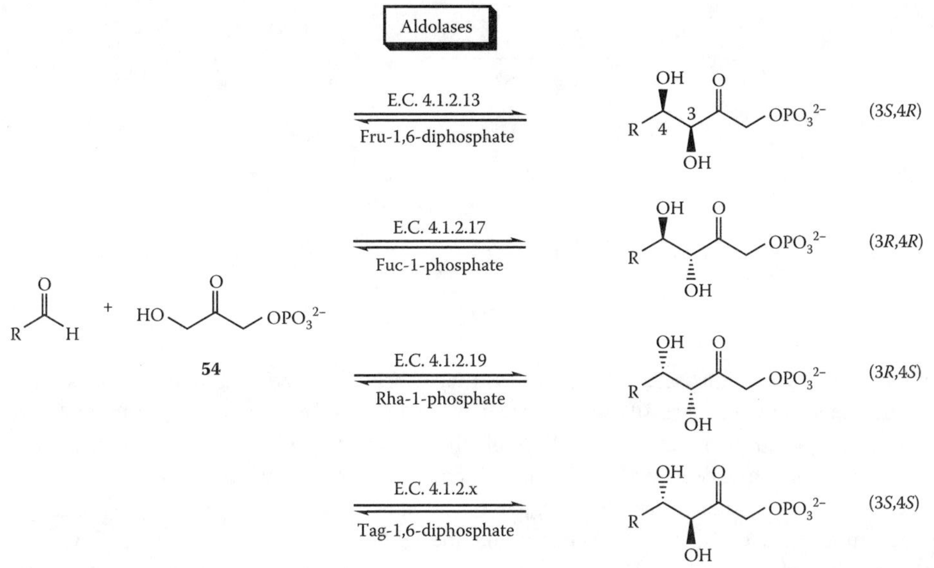

SCHEME 19.32

Although many aldolases have been characterized for research purposes, the four aldolase enzymes described in Scheme 19.32 have not been used commercially to any significant extent. This is likely a result of their availability and the need for dihydroxyacetone phosphate (DHAP) (**54**), the expensive donor substrate required in these aldolase reactions (Scheme 19.32). A number of chemical and enzymatic routes have been described for DHAP synthesis, which could alleviate these concerns.[9,258]

Other aldolases have been described that do not rely on DAHP as a substrate. One such enzyme, 2-deoxyribose-5-phosphate aldolase (E.C. 4.1.2.4), has been cloned into *E. coli,* accepts a broad range of donor and acceptor aldehydes, and has been used to synthesize a number of heterocycles having utility as epithiolone synthons.[259]

The use of reactive immunization to generate catalytic antibodies (or abzymes) that catalyze aldolase reactions has been described, offering additional utility for this synthetically useful transformation.[260] Two such abzymes, 38C2 and 84G3, are available commercially and their respective, diverse activities have been described.[261,262]

Specific reference to the cloning of selected aldolases has been described, and as larger quantities of these biocatalysts become available, process development studies for the synthesis of chiral intermediates will continue to advance.[263]

19.2.4.2 Decarboxylases

A number of decarboxylase enzymes have been described as catalysts for the preparation of chiral synthons, which are difficult to access chemically (see Chapter 2).[264] The amino acid decarboxylases catalyze the pyridoxal phosphate (PLP)-dependent removal of CO_2 from their respective substrates. This reaction has found great industrial utility with one specific enzyme in particular, L-aspartate-β-decarboxylase (E.C. 4.1.1.12) from *Pseudomonas dacunhae*. This biocatalyst, most often used in immobilized whole cells, has been utilized by Tanabe to synthesize L-alanine on an industrial scale (multi-tons) since the mid-1960s (Scheme 19.33).[242,265] Another use for this biocatalyst has been the resolution of racemic aspartic acid to produce L-alanine and D-aspartic acid (Scheme 19.34). The cloning of the L-aspartate-β-decarboxylase from *Alcaligenes faecalis* into *E. coli* offers additional potential to produce both of these amino acids.[266]

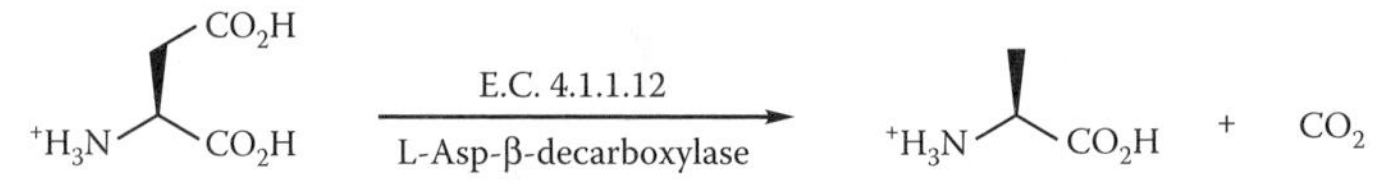

SCHEME 19.33

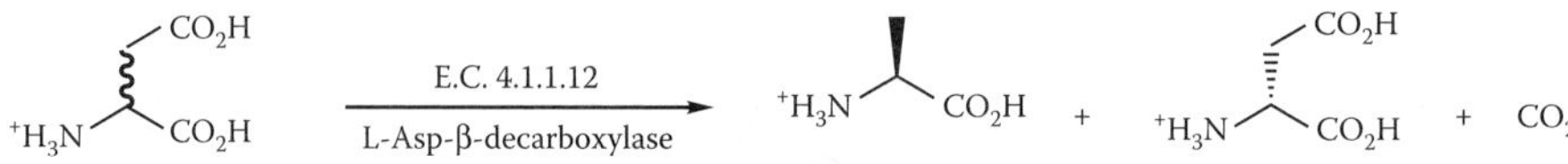

SCHEME 19.34

Numerous other amino acid decarboxylases have been isolated and characterized, and much interest has been shown as a result of the irreversible nature of the reaction with the release of CO_2 as the thermodynamic driving force. Although these enzymes have narrow substrate-specificity profiles, their utility has been widely demonstrated. Additional industrial processes will continue to be developed once other decarboxylases become available. Such biocatalysts would include the aromatic amino acid (E.C. 4.1.1.28), phenylalanine (E.C. 4.1.1.53) and tyrosine (E.C. 4.1.1.25) decarboxylases, which likely could be used to produce derivatives of their respective substrates. These derivatives are finding increased use in the development of peptidomimetic drugs and as possible positron emission tomography imaging agents.[267,268]

The α-keto acid decarboxylases such as pyruvate (E.C. 4.1.1.1) and benzoyl formate (E.C. 4.1.1.7) decarboxylases are a thiamine pyrophosphate (TPP)–dependent group of enzymes, which in addition to nonoxidatively decarboxylating their substrates, catalyze a carboligation reaction forming a C–C bond leading to the formation of α-hydroxy ketones.[269,270] The hydroxy ketone (*R*)-phenylacetylcarbinol (**55**), a precursor to L-ephedrine (**56**), has been synthesized with pyruvate decarboxylase (Scheme 19.35). BASF scientists have made mutations in the pyruvate decarboxylase from *Zymomonas mobilis* to make the enzyme more resistant than the wild-type enzyme to inactivation by acetaldehyde for the preparation of chiral phenylacetylcarbinols.[271]

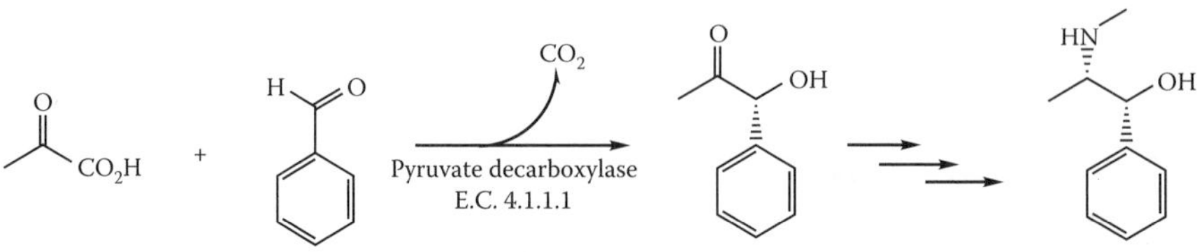

SCHEME 19.35

Benzoyl formate decarboxylase from *Pseudomonas putida* has been used to synthesize chiral 2-hydroxy ketones and bis(α-hydroxy) ketones, which find their use as pharmaceutical intermediates and as new multidentate ligands for asymmetric transition metal catalysis, respectively.[272,273] Combining this decarboxylase activity with AdH has allowed all the stereoisomers of 1-phenylpropane-1,2-diol to be synthesized.[274]

19.2.4.3 Hydroxynitrile Lyases

The hydroxynitrile lyase (HNL) class of enzymes, also referred to as oxynitrilases, consists of enzymes that catalyze the formation of chiral cyanohydrins by the stereospecific addition of hydrogen cyanide (HCN) to aldehydes and ketones (Scheme 19.36).[275–279] These chiral cyanohydrins are versatile synthons, which can be further modified to prepare chiral α-hydroxy acids, α-hydroxy aldehydes and ketones, acyloins, vicinal diols, ethanolamines, and α- and β-amino acids, to name a few.[280] Both (*R*)- and (*S*)-selective HNLs have been isolated, usually from plant sources, where their natural substrates play a role in defense mechanisms of the plant through the release of HCN. In addition to there being HNLs with different stereo-preferences, two different classifications have been defined, based on whether the HNL contains a flavin adenine dinucleotide (FAD) co-factor.

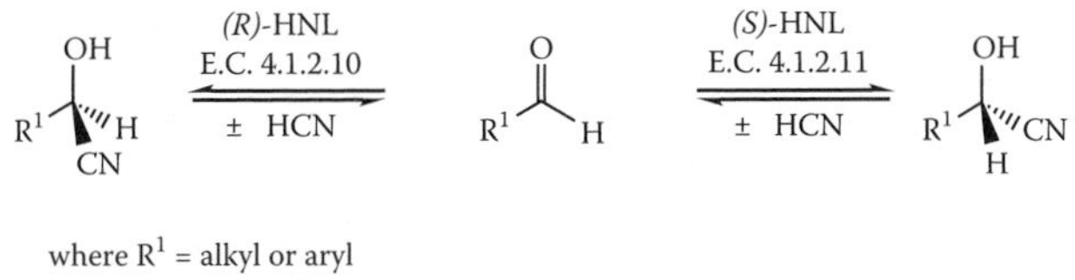

where R^1 = alkyl or aryl

SCHEME 19.36

One of the most common FAD-dependent HNLs is an (*R*)-HNL from *Prunus amygdalus* (*Pa*HNL, from almonds) (E.C. 4.1.2.10). This enzyme has a very broad substrate specificity whereby a wide range of aromatic and aliphatic aldehydes are converted into (*R*)-cyanohydrins with high enantiomeric excess. A common method used to obtain products with high ee and yield requires that the enzyme reaction be conducted in water-saturated organic solvent such as ethyl acetate or diisopropyl ether, which helps to minimize the nonenzymatic addition of HCN in purely aqueous reaction media. A variation of this was described by van Langen and co-workers whereby a two-phase system was used with *Pa*HNL for the synthesis of (*R*)-2-chloromandelic acid, the chiral intermediate used in the antithrombotic agent clopidogrel (Scheme 19.37) (see also Chapter 31).[281]

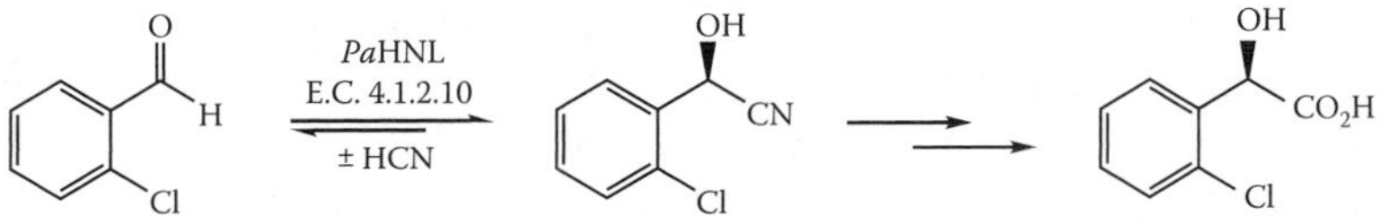

SCHEME 19.37

Although the (*R*)-specific *Pa*HNL has been more readily available than many of the (*S*)-HNLs, this has changed as a result of recombinant DNA technology, which has allowed the cloning and overexpression of a number of these FAD-independent, (*S*)-specific enzymes. The cloned (*S*)-HNLs are from *Hevea brasiliensis* (*Hb*HNL, from rubber tree) (E.C. 4.1.2.39) and *Manihot esculenta* (*Me*HNL, from cassava) (E.C. 4.1.2.37). Another useful (*S*)-HNL is from *Sorghum bicolor* (*Sb*HNL, from millet seedlings) (E.C. 4.1.2.11), which has been cloned but enzyme activity has not been demonstrated, likely because of its glycosylation requirements.[279] In addition, an (*R*)-HNL has been cloned from *Linum usitatissimum* (*Lu*HNL, from flax seedlings).[282] Especially for the cloned

(S)-HNLs, their availability in larger amounts will allow their characterization and use as industrial biocatalysts to increase.

In the references cited earlier, numerous studies on the substrate specificities can be found for both (R)- and (S)-HNLs along with different processing conditions, which make these versatile enzymes amenable to preparing chiral intermediates for a wide variety of uses.

19.2.5 ISOMERASES (E.C. 5.x.x.x)

The isomerase class of biocatalysts represents a small number of enzymes that is mainly composed of the racemases, epimerases, and mutases. In the case of racemases and epimerases, the stereo-chemistry of at least one carbon center is changed, whereas mutases catalyze the transfer of a functional group, such as an amino function, to an adjacent carbon, forming a new stereocenter. One of the most industrially important biotransformations involves an enzyme from this class, D-xylose isomerase (E.C. 5.3.1.5), also known as glucose isomerase. It catalyzes the conversion of D-glucose (57) to D-fructose (58), which is necessary for the production of high-fructose corn syrup (HFCS), a sucrose substitute used by the food and beverage industries (Scheme 19.38). It is one of the highest tonage value enzymes produced in the world and is used in nearly all processes in a cell-free or whole-cell immobilized form.[241,283–285]

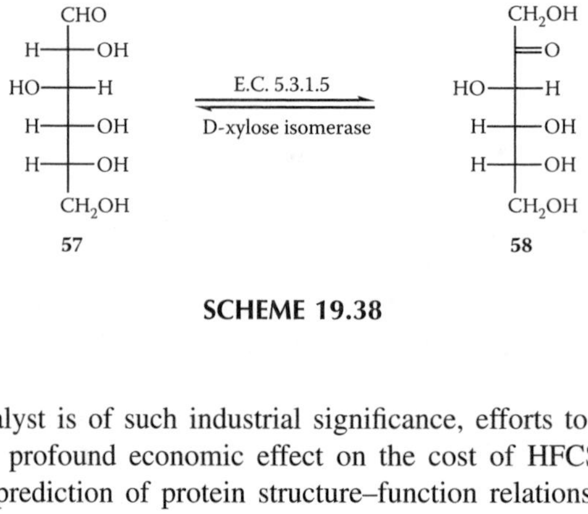

SCHEME 19.38

Because this biocatalyst is of such industrial significance, efforts to redesign it with altered properties could have a profound economic effect on the cost of HFCS. With the advances in molecular biology and prediction of protein structure–function relationships, these studies have been under way for a number of years and include thermal stabilization, alteration of pH and temperature optima, and modifications to substrate specificity.[284,286,287]

Few examples of epimerases exist for the synthesis of chiral fine chemicals. One example is the epimerase-catalyzed conversion of N-acetylglucosamine to N-acetylmannosamine.[288] Another epimerase reaction is used for the preparation of the synthetically useful inositols;[289] D-myo-inositol (59) is converted into D-chiro-inositol (60) by the enzyme D-myo-inositol 1-epimerase (Scheme 19.39), which has been cloned into E. coli from an Agrobacterium sp.[290]

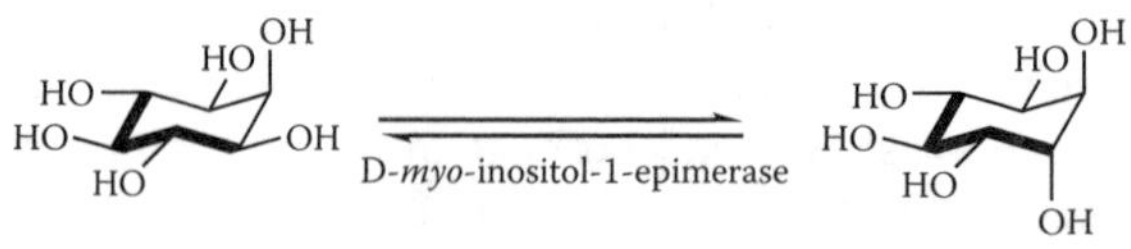

SCHEME 19.39

19.2.5.1 Mutases

Although this class of biocatalyst is currently not used commercially, the potential for these types of enzymes, such as the aminomutases (E.C. 5.4.3.x), has yet to be realized. The general reaction is the conversion of an α-L-amino acid to its corresponding β-L-amino acid with the migration of the 2-amino group to carbon–3 (Scheme 19.40). Some of these biocatalysts require coenzymes such as pyridoxal phosphate and are activated by *S*-adenosylmethionine or use vitamin B_{12} and proceed via a radical mechanism.[291,292] A number of amino acid aminomutases using substrates as L-lysine, L-leucine, D-ornithine, L-alanine, and L-tyrosine have been described.[292-294] An aminomutase using L-phenylalanine as substrate has been identified in the biosynthesis of the taxol sidechain, and through gene cloning the biosynthesis of taxanes and taxane-related compounds have been improved.[295,296] The enediyne antitumor antibiotic C-1027 (lidamycin) isolated from *Streptomyces globisporus* contains an unusual (*S*)-3-chloro-4,5-dihydroxy-β-phenylalanine moiety, which has been determined to be derived from the conversion of L-tyrosine to (*S*)-β-tyrosine by an aminomutase.[297,298] These β-amino acids are finding utility in pharmaceuticals and in the specialty chemical arena and are not easy to produce chemically so that biocatalytic routes are becoming more attractive.[293,299,300]

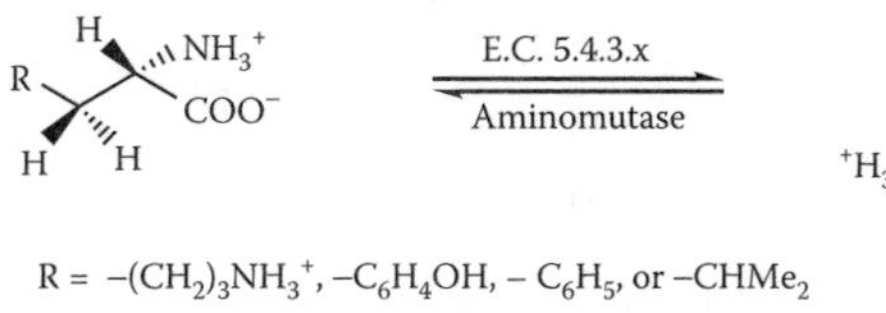

$$R = -(CH_2)_3NH_3{}^+, -C_6H_4OH, -C_6H_5, \text{ or } -CHMe_2$$

SCHEME 19.40

19.3 METABOLIC PATHWAY ENGINEERING

Metabolic pathway engineering is an approach involving more complex types of biotransformations.[301–303] This is accomplished by the genetic manipulation of several, usually sequential, biotransformation reactions, whereby their respective genes are often arranged in clusters. Recombinant DNA techniques are used to clone the genes of interest and overexpress the corresponding enzymes encoded by the cloned genes. This approach has been used to improve production of primary and secondary metabolites, commodity chemicals (e.g., L-phenylalanine, 1,3-propanediol), and drug development candidates.[304,305] Specific examples include the isoprenoids,[306,307] polyketides, and nonribosomal peptides;[308,309] biopolymers;[310] UAAs;[311] exopolysaccharides and B vitamins in lactic acid bacteria;[312] hydrocortisone;[313] starting materials for natural product synthesis (substituted cyclohexadiene-*trans*-diols);[314] and chiral 1,2,4-butanetriols,[315] to name a few. In addition, directed evolution or molecular breeding techniques have been used to improve the desired pathway for the synthesis of specific metabolites. Much of this work is driven to ultimately use renewable substrates as opposed to those derived from petroleum-based feedstocks. Reviews have been published on amino acids, chemicals, and cellular and metabolic engineering.[316–318] A few pertinent examples of this rapidly developing technology are discussed in more detail in the following sections.

19.3.1 AROMATIC AMINO ACID PATHWAY

With the increasing use of L-phenylalanine (L-phe), especially as a result of the artificial sweetener aspartame (**28**), this amino acid and L-tryptophan have become major compounds of focused research. The L-phe pathway has been the subject of intense research, details of which can be found in this book in Chapter 3. Early work focused on the aromatic amino acid pathway itself and the deregulation of the specific genes leading to L-phe synthesis. Subsequent studies were concerned with other pathways that produced precursors to direct carbon flow to the common aromatic pathway, such as phosphenolpyruvate (PEP) and erythrose-4-phosphate (E4P).[319] This has been demonstrated whereby other genes encoding enzymes such as transketolase (*tktA*) and PEP synthase (*pps*) have been introduced into selected bacterial strains, and striking effects have been observed for L-phe synthesis.[320,321] Another gene (*tal*) encoding the enzyme transaldolase was cloned into *E. coli* that showed a significant increase in the production of 3-deoxy-D-arabinoheptulosonate-7-phosphate (DAHP) from glucose.[322] DAHP is the first common intermediate in the aromatic amino acid pathway (see Chapter 3).

Several other important compounds found in the common aromatic amino acid pathway whose overproduction has been studied are shikimic acid (**61**) and, to a lesser extent, quinic acid (**62**) (Scheme 19.41).[323] Both **61** and **62** are naturally occurring, highly functionalized carbocyclic rings with asymmetric centers, which can be used as starting material for the synthesis of GS4104 (**63**), a neuraminidase inhibitor discovered by Gilead Sciences and developed by Roche Pharmaceuticals under the trade name of Tamiflu®.[324,325] Manipulation of the aromatic amino acid pathway in *E. coli* has allowed for numerous strains to be assembled that produce both **61** and **62** as well as other intermediates.[326,327] As reported by Chandran and co-workers, an *E. coli* strain has been constructed that synthesized 87 g/L (0.5M) of **61** in 36% (mol/mol) yield with a maximum productivity of 5.2 gL^{-1}h^{-1}.[328]

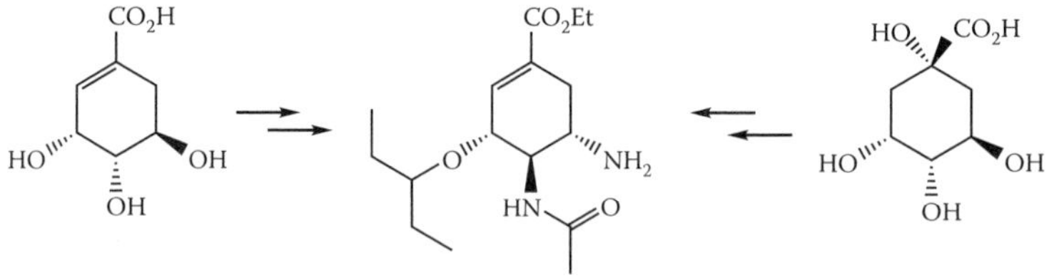

SCHEME 19.41

In addition to the common aromatic pathway in *E. coli* being manipulated to overproduce **61**, *Bacillus* and *Citrobacter* species have also been genetically altered to secrete **61**.[329,330] The levels of **61** produced by these strains, however, have been quite modest compared to that developed for *E. coli* strains.

Other work of Frost and co-workers has focused on the aromatic pathway with respect to the biosynthesis of the quinoid organics benzoquinone and hydroquinone, as well as gallic acid, pyrogallol, and adipic acid production.[331–336] The same principles have been used whereby a cloned gene is overexpressed that allows the recombinant organism to funnel carbon down a specific, selected metabolic pathway. Although these compounds are currently synthesized from petroleum-based feed stocks primarily for economic reasons, environmental concerns and the shortage of nonrenewable starting materials in the future will necessitate other approaches such as these for the production of such commodity chemicals.[332]

19.3.2 POLY-β-HYDROXYALKANOATES

Another example where metabolic pathway engineering has made a dramatic impact is in the biodegradable polymer field. One of the most widely studied polymers in this family is poly-β-hydroxybutyrate (PHB) (**64**). A related member of the poly-β-hydroxyalkanoate (PHA) family commercialized by Imperial Chemical Industries (ICI), which later became Zeneca Bio Products, is a co-polymer consisting of β-hydroxybutyric acid and β-hydroxyvaleric acid. This biodegradable polymer, marketed under the trade name Biopol™, was first used in plastic shampoo bottles by the Wella Corporation.[337] In the early part of 1996, the Biopol product line was purchased by the Monsanto Company from Zeneca. Monsanto terminated production of Biopol at the end of 1998 citing high production costs and a desire to focus on agricultural applications of biotechnology rather than on industrial applications. Then in 2001 Monsanto divested the Biopol assets and technology selling it to Metabolix, Inc. of Cambridge, MA.[338]

The *phbA, phbB,* and *phbC* genes from *Alcaligenes eutrophus (Ralstonia eutrophus)* encoding the biosynthetic enzymes β-ketothiolase, acetoacetyl-CoA reductase (NADPH-dependent), and PHB synthase, respectively, have been cloned into *E. coli* (Scheme 19.42).[339–342] The use of *in vitro* evolution using error-prone polymerase chain reaction has led to enhanced accumulation of PHA in a resultant recombinant strain.[343] Additional studies to enhance the biosynthesis of PHB through the use of metabolic engineering have been discussed.[344]

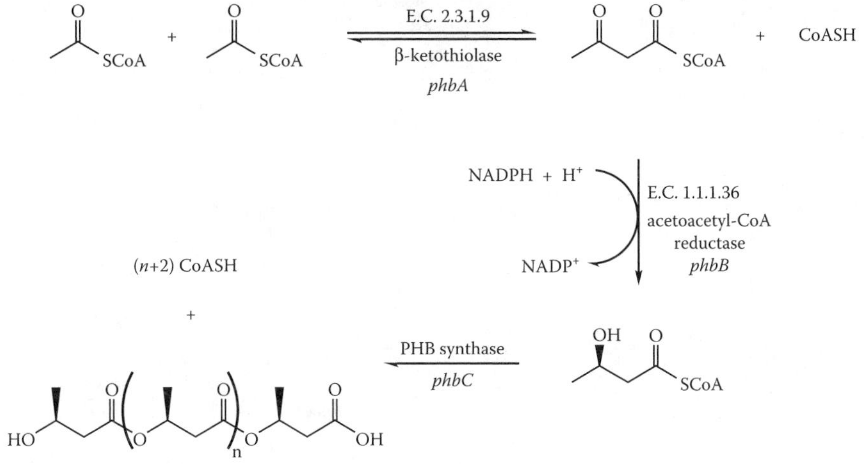

SCHEME 19.42

PHB has been shown to accumulate to levels approaching 90% of the bacterial dry cell weight when these pathway enzymes have been overexpressed.[345] With these efficiencies and with governmental support for developing this technology in plants, it will just be a matter of time before the production of biopolymers such as these can compete economically with the petrochemically derived plastics.[346]

19.3.3 POLYKETIDES AND NONRIBOSOMAL PEPTIDES

A final example of metabolic pathway engineering is based on polyketide and nonribosomal peptide biosynthesis. Polyketides and nonribosomal peptides are complex natural products with numerous chiral centers, which are of substantial economic benefit as pharmaceuticals. These natural products function as antibiotics [erythromycin A (**65**), vancomycin (**66**)], antifungals (rapamycin, amphotericin B), antiparasitics [avermectin A1$_a$ (**67**)], antitumor agents [epothiolone A (**68**), calicheamicin γ_1], and immunosuppressants [FK506 (**69**), cyclosporin A]. Because this exponentially growing and intensely researched field has developed, the reader is directed to review articles for additional details.[347–359] Also with the potential economic benefit to develop the next blockbuster pharmaceutical, a number of patents and patent applications have been published.[360–366]

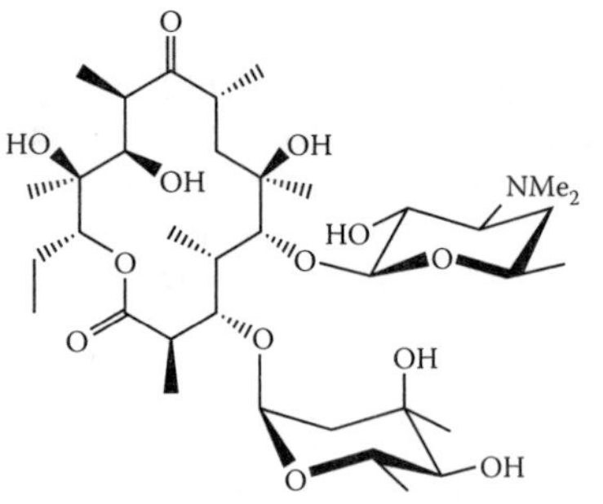

65

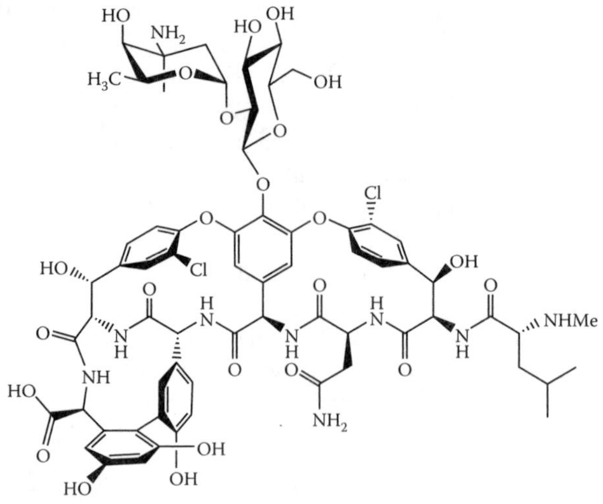

66

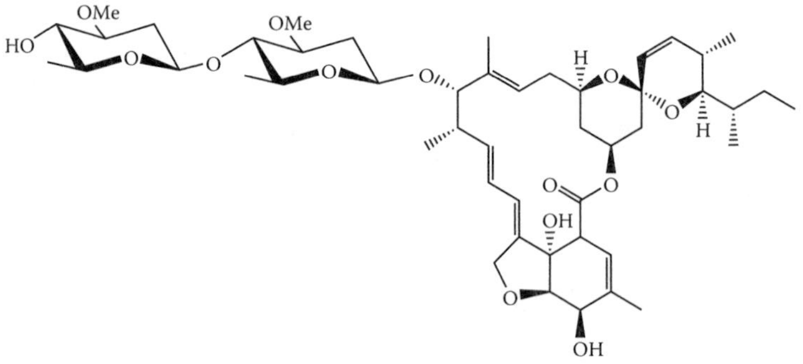

67

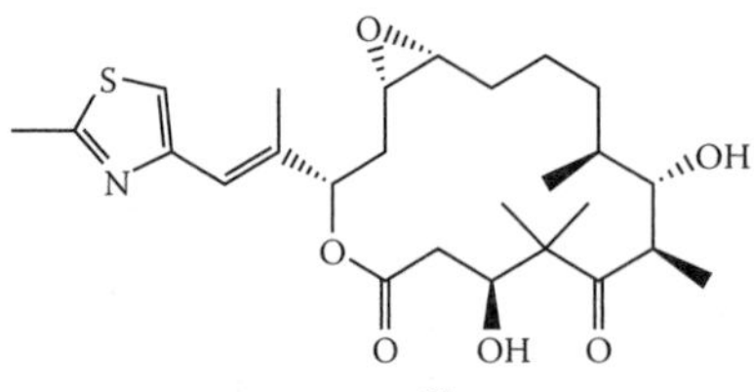

68

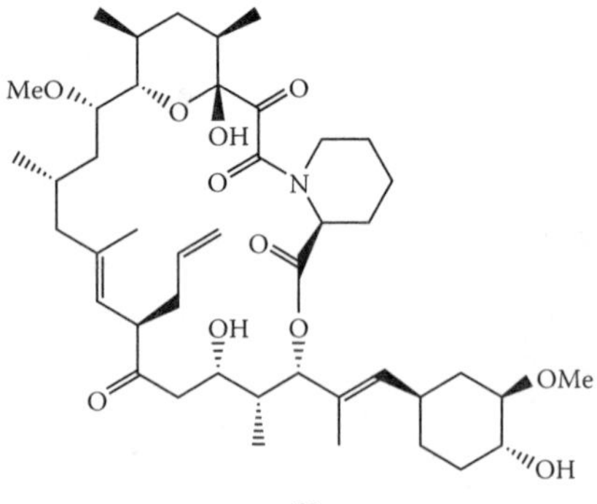

69

The biosynthesis of polyketides is analogous to the formation of long-chain fatty acids catalyzed by the enzyme fatty acid synthase (FAS). These FASs are multi-enzyme complexes that contain numerous enzyme activities. The complexes condense coenzyme A (CoA) thioesters (usually acetyl, propionyl, or malonyl) followed by a ketoreduction, dehydration, and enoylreduction of the β-keto moiety of the elongated carbon chain to form specific fatty acid products. These subsequent enzyme activities may or may not be present in the biosynthesis of polyketides.

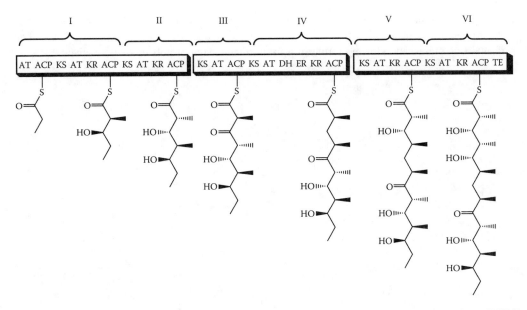

FIGURE 19.4 Modular organization of the six modules (I–VI) of 6-deoxyerythronolide B synthase (DEBS) enzyme as derived from *Saccharopolyspora erythraea*. Enzyme activities are acyltransferases (AT), acyl carrier proteins (ACP), β-ketoacyl-ACP synthases (KS), β-ketoreductases (KR), dehydratases (DH), enoyl reductases (ER), and thioesterases (TE). The TE-catalyzed release of the polyketide chain results in the formation of 6-dEB (**70**).[375,379,383]

Bacterial polyketide synthases (PKSs) are the multifunctional enzyme complexes having multiple enzyme activities (initiation, elongation, elimination, and termination) responsible for polyketide biosynthesis, which have been generally classified into three types based on their enzyme architecture.[367] Type I PKSs have a modular structure (Figure 19.4) and generally produce aliphatic compounds whereby the elongated acyl chain is translocated from upstream to downstream carrier protein domains that contain the tethering thiol group. Type II PKSs are often iterative producing multicyclic aromatic compounds and are characterized such that the elongated acyl chain remains tethered to the same carrier protein while the acyl monomers are on distinct subunits. Type III PKSs are most often found in plants, whereby stilbene synthase (E.C. 2.3.1.95) and chalcone synthase (E.C. 2.3.1.74) are predominate members and lack the use of an acyl carrier protein, using the free CoA esters as substrates directly.[368] A bacterial Type III PKS has been described that produces the UAA precursor 3,5-dihydroxyphenylacetyl-CoA, which is a key intermediate in the biosynthesis of the glycopeptide antibiotic family containing vancomycin (**66**) and teicoplanin.[369]

A similar modular type arrangement of activities is present in the nonribosomal peptide synthetases (NRPS). Several examples have been described that use a hybrid PKS-NRPS system to synthesize epothiolones,[370] rifamycin,[371] and C-1027[372] with reviews on these hybrid systems being described.[373,374]

A model system that has been studied thoroughly and used to develop the modular organization of enzymes has been the PKS deoxyerythronolide B synthase (DEBS) from *Saccharopolyspora erythraea*.[375] This enzyme consists of 6 modules encoding the various enzymes for condensation and further processing activities to produce 6-deoxyerythronolide B (6-dEB) (**70**), the aglycon of erythromycin (**65**). The modular arrangement of enzymes is shown in Figure 19.4. Additional studies have shown that, through manipulation of these enzyme modules, numerous derivatives of polyketides can be synthesized. Through the reprogramming of genes in their respective modules by site-directed mutagenesis, by deletion of certain modules, or by introduction of an unnatural enzymatic activity, the synthesis of polyketide derivatives has been demonstrated.[376–379] Selected derivative structures that have been confirmed are shown by structures **71–73**.[380] In addition, by incubating certain precursors with strains containing designed, mutant DEBSs, various analogues of erythromycin have been synthesized.[381] Purification and characterization of various DEBSs mutants also have been published, allowing further insight into the design of polyketide derivatives.[382]

70

71   72   73

Because these genes are often arranged in clusters, this has allowed a combinatorial biosynthetic approach for the construction of diverse libraries of polyketides and nonribosomal peptides.[384–387] The use of *E. coli* as a production host whereby precursor supply and selected pathway enzymes are modified has resulted in respectable titers of targeted compounds as well as novel derivatives.[388–393]

Despite these complex biochemical systems, much progress has been made to understand and manipulate them at the genetic level. This will ultimately lead to the generation of novel products, which, on screening against selected targets, might allow new polyketides and nonribosomal peptides to be identified with enhanced pharmaceutical properties.

19.4 SCREENING FOR BIOCATALYSTS

If a biocatalyst is not known to exist for a certain biotransformation, then screening for that activity is generally undertaken, especially when the chemical route is too difficult, is too costly, or does not provide the needed chiral purity.[394] Investigation of commercial enzymes would generally be the first step of a screening program prior to embarking on the more labor-intensive testing of fungal or bacterial cultures.[395] Using a microtiter plate format, many process variables such as temperature, pH, solvent, substrate concentration, etc., may be tested simultaneously and optimized prior to scale up. This is especially important for the study of solvents because many substrates are usually not soluble in 100% aqueous solution. A microtiter plate format has also been described for microbes as well as isolated enzymes.[75,396]

In addition, screening for biotransformation activity is important when a "natural" claim is needed, which occurs most notably in the flavor industry.[397] Screening for novel biocatalysts to perform specific biotransformations can require extensive resources, depending on the details of the screening protocol. Therefore, the choice to screen for a biocatalyst can only be justified when the resultant product or biotransfomation exhibits significant commercial potential. A number of literature citations and patents have appeared that describe screening using enzymes and/or cultures for microbial reductions and oxidations, ester synthesis, acylation of a diol, or chiral epoxide hydrolysis.[183,398–402] These are but a few examples of biotransformations involving chiral compounds successfully being carried out.

An example of a project of this nature is the screening for microorganisms to produce aspartame (**28**) from a precursor that is easy synthesized by chemical methods. Microorganisms were screened for their ability to catalyze the *trans*-addition of ammonia across the double bond of *N*-fumaryl-L-phenylalanine methyl ester (FumPM) (**74**) (Scheme 19.43).[403] This is essentially the reaction of a mutated aspartase because the native enzyme has such strict substrate specificity (see Section 19.2.4). Although the literature touts this as a successful screening effort, this process has not been practiced commercially because the yields are extremely low.[404,405]

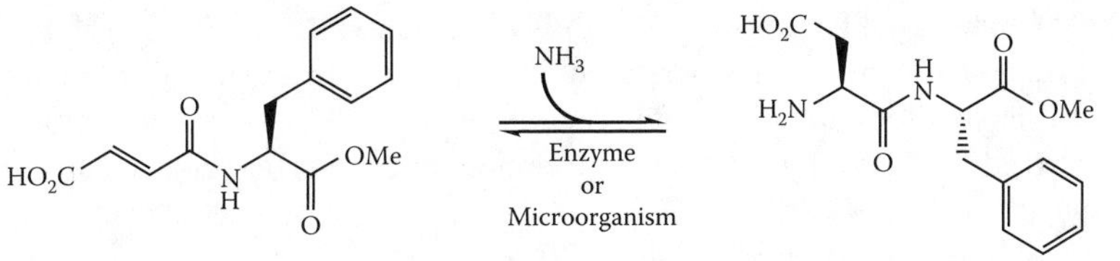

SCHEME 19.43

Another screening example related to the sweetener aspartame **28** was based on finding an enzyme activity that could degrade the cyclized diketopiperazine (DKP), which is bitter and known to form at basic pH from **28**. Using the culture enrichment technique and the aspartame-DKP (**75**) as the sole carbon and nitrogen source, Pantaleone and co-workers isolated a novel cyclic dipeptidase (DKPase) responsible for the hydrolytic activity outlined in Scheme 19.44.[406] The two most likely routes of degradation occur by either the hydrolysis to form the dipeptide via Route A (L-aspartyl-L-phenylalanine, **76**) or Route B (L-phenylalanyl-L-aspartate, **77**). Several cultures were isolated that used Route A with the best culture being identified as *Bacillus circulans*. In addition to removing this bitter component (**75**) via degradation, such a biocatalyst can also be thought of in terms of aspartame synthesis, which would not be dependent on L-phenylalanine whereby hydrolysis of **75** to **76** followed by methylation would lead to **28**.

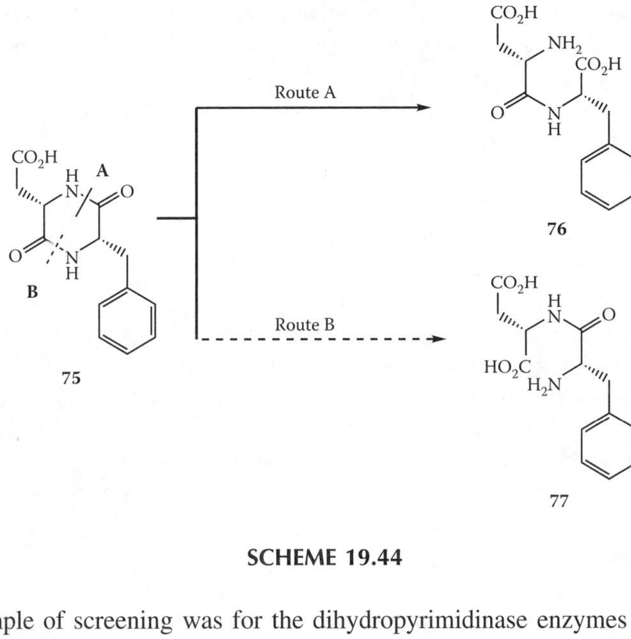

SCHEME 19.44

Another example of screening was for the dihydropyrimidinase enzymes (E.C. 3.5.2.2), also known as hydantoinases, used for the production of either L- or D-amino acids.[407,408] These studies were carried out in the mid- to late 1970s where racemic 5-monosubstituted hydantoins **78** were used as the sole carbon and nitrogen sources with a variety of soil microorganisms. Those microorganisms that grew were then further characterized as having the desired activity. This has resulted in the isolation and characterization of a number of biocatalysts from various microorganisms that catalyze the enantioselective cleavage of hydantoins. Both L- and D-hydantoinase enzymes have been described in the literature, which can be further coupled to either a chemical or enzymatic hydrolysis of the N-carbamoyl-amino acid (Scheme 19.45). In addition, many different substituent R-groups at carbon–5 have been studied for each hydantoinase.[407-409] For industrial purposes, R = p-hydroxyphenyl and phenyl have found most utility, leading to the production of D-p-hydroxyphenylglycine and D-phenylglycine, respectively, which are used as the unnatural sidechains in a number of well-known β-lactam antibiotics. Another commercially important chiral pharmaceutical intermediate synthesized using an hydantoinase in combination with an L-N-carbamoylase is (S)-2-amino-4-phenylbutanoic acid. This UAA, also known as L-homophenylalanine, is used in the synthesis of many hypertensive drugs of the ACE inhibitor class.[410]

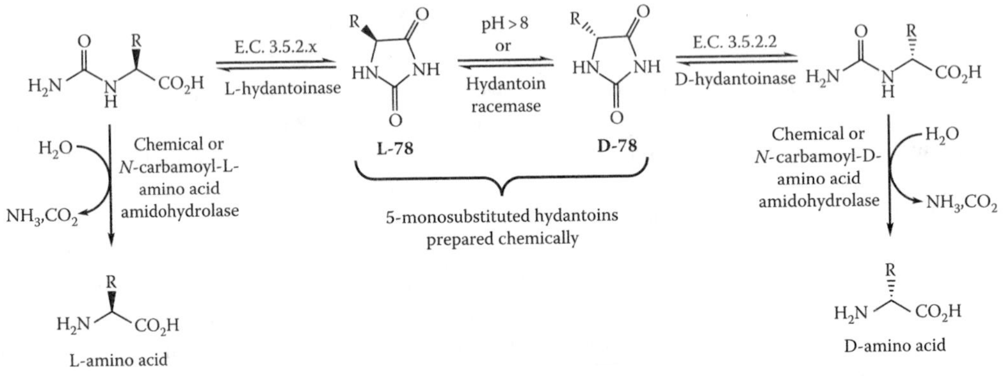

SCHEME 19.45

Two additional hydantoinases have been isolated from thermophilic microorganisms and are being sold commercially in an immobilized form by Roche Applied Science, distributed in the United States by BioCatalytics (Pasadena, CA) for the production of D-amino acids.[411] These purified hydantoinases along with three selected bacterial strains, *Agrobacterium tumefaciens* CIP 67.1, *Pseudomonas fluorescens* CIP 69.13, and *Pseudomonas sp.* ATCC 43648, were tested with a number of ring-substituted D-phenylglycines and shown to be effective at producing these derivatives.[412]

Another screening method has been described for the enzymatic deamidation of (*rac*)-N-carbamoyl amino acid amides to enantiopure N-carbamoyl-L-amino acids.[413]

As a result of the broad use of hydantoinases for the preparation of both proteogenic and nonproteogenic amino acids, studies into their structure and mechanism continue to be pursued.[414-416] Directed evolution methods have been used to invert the enantioselectivity of an hydantoinase for the production of L-methionine,[417] and improvement in catalytic efficiency of an hydantoinase through mutagenesis has been reported.[418]

The reader is referred to the reviews on screening for additional details of this approach to biocatalyst identification.[419-425]

19.5 SUMMARY

As discussed in this chapter, the use of microorganisms or commercially available enzymes to carry out biotransformations is well-established. Numerous routes have been developed to prepare chiral fine chemicals with enzymes from all classes, often in combination with chemical methods, and it is clear that many of the major pharmaceutical companies are investing heavily in this type of research and development. With approaches such as DKR, pathway engineering, and screening for biocatalysts showing such promise, many new biotransformations will undoubtedly be discovered to develop novel, "green" bioprocesses in the future.

REFERENCES

1. Stinson, S. C. *Chem. Eng. News* 1995, *73*, 44.
2. Stinson, S. C. *Chem. Eng. News* 1996, *74*, 35.
3. Martineau, W. *Chiral Chemicals,* The Freedonia Group, Inc.: Cleveland, 1996.
4. Stinson, S. C. *Chem. Eng. News* 2001, *79*, 79.
5. Rouhi, A. M. *Chem. Eng. News* 2002, *80*, 43.
6. Schoemaker, H. E., Mink, D., Wubbolts, M. G. *Science* 2003, *299*, 1694.
7. Stinson, S. C. *Chem. Eng. News* 1997, *75*, 37.
8. Stinson, S. C. *Chem. Eng. News* 1997, *75*, 34.
9. Faber, K. *Biotransformations in Organic Chemistry,* 2nd ed., Springer-Verlag: Berlin, 1995.
10. Blanch, H. W., Clark, D. S. *Applied Biocatalysis,* Marcel Dekker: New York, 1991, Vol. 1.
11. West, S. Chemical Biotransformations. In *Industrial Enzymology,* Godfrey, T., West, S. Eds., Macmillan Press Ltd: London, 1996, p. 155.
12. *Enzymes in Organic Synthesis,* Ciba Foundation Symposium 111, 1984, London.
13. Laane, C., Tramper, J., Lilly, M. D. Biocatalysis in Organic Media. In *Studies in Organic Chemistry,* Elsevier: Amsterdam, 1987, Vol. 29.
14. Roberts, S. M., Wiggins, K., Casy, G., Phythian, S. *Preparative Biotransformations: Whole Cell and Isolated Enzymes in Organic Synthesis,* John Wiley & Sons: Chichester, 1992.
15. Wong, C.-H., Whitesides, G. M. *Enzymes in Synthetic Organic Chemistry,* Elsevier: New York, 1994, Vol. 12.
16. Santaniello, E., Ferraboschi, P., Grisenti, P., Manzocchi, A. *Chem. Rev.* 1992, *92*, 1071.
17. Patel, R. N. *Stereoselective Biocatalysis,* Marcel Dekker, Inc.: New York, 2000.
18. Roberts, S. M., Williamson, N. M. *Curr. Org. Chem.* 1997, *1*, 1.
19. Bornscheuer, U. T., Kazlauskas, R. J. *Hydrolases in Organic Synthesis: Regio- and Stereoselective Biotransformations,* 1st ed., Wiley-VCH: Weinheim, 1999.
20. Drauz, K., Waldmann, H. *Enzyme Catalysis in Organic Synthesis,* Wiley-VCH: Weinheim, 2002.
21. Breuer, M., Ditrich, K., Habicher, T., Hauer, B., Keßeler, M., Stürmer, R., Zelinski, T. *Angew. Chem., Int. Ed.* 2004, *43*, 788.
22. *Enzyme Nomenclature,* Academic Press: New York, 1992.
23. Bairoch, A. *Nucleic Acids Res.* 2000, *28*, 304.
24. White, J. S., White, D. C. *Source Book of Enzymes,* CRC Press: Boca Raton, 1997.
25. Chapter 2: Reduction Reactions. In *Preparative Biotransformations: Whole Cell and Isolated Enzymes in Organic Synthesis,* Roberts, S. M., Wiggins, K., Casy, G., Phythian, S. Eds., John Wiley & Sons: Chichester, 1992.
26. Sheldon, R. A. Chapter 7: Enzymatic Transformations. In *Chirotechnology: Industrial Synthesis of Optically Active Compounds,* Marcel Dekker, Inc.: New York, 1993, p. 205.
27. Wong, C.-H., Whitesides, G. M. Chapter 3: Oxidoreductions. In *Enzymes in Synthetic Organic Chemistry,* Elsevier Science Inc.: Tarrytown, 1994, Vol. 12, p. 131.
28. Turner, N. J. Asymmetric Synthesis Using Enzymes and Whole Cells. In *Advanced Asymmetric Synthesis,* Stephenson, G. R. Ed., Chapman & Hall: London, 1996, p. 260.
29. Hummel, W., Kula, M.-R. *Eur. J. Biochem.* 1989, *184*, 1.
30. Stewart, J. D. *Curr. Opin. Chem. Biol.* 2001, *5*, 120.
31. Matsushita, K., Yakushi, T., Takaki, Y., Toyama, H., Adachi, O. *J. Bacteriol.* 1995, *177*, 6552.
32. Bradshaw, C. W., Wong, C.-H., Shen, G.-J. U.S. Pat. 5,385,833, 1995.
33. Tripp, A. E., Burdette, D. S., Zeikus, J. G., Phillips, R. S. *J. Am. Chem. Soc.* 1998, *120*, 5137.
34. Heiss, C., Phillips, R. S. *J. Chem. Soc., Perkin Trans I* 2000, 2821.
35. Heiss, C., Laivenieks, M., Zeikus, J. G., Phillips, R. S. *Bioorg. Med. Chem.* 2001, *9*, 1659.

36. Gröger, H., Hummel, W., Buchholz, S., Drauz, K., Nguyen, T. V., Hüsken, H., Abokitse, K. *Org. Lett.* 2003, *5*, 173.
37. Gröger, H., Hummel, W., Rollmann, C., Chamouleau, F., Hüsken, H., Werner, H., Wunerlich, C., Abokitse, K., Drauz, K., Buchholz, S. *Tetrahedron* 2004, *60*, 633.
38. Simon, E. S., Plante, R., Whitesides, G. M. *Appl. Biochem. Biotechnol.* 1989, *22*, 169.
39. Kim, M.-J., Kim, J. Y. *J. Chem. Soc., Chem. Commun.* 1991, 326.
40. Bradshaw, C. W., Wong, C.-H., Hummel, W., Kula, M.-R. *Bioorg. Chem.* 1991, *19*, 29.
41. Schmidt, E., Ghisalba, O., Gygas, D., Sedelmeier, G. *J. Biotechnol.* 1992, *24*, 315.
42. Kallwass, H. K. W. *Enzyme Microb. Technol.* 1992, *14*, 28.
43. Krix, G., Bommarius, A. S., Drauz, K., Kottenhahn, M., Schwarm, M., Kula, M.-R. *J. Biotechnol.* 1997, *53*, 29.
44. Hanson, R. L., Bembenek, K. S., Patel, R. N., Szarka, L. J. *Appl. Microbiol. Biotechnol.* 1992, *37*, 599.
45. Casy, G., Lee, T. V. PCT Appl. WO 93/13215, 1993.
46. Tao, J., McGee, K. *Org. Proc. Res. Dev.* 2002, *6*, 520.
47. Goncalves, L. P. B., Antunes, O. A. C., Pinto, G. F., Oestreicher, E. G. *Tetrahedron: Asymmetry* 2000, *11*, 1465.
48. Pantaleone, D. P., Geller, A. M., Taylor, P. P. *J. Mol. Catal. B: Enzymatic* 2001, *11*, 795.
49. Senkpeil, R. F., Pantaleone, D. P., Taylor, P. P. PCT Appl. 02/033110, 2002.
50. Ohshima, T., Soda, K. *Trends Biotechnol.* 1989, *7*, 210.
51. Bommarius, A. S. Reduction of C=N Bonds. In *Enzyme Catalysis in Organic Synthesis*, Drauz, K., Waldmann, H. Eds., Wiley-VCH Verlag: Weinheim, 2002, Vol. 3, p. 1047.
52. Ohshima, T., Soda, K. Stereoselective Biocatalysis: Amino Acid Dehydrogenases and Their Applications. In *Stereoselective Biocatalysis,* Patel, R. N. Ed., Marcel Dekker, Inc.: New York, 2000, p. 877.
53. Galkin, A., Kulakova, L., Yamamoto, H., Tanizawa, K., Tanaka, H., Esaki, N., Soda, K. *J. Ferment. Bioeng.* 1997, *83*, 299.
54. Katsumata, R., Hashimoto, S. U.S. Pat. 5,559,016, 1996.
55. Ohshima, T., Wandrey, C., Conrad, D. *Biotechnol. Bioeng.* 1989, *34*, 394.
56. Kato, Y., Fukumoto, K., Asano, Y. *Appl. Microbiol. Biotechnol.* 1993, *39*, 301.
57. Hummel, W., Schmidt, E., Wandrey, C., Kula, M.-R. *Appl. Microbiol. Biotechnol.* 1986, *25*, 175.
58. Asano, Y., Nakazawa, A. *Agric. Biol. Chem.* 1987, *51*, 2035.
59. Asano, Y., Yamada, A., Kato, Y., Yamaguchi, K., Hibino, Y., Hirai, K., Kondo, K. *J. Org. Chem.* 1990, *55*, 5567.
60. Bommarius, A. S., Schwarm, M., Drauz, K. *Chimia* 2001, *55*, 50.
61. Bommarius, A. S., Schwarm, M., Stingl, K., Kottenhahn, M., Huthmacher, K., Drauz, K. *Tetrahedron: Asymmetry* 1995, *6*, 2851.
62. Patel, R. N., Chu, L., Mueller, R. *Tetrahedron: Asymmetry* 2003, *14*, 3105.
63. Wichmann, R., Wandrey, C., Buckmann, A. F., Kula, M.-R. *Biotechnol. Bioeng.* 1981, *23*, 2789.
64. Kragl, U., Kruse, W., Hummel, W., Wandrey, C. *Biotechnol. Bioeng.* 1996, *52*, 309.
65. Hanson, R. L., Singh, J., Kissick, T. P., Patel, R. N., Szarka, L. J., Mueller, R. H. *Bioorg. Chem.* 1990, *18*, 118.
66. Sutherland, A., Willis, C. L. *Tetrahedron Lett.* 1997, *38*, 1837.
67. Csuk, R., Glänzer, B. I. Yeast-Mediated Stereoselective Biocatalysis. In *Stereoselective Biocatalysis,* Patel, R. N. Ed., Marcel Dekker, Inc.: New York, 2000, p. 527.
68. Nakamura, K., Matsuda, T. Reduction Reactions. In *Enzyme Catalysis in Organic Synthesis,* Drauz, K., Waldmann, H. Eds., Wiley-VCH Verlag: Weinheim, 2002, Vol. 3, p. 991.
69. Ward, O. P., Young, C. S. *Enzyme Microb. Technol.* 1990, *12*, 482.
70. Csuk, R., Glänzer, B. I. *Chem. Rev.* 1991, *91*, 49.
71. Servi, S. *Synthesis* 1990, 1.
72. Ager, D. J., East, M. B. *Asymmetric Synthetic Methodology,* CRC Press, Inc.: Boca Raton, 1996.
73. Chartrain, M., Greasham, R., Moore, J., Reider, P., Robinson, D., Buckland, B. *J. Mol. Catal. B: Enzymatic* 2001, *11*, 503.
74. Nakamura, K., Yamanaka, R., Matsuda, T., Harada, T. *Tetrahedron: Asymmetry* 2003, *14*, 2659.
75. Homann, M. J., Vail, R. B., Previte, E., Tamarez, M., Morgan, B., Dodds, D. R., Zaks, A. *Tetrahedron* 2004, *60*, 789.

76. Anderson, B. A., Hansen, M. M., Harkness, A. R., Henry, C. L., Vicenzi, J. T., Zmijewski, M. J. *J. Am. Chem. Soc.* 1995, *117*, 12358.

77. Vicenzi, J. T., Zmijewski, M. J., Reinhard, M. R., Landen, B. E., Muth, W. L., Marler, P. G. *Enzyme Microb. Technol.* 1997, *20*, 494.

78. Zmijewski, M. J., Vicenzi, J., Landen, B. E., Muth, W., Marler, P., Anderson, B. *Appl. Microbiol. Biotechnol.* 1997, *47*, 162.

79. Reddy, J., Tschaen, D., Shi, Y.-J., Pecore, V., Katz, L., Greasham, R., Chartrain, M. *J. Ferment. Bioeng.* 1996, *81*, 304.

80. Chartrain, M., McNamara, J., Greasham, R. *J. Ferment. Bioeng.* 1996, *82*, 507.

81. Katz, L., King, S., Greasham, R., Chartrain, M. *Enzyme Microb. Technol.* 1996, *19*, 250.

82. Blacker, A. J., Holt, R. A. Development of a Multi-stage Chemical and Biological Process for an Optically Active Intermediate for an Anti-glaucoma Drug. In *Chirality in Industry II,* Collins, A. N., Sheldrake, G. N., Crosby, J. Eds., John Wiley & Sons: Chichester, 1997, p. 245.

83. Hunt, J. R., Carter, A. S., Murrell, J. C., Dalton, H., Hallinan, K. O., Crout, D. H. G., Holt, R. A., Crosby, J. *Biocatal. Biotransform.* 1995, *12*, 159.

84. Hallinan, K. O., Crout, D. H. G., Hunt, J. R., Carter, A. S., Dalton, H., Murrell, J. C., Holt, R. A., Crosby, J. *Biocatal. Biotransform.* 1995, *12*, 179.

85. Sih, C. J., Zhou, B.-N., Gopalan, A. S., Shieh, W.-R., Chen, C.-S., Girdaukas, G., Vanmiddlesworth, F. *Proc. NY Acad. Sci.* 1984, *434*, 186.

86. Sih, C. J. U.S. Pat. 4,710,468, 1987.

87. Bare, G., Jacques, P., Hubert, J. B., Rikir, R., Thonart, P. *Appl. Biochem. Biotechnol.* 1991, *28/29*, 445.

88. Butler, D. E., Le, T. V., Millar, A., Nannings, T. N. U.S. Pat. 5,155,251, 1992.

89. Thottathil, J. K., Pendri, Y., Li, W.-S., Kronenthal, D. R. U.S. Pat. 5,278,313, 1994.

90. Karanewsky, D. S., Badia, M. C., Ciosek, C. P. J., Robl, J. A., Sofia, M. J., Simpkins, L. M., DeLange, B., Harrity, T. W., Biller, S. A., Gordon, E. M. *J. Med. Chem.* 1990, *33*, 2952.

91. Shimizu, S., Kataoka, M., Kita, K. *J. Mol. Catal. B: Enzymatic* 1998, *5*, 321.

92. Kita, K., Fukura, T., Nakase, K.-I., Okamoto, K., Yanase, H., Kataoka, M., Shimizu, S. *Appl. Environ. Microbiol.* 1999, *65*, 5207.

93. Yasohara, Y., Kizaki, N., Hasegawa, J., Takahashi, S., Wada, M., Kataoka, M., Shimizu, S. *Appl. Microbiol. Biotechnol.* 1999, *51*, 847.

94. Yasohara, Y., Kizaki, N., Hasegawa, J., Wada, M., Kataoka, M., Shimizu, S. *Tetrahedron: Asymmetry* 2001, *12*, 1713.

95. Wada, M., Yoshizumi, A., Nakamori, S., Shimizu, S. *Appl. Environ. Microbiol.* 1999, *65*, 4399.

96. Kataoka, M., Kita, K., Wada, M., Yasohara, Y., Hasegawa, J., Shimizu, S. *Appl. Microbiol. Biotechnol.* 2003, *62*, 437.

97. Wada, M., Yoshizumi, A., Noda, Y., Kataoka, M., Shimizu, S., Takagi, H., Nakamori, S. *Appl. Environ. Microbiol.* 2003, *69*, 933.

98. Martzen, M. R., McCraith, S. M., Spinelli, S. L., Torres, F. M., Fields, S., Grayhack, E. J., Phizicky, E. M. *Science* 1999, *286*, 1153.

99. Kaluzna, I., Andrew, A. A., Bonilla, M., Martzen, M. R., Stewart, J. D. *J. Mol. Catal. B: Enzymatic* 2002, *17*, 101.

100. Rodriguez, S., Schroeder, K. T., Kayser, M. M., Stewart, J. D. *J. Org. Chem.* 2000, *65*, 2586.

101. Rodriguez, S., Kayser, M. M., Stewart, J. D. *J. Am. Chem. Soc.* 2001, *123*, 1547.

102. Chartrain, M., Armstrong, J., Katz, L., Keller, J., Mathre, D., Greasham, R. *J. Ferment. Bioeng.* 1995, *80*, 176.

103. Patel, R. N., Banerjee, A., McNamee, C. G., Brzozowsik, D., Hanson, R. L., Szarka, L. J. *Enzyme Microb. Technol.* 1993, *15*, 1014.

104. Patel, R. N., Robinson, R. S., Szarka, L. J., Kloss, J., Thottathil, J. K., Mueller, R. H. *Enzyme Microb. Technol.* 1991, *13*, 906.

105. Patel, R. N., Szarka, L. J., Mueller, R. H. U.S. Pat. 5,559,017, 1996.

106. Patel, R. N., McNamee, C. G., Banerjee, A., Howell, J. M., Robinson, R. S., Szarka, L. J. *Enzyme Microb. Technol.* 1992, *14*, 731.

107. Patel, R. N. *Adv. Appl. Microbiol.* 1997, *43*, 91.

108. Wong, C.-H., Whitesides, G. M. Chapter 5: Synthesis of Glycoside Bonds. In *Enzymes in Synthetic Organic Chemistry,* Elsevier Science Ltd: Tarrytown, New York, 1994, Vol. 12, p. 252.

109. Schirch, L. *Serine Hydroxymethyltransferase,* Vol. 53, John Wiley & Sons: New York, 1982, p. 83.
110. Saeed, A., Young, D. W. *Tetrahedron* 1992, *48,* 2507.
111. Dotzlaf, J. E., Gazak, R. J., Kreuzman, A. J., Kroeff, E. P., Queener, S. W., Vicenzi, J. T., Yeh, J. M., Zock, J. M. Eur. Pat. EP 628,638 A2, 1994.
112. Miller, M. J. Eur. Pat. EP 460,883 A2, 1991.
113. Vassilev, V. P., Uchiyama, T., Kajimoto, T., Wong, C.-H. *Tetrahedron Lett.* 1995, *36,* 4081.
114. Kreuzman, A. J., Zock, J. M., Dotzlaf, J. E., Vicenzi, J. T., Queener, S. W., Yeh, W. K. *J. Indust. Microbiol. Biotechnol.* 1997, *19,* 369.
115. Jackson, B. G., Pedersen, S. W., Fisher, J. W., Misner, J. W., Gardner, J. P., Staszak, M. A., Doecke, C., Rizzo, J., Aikins, J., Farkas, E., Trinkle, K. L., Vicenzi, J., Reinhard, M., Kroeff, E. P., Higginbotham, C. A., Gazak, R. J., Zhang, T. Y. *Tetrahedron* 2000, *56,* 5667.
116. Ura, D., Hashimukai, T., Matsumoto, T., Fukuhara, N. Eur. Pat. EP 421,477 A1, 1990.
117. Hatakeyama, K., Goto, M., Terasawa, M., Yugawa, H. Jap. Pat. JP 09028391, 1997.
118. Nikumaru, S., Fukuhara, N. Jap. Pat. JP 08084593, 1996.
119. Bull, C., Durham, D. R., Gross, A., Kupper, R. J., Walter, J. F. Aspartame Synthesis in a Miscible Organic Solvent. In *Biocatalytic Production of Amino Acids and Derivatives,* Rozzell, J. D., Wagner, F. Eds., Wiley: New York, 1992, p. 241.
120. Chmurny, A. B., Gross, A. T., Kupper, R. J., Roberts, R. L. U.S. Pat. 4,710,583, 1987.
121. Chmurny, A. B., Gross, A. T., Kupper, R. J., Roberts, R. L. U.S. Pat. 4,873,359, 1989.
122. Gross, A. T. U.S. Pat. 5,002,872, 1991.
123. Christen, P., Metzler, D. E. In *Transaminases,* in series *Biochemistry,* Meister, A. Ed., John Wiley & Sons: New York, 1985, Vol. 2.
124. Crump, S. P., Rozzell, J. D. Biocatalytic Production of Amino Acids by Transamination. In *Biocatalytic Production of Amino Acids and Derivatives,* Rozzell, J. D., Wagner, F. Eds., Hanser: Munich, 1992, p. 43.
125. Stirling, D. I., Zeitlin, A. L., Matcham, G. W. U.S. Pat. 4,950,606, 1990.
126. Stirling, D. I. The Use of Aminotransferases for the Production of Chiral Amino Acid and Amines. In *Chirality in Industry,* Collins, A. N., Sheldrake, G. N., Crosby, J. Eds., John Wiley & Sons: New York, 1992, p. 209.
127. Shin, J.-S., Kim, B.-G. *Biotechnol. Bioeng.* 1999, *65,* 206.
128. Yun, H., Yang, Y.-H., Cho, B.-K., Hwang, B.-Y., Kim, B.-G. *Biotechnol. Lett.* 2003, *25,* 809.
129. Cho, B.-K., Cho, H. J., Yun, H., Kim, B.-G. *J. Mol. Catal. B: Enzymatic* 2003, *26,* 273.
130. Cho, B.-K., Cho, H. J., Park, S.-H., Yun, H., Kim, B.-G. *Biotechnol. Bioeng.* 2003, *81,* 783.
131. Patel, R. N., Banerjee, A., Nanduri, V. B., Goldberg, S. L., Johnston, R. M., Hanson, R. L., McNamee, C. G., Brzozowski, D. B., Tully, T. P., Ko, R. Y., LaPorte, T. L., Cazzulino, D. L., Swaminathan, S., Chen, C.-K., Parker, L. W., Venit, J. J. *Enzyme Microb. Technol.* 2000, *27,* 376.
132. Ager, D. J., Li, T., Pantaleone, D. P., Senkpeil, R. F., Taylor, P. P., Fotheringham, I. G. *J. Mol. Catal. B: Enzymatic* 2001, *11,* 199.
133. Tanizawa, K., Asano, S., Masu, Y., Kuramitsu, S., Kagamiyama, H., Tanaka, H., Soda, K. *J. Biol. Chem.* 1989, *264,* 2450.
134. Taylor, P. P., Fotheringham, I. G. *Biochim. Biophys. Acta* 1997, *1350,* 38.
135. Rozzell, J. D. U.S. Pat. 4,826,766, 1989.
136. Ager, D. J., Fotheringham, I. G., Laneman, S. A., Pantaleone, D. P., Taylor, P. P. *Chimica Oggi* 1997, *15 (3/4),* 11.
137. Taylor, P. P., Pantaleone, D. P., Senkpeil, R. F., Fotheringham, I. G. *Trends Biotechnol.* 1998, *16,* 412.
138. Then, J., Bartsch, K., Deger, H.-M., Grabley, S., Marquardt, R. Eur. Pat. EP 248,357 A2, 1987.
139. Meiwes, J., Schudok, M., Kretzschmar, G. *Tetrahedron: Asymmetry* 1997, *8,* 527.
140. Bartsch, K., Schneider, R., Schulz, A. *Appl. Environ. Microbiol.* 1996, *62,* 3794.
141. Kretzschmar, G., Meines, J., Schudok, M., Hamann, P., Lerch, U., Grabley, S. U.S. Pat. 5,480,786, 1996.
142. Godfrey, T., West, S. I. Introduction to Industrial Enzymology. In *Industrial Enzymology,* Godfrey, T., West, S. Eds., Macmillan Press Ltd.: London, 1996, p. 3.
143. Bentley, I. S., Williams, E. C. Starch Conversion. In *Industrial Enzymology,* Godfrey, T., West, S. Eds., Macmillan Press Ltd: London, 1996, p. 339.
144. Faber, K., Orru, R. V. A. Hydrolysis of Epoxides. In *Enzyme Catalysis in Organic Synthesis,* Drauz, K., Waldmann, H. Eds., Wiley-VCH Verlag: Weinheim, 2002, Vol. 2, p. 579.

145. Pace, H. C., Brenner, C. *Genome Biol.* 2001, *2*, 1.
146. Nagasawa, T., Yamada, H. *Pure Appl. Chem.* 1995, *67*, 1241.
147. Fallon, R. D., Stieglitz, B., Turner Jr., I. *Appl. Microbiol. Biotechnol.* 1997, *47*, 156.
148. Crosby, J., Moilliet, J., Parratt, J. S., Turner, N. J. *J. Chem. Soc., Perkin Trans. I* 1994, 1679.
149. Meth-Cohn, O., Wang, M.-X. *J. Chem. Soc., Chem. Commun.* 1997, 1041.
150. DeSantis, G., Zhu, Z., Greenberg, W. A., Wong, K., Chaplin, J., Hanson, S. R., Farwell, B., Nicholson, L. W., Rand, C. L., Weiner, D. P., Robertson, D. E., Burk, M. J. *J. Am. Chem. Soc.* 2002, *124*, 9024.
151. DeSantis, G., Wong, K., Farwell, B., Chatman, K., Zhu, Z., Tomlinson, G., Huang, H., Tan, X., Bibbs, L., Chen, P., Kretz, K., Burk, M. J. *J. Am. Chem. Soc.* 2003, *125*, 11476.
152. Kamphuis, J., Boesten, W. H. J., Kaptein, B., Hermes, H. F. M., Sonke, T., Broxterman, Q. B., Tweel, W. J. J. V. D., Schoemaker, H. E. The Production and Uses of Optically Pure Natural and Unnatural Amino Acids. In *Chirality in Industry,* Collins, A. N., Sheldrake, G. N., Crosby, J. Eds., John Wiley & Sons: Chichester, 1992, p. 187.
153. Powell, L. W. Immobilized Enzymes. In *Industrial Enzymology,* Godfrey, T., West, S. Eds., Macmillan Press LTD: Hampshire, UK, 1996, p. 265.
154. Fité, M., Capellas, M., Benaiges, M. D., Caminal, G., Clapés, P., Alvaro, G. *Biocatal. Biotransform.* 1997, *14*, 317.
155. Youshko, M. I., Rantwijk, F. v., Sheldon, R. A. *Tetrahedron: Asymmetry* 2001, *12*, 3267.
156. Theil, F. *Chem. Rev.* 1995, *95*, 2203.
157. Collins, A. N., Sheldrake, G. N., Crosby, J. *Chirality in Industry II,* John Wiley & Sons: New York, 1997.
158. Wong, C.-H., Whitesides, G. M. Chapter 2: Use of Hydrolytic Enzymes: Amidases, Proteases, Esterases, Lipases, Nitrilases, Phosphatases, Epoxide Hydrolases. In *Enzymes in Synthetic Organic Chemistry,* Elsevier: New York, 1994, p. 41.
159. Gais, H. J., Theil, F. Hydrolysis and Formation of C-O Bonds. In *Enzyme Catalysis in Organic Synthesis,* Drauz, K., Waldmann, H. Eds., Wiley-VCH Verlag: Weinheim, 2002, Vol. 2, p. 335.
160. Bornscheuer, U. T., Kazlauskas, R. J. *Hydrolases in Organic Synthesis: Regio- and Stereoselective Biotransformations,* Wiley-VCH: Weinheim, 1999, p. Chapter 2.
161. Theil, F. *Tetrahedron* 2000, *56*, 2905.
162. Berglund, P. *Biomol. Eng.* 2001, *18*, 13.
163. Secundo, F., Carrea, G. *Chem. Eur. J.* 2003, *9*, 3194.
164. Handfeld, U. *Org. Biomol. Chem.* 2003, *1*, 2405.
165. Kazlauskas, R. J., Weissfloch, A. N. E., Rappaport, A. T., Cuccia, L. A. *J. Org. Chem.* 1991, *56*, 2656.
166. Lemke, K., Lemke, M., Theil, F. *J. Org. Chem.* 1997, *62*, 6268.
167. Provencher, L., Wynn, H., Jones, J. B., Krawczyk, A. R. *Tetrahedron: Asymmetry* 1993, *4*, 2025.
168. Ader, U., Andersch, P., Berger, M., Goergens, U., Seemayer, R., Schneider, M. *Pure Appl. Chem.* 1992, *64*, 1165.
169. Jones, J. B. *Aldrichimica Acta* 1993, *26*, 105.
170. Holzwarth, H.-C., Pleiss, J., Schmid, R. D. *J. Mol. Catal. B: Enzymatic* 1997, *3*, 73.
171. Nishizawa, K., Ohgami, Y., Matsuo, N., Kisida, H., Hirohara, H. *J. Chem. Soc., Perkin Trans. 2* 1997, 1293.
172. Cygler, M., Grochulski, P., Kazlauskas, R. J., Schrag, J. D., Bouthillier, F., Rubin, B., Serriqi, A. N., Gupta, A. K. *J. Am. Chem. Soc.* 1994, *116*, 3180.
173. Toone, E. J., Jones, J. B. *Tetrahedron: Asymmetery* 1991, *2*, 1041.
174. Gascoyne, D. G., Finkbeiner, H. L., Chan, K. P., Gordon, J. L., Stewart, K. R., Kazlauskas, R. J. *J. Org. Chem.* 2001, *66*, 3041.
175. Janes, L. E., Kazlauskas, R. J. *Tetrahedron: Asymmetry* 1997, *8*, 3719.
176. Kazlauskas, R. J. *Trends Biotechnol.* 1994, *12*, 464.
177. Kazlauskas, R. J., Weissfloch, A. N. E. *J. Mol. Catal. B: Enzymatic* 1997, *3*, 65.
178. Zuegg, J., Hönig, H., Schrag, J. D., Cygler, M. *J. Mol. Catal. B: Enyzmatic* 1997, *3*, 83.
179. Pleiss, J., Fischer, M., Peiker, M., Thiele, C., Schmid, R. D. *J. Mol. Catal. B: Enzymatic* 2000, *10*, 491.
180. Ohno, M., Otsuka, M. *Org. Reactions* 1989, *37*, 1.
181. Tamm, C. *Pure Appl. Chem.* 1992, *64*, 1187.
182. Lange, S., Musidlowska, A., Schmidt-Dannert, C., Schmitt, J., Bornscheuer, U. T. *Chem. Biochem.* 2001, *2*, 576.

183. Saksena, A. K., Girijavallabhan, V. M., Pike, R. E., Wang, H., Lovey, R. G., Liu, Y.-T., Ganguly, A. K., Morgan, W. B., Zaks, A. U.S. Pat. 5,403,937, 1995.
184. Patel, R. N., Banerjee, A., Szarka, L. J. *J. Amer. Oil Chem. Soc.* 1996, *73*, 1363.
185. Dodds, D. R., Heinzelman, C., Homann, M., Morgan, W. B., Previte, E., Roehl, R. A., Vail, R., Zaks, A. *Chimica Oggi* 1996, *14*, 9.
186. Roberge, C., Cvetovich, R. J., Amato, J. S., Pecore, V., Hartner, F. W., Greasham, R., Chartrain, M. *J. Ferment. Bioeng.* 1997, *83*, 48.
187. Morgan, B., Dodds, D. R., Zaks, A., Andrews, D. R., Klesse, R. *J. Org. Chem.* 1997, *62*, 7736.
188. Schoffers, E., Golebiowski, A., Johnson, C. R. *Tetrahedron* 1996, *52*, 3769.
189. Egri, G., Fogassy, E., Novák, L., Poppe, L. *Tetrahedron: Asymmetry* 1997, *8*, 547.
190. Carnell, A. J. *J. Mol. Catal. B: Enzymatic* 2002, *19-20*, 83.
191. Klibanov, A. M. *Acc. Chem. Res.* 1990, *23*, 114.
192. Mori, K. *Synlett* 1995, *11*, 1097.
193. Yokomatsu, T., Minowa, T., Murano, T., Shibuya, S. *Tetrahedron* 1998, *54*, 9341.
194. Kirihara, M., Kawasaki, M., Takuwa, T., Kakuda, H., Wakikawa, T., Takeuchi, Y., Kirk, K. L. *Tetrahedron: Asymmetry* 2003, *14*, 1753.
195. Patel, R. N., Robison, R. S., Szarka, L. J. *Appl. Microbiol. Biotechnol.* 1990, *34*, 10.
196. May, O., Verseck, S., Bommarius, A., Drauz, K. *Org. Proc. Res. Dev.* 2002, *6*, 452.
197. Ward, R. S. *Tetrahedron: Asymmetry* 1995, *6*, 1475.
198. Caddick, S., Jenkins, K. *Chem. Soc. Rev.* 1996, 447.
199. Noyori, R., Tokunaga, M., Kitamura, M. *Bull. Chem. Soc. Jpn.* 1995, *68*, 36.
200. Stecher, H., Faber, K. *Synthesis* 1997, 1.
201. Pàmies, O., Bäckvall, J.-E. *Trends Biotechnol.* 2004, *22*, 130.
202. Pellissier, H. *Tetrahedron* 2003, *59*, 8291.
203. Pàmies, O., Bäckvall, J.-E. *Chem. Rev.* 2003, *103*, 3247.
204. Kim, M.-J., Ahn, Y., Park, J. *Curr. Opin. Biotechnol.* 2002, *13*, 578.
205. Schnell, B., Faber, K., Kroutil, W. *Adv. Synth. Catal.* 2003, *345*, 653.
206. Stecher, H., Faber, K. *Synthesis* 1997, 1.
207. Larsson, A. L. E., Persson, B. A., Bäckvall, J.-E. *Angew. Chem., Int. Ed.* 1997, *36*, 1211.
208. Stürmer, R. *Angew. Chem., Int. Ed.* 1997, *36*, 1173.
209. Thuring, J. W. J. F., Klunder, A. J. H., Nefkens, G. H. L., Wegman, M. A., Zwanenburg, B. *Tetrahedron Lett.* 1996, *37*, 4759.
210. Williams, J. M. J., Parker, R. J., Neri, C. Enzymatic Kinetic Resolution. In *Enzyme Catalysis in Organic Synthesis,* Drauz, K., Waldmann, H. Eds., Wiley-VCH Verlag: Weinhein, 2002, Vol. 1, p. 287.
211. Fruton, J. S. *Adv. Enzymol. Relat. Areas Mol. Biol.* 1982, *53*, 239.
212. Chaiken, I. M., Komoriya, A., Ohno, M., Widmer, F. *Appl. Biochem. Biotechnol.* 1982, *7*, 385.
213. Jakubke, H.-D., Kuhl, P., Könnecke, A. *Angew. Chem., Int. Ed.* 1985, *24*, 85.
214. Kullmann, W. *Enzymatic Peptide Synthesis,* CRC Press, Inc.: Boca Raton, 1987.
215. Heiduschka, P., Dittrich, J., Barth, A. *Pharmazie* 1990, *45*, 164.
216. Schellenberger, V., Jakubke, H.-D. *Angew. Chem., Int. Ed.* 1991, *30*, 1437.
217. Jakubke, H.-D. Hydrolysis and Formation of Peptides. In *Enzyme Catalysis in Organic Synthesis,* Drauz, K., Waldmann, H. Eds., Wiley-VCH Verlag: Weinheim, 2002, Vol. 2, p. 800.
218. Bordusa, F. *Chem. Rev.* 2002, *102*, 4817.
219. Breddam, K. *Carlsberg Res. Commun.* 1986, *51*, 83.
220. Lye, G. J., Woodley, J. M. *Trends Biotechnol.* 1999, *17*, 395.
221. Haensler, M., Thust, S., Klossek, P., Ullmann, G. *J. Mol. Catal. B: Enzymatic* 1999, *6*, 95.
222. Erbeldinger, M., Ni, X., Halling, P. J. *Biotechnol. Bioeng.* 1999, *63*, 316.
223. Wehofsky, N., Kirbach, S. W., Haensler, M., Wissmann, J.-D., Bordusa, F. *Org. Lett.* 2000, *2*, 2027.
224. Oyama, K. Enzymatic Synthesis of Aspartame in Organic Solvents, Biocatalysis in Organic Media, 1986, Wageningen, The Netherlands. Published in *Studies in Organic Chemistry,* Laane, C., Tramper, J., Lilly, M. D. Eds., Elsevier, Amsterdam, 1987, Vol. 29, p. 209.
225. Pantaleone, D. P., Dikeman, R. N. Enzymatic Synthesis of Aspartame Precursors. In *Science for the Food Industry of the 21st Century,* Yalpani, M. Ed., ATL Press: Mt. Prospect, 1993, Vol. 1, p. 173.
226. Ager, D. J., Pantaleone, D. P., Henderson, S. A., Katritzky, A. R., Prakash, I., Walters, D. E. *Angew. Chem., Int. Ed.* 1998, *37*, 1803.

227. De Martin, L., Ebert, C., Gardossi, L., Linda, P. *Tetrahedron Lett.* 2001, *42*, 3395.

228. Sears, P., Wong, C.-H. *Biotechnol. Prog.* 1996, *12*, 423.

229. Wehofsky, N., Thust, S., Burmeister, J., Klussmann, S., Bordusa, F. *Angew. Chem., Int. Ed.* 2003, *42*, 677.

230. Imaoka, Y., Kawamoto, T., Ueda, M., Tanaka, A. *Appl. Microbiol. Biotechnol.* 1994, *40*, 653.

231. Van Langen, L. M., Van Rantwijk, F., Svedas, V. K., Sheldon, R. A. *Tetrahedron: Asymmetry* 2000, *11*, 1077.

232. Moree, W. J., Sears, P., Kawashiro, K., Witte, K., Wong, C.-H. *J. Am. Chem. Soc.* 1997, *119*, 3942.

233. Wang, Y.-G., Yakovlevsky, K., Zhang, B., Margolin, A. L. *J. Org. Chem.* 1997, *62*, 3488.

234. Sekizaki, H., Itoh, K., Toyota, E., Tanizawa, K. *Tetrahedron Lett.* 1997, *38*, 1777.

235. Sekizaki, H., Itoh, K., Toyota, E., Tanizawa, K. *Chem. Pharm. Bull.* 1996, *44*, 1585.

236. Sergeeva, M. V., Paradkar, V. M., Dordick, J. S. *Enzyme Microb. Technol.* 1997, *20*, 623.

237. Chen, S.-T., Chen, S.-Y., Wang, K.-T. *J. Org. Chem.* 1992, *57*, 6960.

238. Zhang, X.-z., Wang, X., Chen, S., Fu, X., Wu, X., Li, C. *Enzyme Microb. Technol.* 1996, *19*, 538.

239. Eichhorn, U., Bommarius, A. S., Drauz, K., Jakubke, H.-D. *J. Peptide Sci.* 1997, *3*, 245.

240. van der Werf, M. J., van den Tweel, W. J. J., Kamphuis, J., Hartmans, S., de Bont, J. A. M. *Trends Biotechnol.* 1994, *12*, 95.

241. Gerhartz, W. *Enzymes in Industry,* VCH: Weinheim, 1990.

242. Collins, A. N., Sheldrake, G. N., Crosby, J. *Chirality in Industry,* John Wiley & Sons: New York, 1992.

243. Renard, G., Guilleux, J.-C., Bore, C., Malta-Valette, V., Lerner, D. A. *Biotechnol. Lett.* 1992, *14*, 673.

244. Yanaka, M., Ura, D., Takahashi, A., Fukahara, N. Jap. Pat. JP 06113870 A2, 1994.

245. Liu, W. U.S. Pat. 5,981,239, 1999.

246. Zhao, J.-S., Yang, S.-K. *China J. Chem.* 1995, *13*, 241.

247. Gloge, A., Zoß, J., Kövári, Á., Poppe, L., Rétey, J. *Chem. Eur. J.* 2000, *6*, 3386.

248. Fessner, W.-D., Helaine, V. *Curr. Opin. Biotechnol.* 2001, *12*, 574.

249. Fessner, W.-D. Enzymatic Asymmetric Synthesis Using Aldolases. In *Stereoselective Biocatalysis,* Patel, R. Ed., Marcel Dekker, Inc.: New York, 2000, p. 239.

250. Takayama, S., McGarvey, G. J., Wong, C.-H. *Ann. Rev. Microbiol.* 1997, *51*, 285.

251. Wong, C.-H., Liu, K. C. PCT Appl. WO 92/21655, 1992.

252. Wong, C.-H., Sharpless, K. B. PCT Appl. WO 95/16049, 1995.

253. Bednarski, M. D. Applications of Enzymatic Aldol Reactions in Organic Synthesis. In *Applied Biocatalysis,* Blanch, H. W., Clark, D. S. Eds., Marcel Dekker, Inc.: New York, 1991, Vol. 1, p. 87.

254. Gefflaut, T., Blonski, C., Perie, J., Willson, M. *Prog. Biophys. Mol. Biol.* 1995, *63*, 301.

255. Wang, P. G., Fitz, W., Wong, C.-H. *CHEMTECH* 1995, *25*, 22.

256. Abeles, R. H., Frey, P. A., Jencks, W. P. *Biochemistry,* Jones and Bartlett: Boston, 1992, p. 45.

257. Hupe, D. J. Imine Formation in Enzymatic Reactions. In *Enzyme Mechanisms,* Page, M. I., Williams, A. Eds., The Royal Society of Chemistry: London, 1987, p. 317.

258. Fessner, W.-D., Sinerius, G. *Angew. Chem., Int. Ed.* 1994, *22*, 209.

259. Liu, J., Wong, C.-H. *Angew. Chem., Int. Ed.* 2002, *41*, 1404.

260. Wagner, J., Lerner, R. A., Barbas, C. F., III. *Science* 1995, *270*, 1797.

261. Barbas, C. F., III, Heine, A., Zhong, G., Hoffmann, T., Gramatikova, S., Björnestedt, R., List, B., Anderson, J., Stura, E. A., Wilson, I. A., Lerner, R. A. *Science* 1997, *278*, 2085.

262. Maggiotti, V., Resmini, M., Gouverneur, V. *Angew. Chem., Int. Ed.* 2002, *41*, 1012.

263. Garcia-Junceda, E., Shen, G.-J., Sugai, T., Wong, C.-H. *Bioorg. Med. Chem.* 1995, *3*, 945.

264. Ward, O. P., Baev, M. V. Decarboxylases in Stereoselective Catalysis. In *Stereoselective Biocatalysis,* Patel, R. Ed., Marcel Dekker, Inc.: New York, 2000, p. 267.

265. Calton, G. J. The Enzymatic Preparation of L-Alanine. In *Biocatalytic Production of Amino Acids and Derivatives,* Rozzell, J. D., Wagner, F. Eds., Hanser Publishers: Munich, 1992, p. 59.

266. Chen, C.-C., Chou, T.-L., Lee, C.-Y. *J. Ind. Microbiol. Biotechnol.* 2000, *25*, 132.

267. Ritter, S. K. *Chem. Eng. News* 1995, *73*, 39.

268. DeJesus, O. T. *Drug Dev. Res.* 2003, *59*, 249.

269. Pohl, M., Lingen, B., Müller, M. *Chem. Eur. J.* 2002, *8*, 5288.

270. Iding, H., Siegert, P., Mesch, K., Pohl, M. *Biochim. Biophys. Acta.* 1998, *1385*, 307.

271. Breur, M., Hauer, B., Friedrich, T. PCT Appl. WO 03/020921, 2003.

272. Dünnwald, T., Müller, M. *J. Org. Chem.* 2000, *65*, 8608.

273. Dünnwald, T., Greiner, L., Iding, H., Liese, A., Müller, M., Pohl, M. Eur. Pat. EP 1048737 A2, 2000.

274. Kihumbu, D., Stillger, T., Hummel, W., Liese, A. *Tetrahedron: Asymmetry* 2002, *13*, 1069.

275. Brussee, J., Van der Gen, A. Biocatalysis in the Enantioselective Formation of Chiral Cyanhydrins, Valuable Building Blocks in Organic Synthesis. In *Stereoselective Biocatalysis,* Patel, R. N. Ed., Dekker: New York, 2000, p. 289.

276. Effenberger, F., Forster, S., Wajant, H. *Curr. Opin. Biotechnol.* 2000, *11*, 532.

277. Effenberger, F. Hydroxynitrile Lyases in Stereoselective Synthesis. In *Stereoselective Biocatalysis,* Patel, R. N. Ed., Dekker: New York, 2000, p. 321.

278. Griengl, H., Schwab, H., Fechter, M. *Trends Biotechnol.* 2000, *18*, 252.

279. Schmidt, M., Griengl, H. *Top. Curr. Chem.* 1999, *200*, 193.

280. North, M. *Tetrahedron: Asymmetry* 2003, *14*, 147.

281. van Langen, L. M., van Rantwijk, F., Sheldon, R. A. *Org. Proc. Res. Dev.* 2003, *7*, 828.

282. Trummler, K., Wajant, H. *J. Biol. Chem.* 1997, *272*, 4770.

283. Rotheim, P. *The Enzyme Industry: Specialty and Medical Applications,* Business Communications Company, Inc.: Norwalk, CT, 1994.

284. Bhosale, S. H., Rao, M. B., Deshpande, V. V. *Microbiol. Rev.* 1996, *60*, 280.

285. Pedersen, S. *Bioprocess Technol.* 1993, *16*, 185.

286. Quax, W. J., Luiten, R. G. M., Schuurhuizen, P. W., Mrabet, N. U.S. Pat. 5,376,536, 1994.

287. Starnes, R. L., Kelly, R. M., Brown, S. H. U.S. Pat. 5,219,751, 1993.

288. Kragl, U., Gygax, D., Ghisalba, O., Wandrey, C. *Angew. Chem.* 1991, *103*, 854.

289. Painter, G. F., Eldridge, P. J., Falshaw, A. *Bioorg. Med. Chem.* 2004, *12*, 225.

290. Yamamoto, R., Yamaguchi, M., Kobayashi, A., Machida, K., Isshiki, K., Kanbe, K., Tamamura, T. PCT Appl. WO 02/055715, 2002.

291. Frey, P. A. *Chem. Rev.* 1990, *90*, 1343.

292. Frey, P. A., Reed, G. H. *Adv. Enzymol. Relat. Areas Mol. Biol.* 1992, *66*, 1.

293. Liao, H. H., Gokarn, R. R., Gort, S. J., Jessen, H. J., Selifonova, O. PCT Appl. WO 03/062173, 2003.

294. Chen, H.-P., Wu, S.-H., Lin, Y.-L., Chen, C.-M., Tsay, S.-S. *J. Biol. Chem.* 2001, *276*, 44744.

295. Fleming, P. E., Mocek, U., Floss, H. G. *J. Am. Chem. Soc.* 1993, *115*, 805.

296. Steele, C. L., Chen, Y., Dougherty, B. A., Hofstead, S., Lam, K. S., Li, W., Xing, Z. PCT Appl. WO 03/066871, 2003.

297. Christenson, S. D., Wu, W., Spies, M. A., Shen, B., Toney, M. D. *Biochemistry* 2003, *42*, 12708.

298. Christenson, S. D., Liu, W., Toney, M. D., Shen, B. *J. Am. Chem. Soc.* 2003, *125*, 6062.

299. Cardillo, G., Tomasini, C. *Chem. Soc. Rev.* 1996, *25*, 117.

300. Borman, S. *Chem. Eng. News*, 1997, *75*, 32.

301. Bailey, J. E. *Science* 1991, *252*, 1668.

302. Stephanopoulos, G., Vallino, J. J. *Science* 1991, *252*, 1675.

303. Stephanopoulos, G. *Curr. Opin. Biotechnol.* 1994, *5*, 196.

304. Mijts, B. N., Schmidt-Dannert, C. *Curr. Opin. Biotechnol.* 2003, *14*, 597.

305. Kholsa, C., Keasling, J. D. *Nature Rev. Drug Disc.* 2003, *2*, 1019.

306. Schmidt-Dannert, C., Umeno, D., Arnold, F. H. *Nat. Biotechnol.* 2000, *18*, 750.

307. Misawa, N., Shimada, H. *J. Biotechnol* 1999, *59*, 169.

308. Carreras, C. W., Rembert, P., Khosla, C. *Top. Curr. Chem.* 1997, *188*, 85.

309. Marahiel, M. A., Stachelhaus, T., Mootz, H. D. *Chem. Rev.* 1997, *97*, 2651.

310. Madison, L. L., Huisman, G. *J. Microbiol. Mol. Biol. Rev.* 1999, *63*, 21.

311. Maier, T. H. P. *Nat. Biotechnol.* 2003, *21*, 422.

312. Kleerebezem, M., Hugenholtz, J. *Curr. Opin. Biotechnol.* 2003, *14*, 232.

313. Szczebara, F. M., Chandelier, C., Villeret, C., Masurel, A., Bourot, S., Duport, C., Blanchard, S., Groisillier, A., Testet, E., Costaglioli, P., Cauet, G., Degryse, E., Balbuena, D., Winter, J., Achstetter, T., Spagnoli, R., Pompon, D., Dumas, B. *Nat. Biotechnol.* 2003, *21*, 143.

314. Franke, D., Lorbach, V., Esser, S., Dose, C., Sprenger, G. A., Halfar, M., Thömmes, J., Müller, R., Takors, R., Müller, M. *Chem. Eur. J.* 2003, *9*, 4188.

315. Niu, W., Molefe, M. N., Frost, J. W. *J. Am. Chem. Soc.* 2003, *125*, 12998.

316. Krämer, R. *J. Biotechnol.* 1996, *45*, 1.

317. Chotani, G., Dodge, T., Hsu, A., Kumar, M., LaDuca, R., Trimbur, D., Weyler, W., Sanford, K. *Biochim. Biophys. Acta.* 2000, *1543*, 434.

318. Cameron, D. C., Tong, I.-T. *Appl. Biochem. Biotechnol.* 1993, *38*, 105.
319. Flores, N., Xiao, J., Berry, A., Bolivar, F., Valle, F. *Nat. Biotechnol.* 1996, *14*, 620.
320. Patnaik, R., Liao, J. C. *Appl. Environ. Microbiol.* 1994, *60*, 3903.
321. Patnaik, R., Spitzer, R. G., Liao, J. C. *Biotechnol. Bioeng.* 1995, *46*, 361.
322. Lu, J.-l., Liao, J. C. *Biotechnol. Bioeng.* 1997, *53*, 132.
323. Krämer, M., Bongaerts, J., Bovenberg, R., Kremer, S., Müller, U., Orf, S., Wubbolts, M., Raeven, L. *Metab. Eng.* 2003, *5*, 277.
324. Federspiel, M., Fischer, R., Hennig, M., Mair, H.-J., Oberhauser, T., Rimmler, G., Albiez, T., Bruhin, J., Estermann, H., Gandert, C., Göckel, V., Götzö, S., Hoffmann, U., Huber, G., Janatsch, G., Lauper, S., Röckel-Stäbler, O., Trussardi, R., Zwahlen, A. G. *Org. Proc. Res. Dev.* 1999, *3*, 266.
325. Rohloff, J. C., Kent, K. M., Postich, M. J., Becker, M. W., Chapman, H. H., Kelly, D. E., Lew, W., Louie, M. S., McGee, L. R., Prisbe, E. J., Schultze, L. M., Yu, R. H., Zhang, L. *J. Org. Chem.* 1998, *63*, 4545.
326. Knop, D. R., Draths, K. M., Chandran, S. S., Barker, J. L., von Daeniken, R., Weber, W., Frost, J. W. *J. Am. Chem. Soc.* 2001, *123*, 10173.
327. Draths, K. M., Knop, D. R., Frost, J. W. *J. Am. Chem. Soc.* 1999, *121*, 1603.
328. Chandran, S. S., Yi, J., Draths, K. M., von Daeniken, R., Weber, W., Frost, J. W. *Biotechnol. Prog.* 2003, *19*, 808.
329. Shirai, M., Reiko, M., Satoshi, S., Kosuke, S., Saburo, Y., Katsuhiro, S., Tetsu, Y., Kenichi, O. Eur. Pat. EP 1,092,766, 2001.
330. Iomantas, Y. A. V., Abalakina, E. G., Polanuer, B. M., Yampolskaya, T. A., Bachina, T. A., Kozlov, Y. I. U.S. Pat. 6,436,664, 2002.
331. Draths, K. M., Pompliano, D. L., Conley, D. L., Frost, J. W., Berry, A., Disbrow, G. L., Staversky, R. J., Lievense, J. C. *J. Am. Chem. Soc.* 1992, *114*, 3956.
332. Frost, J. W., Lievense, J. *New J. Chem.* 1994, *18*, 341.
333. Draths, K. M., Ward, T. L., Frost, J. W. *J. Am. Chem. Soc.* 1992, *114*, 9725.
334. Draths, K. M., Frost, J. W. *J. Am. Chem. Soc.* 1994, *116*, 399.
335. Ran, N., Knop, D. R., Draths, K. M., Frost, J. W. *J. Am. Chem. Soc.* 2001, *123*, 10927.
336. Kambourakis, S., Draths, K. M., Frost, J. W. *J. Am. Chem. Soc.* 2000, *122*, 9042.
337. Keeler, R. *R&D Magazine*, 1991, *33*, 46.
338. Poirier, Y., Dennis, D. E., Klomparens, K., Somerville, C. *Science* 1992, *256*, 520.
339. Slater, S. C., Voige, W. H., Dennis, D. E. *J. Bacteriol.* 1988, *170*, 4431.
340. Schubert, P., Steinbüchel, A., Schlegel, H. G. *J. Bacteriol.* 1988, *170*, 5837.
341. Peoples, O. P., Sinskey, A. J. *J. Biol. Chem.* 1989, *264*, 15293.
342. Peoples, O. P., Sinskey, A. J. *J. Biol. Chem.* 1989, *264*, 15298.
343. Doi, Y., Taguchi, S., Kichise, T. PCT Appl. WO 03/050277, 2003.
344. Lee, J.-N., Shin, H.-D., Lee, Y.-H. *Biotechnol. Prog.* 2003, *19*, 1444.
345. Steinbüchel, A., Schlegel, H. G. *Mol. Microbiol.* 1991, *5*, 535.
346. Lee, S. Y. *Trends Biotechnol.* 1996, *14*, 431.
347. Hutchinson, C. R. *Curr. Opin. Microbiol.* 1998, *1*, 319.
348. Hopwood, D. A. *Chem. Rev.* 1997, *97*.
349. Katz, L. *Chem. Rev.* 1997, *97*, 2557.
350. Khosla, C. *Chem. Rev.* 1997, *97*, 2577.
351. Marahiel, M. A., Stachelhaus, T., Mootz, H. D. *Chem. Rev.* 1997, *97*, 2651.
352. Cane, D. E., Walsh, C. T., Khosla, C. *Science* 1998, *282*, 63.
353. Khosla, C. *J. Org. Chem.* 2000, *65*, 8127.
354. Pohl, N. L. *Curr. Opin. Chem. Biol.* 2002, *6*, 773.
355. Reeves, C. D. *Crit. Rev. Biotechnol.* 2003, *23*, 95.
356. Liou, G. F., Khosla, C. *Curr. Opin. Chem. Biol.* 2003, *7*, 279.
357. Donadio, S., Sosio, M. *Comb. Chem. High Throughput Screening* 2003, *6*, 489.
358. Katz, L., Donadio, S. *Ann. Rev. Microbiol.* 1993, *47*, 875.
359. Hutchinson, C. R., Fujii, I. *Ann. Rev. Microbiol.* 1995, *49*, 201.
360. Farnet, C. M., Staffa, A. PCT Appl. WO 03/089641, 2003.
361. Grandi, G., De Ferra, F., Rodrigues, F. Eur. Pat. EP 789078 A2, 1997.
362. Gregory, M. A., Gaisser, S., Petkovic, H., Moss, S. PCT Appl. WO 04/007709, 2004.

363. Kealey, J. T., Dayem, L. C., Santi, D. V. PCT Appl. WO 04/007688, 2004.
364. Marahiel, M. A., Stachelhaus, T., Mootz, H., Konz, D. PCT Appl. WO 00/052152, 2000.
365. Santi, D. V., Xue, Q., Ashley, G. PCT Appl. WO 00/063361, 2000.
366. Santi, D. V., Xue, Q., Ashley, G. U.S. Pat. 6,399,789, 2002.
367. Shen, B. *Curr. Opin. Chem. Biol.* 2003, *7*, 285.
368. Morita, H., Noguchi, H., Schröder, J., Abe, I. *Eur. J. Biochem.* 2001, *268*, 3759.
369. Tseng, C. C., McLoughlin, S. M., Kelleher, N. L., Walsh, C. T. *Biochemistry* 2004, *43*, 970.
370. Walsh, C. T., O'Connor, S. E., Schneider, T. L. *J. Indust. Microbiol. Biotechnol.* 2003, *30*, 448.
371. Admiraal, S. J., Khosla, C., Walsh, C. T. *J. Am. Chem. Soc.* 2003, *125*, 13664.
372. Liu, W., Christenson, S. D., Standage, S., Shen, B. *Science* 2002, *297*, 1170.
373. Cane, D. E., Walsh, C. T. *Chem. Biol.* 1999, *6*, R319.
374. Hutchinson, C. R. *Proc. Natl. Acad. Sci. USA* 2003, *100*, 3010.
375. Donadio, S., Staver, M. J., McAlpine, J. B., Swanson, S. J., Katz, L. *Science* 1991, *252*, 675.
376. Donadio, S., McAlpine, J. B., Sheldon, P. J., Jackson, M., Katz, L. *Proc. Natl. Acad. Sci. USA* 1993, *90*, 7119.
377. Kao, C. M., Luo, G., Katz, L., Cane, D. E., Khosla, C. *J. Am. Chem. Soc.* 1995, *117*, 9105.
378. Pieper, R., Ebert-Khosla, S., Cane, D., Khosla, C. *Biochemistry* 1996, *35*, 2054.
379. Tsukamoto, N., Chuck, J.-A., Luo, G., Kao, C. M., Khosla, C., Cane, D. E. *Biochemistry* 1996, *35*, 15244.
380. Kao, C. M., Luo, G., Katz, L., Cane, D. E., Khosla, C. *J. Am. Chem. Soc.* 1996, *118*, 9184.
381. Jacobsen, J. R., Hutchinson, C. R., Cane, D. E., Khosla, C. *Science* 1997, *277*, 367.
382. Pieper, R., Gokhale, R. S., Luo, G., Cane, D. E., Khosla, C. *Biochemistry* 1997, *36*, 1846.
383. McDaniel, R., Kao, C. M., Fu, H., Hevezi, P., Gustafsson, C., Betlach, M., Ashley, G., Cane, D. E., Khosla, C. *J. Am. Chem. Soc.* 1997, *119*, 4309.
384. Walsh, C. T. *ChemBioChem* 2002, *3*, 124.
385. Chu, D. T. W., Plattner, J. J., Katz, L. *J. Med. Chem.* 1996, *39*, 3858.
386. Khosla, C., Zawada, R. J. X. *Trends Biotechnol.* 1996, *14*, 335.
387. Khosla, C., Caren, R., Kao, C. M., McDaniel, R., Wang, S.-W. *Biotechnol. Bioeng.* 1996, *52*, 122.
388. Pfeifer, B. A., Admiraal, S. J., Gramajo, H., Cane, D. E., Khostla, C. *Science* 2001, *291*, 1790.
389. Lombó, F., Pfeifer, B., Leaf, T., Ou, S., Kim, Y. S., Cane, D. E., Licari, P., Khosla, C. *Biotechnol. Prog.* 2001, *17*, 612.
390. Pfeifer, B., Hu, Z., Licardi, P. J., Khosla, C. *Appl. Environ. Microbiol.* 2002, *68*, 3287.
391. Dayem, L. C., Carney, J. R., Santi, D. V., Pfeifer, B. A., Khosla, C., Kealey, J. T. *Biochemistry* 2002, *41*, 5193.
392. Kennedy, J., Murli, S., Kealey, J. T. *Biochemistry* 2003, *42*, 14342.
393. Kinoshita, K., Pfeifer, B., Khosla, C., Cane, D. E. *Bioorg. Med. Chem. Lett.* 2003, *13*, 3701.
394. Kieslich, K., van der Beek, C. P., de Bont, J. A. M., van den Tweel, W. J. J. *New Frontiers in Screening for Microbial Biocatalysts*, Elsevier: Amsterdam, 1998.
395. Cheetham, P. S. J. *Enzyme Microb. Technol.* 1987, *9*, 194.
396. Stahl, S., Greasham, R., Chartrain, M. *J. Biosci. Bioeng.* 2000, *89*, 367.
397. Cheetham, P. S. J. *Trends Biotechnol.* 1993, *11*, 478.
398. Patel, R. N., Banerjee, A., Szarka, L. J. *J. Am. Oil Chem. Soc.* 1995, *72*, 1247.
399. Lorraine, K., King, S., Greasham, R., Chartrain, M. *Enzyme Microb. Technol.* 1996, *19*, 250.
400. Takahashi, E., Nakamichi, K., Furui, M. *J. Ferment. Bioeng.* 1995, *80*, 247.
401. Linko, Y.-Y., Lamsa, M., Huhtala, A., Rantanen, O. *J. Am. Oil Chem. Soc.* 1995, *72*, 1293.
402. Zhang, J., Reddy, J., Roberge, C., Senanayake, C., Greasham, R., Chartrain, M. *J. Ferment. Bioeng.* 1995, *80*, 244.
403. Chung, W., Goo, Y. M. *Arch. Pharm. Res.* 1988, *11*, 139.
404. Nakayama, A., Torigoe, Y. Jap. Pat. JP 63185395, 1987.
405. Xu, J., Wutong, J., Jin, S., Yao, W. *Zhongguo Yaoke Daxue Xuebao* 1994, *25*, 53.
406. Pantaleone, D. P., Giegel, D. A., Schnell, D. Isolation, Partial Purification and Characterization of a Novel Cyclic Dipeptidase. In *New Frontiers in Screening for Microbial Biocatalysis,* Kieslich, K., van der Beek, C. P., de Bont, J. A. M., van den Tweel, W. J. J. Eds., Elsevier Science: Amsterdam, 1998, Vol. 53, p. 201.

407. Syldatk, C., Muller, R., Siemann, M., Krohn, K., Wagner, F. Microbial and Enzymatic Production of D-Amino Acids from DL-5-Monosubstituted Hydantoins. In *Biocatalytic Production of Amino Acids and Derivatives,* Rozzell, J. D., Wagner, F. Eds., Carl Hanser Verlag: Munich, 1992, p. 75.

408. Syldatk, C., Muller, R., Pietzsch, M., Wagner, F. Microbial and Enzymatic Production of L-Amino Acids from DL-5-Monosubstituted Hydantoins. In *Biocatalytic Production of Amino Acids and Derivatives,* Rozzell, J. D., Wagner, F. Eds., Carl Hanser Verlag: Munich, 1992, p. 129.

409. Pietzsch, M., Syldatk, C. Hydrolysis and Formation of Hydantoins. In *Enzyme Catalysis in Organic Synthesis*, Drauz, K., Waldmann, H. Eds., Wiley-VCH Verlag: Weinheim, 2002, Vol. 2, p. 761.

410. Lo, H.-H., Kao, C.-H., Lee, D.-S., Yang, T.-K., Hsu, W.-H. *Chirality* 2003, *15*, 699.

411. Keil, O., Schneider, M. P., Rasor, J. P. *Tetrahedron: Asymmetry* 1995, *6*, 1257.

412. Garcia, M. J., Azerad, R. *Tetrahedron: Asymmetry* 1997, *8*, 85.

413. Trauthwein, H., May, O., Dingerdissen, U., Buchholz, S., Drauz, K. *Tetrahedron Lett.* 2003, *44*, 3737.

414. Wilms, B., Wiese, A., Syldatk, C., Mattes, R., Altenbuchner, J. *J. Biotechnol.* 2001, *86*, 19.

415. May, O., Verseck, S., Bommarius, A., Drauz, K. *Org. Proc. Res. Dev.* 2002, *6*, 452.

416. Abendroth, J., Niefind, K., May, O., Siemann, M., Syldatk, C., Schomburg, D. *Biochemistry* 2002, *41*, 8589.

417. May, O., Nguyen, P. T., Arnold, F. H. *Nat. Biotechnol.* 2000, *18*, 317.

418. Arnold, F. H., May, O., Drauz, K., Bommarius, A. PCT Appl. WO 00/058449, 2000.

419. Yamada, H. Screening of Novel Enzymes for the Production of Useful Compounds. In *New Frontiers in Screening for Microbial Biocatalysts,* Kieslich, K., van der Beek, C. P., de Bont, J. A. M., van den Tweel, W. J. J. Eds., Elsevier Science: Amsterdam, 1998, Vol. 53, p. 13.

420. Taylor, S. J. C., Holt, K. E., Brown, R. C., Keene, P. A., Taylor, I. N. Choice of Biocatalyst in the Development of Industrial Biotransformations. In *Stereoselective Biocatalysis*, Patel, R. N. Ed., Marcel Dekker, Inc., New York, 2000, p. 397.

421. Asano, Y. *J. Biotechnol.* 2002, *94*, 65.

422. Ogawa, J., Shimizu, S. *Curr. Opin. Biotechnol.* 2002, *13*, 367.

423. Wahler, D., Reymond, J.-L. *Curr. Opin. Chem. Biol.* 2001, *5*, 152.

424. Lorenz, P., Liebeton, K., Niehaus, F., Eck, J. *Curr. Opin. Biotechnol.* 2002, *13*, 572.

425. Jaeger, K. E., Eggert, T., Eipper, A., Reetz, M. T. *Appl. Microbiol. Biotechnol.* 2001, *55*, 519.

20 Combining Enzyme Discovery and Evolution to Develop Biocatalysts

Mark J. Burk, Nelson Barton, Grace DeSantis, William Greenberg, David Weiner, and Lishan Zhao

CONTENTS

20.1 INTRODUCTION

In nature, organisms produce a vast array of complex chemical entities via numerous precisely orchestrated enzyme-mediated reactions. This remarkable ability of nature to perform sophisticated chemical syntheses with unerring selectivity has long been an inspiration to medicinal and process chemists. Enzymes offer exquisitely precise chemo-, regio-, and stereocontrol and can accelerate chemical transformations that are challenging to perform by conventional chemical synthetic methodology. Reactions often proceed at room temperature and under neutral aqueous conditions, in the absence of toxic organic solvents or heavy metal catalysts. In addition, by virtue of enzyme selectivity, biocatalytic routes can obviate synthetic protecting group manipulations. Consequently, the application of biocatalysis for the synthesis of chiral as well as achiral pharmaceutical intermediates has received increased attention. As the number of available enzymes continues to grow, so do the types of chemical transformations that biocatalysts are able to perform reliably *in vitro*. To date, the utility of both enzymes and whole cells in biotransformations has expanded well beyond traditional examples such as bioremediation and the industrial processing of pulp and paper, food, and textiles.

Enzymes often prove to be the catalyst of choice for numerous transformations, and their prowess is particularly noteworthy for the synthesis of chiral molecules. The ability of biocatalysts to impart chirality through conversion of prochiral molecules or by transformation of only one stereoisomer of a racemic mixture stems from the inherent chirality of enzymes. As noted in the introduction to this book (Chapter 1), the chiral drug market is increasing, partly as a result of the need to produce single enantiomers as advocated by the U.S. Food and Drug Administration.[1] The ability to extend the patent life of a drug through a racemic switch also plays a role in this increase. An example of a racemic switch is Astra Zeneca's Esomeprazole, a proton pump inhibitor (see Chapter 31).[2]

20.1.1 BIOCATALYST CATALOGUE

The number of useful biocatalysts has been increasing rapidly, and this growth has been spurred by the development of technology to access previously uncultured organisms and/or the DNA encoding new enzymes with novel properties.[3] There is an enzyme-catalyzed counterpart for many types of chemically derived organic reactions (see Chapter 19). Biocatalysts can perform classes of reactions that are the staple tools of medicinal chemists, including aldol condensations, asymmetric epoxidations, Baeyer-Villiger oxidations, and Claisen rearrangements. Perhaps more importantly, there are enzyme-catalyzed processes known where the chemical equivalent has not been developed or is inherently ineffective. It is exactly in these areas, where chemistry struggles, that biocatalysis can play an important role in organic synthesis.

20.1.2 CHALLENGES TO BIOCATALYST APPLICATION

Despite the successful development of biocatalysts for a variety of important transformations, enzymes historically have suffered from numerous shortcomings that have limited their utility in chemical, agrochemical, and pharmaceutical applications.[4,5] Attributes such as low specific activity, inadequate substrate scope, low or undesired enantiomer specificity, intolerance to organic solvents, and low volumetric productivity were challenges often encountered when attempting to implement biocatalytic processes. However, many of these challenges were related to the deficiencies associated with the available enzymes. As in chemical catalysis, identification and development of an optimum biocatalyst for a given application requires a large number of enzymes (a library) that can be screened for the desired transformation with the specific substrate of interest. In general, large libraries of biocatalysts from a range of enzyme classes have not been available until relatively recently. Many different enzyme classes now may be purchased as off-the-shelf preparations from a variety of enzyme suppliers. Biocatalysis database tools also are available to aid in route design and biocatalyst selection. Should a suitable biocatalyst for a particular transformation not be commercially available, a number of approaches are possible to obtain an appropriate catalyst. Modern discovery programs are better able to fully exploit the diversity present in nature and can be used to rapidly furnish enabling enzymes for a particular application. In addition, when necessary, enzyme discovery may be combined with evolution technologies entailing iterative enzyme mutation and high-throughput screening. The combination of discovery and evolution approaches has proved to be very successful in overcoming many of the challenges that have faced the commercialization of biocatalysis.

20.2 RECENT ADVANCES IN DISCOVERY AND EVOLUTION TECHNOLOGIES

20.2.1 DISCOVERY

Our planet is home to an extraordinarily rich diversity of microorganisms with a vast range of metabolic pathways. Even in a single soil sample there can be thousands of different microbial species, yet only a tiny fraction of this microbial diversity can be cultured under standard laboratory conditions.[6,7] Thus, to fully exploit the available biodiversity in nature, it is necessary to go beyond screening enzymes from just the most easily cultured organisms. To address this need, we have developed two different methods to access natural biodiversity. The first technique involves high-throughput cultivation of organisms through single-cell encapsulation and microcolony selection methods.[8] This technology has allowed the discovery of a vast range of new organisms that could not be cultivated using conventional methods. In addition, we have developed a highly effective culture-independent gene discovery approach, which uses direct cloning of environmentally derived DNA from a large array of biotopes.[9,10] Soil, aquatic, littoral, and marine samples have been collected worldwide from hundreds of microenvironments and processed into large genomic DNA libraries. Environmental DNA (eDNA) libraries comprising oligonucleotide fragments, usually of 1–5 kB in length, are cloned into expression vectors and are expressed in an easily manipulated, compatible host. The process results in libraries having a population of 10^6–10^9 unique clones, each with the potential to express the individual or multiple genes on a discrete eDNA insert. These eDNA libraries are maintained for high-throughput screening efforts that search for particular activities or sequences of interest. We have found that these approaches are often superior to the more traditional methods of screening cultures of discrete organisms for discovery of novel enzymes.

20.2.2 EVOLUTION

20.2.2.1 Single-Gene Methods

Evolution has been shown to be an efficient tool for the improvement of enzymes and pathways. Stochastic approaches to introduce point mutations in genes include error prone polymerase chain reaction (ePCR),[11] the use of low-fidelity (mutator) strains,[12] and chemical mutagens.[13]

More recently, we have introduced GSSM™ technology, which is a unique method for rapid laboratory evolution of proteins whereby each amino acid of a protein is replaced with each of the other 19 commonly occurring proteinogenic amino acids.[14] GSSM technology is used at the genetic level through the use of degenerate primer sets, comprised of either 32 or 64 codon variants, for each amino acid residue of the wild-type (WT) enzyme. Use of standard methods for DNA replication generates a library of genes possessing all codon variations required for site saturation mutagenesis of the original gene.[15] Transformation and expression of the collection of GSSM gene variants in a host organism such as *Escherichia coli* furnish a comprehensive library of single-site enzyme mutants. In contrast to the use of ePCR and the use of low-fidelity (mutator) strains to introduce genetic mutations into genes, GSSM technology offers comprehensive access to all single amino acid genetic variants. Amino acid changes introduced by ePCR or mutator strains are limited to those that result from a single base change in the WT codon, and these methods statistically allow no more than 6 of the possible 19 changes at any given residue in a single round of mutation.

20.2.2.2 Multiple Gene Methods

A number of unique methods for generating chimeric genes from multiple gene families have been reported.[16] However, most of the methods used for DNA reassembly are PCR-based and rely on the extension of cross-annealed random DNA fragments from different parents. Because the number and location of crossover events is dependent on the relative parental sequence identity, the sites of crossover are not easily controlled. Consequently, PCR-based approaches for gene reassembly, including gene shuffling,[17] result in overrepresentation of full-length parental sequences, thereby diminishing the sequence diversity of the chimeric library. The presence of these parental genes can dramatically increase the number of clones that need to be screened to obtain coverage of the library and identify useful variants. Moreover, low sequence diversity results from the required stretches of homology and limits the possible solutions that may be available for a given application.

Ligation-based methods, such as our GeneReassembly™ technology,[18] are very effective for generating highly diverse gene chimeras. Reassembled genes can be generated from designed DNA segments, which allows for precise control over the number and locations of the demarcation sites. Genes can be recombined at any position, even at nonconserved sites. DNA segments with compatible overhangs are joined with the same efficiency, independent of their origin and sequence identity. The efficiency of GeneReassembly technology is, therefore, not dependent on the sequence similarity of the parental genes. Each possible combination event (including WT parent segments) occurs with the same probability, and consequently the resulting library is not contaminated with an abundance of parental genes. The complexity of the reassembled library can be tuned to suit the specific screening strategy. Higher complexity can be achieved by using more parent genes or by increasing the number of segments. The segments can be generated by oligonucleotide synthesis, enzymatic cleavage, or PCR. The use of synthetically prepared oligonucleotides affords the opportunity to optimize codon usage and to add/eliminate specific restriction sites during gene reassembly. As the reassembly sites are known precisely, it is also possible to include knowledge of structural elements in the design of a GeneReassembly procedure.

Biodiversity-based enzyme discovery applied in combination with evolution methods, such as the GSSM and GeneReassembly technologies, provides an effective platform for developing enzymes with new activities that can be used in processes that currently are challenging or impossible to perform with existing methodologies. In the following, we describe several biocatalytic platforms developed at Diversa that highlight the power of this approach.

20.3 NITRILASES

Nitrilases (E.C. 3.5.5.1) promote the mild hydrolytic conversion of organonitriles directly to the corresponding carboxylic acids.[19] However, less than 20 microbially derived nitrilases had been characterized at the start of this work, despite their potential synthetic value. The paucity of enzymes and the limited substrate scope of the handful of enzymes available have limited practical commercial development of nitrilase-catalyzed conversions, except in a few cases. Accordingly, we engaged in a discovery effort centered around exploiting the natural diversity available within our environmental DNA libraries and have discovered and characterized more than 200 new sequence unique nitrilases.[20] All of the newly discovered nitrilases possess the conserved catalytic triad Glu-Lys-Cys that is characteristic for this enzyme class.[19]

The nitrilase collection was evaluated for the production of α-hydroxy acids formed through hydrolysis of cyanohydrins. Since cyanohydrins racemize readily under basic conditions,[21] a dynamic kinetic resolution (DKR) process is possible, permitting access to α-hydroxy acids in 100% theoretical yield. One important application of this type involves commercial production of (R)-mandelic acid from mandelonitrile.[22,23] Mandelic acid and derivatives find broad use as intermediates and resolving agents for production of many pharmaceutical and agricultural products.[24] The nitrilase library was screened for activity and enantioselectivity in the hydrolysis of mandelonitrile to mandelic acid (Scheme 20.1) and several enzymes afforded mandelic acid with >90% ee. One enzyme (nitrilase I) generated (R)-mandelic acid quantitatively with 98% ee. The broad substrate scope of this enzyme was demonstrated on a range of mandelic acid derivatives as well as aromatic and heteroaromatic analogues (Table 20.1). Nitrilase I was effectively applied to a series of ortho-, meta-, and para-substituted mandelonitrile derivatives, and products generally were produced with very high enantioselectivities. Other larger aromatic groups, such as 1-naphthyl and 2-naphthyl, as well as hetero aromatic 3-pyridyl and 3-thienyl analogues also were accommodated within the active site, and hydrolysis with nitrilase I yielded glycolic acid products with high ee. This was the first reported demonstration of a nitrilase that affords such a broad range of mandelic acid derivatives and heteroaromatic analogues with high levels of enantioselectivity. It is noteworthy that for several of these processes the reactions were run in such a manner that product concentrations of up to 2M were achieved.

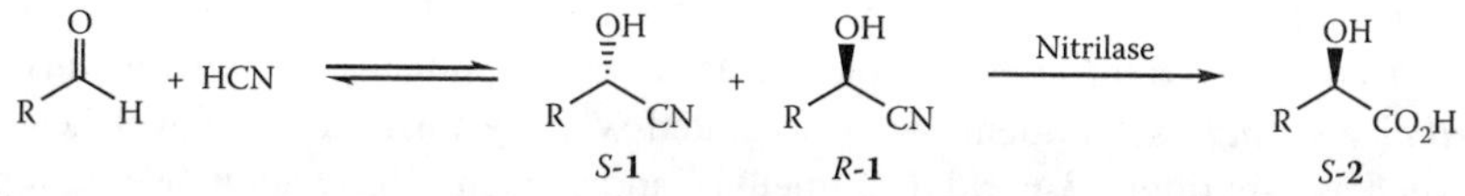

SCHEME 20.1

TABLE 20.1
Nitrilase I-Catalyzed Production of Mandelic Acid Derivatives Under Dynamic Kinetic Resolution Conditions (Scheme 20.1)

Entry	R- Sidechain	ee (%)
1	C_6H_5	98
2	$2\text{-Cl-}C_6H_5$	97
3	$2\text{-Br-}C_6H_5$	96
4	$2\text{-Me-}C_6H_5$	95
5	$3\text{-Cl-}C_6H_5$	98
6	$3\text{-Br-}C_6H_5$	99
7	$4\text{-F-}C_6H_5$	99
8	1-naphthyl	95
9	2-naphthyl	98
10	3-pyridyl	97
11	3-thienyl	95

TABLE 20.2
Nitrilase II-Catalyzed Production of Aryllactic Acid Derivatives and Analogues Under DKR Conditions

Entry	R- Sidechain	ee (%)
1	$\text{-CH}_2\text{-}C_6H_5$	96
2	$\text{-CH}_2\text{-2-Me-}C_6H_5$	95
3	$\text{-CH}_2\text{-2-Br-}C_6H_5$	95
4	$\text{-CH}_2\text{-2-F-}C_6H_5$	91
5	$\text{-CH}_2\text{-3-Me-}C_6H_5$	95
6	$\text{-CH}_2\text{-3-F-}C_6H_5$	99
7	$\text{-CH}_2\text{-1-naphthyl}$	96
8	$\text{-CH}_2\text{-2-pyridyl}$	99
9	$\text{-CH}_2\text{-3-pyridyl}$	97
10	$\text{-CH}_2\text{-2-thienyl}$	96
11	$\text{-CH}_2\text{-3-thienyl}$	97

The nitrilase collection was similarly applied for the production of aryllactic acid derivatives through hydrolysis of the corresponding cyanohydrins. Phenyllactic acid and derivatives can serve as versatile building blocks for the preparation of numerous biologically active compounds.[24] Several enzymes in the nitrilase collection provided phenyllactic acid with high enantiomeric excess (Table 20.2). One enzyme, nitrilase II, was further characterized and afforded (S)-phenyllactic acid (**2**, R = PhCH$_2$), which was isolated in high yield and with 96% ee. This is a significant advance compared to the highest enantioselectivity previously reported for this biocatalytic conversion, which was 75% ee achieved through a whole-cell transformation using a *Pseudomonas* strain.[25] In addition, *ortho*- and *meta*-substituents were tolerated well by nitrilase II. Novel heteroaromatic derivatives, such as 2-pyridyl-, 3-pyridyl, 2-thienyl-, and 3-thienyllactic acids, were prepared with high conversions and enantioselectivities. It is interesting that *para*-substituents significantly lowered reaction rates, with full conversion taking more than 2 weeks under these conditions with nitrilase II. However, it is anticipated that another enzyme in the >200 member collection may be suitable for the transformation of *para*-substituted phenylacetaldehyde cyanohydrin substrates.

The nitrilase collection also was applied in a high-yielding desymmetrization process to access a key intermediate used to manufacture a major cholesterol-lowering drug. Nitrilase-catalyzed hydrolysis of one nitrile group of the prochiral substrate 3-hydroxyglutaronitrile (**3**) can afford (*R*)-4-cyano-3-hydroxybutyric acid (**4**) (Scheme 20.2), which, once esterified to ethyl-(*R*)-4-cyano-3-hydroxybutyrate, is an intermediate used in the commercial scale production of Pfizer's cholesterol-lowering drug, Lipitor™ (see also Chapter 31).[26] Previously reported attempts to use enzymes for this process provided 4-cyano-3-hydroxybutyrate with low enantioselectivity (22% ee) and the undesired *S*-absolute configuration.[27] Screening the nitrilase library uncovered 4 sequence-unique enzymes that generated the requisite (*R*)-4-cyano-3-hydroxybutyrate with quantitative conversion (>95%) and >90% ee. It is interesting that the same screening program also identified several nitrilases that afford the opposite enantiomer, the *S*-product in high enantiomeric excess. Thus, the extensive screen of biodiversity revealed enzymes that provide ready access to either enantiomer with high enantioselectivities, underscoring the advantage of having access to a large and diverse library of nitrilases.[20]

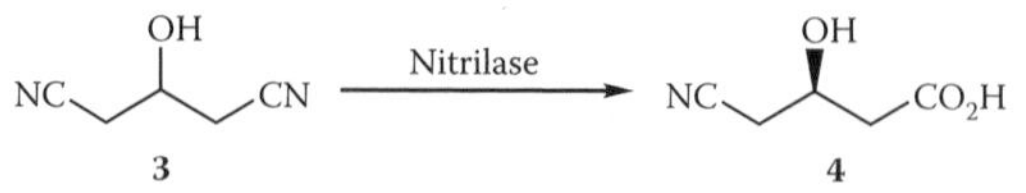

SCHEME 20.2

The most effective nitrilase identified in discovery efforts catalyzed the complete conversion of 3-hydroxyglutaronitrile (HGN) to (*R*)-4-cyano-3-hydroxybutyric acid (**4**) in 24 hours with a product ee of 95% at 100 mM substrate concentration.[20] However, as the substrate concentration was increased, substantially lowered enantiomeric excesses were observed. To develop a practical process, the objective was to identify a nitrilase that would catalyze the hydrolysis with high enantioselectivity at substrate concentrations up to 3M. At substrate concentrations of 0.5M, 1M, 2M, and 3M, enantiomeric excesses were 92.1%, 90.7%, 89.2%, and 87.6%, respectively, determined at reaction completion. Enzyme selectivity and productivity often are compromised at such high substrate concentrations, and in this case, 3M is 33% by volume of organic substrate. To develop a viable process that yielded (*R*)-4-cyano-3-hydroxybutyric acid (**4**) with high enantiomeric excess and high volumetric productivities we applied GSSM technology for the evolution of our lead nitrilase enzyme.

We exploited the degeneracy of the genetic code and made a 32 codon NNK (K= G,T) library containing every singe amino acid variant of our best WT enzyme.[28] The nitrilase GSSM library consisted of 10,528 genetic mutants and was screened completely using a mass spectroscopy assay where one of the two enantiotopic nitrile groups of hydroxyglutaronitrile was selectively labeled with ^{15}N such that enantiomeric products were isotopically differentiated and an ee determination can be made. Amino acid changes at several different residues led to enhanced enantioselectivity over the WT enzyme. Characterization of primary hits showed that residues Ala190 and Phe191 were enantioselectivity "hot spots" with several mutants affording product with higher ee. Each of these up-mutant enzymes was evaluated at molar substrate concentrations, and although many of the variants did not perform well at this higher substrate loading, the serine, histidine, and threonine variants at position 190 exhibited significant ee enhancement. The Ala190His variant was found to be the most selective and most active of the GSSM mutants allowing complete conversion to (*R*)-4-cyano-3-hydroxybutyric acid (**4**) at 3M substrate concentration in >98.5% ee within 15 hours. This is a dramatic improvement relative to WT, which yields **4** in only 88% ee after 24 hours under these conditions. The transformation was performed effectively on a multi-gram scale, and the desired intermediate **4** was generated with a volumetric productivity of 612 gL^{-1} d^{-1} and 98.5% ee using the A190H enzyme.

The unique benefits of GSSM evolution technology were demonstrated through identification and development of a nitrilase that provides practical access to a key commercial intermediate for the Lipitor drug. It is important to recognize that other evolution technologies are unlikely to have provided the desired A190H variant from our WT enzyme. For example, performing ePCR on our WT nitrilase would be statistically unlikely to furnish the improved A190H variant enzyme because two base changes in the Ala codon (GCN) are required to afford a His residue (CAT/C) at position 190. ePCR is a mutagenesis method that leads to single base mutations along a gene; such point mutations statistically would be unlikely to occur in adjacent base positions required for this residue change. Gene recombination methods also would be ineffectual because no currently reported or known nitrilases have a His residue at or near position 190.

20.4 CEPHALOSPORIN C AMIDASE

Semi-synthetic cephalosporins are among the most widely used antibiotics in the world. Many of these antibiotics are synthesized from 7-aminocephalosporanic acid (7-ACA) (**5**), a compound obtained through the deacylation of cephalosporin C (Ceph C) (**6**). Traditionally, this deacylation has been carried out using a chemical process. However, the chemical process involves the use of numerous toxic compounds that generate a costly chemical wastestream. For this reason, an enzymatic route for the production of 7-ACA from Ceph C is very appealing. Currently, enzymatic production of 7-ACA from Ceph C is accomplished using a 2-enzyme process.[29] The first enzyme, D-amino acid oxidase, is used to oxidize Ceph C to a keto acid intermediate that is then decarboxylated by hydrogen peroxide to yield glutaryl-7-ACA. Glutaryl-7-ACA is then deacylated to 7-ACA through the action of the second enzyme, glutaryl-7-ACA acylase. Although this 2-enzyme process currently is being used to manufacture 7-ACA, a single enzyme process using an amidase that directly converts Ceph C to 7-ACA would be much more efficient and therefore highly desirable (Scheme 20.3).

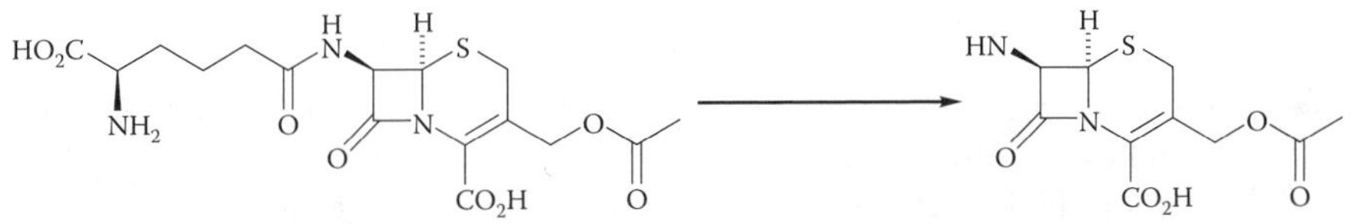

SCHEME 20.3

Previous efforts have failed to identify an enzyme with robust Ceph C amidase activity. Some glutaryl-7-ACA acylases can directly convert Ceph C to 7-ACA, but they do so with very poor efficiency and have not been considered for a single-enzyme manufacturing process.[30-33] Nonetheless, glutaryl-7-ACA acylases with measurable activity on Ceph C are classified as cephalosporin C acylases. Mutagenesis approaches such as ePCR have been used in an attempt to improve the activity of these enzymes on Ceph C, but only marginal improvements in the desired activity have been observed.[31] The crystal structure of one of these Ceph C acylases has been reported and provides some clues regarding their poor activity on Ceph C.[33] The conclusions drawn from this effort support the idea that the poor activity is a result of the inability of these enzymes to accommodate the D-2-aminoadipyl sidechain in the substrate-binding pocket. These results are not particularly surprising given that the initial steps of the 2-enzyme process for 7-ACA production are designed to convert the D-2-aminoadipyl sidechain of Ceph C to the glutaryl sidechain, the preferred substrate for Ceph C acylases. It has been suggested that extensive remodeling of the substrate binding pocket of these Ceph C acylases may be required to significantly improve its ability to accommodate the D-2-aminoadipyl sidechain.[33] Nonetheless, the available structural information has led to the development of new, targeted mutagenesis proposals hypothesized to improve the Ceph C amidase activity of these glutaryl-7-ACA acylases.[34,35]

Another approach to finding a one-enzyme solution is to discover a more active Ceph C amidase in nature, and this is the strategy that we have taken at Diversa. We have developed and implemented a proprietary high-throughput screening platform, known as GigaMatrix™ technology, for the discovery of rare bioactivities from our large collection of environmental libraries.[36,37] GigaMatrix technology uses a 100,000-well microplate with the footprint of a standard 96-well plate. The system uses an optical interface to visualize the plate and possesses an automated recovery system to break out primary hits for further characterization.[36] A Ceph C amidase assay amenable to the GigaMatrix platform was used to rapidly screen environmental libraries containing more than 10^6 members. Using this strategy, 13 novel D-specific amidases were discovered. One of the best enzymes was a unique gene product in a sequence class that is clearly distinct from previously discovered Ceph C/glutaryl-7-ACA acylases.

The objective was to identify an amidase that could efficiently convert Ceph C to 7-ACA at high substrate load with negligible product inhibition, preferably at neutral or slightly acidic pH, because Ceph C is less stable at alkaline pH. The most effective amidase discovered was initially characterized for its ability to convert Ceph C to 7-ACA. These results revealed that the recombinant enzyme possesses favorable pH performance and good substrate conversion at low substrate concentration (10mM, pH 7.5, >95% substrate conversion). This enzyme and several other new amidases are being further evaluated at higher substrate concentrations to identify a lead candidate for a 1-enzyme hydrolysis of Ceph C. Preliminary results indicate that enzyme activity is reduced at higher substrate concentrations. Accordingly, *in vitro* optimization using Diversa's evolution platforms will be implemented for development of an ideal Ceph C amidase for this process. Overall, discovery from nature provided an excellent starting point for this evolution program because Ceph C amidases exhibiting substantial activity even at low concentration were unavailable previously.

20.5 ALDOLASES

The aldol reaction is extremely useful in organic synthesis as a way of generating new carbon–carbon bonds and introducing up to two new stereocenters into a molecule (Scheme 20.4). Controlling the stereochemistry of aldol reactions can be very challenging and has been an area of intense research. Although several elegant asymmetric methods have been developed,[38] there remains a need for additional methods, and in recent years enzyme-catalyzed approaches using aldolases have received attention as alternatives to chemical methodologies.[39]

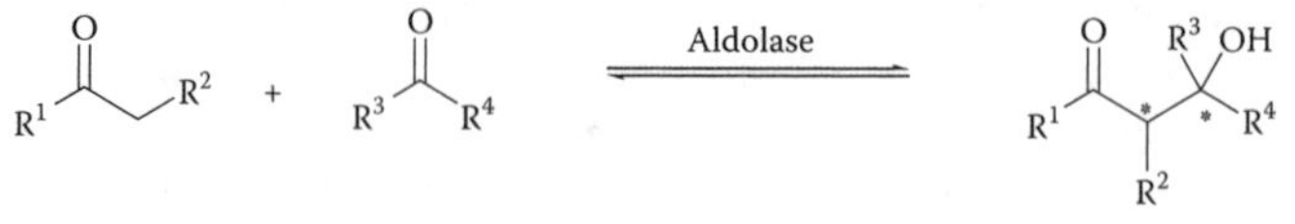

SCHEME 20.4

We have developed a process for synthesis of key intermediates in the preparation of statin cholesterol-lowering drugs using a deoxyribose-5-phosphate aldolase (DERA). Our work builds on the pioneering efforts of Gijsen and Wong,[40] who observed that DERA can catalyze the stereo-selective tandem condensation of two equivalents of acetaldehyde with a series of aldehyde acceptors to afford 6-membered lactol derivatives. Because the DERA-catalyzed reaction is an equilibrium process, the intermediate 4-carbon adduct is reversibly formed under the reaction conditions. The second condensation between this intermediate and a second equivalent of acetaldehyde drives the equilibrium favorably as a result of the stability of the cyclized lactol form of the product (Scheme 20.5). The elegance and power of this approach for statin synthesis was evident; however, serious challenges remained. For example, the reported process required extensive (3-day) reaction times and high enzyme loading (20% wt/wt) and suffered from substrate inhibition.

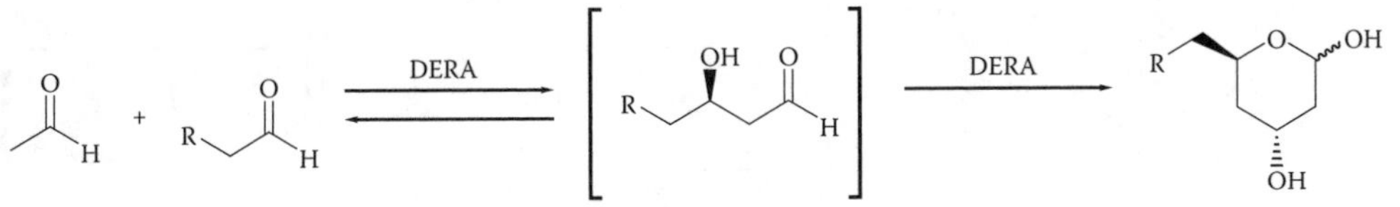

SCHEME 20.5

To make the DERA-catalyzed process commercially attractive, improvements were required in catalyst load, reaction time, and volumetric productivity. We undertook an enzyme discovery program, using a combination of activity- and sequence-based screening, and discovered 15 DERAs that are active in the previously mentioned process. Several of these enzymes had improved catalyst load relative to the benchmark DERA from *E. coli*. In the first step of our process, our new DERA enzymes catalyze the enantioselective tandem aldol reaction of two equivalents of acetaldehyde with one equivalent of chloroacetaldehyde (Scheme 20.6). Thus, in 1 step a 6-carbon lactol with two stereogenic centers is formed from achiral 2-carbon starting materials. In the second step, the lactol is oxidized to the corresponding lactone **7** with sodium hypochlorite in acetic acid, which is crystallized to an exceptionally high level of purity (99.9% ee, 99.8% de).

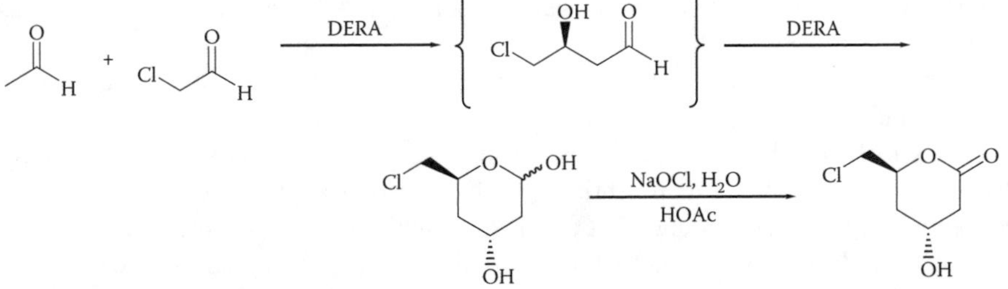

SCHEME 20.6

Lactone **7** can serve as a versatile starting material for the synthesis of a range of HMG-CoA (3-hydroxy-3-methylglutaryl coenzyme A) reductase inhibitors, commonly referred to as statins, which includes Lipitor (**8**) and Crestor™ (**9**) compounds (Scheme 20.7). In a single transformation, the lactone ring can be opened and the chloride can be displaced by either cyanide or hydroxide to access advanced Lipitor or Crestor intermediates, respectively.

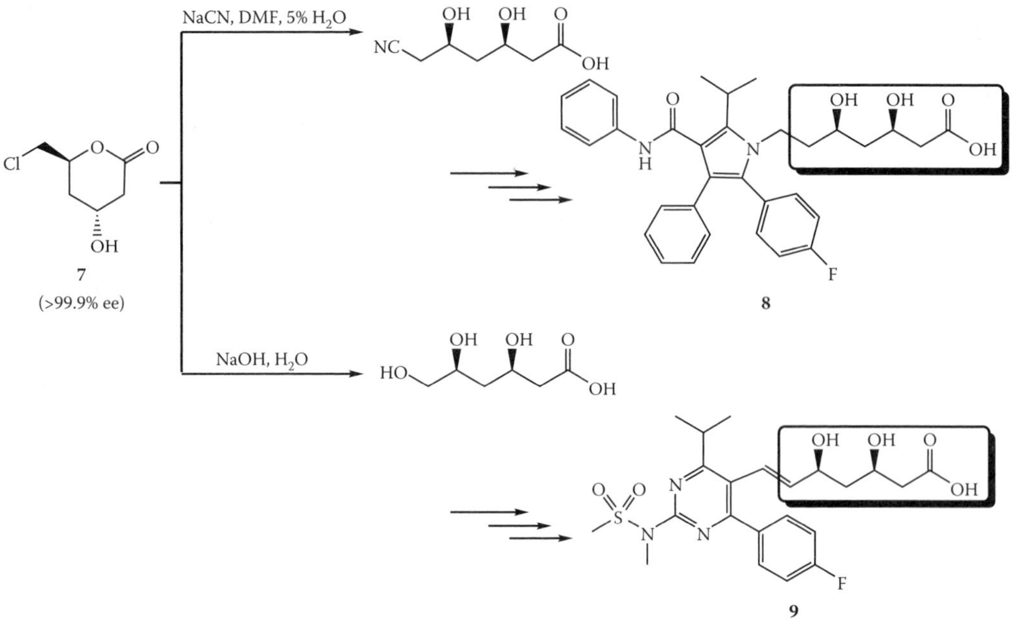

SCHEME 20.7

Overall, we have discovered new DERA enzymes that have allowed the development of a useful process for production of key intermediates for the preparation of statin drugs. Through the use of our best DERA enzyme combined with process development, we have improved catalyst load from 20% w/w to 2% w/w, improved reaction time from several days to 3 hours, and improved volumetric productivity from 12 g/L to 115 g/L, relative to the previously published procedures.

20.6 EPOXIDE HYDROLASES

Epoxide hydrolases (EHs) (E.C. 3.3.2.x) are capable of selectively hydrolyzing racemic or *meso*-epoxides to produce single-enantiomer epoxides or vicinal diols that can serve as key building blocks for the synthesis of a range of pharmaceutical compounds.[41] On a commercial scale, chemical methods for the synthesis of chiral epoxides and diols often suffer from limitations, the most important of which are a lack of substrate scope and the utilization of expensive and sensitive metal catalysts. In contrast, microbial EHs do not require cofactors and, in principle, are not limited to particular classes of epoxides as substrates. Because no single enzyme is likely to function optimally on all types of epoxides, the full potential of EHs in varied applications will only be realized through creation of a library or collection of enzymes with differing substrate scopes. Fewer than 15 microbial epoxide hydrolases had been reported and still fewer had been carefully studied, hampering the development of EHs in the pharmaceutical and chemical industries. To overcome this limitation, Diversa initiated an epoxide hydrolase program. Several high-throughput screening assays were developed and were used to screen eDNA libraries to discover novel EHs. To date, more than 50 novel microbial EHs have been discovered. On examination of the newly discovered sequences, it became apparent that the Diversa EHs are highly diverse and represent a wide range of sequence-space. Thus, we were keen to test whether the sequence-diversity of our EH library extended to useful functional diversity of the catalysts.

The majority of our EHs have been subcloned and expressed in bacterial hosts for screening. To probe the functional diversity of these enzymes, several substrates representing alkyl and aryl as well as terminal and internal epoxides were used to screen the enzyme library. A few examples are shown in Schemes 20.8 (kinetic resolutions of *rac*-epoxides) and 20.9 (desymmetrization of *meso*-epoxides).*

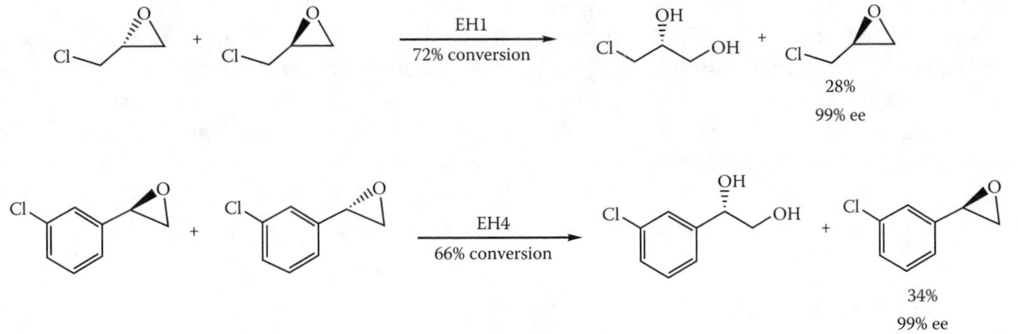

SCHEME 20.8

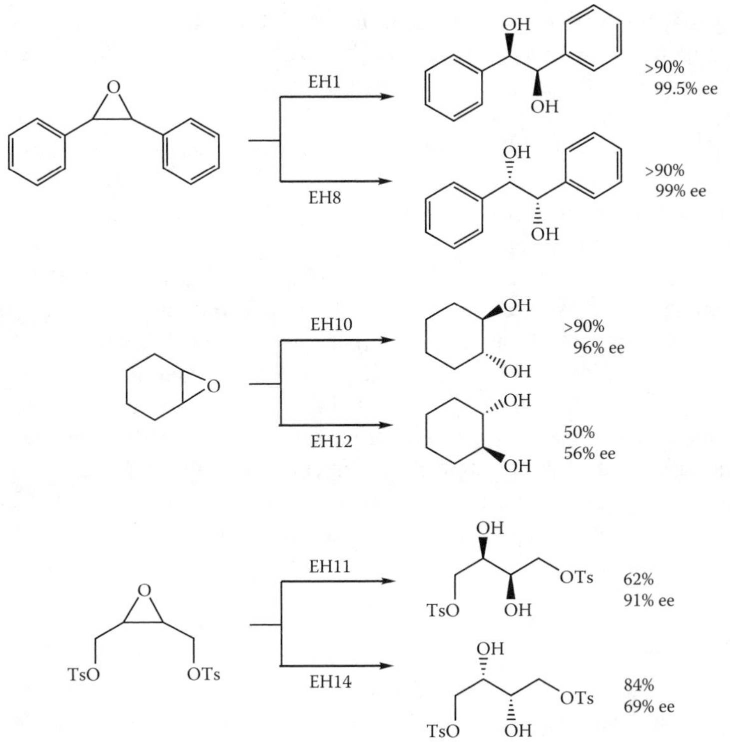

SCHEME 20.9

   Active hits were found for every type of substrate screened, including those for which other known microbial epoxide hydrolases were ineffective. For example, hydrolysis of *cis*-stilbene oxide was not successful with several microbial EHs tested previously.[42,43] By contrast, several of our new enzymes actively hydrolyzed this substrate and exhibited excellent enantioselectivities (>99% ee). It is important to note that these enzymes were found to be capable of selectively hydrolyzing a wide range of *meso*-epoxides, including cyclic and acyclic alkyl- and aryl-substituted substrates.

---

* Results in these schemes are from preliminary enzyme testing and reactions are not optimized. Note that yields are based on chromatographic results, not isolated products.

The corresponding chiral diols were furnished with very high ee's and yields in these processes. In kinetic resolution of terminal epoxides, even with short-chain aliphatic epoxides such as epichlorohydrin, EHs were found with reasonable enantioselectivities (E up to 10) compared with previously reported enzymes that had poor selectivities (E<2).[44] In all cases studied, enzymes were found that had complementary enantioselectivities giving access to either enantiomer of the epoxides in kinetic resolution reactions or either enantiomer of the diols in desymmetrization reactions. It is worth pointing out that the enantioselectivities of promising Diversa EHs may be significantly improved through further reaction engineering or by directed evolution. Given the wide range of sequence divergence and high functional diversity of the enzymes, the Diversa EH library represents a valuable platform for the synthesis of chiral epoxides and diols.

20.7 SUMMARY

Application of biocatalysis in the laboratory and on multi-ton commercial scales is dramatically increasing as a result of scientific advances on several fronts. New methods for mining biodiversity have provided access to a growing collection of novel off-the-shelf enzymes that may be rapidly screened for a given application. As the range of newly discovered enzymes increases, we also are observing an expanding range of transformations that may be catalyzed effectively by biocatalysts. In many cases, enzymatic transformations are being developed for processes where traditional chemical methods perform poorly. Sophisticated and robust evolution technologies have been developed to optimize and tailor the properties of enzymes for a transformation of interest. To efficiently identify enzymes with desirable properties, high-throughput screening tools and assays that greatly expedite the selection process have been developed. Finally, new host organisms are being developed for the vital activity of producing enzymes in a cost-effective manner. The increased understanding of the molecular interactions that govern enzyme-substrate specificity and the availability of burgeoning databases that can serve to guide the appropriate choice of a biocatalyst also are contributing to the growing prominence of biocatalysis in chiral fine chemical production. In this chapter, we have outlined our efforts to develop new technologies that can be applied to biocatalyst development. In particular, we have established a two-pronged strategy for biocatalyst development, which entails enzyme discovery from natural biodiversity combined with effective evolution technologies that allow the honing of enzyme properties for a given application. The examples provided serve as a testament to the power of this approach, and it is anticipated that these robust methods will lead to a wide array of novel transformations for the production of chiral fine chemicals in the future.

ACKNOWLEDGMENTS

The authors would like to thank their respective teams for their important contribution to this work. In addition Jay Short, Dan Robertson, Keith Kretz, Mike Lafferty, Xuqiu Tan, Lisa Bibbs, Pat Simms, Eric Mather, Brian Morgan, Gerhard Fry, and their teams are acknowledged for their contribution to the projects described in this review.

REFERENCES

1. Food and Drug Administration Policy Statement for the Development of New Stereoisomeric Drugs. *Chirality* 1992, *4*, 338.
2. Rouhi, A. M, *Chem. Eng. News* 2003, *81*, 56.
3. Gray, K. A., Richardson, T. H., Robertson, D. E., Swanson, P. E., Subramanian, M. V. *Adv. Appl. Microbiol.* 2003, *52*, 1.
4. Drauz, K., Waldmann, H., Roberts, S. R. Eds. *Enzyme Catalysis in Organic Synthesis,* Wiley-VCH: Weinheim, Germany, 2nd ed., 2002.

5. Schmidt, A., Dordick, J. S., Hauer, B., Kiener, A., Wubbolts, M., Witholt. B. *Nature* 2001, *409*, 258.
6. Pace, N. R. *Science* 1997, *276*, 734.
7. Bohannan, B, J. M., Hughes, J. *Curr. Opin. Microbiol.* 2003, *6*, 282.
8. Zengler, K., Toledo, G., Rappé, M,, James Elkins, J., Mathur, E. J., Short, J. M., Keller, M. *Proc. Natl. Acad. Sci. USA* 2002, *99*, 15681.
9. Short, J. M. *Nature Biotech.* 1997, *15*, 1322.
10. Robertson, D. E., Mathur, E. J., Swanson, R. V., Marrs, B. L., Short, J. M. *SIM News* 1996, *46*, 3.
11. Cadwell, R. C., Joyce G. F. *PCR Methods Appl.* 1992, *2*, 28.
12. Greener, A., Callahan, M., Jerpseth, B. *Methods in Molecular Biology* 1996, *57*, 375.
13. Maniatis, T., Fritsch, E. F., Sambrook, J., In *Molecular Cloning: A Laboratory Manual,* Cold Spring Harbor Laboratory: Cold Spring Harbor, NY, 1982.
14. Short, J. M. U.S. Pat. 6,562,594, 2003.
15. Gray, K. A., Richardson, T. H., Kretz, K., Short, J. M., Bartnek, F., Knowles, R., Kan, L., Swanson, P. E., Robertson, D. E. *Adv. Synth. Catal.* 2001, *343*, 607.
16. Jaeger, K. E., Eggert, T., Eipper, A., Reetz, M. T. *Appl. Microbiol Biotechnol.* 2001, *55*, 519.
17. Zhang, J. H., Dawes, G., Stemmer, W. P. *Proc. Natl. Acad. Sci, USA* 1997, *94*, 4504.
18. Short, J. M., Djavakhishvili, T. D., Frey, G. J. U.S. Pat. 6,361,974, 2002.,
19. Pace, H., Brenner, C. *Genome Biology* 2001, *2*, 0001.1-0001.9.
20. DeSantis, G., Zhu, Z., Greenberg, W. A., Wong, K., Chaplin, J., Hanson, S. R., Farwell, B., Nicholson, L. W., Rand, C. L., Weiner, D. P., Robertson, D. E., Burk, M. J. *J. Am. Chem. Soc.* 2002, *124*, 9024.
21. Inagaki, M., Hiratake, J., Nishioka, T., Oda, J. *J. Org. Chem.* 1992, *57*, 5643.
22. Ress-Loschke, M., Friedrich, T., Hauer, B., Mattes, R., Engels, D. PCT Appl. WO 00/23577, 2000.
23. Yamamoto, K., Oishi, K., Fujimatsu, I., Komatsu, K. *Appl. Environ. Microbiol.* 1991, *57*, 3028.
24. Coppola, G. M., Schuster, H. F. *Chiral α-Hydroxy Acids in Enantioselective Synthesis,* Wiley-VCH: Weinheim, Germany, 1997.
25. Hashimoto, Y., Kobayashi, E., Endo, T., Nishiyama, M., Horinouchi, S. *Biosci. Biotech. Biochem.* 1996, *60*, 1279.
26. Johnson, F., Panella, J. P., Carlson, A. A. *J. Org. Chem.* 1962, *27*, 2241.
27. Beard, T., Cohen, M. A., Parratt, J. S., Turner, N. J. *Tetrahedron: Asymmetry* 1993, *4*, 1085.
28. DeSantis, G., Wong, K., Farwell, B., Chatman, K., Zhu, Z.: Tomlinson, G., Huang, H., Tan, X., Bibbs, L., Chen, P., Kretz, K., Burk, M. J. *J. Am. Chem. Soc.* 2003, *125*, 11476.
29. Monti, D., Carrea, G., Riva, S., Baldaro, E., Frare, G. *Biotechnol. Bioengin.* 2000, *70*, 239.
30. Matsuda, A., Matsuyama, K., Yamamoto, K., Ichikawa, S., Komatsu, K. J. *Bacteriology* 1987, *169*, 5815.
31. Ishii, Y., Saito, Y., Fujimura, T, Sasaki, H., Noguchi, Y., Yamada, H., Niwa, M., Shimomura, K. *Eur. J. Biochem.* 1995, *230*, 773.
32. Li, Y., Chen, J., Jiang, W., Mao, X., Zhao, G., Wang, E. *Eur. J. Biochem.* 1999, *262*, 713.
33. Kim, Y., Yoon, K, Khang, Y., Turley, S., Hol, W. *Structure* 2000, *8*, 1059.
34. Kim, Y, Hol, W. *Chem. Biology* 2001, *8*, 1253.
35. Fritz-Wolf, K., Koller, K., Lange, G., Liesum, A., Sauber, K., Schreuder, H., Aretz, W., Kabsch, W. *Protein Science* 2002, *11*, 92.
36. Lafferty, M., Dycaico, M. *Methods in Enzymology* 2004, *388*, 119.
37. Short, J. M., Keller, M., Lafferty, W. M. PCT Int. Appl. WO 02/31203, 2002.
38. Sakthivel, K., Notz, W., Bui, T., Barbas III, C. F. *J. Am. Chem. Soc.* 2001, *123*, 5260, and references cited therein.
39. Silvestri, M. G., Desantis, G., Mitchell, M., Wong, C.-H. *Topics Stereochem.* 2003, *23*, 267.
40. Gijsen, H. J. M., Wong, C.-H. *J. Am. Chem. Soc.* 1994, *116*, 8422, Gijsen, H.J.M., Wong, C.-H. *J. Am. Chem. Soc.* 1995, *117*, 7585.
41. Archelas, A., Furstoss, R. *Ann.. Rev. Microbiol.* 1997, *51*, 491.
42. Rink, R., Fennema, M., Smids, M., Dehmel, U., Janssen, D. B. *J. Biol. Chem.* 1997, *272*, 14650.
43. Morisseau, C., Archelas, A., Guitton, C., Faucher, D., Furstoss, R., Baratti, J. C. *Eur. J. Biochem.* 1999, *263*, 386.
44. Spelberg, J. H. L., Hylckama Vlieg, J. E. T., Bosma, T., Kellogg, R. M., Janssen, D. B. *Tetrahedron: Asymmetry* 1999, *10*, 2863.

21 One-Pot Synthesis and the Integration of Chemical and Biocatalytic Conversions

Paul A Dalby, Gary J Lye, and John M Woodley

CONTENTS

21.1 INTRODUCTION

Biocatalysis is becoming an established method to assist in the manufacture of synthetic targets and, in particular, chiral fine chemicals.[1-3] The ability of enzymes (whether isolated or within an intact cell) to cause resolution or synthesis of chiral centers is without precedent, and there is consequently a strong drive to implement such processes in industry.

Most fine chemical syntheses involve multiple steps including protection, catalysis, and deprotection operations. In those cases where exquisite enantio- or regioselectivity or specificity is demanded, then biocatalysis will be considered. The mild conditions often used for biocatalysis confer the added advantage of overcoming the need for protection and deprotection in many cases.[4] Nevertheless, the enzymatic process will usually be one or at best a few steps, in an otherwise 10–20-step synthesis. Consequently, biocatalytic reactions are most normally preceded and followed by a chemical conversion. With this in mind, implementation needs to consider the integration of the biocatalytic step with the neighboring operations.

The integration of two unit operations lies at the heart of process engineering. More often in bioprocesses it is the integration of product formation with the following recovery steps that is critical.[5] In the specific case of biocatalytic processes the product recovery is also critical, but in this chapter the focus will be on the integration of the surrounding chemical steps with the biocatalysis.

21.2 EFFECTS OF ONE OPERATION ON ANOTHER

Chemical and biocatalytic steps are frequently incompatible as a result of reagents and media. These effects are, in part, dependent on the order of the operations (i.e., if chemical catalysis precedes or follows biocatalysis). Whether reactions are carried out in a sequential or one-pot mode, the interaction of one step with the following one is critical. Where a chemical step precedes the biocatalytic one, then the following considerations need to be taken into account in the biocatalytic reaction:

- Extreme pH or temperature
- Organic solvent–based medium
- Residual levels of chemical catalyst
- Diluents or reagents present in excess used to drive equilibrium
- High concentrations of the target intermediate (i.e., product of chemistry and reactant of biocatalysis)

Where the biocatalytic step precedes the chemical one, then there are also issues that need to be taken into account in the chemical reaction:

- Water-based medium
- Residual protein levels (from catalyst)
- Low concentrations of intermediates (i.e., product of biocatalysis and reactant of chemistry)
- For whole-cell catalyzed reactions, other extracellular metabolites or cellular debris

Any or all of these conditions may limit productivity, and whichever of the previously listed sequences occur in a particular process, the biocatalytic conversion needs to be selected and designed to cope with these issues. Such issues may be solved by process techniques (using purification or a change of medium/catalyst between the reactions) or potentially by alteration of the properties of the biocatalyst via selection or directed evolution. A further approach is to combine operations in a single-pot operation using process techniques, biocatalyst modification, and a degree of compromise.

21.3 INTEGRATION PRINCIPLES

21.3.1 Rationale for Integration

The use of operations in a single pot may require compromise of process conditions and complex operation. Hence the logic for complete integration should be established ahead of evaluation.

21.3.1.1 Kinetic Considerations

The first group of biocatalytic reactions where integration needs to be considered includes those cases where one (or more) reactant(s) or one (or more) product(s) are damaging to the biocatalyst. They may be toxic to a microbial catalyst, denature an enzyme, or result in unfavorable kinetics; in each case this will demand controlled concentrations in the environment of the catalyst itself. A range of techniques from the use of feeding, an auxiliary phase,[6] or *in situ* product removal[5,7] can be used to help control the concentration. A further approach is to integrate the biocatalytic step with the preceding or following step such that the reactant is generated *in situ* or the product is removed reactively. Clearly such an approach demands that the biocatalysts can tolerate the alternative conditions/reagents of the neighboring operation. In these cases a high rate and therefore product-to-catalyst ratio can be achieved.

21.3.1.2   Thermodynamic Considerations

Many chemical and biocatalytic conversions involve reactions with an unfavorable equilibrium such as condensations.[8] In both cases the Law of Mass Action will apply such that the removal of one species from the reaction mixture will shift the equilibrium position. This is particularly useful for biocatalytic conversions where the alternative approach of using a reactant in excess may have deleterious effects on the biocatalyst.

Likewise, conversions where a reactant is unstable should be linked with the previous step to keep the yield as high as possible, and conversions where the product is unstable should be linked with the following step to keep the yield as high as possible. In addition, reactions requiring deracemization to take advantage of the unreacted enantiomer will also benefit from one-pot operation to improve the yield. In each of these cases it is an increase in the thermodynamic yield of product on substrate rather than any kinetic advantage that is gained.

21.3.1.3   Engineering, Economic, and Environmental Considerations

From a process-engineering standpoint there are further considerations that make integrated one-pot syntheses attractive. The various bases by which potential benefits can be quantified and compared have been defined previously.[5] The yield of individual conversions can be increased by minimizing degradation of unstable intermediate products (as described in Section 21.3.1.1) or by shifting an unfavorable reaction equilibrium (as described in Section 21.3.1.2). Overall process yields can be increased by reducing the total number of process steps and the associated handling losses in each piece of a plant process.[9] In the case of integrated reactions the total number of steps is reduced because it is not necessary to isolate an intermediate product after the initial conversion. Such advantages can be gained in any pair of integrated reactions regardless of whether the conversions are purely chemical, purely biological, or a combination of the two as described here. Integrated processes can also help reduce the overall process time (and associated manpower costs), particularly at large scale where the times taken to fill and empty successive process vessels can be considerable.

Economic and environmental benefits also arise from the adoption of one-pot syntheses. A number of case studies have shown decreased capital and operating costs arising from the use of integrated reaction–separation systems.[10,11] Although such detailed analyses are yet to be presented for integrated reaction systems, the underlying assumptions and calculation procedures are identical, so similar financial savings can be envisaged. The high level of waste generation in the fine chemical and pharmaceutical sectors, estimated to be between 5–50 and 25–100 kg (waste)/kg (product), respectively, is also now forcing companies to consider the E-factor and atom efficiency of synthetic routes.[12] Here biocatalytic reactions can help as a result of the avoidance of protection and deprotection steps, whereas integrated reactions can minimize the amount of waste solvent or contaminated resin generated during the isolation of intermediates.

21.3.2   ONE-POT SYNTHESIS AND ONE-POT PROCESSING

Where chemical and biocatalytic steps are operated in a single reactor, the gains may be considerable. To describe such operations the term "one-pot synthesis" is increasingly being used in the literature. However, close inspection of the details being reported suggests that there are currently a number of interpretations of what is considered a one-pot process. At one extreme is the ideal notion of multiple chemical/biocatalytic reactions occurring simultaneously with no need to isolate intermediate products or change reaction conditions. At the other extreme, various chemical and biocatalytic reactions occur in one pot but there may be some purification of the intermediate product, or change of solvent, between two essentially sequential reactions. Table 21.1 attempts to more precisely define what is meant by "one-pot synthesis" and gives some relevant published examples. It also introduces the terms "sequential one-pot synthesis" and "one-pot processing" to

TABLE 21.1
Definition and Operational Characteristics of One-Pot Syntheses and One-Pot Processing

Process Definition	Operational Characteristics	Examples	Refs
1: ("True") one-pot synthesis	Multiple reactions occurring simultaneously	Domino reactions	13
	Constant reaction conditions (e.g., solvent, pH, T°C)	*In situ* racemization	14
	No isolation of intermediate products		
2: Sequential one-pot synthesis	Reactions occur sequentially	Regioselective enzymatic	15
	Variable reaction conditions (e.g., pH, T°C)	protection and chemical	
	No isolation of intermediate products	acetylation	
3: One-pot processing[a]	Reactions occur sequentially	Lotrafiban production	16
	Variable reaction conditions (e.g., solvent, pH, T°C)		
	Partial purification of intermediates		

[a] Usually requires custom reactor configurations (e.g., stirred reactor with integrated membrane separation or distillation capabilities).

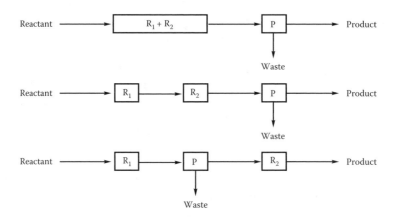

FIGURE 21.1 Flow-sheet options for integration. **1:** ("True") one-pot synthesis, **2:** Sequential one-pot synthesis; **3:** One-pot processing. For definitions see Table 21.1. Where R_1, R_2, and P represent Reaction 1, Reaction 2, and Purification, respectively.

more precisely describe situations in which reaction conditions may be changed between steps or there is some degree of intermediate purification, respectively. Figure 21.1 illustrates the possible flow sheets with these three forms of integration.

Although operation in a single pot has many advantages, it may prove intractable to alter the biocatalyst to perfectly match the chemical environment. In such cases a compromise may need to be reached such that neither the chemical or biocatalytic operations are operating optimally but under a compromised condition. Such conditions must still prove beneficial in relation to operation in separate reactors, each under optimal conditions.

21.3.2.1 Biocatalyst Environment: Types of Chemistry

Choice of process options for the successful integration of biocatalytic and chemical synthesis steps is heavily dependent on the type of chemical reaction used prior to or subsequent to the biocatalytic step and also on the robustness of the biocatalyst to inactivation by components of the chemical step. The use of organic solvent or highly reactive chemical reagents before subsequent conversion using a biocatalyst is usually expected to require the complete purification of the product of the chemical step. There are many examples in the literature where such an approach has been applied, including the use of chromatography after a Grignard reaction in tetrahydrofuran (THF),[17] ozonolysis in dichloromethane and methanol,[18] or after methylation with diazomethane-silica gel in ether.[19] Similarly, the aqueous environment required for a biocatalytic step must often be replaced by extraction of the target compound into an organic phase,[17,18] or simply by evaporation,[20] to allow subsequent water-sensitive chemical reactions to proceed.

Despite the previously mentioned restrictions, examples are emerging in which a closer integration is achieved by using enzymes that can operate in organic solvents or by using chemistry compatible with aqueous solvents. Lipases, for example those from *Candida antartica* and *Pseudomonas sp.*, can be used in organic solvents such as cyclohexane, *t*-BuOH, THF, and even triethylamine, either in free solution or immobilized.[15,21-23] They have been used successfully in one-pot reactions in the presence of acetone cyanohydrin with strongly basic anionic exchange resins, Pd–C, organometallic ruthenium, aluminum, and iridium catalysts.

The use of enzymes in organic solvents can also take advantage of spontaneous air oxidations, rearrangements, or cycloadditions, such as the air-oxidation of catechol produced from phenol by tyrosinase, to form *ortho*-quinone, followed by a Diels-Alder addition to various dienophiles in chloroform.[13] It is interesting that an attempt to allylate the final product with the Lewis acid Sc(CF$_3$SO$_3$)$_3$ and (Allyl)$_4$Sn would not proceed unless the enzyme and excess dienophile were removed first.

As described previously, the conditions of the chemical step can completely deactivate the biocatalyst, but they could also modify the functional group intended for transformation by the enzyme. One-pot aqueous reactions can still be achieved in these cases, but the chemical reagents must be added after completion of the enzyme reaction. For example, in the chemo-enzymatic synthesis of 2′-deoxynucleosides, *Candida antartica* lipase is used in dry THF to regioselectively propionylate 2-deoxy-D-ribose at the primary hydroxyl group. This is followed by the addition of acetic anhydride and pyridine to acetylate the remaining free hydroxyl groups, only after the primary hydroxyl group has been blocked.[15] Although only used for preparative chemistry with an expensive single use of the enzyme, such reports indicate potential integration if the enzyme could be made robust enough for multiple cycles.

Many chemical transformations can be adapted for use in water and are therefore potentially compatible with the preferred conditions for biocatalysis. The possibility of performing one-pot reactions with aqueous chemical steps alongside biocatalysis has opened up the possibility of driving reversible enzyme reactions to completion,[24] enhancing enzyme-substrate binding through reversible modification with arsenates,[25] and achieving deracemizations or dynamic kinetic resolutions (DKRs) with boranes or mild bases.[14,26-28]

The use of aqueous chemical steps in one pot with biocatalysis can often lead to gradual deactivation of the biocatalyst, thus limiting its reusability. For example, although vinyl esters as acyl donors can make an enzymatic transesterification irreversible, the resulting formation of an aldehyde by-product can lead to enzyme deactivation through the formation of a Schiff's base with lysine sidechains.[24]

21.4 PRACTICAL EXAMPLES

In the following section some specific examples of one-pot syntheses are described together with the rationale for adopting such approaches. These are summarized in Table 21.2. The examples have been divided between approaches that are applicable for use in aqueous environments and those that must be used in organic solvents.

21.4.1 CHEMISTRY IN AQUEOUS ENVIRONMENTS

An emerging technique that takes advantage of the compatibility of aqueous chemistry and biocatalysis is the use of *in situ* reactions to promote deracemizations or DKRs.[31] For example, reductive

TABLE 21.2
Summary of Published Approaches to One-Pot Syntheses Showing the Target Classes of Compounds and the Rationale for Process Integration

Reaction/Product	Process Step(s)	Solvent	Chemical Step(s)	Enzyme Step(s)	Refs
Aldol addition with unphosphorylated substrate	One-pot	H_2O	Dihydroxyacetone + $HOAsO_3^{2-}$	Aldolase (FruA) + R-CHO	25
Deracemization of amines	One-pot	H_2O	Ammonia–borane	Preevolved amine oxidase	26
Deracemization of amino-acids	One-pot	H_2O	$NaCNBH_3$ or $NaBH_4$	Amino acid oxidase	2,7,14
Irreversible transesterification using enol esters	One-pot	H_2O	Enol tautomerization	Lipase	24
DKR of thioesters for carboxylates	One-pot	H_2O, MeCN	Trioctylamine	Subtilisin Carlsberg	28
DKR of thioesters for oxo-esters	One-pot	Toluene	Triethylamine	CA lipase + n-BuOH	28
DKR of aldehydes via cyanohydrins for cyanohydrin acetates	One-pot	$(iPr)_2O$	Acetone cyanohydrin + anion exchange resin (OH- form)	Pseudomonas sp. lipase + isopropenyl acetate	29
Enantioselective synthesis of 3-hydroxypiperidin-2-ones	Purification One-pot sequential Purification Purification One-pot	MeOH, CH_2Cl_2 H_2O H_2O Not given Not given Not given	1) O_3 (ozonolysis) 4) CH_2N_2 5) PtO_2, H_2 6) Spontaneous cyclization	2) CR lipase 3) LDH + FDH + NADH	30
Catechol synthesis from glucose-6-phosphate	One-pot processing	AcOH	2) Conc. HI, 70°C, under Ar	1) DOIS +NAD	20
2'-deoxynucleosides	One-pot sequential	Dry THF	2) Ac_2O, pyr	1) CA lipase + propionic anhydride	15
DKR of phenethylamine acetylation	One-pot	Et_3N	Pd–C,H_2, 50°C, under Ar	Immobilized CA lipase + AcOEt	23
DKR of alcohol acetylation (e.g., (R)-1-indanol)	One-pot	t-BuOH	Organometallic Ruthenium catalyst for alcohol racemization, 70°C, under Ar, 1 eq. ketone	CA lipase + ROAc	22
DKR of alcohol acetylation	One-pot	C_6H_{12} or CH_2Cl_2, and vinyl acetate	Al, Ir, or Rh catalysts for secondary alcohol racemization with up to 1 eq. ketone	Pseudomonas sp. lipase or Pseudomonas fl. lipase at 20–80°C	21
Hydroxylation-oxidation-Diels-Alder domino reaction	One-pot One-pot Purification	$CHCl_3$ $CHCl_3$ $CHCl_3$ $CHCl_3$	2) Air oxidation of catechol to ortho-quinone 3) Diels-alder with dienophiles 4) Allylation with Lewis acid $Sc(CF_3SO_3)_3$ and $(Allyl)_4Sn$	1) Tyrosinase for phenol to catechol	13

CA, *Candida antarctica*; CR, *Candida rugosa*; DKR, dynamic kinetic resolution; DOIS, 2-deoxy-cyllo-inosose synthase; LDH, lactate dehydrogenase; THF, tetrahydrofuran.

aminations using borohydride, cyanoborohydride, or amino-boranes have been used in one-pot reactions alongside amine oxidases to deracemize α-amino-acids and α-methylbenzylamine.[14,26,27] A D-amino-acid oxidase was used to form an imine from the corresponding D-amine of a racemic mixture (Scheme 21.1) (see also Chapter 2).[26] The borane then reduced the imine *in situ* to form both the D- and L-amine. In this manner, the original racemic amine can be deracemized to the L-amine in high yield and enantiomeric excess.

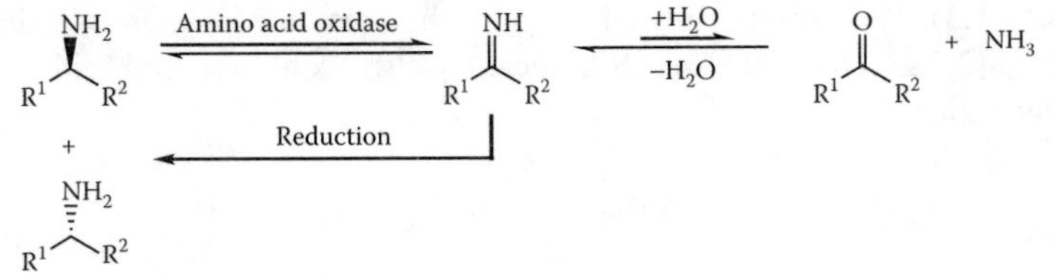

$$NH_2 \xrightarrow{\text{Amino acid oxidase}} NH \underset{-H_2O}{\overset{+H_2O}{\rightleftharpoons}} O + NH_3$$

$$NH_2 \xleftarrow{\text{Reduction}}$$

SCHEME 21.1

DKR of thioesters (Scheme 21.2) with a chiral center at the α-carbon has been achieved in a water/acetonitrile biphasic system by racemization with mild organic bases, such as trioctylamine, coupled to enantioselective hydrolysis of the thioester with subtilisin Carlsberg.[28] Such an approach can be applied to a wide variety of thioesters but not oxoesters, which have less acidic α-protons.

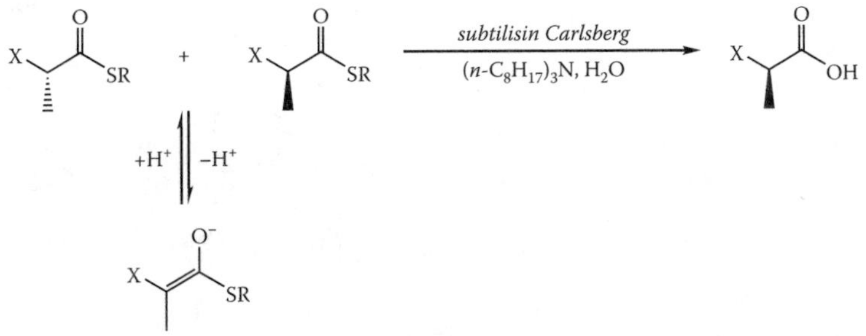

$$X\text{-CO-SR} + X\text{-CO-SR} \xrightarrow[\;(n\text{-}C_8H_{17})_3N,\;H_2O\;]{\text{subtilisin Carlsberg}} X\text{-CO-OH}$$

$$+H^+ \Updownarrow -H^+$$

SCHEME 21.2

21.4.2 CHEMISTRY IN ORGANIC ENVIRONMENTS

The DKR hydrolysis of thioesters described previously has also been extended to transesterifications in toluene, using triethylamine and *Candida antartica* lipase.[28] This general approach can therefore be applied to the resolution of a wide range of both water-soluble and water-insoluble thioesters by selecting an appropriate solvent, base, and enzyme system.

Dinh and co-workers have shown that transition metal and aluminum catalysts can be used for DKR during the enantioselective biotransformation of alcohols to esters.[21] *Pseudomonas fluorescens* lipase catalyzes the transesterification of R-phenethyl alcohol with vinyl acetate in dichloromethane or cyclohexane. The temporary oxidation and reduction of the remaining alcohol using the transition metal catalyst allows the S-enantiomer to be recycled during the reaction. Two of the catalysts, [Rh(cod)Cl]$_2$ and Rh$_2$(OAc)$_4$, achieved respectively 80% ee at 76% conversion and 98% ee at 60% conversion. In principle, high enantioselectivity should be achievable at 100% conversion, and the results achieved led the authors to suggest that the transition metal affects the enzyme selectivity or possibly catalyzes an achiral transesterification.[21] One further drawback of using these transition metal catalysts is the requirement for up to stoichoimetric quantities of the ketone formed during temporary oxidation. A similar requirement for 1-indanone was also obtained by Larsson and co-workers in the esterification of R-1-indanol by DKR using a ruthenium catalyst and *Candida antartica* lipase in *t*-BuOH.[22] The authors also found that a careful choice of ruthenium catalyst for the racemization step had to be made to ensure compatibility with the enzyme and acyl donor. The initially most promising catalyst [(PPh$_3$)$_3$RuCl$_2$] proved to be less active in their presence.

Muller and co-workers have demonstrated the potential of coupling several spontaneous chemical steps to biocatalysis in a one-pot domino reaction to form bicyclo[2.2.2]octenes.[13] A tyrosinase from mushrooms was immobilized on glass beads with the phenol substrate in a mixture of chloroform and a dienophile under air. Tyrosinase can transform a wide variety of phenols to the corresponding catechol, and the presence of air resulted in spontaneous oxidation to the ortho-quinone (Scheme 21.3). The presence of a dienophile then resulted in a Diels-Alder cycloaddition to form the bicyclo[2.2.2]octene product. Significant yields were achieved with a broad range of phenols and dienophiles.

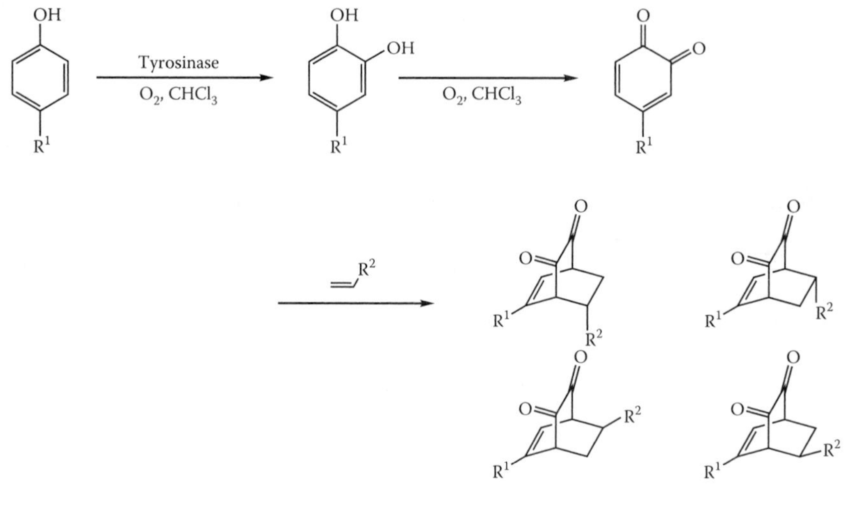

SCHEME 21.3

21.5 SUMMARY AND FUTURE TRENDS

Many of the issues described in this chapter that potentially arise when integrating chemical with biocatalyst transformations can be addressed by modifying the biocatalyst. In particular, directed evolution can be used to modify enzyme properties[32–34] and to some extent the properties of whole cells.[35] The carryover of reagents, solvent, catalysts, and side products formed during chemical transformations may affect subsequent biocatalytic steps. Directed evolution could, therefore, be used to evolve an enzyme that is more soluble and active in the presence of organic solvent, for example in the presence of dimethylformide (DMF)[36] or potentially for an enzyme that is resistant to inactivation through degradation or modification by reactive chemical contaminants. In cases where biological contaminants from a biocatalytic step poison subsequent chemical reactions, directed evolution could potentially be used to minimize these effects. For example, an enzyme evolved for enhanced product selectivity or product yield, or for reduced product inhibition, would minimize the concentration of remaining substrates or sideproducts able to contaminate subsequent less selective chemical steps. Enhanced enzyme immobilization, and, therefore, an improved purity of the biocatalyst would also reduce the potential levels of contaminants in subsequent chemical transformations.

Whether processes include biocatalysis followed by chemistry or vice versa, it would be advantageous to match the solvent composition, pH, and process intensities of both steps. In many cases the solution of choice may be to modify the enzyme to operate most effectively under conditions that favor the chemical transformation (i.e., in an organic solvent or an aqueous-organic solvent biphasic mixture). The engineering principles governing the design of such processes are now well-established,[37] and many such processes are operated commercially.[38] Alternatively, research has identified room temperature ionic liquids as an entirely new class of reaction solvents.[39] These are proving to be very interesting media for both chemical[40,41] and biocatalytic[42–44] conversions and may have a particular role in facilitating integrated chemical and biocatalytic processes. For example, the chemo-enzymatic synthesis of the opioid oxycodone has been reported using a single, biocompatible ionic liquid as the reaction medium.[45]

In this chapter the principles of integration of biocatalytic and chemical steps have been explained. The key factors limiting the successful operation of integrated reaction sequences have also been identified together with genetic and process-engineering solutions that can be used to overcome such limitations. Clearly the industrial implementation of integrated conversion processes will require issues of scale up to be addressed, particularly those related to reactor heterogeneity. Such studies will be crucial if the integration benefits identified in the development laboratory are to be predictably reproduced at manufacturing scale.

REFERENCES

1. Schmid, A., Dordick, J. S., Hauer, B., Wubbolts, M., Witholt, B. *Nature* 2001, *409*, 258.
2. Schmid, A., Hollman, F., Park, J. B., Buhler, B. *Curr. Opin. Biotechnol.* 2002, *13*, 359.
3. Straathof, A. J. J., Panke, S., Schmid, A. *Curr. Opin. Biotechnol.* 2002, *13*, 548.
4. Rozzell, J. D. *Bioorg. Med. Chem.* 1999, *7*, 2253.
5. Lye, G. J., Woodley, J. M. *Trends Biotechnol.* 1999, *17*, 395.
6. Straathof, A. J. J. *Biotechnol. Prog.* 2003, *19*, 755.
7. Stark, D., von Stockar, U. *Adv. Biochem. Eng. Biotechnol.* 2003, *80*, 149.
8. Turner, N. J. *Curr. Opin. Biotechnol.* 2000, *11*, 527.
9. Fish, N. M., Lilly, M. D. *Biotechnol.* 1984, *2*, 623.
10. Hoeks, F. W. J. M. M., Muhle, J., Bohlen, L., Psenicka, I. *Chem. Eng. J.* 1996, *61*, 53.
11. Weuster-Botz, D., Karutz, M., Joksch, B., Schartges, D., Wandy, C. *Appl. Microbiol. Biotechnol.* 1996, *46*, 209.
12. Sheldon, R. A. *Pure Appl. Chem.* 2000, *72*, 1233.
13. Muller, G. H., Lang, A., Seithel, D. R., Waldmann, H. *Chem. Eur. J.* 1998, *4*, 2513.
14. Beard, T. M., Turner, N. J. *J. Chem. Soc., Chem. Commun.* 2002, *2002*, 246.
15. Prasad, A. K., Sorensen, M. D., Parmar, V. S., Wengel, J. *Tetrahedron Lett.* 1995, *36*, 6163.
16. Atkins, R. J., Banks, A., Bellingham, R. K., Breen, G. F., Carey, J. S., Etridge, S. K., Hayes, J. F., Hussain, N., Morgan, D. O., Oxley, P., Passey, S. C., Walsgrove, T. C., Wells, A. S. *Org. Proc. Res. Dev.* 2003, *7*, 663.
17. Macritchie, J. A., Silcock, A., Wills, C. L. *Tetrahedron: Asymmetry* 1997, *8*, 3895.
18. Bhalay, G., Clough, S., McLaren, L., Sutherland, A., Wills, C. L. *J. Chem. Soc., Perkin Trans. I* 2000, 901.
19. Kimura, M., Kuboki, A., Sugai, T. *Tetrahedron: Asymmetry* 2002, *13*, 1059.
20. Kakinuma, K., Nango, E., Kudo, F., Matsushima, Y., Eguchi, T. *Tetrahedron Lett.* 2000, *41*, 1935.
21. Dinh, P. M., Howarth, J. A., Hudnott, A. R., Williams, J. M. *Tetrahedron Lett.* 1996, *37*, 7623.
22. Larsson, A. L. E., Persson, B. A., Backvall, J. E. *Angew. Chem., Int. Ed.* 1997, *36*, 1211.
23. Reetz, M. T., Schimossek, K. *Chimia* 1996, *50*, 668.
24. Faber, K., Riva, S. *Synthesis* 1992, 895.
25. Schoevaart, R., van Rantwijk, F., Sheldon, R. A. *J. Org. Chem.* 2001, *66*, 4559.
26. Alexeeva, M., Enright, A., Dawson, M. J., Mahmoudian, M., Turner, N. J. *Angew. Chem., Int. Ed.* 2002, *41*, 3177.
27. Soda, K., Oikawa, T., Yokoigawa, K. *J. Mol. Catal. B: Enzyme* 2001, *11*, 149.
28. Um, P. J., Drueckhammer, D. G. *J. Am. Chem. Soc.* 1998, *120*, 5605.
29. Inagaki, M., Hiratake, J., Nishioka, T., Oda, J. *J. Am. Chem. Soc.* 1991, *113*, 9360.
30. Gibbs, G., Hateley, M. J., McLaren, L., Welham, M., Willis, C. L. *Tetrahedron Lett.* 1999, *40*, 1069.
31. Turner, N. J. *Curr. Opin. Chem. Biol.* 2004, *8*, 114.
32. Stemmer, W. P. C. *Nature* 1994, *370*, 389.

33. May, O., Nguyen, P. T., Arnold, F. H. *Nature Biotech,* 2000, *18*, 317.
34. Dalby, P. A. *Curr. Opin. Struct. Biol.* 2003, *13*, 500.
35. Patnaik, R., Louie, S., Gavrilovic, V., Perry, K., Stemmer, W. P. C., Ryan, C. M., del Cardayre, S. *Nature Biotech,* 2002, *20*, 707.
36. Moore, J. C., Arnold, F. H. *Nature Biotech,* 1996, *14*, 458.
37. Lye, G. J., Woodley, J. M. Advances in the selection and design of two-liquid biocatalytic reactors. In *Multiphase Bioreactor Design,* Cabral, J. M. S., Mota, M., Tramper, J. Eds., Taylor and Francis: New York, 2001, p. 115.
38. Liese, A., Seelbach, K., Wandrey, C. *Industrial Biotransformations*, Wiley, VCH: Weinheim, 2000.
39. Seddon, K. R. *J. Chem. Technol. Biotechnol.* 1997, *68*, 351.
40. Gordon, C. *App. Catal. A: General* 2001, *222*, 101.
41. Welton, T. *Chem. Rev.* 1999, *99*, 2071.
42. van Rantwijk, F., Madeira Lau, R., Sheldon, R. A. *Trends Biotechnol.* 2003, *21*, 131.
43. Roberts, N. J., Lye, G. J. Application of room temperature ionic liquids in biocatalysis: Opportunities and challenges. In *Ionic Liquids: Industrial Applications for Green Chemistry,* Rogers, R. D., Sheldon, K. R. Eds., American Chemical Society: Washington, D.C., 2002, Vol. ACS Symposium Series 818.
44. Kragl, U., Eckstein, M., Kaftzik, N. *Curr. Opin. Biotechnol.* 2002, *13*, 565.
45. Walker, A. J., Bruce, N. C. *Tetrahedron* 2004, *60*, 561.

22 Substitution Reactions

David J. Ager

CONTENTS

22.1 INTRODUCTION

This chapter covers reactions where a stereogenic sp^3 center is inverted or undergoes a substitution reaction. To control the stereochemical outcome of a substitution reaction, an S$_N$2 mechanism usually has to be used. As a consequence, inversion of configuration is usually observed, although in some reactions two sequential substitutions can occur to give the overall appearance of retention. The use of an S$_N$2 reaction necessitates the establishment of the stereogenic center in the reactant. When an isolated center undergoes this type of reaction, it is usually to correct stereochemistry.

This chapter also includes reactions of epoxides and surrogates, as well as reactions at isolated centers. In the cases of vicinal functional groups, an asymmetric reaction can be used to introduce the vicinal functionalization and generate the stereogenic center, as with an epoxidation reaction.[1]

Another type of substitution reaction is increasing in popularity—the use of an allylic substrate, such as an allyl acetate where the nucleophile is introduced with stereochemical control in the presence of a palladium catalyst and a chiral ligand. Reactions where a chiral anion, be it derived from a chiral heteroatom group, such as a sulfoxide, or an auxiliary, such as Evans's oxazolidinones, are not included in this chapter because the alkyl halide is usually relatively simple and the stereochemical selectivity is derived from the system itself.

22.2 S$_N$2 REACTIONS

Often the purpose of a substitution reaction at a saturated center is to correct stereochemistry. An example is provided in the Sumitomo approach to pyrethroid alcohols where the undesired isomer is inverted in the presence of the required one (Scheme 22.1).[2]

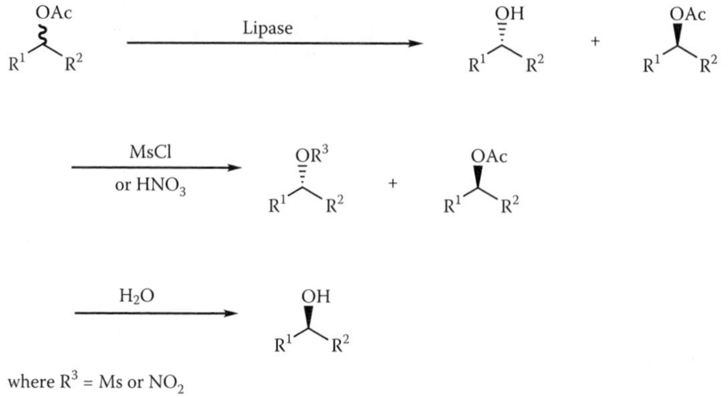

SCHEME 22.1

The phenoxypropionic acid herbicides (Chapter 31) rely on an S_N2 reaction to convert the *S*-2-chloropropionic acid (**1**) to the *R*-2-aryloxy derivative **2**.[3] These compounds are also available from lactic acid by substitution reactions; *R*-lactic acid (**3**) requires two inversions, whereas *S*-lactate (**4**) only requires one (Schemes 22.2 and 22.3).[4]

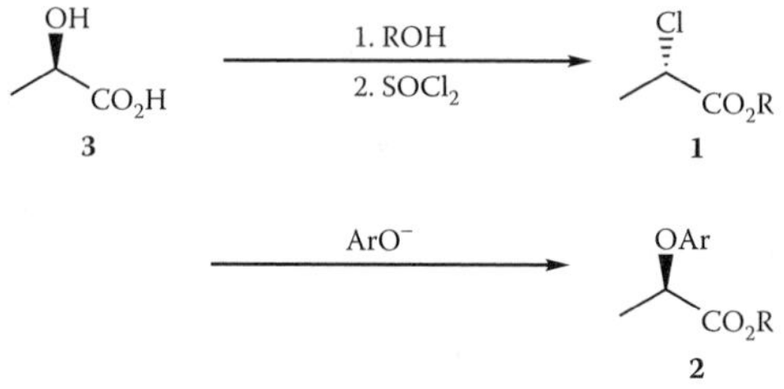

SCHEME 22.2

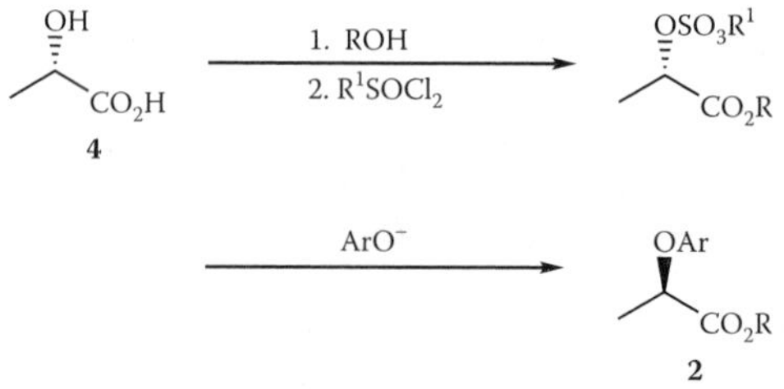

SCHEME 22.3

An additional example of the undesired isomer being inverted is provided in an approach to the pyrethroid, prallethrin (**5**) (Scheme 22.4).[5]

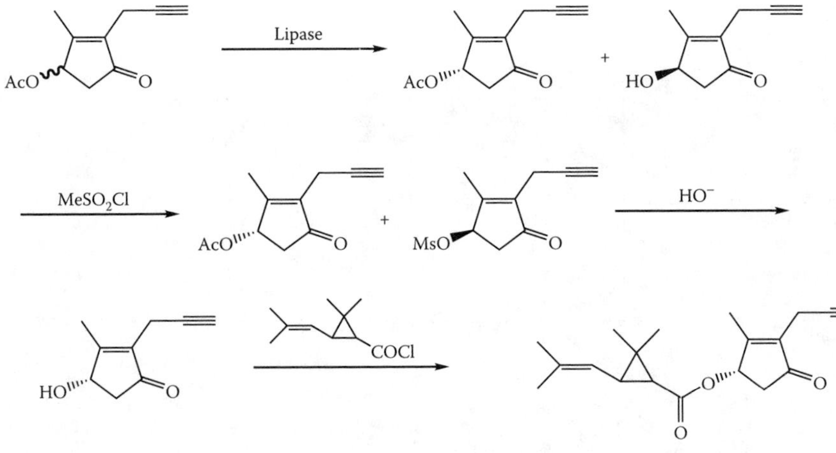

SCHEME 22.4

Reaction conditions must be carefully controlled to ensure that these types of substitution reactions proceed through an S_N2 mechanism and inversion occurs at the stereogenic center. The Mitsunobu reaction is the most common way to bring about inversion by nucleophilic attack, but it is plagued by the formation of large amounts of by-products.[6,7] In some examples, as with hydroxy groups, this is not a trivial undertaking.[1] An example for the inversion of the hydroxy group is with menthol that uses dicyclohexylcarbodiimide to activate the hydroxy group (Scheme 22.5).[8]

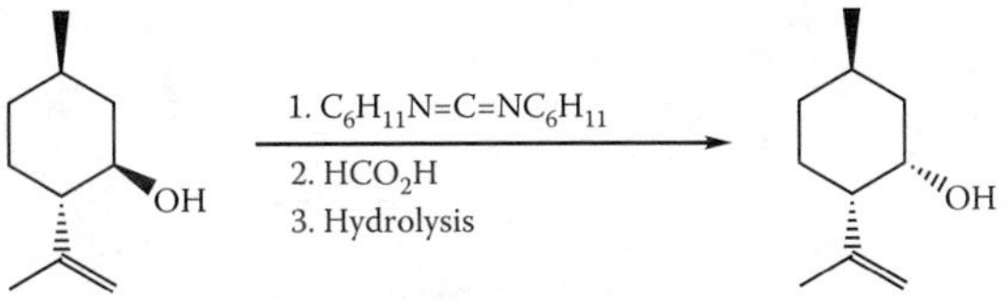

SCHEME 22.5

The use of picolinic and 6-methylpicolinic acids in a Mitsunobu reaction has been advocated because the resultant esters are readily cleaved by copper(II) (Scheme 22.6). The advantage is that ester cleavage occurs under neutral conditions, which minimizes the risk of elimination even with base-sensitive systems.[9]

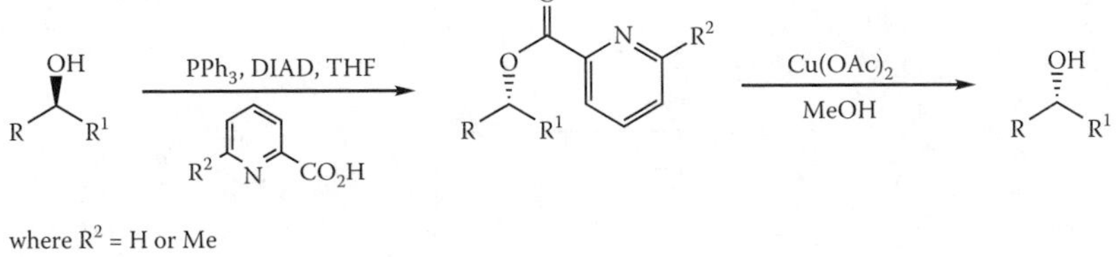

where R^2 = H or Me

SCHEME 22.6

The Mitsunobu reaction of 2,3-dihydroxy esters is regioselective and occurs at the β-position only. Through careful choice of the nucleophile and the reaction sequence, both stereoisomers at this position can be obtained (Scheme 22.7).[10]

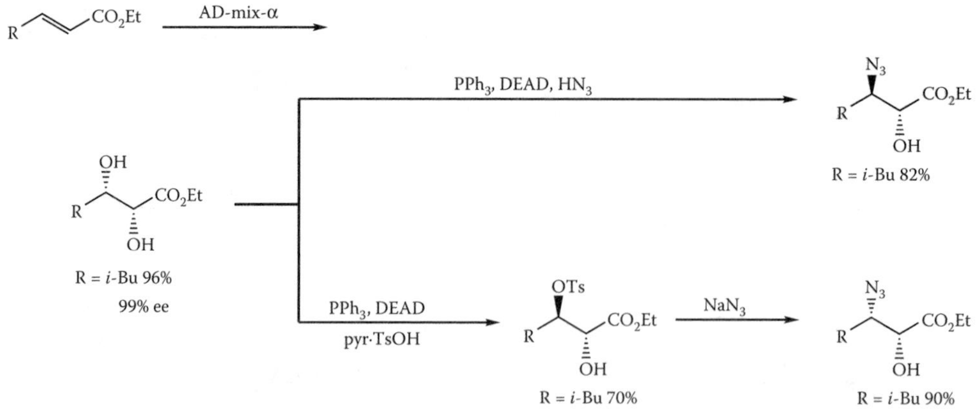

SCHEME 22.7

22.3 EPOXIDE OPENINGS

Although considerable effort has been expended on asymmetric routes to the structural motif within a number of beta-blocker drugs, there has been little financial reward. Through the use of a chiral equivalent of epichlorohydrin (X = Cl), a substitution reaction at the sp³ center followed by epoxide opening allows for entry into this class of drugs. The stereogenic center of the epoxide is retained throughout this sequence (Scheme 22.8).[11]

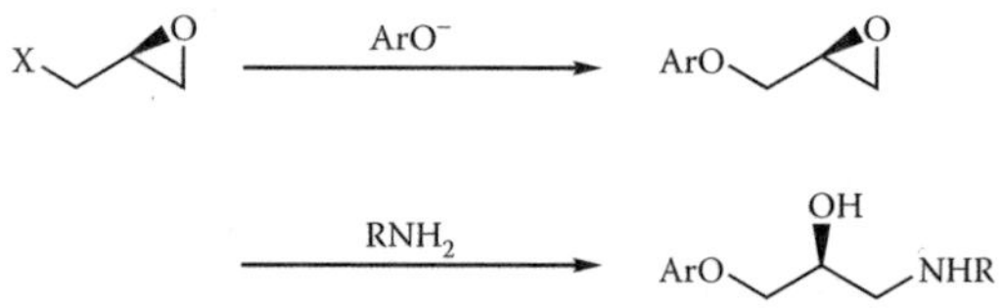

SCHEME 22.8

A considerable amount of work was required to optimize the leaving group and avoid racemization through a Payne rearrangement mechanism.[12] Of course, the Sharpless epoxidation of allyl alcohols is well-known to access these 3-functionalized epoxides.

Manganese complexes provide sufficient induction for synthetic utility for the preparation of unsubstituted epoxides (Scheme 22.9) (Chapter 9).[13–18] The manganese(III)salen complex **6** can also use bleach as the oxidant rather than an iodosylarene.[19,20]

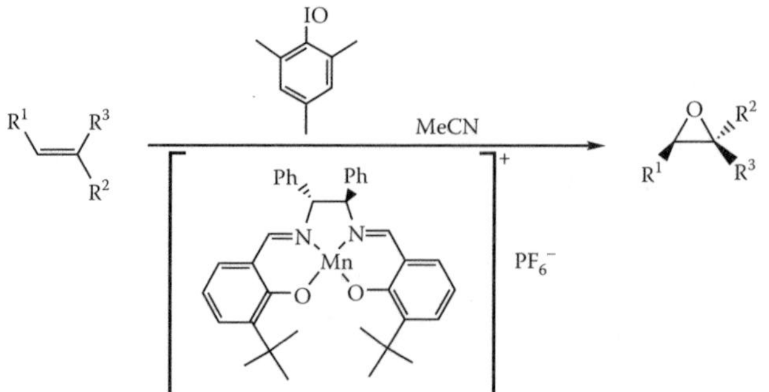

SCHEME 22.9

The regioselective control for the nucleophilic opening of an epoxide in an acyclic system is well-known.[1,20] Under basic conditions, the nucleophile usually attacks the sterically less encumbered site, whereas under acidic conditions, the sterically more hindered site is favored.[21-25] The product invariably contains the functional groups in an *ancat*-disposition, when an S_N2 pathway is followed.[1,20] Epoxides react with a wide variety of nucleophiles.[26]

22.3.1 HIV PROTEASE INHIBITORS

Human immunodeficiency virus (HIV) protease inhibitors have provided the pharmaceutical and fine chemical industry with a number of difficult problems to solve. Most have complex structures with a number of stereogenic centers that make synthetic sequences long. In addition, the dosages of these drugs are relatively high, and large volumes have to be made. Because most may be considered peptidomimetics, amino acids are often used as a starting material. A number of candidates have relied on a homologation, formation of an epoxide, and opening with a nitrogen nucleophile (e.g., Scheme 22.10).[27-35] The recrystallization of the intermediate removes the unwanted diastereoisomer.

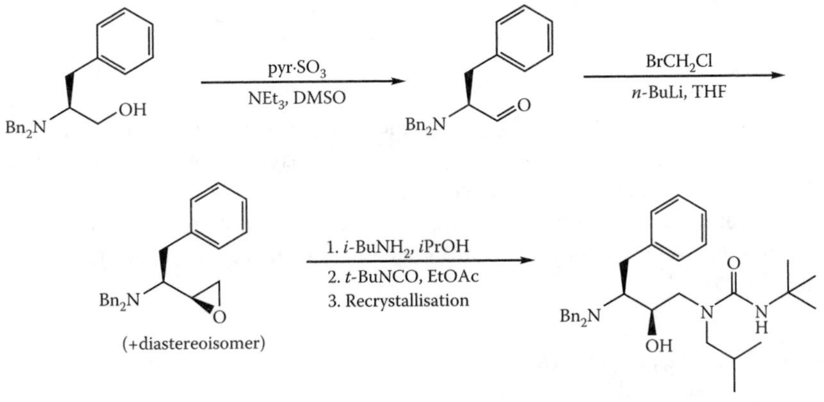

SCHEME 22.10

22.4 CYCLIC SULFATE REACTIONS

Cyclic sulfates provide a useful alternative to epoxides now that it is viable to produce a chiral diol from an alkene. These cyclic compounds are prepared by reaction of the diol with thionyl chloride, followed by ruthenium-catalyzed oxidation of the sulfur (Scheme 22.11).[36] This oxidation has the advantage over previous procedures because it only uses a small amount of the transition metal catalyst.[1,37,38]

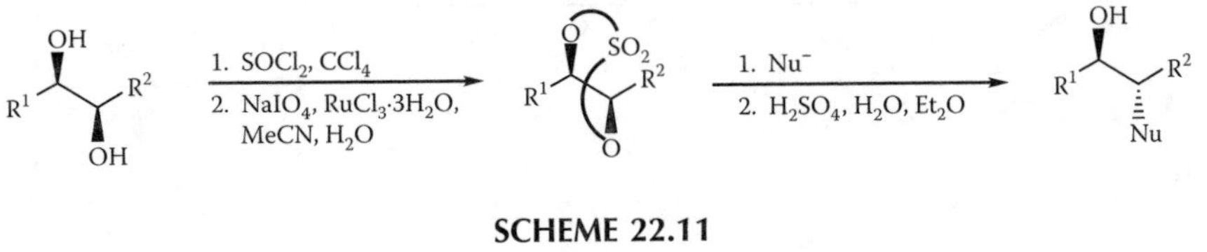

SCHEME 22.11

22.5 IODOLACTONIZATIONS

Intramolecular cyclization is a useful method for the preparation of lactones and cyclic ethers.[39] The most common examples are iodolactonization and iodoetherification, the difference being that the former uses a carboxylic acid derivative as the nucleophile, and the latter relies on a hydroxy group. Thus, butyrolactones are available from γ,δ-unsaturated carboxylic acid derivatives,[1,40,41] whereas unsaturated alcohols lead to cyclic ethers.[42–45] Lactones are also available from a wide variety of nucleophiles such as carbonates,[46] orthoesters,[47] or carbamates,[48,49] which can all be used in place of a carboxylate anion.[49,50]

The intermediate iodonium ion controls the relative stereochemistry of the cyclization. An asymmetric center present within the substrate, therefore, allows for enantioselectivity (*cf.* Scheme 22.12).[51] The reaction is very susceptible to the steric interactions within the transition state.[48,51–53]

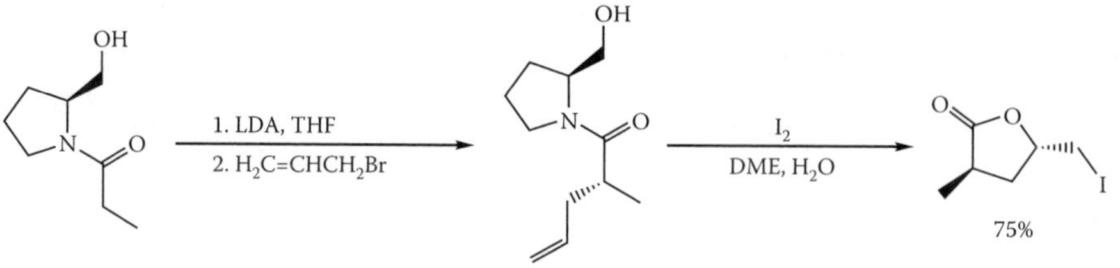

SCHEME 22.12

Variations on this cyclization methodology include the use of bromine, rather than iodine; again, both lactones and ethers can be prepared.[1,47,54–57] An asymmetric bromolactonization procedure affords α-hydroxy acids with good asymmetric induction.[57–62]

A further variation uses selenium or sulfur as the electrophilic species to induce cyclization. The diversity of organoselenium or organosulfur chemistry is then available for further modification of the product.[1,63–75]

22.6 ALLYLIC SUBSTITUTIONS

The allylic class of substitution reactions is gaining in popularity as catalysts become more general and the factors that govern the selectivity are better understood.[76,77] Allylic systems can undergo nucleophilic substitution by either an S_N2 or S_N2' reaction.[78–81] The S_N2 reaction can compete effectively with an S_N2' reaction.

The use of a metal catalyst, such as palladium, also provides for some asymmetric induction when an allylic system is treated with a stabilized anion (Scheme 22.13)[82–103] or with other nucleophiles.[82,87,104–111] This approach also allows for the kinetic resolution of allyl acetates.[104,112]

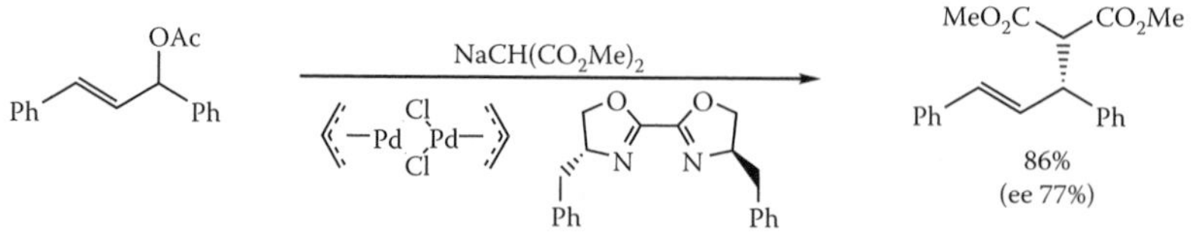

SCHEME 22.13

The mechanism of allylic alkylation reactions has been the subject of significant investigation.[80,113,114] The two major approaches were based on the use of a cyclic substrate and use of chiral substrates and have given rise to the model shown in Scheme 22.14. The model also includes the following general observations:

1. The allyl complex is generally formed with inversion of configuration.
2. If the leaving group is antiperiplanar to the metal, then formation of the π-allyl intermediate usually follows.[115–117]
3. Attack by a "soft" nucleophile leads to inversion of configuration in the second step.[118–127]
4. Attack by a "hard" nucleophile results in retention of configuration in the second step
   because the metal center is attacked first (pathway B).[128–131]

Palladium-catalyzed allylation reactions proceed with net retention of configuration that is the
result of two inversions.[118–120,132–135] Grignard reactions with nickel or copper catalysis proceed with
net inversion.[128,136–138]

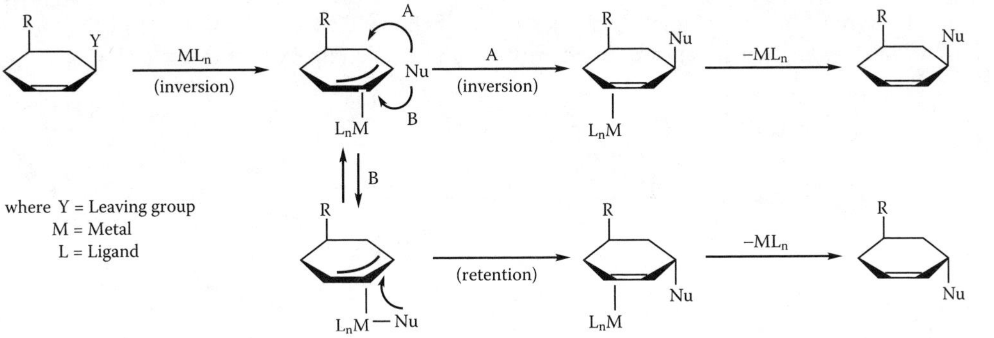

SCHEME 22.14

The methodology suffers from competition between *syn-* and *anti-*mechanisms in acyclic cases.
Indeed, the pathway may depend on the nature of nucleophile and substrate.[78,139] In addition, there
is no clearcut differentiation between "hard" and "soft" nucleophiles in the model described.

Use of a chiral catalyst for allylic substitution reactions allows for desymmetrization reactions
and asymmetric reactions (Scheme 22.15).[140–143]

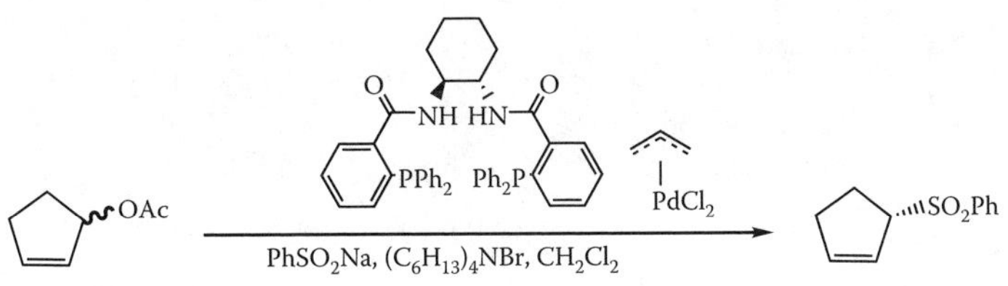

SCHEME 22.15

When the allylic substituent is the source of chirality, then palladium-catalyzed substitutions
can lead to good asymmetric induction (Schemes 16 and 17).[109] In this example, the geometry of
the double bond in the substrate controls the stereochemical outcome of the reaction.

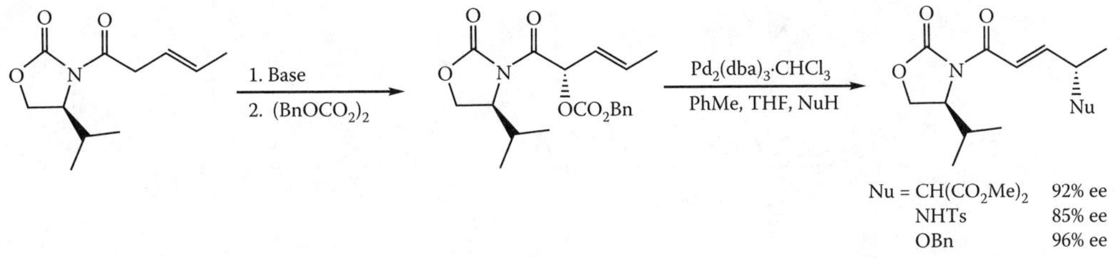

Nu = CH(CO₂Me)₂	92% ee
NHTs	85% ee
OBn	96% ee

SCHEME 22.16

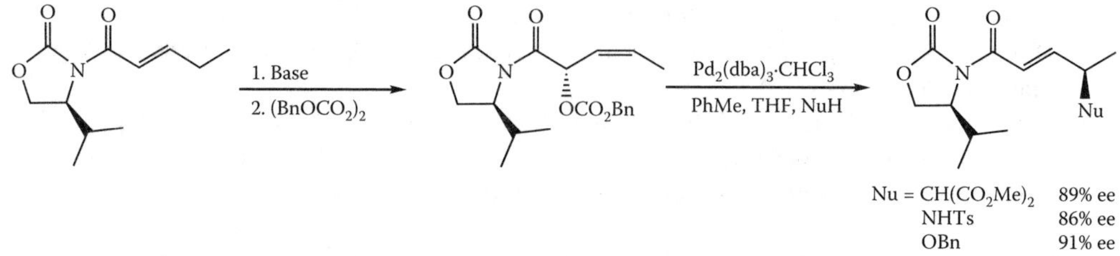

SCHEME 22.17

1-(2-Naphthyl)ethyl acetate (**7**) has been used as a substrate for palladium-catalyzed nucleophilic substitutions in the presence of chiral ligands (Scheme 22.18). The presence of the unsaturation from the aromatic ring allows the intermediate π-allyl complex to form.[144]

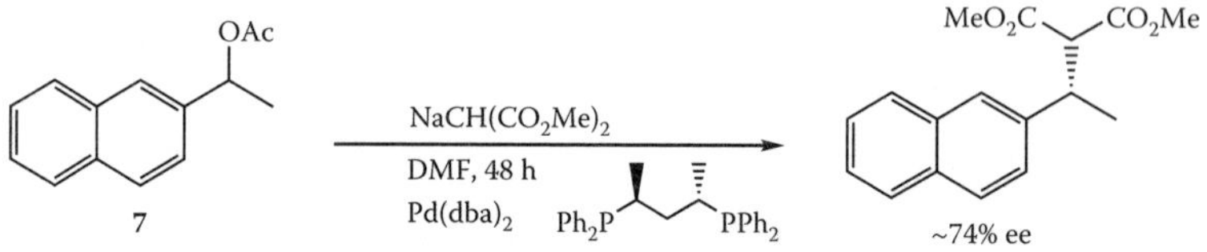

SCHEME 22.18

Amino acids can be accessed by a palladium-catalyzed allylic substitution reaction (Scheme 22.19).[145] Some ligand screening is necessary to obtain high selectivity in a specific case. Indeed, a wide range of ligands has been proposed for this and analogous reactions.[145–148]

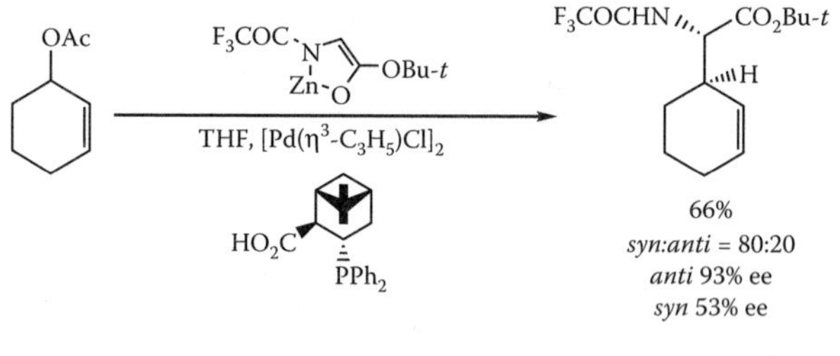

SCHEME 22.19

An alternative amino acid synthesis that leads to γ-amino acids is outlined in Scheme 22.20.[149]

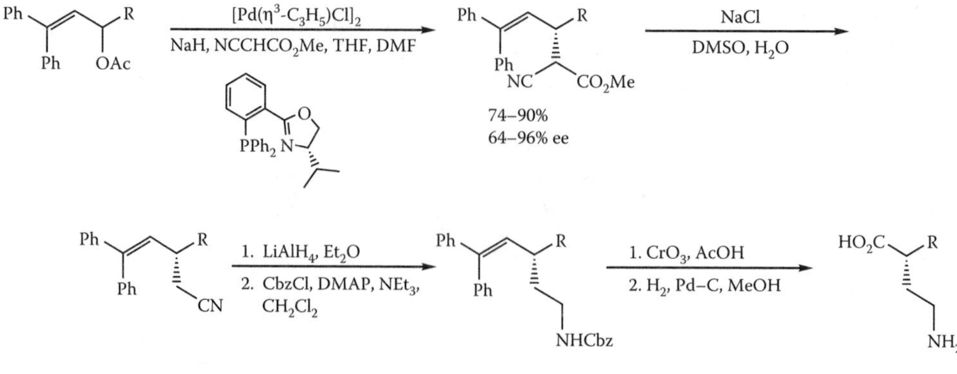

SCHEME 22.20

A useful transformation that leads to chiral butenolides, in their own right useful chiral building blocks, is the reaction of the butenolide **8** with a nucleophile, such as a phenol, in the presence of a palladium catalyst (Scheme 22.21).[150]

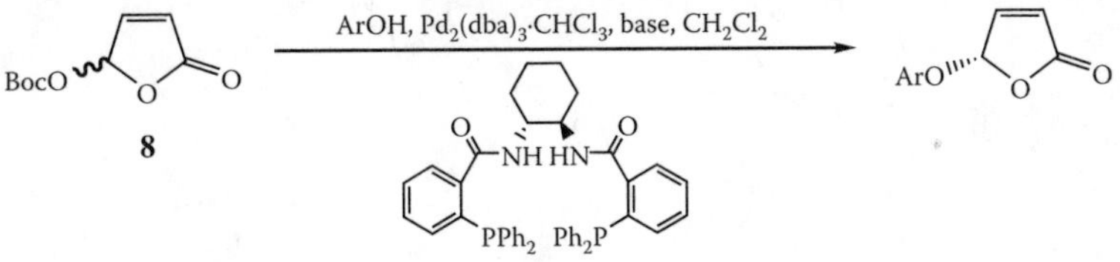

SCHEME 22.21

For an example of an allylic substitution in a nonsteroidal antiinflammatory drug synthesis, see Chapter 11, Section 11.3.5.

Metals other than palladium and molybdenum can be used for allylic substitution reactions. For example, nickel in the presence of the oxazolinylferrocenylphosphine **9** provides good asymmetric induction for the reaction of a Grignard reagent with allylic electrophilic systems such as acetates.[151]

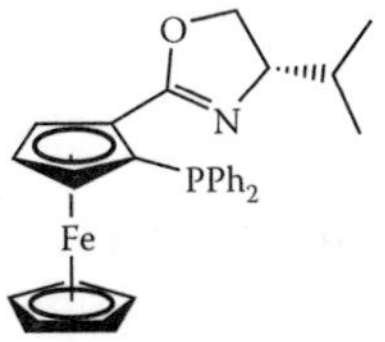

9

An example of an S_N2' substitution reaction with a mixed zinc-copper reagent is shown in Scheme 22.22. The product **9** was used to prepare the bicyclic enone **10.** In this example the starting alcohol gave rise to the asymmetric induction. The substitution was *anti*-selective.[152]

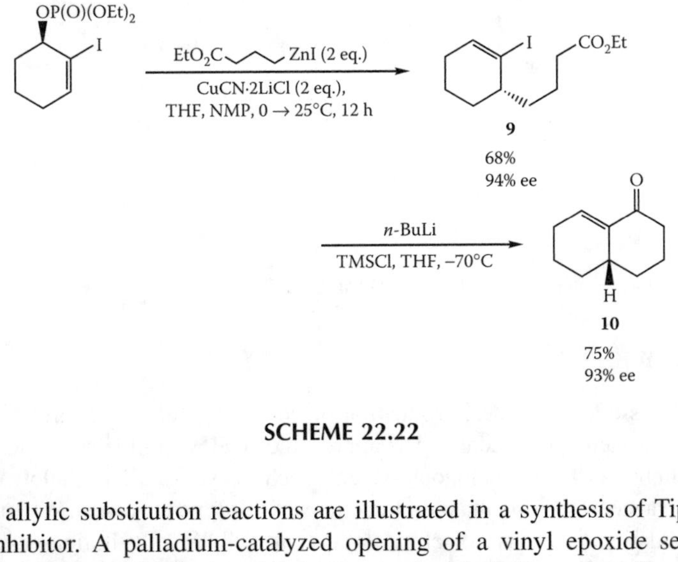

SCHEME 22.22

The various allylic substitution reactions are illustrated in a synthesis of Tipranavir (**11**), an HIV protease inhibitor. A palladium-catalyzed opening of a vinyl epoxide set the quaternary stereogenic center (Scheme 22.23). A molybdenum-catalyzed allylic nuclophilic displacement was used to access the benzylioc stereogenic center (Scheme 22.24).[153]

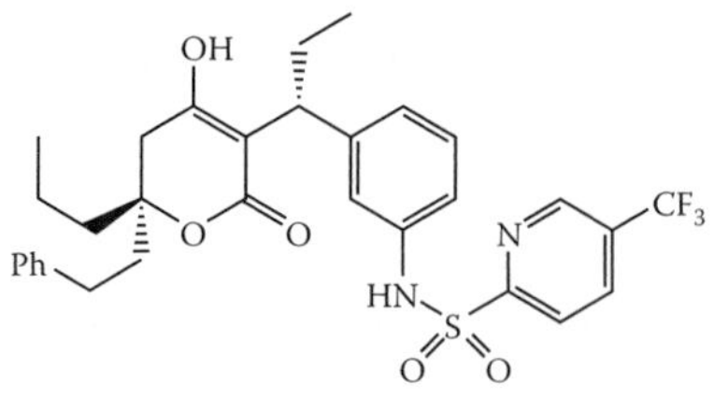

11

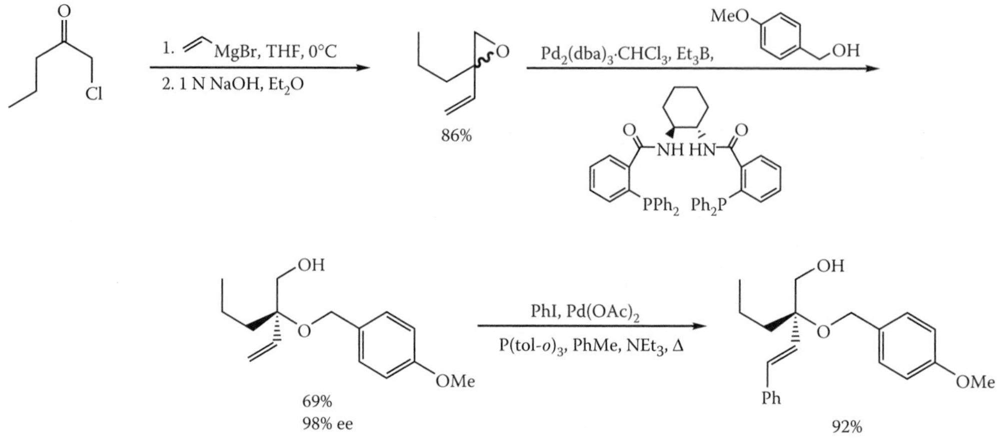

SCHEME 22.23

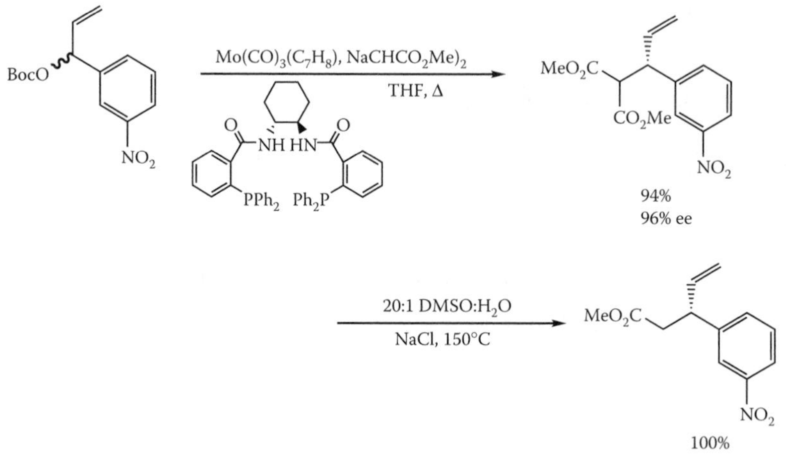

SCHEME 22.24

22.7 SUMMARY

Substitution reactions allow for the introduction or change of functional groups but rely on the prior formation of the stereogenic center. The approach can allow for the correction of stereochemistry. Reactions of epoxides, and analogous systems such as cyclic sulfates, allow for 1,2-functionality to be set up in a stereospecific manner. Reactions of this type have been key to the applications of asymmetric oxidations. The use of chiral ligands for allylic substitutions does allow for the introduction of a new stereogenic center. With efficient catalysts now identified, it is surely just a matter of time before this methodology is used at scale.

REFERENCES

1. Ager, D. J., East, M. B. *Asymmetric Synthetic Methodology*, CRC Press: Boca Raton, 1995.
2. Danda, H., Nagatomi, T., Maehara, A., Umemura, T. *Tetrahedron* 1991, *47*, 8701.
3. Sheldon, R. A. *Chirotechnology: Industrial synthesis of optically active compounds*, Marcel Dekker: New York, 1993.
4. Gras, G. Ger. Pat. DE 3,024,265, 1979.
5. Hirohara, H., Mitsuda, S., Ando, E., Komaki, R. In *Biocatalysts in Organic Synthesis,* Tramper, J., Van der Plas, H. C., Linko, P. Eds., Elsevier: Amsterdam, 1985, p. 119.
6. Mitsunobu, O. *Synthesis* 1981, 1.
7. Dembinski, R. *Eur. J. Org. Chem.* 2004, 2763.
8. Kaulen, J. Ger. Pat. DE 3511210, 1986.
9. Sammakia, T., Jacobs, J. S. *Tetrahedron Lett.* 1999, *40*, 2685.
10. Ko, S. Y. *J. Org. Chem.* 2002, *67*, 2689.
11. Sheldon, R. A. *Chirotechnology: Industrial Synthesis of Optically Active Compounds*, Marcel Dekker, Inc.: New York, 1993.
12. Klunder, J. M., Onami, T., Sharpless, K. B. *J. Org. Chem.* 1989, *54*, 1295.
13. Zhang, W., Loebach, J. L., Wilson, S. R., Jacobsen, E. N. *J. Am. Chem. Soc.* 1990, *112*, 2801.
14. Irie, R., Noda, K., Ito, Y., Matsumoto, N., Katsuki, T. *Tetrahedron Lett.* 1990, *31*, 7345.
15. Zhang, W., Loebach, J. L., Wilson, S. R., Jacobsen, E. J. *J. Am. Chem. Soc.* 1990, *112*, 2801.
16. Van Draanen, N. A., Arseniyadis, S., Crimmins, M. T., Heathcock, C. H. *J. Org. Chem.* 1991, *56*, 2499.
17. Okamoto, Y., Still, W. C. *Tetrahedron Lett.* 1988, *29*, 971.
18. Schwenkreis, T., Berkessel, A. *Tetrahedron Lett.* 1993, *34*, 4785.
19. Zhang, W., Jacobsen, E. N. *J. Org. Chem.* 1991, *56*, 2296.
20. Deng, L., Jacobsen, E. N. *J. Org. Chem.* 1992, *57*, 4320.
21. Parker, R. E., Isaacs, N. S. *Chem. Rev.* 1959, *59*, 737.
22. Boireau, G., Abenhaim, D., Bernardon, C., Henry-Basch, E., Sabourault, B. *Tetrahedron Lett.* 1975, 2521.
23. Schauder, J. R., Krief, A. *Tetrahedron Lett.* 1982, *23*, 4389.
24. Tanaka, T., Inoue, T., Kamei, K., Murakami, K., Iwata, C. *J. Chem. Soc., Chem. Commun.* 1990, 906.
25. Bartók, M., Láng, K. L. *Chem. Heterocycl. Comp.* 1984, *42 (pt.3)*, 1.
26. Williams, N. R. *Adv. Carbohydr. Chem. Biochem.* 1970, *25*, 109.
27. Bennett, F., Patel, N. M., Girijavallabhan, V. M., Ganguly, A. K. *Synlett* 1993, 703.
28. Liu, C., Ng, J. S., Behling, J. R., Yen, C. Y., Campbell, A. L., Fuzail, K. S., Yonan, E. E., Mehrota, D. V. *Org. Proc. Res. Dev.* 1997, *1*, 45.
29. Ng, J. S., Przybyla, C. A., Liu, C., Yen, J. C., Muellner, F. W., Weyker, C. L. *Tetrahedron* 1995, *51*, 6397.
30. Ng, J. S., Przybyla, C. A., Mueller, R. A., Vazquez, M. L., Getman, D. P. U.S. Pat. 5,583,238, 1996.
31. Vazquez, M. L., Bryant, M. L., Clare, M., DeCresenzo, G. A., Doherty, E. M., Freskos, J. N., Getman, D. P., Houseman, K. A., Julian, J. A., Kocan, G. P., Mueller, R. A., Shieh, H.-S., Stallings, W. C., Stegeman, R. A., Talley, J. J. *J. Med. Chem.* 1995, *38*, 581.
32. Ghosh, A. K., Hussain, K. A., Fidanze, S. *J. Org. Chem.* 1997, *62*, 6080.
33. Ahmad, S., Spergel, S. H., Barrish, J. C., DiMarco, J., Gougoutas, J. *Tetrahedron: Asymmetry* 1995, *6*, 2893.
34. Askin, D., Wallace, M. A., Vacca, J. P., Reamer, R. A., Volante, R. P., Shinkai, I. *J. Org. Chem.* 1992, *57*, 2771.
35. Sham, H. L., Betebenner, D. A., Zhao, C., Wideburg, N. E., Saldivar, A., Kempf, D. J., Plattner, J. J., Norbeck, D. W. *J. Chem. Soc., Chem. Commun.* 1993, 1052.
36. Gao, Y., Sharpless, K. B. *J. Am. Chem. Soc.* 1988, *110*, 7538.
37. Denmark, S. E. *J. Org. Chem.* 1981, *46*, 3144.
38. Lowe, G., Salamone, S. J. *J. Chem. Soc., Chem. Commun.* 1983, 1392.
39. Cardillo, G., Orena, M. *Tetrahedron* 1990, *46*, 3321.
40. Bartlett, P. A. *Tetrahedron* 1980, *36*, 3.
41. Bartlett, P. A., Richardson, D. P., Myerson, J. *Tetrahedron* 1984, *40*, 2317.
42. Williams, D. R., White, F. H. *Tetrahedron Lett.* 1986, *27*, 2195.
43. Baldwin, S. W., McIver, J. M. *J. Org. Chem.* 1987, *52*, 322.

44. Labelle, M., Morton, H. E., Guidon, Y., Springer, J. P. *J. Am. Chem. Soc.* 1988, *110*, 4533.
45. Kang, S. H., Lee, S. B. *Tetrahedron Lett.* 1993, *34*, 7579.
46. Bongini, A., Cardillo, G., Orena, M., Porzi, G., Sandri, S. *J. Org. Chem.* 1982, *47*, 4626.
47. Williams, D. R., Harigaya, Y., Moore, J. L., D'asa, A. *J. Am. Chem. Soc.* 1984, *106*, 2641.
48. Hirama, M., Uei, M. *Tetrahedron Lett.* 1982, *23*, 5307.
49. Kobayashi, S., Isobe, T., Ohno, M. *Tetrahedron Lett.* 1984, *25*, 5079.
50. Friesen, R. W. *Tetrahedron Lett.* 1990, *31*, 4249.
51. Najdi, S., Reichlin, D., Kurth, M. J. *J. Org. Chem.* 1990, *55*, 6241.
52. Kurth, M. J., Beard, R. L., Olmstead, M., Macmillan, J. G. *J. Am. Chem. Soc.* 1989, *111*, 3712.
53. Fuji, K., Node, M., Naniwa, Y., Kawabata, T. *Tetrahedron Lett.* 1990, *31*, 3175.
54. Fukuyama, T., Wang, C.-L. J., Kishi, Y. *J. Am. Chem. Soc.* 1979, *101*, 260.
55. Al-Dulayymi, J., Baird, M. S. *Tetrahedron Lett.* 1989, *30*, 253.
56. Jung, M. E., Lew, W. *J. Org. Chem.* 1991, *56*, 1347.
57. Jew, S.-s., Terashima, S., Koga, K. *Tetrahedron* 1979, *35*, 2337.
58. Terashima, S., Jew, S.-S. *Tetrahedron Lett.* 1977, 1005.
59. Terashima, S., Jew, S.-S., Koga, K. *Chemistry Lett.* 1977, 1109.
60. Terashima, S., Jew, S.-S., Koga, K. *Tetrahedron Lett.* 1977, 4507.
61. Terashima, S., Jew, S.-S., Koga, K. *Tetrahedron Lett.* 1978, 4937.
62. Jew, S.-S., Terashima, S., Koga, K. *Chem. Pharm. Bull.* 1979, *27*, 2351.
63. Nicolaou, K. C. *Tetrahedron* 1981, *37*, 4097.
64. Nicolaou, K. C., Barnette, W. E. *J. Chem. Soc., Chem. Commun.* 1977, 331.
65. Murata, S., Suzuki, T. *Tetrahedron Lett.* 1987, *28*, 4415.
66. Clive, D. L. J., Chittattu, G. *J. Chem. Soc., Chem. Commun.* 1977, 484.
67. Tiecco, M., Testaferri, L., Tingoli, M., Bartoli, D. *Tetrahedron* 1989, *45*, 6819.
68. Nicolaou, K. C., Sipio, W. J., Magolda, R. J., Claremon, D. A. *J. Chem. Soc., Chem. Commun.* 1979, 83.
69. Nicolaou, K. C., Seitz, S. P., Sipio, W. J., Blount, J. F. *J. Am. Chem. Soc.* 1979, *101*, 3884.
70. Bartlett, P. A., Holm, K. H., Morimoto, A. *J. Org. Chem.* 1985, *50*, 5179.
71. Nicolaou, K. C., Lysenko, Z. *J. Chem. Soc., Chem. Commun.* 1977, 293.
72. Nicolaou, K. C., Barnette, W. E., Magolda, R. L. *J. Am. Chem. Soc.* 1981, *103*, 3480.
73. Nicolaou, K. C., Barnette, W. E., Magolda, R. L. *J. Am. Chem. Soc.* 1981, *103*, 3486.
74. Nicolaou, K. C., Barnette, W. E., Magolda, R. L. *J. Am. Chem. Soc.* 1978, *100*, 2567.
75. Nicolaou, K. C., Claremon, D. A., Barnette, W. E., Seitz, S. P. *J. Am. Chem. Soc.* 1979, *101*, 3704.
76. Acemoglu, L., Williams, J. M. J. In *Handbook of Organopalladium Chemistry for Organic Synthesis,* Negishi, E.-I. Ed., John Wiley & Sons, Inc.: Hoboken, NJ, 2002, Vol. 2, p. 1945ff.
77. Pfaltz, A., Lautens, M. In *Comprehensive Asymmetric Catalysis,* Jacobsen, E. N., Pfaltz, A., Yamamoto, H. Eds., Springer-Verlag: Berlin, 1999, Vol. 2, p. 833.
78. Magid, R. M. *Tetrahedron* 1980, *36*, 1901.
79. Tsuji, J. *Pure Appl. Chem.* 1989, *61*, 1673.
80. Consiglio, G., Waymouth, R. M. *Chem. Rev.* 1989, *89*, 257.
81. Uguen, D. *Tetrahedron Lett.* 1984, *25*, 541.
82. Trost, B. M. *Acc. Chem. Res.* 1980, *13*, 385.
83. Muller, D., Umbricht, G., Weber, B., Pfaltz, A. *Helv. Chim. Acta* 1991, *74*, 232.
84. Leutenegger, U., Umbricht, G., Fahrni, C., von Matt, P., Pfaltz, A. *Tetrahedron* 1992, *48*, 2143.
85. von Matt, P., Pfaltz, A. *Angew. Chem., Int. Ed.* 1993, *32*, 566.
86. Reiser, O. *Angew. Chem., Int. Ed.* 1993, *32*, 547.
87. Trost, B. M., Van Vranken, D. L. *Chem. Rev.* 1996, *96*, 395.
88. Nakano, H., Okuyama, Y., Yanagida, M., Hongo, H. *J. Org. Chem.* 2001, *66*, 620.
89. Li, Z.-P., RTang, F.-Y., Xu, H.-D., Wu, X.-Y., Zhou, Q.-L., Chan, A. S. C. *J. Mol. Catal. A: Chemical* 2003, *193*, 89.
90. Kodama, H., Taiji, T., Ohta, T., Furukawa, I. *Tetrahedron: Asymmetry* 2000, *11*, 4009.
91. Vasse, J.-L., Stranne, R., Zalubovskis, R., Gayet, C., Moberg, C. *J. Org. Chem.* 2003, *68*, 3258.
92. Rassias, G. A., Page, P. C. B., Reigner, S., Christie, S. D. R. *Synlett* 2000, 379.
93. Nakano, H., Okuyama, Y., Hongo, H. *Tetrahedron Lett.* 2000, *41*, 4615.
94. Chelucci, G. *Tetrahedron: Asymmetry* 1997, *8*, 2667.
95. Saitoh, A., Achiwa, K., Tanaka, K., Morimoto, T. *J. Org. Chem.* 2000, *65*, 4227.

441

96. Lee, S.-g., Lee, S. H., Song, C. E., Chung, B. Y. *Tetrahedron: Asymmetry* 1999, *10*, 1795.
97. Hiroi, K., Suzuki, Y., Kawagishi, R. *Tetrahedron Lett.* 1999, *40*, 715.
98. Yonehara, K., Hashizume, T., Mori, K., Ohe, K., Uemaru, S. *J. Org. Chem.* 1999, *64*, 9374.
99. Hayashi, T. *J. Organometal. Chem.* 1999, *576*, 195.
100. Helmchen, G. *J. Organometal. Chem.* 1999, *576*, 203.
101. Enders, D., Peters, R., Runsink, J., Bats, J. W. *Org. Lett.* 1999, *1*, 1863.
102. Tye, H., Smyth, D., Eldred, C., Wills, M. *J. Chem. Soc., Chem. Commun.* 1997, 1053.
103. Page, P. C. B., Heaney, H., Reignier, S., Rassias, G. A. *Synlett* 2003, 22.
104. Sawamura, M., Ito, Y. *Chem. Rev.* 1992, *92*, 857.
105. Trost, B. M. *Pure Appl. Chem.* 1979, *51*, 787.
106. Trost, B. M. *Tetrahedron* 1977, *33*, 2615.
107. Hayashi, T. *Pure Appl. Chem.* 1988, *60*, 7.
108. Ohta, T., Sasayama, H., Nakajima, O., Kurahashi, N., Fujii, T., Furukawa, I. *Tetrahedron: Asymmetry* 2003, *14*, 537.
109. Sugiura, M., Yagi, Y., Wei, S.-Y., Nakai, T. *Tetrahedron Lett.* 1998, *39*, 4351.
110. Trost, B. M., Radinov, R. *J. Am. Chem. Soc.* 1997, *119*, 5962.
111. Hamada, Y., Seto, N., Takayanagi, Y., Nakano, T., Hara, O. *Tetrahedron Lett.* 1999, *40*, 7791.
112. Hayashi, T., Yamamoto, A., Ito, Y. *J. Chem. Soc., Chem. Commun.* 1986, 1090.
113. Trost, B. M., Van Vranken, D. L., Bingel, C. *J. Am. Chem. Soc.* 1992, *114*, 9327.
114. Clyne, D. S., Mermet-Bouvier, Y. C., Nomura, N., RajanBabu, T. V. *J. Org. Chem.* 1999, *64*, 7601.
115. Fiaud, J.-C., Aribi-Zouioneche, L. *J. Chem. Soc., Chem. Commun.* 1986, 390.
116. Fiaud, J.-C., Legros, J.-Y. *J. Org. Chem.* 1987, *52*, 1907.
117. Faller, J. W., Linebarrier, D. *Organometallics* 1988, *7*, 1670.
118. Trost, B. M., Verhoeven, T. R. *J. Org. Chem.* 1976, *41*, 3215.
119. Trost, B. M., Verhoeven, T. R. *J. Am. Chem. Soc.* 1980, *102*, 4730.
120. Trost, B. M., Weber, L. *J. Am. Chem. Soc.* 1975, *97*, 1611.
121. Keinan, E., Sahai, M., Roth, Z., Nudelman, A., Herzig, J. *J. Org. Chem.* 1985, *50*, 3558.
122. Trost, B. M., Keinan, E. *J. Am. Chem. Soc.* 1978, *100*, 7779.
123. Fiaud, J.-C. *J. Chem. Soc., Chem. Commun.* 1983, 1055.
124. Auburn, P. R., Whelan, J., Bosnich, B. *J. Chem. Soc., Chem. Commun.* 1986, 186.
125. Trost, B. M., Schmuff, N. R. *J. Am. Chem. Soc.* 1985, *107*, 396.
126. Trost, B. M., Verhoeven, T. R., Fortunak, J. M. *Tetrahedron Lett.* 1979, 2301.
127. Bäckvall, J.-E., Nordberg, R. E. *J. Am. Chem. Soc.* 1981, *103*, 4959.
128. Consiglio, G., Morandini, F., Piccolo, O. *J. Am. Chem. Soc.* 1981, *103*, 1845.
129. Sheffy, F. K., Stille, J. K. *J. Am. Chem. Soc.* 1983, *105*, 7173.
130. Matsushita, H., Negishi, E. *J. Chem. Soc., Chem. Commun.* 1982, 150.
131. Hayashi, T., Yamamoto, A., Hagihara, T. *J. Org. Chem.* 1986, *51*, 723.
132. Trost, B. M., Schmuff, N. R., Miller, M. J. *J. Am. Chem. Soc.* 1980, *102*, 5979.
133. Trost, B. M., Strege, P. E. *J. Am. Chem. Soc.* 1977, *99*, 1649.
134. Tanikaga, R., Jun, T. X., Kaji, A. *J. Chem. Soc., Perkin Trans. I* 1990, 1185.
135. Mackenzie, P. B., Whelan, J., Bosnich, B. *J. Am. Chem. Soc.* 1985, *107*, 2046.
136. Felkin, H., Joly-Goudket, M., Davies, S. G. *Tetrahedron Lett.* 1981, *22*, 1157.
137. Gendreau, Y., Normant, J. F. *Tetrahedron Lett.* 1979, 1617.
138. Consiglio, G., Morandini, F., Piccolo, O. *J. Chem. Soc., Chem. Commun.* 1983, 112.
139. Hayashi, T., Yamamoto, A., Hayihara, T., Ito, Y. *Tetrahedron Lett.* 1986, *27*, 191.
140. Trost, B. M., Organ, M. G., O'Doherty, G. A. *J. Am. Chem. Soc.* 1995, *117*, 9662.
141. Trost, B. M., Lee, C. B., Weiss, J. M. *J. Am. Chem. Soc.* 1995, *117*, 7247.
142. Trost, B. M. *Pure Appl. Chem.* 1996, *68*, 779.
143. Trost, B. M. *Acc. Chem. Res.* 1996, *29*, 355.
144. Legros, J.-Y., Boutros, A., Fiaud, J.-C., Toffano, M. *J. Mol. Catal. A: Chemical* 2003, *196*, 21.
145. Weiß, T. D., Helmchen, G., Kazmaier, U. *J. Chem. Soc., Chem. Commun.* 2002, 1270.
146. Trost, B. M., Bunt, R. C. *Angew. Chem., Int. Ed.* 1996, *35*, 99.
147. Gilbertson, S. R., Xie, D., Fu, Z. *J. Org. Chem.* 2001, *66*, 7240.
148. Helmchen, G., Kudis, S., Sennhenn, P., Steinhagen, H. *Pure Appl. Chem.* 1997, *69*, 513.
149. Martin, C. J., Rawson, D. J., Williams, J. M. J. *Tetrahedron: Asymmetry* 1998, *9*, 3723.

150. Trost, B. M., Toste, F. D. *J. Am. Chem. Soc.* 2003, *125*, 3090.
151. Chung, K.-G., Miyake, Y., Uemura, S. *J. Chem. Soc., Perkin Trans. I* 2000, 2725.
152. Calaza, M. I., Hupe, E., Knochel, P. *Org. Lett.* 2003, *5*, 1059.
153. Trost, B. M., Andersen, N. G. *J. Am. Chem. Soc.* 2003, *124*, 14320.

23 Industrial Applications of Chiral Auxiliaries

David R. Schaad

CONTENTS

23.1   INTRODUCTION

The use of chiral auxiliaries has traditionally been in the academic and small-scale synthetic arenas. This, in part, has been a result of the infancy of chiral auxiliaries themselves in organic synthesis. Many of these chiral substrates have been developed only in the last 10–20 years, and their synthetic utility is only now being realized. Additionally, many chiral auxiliaries are difficult to prepare and handle at large scale. Only recently has their been considerable effort to produce large quantities of the auxiliaries. Another important consideration is the cost of the auxiliary compared to alternate methods, such as resolution. The experimental reaction conditions for some of the more common chiral auxiliaries often call for specialized equipment, extreme temperature conditions, or both. These factors, to date, have combined to limit the use of chiral auxiliaries on an industrial scale.

This chapter will attempt to summarize the application of chiral auxiliaries at a scale important to the chiral fine chemical business. In addition, and as important, there are numerous cases where a chiral auxiliary has been used in the early stages of pharmaceutical development and the potential exists for this approach to be used in future work at larger scale. Last, there exists a limited number of cases where the auxiliary exists as part of an active drug. The known industrial applications, the potential applications, and the presence of auxiliaries in active drugs will all be addressed.

23.1.1   GENERAL CHIRAL AUXILIARY REVIEW

By definition, a chiral auxiliary differs from a chiral template in that the auxiliary is capable of being recycled after the desired asymmetric reaction. Hence, chiral templates will not be included in this chapter. The uses of a chiral moiety as a ligand for a reagent have also been excluded.

A number of reviews exist on the formation and uses of chiral auxiliaries.[1–7] Some of these auxiliaries, such as oxazolines, can be used on their own or incorporated into chiral ligands for asymmetric transition metal-catalyzed synthesis.[7]

Many chiral auxiliaries are derived from 1,2-amino alcohols.[7] These include oxazolidinones (**1**),[7–9] oxazolines (**2**),[10,11] bis-oxazolines (**3**),[12,13] oxazinones (**4**),[14] and oxazaborolidines (**5**).[15–17] Even the 1,2-amino alcohol itself can be used as a chiral auxiliary.[18–22] Other chiral auxiliaries examples include camphorsultams (**6**),[23] piperazinediones (**7**),[24] SAMP [(S)-1-amino-2-methoxy-methylpyrrolidine] (**8**) and RAMP (*ent*-**8**),[25] chiral boranes such as isopinocampheylborane (**9**),[26] and tartaric acid esters (**10**). For examples of terpenes as chiral auxiliaries, see Chapter 5. Some of these auxiliaries have been used as ligands in reagents (e.g., Chapters 17 and 24), such as **3** and **5**, whereas others have only been used at laboratory scale (e.g., **6** and **7**). It should be noted that some auxiliaries may be used to synthesize starting materials, such as an unnatural amino acid, for a drug synthesis, and these may not have been reported in the primary literature.

23.2 CHIRAL AUXILIARY STRUCTURES IN PHARMACEUTICALS

Chiral auxiliaries can serve as a chiral template, and the structural motif can be incorporated into an active pharmaceutical. Although the chemistry of the chiral unit can be exploited, this is not a true application of an auxiliary as the unit becomes part of the final molecule.

The oxazolidinone-based compounds are the most common (Figure 23.1), and examples include antibacterial,[27,28] adhesion receptor antagonists,[29,30] tumor necrosis factor inhibitors,[31] platelet aggregation inhibitors,[32] antimigraine drugs,[33] and monoamine oxidase inhibitors.[34,35] In one case, an oxazolidinone has been used as a transdermal drug delivery agent.[36]

Oxazoles and isooxazoles are also found in many active drugs (Figure 23.2). Although not as prevalent compared to the oxazolidinones, examples include antiinflammatories[37,38] and nitric oxide–associated diseases.[39]

Although this application area is intriguing, the focus of this chapter remains on the broader application of chiral auxiliaries for asymmetric synthesis.

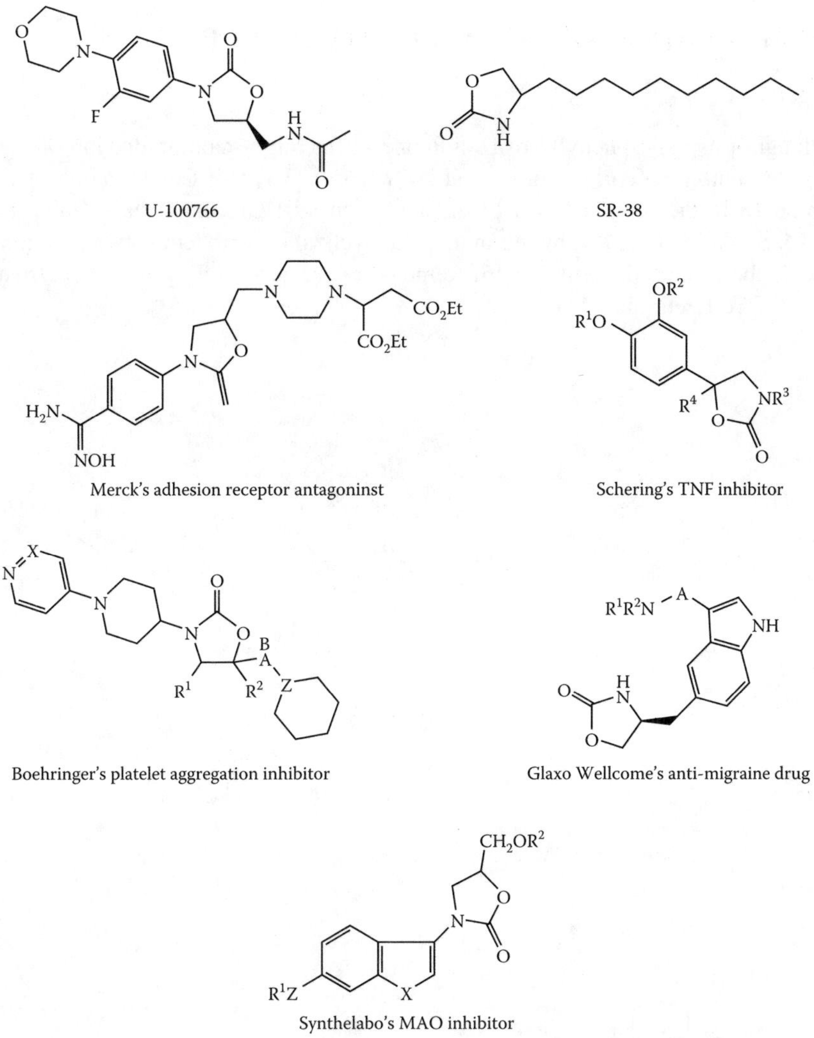

U-100766

SR-38

Merck's adhesion receptor antagoninst

Schering's TNF inhibitor

Boehringer's platelet aggregation inhibitor

Glaxo Wellcome's anti-migraine drug

Synthelabo's MAO inhibitor

FIGURE 23.1 Examples of oxazolidinones as part of the structures of active pharmaceuticals.

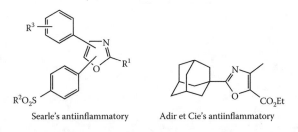

Searle's antiinflammatory

Adir et Cie's antiinflammatory

Takeda NO production inhibitor

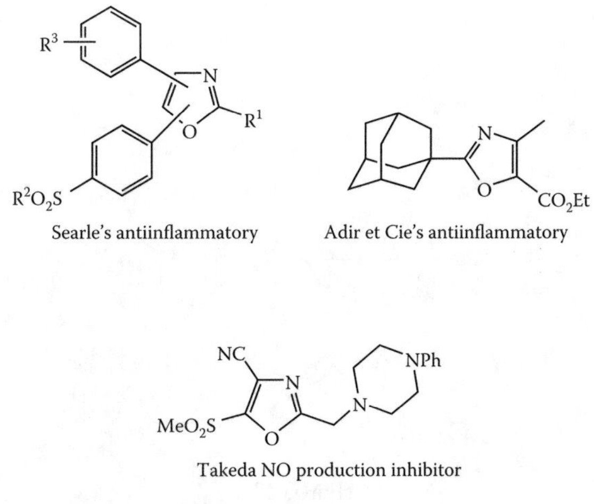

FIGURE 23.2 Examples of oxazoles and isoxazoles in active pharmaceuticals.

23.3 THE APPLICATION OF CHIRAL AUXILIARIES IN INDUSTRY

23.3.1 EZETIMIBE

Schering-Plough uses (*S*)-4-phenyl-2-oxazolidinone in the large-scale production of their choles-terol absorption inhibitor Zetia (ezetimibe) (**11**) (Scheme 23.1).[40–42] Condensation of the alcohol **12** with imine **13** in the presence of a Lewis acid such as TiCl$_4$ and tertiary amine base yields compound **14**. Silylation followed by intramolecular cyclization with tetrabutylammonium fluoride (TBAF) yields the protected ezetimibe **15**. Removal of the protecting groups is carried out with weak acid to afford Ezetimibe (**11**).

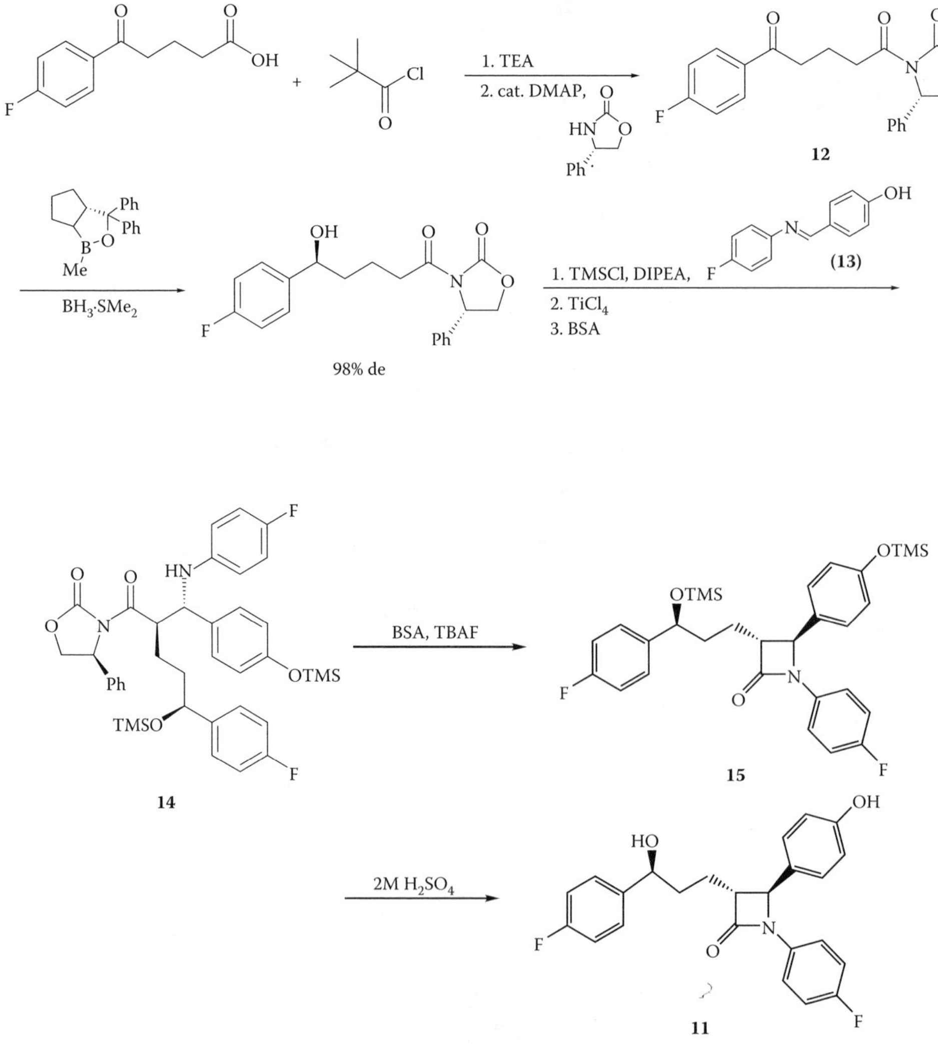

SCHEME 23.1

23.3.2 PNP405

The (*R*)-4-phenyl-2-oxazolidinone auxiliary was used in the process development and scale up of Novartis' purine nucleoside phosphorylase inhibitor PNP405 (**16**) (Scheme 23.2).[43] Asymmetric alkylation of **17** with bromoacetonitrile provided a 7:1 diastereoisomeric ratio of crude **18**. Recrystallization afforded **18** in 80% yield and >99% de. Simple addition of sodium borohydride in tetrahydrofuran (THF)-water at room temperature[44] resulted in the desired γ-cyano alcohol **19** and recovery of the auxiliary.

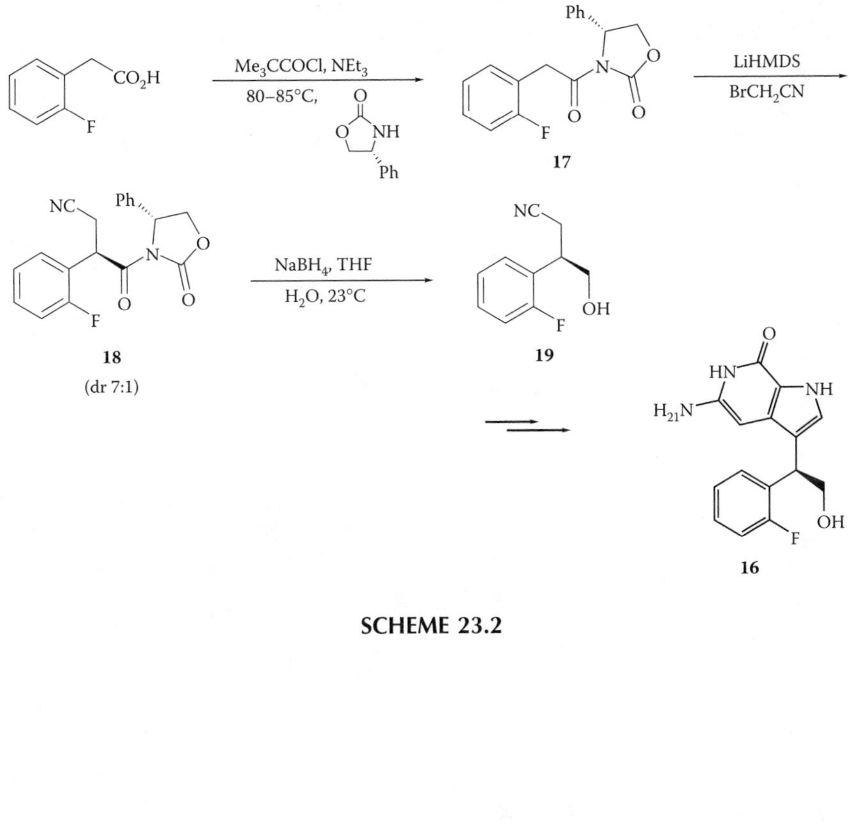

SCHEME 23.2

23.3.3 Zambon Naproxen Process

Tartaric acid has been used as a chiral auxiliary in a patented route to (*S*)-naproxen (**20**) (Scheme 23.3).[45–49] In initial studies, the acetal **21** was used to allow a stereoselective bromination that resulted in a 91:9 ratio of the (*RRS*)- and (*RRR*)-bromo derivatives **22** and **23**. The bromo acetal diesters could be completely separated. Debromination of **22**, followed by acid hydrolysis, led to formation of (*S*)-naproxen (**20**) in 80% yield, >99% ee, and recovery of the auxiliary. Conversely, debromination and hydrolysis of **23** gave only 12% yield of (*R*)-naproxen and 86% ee. In this case, the hydroxy acetal **24** was the major product (68%). However, the auxiliary was recovered in enantiomerically pure form.

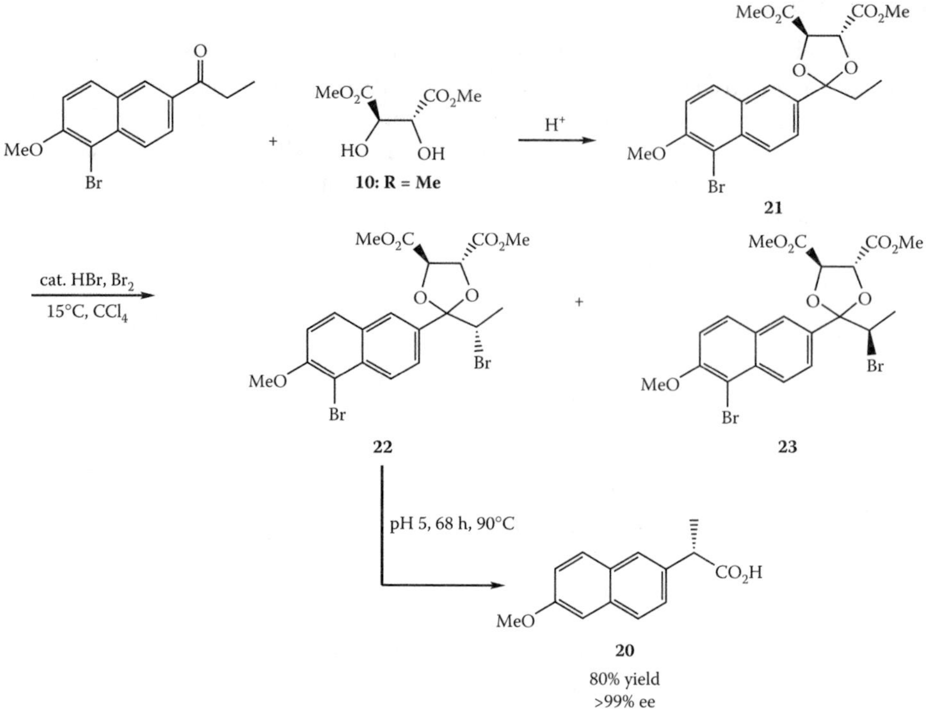

SCHEME 23.3

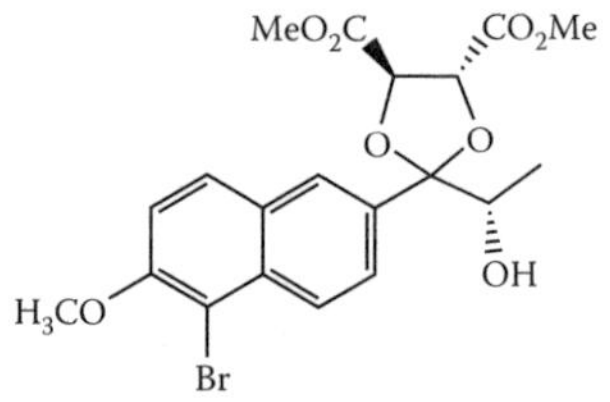

23.3.4 PARKE-DAVIS ROUTE TO CI-1008

An anticonvulsant, CI-1008, also called Pregabalin (**25**), was developed using the Evans' oxazolidinone auxiliary **26** derived from (*1S,2R*)-(–)-norephedrine.[50] The route was scaled up to produce several hundred kilogram quantities at production and pilot plant scale and provided early clinical trial material (Scheme 23.4).

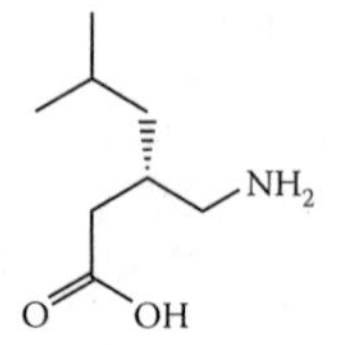

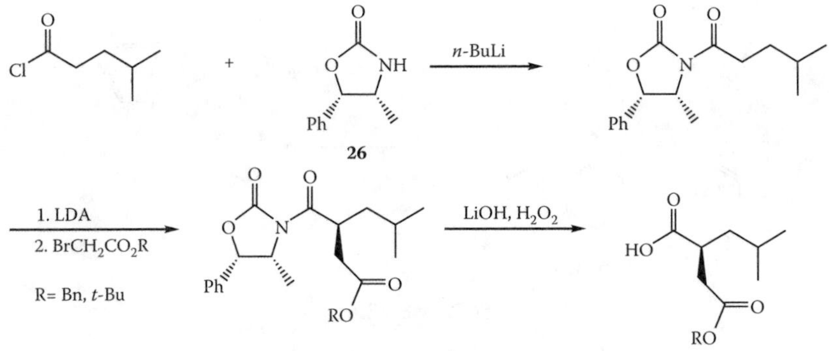

SCHEME 23.4

It was found that the acylation step could be run efficiently at 0°C. However, it was observed that even a slight excess of *n*-BuLi (<5%) resulted in a 15% yield loss at the acylation step, accounted for by the epimerization of the carbon adjacent to the phenyl ring on the oxazolidinone. It was noticed that a red color resulted when excess *n*-BuLi was present as a result of the formation of a dianionic species. Thus, stopping the addition when only the characteristic yellow color was present maintained the normally high yields (95%) for this acylation.

The alkylation step could be run as high as −35°C, but higher than this temperature a side product formed. This was a result of decomposition of the lithium enolate through a ketene that underwent alkylation with *tert*-butyl bromoacetate (Scheme 23.5).

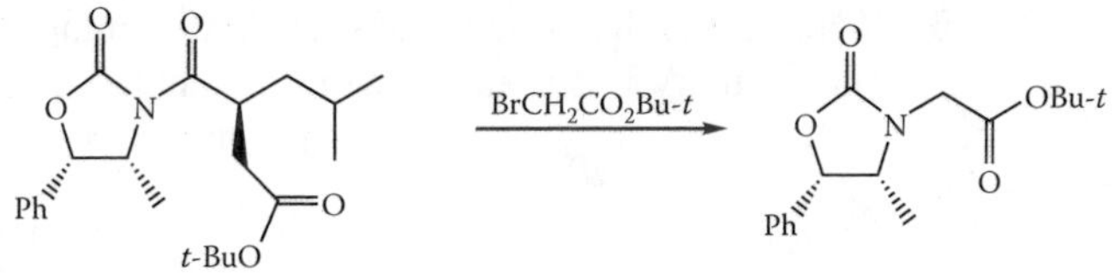

SCHEME 23.5

Overall recovery of the auxiliary was 64%. On cost estimates for scale up work it was found that the previously mentioned chiral auxiliary route exceeded the desired cost by a factor of six, primarily as a result of the cost of the chiral oxazolidinone and a yield of 33% over 10 steps. The route was ultimately replaced with a classical resolution protocol using mandelic acid, and this has been superseded by asymmetric approaches (see Chapter 12).

23.4 POTENTIAL APPLICATIONS OF CHIRAL AUXILIARIES

There are numerous examples where the potential for the application of chiral auxiliaries, once again primarily oxazolidinones, exists for an industrial setting.

The synthesis of calcium channel blockers of the diltiazem group, effective in lowering blood pressure, has been reported by Hoffmann-LaRoche (Scheme 23.6).[51] A screening of several auxiliaries yielded acceptable ee, but it was found that the use of (*1R,2S*)-2-phenylcyclohexanol (**27**) gave the required diastereoisomer **28** as the major isomer.[52–54] The intermediate could be isolated readily, and the auxiliary could be recycled simply by base hydrolysis. Multi-kilogram quantities of enantiomerically pure naltiazem (**29**) and diltiazem (**30**) were produced by this method.

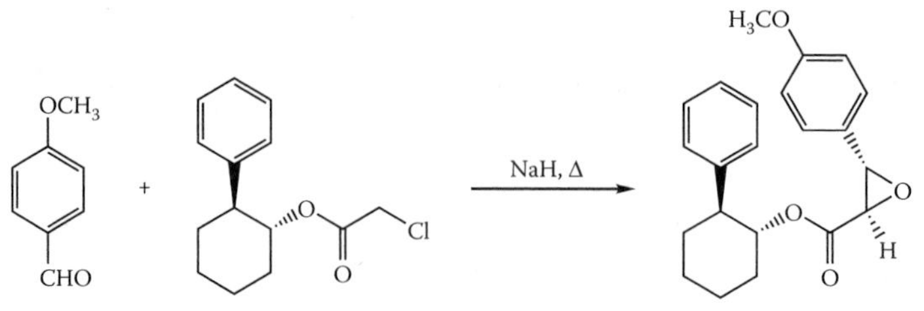

SCHEME 23.6

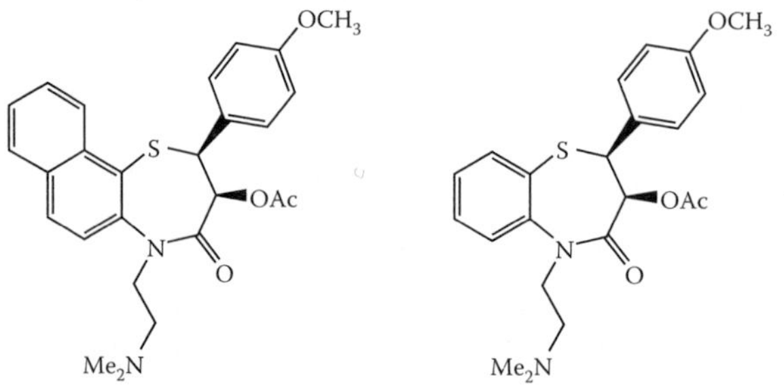

Alternatively, diltiazem (**30**) has been prepared using the Evans' auxiliary derivative **31** derived from L-valine (Scheme 23.7).[55] After dehydration of the adduct from the condensation of **31** with anisaldehyde through the mesylate, the enol ether was formed with a *Z:E* ratio of 4:1. This imide was then treated with 2-aminothiophenol in the presence of 0.1 equiv. 2-aminothiophenoxide with no change in the isomer ratio. The auxiliary was removed with trimethylaluminum, with concomitant formation of the lactam. After separation by crystallization, the correct diastereoisomer was converted to diltiazem in >99%ee.

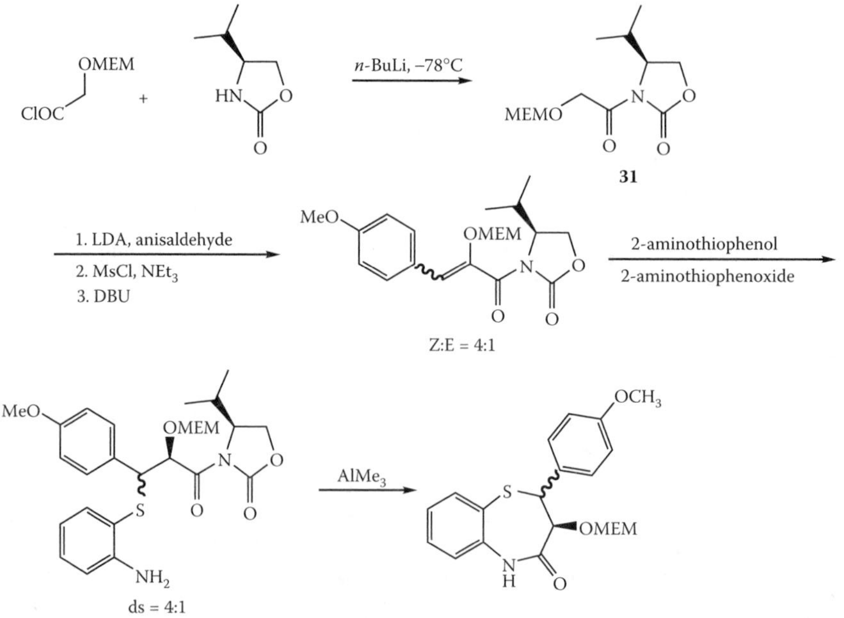

SCHEME 23.7

Lilly has used an Evans' auxiliary in its synthesis of LY309887 (**32**), a dideazafolate antitumor agent (Scheme 23.8).[2] The de of the key step, reaction of the titanium-derived enolate, was >98%. It was noted that the auxiliary could be recycled after cleavage with lithium borohydride.

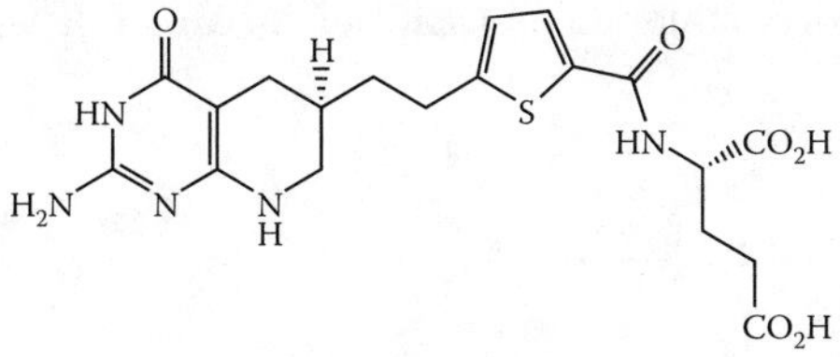

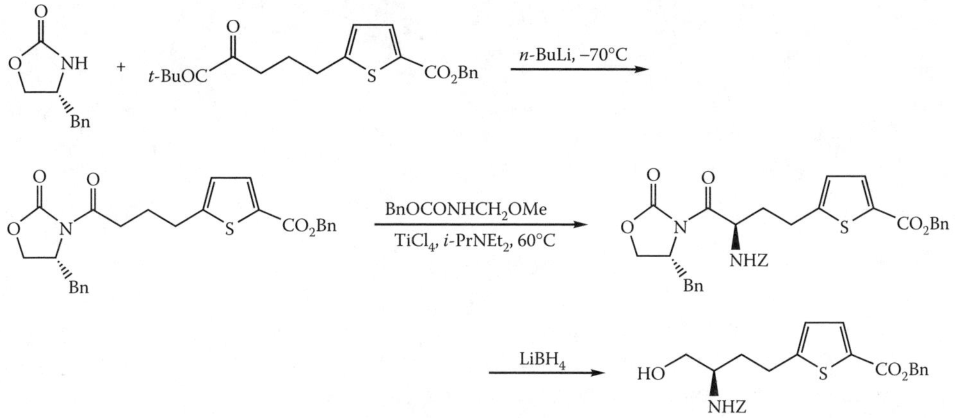

SCHEME 23.8

Amides of (*S*)-lactic acid have been used as chiral auxiliaries in the dynamic kinetic resolution of racemic ibuprofen (Scheme 23.9).[56] The therapeutically effective (*S*)-isomer **33** was obtained in 80% yield, with complete recovery of the pyrrolidine-derived (*S*)-lactamide auxiliary **34**.

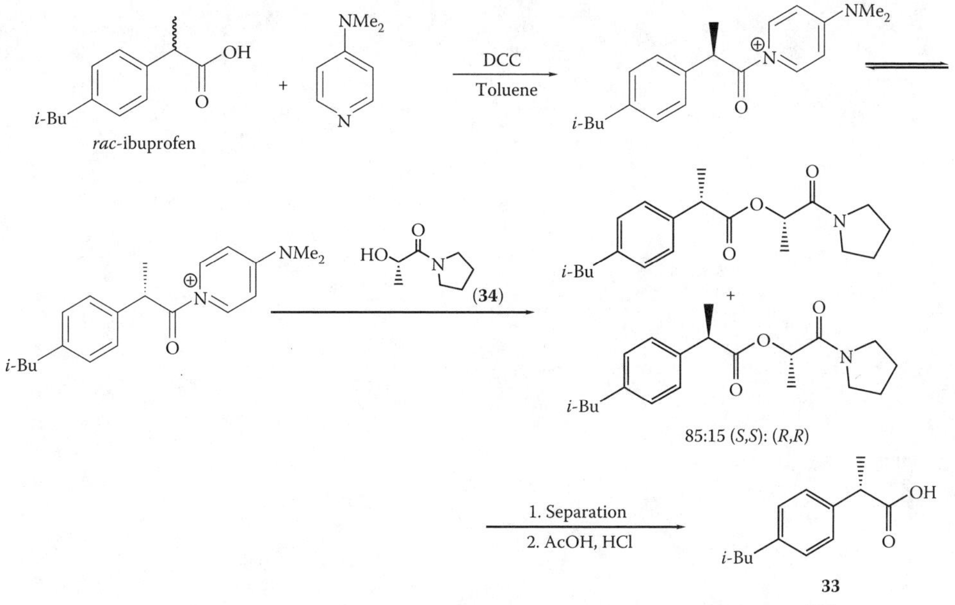

SCHEME 23.9

Merck and Banyu process research and development teams have demonstrated the use of chiral auxiliaries for the production of an endothelin receptor antagonist.[57] In one step, (1*R*, 2*S*)-*cis*-aminoindanol is used as the chiral auxiliary in the alkylation of the enolate derived from the propionyl amide **35** with benzyl chloride **36** (Scheme 23.10) (*cf.* Chapter 24).

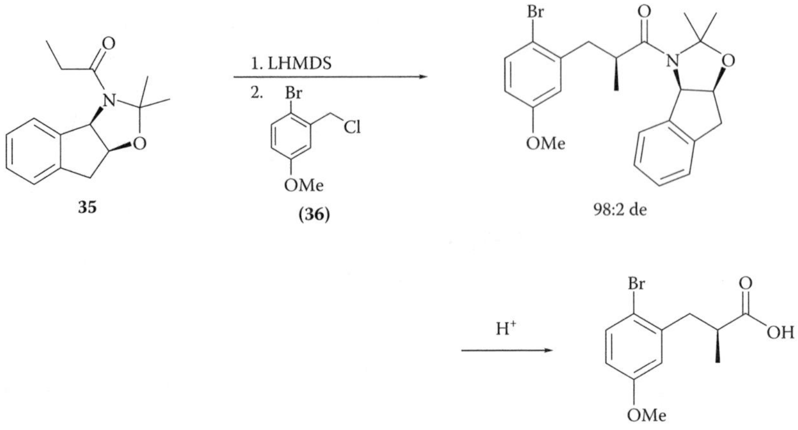

SCHEME 23.10

In a separate step of the process, either (1*S*, 2*S*)-(+)-pseudoephedrine (**37**) or *N*-methyl-(1*S*, 2*R*)-*cis*-1-amino-2-indanol (**38**) was successfully used as auxiliaries for the chiral Michael addition of an aryllithium to give compound **39** (Scheme 23.11).[57]

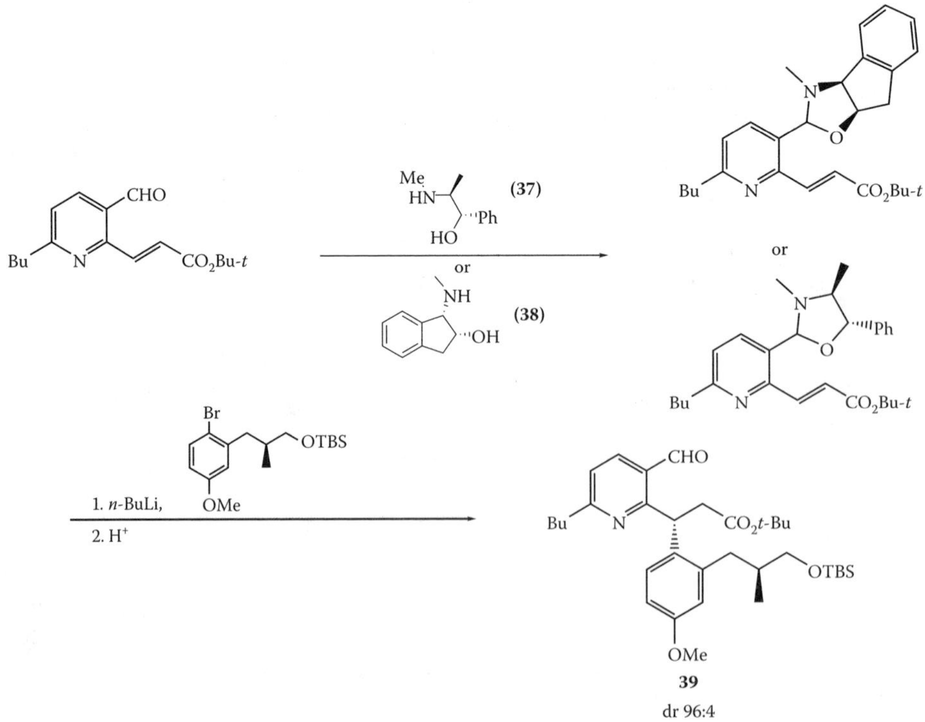

SCHEME 23.11

Bio-Mega/Boehringer Ingelheim have used an oxazolidinone auxiliary in the synthesis of renin inhibitors (e.g., **40**) for the treatment of hypertension and congestive heart failure.[58] A multi-step derivitization of the oxazolidinone **41** from reaction of (*S*)-4-(1-methylethyl)-2-oxazolidinone and 4-bromo-4-pentenoic acid yielded the desired compound.

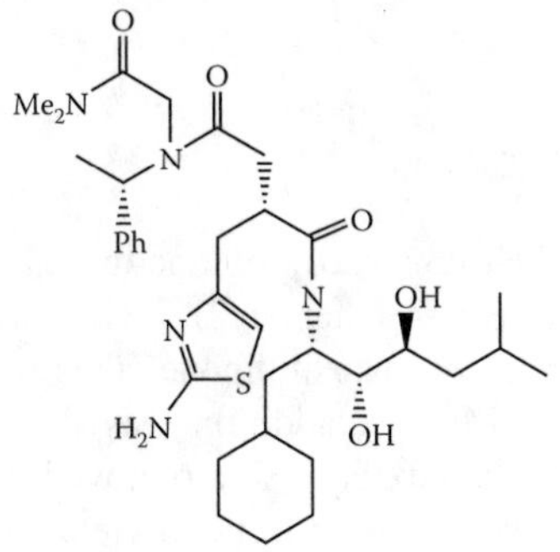

40 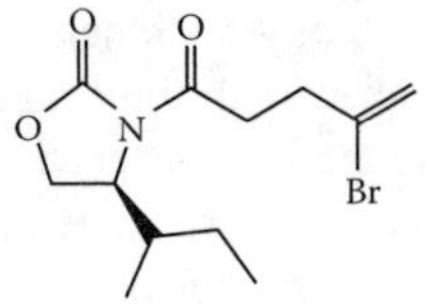   **41**

The optically active alkaloid Bao Gong Teng A (**42**), used to treat glaucoma, has been prepared with methyl (*S*)-lactate as the auxiliary. The key step was the asymmetric 1,3-dipolar cycloaddition of a 3-hydroxypyridinium betaine with the acrylate **43** where the auxiliary is attached directly to the dienophile (Scheme 23.12). The auxiliary was removed, after several steps, yielding the desired product in optically pure form but in only 9% overall yield.[59,60]

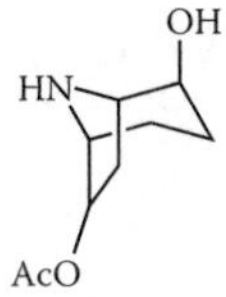

42

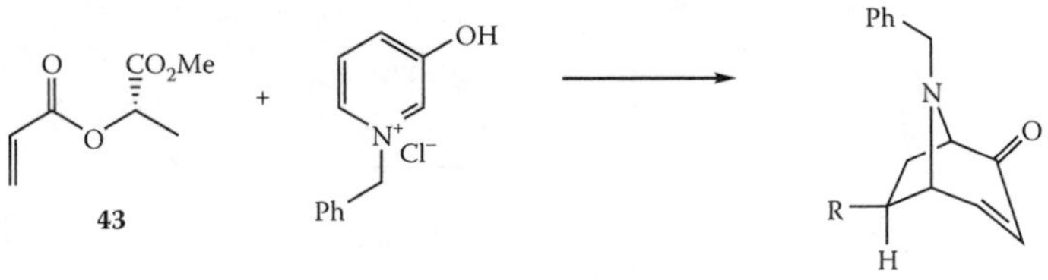

43

SCHEME 23.12

(*S*)-Proline has been used in the synthesis of (*R*)-etomoxir (**44**), a powerful hypoglycemic agent (Scheme 23.13).[61] A proline derivative was used for the diastereoselective bromolactonization step. After hydrolysis of the auxiliary and further functional group transformations, the desired product was obtained in an overall yield of 45%.

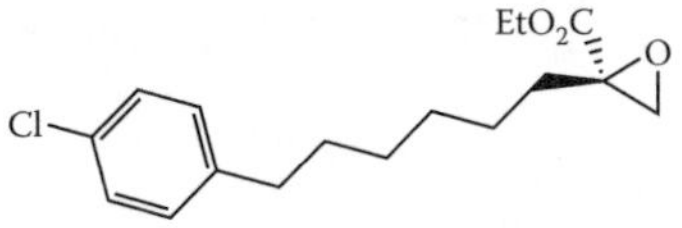

44

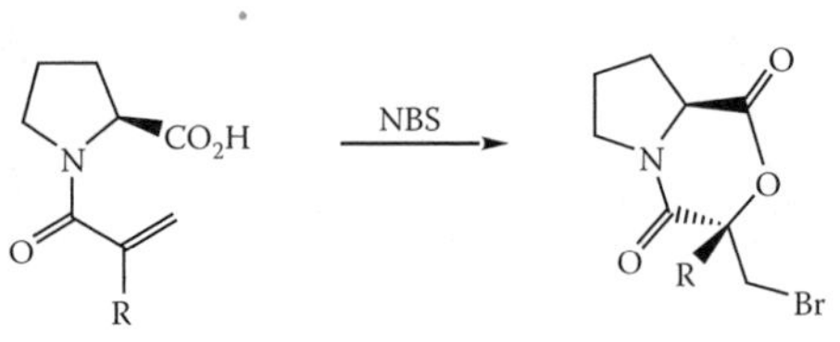

R= -(CH$_2$)$_6$OBn

SCHEME 23.13

Merck has used the L-phenylalanine-derived Evans' oxazolidinone to make matrix metallopro-
teinase inhibitors such as **45** (Scheme 23.14) (see also Chapter 2).[62] The mixed anhydride of 4-
butyric acid was reacted with the lithium anion of the oxazolidinone. This was enolized with the
standard titanium reagents. An enantioselective Michael addition was then carried out by the
addition of *t*-butyl acrylate at low temperature. The auxiliary was removed with LiOH/peroxides
to give the acid, which was further derivatized over multiple steps to yield the desired drug.

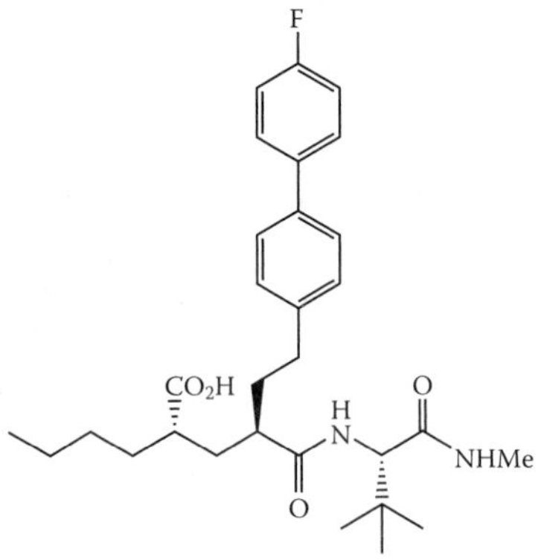

45

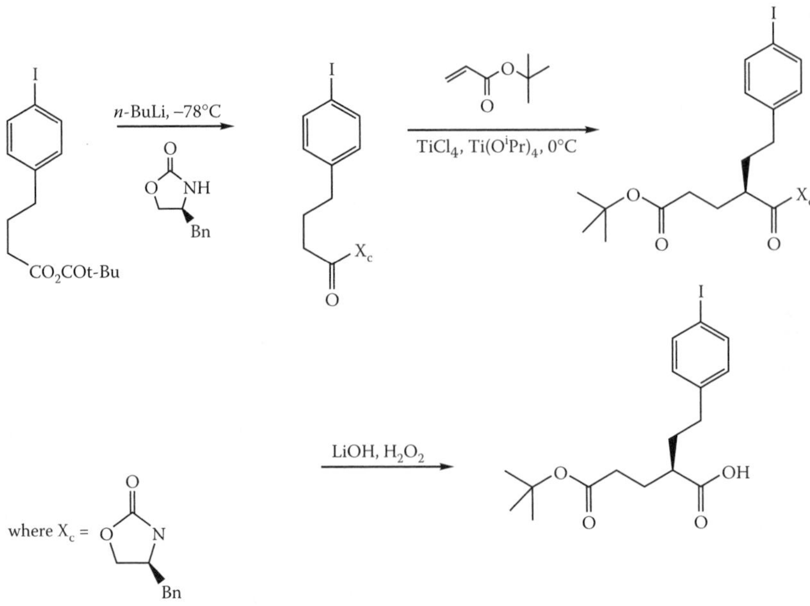

SCHEME 23.14

23.5 SUMMARY

The use of chiral auxiliaries has seen, and will continue to see, an ever-increasing use in an industrial setting. Heretofore, the use of such auxiliaries has been limited by their price, availability, and novelty to large-scale production. The synthetic utility of many of the auxiliaries, particularly the oxazolidinones, is just now being realized. Several of these auxiliaries are just now becoming commercially available, which in turn should make them more competitive with alternate production methods currently in practice. With the demand for producing single-enantiomer pharmaceuticals continuing to increase, the future industrial use of chiral auxiliaries appears promising.

REFERENCES

1. Gao, Y., Boschetti, E., Guerrier, L. *Ann. Pharm. Rev.* 1994, *52*, 184.
2. Barnett, C. J., Wilson, T.M., Evans, D. A., Somers, T. C. *Tetrahedron Lett.* 1997, *38*, 735.
3. Blaser, H.-U. *Chem. Rev.* 1992, *92*, 935.
4. Meyers, A. I. in *Stereocontrolled Organic Synthesis,* Trost, B. M. Ed., Blackwell: Oxford, 1994, p. 144.
5. Foskolos, G. *Pharmakeutike* 1995, *8*, 155.
6. Marchand-Brynaert, J. *Belg. Chim. Nouv.* 1991, *8*, 155.
7. Ager, D. J., Prakash, I., Schaad, D. R. *Chem. Rev.* 1996, *96*, 835.
8. Evans, D. A. *Aldrichimica Acta* 1982, *15*, 23.
9. Swern, D., Dyen, M. E. *Chem. Rev.* 1967, *67*, 197.
10. Gant, T. G., Meyers, A. I. *Tetrahedron* 1994, *50*, 2297.
11. Reuman, M., Meyers, A. I. *Tetrahedron* 1985, *41*, 837.
12. Bolm, C. *Angew. Chem., Int. Ed.* 1991, *30*, 542.
13. Pfaltz, A. *Acc. Chem. Res.* 1993, *26*, 339.
14. Williams, R. M. *Aldrichimica Acta* 1992, *25*, 11.
15. Corey, E. J. U.S. Pat. 4,943,635, 1990.
16. Itsuno, S., Sakurai, Y., Ito, K., Hirao, A., Nakahama, S. *Bull. Chem. Soc., Jpn.* 1987, *60*, 395.
17. Itsuno, S., Nakano, M., Miyazaki, K., Masuda, H., Ito, K., Hirao, A., Nakahama, S. *J. Chem. Soc., Perkin Trans. I* 1985, 2039.
18. Meyers, A. G., Gleason, J. L., Yoon, T. *J. Am. Chem. Soc.* 1995, *117*, 8488.
19. Meyers, A. G., Yang, B. H., Chen, H., McKinistry, L., Kopecky, D., Gleason, J. L. *J. Am. Chem. Soc.* 1997, *119*, 6496.
20. Meyers, A. G., Yang, B. H., Chen, H., Gleason, G. L. *J. Am. Chem. Soc.* 1994, *116*, 9361.
21. Larcheveque, M., Ignatova, E., Cuvigny, T. *Tetrahedron Lett.* 1978, *19*, 3961.
22. Meyers, A. G., Gleason, J. L., Yoon, T., Kung, D. W. *J. Am. Chem. Soc.* 1997, *119*, 656.
23. Oppolzer, W., Moretti, R., Thomi, S. *Tetrahedron Lett.* 1989, *30*, 5603.
24. Schoellkopf, U. *Pure Appl. Chem.* 1983, *55*, 1799.
25. Enders, D., Papadopoulos, K., Rendenbach, B. E. M. *Tetrahedron Lett.* 1986, *27*, 3491.
26. Brown, H. C., Singaram, B. *J. Am. Chem. Soc.* 1984, *106*, 1797.
27. Brickner, S. J., Hutchinson, D. K., Barbachyn, M. R., Manninen, P. R., Ulanowicz, D. A., Garmon, S. A., Grega, K. C., Hendges, S. K., Toops, D. S., Ford, C. W., Zurenko, G. E. *J. Med. Chem.* 1996, *39*, 673.
28. Barbachyn, M. R., Hutchinson, D. K., Brickner, S. J., Cynamon, M. H., Kilburn, J. O., Klemens, S. P., Glickman, S. E., Grega, K. C., Hendges, S. K., Toops, D. S., Ford, C. W., Zurenko, G. E. *J. Med. Chem.* 1996, *39*, 680.
29. Gante, J., Juraszyk, H., Raddatz, P., Wurziger, H., Bernotat-Danielowski, S., Melzer, G. Eur. Pat. EP 96-101746, 1996.
30. Gante, J., Juraszyk, H., Raddatz, P., Wurziger, H., Bernotat-Danielowski, S., Melzer, G. Eur. Pat. EP 96-106423, 1996.
31. Graf, H., Wachtel, H., Schneider, H., Faulds, D., Perez, D., Dinter, H. PCT Int. Appl. WO 96-DE257, 1996.
32. Tsaklakidis, C., Schaefer, W., Doerge, L., Esswein, A., Friebe, W.-G. Ger. Offen. DE 95-19524765, 1995.

33. Patel, R. PCT Int. Appl. WO 96-GB1886, 1996.
34. Jegham, S., Puech, F., Burnier, P., Berthon, D. PCT Int. Appl. WO 96-FR1732, 1996.
35. Jegham, S., Puech, F., Burnier, P., Berthon, D., Leclerc, O. PCT Int. Appl. WO 96-FR1511, 1996.
36. Rajadhyaksha, V. J., Pfister, W. R. *Drug Cosmetic Ind.* 1996, *157 (3)*, 36.
37. de Nanteuil, G., Vincetn, M., Lila, C., Bonnet, J., Fradin, A. Eur. Pat. EP 93-402967, 1993.
38. Talley, J. J., Bertenshaw, S., Rogier, D. J., Graneto, M., Brown, D. L., Devadas, D., Lu, H.-F., Sikorski, J. A. PCT Int. Appl. WO 96-US6992, 1996.
39. Sugihara, Y., Uchibayashi, N., Matsumura, K., Nozaki, Y., Ichimori, Y. PCT Int. Appl. WO 9724340, 1997.
40. Thiruvengadam, T. K., Fu, X., Tann, C.-H., McAllister, T. L., Chiu, J. S., Colon, C. U.S. Pat. 6,207,822, 2001
41. Wu, G., Wong, Y., Chen, X., Ding, Z. *J. Org. Chem.* 1999, *64*, 3714
42. Rosenblum, S. R., Huynh, T., Afonso, A., Davis, H. R., Yumibe, N., Clader, J. W., Burnett, D. A. *J. Med. Chem.* 1998, *41*, 973.
43. Prashad, M., Har, D., Chen, L., Kim, H.-Y., Repic, O., Blacklock, T. J. *J. Org. Chem.* 2002, *67*, 6612
44. Prashad, M., Har, D., Kim. H.-Y., Repic, O. *Tetrahedron Lett.* 1998, *38*, 7067.
45. Castaldi, G., Giordano, C. *Synthesis* 1987, 1039.
46. Crosby, J. *Tetrahedron* 1991, *47*, 4789.
47. Giordano, C., Castaldi, G., Cavicchioli, S. U.S. Pat. 4,888,433, 1989.
48. Giordano, C., Castaldi, G., Cavicchioli, S., Villa, M. G. *Tetrahedron* 1989, *45*, 4243.
49. Giordano, C., Castaldi, G., Uggeri, F., Cavicchioli, S. U.S. Pat. 4,855,464, 1989.
50. Hoekstra, M. S., Sobieray, D. M., Schwindt, M. A., Mulhern, T. A., Grote, T. M., Huckabee, B. K., Hendrickson, V. S., Franklin, L. C., Granger, E. J., Karrick, G. L. *Org. Proc. Res. Dev.* 1997, *1*, 26.
51. Schwartz, A., Madan, P. B., Mohacsi, E., O'Brien, J. P., Todaro, L. J., Coffen, D. L. *J. Org. Chem.* 1992, *57*, 851.
52. Oppolzer, W., Chapuis, C., Kelly, M. J. *Helv. Chim. Acta.* 1983, *66*, 2358.
53. Helmchen, G., Selim, A., Dorsch, D., Taufer, I. *Tetrahedron Lett.* 1983, *24*, 3213.
54. Whitesell, J. K., Chen, H. C., Lawrence, R. M. *J. Org. Chem.* 1986, *51*, 4663.
55. Miyata, O., Shinada, T., Ninomiya, N., Naito, T. *Tetrahedron* 1997, *53*, 2421.
56. Ammazzalorso, A., Amorosso, R., Bettoni, G., De Filippis, B., Giampietro, L., Pierini, M., Tricca, M. L. *Tetrahedron Lett.* 2002, *43*, 4325.
57. Song, Z. J., Zhao, M., Desmond, R., Devine, P., Tschaen, D. M., Tillyer, R., Frey, L., Heid, R., Xu, F., Foster, B., Li, J., Reamer, R., Volante, R., Grabowski, E. J. J., Dolling, U. H., Reider, P. J., Okada, S., Kato, Y., Mano, E. *J. Org. Chem.* 1999, *64*, 9658.
58. Lavallee, P., Simoneau, B. U.S. Pat. 5,541,163, 1993.
59. Pham, V. C., Charlton, J. L. *J. Org. Chem.* 1995, *60*, 8051.
60. Jung, M. E., Longmei, Z., Tangsheng, P., Huiyan, Z., Yan, L., Jingyu, S. *J. Org. Chem.* 1992, *57*, 3528.
61. Jew, S., Kim, H., Jeong, B., Park, H. *Tetrahedron: Asymmetry* 1997, *8*, 1187.
62. Esser, C. K., Bugianesi, R. L., Caldwell, C. G., Chapman, K. T., Durette, P. L., Girotra, N. N., Kopka, I. E., Lanza, T. J., Levorse, D. A., MacCoss, M., Owens, K. A., Ponpipom, M. M., Simeone, J. P., Harrison, R. K., Niedzwiecki, L., Becker, J. W., I., M. A., Axel, M. G., Christen, A. J., McDonnell, J., Moore, V. L., Olszewski, J. M., Saphos, C., Visco, D. M., Shen, F., Colletti, A., Krieter, P. A., Hagmann, W. K. *J. Med. Chem.* 1997, *40*, 1026.

24 The Role and Importance of *cis*-1-Amino-2-Indanol in Biological Systems and Asymmetric Processes

Chris H. Senanayake, Dhileepkumar Krishnamurthy, and Isabelle Gallou

CONTENTS

24.1 INTRODUCTION

The significance of asymmetric synthesis has long been appreciated by the scientific community. Indeed, natural products generally exist as a single enantiomer as a consequence of the inherent chirality of the enzymes that produce them. Furthermore, enzymes, receptors and other binding sites in biological systems recognize compounds with a specific chirality. Two enantiomers of a chiral molecule can display dramatically different biological activities, to the point that one enantiomer is an active drug and the other exhibits fatal toxicity. Therefore, the development of new asymmetric synthetic methods for preparation of enantiopure materials has become of critical importance.[1,2]

In the search for easily accessible chiral templates for asymmetric transformations, chiral amino alcohols derived from α-amino acids have been identified as versatile reagents for the generation of enantiopure compounds and have been used in numerous efficient auxiliary-directed and catalytic asymmetric reactions. Amino alcohols that are not derived from the chiral pool sometimes offer advantages in terms of structural diversity and conformational properties. Among those, *cis*-1-amino-2-indanol (**1**) played a crucial role in the development of biological systems. Since its discovery as a valuable human immunodeficiency virus protease (HIV-PR) inhibitor ligand, many asymmetric methodologies have emerged that use **1** as a constrained phenyl glycinol surrogate. The rigid skeleton of auxiliaries and catalysts derived from *cis*-aminoindanol is the key element in attaining high selectivities. Because both enantiomers of **1** are readily accessible and now commercially available, both enantiomers of a target molecule can be easily prepared.[3–5]

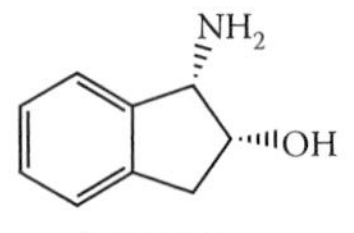

(1*S*,2*R*)-**1**

This chapter focuses on the importance of *cis*-1-amino-2-indanol as a chiral template in the development of new methodologies for the asymmetric synthesis of organic compounds. Other chiral auxiliaries are discussed in Chapter 23. The use of the amino alcohol **1** in resolutions is discussed in Chapter 8, whereas applications of **1** as a ligand are in Chapter 17.

24.2 *CIS*-1-AMINO-2-INDANOL IN BIOLOGICAL TARGETS

In the course of structure-activity relationship studies, the incorporation of *cis*-aminoindanol as chiral amino alcohol P_2' ligand in an HIV-PR inhibitor series led to the discovery (at Merck) of potent inhibitor L-685,434 (**2**). This compound displayed enzyme-inhibitory potency (IC_{50}) of 0.3 nM and antiviral potency (CIC_{95}) of 400 nM. Subsequent crystallographic experiments of enzyme-inhibitor complexes revealed that the indane hydroxyl group acted as carbonyl surrogate, which seemed to indicate that conformationally constrained β-hydroxy amines could be generally useful as amino acid replacements. However, L-685,434 suffered from aqueous insolubility (Figure 24.1).[6]

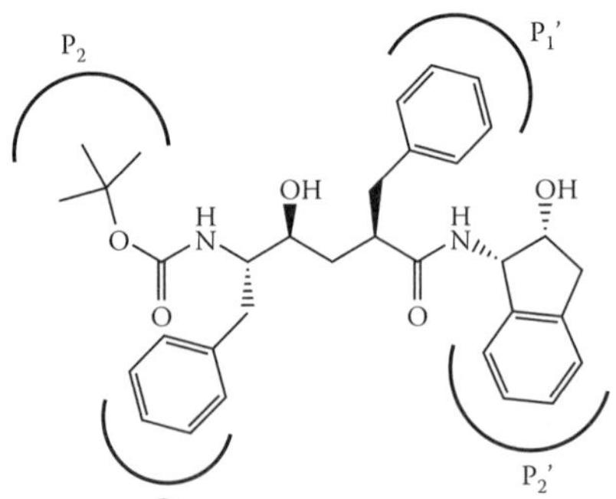

FIGURE 24.1 Structure of human immunodeficiency virus protease (HIV-PR) inhibitor L-685,434 (**2**).

TABLE 24.1
Enzyme Inhibitory and Antiviral Potencies
of Selected HIV-PR[a] Inhibitors

Inhibitor	IC_{50} (nM)	CIC_{95} (nM)	Reference
L-685,434 (**2**)	0.3	400	6
L-689,502 (**3**)	0.45	6–50	7
L-693,549 (**4**)	0.1	(MIC_{100} = 25–50)	9
L-700,417 (**5**)	0.67	100	10
6	110	—	12
7	(K_i = 220 nM)	—	13
8	<0.03	3	15
11	7.6	400	16
Indinavir (**9**)	0.35	25–100	16
L-748,496 (**17**)	0.12	6–12	25
MK-944a (**18**)	0.18	29	26
19	47	—	27
20	3	—	28
21	(K_i = 0.4 nM)	—	29
22	(K_i = 0.5 nM)	—	30

[a] HIV-PR, human immunodeficiency virus protease.

The physical properties of **2** were modified by introduction of polar substituents to improve both antiviral potency and hydrophilicity. These studies led to the discovery of L-689,502 (**3**) and L-693,549 (**4**), each bearing a polar, hydrophilic substituent at the *para* position of the P_1' phenyl ring.[7–9] Both compounds indeed displayed improved solubilities and antiviral potencies (Table 24.1). An inhibitor with *pseudo-C_2*-symmetry, L-700,417 (**5**) was designed by rotation of the C-terminal half of **1** around the central hydroxyl-bearing carbon (Figure 24.2).[10] Askin and co-workers reported a concise and practical synthesis of compounds **2–5** by diastereoselective alkylation of a chiral amide enolate derived from (1*S*,2*R*)-aminoindanol.[11] This strategy, which efficiently used the *cis*-aminoindanol platform as chiral auxiliary, is fully detailed later in this chapter.

Further modification on **2** at P_1, P_1', and P_2' led to several hydroxy ethylene isosteres (e.g., **6** and **7**),[12–14] which displayed poorer enzyme inhibitory and antiviral potencies than **2** (Table 24.1). Incorporation of the constrained 3-*S*-tetrahydrofuranyl urethane in place of the *N*-terminus acyclic *tert*-butyl urethane in **2** resulted in a remarkable 10-fold increase in enzyme inhibitory potency (IC_{50} <0.03 nM) and more than 100-fold increase in antiviral potency (CIC_{95} = 3 nM).[15] Although very potent, the optimized inhibitors of the hydroxy ethylene isostere series still lacked adequate aqueous solubility and acceptable pharmacokinetic profile (Figure 24.2).

In 1994 Merck researchers disclosed the design and pharmacologic properties of orally bio-available HIV-PR inhibitor L-735,524 (**9**), which eventually became known as therapeutic agent Indinavir and was commercialized as Crixivan®.[16–22] It was hypothesized that incorporation of a basic amine, present in the potent HIV-PR inhibitor Ro 31-8959 (**10**, Saquinavir®), into the backbone of the L-685,434 series would improve the bioavailability of these compounds. Replacement of P_2/P_1 ligand of **2** (*N*-Boc and phenyl moieties) by the P_2'/P_1' ligand of **10** (decahydroisoquinoline *tert*-butyl amide) was envisioned to enhance aqueous solubility of the series. Despite the high enzyme-inhibitory

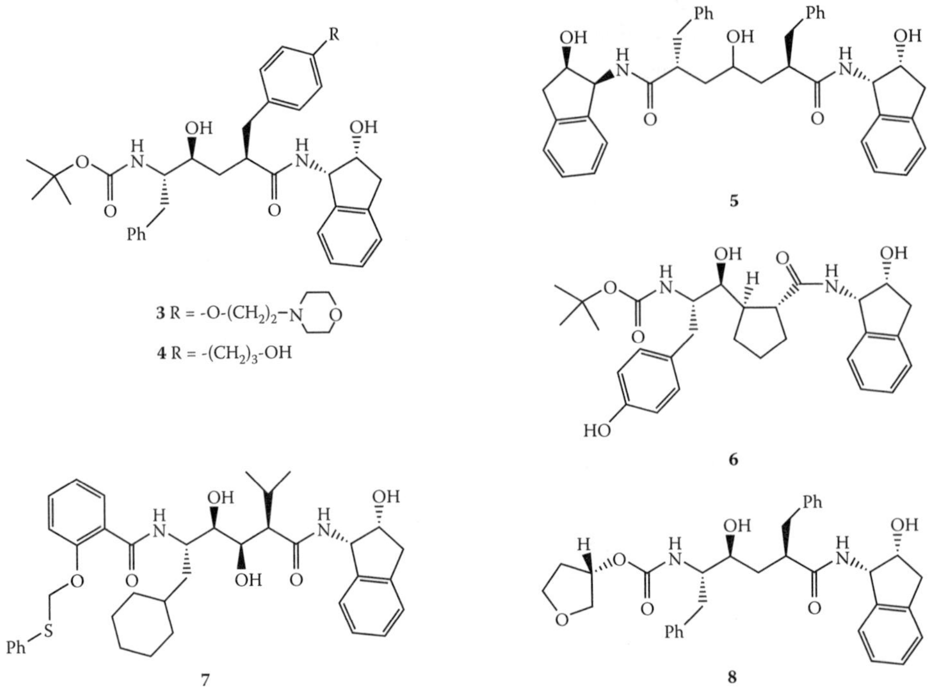

FIGURE 24.2 Human immunodeficiency virus protease (HIV-PR) inhibitors **3–8**.

potency of the resulting **11** ($IC_{50} = 7.6$ nM) and its favorable pharmacokinetic profile, this novel inhibitor displayed low antiviral potency ($CIC_{95} = 400$ nM). Systematic modification of the decahydroisoquinoline ring by 3-pyridylmethylpiperazine finally led to the discovery of L-735,524, which possessed high enzyme-inhibitory and antiviral potencies ($IC_{50} = 0.35$ nM, $CIC_{95} = 25$–100 nM) as well as a good pharmacokinetic profile. Improvement in formulation with the sulfate salt of **9** gave increased aqueous solubility and oral bioavailability (Figure 24.3).[16]

The industrial production of Crixivan® (**9**·H_2SO_4) took advantage of the chirality of (1*S*,2*R*)-aminoindanol to set the two central chiral centers of **9** by an efficient diastereoselective alkylation-epoxidation sequence.[17] The lithium enolate of **12** reacted with allyl bromide to give **13** in 94% yield and 96:4 diastereoselective ratio. Treatment of a mixture of olefin **13** and *N*-chlorosuccinimide in isopropyl acetate–aqueous sodium carbonate with an aqueous solution of sodium iodide led to the desired iodohydrin in 92% yield and 97:3 diastereoselectivity. The resulting compound was converted to the epoxide **14** in quantitative yield. Epoxide opening with piperazine **15** in refluxing methanol followed by Boc-removal gave **16** in 94% yield. Finally, treatment of piperazine derivative **16** with 3-picolyl chloride in sulfuric acid afforded Indinavir sulfate in 75% yield from epoxide **14** and 56% yield for the overall process (Scheme 24.1).[17–22]

Continuing efforts[23,24] in the search for more effective protease inhibitors containing amino-indanol as P_2' ligand led to the discovery of the Indinavir backup candidate L-748,496 (**17**). Replacement of the P_1' benzyl moiety of **9** by a pyridyl methyl group improved both aqueous solubility and oral bioavailability of compounds with a highly lipophilic P_3 ligands such as [3,2*b*]-thienothiophene. HIV-PR inhibitor **17** was revealed to be 3 times as potent as Indinavir against HIV-1 ($IC_{50} = 0.12$ nM) and in cell culture and displayed excellent pharmacokinetic properties.[25] However, this compound did not show improved pharmacokinetic half-life over Indinavir in several

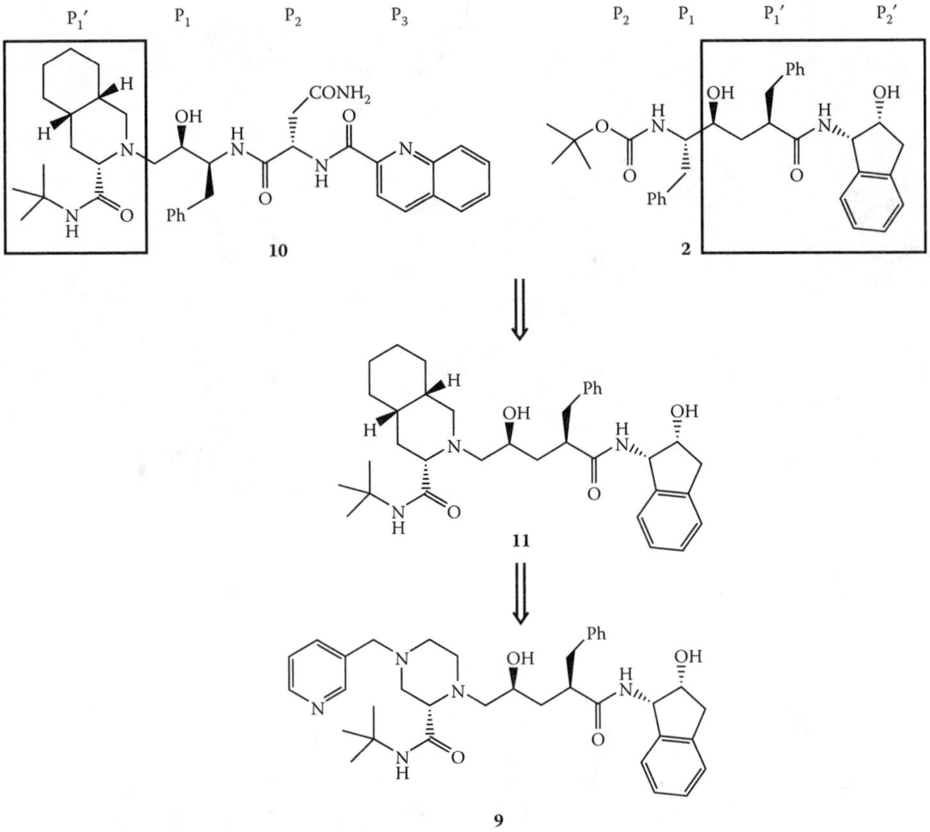

FIGURE 24.3 Design concept of Indinavir.

animal models. Benzofuran derivative **18**, despite its low aqueous solubility, was later identified as a more suitable backup candidate because it was found to be slightly more effective in cell culture assays than Indinavir in suppressing the spread of HIV-1 infection (Figure 24.4, Table 24.1).[26]

In 1996 researchers at Sandoz disclosed the incorporation of *cis*-aminoindanol in their lead compound for HIV-PR inhibition.[27] The most potent compound of their new series (**19**) only showed an IC_{50} of 47 nM and had nonexistent oral bioavailability (Figure 24.4, Table 24.1).[27]

Research groups at Glaxo-Wellcome and Vertex have disclosed a novel series of HIV-PR inhibitors containing a conformationally constrained P_1/P_2 heterocycle.[28] Preliminary studies suggested that a 5-membered ring scaffold may be superior to somewhat more flexible 6-membered ring systems such as pyrimidines. Ligand screening led to compound **20**, described as roughly equipotent to Indinavir (Figure 24.4, Table 24.1).[28]

Finally, Samuelsson and co-workers designed a new C_2-symmetric HIV-PR inhibitor, on the basis of X-ray crystal structures, which indicated that the HIV protease existed as a C_2-symmetric dimer. The peptidomimetic scaffold of this new class of inhibitors was based on D-mannitol and duplication of the C-terminus. Compounds **21** and **22** were rapidly synthesized and displayed comparable enzyme-inhibitory and antiviral potencies to Indinavir (Figure 24.4, Table 24.1).[29–32]

Researchers at DuPont also took advantage of the *cis*-aminoindanol moiety to develop a new series of potent, selective, and orally bioavailable aggrecanase inhibitors.[33] Aggrecanase is mainly

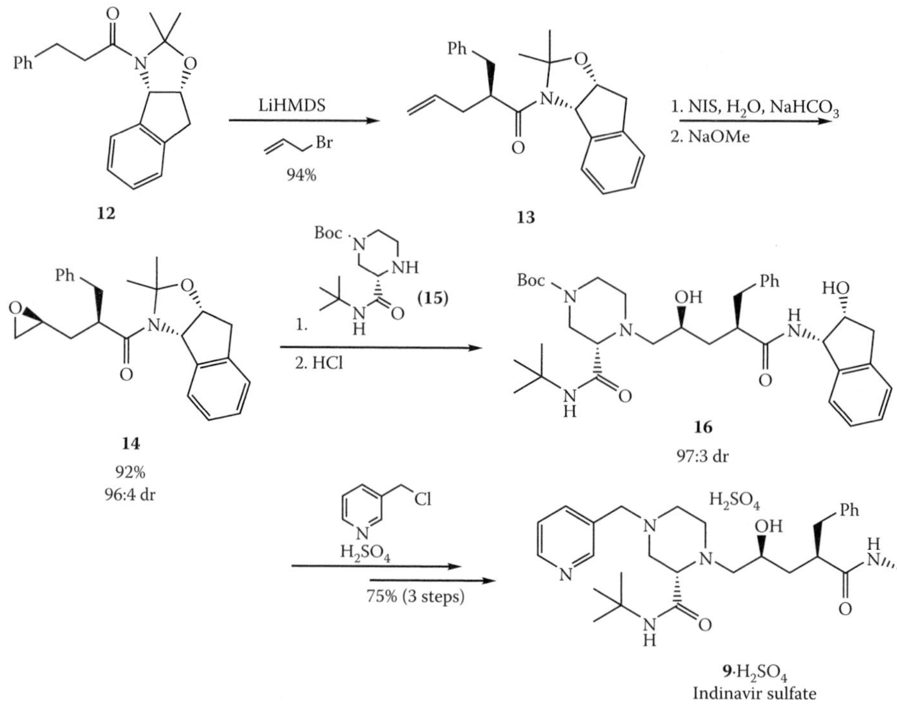

SCHEME 24.1

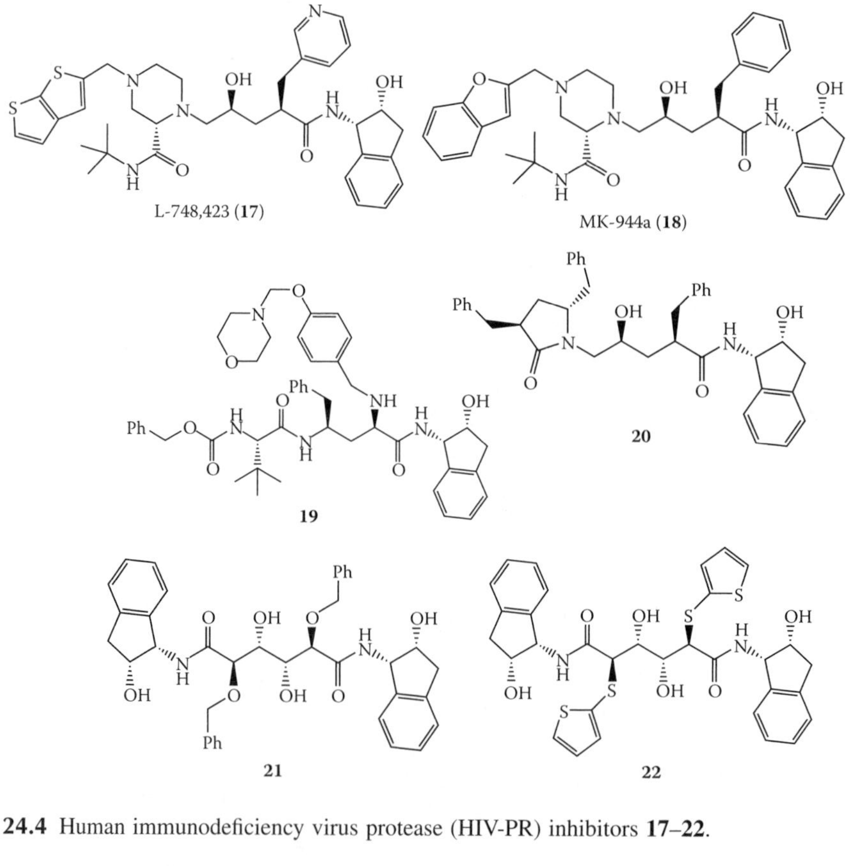

FIGURE 24.4 Human immunodeficiency virus protease (HIV-PR) inhibitors **17–22**.

responsible for a degenerative joint disease. Preliminary studies identified **23** as a very potent and selective aggrecanase inhibitor (IC_{50} = 12 nM), with excellent oral bioavailability and pharmaco-kinetic profile.[34] Subsequent work at Bristol-Myers-Squibb led to the discovery of **24**, in which the potentially metabolically labile hydroxyphenyl has been replaced by a biphenyl group. This new inhibitor displayed remarkably increased potency (IC_{50} = 1.5 nM) and selectivity over **23** and represented a fresh lead in development of new medication for degenerative joint disease (Figure 24.5).[35]

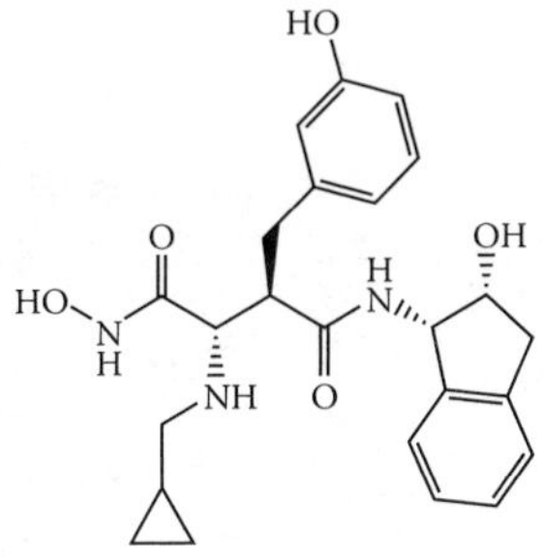

FIGURE 24.5 Structures of aggrecanase inhibitors.

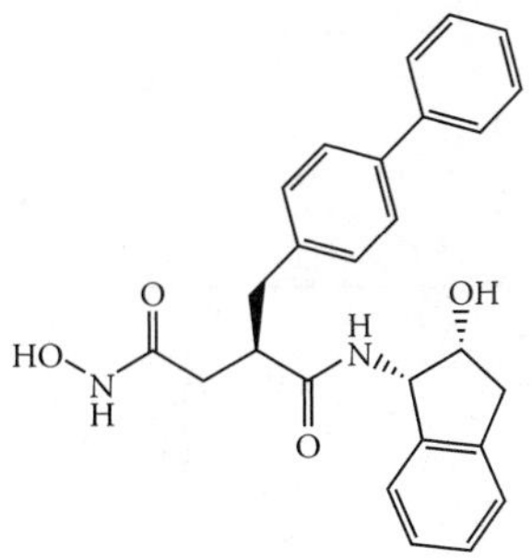

24.3 PRACTICAL SYNTHESIS OF ENANTIOPURE *CIS*-1-AMINO-2-INDANOL

Until recently, syntheses of enantiopure *cis*-aminoindanols relied either on starting materials from the chiral pool or on chemical or enzymatic resolution of racemic intermediates.[7,36–41] Advances in catalytic asymmetric epoxidation (AE) and asymmetric dihydroxylation (AD) of prochiral olefins[42–48] enabled the development of truly practical asymmetric syntheses of *cis*-1-amino-2-indanols (see Chapter 9).[49–54]

Research groups at Sepracor[53,54] and Merck[50–52] independently developed similar strategies to access (1*S*)-amino-(2*R*)-indanol. Both processes used Jacobsen's Mn-(salen) catalyst (MnLCl, **28**)[42–44,55] for indene epoxidation, followed by chirality transfer of the C–O bond of indene oxide **26** to obtain enantiopure (1*S*)-amino-(2*R*)-indanol (Scheme 24.2).

The Sepracor group demonstrated that (1*R*,2*S*)-indene oxide **26** could be prepared from readily available indene **25** in the presence of 1.5 mol% of (*R*,*R*)-MnLCl and 13% NaOCl in dichloromethane in 83% yield and 84% ee (Scheme 24.2). Chiral indene oxide **26** was then subjected to nucleophilic opening with ammonia to provide *trans*-aminoindanol, which was transformed without isolation to its benzamide under the Schotten-Baumann condition (83% ee, >99.5% ee after recrystallization). The optically pure *trans*-benzamidoindane was then converted to the optically pure benzaoxazoline **27** by exposure to 80% H_2SO_4, followed by addition of water to give *cis*-1-amino-2-indanol.[53,54] The sequence was demonstrated on multi-kilogram scale to prepare optically pure (1*R*, 2*S*)-**1** in 40% yield from indene.

A complementary approach to the synthesis (1*R*,2*S*)-**1** has been developed by Merck using (*S*,*S*)-MnLCl catalyst in a hypochlorite medium to provide (1*S*, 2*R*)-indene oxide **26**. This intermediate was converted without isolation to *cis*-aminoindanol in a stereoselective and regioselective manner using Ritter technology (Scheme 24.2). Several key issues, detailed later in this chapter, have been addressed and resolved in both Jacobsen's AE[52] and Ritter technology[50] to develop a reproducible and practical large-scale process for the synthesis of enantiopure *cis*-aminoindanol.

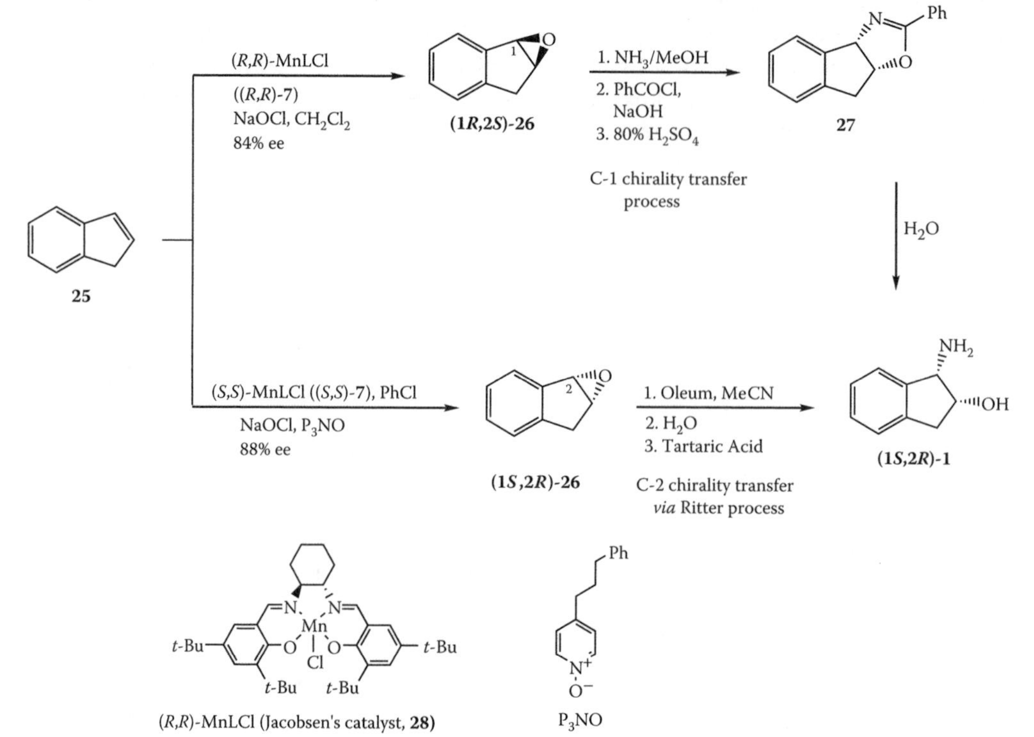

SCHEME 24.2

24.3.1 ASYMMETRIC EPOXIDATION OF INDENE

Chiral manganese-salen complexes have proved to be effective catalysts in the asymmetric epoxidation of unfunctionalized olefins.[42–48] In these salen systems, addition of appropriate N-oxides both activates and stabilizes the catalyst systems.[56,57] It was illustrated that the addition of an axial ligand, such as commercially available phenyl propyl pyridine N-oxide (P$_3$NO),[55] to the (S,S)-MnLCl-NaOCl-PhCl system resulted in a highly activated and stabilized catalyst for indene epoxidation.[52] Furthermore, the catalyst loading could be reduced to less than 0.4 mol%. Mechanistic studies indicated that the active catalyst was Mn(V)-oxo **29** species[52] and that hypochlorous acid (HOCl) was the true oxidant.[58] The slow step of the epoxidation process was identified as the oxidation of Mn(III) species **30** to Mn(V)-oxo **29** (Scheme 24.3).[58] During the development of the epoxidation of indene, the Merck group observed that NaOCl decomposed throughout the course of the reaction, which presented problems with reagent stability and stoichiometry. A secondary oxidation process was identified to produce isonicotinic acid and benzoic acid via benzylic oxidation of P$_3$NO.[52] Decomposition occurred as a result of the insufficient hydroxide amount in the hypochlorite. Based on the equilibrium equations for HOCl and water dissociation, HOCl concentration is inversely proportionate to the hydroxyl ion concentration (Scheme 24.3).[58] Furthermore, the hydroxyl ion concentration can be reduced by acid–base reaction with the carboxylic acid generated from the secondary oxidation of P$_3$NO. Proper adjustment of the hydroxyl ion concentration of commercial 2M NaOCl from 0.03–0.18M to 0.3M stabilized the hypochlorite and minimized the secondary oxidation. This reagent mixture was demonstrated on a multi-kilogram scale to afford indene oxide in 89% yield with an optical purity of 88% ee.[52]

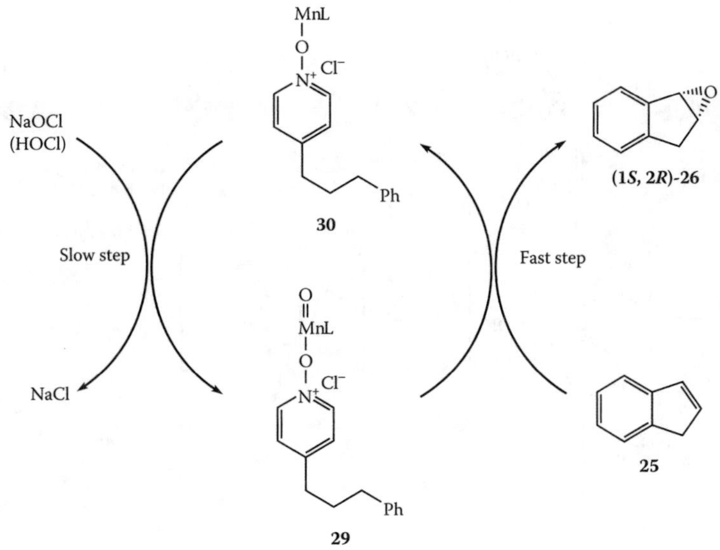

$$HOCl \xrightleftharpoons{K_{eq}} H^+ + OCl^- \qquad \textbf{eq 1}$$

$$H_2O \xrightleftharpoons{K_w} H^+ + OH^- \qquad \textbf{eq 2}$$

$$[HOCl] = \frac{K_{eq}\,[H_2O]\,[OCl^-]}{K_w\,[OH^-]} \qquad \textbf{eq 3}$$

SCHEME 24.3

24.3.2 RITTER-TYPE PROCESS FOR *CIS*-1-AMINO-2-INDANOL SYNTHESIS

Styrene oxide has been shown to give poor yields of regioisomeric oxazolines when exposed to the Ritter reaction conditions.[59,60] When indene oxide **26** was subjected to the Ritter condition (acetonitrile, 97% H_2SO_4), methyl oxazoline **35** was formed as the major product with moderate yield.[50] Low temperature nuclear magnetic resonance studies revealed important mechanistic aspects of the Ritter process (Scheme 24.4). At –40°C, epoxide **26** led to a 1:1 mixture of methyl oxazoline **35** and sulfate **32**. On warming the reaction to 22°C, the sulfate converted to the oxazoline. The proposed mechanism for *syn*-selective oxazoline formation is an acid-induced ring opening of indene oxide to produce nitrilium species **34** via C-1 carbenium ion **31**, which undergoes a thermodynamically driven equilibration process for the formation of the *cis*-5,5-ring derived oxazoline. In this fascinating Ritter process two roadblocks for the product formation were noted: a) polymerization via carbenium ion, and b) the hydrogen shift process from the recipient carbenium ion to form 2-indanone (>12%). The by-product formation can be suppressed by stabilizing the carbenium ion using a catalytic amount of sulfur trioxide in the Ritter mixture. Sulfur trioxide captured the epoxide to form sulfate intermediate, which then underwent product formation (Scheme 24.5). In addition, the chirality of the epoxide was effectively transferred from the C–2 to the C–1 position of the amino alcohol. By using the Ritter acid as an oleum (21% SO_3- H_2SO_4), a highly practical and cost-effective process was developed for the conversion of chiral indene oxide **26** to chiral *cis*-1-amino-2-indanol **1** (>80%, yield).[17,50]

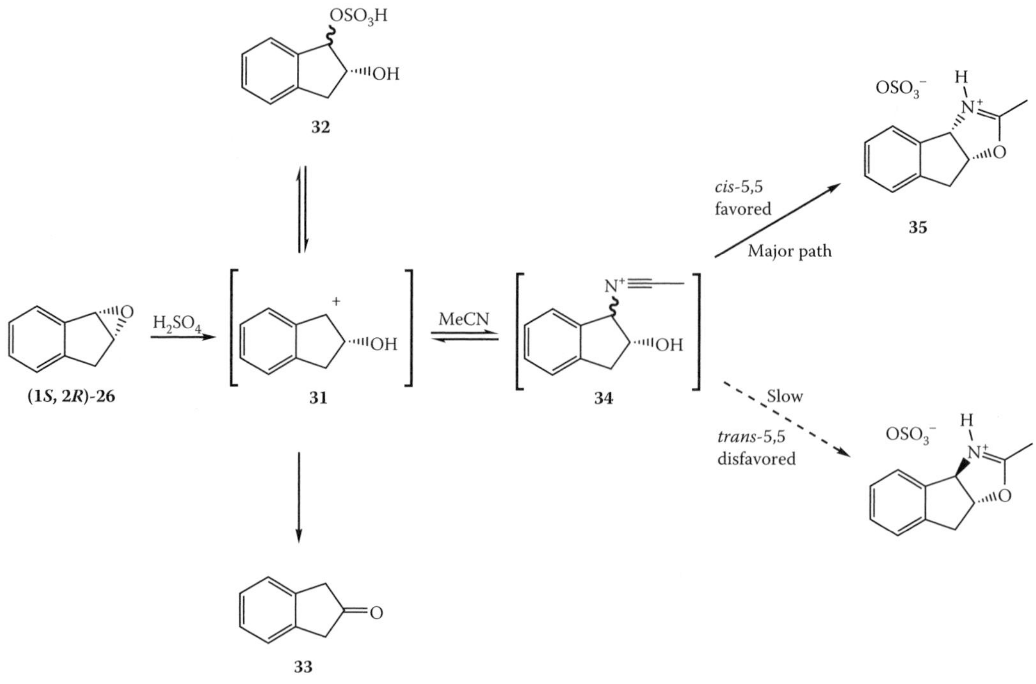

SCHEME 24.4

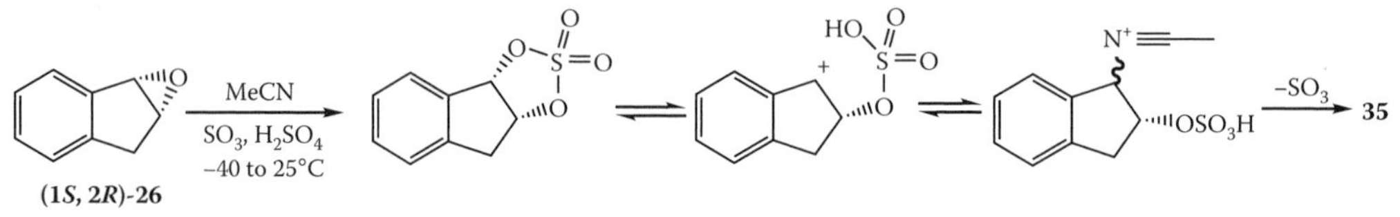

SCHEME 24.5

24.4 *CIS*-1-AMINO-2-INDANOL–DERIVED CHIRAL AUXILIARIES

Conformationally constrained *cis*-aminoindanols and its derivatives have been used as chiral auxiliaries in a number of asymmetric reactions. The availability of both *cis* enantiomers, the high levels of asymmetric induction attained, and the ease of recovery are all assets to the development of efficient and practical processes with those auxiliaries.

24.4.1 ALDOL CONDENSATION

In 1992 Ghosh and co-workers provided the first example of the utility of rigid *cis*-1-amino-2-indanol–derived oxazolidinone **36** as the chiral auxiliary in the asymmetric *syn*-aldol reaction.[60,61] Aldol condensation of the boron enolate of **37** with various aldehydes proceeded with complete diastereofacial selectivity. Effective removal and recovery of the chiral auxiliary was carried out under mild hydrolysis conditions (Scheme 24.6). As both enantiomers of the chiral auxiliary were readily available, both enantiomers of the *syn*-aldol could be prepared with equal asymmetric induction.

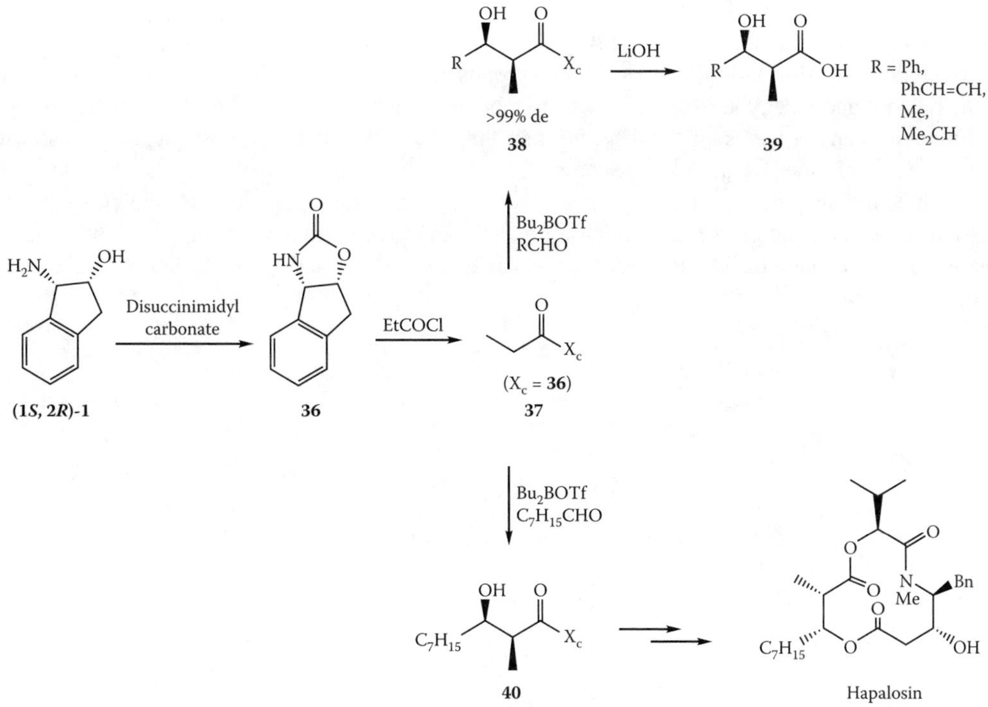

SCHEME 24.6

Ghosh also took advantage of the C–2 hydroxyl moiety of aminoindanols as a handle in the aldol reaction. Chiral sulfonamide **41** was *O*-acylated to give ester **42**. The titanium enolate of ester **42** was formed as a single isomer and added to a solution of aldehyde, precomplexed with titanium tetrachloride, to yield the *anti*-aldol product **43** in excellent diastereoselectivities.[63] One additional advantage of the ester-derived chiral auxiliaries was their ease of removal under mild conditions. Thus, hydrolysis of **43** afforded *anti*-α-methyl-β-hydroxy acid **44** as a pure enantiomer and *cis*-1-*p*-tolylsulfonamido-2-indanol was recovered without loss of optical purity (Scheme 24.7).[63]

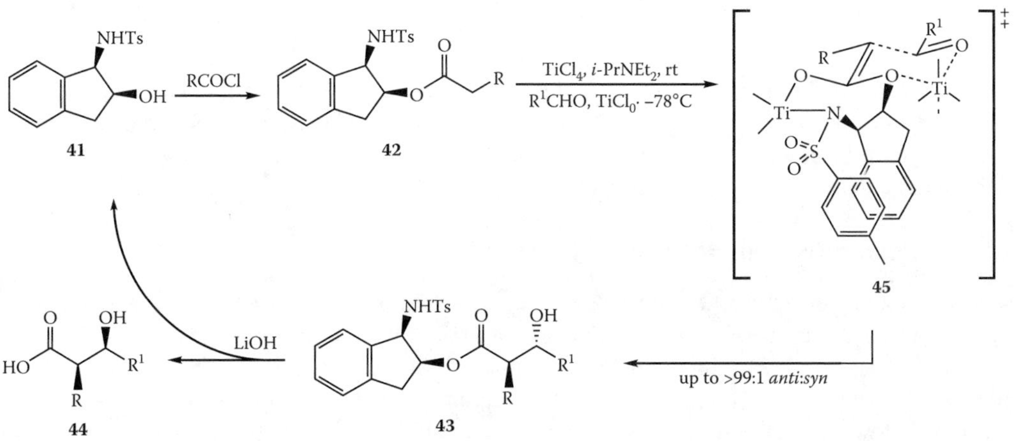

SCHEME 24.7

The stereochemical outcome was rationalized by a Zimmerman-Traxler type transition state **45**.[64] Assuming the titanium enolate of **42** has a Z-geometry and forms a 7-membered metallacycle with a chairlike conformation, a model can be proposed where a second titanium metal coordinates to the indanol and aldehyde oxygens in a 6-membered chairlike conformation. The involvement of two titanium centers was supported by the fact that aldehydes that were not precomplexed with titanium tetrachloride did not react (Scheme 24.7).[63] Ghosh and co-workers further hypothesized that a chelating substituent on the aldehyde would alter the transition state **46** and consequently the stereochemical outcome of the condensation, leading to *syn*-aldol products **47**.[64] Indeed, reaction of the titanium enolate of **42** with bidentate oxyaldehydes proceeded with excellent *syn*-diastereo-selectivity (Scheme 24.8).[65]

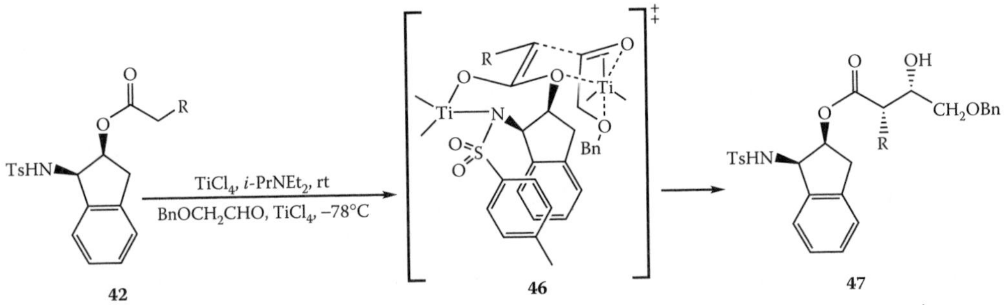

SCHEME 24.8

Thus, with proper choice of chiral template and aldehyde, all four possible *syn-* and *anti*-aldol products can be prepared with predictable stereochemistry. Both boron and titanium enolate meth-odologies have been successfully applied to the stereoselective syntheses of several biologically active compounds[66–69] and natural product synthons.[61,70,71]

24.4.2 ALKYLATION AND HOMOALDOL REACTIONS

A concise and practical synthesis of HIV-1 protease inhibitor **2** was developed by Askin and co-workers, using the rigid tricyclic aminoindanol acetonide as a chiral platform.[11] The diastereo-selective alkylation of (Z)-lithium enolate of amide **48** with amino epoxide **49** gave intermediate **50** in >90% yield and >98% de (Scheme 24.9). Lithium carbamate salt of **49** presumably activated the epoxide toward electrophilic epoxide opening, and alkylation of the (Z)-enolate occurred from the less hindered β-face. The amino alcohol was deprotected by treatment with camphorsulfonic acid and gave **2** in good yield.

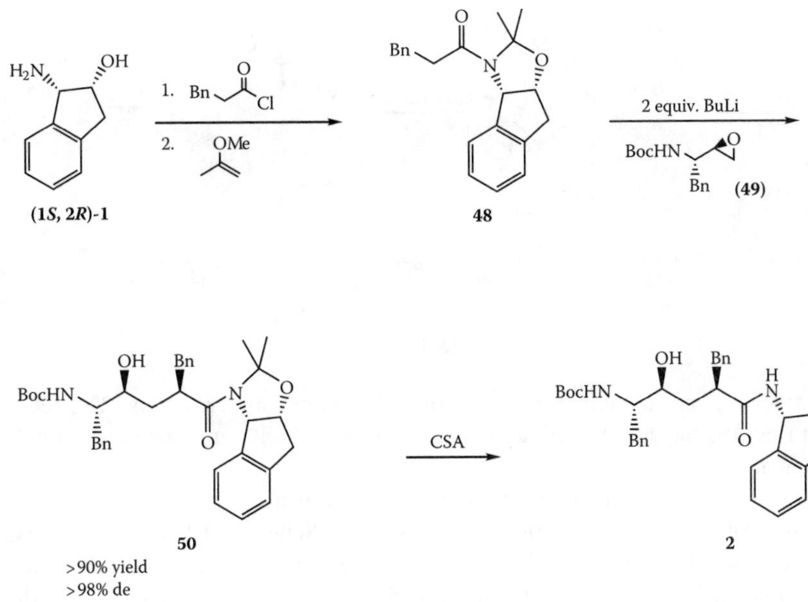

SCHEME 24.9

This alkylation strategy has been successfully implemented to the diastereoselective synthesis of a number of biologically active compounds,[17,19,32,72,73] including the orally active HIV protease inhibitor Crixivan® (>95% de)[17,19] and nucleoside antibiotic (+)-sinefungin (**51**) (>99% de).[72] The C–6 amine stereochemistry of (+)-sinefungin was set by a highly diastereoselective allylation of (1*S*,2*R*)-1-amino-2-indanol–derived oxazolidinone **52** (Scheme 24.10).

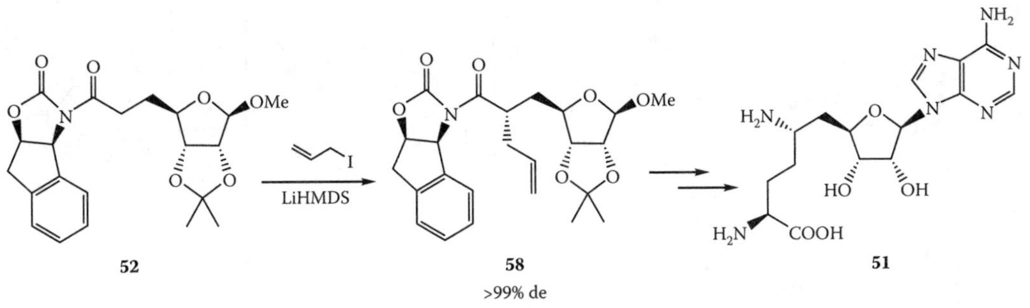

SCHEME 24.10

The low reactivity of glycine enolate with unactivated alkyl halides to form α-amino acids could be overcome by stabilizing the nucleophile using *cis*-aminoindanol–derived hippuric acid **53**. This key substrate was readily prepared from commercially available azalactone **54** by a one-pot operation (85% yield, 2 steps). The lithium enolate of amide acetonide **53** with a wide range of alkyl halides proceeded in moderate yields (>60%) and excellent diastereoselectivities (>95% de). Assuming that lithium halide would facilitate the dissociation of the amide enolate from the aggregated state and thus enhance its reactivity, 4 equivalents of lithium chloride were used as additive and resulted in a 25% increase in yield (Scheme 24.11). Reactions with secondary halides

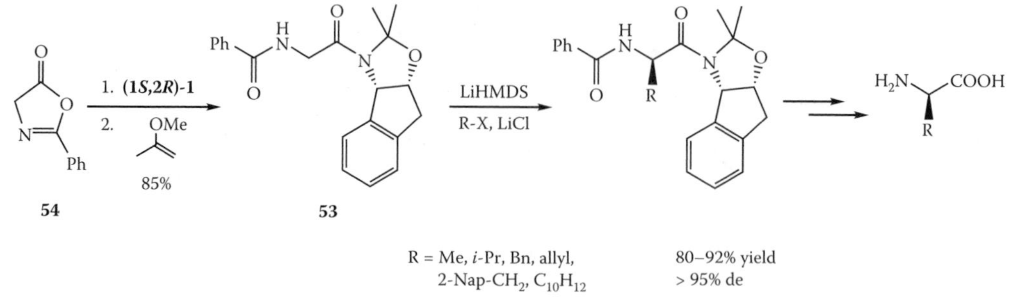

SCHEME 24.11

remained low yielding, presumably because of competing elimination reactions. The benzoyl amide could be removed and the chiral auxiliary could be recovered without epimerization of the amino acid chiral center.[74]

The methodology was extended by Askin to the asymmetric syntheses of *syn*- and *anti*-2,4-substituted-γ-butyrolactones in a stereoselective fashion[20] (Scheme 24.12). Amide **48** was diastereoselectively allylated (94% yield, 92% de), and olefin **55** was then subjected to Yoshida's unbuffered condition (I$_2$/THF/H$_2$O) to give iodolactone **56** in 97:3 *anti:syn* ratio.[75] The efficient 1,3-chirality transfer is consistent with a highly ordered imidate transition state **57** wherein the A$_{1,3}$ strain between the iminium indanol substituent and the benzyl group shifts the latter in a *pseudo-axial* orientation, which results in the high preference for the formation of thermodynamically less stable *anti*-2,4-substituted-lactone **56**. Pro-(2*S*)-diastereomer **55** was prepared in high yields with excellent diastereoselectivity by reversal of the order of introduction of the benzyl and allyl groups. It is interesting that exposure to the buffered iodohydrin process (NIS, H$_2$O, NaHCO$_3$) resulted in the formation of the 2,4-*syn*-product **58** with outstanding selectivity (97% de). Epoxide formation in a basic medium followed by acid-mediated lactonization gave *syn*-2,4-disubstituted-γ-butyrolactone **59**.[20]

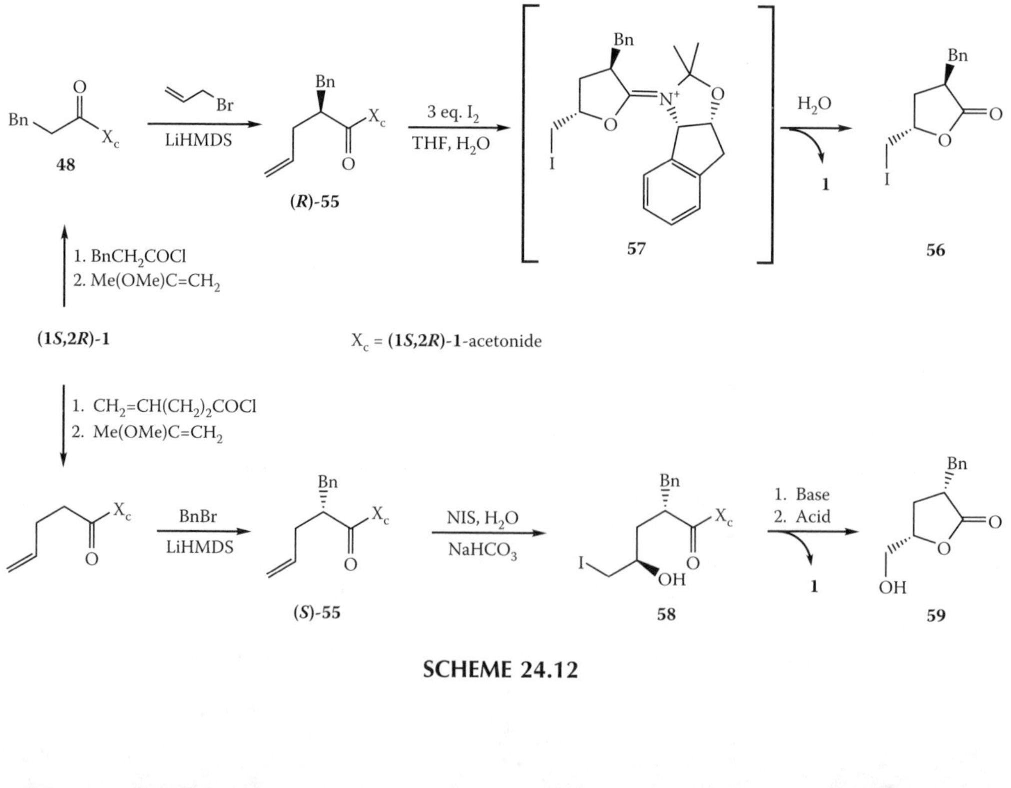

SCHEME 24.12

In another example, Armstrong and McWilliams have shown aminoindanol acetonide to be a powerful chiral auxiliary for stereocontrolled 1,6-asymmetric induction.[76,77] They showcased this process in an elegant tandem 1,2-migration–homoaldol protocol for the synthesis of highly functionalized *syn*-2,4-γ-butyrolactones (Scheme 24.13).[77] Lithium enolate **60** was reacted with bis(iodomethyl)zinc in the presence of lithium benzylalkoxide. The higher order alkoxyzincate **61** obtained is thought to undergo a 1,2-migration to give alkoxyzincate **62**. The stereoselective migration, which proceeded in a remarkable >98% de, set the absolute configuration of the α-center. Zinc homoenolate **62** was then transmetalated with (*i*-PrO)TiCl₃ and subjected to homoaldol reaction with *N*-(*tert*-butoxycarbonyl)phenylalaninal. Homoaldol **63** was produced in 59% yield and >98% de for the overall 2-step transformation. Treatment of γ-hydroxyamide **63** with *p*-toluenesulfonic acid induced cyclization to lactone **64** in good yield. Furthermore, pure (1*S*,2*R*)-*cis*-aminoindanol crystallized from the reaction mixture as the *p*-toluenesulfonate salt and was recovered by simple filtration. Thus, *cis*-aminoindanol acetonide as the single chiral-controlling element was responsible for the asymmetric induction in the two disparate transformations.

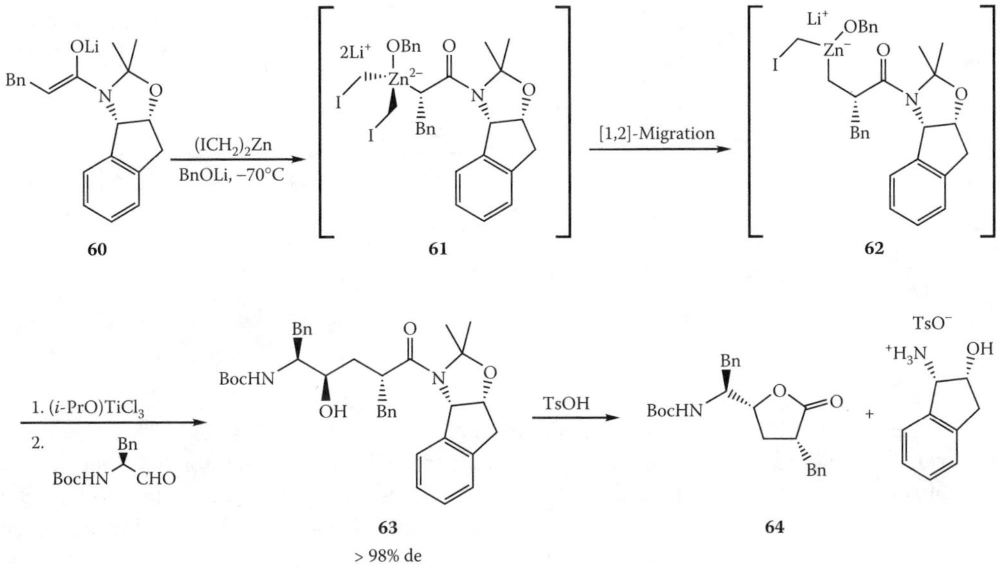

SCHEME 24.13

24.4.3 Conjugate Addition

The asymmetric synthesis of endothelin receptor antagonist **65** by scientists at Merck took advantage of *cis*-aminoindanol as a chiral auxiliary, both for stereoselective alkylation and chiral Michael addition.[78] Their approach to the conjugate addition was based on the reaction reported by Alexakis and co-workers[79] and work by Frey and co-workers,[80] where the stereoselectivity of the 1,4-addition of lithium diphenylcuprate or aryl lithium to α,β-unsaturated esters was controlled by a neighboring chiral auxiliary. Aldehyde **66** was treated with (1*S*,2*R*)-*N*-methyl-1-amino-2-indanol to yield the *N*,*O*-acetal **67** quantitatively. The aryllithium generated from **68** was added at low temperature to Michael acceptor **67** followed by acidic workup. Aldehyde **69** was obtained in 92% overall yield from **66** and 92% de. In comparison, conjugate addition to dimethyl acetal of **66** using external chiral additives led to diastereoselectivities that would not exceed 67% de (Scheme 24.14).

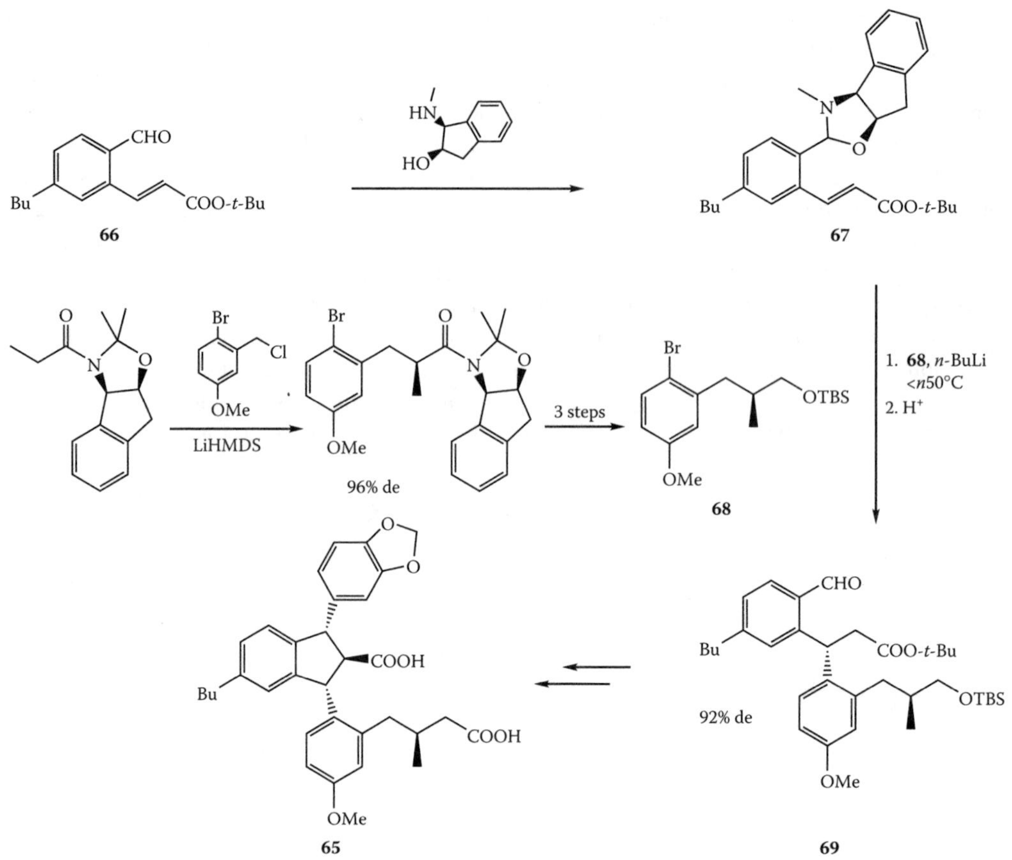

SCHEME 24.14

The alternative strategy of using *cis*-aminoindanol as a chiral auxiliary on the Michael donor has also been explored.[81] Chiral amide enolates were reacted with α,β-unsaturated ester **70**, and the resultant adducts were reduced and cyclized to δ-lactones **73** to determine the facial selectivity on the Michael acceptor. It is interesting that protected amino alcohol **71** did not lead to significant diastereofacial discrimination, whereas **72** afforded lactone **73** with high 4-(*S*)-selectivity (Scheme 24.15).

24.4.4 DIELS-ALDER REACTION

Researchers at Merck have demonstrated the utility of *cis*-aminoindanol–derived oxazolidinone **36** as a chiral auxiliary in the asymmetric Diels-Alder reaction (see Chapter 26).[82] Pioneering work by Evans[83,84] has shown that very high levels of diastereofacial discrimination could be achieved in the Diels-Alder reaction of isoprene with oxazolidinone **75** (88% de). However, much lower levels of selectivity were observed with phenyl glycinol derivative **76** (35% de).[83,84] The Merck group hypothesized that the low level of selectivity of **76** resulted from the rotationally labile phenyl moiety, which could be more or less sterically demanding depending on its conformation. Conformationally constrained phenylglycinol analogue **77** was subjected to Evans' Diels-Alder reaction conditions (1.4 equiv. Et₂AlCl, –15°C, CH₂Cl₂) and gave excellent *endo*- and diastereoselectivities (93% de). Homologous 6- and 7-membered ring-containing systems **78** and **79** led to low levels of asymmetric induction (30–35% de), thus demonstrating the importance of the rigidity of the aminoindanol platform (Scheme 24.16).[91]

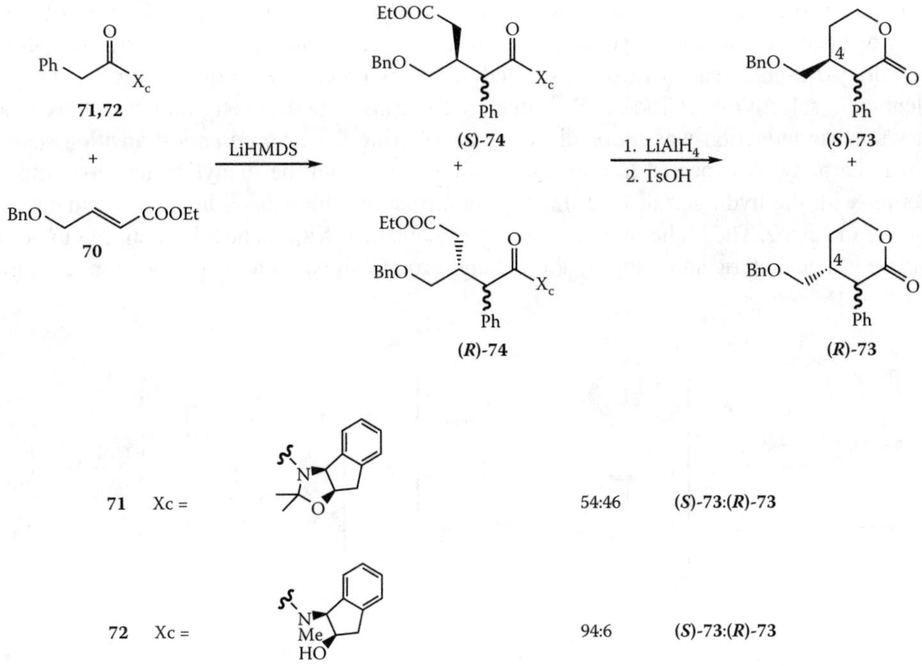

SCHEME 24.15

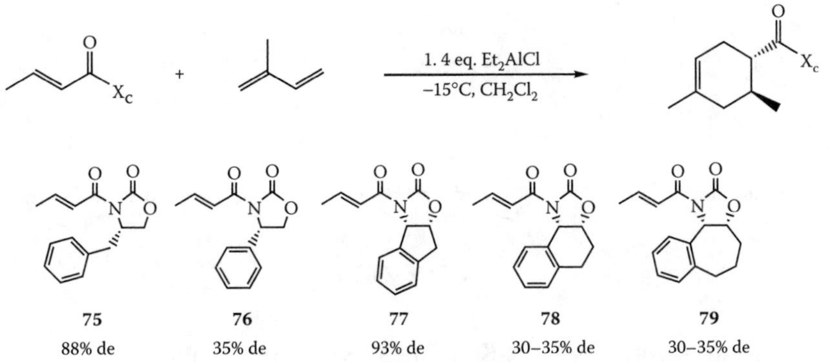

SCHEME 24.16

Ghosh has also shown that *cis*-1-(arylsulfonamido)-2-indanols could be used as excellent chiral auxiliaries in Diels-Alder reaction.[85] Reaction of **80** with cyclopentadiene using a variety of Lewis acids led to **81** in good yield, with *endo:exo* ratios superior to 99:1 and diastereoselectivities ranging from 72% to 92%. The best selectivities were observed using titanium tetrachloride as the Lewis acid. The high degree of diastereoselection was presumed to result from effective metal chelation (Scheme 24.17).

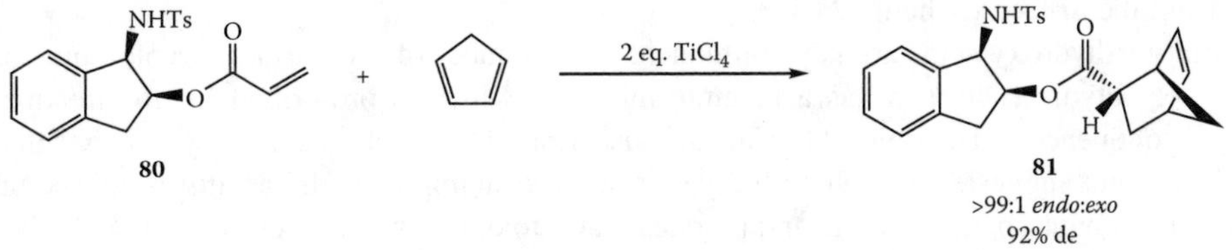

SCHEME 24.17

In contrast to the previous examples, Rawal has studied the advantages of linking the chiral auxiliary to the diene moiety.[86] Cycloadditions of readily accessible and stable 1-(2-oxazolidinon-3-yl)-3-siloxy-1,3-butadiene with α-substituted acroleins proceeded in quantitative yields and in excellent *endo*-selectivities. Crude cycloadducts were transformed directly into cyclohexenone **82**. The asymmetric induction was rationalized by considering the favored *endo*-transition state. The carbamate carbonyl is expected to adopt a position away from the dienyl moiety to avoid steric interaction with the hydrogen at C–2. In this conformation, the oxazolidinone substituent blocks one face of the diene. The higher diastereoselectivity obtained with diene **83** is believed to originate from more efficient steric shielding by the indanyl group compared to isopropyl or phenyl groups (Scheme 24.18).

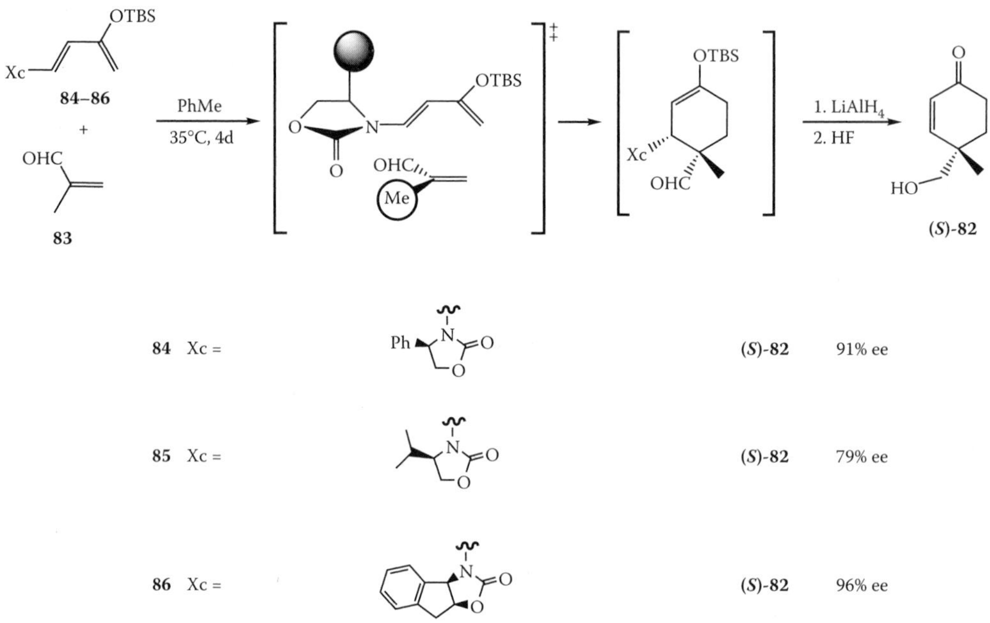

SCHEME 24.18

24.4.5 ASYMMETRIC ADDITIONS TO α-KETOESTERS AND α-KETOAMIDES

Secondary α-hydroxy acids were readily accessed by reduction of α-ketoesters bearing *cis*-1-arylsulfonamido-2-indanol derivatives as a chiral auxiliary in high yields and good to excellent diastereoselectivities (from 4:1 to >99:1) using bulky alkyl hydrides.[87] Indeed, reduction of **87** using L-selectride and zinc chloride afforded α-hydroxy ester **88** in 96% yield and >99:1 diastereoselectivity. Hydrolysis under mild conditions released the chiral auxiliary and produced essentially optically pure α-hydroxy acids **89**. The high degree of stereoselection was attributed to metal chelation of the carbonyl oxygens, thus locking **87** in an s-*cis* conformation. The vicinal toluene sulfonamide most probably shields the *Re*-face, which consequently leads to the preferential hydride attack from the *Si*-face (Scheme 24.19).

Tertiary α-hydroxy acids are key components of a number of muscarinic receptor antagonists, such as oxybutynin (Ditropan® as a racemic mixture), which is prescribed for the treatment of urinary incontinence and exhibits classical antimuscarinic side effects, such as dry mouth.[88] Because biological results suggested that (*S*)-oxybutynin displayed an improved therapeutic profile compared to its racemic counterpart, a general and practical entry to optically pure key subunit (*S*)-**90** became of primary importance.[89] In their investigations, researchers at Sepracor considered the advantage

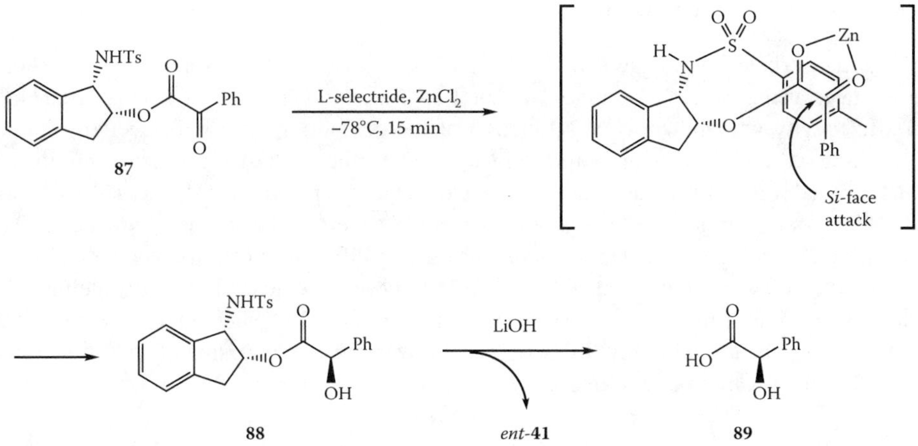

SCHEME 24.19

of using the C–1 amine or C–2 alcohol of *cis*-aminoindanol as a chiral handle. Hydroxy acid (*S*)-**90** was generated indifferently by diastereoselective cyclohexyl or phenyl Grignard addition to the appropriate ketoester or ketoamide, followed by removal of aminoindanol unit (Scheme 24.20).[90] Phenyl Grignard addition to ketoamide **91** and ketoester **93** proceeded via magnesium coordination and led to high diastereoselectivities. In the case of cyclohexyl Grignard addition, zinc chloride was a necessary additive to achieve good degrees of diastereoselection. It is interesting that *N*-tosyl-derived ketoesters provided the most expedient avenue to the preparation of optically pure (*S*)-acid **90**.

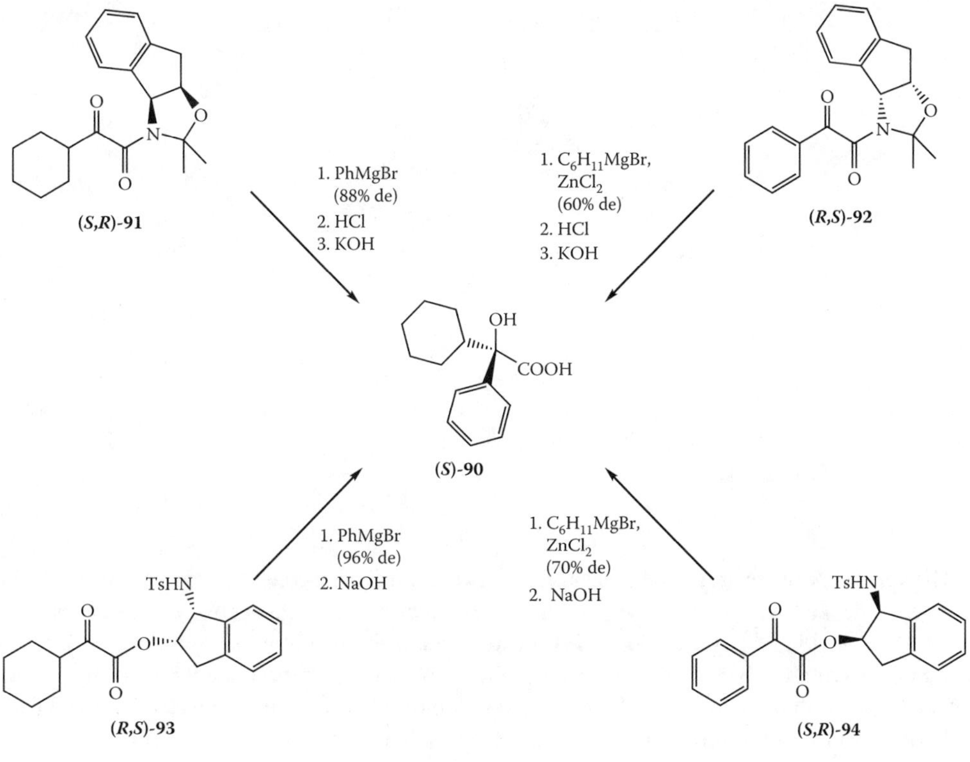

SCHEME 24.20

24.4.6 ASYMMETRIC REARRANGEMENTS

Kress and co-workers investigated the stereoselective [2,3]-Wittig rearrangement of α-allyloxy amide enolates using (1S,2R)-1-amino-2-indanol as the chiral auxiliary (Scheme 24.21).[91] The diastereoselectivity was shown to be a function of the counterion with *syn*-selectivity increasing from K<Na<Li<Zr. Although zirconium enolates led to the highest diastereoselectivities (94:6 *syn:anti*), low conversion was an issue (73% yield). Optimal results were observed using a LiH-MDS–HMPA combination, and **95** was obtained in 97% yield and 89:11 *syn:anti* selectivity. The scope of this methodology was extended to olefins bearing different substitution patterns. In general, *trans*-substituted olefins exhibited excellent *syn*-diastereoselectivities, whereas *cis*-olefins afforded low diastereofacial discrimination (1:2 *syn:anti*). Unsubstituted allyl ether led to diastereoselectivity greater than 98%. The utility of this process was demonstrated by conversion of **95** to functionalized acyclic and cyclic α-amino esters **96** and **97**.

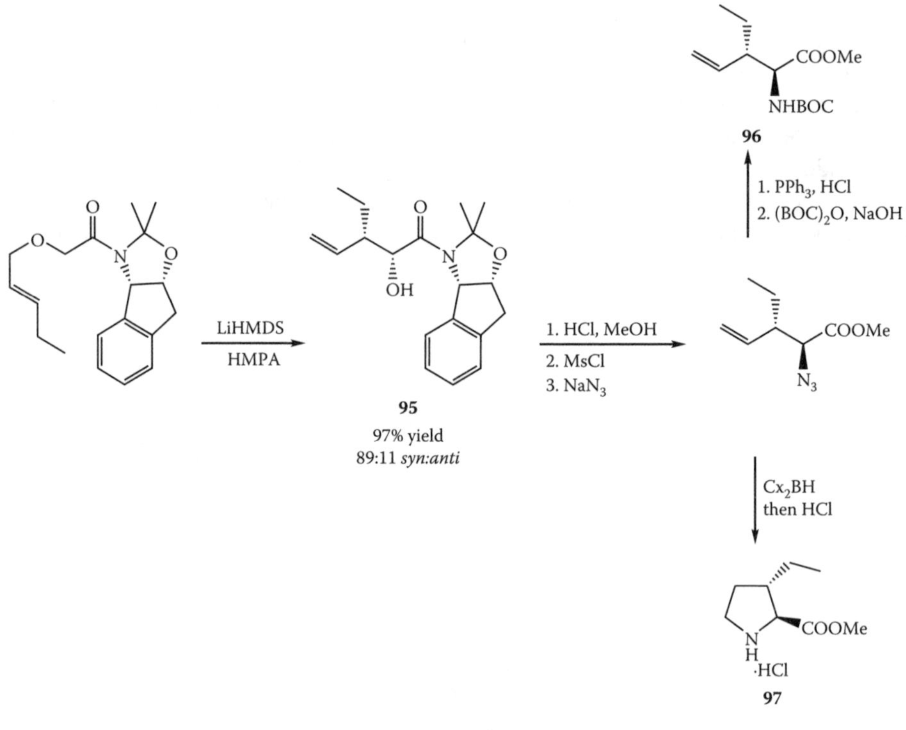

SCHEME 24.21

The versatility of *cis*-aminoindanol as chiral auxiliary has been considered in various Claisen[92,93] rearrangements and was found to be particularly efficient in the 6-azaelectrocyclization reaction.[93] Indeed, the reaction of (E)-3-carbonyl-2,4,6-trienal **98** with enantiopure *cis*-aminoindanol **1** proceeded under remarkably mild conditions to produce pentacyclic piperidine **99** as a single isomer. The reaction was thought to proceed via isomerization of dihydropyridine intermediate **100** toward the thermodynamically more stable aminoacetal **99** (Scheme 24.22).

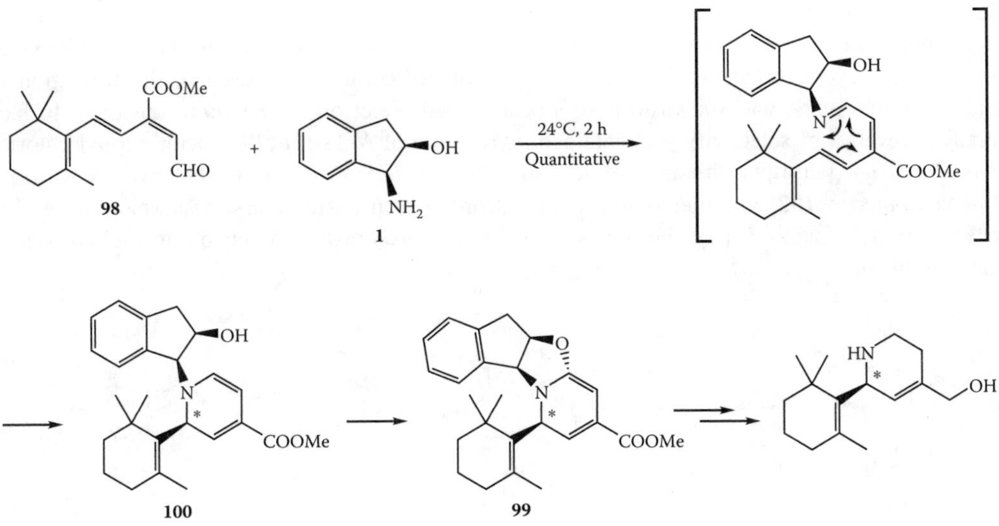

SCHEME 24.22

24.4.7 ELECTROPHILIC AMINATION

A novel asymmetric synthesis of α-amino acids via electrophilic amination has been demonstrated by Zheng and Armstrong and co-workers.[94] No "+NHBoc" was observed when lithium *tert*-butyl-*N*-tosyloxycarbamate (LiBTOC) was reacted with zinc and lithium enolates of **48**. Transmetallation of the lithium enolate with copper(I) cyanide was necessary to generate a reactive amide cuprate, which then added efficiently to the electrophile. The electrophilic amination of chiral cuprates with LiBTOC provided an expedient approach to α-amino acids with predictable absolute configuration in high enantiomeric purity and good yield (Scheme 24.23).

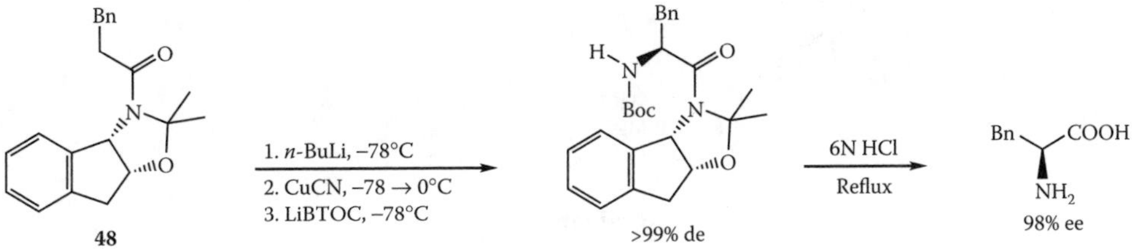

SCHEME 24.23

24.4.8 PRACTICAL SYNTHESIS OF ENANTIOPURE SULFINAMIDES AND SULFOXIDES

Enantiopure sulfinamides have proved a powerful tool for the asymmetric synthesis of chiral amines[95–102] and were efficiently used in the preparation of natural products and potential drugs such as (*R*)-didesmethylsibutramine.[99] A general method for the modular synthesis of optically pure aryl and tertiary alkyl sulfinamide auxiliaries was developed by Senanayake and co-workers from *N*-sulfonyl-1,2,3-oxathiazolidine-2-oxides.[103] Reaction of *N*-Mes-(1*R*,2*S*)-**1**, containing the conformationally constrained indane platform and an arylsulfonyl group as *N*-activator, with thionyl chloride and 3,5-lutidine in THF provided oxathiazolidine oxide **101** in excellent 97:3 *endo:exo* selectivity. The activated N–S bond was cleaved chemoselectively with a variety of organometallic reagents with inversion of configuration at the sulfur atom. Subsequent mild displacement of the O–S bond with a nitrogen nucleophile also proceeded with inversion of configuration at the sulfur atom and led to enantiopure sulfinamides **102** in good overall yield. Enantiopure aminoindanol **103** was recovered and readily recycled (Scheme 24.24). The base–solvent combination used in the formation of oxathiazolidine oxide was shown to have a pronounced effect on the *endo:exo* outcome. In fact, complete reversal of selectivity was obtained by reaction of *N*-Ts-(1*R*,2*S*)-**1** with thionyl chloride using 2,6-di-*tert*-butyl-pyridine in THF, leading to a 2:98 *endo:exo* ratio, which gave access to the (*S*) enantiomer of **102**. The methodology thus afforded both enantiomers of sulfinamides **102** in 99% ee from 1 enantiomer of the indane platform and was proved amenable to multi-kilogram scale production.

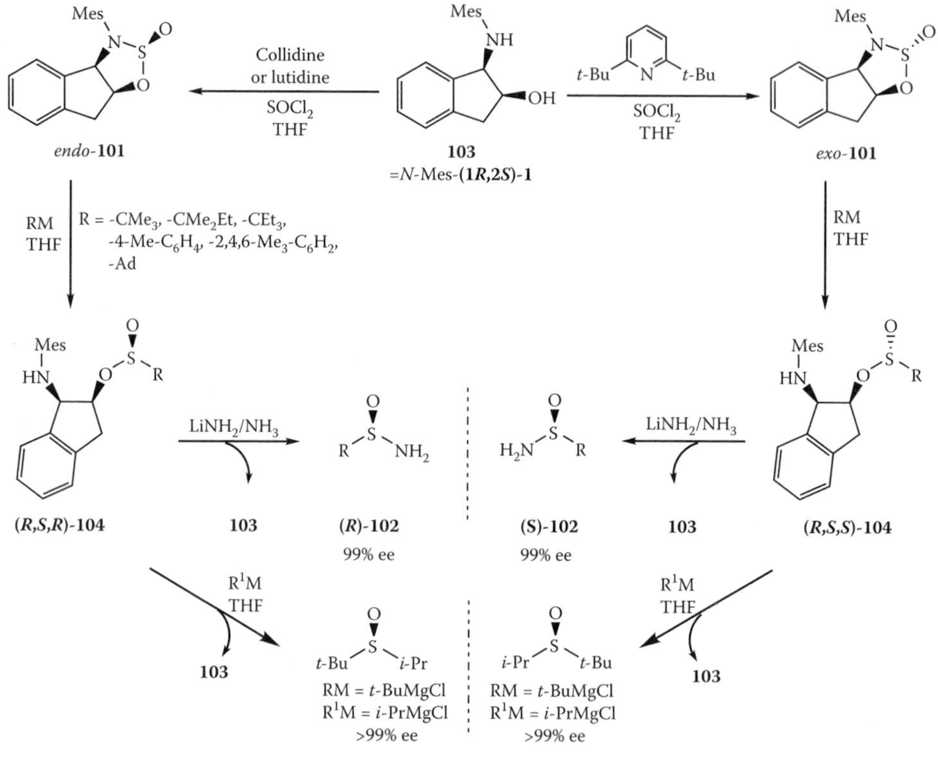

SCHEME 24.24

The same strategy was successfully applied to the synthesis of enantiopure sulfoxides, which are often used as chiral controllers for C–C bond formations or as ligands in catalytic asymmetric processes.[104] The O–S bond of **104** was displaced by a carbon nucleophile to give enantiopure sulfoxides (Scheme 24.24). Indeed, treatment of individual diastereoisomers of **104** (R = *t*-Bu) with isopropylmagnesium chloride provided the corresponding enantiomers of *t*-BuS(O)-*i*-Pr in excellent yield and optical purity, with remarkable recovery of enantiopure **103**. In the course of the study, it was discovered that the S–N bond of **101** could also be displaced with mild reagents such as organozincs of sterically congested halides (R = Ad) to give sulfinate intermediate **104**. In addition to alkyl–alkyl chiral sulfoxides, this powerful process gave access to either enantiomers of alkyl–aryl and aryl–aryl sulfoxides. Finally, *tert*-butyl-(*tert*-butylsulfinyl)acetate and diethyl-(*tert*-butylsulfi-nyl)-methyl-phosphonate were generated by addition of the corresponding lithium reagents to the *tert*-butylsulfinate, which demonstrated the ability of this methodology to lead to novel structures and possibly to new biological targets (Table 24.2).[104]

TABLE 24.2
Enantiopure Sulfoxides Formation by Double Nucleophilic
Displacement of Oxathiazolidine 101

RM	R¹M	Sulfoxide	Yield (%)	% ee
t-BuMgCl	*i*-BuMgCl		90	99.5 (*S*)
AdZnBr	*n*-BuMgCl		93	99 (*S*)
t-BuMgCl	PhMgCl		77	99 (*R*)
p-tolylMgBr	EtMgCl, CuBr.SMe₂		89	95 (*S*)
EtMgCl, CuBr.SMe₂	*p*-tolylMgBr		83	90 (*S*)
t-BuMgCl	CH2=C(OLi)O-*t*-Bu		93	99 (*R*)
t-BuMgCl	LiCH₂P(O)(OEt)₂		93	99 (*S*)
LiCH₂P(O)(OEt)₂	*t*-BuMgCl		83	98 (*R*)

24.5 RELATED AMINO ALCOHOLS AND THEIR APPLICATIONS IN ASYMMETRIC SYNTHESIS

24.5.1 *CIS*-2-AMINO-3,3-DIMETHYL-1-INDANOL

Saigo and co-workers designed and demonstrated the general utility of *cis*-2-amino-3,3-dimethyl-1-indanol as a chiral auxiliary in a variety of carbon–carbon and carbon–heteroatom bond formation reactions and as ligands in several catalytic processes.[105–109] Racemic *cis*-2-amino-3,3-dimethyl-1-indanol **105** was prepared in 3 steps from 3,3-dimethyl-1-indanone by oxime formation followed by sequential reduction of the keto and imino functional groups (Scheme 24.25). The resolution of **105** was performed using (*S*)-mandelic acid. Salt (+)-**106** crystallized from ethanol and recrystallization followed by treatment with an alkaline solution gave optically pure (1*R*,2*S*)-**105** in 35% yield. The enantiomeric (1*S*,2*R*)-**105** was obtained in 37% yield from the crystallization–recrystallization filtrates by successive treatment with base and (*R*)-mandelic acid, followed by filtration of the crystalline (−)-**106** and treatment of the salt with alkali.[105,106]

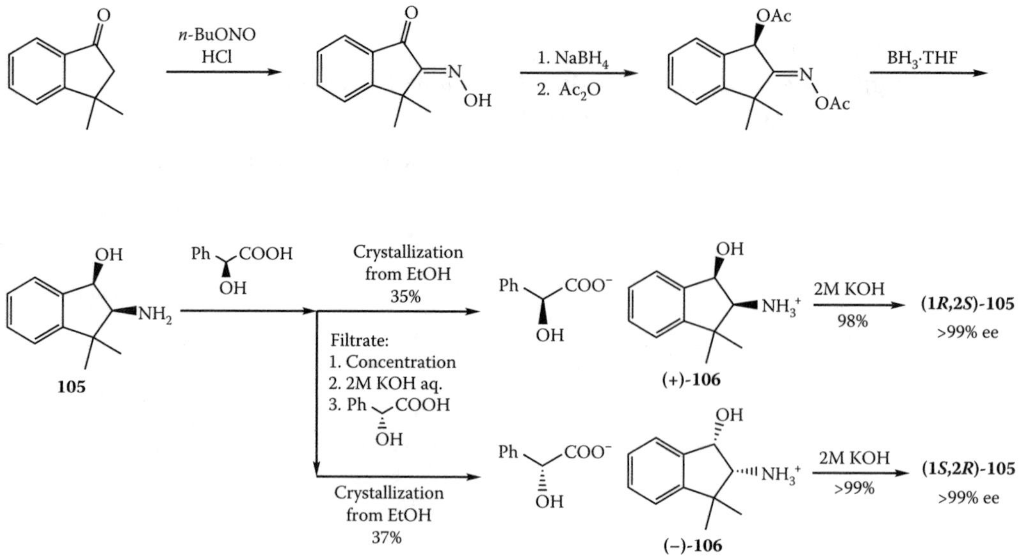

SCHEME 24.25

The diastereoselective alkylation of *N*-acyloxazolidinones enolates was examined first. Lithium enolates of **107** were reacted with a variety of alkyl halides, and alkylation products were formed with excellent diastereoselectivities (94–99% de). Hydrolysis gave optically pure carboxylic acids, and the chiral auxiliary was recovered for reuse almost quantitatively.[105,106] Highly diastereoselective bromination was also achieved by reaction of the boron enolate of **107** with *N*-bromosuccinimide (NBS) (98% de). Optically pure amino acids could be accessed by simple synthetic transformations (Scheme 24.26).[106]

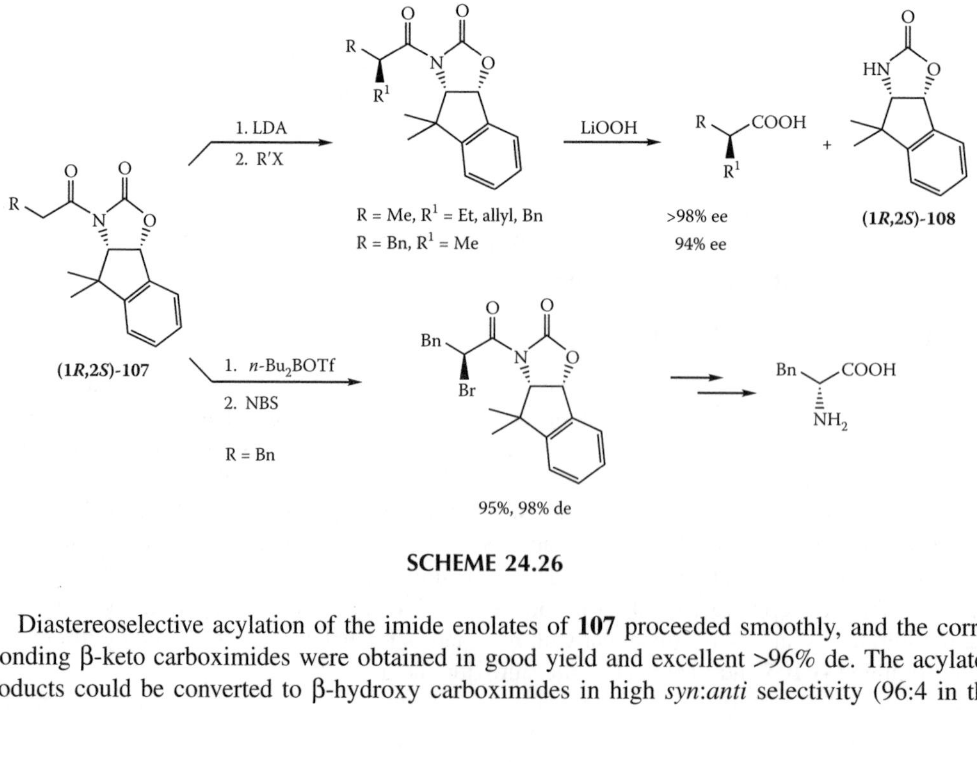

SCHEME 24.26

Diastereoselective acylation of the imide enolates of **107** proceeded smoothly, and the corresponding β-keto carboximides were obtained in good yield and excellent >96% de. The acylated products could be converted to β-hydroxy carboximides in high *syn:anti* selectivity (96:4 in the case of R = Ph) by treatment with zinc borohydride, and recrystallization afforded optically pure *syn*-products. Hydrolysis followed by methylation led to chiral β-hydroxy esters in good yield and >99% ee (Scheme 24.27). In addition, reaction of the sodium enolate of **107** with 2-(*p*-toluene-sulfonyl)-3-phenyloxyaziridine followed by acidic quenching led to the hydroxylated product with high diastereoselectivity (86% de). Treatment with magnesium methoxide gave (*R*)-methyl mandelate in 86% ee, and the chiral auxiliary **108** was recovered for reuse (Scheme 24.27).[106]

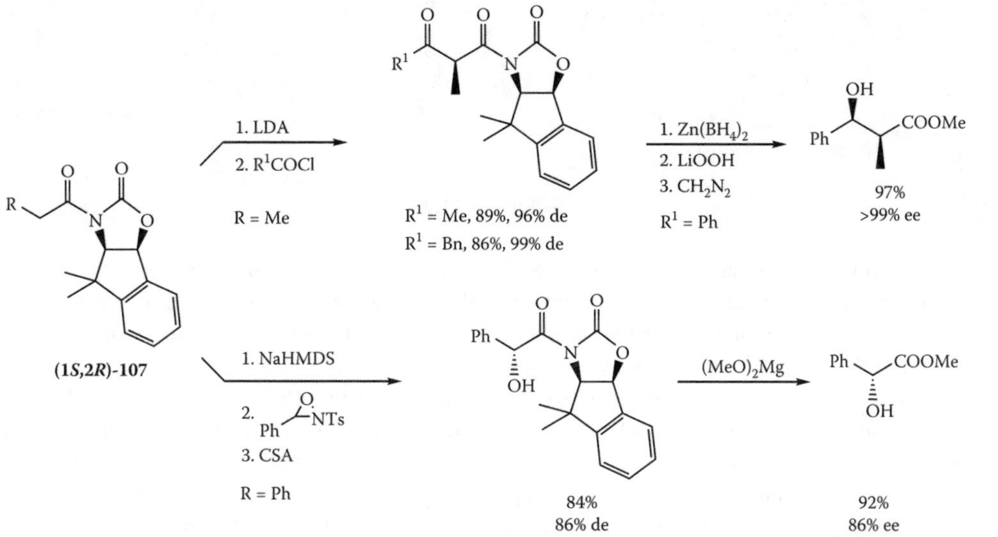

SCHEME 24.27

The stereochemical outcome of these electrophilic additions is consistent with a transition state in which the metal chelates the oxazolidinone carbonyl and the enolate oxygen. Reaction with an electrophile would, therefore, occur at the less hindered diastereotopic face of the (*Z*)-enolate, away from the shielding methyl groups of the auxiliary (Figure 24.6). Because both enantiomers of oxazolidinone **108** are equally available, the direction of the asymmetric induction can be controlled by proper choice of the absolute stereochemistry of the chiral auxiliary.[106]

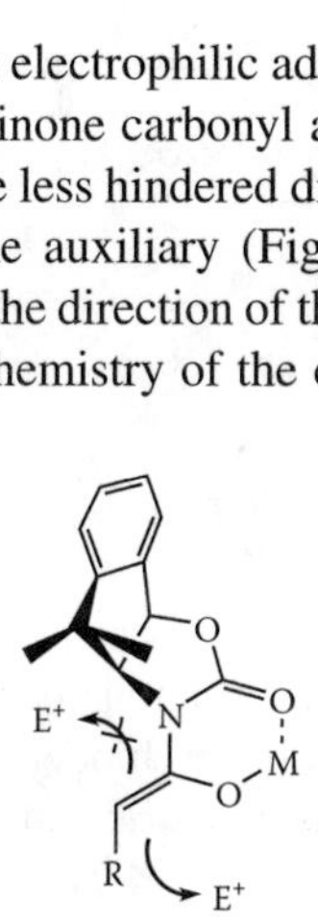

FIGURE 24.6 Chelating model for electrophilic additions.

Saigo and co-workers reasoned that, by analogy, high levels of diastereofacial discrimination could be achieved in the Lewis acid–mediated Diels-Alder reaction of dienes with oxazolidinone **108**-derived dienophiles. Indeed, excellent regioselectivities (*endo:exo*) and diastereoselectivities were reached in the Diels-Alder reaction of **109** with cyclic and acyclic dienes using Et$_2$AlCl as the activator (Scheme 24.28).[107] The selectivities obtained actually surpassed those reported with *cis*-1-amino-2-indanol **1** as the chiral auxiliary (93% de) (see Scheme 24.16).[82] The additional bulk introduced in the rigid backbone by the vicinal methyl groups proved an asset to improved stereocontrol.

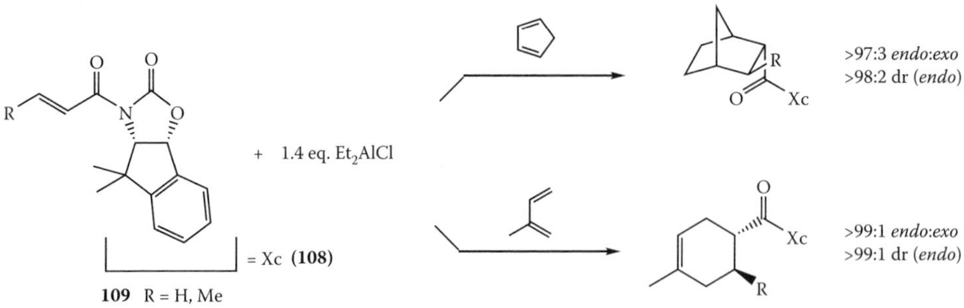

SCHEME 24.28

cis-2-Amino-3,3-dimethyl-1-indanol was also used as chiral auxiliary for the asymmetric Ireland-Claisen rearrangement of allyl carboxylates. Preliminary studies on the racemic system demonstrated the potential of the process as unoptimized reaction conditions (NaHMDS, TMSCl in THF) led to the rearranged products in a promising 84:16 diastereoselectivity. The reaction solvent was shown to affect the selectivity, and slightly improved results (87:13) were obtained in Et₂O. It also appeared that the diastereoselectivity depended on the bulkiness of the silane used. Smaller silylating agents such as dimethylsilyl chloride (DMSCl) gave better selectivities (93:7). The Ireland-Claisen rearrangement of several enantiopure allyl carboxylates was carried out, and the corresponding products were obtained with high diastereoselectivity (Scheme 24.29).[108]

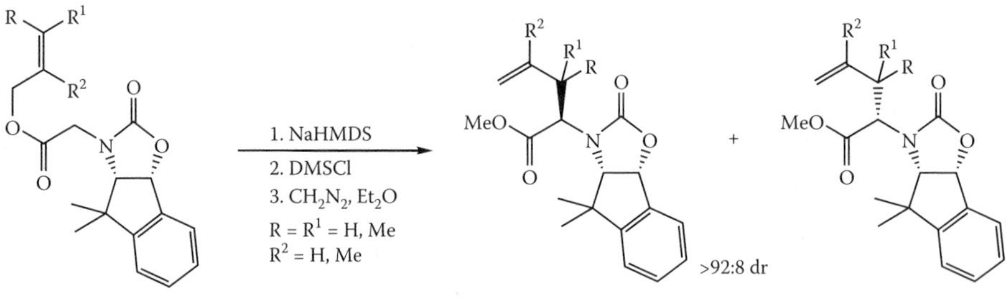

SCHEME 24.29

A six-membered chairlike transition state was proposed in which the (*E*)-ketene silyl acetal is placed in the opposite direction to the oxazolidinone carbonyl to avoid dipole moment repulsion, with the allylic moiety oriented away from the hindered indane backbone. This transition state would account for the direction of the asymmetric induction and also explain the better results observed with smaller silylating groups (Figure 24.7).[108]

24.5.2 *cis*-1-Amino-2-Hydroxy-1,2,3,4-Tetrahydronaphthalene

Another promising conformationally constrained amino alcohol is *cis*-1-amino-2-hydroxy-1,2,3,4-tetrahydronaphthalene. Early studies using the tetrahydronaphthol backbone as chiral auxiliary or as a ligand in catalysis (see Section 17.8.3) often showed that the relative flexibility of the 6-membered ring was highly detrimental to the achievement of good levels of asymmetric induction, such as in the Diels-Alder reaction[82] or in the addition of diethyl zinc to benzaldehyde.[110] However, it led to similar or slightly improved selectivities compared to the more rigid indanol platform in the borane reduction of carbonyls[40,63,64] and the addition of Grignard reagents to ketones.[65]

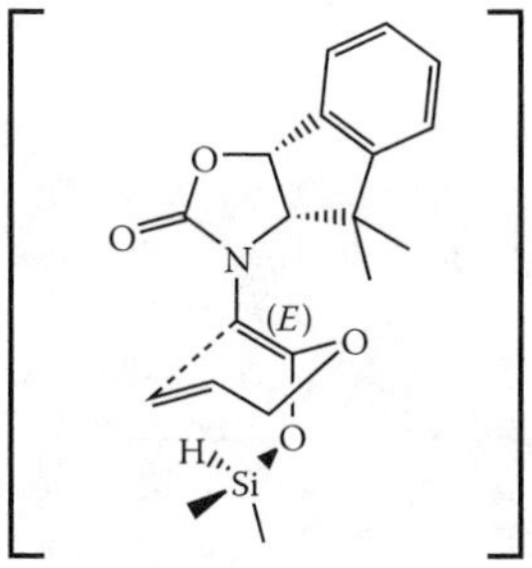

FIGURE 24.7 Transition state with ketene silyl acetals.

24.6 SUMMARY

Since the discovery of *cis*-1-amino-2-indanol as a ligand for HIV-protease inhibitors and the development of a practical industrial process for the synthesis of either *cis*-isomers in enantiopure form, the remarkable properties of the rigid indane platform have been extensively used in an ever-increasing number of asymmetric methodologies.

cis-1-Amino-2-indanol–based chiral auxiliaries have proved effective in various carbon–carbon and carbon–heteroatom bond formations, especially in aldol and Diels-Alder reactions. Examples of the utility of these chiral auxiliaries include the syntheses of α-amino acids, γ-lactones, sulfina-mides, sulfoxides, natural products, and pharmaceutical drugs in a highly stereoselective fashion. The industrial process for Crixivan is a particularly striking demonstration of this concept. The availability of optically active *cis*-1-amino-2-indanol, the high levels of asymmetry it induces, and its easy recovery and recycling have proved remarkable assets to the development of practical and economical large-scale processes.

REFERENCES AND NOTES

1. For examples of asymmetric processes for drug synthesis, see Senanayake, C. H., Krishnamurthy, D. *Current Opinion in Drug Discovery & Development* 1999, *2*, 590.
2. For examples of asymmetric processes for drug synthesis, see Senanayake, C. H., Pflum, D. A. *Chemica Oggi* 1999, *9*, 21.
3. For previous review on the subject of *cis*-1-amino-2-indanol, see Ghosh, A. K., Fidanze, S., Senan-ayake, C. H. *Synthesis* 1998, 937.
4. For previous review on the subject of *cis*-1-amino-2-indanol, see Senanayake, C. H. *Aldrichimica Acta* 1998, *31*, 3.
5. Senanayake, C. H., Jacobsen, E. N. Chiral (Salen)Mn(III) Complexes in Asymmetric Epoxidations: Practical Synthesis of *Cis*-Aminoindanol and Its Application to Enantiopure Drug Synthesis, *Process Chemistry in the Pharmaceutical Industry*, Gadamasetti, K. G. Marcel Dekker: New York, 1999, Chapter 18, 327.
6. Lyle, T. A., Wiscount, C. M., Guare, J. P., Thompson, W. J., Anderson, P. S., Darke, P. L., Zugay, J. A., Emini, E. A., Schleif, W. A., Quintero, J. C., Dixon, R. A. F., Sigal, I. S., Huff, J. R. *J. Med. Chem.* 1991, *34*, 1228.
7. Thompson, W. J., Fitzgerald, P. M. D., Holloway, M. K., Emini, E. A., Darke, P. L., McKeever, B. M., Schleif, W. A., Quintero, J. C., Zugay, J. A., Tucker, T. J., Schwering, J. E., Homnick, C. F., Nunberg, J., Springer, J. P., Huff, J. R. *J. Med. Chem.* 1992, *35*, 1685.
8. Hungate, R. W., Chen, J. L., Starbuck, K. E., Vacca, J. P., McDaniel, S. L., Levin, R. B., Dorsey, B., D., Guare, J. P., Holloway, M. K., Whitter, W., Darke, P. L., Zugay, J. A., Schleif, W. A., Emini, E. A., Quintero, J. C., Lin, J. H., Chen, I.-W., Anderson, P. S., Huff, J. R. *Bioorg. Med. Chem.* 1994, *2*, 859.
9. Young, S. D., Payne, L. S., Thompson, W. J., Gaffin, N., Lyle, T. A., Britcher, S. F., Graham, S. L., Schultz, T. H., Deana, A. A., Darke, P. L., Zugay, J. A., Schleif, W. A., Quintero, J. C., Emini, E. A., Anderson, P. S., Huff, J. R. *J. Med. Chem.* 1992, *35*, 1702.

10. Bone, R., Vacca, J. P., Anderson, P. S., Holloway, M. K. *J. Am. Chem. Soc.* 1991, *113*, 9382.

11. Askin, D., Wallace, M. A., Vacca, J. P., Reamer, R. A., Volante, R. P., Shinkai, I. *J. Org. Chem.* 1992, *57*, 2771.

12. Thompson, W. J., Ball, R. G., Darke, P. L., Zugay, J. A., Thies, J. E. *Tetrahedron Lett.* 1992, *33*, 2957.

13. Thaisrivongs, S., Turner, S. R., Strohbach, J. W., TenBrink, R. E., Tarpley, W. G., McQuade, T. J., Heinrikson, R. L., Tomasselli, A. G., Hui, J. O., Howe, W. J. *J. Med. Chem.* 1993, *36*, 941.

14. Dorsey, B. D., Plzak, K. J., Ball, R. G. *Tetrahedron Lett.* 1993, *34*, 1851.

15. Ghosh, A. K., Thompson, W. J., McKee, S. P., Duong, T. T., Lyle, T. A., Chen, J. C., Darke, P. L., Zugay, J. A., Emini, E. A., Schleif, W. A., Huff, J. R., Anderson, P. S. *J. Med. Chem.* 1993, *36*, 292.

16. Dorsey, B. D., Levin, R. B., McDaniel, S. L., Vacca, J. P., Guare, J. P., Darke, P. L., Zugay, J. A., Emini, E. A., Schleif, W. A., Quintero, J. C., Lin, J. H., Chen, I.-W., Holloway, M. K., Fitzgerald, P. M. D., Axel, M. G., Ostovic, D., Anderson, P. S., Huff, J. R. *J. Med. Chem.* 1994, *37*, 3443 and references therein.

17. Reider, P. J. *Chimia* 1997, *51*, 306.

18. Askin, D., Eng, K. K., Rossen, K., Purick, R. M., Wells, K. M., Volante, R. P., Reider, P. J. *Tetrahedron Lett.* 1994, *35*, 673.

19. Maligres, P. E., Upadhyay, V., Rossen, K., Cianciosi, S. J., Purick, R. M., Eng, K. K., Reamer, R. A., Askin, D., Volante, R. P., Reider, P. J. *Tetrahedron Lett.* 1995, *36*, 2195.

20. Maligres, P. E., Weissman, S. A., Upadhyay, V., Cianciosi, S. J., Reamer, R. A., Purick, R. M., Sager, J., Rossen, K., Eng, K. K., Askin, D., Volante, R. P., Reider, P. J. *Tetrahedron* 1996, *52*, 3327.

21. Rossen, K., Volante, R. P., Reider, P. J. *Tetrahedron Lett.* 1997, *38*, 777.

22. Rossen, K., Weissman, S. A., Sager, J., Reamer, R. A., Askin, D., Volante, R. P., Reider, P. J. *Tetrahedron Lett.* 1995, *36*, 6419.

23. Coburn, C. A., Young, M. B., Hungate, R. W., Isaacs, R. C. A., Vacca, J. P., Huff, J. R. *Bioorg. Med. Chem. Lett.* 1996, *6*, 1937.

24. Tata, J. R., Charest, M., Lu, Z., Raghavan, S., Huening, T., Rano, T. PCT Appl. WO 01/05230, 2001.

25. Dorsey, B. D., McDaniel, S. L., Levin, R. B., Vacca, J. P., Darke, P. L., Zugay, J. A., Emini, E. A., Schleif, W. A., Lin, J. H., Chen, I.-W., Holloway, M. K., Anderson, P. S., Huff, J. R. *Bioorg. Med. Chem. Lett.* 1994, *4*, 2769; Dorsey, B. D., McDaniel, S. L., Levin, R. B., Michelson, S. R., Vacca, J. P., Darke, P. L., Zugay, J. A., Emini, E. A., Schleif, W. A., Lin, J. H., Chen, I.-W., Holloway, M. K., Anderson, P. S., Huff, J. R. *Bioorg. Med. Chem. Lett.* 1995, *5*, 773.

26. Dorsey, B. D., McDonough, C., McDaniel, S. L., Levin, R. B., Newton, C. L., Hoffman, J. M., Darke, P. L., Zugay-Murphy, J. A., Emini, E. A., Schleif, W. A., Olsen, D. B., Stahlhut, M. W., Rutkowski, C. A., Kuo, L. C., Lin, J. H., Chen, I.-W., Michelson, S. R., Holloway, M. K., Huff, J. R., Vacca, J. P. *J. Med. Chem.* 2000, *43*, 3386.

27. Lehr, P., Billich, A., Charpiot, B., Ettmayer, P., Scholz, D., Rosenwirth, B., Gstach, H. *J. Med. Chem.* 1996, *39*, 2060.

28. Spaltenstein, A., Almond, M. R., Bock, W. J., Cleary, D. G., Furfine, E. S., Hazen, R. J., Kazmierski, W. M., Salituro, F. G., Tung, R. D., Wright, L. L. *Bioorg. Med. Chem. Lett.* 2000, *10*, 1159.

29. Alterman, M., Björsne, M., Mühlman, A., Classon, B., Kvarnström, I., Danielson, H., Markgren, P.-O., Nillroth, U., Unge, T., Hallberg, A., Samuelsson, B. *J. Med. Chem.* 1998, *41*, 3782.

30. Mühlman, A., Classon, B., Hallberg, A., Samuelsson, B. *J. Med. Chem.* 2001, *44*, 3402.

31. Oscarsson, K., Classon, B., Kvarnström, I., Hallberg, A., Samuelsson, B. *Can. J. Chem.* 2000, *78*, 829.

32. Mühlman, A., Lindberg, J., Classon, B., Unge, T., Hallberg, A., Samuelsson, B. *J. Med. Chem.* 2001, *44*, 3407.

33. Yao, W., Decicco, C. P. PCT Appl. WO 99/09000, 1999.

34. Yao, W., Wasserman, Z. R., Chao, M., Reddy, G., Shi, E., Liu, R.-Q., Covington, M. B., Arner, E. C., Pratta, M. A., Tortorella, M., Magolda, R. L., Newton, R., Qian, M., Ribadeneira, M. D., Christ, D., Wexler, R. R., Decicco, C. P. *J. Med. Chem.* 2001, *44*, 3347.

35. Yao, W., Chao, M., Wasserman, Z. R., Liu, R.-Q., Covington, M. B., Newton, R., Christ, D., Wexler, R. R., Decicco, C. P. *Bioorg. Med. Chem. Lett.* 2002, *12*, 101.

36. Didier, E., Loubinoux, B., Ramos Tombo, G. M., Rihs, G. *Tetrahedron Lett.* 1991, *47*, 4941.

37. Boyd, D. R., Sharma, N. D., Bowers, N. I., Goodrich, P. A., Groocock, M. R., Blacker, A. J., Clarke, D. A., Howard, T., Dalton, H. *Tetrahedron: Asymmetry* 1996, *7*, 1559.

38. Takahashi, M., Koike, R., Ogasawara, K. *Chem. Pharm. Bull.* 1995, *43*, 1585.

39. Takahashi, M., Ogasawara, K. *Synthesis* 1996, 954.
40. Ghosh, A. K., Kincaid, J. F., Haske, M. G. *Synthesis* 1997, 541.
41. Kajiro, H., Mitamura, S., Mori, A., Hiyama, T. *Bull. Chem. Soc. Jpn.* 1999, *72*, 1093.
42. Jacobsen, E. N. Asymmetric Catalytic Epoxidation of Unfunctionalized Olefins, *Catalytic Asymmetric Synthesis*, Ojima, I. VCH: New York, 1994, Chapter 4.2, 159 and references therein.
43. Katsuki, T. Asymmetric Epoxidation of Unfunctionalized Olefins and Related Reactions, *Catalytic Asymmetric Synthesis*, Ojima, I. VCH: New York, 2000, Chapter 6B, 287 and references therein.
44. Jacobsen, E. N., Zhang, W, Muci, A. R., Ecker, J. R., Deng, L. *J. Am. Chem. Soc.* 1991, *113*, 7063.
45. Halterman, R. L., Jan, S.-T. *J. Org. Chem.* 1991, 56, 5253.
46. Hirama, M., Oishi, T., Itô, S. *J. Chem. Soc., Chem. Commun.* 1989, 665.
47. Hanessian, S., Meffre, P., Girard, M., Beaudoin, S., Sancéau, J.-Y., Bennani, Y. *J. Org. Chem.* 1993, *58*, 1991.
48. Sasaki, H., Irie, R., Hamada, T., Suzuki, K., Katsuki, T. *Tetrahedron* 1994, *50,* 11827.
49. Lakshman, M. K., Zajc, B. *Tetrahedron Lett.* 1996, *37*, 2529.
50. Senanayake, C. H., Roberts, F. E., DiMichele, L. M., Ryan, K. M., Liu, J., Fredenburgh, L. E., Foster, B. S., Douglas, A. W., Larsen, R. D., Verhoeven, T. R., Reider, P. J. *Tetrahedron Lett.* 1995, *36*, 3993.
51. Senanayake, C. H., DiMichele, L. M., Liu, J., Fredenburgh, L. E., Ryan, K. M., Roberts, F. E., Larsen, R. D., Verhoeven, T. R., Reider, P. J. *Tetrahedron Lett.* 1995, *36*, 7615.
52. Senanayake, C. H., Smith, G. B., Ryan, K. M., Fredenburgh, L. E., Liu, J., Roberts, F. E., Hughes, D. L., Larsen, R. D., Verhoeven, T. R., Reider, P. J. *Tetrahedron Lett.* 1996, *37*, 3271.
53. Gao, Y., Hong, Y., Nie, X., Bakale, R. P., Feinberg, R. R., Zepp, C. M. U.S. Pat. 5,599,985, 1997.
54. Gao, Y., Hong, Y., Nie, X., Bakale, R. P., Feinberg, R. R., Zepp, C. M. U.S. Pat. 5,616,808, 1997.
55. Available from Aldrich Chemical Co., Inc., Multi-kilogram quantities are available from ChiRex, Newcastle, England.
56. Srinivasan, K., Michaud, P., Kochi, J. K. *J. Am. Chem. Soc.* 1986, *108*, 2309.
57. Larrow, J. F., Jacobsen, E. N. *J. Am. Chem. Soc.* 1994, *116*, 12129 and references therein.
58. Hughes, D. L., Smith, G. B., Liu, J., Dezeny, G. C., Senanayake, C. H., Larsen, R. D., Verhoeven, T. R., Reider, P. J. *J. Org. Chem.* 1997, *62*, 2222.
59. Ritter, J. J., Minieri, P. P. *J. Am. Chem. Soc.* 1948, *70*, 4045.
60. For a review on the Ritter reaction, see Bishop, R. *Comprehensive Org. Syn.* 1991, *6*, 261.
61. Ghosh, A. K., Duong, T. T., McKee, S. P. *J. Chem. Soc., Chem. Commun.* 1992, 1673.
62. Evans, D. A. *Aldrichimica Acta* 1982, *15*, 23.
63. Ghosh, A. K., Onishi, M. *J. Am. Chem. Soc.* 1996, *118*, 2527.
64. Zimmerman, H. E., Traxler, M. D. *J. Am. Chem. Soc.* 1957, *79*, 1920.
65. Ghosh, A. K., Fidanze, S., Onishi, M., Hussain, K. A. *Tetrahedron Lett.* 1997, *38*, 7171.
66. Ghosh, A. K., Liu, W., Xu, Y., Chen, Z. *Angew. Chem., Int. Ed.* 1996, *35*, 74.
67. Ghosh, A. K., Liu, C. *J. Am. Chem. Soc.* 2003, *125*, 2374.
68. Ghosh, A. K., Bischoff, A. *Org. Lett.* 2000, *2*, 1573.
69. Ghosh, A. K., Fidanze, S. *Org. Lett.* 2000, *2*, 2405.
70. Ghosh, A. K., Hussain, K. A., Fidanze, S. *J. Org. Chem.* 1997, *62*, 6080.
71. Ghosh, A. K., Fidanze, S. *J. Org. Chem.* 1998, *63*, 6146.
72. Ghosh, A. K., Liu, W. *J. Org. Chem.* 1996, *61*, 6175.
73. Myers, A. G., Barbay, J. K., Zhong, B. *J. Am. Chem. Soc.* 2001, *123*, 7207.
74. Lee, J., Choi, W.-B., Lynch, J. E., Volante, R. P., Reider, P. J. *Tetrahedron Lett.* 1998, *39*, 3679.
75. Tamaru, Y., Mizutani, M., Furukawa, Y., Kawamura, S.-I., Yoshida, Z.-I., Yanagi, K., Minobe, M. *J. Am. Chem. Soc.* 1984, *106*, 1079.
76. Armstrong, J. D. III, Hartner, F. W. Jr., DeCamp, A. E., Volante, R. P., Shinkai, I. *Tetrahedron Lett.* 1992, *33*, 6599.
77. McWilliams, J. C., Armstrong, J. D. III, Zheng, N., Bhupathy, M., Volante, R. P., Reider, P. J. *J. Am. Chem. Soc.* 1996, *118*, 11970.
78. Song, Z. J., Zhao, M., Desmond, R., Devine, P., Tschaen, D. M., Tillyer, R., Frey, L., Heid, R., Xu, F., Foster, B., Li, J., Reamer, R., Volante, R., Grabowski, E. J. J., Dolling, U. H., Reider P. J., Okada, S., Kato, Y., Mano, E. *J. Org. Chem.* 1999, *64*, 9658.
79. Alexakis, A., Sedrani, R., Mangeney, P., Normant, J. F. *Tetrahedron Lett.* 1988, *29*, 4411.

80. Frey, L. F., Tillyer, R. D., Caille, A.-S., Tschaen, D. M., Dolling, U.-H., Grabowski, E. J. J., Reider, P. J. *J. Org. Chem.* 1998, *63*, 3120.
81. Smitrovich, J. H., Boice, G. N., Qu, C., DiMichele, L., Nelson, T. D., Huffman, M. A., Murry, J., McNamara, J., Reider, P. J. *Org. Lett.* 2002, *4*, 1963.
82. Davies, I. W., Senanayake, C. H., Castonguay, L., Larsen, R. D., Verhoeven, T. R., Reider, P. J. *Tetrahedron Lett.* 1995, *36*, 7619.
83. Evans, D. A., Chapman, K. T., Bisaha, J. *J. Am. Chem. Soc.* 1988, *110*, 1238.
84. Evans, D. A., Chapman, K. T., Hung, D. T., Kawaguchi, A. T. *Angew. Chem., Int. Ed.* 1987, *26*, 1184.
85. Ghosh, A. K., Mathivanan, P. *Tetrahedron: Asymmetry* 1996, *7*, 375.
86. Janey, J. M., Iwana, T., Kozmin, S. A., Rawal, V. H. *J. Org. Chem.* 2000, *65*, 9059.
87. Ghosh, A. K., Chen, Y. *Tetrahedron Lett.* 1995, *36*, 6811.
88. Yaker, Y. E., Goa, K. L., Fitton, A. *Drug & Aging* 1995, *6*, 243.
89. Grover, P. T., Bhongle, N. N., Wald, S. A., Senanayake, C. H. *J. Org. Chem.* 2000, *65*, 6283.
90. Senanayake, C. H., Fang, K., Grover, P., Bakale, R. P., Vandenbossche, C. P., Wald, S. A. *Tetrahedron Lett.* 1999, *40*, 819.
91. Kress, M. H., Yang, C., Yasuda, N., Grabowski, E. J. J. *Tetrahedron Lett.* 1997, *38*, 2633.
92. Mulder, J. A., Hsung, R. P., Frederick, M. O., Tracey, M. R., Zificsak, C. A *Org. Lett.* 2002, *4*, 1383.
93. Tanaka, K, Katsumura, S. *J. Am. Chem. Soc.* 2002, *124*, 9660.
94. Zheng, N., Armstrong, J. D. III, McWilliams, J. C., Volante, R. P. *Tetrahedron Lett.* 1997, *38*, 2817 and references therein.
95. Davis, F. A., Zhou, P., Chen, B.-C. *Chem. Soc. Rev.* 1998, *27*, 13.
96. Ellman, J. A., Owens, T. D., Tang, T. P. *Acc. Chem. Rev.* 2002, *35*, 984 and references therein.
97. Weix, D. J., Ellman, J. A. *Org. Lett.* 2003, *5*, 1317.
98. Prakash, G. K. S., Mandal, M., Olah, G. A. *Org. Lett.* 2001, *3*, 2847.
99. Han, Z., Krishnamurthy, D., Pflum, D., Grover, P., Wald, S. A., Senanayake, C. H. *Org. Lett.* 2002, *4*, 4025.
100. Davis, F. A., Prasad, K. R., Nolt, M. B., Wu, Y. *Org. Lett.* 2003, *5*, 925.
101. Pflum, D. A., Krishnamurthy, D., Han, Z., Wald, S. A., Senanayake, C. H. *Tetrahedron Lett.* 2002, *43*, 923.
102. Han, Z., Krishnamurthy, D., Grover, P., Fang, Q. K., Pflum, D. A., Senanayake, C. H. *Tetrahedron Lett.* 2003, *44*, 4195.
103. Han, Z., Krishnamurthy, D., Grover, P., Fang, Q. K., Senanayake, C. H. *J. Am. Chem. Soc.* 2002, *124*, 7880.
104. Han, Z., Krishnamurthy, D., Grover, P., Wilkinson, H. S., Fang, Q. K., Su, X., Lu, Z.-H., Magiera, D., Senanayake, C. H. *Angew. Chem., Int. Ed.* 2003, *42*, 2032 and references therein.
105. Sudo, A., Saigo, K. *Tetrahedron: Asymmetry* 1995, *6*, 2153, Sudo, A., Saigo, K. *Tetrahedron: Asymmetry* 1996, *7*, 619.
106. Sudo, A., Saigo, K. *Tetrahedron: Asymmetry* 1996, *7*, 2939 and references therein.
107. Sudo, A., Saigo, K. *Chem. Lett.* 1997, 97.
108. Matsui, S., Oka, Y., Hashimoto, Y., Saigo, K. *Enantiomer* 2000, 5, 105.
109. Sudo, A., Saigo, K. *J. Org. Chem.* 1997, *62*, 5508.
110. Bellucci, C. M., Bergamini, A., Cozzi, P. G., Papa, A., Tagliavini, E., Umani-Ronchi, A. *Tetrahedron: Asymmetry* 1997, *8*, 895.

25 Enantiopure Amines by Chirality Transfer Using (*R*)-Phenylglycine Amide

Ben de Lange, Wilhelmus H. J. Boesten, Marcel van der Sluis,
Patrick G. H. Uiterweerd, Henk L. M. Elsenberg,
Richard M. Kellogg, and Quirinus B. Broxterman

CONTENTS

25.1 INTRODUCTION

Enantiopure amines are present as key functional elements in a large majority of pharmaceutical drugs and drug candidates. Although in recent years great progress in asymmetric synthesis has been realized based on homogeneous catalysis,[1] biocatalysis,[2] and resolution processes combined with efficient racemization of the undesired enantiomer,[3] cost-effective synthesis of enantiopure amines continues to be a challenge for the fine chemical industry. The efficiency of these processes depends on, among other factors, the availability of a (chiral) starting material, reagents, auxiliaries, and catalysts. Furthermore, the presence of an efficient recycling process for expensive chiral auxiliaries and catalysts can be of decisive importance for an economically feasible process.

For the preparation of enantiopure amines, diastereoselective synthesis using a chiral auxiliary can be a viable approach. In this concept, in the first step a chiral intermediate is formed by reaction of a prochiral substrate with the chirality transfer agent. The key second step is a diastereoselective reaction. This is followed by cleavage of the chiral auxiliary to give the product amine. This concept is illustrated in Figure 25.1.

The synthesis of enantiopure amino-functionalized compounds such as α- and β-amino acids or nonfunctionalized amines can be envisaged by the use of aldehydes, ketones, α- or β-keto acids, or derivatives thereof as substrates for imine formation followed by, for example, diastereoselective Strecker reactions, reductions, or organometallic addition reactions. In the literature, diastereoselective syntheses based on a large variety of chiral auxiliaries, such as α-arylethylamines,[4]

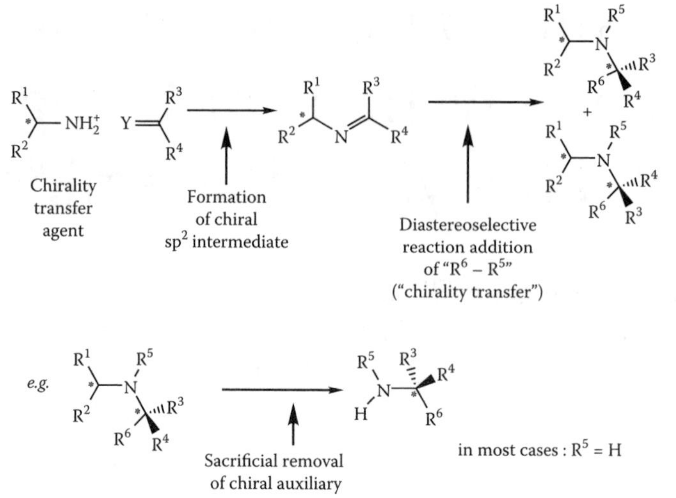

FIGURE 25.1 Diastereoselective chirality transfer reactions.

β-amino alcohols and derivatives,[5] amino diols,[6] sugar derivatives,[7] and sulfinates[8] have been reported (see Chapter 23). However, in the production of most pharmaceuticals, the drawback of most chiral auxiliaries can be cost, availability, or a combination of both. Auxiliaries are used in stoichiometric amounts and, when used as a template, which is in a sacrificial mode, they are, in principle, lost during the conversion. Even when the auxiliaries have the potential to be recycled, this can be difficult in a pharmaceutical application because of good manufacturing practice requirements (see Chapters 1 and 23). Another drawback is the need for removal of some of these auxiliaries by procedures unsuitable for large-scale preparations. Examples include oxidation with Pb(OAc)$_4$ or treatment with HIO$_4$/MeNH$_2$.[9] Furthermore, purification by chromatography in a separate step may be needed to obtain diastereoisomerically pure compounds, and this can lead to unacceptable losses, difficulties in scaling up, and additional costs.

Enantiopure (R)-phenylglycine amide [abbreviated (R)-PGA] (1) has become readily accessible at DSM as a result of its application on industrial scale as a key intermediate in enzymatic routes for the preparation of β-lactam antibiotics (Scheme 25.1).[10]

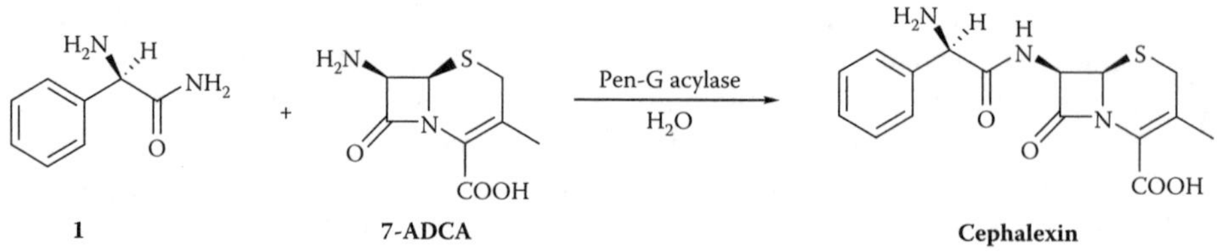

SCHEME 25.1

Either (S)-specific aminopeptidase catalyzed hydrolysis of racemic PGA[11] or crystallization-induced asymmetric transformation of racemic PGA with (S)-mandelic acid as resolving agent[12] can be used to prepare (R)-PGA. As a result of its ready availability on large scale within DSM, we envisaged the application of (R)-PGA for the production of enantiomerically pure amine functionalized compounds using the chirality transfer concept. Obviously, (S)-phenylglycine amide is also available and can be used for the preparation of the opposite enantiomer of the amines described.

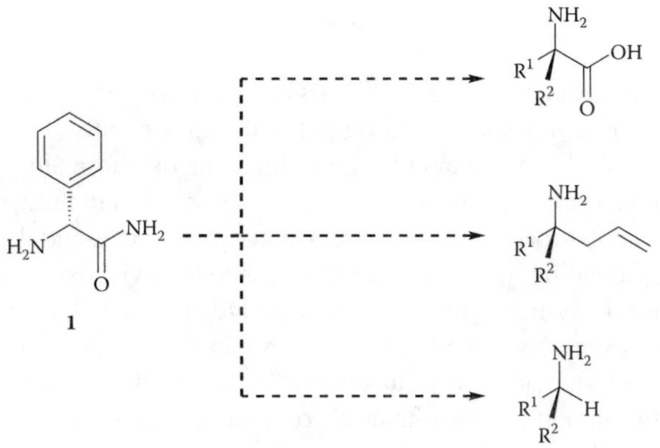

FIGURE 25.2 Chirality transfer routes of (R)-PGA [(R)-phenylglycine amide] toward α-amino acids, homoallylamines, and amines.

The anticipated facile removal of the template by catalytic hydrogenolysis (although the selectivity when two benzyl groups are present around the nitrogen atom may need to be determined) and the crystalline nature of PGA-containing intermediates are advantages that arise from the presence of the benzyl and amide groups, respectively. These properties can allow for easy enrichment of diastereoisomers in cases where the diastereoselectivity is not absolute. The slightly acidic α-H of (R)-PGA causes it to be somewhat sensitive toward racemization and is another aspect to address. Prolonged high temperatures, long reaction times, and/or basic conditions need to be avoided.

In this chapter, recent applications of (R)-phenylglycine amide (**1**) in asymmetric synthesis are presented (Figure 25.2). The first section deals with diastereoselective Strecker reactions for the preparation of α-amino acids and derivatives, whereas the second section focuses on diastereoselective allylation of imines for preparation of enantiomerically pure homoallylamines. This latter class of compounds is a well-known intermediate for the synthesis of, for example, many types of amines, amino alcohols, and β-amino acids. The final section describes reduction of imines providing enantiomerically pure amines. (S)-3,3-Dimethyl-2-butylamine and (S)-1-aminoindane will be presented as leading examples. The results described in this chapter originate from a long-standing cooperation in the field of chiral technology development between DSM Pharma Chemicals and Syncom B.V.

25.2 ENANTIOPURE α-AMINO ACIDS PREPARED BY MEANS OF DIASTEREOSELECTIVE STRECKER REACTIONS

Industrial production of α-amino acids via the Strecker reaction is one of the most general methods to obtain these compounds in a cost-effective way because it makes use of inexpensive and easily accessible starting materials (Chapter 2).[13] The amino nitrile derived from the Strecker reaction is usually hydrolyzed to the racemic amino acid amide or the racemic amino acid followed by resolution processes.[14] This leads to a maximum yield of 50% if the unwanted enantiomer is not racemized.[2] In principle, asymmetric synthesis approaches leading to a maximum yield of 100% of a single enantiomer are more advantageous.[15] Substantial progress in catalytic asymmetric Strecker reactions has been reported using sophisticated, and in many cases complex, chiral catalysts.[16] In the examples reported so far, N-protection of the amino donor is needed, which leads to an additional step for removal of this protective group. Economically feasible processes will require either highly efficient recycling of the ligand and/or the metal used in the catalysis or very high substrate-to-catalyst ratios. Therefore, application of a cost-efficient, readily available, chiral auxiliary in a diastereoselective Strecker reaction can have economic advantages over the use of asymmetric catalytic Strecker approaches.

Diastereoselective Strecker syntheses with varying diastereoselectivities have been reported.[4] In many cases, the α-amino nitriles need to be purified in a separate step to obtain diastereoisomerically pure compounds by, for example, crystallization or chromatography. Crystallization-induced asymmetric transformation would be an interesting solution to these problems (see Chapter 7).[17,18] In this crystallization process, one diastereoisomer precipitates and the other one epimerizes in solution via the corresponding imine; this can lead both to high yield and high diastereoselectivity in a single step. To prove the principle of this concept, pivaldehyde and 3,4-dimethoxyphenylacetone were chosen as typical examples to be tested in the Strecker reaction with (R)-PGA and HCN.[19] Use of these starting materials lead to enantiomerically enriched tert-leucine and α-methyl-dopa, respectively; these are two important nonproteogenic α-amino acids with pharmaceutical applications.[20]

The asymmetric Strecker reaction of (R)-PGA (1), pivaldehyde (2), and HCN generated in situ[21] from NaCN/AcOH in methanol as solvent gave the (R,S)-3 and (R,R)-3 aminonitriles in 80% yield. However, the diastereoisomeric excess was only 30%. It was found that by performing the reaction in water and heating for 24 hours at 70°C, the de of aminonitrile (R,S)-3 could be increased to >98%. The isolated yield was 93%.

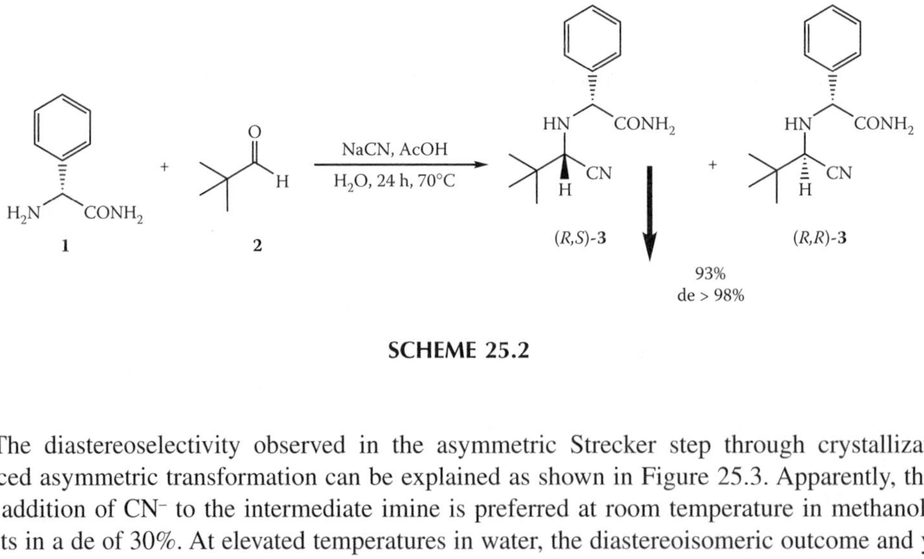

SCHEME 25.2

The diastereoselectivity observed in the asymmetric Strecker step through crystallization-induced asymmetric transformation can be explained as shown in Figure 25.3. Apparently, the re-face addition of CN⁻ to the intermediate imine is preferred at room temperature in methanol and results in a de of 30%. At elevated temperatures in water, the diastereoisomeric outcome and yield of the process are controlled by the reversible nature of the conversion of the amino nitriles into the intermediate imine and by the difference in solubility between both diastereoisomers under the conditions applied.[22,23]

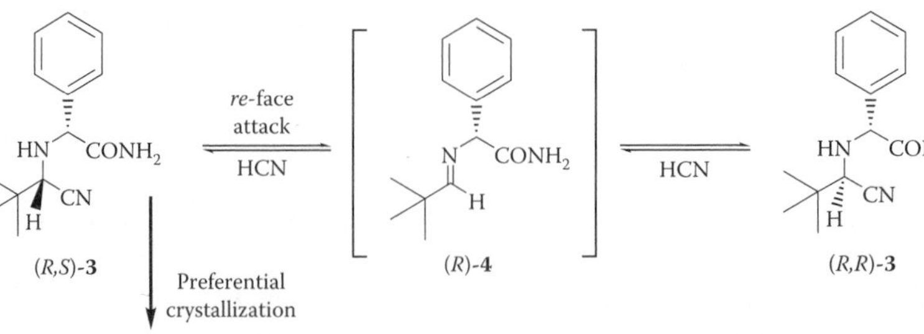

FIGURE 25.3 Crystallization-induced asymmetric transformation of amino nitrile **3**.

One of the possibilities to convert the amino nitrile (*R*,*S*)-**3** to (*S*)-*tert*-leucine (**7**) is shown in the reaction sequence of Scheme 25.3. Hydrolysis of the aminonitrile (*R*,*S*)-**3** to the diamide **5** was performed in concentrated sulfuric acid. Removal of the phenylacetamide group using H_2 and Pd–C gave (*S*)-*tert*-leucine amide (**6**). Acidic hydrolysis of the amide **6** yielded (*S*)-*tert*-leucine (**7**) in 73% overall yield for the 3 steps and an enantiomeric excess of >98%. The overall yield for this nonoptimized protocol is 66% based on pivaldehyde. Obviously, other routes to convert the amino nitrile to the amino acid can be envisaged. As an example, by heating the diamide in sulfuric acid, after dilution with water, the diacid can be obtained that, after hydrogenolysis, affords the amino acid in 2 steps from the amino nitrile **3**.

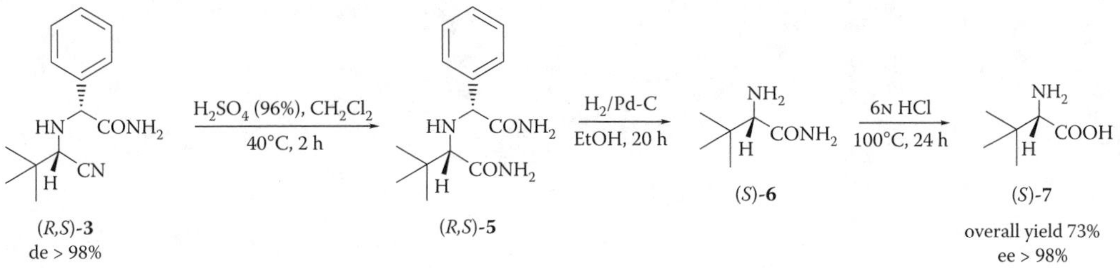

SCHEME 25.3

In a second example using 3,4-dimethoxyphenylacetone (**8**) and (*R*)-PGA (**1**), it was found that an equilibrium composition of 55:45 exists between the two diastereoisomers (*R*,*S*)-**9** and (*R*,*R*)-**9** in methanol as solvent. By fine-tuning the reaction, it was found that in a mixture of methanol and water (6:1), almost diastereoisomerically pure (de >98%) amino nitrile (*R*,*S*)-**9** precipitated in 76% isolated yield (Scheme 25.4). Clearly, in this case a crystallization-induced asymmetric transformation has also occurred.

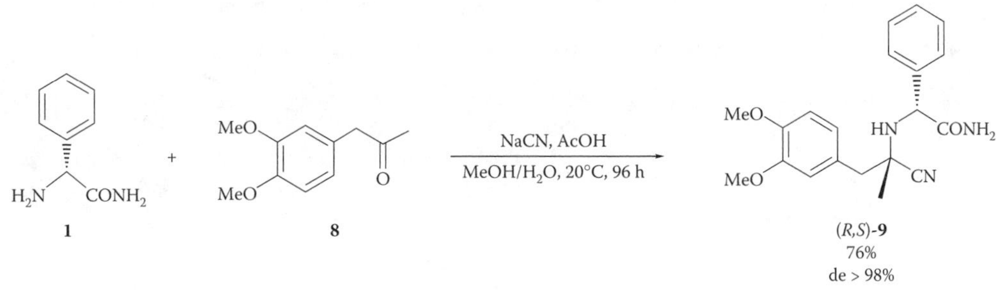

SCHEME 25.4

Several other amino nitriles were obtained as crystalline materials from aqueous methanol mixtures, including those derived from acetophenone, isobutyraldehyde, and 3-methyl-2-butanone. For each substrate, conditions had to be optimized to obtain diastereoisomerically pure amino nitriles.

25.3 DIASTEREOSELECTIVE ALLYLATION REACTIONS

Chiral homoallylamines are valuable intermediates for the preparation of compounds such as amines, β-amino acids, 1-amino-3,4-epoxides, and 1,3-amino alcohols.[24] High 1,3-asymmetric induction has been achieved during the allylation of imines derived from chiral auxiliaries such as β-amino alcohols and α-amino acid esters.[25] Drawbacks of these methods are the limited availability

TABLE 25.1
Diasteroselective Allylation Reactions of (R)-PGA[a] (1)
with Several Aldehydes (Scheme 25.5)

R¹	Yield of 10 (%)	de of 11 (%)	Yield of 11 (%)
Phenyl	84	>98	93
3-Piperonyl	90	>98	81
4-HO-Phenyl	87	>98	90
2,5-(MeO)₂-C₆H₃	90	>98	93
3-Pyridyl	89	>98	43
2-Furyl	80	90	88
2-Thiophene	95	>98	90
t-Butyl	95	>98	89
i-Propyl	86	98	77
i-Butyl	95	>98	94
Me	91	88	95

[a] (R)-PGA, (R)-phenylglycine amide.

and/or the need of removal of these auxiliaries by procedures unsuitable for large-scale preparations, such as oxidation with Pb(OAc)₄ or treatment with HIO₄/MeNH₂. As in the case of diastereoselective Strecker reactions, we applied (R)-PGA as a chiral-inducing agent in diastereoselective allylation reactions.[26]

(R)-Imines **10** were obtained easily in yields >85% by stirring a mixture of (R)-PGA (**1**) with the corresponding aldehyde (R¹CHO) in, for example, dichloromethane at room temperature. The addition of allylzinc bromide to the imines **10** in THF at 0°C provided the (R,R)-PGA-homoallyl-amines **11** in yields up to 94% with de >98% in most cases (Scheme 25.5 and Table 25.1).

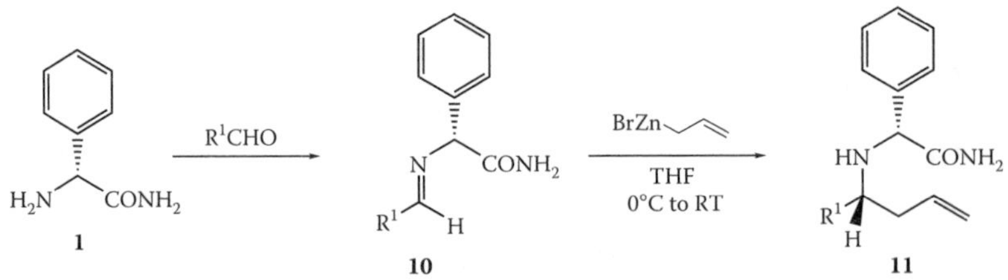

SCHEME 25.5

Functionalized allylzinc reagents, such as crotylzinc bromide[27] and methallylzinc bromide, also gave the analogous addition products in >95% yield and de >98%. The high selectivity can be rationalized in a model assuming chelation of the zinc atom between the two heteroatoms of the (R)-phenylglycine amide-imine to form a 5-membered intermediate. Furthermore, a 6-membered chairlike transition state is formed from the allylic system and the C = N double bond of the imine. Then, the re-face 1,2-addition proceeds in a concerted fashion by an allylic rearrangement (Figure 25.4). In accord with this conclusion, crotylzinc bromide provides the methallyl-substituted product.

These diastereomerically and enantiomerically enriched compounds are useful as intermediates for a variety of interesting building blocks such as 1-aminobutanes, nonprotected homoallylamines, β-amino acids, and γ-amino alcohols (Figure 25.5). The conversion to 1-aminobutanes and homo-allylamines will be described.

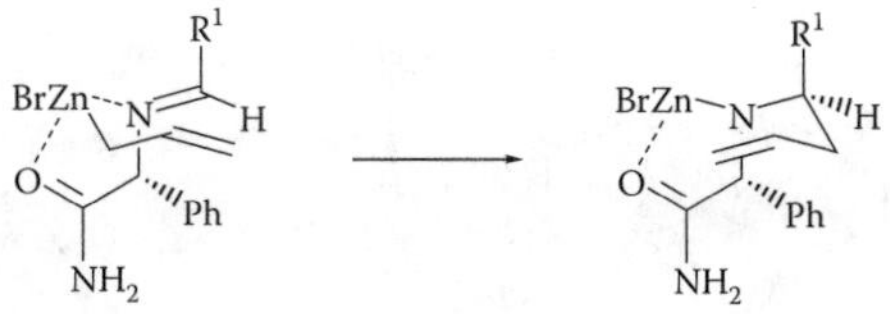

FIGURE 25.4 Chelation controlled addition of allylzinc bromide to (R)-PGA [(R)-phenylglycine amide] imines.

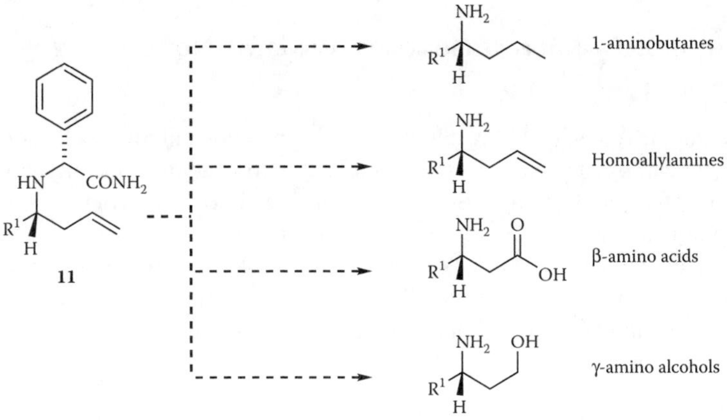

FIGURE 25.5 Conversions of phenylglycine-protected homoallylamines.

25.3.1 SYNTHESIS OF ENANTIOPURE 1-AMINOBUTANES FROM PGA-HOMOALLYLAMINES

Catalytic hydrogenation of PGA-homoallylamines simultaneously reduced the double bond and removed the chiral auxiliary in one step. Some typical examples of enantiomerically pure (R)-aminobutanes **12** obtained are shown in Scheme 25.6. The nonoptimized yields varied between 49% and 88% with ee values of 94% to >98%. The high enantiomeric excesses of these chiral amines are in agreement with the equally high diastereoselectivity of the allylation reaction and lack of racemization of the phenylglycine amide moiety in any of the steps. Enantiomerically pure chiral (R)-α-propylpiperonylamine **12c** is an important building block of the human leukocyte elastase inhibitor L-694,458 (**13**).[28]

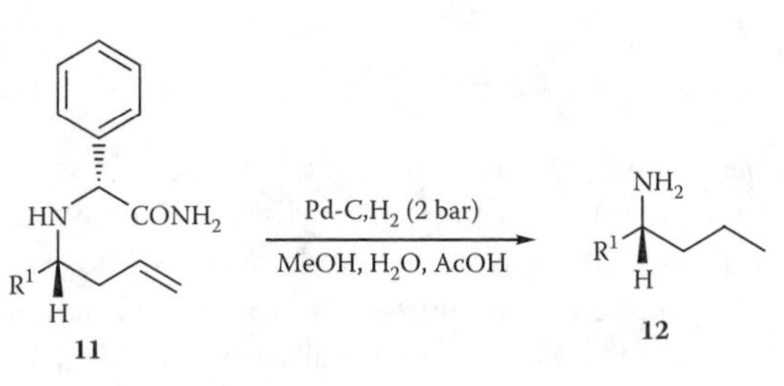

SCHEME 25.6

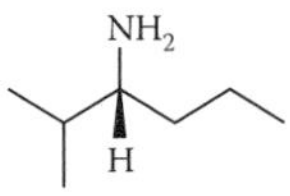

25.3.2 SYNTHESIS OF ENANTIOPURE HOMOALLYLAMINES VIA NONREDUCTIVE REMOVAL OF (*R*)-PGA

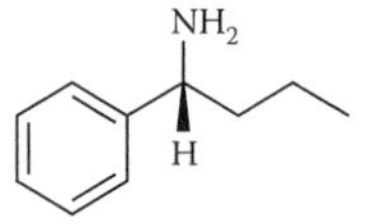

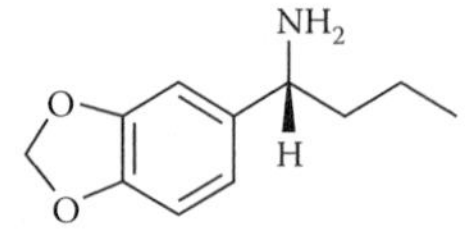

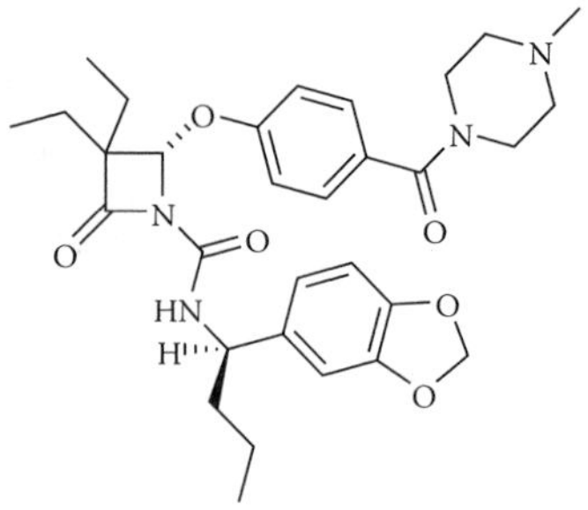

Chiral homoallylamines are valuable synthons for the preparation of biologically active components including β-amino carboxylic acids or esters, obtained by oxidation of the allylic functionality.[1,29] Because removal of the chiral auxiliary by hydrogenation leads to the loss of the allylic functionality, we developed alternative routes for the conversion of the adduct into the unprotected homoallylamines. As a typical example, (*R*,*R*)-PGA-homoallylamine derived from isobutyraldehyde **11i** was used to develop the so-called "*retro*-Strecker" and the "decarbonylation" method for the conversion of (*R*)-phenylglycine amide protected homoallylamines into *N*-benzylidene protected homoallylamines **15** (Scheme 25.7).

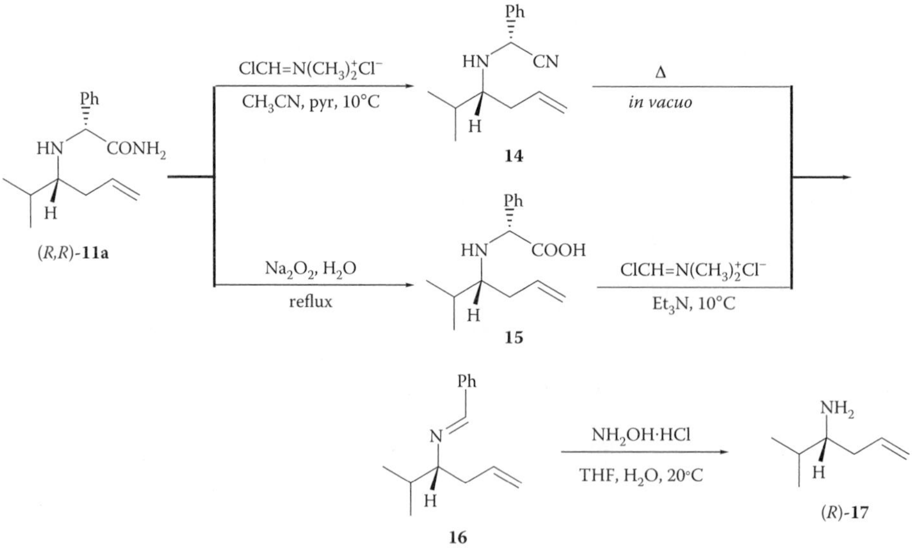

SCHEME 25.7

The first method is based on the facile elimination of HCN in a *retro*-Strecker fashion. The PGA-homoallylamine **11a** was converted into the nitrile **14** by dehydration of the amide moiety using Vilsmeier's reagent. Loss of HCN from the crude nitrile **14** occurred on heating *in vacuo* to yield the homoallylamine **16** masked as the imine. To prove the principle, hydrolysis by treatment with hydroxylamine hydrochloride[30] in aqueous tetrahydrofuran (THF) gave the homoallylamine **17** with an ee of 94%. The overall yield of the 3 steps is approximately 80%. Obviously, other methodologies can also be used to generate the desired unsaturated enantiopure amine **17**.

The second method, referred to as the "decarbonylation" reaction, concerns a base-catalyzed elimination of HCl and CO from an acid chloride.[31] The amide moiety was converted into the amino acid **15** by reaction with Na_2O_2 in water.[32] Alternatively, amides can be converted into the corresponding carboxylic acid by treatment with concentrated HCl.[33] However, this method produced lower yields as a result of some decomposition. The carboxylic acid **15** was treated with the Vilsmeier reagent and triethylamine furnishing the imine **16** via decarbonylation. The imine **16** was then converted, in a similar manner to the "*retro*-Strecker" method, to the unprotected homoallylamine **17** in 75% overall yield and an ee of 98%.

Conversions of PGA-functionalized homoallylamines into β-amino acid derivatives[34] and aminoalcohols are being investigated.

25.4 DIASTEREOSELECTIVE HYDROGENATIONS OF (*R*)-PGA-IMINES

Industrial procedures for the preparation of enantiopure amines currently involve, for example, separation of enantiomers by crystallization of diastereoisomeric salts. Major drawbacks of this approach are the handling of substantial amounts of solids and the need to develop recycle loops for the undesired enantiomer and auxiliary. Enzymatic resolution processes, for example by lipase-mediated acylation combined with efficient racemization and recycling of the unwanted enantiomer, have significantly improved this resolution concept.[35] Furthermore, extensive progress has been made in the development of asymmetric hydrogenation approaches[36] (see Chapter 12) or biocatalytic transaminase processes (see Chapters 3 and 19) for the synthesis of optically active amines. Also, routes based on diastereoselective synthesis using diverse chiral auxiliaries—with varying results in selectivities—have been described extensively.[4,8a,37]

In this section, we describe as typical examples the application of (*R*)-PGA as a chiral vehicle in the preparation of (*S*)-1-aminoindane[38] (see also Chapter 24) and (*S*)-3,3-dimethyl-2-butylamine. (*S*)-1-Aminoindane is present as a key structural element (or with additional functionalization) in therapeutic agents under clinical investigation [e.g., Rasagiline mesylate, (**18**) for the treatment of Parkinson's disease,[39] and Irindalone (**19**), which displays potent antihypertensive activity].[40]

18

19

The typical technologies used for the preparation of amines have also been used for the synthesis of optically pure (*R*)- or (*S*)-1-aminoindane. For example, resolution approaches include the diastereoisomeric salt formation of racemic *N*-benzyl-1-aminoindane with (*S*)-mandelic acid[41] or (*R*,*R*)-tartaric acid,[42] which resulted in, after hydrogenation, (*R*)-1-aminoindane with >99% ee. Also, resolutions that use enzymatic acylation concepts have been described.[43,44] The maximum theoretic yield of 50% is a clear limitation of these methods. Asymmetric synthetic approaches to chiral 1-aminoindanes have been described, including enantioselective hydrosilylation of 1-indanoxime[45,46] and hydroboration of indene.[47] However, ee values were low to moderate.

Likewise, chiral templates such as (*R*)-phenylethylamine[37] and (*R*)-phenylglycinol[48] have been used. Use of (*R*)-phenylethylamine[37] provided (*R*)-1-aminoindane from 1-indanone in 3 steps, in an overall yield of 5% and >99% ee. The low yield is caused by the nonselective removal of the chiral auxiliary in the final step. Similarly, (*R*)-phenylglycinol gave (*S*)-1-aminoindane from 1-indanone in 39% overall yield. Drawbacks in this route are the limited availability of this expensive chiral auxiliary; the use of chromatography to obtain diastereoisomerically pure intermediates; and auxiliary removal by procedures unsuitable for large-scale preparations, such as oxidation with Pb(OAc)$_4$. Because of the limited availability of routes toward enantiomerically pure aminoindanes, it was reasoned that chirality transfer using (R)-PGA might be a good alternative, and a 3-step approach was studied: imine formation, diastereoselective reduction, and removal of the chiral auxiliary.[49]

The imine **20** from (R)-PGA **1** with 1-indanone **21** was obtained in 90% yield and ee >98% after boiling in i-PrOAc for 6 hours in the presence of a catalytic amount of p-TsOH·H$_2$O (Scheme 25.8). Prolonged heating under reflux of the mixture resulted in partial racemization of the chiral transfer agent.

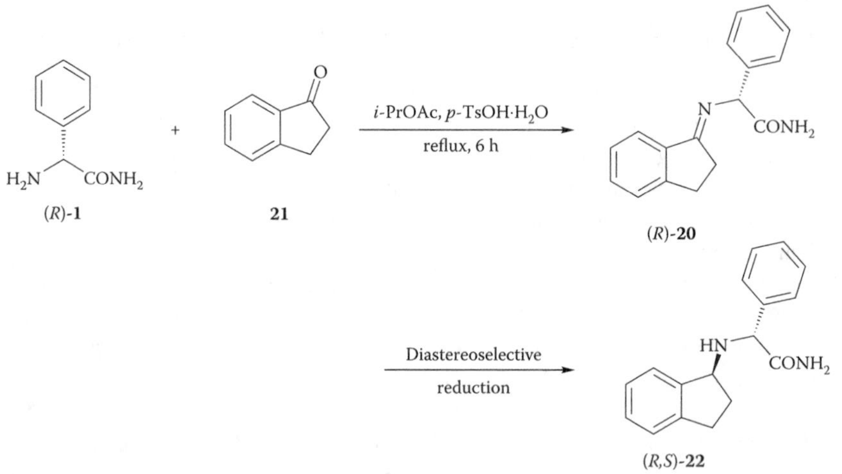

SCHEME 25.8

The diastereoselective reduction of the imine **20** is the second and key step in the preparation of (S)-1-aminoindane. The results of reduction with NaBH$_4$ and H$_2$ with various heterogeneous catalysts are given in Table 25.2.

The highest diastereoselectivity was found with Raney nickel as catalyst (Entry 2). A disadvantage of this approach was the necessity to use large amounts of Raney nickel (100 w/w%) to ensure complete conversion; this would lead to handling problems on production scale. Reductions with less Raney nickel or in other solvents gave either incomplete conversions or lower de. In general, nearly complete conversions using the Pd, Pt, and Pd/Pt mixed catalysts (10 w/w%) in i-PrOAc were obtained. However, the de of the product was consistently lower than in the case in which Raney nickel was used. Therefore, we decided to focus on the "second best," namely Pd-C, and sought to optimize this reduction step in terms of catalyst loading and type of solvent. Argonaut's parallel pressure reactor, the Endeavor, was used for this purpose.[50] A series of experiments with 5% Pd-C catalyst was conducted to explore the effect of variation of the solvent. The results are summarized in Table 25.3.

It is interesting that a clear correlation was found between solvent polarity (the dielectric constant ε) and de. Toluene is the most suitable solvent (Entry 1): (R,S)-**22** was obtained with a de of 90% (which is comparable to the RaNi reductions) but with only 8 w/w% 5% Pd/C. The reduction step was further optimized. The best conditions were found at 4 w/w% 5% Pd/C and 25 w/w% concentration of imine **20** in toluene, which ensured still fast reaction times (<5 h), complete conversion, and high de values. These reduction conditions have been tested successfully on a several hundred gram scale.

TABLE 25.2
The Effect of Type of Catalyst on the Hydrogenation of (*R*)-20 in *i*-PrOAc

Entry	Reducing Agent	Loading (w/w%)	Conditions	Time (h)	Conversion (%)	de[a] (%)
1	NaBH$_4$	2 eq	20°C, EtOH	2	>99	18
2	Ra-Ni[b]	100	50°C, 3.5 bar H$_2$	8	>99	88
3	5% Pd-C[c]	10	25°C, 3.5 bar H$_2$	5	>99	80
4	5% Pt-C[d]	10	25°C, 3.5 bar H$_2$	2	98	64
5	1% Pt-C[e]	10	25°C, 3.5 bar H$_2$	>10	98	68
6	8% Pd/2% Pt-C[f]	10	25°C, 3.5 bar H$_2$	5	>99	76

[a] The de was determined by ^{1}H NMR analysis on the basis of the relative integration of the ^{1}H-aminoindane triplet of (*R,S*) at 4.34 ppm and (*R,R*) at 4.19 ppm. The de was confirmed by high-performance liquid chromatography analysis.
[b] Doduco ACTIMET "S," charge 768-18071.
[c] Engelhard ESCAT 142, moist-reduced, 52% moisture.
[d] Aldrich, dry.
[e] Engelhard ESCAT 261, moist-reduced, 51% moisture.
[f] Johnson-Matthey, 1130/8, batch 10, type 464, 51% moisture.

TABLE 25.3
The Effect of Type of Solvent on the Hydrogenation of (*R*)-20 with 5% Pd-C[a] (8 w/w%) at 25°C, 3.5 bar H$_2$

Entry	Solvent	ε	Time (h)	Conversion (%)	de (%)
1	Toluene	2.38	3	>99	90
2	MTBE	4.5	>10	74	86
3	*i*-PrOAc	5[b]	6	98	80
4	THF	7.58	2	97	68
5	*n*-Pentanol	13.9	6	98	58
6	EtOH	24.55	2	97	42

[a] Engelhard ESCAT 142, moist-reduced, 52% moisture.
[b] Not found in the literature. The value given here is estimated from the values of 4.63 for *i*-pentyl acetate and 6.68 for MeOAc.

Furthermore, it was found that a single recrystallization of the HCl-salt of the PGA-aminoindane provided (*R,S*)-**22** in 84% overall yield based on imine **20** and a de >98%. This high substrate concentration and low Pd loadings (the Pd catalyst might be recycled, a conjecture not yet proved), combined with the high diastereoselectivity and easy enrichment via crystallization, make this a scalable procedure.

Catalytic hydrogenation of (*R,S*)-**22** was unsatisfactory for removal of the chiral auxiliary. The regioselectivity of the *N*-benzylation step (5% Pd-C, 5 bar H$_2$, 40°C, EtOH, AcOH) gave a (*S*)-1-amino-indane/(*R*)-PGA ratio of 4:6. This means that 60% of the key intermediate was converted to the undesired product. Therefore, the "*retro*-Strecker" method, analogous to the homoallylamines case, was applied as an alternative. The initially 3-step method could be improved to an efficient one-pot procedure (Scheme 25.9).

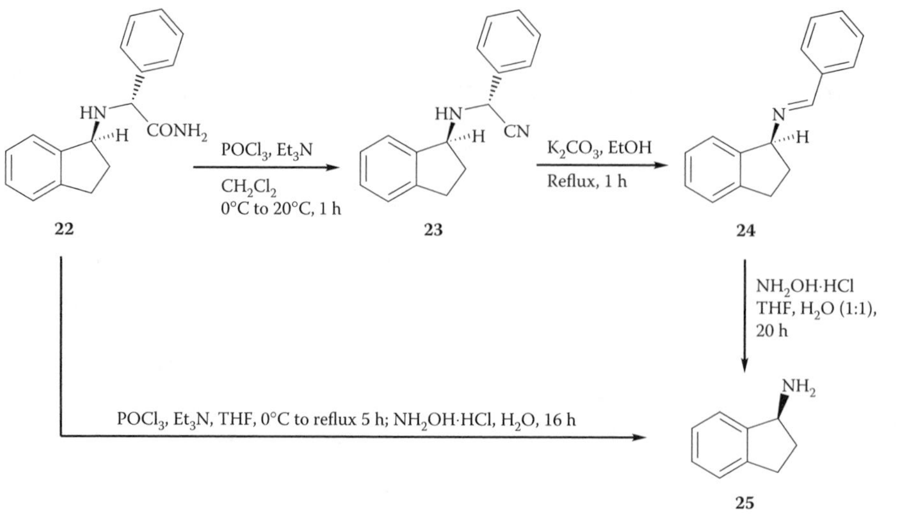

SCHEME 25.9

Dehydration of the amide **22** with POCl₃/Et₃N to the nitrile **23** proved to be less time-consuming and cleaner and gave improvement of yields relative to the Vilsmeier procedure previously used. Subsequently, the mixture was heated at reflux to effect the *retro*-Strecker reaction. Finally, addition of an aqueous solution of NH₂OH·HCl resulted in hydrolysis of the intermediate imine **24** to (S)-1-aminoindane (**25**). Instead of using NH₂OH·HCl, the hydrolysis can also be accomplished by addition of an excess of 30% HCl followed by heating under reflux. Thus, (S)-1-aminoindane (**25**) with an ee of 96% was prepared in 58% overall yield from (R)-PGA (**1**) and 1-indanone by means of an effective 3-step procedure.

As an additional example, we investigated the synthesis of (S)-3,3-dimethyl-2-butylamine (**26**) using the 3-step approach (Scheme 25.10). In this case imine formation proved to be more difficult. Reaction of (R)-PGA **1** and ketone **27** in boiling *i*-PrOAc was slow, and a low conversion was observed. The imine was obtained in 97% yield as a white solid by heating under reflux in a mixture of toluene and cyclohexane for 20 hours. Unfortunately, the (R)-PGA moiety was partly racemized into a mixture of 85% (R)-**28** and 15% (S)-**28**.

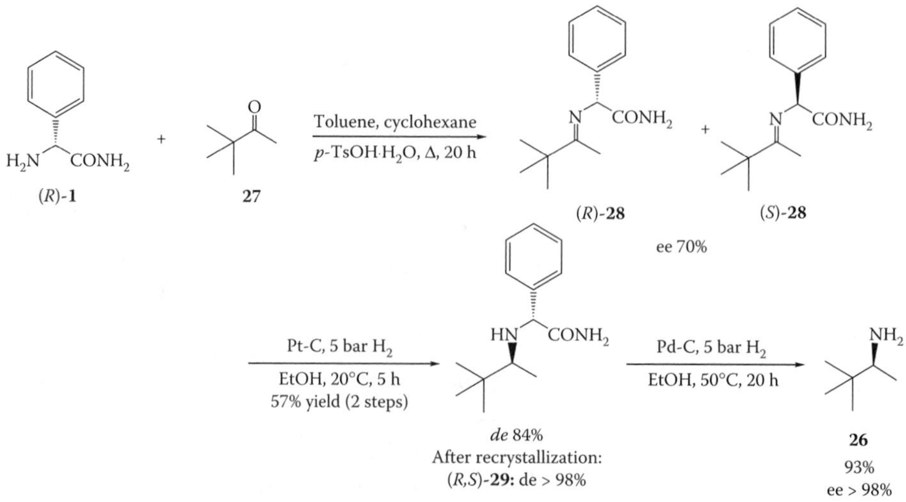

SCHEME 25.10

TABLE 25.4
Diastereoselective Reductions of (R)-28/(S)-28;
Ratio 85:15

Entry	Reductant	Solvent	Conditions	de (%)
1	NaBH$_4$	MeOH	0°C, 1 h	80
2	RaNi	EtOH	20°C, 1 bar H$_2$, 72 h	88[a]
3	5% Pt/C	MeOH	20°C, 5 bar H$_2$, 5 h	84
4	5% Pd/C	MeOH	20°C, 5 bar H$_2$, 5 h	80

[a] Conversion ca. 75% (entries 1, 3 and 4: conversions quantitative).

The solid thus obtained was subjected to various diastereoselective reduction conditions. The results are summarized in Table 25.4.

High selectivities were obtained in case of NaBH$_4$ reduction as well as heterogeneous reduction. Crystallization from hexane afforded diastereoisomerically pure (R,S)-**29** in an overall yield of 57% for the 2 steps. This clearly demonstrates that (as found also for (S)-1-aminoindane) the key asymmetric step does not always need to deliver 100% selectivity, as long as a good purification method is available. The yield can, of course, be drastically improved by prevention of racemization in the imine-formation step and increasing the selectivity in the reduction step, as described for (S)-1-aminoindane.

Removal of the chiral-inducing agent was accomplished in a straightforward fashion by means of hydrogenolysis using H$_2$ and 5% Pd-C in ethanol to give the amine **27** as its HCl salt in 93% yield (ee >98%). The total reaction sequence from (R)-PGA (**1**) and 3,3-dimethyl-2-butanone (**27**) has been carried out to give a multi 10-g sample of (S)-3,3-dimethylbutylamine (**26**) as its HCl salt with an overall yield of 53% based on **1**.

More applications of this chirality transfer approach to enantiomerically pure amines using (R)- or (S)-phenylglycine amide are under investigation.

25.5 SUMMARY

In summary, (R)-phenylglycine amide (**1**)—available on a large scale within DSM—is an excellent chiral-inducing agent for application in diastereoselective synthesis. Diastereoselective reactions with imines of (R)-PGA, for example Strecker reactions, combined with a crystallization-induced asymmetric transformation, allylation, or reductions proceed with high to excellent de values. Also, it is clear that the key asymmetric step does not have to be completely 100% selective because crystallization is an efficient purification method for the PGA-functionalized diastereoisomers. Furthermore, after performing its role, the chiral template is conveniently removed under either reductive or nonreductive conditions. Obviously, (S)-phenylglycine amide is also accessible and can be used for the preparation of the opposite isomer of the products described. Based on the results obtained, we believe that in several cases "old-fashioned" diastereoselective transformations can compete with advanced asymmetric catalysis. Currently, we are further extending this chirality transfer technology using PGA toward routes for enantiopure β-amino acid derivatives and highly functionalized amines.

REFERENCES AND NOTES

1. a) Jacobsen, E. N., Pfaltz, A., Yamamoto, H. Eds. *Comprehensive Asymmetric Catalysis*, Springer-Verlag: Berlin, 1999; b) Van den Berg, M., Minnaard, A. J., Schudde, E. P., Van Esch, J., De Vries A. H. M., De Vries J. G., Feringa, B. L. *J. Am. Chem. Soc.* 2000, *122*, 11539.

2. Patel, R., Ed. *Stereoselective Biocatalysis,* Dekker: New York, 2000.

3. Eliel, E. L., Wilen, S. H., Mander, L. W. Eds. *Stereochemistry of Organic Compounds*, John Wiley & Sons: New York, 1994, Chapter 7, p. 297.

4. For a review on the use of α-methylbenzylamine, see Juaristi, E., Leon-Romo, J. L., Reyes, A., Escalante, J. *Tetrahedron: Asymmetry* 1999, *10*, 2441.

5. a) Dave, R. H., Hosangadi, B. D. *Tetrahedron* 1999, *55*, 11295; b) Ma, D., Tian, H., Zou, G. *J. Org. Chem.* 1999, *64*, 120; c) Chakraborty, T. K., Hussain, K. A., Reddy, G. V. *Tetrahedron* 1995, *51*, 9179.

6. Weinges, K., Brachmann, H., Stahnecker, P., Rodewald, H., Nixdorf, M., Irngarter H. *Liebigs Ann. Chem.* 1985, 566.

7. Kunz, H., Sager, W., Schanzenbach, D., Decker, M. *Liebigs Ann. Chem.* 1991, 649.

8. a) Ellman, J. A., Owens, T. D., Tang, P. D. *Acc. Chem. Res.* 2002, *35*, 984; b) Davis, F. A., Fanelli, D. L. *J. Org. Chem.* 1998, *63*, 1981.

9. Wright, D. L., Schulte II, J. P., Page, M. A. *Org. Lett.* 2000, *2*, 1847.

10. Bruggink, A., Roos, E. C., de Vroom, E. *Org. Proc. Res. Dev.* 1998, *2*, 128.

11. Sonke, T., Kaptein, B., Boesten, W. H. J., Broxterman, Q. B., Schoemaker, H. E., Kamphuis, J., Formaggio, F., Toniolo, C., Rutjes, F. P. J. T. in ref 2, Chapter 2.

12. a) Boesten, W. H. J. Eur. Pat. Appl. EP 442584, 1991; b) Boesten, W. H. J. Eur. Pat. Appl. EP 442585, 1991.

13. a) Kunz, H. in *Houben-Weyl: Stereoselective Synthesis*, Helmchen, G., Hoffmann, R. W., Mulzer, J., Schaumann, E., Eds., Vol. E 21, D.1.4.4., Thieme Verlag, Stuttgart, 1995, 1931; b) Shafran, Y. M., Bakulev, V. A., Mokrushin, V. S. *Russ. Chem. Rev.* 1989, *58*, 148; c) Strecker, A. *Ann. Chem. Pharm.* 1850, *75*, 27.

14. For reviews see a) Patel, R., Ed. *Stereoselective Biocatalysis,* Dekker: New York, 2000, Chapter 2; b) Drauz, K., Waldmann, H. In *Enzyme Catalysis in Organic Synthesis, Volume 1*, VCH: Weinheim, 1995, p. 393.

15. a) Calmes, M., Daunis, J. *Amino Acids*, 1999, *16*, 215; b) Cativiela, C., Díaz-de-Villegas, M. D. *Tetrahedron: Asymmetry* 1998, *9*, 3517.

16. For reviews on catalytic asymmetric Strecker reactions, see a) Gröger, H. *Chem. Rev.* 2002, *103*, 2795; b) Yet, L. *Angew. Chem., Int. Ed.* 2001, *40*, 875.

17. Few examples of crystallization-induced asymmetric transformations in Strecker reactions based on arylalkyl-methyl ketones have been reported: a) Weinges, K., Gries, K., Stemmle, B., Schrank, W. *Chem. Ber.* 1977, *110*, 2098; b) Weinges, K., Klotz, K-P., Droste, H. *Chem. Ber.* 1980, *113*, 710.

18. For a broad discussion of crystallization-induced asymmetric transformation, see Vedejs, E., Chapman, R. W., Lin, S., Muller, M., Powell, D. R. *J. Am. Chem. Soc.* 2000, *122*, 3047 and references cited therein.

19. a) Boesten, W. H. J., Seerden, J.-P. G., de Lange, B., Dielemans, H. J. A., Elsenberg, H. L. M., Kaptein, B., Moody, H. M., Kellogg, R. M., Broxterman, Q. B. *Org. Lett.* 2001, *3*, 1121; b) Boesten, W. H. J., Seerden, J.-P. G., de Lange, B., Dielemans, H. J. A., Elsenberg, H. L. M., Kaptein, B., Moody, H. M., Kellogg, R. M., Broxterman, Q. B. PCT Int. Appl. WO 01 42,173, 2001.

20. Bommarius, A. S., Schwarm, M., Stingl, K., Kottenhahn, M., Huthmacher, K., Drauz, K. *Tetrahedron: Asymmetry,* 1995, *6*, 2851.

21. HCN can be used at large scale.

22. For example, in the case of phenylacetone, it was found that in solution the initially formed minor isomer preferentially precipitated under crystallization conditions.

23. See for a discussion of asymmetric transformation of α-amino nitriles with mandelic acid: Hassan, N. A., Bayer, E., Jochims, J. C. *J. Chem. Soc., Perkin Trans. I* 1998, 3747.

24. Laschat, S., Kunz, H. *J. Org. Chem* 1991, *56*, 5883.

25. a) Alvaro, G., Martelli, G., Savoia, D. *J. Chem. Soc., Perkin Trans. I* 1998, *1*, 777; b) Razavi, H., Polt, R. *J. Org. Chem.* 2000, *65*, 5693.

26. Van der Sluis, M., Dalmolen J., de Lange, B., Kaptein, B., Kellogg, R. M., Broxterman, Q. B. *Org. Lett.* 2001, *3*, 3943, Van der Sluis, M., Dalmolen J., de Lange, B., Kaptein, B., Kellogg, R. M., Broxterman, Q. B. PCT Int. Appl. WO 01 90,04, 2001.

27. Two isomers in a ratio of 1:1.3 were obtained for the additional methyl group present in the crotyl moiety.

28. Cvetovich, R. J., Chartrain, M., Hartner, Jr, F. W., Roberge, C., Amato, J. S., Grabowski, E. J. J. *J. Org. Chem.* 1996, *61*, 6575.

29. a) Moody, C. J., Hunt, J. C. A. *Synlett* 1998, 733; b) Polniaszek, R. P., Belmont, S. E., Alvarez, R. *J. Org. Chem.* 1990, *55*, 215; c) Laschat, S., Kunz, H. *Synlett* 1990, 51; d) Hua, D. H., Miao, S. W. M., Chen, J. S., Iguchi, S. *J. Org. Chem.* 1991, *56*, 4.

30. Logers, M., Overman, L. E., Welmaker, G. S. *J. Am. Chem. Soc.* 1995, *117*, 9139.

31. Decarbonylation has been reported in the reaction of α-anilino-α,α-diphenylacetic acid with *p*-toluenesulfonylchloride and pyridine furnishing CO and the imine of benzophenone and aniline: Sheenan, J. C., Frankenfeld, J. W. *J. Org. Chem.* 1962, *27*, 628.

32. Vaughn, H. L., Robbins, M. D. *J. Org. Chem.* 1975, *40*, 1187.

33. Houben Weyl series, *Methoden der Organischen Chemie*, Georg Thieme Verlag, 1985, Band E5, p. 259.

34. For a review on recent advances in the synthesis of β-amino acids, see Liu, M., Sibi, M. P. *Tetrahedron*, 2002, *58*, 7991.

35. Ladner, W. E., Ditrich, K. *Chimica Oggi* 1999, *July/August*, 51.

36. For a review on selective hydrogenation for fine chemicals: recent trends and new developments, see Blaser, H-.U., Malan, C., Pugin, B., Spindler, F., Steiner, H., Studer, M. *Adv. Synth. Catal.* 2003, *345*, 103.

37. Gutman, A. L., Etinger, M., Nisnevich, G., Polyak, F. *Tetrahedron: Asymmetry* 1998, *9*, 4369.

38. Using (*S*)-PGA as the chiral auxiliary would yield (*R*)-1-aminoindane or, (*R*)-3,3-dimethyl-2-butylamine, respectively.

39. Graul, A., Castaner, J. *Drugs of the Future*, 1998, *23*, 903.

40. Bøgesø, K. P., Arnt, J., Boeck, V., Christensen, A.V., Hyttel, J., Gundertofte Jensen, K. *J. Med. Chem.* 1988, *31*, 2247.

41. a) Lidor, R., Bahar, E. W.O. Patent 96/21640, 1996; b) Lidor, R., Bahar, E., Zairi, O., Atili, G., Amster, D. *Org. Prep. Proced. Int.* 1997, *29*, 701.

42. Gutman, A. L., Zaltzman, I., Ponomarev, V., Sotrihin, M., Nisnevich, G. W.O. Patent 02/068376 A1, 2002.

43. a) Gutman, A. L., Meyer, E., Kalerin, E., Polyak, F., Sterling, J. *Biotechnol. Bioeng.* 1992, *40*, 760; b) Gutman, A. L., Shkolnik, E., Meyer, E., Polyak, F., Brenner, D., Boltanski, A. *Ann. N. Y. Acad. Sci.* 1996, *799*, 620.

44. For other enzymatic approaches, see a) Stirling, D. I., Matcham, G. W., Zeitlin, A. L. U.S. Patent 5,300,437, 1994; b) Shin, J.-S., Kim, B.-G., Shin, D.-H. *Enzyme Microbial Technol.* 2001, *29*, 232; c) Laumen, K., Brunella, A., Graf, M., Kittelmann, M., Walser, P., Ghisalba, O. *Pharmacochem. Lib.* 1998, *29*, 17.

45. Brunner, H., Becker, R., Gauder, S. *Organometallics* 1986, *5*, 739.

46. Takei, I., Nishibayashi, Y., Ishii, Y., Mizobe, Y., Uemura, S., Hidai, M. *J. Chem. Soc., Chem. Commun.* 2001, 2360.

47. Fernandez, E., Maeda, K., Hooper, M.W., Brown, J. M. *Chem. Eur. J.* 2000, *6*, 1840.

48. Stalker, R. A., Munsch, T. E., Tran, J. D., Nie, X., Warmuth, R., Beatty, A., Aakeröy, C. B. *Tetrahedron* 2002, *58*, 4837.

49. Uiterweerd, P. G. H., van der Sluis, M., Kaptein, B., de Lange, B., Kellogg, R. M., Broxterman, Q.B. *Tetrahedron Asymmetry*, 2003, *14*, 3479.

50. For a review on the power of High Throughput Experimentation (HTE) in Homogeneous Catalysis Research for Fine Chemicals, see de Vries, J. G., de Vries, A. H. M. *Eur. J. Org. Chem.* 2003, 799.

26 Pericyclic Reactions

Michael B. East

CONTENTS

26.1 INTRODUCTION

The literature abounds with examples of pericyclic (and pericyclic-type) reactions that have been used to synthesize a broad range of target molecules including natural products.[1-3] (Some transformations do not proceed in a concerted manner, but follow a stepwise mechanism. These reactions have been included where the rules governing pericyclic reactions can also predict the stereochemical outcome.) Although few of these reactions have been scaled up outside of the laboratory, their potential to provide a complex target molecule with a minimal number of transformations is unparalleled in the organic chemists' repertoire. For example, in the Diels-Alder reaction, up to six stereogenic centers can be controlled in a single reaction.[4] Recent progress, especially with regard to asymmetric catalysts, should encourage the use of these reactions at scale.

This chapter has been organized into sections based on the major reaction types. Hence, the hetero Diels-Alder reaction has been included as a subsection of the Diels-Alder reaction. Although there are numerous reports of asymmetric pericyclic reactions, our discussion concentrates on reactions that are most likely to allow for scale up.

26.2 THE DIELS-ALDER REACTION

The Diels-Alder reaction, now more than 60 years old, has regained new prominence with the ability to use asymmetric technology to control relative and absolute stereochemistry while creating two new carbon–carbon bonds.[1–3,5,6] As a result of the diversity of dienes and dienophiles, the application of asymmetric methodology to the Diels-Alder reaction has lagged behind other areas such as aldol reactions.[3] However, considerable advances have been made in recent years through the use of chiral dienophiles and catalysts.[3,6–21]

26.2.1 REGIOCHEMISTRY AND STEREOCHEMISTRY

Cycloadditions that involve two unsymmetric reactants can lead to regioisomers. The regioselectivity of these adducts can be predicted with a high degree of success through the use of frontier molecular orbital theory.[22–25] The *ortho* product (this nomenclature follows the analogy of disubstituted aromatic systems) is usually the preferred isomer from 1-substituted dienes, whereas 2-substituted dienes provide the *para* isomer as the major adduct. However, when a Lewis acid is used as a catalyst in the reaction, the ratio of these isomers can alter dramatically and, occasionally, can be reversed.[22]

In addition, Diels-Alder adducts are formed through two types of approaches that lead to *endo* or *exo* isomers. The *endo* isomer is usually favored over the *exo* isomer, although the *exo* isomer is generally the thermodynamically preferred product. This is known as the Alder, or *endo*, rule and can be attributed to the additional stability gained by secondary molecular orbital overlap during the cycloaddition.[22,26,27] Again, the use of a Lewis acid catalyst can alter the *endo/exo* ratio and has even been shown to give the thermodynamic *exo* adduct as the major product.[28]

Complete stereoselectivity occurs in the Diels-Alder reaction through *syn* addition of the dienophile to the diene. Hence, the reaction of dimethyl fumarate and dimethyl maleate with cyclopentadiene yields the *trans* and *cis* adducts, respectively (Scheme 26.1).[29]

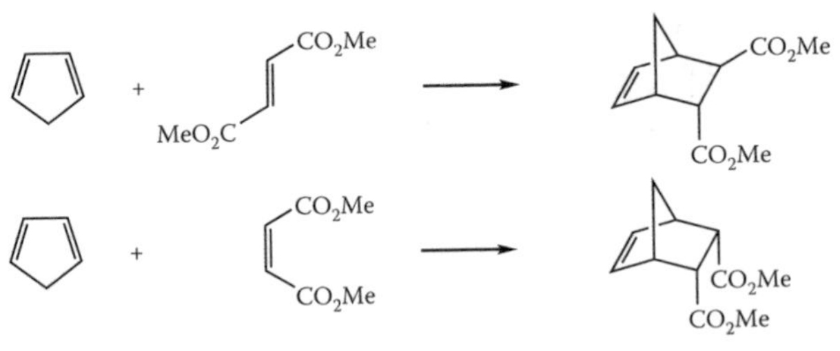

SCHEME 26.1

The use of maleic anhydride provides a *cis*-dienophile that can react with cyclopentadiene in a classic Diels-Alder reaction. The synthesis of tandospirone (**1**) takes advantage of the stereochemical control to set the relative configuration at four stereogenic centers.[30–32]

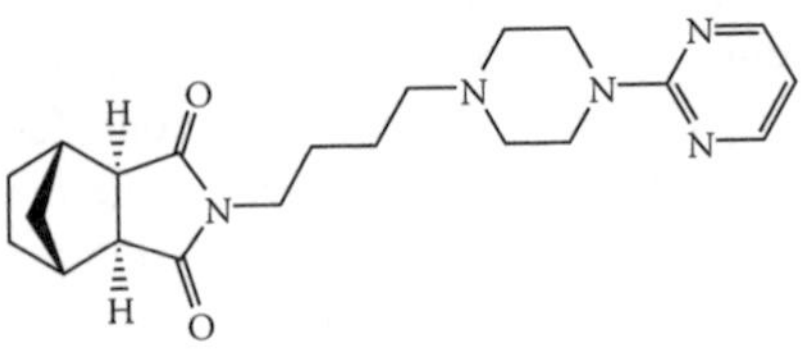

1

26.2.2 CATALYSTS

The financial advantages of using catalysts on an industrial scale are obvious. In addition to increased regiochemistry and stereochemistry, the addition of a Lewis acid catalyst often allows for a dramatic increase in the rate of Diels-Alder reactions; this has been attributed to changes in energy of molecular orbitals when complexed, decreasing the activation energy.[22,33] Although highly diastereoselective thermal Diels-Alder cycloadditions are known,[34] the addition of a Lewis acid appears to be essential to maximize selectivity for the vast majority of Diels-Alder reactions.[22,35–37] The use of a catalyst allows the reaction to be run at lower temperatures and, thus, allows for greater differentiation of the diastereomeric transition states.

A large number of Lewis acids have been advocated for the catalysis of the Diels-Alder reaction.[3] The vast majority are based on zinc, aluminum, titanium, and boron derivatives. It should be noted that the use of some metals, such as zinc, causes significant disposal problems in wastewater streams and can add significantly to costs. Polymerization can be minimized by use of Lewis acids where strongly electron-withdrawing substituents, such as chloride, have been replaced by more moderate ligands, such as alkoxy, or weaker Lewis acids such as lanthanide complexes.[37–45] Although these changes allow for the preservation of fragile functional groups and avoid substrate decomposition, they also compromise the Lewis acidity and the catalyst may be less effective. In addition to Lewis acids, proteins, antibodies, enzymes, and radicals have been found to catalyze the Diels-Alder cycloaddition.[46]

26.2.2.1 Chiral Lewis Acids

The discovery that chiral Lewis acids can catalyze the asymmetric Diels-Alder reaction is a major milestone for the scale up and practice of this reaction on an industrial scale. The use of such a catalyst obviates the need for a chiral auxiliary on the diene or dienophile. The vast majority of chiral auxiliaries that have been used in the Diels-Alder reaction are either not commercially available or are expensive. In addition, the chemical steps needed to attach and remove the chiral auxiliary increase the cost and complexity of the synthesis. Chiral catalysts may also be recovered or recycled, further decreasing cost.[47] Research in this area is very active, and catalysts based on a number of metals (Table 26.1) have shown encouraging asymmetric induction.[21] Our understanding of the role these catalysts play in the asymmetric induction of Diels-Alder reactions is increasing, and more general reagents should appear.[27,48–54]

The reaction of methacrolein with cyclopentadiene catalyzed by a chiral menthoxyaluminum complex gives adducts with ee's of up to 72%, but with other dienophiles little, if any, induction was noted.[94,95] A chiral cyclic amido aluminum complex **2** catalyzes the cycloaddition of cyclopentadiene with the *trans*-crotyl derivative **3** in good yield and enantioselectivity (Scheme 26.2).[47] This chiral catalyst can also be easily recovered.

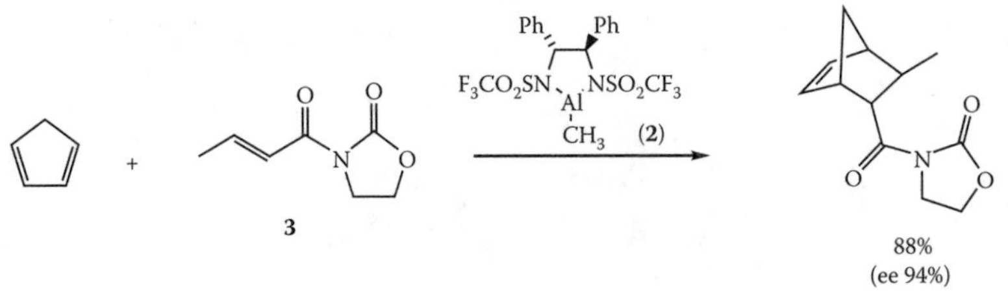

SCHEME 26.2

High ee's have been observed in analogous reactions with titanium catalysts derived from tartaric acid,[56,81,90] and bisoxazolines.[17,70,96,98,99] The development of a robust air-stable catalyst that

TABLE 26.1
Chiral Catalysts Used in the Diels-Alder Reaction

Ligand system	Metal	Reference	Ligand system	Metal	Reference
	Ti	55		Al, B, Ti	11,12,56–62
	Al	47,53,63,64		Al, B, Ti	13,52,65–69
	Al, Sn, Ti	59		Ti	14
	Cu, Fe, Mg	8,70–72		Cu,Cr	73–78
	Cu	16,79		Cr	15
	Al	80		Al, Sn, Ti	59,81–85
	Ti	86–88		Al, B, Sn, Ti	56,59,89–91
	B	52,92,93		Al	59

TABLE 26.1 (continued)
Chiral Catalysts Used in the Diels-Alder Reaction

	Al	19	R*OMCl$_2$	Al	94,95
	Co	96		B	97

Ph Ph O M O (first structure)

Ph$_2$P PPh$_2$ (second structure)

MCl$_2$ (third structure)

results in high asymmetric induction in hetero Diels-Alder reactions offers promise for scale up. Although the rate of reaction slows up to fourfold after five recycles, the selectivity and enantioselectivity are maintained.[72]

The vast majority of successful chiral catalysts to date are based on tartaric acid, BINOL, or oxazolidinone derivatives (Table 26.1). Because derivatives of both of these compounds are commercially available, scale up should not present a problem. If the observed asymmetric induction is found to be low with catalysts based on tartaric acid or oxazolidinones, the sterically hindered titanium BINAP-type complexes should allow for increased selectivity. In addition, nontoxic metal counterions, such as iron and aluminum, do not appear to compromise the asymmetric induction.

26.2.2.2 Chiral Bases

In addition to Lewis acid catalysts, chiral bases catalyze the Diels-Alder reaction.[100,101] Although the use of bases as catalysts does show promise, especially for acid-sensitive functionality, the current level of asymmetric induction may not be acceptable for scale up. In addition, only a limited number of dienes and dienophiles have been shown to undergo the chiral-base catalyzed Diels-Alder reaction.

26.2.2.3 Biological Catalysts

Although the use of enzymes, as catalysts in the Diels-Alder reaction, has not been particularly successful,[46,102,103] abzymes can be used to control regioselectivity and stereoselectivity in the Diels-Alder reaction.[46,104] Biological catalysts have not been used on significant scale in the Diels-Alder reaction. Further, the use of abzymes would require a large investment initially, and any subtle change in the substituents on either the diene or dienophile may dramatically reduce asymmetric induction and yield.

26.2.3 CHIRAL DIENOPHILES

Prior to the advent of chiral catalysts, the most widely used approach to chiral adducts has involved the use of chiral dienophiles and, of these, acrylates, derived from a chiral alcohol or amine and acryloyl chloride, are the most common (Table 26.2).[3,4]

TABLE 26.2
Chiral Dienophiles Used in the Diels-Alder Reaction

Dienophile	Catalyst	Reference	Dienophile	Catalyst	Reference
	TiCl$_4$ EtAlCl$_2$ Me$_2$AlCl	6,105,106		TiCl$_2$(OiPr)$_2$	107
	TiCl$_2$(OiPr)$_2$	39,40		Et$_2$AlCl BF$_3$·OEt$_2$ TiCl$_4$ TiCl$_2$(OiPr)$_2$	108–110
	–	111–113		MeAlCl$_2$	114
	ZnCl$_2$	115		Me$_2$AlCl EtAlCl$_2$	38,116
	TiCl$_4$ Et$_2$AlCl Clay	40,107,117-126		Et$_2$AlCl iBu$_2$AlCl AlCl$_3$	127–129
	ZnX$_2$	130,131		ZnX$_2$ MgBr$_2$·OEt$_2$ SiO$_2$ LiClO$_4$ Et$_2$AlCl BF$_3$·OEt$_2$ Eu(fod)$_3$ TiCl$_4$	7,34,132–146
	–	147		– Et$_3$Al EtAlCl$_2$ Et$_2$AlCl Cp$_2$TiCl$_2$	131,148–152
	ZnCl$_2$ BF$_3$·OEt$_2$ SnCl$_4$ Et$_2$AlCl	153		–	154
	–	155	where X = O, NAc	–	156–159

TABLE 26.2 (continued)
Chiral Dienophiles Used in the Diels-Alder Reaction

(structure)	Et₂AlCl	160	(structure) R*O	–	161,162
(structure)	BF₃·OEt₂	10	(structure) MenO₂C···CO₂Men	AlCl₃	163,164
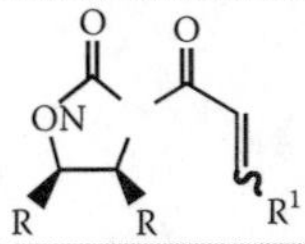 (structure)	Et₂AlCl Me₂AlCl	120,165–168	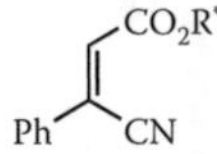 (structure) Ph···CN, CO₂R*	–	169
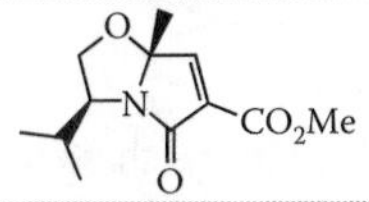 (structure) CO₂Me	ZnCl₂	170	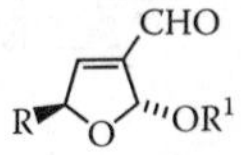 (structure) CHO, R···OR¹	Eu(fod)₃	171
(structure)	– Eu(fod)₃ t-BuMe₂SiOTf	172,173	(structure)	TiCl₄	174–176
(structure)	TiCl₄	174–176	(structure) R¹ tBu, OH	ZnCl₂ Ti(OiPr)₄ BF₃·OEt₂	177
(structure) OEt, NR₂*	BF₄⁻·OEt₃⁺	178	(structure) Ph₃CO	ZnCl₂ TiCl₂(OiPr)₂ BF₃·OEt₂ Et₂AlCl EtAlCl₂ iBu₂AlCl TiCl₄	179
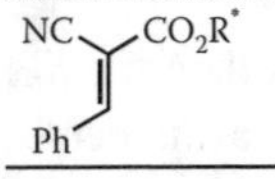 (structure) NC, Ph, CO₂R*	EtAlCl₂ TiCl₄	180			

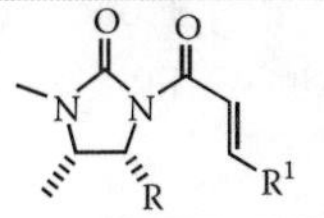

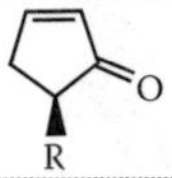

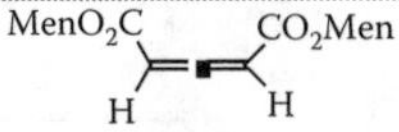

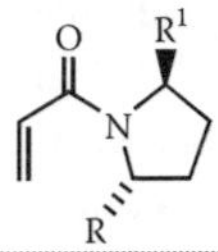

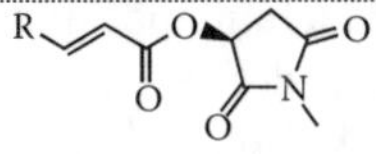

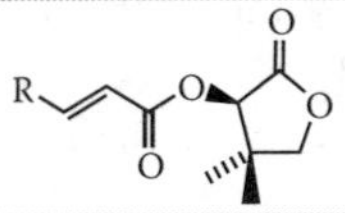

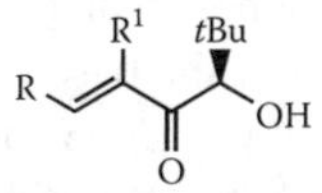

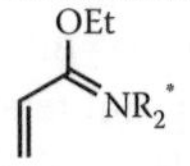

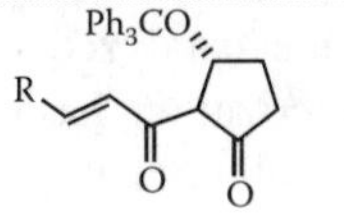

[a]The major products are *endo*.

The use of a chiral auxiliary or group on the dienophile can provide for face selectivity. High *endo–exo* selectivities have been achieved with bicyclic adducts, together with high asymmetric induction. Chiral dienophiles can be classified as either type I or type II reagents (Figure 26.1).[4]

FIGURE 26.1 Type I and II dienophiles for the Diels-Alder reaction.

Type I reagents, such as chiral acrylates, incorporate a chiral group in a simple and straightforward manner. Type II reagents, where the chiral group is one atom closer to the double bond, are more difficult to synthesize and recycle. In addition, significant stereoselection is only observed in the presence of a Lewis acid (*vide infra*); this is especially true of type I reagents.[22,35–37] However, for type II dienophiles such as the sulfoxide **4**, thermal Diels-Alder conditions can provide high asymmetric induction and excellent overall yields (e.g., Scheme 26.3).[132]

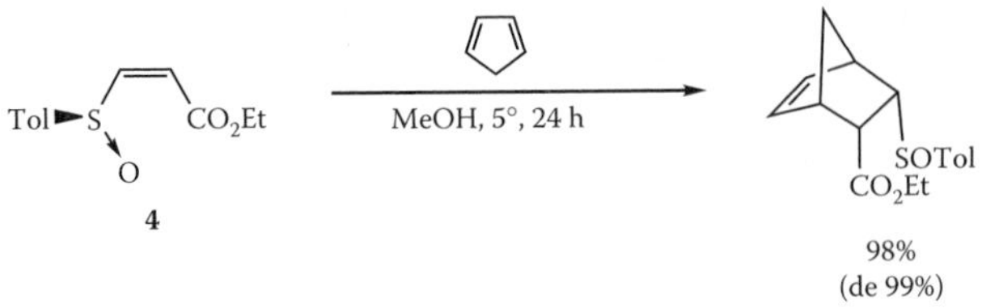

SCHEME 26.3

Although a large number of chiral dienophiles have been developed (Table 26.2), their ability to provide high asymmetric induction appears to be limited to specific dienes. However, there are some dienophiles that tolerate a wider variety of dienes including menthol derivatives,[117,118] camphor derivatives,[6,39,40,105,107–113,181,182] and oxazolidinones.[120,165,183,184] It should be noted that even these auxiliaries would require an efficient recycle protocol for economic scale up. One exception is the use of sacrificial chiral oxazolidinones, which are relatively inexpensive. This approach has been used in the large-scale preparation of the base cyclohexane unit of Ceralure B$_1$.[168] A procedure has been developed for the preparation of (*1S,2S*)-5-norbornene-2-carboxylic acid where the D-pantalactone auxiliary can be recycled efficiently.[185,186]

The ethyl aluminum dichloride–catalyzed synthesis of (*R*)-(+)-cyclohex-3-enecarboxylic acid, using galvinoxyl to inhibit polymerization, has been successfully scaled up to the kilogram level.[106] An improved synthesis of the chiral auxiliary, *N*-acryloylbornane-10,2-sultam, was also described together with a recycle protocol.

26.2.4 CHIRAL DIENES

Chiral dienes have proved to be less popular in asymmetric Diels-Alder reactions than their chiral dienophile counterparts. This is primarily a result of the problem of designing a molecule that incorporates a chiral moiety, such as the formation of a chiral isoprenyl ether or vinyl ketene acetal.[187–190] In addition, diastereoselectivities often are not high,[54,191–199] as illustrated by the cycloaddition of the chiral butadiene **5** with acrolein (Scheme 26.4). Improved stereoselection is observed through the use of double asymmetric induction, although this is a somewhat wasteful protocol.[35,54,177,200]

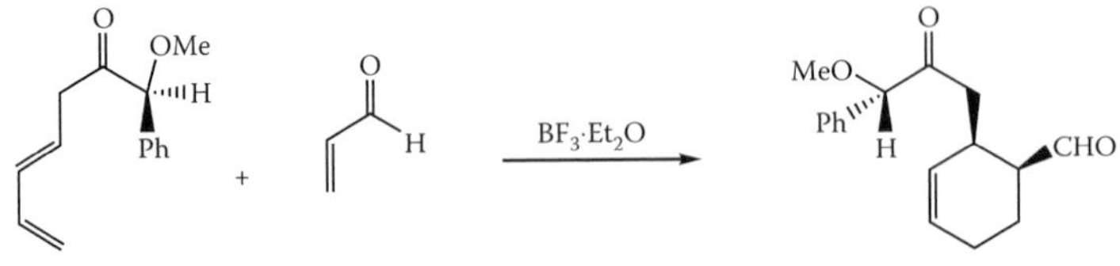

SCHEME 26.4

Sugar-substituted dienes are readily accessible and, although asymmetric induction may be low, the desired product is often easily isolated by crystallization.[192,194,195,201,202] Chiral dienophiles are usually more difficult to prepare, and it is often difficult to recycle the chiral auxiliary. Hence, Diels-Alder reactions involving chiral dienophiles should be avoided unless the chiral auxiliary becomes incorporated into the target molecule.

26.2.5 COOPERATIVE BLOCKING GROUPS

The use of a chiral fumarate ester allows for asymmetric induction irrespective to the approach of the dienophile to the diene. In particular, dimenthyl fumarate (**6**) has been advocated for large scale because of its ready availability, low cost, excellent yields, and high asymmetric induction.[35,128,129,203–208] Although other more exotic chiral auxiliaries may be used,[32] the use of **5** coupled with a homogeneous Lewis acid catalyst at low temperatures allows for remarkably high diastereoselectivity with a number of dienes (Scheme 26.5).[125,164,209]

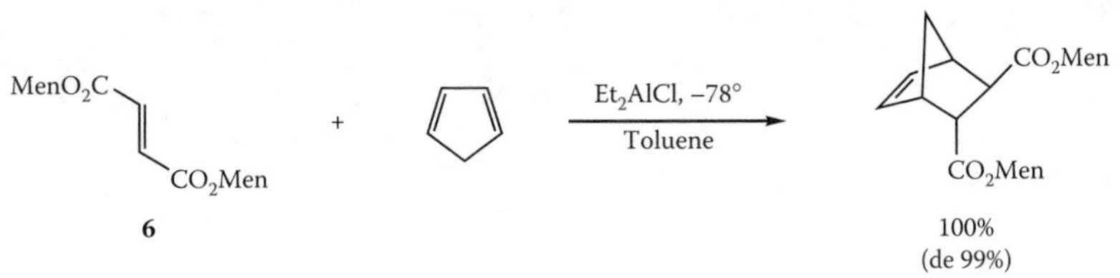

SCHEME 26.5

26.2.6 SOLVENT EFFECTS

The choice of solvent has had little, if any, influence on the majority of Diels-Alder reactions.[210,211] Although the addition of a Lewis acid might be expected to show more solvent dependence, generally there appears to be little effect on asymmetric induction.[118,129] However, a dramatic effect of solvent polarity has been observed for chiral metallocene triflate complexes.[212] The use of polar solvents, such as nitromethane and nitropropane, leads to a significant improvement in the catalytic properties of a copper Lewis acid complex in the hetero Diels-Alder reaction of glyoxylate esters with dienes.[213]

26.2.7 INVERSE ELECTRON DEMAND DIELS-ALDER REACTIONS

The asymmetric cycloaddition of electron-rich dienophiles with electron poor dienes, although discovered relatively recently, has provided some encouraging results (Scheme 26.6).[214–223] It should be noted that a multi-step synthesis is often required to synthesize the diene[215,220] and a recycle protocol for the chiral auxiliary may not be possible.[214,218,220]

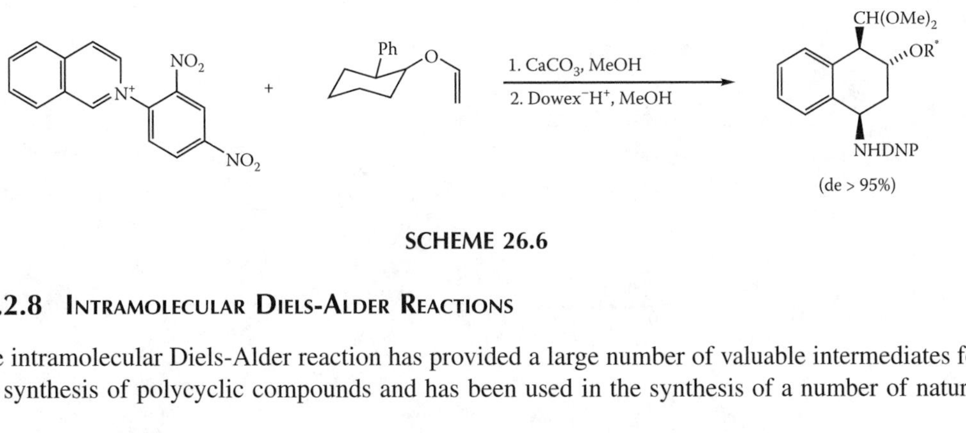

SCHEME 26.6

26.2.8 INTRAMOLECULAR DIELS-ALDER REACTIONS

The intramolecular Diels-Alder reaction has provided a large number of valuable intermediates for the synthesis of polycyclic compounds and has been used in the synthesis of a number of natural products.[35,224–228] A distinct advantage of this reaction is the ability to construct the reactant to provide for either *endo* or *exo* attack, which allows for excellent stereoselection.[225,229–234]

The simplicity and power of this reaction is illustrated by the spontaneous cyclization of the derivative **7**, formed from the condensation of (*R*)-citronellal (**8**) and 5-*n*-pentyl-1,3-cyclohexanedione, which occurs with complete stereochemical control (Scheme 26.7).[235]

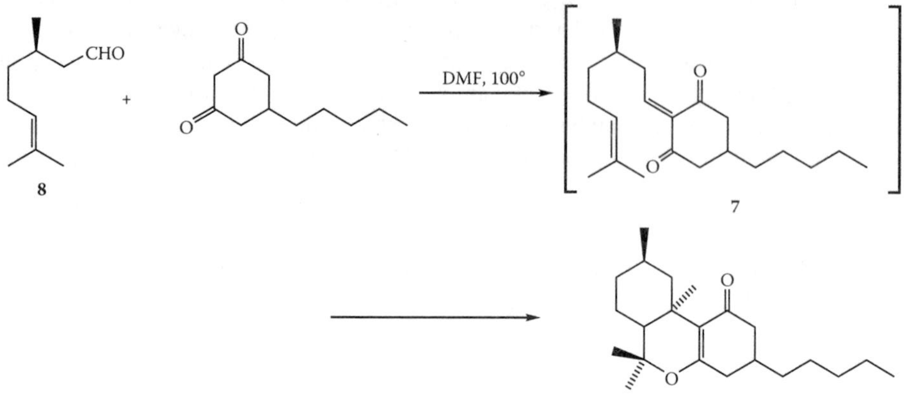

SCHEME 26.7

Lewis acid catalysts have been used in this approach.[225,236–238] There are also a number of examples of hetero intramolecular Diels-Alder reactions.[225,238–241]

26.2.9 HETERO DIELS-ALDER REACTIONS

Although the reactions of carbonyl compounds with dienes follow a stepwise addition, the overall stereochemical outcome mimics that of the Diels-Alder reaction.[1,2,5,242–244] The products of the hetero Diels-Alder reaction have been used as substrates for a wide variety of transformations and natural products syntheses.[242,245–258] Many hetero Diels-Alder reactions use chiral auxiliaries or catalysts that have been successful for the Diels-Alder reaction.

As with the Diels-Alder reaction, the addition of a Lewis acid appears to be essential to maximize selectivity.[259–263] Again, it should be noted that the use of some metals, such as zinc, causes significant disposal problems in wastewater streams and can add significantly to costs. The type of Lewis acid used can determine the relative stereochemical outcome of the reaction (Scheme 26.8).[259,264]

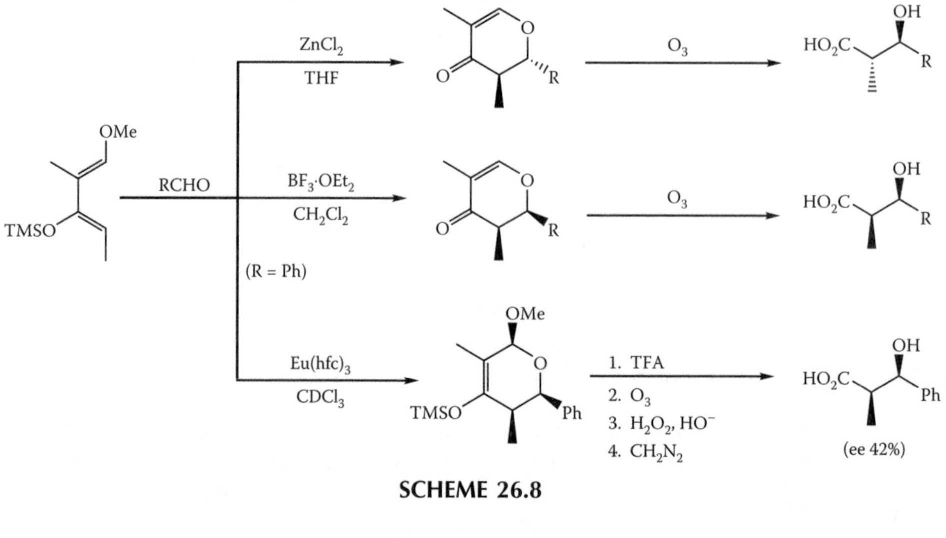

SCHEME 26.8

Chiral catalysts do provide high degrees of asymmetric induction.[61,265–272] Double asymmetric induction was observed in the synthesis of homochiral pyranose derivatives using a chiral auxiliary–chiral catalyst combination.[273,274]

The imine of either *R*- or *S*-(α)-methylbenzylamine and ethylglyoxalate has been reacted with Danishefsky's diene to prepare a piperidine derivative.[275] It is fortunate that the desired isomer crystallized out. This imine has also been reacted with cyclopentadiene to generate 2-azabicyclo[2.2.1]heptane derivative at scale in moderate yields.[276,277] In both cases the chiral auxiliary was removed by hydrogenation.

A hetero-Diels-Alder reaction has been used to prepare racemic 2-ethoxycarbonyl-3,6-dihydro-2H-pyran (**9**). This ester **9** was resolved by *Bacillus lentus* protease to provide the *R*-isomer. Reduction, protection, and ozonolysis provided the bis-mesylate **10** (Scheme 26.9), a key intermediate in the synthesis of the PKC (protein kinase C) inhibitor LY333531 (**11**).[278] This resolution approach was used because it was more efficient than an asymmetric Diels-Alder reaction.

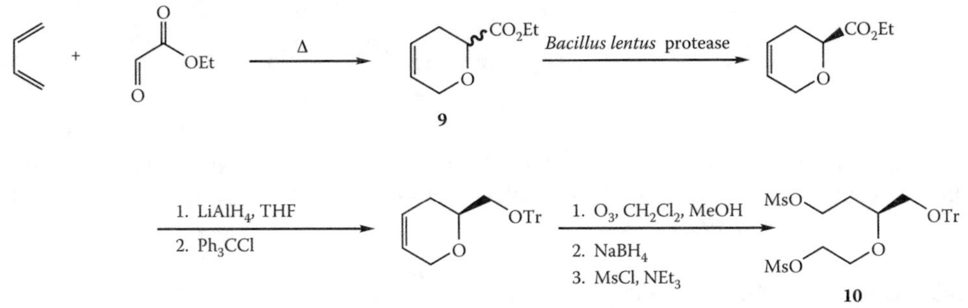

SCHEME 26.9

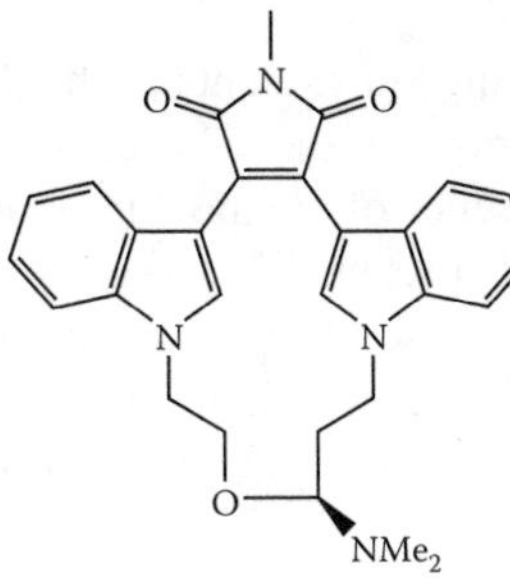

11

26.3 CLAISEN-TYPE REARRANGEMENTS

The Claisen rearrangement has been instrumental in the synthesis of a number of natural products.[279–289] Many useful derivatives have been prepared using the Claisen-type rearrangement including enol ethers,[290] amides,[291–293] esters and orthoesters,[294–296] acids,[297,298] oxazolines,[299] ketene acetals,[300,301] and thioesters.[302] Many of these variants use a cyclic primer to control relative and absolute stereochemistry. The Claisen and oxy Cope provide the best candidates for scale up as a result of the irreversible nature of these reactions.

26.3.1 Cope Rearrangement

The [3,3]-sigmatropic rearrangement of a 1,5-hexanediene is known as the Cope rearrangement and usually proceeds through a chair transition state. Generally, a large substituent at C–3 (or C–4) prefers to adopt an equatorial-like confirmation.[303,304] As the reaction is concerted, chirality at C–3 (or C–4) is transferred to the new chiral center at C–1 (or C–6). The reaction can be catalyzed by transition metals.[305] The use of a palladium catalyst allows for the reaction to be conducted at room temperature instead of extremely high temperatures (Scheme 26.10).[306,307]

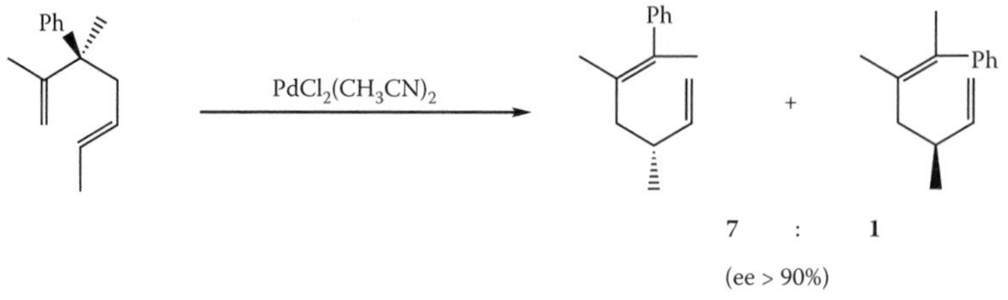

SCHEME 26.10

The Cope reaction is reversible, but the introduction of an oxygen atom (the oxy-Cope and anionic oxy-Cope rearrangements) results in the formation of an enolate whose stability drives the reaction to completion.[308,309] Because the enolate is formed regioselectively, it can be used for further transformations *in situ*.[310–312]

26.3.2 Claisen Rearrangement

The [3,3]-sigmatropic rearrangement of allyl vinyl ethers is the Claisen rearrangement and is mechanistically analogous to the Cope rearrangement. Because the product of the reaction is a carbonyl compound, the rearrangement usually goes to completion. The use of a large substituent, such as a bulky silyl group, that can occupy the equatorial position allows for good stereoselectivity (*cf.* Scheme 26.11).[313]

The presence of π-electron–donating substituents at the 2-position of the vinyl portion of the ether allows for significant acceleration of the Claisen rearrangement.[314–318] Aliphatic Claisen rearrangements can proceed in the presence of organoaluminum compounds,[286,319,320] although other Lewis acids have failed to show reactivity.[286,321–324] Useful levels of (Z)-stereoselection and asymmetric induction have been obtained by use of bulky chiral organoaluminum Lewis acids.[325–327]

The Claisen reaction can be catalyzed by palladium;[328] thus, the [3,3]-sigmatropic rearrangement of a bulky allylsilane derivative **12** proceeds with high selectivity (Scheme 26.11).[329–331]

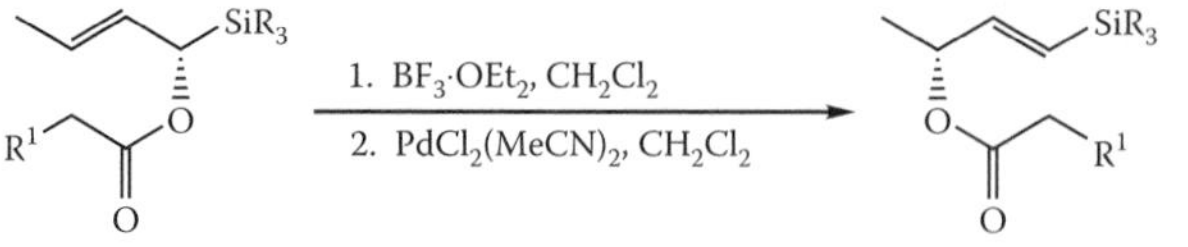

SCHEME 26.11

26.3.3 Ester Enolate Claisen Rearrangement

The most synthetically useful Claisen rearrangement, the ester enolate reaction of allyl esters, requires only relatively mild reaction conditions and is most amenable to scale up.[297,332] The geometry of the initially formed enol ether substrate is controlled by the choice of solvent system (Scheme 26.12).[298,333–335] Thus, the methodology provides a useful alternative to an aldol approach.

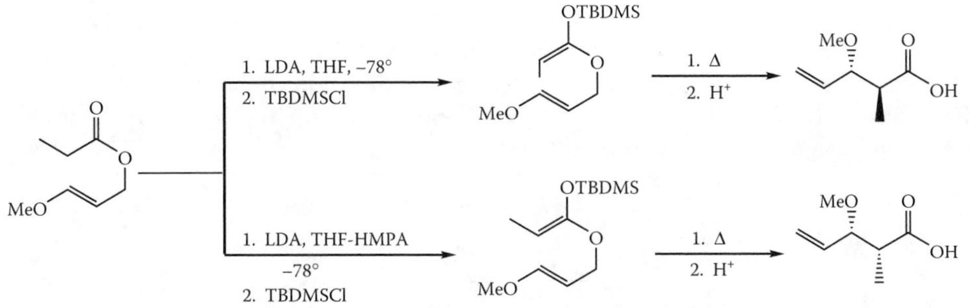

SCHEME 26.12

The Claisen ester enolate reaction has proved to be extremely useful in the synthesis of a large number of natural products.[3] In addition, the rearrangement has been extended to allow the preparation of useful intermediates such as α-alkoxy esters,[88,329,331,336–343] α-phenylthio esters,[339,344,345] α- and β-amino acids,[340,346–350] α-fluoro esters,[351] cycloalkenes,[352,353] tetronic acids,[354] and dihydropyrans.[355–357]

The use of a heteroatom α to the ester carbonyl group allows for the formation of a chelate with the metal counterion; hence, the geometry of the ester enolate can be assured.[336–338,358,359] This approach was used in the rearrangement of the glycine allylic esters **13** to γ,δ-unsaturated amino acids in good yields and excellent diastereoselectivity (Scheme 26.13).[358] The enantioselectivity could be reversed by using quinidine instead of quinine.

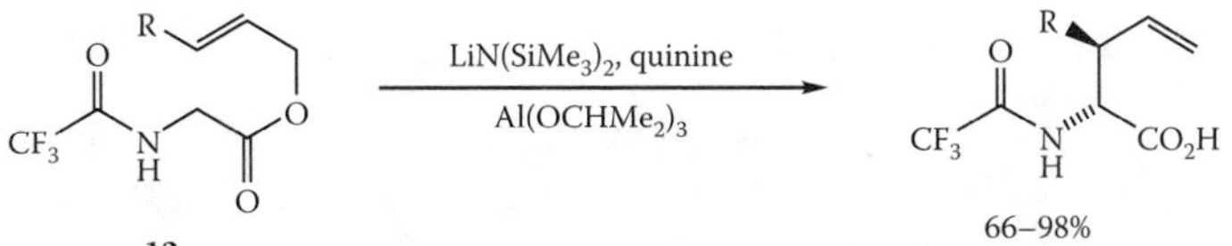

13

66–98%
(de 98%)
(ee 79–90%)

SCHEME 26.13

26.4 THE ENE REACTION

The ene reaction,[3,6,360–365] the addition of a carbon–carbon or carbon–oxygen double bond with concomitant transfer of an allylic hydrogen, can allow for chirality transfer.[366–369] The reaction has similarities to the Diels-Alder reaction in that a σ-bond is formed at the expense of a π-bond. In addition, the use of a Lewis acid as a catalyst allows for control of the relative stereochemistry (Scheme 26.14).[370–372] Large-scale reactions will be complicated by the need to use either high temperatures or Lewis acids. In addition, thermal and Friedel-Crafts–type degradation products may be problematic with the use of these conditions.[361,373]

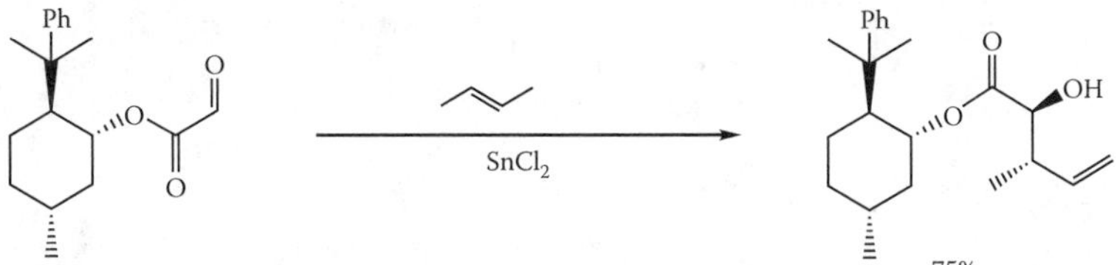

75%
(de 95%)

SCHEME 26.14

Lewis acid–catalyzed reactions can be conducted at much lower temperatures than the uncatalyzed reaction, which can require temperatures in excess of 200°C.[6,272,360,363,370,373–389]

The methodology allows for a selective preparation of cyclic compounds[361,390–394] as well as acyclic ones (Scheme 26.15).[363,383,395–399] Stereochemical control for acyclic reactions can be increased with the use of a cyclic primer.[369,372,375,400–409]

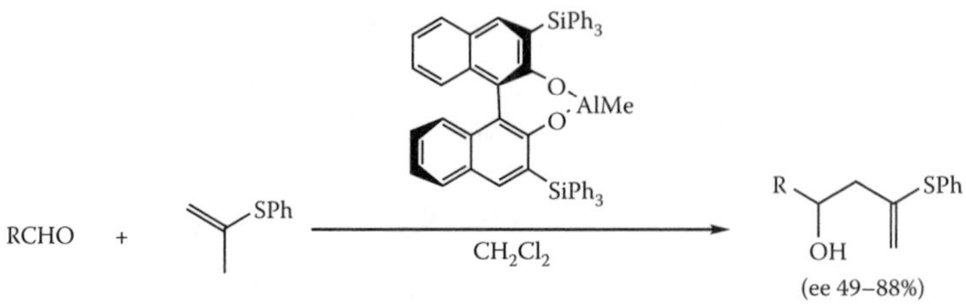

SCHEME 26.15

Intramolecular ene reactions can be highly diastereoselective.[401,410–416] An example is provided by the synthesis of the corynantheine derivative **14** (Scheme 26.16).[410]

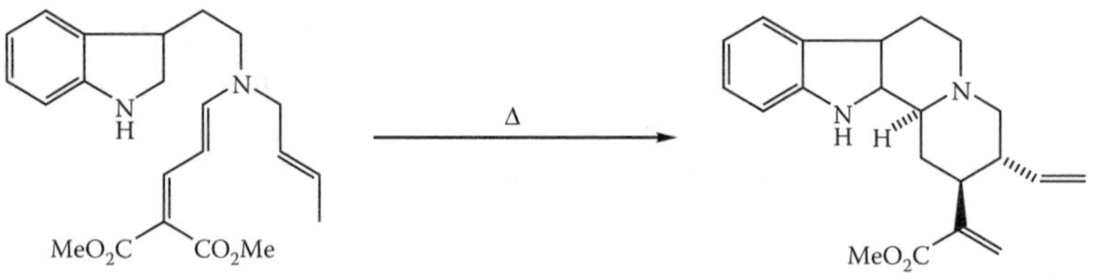

SCHEME 26.16

26.5 DIPOLAR CYCLOADDITIONS

Dipolar cycloadditions, closely related to the Diels-Alder reaction,[1,2] result in the synthesis of a five-membered adduct including cyclopentane derivatives.[417–426]

The cycloaddition of a nitrile oxide with a chiral allylic ether affords an isoxazoline with selectivity for the *pref*-isomer. This selectivity increases with the size of the alkyl substituent and is insensitive to the size of the allyl oxygen substituent. However, allyl alcohols tend to form the *parf*-isomer preferentially, although the selectivity is often low.[427–434] The product of dipolar cycloadditions based on nitrile oxides, the isoxazoline moiety, can be converted into a large variety of functional groups under relatively mild conditions.[3] Among other products, the addition can be used to prepare β-hydroxy ketones (Scheme 26.17).[435] The isoxazoline moiety can be used to control the relative stereochemistry through chelation control.[436,437]

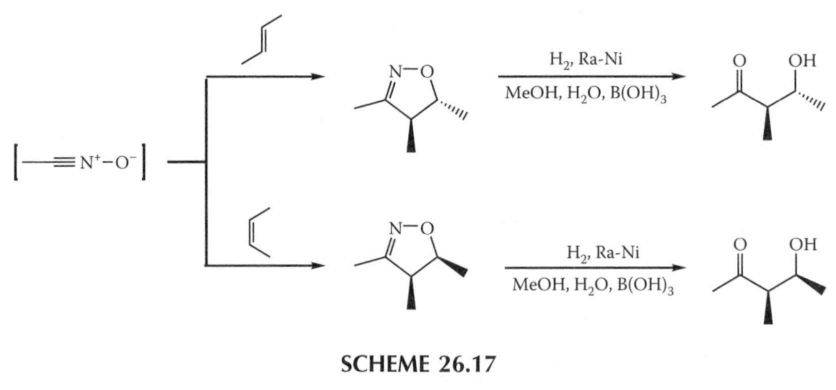

SCHEME 26.17

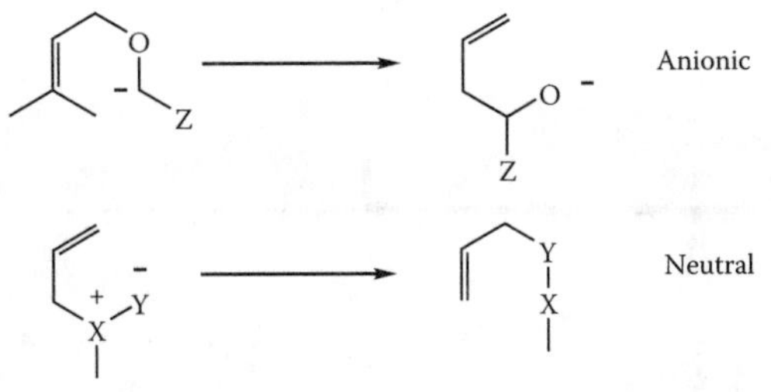

FIGURE 26.2 Anionic and neutral [2,3]-sigmatropic rearrangements.

26.6 [2,3]-SIGMATROPIC REARRANGEMENTS

There are two classes of [2,3]-sigmatropic rearrangements: anionic and neutral (Figure 26.2).[438,439] The Wittig rearrangement provides the most useful example of anionic rearrangements, whereas the Evans rearrangement is the most important example of neutral rearrangements.[438,440] Numerous other examples such as allylic sulfenates, amine oxides, and selenoxides are known, but the electronic and stereochemical arguments used for the two major reactions allow similarities and extrapolations to be drawn with confidence.[438] The [2,3]-rearrangement often needs to be conducted at low temperature to minimize competition from a [3,3]-rearrangement or nonconcerted process adding to the problem of scale up.[441–443]

26.6.1 Wittig Rearrangement

The Wittig rearrangement is primarily used in the transformation of an allylic ether to an α-allyl alcohol (Scheme 26.18).[444,445] The transition-state geometry plays an important role to determine the reaction outcome that, in turn, is dependent on the stereochemistry of the double bond (Figure 26.3).[438,442,444,446–454]

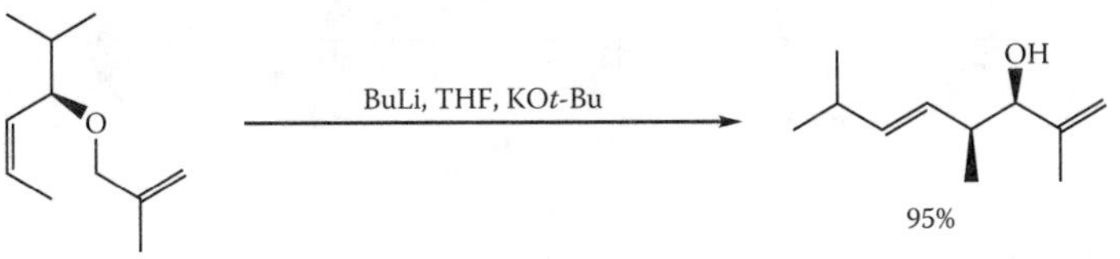

SCHEME 26.18

The Wittig rearrangement has been used in the synthesis of a wide range of compounds that rely on this versatile protocol,[455] ranging from large antibiotics,[456,457] to dihydrofurans,[458,459] to steroid derivatives.[460–462] However, the use of strongly basic conditions is the major hindrance to the implementation of this reaction at scale.

26.6.2 Evans Rearrangement

The other major [2,3]-rearrangement, the Evans rearrangement,[463] can be extended to synthesize functionalized allyl alcohols.[464,465] The Evans rearrangement involves the rearrangement of an allylic sulfoxide to an allyl alcohol.[463] The reverse reaction is also possible: the conversion of an allyl alcohol to the analogous sulfoxide. The "push–pull" nature of the two rearrangements (Scheme 26.19), coupled with the transfer of chirality, provides a method to invert an allyl alcohol together with the isomerization of the alkene (Scheme 26.20).[438,466]

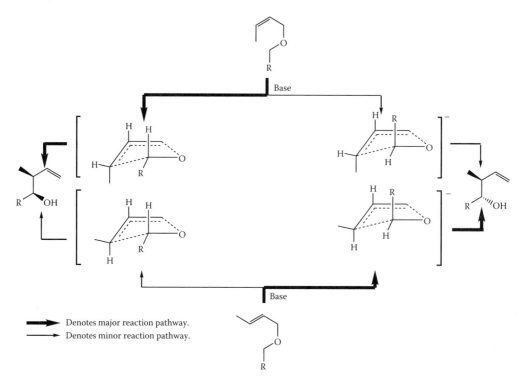

FIGURE 26.3 Reaction pathways for the Wittig rearrangement.

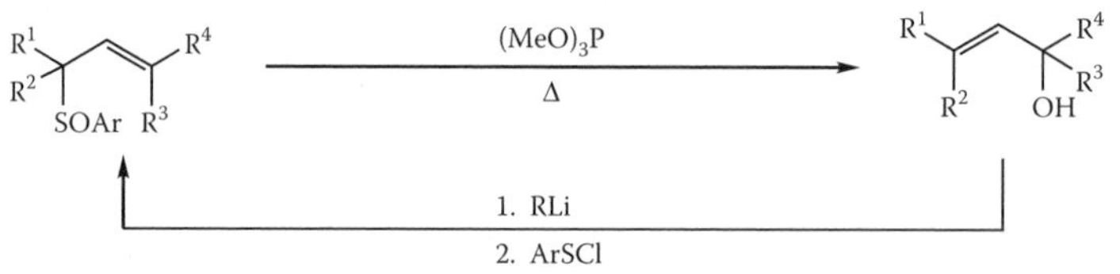

SCHEME 26.19

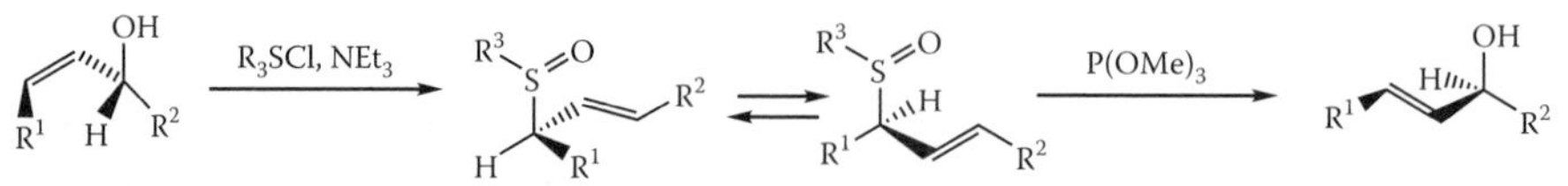

SCHEME 26.20

The Evans rearrangement can be driven to completion by the addition of a thiophile, such as trimethylphosphite (Scheme 26.19).[440,467–470] This strategy allows the chemistry of the allyl phenyl sulfoxide, or other sulfur precursor, to be exploited before the allyl alcohol is unmasked.[463,471–474] The addition of phenylsulfenyl chloride to an alkene, followed by the elimination of hydrogen chloride and subsequent rearrangement, provides a useful synthesis of allyl alcohols.[473,475] The [2,3]-Evans sigmatropic rearrangement is concerted and allows for stereochemical transfer.[476,477] The reverse reaction, formation of the allyl sulfoxide, results from the treatment of an allyl alcohol using a base followed by arylsulfenyl chloride to produce the allyl sulfoxide.[478,479]

26.7 OTHER PERICYCLIC REACTIONS

Although photochemical cycloadditions can prove difficult to scale up, they do offer access to cycloadducts not directly accessible by other methods.[480,481] The initial studies into asymmetric synthesis by photochemical means were in the solid phase or organized assemblies, and few examples were known in solution.[482,483] Circular polarized light, a chiral agent, has the potential to induce asymmetric synthesis, although useful ee's have not yet been obtained.[481]

The condensation of an imine with a Reformatsky-type reagent and tandem reactions can result in asymmetric induction.[3,207,484–487] The reaction of a ketene with an electron-rich alkene results in a [2+2] cycloaddition, although other systems can also be used.[488–490] The stereochemistry of the adduct is *cis,* and functionalized ketenes can also be used. The ketene can be generated *in situ* (Scheme 26.21).[491]

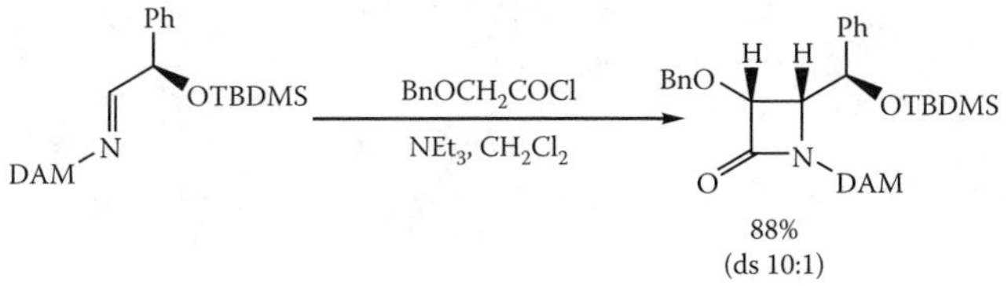

SCHEME 26.21

The cycloaddition of ketene with chloral in the presence of catalytic quantities of quinidine or quinine leads to the oxetanones in high optical and chemical yield (Scheme 26.22). This reaction is practiced on an industrial scale with the chiral building blocks malic and citramalic acids being formed by hydrolysis.[492]

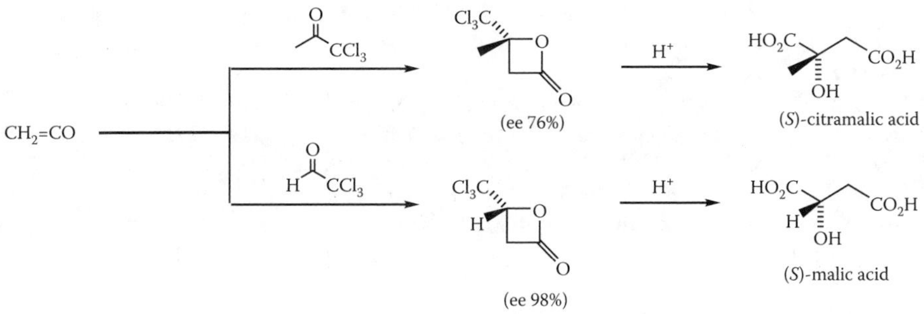

SCHEME 26.22

26.8 SUMMARY

Many types of pericyclic and cycloaddition reactions have been documented. Although there are no general guidelines for the asymmetric preparation, reagents such as chiral catalysts are providing more general routes. Many of the reactions discussed rely on the use of low temperature. Although it is expensive to conduct low temperature reactions on an industrial scale, reactions that need temperatures of down to −105°C can be conducted. It should be noted that at such temperatures only stainless steel vessels, which require neutral or basic conditions, can be used at these extreme temperatures. In addition, reactions that involve the use of a metal can cause contamination problems in wastewater or the product.

Other sections of this book should be accessed because many chiral auxiliaries and catalysts appear to cross over from one type of pericyclic reaction to another. It should also be noted that, although a large number of pericyclic reactions are known, large differences in asymmetric induction can occur with only subtle changes in reagents. This is especially true of the Diels-Alder reaction.

The Claisen rearrangement, Cope rearrangement, and associated variants are powerful tools that can be used to create a number of new chiral centers in an expeditious manner, but the use of heavy metals, such as mercury, should be avoided. Of these reactions, the Ireland-Claisen ester enolate reaction provides the most versatile synthetic pathway with minimal scale up problems.

Our understanding of chiral catalyst has greatly increased in recent years, and we are beginning to see catalysts that are less substrate dependent. The advantages of chiral catalysts on an industrial scale are obvious. With the enormous potential of pericyclic reactions, particularly because more than one carbon–carbon bond and a number of stereogenic centers can all be created at the same time, it can only be a matter of time before they become a key weapon in the process scale up strategy to chiral compounds.

REFERENCES

1. Desimoni, G., Tacconi, G., Barco, A., Polinni, G. P. In *Natural Product Synthesis Through Pericyclic Reactions*, American Chemical Society: Washington, D. C., 1983, p. 119.
2. Kocovsky, P., Turecek, F., Hajicek, J. In *Synthesis of Natural Products: Problems of Stereoselection*, CRC Press: Boca Raton, 1986, Vol. 1.
3. Ager, D. J., East, M. B. *Asymmetric Synthetic Methodology*, CRC: Boca Raton, 1995.
4. Masamune, S., Choy, W., Petersen, J. S., Sita, L. R. *Angew. Chem., Int. Ed.* 1985, *24*, 1.
5. Danishefsky, S. *Acc. Chem. Res.* 1981, *14*, 400.
6. Oppolzer, W. *Angew. Chem., Int. Ed.* 1984, *23*, 876.
7. Alonso, I., Carretero, J. C., Ruano, J. L. G. *J. Org. Chem.* 1993, 3231.
8. Evans, D. A., Lectka, T., Miller, S. J. *J. Am. Chem. Soc.* 1993, *115*, 6460.
9. Carretero, C. J. C., Ruano, G. J. L., Lorente, A., Yuste, F. *Tetrahedron: Asymmetry* 1993, *4*, 177.
10. Hoffmann, H., Bolte, M., Berger, B., Hoppe, D. *Tetrahedron Lett.* 1993, *34*, 6537.
11. Bao, J., Wulff, W. D., Rheongold, A. L. *J. Am. Chem. Soc.* 1993, *115*, 3814.
12. Kobayashi, S., Hachiya, I., Ishitani, H., Araki, M. *Tetrahedron Lett.* 1993, *34*, 2581.
13. Corey, E. J., Loh, T.-P. *Tetrahedron Lett.* 1993, *34*, 3979.
14. Corey, E. J., Roper, T. D., Ishihara, K., Sarakinos, G. *Tetrahedron Lett.* 1993, *34*, 8399.
15. Uemura, M., Hayashi, Y., Hayashi, Y. *Tetrahedron Asymmetry* 1993, *4*, 2291.
16. Evans, D. A., Lectka, T., Miller, S. J. *Tetrahedron Lett.* 1993, *34*, 7027.
17. Nakanishi, S., Kumeta, K., Sawai, Y., Takata, T. *J. Organomet. Chem.* 1996, *515*, 99.
18. Arai, Y., Masuda, T., Tsutomu, Y., Shiro, M. *J. Chem. Soc., Perkin Trans. 1* 1996, 759.
19. Bao, J., Wulff, W. D., Rheingold, A. L. *J. Am. Chem. Soc.* 1993, *115*, 3814.
20. Posner, G. H., Carry, J. C., Anjeh, T. E. W., French, A. N. *J. Org. Chem.* 1992, *57*, 7012.
21. Kagan, H. B., Riant, O. *Chem. Rev.* 1992, *92*, 1007.
22. Fleming, I. *Frontier Orbitals and Organic Chemical Reactions*, Wiley: New York, 1976.
23. Houk, K. N. *Acc. Chem. Res.* 1975, *8*, 361.
24. Fringuelli, F., Taticchi, A. *Dienes in the Diels-Alder Reaction*, Wiley: New York, 1990.
25. Yamauchi, M., Honda, Y., Matsuki, N., Watanabe, T., Date, K., Hiramatsu, H. *J. Org. Chem.* 1996, *61*, 2719.
26. Houk, K. N., Strozier, R. W. *J. Am. Chem. Soc.* 1973, *95*, 4094.
27. Chao, T.-M., Baker, J., Hehre, W. J., Kahn, S. D. *Pure Appl. Chem.* 1991, *63*, 283.
28. Brion, F. *Tetrahedron Lett.* 1982, *23*, 5299.
29. Sauer, J., Wuest, H., Mielert, A. *Chem. Ber.* 1964, *97*, 3183.
30. Wu, Y.-H., Rayburn, J. W., Allen, L. E., Ferguson, H. C., Kissel, J. W. *J. Med. Chem.* 1972, *15*, 477.
31. Fuji, A., Yukio, A., Kikuo, I. Eur. Pat. EP 82402, 1983.
32. Ishizumi, K., Antoku, F., Asami, Y. U.S. Pat. 4,543,355, 1985.
33. Nelson, D. J. *J. Org. Chem.* 1986, *51*, 3185.
34. Koizumi, T., Hakamada, I., Yoshii, E. *Tetrahedron Lett.* 1984, *25*, 87.
35. Paquette, L. A. In *Asymmetric Synthesis,* Morrison, J. D. Ed., Academic Press: Orlando, 1984, Vol. 3, p. 455.

36. Poll, T., Metter, J. O., Helmchen, G. *Angew. Chem., Int. Ed.* 1985, *24*, 112.
37. Yamamoto, Y., Suzuki, I. *J. Org. Chem.* 1993, *58*, 4783.
38. Oppolzer, W., Wills, M., Kelly, M. J., Signer, M., Blagg, J. *Tetrahedron Lett.* 1990, *31*, 5015.
39. Oppolzer, W., Chopius, C. *Tetrahedron Lett.* 1983, *24*, 4665.
40. Oppolzer, W., Chapius, C., Dao, G. M., Reichlin, D., Godel, T. *Tetrahedron Lett.* 1982, *23*, 4781.
41. Bednarski, M., Danishefsky, S. *J. Am. Chem. Soc.* 1983, *105*, 3716.
42. Bonnesen, P. V., Puckett, C. L., Honeychuck, R. V., Hersh, W. H. *J. Am. Chem. Soc.* 1989, *111*, 6070.
43. Grieco, P. A., Nunes, J. J., Gaul, M. D. *J. Am. Chem. Soc.* 1990, *112*, 4595.
44. Smith, D. A., Houk, K. N. *Tetrahedron Lett.* 1991, *32*, 1549.
45. Danishefsky, S., Bednarski, M. *Tetrahedron Lett.* 1984, *25*, 721.
46. Pindur, U., Lutz, G., Otto, C. *Chem. Rev.* 1993, *93*, 741.
47. Corey, E. J., Imwinkelried, R., Pikul, S., Xiang, Y. B. *J. Am. Chem. Soc.* 1989, *111*, 5493.
48. Gothelf, K. V., Jorgensen, K. A. *J. Org. Chem.* 1995, *60*, 6847.
49. Gothelf, K. V., Hazell, R. G., Jorgensen, K. A. *J. Am. Chem. Soc.* 1995, *117*, 4435.
50. de Pascual-Teresa, B., Gonzalez, J., Asensio, A., Houk, K. N. *J. Am. Chem. Soc.* 1995, *117*, 4347.
51. Haase, C., Sarko, C. R., Dimare, M. *J. Org. Chem.* 1995, *60*, 1777.
52. Corey, E. J., Loh, T.-P., Roper, T. D., Azimioara, M. D., Noe, M. C. *J. Am. Chem. Soc.* 1992, *114*, 8290.
53. Corey, E. J., Sarshar, S. *J. Am. Chem. Soc.* 1992, *114*, 7938.
54. Trost, B. M., O'Krongly, D., Belletire, J. L. *J. Am. Chem. Soc.* 1980, *102*, 7595.
55. Maruoka, K., Murase, N., Yamamoto, H. *J. Org. Chem.* 1993, *58*, 2938.
56. Yamamoto, H., Maruoka, K., Furuta, K., Naruse, Y. *Pure Appl. Chem.* 1989, *61*, 419.
57. Terada, M., Mikami, K., Nakai, T. *Tetrahedron Lett.* 1991, *32*, 935.
58. Kaufmann, D., Boese, R. *Angew. Chem., Int. Ed.* 1990, *29*, 545.
59. Ketter, A., Glahsl, G., Herrmann, R. *J. Chem. Research (S)* 1990, 278.
60. Hattori, K., Yamamoto, H. *J. Org. Chem.* 1992, *57*, 3264.
61. Motoyama, Y., Terada, M., Mikami, K. *Synlett* 1995, 967.
62. Kobayashi, S., Araki, M., Hachiya, I. *J. Org. Chem.* 1994, *59*, 3758.
63. Corey, E. J., Sarshar, S., Lee, D.-H. *J. Am. Chem. Soc.* 1994, *116*, 12089.
64. Corey, E. J. *Tetrahedron Lett.* 1991, *32*, 7517.
65. Takasu, M., Yamamoto, H. *Synlett* 1990, 194.
66. Seerden, J.-P. G., Scheeren, H. W. *Tetrahedron Lett.* 1993, *34*, 2669.
67. Sartor, D., Saffrich, J., Helmchen, G. *Synlett* 1990, 197.
68. Sartor, D., Saffrich, J., Helmchen, G., Richards, C. J., Lambert, H. *Tetrahedron: Asymmetry* 1991, *2*, 639.
69. Corey, E. J., Loh, T. P. *J. Am. Chem. Soc.* 1991, *113*, 8966.
70. Corey, E. J., Imai, N., Zhang, H.-Y. *J. Am. Chem. Soc.* 1991, *113*, 728.
71. Corey, E. J., Ishihara, K. *Tetrahedron Lett.* 1992, *33*, 6807.
72. Evans, D. A., Johnson, J. S., Olhava, E. J. *J. Am. Chem. Soc.*, 2000, *122*, 1635.
73. Davies, I. W., Senanayake, C. H., Larsen, R. D., Verhoeven, T. R., Reider, P. J. *Tetrahedron Lett.* 1996, *37*, 1725.
74. Thompson, C. F., Januson, T. F., Jacobsen, E. N. *J. Am. Chem. Soc.* 2001, *123*, 9974.
75. Thompson, C. F., Januson, T. F., Jacobsen, E. N. *J. Am. Chem. Soc.* 2000, *122*, 10482.
76. Chavez, D. E., Jacobsen, E. N. *Angew. Chem., Int. Ed.* 2001, *40*, 3667.
77. Dossiter, A. G., Januson, T. F., Jacobsen, E. N. *Angew. Chem., Int. Ed.* 1999, *38*, 2398.
78. Liu, P., Jacobsen, E. N. *J. Am. Chem. Soc.*, 2001, *123*, 10772.
79. Knol, J., Meetama, A., Feringa, B. L. *Tetrahedron: Asymmetry* 1995, *6*, 1069.
80. Rebiere, F., Riant, O., Kagan, H. B. *Tetrahedron: Asymmetry* 1990, *1*, 199.
81. Narasaka, K., Iwasawa, N., Inoue, M., Yamada, T., Nakashima, M., Sugimori, J. *J. Am. Chem. Soc.* 1989, *111*, 5340.
82. Corey, E. J. *Tetrahedron Lett.* 1991, *32*, 6289.
83. Narasaka, K., Tanaka, H., Kanai, F. *Bull. Chem. Soc., Jpn* 1991, *64*, 387.
84. Engler, T. A., Letavic, M. A., Takusagawa, F. *Tetrahedron Lett.* 1992, *33*, 6731.
85. Cativiela, C., Lopez, P., Mayoral, J. A. *Tetrahedron: Asymmetry* 1991, *2*, 1295.
86. Devine, P. N., Oh, T. *J. Org. Chem.* 1992, *57*, 396.
87. Devine, P. N., Oh, T. *Tetrahedron Lett.* 1991, *32*, 883.

88. Oh, T., Wrobel, Z., Devine, P. N. *Synlett* 1992, 81.
89. Furuta, K., Shimizu, S., Miwa, Y., Yamamoto, H. J. *Org. Chem.* 1989, *54*, 1481.
90. Furuta, K., Miwa, Y., Iwanaga, K., Yamamoto, H. *J. Am. Chem. Soc.* 1988, *110*, 6254.
91. Furuta, K., Kanematsu, A., Yamamoto, H., Takaoka, S. *Tetrahedron Lett.* 1989, *30*, 7231.
92. Ishihara, K., Gao, Q., Yamamoto, H. *J. Org. Chem.* 1993, *58*, 6917.
93. Gao, Q., Maruyama, T., Mouri, M., Yamamoto, H. *J. Org. Chem.* 1992, *57*, 1951.
94. Takemura, H., Komeshima, N., Takashito, I., Hashimoto, S.-I., Ikota, N., Tomioka, K., Koga, K. *Tetrahedron Lett.* 1987, *28*, 5687.
95. Hashimoto, S.-I., Komeshima, N., Koga, K. *J. Chem. Soc., Chem. Commun.* 1979, 437.
96. Lautens, M., Lautens, J. C., Smith, A. C. *J. Am. Chem. Soc.* 1990, *112*, 5627.
97. Hawkins, J. M., Loren, S. *J. Am. Chem. Soc.* 1991, *113*, 7794.
98. Kuendig, E. P., Bourdin, B., Bernardinelli, G. *Angew. Chem.* 1994, *106*, 1931.
99. Wada, E., Pei, W., Kanemasa, S. *Chem. Lett.* 1994, 2345.
100. Riant, O., Kagan, H. B. *Tetrahedron Lett.* 1989, *30*, 7403.
101. Okamura, H., Nakamura, Y., Iwagawa, T., Nakatani, M. *Chem. Lett.* 1996, 193.
102. Van der Eycken, J., Vandewalle, M., Heinemann, G., Laumen, K., Schneider, M. P., Kredel, J., Sauer, J. *J. Chem. Soc., Chem. Commun.* 1989, 306.
103. Janssen, A. J. M., Klunder, A. J. H., Zwanenburg, B. *Tetrahedron Lett.* 1990, *31*, 7219.
104. Braisted, A. C., Schultz, P. G. *J. Am. Chem. Soc.* 1990, *112*, 7430.
105. Oppolzer, W., Rodriguez, I., Blagg, J., Bernardinelli, G. *Helv. Chim. Acta* 1989, *72*, 123.
106. Thom, C., Kocienski, P., Jarowicki, K. *Synthesis* 1993, 475.
107. Oppolzer, W., Chapuis, C., Bernardinelli, G. *Tetrahedron Lett.* 1984, *25*, 5885.
108. Tanaka, K., Uno, H., Osuga, H., Suzuki, H. *Tetrahedron: Asymmetry* 1993, *4*, 629.
109. Banks, M. R., Blake, A. J., Alexander, J., Cadogan, J. I. G., Doyle, A. A., Gosney, I., Hodgson, P. K. G., Thorburn, P. *Tetrahedron* 1996, *52*, 4079.
110. Palomo, C., Berree, F., Linden, A., Villalgordo, J. M., Facultad, Q. *J. Chem. Soc., Chem. Commun.* 1994, 1861.
111. Langlois, Y., Pouilhes, A. *Tetrahedron: Asymmetry* 1991, *2*, 1223.
112. Kouklovsky, C., Pouilhes, A., Langlois, Y. *J. Am. Chem. Soc.* 1990, *112*, 6672.
113. Pouilhes, A., Uriarte, E., Kouklovsky, C., Langlois, N., Langlois, Y., Chiaroni, A., Riche, C. *Tetrahedron Lett.* 1989, *30*, 1395.
114. Boeckman, R. K., Nelson, S. G., Gaul, M. D. *J. Am. Chem. Soc.* 1992, *114*, 2258.
115. Arai, Y., Matsui, M., Koizumi, T., Shiro, M. *J. Org. Chem.* 1991, *56*, 1983.
116. Oppolzer, W., Seletsky, B. M., Bernardinelli, G. *Tetrahedron Lett.* 1994, *35*, 3509.
117. Oppolzer, W., Kurth, M., Reichlin, D., Chapuis, C., Mohnhaupt, M., Moffatt, F. *Helv. Chim. Acta* 1981, *64*, 2802.
118. Oppolzer, W., Kurth, M., Reichlin, D., Moffatt, F. *Tetrahedron Lett.* 1981, *22*, 2545.
119. Oppolzer, W., Chapuis, C., Kelly, M. J. *Helv. Chim. Acta* 1983, *66*, 2358.
120. Evans, D. A., Chapman, K. T., Bisaha, J. *J. Am. Chem. Soc.* 1984, *106*, 4261.
121. Corey, E. J., Ensley, H. E. *J. Am. Chem. Soc.* 1975, *97*, 6908.
122. Cativiela, C., Figueras, F., Fraile, J. M., Garcia, J. I., Mayoral, J. A. *Tetrahedron: Asymmetry* 1991, *2*, 953.
123. Cativiela, C. L., P., Mayoral, J. A. *Tetrahedron: Asymmetry* 1990, *1*, 61.
124. Gras, J. L., Poncet, A., Nouguier, R. *Tetrahedron Lett.* 1992, *33*, 3323.
125. Takayama, H., Iyobe, A., Koizumi, T. *J. Chem. Soc., Chem. Commun.* 1986, 771.
126. Ohkata, K., Kubo, T., Miyamoto, K., Ono, M., Yamamoto, J., Akiba, K.-Y. *Heterocycles* 1994, *38*, 1483.
127. Waldmann, H., Drager, M. *Tetrahedron Lett.* 1989, *30*, 4227.
128. Walborsky, H. M., Barash, L., Davis, T. C. *Tetrahedron* 1963, *19*, 2333.
129. Furuta, K., Iwanaga, K., Yamamoto, H. *Tetrahedron Lett.* 1986, *27*, 4507.
130. Reymond, J.-L., Vogel, P. *J. Chem. Soc., Chem. Commun.* 1990, 1070.
131. Chen, Y., Vogel, P. *Tetrahedron Lett.* 1992, *33*, 4917.
132. Ronan, B., Kagan, H. B. *Tetrahedron: Asymmetry* 1991, *2*, 75.
133. Lopez, R., Carretero, J. C. *Tetrahedron: Asymmetry* 1991, *2*, 93.
134. Fuji, K., Tanaka, K., Abe, H., Itoh, A., Node, M., Taga, T., Miwa, Y., Shiro, M. *Tetrahedron: Asymmetry* 1991, *2*, 179.

135. Alonso, I., Carretero, J. C., Garcia Ruano, J. L. *Tetrahedron Lett.* 1991, *32*, 947.
136. Fuji, K., Tanaka, K., Abe, H., Matsumoto, K., Taga, T., Miwa, Y. *Tetrahedron: Asymmetry* 1992, *3*, 609.
137. Arai, Y. K., S.-i, Takeuchi, Y., Koizumi, T. *Tetrahedron Lett.* 1985, *26*, 6205.
138. Arai, Y. M., M., Koizumi, T, J. *Chem. Soc. Perkin Trans. I* 1990, 1233.
139. Takahashi, T., Kotsubo, H., Iyobe, A., Namiki, T., Koizumi, T. *J. Chem. Soc., Perkin Trans. I* 1990, 3065.
140. Takahashi, T., Iyobe, A., Arai, Y., Koizumi, T. *Synthesis* 1989, 189.
141. Takayama, H., Hayashi, K., Koizumi, T. *Tetrahedron Lett.* 1986, *27*, 5509.
142. Takayama, H., Iyobe, A., Koizumi, T. *Chem. Pharm. Bull.* 1987, *35*, 433.
143. Alonso, I. C., J. C., Ruano, J. L. G. *Tetrahedron Lett.* 1989, *30*, 3853.
144. Alonso, I., Cid, M. B., Carretero, J. C., Ruano, J. L., Hoyos, M. A. *Tetrahedron: Asymmetry* 1991, *2*, 1193.
145. Carreno, M. C. R., J. L. G., Urbano, A. *Tetrahedron Lett.* 1989, *30*, 4003.
146. Arai, Y., Koizumi, T. *Sulfur Rep.* 1993, *15*, 41.
147. Serrano, J. A., Caceres, L. E., Roman, E. *J. Chem. Soc., Perkin Trans. I* 1992, 941.
148. Kim, K. S., Cho, I. H., Joo, Y. H., Yoo, I. J., Song, J. H., H., K. J. *Tetrahedron Lett.* 1992, *33*, 4029.
149. Jurczak, J., Tkacz, M. *Synthesis* 1979, 42.
150. Horton, D., Machinami, T., Takagi, Y. *Carbohydrate Res.* 1983, *121*, 135.
151. Horton, D., Machinami, T. *J. Chem. Soc., Chem. Commun.* 1981, 88.
152. Franck, R. W. J., T. V., Olejniczak, K., Blount, J. F. *J. Am. Chem. Soc.* 1982, *104*, 1106.
153. Liu, H.-J., Chew, S. Y., Browne, E. N. C. *Tetrahedron Lett.* 1991, *32*, 2005.
154. Waldner, A. *Tetrahedron Lett.* 1989, *30*, 3061.
155. Sato, M., Orii, C., Sakaki, J.-I., Kaneko, C. *J. Chem. Soc., Chem. Commun.* 1989, 1435.
156. Roush, W. R., Brown, B. B. *Tetrahedron Lett.* 1989, *30*, 7309.
157. Roush, W. R., Essenfeld, A. P., Warmus, J. S., Brown, B. B. *Tetrahedron Lett.* 1989, *30*, 7305.
158. Mattay, J. M., J., Maas, G. *Chem. Ber.* 1989, *122*, 327.
159. Kneer, G. M., J., Raabe, G., Kruger, C., Lauterwein, J. *Synthesis* 1990, 599.
160. Jensen, K. N., Roos, G. H. P. *Tetrahedron: Asymmetry* 1992, *3*, 1553.
161. De Jong, J. C., van Bolhuis, F., Feringa, B. L. *Tetrahedron: Asymmetry* 1991, *2*, 1247.
162. De Jong, J. C., Jansen, J. F. G. A., Feringa, B. L. *Tetrahedron Lett.* 1990, *31*, 3047.
163. Aso, M., Ikeda, I., Kawabe, T., Shiro, M., Kanematsu, K. *Tetrahedron Lett.* 1992, *33*, 5789.
164. Ikeda, I., Honda, K., Osawa, E., Shiro, M., Aso, M., Kanematsu, K. *J. Org. Chem.* 1996, *61*, 2031.
165. Evans, D. A., Chapman, K. T., Bisaha, J. *J. Am. Chem. Soc.* 1988, *110*, 1238.
166. Kimura, K., Murata, K., Otsuka, K., Ishizuka, T., Haratake, M., Kunieda, T. *Tetrahedron Lett.* 1992, *33*, 4461.
167. Chapuis, C., Bauer, T., Jezewski, A., Jurczak, J. *Pol. J. Chem.* 1994, *68*, 2323.
168. Raw, A. S., Jang, E. B. *Tetrahedron* 2000, *56*, 3285.
169. Cativiela, C., Mayoral, J. A., Avenoza, A., Peregrina, J. M., Lahoz, F. J., Gimeno, S. *J. Org. Chem.* 1992, *57*, 4664.
170. Meyers, A. I., Busacca, C. A. *Tetrahedron Lett.* 1989, *30*, 6977.
171. Rehnberg, N., Sundin, A., Magnusson, G. *J. Org. Chem.* 1990, *55*, 5477.
172. Lamy-Schelkens, H. G., L. *Tetrahedron Lett.* 1989, *30*, 5891.
173. Waldmann, H., Dräger, M. *Liebigs Ann. Chem.* 1990, 681.
174. Linz, G., Weetmen, J., Hady, A. F. A., Helmchan, G. *Tetrahedron Lett.* 1989, *30*, 5599.
175. Helmchen, G. H., A. F. A., Hartmann, H., Karge, R., Krotz, A., Sartor, K., Urmann, M. *Pure Appl. Chem.* 1989, *61*, 409.
176. Poll, T. H., A. F. A., Karge, R., Linz, G., Weetman, J., Helmchen, G. *Tetrahedron Lett.* 1989, *30*, 5595.
177. Masamune, S., Reed, L. A. I., Davis, J. T., Choy, W. *J. Org. Chem.* 1983, *48*, 4441.
178. Jung, M. E., Vaccaro, W. D., Buszek, K. R. *Tetrahedron Lett.* 1989, *30*, 1893.
179. Tomioka, K. H., N., Suenaga, T., Koga, K. *J. Chem. Soc., Perkin Transactions I* 1990, 426.
180. Avenoza, A. C., C., Mayoral, J. A., Peregrina, J. M., Sinou, D. *Tetrahedron: Asymmetry* 1990, *1*, 765.
181. Yang, T.-K., Chen, C.-J., Lee, D.-S., Jong, T.-T., Jiang, Y.-Z., Mi, A.-Q. *Tetrahedron: Asymmetry* 1996, *7*, 57.
182. Oppolzer, W., Chapuis, C., Dupuis, D., Guo, M. *Helv. Chim. Acta* 1985, *68*, 2100.
183. Evans, D. A., Chapman, K. T., Bisaha, J. *Tetrahedron Lett.* 1984, *25*, 4071.

184. Evans, D. A., Chapman, K. T., Hung, D. T., Hawaguchi, A. T. *Angew. Chem., Int. Ed.* 1987, *26*, 1184.
185. Chang, H., Zhou, L., McCargar, R. D., Mahmud, T., Hirst, I. *Org. Proc. Res. Dev.* 1999, *3*, 289.
186. Poll, T., Sobczak, A., Hartmann, A., Helmchen, G. *Tetrahedron Lett.* 1985, *26*, 3095.
187. Thiem, R., Rotscheidt, K., Breitmaier, E. *Synthesis* 1989, 836.
188. Konopelski, J. P., Boehler, M. A. *J. Am. Chem. Soc.* 1989, *111*, 4515.
189. Rieger, R., Btreitmaier, E. *Synthesis* 1990, 697.
190. Lyssikatos, J. P., Bednarski, M. D. *Synlett* 1990, 230.
191. McDougal, P. G., Jump, J. M., Rojas, C., Rico, J. G. *Tetrahedron Lett.* 1989, *30*, 3897.
192. Beagley, B., Larsen, D. S., Pritchard, R. G., Stoodley, R. J. *J. Chem. Soc., Perkin Trans. I* 1990, 3113.
193. Giuliano, R. M., Jordan, A. D. J., Gauthier, A. D., Hoogstein, K. *J. Org. Chem.* 1993, *58*, 4979.
194. Larsen, D. S., Stoodley, R. J. *J. Chem. Soc., Perkin Trans. I* 1989, 1841.
195. Gupta, R. C., Larsen, D. S., Stoodley, R. J., Slawin, A. M. Z., Williams, D. J. *J. Chem. Soc., Perkin Trans. I* 1989, 739.
196. Bird, C. W., Lewis, A. *Tetrahedron Lett.* 1989, *30*, 6227.
197. Bloch, R., Chaptal-Gradoz, N. *Tetrahedron Lett.* 1992, *33*, 6147.
198. Kozikowski, A. P., Nieduzak, T. R. *Tetrahedron Lett.* 1986, *27*, 819.
199. Defoin, A., Pires, J., Tissot, I., Tschamber, T., Bur, D., Zehnder, M., Streith, J. *Tetrahedron: Asymmetry* 1991, *2*, 1209.
200. Masamune, S., Reed, L. A. I., Choy, W. *J. Org. Chem.* 1983, *48*, 1137.
201. Gupta, R. C., Raynor, C. M., Stoodley, R. J., Slawin, A. M. Z., Williams, D. J. *J. Chem. Soc., Perkin Trans. I* 1988, 1773.
202. Larsen, D. S., Stoodley, R. J. *J. Chem. Soc., Perkin Trans. I* 1990, 1339.
203. Corey, E. J. *Pure Appl. Chem.* 1990, *62*, 1209.
204. Tolbert, L. M., Ali, M. B. *J. Am. Chem. Soc.* 1981, *103*, 2104.
205. Walborsky, H. M., Barash, L., Davis, T. C. *J. Org. Chem.* 1961, *26*, 4778.
206. Misumi, A., Iwanaga, K., Furuta, K., Yamamoto, H. *J. Am. Chem. Soc.* 1985, *107*, 3343.
207. Tolbert, L. M., Ali, M. B. *J. Am. Chem. Soc.* 1982, *104*, 1742.
208. Tolbert, L. M., Ali, M. B. *J. Am. Chem. Soc.* 1984, *106*, 3806.
209. Charlson, J. L., Chee, G., McColeman, H. *Can. J. Chem.* 1995, *73*, 1454.
210. Reichardt, C. *Solvents and Solvent Effects in Organic Chemistry*, VCH: Weinheim, 1988.
211. Sauer, J., Sustmann, R. *Angew. Chem., Int. Eng. Ed.* 1980, *19*, 779.
212. Jaquith, J. B., Guan, J., Wang, S., Collins, S. *Organometallics* 1995, *14*, 1079.
213. Johannsen, M., Joergensen, K. A. *Tetrahedron* 1996, *52*, 7321.
214. Choudhury, A., Franck, R. W., Gupta, R. B. *Tetrahedron Lett.* 1989, *30*, 4921.
215. Posner, G. H., Wettlaufer, D. G. *Tetrahedron Lett.* 1986, *27*, 667.
216. Arnold, T., Reissig, H.-U. *Synlett* 1990, 514.
217. Boger, D. L., Corbett, W. L., Curran, T. T., Kasper, A. M. *J. Am. Chem. Soc.* 1991, *113*, 1713.
218. Backvall, J.-E., Rise, F. *Tetrahedron Lett.* 1989, *30*, 5347.
219. Mattay, J., Kneer, G., Mertes, J. *Synlett* 1990, 145.
220. Posner, G. H., Wettlaufer, D. G. *J. Am. Chem. Soc.* 1986, *108*, 7373.
221. Greene, A. E., Charbonnier, F., Luche, M. J., Moyano, A. *J. Am. Chem. Soc.* 1987, *109*, 4752.
222. Boger, D. L., Mullican, M. D. *Tetrahedron Lett.* 1982, *23*, 4551.
223. Marko, I. E., Evans, G. R., Seres, P., Chelle, I., Janousek, Z. *Pure Appl. Chem.* 1996, *68*, 113.
224. Ihara, M., Katsumata, A., Egashira, M., Suzuki, S., Tokunaga, Y., Fukumoto, K. *J. Org. Chem.* 1995, *60*, 5560.
225. Fallis, A. G. *Can. J. Chem.* 1984, *62*, 183.
226. Takano, S. *Pure Appl. Chem.* 1987, *59*, 353.
227. Brieger, G., Bennett, J. N. *Chem. Rev.* 1980, *80*, 63.
228. Jung, M. E., Zimmerman, C. N., Lowen, G. T., Khan, S. I. *Tetrahedron Lett.* 1993, *34*, 4453.
229. Coe, J. W., Roush, W. R. *J. Org. Chem.* 1989, *54*, 915.
230. Roush, W. R., Kageyama, M., Riva, R., Brown, B. B., Warmus, J. S., Moriarty, K. J. *J. Org. Chem.* 1991, *56*, 1192.
231. Pyne, S. P., Spellmeyer, D. C., Chen, S., Fuchs, P. L. *J. Am. Chem. Soc.* 1982, *104*, 5728.
232. Pyne, S. G., Hensel, M. J., Fuchs, P. L. *J. Am. Chem. Soc.* 1982, *104*, 5719.
233. Gibbs, R. A., Bartels, K., Lee, R. W. K., Okamura, W. H. *J. Am. Chem. Soc.* 1989, *111*, 3717.

234. Davidson, A. H., Moloney, B. A. *J. Chem. Soc., Chem. Commun.* 1989, 445.
235. Tietze, L.-F., von Kiedrowski, G., Harms, K., Clegg, W., Sheldrick, G. *Angew. Chem., Int. Ed.* 1980, *19*, 134.
236. Bailey, M. S., Brisdon, B. J., Brown, D. W., Stark, K. M. *Tetrahedron Lett.* 1983, *24*, 3037.
237. Roush, W. R., Gillis, H. R., Ko, A. I. *J. Am. Chem. Soc.* 1982, *104*, 2269.
238. Levin, J. I. *Tetrahedron Lett.* 1989, *30*, 2355.
239. Sisko, J., Weinreb, S. M. *Tetrahedron Lett.* 1989, *30*, 3037.
240. Uyehara, T., Suzuki, I., Yamamoto, Y. *Tetrahedron Lett.* 1990, *31*, 3753.
241. Annunziata, R., Cinquini, M., Cozzi, F., Raimondi, L. *Tetrahedron Lett.* 1989, *30*, 5013.
242. Danishefsky, S. J. *Aldrichimica Acta* 1986, *19*, 59.
243. Kametani, T., Chu, S.-D., Honda, T. *Heterocycles* 1987, *25*, 241.
244. Petrzilka, M., Grayson, J. I. *Synthesis* 1981, 753.
245. Danishefsky, S., DeNinno, M. P., Phillips, G. B., Zelle, R. E., Lartey, P. A. *Tetrahedron* 1986, *42*, 2809.
246. Fraser-Reid, B., Rahman, M. A., Kelly, D. R., Srivastava, R. M. *J. Org. Chem.* 1984, *49*, 1835.
247. Wender, P. A. *J. Am. Chem. Soc.* 1987, *109*, 4390.
248. Danishefsky, S. J., Pearson, W. H., Harvey, D. F., Maring, C. J., Springer, J. P. *J. Am. Chem. Soc.* 1985, *107*, 1256.
249. Danishefsky, S. J., Myles, D. C., Harvey, D. F. *J. Am. Chem. Soc.* 1987, *109*, 862.
250. Danishefsky, S. J., Armistead, D. M., Wincott, F. E., Selnick, H. G., Hungate, R. *J. Am. Chem. Soc.* 1987, *109*, 8117.
251. Egbertson, M., Danishefsky, S. J. *J. Org. Chem.* 1989, *54*, 11.
252. Danishefsky, S. J., Selnick, H. G., Armistead, D. M., Wincott, F. E. *J. Am. Chem. Soc.* 1987, *109*, 8119.
253. Danishefsky, S. J., Armistead, D. M., Wincott, F. E., Selnick, H. G., Hungate, R. *J. Am. Chem. Soc.* 1989, *111*, 2967.
254. Tietze, L. F., Schnieder, C. *J. Org. Chem.* 1991, *56*, 2476.
255. Danishefsky, S. J., DeNinno, M. P. In *Trends in Synthetic Carbohydrate Chemistry,* Horton, D., Hawkins, L. D. Eds., American Chemical Society: Washington, D. C., 1989, p. 160.
256. Danishefsky, S. J., DeNinno, M. P., Audia, J. E., Schulte, G. In *Trends in Synthetic Carbohydrate Chemistry,* Horton, D., Hawkins, L. D. Eds., American Chemical Society: Washington, D. C., 1989, p. 176.
257. Danishefsky, S. J., DeNinno, M. P. *Angew. Chem., Int. Ed.* 1987, *26*, 15.
258. Golebiowski, A., Jurczak, J. *J. Chem. Soc., Chem. Commun.* 1989, 263.
259. Danishefsky, S., Larson, E. R., Askin, D. *J. Am. Chem. Soc.* 1982, *104*, 6457.
260. Belanger, J., Landry, N. L., Pare, J. R. J., Jankowski, K. *J. Org. Chem.* 1982, *47*, 3649.
261. Danishefsky, S., Bednarski, M., Izawa, T., Maring, C. *J. Org. Chem.* 1984, *49*, 2290.
262. Danishefsky, S., Kerwin, J. F. J., Kobayashi, S. *J. Am. Chem. Soc.* 1982, *104*, 358.
263. Larson, E. R., Danishefsky, S. *J. Am. Chem. Soc.* 1982, *104*, 6458.
264. Danishefsky, S. J., Pearson, W. H., Harvey, D. F. *J. Am. Chem. Soc.* 1984, *106*, 2456.
265. Maruoka, K., Yamamoto, H. *J. Am. Chem. Soc.* 1989, *111*, 789.
266. Maruoka, K., Itoh, T., Shirasaka, T., Yamamoto, H. *J. Am. Chem. Soc.* 1988, *110*, 310.
267. Wada, E., Yasuoka, H., Kanemasa, S. *Chem. Lett.* 1994, 1637.
268. Faller, J. W., Smart, C. J. *Tetrahedron Lett.* 1989, *30*, 1189.
269. Katagiri, N., Makino, M., Tamura, T., Takahiro, K., Kaneko, C. *Chem. Pharm. Bull.* 1996, *44*, 850.
270. Mikami, K., Kotera, O., Motoyama, Y., Sakaguchi, H. *Synlett* 1995, 975.
271. Inanaga, J., Sugimoto, Y., Hanamoto, T. *New J. Chem.* 1995, *19*, 707.
272. Johannsen, M., Joergensen, K. A. *J. Org. Chem.* 1995, *60*, 5757.
273. Bednarski, M., Danishefsky, S. *J. Am. Chem. Soc.* 1983, *105*, 6968.
274. Page, P. C. B., Prodger, J. C. *Synlett* 1991, 84.
275. Tararov, V. I., Kadyrov, R., Kadyrova, Z., Dubrovina, N., Borner, A. *Tetrahedron: Asymmetry.* 2002, *13*, 25.
276. Lau, J. F., Hansen, T. K., Kilburn, J. P., Frydenvang, K., Holsworth, D. D., Ge, Y., Uyeda, R. T., Judge, L. M., Andersen, H. S. *Tetrahedron* 2002, *58*, 7339.
277. Ekegren, J. K., Modin, S. A., Alonso, D. A., Andersson, P. G. *Tetrahedron: Asymmetry.* 2002, *13*, 447.
278. Caille, J.-C., Govindan, C. K., Junga, H., Lalonde, J., Yao, Y. *Org. Proc. Res. Dev.* 2002, *6*, 471.
279. Ziegler, F. E. *Acc. Chem. Res.* 1977, *10*, 227.

280. Ziegler, F. E. *Chem. Rev.* 1988, *88*, 1423.
281. Rhoads, S. J., Rawlins, N. R. *Org. React.* 1975, *22*, 1.
282. Bennett, G. B. *Synthesis* 1977, *589*.
283. Blechert, S. *Synthesis* 1989, 71.
284. Tarbell, D. S. *Org. React.* 1944, *2*, 1.
285. Tarbell, D. S. *Chem. Rev.* 1940, *27*, 495.
286. Takai, K., Mori, I., Oshima, K., Nozaki, H. *Tetrahedron Lett.* 1981, *22*, 3985.
287. Hansen, H.-J., Schmid, H. *Tetrahedron* 1974, *30*, 1959.
288. Ferrier, R. J., Vethavijasar, N. *J. Chem. Soc., Perkin Trans. I* 1973, 1791.
289. Mandai, T., Ueda, M., Hasegawa, S.-i., Kawada, M., Tsuji, J. *Tetrahedron Lett.* 1990, *31*, 4041.
290. Reetz, M. T., Gansäuer, A. *Tetrahedron* 1993, *49*, 6025.
291. Felix, D., Gschwend-Steen, D., Wick, A. E., Eschenmoser, A. *Helv. Chim. Acta* 1969, *52*, 1030.
292. Ito, S., Tsunoda, T. *Pure Appl. Chem.* 1990, *62*, 1405.
293. Nubbemeyer, U. *Synthesis* 1993, 1120.
294. Jones, G. B., Huber, R. S., Chau, S. *Tetrahedron* 1993, *49*, 369.
295. Mulzer, J., Scharp, M. *Synthesis* 1993, 615.
296. Bao, R., Valverde, S., Herradón, B. *Synlett* 1992, 217.
297. Ireland, R. E., Mueller, R. H. *J. Am. Chem. Soc.* 1972, *94*, 5897.
298. Ireland, R. E., Willard, A. K. *Tetrahedron Lett.* 1975, 3975.
299. Kurth, M., Soares, C. J. *Tetrahedron Lett.* 1987, *28*, 1031.
300. Funk, R. L., Abelman, M. M., Munger, J. D. *Tetrahedron* 1986, *42*, 2831.
301. Nubbemeyer, U., Öhrlein, R., Gonda, J., Ernst, B., Bellus, D. *Angew. Chem., Int. Ed.* 1991, *30*, 1465.
302. Beslin, P., Perrio, S. *Tetrahedron* 1993, *49*, 3131.
303. Dewar, M. J. S., Jie, C. *Acc. Chem. Res.* 1992, *25*, 537.
304. Hill, R. K., Gilman, N. W. *J. Chem. Soc., Chem. Commun.* 1967, 617.
305. Lutz, R. P. *Chem. Rev.* 1984, *84*, 205.
306. Overman, L. E., Jacobsen, E. J. *J. Am. Chem. Soc.* 1982, *104*, 7225.
307. Overman, L. E., Knoll, F. M. *J. Am. Chem. Soc.* 1980, *102*, 865.
308. Paquette, L. A. *Angew. Chem., Int. Ed.* 1990, *29*, 609.
309. Paquette, L. A., Maynard, G. D. *Angew. Chem., Int. Ed.* 1991, *30*, 1368.
310. Paquette, L. A., Oplinger, J. A. *Tetrahedron* 1989, *45*, 107.
311. Wender, P. A., Sieburth, S. M., Petraitis, J. J., Singh, S. K. *Tetrahedron* 1981, *37*, 3967.
312. Uma, R., Rajogopalan, K., Swaminathan, S. *Tetrahedron* 1986, *42*, 2757.
313. Mikami, K., Maeda, T., Kishi, N., Nakai, T. *Tetrahedron Lett.* 1984, *25*, 5151.
314. Koreeda, M., Luengo, J. I. *J. Am. Chem. Soc.* 1985, *107*, 5572.
315. Denmark, S. E., Harmata, M. A. *J. Am. Chem. Soc.* 1982, *104*, 4972.
316. Denmark, S. E., Harmata, M. A., White, K. S. *J. Am. Chem. Soc.* 1989, *111*, 8878.
317. Denmark, S. E., Rajendra, G., Marlin, J. E. *Tetrahedron Lett.* 1989, *30*, 2469.
318. Denmark, S. E., Marlin, J. E. *J. Org. Chem.* 1991, *56*, 1003.
319. Nonoshita, K., Banno, H., Maruoka, K., Yamamoto, H. *J. Am. Chem. Soc.* 1990, *112*, 316.
320. Maruoka, K., Banno, H., Nonoshita, K., Yamamoto, H. *Tetrahedron Lett.* 1989, *30*, 1265.
321. van der Baan, J. L., Bickelhaupt, F. *Tetrahedron Lett.* 1986, *27*, 6267.
322. Hill, R. K., Khatri, H. N. *Tetrahedron Lett.* 1978, 4337.
323. Maruoka, K., Banno, H., Yamamoto, H. *J. Am. Chem. Soc.* 1990, *112*, 7791.
324. Yamamoto, H., Maruoka, K. *Pure Appl. Chem.* 1990, *62*, 2063.
325. Masse, C. E., Panek, J. S. *Chemtracts: Org. Chem.* 1995, *8*, 242.
326. Maruoka, K., Saito, S., Yamamoto, H. *J. Am. Chem. Soc.* 1995, *117*, 1165.
327. Boeckman, R. K. J., Need, M. J., Gaul, M. D. *Tetrahedron Lett.* 1995, *36*, 803.
328. Sugiura, M., Yanagisawa, M., Nakai, T. *Synlett* 1995, 447.
320. Panek, J. S., Sparks, M. A. *J. Org. Chem.* 1990, *55*, 5564.
330. Panek, J. S., Clark, T. D. *J. Org. Chem.* 1992, *57*, 4323.
331. Sparks, M. A., Panek, J. S. *J. Org. Chem.* 1991, *56*, 3431.
332. Pereira, S., Srebnik, M. *Aldrichimica Acta* 1993, *26*, 17.
333. Ireland, R. E., Wilcox, C. S. *Tetrahedron Lett.* 1977, 2839.
334. Ireland, R. E., Wipf, P., Armstrong, J. D. *J. Org. Chem.* 1991, *56*, 650.

335. Narula, A. S. *Tetrahedron Lett.* 1981, *22*, 2017.
336. Sato, T., Tajima, K., Fujia, T. *Tetrahedron Lett.* 1983, *24*, 729.
337. Fujisawa, T., Kohama, H., Tajima, K., Sato, T. *Tetrahedron Lett.* 1984, *25*, 5155.
338. Kallmerten, J., Gould, T. J. *Tetrahedron Lett.* 1983, *24*, 5177.
339. Ager, D. J., Cookson, R. C. *Tetrahedron Lett.* 1982, *23*, 3419.
340. Tsunoda, T., Tatsuki, S., Shiraishi, Y., Akasaka, M., Itô, S. *Tetrahedron Lett.* 1993, *34*, 3297.
341. Tadano, K.-i., Minami, M., Ogawa, S. *J. Org. Chem.* 1990, *55*, 2108.
342. Burke, S. D., Pacofsky, G. J. *Tetrahedron Lett.* 1986, *27*, 445.
343. Bartlett, P. A., Tanzella, D. J., Barstow, J. F. *J. Org. Chem.* 1982, *47*, 3941.
344. Lythgoe, B., Milner, J. R., Tideswell, J. *Tetrahedron Lett.* 1975, 2593.
345. Richardson, S. K., Sabol, M. R., Watt, D. S. *Synth. Commun.* 1989, *19*, 359.
346. Bartlett, P. A., Holm, K. H., Morimoto, A. *J. Org. Chem.* 1985, *50*, 5179.
347. Bartlett, P. A., Barstow, J. F. *Tetrahedron Lett.* 1982, *23*, 623.
348. Bartlett, P. A., Barstow, J. F. *J. Org. Chem.* 1982, *47*, 3933.
349. Baumann, H., Duthaler, R. O. *Helv. Chim. Acta* 1988, *71*, 1025.
350. Dell, C. P., Khan, K. M., Knight, D. W. *J. Chem. Soc., Perkin Trans. I* 1994, 341.
351. Welch, J. T., Plummer, J. S., Chou, T.-S. *J. Org. Chem.* 1991, *56*, 353.
352. Danishefsky, S., Tsuzuki, K. *J. Am. Chem. Soc.* 1980, *102*, 6891.
353. Danishefsky, S., Funk, R. L., Kerwin, J. F. *J. Am. Chem. Soc.* 1980, *102*, 6889.
354. Brandage, S., Flodman, L., Norberg, A. *J. Org. Chem.* 1984, *49*, 927.
355. Ireland, R. E., Wuts, P. G. M., Ernst, B. *J. Am. Chem. Soc.* 1981, *103*, 3205.
356. Burke, S. D., Cobb, J. E. *Tetrahedron Lett.* 1986, *27*, 4237.
357. Ireland, R. E., Thaisrivongs, S., Vanier, N., Wilcox, C. S. *J. Org. Chem.* 1980, *45*, 48.
358. Kazmaier, U., Krebs, A. *Angew. Chem., Int. Ed.* 1995, *34*, 2012.
359. Kazmaier, U., Maier, S. *J. Chem. Soc., Chem. Commun.* 1995, 1991.
360. Snider, B. B. *Acc. Chem. Res.* 1980, *13*, 426.
361. Oppolzer, W. *Pure Appl. Chem.* 1981, *53*, 1181.
362. Oppolzer, W. *Angew. Chem., Int. Ed.* 1989, *28*, 38.
363. Mikami, K., Shimizu, M. *Chem. Rev.* 1992, *92*, 1021.
364. Thomas, B. E., Houk, K. N. *J. Am. Chem. Soc.* 1993, *115*, 790.
365. Hoffmann, H. M. R. *Angew. Chem., Int. Ed.* 1969, *8*, 556.
366. Lehmkuhl, H. *Bull. Soc. Chim., Fr.* 1981, 87.
367. Oppolzer, W., Jacobsen, E. J. *Tetrahedron Lett.* 1986, *27*, 1141.
368. Oppolzer, W., Robbiani, C., Battig, K. *Tetrahedron* 1984, *40*, 1391.
369. Oppolzer, W., Begley, T., Ashcroft, A. *Tetrahedron Lett.* 1984, *25*, 825.
370. Mikami, K., Shimizu, M., Nakai, T. *J. Org. Chem.* 1991, *56*, 2952.
371. Whitesell, J. K., Deyo, D., Bhattacharya, A. *J. Chem. Soc., Chem. Commun.* 1983, 802.
372. Whitesell, J. K., Bhattacharya, A., Aguilar, D. A., Henke, K. *J. Chem. Soc., Chem. Commun.* 1982, 989.
373. Terada, M., Sayo, N., Mikami, K. *Synlett* 1995, 411.
374. Shimizu, M., Mikami, K. *J. Org. Chem.* 1992, *57*, 6105.
375. Mikami, K., Loh, T.-P., Nakai, T. *Tetrahedron Lett.* 1988, *29*, 6305.
376. Houston, T. A., Tanaka, Y., Koreeda, M. *J. Org. Chem.* 1993, *58*, 4287.
377. Mikami, K., Terada, M., Nakai, T. *J. Am. Chem. Soc.* 1989, *111*, 1940.
378. Mikami, K., Terada, M., Nakai, T. *J. Am. Chem. Soc.* 1990, *112*, 3949.
379. Mikami, K., Yajima, T., Takasaki, T., Matsukawa, S., Terada, M., Uchimaru, T., Maruta, M. *Tetrahedron* 1996, *52*, 85.
380. Terada, M., Mikami, K. *J. Chem. Soc., Chem. Commun.* 1995, 2391.
381. Kitamoto, D., Imma, H., Nakai, T. *Tetrahedron Lett.* 1995, *36*, 1861.
382. Mikami, K., Motoyama, Y., Terada, M. *Inorg. Chim. Acta* 1994, *222*, 71.
383. Mikami, K., Matsukawa, S. *Tetrahedron Lett.* 1994, *35*, 3133.
384. Terada, M., Mikami, K. *J. Chem. Soc., Chem. Commun.* 1994, 833.
385. Mikami, K., Terada, M., Narisawa, S., Nakai, T. *Org. Synth.* 1993, *71*, 14.
386. Terada, M., Matsukawa, S., Mikami, K. *J. Chem. Soc., Chem. Commun.* 1993, 327.
387. Van der Meer, F. T., Feringa, B. L. *Tetrahedron Lett.* 1992, *33*, 6695.
388. Mikami, K. *Pure Appl. Chem.* 1996, *68*, 639.

389. Faller, J. W., Liu, X. *Tetrahedron Lett.* 1996, *37*, 3449.
390. Sarko, T. K., Gosh, S. H., Subba Rao, P. S. V., Satapathi, T. K. *Tetrahedron Lett.* 1990, *31*, 3461.
391. Oppolzer, W., Keller, T. H., Kuo, D. L., Pachinger, W. *Tetrahedron Lett.* 1990, *31*, 1265.
392. Oppolzer, W. *Pure Appl. Chem.* 1990, *62*, 1941.
393. Takacs, J. M., Anderson, L. G., Greswell, M. W., Takacs, B. E. *Tetrahedron Lett.* 1987, *28*, 5627.
394. Mikami, K., Takahashi, K., Nakai, T. *Tetrahedron Lett.* 1989, *30*, 357.
395. Mikami, K., Terada, M., Narisawa, S., Nakai, T. *Synlett* 1992, 255.
396. Tanino, K., Nakamura, T., Kuwajima, I. *Tetrahedron Lett.* 1990, *31*, 2165.
397. Snider, B. B., Cartaya-Mari, C. P. *J. Org. Chem.* 1984, *49*, 1688.
398. Maruoka, K., Hoshino, Y., Shirasaka, T., Yamamoto, H. *Tetrahedron Lett.* 1988, *29*, 3967.
399. Mikami, K., Narisawa, S., Shimizu, M., Terada, M. *J. Am. Chem. Soc.* 1992, *114*, 6566.
400. Kim, B. H., Lee, J. Y., Kim, K., Whang, D. *Tetrahedron: Asymmetry* 1991, *2*, 27.
401. Oppolzer, W., Pitteloud, R. *J. Am. Chem. Soc.* 1982, *104*, 6478.
402. Mikami, K., Kaneko, M., Loh, T.-P., Terada, M., Nakai, T. *Tetrahedron Lett.* 1990, *31*, 3909.
403. Benner, J. P., Gill, G. B., Parrott, S. J., Wallace, B., Belay, M. J. *J. Chem. Soc., Perkin Trans. I* 1984, 314.
404. Whitesell, J. K., Nabona, K., Deyo, D. *J. Org. Chem.* 1989, *54*, 2258.
405. Whitesell, J. K., Carpenter, J. F., Yaser, H. K., Machajewski, T. *J. Am. Chem. Soc.* 1990, *112*, 7653.
406. Whitesell, J. K., Bhattacharya, A., Buchanan, C. M., Chen, H. H., Deyo, D., James, D., Liu, C.-L., Minton, M. A. *Tetrahedron* 1986, *40*, 2993.
407. Snider, B. B., Rodini, D. J. *Tetrahedron Lett.* 1980, *21*, 1815.
408. Snider, B. B., Rodini, D. J., Cann, R. S. E., Sealfon, S. *J. Am. Chem. Soc.* 1979, *101*, 5283.
409. Snider, B. B., van Straten, J. W. *J. Org. Chem.* 1979, *44*, 3567.
410. Tietze, L. F., Wichmann, J. *Angew. Chem., Int. Ed.* 1992, *31*, 1079.
411. Duncia, J. V., Lansbury, P. T., Miller, T., Snider, B. B. *J. Am. Chem. Soc.* 1982, *104*, 1930.
412. Snider, B. B., Duncia, J. V. *J. Org. Chem.* 1981, *46*, 3223.
413. Snider, B. B., Duncia, J. V. *J. Am. Chem. Soc.* 1980, *102*, 5926.
414. Hiroi, K., Umemura, M. *Tetrahedron Lett.* 1992, *33*, 3343.
415. Narasaka, K., Yujiro, S., Yamada, J. *Isr. J. Chem.* 1991, *31*, 261.
416. Murase, N., Ooi, T., Yamamoto, H. *Synlett* 1991, 857.
417. Trost, B. M., King, S., Nanninga, T. N. *Chemistry Lett.* 1987, 15.
418. Trost, B. M., Mignani, S. M. *Tetrahedron Lett.* 1985, *26*, 6313.
419. Trost, B. M., Lynch, J., Renaut, P., Steinman, D. H. *J. Am. Chem. Soc.* 1986, *108*, 284.
420. Trost, B. M., Bonk, P. J. *J. Am. Chem. Soc.* 1985, *107*, 1778.
421. Trost, B. M., Miller, M. L. *J. Am. Chem. Soc.* 1988, *110*, 3687.
422. Trost, B. M., Seoane, P., Mignani, S., Acemoglu, M. *J. Org. Chem.* 1989, *54*, 7487.
423. Trost, B. M., Acemoglu, M. *Tetrahedron Lett.* 1989, *30*, 1495.
424. Molander, G. A., Andrews, S. W. *Tetrahedron Lett.* 1989, *30*, 2351.
425. Trost, B. M. *Angew. Chem., Int. Ed.* 1986, *25*, 1.
426. Horiguchi, Y., Suchiro, I., Sasaki, A., Kuwajima, I. *Tetrahedron Lett.* 1993, *34*, 6077.
427. Houk, K. N., Moses, R. S., Wu, Y.-D., Rondan, N. G., Jager, V., Schohe, R., Fronczek, F. R. *J. Am. Chem. Soc.* 1984, *106*, 3880.
428. Kozikowski, A. P., Ghosh, A. K. *J. Org. Chem.* 1984, *49*, 2762.
429. Vasella, A. *Helv. Chim. Acta* 1977, *60*, 426.
430. Kozikowski, A. P., Ghosh, A. K. *J. Am. Chem. Soc.* 1982, *104*, 5788.
431. Huber, R., Vasella, A. *Tetrahedron* 1990, *46*, 33.
432. Kanemasa, S., Tsuruoka, T., Wada, E. *Tetrahedron Lett.* 1993, *34*, 87.
433. Hoffmann, R. W. *Chem. Rev.* 1989, *89*, 1841.
434. Curran, D. P., Kim, B. H. *Synthesis* 1990, 312.
435. Curran, D. P. *J. Am. Chem. Soc.* 1983, *105*, 5826.
436. Annunziata, R., Cinquini, M., Cozzi, F., Restelli, A. *J. Chem. Soc., Chem. Commun.* 1984, 1253.
437. Kamimura, A., Marumo, S. *Tetrahedron Lett.* 1990, *31*, 5053.
438. Hoffmann, R. W. *Angew. Chem., Int. Ed.* 1979, *18*, 563.
439. Tomooka, K., Watanabe, M., Nakai, T. *Tetrahedron Lett.* 1990, *31*, 7353.
440. Evans, D. A., Andrews, G. C., Sims, C. L. *J. Am. Chem. Soc.* 1971, *93*, 4956.

441. Baldwin, J. E., DeBernardis, J., Patrick, J. E. *Tetrahedron Lett.* 1970, 353.
442. Baldwin, J. E., Patrick, J. E. *J. Am. Chem. Soc.* 1971, *93*, 3556.
443. Rautenstrauch, V. *J. Chem. Soc., Chem. Commun.* 1970, 4.
444. Felkin, H., Frajerman, C. *Tetrahedron Lett.* 1977, 3485.
445. Tsai, D. J.-S., Midland, M. M. *J. Am. Chem. Soc.* 1985, *107*, 3915.
446. Marshall, J. A., Jenson, T. M. *J. Org. Chem.* 1984, *49*, 1707.
447. Still, W. C., Mitra, A. *J. Am. Chem. Soc.* 1978, *100*, 1927.
448. Mikami, K., Kimura, Y., Kishi, N., Nakai, T. *J. Org. Chem.* 1983, *48*, 279.
449. Mikami, K., Fujimoto, K., Kasuga, T., Nakai, T. *Tetrahedron Lett.* 1984, *25*, 6011.
450. Mikami, K., Takahashi, O., Tabei, T., Nakai, T. *Tetrahedron Lett.* 1986, *27*, 4511.
451. Mikami, K., Maeda, T., Nakai, T. *Tetrahedron Lett.* 1986, *27*, 4189.
452. Rautenstrauch, V., Buchi, G., Wuest, H. *J. Am. Chem. Soc.* 1974, *96*, 2576.
453. Hoffmann, R., Brückner, R. *Angew. Chem., Int. Ed.* 1992, *31*, 647.
454. Antoniotti, P., Tonachini, G. *J. Org. Chem.* 1993, *58*, 3622.
455. Nakai, T., Mikami, K. *Org. React.* 1994, *46*, 105.
456. Balestra, M., Wittman, M. D., Kallmerten, J. *Tetrahedron Lett.* 1988, *29*, 6905.
457. Bruckner, R. *Tetrahedron Lett.* 1988, *29*, 5747.
458. Marshall, J. A., Robinson, E. D., Zapata, A. *J. Org. Chem.* 1989, *54*, 5854.
459. Marshall, J. A., Wang, X.-j. *J. Org. Chem.* 1990, *55*, 2995.
460. Mikami, K., Kawamoto, K., Nakai, T. *Tetrahedron Lett.* 1985, *26*, 5799.
461. Midland, M. M., Kwon, Y. C. *Tetrahedron Lett.* 1985, *26*, 5021.
462. Fujimoto, Y., Ohhana, M., Terasawa, T., Ikekawa, N. *Tetrahedron Lett.* 1985, *26*, 3239.
463. Evans, D. A., Andrews, G. C. *Acc. Chem. Res.* 1974, *7*, 147.
464. Burgess, K., Henderson, I. *Tetrahedron Lett.* 1989, *30*, 4325.
465. Sato, T., Otera, J., Nozaki, H. *J. Org. Chem.* 1989, *54*, 2779.
466. Goldmann, S., Hoffmann, R. W., Maak, N., Geueke, K. *J. Chem. Ber.* 1980, *113*, 831.
467. Grieco, P. A. *J. Chem. Soc., Chem. Commun.* 1972, 702.
468. Evans, D. A., Andrews, G. C. *J. Am. Chem. Soc.* 1972, *94*, 3672.
469. Babler, J. H., Haack, R. A. *J. Org. Chem.* 1982, *47*, 4801.
470. Hoffmann, R. W., Goldmann, S., Maak, N., Gerlach, R., Frickel, F., Steinbach, G. *Chem. Ber.* 1980, *113*, 819.
471. Ogiso, A., Kitazawa, E., Kurabayashi, M., Sato, A., Takahashi, S., Noguchi, H., Kuwano, H., Kobayashi, S., Mishima, H. *Chem. Pharm. Bull.* 1978, *26*, 3117.
472. Evans, D. A., Andrews, G. C., Fujimoto, T. T., Wells, D. *Tetrahedron Lett.* 1973, 1389.
473. Masaki, Y., Hashimoto, K., Sakuma, K., Kaji, K. *J. Chem. Soc., Chem. Commun.* 1979, 855.
474. Evans, D. A., Andrews, G. C., Fujimoto, T. T., Wells, D. *Tetrahedron Lett.* 1973, 1385.
475. Masaki, Y., Hashimoto, K., Kaji, K. *Tetrahedron Lett.* 1978, 4539.
476. Kojima, K., Koyama, K., Ameniya, S. *Tetrahedron* 1985, *41*, 4449.
477. Miller, J. G., Kurz, W., Untch, K. G., Stork, G. *J. Am. Chem. Soc.* 1974, *96*, 6774.
478. Bickart, P., Carson, F. W., Jacobus, J., Miller, E. G., Mislow, K. *J. Am. Chem. Soc.* 1968, *90*, 4869.
479. Baudin, J.-B., Julai, S. A. *Tetrahedron Lett.* 1989, *30*, 1963.
480. Leblanc, Y., Fitzsimmons, B. J. *Tetrahedron Lett.* 1989, *30*, 2889.
481. Inoue, Y. *Chem. Rev.* 1992, *92*, 741.
482. Kagan, H. B., Fiaud, J. C. *Top. Stereochem.* 1978, *10*, 175.
483. Kaupp, G., Haak, M. *Angew. Chem., Int. Ed.* 1993, *32*, 694.
484. Sayo, N., Kimara, Y., Nakai, T. *Tetrahedron Lett.* 1982, *23*, 3931.
485. Koch, H., Runsink, J., Scharf, H.-D. *Tetrahedron Lett.* 1983, *24*, 3217.
486. Zamojski, A., Jarosz, S. *Tetrahedron* 1982, *38*, 1447.
487. Zamojski, A., Jarosz, S. *Tetrahedron* 1982, *38*, 1453.
488. Engler, T. A., Ali, M. H., Velde, D. V. *Tetrahedron Lett.* 1989, *30*, 1761.
489. Marouka, K., Concepcion, A. B., Yamamoto, H. *Synlett* 1992, 31.
490. Müller, F., Mattay, J. *Chem. Rev.* 1993, *93*, 99.
491. Kobayashi, Y., Takemoto, Y., Ito, Y., Terashima, S. *Tetrahedron Lett.* 1990, *31*, 3031.
492. Wynberg, H., Staring, E. G. J. *J. Am. Chem. Soc.* 1982, *104*, 166.

27 Asymmetric Free-Radical Reductions Mediated by Chiral Stannanes, Germanes, and Silanes

Jens Beckmann, Dainis Dakternieks, and Carl H. Schiesser

CONTENTS

27.1 INTRODUCTION

Free-radical technology had to come a long way before it gained its current well-deserved recognition among synthetic chemists. Traditionally, free-radical chemistry has often been neglected for rational synthesis of fine chemicals because of the perception that radicals are uncontrollable and unselective. However, this negative view was revised with the adoption of tributyltin hydride in organic synthesis after it was recognized that reactions involving free radicals can indeed occur with a high degree of regioselectivity and diastereoselectivity. Nevertheless, the first examples of radical reactions that proceed with genuine enantiocontrol were reported only in the mid 1990s.[1–5]

The majority of applications that involve tributyltin hydrides are chain processes under reducing conditions, and from these, selective halogen abstractions from C–Hal bonds are by far the most prominent.[6-8] The general mechanism of such a reaction is outlined in Scheme 27.1. After the reaction has been initiated, the first step of the radical chain mechanism involves the abstraction of a halogen atom from the alkyl halide **A** by a triorganotin radical to give rise to a triorganotin halide and a planar (sp^2 configured) carbon radical **B**, which subsequently reacts with the triorganotin hydride producing the product **C** and a new triorganotin radical, which participates in a new cycle.

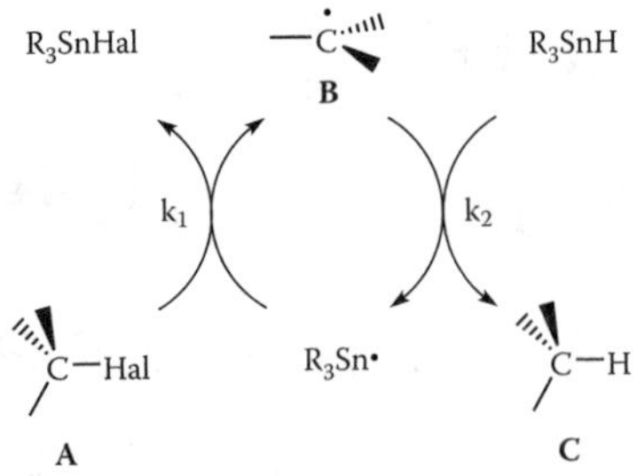

$$R_3SnHal \qquad -\overset{\bullet}{\underset{B}{C}}^{\text{\tiny\textbackslash\textbackslash\textbackslash}} \qquad R_3SnH$$

$$k_1 \qquad\qquad k_2$$

$$\underset{A}{\overset{\text{\tiny\textbackslash\textbackslash}}{C}-Hal} \qquad R_3Sn^{\bullet} \qquad \underset{C}{\overset{\text{\tiny\textbackslash\textbackslash}}{C}-H}$$

SCHEME 27.1

If the first reaction step in Scheme 27.1 affords a prochiral radical **B**, then, in principle, the second step will produce a chiral product **C** (provided all three substituents of **B** are different and none are hydrogen) and the stereochemical outcome may be controlled by the stereochemistry of the organotin hydride. Optically inactive organotin hydrides (e.g., Bu₃SnH) will attack the enantiotopic faces of the planar radical with equal probability, and the product will be racemic. However, when the organotin hydride is chiral, the transition states involved in the second reaction step are diastereotopic, and consequently the attack by the organotin hydride at the *Re*- and *Si*-faces of the planar radical will involve different activation energies (Figure 27.1). If the difference of the activation energies is sufficiently high and appropriate reaction conditions are applied (e.g., low temperatures), effective discrimination of the enantiotopic faces of the planar radical by the chiral organotin hydride is possible and one enantiomer of **C** will be preferred over the other.

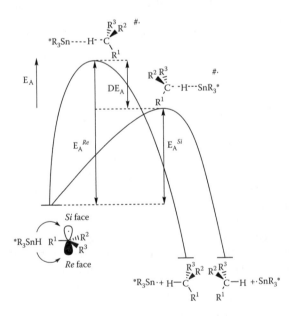

FIGURE 27.1 Diastereotopic transition states involved in the reaction of a prochiral carbon-centered radical with a chiral organotin hydride.

27.2 STOICHIOMETRIC FREE-RADICAL REDUCTIONS

As a result of the limited configurational stability of optically active organotin compounds, in which the chirality is on the tin atom, most advances in enantioselective free-radical reductions involve organostannanes where the elements of chirality are contained in the organic substituents. Selected

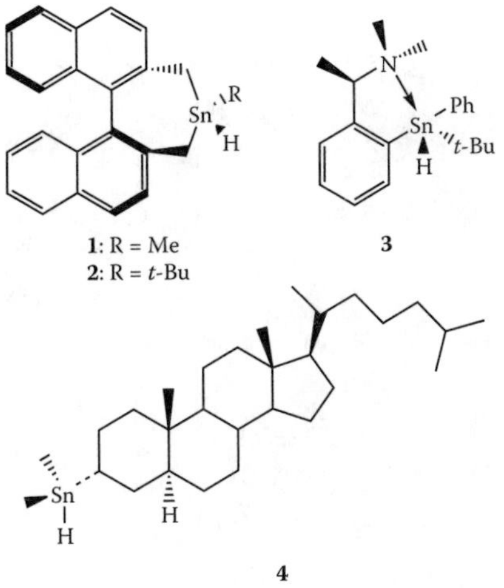

FIGURE 27.2 Examples of C-centered optically active organotin hydrides.

early examples based on chiral 1,1′-binaphthyl-,[9,10] 2-[(1-dimethylaminoalkyl)phenyl]-,[11,12] and cholestannyl groups[13] are shown in Figure 27.2.

Feasibility studies of **1–4** in free-radical reductions involving α-bromo ketones and esters demonstrated enantioselectivities of up to 51% ee. Although these results supply proof of concept, the aim to provide a widely applicable synthetic reagent required significant improvement of ee values and access to less costly chiral-reducing agents. To address this problem, chiral organostannanes derived from menthol were developed.[14–24] Menthol, a naturally occurring terpene alcohol, provides a convenient source of chirality, where both enantiomers are inexpensive and readily available on scale (see Chapter 5). Up to three menthyl groups can be attached to a tin atom using the configurationally stable Grignard reagents derived from the corresponding menthyl chloride enantiomers.[25] The configurations of the respective menthyl groups are very similar to those of the parent alcohols, (−)- and (+)-menthol. This technology enables the cost-effective production of a variety of tailor-made menthyltin hydrides; a selection explored by Chirogen is depicted in Figure 27.3.

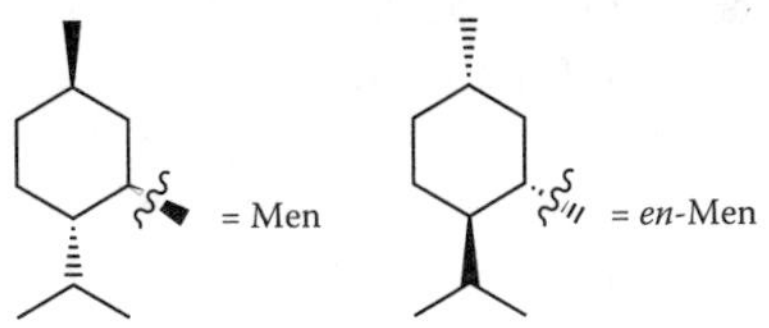

MenPh$_2$SnH	Men$_2$PhSnH	Men$_3$SnH
5	**6**	**7**
en-MenPh$_2$SnH	*en*-Men$_2$PhSnH	*en*-Men$_3$SnH
5′	**6′**	**7′**

FIGURE 27.3 Examples of menthyltin hydrides evaluated by Chirogen for enantioselective free-radical reductions.

For α-bromo ketones and related compounds, enhancement of the enantioselectivities has been achieved by the addition of simple Lewis acids (e.g., BF$_3$, Cp$_2$TiCl$_2$, or magnesium salts, which presumably form Lewis acid–Lewis base complexes with the carbonyl functions of the substrates).[26,27] The stereochemical effect of some Lewis acid additives on the reduction of a racemic α-bromo ester, namely ethyl 2-bromo-2-phenylpropanoate, is summarized in Scheme 27.2.

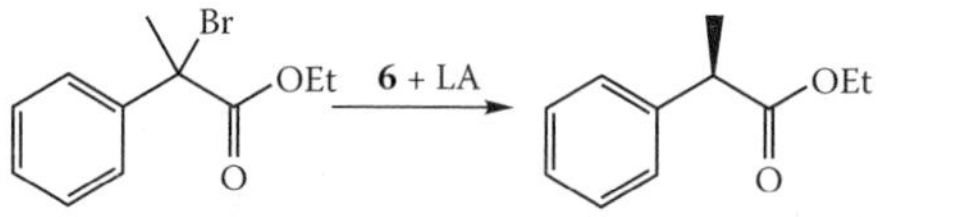

LA = none: 2 % ee
LA = BF$_3$:   32 % ee
LA = Cp$_2$ZrCl$_2$:   36 % ee
LA = MgBr$_2$:   92 % ee
LA = MgI$_2$:   96% ee
LA = Mg(OTf)$_2$:   98% ee

the (S)-enantiomer dominates in these examples

SCHEME 27.2

27.2.1 Scope and Limitations

The versatility of organotin hydrides lies in their ability to reduce with a high degree of selectivity a variety of functional groups, including halides, sulfides, selenides, thionocarbonates, and dithio-carbonates; nitro groups; sulfoxide, sulfate, and sulfonyl moieties, and others.[6–8,28] Although the direct reduction of alcohols is not possible, appropriate derivatization to thiocarbonates or dithio-carbonates make them susceptible to radical transformations. These reactions are commonly referred to as Barton-McCombie reductions.[6–8,28] One of the most striking attributes of triorganotin hydrides, regardless of whether they are chiral, is their complementary nature to alternative reducing agents, such as NaBH$_4$, performing under polar reaction conditions. For instance, triorganotin hydrides selectively reduce halogen atoms in the presence of carbonyl functions or alcohol groups.[6–8,28] Triorganotin hydrides are generally applicable for sensitive molecules possessing a variety of functional groups that may decompose or epimerize under ionic reaction conditions. The use of triorganotin hydrides is particularly attractive where polar reactions suffer from steric hindrance that requires forcing reaction conditions. In these cases, free-radical intermediates usually possess a greater accessibility to encumbered sites because they are neutral and only weakly solvated. Besides application in radical reductions, triorganotin hydrides are also useful for initiating radical processes that involve selective additions to double or triple bonds (hydrostannylations) and proceed with the subsequent rearrangements of the radical intermediates. A variety of elegant tandem and cascade reactions have been reported that have been initiated by triorganotin hydrides.[28] Limitation to the applicability of triorganotin hydrides are encountered whenever functional groups susceptible to radical processes, such as nitro groups, may compete with the intended reaction or whenever radical inhibitors are present.

27.2.2 Examples Relevant to the Fine Chemical Industry

Prime targets for free-radical reductions are precursors possessing activated C–Hal bonds in α-position to carbonyl functions. These cover a vast range of compound classes, such as α-halogenated ketones, esters, amino acids, α-hydroxy acids, etc. In general, these compounds are accessible by halogenation of the appropriate parent compound using standard reagents, such as bromine or N-bromosuccinimide (NBS).[29] Thus, a two-step sequence involving the halogenation of a chiral (racemic) carbon atom situated in the α-position to a carbonyl function followed by an enantio-selective free-radical reduction produces exclusively the desired enantiomer. This reaction sequence effectively provides a chemical deracemization alternative to classic resolution techniques. An illustrative example is the deracemization of methyl N-TFA phenylglycinate (Scheme 27.3) (see Chapter 25).[29]

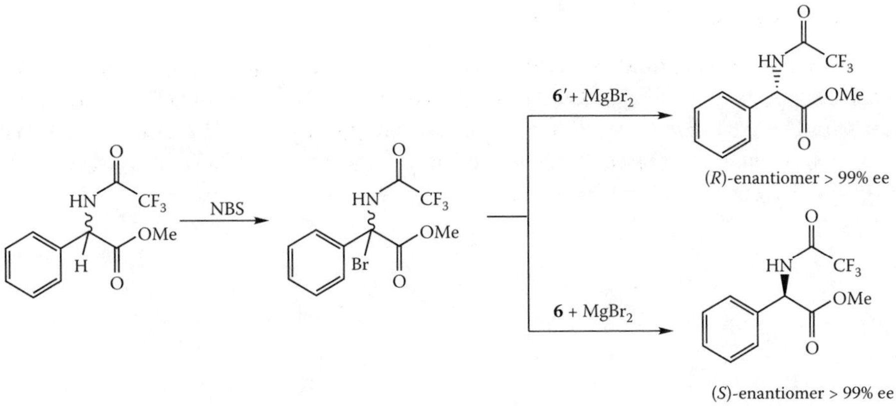

SCHEME 27.3

Therefore, the methyl *N*-TFA phenylglycinate is brominated using NBS and subsequently reduced with either Men$_2$PhSnH (**6**) or *ent*-Men$_2$PhSnH (**6′**) in the presence of MgBr$_2$ to produce the same material, methyl *N*-TFA phenylglycinate, as either the *R*- or *S*-enantiomer with ee's greater than 99%.[29] The stereochemistry of the product can be easily selected by choosing the parity of the reducing agent (e.g., **6** or **6′**). Compared to classic resolution technologies where the maximum yield of the desired enantiomer isolated from racemic mixture cannot exceed 50%, this chemical deracemization could provide yields close to 100%. Furthermore, this chemical deracemization can switch between the antipodes allowing the conversion of the (*S*)-enantiomer into the (*R*)-form and vice versa. In a similar way, both enantiomers of benzyl *N*-TFA *tert*-leucinate as well as ethyl esters of naproxen and ibuprofen, two nonsteroidal antiinflammatory drugs, have been deracemized with enantioselectivities in excess of 96% ee (Figure 27.4).[29]

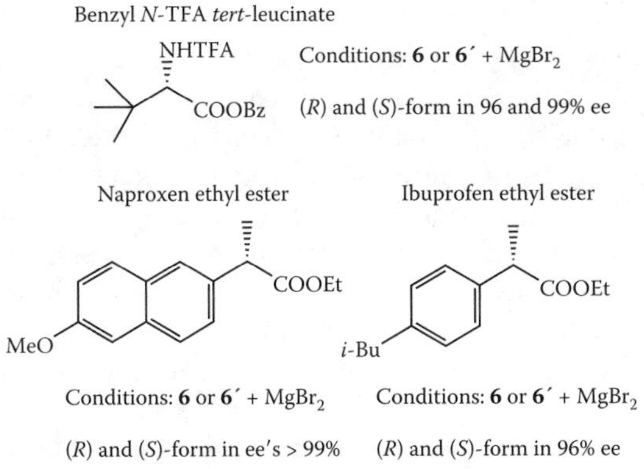

FIGURE 27.4 Deracemization of protected forms of *tert*-leucine, naproxen, and ibuprofen.

27.3 STRATEGIES FOR THE AVOIDANCE OF TIN WASTE

Despite the unique synthetic possibilities offered by organotin reagents, their use in the chemical industry has steadily declined in recent years as a result of perceived toxicity and environmental concerns associated with disposal of tin wastes.[30,31] Consequently, appreciable efforts have been made to develop heavy metal-free alternatives based on silicon, germanium, or phosphorus chemistry. Other strategies have been applied to immobilize the tin reagents or to reduce their usage to less than stoichiometric amounts.

27.3.1 IMMOBILIZATION OF TIN REAGENTS

The immobilization of nonchiral organotin hydrides has been addressed in a number of works. Various solid supports (e.g., silica, alumina, and polystyrene) and different ways to covalently bind organotin hydrides have been used.[32] Chirogen has developed immobilized menthyltin hydrides attached to silica and polystyrene supports, as shown in Figure 27.5. Both systems have been developed to the proof-of-concept stage.

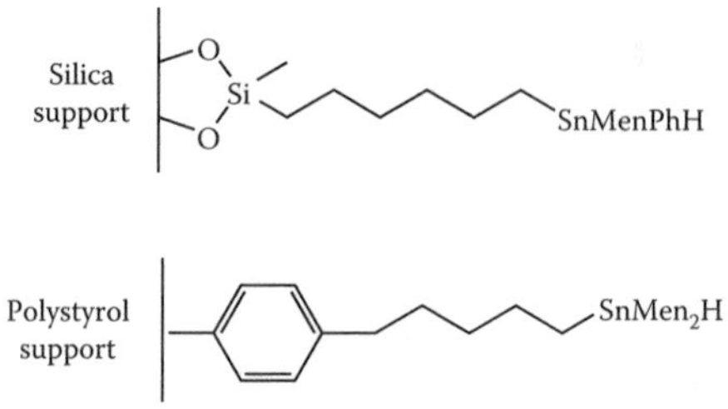

FIGURE 27.5 Immobilized menthyltin hydrides at silica and polystyrol supports.

Although these systems are feasible for enantioselective free-radical reductions, the enantioselectivities achieved are generally lower than those obtained in homogeneous reactions. As in the case for most immobilized reagents, the loading of active tin sites on the surface of the support materials is rather low, especially for silica-supported systems. One associated problem is the relatively low concentration of radical donors, which reduces the radical delivery rate to levels that render the propagation of the radical chain difficult. Other drawbacks involve temperature control issues as a result of the heterogeneous nature of the reactions as well as leaching of the organotin moieties from the carrier support.

27.3.2 REDUCTIONS CATALYTIC IN TIN

An intriguing solution to reduction of tin waste levels is the use of organostannanes (e.g., Bu_3SnH) in catalytic amounts in the presence of a coreductant, such as $(MeHSiO)_n$ or $PhSiH_3$, which regenerates the organostannane but does not reduce the substrate directly.[33,34] Achiral systems were optimized to use as little as 5% of the stoichiometrically required amount of organostannane. Chirogen has developed a chiral version of this catalytic cycle for the use of its menthyltin hydrides (e.g., compounds **5–7**) using borane as a coreductant. One advantage of this technology is that borane can cleave one phenyl group in phenyl-substituted organotin compounds to afford organotin hydrides.[35] Thus, the use of an excess of borane with a small amount of an appropriate triorganophenyltin compound (e.g., Men_3SnPh) provides an effective catalytic system for (enantioselective) free-radical reductions. It is important to note that borane does not undergo free-radical reductions with C–Hal bonds of the substrate molecules. The basis of this technology is illustrated in Scheme 27.4.

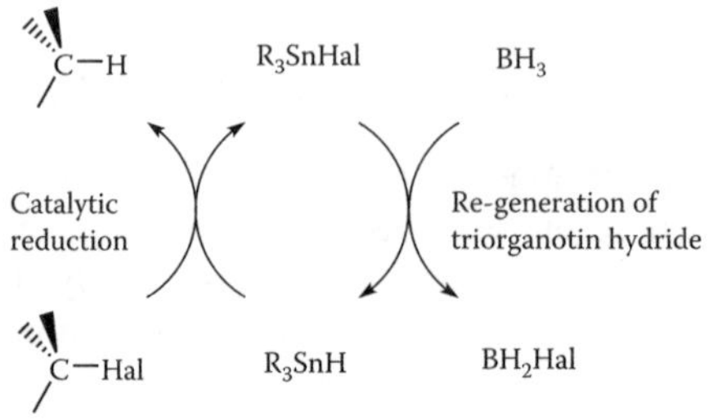

SCHEME 27.4

Large amounts of borane can be prepared conveniently by the reaction of $NaBH_4$ with dimethyl sulfate. Excess of borane and associated by-products may be hydrolyzed after the reduction to yield environmentally compatible boronic acid, $B(OH)_3$.

27.3.3 Reducing Agents Based on Germanium and Silicon

Triorganotin hydrides are good reagents for free-radical reductions because of the lability of the Sn–H bond and because of their ability to undergo homolytic bond cleavages for the delivery of hydrogen radicals at acceptable rates. Compared to this, the M–H bonds in organosilanes and organogermanes, R_3MH (M = Si, Ge), are kinetically more stable and the hydrogen donor ability is much lower when R represents a simple alkyl or aryl group. Whereas organogermanes, R_3GeH (R = alkyl, aryl), may be considered as borderline cases, the hydrogen delivery rate toward common organic radicals is generally too slow with organosilanes, R_3SiH (R = alkyl, aryl).[36,37] However, the Si–H bond can be effectively activated by appropriate substituents, such as triorganosilyl- and organothio groups. Thus, tris(trimethylsilyl)silane, $(Me_3Si)_3SiH$, and tris(methylthio)silane, $(MeS)_3SiH$, undergo free-radical reductions and are currently the most successful metal-free alternative to replace tributyltin hydride.[38,39] Organosilicon compounds are generally considered to be environmentally friendly because their ultimate degradation product is silica. Chirogen has developed chiral organogermanes and organosilanes for the purpose of enantioselective free-radical reductions; examples are depicted in Figure 27.6.

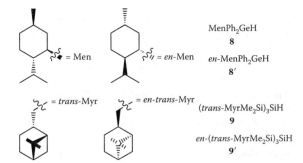

FIGURE 27.6 Heavy metal–free chiral-reducing agents based on Ge and Si chemistry.

The chiral organogermanes, $MenPh_2GeH$ (**8**) and en-$MenPh_2GeH$ (**8′**) are again based on the availability of (+)- and (–)-menthol, whereas the chiral tetrasilanes, ($trans$-$MyrMe_2Si)_3SiH$ (**9**) and en-($trans$-$MyrMe_2Si)_3SiH$ (**9′**) are derivatives of (+)- and (–)-β-pinene (see Chapter 5). The reagents **8** and **9** have been used successfully in free-radical reductions of α-brominated methyl N-TFA phenylglycinate and benzyl N-TFA $tert$-leucinate, respectively, producing enantioselectivies of 99% ee in both cases. Besides environmental reasons, chiral organogermanes and organosilanes offer another advantage—namely, they produce excellent enantioselectivities even at comparatively high temperatures up to –20°C. This fundamental difference to organostannanes, which generally provide best results at –78°C, is attributed to the closer proximity of the organosilane with the radical substrate at the diastereotropic transition state of the hydrogen transfer, which allows a more effective transfer of chiral information.[40]

27.4 SUMMARY

Free-radical reductions mediated by chiral stannanes, germanes, and silanes may occur with enantio-selectivities in excess of 99% ee. Owing to the involvement of radical intermediates and the mild reaction conditions, this process is applicable for a large variety of simple or even complex target molecules that are incompatible to asymmetric reductions that require ionic reaction conditions.

REFERENCES

1. Curran, D. P., Porter, N. A., Giese, B. *Stereochemistry or Radical Reactions. Concepts, Guidelines, and Synthetic Applications,* VCH: Weinheim, 1996.
2. Sibi, M. P., Porter, N. A. *Acc. Chem. Res.* 1999, *32*, 163.
3. Renaud, P., Sibi, M., Eds. *Radicals in Organic Synthesis,* Wiley-VCH: Weinheim, 2001.
4. Bar, G., Parsons, A. F. *Chem. Soc. Rev.* 2003, *32*, 251.
5. Sibi, M. P., Manyem, S., Zimmerman, J. *Chem. Rev.* 2003, *103*, 3263.
6. Neumann, W. P. *Synthesis* 1987, 665.
7. Curran, C. P. *Synthesis* 1988, 417.
8. Curran, C. P. *Synthesis* 1988, 489.
9. Nanni, D., Curran, D. P. *Tetrahedron: Asymmetry* 1996, *7*, 2417.
10. Blumenstein, M., Schwarzkopf, K., Metzger, J. O. *Angew. Chem., Int. Ed.* 1997, *36*, 235.
11. Schwarzkopf, K., Metzger, J. O., Saak, W., Pohl, S. *Chem. Ber.* 1997, *130*, 1539.
12. Schwarzkopf, K., Blumenstein, M., Hayen, A., Metzger, J. O. *Eur. J. Org. Chem.* 1998, 177.
13. Skidmore, M. A., Schiesser, C. H. *Phosphorous, Sulfur, Silicon* 1999, *150-151*, 177.
14. Schumann, H., Wassermann, B. C. *J. Organomet. Chem.* 1989, *365*, C1.
15. Schumann, H., Wassermann, B. C., Hahn, F. E. *Organometallics* 1992, *11*, 2803.
16. Schumann, H., Wassermann, B. C., Pickardt, J. *Organometallics* 1993, *12*, 3051.
17. Podesta, J. C., Radivoy, G. E. *Organometallics* 1994, *13*, 3364.
18. Podesta, J. C., Chopa, A. B., Radivoy, G. E., Vitale, C. A. *J. Organomet. Chem.* 1995, *494*, 11.
19. Lucas, C., Santini, C. C., Prinz, M., Cordonnier, M.-A., Basset, J.-M., Connil, M.-F., Jousseaume, B. *J. Organomet. Chem.* 1996, *520*, 101.
20. Vitale, C. A., Podesta, J. C. *J. Chem. Soc., Perkin Trans. I* 1996, 2407.
21. de Mallmann, A., Lot, O., Perrier, N., Lefebvre, F., Santini, C., Basset, J. M. *Organometallics* 1998, *17*, 1031.
22. Dakternieks, D., Dunn, K., Henry, D. J., Schiesser, C. H., Tiekink, E. R. T. *Organometallics* 1999, *18*, 3342.
23. Dakternieks, D., Dunn, K., Henry, D. J., Schiesser, C. H., Tiekink, E. R. T. *J. Chem. Soc., Dalton Trans.* 2000, 3693.
24. Dakternieks, D., Dunn, K., Henry, D. J., Schiesser, C. H., Tiekink, E. R. T. *J. Organomet. Chem.* 2000, *605*, 209.
25. Hoffmann, R. W. *Chem. Soc. Rev.* 2003, *32*, 225.
26. Renaud, P., Gerster, M. *Angew. Chem., Int. Ed.* 1998, *37*, 2562.
27. Dakternieks, D., Dunn, K., Perchyonok, T., Schiesser, C. H. *J. Chem. Soc., Chem. Commun.* 1999, 1665.
28. Carland, M. W., Schiesser, C. H. In *The Chemistry of Organic Germanium, Tin and Lead Compounds,* Vol. 2, Rappoport, Z., Ed., Wiley & Sons: Chichester, 2002, 1401.
29. Dakternieks, D., Perchyonok, T., Schiesser, C. H. *Tetrahedron: Asymmetry* 2003, *14*, 3057.
30. Baguley, P. A., Walton, J. C. *Angew. Chem., Int. Ed.* 1998, *37*, 3072.
31. Studer, A., Amrein, S. *Synthesis* 2002, 835.
32. Jurkschat, K., Mehring, M. In *The Chemistry of Organic Germanium, Tin and Lead Compounds,* Vol. 2, Rappoport, Z., Ed. Wiley & Sons: Chichester, 2002, 1543.
33. Lopez, R. M., Hays, D. S., Fu, G. C. *J. Am. Chem. Soc.* 1997, *119*, 6949.
34. Hays, D. S., Fu, G. C. *J. Org. Chem.* 1998, *63*, 2796.
35. Faraoni, M. B., Koll, L. C., Mandolesi, S. D. Zuniga, A. E., Podesta, J. C. *J. Organomet. Chem.* 2000, *613*, 236.
36. Chatgilialoglu, C., Ballestri, M. *Organometallics* 1995, *14*, 5017.
37. Chatgilialoglu, C., Timokhin, V. I., Ballestri, M. *J. Org. Chem.* 1998, *63*, 1327.
38. Chatgilialoglu, C. *Chem. Rev.* 1995, *95*, 1229.
39. Chatgilialoglu, C., Schiesser, C. H. In *The Chemistry of Organic Silicon Compounds,* Vol. 3, Rappaport, Z. Apeloig, Y., Eds. Wiley & Sons: New York, 2001, 341.
40. Dakternieks, D., Henry, D. J., Schiesser, C. H. *J. Chem. Soc., Perkin Trans 2,* 1998, 591.

28 Metathesis Reactions

David R. Allen

CONTENTS

28.1 INTRODUCTION

Olefin metathesis, the rearrangement of carbon–carbon bonds in the presence of a metal carbene, has become an integral tool for the synthetic organic chemist. The simplistic nature in the execution of this chemistry lends itself well to use on a large scale. The reaction offers alternatives to other alkene synthesis, such as the Wittig reaction with the large amounts of by-products produced, and can tolerate a wide variety of functionality. With the emergence of several well-defined, highly active catalysts, the deployment of this type of chemistry on an industrial scale will continue to expand over the coming years.

Although the metathesis reaction does not involve stereogenic centers, chiral catalysts are available that can differentiate between enantiomers of substrates (*vide infra*). Although the reaction has found widespread usage in the polymer industry, this chapter will be confined to fine chemical applications.

28.2 REACTION TYPES

In general, there are three modes of olefin metathesis: ring closing metathesis (RCM),[1-4] ring opening metathesis polymerization (ROMP),[5,6] and cross metathesis (CM).[7-9] Although all three have industrial applications, the main use of olefin metathesis for fine chemicals production lies in the modes of CM and RCM (Scheme 28.1).

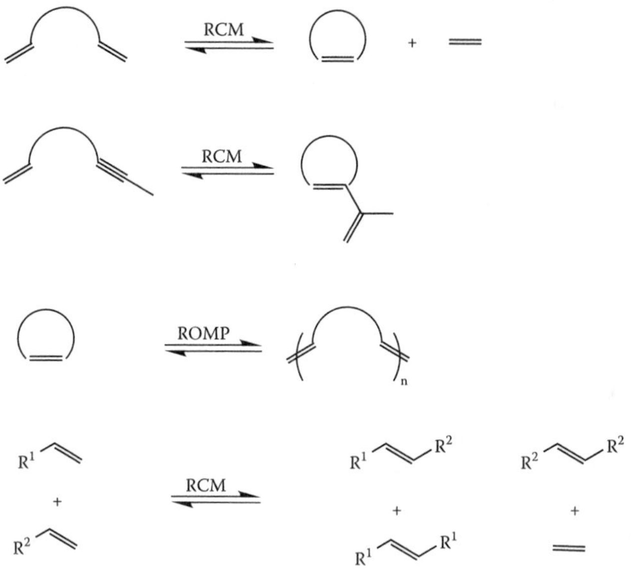

SCHEME 28.1

The metal carbene catalysts used in these reaction fall into four broad categories: titanium,[10-12] tungsten,[13-15] molybdenum,[16-18] and ruthenium.[19-21] Early researchers used the titanium- and tungsten-based catalyst, but these systems lacked the functional group tolerance and stability required for broad academic use. The development of the molybdenum- and ruthenium-based catalysts provided an acceptable blend of good functional group tolerance, chemical stability, and reactivity to allow for widespread use in organic synthesis (Figure 28.1).[22-28]

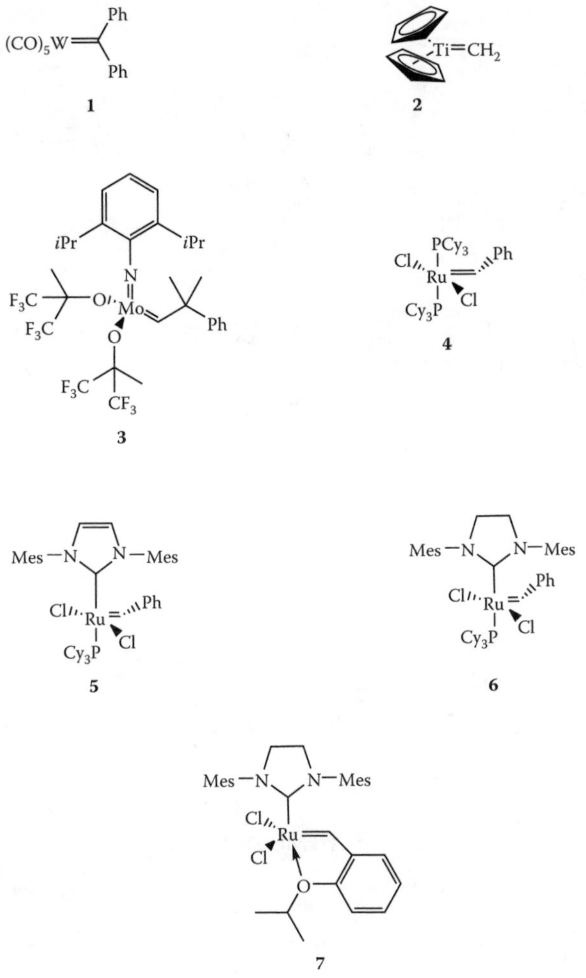

FIGURE 28.1 Structures of some metathesis catalysts.

28.3 MECHANISM

The mechanism for these metathesis catalysts appears to be the same regardless of mode of metathesis or metal center. The first step in the mechanism is the loss of a phosphine ligand to generate a 14 electron species **8**.[29] This 4-coordinate intermediate is then trapped by an olefin followed by formation of a metallocyclobutane **9** or **10**. This species collapses to either give product or unchanged starting material (Scheme 28.2).[30,31] In some cases, the regiochemistry can be controlled by the substrate itself or by the steric requirements of the ligands. In many cases, however, formation of **9** and **10** compete.

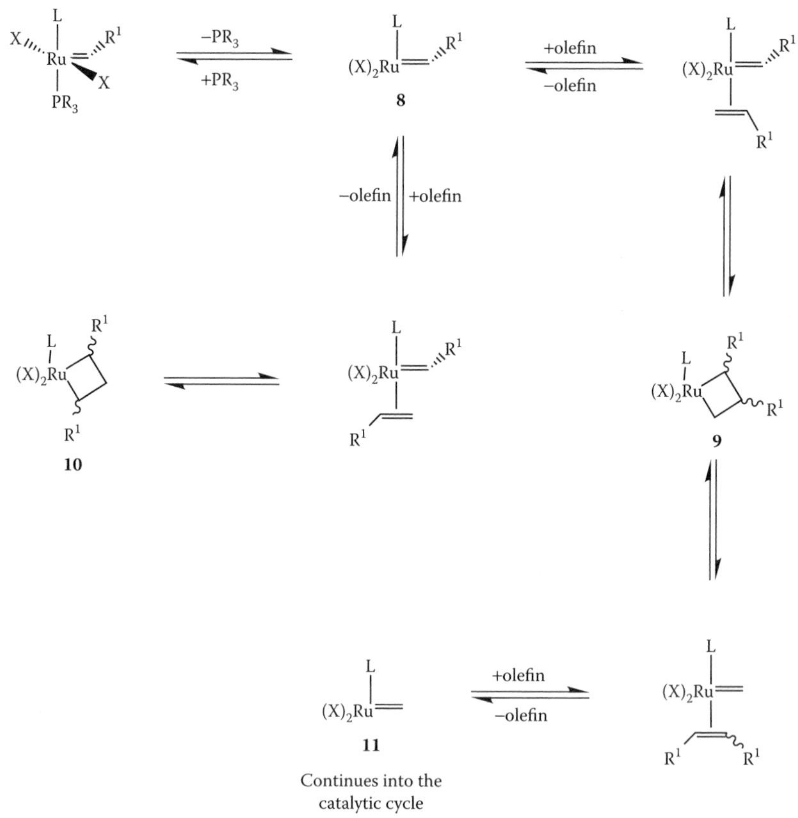

SCHEME 28.2

It is worthy to note here that the methylidene complex **11** is a poor initiator for olefin metathesis reactions at room temperature. Although this complex can undergo multiple catalytic turnovers, if it is intercepted by free phosphine ligand, it becomes incapable of reentering the metathesis catalytic cycle.[32]

28.4 RING CLOSING METATHESIS

Because the metathesis reactions all involve the same mechanism, this and the following sections provide representative examples of the reaction types as they have been used in the fine chemical industry.

Merck has used a diastereoselective double RCM in the synthesis of NK-1 receptor antagonists (Scheme 28.3).[33] Previously reported studies by the same group showed that amino acid–derived tetraene **12** smoothly underwent ring closure in a diastereoselective fashion to give the corresponding spirocyclic compounds **13** in excellent diastereoselectivity.[34] An extension of this work to the tetraene system derived from phenylglycine gave the desired spirocycle in 86% yield and 70% ds.

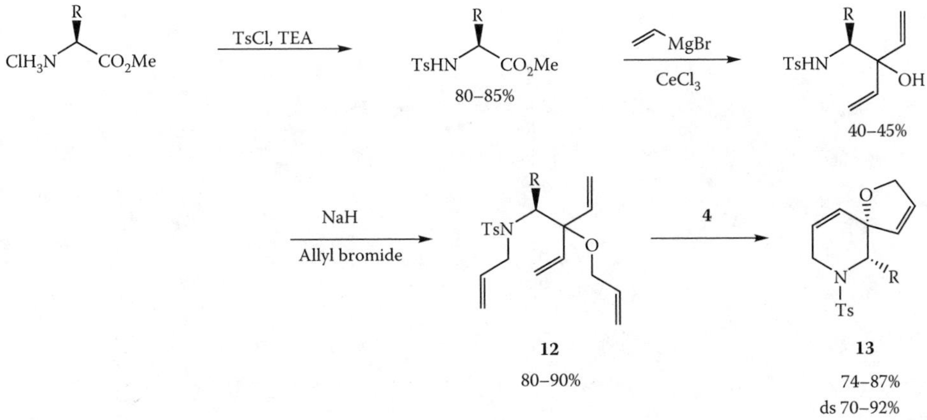

SCHEME 28.3

The predominant reaction pathway appears to be the formation of the dihydrofuran ring. This is followed by the piperidine ring closure. The evidence for this is the detection of only one of the two possible piperidine diene diastereoisomers along with both dihydrofuran diene diastereoisomers when the reaction was prematurely quenched (Scheme 28.4).[33]

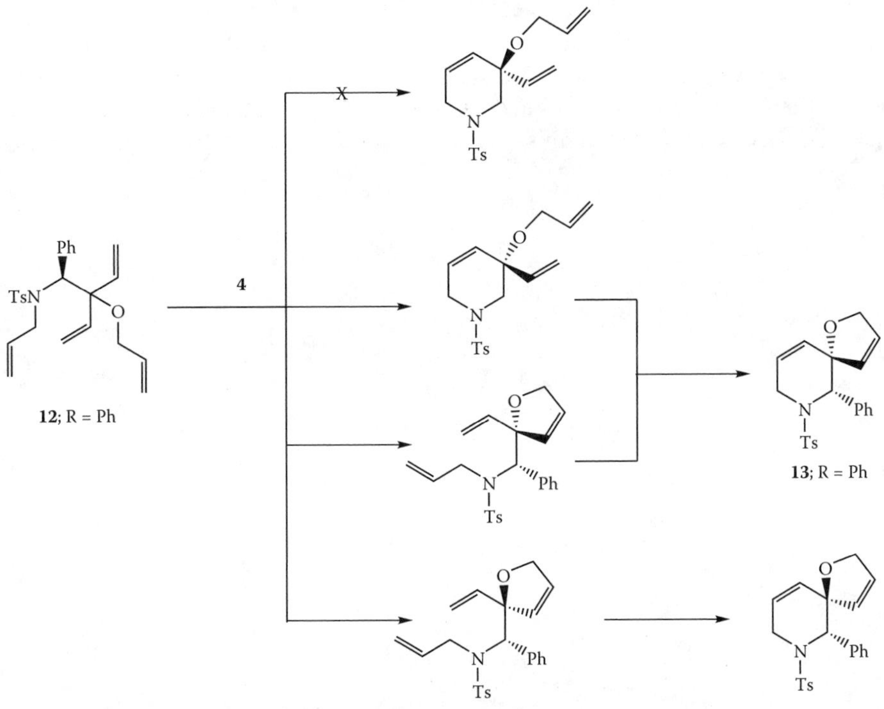

SCHEME 28.4

With the spirocycle in hand, a regioselective and stereoselective Heck reaction was used to install the substituted aromatic portion of the molecule (Scheme 28.5). It is interesting that the addition of water to the reaction provided the high selectivity. The remaining olefin was reduced over Pearlman's catalyst, and the protecting groups were removed to provide the desired NK-1 antagonist **14** in 7 overall steps.[33]

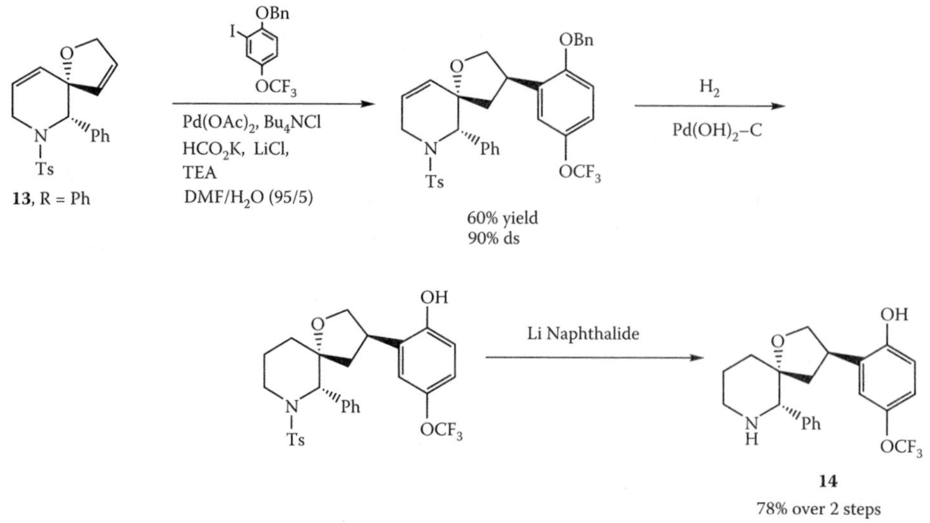

SCHEME 28.5

Other groups have used RCM to make much larger ring systems. Researchers at GSK have made human Cathespin K inhibitors **15** and **16** using RCM to synthesize a key azepanone pharmacophore **17** (Scheme 28.6).[35,36] In this case a 7-membered ring was formed by the metathesis reaction.

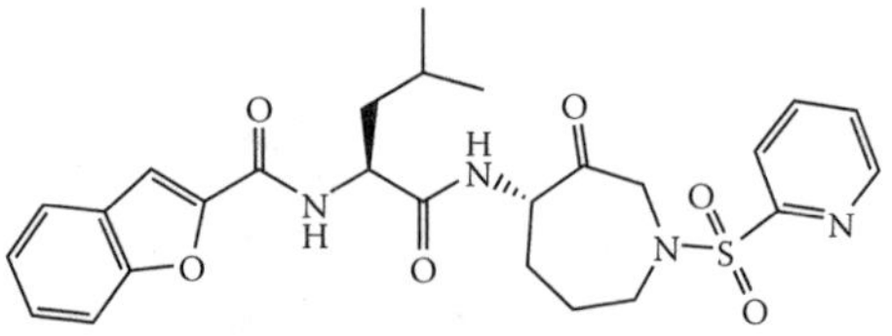

15

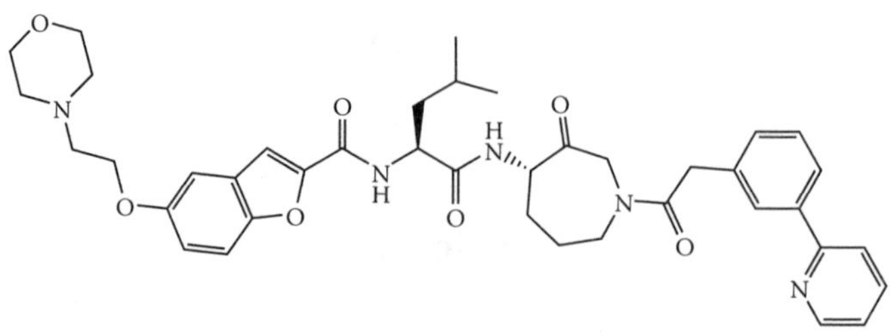

16

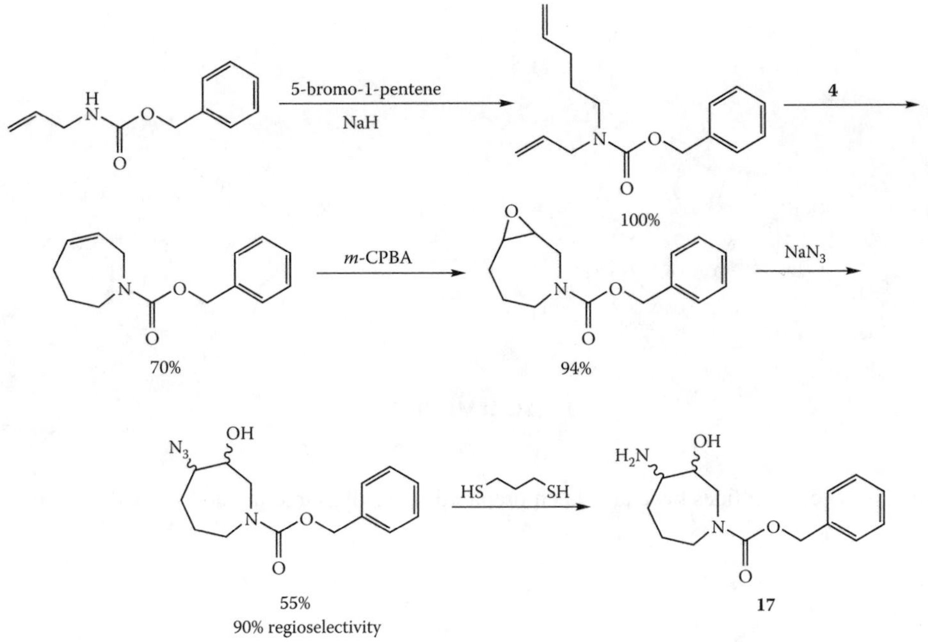

SCHEME 28.6

Another group at Boehringer Ingelheim has used RCM to prepare macrocyclic peptides active against the hepatitis C virus.[37–39] The metathesis reactions were used to form 15-membered rings and the *trans*-alkene (Schemes 28.7 and 28.8). The length of the sidechains determined the position of the unsaturation in the macrocycle.

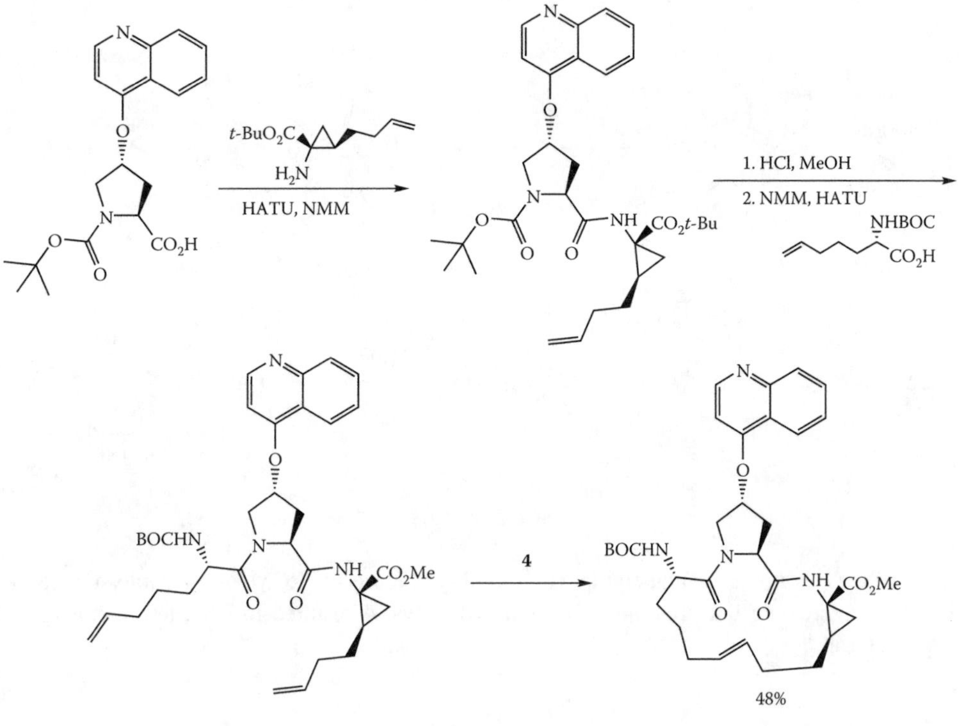

SCHEME 28.7

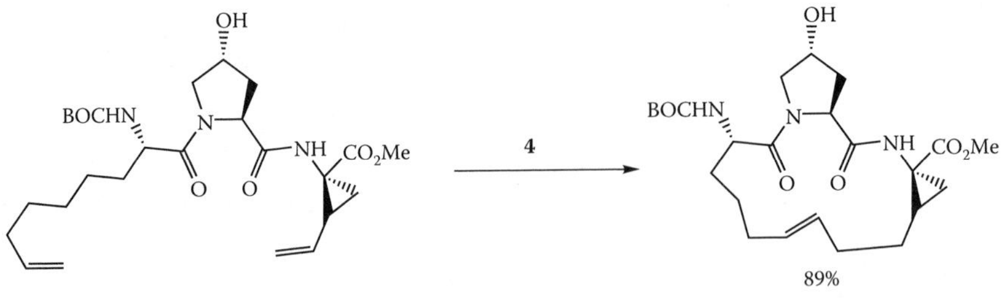

SCHEME 28.8

Smaller cyclic peptides have also been prepared using the same methodology (Scheme 28.9).[40]

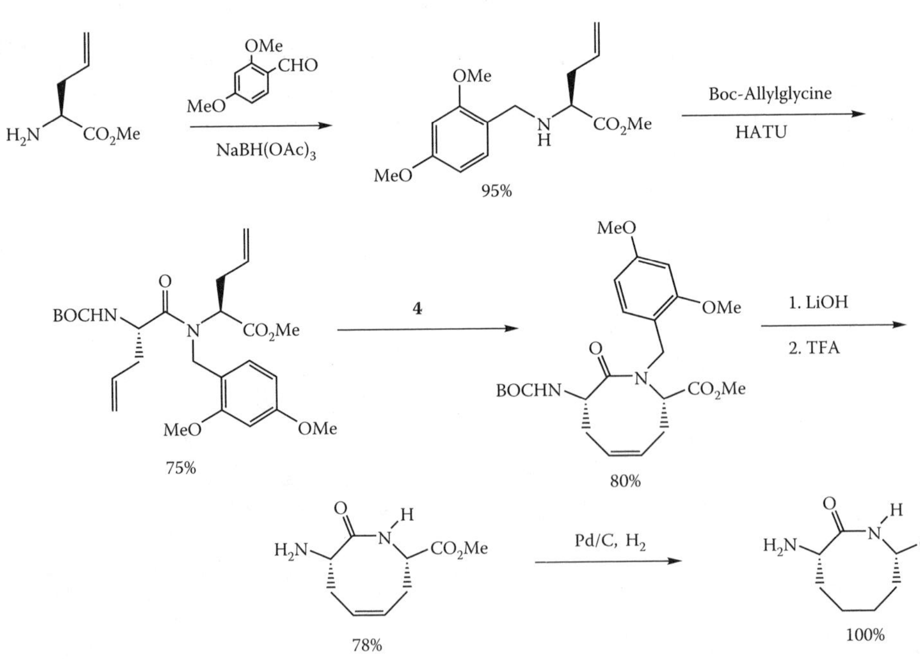

SCHEME 28.9

In a related example, pentapeptide **18** smoothly undergoes RCM using catalyst **5** (Scheme 28.10). It is interesting that treatment of **18** with catalyst **4** resulted in complete recovery of the starting material.[41]

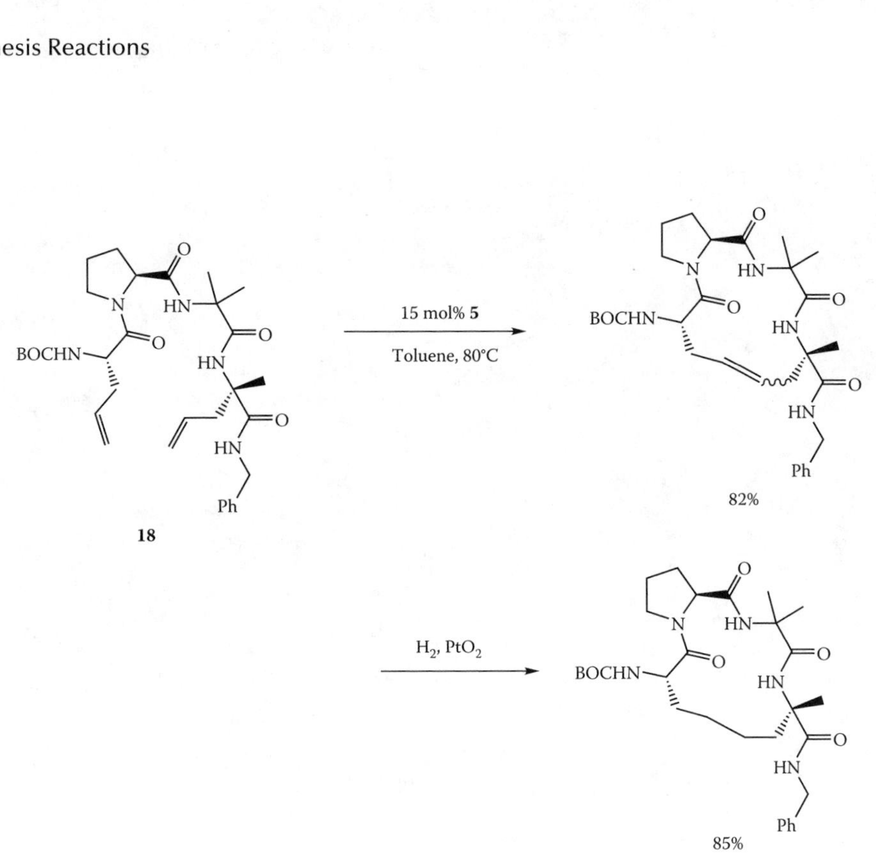

SCHEME 28.10

The required α,α-disubstituted amino acid for **18** was prepared by an enzymatic resolution of the corresponding racemic amino amide (Scheme 28.11) (see also Chapters 2 and 19).[42]

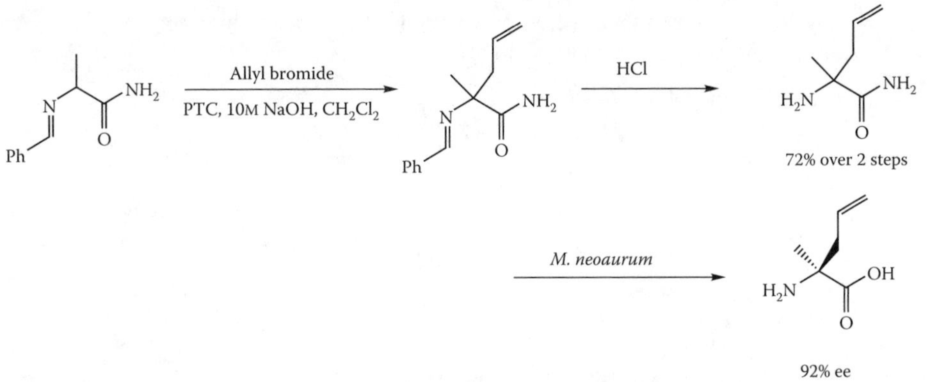

SCHEME 28.11

Substituted pipecolic acid derivatives can be accessed from a suitably protected allylglycine derivative by first use of a palladium-catalyzed *N,O*-acetal formation followed by RCM.[43] Treatment of **19** with boron trifluoride etherate followed by a variety of nucleophiles formed the corresponding substituted products **20** and **21** (Scheme 28.12).

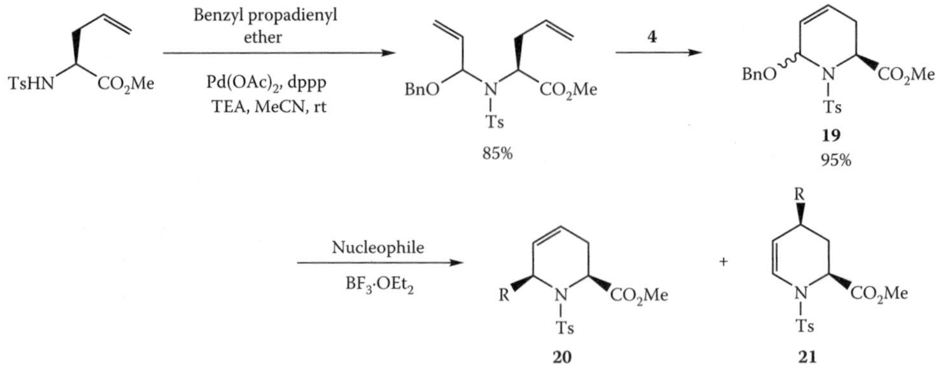

R	20:21	Yield
Allyl	90:10	62
CN	100:0	70

SCHEME 28.12

RCM of enamides **22** using catalyst **4** or **5** produces 5- or 6-membered rings (Scheme 28.13).[44]

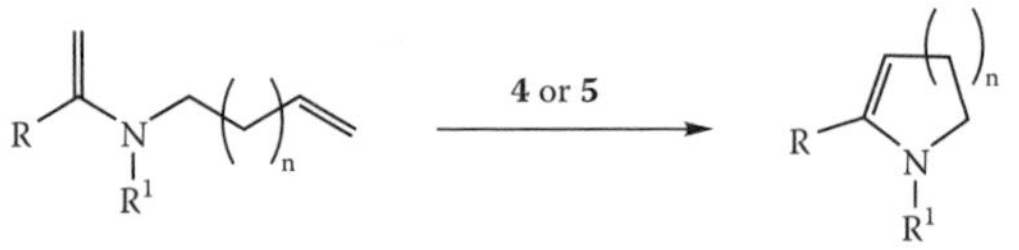

R = H or Me
R^1 = Ts, Bz, CO_2Et
n = **1–3**

SCHEME 28.13

It is worthy to note that attempted RCM to produce a 7-membered ring was unsuccessful. Instead, isomerization of the terminal olefin to an internal olefin followed by RCM gave the substituted piperidine (Scheme 28.14).[44]

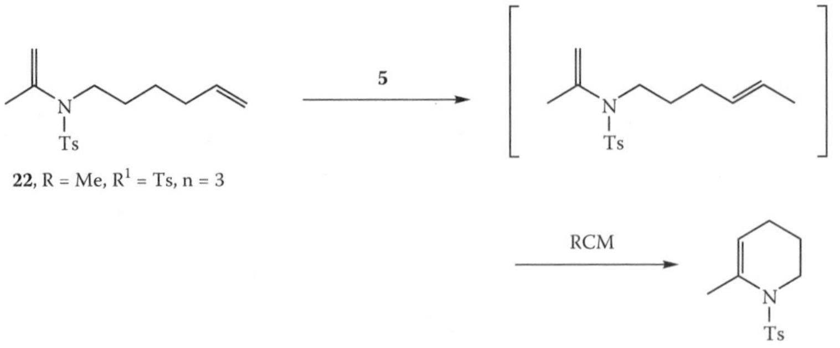

22, R = Me, R^1 = Ts, n = 3

SCHEME 28.14

An intersecting RCM is the alkyne variant called ring closing alkyne metathesis (RCAM). In this case, two alkynes, usually methyl terminated, are converted to the cycloalkyne. Unlike RCM, RCAM does not produce geometric isomers. Using this methodology, several constrained diaminosuberic analogues can be produced (Scheme 28.15).[45]

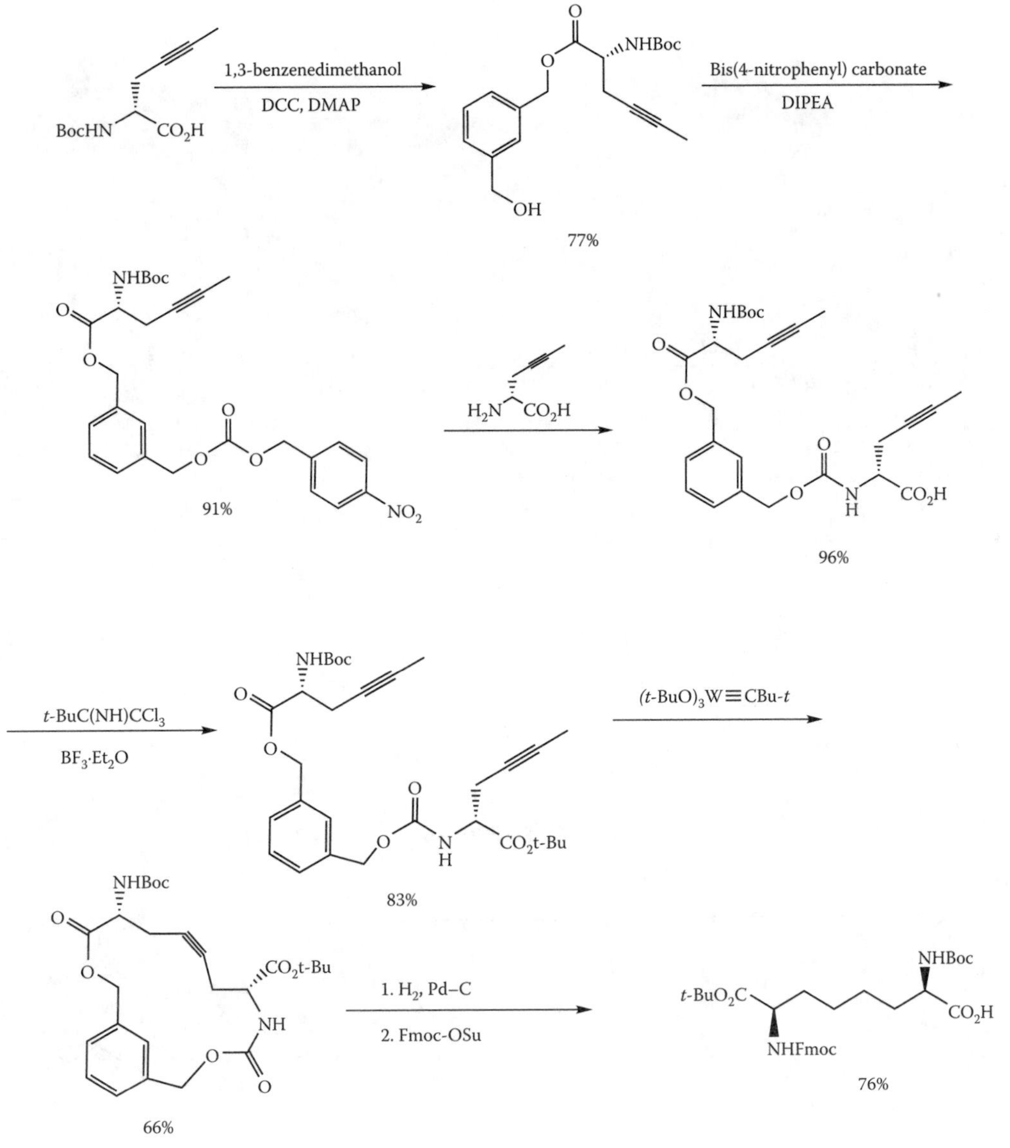

SCHEME 28.15

The required chiral hexynoic amino acid for the sequence in Scheme 28.15 was prepared by an enzymatic resolution approach (Scheme 28.16).[45]

28.5 CROSS METATHESIS

Olefin metathesis can be very useful in the CM mode, as shown in the synthesis of insect pheremones.[46] In the first example for the synthesis of the peach twig borer pheromone **23**, an excess of 1-hexene was used to increase the yield of the desired product. However, both of the other products could be recycled (Scheme 28.17). In the second example, CM was used to change the ester groups of meadowfoam oil (**24**) through cleavage of the alkenes rather than ester bonds. The sequence resulted in the synthesis of the mosquito pheromone (**25**) (Scheme 28.18).

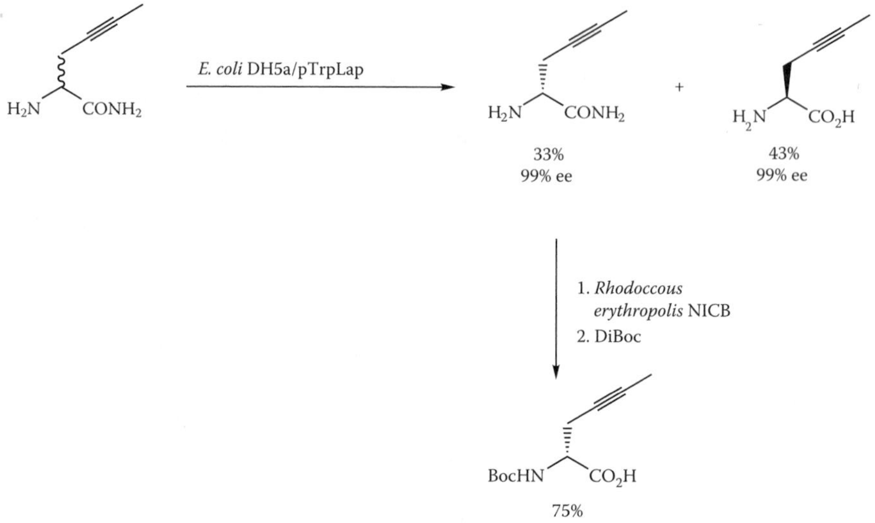

SCHEME 28.16

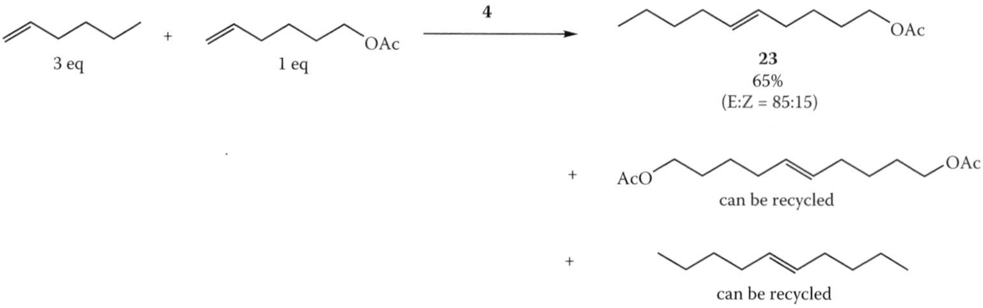

SCHEME 28.17

Other pheromones prepared using CM include 11-tetradecenyl acetate (Omnivorous Leafroller), 8,10-dodecadienol (Codling Moth), 9-tetradecenyl formate (Diamondback Moth),[47] 9,11-hexadecadienal (Pecan Nut Casebearer),[48] and 4-tridecenyl acetate (Tomato Pinworm).[49]

Alternatively, CM can be used in tandem with other catalytic methods to rapidly produce complex molecules. In the example shown in Scheme 28.19 for the synthesis of (–)-ketorolac (**26**), CM produces the starting material for the next catalytic reaction.[46,50]

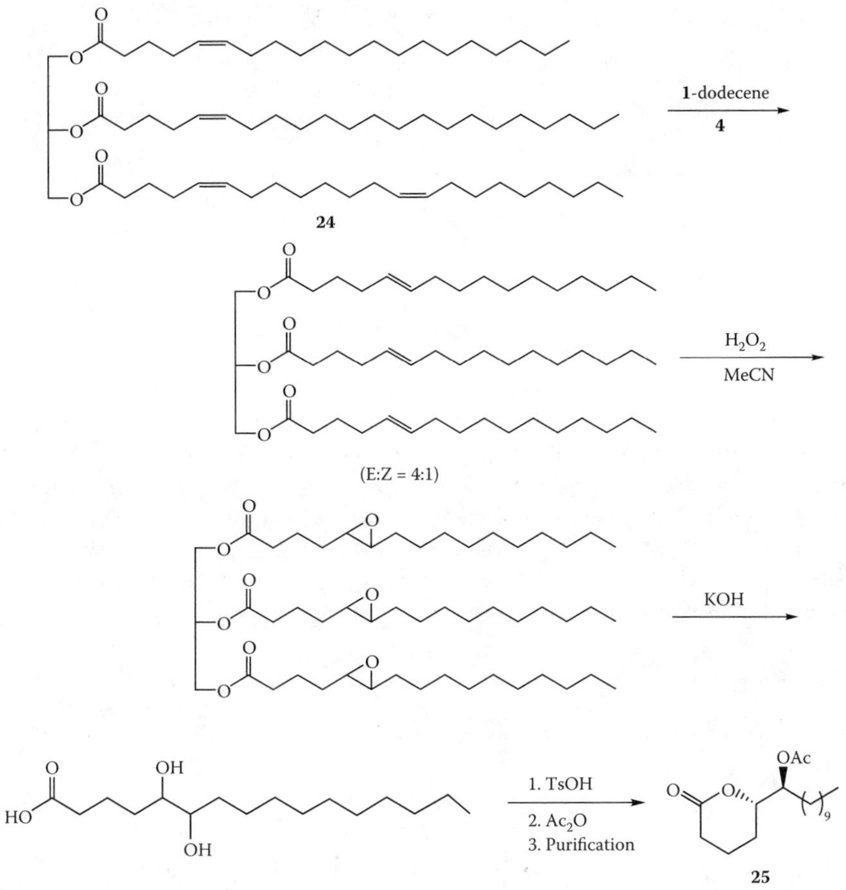

SCHEME 28.18

28.6 CHIRAL METATHESIS

Enantiomerically pure olefin metathesis catalysts can be used to promote reactions whereby simple achiral substrates are transformed into more complex chiral molecules. Much like their achiral counterparts, chiral olefin metathesis catalysts can be used in three distinct fashions. Chiral metathesis reactions have been reviewed.[51–53]

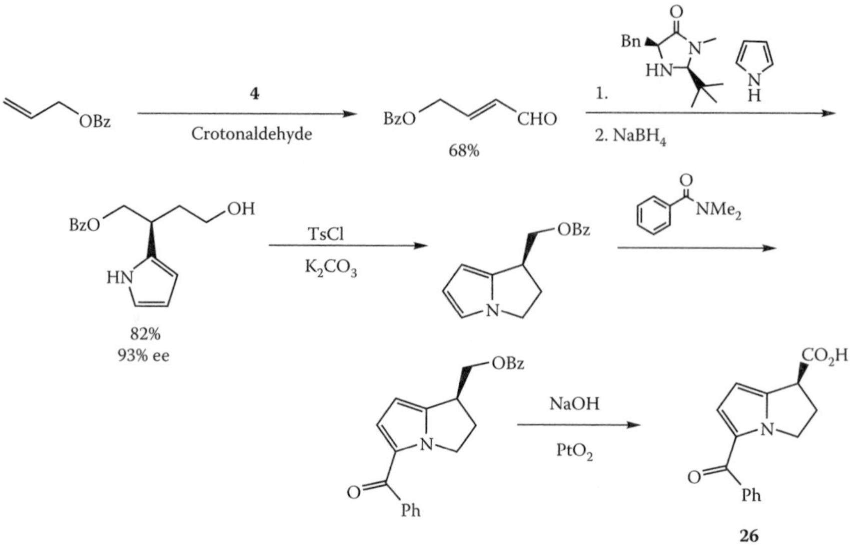

SCHEME 28.19

A chiral catalyst can differentiate between the two enantiomers of a substrate to provide nonracemic products.[54] The original catalyst **27** has been modified to the BINOL derivative **29** because this has advantages for the formation of 6-membered rings.[55] The THF (tetrahydrofuran) plays a key role and enhances enantioselectivity.[56] The catalyst **30** is useful for asymmetric ring closing metathesis (ARCM) and cross-metathesis reactions that form 5-membered rings.[57] The reactions are not always limited to molybdenum as the metal; tungsten has been used with success.[58] The catalysts are also amenable to being used on a polymer support.[59]

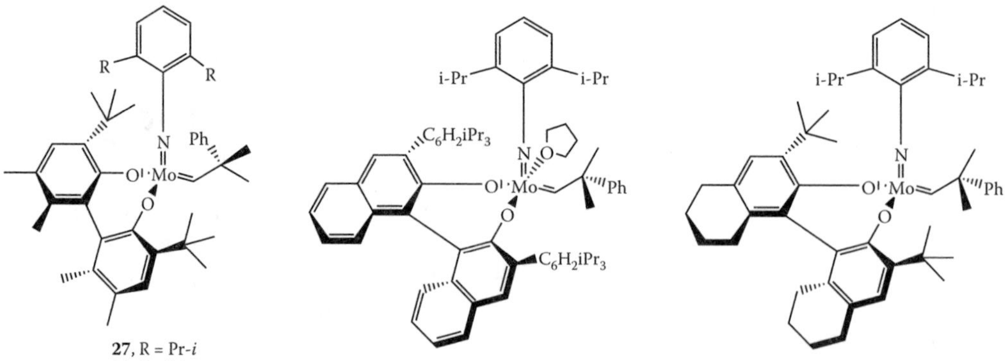

Kinetic resolution of racemic dienes, such as **31**, can be accomplished by ARCM using a chiral catalyst (Figure 28.2).[51,54,60]

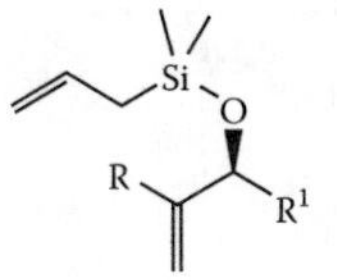

31

R	R^1	K$_{rel}$	Catalyst
Me	C$_5$H$_{11}$	25	29
Me	C$_6$H$_{12}$	14	28
H	C$_6$H$_{12}$	25	27

FIGURE 28.2 Substrates for asymmetric ring closing metathesis.

These chiral catalysts can also be used to desymmetrize *meso* substrates to produce enantio pure 5-, 6-, and 7-membered rings (Scheme 28.20).[57,61–64]

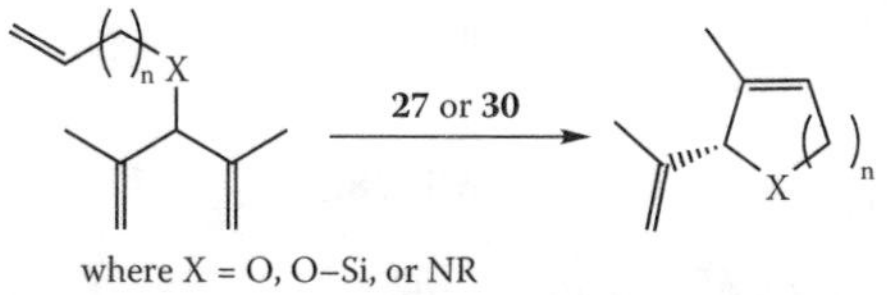

where X = O, O–Si, or NR

SCHEME 28.20

An elegant application of this methodology was demonstrated by Burke and colleagues in a short synthesis of *endo*-brevicomin (**31**) (Scheme 28.21).[65]

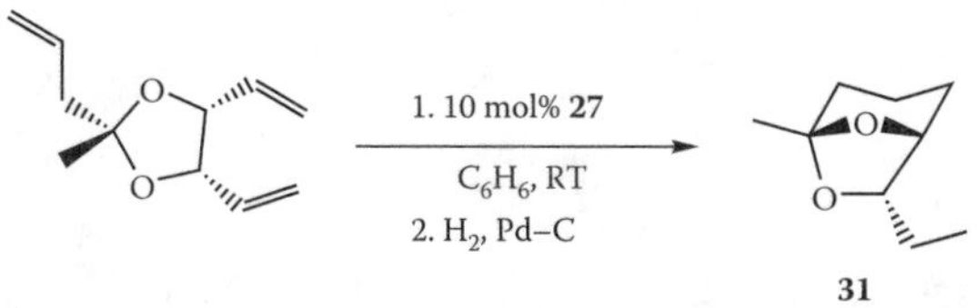

31
78%
59% ee

SCHEME 28.21

The asymmetric reactions can also be coupled with achiral metathesis reactions as in, for example, AROM/RCM or AROM/CM reactions.[66,67] Examples of the latter are given in Scheme 28.22.[68–70]

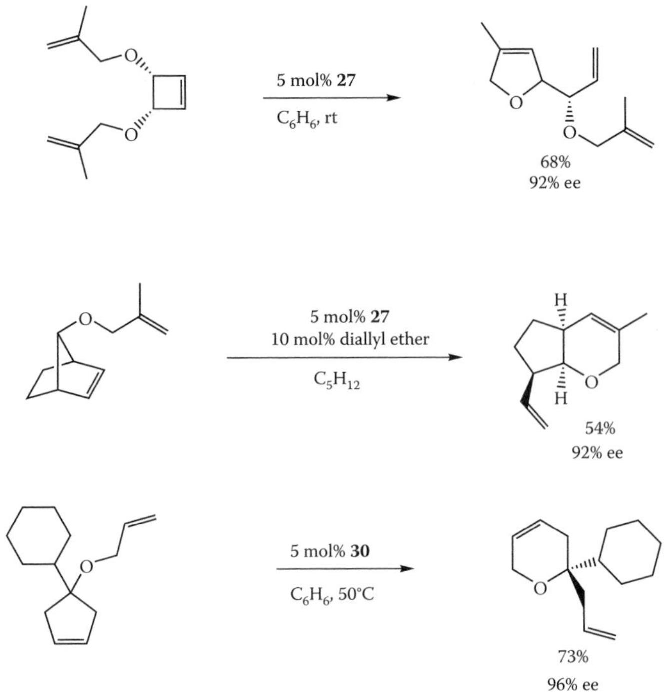

SCHEME 28.22

This methodology has been applied to the synthesis of the dihydropyran portion **32** of tipranavir (**33**), an anti-human immunodeficiency virus compound (Scheme 28.23).[56,70] The dihydropyran **32** can be accessed from the cyclopentene **34** by what is formally a tandem AROM/RCM reaction.

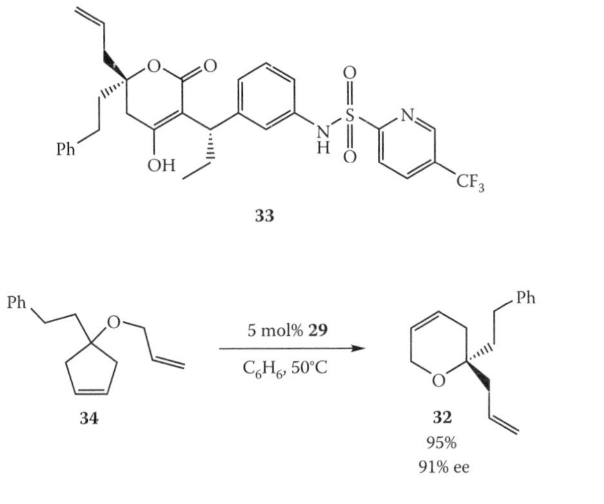

33

34

32
95%
91% ee

SCHEME 28.23

Asymmetric ring opening metathesis reactions can also be performed.[67,66] In these cases, it is common for the reaction to be coupled with a cross metathesis or other reaction to avoid polymerization of the product.[51] The catalysts are constantly being improved to provide higher selectivities and faster reactions. The catalyst **35** is recyclable and air-stable (Scheme 28.24).[60,71]

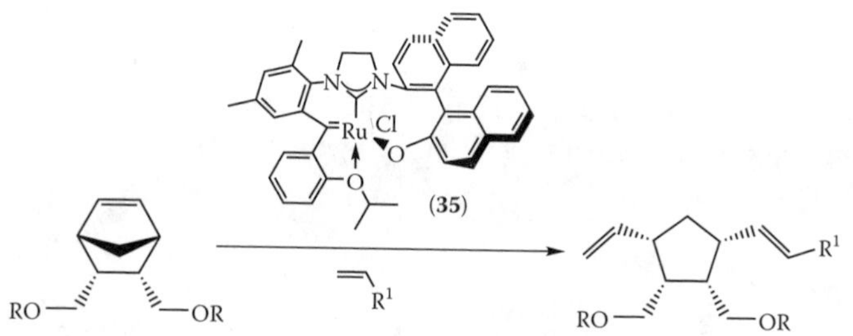

SCHEME 28.24

28.7 TANDEM CATALYSIS

Grubbs and coworkers have shown that **4** or **6** can be used as a hydrogenation or dehydrogenation catalyst. This dual use was demonstrated in the synthesis of (*R*)-(−)-muscone (**36**) (Scheme 28.25).[72,73]

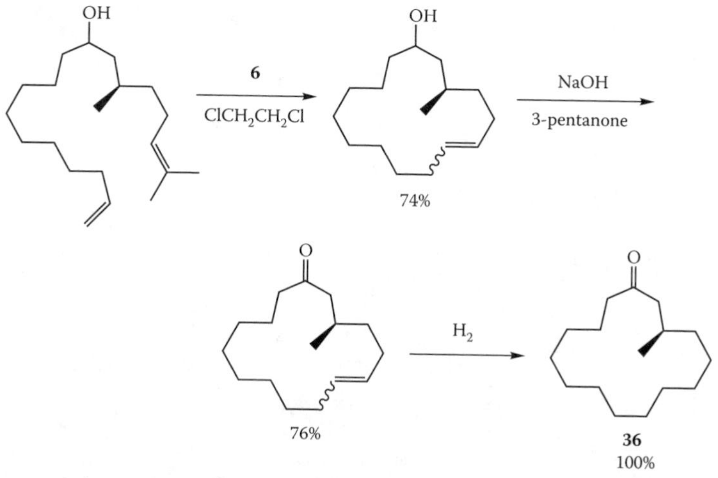

SCHEME 28.25

28.8 RUTHENIUM REMOVAL

In some instances removal of ruthenium from the desired product has proved difficult. To overcome this problem, there are several reported ways to remove the ruthenium catalyst after the olefin metathesis reaction is complete. The first and most common is addition of a modifier or adsorbent followed by chromatography. The additives include dimethyl sulfoxide (DMSO), triphenylphosphine oxide,[74] or activated carbon.[75] Although this is generally not suitable for commercial production, it can be quite effective. A convenient alternative is the use of Brockmann I basic alumina, followed by simple filtration through a bed of filter agent such as activated carbon. This method is capable of reducing ruthenium levels down to less than 20 ppm.[76]

Several chemical techniques have also been developed, such as simple oxidation using hydrogen peroxide or lead tetraacetate.[77] Both these protocols suffer from harsh reaction conditions and additional toxicity issues. A more functional group-tolerant method is the use of tris(hydroxymethyl)phosphine to coordinate the ruthenium.[78] This complex is water soluble and can be washed from the product by several simple water washes. The main drawback to this method is the large excess of phosphine needed (25 mole equiv./mole of Ru).[46]

Rather than modifying the workup conditions, other groups have changed the reaction solvent or catalyst structure. The use of ionic liquids as a reaction solvent allows for easy catalyst and product separation as well as catalyst recycling.[79] Alternatively, others have attached the various metathesis catalysts onto solid supports,[80–82] which again allows for easy workup and catalyst recycle.

28.9 SUMMARY

In just a short time, olefin metathesis has become an important tool to the synthetic organic chemist. The large-scale use of this chemistry has already been seen in the polymer and fragrance industries. As drug candidates move through the development pipeline, the commercial application of this chemistry probably will be put into practice. The applications of the asymmetric catalysts allow for an efficient coupling of two reactions with the same catalyst and reaction conditions.

REFERENCES

1. Grubbs, R. H., Miller, S. J., Fu, G. C. *Acc. Chem. Res.* 1995, *28*, 446.
2. Tsuji, J., Hashiguchi, S. *Tetrahedron Lett.* 1980, *21*, 2955.
3. Armstrong, S. K. *J. Chem. Soc. Perkin Trans. 1* 1998, 371.
4. Furstner, A., Langemann, K. *Synthesis* 1997, 792.
5. Schuster, M., Blechert, S. *Angew. Chem., Int. Ed.* 1997, *37*, 2036.
6. Furstner, A. *Angew. Chem., Int. Ed.* 2000, *39*, 3012.
7. Brummer, O., Ruckert, A., Blechert, S. *Chem. Eur. J.* 1997, *3*, 441.
8. Blackwell, H. E., O'Leary, D. J., Chatterjee, A. K., Washenfelder, R. A., Bussmann, D. A., Grubbs, R. H. *J. Am. Chem. Soc.* 2000, *122*, 58.
9. Blechert, S. *Pure Appl. Chem.* 1999, *71*, 1393.
10. Nicolau, K. C., Postema, M. H. D., Yue, E. W., Nadin, A. *J. Am. Chem. Soc.* 1996, *118*, 100.
11. Stille, J. R., Grubbs, R. H. *J. Am. Chem. Soc.* 1986, *108*, 855.
12. Stille, J. R., Santarsiero, B. D., Grubbs, R. H. *J. Org. Chem.* 1990, *55*, 843.
13. McGinnis, J., Katz, T. J., Hurwitz, S. *J. Am. Chem. Soc.* 1976, *98*, 605.
14. Lee, S. J., McGinnis, J., Katz, T. J. *J. Am. Chem. Soc.* 1976, *98*, 7818.
15. Casey, C. P., Burkhardt, T. J. *J. Am. Chem. Soc.* 1973, *95*, 5833.
16. Schrock, R. R., Murdzek, J. S., Bazan, G. C., Robbins, J., DiMare, M., O'Regan, M. *J. Am. Chem. Soc.* 1990, *112*, 3875.
17. Bazan, G. C., Khosravi, E., Schrock, R. R., Feast, W. J., Gibson, V. C., O'Regan, M. B., Thomas, J. K., Davis, W. M. *J. Am. Chem. Soc.* 1990, *112*, 8378.
18. Bazan, G. C., Oskam, J. H., Cho, H.-N., Park, L. Y., Schrock, R. R. *J. Am. Chem. Soc,* 1991, *113*, 6899.
19. Grubbs, R. H., Chang, S. *Tetrahedron* 1998, *54*, 4413.
20. Trinka, T. M., Grubbs, R. H. *Acc. Chem. Res.* 2001, *34*, 18.
21. Nguyen, S. T., Johnson, L. K., Grubbs, R. H. *J. Am. Chem. Soc.* 1992, *114*, 3974.
22. Fu, G. C., Nguyen, S. T., Grubbs, R. H. *J. Am. Chem. Soc.* 1993, *115*, 9856.
23. Nguyen, S. T., Grubbs, R. H., Ziller, J. W. *J. Am. Chem. Soc.* 1993, *115*, 9858.
24. Wu, Z., Nguyen, S. T., Grubbs, R. H., Ziller, J. W. *J. Am. Chem. Soc.* 1995, *117*, 5503.
25. Nguyen, S. T., Grubbs, R. H. *J. Organometal. Chem.* 1995, *497*, 195.
26. Lynn, D. M., Kanaoka, S., Grubbs, R. H. *J. Am. Chem. Soc.* 1996, *118*, 784.
27. Schwab, P., France, M. B., Ziller, J. W., Grubbs, R. H. *Angew. Chem., Int. Ed.* 1995, *34*, 2039.
28. Schwab, P., Grubbs, R. H., Ziller, J. W. *J. Am. Chem. Soc.* 1996, *118*, 100.
29. Love, J. A., Sanford, M. S., Day, M. W., Grubbs, R. H. *J. Am. Chem. Soc.* 2003, *125*, 10103.
30. Dias, E. L., Nguyen, S. T., Grubbs, R. H. *J. Am. Chem. Soc.* 1997, *119*, 3887.
31. Love, J. A., Morgan, J. P., Trnka, T. M., Grubbs, R. H. *Angew. Chem., Int. Ed.* 2002, *41*, 4035.
32. Sanford, M. S., Love, J. A., Grubbs, R. H. *J. Am. Chem. Soc.* 2001, *123*, 6543.
33. Wallace, D. J., Goodman, J. M., Kennedy, D. J., Davies, A. J., Cowden, C. J., Ashwood, M. S., Cottrell, I. F., Dolling, U.-H., Reider, P. *J. Org. Lett.* 2001, *3*, 671.

34. Wallace, D. J., Bulger, P. G., Kennedy, D. J., Ashwood, M. S., Cottrell, I. F., Dolling, U.-H. *Synlett* 2001, 357.

35. Marquis, R. W., Ru, Y., LoCastro, S. M., Zeng, J., Yamashita, D. S., Oh, H.-J., Erhard, K. F., Davis, L. D., Tomaszek, T. A., Tew, D., Salyers, K., Proksch, J., Ward, K., Smith, B., Levy, M., Cummings, M. D., Haltiwanger, R. C., Trescher, G., Wang, B., Hemling, M. E., Quinn, C. J., Cheng, H.-Y., Lin, F., Smith, W. W., Janson, C. A., Zhao, B., McQueney, M. S., D'Alessio, K., Lee, C.-P., Marzulli, A., Dodds, R. A., Blake, S., Hwang, S.-M., James, I. E., Gress, C. J., Bradley, B. R., Lark, M. W., Gowen, M., Veber, D. F. *J. Med. Chem.* 2001, *44*, 1380.

36. Marquis, R. W., Yu, R., Veber, D. F., Cummings, D. M., Thompson, S. K., Yamashita, D. S. U.S. Pat. 2003/144175 A1, 2003.

37. Tsantrizos, Y. S., Cameron, D. R., Faucher, A.-M., Ghiro, E., Goudreau, N., Halmos, T., Llinas-Brunet, M. U.S. Pat. 6,608,027 B1, 2003.

38. Llinas-Brunet, M., Gorys, V. J. PCT Pat. Appl. WO 03064455 A2, 2003.

39. Tsantrizos, Y. S., Cameron, D. R., Faucher, A.-M., Goudreau, N., Halmos, T., Llinas-Brunet, M. PCT Pat. Appl. WO 0059929, 2000.

40. Creighton, C. J., Reitz, A. B. *Org. Lett.* 2001, *3*, 893.

41. Kaptein, B., Broxterman, Q. B., Schoemaker, H. E., Rutjes, F. P. J. T., Veerman, J. J. N., Kamphuis, J., Peggion, C., Formaggio, F., Toniolo, C. *Tetrahedron* 2001, *57*, 6567.

42. Kaptein, B., Boesten, W. H. J., Broxterman, Q. B., Schoemaker, H. E., Kamphuis, J. *Tetrahedron Lett.* 1992, 6007.

43. Tjen, K. C. M. F., Kinderman, S. S., Schoemaker, H. E., Hiemstra, H., Rutjes, F. P. J. T. *J. Chem. Soc., Chem. Comm.* 2000, 699.

44. Kinderman, S. S., van Maarseveen, J. H., Schoemaker, H. E., Hiemstra, H., Rutjes, F. P. J. T. *Org. Lett.* 2001, *3*, 2045.

45. Aguilera, B., Wolf, L. B., Nieczypor, P., Rutjes, F. P. J. T., Overkleeft, H. S., van Hest, J. C. M., Schoemaker, H. E., Wang, B., Mol, J. C., Furstner, A., Overhand, M., van der Marel, G. A., van Boom, J. H. *J. Org. Chem.* 2001, *66*, 3584.

46. Pederson, R. L., Fellows, I. M., Ung, T. A., Ishihara, H., Hajela, S. P. *Adv. Synth. Catal.* 2002, *344*, 728.

47. Pederson, R. L., Grubbs, R. H. U.S. Pat. 6,215,019 B1, 2001.

48. Pederson, R. L., Grubbs, R. H. PCT Pat. Appl. WO 0136368 A2, 2001.

49. Pederson, R. L., Grubbs, R. H. U.S. Pat. 2002/0022741 A1, 2002.

50. MacMillan, D. W. C. PCT Pat. Appl. WO 03002491 A2, 2003.

51. Schrock, R. R., Hoveyda, A. H. *Angew. Chem., Int. Ed.* 2003, *42*, 4592.

52. Hoveyda, A. H., Gillingham, D. G., van Veldhuizen, J. J., Kataoka, O., Garber, S. B., Kingsbury, J. S., Harrity, J. P. A. *Org. Bio. Chem.* 2004, *2*, 8.

53. Hoveyda, A. H., Schrock, R. R. *Chem. Eur. J.* 2001, *7*, 945.

54. Alexander, J. B., La, D. S., Cefalo, D. R., Hoveyda, A. H., Schrock, R. R. *J. Am. Chem. Soc.* 1998, *120*, 4041.

55. Zhu, S. S., Cefalo, D. R., La, D. S., Jamieson, J. Y., David, W. M., Hoveyda, A. H., Schrock, R. R. *J. Am. Chem. Soc.* 1999, *121*, 8251.

56. Teng, X., Cefalo, D. R., Schrock, R. R., Hoveyda, A. H. *J. Am. Chem. Soc.* 2002, *124*, 10779.

57. Aeilts, S. L., Cefalo, D. R., Bonitatebus, P. J., Houser, J. H., Hoveyda, A. H., Schrock, R. S. *Angew. Chem., Int. Ed.* 2001, *40*, 1452.

58. Tsang, W. C. P., Hultzsch, K. C., Alexander, J. B., Bonitatebas, P. J., Schrock, R. R., Hoveyda, A. H. *J. Am. Chem. Soc.* 2003, *125*, 2652.

59. Hultzsch, K. C., Jernelius, J. A., Hoveyda, A. H., Schrock, R. R. *Angew. Chem., Int. Ed.* 2002, *41*, 589.

60. Van Veldhuizen, J. J., Garber, S. B., Kingsbury, J. S., Hoveyda, A. H. *J. Am. Chem. Soc.* 2002, *124*, 4954.

61. Dolman, S. J., Sattely, E. S., Hoveyda, A. H., Schrock, R. R. *J. Am. Chem. Soc.* 2002, *124*, 6991.

62. Kiely, A. F., Jernelius, J. A., Schrock, R. R., Hoveyda, A. H. *J. Am. Chem. Soc.* 2002, *124*, 2868.

63. La, D. S., Alexander, J. B., Cefalo, D. R., Graf, D. D., Hoveyda, A. H., Schrock, R. R. *J. Am. Chem. Soc.* 1998, *120*, 9720.

64. Dolman, S. J., Schrock, R. R., Hoveyda, A. H. *Org. Lett.* 2003, *5*, 4899.

65. Burke, S. D., Muller, N., Beaudry, C. M. *Org. Lett.* 1999, *1*, 1827.

66. La, D. S., Sattely, E. S., Ford, J. G., Schrock, R. R., Hoveyda, A. H. *J. Am. Chem. Soc.* 2001, *123*, 7767.

67. La, D. S., Ford, J. G., Sattely, E. S., Bonitatebus, P. J., Schrock, R. R., Hoveyda, A. H. *J. Am. Chem. Soc.* 1999, *121*, 11603.
68. Weatherhead, G. S., Ford, J. G., Alexanian, E. J., Schrock, R. R., Hoveyda, A. H. *J. Am. Chem. Soc.* 2000, *122*, 1828.
69. Harrity, J. P. A., La, D. S., Cefalo, D. R., Visser, M. S., Hoveyda, A. H. *J. Am. Chem. Soc.* 1998, *120*, 2343.
70. Cefalo, D., Kiely, A. F., Wuchrer, M., Jamieson, J. Y., Schrock, R. R., Hoveyda, A. H. *J. Am. Chem. Soc.* 2001, *123*, 3139.
71. van Veldhuizen, J. J., Gillingham, D. G., Garber, S. B., Kataoka, O., Hoveyda, A. H. *J. Am. Chem. Soc.* 2003, *125*, 12502.
72. Louie, J., Bielawski, C. W., Grubbs, R. H. *J. Am. Chem. Soc.* 2001, *123*, 11312.
73. Kamat, V. P., Hagiwara, H., Katsumi, T., Hoshi, T., Suzuki, T., Ando, M. *Tetrahedron* 2000, *56*, 4397.
74. Ahn, Y. M., Yang, K., Georg, G. I. *Org. Lett.* 2001, *3*, 1411.
75. Cho, J. H., Kim, B. M. *Org. Lett.* 2003, *5*, 531.
76. Conde, J. J. PCT Pat. Appl. WO 03026770 A1, 2003.
77. Paquette, L. A., Schloss, J. D., Efremov, I., Fabris, F., Gallou, F., Mendez-Andino, J., Yang, J. *Org. Lett.* 2000, *2*, 1259.
78. Maynard, H. D., Grubbs, R. H. *Tetrahedron Lett.* 1999, *40*, 4137.
79. Buijsman, R. C., van Vuuren, E., Sterrenburg, J. G. *Org. Lett.* 2001, *3*, 3785.
80. Yao, Q. *Angew. Chem., Int. Ed.* 2000, *39*, 3896.
81. Schurer, S. C., Gessler, S., Buschmann, N., Blechert, S. *Angew. Chem., Int. Ed.* 2000, *39*, 3898.
82. Ahmed, M., Barrett, A. G. M., Braddock, D. C., Cramp, S. M., Procopiou, P. A. *Tetrahedron Lett.* 1999, *40*, 8657.

29 Synthesis of Homochiral Compounds: A Small Company's Role*

Karen Etherington, Ed Irving, Feodor Scheinmann, and Basil Wakefield

CONTENTS

29.1 INTRODUCTION

A small chemical company may occupy a small, specialized niche, or it may provide services covering a wide area. Ultrafine falls into the latter category and in many ways may be compared with the chemical research and development department of a large company. Because it is so much smaller, emphasis must be placed on versatility as well as high scientific expertise.

Many of the custom synthesis and research and development contracts tackled at Ultrafine have required the preparation of homochiral compounds, so it has been necessary to build up expertise in the whole range of available methodology. Some of the most interesting examples remain confidential, but the selection that follows illustrates how some of the challenges were tackled.

29.2 CLASSICAL RESOLUTION

29.2.1 2-CHLOROPROPIONIC ACID

A number of herbicides are 2-aryloxypropionic acids (see Chapter 31). These herbicides were originally used as racemates, but in the agrochemical field, as in the pharmaceutical field, there

---

* Ultrafine (UFC Ltd.) is now part of the Sigma-Aldrich Corporation operating within the SAFC division.

has been an increasing requirement for homochiral active enantiomers. We were approached by a manufacturer of a key intermediate for these herbicides, 2-chloropropionic acid (**1**). Because the manufacturing process was already in place, a resolution rather than an asymmetric synthesis was needed. The process should give as high a yield as possible of the required enantiomer; the resolving agent should be cheap and/or recyclable; and it should be possible to recover, racemize, and then recycle the "unwanted" 2-chloropropionic acid.

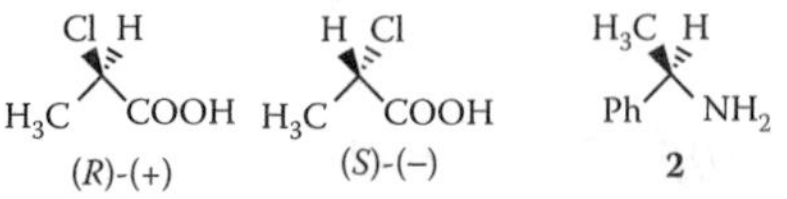

The (*S*)-(–)-isomer (**1b**) was the one required for synthesis of the herbicides. Reaction of racemic 2-chloropropionic acid with one equivalent of (*R*)-(+)-α-phenylethylamine (**2**) gave a quantitative yield of the mixture of diastereoisomeric salts, which after 4 recrystallizations gave the required diastereoisomer in 20% yield [i.e., 40% based on the (*S*)-(–)-isomer present] with a de of 88% (determined by nuclear magnetic resonance, or NMR). The free acid (**1b**) was obtained quantitatively and without racemization on acidification of the salt, and the (*R*)-(+)-α-phenylethylamine was also recovered without racemization in 89% yield.

Some preliminary experiments were performed on the reaction of racemic 2-chloropropionic acid with half an equivalent of (*R*)-(+)-α-phenylethylamine (**2**). These gave the required diastereoisomeric salt with de *ca* 80% in 39% yield based on **2** after only 1 recrystallization.

Our methodology was regarded by the client as well-suited to scale up. However, although the resolution was then, almost 20 years ago, regarded as an efficient example of its kind, it is noteworthy that it is now conducted on a large scale by an enzymatic method (see Chapter 31).[1]

29.2.2 Anatoxin a

Anatoxin-a (**3**) is a powerful neurotoxin (inhibitor of acetylcholine esterase) found in freshwater blue–green algae. The compound was required as an analytical standard and also for development of an immunoassay. It was synthesized by the 8-step sequence summarized in Scheme 29.1. The key intermediate **4** was resolved by separation of the dibenzoyl tartrates, and the remaining steps then gave both (+)- and (–)-anatoxin-a.[2] The efficiency of the resolution was monitored by formation of the BocAla derivatives, which were distinguishable by NMR; an ee of >98% was achieved even before recrystallization of the salts. It is also noteworthy that the published absolute configuration of the intermediate **5**[3] was shown to be in error by X-ray crystallography.

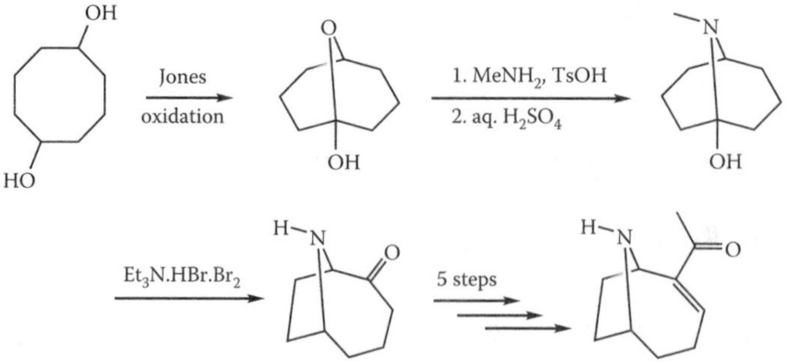

SCHEME 29.1

5

29.2.3 CHROMATOGRAPHY

Separation by chromatography, using a chiral stationary phase, may be regarded as a modern form of classical resolution. Owing to the advances in chiral chromatography over the last 20 years, in particular the wide range of chiral columns now available, this has now become a routinely used analytical technique. Advances have also been made in the methods of detection, and for high-performance liquid chromatography (HPLC) the advent of circular dichroism-based detectors for the analysis of chiral compounds in biological matrices has further enhanced its usefulness. Furthermore, in contrast to the limited enantiomeric purity data that can be obtained by measuring optical rotations, chiral HPLC allows for the accurate determination of the enantiomeric excess.

Although classical column chromatography over chiral stationary phases has been used for several years, HPLC has until recently been regarded mainly as an analytical tool. However, methods have now been developed and many papers have been published on the various applications of particular stationary phases in preparative HPLC, which may be the method of choice when the preparation of small amounts of material are required for screening purposes, for use in biological assays, or as standards in purity assays. The use of chromatographic techniques may be quicker and less risky than custom synthesis and resolution and can guarantee material to a defined specification.

The majority of preparative separations undertaken at Ultrafine have used chiral stationary phases based on either cellulose or amylose derivatives. In one project 500 mg of salts of both the (R,R)- and (S,S)-Formoterol (6) enantiomers were prepared using an OJ column with a resulting ee of >97%.[4,5] A loading of 200 mg/mL on a semipreparative column was achievable without loss of resolution or purity relative to the racemate.

6

Cardiovascular drugs is another area where chiral chromatography has been exploited.[4,6] For example, the enantiomers of the widely used anticoagulant warfarin have been determined in plasma using a wide range of chiral columns and conditions. They have also been separated on a preparative scale using supercritical fluid chromatography, which has the advantage of a reduction in solvent waste and ease of isolation of the prepared enantiomers. The current major disadvantage is the complexity of the equipment required.[7]

Advances in preparative enantioseparation by simulated moving bed (SMB) chromatography have occurred in the last 10 years. SMB was invented in the 1960s and was used by the petrochemical and sugar industries. Now with the improvements in stationary phases and hardware it is an option for the large-scale preparation of enantiomerically pure material. The majority of the latest published data are using either amylose- or cellulose-based phases because of their selectivity. There are now examples in the literature of the commercial separation on the multi-ton scale.[8]

29.3   THE CHIRAL POOL

29.3.1   A KEY LEUKOTRIENE INTERMEDIATE

The predecessor of Ultrafine was the Fine Chemicals Unit of Salford University Industrial Centre, which was set up partly to exploit a new route to prostaglandins.[9] When the independent company was founded, it made sense to offer other eicosanoids. A key intermediate for the synthesis of leukotriene A_4 (LTA$_4$) (**7**), and thence LTC$_4$, LTD$_4$, and LTE$_4$, was the epoxyalcohol (**10**), whose synthesis from 2-deoxy-D-ribose (**8**) (Scheme 29.2) had been reported.[10]

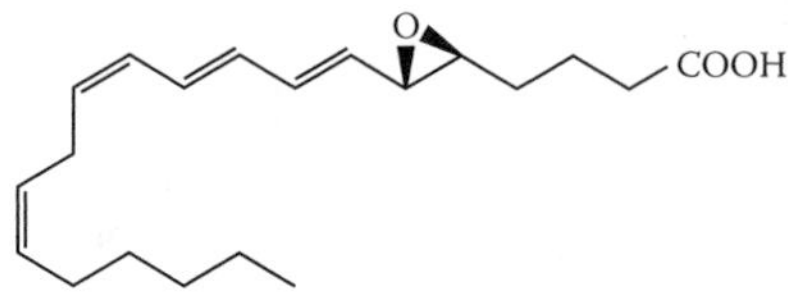

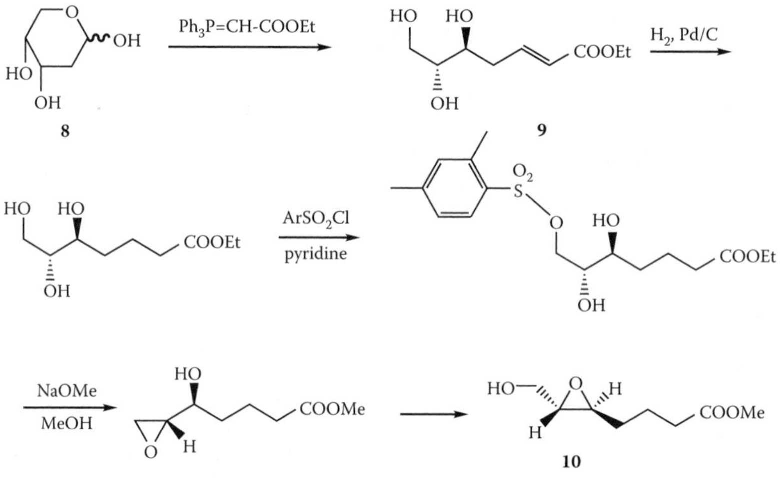

SCHEME 29.2

This route was successfully used for the synthesis of the desired leukotrienes, but it was later modified and improved as outlined in Scheme 29.3; this modified route was also used for the synthesis of [8,9,10,11-^{13}C$_4$]-leukotriene A$_4$ methyl ester.

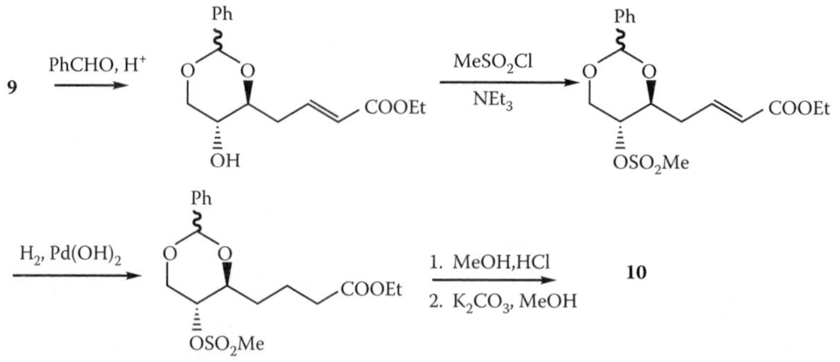

SCHEME 29.3

29.4 ENZYME-CATALYZED KINETIC RESOLUTION

29.4.1 SYNTHESIS OF LEUKOTRIENE B$_4$ (LTB$_4$)

One way in which a small company can keep abreast of new developments is to collaborate with universities. Moreover, such collaboration helps to motivate staff (and publication of results provides advertising). One fruit of such a collaboration was a synthesis of LTB$_4$, in which chiral moieties of the molecule are derived from the enantiomers of a common intermediate (**11**).[11] Several routes have been devised for enzyme-catalyzed kinetic resolution of bicyclo[3.2.0]heptenones.[12] An efficient one that was used for the synthesis of LTB$_4$ (**12**) is shown in Scheme 29.4.

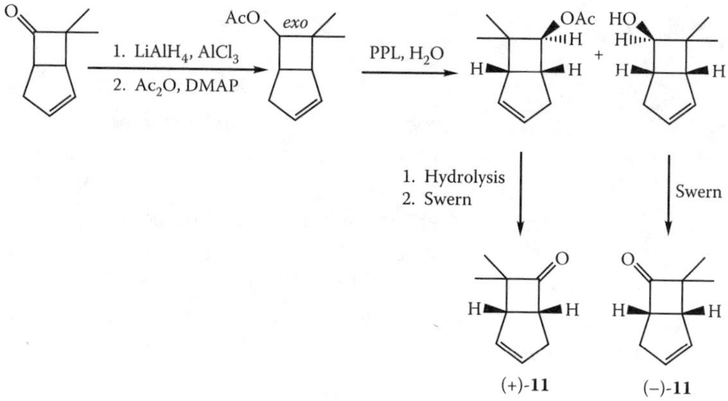

1. LiAlH$_4$, AlCl$_3$
2. Ac$_2$O, DMAP

PPL, H$_2$O

1. Hydrolysis
2. Swern

Swern

(+)-**11**   (−)-**11**

SCHEME 29.4

The target molecule was then constructed from the two enantiomers, (+)-(**11**) and (−)-(**11**), as shown in Scheme 29.5.

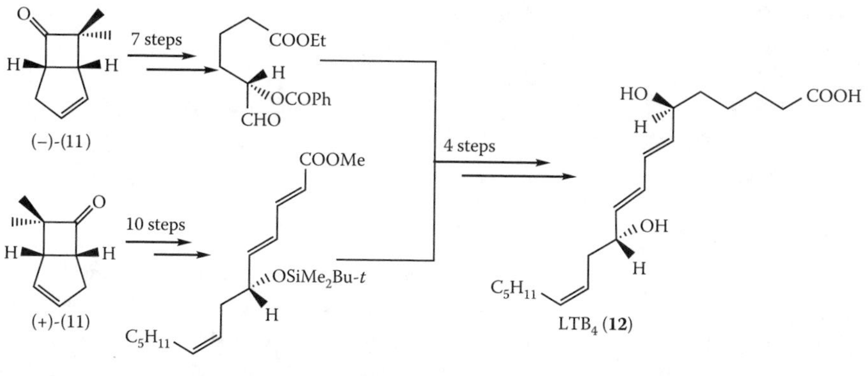

(−)-(**11**)   7 steps

(+)-(**11**)   10 steps

4 steps

LTB$_4$ (**12**)

SCHEME 29.5

29.4.2 THE METHADONE STORY

Racemic methadone (**13**) continues to be used as a maintenance drug in the treatment of addiction to heroin (**14**).[13] Methadone has also been used in treating severe pain.[14] The value in using oral racemic methadone is that it also helps to combat the spread of human immunodeficiency virus by reducing injection of heroin.[13] Under medical supervision, the addict can lead a more stable life, but there is a temptation to remain on racemic methadone to avoid the withdrawal symptoms known as cold turkey.[13] The illegal use of methadone taken together with other drugs such as benzodiazepines and alcohol has led to fatalities, although methadone overdose can be treated by administration of naltrexone (**15**).[15] It is known that levomethadone [(*R*)-(–)-methadone] is the active principle[14b] and, therefore, by racemate switching it should be possible to reduce the given dose by half.

The preparation of homochiral compounds by formation and separation of diastereoisomers or by kinetic resolution of racemates, at or near the end of a total synthesis, has been a method of choice. This avoids the possibility of racemization should chirality be introduced earlier. However, the costs are high because only half by weight of the homochiral compound is theoretically possible from the racemate unless the optical antipode can also be easily inverted to the desired product. Indeed, previous methods for producing levomethadone based on the classical resolution at the end, or at the penultimate stage of the synthesis, were costly and not very effective. Levomethadone hydrochloride has previously been marketed as *L*-Polamidon and Levadone[16] but was subsequently withdrawn because of the high cost of production.

Resolution of a cheap racemate at the start of a synthesis is economically advantageous if it can be demonstrated that chirality is not lost in the subsequent stages. Moreover, optical purity can be enhanced by purification of intermediates during the total synthesis. This strategy of resolution at the beginning has been applied to a new synthesis of optically active methadones and levo-α-acetylmethadol (LAAM) (**16**).

29.4.2.1 Early Methods for Producing Chiral Methadones

The previous methods of preparing levomethadone were based on separation of the D-tartrate salts of racemic methadone and also separation of the diastereoisomeric tartrates of the intermediate at the penultimate stage, namely the nitrile **17**.[17–19] Diastereoisomeric *d*-α-bromocamphor-10-sulfonate salts have also been separated.[19] However, we and others have noted that there were difficulties in the resolution of racemic methadone by preparation of diastereoisomeric salts.[20]

Synthesis of single-enantiomer methadones has also been achieved using the chiral pool approach. Thus, Barnett and Smirz obtained (*S*)-(+)-1-dimethylamino-2-propanol (**18**) from (–)-

ethyl L-lactate.[21] From this intermediate, *d*-methadone can be synthesized. However, it would be impractical to synthesize *l*-methadone by this method because (+)-ethyl D-lactate is approximately 450 times more expensive than the L-isomer.[22]

29.4.2.2 Lipase-Catalyzed Resolution of (*R,S*)–Dimethylaminopropan-2-ol

An alternative approach for the synthesis of levomethadone uses lipase-catalyzed resolution of a cheap[23] starting material, (*R,S*)-dimethylaminopropan-2-ol (**18**) (Scheme 29.6), and demonstration that configurational integrity is maintained during the subsequent steps of the total synthesis. Our approach has also led to the synthesis of *d*-methadone and of LAAM **16**.

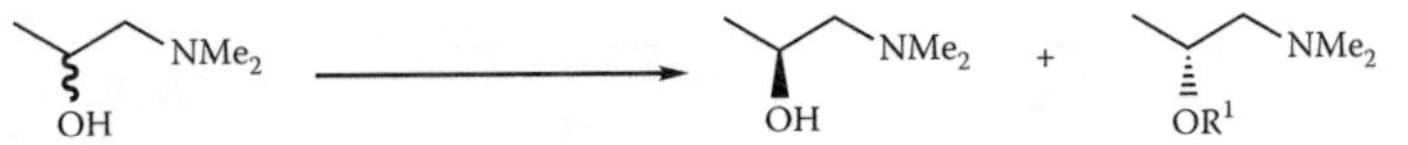

SCHEME 29.6

(*R,S*)-1-Dimethylaminopropan-2-ol (**18**) is an attractive substrate for lipase-catalyzed resolution by transesterification (Scheme 29.6) because of the difference in the size of the two groups attached to the stereogenic center. Thus, it should be possible to gain access to both (*R*) and (*S*)-isomers of the amino alcohol and provide syntheses of both (*R*) and (*S*)-methadone. The initial resolution of (*R,S*)-1-dimethylaminopropan-2-ol (**18**) was carried out on small scale using Lipase PS (Amano) from *Pseudomonas cepacia* with vinyl acetate as both acyl donor and solvent. After 2 days the lipase was removed by filtration and the acetate was isolated by chromatography and analyzed for optical purity by chiral shift reagent NMR spectroscopy [Eu(hfc)$_3$].[24] The initial experiment gave 60% conversion to the (*R*)-ester **19** (R^1 = Ac) with 46% ee. The optical purity could be improved by removing the lipase after 16 hours of reaction time; this gave conversion to the ester of 33% with 88% ee. From these initial experiments we were not able to isolate the unreacted (*S*)-alcohol (*S*)-**18** by chromatography.

To achieve better results a series of lipases was screened using the Chirazyme® screening kit and analysis by gas chromatography with a chiral stationary phase. *Candida antartica* Lipase B, available as Novozyme® 435, was chosen for further development with the acyl transfer agent vinyl propionate.

Racemic 1-dimethylaminopropan-2-ol (**18**) was acylated with propanoyl chloride, and the reaction product was analyzed by gas chromatography (GC). The resultant (*R*)- and (*S*)-esters **19** (R^1 = COEt) were resolved by GC using an α-cyclodextrin column. Reaction of the racemic alcohol **18** was then carried out in the presence of Novozyme 435 and vinyl propanoate, and the reaction was followed by GC. After 4 hours the reaction was approximately 2% complete and the ee of the propanoate ester **19** (R^1 = COEt) was 95.9%. After 88 hours the conversion had reached 50% and the ee was still 95.6%. The remarkable specificity of Novozyme 435 for the (*R*)-amino alcohol (*R*)-**18** was evident because even after 3.5 days reaction time, only a very small amount of the (*S*)-ester was detected. The reaction could be scaled up; thus, 1 kg of the racemic amino alcohol **18** was treated with vinyl propanoate (0.5 equivalents) and Novozyme 435 (3% by weight). After 3 days, both optically active products were isolated by distillation at reduced pressure. The (*S*)-amino alcohol (*S*)-**18** was recovered in 45% yield, which compared favorably with a yield of 32% for resolution on a small scale. The (*R*)-propanoate **19** (R^1 = COEt) distilled as a colorless oil in 36% yield — slightly higher than that obtained from the small-scale resolution. The overall recovery was 81% from the scaled up reaction.

The resulting products were then converted into both (*R*) and (*S*)-methadone, following largely the literature procedures with careful monitoring of the optical activity for each stage of the synthesis (Scheme 29.7).

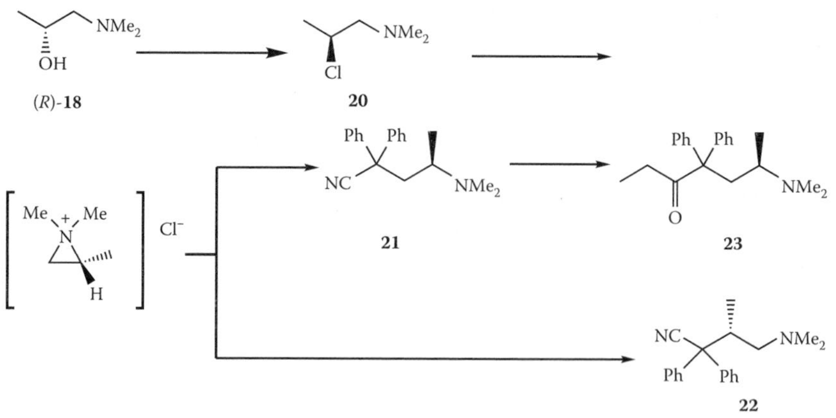

SCHEME 29.7

Thus (R)-1-dimethylamino-2-propyl propanoate was hydrolyzed in methanol using the Zemplen procedure with a catalytic amount of freshly prepared sodium methoxide, and the resultant alcohol (R)-**18** was then treated with thionyl chloride in chloroform to give (S)-(+)-1-dimethylamino-2-chloropropane **20** as the hydrochloride salt, with [α]$_D$ +65.9° after 3 recrystallizations [lit. value[21] –65° for the (R) isomer]. Thus, the chlorination had occurred with total inversion of stereochemistry.[21] Reaction of product **20** with the sodium salt of diphenylacetonitrile in the presence of 18-crown-6 phase transfer catalyst favored the formation of the desired aminonitrile **21** over the unwanted[25] isomeric product **22**. The (R)-**21** had [α]$_D$ –50.2°, which compares favorably to the value of +49° previously reported for the (S)-isomer.[21] Conversion of the nitrile (**21**) through to methadone was achieved by the Grignard reaction with ethylmagnesium bromide[26] to give (R)-(–)-methadone (levomethadone) (**23**), isolated as its hydrochloride salt in >99% ee, as shown by chiral HPLC, [α]$_D$ –136°. Synthesis of d-methadone [(S)-(+)-methadone] was carried out analogously to give (S)-methadone hydrochloride having a rotation of [α]$_D$ +136°.

29.4.2.3 Levo-α-Acetylmethadol

Levo-α-acetylmethadol (**16**), marketed as Orlaam and also known as LAAM, was approved in 1993 by the U.S. Food and Drug Administration for treatment of drug addiction.[27] The advantage of using LAAM is that it is effective for 48–72 hours after the oral dose is taken, compared to 24 hours for racemic methadone.[27] However, although LAAM continues to be used in the United States, the European Medicines Agency (EMEA) has withdrawn marketing authorization for Orlaam pending further risk/benefit reassessment owing to reported cardiac disorders. The switch to methadone or detoxification is currently advised.[28]

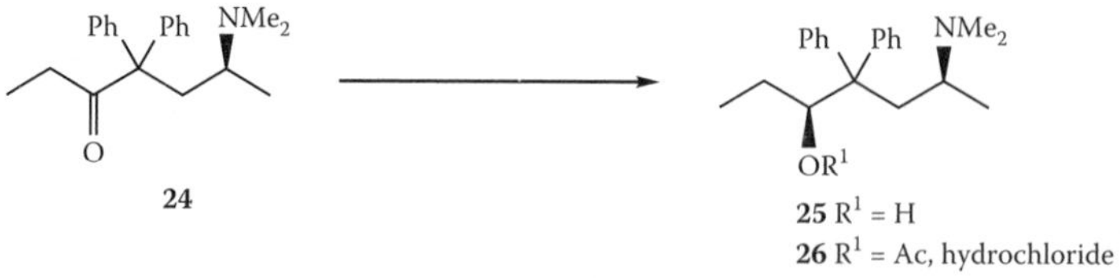

SCHEME 29.8

Previously, LAAM has been prepared from d-methadone (**24**) (Scheme 29.8) via catalytic reduction using hydrogen or, alternatively, using sodium and propanol.[29] However, the most convenient method for the reduction was using sodium borohydride in the presence of cerium(III) chloride. The resultant methadol **25** was obtained in quantitative yield and was acetylated using acetyl chloride. LAAM was isolated as the hydrochloride salt **26** with $[\alpha]_D$–60.6°, in accord with the literature value of –59°.[30] Full experimental details of this work have been reported.[31]

29.4.3 ENZYMATIC GENERATION AND *IN SITU* SCREENING OF DYNAMIC COMBINATORIAL LIBRARIES

Dynamic combinatorial chemistry (DCC) is a rapidly emerging field that offers a possible alternative to the approach of traditional combinatorial chemistry (CC).[32] Whereas CC involves the use of irreversible reactions to efficiently generate static libraries of related compounds, DCC relies on the use of reversible reactions to generate dynamic mixtures. The binding of one member of the dynamic library to a molecular trap (such as the binding site of a protein) is expected to perturb the library in favor of the formation of that member (Figure 29.1).

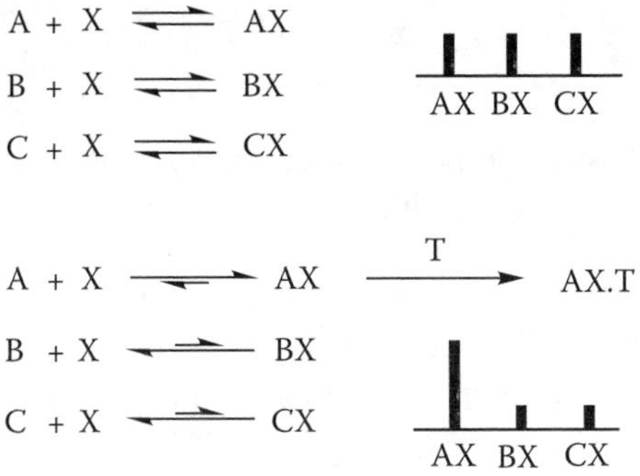

FIGURE 29.1 The dynamic combinatorial chemistry (DCC) concept: reversible reactions performed with a limiting amount of **X** generate a mixture of compounds **AX, BX,** and **CX.** The binding of **AX** to molecular trap **T** causes perturbation of the equilibria involving **A** and **X** to give overall amplification of **AX** at the expense of the other library members.

Comparison of the "perturbed" library with that generated in the absence of the trap should indicate which members of the library are interacting with the trap, which effectively offers *in situ* screening of the combinatorial library.

The DCC concept has already been proved by several research groups, including those of Lehn and Sanders.[32,33] However, significant experimental challenges remain before the method may be considered a practical complement to traditional CC. In particular, the *in situ* screening, which is an attractive feature of DCC, demands the use of conditions amenable both to the library formation and trapping stages. If the trap is envisaged as a protein or other biomolecule, the system is likely to require aqueous and near-physiologic conditions, where few covalent bond-forming reactions are compatible. To our knowledge no DCC experiment involving the formation of carbon–carbon bonds under physiologic conditions has been performed.[34]

Enzyme-catalyzed reactions, which are characteristically reversible under physiologic conditions, are ideally suited to the generation of dynamic combinatorial libraries. Many enzymes with broad specificity (required for library diversity) are already commercially available, and the application of modern techniques in directed evolution may be expected to increase their number.

In considering the application of enzyme catalysis to DCC, we were encouraged by the thermodynamic resolution of a dynamic mixture of aldol products by Whitesides and co-workers through the use of a broad-specificity aldolase to lead to reversible formation of carbon–carbon bonds under mild conditions.[35] For the current investigation[36] we chose a related enzyme, *N*-acetylneuraminic acid aldolase (NANA aldolase, EC 4.1.3.3), which catalyzes the cleavage of *N*-acetylneuraminic acid (sialic acid, **27a**) to *N*-acetylmannosamine (ManNAc, **28a**), and sodium pyruvate **29**; in the presence of excess sodium pyruvate, aldol products **27a–c** are generated from the respective substrates **28a–c** (Scheme 29.9). Thus, the equilibrium may be driven toward the formation of an aldol product in which the enzyme will accept a range of reducing sugars as the electrophilic component.

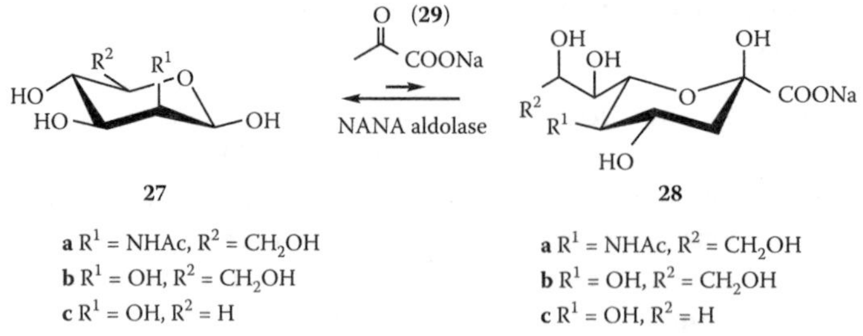

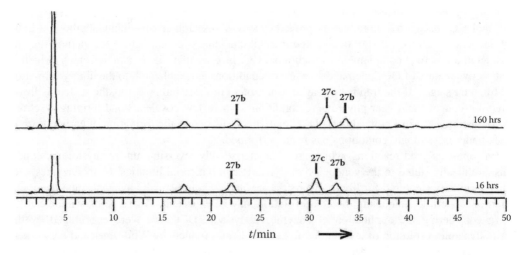

SCHEME 29.9

We succeeded in the generation of a small dynamic library of aldol products **27a**, **27b** (ketode-oxynonulosonic acid, KDN), and **27c** (ketodeoxyoctulosonic acid, KDO) through the action of NANA aldolase on a mixture of the corresponding substrates **28a**, **28b** (D-mannose), and **28c** (D-lyxose). Wheat germ agglutinin (WGA), a well-studied and readily available plant lectin,[37] was chosen as the molecular trap. WGA is known to specifically bind sialic acid with modest (mM) affinity, with the diequatorial C–4 hydroxy and C–5 acetamido groups of sialic acid forming the primary recognition motif. We thus expected amplification of sialic acid to occur when the dynamic library containing sialic acid was generated in the presence of WGA.

Generation of the dynamic library proved straightforward; a mixture containing equimolar amounts of the three substrates **28a–c** and two equivalents of sodium pyruvate was incubated in the presence of NANA aldolase (Figures 29.2 and 29.3). Aliquots of the incubation mixture were withdrawn at intervals and analyzed by ion-exchange HPLC.

The three aldol products gave reproducible peak areas. As expected, the use of a small excess of sodium pyruvate resulted in low conversion (15–40%) to the aldol products[38] and gave 2–5 mM concentrations of each.

FIGURE 29.2 High-performance liquid chromatography traces showing the change in concentration of sialic acid (**27a**), ketodeoxynonulosonic acid (KDN) (**27b**), and ketodeoxyoctulosonic acid (KDO) (**27c**) over time for a control incubation.

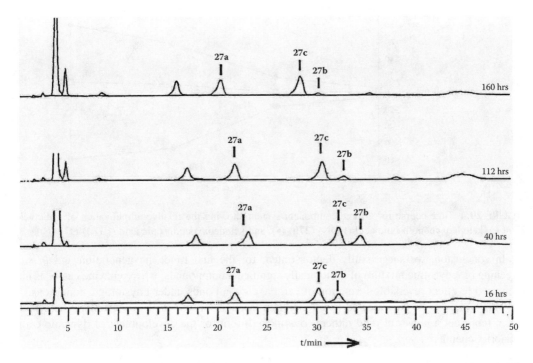

FIGURE 29.3 High-performance liquid chromatography traces showing the change in concentration of sialic acid (**27a**), ketodeoxynonulosonic acid (KDN) (**27b**), and ketodeoxyoctulosonic acid (KDO) (**27c**) over time for incubation with wheat germ agglutinin (WGA).

It was evident that the aldol reaction had reached equilibrium after 16 hours incubation because the product distribution changed very little after this time. Evidence that the enzyme does indeed catalyze both aldol formation and cleavage on the timescale used was demonstrated by reequilibrating a mixture of sialic acid and D-mannose in the presence of the NANA aldolase. After incubation overnight the mixture contained, in addition to the initial components, KDN and sodium pyruvate, which could only arise through a retro-aldol cleavage of sialic acid followed by aldol formation of KDN.

When the incubations were performed in the presence of WGA (sufficient lectin was added to provide at least one equivalent of binding sites, based on the amount of sialic acid formed in the control experiments) the product distribution was observed to change dramatically over time (Figure 29.3).

The system appears initially to approach a similar distribution to that observed in the control incubation, but thereafter the relative concentration of sialic acid is seen to increase, with a corresponding decrease in the relative concentration of KDN.

A plot of the percentage amplification in the relative concentration (as estimated from relative peak area) of each aldol product over controls is shown in Figure 29.4. The peak assigned to sialic acid contributed approximately 40% of the total aldol product peak area after 160 hours in the presence of WGA but only 22% of the total aldol product peak area in the control mixture after the same time. This difference corresponds to a relative amplification of about 80%. Conversely, the contribution of KDN to the total aldol product peak area after 160 hours was 31% in the control incubation but only 6% in the presence of WGA; this difference corresponds to a relative suppression of 80%. The observation that KDN and KDO were not suppressed to the same extent might be a consequence of WGA having a weak binding affinity for KDO.

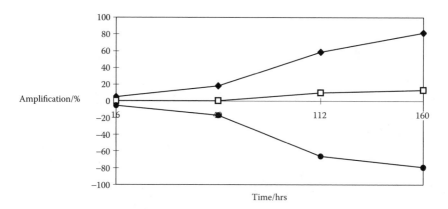

FIGURE 29.4 Time course for a three-component system showing the relative amplification of sialic acid **27a** (♦), ketodeoxynonulosonic acid (KDN) (**27b**) (●), and ketodeoxyoctulosonic acid (KDO) (**27c**) (□).

In conclusion, we successfully demonstrated, for the first time, the generation and *in situ* screening of a dynamic mixture of biologically significant compounds, where enzyme catalysis has been used to effect reversible formation of carbon–carbon bonds under physiologic conditions.

This work can be extended to larger libraries through the use of broad-specificity enzymes, which represent a powerful (and hitherto overlooked) tool for the development of dynamic combinatorial chemistry.

ACKNOWLEDGMENT

We thank Andrew Stachulski for his help in the preparation of this chapter in the first edition.

REFERENCES AND NOTES

1. Taylor, S. C. In *Opportunities in Biotransformations,* Coppang, L. G., Martin, R.E., Pickett. J. A., Bucke, C., Bunch, A. W., Eds. Elsevier: London, 1990, p. 170.
2. Ferguson, J. R., Lumbard, K. W., Scheinmann, F., Stachulski, A. V., Stjernlöf, P., Sundell. S. *Tetrahedron Lett.* 1995, *36*, 8867.
3. Stjernlöf, P., Trogen, L., Anderson, A. *Acta Chem. Scand.* 1989, *43*, 917.
4. *cf.* Bojarski, J. *J. Biochem. Biophys. Methods* 2002, *54*, 197–220.
5. *cf.* Hett, R., Fang, Q. K., Gao, Y., Hong, Y., Butler, H. T., Nie, X., Wald, S. A. *Tetrahedron Lett.* 1997, *38*, 1125–1128.
6. Okamoto, Y., Kaida, Y. *J. Chromatography, A,* 1994, *666*, 403–419.
7. Williams, K. L., Sander, L. C. *J. Chromatography, A,* 1997, *906*, 149–158
8. Schulte M., Strube, J, *J. Chromatography, A,* 2001, *906*, 399–416.
9. Hart, T. W., Metcalfe, D. A., Scheinmann, F. *J. Chem. Soc. Chem., Comm.* 1979, 156.
10. Rokach, D., Zanboni, R., Lau, C., Guindon, Y. *Tetrahedron Lett.* 1981, *22*, 2759.
11. Cotterill, I. C., Jaouhari, R., Dorman, G., Roberts, S. M., Scheinmann, F., Wakefield, B. J. *J. Chem. Soc. Perkin Trans. I* 1991, 2505.
12. Butt, S., Davies, H. G., Dawson, M. J., Lawrence, G. C., Leaver, J., Roberts, S. M., Turner, M. K., Wakefield, B. J., Wall, W. F., Winders, J. A. *Tetrahedron Lett.* 1985, *26*, 5077.
13. Grant, K. *Scotland on Sunday,* 26 August, 2001, p. 13.
14. a) Morley, J. S., Makin, M. K. *Pain Reviews* 1998, *5*, 51; b) Casy, A.F., Parfitt, R.T. *Opioid Analgesics*, Plenum Press: New York, 1986, p. 303.
15. a) Wolff, K., Rostami-Hodjegan, A., Shires, S., Hay, A.W.M., Freely, M., Calvert, R., Raistrick, D., Tucker, G.T. *Brit. J. Pharmacol.* 1997, *44*, 325 and references quoted therein; b) *Drugs Misuse and Dependence,* Department of Health (UK), 1999, p. 104.

16. *Merck Index, 12th Edition*, Merck & Co., Inc., New Jersey, 1996, p. 1015, monograph no. 6008.
17. Brode, W. R., Hill, M. W. *J. Org. Chem.* 1948, *13*, 191.
18. Larsen, A. A., Tullar, B. F., Elpern, B., Buck, J. S. *J. Am. Chem. Soc.* 1948, *70*, 4194.
19. Howe, E. E., Sletzinger, M. *J. Am. Chem. Soc.* 1949, *71*, 2935.
20. a) Walton, E., Ofner, P., Thorp, R. H. *J. Chem. Soc.* 1949, 648; b) D. Adams, unpublished work.
21. Barnett, C. J., Smirz, J. C. *J. Org. Chem.* 1976, *41*, 710.
22. *Fluka Laboratory Chemicals*, 2001/2002, p. 643, catalogue no. 77637 and 69769.
23. *Fluka Laboratory Chemicals*, 2001/2002, p. 509, catalogue no. 39300.
24. Goering, H., Eikenberry, H., Koermer, G., Lattimer, C. *J. Am. Chem. Soc.* 1974, *96*, 1493.
25. Poupaert, J. H., Van der Jeugd, P., Gerardy, B. M., Claesen, M., Dumont, P. *J. Chem. Res. (S)*, 1981, 192, *J. Chem. Res. (M)*, 1981, 2484.
26. Bockmuehl, V. M., Ehrhart, G. *Liebigs Ann.* 1949, *561*, 52.
27. *LAAM in the Treatment of Opiate Addiction, Chapter 1*, The National Clearing House for Alcohol and Drug Information, www.health.org/govpubs/bkd170/22c.htm.
28. *European Agency for the Evaluation for the Evaluation of Medicinal Products*, 19 April 2001, EMEA/8776/01, www.emea.eu.int/.
29. May, E. L., Mosettig, E. *J. Org. Chem.* 1948, *13*, 459 and 663.
30. Pohland, A., Marshall, F. J., Carney, T. P. *J. Am. Chem. Soc.* 1949, *71*, 460.
31. Hull, J. D., Scheinmann, F., Turner, N. J. *Tetrahedron: Asymmetry* 2003, *14*, 567.
32. For recent reviews, see Otto, S., Furlan, R. L. E., Sanders, J. K. M. *Drug Discovery Dev.* 2002, *7*, 117; Cousins, G. R. L., Poulsen, S. A., Sanders J. K. M. *Curr. Opin. Chem. Biol.* 2000, *4*, 270; Huc, I., Lehn, J.-M. *Proc. Natl. Acad. Sci. USA* 1997, *94*, 2106; Lehn, J.-M. *Chem. Eur. J.* 1999, *5*, 2455.
33. a) Ramström, O., Lehn, J.-M. *ChemBioChem* 2000, *1*, 41; Huc, I., Lehn, J.-M. *Proc. Natl. Acad. Sci. USA* 1997, *94*, 2106; b) Furlan, R. L. E., Ng, Y. F., Otto, S., Sanders, J. K. M. *J. Am. Chem. Soc.* 2001, *123*, 8876; c) Hochgürtel, M., Kroth, H., Piecha, D., Hofmann, M.W., Nicolaou, K. C., Krause, S., Schaaf, G., Sonnemoser, G., Eliseev, A. V. *Proc. Natl. Acad. Sci. USA* 2002, *99*, 3382.
34. A library of vancomycin analogues has been prepared using ring closing metathesis reactions in aqueous solution (Nicolaou, K. C., Hughes, R., Cho, S. Y., Wissinger, N., Labischinski, H., Endermann, R. *Chem. Eur. J.* 2001, *7*, 3824); a kinetic template effect was observed when the library was prepared in the presence of a dipeptide trap.
35. Bednarski, M. D., Simon, E. S., Bischofberger, N., Fessner, W. D., Kim, M. J., Lees, W., Saito, T., Waldmann, H., Whitesides, G. M. *J. Am. Chem. Soc.* 1989, *111*, 627.
36. Lins, R. J., Flitsch, S. L., Turner, N. J., Irving, E., Brown, S. A. *Angew. Chem., Int. Ed.* 2002, *41*, 3405.
37. Peters, B. P., Ebisu, S., Goldstein, I. J., Flashner, M. *Biochemistry* 1979, *18*, 5505.
38. As determined by HPLC analysis of standard solutions of synthetic sialic acid. The control mixture contained D-mannose and ManNAc (0.5 μmol each), sialic acid (0.13 μmol), and NANA aldolase in buffer solution (0.2 U), total volume: 40 μL.

30 Commercial Synthesis of the Antiglaucoma Prostanoid Travoprost

Martin Fox, Mark Jackson, Ian C. Lennon, and Raymond McCague

CONTENTS

30.1 INTRODUCTION

Travoprost is a prostaglandin analogue (prostanoid) developed by Alcon Laboratories in commercial use for the treatment of glaucoma and ocular hypertension, where it is a highly potent compound administered as eye drops.[1] Important in its overall development was the construction of a suitable process for its commercial manufacture, and in this regard an overriding feature of the molecule is the presence of five stereogenic centers. There are four around the cyclopentane ring and one in a sidechain. In addition there are two olefin functions, one with *trans-* and one with *cis-*geometry. Adequate control of the stereochemistry of the molecule was consequently of paramount importance in the design and development of a synthesis. This chapter discusses the considerations that we needed in the synthetic design and to eventually reach a robust supply position for the active pharmaceutical ingredient.

30.2 CHOICES OF SYNTHETIC APPROACH

At the commencement of the project there were several routes available for the synthetic construction of prostanoid compounds; our challenge was which to choose. For this, we were mindful of the stereochemical features of the molecule and the need for a full stereochemical characterization of the active pharmaceutical ingredient (API) produced. This was required in the regulatory submissions,

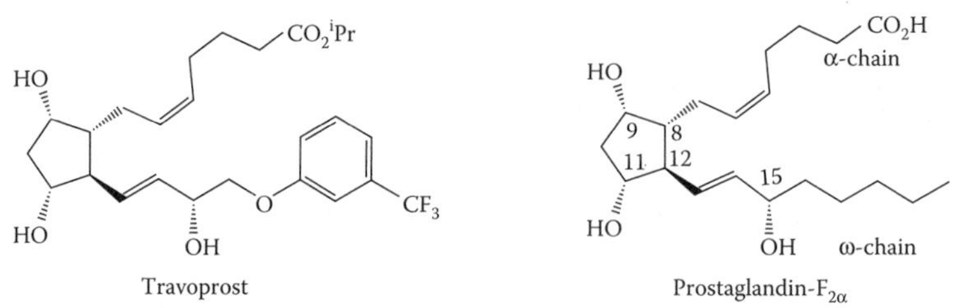

FIGURE 30.1 Travoprost and prostaglandin-$F_{2\alpha}$.

the approval of which was necessary before the pharmaceutical agent could be sold on the market. In particular, if the synthesis produced a stereoisomeric contaminant that was difficult to remove, then the necessary efforts to remove the contaminant might render the synthesis impractical. Therefore, in the following discussion, various literature routes are considered with particular regard to the control of the stereocenters in the molecule.

Travoprost is related to prostaglandin $F_{2\alpha}$, shown in Figure 30.1, where the F refers to the series of prostaglandins containing a cyclopentane-1,3-diol, the 2 refers to the series having a *cis*-olefin function in the carboxylate-bearing sidechain, and the α- refers to the configuration of the hydroxyl function at C–9. Clearly, travoprost has these features; therefore in principle any synthesis of prostaglandin $F_{2\alpha}$ could be adapted to make travoprost.

30.2.1 COREY LACTONE APPROACH TO PROSTAGLANDINS

Around 35 years ago the need for synthetic approaches to prostaglandins and their analogues was recognized because it is not possible to obtain useful amounts from natural sources. This challenge prompted extensive synthetic chemical research and the discovery of many routes to access the compounds, which contributed significantly to the development of the science of organic synthesis. The earliest routes were cumbersome and only provided access to the racemic prostaglandins,[2] but for prostaglandin-E_1 it was soon demonstrated that bioactivity was present only in the natural enantiomer.[3] A landmark in chemical synthesis was the identification by Corey of a versatile bicyclic synthon from which a wide range of prostaglandins and analogues could be made; it is known (in variously protected forms) as Corey lactone (see Scheme 30.1).[4] Its single enantiomer was initially

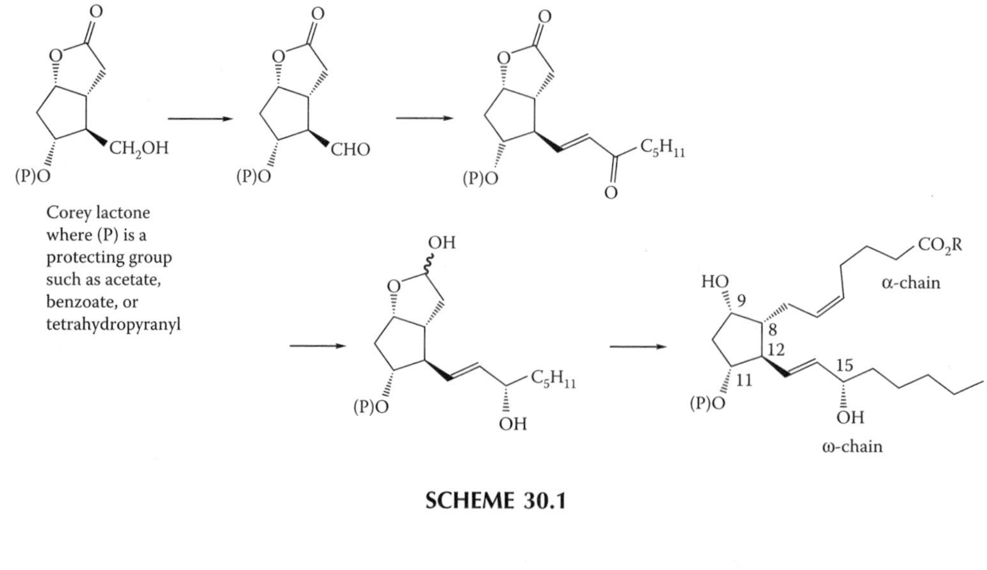

SCHEME 30.1

accessed by resolution of an intermediate hydroxy acid with ephedrine, thereby enabling single-enantiomer prostaglandin derivatives to be synthesized routinely. In the typical route, as shown in Scheme 30.1, the chain from C–13 to C–20 (also known as the ω-chain) is attached via a Horner-Emmons reaction of the 13-carboxaldehyde with a β-keto-phosphonate providing the necessary *trans*-stereochemistry of the olefin. The remaining chain C–1 to C–7 (the α-chain) is then added last with a Wittig reagent to give predominantly the natural *cis*-olefin stereochemistry. Alternatively, it is possible to arrange the synthesis so that the chains are introduced in the opposite order.[5]

Although four of the five stereogenic centers of the prostaglandin are defined by the Corey lactone, chemoselective and stereoselective reduction of the enone to establish the C–15 allylic alcohol is challenging. In the early work[4] the enone was reduced to a diastereoisomeric mixture, which was separated by preparative chromatography and the unwanted isomer was recycled back to the enone by oxidation. Subsequently chiral reagents were used for the reduction. For instance, a trialkylborane derived from thexylborane and (+)-limonene was treated with *t*-butyllithium to provide a borohydride, which reduced the enone at –120°C giving a 4.5:1 mixture of the C–15 epimers.[6] However, this reduction posed a problem, which prompted Corey to design an alternative synthesis via epoxide and sulfur intermediates 20 years later.[7]

Although the earliest syntheses of the Corey lactone, using a cyclopentadienyl-thallium species, were not very attractive, it was soon found that the lactone could be usefully generated by a Baeyer-Villiger reaction on a bicycloheptenone.[8] In turn, this derives from an easy [2 + 2] cycloaddition reaction between dichloroketene and cyclopentadiene (Scheme 30.2).[9] Additionally, an intermediate of the lactone could be resolved by way of the α-methylbenzylamine salt of its opened hydroxyacid, bringing the resolution step earlier in the synthesis than previously.[10]

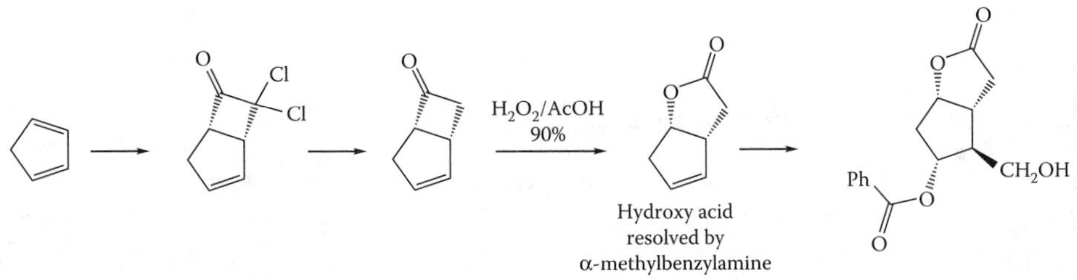

Hydroxy acid
resolved by
α-methylbenzylamine

SCHEME 30.2

30.2.2 Newton-Roberts (tricycloheptanone) Routes to Prostaglandins

Most of the previously mentioned work by Corey was published between 1968 and 1973. A few years later in 1977 an intriguing route, also starting from the dichloroketene cycloaddition–derived bicyclo[3.2.0]hept-2-en-6-one, was devised by Newton and Roberts.[11] An outline of this route is shown in Scheme 30.3. Formation of the bromohydrin gives essentially a single product that establishes the hydroxyl function at C–9 (of the prostaglandin), which after protection can be cyclized to a strained tricyclic ketone intermediate. Reaction of this intermediate with a cuprate reagent led to a facile and completely regiocontrolled opening of the cyclopropane ring.[11] From a stereochemical point of view there are two very interesting features of this approach. The first is that it requires the opposite enantiomer of the bicyclo[3.2.0]hept-2-en-6-one intermediate as needed for the Corey lactone because the overall sequence of steps constitutes an overall inversion of stereochemistry. For comparison, in Scheme 30.3 the structures of the intermediates are drawn with the cyclopentane ring in the same orientation throughout. As a curiosity, it should be noted in the Newton-Roberts approach the silyl-protected OH of the synthon ultimately provides the 9-hydroxyl of the prostaglandin and the lactone provides the 11-hydroxyl, whereas from the Corey lactone the benzoate group ultimately provides the 11-hydroxyl and the lactone provides the 9-hydroxyl. Thus, there is useful complementarity between the two approaches.

It is the second feature of the Newton-Roberts approach that is valuable from a stereocontrol point of view. In the Corey lactone approach, the synthon itself already contains C–13 of the ω-sidechain, which enforces the use of a reagent containing a C-15 carbonyl, making control of the C–15 hydroxyl stereochemistry difficult, whereas in the tricycloheptanone approach the entire sidechain (C–13 to C–20) is added in one piece, which means that this sidechain can be readily introduced in the form of a reagent that has this sidechain stereochemistry already in place. Following the addition of the ω-sidechain and a Baeyer-Villiger oxidation to the lactone, the later steps to add the α-chain are essentially identical to the Corey lactone route because the lactones are regioisomeric.

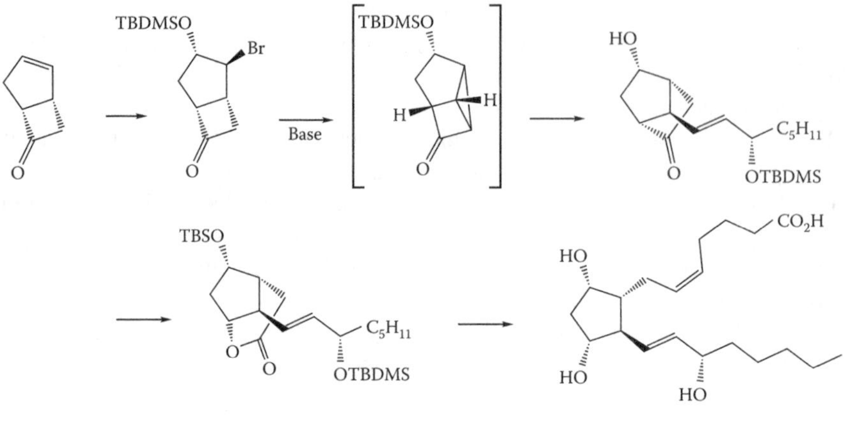

SCHEME 30.3

30.2.3 CYCLOPENTENONE-INTERMEDIATE ROUTES TO PROSTAGLANDINS

A further versatile set of routes to synthesize prostaglandins is based on conjugate additions to the olefin function of cyclopentenone derivatives. Key intermediates used are shown in Scheme 30.4. In the earliest work of this type in the early 1970s[12] an enone synthon was used already bearing the α-chain in its saturated form [i.e., as –(CH$_2$)$_6$CO$_2$Me], and the ω-chain then was added through an appropriate cuprate reagent (Scheme 30.4). As with the Newton-Roberts approach there is again the advantage that the ω-chain can be introduced complete with the requisite chirality at C–15. A nice feature is that the intermediate, which has only the one stereocenter (at what will become the C–11 hydroxyl), can be resolved through biocatalysis with a lipase and an enol acetate, and all the material can be converged to the required enantiomer by way of a Mitsunobu inversion on the unwanted isomer.[13] In a complementary synthesis, and perhaps the first enone-based approach to a fully featured prostaglandin, Stork described a route where the ω-chain was introduced first and an exocyclic methylene was used as a handle on which to add the α-chain.[14]

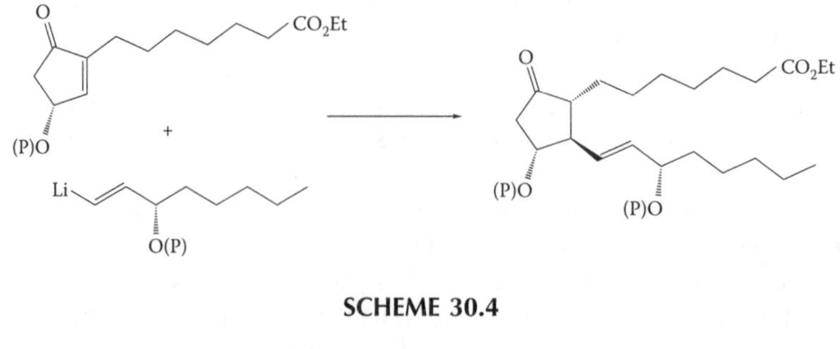

SCHEME 30.4

Conceptually, the most elegant approach in using an enone synthon is a 3-component coupling approach wherein one sidechain is introduced by a conjugate addition onto the enone and then the other sidechain is introduced by way of an alkylation onto the carbon of the formed enolate. This was recognized by Syntex workers in 1974, although only simplified compounds missing the 11-hydroxy were described.[15] Several years later Noyori was successful in bringing this approach into practice for fully functionalized prostanoids.[16,17] A particularly elegant feature is that although the enone synthon contains only the one stereocenter (for C–11), the propensity for *trans*-geometry of vicinal substituents means that the correct C–12 and C–8 configurations are obtained directly (Scheme 30.5). The starting material for this approach is readily accessed from a cyclopentene-1,4-diol. Its *meso*-diacetate is enzymatically desymmetrized, the liberated hydroxyl is protected, the other acetate is removed, and the alcohol is oxidized to the ketone.[18] However, there is a problem with the 3-component coupling approach. The enolate formed by the addition tends to equilibrate with the opposite regioisomer (toward C–10) and if it does so, the C–11 oxygen eliminates. Ways have been found to alleviate this problem by switching the metal of the enolate to tin or zinc[19] or by using a particularly reactive alkynyl triflate electrophile.[20] Another approach was to start with a synthon with a dioxolane attached between the 10- and 11-positions, which prevents the enolate equilibration but requires an additional step to reduce the unwanted oxygen later.[21]

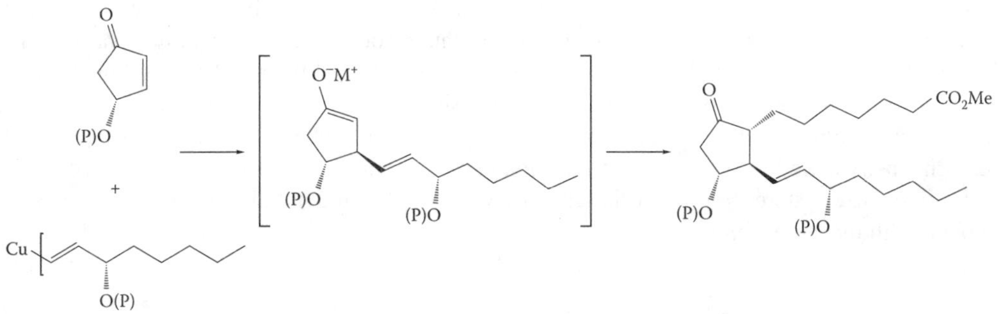

SCHEME 30.5

Noyori's 3-component coupling approach leads directly to the prostaglandin E series that has the ketone function at C–9. For the prostaglandin F series it is necessary to effect a stereoselective reduction of this ketone function. It is claimed that the bulky hydride reagent L-Selectride gives complete formation of the 9α selectivity through addition from the least-hindered side, as dictated by the adjacent α-chain at position 8.[22] It is interesting that the chiral-reducing agent BINAL-H, an aluminum complex of binaphthol, as well as giving high α:β stereochemistry preference, reduces one enantiomer 130-fold faster than the other enantiomer.[22] In principle, that might be used to kinetically discard any unwanted isomer contaminant reaching that far into the synthesis (either the enantiomer, or just unwanted epimer at C–8).

An evolution from Noyori's approach leads to one where a diethylaminomethyl substituent is used to give a one-carbon start to the α-chain. From this synthon there is much flexibility because the ω- and α-chains can be added in either order, and synthesis of the single isomer synthon can start with a Sharpless epoxidation. This route is shown in Scheme 30.6.[23] For completeness, we note that there is another variant of the general enone approach identified by Danishefsky, in which the C–9 and C–11 carbon oxidation states are inverted (ketone at what is to become C–11) where the α-chain is added first as a nucleophile, and then the ω-chain is added second as an electrophile.[24]

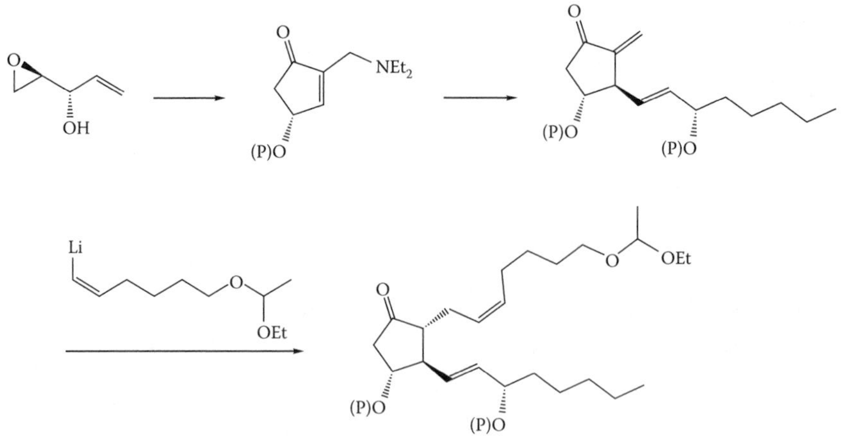

SCHEME 30.6

30.3 CHOICE OF ROUTE FOR TRAVOPROST SYNTHESIS

Mindful of the various options available for the synthesis of travoprost by reference to the known chemical syntheses of the natural prostaglandins, our main concern, as stated earlier, was to ensure rigorous stereocontrol to simplify product-purification aspects and the regulatory submission.

With regard to the stereochemistry, the three basic approaches are illustrated in Figure 30.2. In this figure a tick symbol (✓) represents a stereogenic center where we can be assured that the stereochemistry is rigorously defined. A question mark symbol (?) means that we can be concerned of the rigor of the stereocontrol in the chemistry, and a frown symbol (☹) means we expect a problem with the stereocontrol.

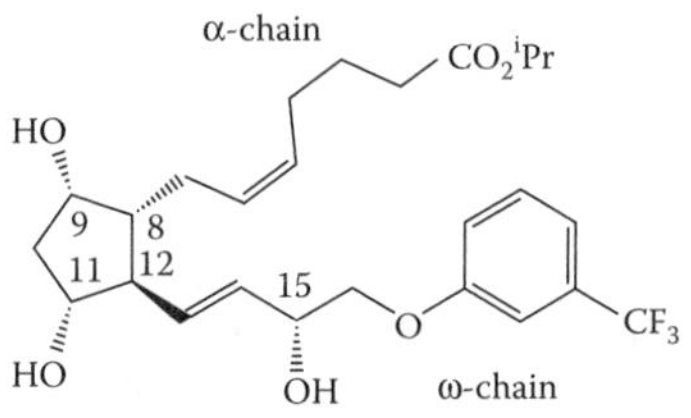

Approach	Integrity of Stereocenter				
	8	9	11	12	15
Corey lactone	✓	✓	✓	?	☹
Newton/Roberts	✓	✓	✓	✓	✓
Enone addition	?	☹	✓	?	✓

FIGURE 30.2 Stereochemical attributes for prostanoid routes.

With the Corey lactone four of the stereocenters (8,9,11,12) are defined in the lactone synthon, although there could be a slight concern with the integrity of the stereocenter at position 12 because as an aldehyde intermediate there is some chance of epimerization at that center via its enol. The main problem with this approach is the control of the stereocenter at position 15, which has to be introduced late into the process. Inevitably, there is some diastereoisomer formed, which can be difficult to remove and therefore is an aspect that would require significant attention to satisfy the regulatory demands for process stereocontrol in a route starting with Corey lactone.

The Newton-Roberts approach, however, defines absolutely all the stereocenters. Four are held together in the tricyclic ketone intermediate manufactured in single enantiomer form, whereas because the ω-sidechain component used has an extra carbon atom relative to the synthon required in the Corey lactone route, there is opportunity to predefine this center rigorously as the vinylic alcohol. There is a matter of regioisomerism in this approach as a result of the selectivity of the Baeyer-Villiger oxidation, which is discussed later, although the regioisomer is fairly easy to remove.

In the enone addition approach, as exemplified by the work of Noyori, there is potentially the least rigorous control of selectivity around the cyclopentane ring because the cyclopentenone synthon used only has one stereocenter defined (for position 11), and the selectivity of a chain of subsequent reactions is relied on. Although almost complete selectivity for formation of the vicinal *trans*-relationship at position 12 may be expected, process control would be needed to verify this. Any imperfection at position 12 will relay a corresponding incorrect configuration at positions 8 and then 9. In particular there is much concern for position 9, which derives from a diastereoselective ketone reduction and which is far removed from the original controlling stereocenter. However, the ω-sidechain can be introduced with its stereocenter rigorously controlled as for the Newton-Roberts approach. Relative to the others the Noyori route could be preferred for the prostaglandin-E series (9-keto) where there is no stereocenter at C–9.

Aside from the stereogenic (chiral) centers, travoprost contains olefinic functions in the α- and ω-chains that need to be of *cis*- and *trans*-configuration, respectively. The *trans*-ω-chain olefin configuration seems to be well-controlled either via the vinyl halide (Newton-Roberts and enone approaches) or via the phosphonate coupling (Corey lactone approach). For the *cis*-α-chain, the Wittig coupling approach also produces some *trans*-isomer by-product, but this is well-accepted in synthetic prostanoids. The alternative is via a Lindlar hydrogenation of the alkyne, but in addition to the *trans*-isomer, overreduced saturated compound (or underreduced alkyne) could be a problematic contaminant. All in all there is nothing to choose between the routes on the basis of olefin configuration.

Thus, in respect to the five stereogenic centers, our analysis showed the Newton-Roberts approach to give the most rigorous control of the stereochemistry; hence it is this route that we chose, initially for exploration and subsequently for development and manufacture. This route was also well-aligned with the available technology base developed within Chirotech. We were confident that we would be able to develop an enzymatic resolution approach for the ω-chain, scale up the classical resolution approach to the bicyclo[3.2.0]hept-2-en-6-one synthon, and finally develop the cuprate addition for large-scale operation. Moreover, we were comfortable that the route could be developed without any issue of third-party intellectual property.

30.4   COMMERCIAL SYNTHESIS OF TRAVOPROST

The initial retrosynthetic analysis for a route to travoprost, using the Newton-Roberts route, provided three key synthons: a) the α-chain was derived from the commercially available Wittig salt, (4-carboxybutyl)triphenylphosphonium bromide; b) the ω-chain required a route to the single enantiomer mixed cuprate reagent, most conveniently derived from the propargylic alcohol via the *trans*-vinyl iodide, to be developed and c) the cyclopentane core could be derived from the single enantiomer protected bromohydrin (Figure 30.3).[11c,25]

Our first goal was to develop a scaleable route to the single enantiomer propargylic alcohol. The racemic alcohol was readily produced on a multi-kilogram scale by alkylation of the phenol with inexpensive bromoacetaldehyde diethyl acetal, hydrolysis to the aldehyde, and addition of the acetylenic Grignard reagent. With this material at hand we developed an efficient bioresolution process that esterified the alcohol in the presence of Chirazyme L9 (Scheme 30.7).[26] At this point

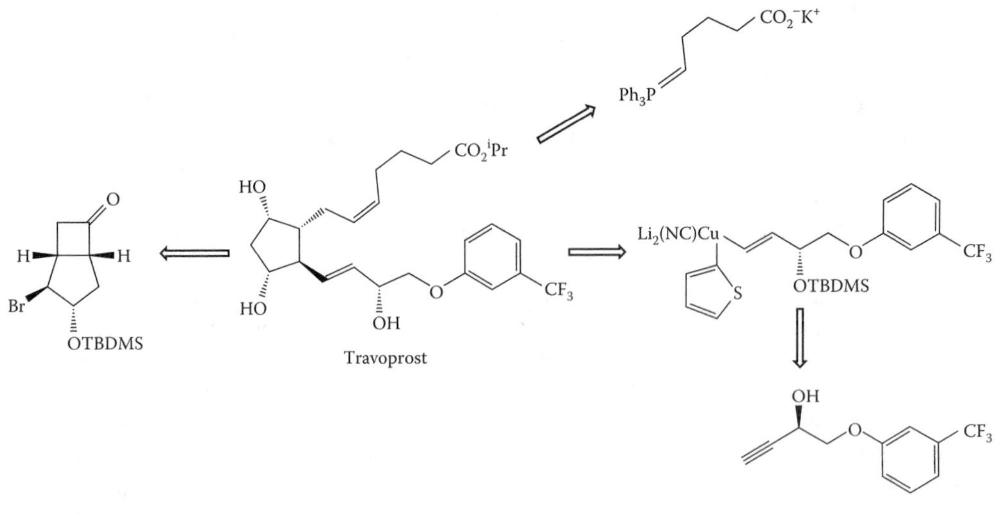

FIGURE 30.3 Retrosynthetic analysis for travoprost.

we had good yield of the desired butyrate ester in >98% ee and we could form a hydrogen sulfate ester of the off-isomer alcohol to afford separation. This would provide the desired compound in a maximum of 50% yield. By forming the mesylate and treating the mixture of ester and mesylate with butyric acid, we could obtain the desired enantiomer of butyrate ester in approximately 90% ee. A Chirazyme L2-mediated ester hydrolysis, protection of the free alcohol, and removal of the residual butyrate ester afforded the desired silyl-protected acetylenic alcohol in 99.5% ee and 65% yield from the racemate.[26] To support the ongoing manufacture of travoprost, in excess of 50 kg of this synthon has been manufactured by methods based on this approach.

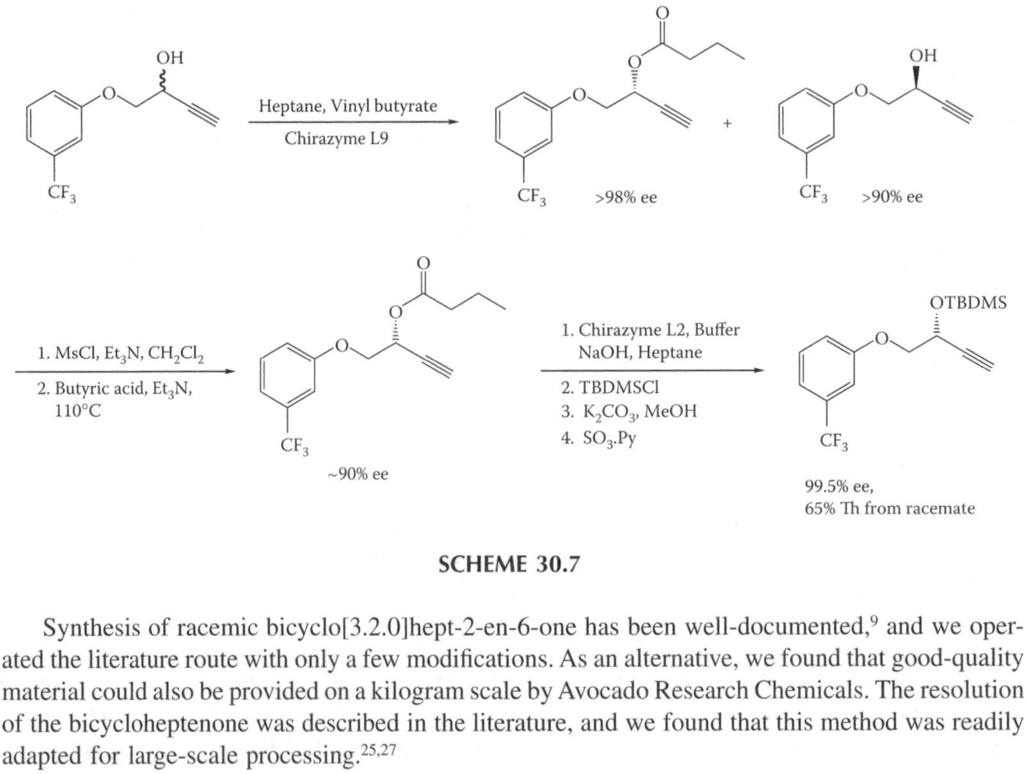

SCHEME 30.7

Synthesis of racemic bicyclo[3.2.0]hept-2-en-6-one has been well-documented,[9] and we operated the literature route with only a few modifications. As an alternative, we found that good-quality material could also be provided on a kilogram scale by Avocado Research Chemicals. The resolution of the bicycloheptenone was described in the literature, and we found that this method was readily adapted for large-scale processing.[25,27]

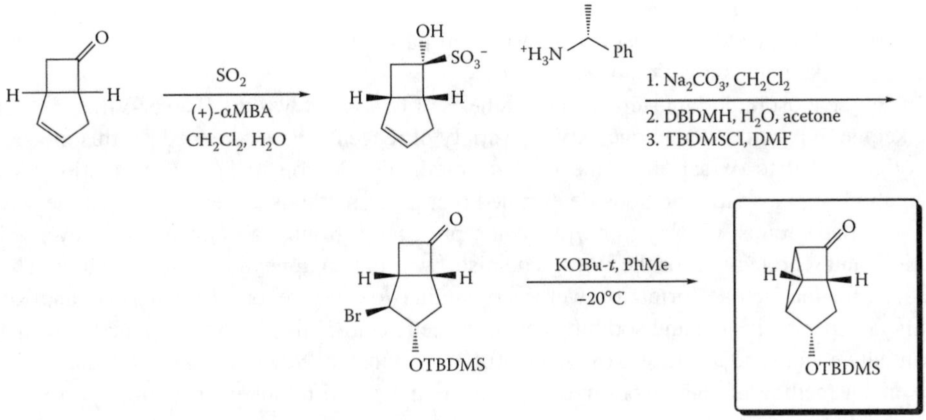

SCHEME 30.8

The resolution was carried out by forming an α-methylbenzylamine salt of the α-hydroxysulfonic acid derivative of the bicycloheptenone (Scheme 30.8), recrystallizing twice, cracking the salt, and isolating the single enantiomer product (99% ee). Formation of the bromohydrin was extremely selective to give the desired product in >90% regio and facial selectivity. The bromohydrin could be recrystallized, but this enriched the product in the hydantoin by-product from the bromination reagent. Protecting the hydroxyl with the *tert*-butyldimethylsilyl group afforded a much more crystalline product. A single recrystallization gave the desired product in high purity and a 38% yield from the salt. We have manufactured in excess of 70 kg of this product to support travoprost manufacturing campaigns. The highly reactive and unstable tricycle was obtained by treatment of the silyl-protected bromohydrin with KOBu-*t* in toluene at –20°C, and was used as required.

The required *trans*-vinyl iodide was made from the acetylene using an *in situ* variant of Schwartz's reagent, as described by Negishi and co-workers.[28] Zirconocene dichloride is readily available, and the procedure was found to be reproducible on scale, as compared to the use of commercial Schwartz's reagent. With the desired crude vinyl iodide in hand we formed the key mixed cuprate reagent by lithiation of the vinyl iodide and treatment with freshly prepared 2-thienylcyanocuprate (Scheme 30.9). Use of commercial cuprate gave unsatisfactory and irreproducible results; we felt, therefore, it was more prudent to make the reagent ourselves.[27]

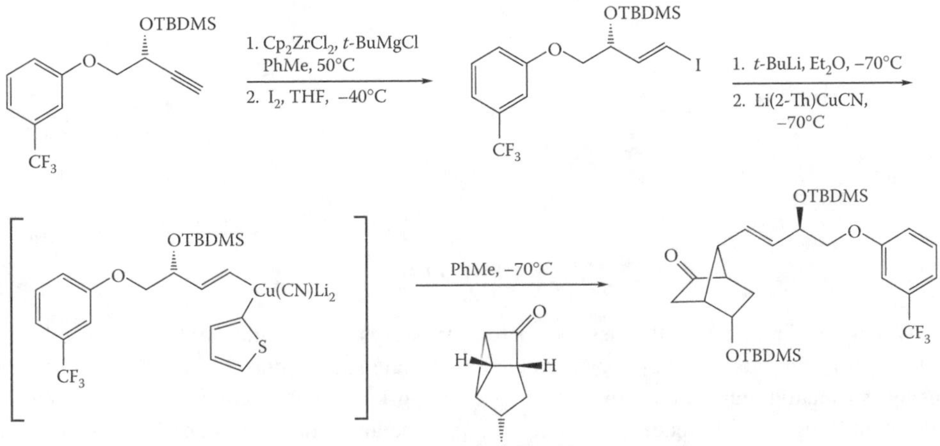

SCHEME 30.9

Reaction of the cuprate with a toluene solution of the tricycle at −70°C afforded the desired ketone in very high yield (60–80%). This was semi-purified by passing through a short silica plug to provide the ketone as a yellow-waxy solid.

One of the most important steps in the synthesis of travoprost was the Baeyer-Villiger oxidation of the ketone to provide the lactone. A wide variety of reagents were screened for this process, but most of these led to oxidation of the ω-chain olefin in addition to lactone formation. A short investigation of enzymatic methods determined that these methods were not compatible with the silyl protecting groups. Clearly, removal of the protecting groups, an enzymatic Baeyer-Villiger oxidation, and reprotection for the Wittig chemistry was not an appealing route to follow. The best conditions for the lactone formation were very similar to those reported by Newton and Roberts for PGF$_{2\alpha}$,[11a] peracetic acid and sodium acetate in acetic acid. This gave the regioisomeric lactones in quantitative yield and a 3:1 ratio (Scheme 30.10). Fortunately Newton and Roberts had described a convenient method of selectively hydrolyzing the unwanted minor isomer to a hydroxy acid,[29] which could be removed by passing the mixture down a short plug of silica gel and recrystallizing. The selectivity of this hydrolysis process is attributed to the severe congestion within the tetrahedral transition state of the minor isomer, leading to faster hydrolysis. It was essential to monitor the hydrolysis reaction to ensure that all the minor isomer had been converted into the hydroxy acid because removal of the regioisomeric lactone by crystallization methods proved to be impractical.

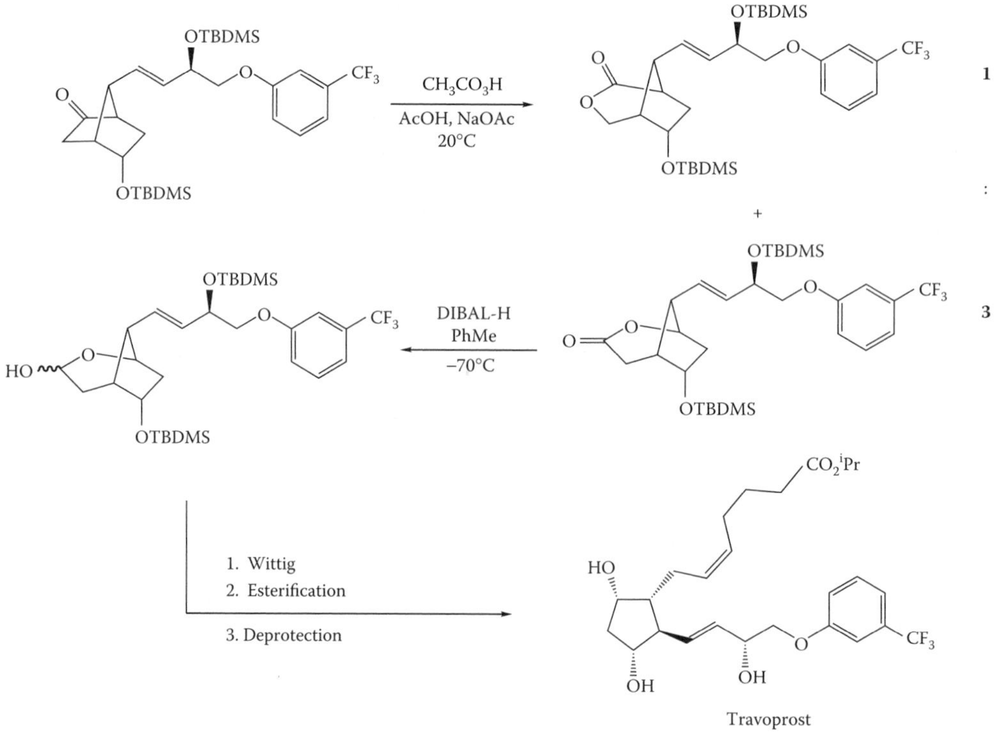

SCHEME 30.10

It was a key finding that the desired lactone was crystalline and could be obtained in high chemical purity. In most syntheses of prostanoids the late stage, protected intermediates are oils, requiring chromatography for purification. By carrying out a recrystallization of the lactone we could obtain a high purity, readily characterized intermediate only four steps from the end of the synthesis. Having a controlled specification at this stage ensured that travoprost of reproducible purity was produced in each manufacturing campaign. In addition, we were able to obtain a crystal structure[27] that confirmed the relative configuration of the lactone intermediate and was useful for travoprost regulatory filings.

The synthesis of travoprost was completed using standard chemistry that has been used on a multitude of prostanoids. The lactone that comes from this route is a regioisomer of that obtained from the Corey lactone route, so it is not surprising that similar chemistry can be used. The lactone was reduced to a lactol and a *cis*-selective Wittig reaction, using a commercially available phosphonium salt, gave an intermediate with both sidechains attached (Scheme 30.10). Through careful temperature control we could limit the amount of α-chain *trans*-olefin to less than 3%, which is a quantity that is allowed in the final specification of prostanoid pharmaceuticals. Silyl migration between the C–11 and C–9 hydroxyls was also observed, but this was of no consequence because after esterification and deprotection it is irrelevant which of the hydroxyls were silylated. As there were free hydroxyl groups in the penultimate intermediate, we used very mild alkylation conditions (DBU, isopropyl iodide) for esterification.[27] Silyl deprotection was achieved using aqueous hydrochloric acid in isopropanol.

Finally, crude travoprost was purified using a Biotage Prep 150-L chromatography unit. The previous four steps (DIBAL-H reduction, Wittig, esterification, and deprotection) were extremely high yielding and clean. The lactone was of exceptionally high purity, so the final bulk active purification was relatively straightforward (Scheme 30.10).

In conclusion, we have shown that the Newton-Roberts tricycle rearrangement route[11] to prostaglandin analogues is suitable for the kilogram manufacture of active pharmaceuticals such as travoprost. The total number of synthetic steps was 22 and the longest linear sequence was 16 steps, providing travoprost in yields from 4% to 7%. Key features of this approach are the rigorous control of all 5 stereocenters, the use of bioresolution for the single enantiomer ω-chain, the use of classical resolution of the cyclopentane core, and the identification of a late-stage crystalline compound to ensure good-quality final product. This route, with some minor modifications that cannot be disclosed here, is the basis of our manufacturing route to travoprost on a low-kilogram scale.

30.5 PROCESS-RELATED IMPURITIES

In a multi-step synthesis of a complex molecule with 5 chiral centers and 2 stereodefined olefins a large number of process-related impurities could be formed. It was important to rationalize which impurities could be realistically expected from the synthetic pathway. There are many obvious impurities that could be readily accessed using the route to travoprost, such as the free acid and the ethyl ester. Other potential impurities are shown in Figure 30.4. We were concerned that the ethyl ester could be formed by transesterification of the isopropyl ester on the Biotage column, but we later found out that it was formed from any quantities of free acid in the crude mixture reacting with the ethyl acetate solvent on the column. Hence it was important to control the esterification reaction to achieve a high conversion. A mixture containing all 7 possible acetates was prepared by reaction of travoprost with acetic anhydride in pyridine. The acetates were also observed on heating an ethyl acetate solution of travoprost in the presence of silica, but prolonged heating was required; thus we were able to prove that the acetates were not formed in the final chromatography. Furthermore, no ethyl ester was observed under these conditions.

The 15-epi diastereoisomer was readily synthesized by using the (*S*)-enantiomer of the acetylenic alcohol and completing the prostanoid synthesis. The epoxide impurity was made by epoxidizing the lactone intermediate with mCPBA, giving a 3:1 mixture of epoxides and carrying this through the synthesis. It is interesting that the final epoxide was only a single diastereoisomer and was identical to a sample isolated from crude travoprost.

Other potential impurities could be generated from the regioisomeric lactone if it was not rigorously removed from the desired isomer. We may expect to see the ring opened regioisomeric

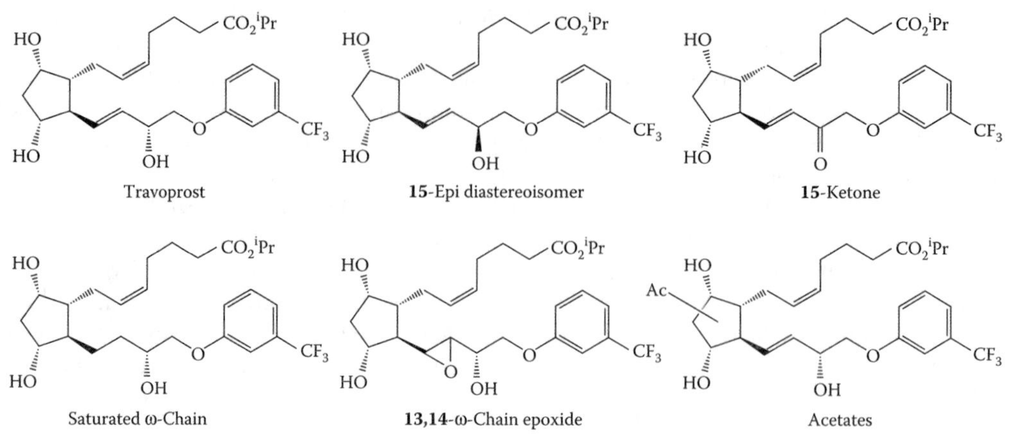

FIGURE 30.4 Proposed process-related impurities.

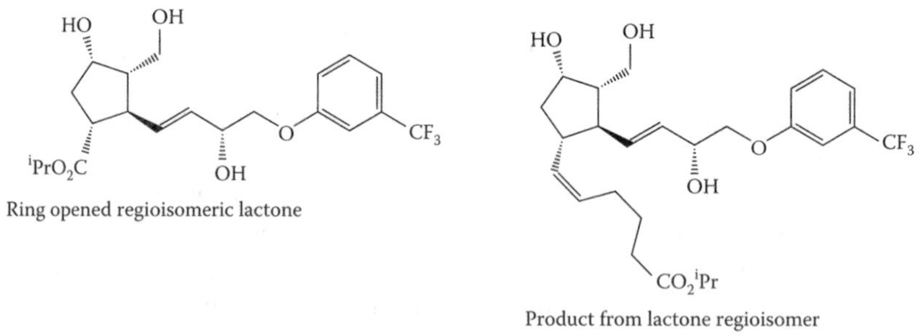

FIGURE 30.5 Process-related impurities from the regioisomeric lactone.

lactone if the hydroxy acid was not removed and the regioisomeric prostanoid (Figure 30.5) if the other lactone carried through the synthesis. By careful control of the synthesis of the lactone and introducing a tight specification, we were able to ensure that these impurities did not appear in the final active ingredient.

The final impurity of interest was the enantiomer of travoprost. During our early-stage synthesis of toxicology and clinical trial material no enantiomeric excess assay was available. Given the convergency in the coupling reaction, we argued that there is no need to establish an enantiomeric excess assay for the coupled intermediate (and subsequently for travoprost itself) because the amount of the enantiomer (with all the stereocenters of opposite configuration) will be vanishingly small. Hence, as in Figure 30.6, if the starting components of the coupling reaction are each 98% ee (i.e., a 99:1 mixture of major and minor enantiomers), then in theory the composition of the coupled product will be 98.01% of the required isomer, 0.99% each of a pair of diastereoisomers, and only 0.01% of the enantiomer (i.e., an enantiomeric excess of 99.98%).

Although it is clear that an assay for, and means to remove, the diastereoisomer is essential, we would argue that there is no need to develop an assay for the enantiomer of travoprost. It also is unnecessary to find a method for removing it, provided we could assure that the starting components are controlled to be of sufficiently high enantiomeric excess. However, this argument relies on a statistical coupling between the components. In principle it is possible that an enantiomer of one component couples with different rates to the enantiomers of the other component. In theory, if there were preferential coupling between the pair that leads to travoprost, there would be preferential formation of the unwanted enantiomer relative to the diastereoisomer and the development of an assay for enantiomer would then be necessary. We therefore needed to prove that

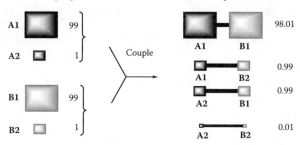

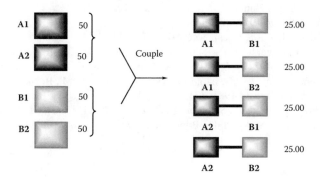

FIGURE 30.6 Enantiomeric excess determination for travoprost.

FIGURE 30.7 Test reaction between racemic components.

there was no such bias in the coupling reaction. Conveniently, this was carried out by coupling together the racemates and determining the diastereoisomeric excess of the resultant mixture.

Statistically, if there is no diastereoisomeric bias in the coupling, each of the 4 possible isomeric components should be present in equal amounts (i.e., 25% of each) and the ratio of diasteroisomers (each of which will be racemic in any case) will be 1:1 (Figure 30.7). On performing the coupling of racemates we observed an actual ratio of 1.2:1.0. Although there is a small steric bias, it is insufficient to challenge our argument. Moreover, the isomer formed in preference was the unwanted diastereosiomer, so that would lead to a depletion of the unwanted enantiomer relative to a completely statistical distribution of coupling products, and at this ratio of diastereoisomers in the racemate-to-racemate coupling, a corresponding enantiomeric excess of 99.985% can be calculated on combining 98% ee components. Thus, we demonstrated that travoprost is produced with almost none of the undesired enantiomer because we routinely achieve >99% ee for the acetylenic alcohol and 99% ee for the bicylcoheptenone intermediate.

ACKNOWLEDGMENTS

We would like to acknowledge the support and encouragement of Evan Kyba and Mark DuPriest of Alcon Laboratories during the development and manufacturing phases of the travoprost project. In addition, we would like to thank all the members of the travoprost team who worked extremely hard to make this manufacturing process a reality. In particular, John Andrews for project management; Graham Ruecroft, Alex Cantrill, and Shaun Jones (process development, Cambridge); Dan Henton and Chris Goralski (process development, Midland); the analytical team in Cambridge, UK; and all members of the manufacturing team in Mirfield, U.K.

REFERENCES

1. Klimko, P. G., Bishop, J., Desantis Jr., L., Sallee, V. L. Eur. Pat. 0639563, 1995, *Chem. Abstr.* 1995, *122*, 290579.
2. a) Corey, E. J., Andersen, N. H., Carlson, R. M., Paust, J., Vedejs, E., Vlattas, I., Winter, R. E. K. *J. Am. Chem. Soc.* 1968, *90*, 3245; b) Schneider, W. P., Axen, U., Lincoln, F. H., Pike, J. E., Thompson, J. L. *J. Am. Chem. Soc.* 1968, *90*, 5895.
3. Corey, E. J., Vlattas, I., Harding, K. *J. Am. Chem. Soc.* 1969, *91*, 535.
4. a) Corey, E. J., Weinshenker, N. M., Schaaf, T. K., Huber, W. *J. Am. Chem. Soc.* 1969, *91*, 5675; b) Corey, E. J., Schaaf, T. K., Huber, W., Koelliker, U., Weinshenker, N. M. *J. Am. Chem. Soc.* 1970, *92*, 397.
5. Schaaf, T. K., Corey, E. J. *J. Org. Chem.* 1972, *37*, 2921.
6. Corey, E. J., Albonico, S. M., Koelliker, U., Schaaf, T. K., Varma, R. K. *J. Am. Chem. Soc.* 1971, *93*, 1491.
7. Bansal, R., Cooper, G. F., Corey, E. J. *J. Org. Chem.* 1991, *56*, 1329.
8. Corey, E. J., Arnold, Z., Hutton, J. *Tetrahedron Lett.* 1970, 307.
9. Grieco, P. *J. Org. Chem.* 1972, *37*, 2363.
10. Corey, E. J., Mann, J. *J. Am. Chem. Soc.* 1973, *95*, 6832.
11. a) Dimsdale, M. J., Newton, R. F., Rainey, D. K., Webb, C. F., Lee, T. V., Roberts, S. M. *J. Chem. Soc., Chem. Commun.* 1977, 716; b) Cave, R. J., Newton, R. F., Reynolds, D. P., Roberts, S. M. *J. Chem. Soc., Perkin Trans. I* 1981, 646; c) Davies, J., Roberts, S. M., Reynolds, D. P., Newton, R. F. *J. Chem. Soc., Perkin Trans. I* 1981, 1317.
12. a) Sih, C. J., Price, P., Sood, R., Salomon, R. G., Peruzzotti, G., Casey, M. *J. Am. Chem. Soc.* 1972, *94*, 3643; b) Alvarez, F. S., Wren, D., Prince, A. *J. Am. Chem. Soc.* 1972, *94*, 7823; c) Kluge, A. F., Untch, K. G., Fried, J. H. *J. Am. Chem. Soc.* 1972, *94*, 9256.
13. Babiak, K. A., Ng, J. S., Dygos, J. H., Weyker, C. L., Wang, Y.-F., Wong, C.-H. *J. Org. Chem.* 1990, *55*, 3377.
14. Stork, G., Isobe, M. *J. Am. Chem. Soc.* 1975, *97*, 4745.
15. Patterson, Jr. J. W., Fried, J. H. *J. Org. Chem.* 1974, *39*, 2506.
16. Suzuki, M., Kawagishi, T., Suzuki, T., Noyori, R. *Tetrahedron Lett.* 1982, *23*, 4057.
17. Noyori, R., Suzuki, M. *Angew. Chem., Int. Ed.* 1984, *23*, 847.
18. a) Laumen, K., Schneider, M. *Tetrahedron Lett.* 1984, *25*, 5875; b) Deardorff, D. R., Matthews, A. J., McMeekin, D. S., Craney, C. L. *Tetrahedron Lett.* 1986, *27*, 1255.
19. a) Suzuki, M., Yanagisawa, A., Noyori, R. *J. Am. Chem. Soc.* 1985, *107*, 3348; b) Morita, Y., Suzuki, M., Noyori, R. *J. Org. Chem.* 1989, *54*, 1785.
20. Gooding, O. W. *J. Org. Chem.* 1990, *55*, 4209.
21. Johnson, C. R., Penning, T. D. *J. Am. Chem. Soc.* 1986, *108*, 5655.
22. Suzuki, M., Yanagisawa, A., Noyori, R. *J. Am. Chem. Soc.* 1988, *110*, 4718.
23. Okamoto, S., Kobayashi, Y., Kato, H., Hori, K., Takahashi, T., Tsuji, J., Sato, F. *J. Org. Chem.* 1988, *53*, 5590.
24. Danishefsky, S. J., Cabal. M. P., Chow, K. *J. Am. Chem. Soc.* 1989, *111*, 3456.
25. Wallis, C. J. Eur. Pat. 0074856, 1983, *Chem. Abstr.* 1984, *99*, 139627
26. Fox, M. E., Jackson, M., Lennon, I. C., McCague, R., Parratt, J. S. *Adv. Synth. Catal.* 2002, *344*, 50.

27. Boulton, L. T., Brick, D., Fox, M. E., Jackson, M., Lennon, I. C., McCague, R., Parkin, N., Rhodes, D., Ruecroft, G. *Org. Proc. Res. Dev.* 2002, *6*, 138.
28. Swanson, D. R., Nguyen, T., Noda, Y., Negishi, E. *J. Org. Chem.* 1991, *56*, 2590.
29. Newton, R. F., Roberts, S. M. *Tetrahedron* 1980, *36*, 2163.

31 Synthesis of Large Volume Products

David J. Ager and Colin R. Bayley

CONTENTS

31.1 INTRODUCTION

The purpose of this chapter is to discuss the syntheses of top-selling chiral compounds so that a perspective may be obtained about the merits of the various asymmetric approaches discussed elsewhere in this book. Of course, the majority of these drugs are mature, which means that a considerable amount of time and money has been expended to reduce costs and optimize the syntheses. In addition, asymmetric synthesis is a new approach and, although the potential may exist today to use an asymmetric oxidation, this was not a serious option 20 years ago. To help with the development of asymmetric synthesis, a discussion has been included on the difference between the top-selling drugs discussed in the first edition of this book and in the current one. In addition to pharmaceuticals, some large volume products in the food and agricultural areas are also discussed.

31.2 PHARMACEUTICALS

The top ten best-selling single enantiomer drugs in 1996 are given in Table 31.1.[1] Because we are only concerned with small molecules, peptide products such as insulin and erythropoietin have been ignored because they are outside of the realms of chemical approaches.

The top-selling chiral drugs for 2002 are shown in Table 31.2.

TABLE 31.1
Top Ten Best-Selling Single Enantiomer Drugs in 1996[a]

Drug	Sales ($B)
Enalapril	2.9
Simvastatin	2.8
Pravastatin	2.4
Amoxicillin	1.9
Lisinopril	1.7
Diltiazem	1.6
Captopril	1.3
Sertraline	1.3
Lovastatin	1.3
Cefaclor	1.2

[a] The data was provided by Technology Catalysts International and is gratefully acknowledged.

TABLE 31.2
Top Ten Best-Selling Single Enantiomer Drugs in 2002[a]

Drug	Sales ($B)
Lipitor	8.0
Simvastatin	5.6
Pravastatin	4.0
Paroxetine	3.3
Clopidogrel	2.9
Sertraline	2.7
Seretride[b]	2.4
Esomeprazole	2.0
Augmentin[c]	1.8
Valsartan	1.7

[a] The data was provided by PharQuest and is gratefully acknowledged.
[b] This is a mixture of fluticasone propionate and salmeterol xinafoate.
[c] This is a mixture of amoxicillin and potassium clavulanate.

Some drugs are sold as mixtures with other active ingredients; this makes calculation of the total sales for a specific component difficult. As an example, atorvastatin is marketed as Lipitor but is also a component of Caduet where an achiral component, amlodipine, is also an active ingredient. In addition, many Japanese companies do not report specific sales of drugs. As with the first edition of this book, large biological molecules, such as Erypo, also known as epoetin and erythropoietin, have not been included. It will be seen that cholesterol-lowering drugs are the big market leaders. Compared to 1996, angiotensin-converting enzyme inhibitors have fallen from the chart. This is partly the result of many of them coming to the end of their patent life time and becoming generic.

31.2.1 LIPITOR

Lipitor (1)[2] is a cholesterol-lowering agent whose mode of action is as an HMG-CoA (3-hydroxy-3-methylglutaryl coenzyme A) reductase inhibitor. It has been marketed by Pfizer and is sold as the calcium salt.

1

The synthesis of this top-selling drug is relatively straightforward because the stereochemistry is contained in a single sidechain.[3–8] One chiral synthon is used to induce the second stereogenic center (Scheme 31.1). The hydroxy nitrile **2** is purchased. Ester enolate chemistry is used to chain extend and, because of the price of **2**, 4 equivalents of the enolate of *tert*-butyl acetate are used to ensure high conversion. In earlier versions of the synthesis, a separate base was used to cause the nonproductive deprotonations. The hydroxy group controls the stereochemical course during the reduction. The small amount of the off-isomer produced after ketal formation is removed by recrystallization of the TBIN (**3**). The boron has to be removed prior to the ketal formation; this is achieved by azeotropic removal of the trimethyl borate with methanol. The resultant amine, TBIA (**4**), is not isolated but is used directly in the final coupling reaction (*vide infra*). (Much of the data for the synthesis, marketing, and therapeutic actions of the pharmaceutical compounds in this chapter has been obtained from the PharQuest database.

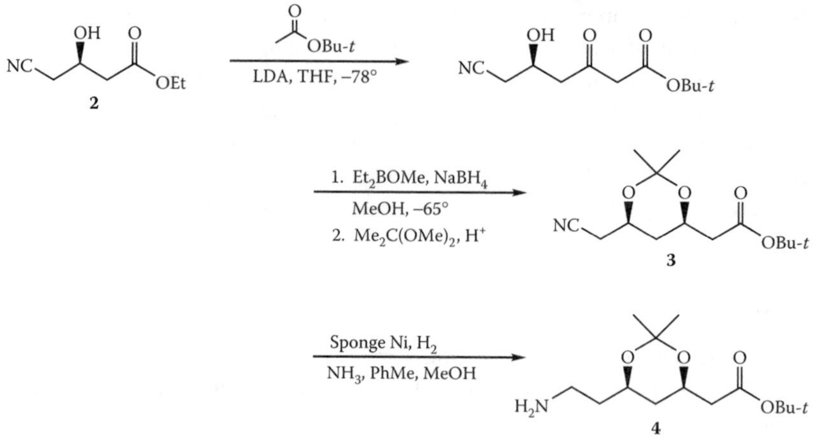

SCHEME 31.1

A Stetter reaction is used to form the other part of the molecule **5** (Scheme 31.2).

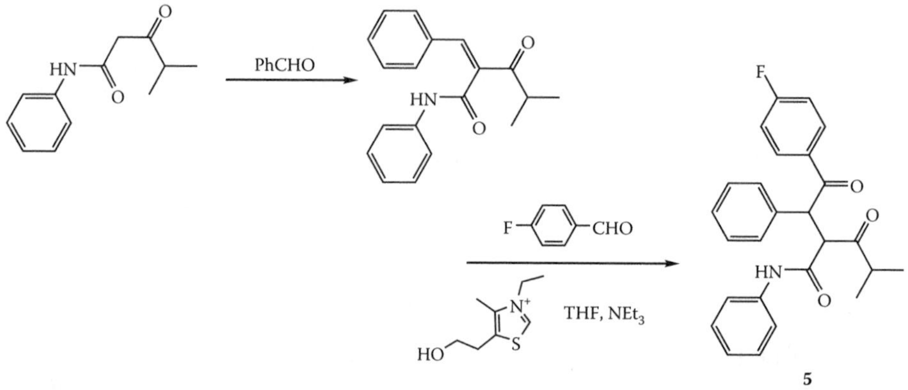

SCHEME 31.2

The TBIA (**4**) is then used in a Paal-Knorr condensation to provide the pyrrole (Scheme 31.3). This reaction is slow. Deprotection, saponification, and cyclization then provide the atorvastatin lactone (**6**). Conversion to the final calcium salt form of **1** is achieved by base treatment and addition of calcium (Scheme 31.3).[9]

The key step with regard to stereoselection lies with the synthesis of the hydroxy nitrile. A number of syntheses have been reported, and different suppliers use different approaches. One is

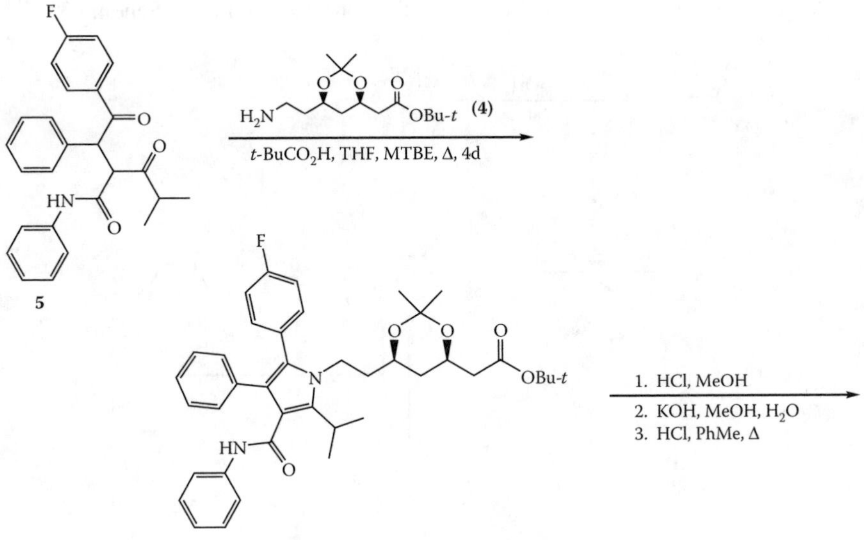

SCHEME 31.3

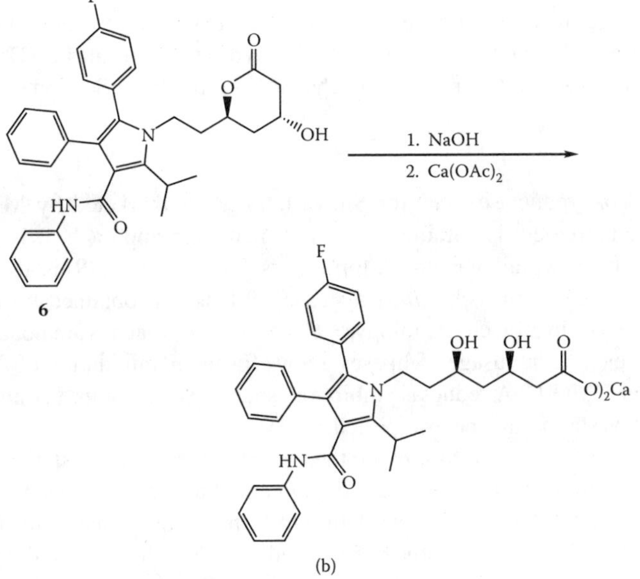

an asymmetric reduction of the readily available chloro ketone **7** by chemical or biological means (Scheme 31.4) (see also Chapters 12, 19, and 20).[10–34] However, subsequent displacement of the halide by cyanide can be problematic.[7,32,34–51] The alternative of cyanide reaction with the epoxide is also a challenge. Resolution methods are also available,[52] as are reductions of the keto nitrile.[53]

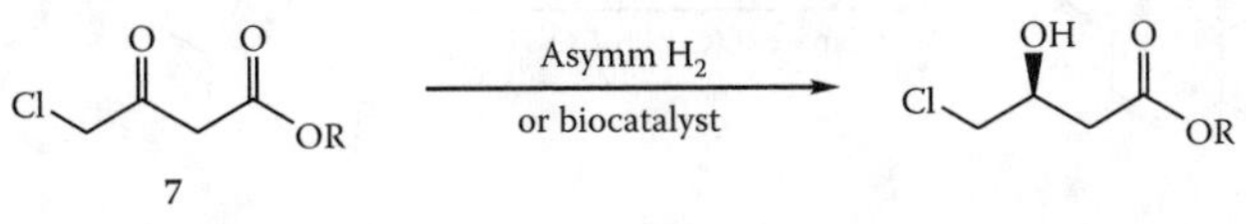

SCHEME 31.4

An alternative has been to use L-malic acid as the source of chirality (Scheme 31.5).[54,55]

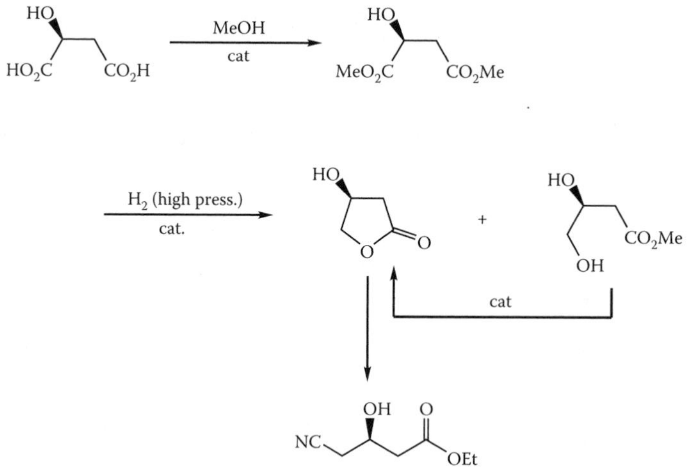

SCHEME 31.5

Other approaches have also been developed and are discussed elsewhere, such as an aldol approach (Chapters 19 and 20) and the hydrolysis of 3-hydroxyglutaronitrile (Chapter 20).[56] In addition the chemistry has some overlap to approaches to carnitine (see Section 31.3.3.).

31.2.2 SIMVASTATIN

The high-value HMG-CoA reductase inhibitor Simvastatin (**8**) is marketed by Merck under the name Zocor. The active ingredient is obtained from a fermentation approach. It is very similar in structure to lovastatin, which has fallen from the top-sellers list. Lovastatin (**9**) is also a cholesterol-reducing drug that is isolated from *Aspergillus terreus*.[57–60] It is still obtained by fermentation,[61] and with the current advances in molecular biology,[62–64] chemical approaches are not able to compete in a cost-effective manner.[65–67] The usage of lipases allows for the manipulation of the butyric acid sidechain to access other HMG-CoA reductase inhibitors such as simvastatin.[68] A number of routes to various portions of lovastatin have been reported.[69]

As mentioned, fermentation is used to provide the core structure of simvastatin, and hydrolysis is used to remove the 2-methylbutyrate sidechain; the dimethylbutyric acid sidechain is introduced by coupling (Scheme 31.6).[70–73] Chemical methylation to form the quaternary dimethylated center in the sidechain has been achieved on synthetic intermediates.[74,75] The chemical approach has to differentiate two hydroxy groups, and this requires protection–deprotection steps.[76–86]

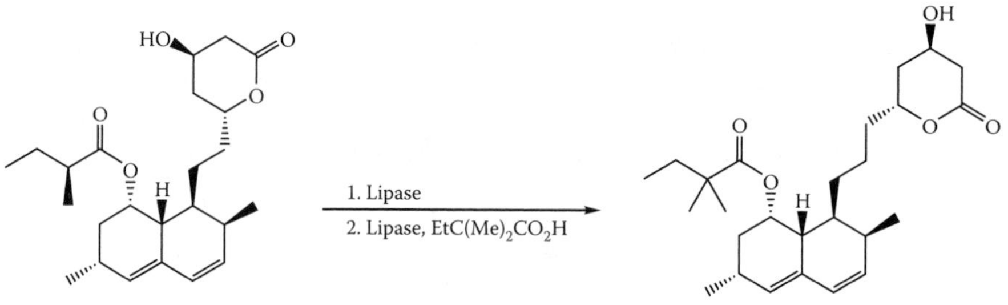

SCHEME 31.6

31.2.3 PRAVASTATIN

Pravastatin (**10**) is another HMG-CoA reductase for the inhibition of cholesterol biosynthesis; it is marketed by Sanyo and Bristol Myers Squibb under the trade names Mevalotin and Pravachol.[87] It has a close structural relationship to lovastatin and simvastatin. It is produced by a two-step sequence. First, mevastatin (**11**), also known as ML-236B or compactin, is prepared by fermentation of *Penicillium citrinum*;[88] it is then enzymatically hydroxylated to produce **11** (Scheme 31.7).[88–101]

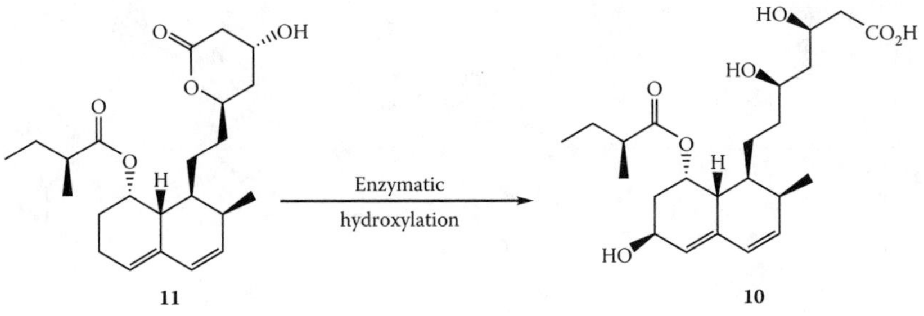

SCHEME 31.7

For an alternative synthesis, see Section 11.3.8.

31.2.4 CLOPIDOGREL

Clopidogrel (**12**) is marketed by Sanofi-Synthlebo and BMS as the hydrogen sulfate salt under the trade name Plavix. It is a platelet aggregation inhibitor that is used as an antithrombotic agent for the reduction of cardiovascular events. The synthesis is straightforward (Scheme 31.8).[102–107] The asymmetry is obtained by a classical resolution.[103] It is possible to recycle the undesired isomer.[108]

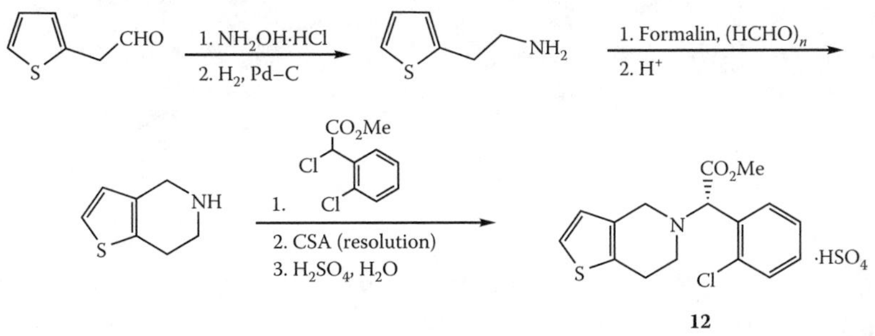

SCHEME 31.8

Alternative syntheses have been proposed, in particular with *o*-chlorophenylglycine providing the stereogenic center.[109–112]

31.2.5 PAROXETINE

Paroxetine (**13**) is a selective serotonin reuptake inhibitor developed by Novo Nordisk's subsidiary, Ferrosan, and licensed to SmithKline Beecham (SB) (now GlaxoSmithKline).[113] It is sold under the name Paxil, as the hydrochloride salt, and it is used to treat depression, anxiety, and obsessive compulsive disorder. The synthesis is relatively straightforward and involves a resolution that also is used to separate the desired, major *trans*-isomer (Scheme 31.9).[114–118] Base is used to equilibrate the substituents of the *cis*-ester **14** to the thermodynamically more stable *trans*-configuration prior to reduction.

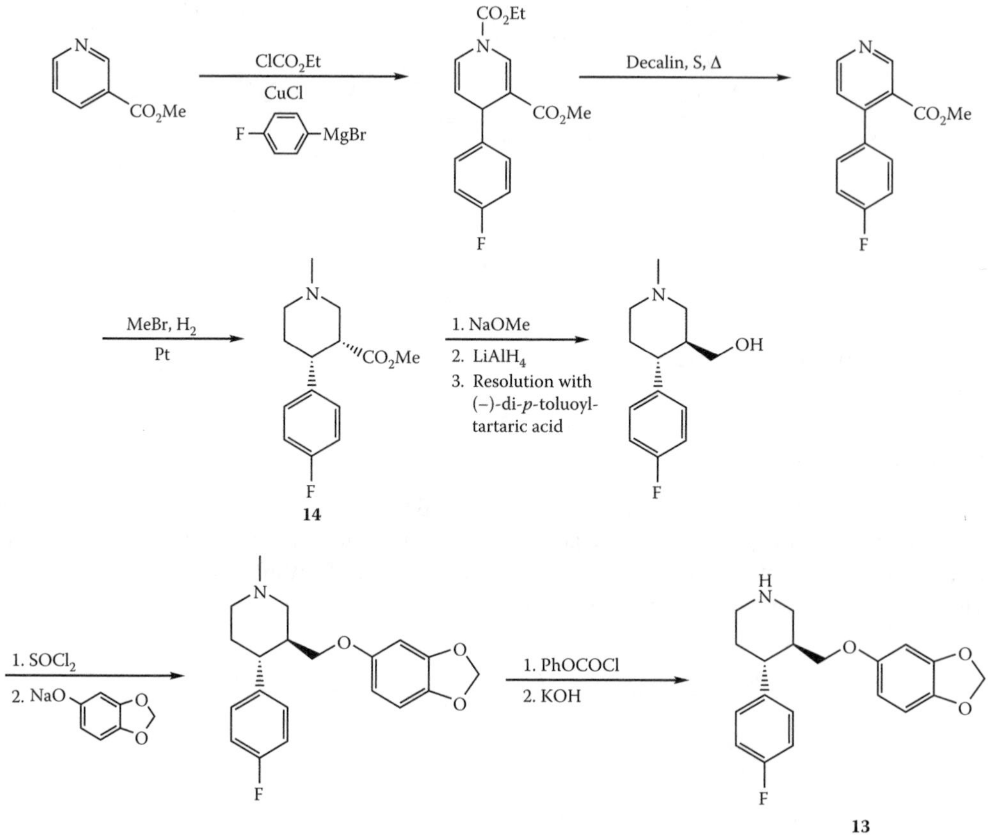

SCHEME 31.9

An alternative approach that alleviates the demethylation step and avoids the use of the fluoro-Grignard reagent is shown in Scheme 31.10. The resultant alcohol **15** of this sequence, after a resolution, can be reacted with thionyl chloride and then coupled with the substituted sodium phenoxide to give **13** (*cf.* Scheme 31.9).[119]

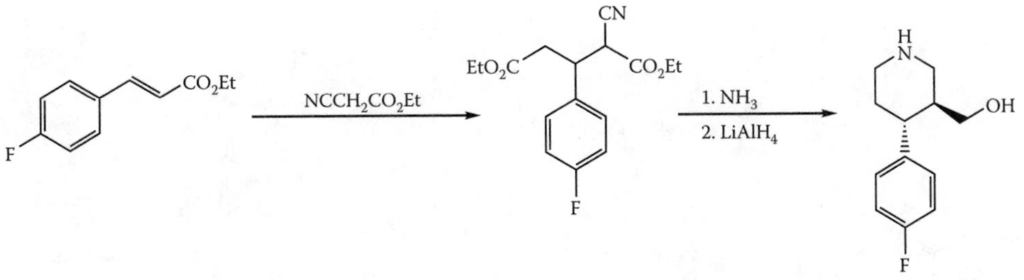

SCHEME 31.10

Alternative syntheses have been proposed,[120–134] including enzymatic resolutions[135–141] and other asymmetric approaches.[142–147]

31.2.6 SERTRALINE

Sertraline (**16**) is an antidepressant that inhibits the uptake of serotonin in the central nervous system.[148] It is marketed by Pfizer under the name Zoloft.[149] One methodology that can be used relies on an asymmetric reduction (see also Chapter 16) (Scheme 31.11).[150–156] The lactone can be used as a chiral starting material for the Friedel-Crafts reaction. The established stereogenic center in the tetralone controls the reduction of the imine.[157] The alternative is a resolution approach (Scheme 31.12).[148,153,158–161]

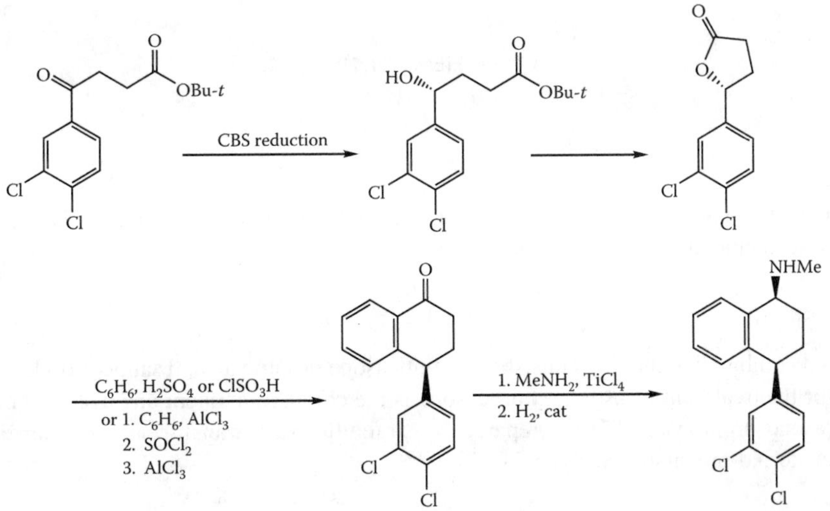

SCHEME 31.11

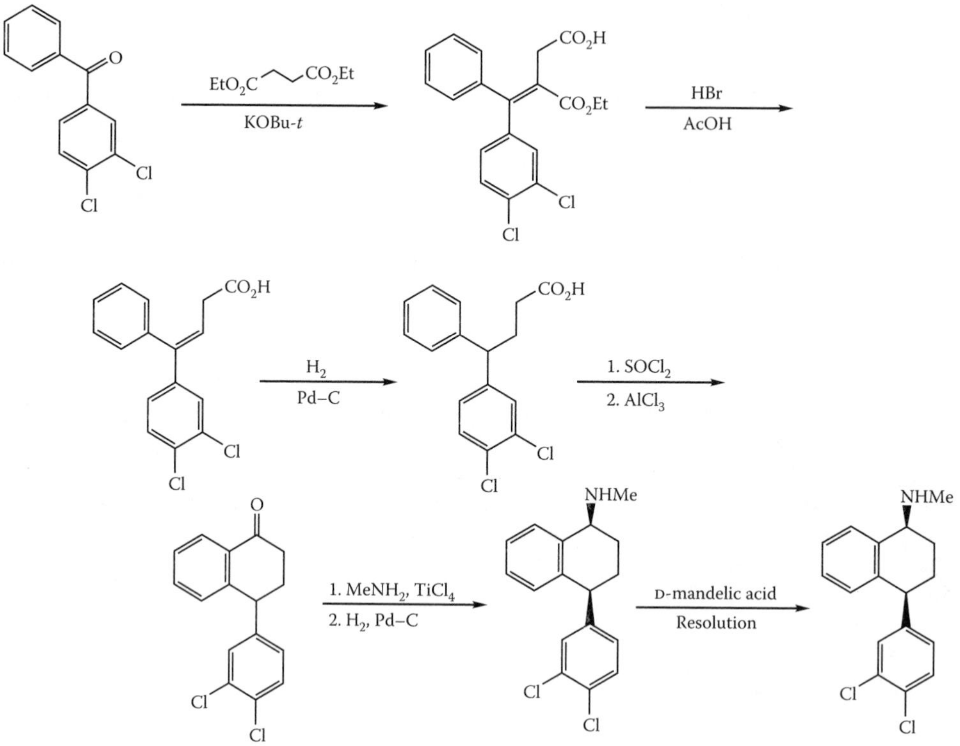

SCHEME 31.12

Alternative syntheses have been proposed, including the interconversion of isomers,[162] by a number of academic and process groups.[163–178]

31.2.7 SERETIDE

Seretide is a combination therapy consisting of fluticasone propionate and salmeterol. It is marketed by GSK for the treatment of asthma. Fluticasone is the chiral component and a steroid derivative.

Fluticasone propionate (**17**) is prepared by straightforward transformations from a suitably substituted steroid precursor (Scheme 31.13).[179,180]

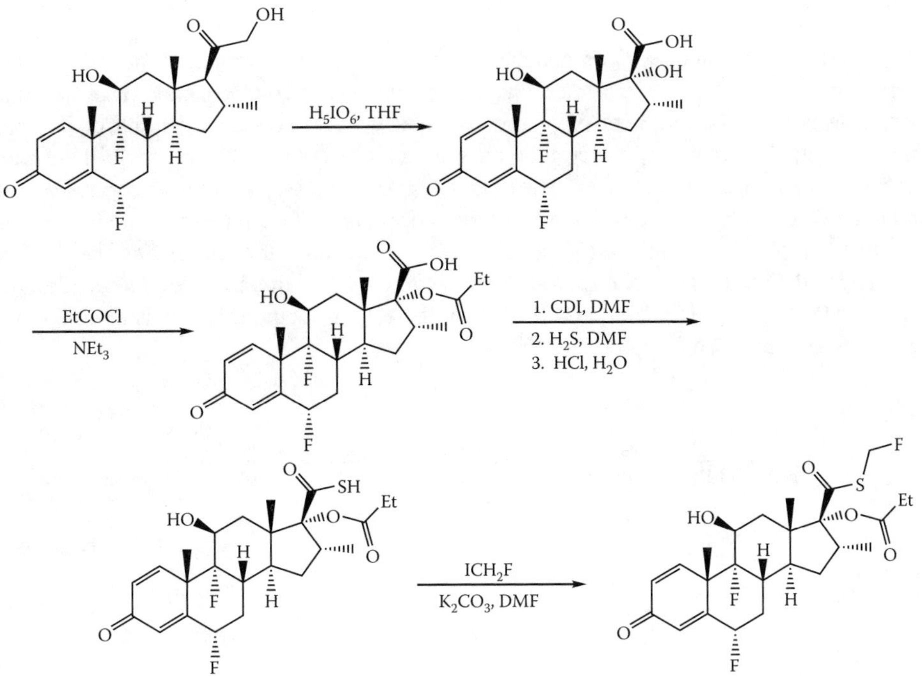

SCHEME 31.13

Although salmeterol (**18**) does contain a stereogenic center, it is used as the racemic mixture; its synthesis has been included for completeness (Scheme 31.14).[181–183]

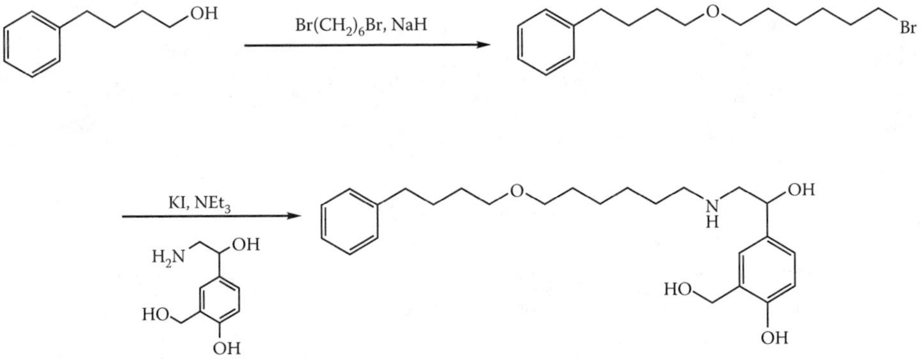

SCHEME 31.14

Other synthetic approaches have been proposed.[184–186] Asymmetric syntheses have been reported.[187,188]

31.2.8 Esomeprazole

AstraZeneca (formerly Astra) has launched the proton-pump inhibitor esomeprazole (**19**) (as Nexium) as a treatment for peptic ulcer, gastroesophageal reflux disease, duodenal ulcer, and esophagitis. Esomeprazole is the (*S*)-enantiomer of omeprazole and was developed as a result of its improved pharmokinetic profile and better potency after oral dosing than (*R*)-form of omeprazole or the racemate. The dosage is higher than would be expected for a simple chiral switch. The stereogenic center is at sulfur. Detailed accounts of the development of the process have been published.[189,190]

The first samples were obtained by resolution.[191–193] Because the last step in the synthesis of the racemate is the oxidation of the sulfide to the sulfoxide,[194–198] this has been modified to provide the *S*-isomer (Scheme 31.15). This is achieved by the Kagan method, which is a variation of the Sharpless epoxidation (see Chapter 9).[199,200]

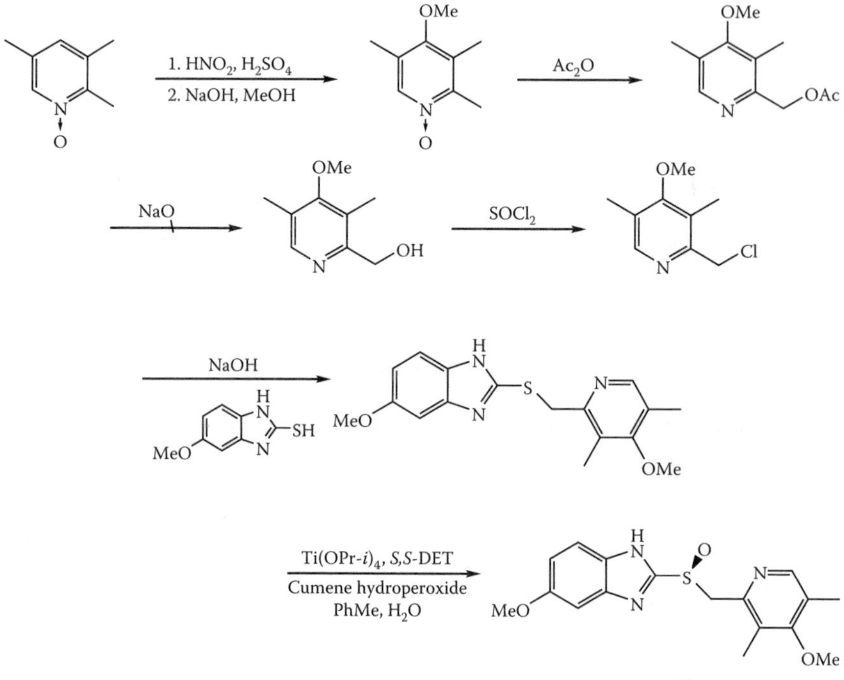

SCHEME 31.15

Alternative approaches include the separation of the isomers,[201,202] biological oxidation,[203] and biological resolution by reduction of one of the sulfoxide isomers.[204]

31.2.9 Augmentin

Augmentin is a combination therapy where the bacterial resistance to penicillins is reduced by the use of the β-lactamase inhibitor clavulanic acid (**20**). It originated from Beechams and is now marketed by GlaxoSmithKline (GSK).

Clavulanic acid is a fermentation product obtained from *Streptomyces clavuligerus*.[205,206]

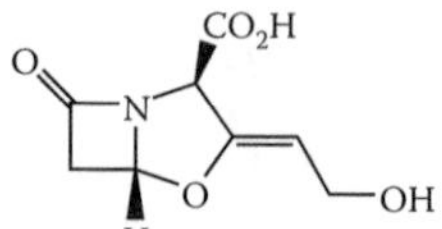

20

Amoxicillin (**21**) is a semi-synthetic penicillin antibiotic. The penicillin portion is derived from fermentation of either penicillin-V or penicillin-G, and then the sidechain is removed to afford 6-APA. This transformation can be done chemically.[69,207] The alternative, which is growing in importance, is to perform an enzymatic cleavage under mild conditions.[208] The D-*p*-hydroxyphenylglycine is then attached as the new sidechain; chemical and enzymatic methods are available to achieve this (Scheme 31.16).[209–215] The phenylglycine amino acid is obtained by a resolution (Chapters 2, 7, and 25) or by enzymatic hydrolysis of a hydantoin (Chapter 2, 6, and 19).[216–220]

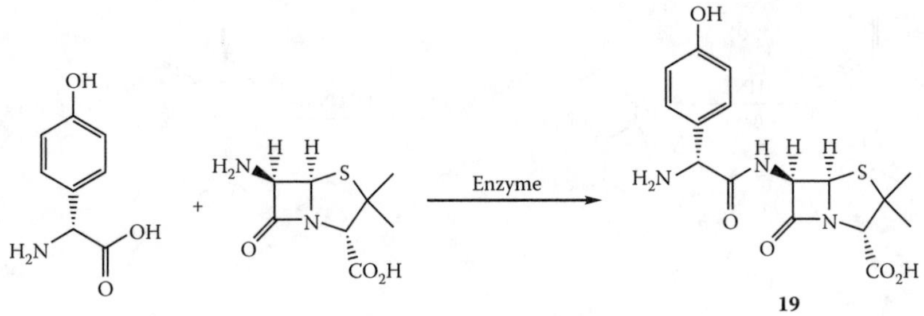

SCHEME 31.16

31.2.10 VALSARTAN

Valsartan (**22**) is an orally active, angiotensin II antagonist that is marketed under the name Diovan by Novartis for hypertension. Novartis has also developed a combination therapy, valsartan plus hydrochlorothiazide, for the second-line therapy of hypertension. The synthesis is straightforward from a stereochemical viewpoint because a chiral pool synthesis is used. The stereogenic center is derived from L-valine (Scheme 31.17).[221–223]

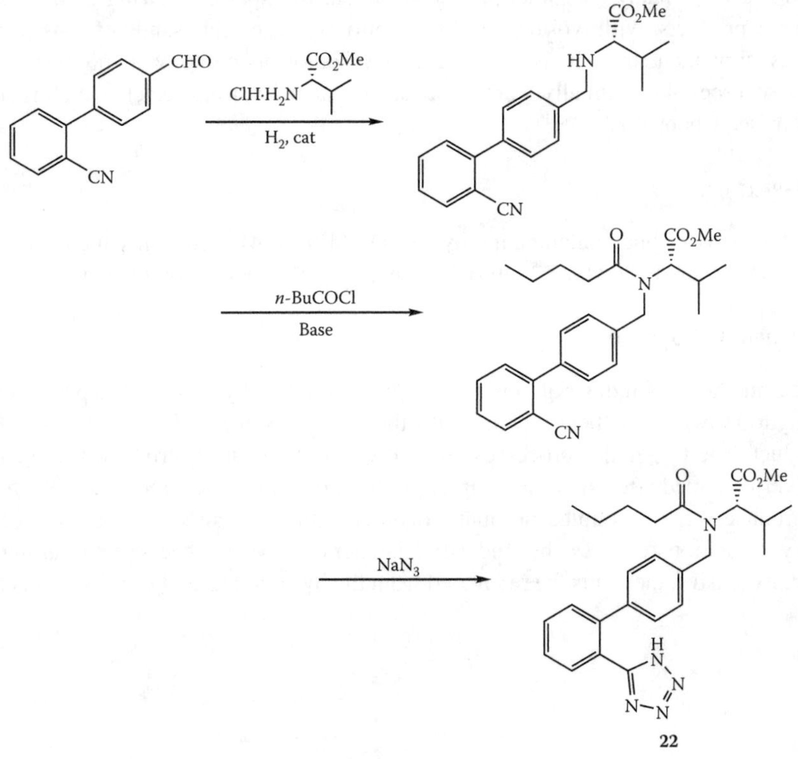

SCHEME 31.17

31.3 FOOD INGREDIENTS

31.3.1 MENTHOL

Although menthol (**23**) is a terpene available from natural sources (Chapter 5), asymmetric synthesis by the Takasago method now accounts for a substantial portion of the market. This synthesis is discussed in detail in Chapter 12. The key step is the asymmetric isomerization of an imine to an allyl amine (Scheme 31.18).[224,225]

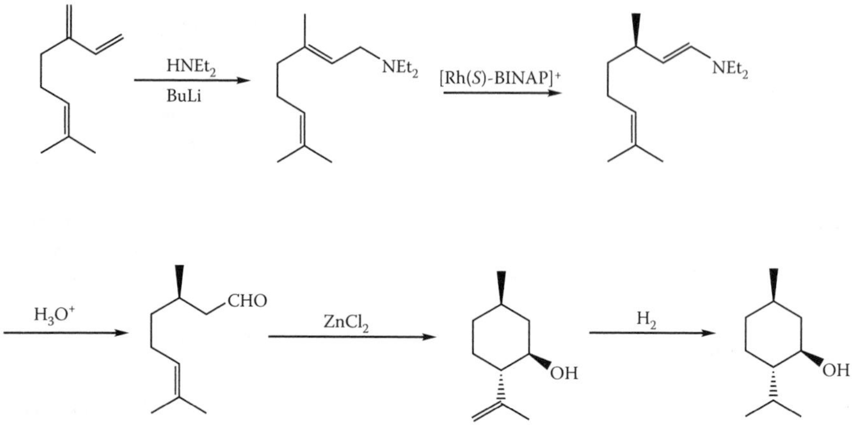

SCHEME 31.18

Menthol is used in many consumer products, such as toothpaste, chewing gum, cigarettes, and pharmaceutical products, with worldwide consumption of many thousands of tons per year.[225,226] Takasago has implemented their asymmetric isomerization technology to produce a variety of optically active terpenoids from allyl amines at various manufacturing scales, which is summarized in Table 31.3 (see Chapter 12).[225,226]

31.3.2 ASPARTAME

Aspartame (L-aspartyl-L-phenylalanine methyl ester) (APM) (**24**) is currently the most widely used nonnutritive sweetener worldwide,[227] although this position is being challenged.

31.3.2.1 Chemical Synthesis

The chemical methods of industrial significance involve the dehydration of aspartic acid to form an acid anhydride, which is then coupled with the phenylalanine or its methyl ester to give the desired product. The two major processes are known as the Z and F processes (Schemes 19 and 20, respectively), named after the protecting group used on the aspartyl moiety.[228–231] Both of these processes produce some β-coupled products together with the desired α-aspartame (**24**), but the selective crystallization removes the undesired isomers. However, because the amino acid raw materials are expensive, they must be recovered from the by-products and waste streams for recycle.

TABLE 31.3
Optically Active Terpenoids Produced by Asymmetric Isomerization of Allylamines (see references 225 and 226)

Name	Structure	Use	Production (tons/year)
(+)-Citronellal		Intermediate	1,500
(−)-Isopulegol		Intermediate	1,100
(−)-Menthol		Pharmaceuticals Tobacco Household products	1,000
(−)-Citronellol		Fragrances	20
(+)-Citronellol		Fragrances	20
(−)-7-Hydroxy-citronellal		Fragrances	40
S-7-Methoxy-citronellal		Insect growth regulator	10
S-3,7-Dimethyl-1-octanal		Insect growth regulator	7

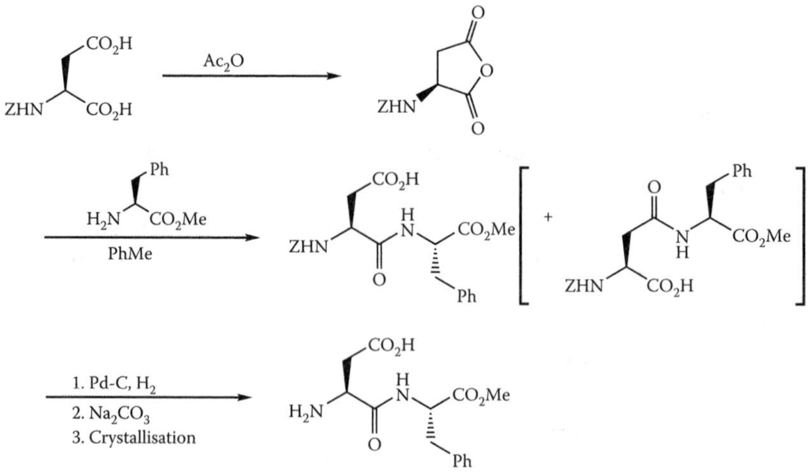

SCHEME 31.19

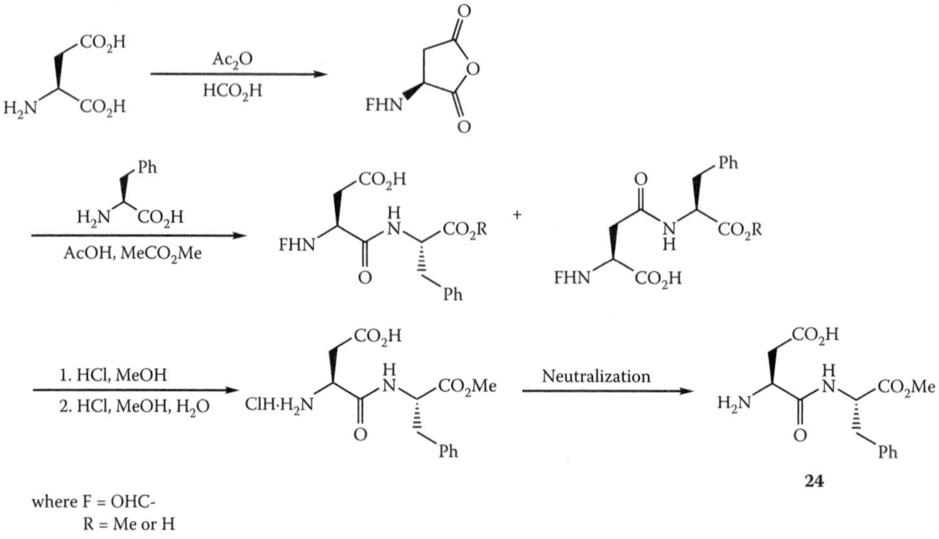

where F = OHC-
R = Me or H

SCHEME 31.20

31.3.2.2 Enzymatic Synthesis

The enzymatic synthesis approaches are discussed in more detail in Chapter 19. A protease can be used to catalyze the synthesis of a peptide bond (Scheme 31.21). When the stoichiometry of the reactions is such that two moles of phenylalanine methyl ester are used with one mole of Z-Asp, the Z-APM•PM product precipitates and shifts the equilibrium to >95% conversion.[232] This is the basis of the commercial TOSOH process operated by Holland Sweetener that uses thermolysin.[233] One significant variation has been the use of racemic PM instead of the L-isomer. Because the enzyme will only recognize the L-PM isomer to form the peptide bond, the unreacted D-PM isomer forms a salt and then, after acidification, the D-PM can be chemically racemized and recycled.

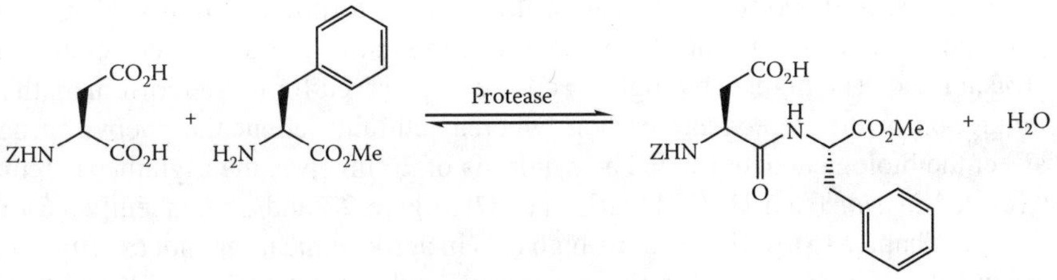

SCHEME 31.21

31.3.3 CARNITINE

The food additive carnitine (**25**) is worthy of mention because it has a very close structural relationship to some beta blockers. It is also called vitamin B$_T$. There are many potential routes to this compound, including an asymmetric hydrogenation method.[234-236] The method is closely related to that used for Lipitor (Section 31.2.1). Reduction of 4-chloro-3-oxobutyrate provides the desired alcohol isomer. Ester hydrolysis and reaction with triethylamine affords **25**. There are two other major approaches; one relies on an asymmetric microbial oxidation (Scheme 31.22).[237]

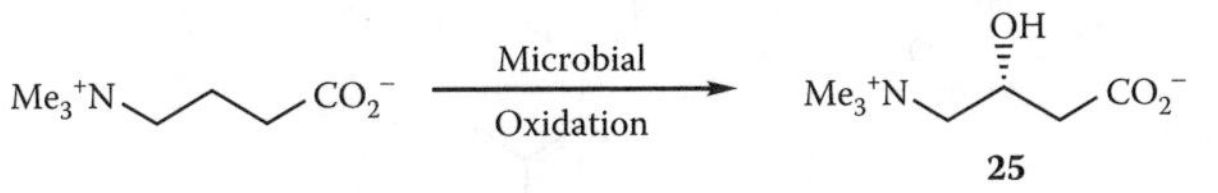

SCHEME 31.22

The other approach starts from epichlorohydrin and involves a classical resolution of the amide. The "wrong" amide is also converted to the desired isomer by a dehydration–rehydration sequence that uses hydrolases (Scheme 31.23).[238,239] A biological variation is to hydrolyze the 4-butyrobetaine as the stereodifferentiating step.[240,241]

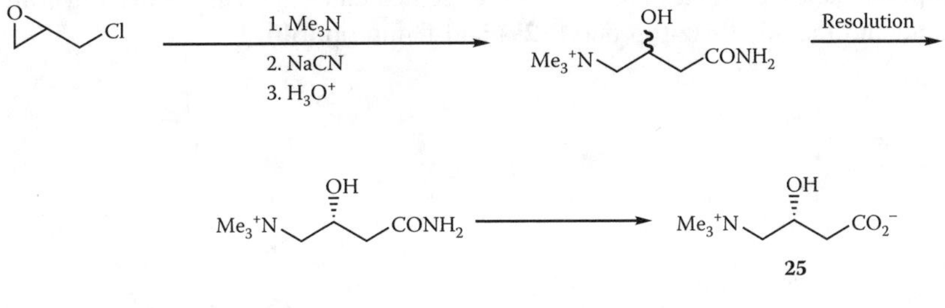

SCHEME 31.23

31.4 AGRICULTURAL PRODUCTS

The primary driving force with agricultural product compounds is cost. This is reflected in the relatively small number of chiral compounds that are sold at scale.

31.4.1 METOLACHLOR

The reduction of metolachlor (**26**) is described in detail elsewhere (Chapters 12 and 15). This compound is sold as an enantio-enriched compound (~80% ee) rather than the pure enantiomer because of economic constraints. Metolachlor (**26**) is a pesticide sold since March 1997 as the enantiomerically enriched form and sold under the trade name DUAL MAGNUM™.[242-244] Since its introduction in 1978, >20,000 tons per year of the racemic form have been sold.[244] Metolachlor exists as 4 diastereoisomers, a result of stereogenic carbon atom and restricted rotation about the phenyl-nitrogen bond. The highest biological activity is observed for diastereoisomers that contain the *S*-configuration at the stereogenic carbon, whereas chirality about the phenyl-nitrogen bond does not affect the biological activity.[242] The synthesis of **26** involves the asymmetric reduction by a catalyst formed *in situ* from [Ir(COD)Cl]$_2$ and **27** to give **28** and subsequently **26** in 80% ee (Scheme 31.24) (Chapter 15).[242] Fortunately, high ee's in agrochemicals are not as critical compared to pharmaceuticals, and the usage of the enantiomerically enriched herbicide leads to a 40% reduction in the environmental loading.[244]

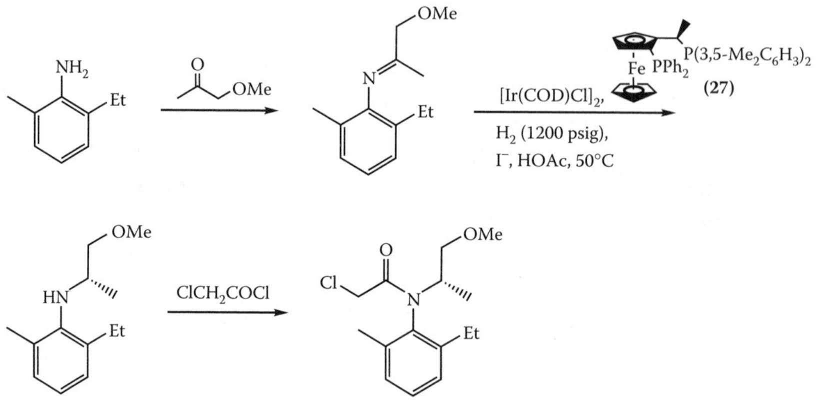

SCHEME 31.24

31.4.2 Phenoxypropionic Acid Herbicides

Phenoxypropionic acid herbicides contain a stereogenic center that can be derived from 2-chloro-propionic acid and include fluazifop-butyl (**29**) and flamprop (**30**).[245]

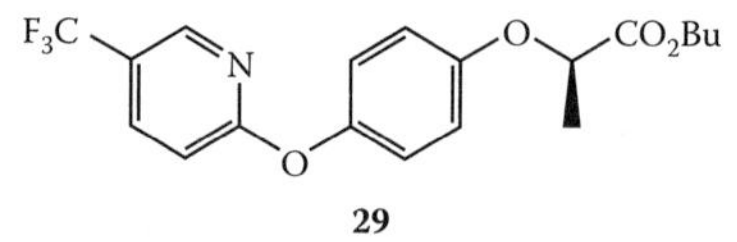

29

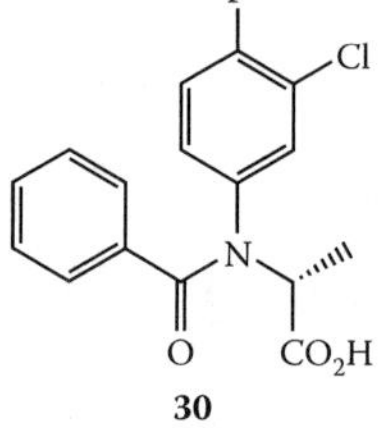

30

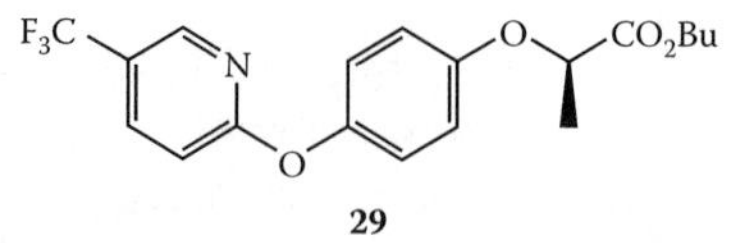

29

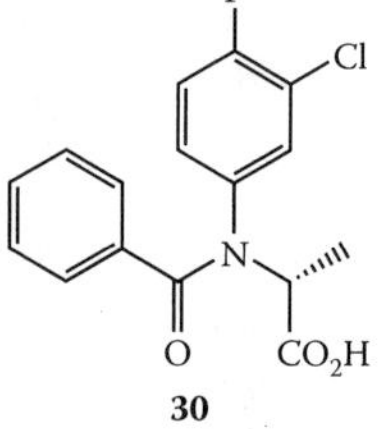

30

A number of enzymatic resolutions have been investigated to the chiral halo ester, including esterification, transesterification, and aminolysis.[69,246–250] Zeneca (ICI) has commercialized the approach that relies on a dehalogenase (Scheme 31.25).[69,245,248,249]

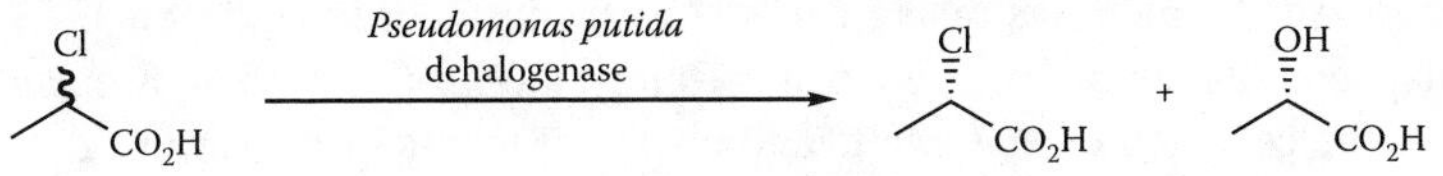

SCHEME 31.25

An alternative approach relies on a lipase method, but excess substrate has to be present to ensure good ee (Scheme 31.26).[250]

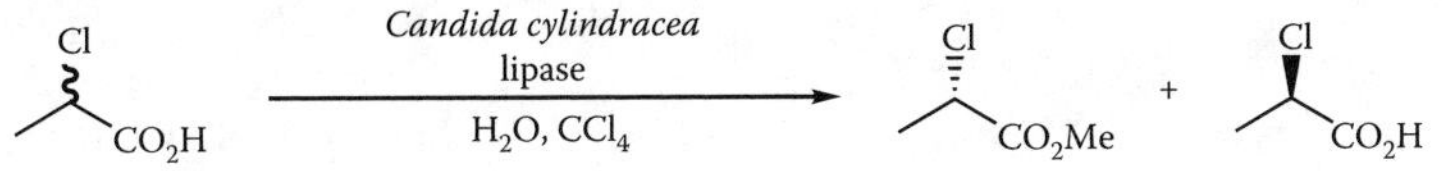

SCHEME 31.26

31.4.3 PYRETHROIDS

Pyrethroids are commercially important insecticides that usually contain a cyclopropyl unit that is *cis*-substituted and a cyanohydrin derivative. They are usually sold as a mixture of isomers. However, asymmetric routes have been developed, especially because these compounds are related to chrysanthemic esters (Chapter 12).[251] The pyrethroids can be resolved through salt formation or by enzymatic hydrolysis.[252]

Although the methodology does not apply to others, one of the members of this family, deltamethrin (**31**), is made at scale by a selective crystallization. In the presence of a catalytic amount of base, only one of the isomers crystallizes from isopropanol (Scheme 31.27).[253]

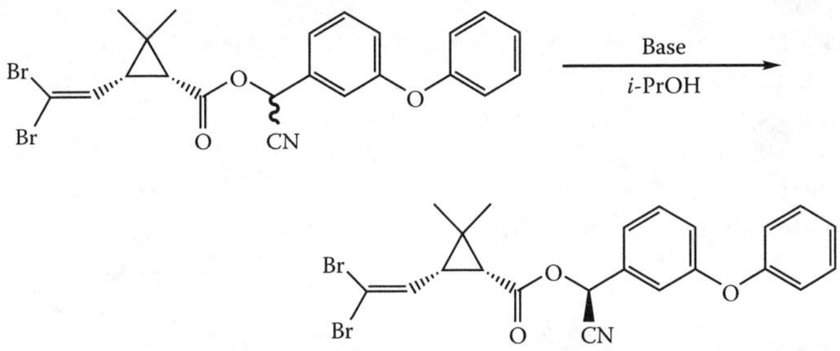

SCHEME 31.27

31.5 SUMMARY

Only a small number of examples of chiral compounds that are produced at high volume have been given in this chapter. The comparison to the first edition's top ten pharmaceutical list shows that asymmetric synthesis is now beginning to play an important role in the synthesis of drugs. Although fermentation and resolution approaches are still in abundance, the drugs that are prepared by these methods are mature and nearing the end of their patent lives. There is no reason to doubt that asymmetric synthetic methods and dynamic kinetic resolutions will continue to gain ground with the drugs in development that will be tomorrow's blockbusters. It will be interesting to see how the chemistry of large volume drugs, such as the 'pril family, changes as these compounds become generic. With so many different approaches, and with the structural diversity of drug candidates, there is a place for all the methods discussed in the chapter and, indeed, this book in the process chemist's arsenal. The one area where we can certainly look forward to significant changes in the next few years is the coupling of enzymatic and chemical reactions to achieve dynamic resolution systems.

REFERENCES

1. Ager, D. J. In *Handbook of Chiral Chemicals,* Ager, D. J. Ed., Marcel Dekker, Inc.: New York, 1999, p. 33.
2. Roth, B. D. U.S. Pat. 4681893, 1987.
3. Butler, D. E., Dejong, R. L., Nelson, J. D., Pamment, M. G., Stuk, T. L. PCT Int. Appl. WO 0255519, 2002.
4. Butler, D. E., Le, T. V., Nanninga, T. N. U.S. Pat. 5298627, 1994.
5. Browser, P. L., Butler, D. E., Deering, C. F., Le, T. V., Millar, A., Nanninga, T. N., Roth, B. D. *Tetrahedron Lett.* 1992, *33*, 2279.
6. Roth, B. D. Eur. Pat. EP 409281, 1991.
7. Butler, D. E., Le, T. V., Millar, A., Nanninga, T. N. U.S. Pat. 5155251, 1992.
8. Baumann, K. L., Butler, D. E., Deering, C. F., Mennen, K. E., Millar, A., Nanninga, T. N., Palmer, C. W., Roth, B. D. *Tetrahedron Lett.* 1992, *33*, 2283.
9. Briggs, C. A., Jennings, R. A., Wade, R. A., Harasawa, K., Ichikawa, S., Minohara, K., Nakagawa, S. PCT Pat. Appl. WO 9703959, 1997.
10. Procter, L. D., Warr, A. J. PCT Int. Appl. WO 03097569, 2003.
11. Ager, D. J., Laneman, S. A. *Tetrahedron: Asymmetry* 1997, *8*, 3327.
12. Zhou, Y.-G., Tang, W., Wang, W.-B., Li, W., Zhang, X. *J. Am. Chem. Soc.* 2002, *124*, 4952.
13. Ema, T., Moriya, H., Kofukuda, T., Ishida, T., Maehara, K., Utaka, M., Sakai, T. *J. Org. Chem.* 2001, *66*, 8682.
14. Yasohara, Y., Kizaki, N., Hasegawa, J., Wada, M., Kataoka, M., Shimizu, S. *Tetrahedron: Asymmetry* 2001, *12*, 1713.
15. Nakai, T., Morikawa, S., Kizaki, N., Yasohara, Y. PCT Int. Appl. WO 03004653, 2003.
16. Genet, J.-P., Marinetti, A., Michaud, G., Bulliard, M. PCT Int. Appl. WO 02040492, 2002.
17. Oehrlein, R., Baisch, G. PCT Int. Appl. WO 02040438, 2002.
18. Yamamoto, Y., Matsuyama, A., Kobatashi, Y. *Biosci. Biotech. Biochem.* 2002, *66*, 481.
19. Bulliard, M., Laboue, B., Roussiasse, S. PCT Int. Appl. WO 02012253, 2002.
20. Saito, T., Matsumura, K., Yokozawa, T., Sayo, N. Eur. Pat. EP 1176135, 2002.
21. Kizaki, N., Yasohara, Y., Hasegawa, J., Wada, M., Kataoka, M., Shimizu, S. *Appl. Microbiol. Biotech.* 2001, *55*, 590.
22. Nishiyama, A., Inoue, K. PCT Int. Appl. WO 00075099, 2000.
23. Yamamoto, Y. Eur. Pat. EP 955375, 1999.
24. Yasohara, Y., Kizaki, N., Hasegawa, J., Takahashi, S., Wada, M., Kataoka, M., Shimizu, S. *Appl. Microbiol. Biotech.* 1999, *51*, 847.
25. Dahl, A. C., Madsen, J. O. *Tetrahedron: Asymmetry* 1998, *9*, 4395.
26. Shimizu, S., Kataoka, M., Kita, K. *Ann. NY Acad. Sci.* 1998, *864*, 87.
27. Yasohara, Y., Kizaki, N., Hasegawa, J., Wada, M., Shimizu, S., Kataoka, M., Yamamoto, K., Kawabata, H., Kita, K. PCT Int. Appl. WO 9835025, 1998.
28. Shimizu, S., Kataoka, M., Kita, K. *J. Mol. Catal. (B): Enzymatic* 1998, *5*, 321.
29. Matsuyama, A., Tomita, A., Kobayashi, Y. Eur. Pat. EP 606899, 1994.
30. Matsuyama, A., Yamamoto, H., Kobayashi, Y. *Org. Proc. Res. Dev.* 2002, *6*, 558.
31. Oonishi, I., Shimaoka, M., Kira, I., Nakazawa, M. Jap. Pat. 06038776, 1994.
32. Mitsushashi, S., Sakurai, K., Kumobayashi, H. Eur. Pat. EP 573184, 1993.

33. Aragozzini, F., Valenti, R., Santaniello, E., Ferraboschi, P., Grisenti, P. *Biocatalysis* 1992, *5*, 325.
34. Wakita, R., Asako, H., Matsumura, K., Shimizu, M. Jap. Pat. 200303193, 2003.
35. Procter, L. D., Warr, A. J. PCT Int. Appl. WO 0309758, 2003.
36. Matsuda, H., Egashira, T., Shibata, T., Minagawa, M. Jap. Pat. 200314690, 2003.
37. Egashira, T., Minagawa, M., Matsuda, H. Jap. Pat. 2003096041, 2003.
38. Suzuki, T., Matsuda, H., Kasai, T. Jap. Pat. 2002220367, 2002.
39. Suzuki, T. Jap. Pat. 2002220369, 2002.
40. Suzuki, T. Jap. Pat. 2002220368, 2002.
41. Shimoshige, T., Hayashi, T., Kumasawa, Y., Watanabe, Y. Jap. Pat. 2002080442, 2002.
42. Kasai, T., Dejima, E. Jap. Pat. 2001335554, 2001.
43. Kasai, T., Dejima, E. Jap. Pat. 2001302607, 2001.
44. Kasai, T., Dejima, E. Jap. Pat. 2001335553, 2001.
45. Matsuda, H., Shibata, T., Hashimoto, H., Kitai, M. Jap. Pat. 10231278, 1998.
46. Matsuda, H., Shibata, T., Tsuchisada, H. Jap. Pat. 2001122841, 2001.
47. Matsuda, H., Shibata, T., Tsuchisada, H. Jap. Pat. 2001122840, 2001.
48. Matsuda, H., Shibata, T., Tsuchisada, H. Jap. Pat. 2001122839, 2001.
49. Matsuda, H., Shibata, T., Tsuchisada, H. Jap. Pat. 2001122838, 2001.
50. Matsuda, H., Shibata, T., Tsuchisada, H. Jap. Pat. 2000212150, 2000.
51. Dotei, H., Koga, T., Matsuda, H. Jap. Pat. 2000212151, 2000.
52. Kasai, N., Suzuki, T. *Adv. Synth. Catal.* 2003, *345*, 437.
53. Wakita, R., Ito, N. Eur. Pat. EP 1201647, 2002.
54. Kwak, B.-S. *Spec. Chem. Magazine* 2002, *22(5)*, 17.
55. Lee, B. N., Jung, I. S., Jang, E. J. DE Pat. 10130740, 2002.
56. Burk, M., Desantis, G., Morgan, B., Zhu, Z. PCT Int. Appl. WO 03106415, 2003.
57. Albers-Schonberg, G., Monaghan, R. L., Alberts, A. W., Hoffman, C. H. U.S. Pat. 4,342,767, 1982.
58. Moore, R. N., Bigam, G., Chan, J. K., Hogg, A. M., Nakashima, T. T., Vederas, J. C. *J. Am. Chem. Soc.* 1985, *107*, 3694.
59. Witter, D. J., Vederas, J. C. *J. Org. Chem.* 1996, *61*, 2613.
60. Yoshida, Y., Witter, D. J., Liu, Y., Vederas, J. C. *J. Am. Chem. Soc.* 1994, *116*, 2693.
61. Gbewonyo, K., Hunt, G., Buckland, B. *Bioprocess Eng* 1992, *8*, 1.
62. Dahiya, J. S. Eur. Pat. EP 556699, 1993.
63. Vinci, V. A., Conder, M. J., Mcada, P. C., Reeves, C. D., Rambosek, J., Davis, C. R. PCT Int. Appl. WO 9512661, 1995.
64. Hutchinson, R. G., Kennedy, J., Park, C. PCT Int. Appl. WO 0037629, 2000.
65. Beck, G., Jendralla, H., Kesseler, K. *Synthesis* 1995, 1014.
66. Hiyama, T. *Pure Appl. Chem.* 1996, *68*, 609.
67. Rosen, T., Heathcock, C. H. *Tetrahedron* 1986, *42*, 4909.
68. Carta, G., Conder, M. J., Gainer, J. L., Stieberg, R. W., Vinci, V. A., Weber, T. W. PCT Int. Appl. WO 9426920, 1994.
69. Crosby, J. In *Chirality in Industry,* Collins, A. N., Sheldrake, G. N., Crosby, J. Eds., Wiley: New York, 1992, p. 1.
70. Conder, M. J., Cianciosi, S. J., Cover, W. H., Dabora, R. L., Pisk, E. T. S., R. W., Tehlewitz, B., Tewalt, G. L. U.S. Pat. 5,223,415, 1993.
71. Conder, M. J., Daborah, R. L., Lein, J., Tewalt, G. L. U.K. Pat., 1992.
72. Dabora, R. L., Tewalt, G. L. U.S. Pat. 5,159,104, 1992.
73. Schimmel, T. G., Borneman, W. S., Conder, M. J. *Appl. Environment. Microbiol.* 1997, *63*, 1307.
74. Kubela, R., Radhakrishnan, J. U.S. Pat. 5,393,893, 1995.
75. Askin, D., Verhoeven, T. R., Liu, T. M. H., Shinkai, I. *J. Org. Chem.* 1991, *56*, 4919.
76. Thaper, R. K., Kumar, Y., Kumar, S. M. D., Misra, S., Khanna, J. M. *Org. Proc. Res. Dev.*, 1999, *3*, 476.
77. Dabak, K., Adiyaman, M. *Helv. Chim. Acta* 2003, *86*, 673.
78. Galeazzi, E., Garcia, G. A., Lora, F., Lopez, G., Martinez, O., Tisselli, E., Trejo, A. U.S. Pat. 6,472,542, 2002.
79. Lee, K. H., Kim, J. W., Choi, K. D., Bae, H. PCT Int. Appl. WO 0172734, 2001.
80. Lee, J., Ha, T., Park, C., Lee, H., Lee, G., Chang, Y. PCT Int. Appl. WO 0357684, 2003.
81. Hong, C. I., Kim, J. W., Shin, K. J., Kang, T. W., Cho, D. O. PCT Int. Appl. WO 0145484, 2001.

82. Taoka, N., Inoue, K. PCT Int. Appl. WO 0034264, 2000.
83. Karimian, K., Tan, T. F., Tao, Y., Li, Y., Doucetti, G. PCT Int. Appl. WO 9965892, 1999.
84. Van Dalen, F., Lemmens, J. M., Van Helvoirt, G. A. P., Peters, T. H. A., Picha, F. PCT Int. Appl. WO 9945003, 1999.
85. Vries, T. R., Wijnberg, H., Faber, W. S., Kalkman-Agayn, V. I., Sibeyn, M. PCT Int. Appl. WO 9832751, 1998.
86. Murthy, K. S. K., Home, S. E., Weeratunga, G., Young, S. PCT Int. Appl. WO 9812188, 1998.
87. Terahara, A., Tanaka, M. DE Pat. 3123499, 1981.
88. Serizawa, N. *Biotechnol. Ann. Rev.* 1996, *2*, 373.
89. Hosobuchi, M., Kurosawa, K., Yoshida, H. *Biotechnol. Bioeng.* 1993, *42*, 815.
90. Garuraja, R., Goel, A., Sridharan, M., Melarkode, R. S., Kulkarni, M., Poomaprajna, A., Sathyanathan, D., Ganesh, S., Suryanarayan, S. PCT Int. Appl. WO 0327302, 2003.
91. Jekkel, A., Ambrus, G., Ilkoy, E., Horvath, I., Konya, A., Szabo, I. M., Nagy, Z., Horvath, G., Mozes, J., Barta, I., Somogyl, G., Salat, J., Boros, S. PCT Int. Appl. WO 0104340, 2001.
92. Jekkel, A., Ambrus, G., Ilkoy, E., Horvath, I., Konya, A., Szabo, I. M., Nagy, Z., Horvath, G., Mozes, J., Barta, I., Somogyl, G., Salat, J., Boros, S. PCT Int. Appl. WO 0103647, 2001.
93. Matsuo, N., Negishi, A., Negishi, Y. Jap. Pat. 2003093045, 2003.
94. Matsuoka, T., Miyakoshi, S., Tanzawa, K., Nakahara, K., Hosobuchi, M., Serizawa, N. *Eur. J. Biochem.* 1989, *184*, 797.
95. Lee, J.-K., Park, J.-W., Seo, D.-J., Lee, S.-C., Kim, J.-Y. PCT Int. Appl. WO 9845410, 1998.
96. Chung, K.-J., Lee, J.-K., Park, J.-W., Seo, D.-J., Lee, S.-C. PCT Int. Appl. WO 9806867, 1998.
97. Nobufusa, S., Ichiro, W. Eur. Pat. EP 776974, 1997.
98. Demain, A. L., Peng, Y., Yashphe, J., David, J. U.S. Pat. 5,942,423.
99. Sibejn, M., Bouman, J., de Pater, R. M., Purmer, C. F. U.S. Pat. 6,268,186, 2001.
100. Serizawa, N., Watanabe, I. U.S. Pat. 5,830,695, 1998.
101. Davis, B. L., Cino, P. M., Szarka, L. U.S. Pat. 6,043,064, 2000.
102. Bakonyi, M., Csatari Nagy, M., Molnar, L., Makovi, Z., Jobb, P., Bal, T. PCT Int. Appl. WO 9851681, 1998.
103. Aubert, D., Ferrand, C., Maffrand, J.-P. U.S. Pat. 4,529,596, 1985.
104. Badore, A., Fréhel, D. U.S. Pat. 4,847,265, 1989.
105. Bakonyi, M., Csatari Nagy, M., Moinar, L., Gajary, A., Alattyani, E. PCT Int. Appl. WO 9851689, 1998.
106. Heymes, A., Castro, B., Bakonyi, M., Csatari Nagy, M., Molnar, L. PCT Int. Appl. WO 9851682, 1998.
107. Castro, B., Dormay, J.-R., Previero, A. PCT Int. Appl. WO 9838322, 1998.
108. Bakonyi, M., Bai, T., Dombrady, Z., Gasper, K., Sapic, A. PCT Int. Appl. WO 0027840, 2000.
109. Pandey, B., Lohray, V. B., Lohray, B. B. PCT Int. Appl. WO 0259128, 2002.
110. Home, S. E., Weeratunga, G., Comanita, B. M., Nagureddy, J. R., McConachie, L. K. PCT Int. Appl. WO 0304502, 2003.
111. Balint, J., Csatarine Nagy, M., Dombrady, Z., Fogansy, E., Gajary, A., Suba, C. PCT Int. Appl. WO 0300636, 2003.
112. Bousquet, A., Musolino, A. PCT Int. Appl. WO 9918110, 1999.
113. Barnes, R. D., Wood-Kaczmar, M. W., Richardson, J. E., Lynch, I. R., Buxton, P. C., Curzons, A. D. Eur. Pat. EP 2233403, 1987.
114. Lucas, E. U.S. Appl. 2002137938, 2002.
115. Borrett, G. Eur. Pat. EP 219934, 1986.
116. Borrett, G. U.S. Pat. 4,861,893, 1989.
117. Christensen, J. A., Squires, R. F. U.S. Pat. 4,007,196, 1977.
118. Lunbeck, J. M., Everland, P., Treppendahl, S., Jakobsen, P. Eur. Pat. EP 374674, 1990.
119. Yu, M. S., Jacewicz, V. W., Shapiro, E. PCT Int. Appl. WO 9853824, 1998.
120. Ennis, D., Lathbury, D. PCT Int. Appl. WO 0228834, 2002.
121. Foguet, R., Ramental, J., Petschen, I., Sallares, J., Camps, F. X., Raga, M. M., Castello, J. M., Armengol, M. P., Fernandez-Cano, D. PCT Int. Appl. WO 0253537, 2002.
122. Ward, N. PCT Int. Appl. WO 0233870, 2002.
123. Borrett, G. T., Fedouloff, M., Hughes, M. J., Share, A. C., Strochan, J. B., Szeto, P., Voyle, M. PCT Int. Appl. WO 0206275, 2002.

124. Froggett, J., Riley, D., Turner, A. PCT Int. Appl. WO 0218338, 2002.
125. Riley, D. PCT Int. Appl. WO 0218337, 2002.
126. Kriedl, J., Czibula, L., Nemes, A., Juhasz, I. D., Papp, E. W., Bagdy, J. N., Dobay, L., Hegedus, I., Harsanyi, K., Borza, I. U.S. Pat. 6,657,062, 2003.
127. Crowe, D., Ward, N., Wells, A. S. PCT Int. Appl. WO 0129032, 2001.
128. Borrett, G. T., Crowe, D., Ward, N., Wells, A. S. PCT Int. Appl. WO 0129031, 2001.
129. Crowe, D., Jones, D. A., Ward, N. PCT Int. Appl. WO 0117966, 2001.
130. Crowe, D., Jones, D. A. PCT Int. Appl. WO 0114369, 2001.
131. Crowe, D., Jones, D. A. PCT Int. Appl. WO 0114335, 2001.
132. Rossi, R., Turchetta, S., Donnarumma, M. PCT Int. Appl. WO 0050422, 2000.
133. Wang, S.-Z., Matsumura, Y. Eur. Pat. EP 810225, 1997.
134. Sugi, K., Itaya, N., Katsura, T., Igi, M., Yamazaki, S., Ishibashi, T., Yamaoka, T., Kawada, Y., Tagami, Y. Eur. Pat. EP 812827, 1997.
135. Bayoud, J. M., Sanchez, P. V., V., G. S., Brieva C, R., De Gonzalez Calvo, G. U.S. Appl. 2003018048, 2003.
136. Palmero, J. M., Fernández-Lorente, G., Mateo, C., Fernández-Lafuente, R., Guisan, J. M. *Tetrahedron: Asymmetry* 2002, *13*, 2375.
137. de Gonzalez, G., Brieva, R., Sánchez, V. M., Bayod, M., Gotor, V. *J. Org. Chem.* 2003, *68*, 3333.
138. Gledhill, L., Kele, C. M. PCT Int. Appl. WO 980255, 1998.
139. Guisan Seijas, J. M., Fernandez Lorente, G., Fernandez-Lafuente, R., Mateo Gonzalez, C., Ceinos Rodriguez, C., Dalmases Barjoan, P., de Ramon Arnat, E. ES Pat. 2161167, 2001.
140. Yu, M. S., Lantos, I., Peng, Z. Q., Yu, J., Cacchio, T. *Tetrahedron Lett.* 2000, *41*, 5647.
141. Zepp, C. M., Gao, Y., Heifner, D. L. U.S. Pat. 5,258,517, 1993.
142. Amat, M., Bosch, J., Hidalgo, J., Canto, M., Perez, M., Llor, N., Molins, E., Miravitlles, C., Orozco, M., Luque, J. *J. Org. Chem.* 2000, *65*, 3074.
143. Senda, T., Ogasawara, M., Hayashi, T. *J. Org. Chem.* 2001, *66*, 6852.
144. Liu, L. T., Hing, P.-C., Huang, H.-L., Chen, S.-F., Wang, C.-L. J., Wen, Y.-S. *Tetrahedron: Asymmetry* 2001, *12*, 419.
145. Adger, B. M., Potter, G. A., Fox, M. E. PCT Int. Appl. WO 9724323, 1997.
146. Murthy, K. S. K., Rey, A. W. PCT Int. Appl. WO 9907680, 1999.
147. Brennan, J. P. PCT Int. Appl. WO 9852920, 1998.
148. Welch, W. M., Harbert, C. A., Koe, B. K., Kraska, A. R. U.S. Pat. 4,536,518, 1985.
149. Welch, W. M., Harbert, C. A., Koe, B. K., Kraska, A. R. Eur. Pat. EP 30081, 1981.
150. Quallich, G. J., Woodall, T. M. *Tetrahedron* 1992, *48*, 10239.
151. Caille, J.-C., Bulliard, M., Laboue, B. In *Chirality in Industry,* Collins, A. N., Sheldrake, G. N., Crosby, J. Eds., Wiley: New York, 1992, p. 391.
152. Quallich, G. J. PCT Int. Appl. WO 9515299, 1995.
153. Quallich, G. J., Williams, M. T., Friedmann, R. C. *J. Org. Chem.* 1990, *55*, 4971.
154. Quallich, G. J., Woodall, T. M. *Tetrahedron Lett.* 1993, *34*, 785.
155. Quallich, G. J., Woodall, T. M. *Synlett* 1993, 929.
156. Quallich, G. J. PCT Int. Appl. WO 9312062, 1993.
157. Quallich, G. J. Eur. Pat. EP 1059287, 2000.
158. Williams, M., Quallich, G. J. *Chem. Ind. (London)* 1990, 315.
159. Quallich, G. J., Williams, M. T. U.S. Pat. 4,839,104, 1988.
160. Williams, M. T. PCT Int. Appl. WO 9301162, 1993.
161. Quallich, G. J., Williams, M. T. U.S. Pat. 4,777,288, 1988.
162. Braish, T. F. U.S. Pat. 5,082,970, 1992.
163. Aronhime, J., Mendelovici, M., Nidam, T., Singer, C. PCT Int. Appl. WO 0145692, 2001.
164. Laitinen, L., Pietikaeinen, P. PCT Int. Appl. WO 0296860, 2002.
165. Mendelovich, M., Nidam, T., Pilarsky, G., Gershon, N. PCT Int. Appl. WO 0168566, 2001.
166. Fischer, E., Treppendahl, S. P., Pedersen, S. B. PCT Int. Appl. WO 0130742, 2001.
167. Jadav, K. J., Chitturi, T. R., Thennati, R. PCT Int. Appl. WO 0149638, 2001.
168. Thommen, M., Hafner, A., Brunner, F., Kimer, H.-J., Kolly, R. PCT Int. Appl. WO 0136378, 2001.
169. Thommen, M., Hafner, A., Kolly, R., Kimer, H.-J., Brunner, F. PCT Int. Appl. WO 0136377, 2001.
170. Barbieri, C., Caruso, E., D'Arrigo, P., Fantoni, G. P., Servi, S. *Tetrahedron: Asymmetry* 1999, *10*, 3931.

171. Chandrasekhar, S., Reddy, M. V. *Tetrahedron* 2000, *56*, 1111.
172. Yun, J., Buchwald, S. L. *J. Org. Chem.* 2000, *65*, 767.
173. Davies, H. M. L., Stafford, D. G., Hansen, T. *Org. Lett.* 2000, *2*, 233.
174. Shum, S. P., Pastor, S. D., Odorisio, P. A. PCT Int. Appl. WO 9955686, 1999.
175. Lautens, M., Rovis, T. *Tetrahedron* 1999, *55*, 8967.
176. Chen, C.-Y., Reamer, R. A. *Org. Lett.* 1999, *1*, 293.
177. Kotay, N. P., Barkoczy, J., Simig, G., Sztuhar, I., Balazs, L., Doman, I., Greff, Z., Ratkai, Z., Seres, P., Clementis, G., Karancsi, T., Ladanyi, L. PCT Int. Appl. WO 9815516, 1998.
178. Corey, E. J., Gant, T. G. *Tetrahedron Lett.* 1994, *35*, 5373.
179. Phillips, G. H., Bailey, E. J., Bain, B. M., Borella, R. A., Buckton, J. B., Clark, J. C., Doherty, A. E., English, A. F., Fazakerley, H., Laing, S. B., Lane-Allman, E., Robinson, J. D., Sandford, P. E., Sharratt, P. J., Steeples, I. P., Stonehouse, R. D., Williamson, C. *J. Med. Chem.* 1994, *37*, 3717.
180. Phillips, G. H., Bain, B. M., Steeples, I. P., Williamson, C. U.S. Pat. 4,335,121, 1982.
181. Skidmore, I. F., Lunts, L. H. C., Finch, H., Naylor, A. U.S. Pat. 5,126,375, 1992.
182. Skidmore, I. F., Lunts, L. H. C., Finch, H., Naylor, A. DE Pat. 3414752, 1984.
183. Finch, H., Naylor, A., Skidmore, I. F., Lunts, L. H. C. U.S. Pat. 4,992,474, 1991.
184. Bessa Bellmunt, J., Dalmases Barjoan, P., Marquillas Olondriz, F. PCT Int. Appl. WO 0018722, 2000.
185. Sallmann, A., Gschwind, H.-P., Francotte, E. PCT Int. Appl. WO 9515953, 1995.
186. Ariza Aranda, J., Serra Masia, J., Monserrat Vidal, C. ES Pat. 2065269, 1995.
187. Goswami, J., Bezbaruah, R. L., Goswami, A., Borthakur, N. *Tetrahedron: Asymmetry* 2002, *12*, 3343.
188. Hett, R., Stare, R., Helquist, P. *Tetrahedron Lett.* 1994, *35*, 9375.
189. Federsel, H.-J. *Chirality* 2003, *15*, S128.
190. Federsel, H.-J., Larsson, M. An Innovative Asymmetric Sulfoxide Oxidation: The Process Development History behind the Antiulcer Agent Esomeprazole. In *Asymmetric Catalysis on Industrial Scale*, Blaser, H.-U., Schmidt, E., Eds., Wiley-VCH: Weinheim, 2004, p. 413.
191. Von Unge, S. PCT Int. Appl. WO 9702261, 1997.
192. Lindberg, P. L., von Unge, S. PCT Int. Appl. WO 9427988, 1994.
193. Lindberg, P. L., von Unge, S. U.S. Pat. 5,714,504, 1998.
194. Brändström, A. E. PCT Int. Appl. WO 9118895, 1991.
195. Gustavsson, A., Källström, Å. U.S. Pat. 5,958,955, 1999.
196. Brändström, A., Lamm, B. R., Hassle, A. U.S. Pat. 4,544,750, 1985.
197. Brändström, A., Lamm, B. R., Hassle, A. U.S. Pat. 4,620,008, 1986.
198. Junggren, U. K., Hassle, A. U.S. Pat. 4,255,431, 1981.
199. Cotton, H., Elebring, T., Larsson, M., Li, L., Sorensen, H., von Unge, S. *Tetrahedron: Asymmetry* 2000, *11*, 3819.
200. Nilsson, M. PCT Int. Appl. WO 0044744, 2000.
201. Broeckx, R. L. M., de Smaele, D., Leurs, S. M. H. PCT Int. Appl. WO 0308406, 2003.
202. Andersson, S., Juza, M. PCT Int. Appl. WO 0351867, 2003.
203. Holt, R., Lindberg, P., Reeve, C., Taylor, S. PCT Int. Appl. WO 9617076, 1996.
204. Graham, D., Holt, R., Lindberg, P., Taylor, S. PCT Int. Appl. WO 9617077, 1996.
205. Fleming, I. D., Noble, D., Noble, H. M., Wall, W. F. U.S. Pat. 4,367,175, 1983.
206. Cole, M., Howarth, T. T., Reading, C. U.S. Pat. 4,529,720, 1985.
207. Gupta, N., Eisberg, N. *Performance Chem.* 1991, 19.
208. Van der Does, T., Kuipers, R. H., Diender, M., Den Hollander, J., Straathof, A., Van der Wielen, L., Heijnen, J. J. PCT Int. Appl. WO 0218618, 2002.
209. Boesten, W. H. J., van Dooren, T. J., Smeets, J. C. M. PCT Int. Appl. WO 9623897, 1996.
210. Clausen, K., Dekkers, R. M. PCT Int. Appl. WO 9602663, 1996.
211. Kaasgaard, S. G., Veitland, U. PCT Int. Appl. WO 9201061, 1992.
212. Grossman, J. H., Hardcastle, G. A. U.S. Pat. 3,980,637, 1976.
213. Nayler, J. H. C., Long, A. A. W., Smith, H., Taylor, T., Ward, N. *J. Chem. Soc. (C)* 1971, 1920.
214. Van Dooren, T. J. G. M., Moody, H. M., Smeets, J. C. M. PCT Int. Appl. WO 9920786, 1999.
215. Kondo, E., Mitsugi, T. U.S. Pat. 4,073,687, 1978.
216. Williams, R. M. *Synthesis of Optically Active α-Amino Acids*, Pergamon Press: Oxford, 1989.
217. Yamada, H., Takahashi, S., Yoshiaki, K., Kumagai, H. *J. Ferment. Technol.* 1978, *56*, 484.

218. Sheldon, R. A., Hulshof, L. A., Bruggink, A., Leusen, F. J. J., van der Haest, A. D., Wijnberg, H. *Chimica Oggi* 1991, *9*, 23.
219. Yamada, H., Takahashi, S., Yoneka, K., Amagasaki, H. Ger. Pat. 2,757,980, 1991.
220. Drauz, K., Kottenhahn, M., Makryaleas, K., Klenk, H., Bernd, M. *Angew. Chem., Int. Ed.* 1991, *30*, 712.
221. Buhlmayer, P. PCT Int. Appl. WO 9730036, 1997.
222. Buhlmayer, P., Ostermayer, F., Schmidlin, T. Eur. Pat. EP 443983, 1991.
223. Buhlmayer, P., Furet, P., Cricione, L., de Gasparo, M., Whitebread, S., Schmidlin, T., Lattmann, R., Wood, J. *Bioorg. Med. Chem. Lett.* 1994, *4*, 29.
224. Noyori, R., Takaya, H. *Acc. Chem. Res.* 1990, *23*, 345.
225. Akutagawa, S., Tani, K. Asymmetric Isomerization of Allylamines. In *Catalytic Asymmetric Synthesis*, Ojima, I. Ed., VCH Publishers, Inc.: New York, NY, 1993, p. 41.
226. Noyori, R., Hasiguchi, S. Yamano, T. Asymmetric Synthesis. In *Applied Homogeneous Catalysis with Organic Compounds*, Herrmann, B. C. W. A. Ed., Wiley-VCH, Weinheim, 2002, Vol. 1, p. 557.
227. Ager, D. J., Pantaleone, D. P., Henderson, S. A., Katritzky, A. R., Prakash, I., Walters, D. E. *Angew. Chem., Int. Ed.* 1998, *37*, 1803.
228. Ariyoshi, Y., Nagano, M., Sato, N., Shimizu, A., Kirimura, J. U.S. Pat. 3,786.039, 1974.
229. Hill, J. B., Gelman, Y. U.S. Pat. 4,946,988, 1990.
230. Boesten, W. H. J. U.S. Pat. 3,879,372, 1975.
231. Hill, J. B., Gelman, Y., Dryden, H. L., Erickson, R., Hsu, K., Johnson, M. R. U.S. Pat. 5,053,532, 1991.
232. Isowa, Y., Ohmori, M., Mori, K., Ichikawa, T., Nonaka, Y., Kihara, K., Oyama, K., Satoh, H., Nishimura, S. U.S. Pat. 4,165,311, 1979.
233. Oyama, K. In *Chirality in Industry*, Collins, A. N., Sheldrake, G. N., Crosby, J. Eds., Wiley: New York, 1992, p. 237.
234. Genet, J. P., Pinel, C., Ratovelomanana-Vidal, V., Mallart, S., Pfister, X., Bischoff, L., De Andrade, M. C. C., Darses, S., Galopin, C., Laffitte, J. A. *Tetrahedron: Asymmetry* 1994, *5*, 675.
235. Kitamura, M., Ohkuma, T., Takaya, H., Noyori, R. *Tetrahedron Lett.* 1988, *29*, 1555.
236. Tinti, M. O., Piccollo, O., Bonifacio, F., Cressini, C., Pereo, S. PCT Int. Appl. WO 0029370, 2000.
237. Kulla, H. G. *Chimia* 1991, *45*, 81.
238. Jung, H., Jung, K., Kieber, H. P. Eur. Pat. 320460, 1989.
239. Giannessi, F., Bolognia, M. L., De Angelis, F. CA Pat. 2111898, 1994.
240. Tscherry, B., Bornscheuer, U., Musidlowska, A., Werlen, J., Zimmerman, T. PCT Int. Appl. WO 02061094, 2002.
241. Held, U., Siebricht, S. *Spec. Chem. Magazine* 2003, *23(6)*, 28.
242. Spindler, F., Pugin, B., Jalett, H.-P., Buser, H.-P., Pittelkow, U., Blaser, H.-U. Catalysis of Organic Reactions. In *Chem. Ind.*, Malz, J. Ed., Dekker: New York, 1996, *Vol. 68*, p. 153.
243. Stinson, S. C. *Chem. & Eng. News* 1997, 34.
244. Blaser, H.-U., Spindler, F. *Chimia* 1997, *51*, 297.
245. Holt, R. A. In *Chirality in Industry II*, Collins, A. N., Sheldrake, G. N., Crosby, J. Eds., Wiley: New York, 1997, p. 113.
246. Bodnar, J., Gubicza, L., Szabo, L.-P. *J. Mol. Catal.* 1990, *61*, 353.
247. Brieva, R., Rebolledo, F., Gotor, V. *J. Chem. Soc., Chem. Commun.* 1990, 1386.
248. Taylor, S. C. In *Chirality in Industry II*, Collins, A. N., Sheldrake, G. N., Crosby, J. Eds., Wiley: New York, 1997, p. 207.
249. Taylor, S. C. U.S. Pat. 4,758,518, 1988.
250. Dahod, S. K., Siuta-Mangano, P. *Biotechnol. Bioeng.* 1987, *30*, 995.
251. Elliot, M. *Pestic. Sci.* 1989, *27*, 337.
252. Schneider, M., Engel, N., Boensmann, H. *Angew. Chem., Int. Ed.* 1984, *23*, 52.
253. Sheldon, R. A. *Chirotechnology: Industrial Synthesis of Optically Active Compounds*, Marcel Dekker: New York, 1993.

Index